Principles of
Human Anatomy

Fourth Edition

Principles of Human Anatomy

Gerard J. Tortora

Bergen Community College

HARPER & ROW, PUBLISHERS, New York
Cambridge, Philadelphia, San Francisco,
London, Mexico City, São Paulo, Singapore, Sydney

1817

Sponsoring Editor: Claudia M. Wilson
Project Editor: Nora Helfgott
Text Design Adaptation and Cover Design: Caliber Design Planning, Inc.
Text Art: Vantage Art, Inc.
Photo Research: Mira Schachne
Production: Kewal K. Sharma
Compositor: Kingsport Press
Printer and Binder: Kingsport Press

Principles of Human Anatomy, Fourth Edition

Copyright © 1986 by Biological Sciences Textbooks, Inc.

Library of Congress Cataloging in Publication Data

Tortora, Gerard J.
  Principles of human anatomy.

  Bibliography: p.
  Includes index.
  1. Human physiology. 2. Anatomy, Human. I. Title.
[DNLM: 1. Anatomy. 2. Physiology. QS 4 T712pa]
QP34.5.T69 1986   612   85–8545
ISBN 0–06–046623–5
85 86 87 88 9 8 7 6 5 4 3 2 1

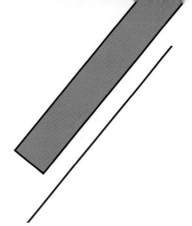

Contents in Brief

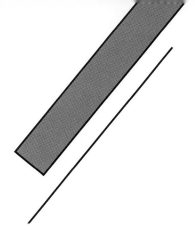

Contents in Detail

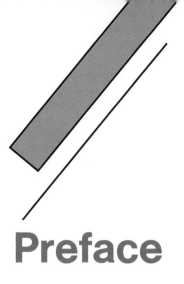

Preface

AUDIENCE

Designed for the introductory course in human anatomy, the fourth edition of *Principles of Human Anatomy* assumes no previous study of the human body. The text is geared to students in health-oriented, medical, and biological programs. Among the students specifically served by this text are those pursuing careers as nurses, medical assistants, physicians' assistants, medical laboratory technologists, radiological technologists, respiratory therapists, dental hygienists, physical therapists, morticians, and medical record keepers. However, because of the scope of the book, *Principles of Human Anatomy,* Fourth Edition, is also useful for students in the biological sciences, science technology, liberal arts, physical education, and in premedical, predental, and prechiropractic programs.

OBJECTIVES

The objectives of *Principles of Human Anatomy* remain unchanged in this fourth edition. Because human anatomy is such a large and complex body of knowledge to present in an introductory course, the first objective is to concentrate on data and unified concepts that contribute to a basic understanding of the structure of the human body. Data unessential to this objective have been minimized. The second objective is to present essential technical vocabulary and important, but difficult, concepts at a reading level that can be handled by the average student. Easy-to-comprehend explanations of terms and step-by-step development of concepts are presented to meet this objective.

THEMES

The fourth edition of this textbook departs from the approaches of most other anatomy texts in that somewhat more emphasis is given to physiology and applications to health. The basic content of the book is anatomy, but *because structure and function are so inseparably related,* it makes little sense to present students with anatomical detail without relating anatomy to some function. The discussion of function gives the students a better understanding of anatomical concepts.

An understanding of structure and function may be enhanced by considering variations from the normal, or the defects and disorders observed in clinical situations. In turn, such conditions are better understood once the context of normal anatomy has been established. Disorders are treated under the special heading **Applications to Health** at the ends of chapters and **Clinical Applications** integrated throughout the text.

ORGANIZATION

The book is organized by systems rather than by regions. The first chapter introduces the levels of structural organization, organ-systems, stuctural plan, anatomical position, regional names, directional terms, planes and sections, and cavities of the human body. Radiographic anatomy and units of measurement are also included. The chapter has been improved by the inclusion of sections dealing with life processes characteristic of humans, autopsy, the dynamic spatial reconstructor (DSR), and magnetic resonance imaging (MRI).

The chapter dealing with the cellular level of organization describes a generalized animal cell to demonstrate the basic structural features and functions of cells. It includes updated material on the structure, chemistry, and functions of plasma membranes; active transport; ribosomes; lysosomes; cell division; and cells and cancer. New to the chapter are discussions on microtomography, receptor-mediated endocytosis, nucleosomes, peroxisomes, the cellular cytoskeleton, periods of interphase, and cancer. Specific changes associated with aging are no longer consolidated in the chapter; they are now found throughout the text as part of the respective chapters. The tissue level of organization is presented through descriptions of the structure, functions, and locations of the principal kinds of tissues. The histology of bone, muscle, nervous tissue, and blood is now considered along with epithelial and connective tissues. Also new to the chapter are sections dealing with suction lipectomy and artificial ligaments.

The discussion of the organ and organ-system levels of organization begins with the chapter on the integumentary system. This chapter includes new material on collagen implants, chemical exfoliation, common baldness, and the Lund-Browder method for estimating the extent of skin burns. As in all the chapters dealing with organ-systems, there is some reference to physiology, disorders, and relevant medical terminology. Chapters dealing with organ-systems also contain discussions of the effects of aging and developmental anatomy.

In the osseous tissue chapter, there is a revised section on the application of pulsating electromagnetic fields (PEMFs) to fracture repair. The chapter on the axial skeleton has been updated to include sections on cranial fossae, gravity inversion, laminectomy, caudal anesthesia, marrow biopsy, sternal puncture, and fractured ribs. A section on shinsplints has been added to the chapter on the appendicular skeleton. New to the chapter on articulations are discussions of arthroscopy, sports injuries, and DMSO and joint disorders.

The muscular system is analyzed through a study of muscle tissue and the locations and actions of the principal skeletal muscles. The chapter on muscle tissue contains revised discussions of the ultrastructure of skeletal muscle, the neuromuscular junction, and fast and slow muscle twitches and new sections on muscle-building anabolic steroids and rigor mortis. The chapter on skeletal muscles has a revised section on group actions of muscles. New to the chapter are an exhibit of the muscles of the floor of the oral cavity and sections dealing with sports injuries.

A major organizational change in this edition is the placement of the chapter on surface anatomy immediately following the muscular system, rather than at the end of the book. However, because of the way the chapter is structured, that is, separate exhibits of various regions of the body with accompanying art, surface anatomy can be studied *anywhere* in the book. The approach used to study surface anatomy is left to the option of the instructor and student. Many new surface anatomy features have been added.

The student is next introduced to the cardiovascular and lymphatic systems. Blood vessels have been placed in a separate chapter; they are no longer considered together with the heart. The chapter on blood contains newly added material on blood substitutes, multiple myeloma, bone marrow transplantation, complete blood count (CBC), and apheresis. The chapter on the heart contains new sections on cardiac tamponade, sinus node dysfunction, the artificial heart, coronary artery disease (CAD), coronary artery spasm, congestive heart failure

(CHF), and cor pulmonale (CP). New to the chapter on blood vessels are discussions of digital subtraction angiography (DSA), the histology of capillaries, blood reservoirs, and deep venous thrombosis (DVT). Flow diagrams indicating arterial distribution and venous drainage for the major regions of the body have also been added. The chapter on the lymphatic system contains new sections on the ultrastructure of lymph capillaries, hypersensitivity (allergy), tissue rejection, autoimmune diseases, and acquired immune deficiency syndrome (AIDS).

The next major area of emphasis is the nervous system. Students are introduced to the structure and function of nervous tissue, the spinal cord and spinal nerves, the brain and cranial nerves, and the autonomic nervous system. The chapter on nervous tissue contains a new section on the organization of neurons into neuronal pools. The chapter on the spinal cord contains revised discussions of nerve damage to plexuses, spinal cord injury, and shingles. New to the chapter on the brain is material on the blood-cerebrospinal fluid barrier, cavitron ultrasonic surgical aspirator (CUSA), brain injuries, brain tumors, positron emission tomography (PET), amyotrophic lateral sclerosis (ALS), brain electrical activity mapping (BEAM), rabies, Reye's syndrome (RS), and Alzheimer's disease (AD). The autonomic nervous system chapter now contains discussions of Raynaud's disease and nicotinic, muscarinic, alpha, and beta receptors.

Sensory structures are considered next. New to this chapter are sections on characteristics of sensations, classification of receptors, histology of muscle spindles and tendon organs, somatosensory cortex, motor cortex, corneal transplantation, radial keratotomy, senile macular degeneration (SMD), cochlear implants, and deafness.

Attention is then turned to the endocrine system. All hormone terminology has been updated, and disorders related to endocrine glands have now been integrated throughout the chapter as clinical applications. New to the chapter are sections dealing with control of hormonal secretions.

The chapter on the respiratory system contains new sections on snoring, nebulization, respiratory failure, bronchitis, pulmonary embolism (PE), and pulmonary edema.

New topics added to the chapter on the digestive system include gastroscopy, liver biopsy, achalasia, heartburn, vomiting, jaundice, mumps, hepatitis, and bulimia.

The urinary system chapter has been expanded by new discussions of extracorporeal shock wave lithotripsy (ESWL) and percutaneous ultrasonic lithotripsy (PUL).

The sections dealing with spermatogenesis, oogenesis, sexual intercourse, and birth control have been moved from the developmental anatomy chapter into the chapter on the reproductive systems. The chapter on the reproductive systems has been further strengthened by the addition of material on the structure of the inguinal canal and inguinal hernia, semen analysis, artificial insemination, premenstrual syndrome (PMS), and toxic shock syndrome (TSS).

The chapter on developmental anatomy now contains new sections on twinning, embryo transfer, and chorionic villi sampling (CVS).

SPECIAL FEATURES

As in previous editions, the book contains numerous learning aids. Users of the book have cited the pedagogical aids as one of the book's many strengths. All of the tested and successful learning aids of previous editions have been kept in the fourth edition, and several new ones have been added. These special features are:

1. Student Objectives. Each chapter opens with a comprehensive list of student objectives. Each objective describes a knowledge or skill students should acquire while studying the chapter. (See **Note to the Student** for an explanation of how the objectives can be used.)

2. Study Outline. A study outline at the end of each chapter provides a brief summary of major topics. This section consolidates the essential points covered in the chapter so that students can recall and relate the points to one another. Page numbers have been added to the outlines so that topics within chapters can be located more easily.

3. Review Questions. Review questions at the end of each chapter provide a check to see if the objectives stated at the beginning of the chapter have been met. After answering the questions, students should reread the objectives to determine whether they have met the goals.

4. Exhibits. Health-science students are generally expected to learn a great deal about the anatomy of certain organ systems, specifically, skeletal muscles, articulations, blood vessels, lymph nodes, and nerves. To avoid interrupting the discussion of concepts and to organize the data, anatomical details have been presented in tabular form in exhibits, most of which are accompanied by illustrations. This method allows a clearer statement of general concepts in the narrative and organizes the specific details to be learned. New summary exhibits have been added to the fourth edition.

5. Applications to Health. Abnormalities of structure or function are grouped at the end of appropriate chapters in sections entitled "Applications to Health." These sections provide a review of normal body processes as well as demonstrations of the importance of the study of anatomy and physiology to a career in any of the health fields. All disorders have been updated, and many new ones have been added.

6. Phonetic Pronunciations. Throughout the text, phonetic pronunciations are provided in parentheses for selected terms. These pronunciations are given at the point where the terms are introduced and are repeated in the glossary. The **Note to the Student** explains the pronunciation key.

7. Key Medical Terms. Glossaries of selected medical terms appear at the end of appropriate chapters; these listings are entitled "Key Medical Terms." All of the glossaries have been revised for the fourth edition.

8. Clinical Applications. Throughout the text, clinical applications are integrated within the narrative. Many new ones have been added.

9. Line Art. The anatomically accurate line drawings in the book are large so that details are easily seen. In the fourth edition, many new illustrations have been added and full color is used *throughout* to differentiate structures and regions.

10. Photographs. The photographs amplify the narrative and the line drawings. Numerous photomicrographs (in full color), newly added scanning electron micrographs, and transmission electron micrographs enhance the histological discussions. Color photographs of specimens clarify gross anatomy discussions and have been added throughout. Photographs of regional dissections, most in full color, have also been added to the fourth edition.

11. Appendixes. Two new appendixes have been added to the fourth edition. Appendix A, "Abbreviations," is an alphabetical list of commonly encountered medical abbreviations. Appendix B, "Eponyms Used in This Text," is an alphabetical list of commonly encountered eponyms and the corresponding current terminology. Eponyms are cited in the text in parentheses immediately following the preferred current terms.

12. Selected Readings. The revised bibliography lists current references for instructors and students.

13. Glossary. The comprehensive glossary of terms has been updated for the fourth edition.

SUPPLEMENTARY MATERIALS

The following supplementary items are available to accompany the fourth edition of *Principles of Human Anatomy*.

1. Learning Guide. Written by Kathleen S. Prezbindowski and Gerard J. Tortora, the *Learning Guide* is designed to help students *learn* anatomy. Rather than asking for mere repetition of the text material, the *Learning Guide* requires students to perform activities such as labeling a drawing or coloring parts of a diagram, as well as answering multiple-choice, matching, and fill-in questions. The book's emphasis is on *using* new information to learn it. To enhance the effectiveness of the exercises, answers are provided for key-concept exercises so that the student has immediate feedback. A Mastery Test at the end of each chapter provides the student with a means of evaluating his or her learning of the chapter material and also gives practice for classroom testing situations.

2. Instructor's Manual. Each chapter in the complimentary *Instructor's Manual,* prepared by Professor Tortora, consists of a chapter overview, a list of instructional concepts relating to the chapter, suggested problem-solving essay questions, and lists of audiovisual materials relating to the chapter topic.

3. Transparencies. Seventy-five full-color transparencies are available. Illustrations from the text that are often shown and discussed in class are the subjects for the transparencies, and special care has been taken to make them clear and usable as overhead projections.

ACKNOWLEDGMENTS

I wish to thank the following people for their contribution to the fourth edition of *Principles of Human Anatomy:* Burton R. Buckley, Northwestern State University of Louisiana; Gerald Collier, San Diego State University; Harry N. Cunningham, Penn State University; Thomas G. Froiland, Northern Michigan University; Robert F. Green, Ventura College; George Harrett-Aigla, City College of San Francisco; Charles K. Levy, Boston University; Suzanne L. Ray, City College of San Francisco; Gary S. Resnick, Saddleback College; Allan Scholz, Eastern Washington University; Lawrence D. Wilkin, University of Iowa; and Louis J. Zanella, Rhode Island Junior College.

Continuing thanks go to the many instructors and students who were kind enough to send me their suggestions for improvement and those people who worked on the three previous editions. Gratitude is also extended for the contributions of the individuals and organizations whose names appear with the photographs in the text. Special thanks to Dr. Michael H. Ross of the University of Florida for providing the excellent color photomicrographs and Dr. Michael C. Kennedy of Hahnemann Medical College for his help in reviewing the illustration program. Finally, for typing drafts of the manuscript and numerous other clerical duties associated with the task of putting together a textbook, thanks to Geraldine C. Tortora.

As the acknowledgments indicate, the participation of many individuals of diverse talent and expertise is required in the production of a textbook of this scope and complexity. For this reason, readers and users of the fourth edition are invited to send their reactions and suggestions to me so that plans can be formulated for subsequent editions.

Gerard J. Tortora

Biology Department
Bergen Community College
400 Paramus Road
Paramus, N.J. 07652

Note to the Student

At the beginning of each chapter is a listing of **Student Objectives.** Before you read the chapter, please read the objectives carefully. Each objective is a statement of a skill or knowledge that you should acquire. To meet these objectives, you will have to perform several activities. Obviously, you must read the chapter carefully. If there are sections of the chapter that you do not understand after one reading, you should reread those sections before continuing. In conjunction with your reading, pay particular attention to the figures and exhibits; they have been carefully coordinated with the textual narrative.

At the end of each chapter are two and sometimes three other learning guides that you may find useful. The first, **Study Outline,** is a concise summary of important topics discussed in the chapter. This section is designed to consolidate the essential points covered in the chapter, so that you may recall and relate them to one another. The second guide, **Review Questions,** is a series of questions designed specifically to help you master the objectives. A third aid, **Key Medical Terms,** appears in some chapters. This is a listing of terms designed to build your medical vocabulary. After you have answered the review questions, you should return to the beginning of the chapter and reread the objectives to determine whether you have achieved the goals.

As a further aid, we have included pronunciations for many terms that may be new to you. These appear in parentheses immediately following the new words, and they are repeated in the glossary of terms at the back of the book. (Of course, since there will always be some conflict among medical personnel and dictionaries about pronunciation, you will come across variations in different sources.) Look at the words carefully and say them out loud several times. Learning to pronounce a new word will help you remember it and make it a useful part of your medical vocabulary. Take a few minutes now to read the following pronunciation key, so it will be familar as you encounter new words. The key is repeated at the beginning of the glossary of terms.

PRONUNCIATION KEY

1. The strongest accented syllable appears in capital letters, for example, bilateral (bī-LAT-er-al) and diagnosis (dī-ag-NŌ-sis).

2. If there is a secondary accent, it is noted by a single quote mark('), for example, constitution (kon'-sti-TOO-shun) and physiology (fiz'-ē-OL-ō-jē). Any additional secondary accents are also noted by a single quote mark, for example, decarboxylation (dē'-kar-bok'-si-LĀ-shun).

3. Vowels marked with a line above the letter are pronounced with the long sound as in the following common words.

ā as in *māke*
ē as in *bē*
ī as in *īvy*
ō as in *pōle*

4. Vowels not so marked are pronounced with the short sound as in the following words.

e as in *bet*
i as in *sip*
o as in *not*
u as in *bud*

5. Other phonetic symbols are used to indicate the following sounds:

a as in *above*
oo as in *sue*
yoo as in *cute*
oy as in *oil*

1

An Introduction to the Human Body

Student Objectives

Define anatomy, with its subdivisions, and physiology.

Define each of the following levels of structural organization that make up the human body: chemical, cellular, tissue, organ, system, and organismic.

Identify the principal systems of the human body, list the representative organs of each system, and describe the function of each system.

List and define several important life processes of humans.

Describe the general anatomical characteristics of the structural plan of the human body.

Define the anatomical position.

Compare common and anatomical terms used to describe various regions of the human body.

Define directional terms used in association with the human body.

Define the common anatomical planes that may be passed through the human body.

Distinguish a cross section, frontal section, and midsagittal section.

List by name and location the principal body cavities and the organs contained within them.

Describe how the abdominopelvic cavity is divided into nine regions and four quadrants.

Define the relationship of radiographic anatomy to the diagnosis of disease.

Contrast the principles employed in conventional radiography and in computed tomography (CT) scanning.

Point out the differences between a roentgenogram and a computed tomography (CT) scan.

Explain how the dynamic spatial reconstructor (DSR) operates and its diagnostic importance in radiographic anatomy.

Describe the principle of magnetic resonance imaging (MRI) and its diagnostic importance.

Define the common metric units of length, mass, and volume, and their U.S. equivalents, that are used in measuring the human body.

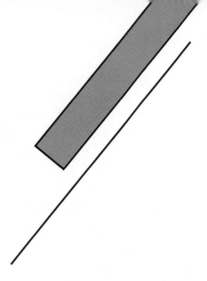

You are about to begin a study of the human body in order to learn how your body is organized and how it functions. The study of the human body involves many branches of science. Each contributes to a comprehensive understanding of how your body normally works and what happens when it is injured, diseased, or placed under stress.

ANATOMY AND PHYSIOLOGY DEFINED

Two branches of science that will help you understand your body parts and functions are anatomy and physiology. **Anatomy** (a-NAT-ō-mē; *anatome* = to dissect) refers to the study of *structure* and the relationships among structures. Anatomy is a broad science, and the study of structure becomes more meaningful when specific aspects of the science are considered. **Surface anatomy** is the study of the form (morphology) and markings of the surface of the body. **Gross (macroscopic) anatomy** deals with structures that can be studied without a microscope. **Systemic (systematic) anatomy** covers specific systems of the body, such as the system of nerves, spinal cord, and brain or the system of heart, blood vessels, and blood. **Regional anatomy** deals with a specific region of the body, such as the head, neck, chest, or abdomen. **Developmental anatomy** is the study of development from the fertilized egg to the adult form. **Embryology** is generally restricted to the study of development from the fertilized egg through the eighth week in utero. Other branches of anatomy are **pathological** (*patho* = disease) **anatomy,** the study of structural changes caused by disease; **histology** (*histo* = tissue), the microscopic study of the structure of tissues; and **cytology** (*cyt* = cell), the study of cells. **Radiographic anatomy,** the study of the structure of the body using radiographic (x-ray) techniques, is discussed in detail later in the chapter.

Whereas anatomy and its branches deal with structures of the body, **physiology** (fiz'-ē-OL-ō-jē) deals with *functions* of the body parts, that is, how the body parts work. However, the two sciences cannot be completely sepa-

rated. Since each part of the body is custom-modeled to carry out a particular set of functions, structure and function are studied together. The structure of a part often determines the functions it will perform, and, in turn, body functions often influence the size, shape, and health of the structures.

LEVELS OF STRUCTURAL ORGANIZATION

The human body consists of several levels of structural organization that are associated with one another in several ways. The lowest level of organization, the **chemical level,** includes all chemical substances essential for maintaining life. All these chemicals are made up of atoms joined together in various ways (Figure 1-1).

The chemicals, in turn, are put together to form the next higher level of organization: the **cellular level. Cells** are the basic structural and functional units of an organism. Among the many kinds of cells in your body are muscle cells, nerve cells, and blood cells. Figure 1-1 shows several isolated cells from the lining of the stomach. Each has a different structure, and each performs a different function.

The next higher level of structural organization is the **tissue level. Tissues** are made up of groups of similarly specialized cells and their intercellular material that perform certain special functions. When the isolated cells shown in Figure 1-1 are joined together, they form a tissue called epithelium, which lines the stomach. Each cell in the tissue has a specific function. Mucous cells produce mucus, a secretion that lubricates food as it passes through the stomach. Parietal cells produce acid in the stomach. Chief cells produce enzymes needed to digest proteins. Other examples of tissues in your body are muscle tissue, connective tissue, and nervous tissue.

In many places in the body, different kinds of tissues are joined together to form an even higher level of organization: the **organ level. Organs** are structures of definite form and function composed of two or more different

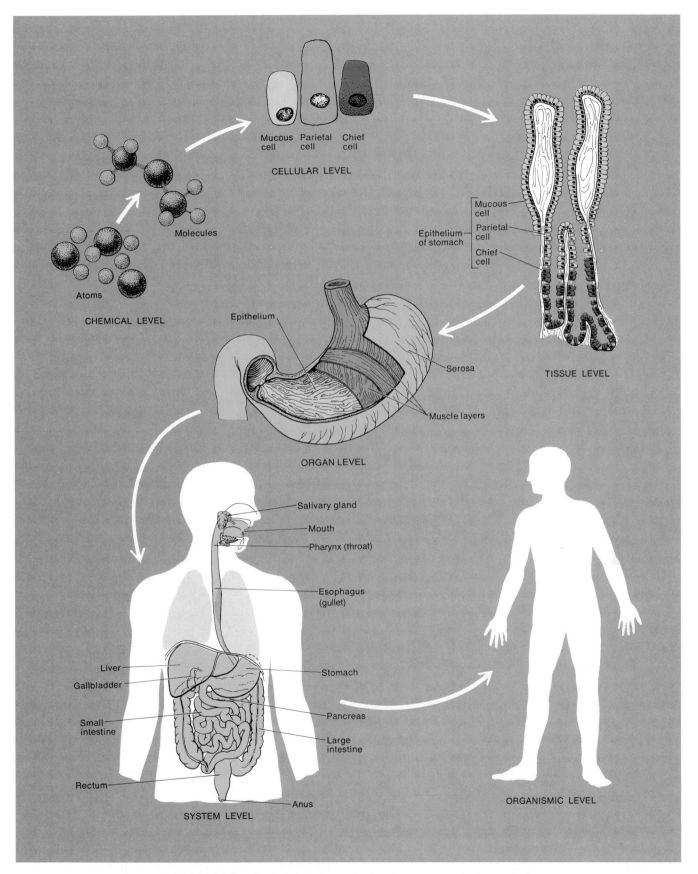

FIGURE 1-1 Levels of structural organization that compose the human body.

tissues. Organs usually have a recognizable shape. Examples of organs are the heart, liver, lungs, brain, and stomach. Figure 1-1 shows three of the tissues that make up the stomach. The serous layer (also called the serosa) around the outside protects the stomach and reduces friction when the stomach moves and rubs against other organs. The muscle tissue layers of the stomach contract to mix food and pass it on to the next digestive organ. The epithelial tissue layer lining the stomach produces mucus, acid, and enzymes.

The next higher level of structural organization in the body is the **system level. A system** consists of an association of organs that have a common function. The digestive system, which functions in the breakdown of food, is composed of the mouth, saliva-producing glands called salivary glands, pharynx (throat), esophagus (gullet), stomach, small intestine, large intestine, rectum, liver, gallbladder, and pancreas.

The highest level of organization is the **organismic level.** All the parts of the body functioning with one another constitute the total **organism**—one living individual.

In the chapters that follow, you will examine the anatomy and physiology of the major body systems. Exhibit 1-1 illustrates these systems, their representative organs, and their general functions. Unless otherwise stated, male and female bodies are similar. The systems are presented in the exhibit in the order in which they are discussed in later chapters.

LIFE PROCESSES

All living forms have certain characteristics that distinguish them from nonliving things. Following are several of the more important life processes of humans:

1. Metabolism. Metabolism is the sum of all the chemical processes that occur in the body. One phase of metabolism, called **catabolism,** provides us with energy needed to sustain life. The other phase of metabolism, called **anabolism,** uses the energy to make various substances that form the body's structural and functional components. Among the many processes contributing to metabolism are: *ingestion,* the taking in of foods; *digestion,* the breaking down of foods into simpler forms that can be used by cells; *absorption,* the uptake of substances by cells; *assimilation,* the buildup of absorbed substances into different substances required by cells; *respiration,* the generation of energy, usually in the presence of oxygen with the release of carbon dioxide; *secretion,* the production and release of a useful substance by cells; and *excretion,* the elimination of wastes produced as a result of metabolism.

2. Excitability. Excitability is our ability to sense changes within and around us. We do this by responding to stimuli (changes in the environment) such as light,

pressure, heat, noises, chemicals, and pain. We constantly respond to our environment to make adjustments that maintain health.

3. Conductivity. Conductivity refers to the ability to carry the effect of a stimulus from one part of a cell to another. This characteristic is highly developed in nerve cells and is developed to a great extent in muscle cells.

4. Contractility. Contractility is the capacity of cells to undergo shortening and change form. Muscle cells exhibit a high degree of contractility.

5. Growth. Growth refers to an increase in size from the inside out. It may be due to an increase in the number of cells or an increase in the size of existing cells.

6. Differentiation. Differentiation is the basic mechanism whereby unspecialized cells change to specialized cells. Specialized cells have structural and functional characteristics that differ from cells from which they originated. Through differentiation, a fertilized egg develops into an embryo, a zygote, and later an organism consisting of a variety of diversified cells.

7. Reproduction. Reproduction refers to either the formation of new cells for growth, repair, or replacement, or the production of a new individual. Through reproduction, life is transmitted from one generation to the next.

STRUCTURAL PLAN

The human body has certain general **anatomical characteristics** that will help you to understand its overall structural plan. For example, humans have a **backbone (vertebral column),** a characteristic that places them in a large group of organisms called *vertebrates.* Another characteristic of the body's organization is that it basically resembles a **tube within a tube** construction. The outer tube is formed by the body wall; the inner tube is the digestive tract. Moreover, humans are for the most part **bilaterally symmetrical**—externally the left and right sides of the body are mirror images.

ANATOMICAL POSITION AND REGIONAL NAMES

In all anatomical texts and charts, descriptions of any region or part of the human body assume that the body is in a specific position, called the **anatomical position,** so that directional terms are clear and any part can be related to any other part. In the anatomical position, the subject is standing erect facing the observer, the upper extremities are placed at the sides, and the palms of the hands are turned forward (Figure 1-2). Once the body is in the anatomical position, it is easier to visualize and understand how it is organized into various regions. The common and anatomical terms, in parentheses, of the principal body regions are also presented in Figure 1-2.

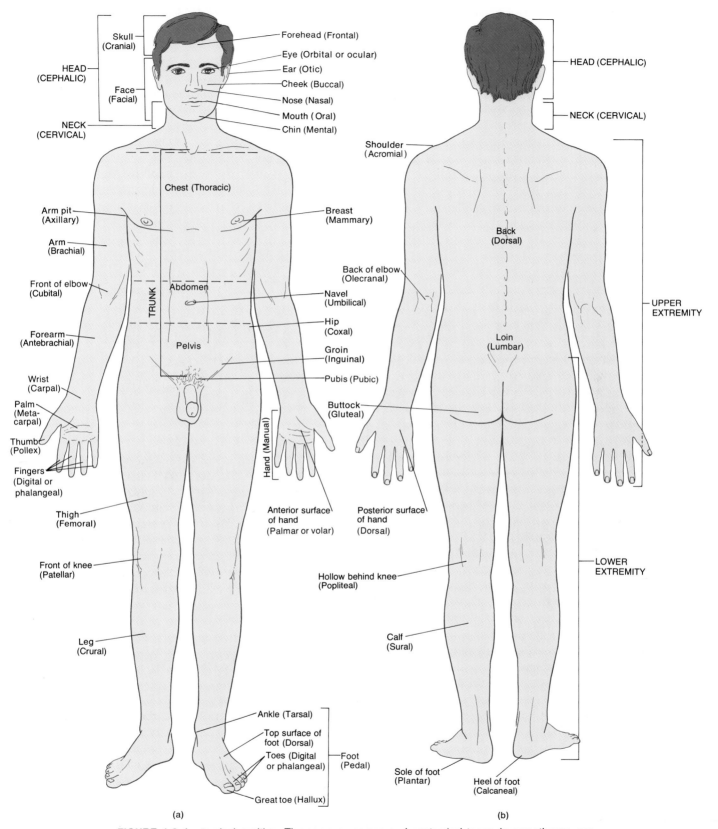

FIGURE 1-2 Anatomical position. The common names and anatomical terms, in parentheses, are indicated for many of the regions of the body. (a) Anterior view. (b) Posterior view.

EXHIBIT 1-1   PRINCIPAL SYSTEMS OF HUMAN BODY: REPRESENTATIVE ORGANS AND FUNCTIONS

1. Integumentary

Definition: The skin and structures derived from it, such as hair, nails, and sweat and oil glands.

Function: Helps regulate body temperature, protects the body, eliminates wastes, synthesizes vitamin D, and receives certain stimuli such as temperature, pressure, and pain.

2. Skeletal

Definition: All the bones of the body, their associated cartilages, and the joints of the body.

Function: Supports and protects the body, provides leverage, produces blood cells, and stores minerals.

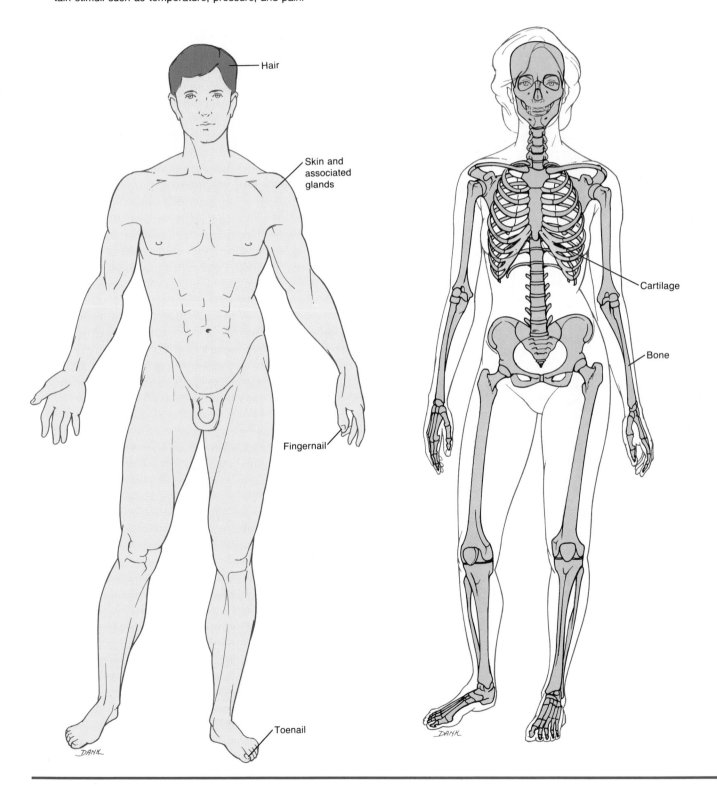

3. Muscular

Definition: All the muscle tissue of the body, including skeletal (shown in the illustration), visceral, and cardiac.

Function: Participates in bringing about movement, maintains posture, and produces heat.

4. Cardiovascular

Definition: Blood, heart, and blood vessels.

Function: Distributes oxygen and nutrients to cells, carries carbon dioxide and wastes from cells, maintains the acid–base balance of the body, protects against disease, prevents hemorrhage by forming blood clots, and helps regulate body temperature.

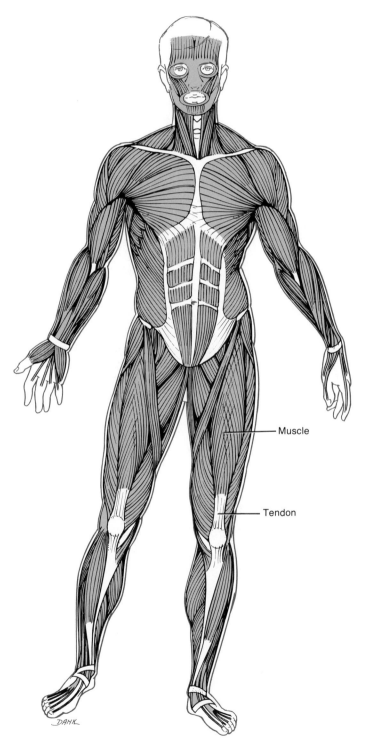

Muscle

Tendon

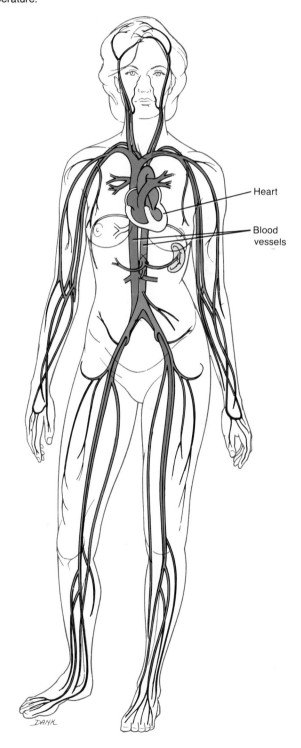

Heart

Blood vessels

EXHIBIT 1-1   (*Continued*)

5. Lymphatic

Definition: Lymph, lymph nodes, lymph vessels, and lymph glands, such as the spleen, thymus gland, and tonsils.

Function: Returns proteins and plasma to the cardiovascular system, transports fats from the digestive system to the cardiovascular system, filters the blood, produces white blood cells, and protects against disease.

6. Nervous

Definition: Brain, spinal cord, nerves, and sense organs, such as the eye and ear.

Function: Regulates body activities through nerve impulses.

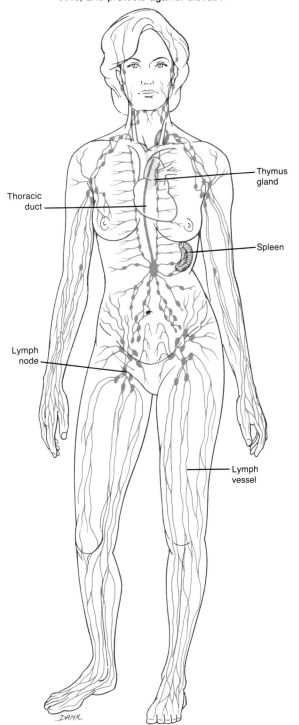

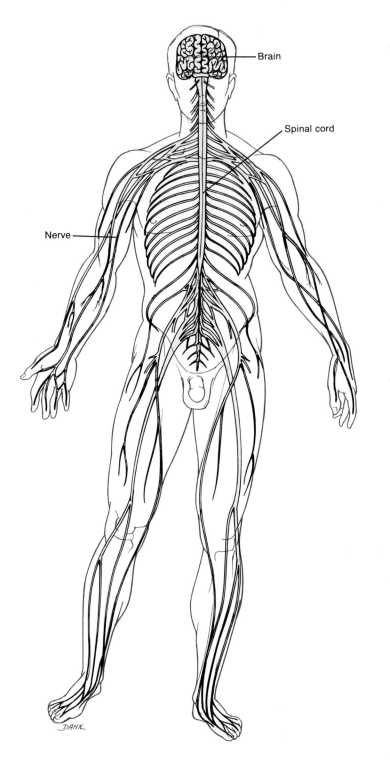

7. Endocrine

Definition: All glands that produce hormones.

Function: Regulates body activities through hormones transported by the cardiovascular system.

8. Respiratory

Definition: The lungs and a series of passageways leading into and out of them.

Function: Supplies oxygen, eliminates carbon dioxide, and helps regulate the acid–base balance of the body.

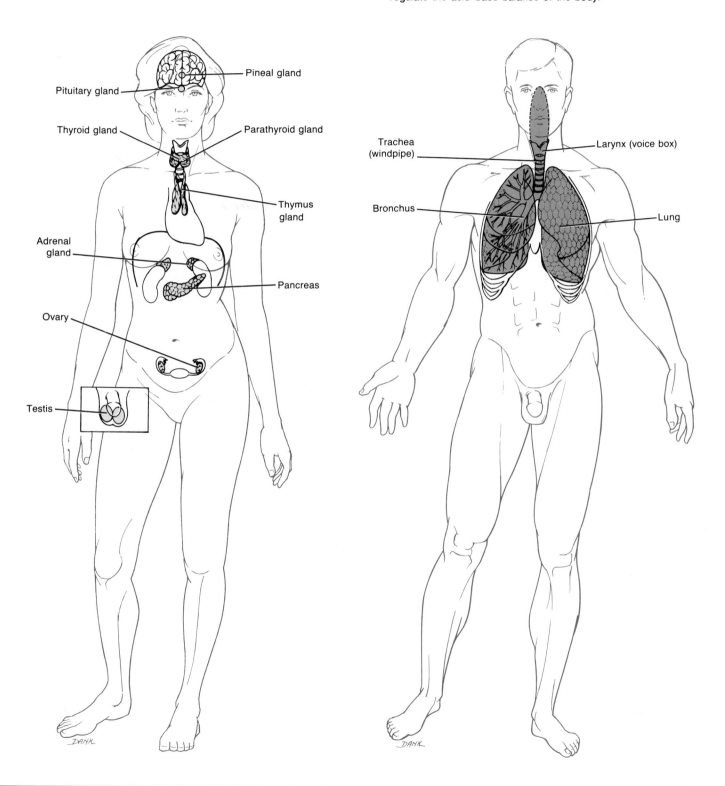

EXHIBIT 1-1 (*Continued*)

9. Digestive

Definition: A long tube and associated organs such as the salivary glands, liver, gallbladder, and pancreas.

Function: Performs the physical and chemical breakdown of food for use by cells and eliminates solid wastes.

10. Urinary

Definition: Organs that produce, collect, and eliminate urine.

Function: Regulates the chemical composition of blood, eliminates wastes, regulates fluid and electrolyte balance and volume, and helps maintain the acid–base balance of the body.

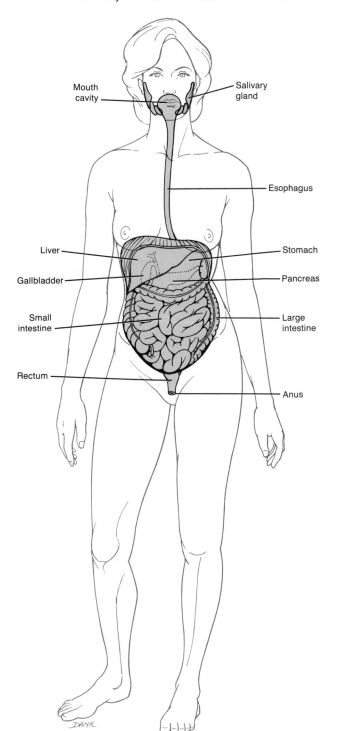

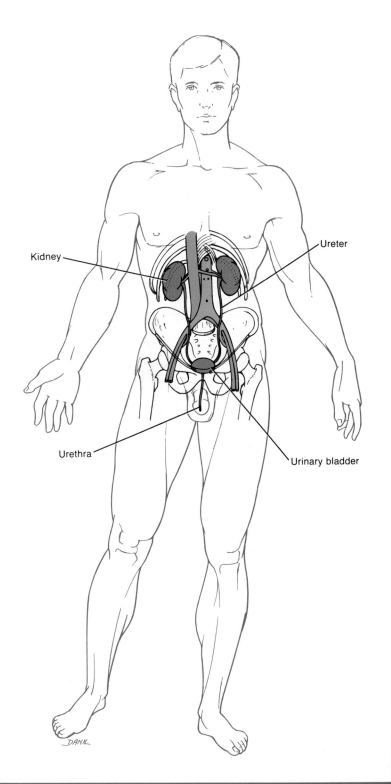

11. Reproductive

Definition: Organs (testes and ovaries) that produce reproductive cells (sperm and ova) and organs that transport and store reproductive cells.

Function: Reproduces the organism.

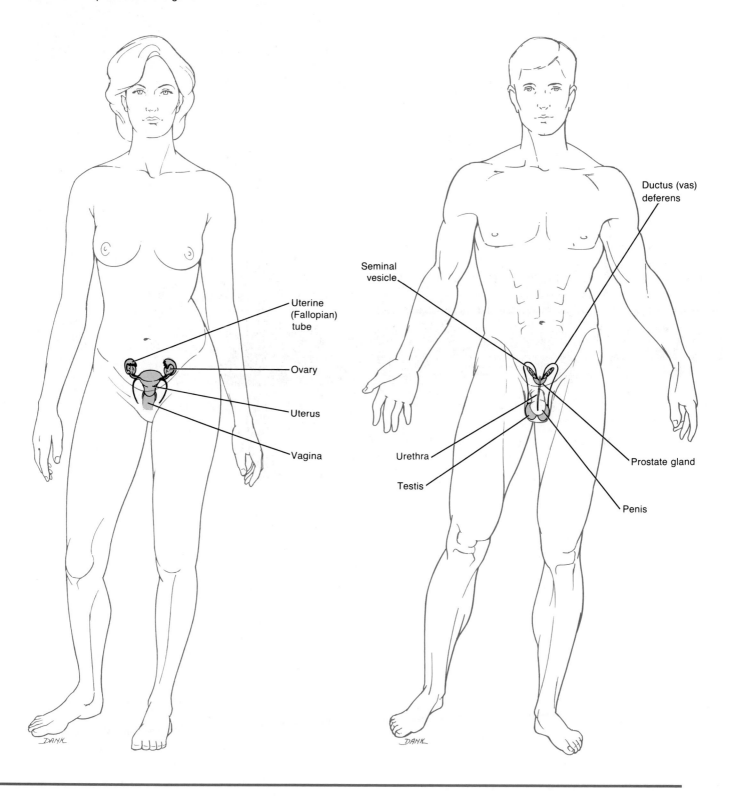

Uterine (Fallopian) tube

Ovary

Uterus

Vagina

Ductus (vas) deferens

Seminal vesicle

Urethra

Testis

Prostate gland

Penis

EXHIBIT 1-2 DIRECTIONAL TERMS

TERM	DEFINITION	EXAMPLE
Superior (cephalad or **craniad)**	Toward the head or the upper part of a structure; generally refers to structures in the trunk.	The heart is superior to the liver.
Inferior (caudad)	Away from the head or toward the lower part of a structure; generally refers to structures in the trunk.	The stomach is inferior to the lungs.
Anterior (ventral)	Nearer to or at the front of the body.	The sternum is anterior to the heart.
Posterior (dorsal)	Nearer to or at the back of the body.	The esophagus is posterior to the trachea.
Medial	Nearer the midline of the body or a structure.	The ulna is on the medial side of the forearm.
Lateral	Farther from the midline of the body or a structure.	The ascending colon of the large intestine is lateral to the urinary bladder.
Intermediate	Between two structures, one of which is medial and the other lateral.	The ring finger is intermediate between the little (medial) and middle (lateral) fingers.
Ipsilateral	On the same side of the body.	The gallbladder and ascending colon of the large intestine are ipsilateral.
Contralateral	On the opposite side of the body.	The ascending and descending colons of the large intestine are contralateral.
Proximal	Nearer the attachment of an extremity to the trunk or a structure; nearer to the point of origin.	The humerus is proximal to the radius.
Distal	Farther from the attachment of an extremity to the trunk or a structure; farther from the point of origin.	The phalanges are distal to the carpals.
Superficial	Toward or on the surface of the body.	The muscles of the thoracic wall are superficial to the viscera in the thoracic cavity (see Figure 1-7b).
Deep	Away from the surface of the body.	The muscles of the arm are deep to the skin of the arm.
Parietal	Pertaining to the outer wall of a body cavity.	The parietal pleura forms the outer layer of the pleural sacs that surround the lungs (see Figure 1-7b).
Visceral	Pertaining to the covering of an organ (viscus).	The visceral pleura forms the inner layer of the pleural sacs and covers the external surface of the lungs (see Figure 1-7b).

DIRECTIONAL TERMS

In order to explain exactly where various body structures are located relative to each other, anatomists use certain **directional terms.** If you want to point out the sternum (breastbone) to someone who knows where the clavicle (collarbone) is, you can say that the sternum is inferior (farther away from the head) and medial (toward the middle of the body) to the clavicle. As you can see, using the terms *inferior* and *medial* avoids a great deal of complicated description. Many directional terms are defined in Exhibit 1-2. The parts of the body used in the examples are labeled in Figure 1-3. Studying the exhibit and the figure together should make clear to you many of the directional relationships among various body parts.

PLANES AND SECTIONS

The structural plan of the human body may also be discussed with respect to **planes** (imaginary flat surfaces) that pass through it. Several of the commonly used planes are illustrated in Figure 1-4. A **midsagittal (median) plane** is a vertical plane that passes through the midline of the body and divides the body or organs into equal right and left sides. A **sagittal (SAJ-i-tal) plane,** also called a **parasagittal** (*para* = near) **plane,** is a plane parallel to a midsagittal plane that divides the body or organs into unequal left and right portions. A **frontal (coronal; kō-RŌ-nal) plane** is a plane at a right angle to a midsagittal (or sagittal) plane that divides the body or organs into anterior and posterior portions. Finally, a **horizontal**

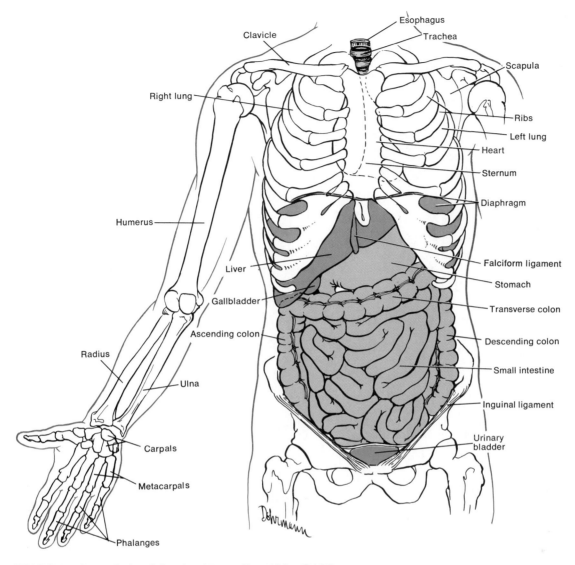

FIGURE 1-3 Anatomical and directional terms. By studying Exhibit 1-2 with this figure, you should gain an understanding of the meanings of the terms *superior, inferior, anterior, posterior, medial, lateral, intermediate, ipsilateral, contralateral, proximal,* and *distal.*

(transverse) plane is a plane that is parallel to the ground, that is, at a right angle to the midsagittal, sagittal, and frontal planes. It divides the body or organs into superior and inferior portions.

When you study a body structure, you will often view it in **section,** that is, look at the flat surface resulting from a cut made through the three-dimensional organ. It is important to know the plane of the section so that you can understand the anatomical relationship of one part to another. Figure 1-5 indicates how three different sections—a **cross section,** a **frontal section,** and a **midsagittal section**—are made through different parts of the brain.

BODY CAVITIES

Spaces within the body that contain internal organs are called **body cavities.** Specific cavities may be distinguished if the body is divided into right and left halves. Figure 1-6 shows the two principal body cavities. The **dorsal body cavity** is located near the posterior (dorsal) surface of the body. It is further subdivided into a **cranial cavity,** which is a bony cavity formed by the cranial (skull) bones and contains the brain, and a **vertebral (spinal) canal,** which is a bony cavity formed by the vertebrae of the backbone and contains the spinal cord and the beginnings of spinal nerves.

The other principal body cavity is the **ventral body cavity.** This cavity, also known as the **coelom** (SĒ-lōm), is located on the anterior (ventral) aspect of the body. Its walls are composed of skin, connective tissue, bone, muscles, and serous membranes. The organs inside the ventral body cavity are called **viscera** (VIS-er-a). Like the dorsal body cavity, the ventral body cavity has two princi-

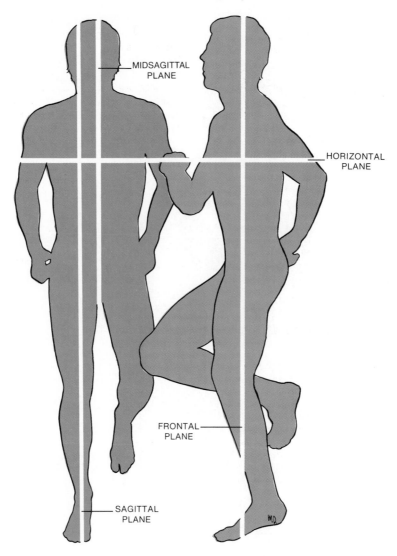

FIGURE 1-4 Planes of the human body.

pal subdivisions—an upper portion, called the **thoracic** (thō-RAS-ik) **cavity** (or chest cavity), and a lower portion, called the **abdominopelvic** (ab-dom'-i-nō-PEL-vik) **cavity.** The anatomical landmark that divides the ventral body cavity into the thoracic and abdominopelvic cavities is the muscular diaphragm.

The thoracic cavity contains several divisions. There are two **pleural cavities** (Figure 1-7). Each is a small potential space between the visceral pleura and parietal pleura, the membranes covering the lungs. The pleural cavities contain a fluid. The **mediastinum** (mē'-dē-as-TĪ-num) is the space, actually a mass of tissue, between the pleurae of the lungs that extends from the sternum to the vertebral column (Figure 1-7a, b). The mediastinum includes all of the contents of the thoracic cavity, except the lungs themselves. The **pericardial** (per'-ē-KAR-dē-al; *peri* = around; *cardi* = heart) **cavity** is a small potential space between the visceral pericardium and parietal pericardium, the membranes covering the heart (Figure 1-7b). The cavity contains a fluid.

The abdominopelvic cavity, as the name suggests, is divided into two portions, although no wall separates them (see Figure 1-6). The upper portion, the **abdominal cavity,** contains the stomach, spleen, liver, gallbladder, pancreas, small intestine, and most of the large intestine. The lower portion, the **pelvic cavity,** contains the urinary bladder, sigmoid colon, rectum, and the internal male or female reproductive organs. One way to mark the division between the abdominal and pelvic cavities is to draw an imaginary line from the symphysis pubis (anterior joint between hipbones) to the superior border of the sacrum (sacral promontory).

ABDOMINOPELVIC REGIONS

To describe the location of organs easily, the abdominopelvic cavity may be divided into the **nine regions** shown in Figure 1-8. Although some unfamiliar terms are used in describing the nine regions and their contents, follow

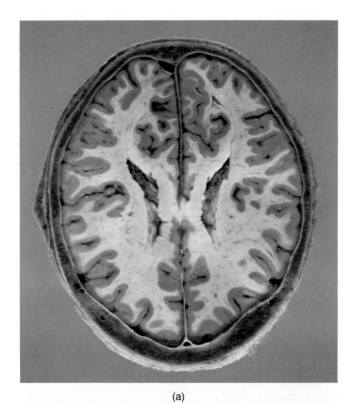

(a)

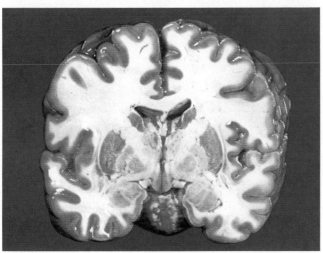

(b)

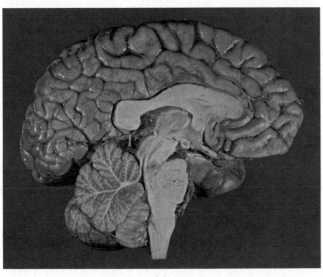

(c)

FIGURE 1-5 Sections through different parts of the brain. (a) Cross section through the brain. (Courtesy of Stephen A. Kieffer and E. Robert Heitzman, *An Atlas of Cross-Sectional Anatomy,* Harper & Row, Publishers, Inc., New York, 1979.) (b) Frontal section through the brain. (Courtesy of C. Yokochi and J. W. Rohen, *Photographic Anatomy of the Human Body,* 2nd ed., 1979, IGAKU-SHOIN, Ltd., Tokyo, New York.) (c) Midsagittal section through the brain. (Courtesy of C. Yokochi and J. W. Rohen, *Photographic Anatomy of the Human Body,* 2nd ed., 1979, IGAKU-SHOIN, Ltd., Tokyo, New York.)

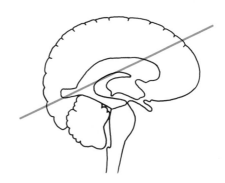

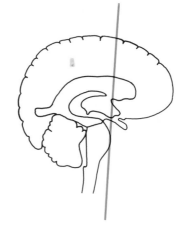

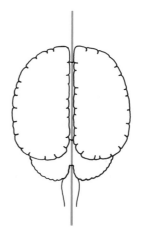

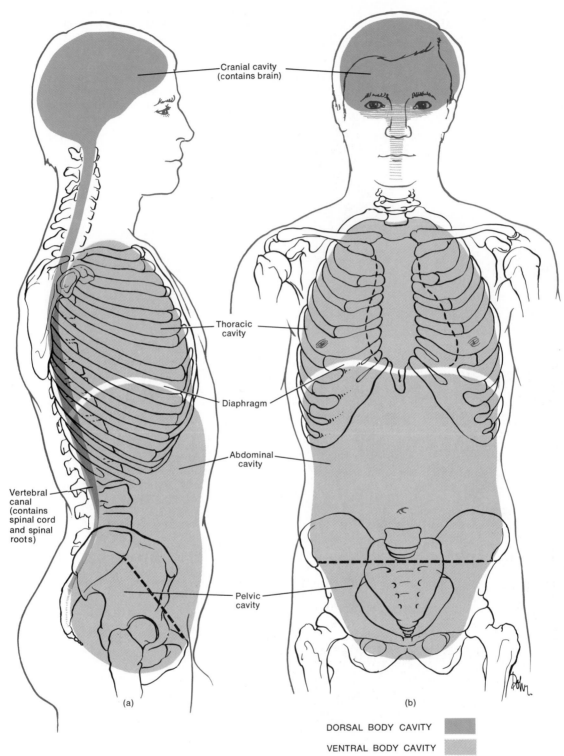

Cranial cavity
(contains brain)

Thoracic
cavity

Diaphragm

Abdominal
cavity

Vertebral
canal
(contains
spinal cord
and spinal
roots)

Pelvic
cavity

(a)

(b)

DORSAL BODY CAVITY

VENTRAL BODY CAVITY

FIGURE 1-6 Body cavities. (a) Location of the dorsal and ventral body cavities. (b) Subdivisions of the ventral body cavity.

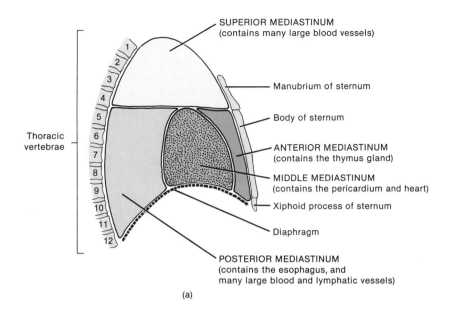

(a)

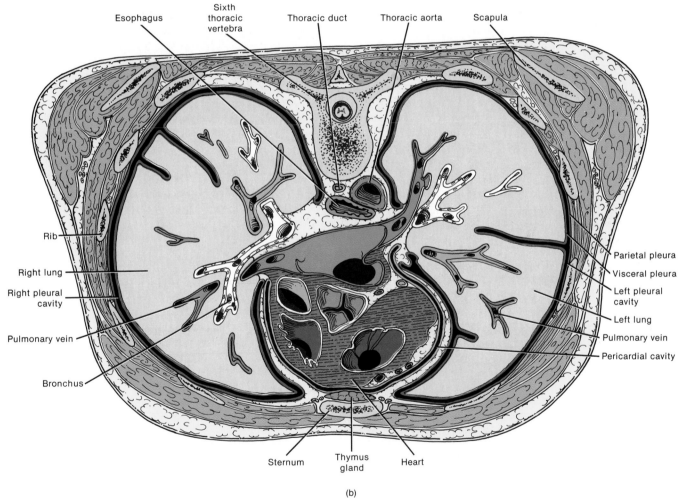

(b)

FIGURE 1-7 Mediastinum. The mediastinum is the space between the pleurae of the lungs that extends from the sternum to the vertebral column. (a) Subdivisions of the mediastinum seen in right lateral view. (b) Mediastinum seen in a cross section of the thorax. Some of the structures shown and labeled may be unfamiliar to you now. However, they are discussed in detail in later chapters.

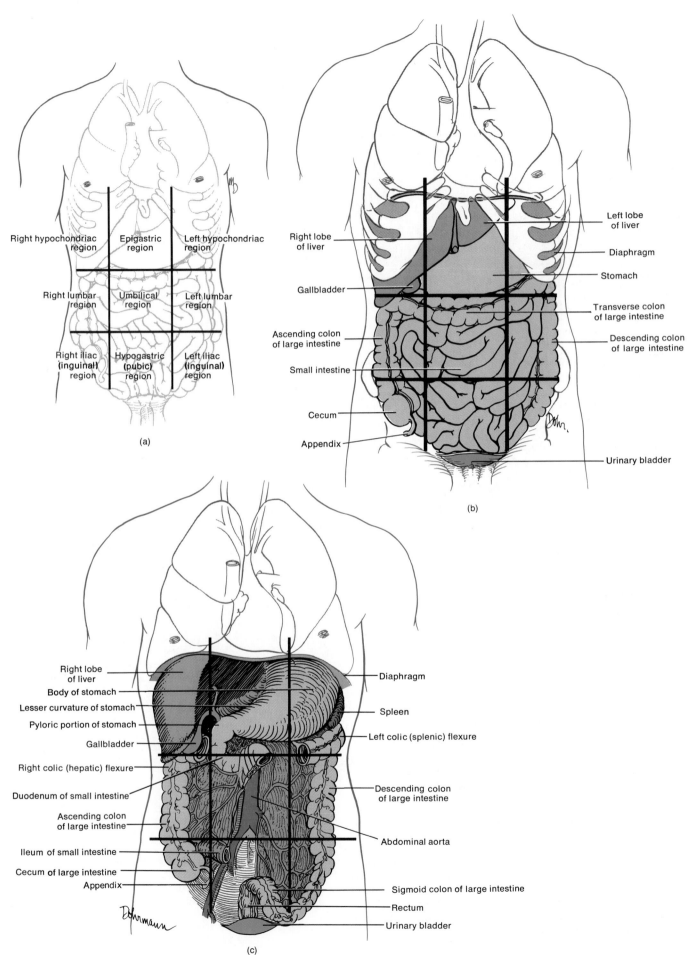

Right hypochondriac region

Epigastric region

Left hypochondriac region

Right lumbar region

Umbilical region

Left lumbar region

Right iliac (inguinal) region

Hypogastric (pubic) region

Left iliac (inguinal) region

(a)

Right lobe of liver

Gallbladder

Ascending colon of large intestine

Small intestine

Cecum

Appendix

Left lobe of liver

Diaphragm

Stomach

Transverse colon of large intestine

Descending colon of large intestine

Urinary bladder

(b)

Right lobe of liver

Body of stomach

Lesser curvature of stomach

Pyloric portion of stomach

Gallbladder

Right colic (hepatic) flexure

Duodenum of small intestine

Ascending colon of large intestine

Ileum of small intestine

Cecum of large intestine

Appendix

Diaphragm

Spleen

Left colic (splenic) flexure

Descending colon of large intestine

Abdominal aorta

Sigmoid colon of large intestine

Rectum

Urinary bladder

(c)

18

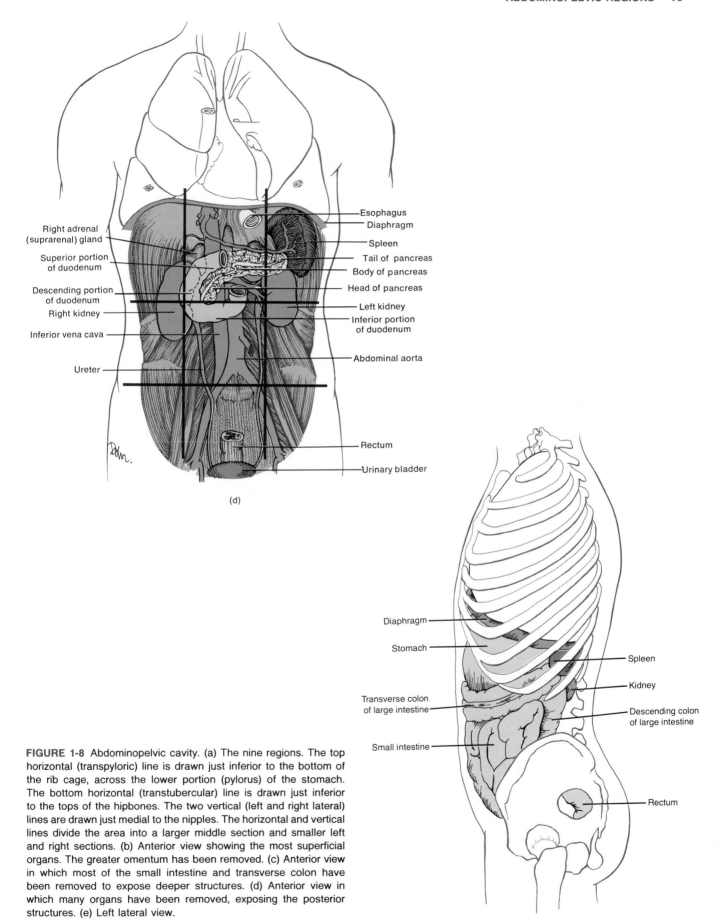

(d)

Right adrenal
(suprarenal) gland

Superior portion
of duodenum

Descending portion
of duodenum

Right kidney

Inferior vena cava

Ureter

Esophagus
Diaphragm
Spleen
Tail of pancreas
Body of pancreas
Head of pancreas
Left kidney
Inferior portion
of duodenum
Abdominal aorta

Rectum
Urinary bladder

Diaphragm
Stomach

Transverse colon
of large intestine

Small intestine

Spleen
Kidney
Descending colon
of large intestine

Rectum

(e)

FIGURE 1-8 Abdominopelvic cavity. (a) The nine regions. The top horizontal (transpyloric) line is drawn just inferior to the bottom of the rib cage, across the lower portion (pylorus) of the stomach. The bottom horizontal (transtubercular) line is drawn just inferior to the tops of the hipbones. The two vertical (left and right lateral) lines are drawn just medial to the nipples. The horizontal and vertical lines divide the area into a larger middle section and smaller left and right sections. (b) Anterior view showing the most superficial organs. The greater omentum has been removed. (c) Anterior view in which most of the small intestine and transverse colon have been removed to expose deeper structures. (d) Anterior view in which many organs have been removed, exposing the posterior structures. (e) Left lateral view.

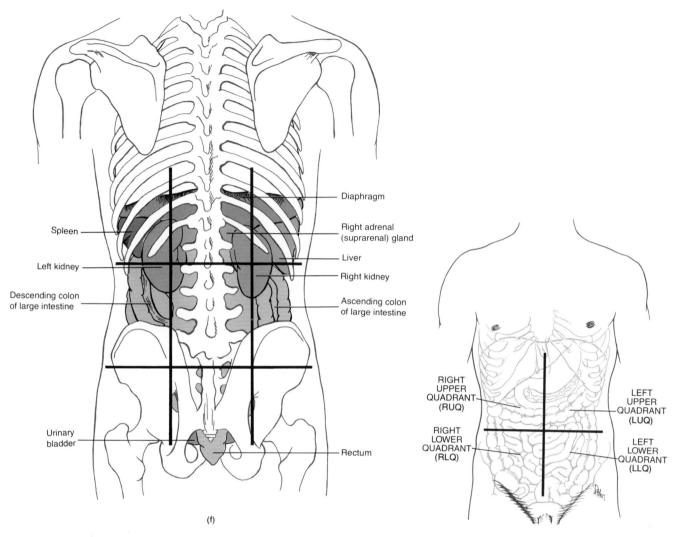

(f)

FIGURE 1-8 (*Continued*) Abdominopelvic cavity. (f) Posterior view.

FIGURE 1-9 Four quadrants of the abdominopelvic cavity. The two lines intersect at right angles at the umbilicus.

the descriptions as well as you can. When the organs are studied in detail in later chapters, they will have more meaning.

The **epigastric** (*epi* = above; *gaster* = stomach) **region** contains the left lobe and medial part of the right lobe of the liver, the pyloric part and lesser curvature of the stomach, the superior and descending portions of the duodenum, the body and superior part of the head of the pancreas, and the two adrenal (suprarenal) glands. The **right hypochondriac** (*hypo* = under; *chondro* = cartilage of ribs) **region** contains the right lobe of the liver, the gallbladder, and the superior third of the right kidney. The **left hypochondriac region** contains the body and fundus of the stomach, the spleen, the left colic (splenic) flexure, the superior two-thirds of the left kidney, and the tail of the pancreas. The **umbilical region** contains the middle of the transverse colon, the inferior part of

the duodenum, the jejunum, the ileum, the hilar regions of the kidneys, and the bifurcations (branching) of the abdominal aorta and inferior vena cava. The **right lumbar** (*lumbus* = loin) **region** contains the superior part of the cecum, the ascending colon, the right colic (hepatic) flexure, the inferior lateral portion of the right kidney, and the small intestine. The **left lumbar region** contains the descending colon, the inferior third of the left kidney, and the small intestine. The **hypogastric (pubic) region** contains the urinary bladder when full, the small intestine, and part of the sigmoid colon. The **right iliac (right inguinal) region** contains the lower end of the cecum, the appendix, and the small intestine. The term *iliacus* refers to the superior portion of the pelvic bone (hipbone). The **left iliac (left inguinal) region** contains the junction of the descending and sigmoid parts of the colon and the small intestine.

ABDOMINOPELVIC QUADRANTS

The abdominopelvic cavity may be divided more simply into **quadrants** (*quad* = four). These are shown in Figure 1-9. In this method, frequently used by clinicians, a horizontal line and a vertical line are passed through the umbilicus. These two lines divide the abdomen into a **right upper quadrant (RUQ), left upper quadrant (LUQ), right lower quadrant (RLQ),** and **left lower quadrant (LLQ).** Whereas the nine-region designation is more widely used for anatomical studies, the quadrant designation is better suited for locating the site of an abdominopelvic pain, tumor, or other abnormality.

CLINICAL APPLICATION

To determine the cause of death accurately, it is necessary to perform an **autopsy** (AW-top-sē; *auto* = self; *opsis* = to view), that is, a postmortem examination of the body. In addition, an autopsy can also be used to uncover the existence of diseases not detected during life, support the accuracy of diagnostic tests, determine the effectiveness and side effects of drugs, analyze the effects of environmental influences on the body, and educate medical students.

A typical autopsy consists of three principal phases of examination. The first phase is examination of the exterior of the body for the presence of wounds, scars, tumors, or other abnormalities. The second phase includes the dissection and gross examination of the major body organs. The third phase of autopsy consists of microscopic examination of tissues from organs to ascertain any pathology. Depending on the circumstances, techniques may also be used to detect and recover microbes and determine the presence of foreign substances in the body.

The autopsy itself begins with a Y-shaped incision to expose the thoracic, abdominal, and pelvic viscera. The upper parts of the Y begin in front of each shoulder, extend inferior to the nipples, and join just inferior to the sternum. The incision is then continued down the middle of the abdominopelvic wall to the symphysis pubis. The sternum is removed by cutting the ribs to expose the thoracic viscera, and the tissues of the abdominopelvic wall are folded back to expose the contents of the abdominopelvic cavity. Unless the information is of special clinical significance, the face and extremities are not usually dissected.

An autopsy usually lasts two to four hours, depending on its comprehensiveness. When the procedure is completed, the organs are returned to the body, except for those that are donated or used to extract a particular pharmacologic substance, and all incisions are sewn up.

RADIOGRAPHIC ANATOMY

A specialized branch of anatomy that is essential for the diagnosis of many disorders is **radiographic** (*radius* = ray) **anatomy,** or **radiology,** which includes the uses of several radiographic techniques (x rays).

CONVENTIONAL RADIOGRAPHY

The most common and familiar type of radiographic anatomy employs the use of a single barrage of x rays. The x rays pass through the body and expose an x-ray film, producing a photographic image called a **roentgenogram** (RENT-gen-ō-gram). A roentgenogram provides a two-dimensional shadow image of the interior of the body (Figure 1-10a). As valuable as they are in diagnosis, conventional x rays compress the body image onto a flat sheet of film, often resulting in an overlap of organs and tissues that could make diagnosis difficult. Moreover, x rays do not always differentiate between subtle differences in tissue density.

COMPUTED TOMOGRAPHY (CT) SCANNING

These diagnostic difficulties have been virtually eliminated by the use of an x-ray technique called **computed tomography (CT) scanning.** First introduced in 1971, CT scanning combines the principles of x-ray and advanced computer technologies. An x-ray source moves in an arc around the part of the body being scanned and repeatedly sends out x-ray beams. As the beams pass through the body, the tissues absorb small amounts of radiation, depending on their densities. Once the beams pass through, they are converted by light-sensitive crystal detectors to electronic signals that are transmitted to the scanner's computer.

Through mathematical analysis of the difference between the total radiation emitted and the amount striking the detectors, the computer reconstructs what happens to the x-ray beams. It determines where and how much of the beams are absorbed as they pass through a tissue. And, like slicing an orange and turning the slice to see the pits, the scanner's computer rotates the mathematical information 90°. It then projects an image, called a **CT scan,** onto a television screen called the physician's console. The CT scan provides a very accurate cross-sectional picture of any area of the body (Figure 1-10b). A series of scans permits a physician to examine layer after layer of a patient's tissues. Additional scans taken from other angles can be used to pinpoint extremely small abnormalities in the tissue.

The CT scan permits a significant differentiation of body parts that was never possible with conventional x rays. The entire CT scanning process takes only seconds, it is completely painless, and the x-ray dose is equal to or less than that of many other diagnostic procedures.

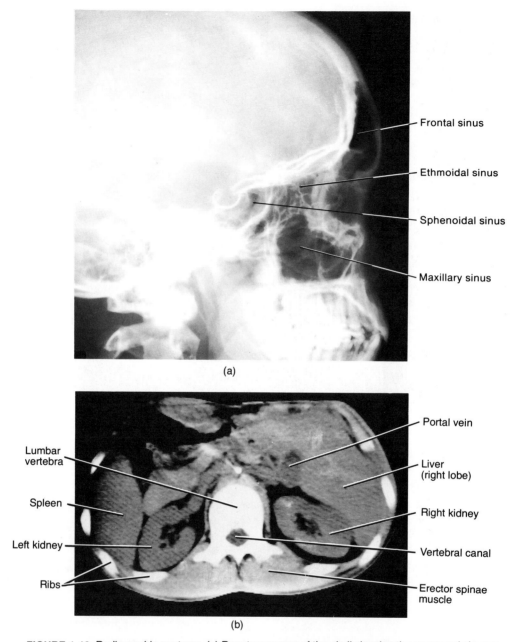

(a)

(b)

FIGURE 1-10 Radiographic anatomy. (a) Roentgenogram of the skull showing the paranasal sinuses. (Courtesy of Eastman Kodak Company.) (b) CT scan of the abdomen. (Courtesy of General Electric Company.)

In subsequent chapters, references will be made to roentgenograms and CT scans. They will have more meaning to you once you have learned more about the anatomy of the body.

DYNAMIC SPATIAL RECONSTRUCTOR (DSR)

A recent development in radiographic anatomy is a highly sophisticated x-ray machine called the **dynamic spatial reconstructor (DSR).** It is the most complex and versatile medical instrument built to date. It resembles a giant pencil sharpener, weighs over 15 tons, and has the ability to construct *moving, three-dimensional,* life-size images

of all or part of an internal organ from any view desired. The image produced by the DSR can be rotated and tipped, and like an electronic knife, the DSR can pictorially slice open an organ and expose its interior (Figure 1-11). The instrument also provides enlargements and stop-action, replay, high-speed, and slow-motion viewing.

During a DSR scan, 28 x-ray guns revolve around the patient, each firing beams of x rays 60 times per second. In a period of 5 seconds, the DSR can produce an amazing 75,000 cross sections. In the same period of time, a CT scanner can make only one section. The level of radiation exposure of the DSR is about double that for a chest roentgenogram.

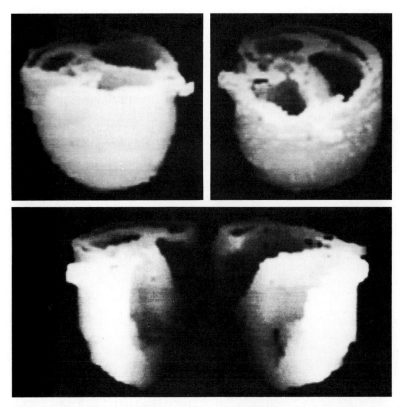

FIGURE 1-11 Dynamic spatial reconstructor (DSR) display of a living dog's heart. One of the unique properties of the DSR is its ability to act like an electronic knife by "dissecting" various organs, as shown in the photos. (Courtesy of Robb, R. A., L. D. Harris, and E. L. Ritman. Computerized x-ray reconstruction tomography in stereometric analysis of cardiovascular dynamics. SPIE—Applications of Optics in Medicine and Biology 89:69–82, 1976.)

The DSR was designed to provide three-dimensional imaging of the heart, lungs, and circulation. It can be used to measure the volumes and movements of the heart and lungs and other internal organs, to detect cancer and heart defects, and to measure tissue damage following a heart attack, stroke, or other disease.

MAGNETIC RESONANCE IMAGING (MRI)

A new procedure in the arsenal for diagnosing disease is **magnetic resonance imaging (MRI),** formerly called **nuclear magnetic resonance (NMR).** It focuses on the nuclei of atoms of a single element in a tissue at a time and determines if the nuclei behave normally in response to an external force such as magnetism. In most studies to date, MRI of hydrogen nuclei has been popular because of the body's large water content. The part of the body to be studied, ranging from a finger to the entire body, is placed in the scanner, exposing the nuclei to a uniform magnetic field (Figure 1-12a) and causing the nuclei to line up in the direction of the magnetic field. Then the aligned nuclei are exposed to a brief burst of an alternating magnetic field at a 90° angle to the first magnetic field. When the magnetic field is cut off, the energy absorbed by the nuclei becomes a small electrical voltage. The voltage is picked up by detectors and relayed to a computer for analysis. The computerized reconstruction shows the density and energy loss (voltage) of the nuclei of a particular element and somewhat resembles a CT scan (Figure 1-12b). The reconstructions can be displayed in color and as two- or three-dimensional images and indicate a biochemical blueprint of cellular activity.

The diagnostic advantage of MRI is that in addition to providing images of diseased organs and tissues, it also provides information about what chemicals are present in the organ and tissue. Such an analysis can indicate a particular disease is in progress, even before symptoms occur. MRI also offers the advantages of being noninvasive (not involving puncture or incision of the skin or insertion of an instrument or foreign material into the body), not utilizing radiation, and gathering biochemical information without time-consuming chemical analyses.

In clinical and preclinical trials, MRI studies have confirmed the findings of other diagnostic techniques such as CT scans and, in some cases, have added more detail. Thus, MRI has proved useful in identifying existing pathologies. Scientists hope that MRI can also be used to perform a biopsy on tumors without an operation, assess mental disorders, measure blood flow, study the evolution

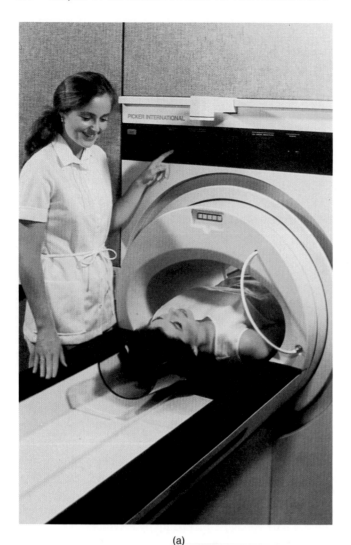

(a)

of hematomas and infarctions in the brain following a stroke, identify the potential for developing a stroke, assess treatment for conditions such as heart disease and stroke, monitor the progress and treatment of a disease, study the effects of toxic drugs on tissues, measure intracellular pH, study metabolism, and determine how donor organs from cadavers may function after transplantation.

MEASURING THE HUMAN BODY

An important aspect of describing the body and understanding how it works is **measurement**—what are the dimensions of an organ, how much does it weigh, how long does it take for a physiological event to occur. Such questions also have clinical importance, for example, in determining how much of a given medication should be administered. As you will see, measurements involving time, weight, temperature, size, length, and volume are a routine part of your studies in a medical science program.

Whenever you come across a measurement in the text, the measurement will be given in metric units. The metric system is standardly used in sciences. To help you compare the metric unit to a familiar unit, the approximate U.S. equivalent will also be given in parentheses directly after the metric unit. For example, you might be told that the length of a particular part of the body is 2.54 cm (1 inch).

As a first step in helping you understand the relationships of the metric system to the U.S. system of measure-

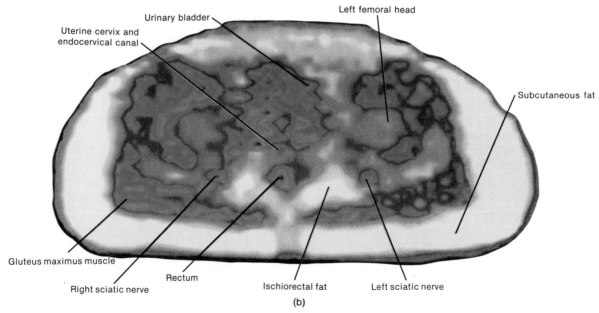

(b)

FIGURE 1-12 Magnetic resonance imaging (MRI). (a) Photograph of a Picker International MRI system. (Courtesy of W. C. Doran, Picker International, Inc., Highland Heights, Ohio.) (b) Image of the female pelvis produced by MRI. (Courtesy of John Cassese, Director of Marketing, Fonar Corporation, Melville, New York.)

EXHIBIT 1-3 METRIC UNITS OF LENGTH AND SOME U.S. EQUIVALENTS

METRIC UNIT (ABBREVIATION)	MEANING OF PREFIX	EQUIVALENT IN METERS	U.S. EQUIVALENT
1 kilometer (km)	kilo = 1,000	1,000 m	3,280.84 ft; 0.62 mi 1 mi = 1.61 km
1 hectometer (hm)	hecto = 100	100 m	328 ft
1 dekameter (dam)	deka = 10	10 m	3.28 ft
1 meter (m)	Standard unit of length		39.37 in; 32.8 ft; 1.09 yd
1 decimeter (dm)	deci = $\frac{1}{10}$	0.1 m	3.94 in
1 centimeter (cm)	centi = $\frac{1}{100}$	0.01 m	0.394 in 1 in = 2.54 cm
1 millimeter (mm)	milli = $\frac{1}{1,000}$	0.001 m = $\frac{1}{10}$ cm	0.0394 in
1 micrometer (μm) [formerly **micron** (μ)]	micro = $\frac{1}{1,000,000}$	0.000,001 m = $\frac{1}{10,000}$ cm	3.94×10^{-5} in
1 nanometer (nm) [formerly **millimicron** (mμ)]	nano = $\frac{1}{1,000,000,000}$	0.000,000,001 m = $\frac{1}{10,000,000}$ cm	3.94×10^{-8} in
1 angstrom (Å)		0.000,000,000,1 m = $\frac{1}{100,000,000}$ cm	3.94×10^{-9} in

EXHIBIT 1-4 METRIC UNITS OF MASS AND SOME U.S. EQUIVALENTS

METRIC UNIT (ABBREVIATION)	EQUIVALENT IN GRAMS	U.S. EQUIVALENT
1 kilogram (kg)	1,000 g	2.205 lb; 1 ton = 907 kg
1 hectogram (hg)	100 g	
1 dekagram (dag)	10 g	0.0353 oz
1 gram (g)	1 g	1 lb = 453.6 1 oz = 28.35 g
1 decigram (dg)	0.1 g	
1 centigram (cg)	0.01 g	
1 milligram (mg)	0.001 g	0.015 gr
1 microgram (gmg)	0.000,001 g	

EXHIBIT 1-5 METRIC UNITS OF VOLUME AND SOME U.S. EQUIVALENTS

METRIC UNIT (ABBREVIATION)	METRIC EQUIVALENT	U.S. EQUIVALENT
1 liter (l)	1,000 ml	33.81 fl oz or 1.057 qt; 946 ml = 1 qt
1 milliliter (ml)	0.001 l	0.0338 fl oz; 30 ml = 1 fl oz 5 ml = 1 teaspoon
1 cubic centimeter (cm³)	0.999972 ml	0.0338 fl oz

ment, three exhibits have been prepared. Exhibit 1-3 contains metric units of length and some U.S. equivalents. Exhibit 1-4 contains metric units of mass and some U.S. equivalents. Exhibit 1-5 contains metric units of volume and some U.S. equivalents. Carefully examine the exhibits. Even if you do not learn all the metric units and their equivalents at this point, you can refer back to the exhibits later.

STUDY OUTLINE

Anatomy and Physiology Defined (p. 2)
1. Anatomy is the study of structure and the relationship among structures.
2. Subdivisions of anatomy include surface anatomy (form and markings of surface features), gross anatomy (macroscopic), systemic, or systematic, anatomy (systems), regional anatomy (regions), developmental anatomy (development from fertilization to adulthood), embryology (development from fertilized egg through eighth week in utero), pathological anatomy (disease), histology (tissues), cytology (cells), and radiographic anatomy (x rays).
3. Physiology is the study of how body structures function.

Levels of Structural Organization (p. 2)

1. The human body consists of several levels of structural organization; among these are the chemical, cellular, tissue, organ, system, and organismic levels.
2. Cells are the basic structural and functional units of an organism.
3. Tissues consist of groups of similarly specialized cells and their intercellular material that perform certain special functions.
4. Organs are structures of definite form and function composed of two or more different tissues.
5. Systems consist of associations of organs that have a common function.
6. The human organism is a collection of structurally and functionally integrated systems.
7. The systems of the human body are the integumentary, skeletal, muscular, nervous, endocrine, cardiovascular, lymphatic, respiratory, digestive, urinary, and reproductive systems (see Exhibit 1-1).

Life Processes (p. 4)

1. All living forms have certain characteristics that distinguish them from nonliving things.
2. Among the life processes in humans are metabolism, excitability, conductivity, contractility, growth, differentiation, and reproduction.

Structural Plan (p. 4)

1. The human body has certain general characteristics.
2. Among the characteristics are a backbone, a tube within a tube organization, and bilateral symmetry.

Anatomical Position and Anatomical Names (p. 4)

1. When in the anatomical position, the subject stands erect facing the observer, the upper extremities are placed at the sides, and the palms of the hands are turned forward.
2. Regional names are terms given to specific regions of the body for reference. Examples of regional names include cranial (skull), thoracic (chest), brachial (arm), patellar (knee), cephalic (head), and gluteal (buttock).

Directional Terms (p. 12)

1. Directional terms indicate the relationship of one part of the body to another.
2. Commonly used directional terms are superior (toward the head or upper part of a structure), inferior (away from the head or toward the lower part of a structure), anterior (near or at the front of the body), posterior (near or at the back of the body), medial (nearer the midline of the body or a structure), intermediate (between a medial and lateral structure), ipsilateral (on the same side of the body), contralateral (on the opposite side of the body), proximal (nearer the attachment of an extremity to the trunk or a structure), distal (farther from the attachment of an extremity to the trunk or a structure), superficial (toward or on the surface of the body), deep (away from the surface of the body), parietal (pertaining to the outer wall of a body cavity), and visceral (pertaining to the covering of an organ).

Planes and Sections (p. 12)

1. Planes are imaginary flat surfaces that are used to divide the body or organs into definite areas. A midsagittal (median) plane is a vertical plane through the midline of the body that divides the body or organs into equal right and left sides; a sagittal (parasagittal) plane is a plane parallel to the midsagittal plane that divides the body or organs into unequal right and left sides; a frontal (coronal) plane is a plane at a right angle to a midsagittal (or sagittal) plane that divides the body or organs into anterior and posterior portions; and a horizontal (transverse) plane is a plane parallel to the ground and at a right angle to the midsagittal, sagittal, and frontal planes that divides the body or organs into superior and inferior portions.
2. Sections are flat surfaces resulting from cuts through body structures. They are named according to the plane on which the cut is made and include cross sections, frontal sections, and midsagittal sections.

Body Cavities (p. 13)

1. Spaces in the body that contain internal organs are called cavities.
2. The dorsal and ventral cavities are the two principal body cavities. The dorsal cavity contains the brain and spinal cord. The organs of the ventral cavity are collectively called viscera.
3. The dorsal cavity is subdivided into the cranial cavity, which contains the brain, and the vertebral, or spinal, canal, which contains the spinal cord and beginnings of spinal nerves.
4. The ventral body cavity is subdivided by the diaphragm into an upper thoracic cavity and a lower abdominopelvic cavity.
5. The thoracic cavity contains two pleural cavities and a mediastinum, which includes the pericardial cavity.
6. The mediastinum is a mass of tissue between the pleurae of the lungs that extends from the sternum to the vertebral column; it contains all contents of the thoracic cavity, except the lungs.
7. The abdominopelvic cavity is divided into a superior abdominal and an inferior pelvic cavity by an imaginary line extending from the symphysis pubis to the sacral promontory.
8. Viscera of the abdominal cavity include the stomach, spleen, pancreas, liver, gallbladder, small intestine, and most of the large intestine.
9. Viscera of the pelvic cavity include the urinary bladder, sigmoid colon, rectum, and internal female and male reproductive structures.

Abdominopelvic Regions (p. 14)

1. To describe the location of organs easily, the abdominopelvic cavity may be divided into nine regions by drawing four imaginary lines (left lateral, right lateral, transpyloric, and transtubercular).
2. The names of the nine abdominopelvic regions are epigastric, right hypochondriac, left hypochondriac, umbilical, right lumbar, left lumbar, hypogastric (pubic), right iliac (inguinal), and left iliac (inguinal).

Abdominopelvic Quadrants (p. 21)

1. To locate the site of an abdominopelvic abnormality in clinical studies, the abdominopelvic cavity may be divided into quadrants by passing imaginary horizontal and vertical lines through the umbilicus.
2. The names of the four abdominopelvic quadrants are right upper quadrant (RUQ), left upper quadrant (LUQ), right lower quadrant (RLQ), and left lower quadrant (LLQ).

Radiographic Anatomy (p. 21)

1. Radiographic anatomy is a specialized branch of anatomy that makes use of x rays.
2. The value of radiographic anatomy is its application in the diagnosis of disease.

Conventional Radiography

1. Conventional radiography uses a single barrage of x rays.
2. The photographic two-dimensional image produced is called a roentgenogram.
3. Conventional radiography has several diagnostic drawbacks, including overlapping of organs and tissues and inability to differentiate subtle differences in tissue density.

Computed Tomography (CT) Scanning

1. CT scanning combines the principles of x-ray and advanced computer technologies.
2. The image produced, called a CT scan, provides a very accurate cross-sectional picture of any area of the body.

Dynamic Spatial Reconstructor (DSR)

1. The DSR is a highly sophisticated x-ray machine that can produce moving, three-dimensional images of different organs of the body,
2. The DSR was designed to provide imaging of the heart, lungs, and circulation.

Magnetic Resonance Imaging (MRI) (p. 23)

1. Magnetic resonance imaging (MRI) is based on the reaction of atomic nuclei to magnetism.
2. MRI can identify existing pathologies, evaluate drug therapy, measure metabolism, and assess the potential for certain diseases.

Measuring the Human Body (p. 24)

1. Various kinds of measurements are important in understanding the human body.
2. Examples of such measurements include organ size, body weight, and amount of medication to be administered.
3. Measurements in this book are given in metric units followed by the approximate U.S. equivalents in parentheses.
4. Metric units of length may be reviewed in Exhibit 1-3, metric units of mass in Exhibit 1-4, and metric units of volume in Exhibit 1-5.

REVIEW QUESTIONS

1. Define anatomy. List and define the various subdivisions of anatomy. Define physiology.
2. Construct a diagram to illustrate the levels of structural organization that characterize the human body. Be sure to define each level.
3. Using Exhibit 1-1 as a guide, outline the function of each system of the body, and list several organs that compose each system.
4. Describe the important life processes in humans.
5. What does bilateral symmetry mean? Why is the body considered to be a tube within a tube? What is a vertebrate?
6. Define the anatomical position. Why is the anatomical position used?
7. Review Figure 1-2. See if you can locate each region on your own body and give both its common and its anatomical name.
8. What is a directional term? Why are these terms important? Can you use each of the directional terms listed in Exhibit 1-2 in a complete sentence?
9. Name the various planes that may be passed through the body. Describe how each plane divides the body.
10. What is meant by the phrase "a part of the body has been sectioned"?
11. Define a body cavity. List the body cavities discussed, and tell which major organs are located in each. What landmarks separate the various body cavities from one another?
12. Name and locate the nine regions of the abdominopelvic area. List the organs, or parts of organs, in each.
13. Describe how the abdominopelvic cavity is divided into four quadrants and name each quadrant.
14. Why is an autopsy performed? Describe the basic procedure.
15. Explain the principle of computed tomography (CT) scanning. Contrast it with conventional radiography in terms of principle and diagnostic value.
16. Distinguish between a CT scan and a roentgenogram.
17. Explain the principle and clinical application of the dynamic spatial reconstructor (DSR).
18. Explain the principle of magnetic resonance imaging (MRI). How is MRI useful in the diagnosis of disease?
19. Convert the following lengths:
    a. A bacterial cell measures 100 μm in length. Give its length in nanometers.
    b. How many meters are in 1 mi?
    c. If a road sign reads 35 km/hour, what would your speedometer have to read to obey the sign?
    d. How many millimeters are in 1 km? In 1 in?
    e. A person's arm measures 2 ft in length. How many centimeters is this?
    f. Convert 0.40 m to millimeters.
    g. How many millimeters are in 5 in?
    h. If you ran 295.2 ft, how many meters would you have run?
    i. If the distance to the moon is 239,000 mi, what is it in meters?
20. Solve the following conversions of mass.
    a. Calculate the milligrams in 0.4 kg and in 1 lb.
    b. If a bottle contains 1.42 g, how many centigrams does it contain?
    c. The indicated dosage of a certain drug is 50 μg. How many milligrams is this?
    d. If you weigh 110 lb, how many kilograms do you weigh?
    e. How many centigrams are in 1 g?
21. Convert the following volumes.
    a. If you excrete 1,200 ml of urine in a day, how many liters is this?
    b. How many milliliters are in 2 l?
    c. Convert 2 pt to milliliters.
    d. If you remove 15 cm^3 of blood from a patient, how many milliliters have you removed?

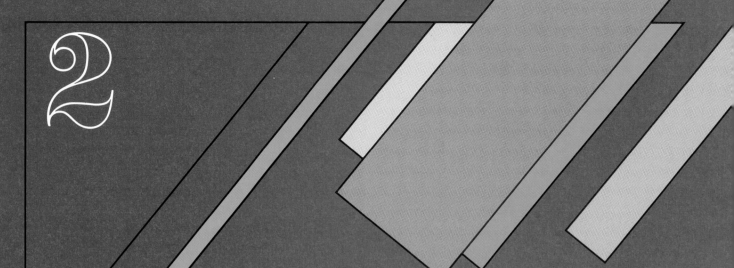

2

Cells

Student Objectives

Define a cell and list a cell's generalized parts.

Explain the structure and molecular organization of the plasma membrane.

Describe how materials move across plasma membranes.

Define the structure and function of several modified plasma membranes.

Describe the chemical composition and functions of cytoplasm.

Describe the structure and two general functions of a cell nucleus.

Define the structure and function of ribosomes.

Distinguish between agranular and granular endoplasmic reticulum (ER) with regard to structure and function.

Describe the structure and functions of the Golgi complex.

Discuss the structure and function of mitochondria as "powerhouses of the cell."

Explain why a lysosome in a cell is called a "suicide packet."

Discuss the structure and role of peroxisomes.

Distinguish between the structure and function of microfilaments and microtubules as components of the cytoskeleton.

Discuss the structure and function of centrioles in cellular reproduction.

Differentiate between cilia and flagella in terms of function.

Define a cell inclusion and give several examples.

Define extracellular material and give several examples.

Describe the principal events of interphase, the stage between cell divisions.

Discuss the stages and events involved in cell division.

Describe the relationship of cancer to cells.

Explain the relationship of aging to cells.

Define key medical terms associated with cells.

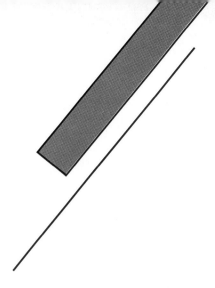

The study of the body at the cellular level of organization is important because many activities essential to life occur in cells and many disease processes originate there. A **cell** may be defined as the basic, living, structural, and functional unit of the body and, in fact, of all organisms. **Cytology** (*cyt* = cell; *logos* = study of) is the branch of science concerned with the study of cells. This chapter concentrates on the structure, functions, and reproduction of cells.

A series of illustrations accompanies each cell structure that you study. A diagram of a generalized animal cell shows the location of the structure within the cell. An electron micrograph shows the actual appearance of the structure.* A diagram of the electron micrograph clarifies some of the small details by exaggerating their outlines. Finally, an enlarged diagram of the structure shows the details that make up the structure.

CLINICAL APPLICATION

Despite the enormous advantages provided by electron microscopy in helping scientists understand detailed cellular structure, one drawback is the fact that specimens must be killed, since they are placed in a vacuum. This problem is now overcome by a process called **microtomography,** in which the principles of electron microscopy and computed tomography (CT) scanning are combined. Microtomography produces high-magnification, three-dimensional images of *living* cells.

The potential applications of microtomography include studying how normal and cancerous cells move, grow, and reproduce. It will also be used to follow the changes that occur as an embryo progresses through its developmental sequences. The technique will also enable scientists to observe the microscopic effects of drugs on living cells and how cancer-causing substances alter living cells.

---

* An **electron micrograph (EM)** is a photograph taken with an electron microscope. The electron microscope can magnify an object more than 200,000 times. In comparison, the light microscope that you probably use in your laboratory magnifies objects up to 1,000 times their size.

GENERALIZED ANIMAL CELL

A **generalized animal cell** is a composite of many different cells in the body. Examine the generalized cell illustrated in Figure 2-1, but keep in mind that *no such single cell actually exists.*

For convenience, we can divide the generalized cell into four principal parts:

1. Plasma (cell) membrane. The outer, limiting membrane separating the cell's internal parts from the extracellular fluid and external environment.

2. Cytoplasm. The substance between the nucleus and the plasma membrane.

3. Organelles. Permanent structures with characteristic morphology that are highly specialized for specific cellular activities.

4. Inclusions. The secretions and storage areas of cells.

Extracellular materials, which are substances external to the cell surface, will also be examined in connection with cells.

PLASMA (CELL) MEMBRANE

The exceedingly thin structure that separates one cell from other cells and from the external environment is called the **plasma (cell) membrane.** Electron microscopy studies have shown that the plasma membrane ranges from 65 to 100 angstroms (Å) in thickness, a dimension below light-microscope limits.†

CHEMISTRY AND STRUCTURE

Proteins are the major component of nearly all plasma membranes, comprising from 50 to 70 percent by weight. Next in abundance are phospholipids. Other constituents in lesser amounts include cholesterol, water, carbohydrates, and ions. Recent studies of membrane structure

---

† One **angstrom,** usually written as 1 Å, = 0.000,000,000,1 m ($\frac{1}{250,000,000}$ in). Another microscopic unit of measurement is the **micrometer (μm),** which is equal to 0.000,001 m ($\frac{1}{25,000}$ in).

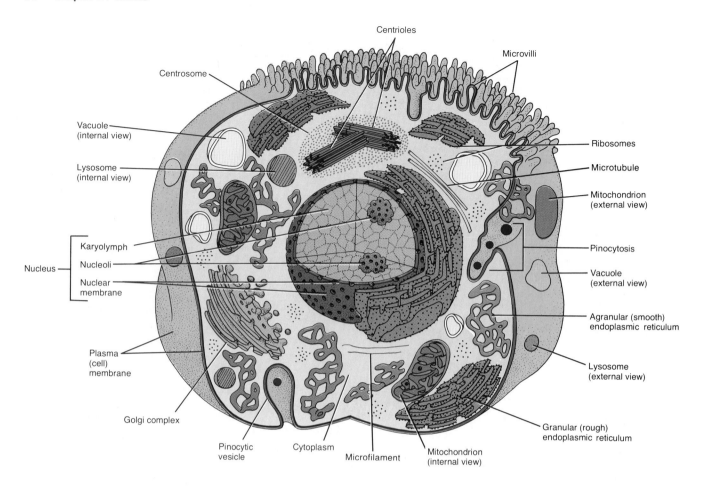

FIGURE 2-1 Generalized animal cell based on electron microscope studies.

suggest a new concept regarding the arrangement of the various molecules. This concept is referred to as the **fluid mosaic model** (Figure 2-2b).

The phospholipid molecules are arranged in two parallel rows, forming a **phospholipid bilayer.** A phospholipid molecule consists of a polar, phosphate-containing "head" which mixes with water and nonpolar fatty acid "tails" that do not mix with water. The molecules are oriented in the bilayer so that "heads" face outward on either side and the "tails" face each other in the membrane's interior.

The membrane proteins are classified into two categories: integral and peripheral. **Integral proteins** are embedded in the phospholipid bilayer among the fatty acid "tails." Some of the integral proteins lie at or near the inner and outer membrane surfaces; others penetrate the membrane completely. Since the phospholipid bilayer is somewhat fluid and flexible and the integral proteins have been observed moving from one location to another in the membrane, the relationship has been compared to icebergs (proteins) floating in the sea (phospholipid bilayer). The subunits of some integral proteins form minute channels through which substances can be transported into and out of the cell (described shortly). Other integral

proteins carry branching chains of carbohydrates. Such combinations of carbohydrate and protein, called *glycoproteins,* provide receptor sites that enable a cell to recognize other cells of its own kind so that they can associate to form a tissue, to recognize and respond to foreign cells that might be potentially dangerous, and to recognize and attach to hormones, nutrients, and other chemicals. Red blood cells also have glycoprotein receptors that prevent them from clumping and producing unwanted blood clots. Diabetes mellitus is a disease believed to be caused by faulty cell receptor sites.

Peripheral proteins are loosely bound to the membrane surface and easily separated from it. Far less is known about them than integral proteins and their functions are not yet completely understood. For example, some peripheral proteins are believed to serve as enzymes that catalyze cellular reactions. An example is cytochrome c, involved in cellular respiration. Other peripheral proteins, such as spectrin of red blood cells, are believed to have a mechanical function by serving as a scaffolding to support the plasma membrane. It is also believed that peripheral proteins may assume a role in changes in membrane shape during such processes as cell division, locomotion, and ingestion.

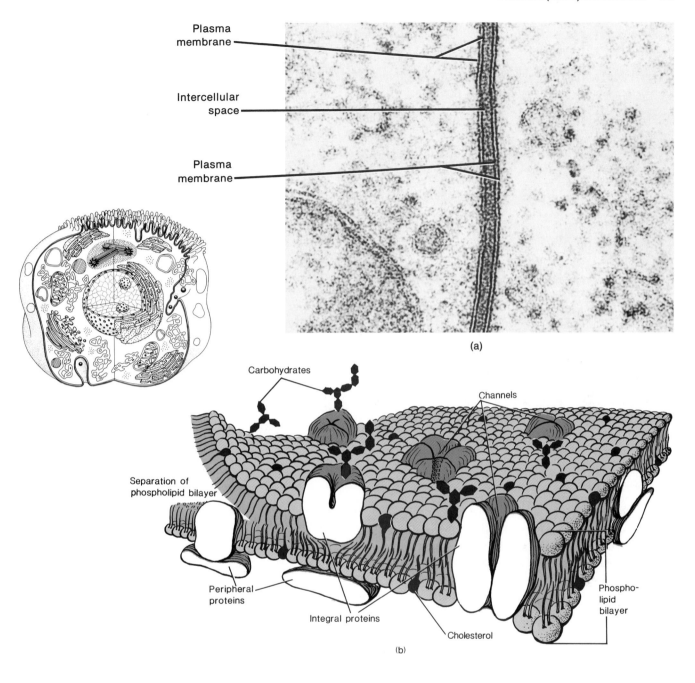

(a)

(b)

FIGURE 2-2 Plasma membrane. (a) Electron micrograph of portions of two plasma membranes separated by an intercellular space at a magnification of 200,000×. (Copyright © Dr. John Fawcett, Photo Researchers.) (b) Enlargement of the plasma membrane showing the latest concept of the relationship of the phospholipid bilayer and protein molecules. The separation of the bilayer is for illustrative purposes only.

FUNCTIONS

Based on the discussion of the chemistry and structure of the plasma membrane, we can now describe its several important functions. First, the plasma membrane provides a flexible boundary that encloses the cellular contents and separates them from the external environment. Second, the membrane facilitates contact with other body cells or with foreign cells or substances. Third, the mem-

brane provides receptors for chemicals such as hormones, neurotransmitters, enzymes, nutrients, and antibodies. Fourth, the plasma membrane mediates the entrance and exit of materials. The ability of a plasma membrane to permit certain substances to enter and exit, but to restrict the passage of others, is called **selective permeability.** Let us look at this mechanism in a bit more detail.

A membrane is said to be *permeable* to a substance if it allows the free passage of that substance. Although plasma membranes are not actually freely permeable to any substance, they do permit some substances to pass more readily than others. For example, water passes more readily than most other substances. The permeability of a plasma membrane appears to be a function of several factors.

1. Size of molecules. Large molecules cannot pass through the plasma membrane. Water and amino acids are small molecules and can enter and exit the cell easily. However, most proteins, which consist of many amino acids linked together, seem to be too large to pass through the membrane. Many scientists believe that the giant-sized molecules do not enter the cell because they are larger than the minute channels in the integral proteins.

2. Solubility in lipids. Substances that dissolve easily in lipids pass through the membrane more readily than other substances, since a major part of the plasma membrane consists of lipid molecules. Examples of lipid-soluble substances are oxygen, carbon dioxide, and steroid hormones.

3. Charge on ions. The charge of an ion attempting to cross the plasma membrane can determine how easily the ion enters or leaves the cell. The protein portion of the membrane is capable of ionization. If an ion has a charge opposite that of the membrane, it is attracted to the membrane and passes through more readily. If the ion attempting to cross the membrane has the same charge as the membrane, it is repelled by the membrane and its passage is restricted. This phenomenon conforms to the rule of physics that opposite charges attract, whereas like charges repel each other.

4. Presence of carrier molecules. Some integral proteins called carriers are capable of attracting and transporting substances across the membrane regardless of size, ability to dissolve in lipids, or membrane charge. Their purpose is to modify membrane permeability. The mechanism by which carriers do this will be described shortly.

MOVEMENT OF MATERIALS ACROSS PLASMA MEMBRANES

The mechanisms whereby substances move across the plasma membrane are essential to the life of the cell. Certain substances, for example, must move into the cell to support life, whereas waste materials or harmful substances must be moved out. Plasma membranes mediate the movements of such materials. The processes involved in these movements may be classed as either passive or active. In **passive processes,** substances move across plasma membranes without assistance from the cell. Their movement involves the kinetic energy of individual molecules. The substances move on their own down a concentration gradient from an area where their concentration is high to an area where their concentration is low. The substances may also be forced across the plasma membrane by pressure from an area where the pressure is

high to an area where it is low. In **active processes,** the cell contributes energy in moving the substance across the membrane, since the substance moves against a concentration gradient.

Passive Processes

• *Diffusion*   A passive process called **diffusion** occurs when there is a *net* or greater movement of molecules or ions from a region of high concentration to a region of low concentration. The movement from high to low concentration continues until the molecules are evenly distributed. At this point, they move in both directions at an equal rate. This point of even distribution is called *equilibrium.* The difference between high and low concentrations is called the *concentration gradient.* Molecules moving from the high-concentration area to the low-concentration area are said to move *down* or *with* the concentration gradient. A good example of diffusion in the body is the movement of oxygen from the blood into the cells and the movement of carbon dioxide from the cells back into the blood. Diffusion ensures that cells receive adequate amounts of oxygen and eliminate carbon dioxide as part of their normal metabolism. Large and small lipid-soluble molecules pass through the phospholipid layer of the plasma membrane by diffusion. Small molecules that are not lipid-soluble, such as certain ions (sodium, potassium, chloride), are able to diffuse through channels formed by integral proteins in the membrane.

• *Facilitated Diffusion*   Another type of diffusion through a selectively permeable membrane occurs by a process called **facilitated diffusion.** This process is accomplished with the assistance of integral proteins in the membrane that serve as carriers. Although some chemical substances are large molecules and insoluble in lipids, they can still pass through the plasma membrane. Among these are different sugars, especially glucose. In the process of facilitated diffusion, it is believed that glucose is picked up by a carrier. The combined glucose-carrier is soluble in the phospholipid bilayer of the membrane, and the carrier moves the glucose to the inside of the membrane and then inside the cell. The carrier makes the glucose soluble in the phospholipid bilayer of the membrane so it can pass through the membrane. By itself, glucose is insoluble and cannot penetrate the membrane. In the process of facilitated diffusion, the cell does not expend energy, and the movement of the substance is from a region of its high concentration to a region of its low concentration.

• *Osmosis*   Another passive process by which materials move across membranes is **osmosis.** It is the net movement of water molecules through a selectively permeable membrane from an area of high water concentration to an area of low water concentration. The water molecules pass through channels in integrated proteins in the membrane. Water moves between various compartments of the body by osmosis.

● *Isotonic, Hypotonic, and Hypertonic Solutions* Osmosis may also be understood by considering the effects of different water concentrations on red blood cells. If the normal shape of a red blood cell is to be maintained, the cell must be placed in an **isotonic solution.** This is a solution in which the total concentrations of water molecules and solute molecules are the same on both sides of the selectively permeable cell membrane. The concentrations of water and solute in the extracellular fluid outside the red blood cell must be the same as the concentration of the intracellular fluid. Under ordinary circumstances, a 0.85 percent NaCl solution is isotonic for red blood cells. In this condition, water molecules enter and exit the cell at the same rate, allowing the cell to maintain its normal shape.

A different situation results if red blood cells are placed in a solution that has a lower concentration of solutes and, therefore, a higher concentration of water. This is called a **hypotonic solution.** In this condition, water molecules enter the cells faster than they can leave, causing the red blood cells to swell and eventually burst. The rupture of red blood cells in this manner is called **hemolysis** (hē-MOL-i-sis), or **laking.** Distilled water is a strongly hypotonic solution.

A **hypertonic solution** has a higher concentration of solutes and a lower concentration of water than the red blood cells. One example of a hypertonic solution is a 10 percent NaCl solution. In such a solution, water molecules move out of the cells faster than they can enter. This situation causes the cells to shrink. The shrinkage of red blood cells in this manner is called **crenation** (krē-NĀ-shun). Red blood cells may be greatly impaired or destroyed if placed in solutions that deviate significantly from the isotonic state.

● *Filtration* A third passive process involved in moving materials in and out of cells is **filtration.** This process involves the movement of solvents such as water and dissolved substances such as sugar across a selectively permeable membrane by gravity or mechanical pressure, usually hydrostatic (water) pressure. Such a movement is always from an area of higher pressure to an area of lower pressure and continues as long as a pressure difference exists. Most small- to medium-sized molecules can be forced through a cell membrane.

An example of filtration occurs in the kidneys, where the blood pressure supplied by the heart forces water and small molecules like urea through thin cell membranes of tiny blood vessels and into the kidney tubules. In this basic process, protein molecules are retained by the body, since they are too large to be forced through the cell membranes of the blood vessels. The molecules of harmful substances such as urea are small enough to be forced through and eliminated, however.

Active Processes

When cells actively participate in moving substances across membranes, they must expend energy. Cells can even move substances against a concentration gradient. The active processes considered here are active transport and endocytosis (phagocytosis, pinocytosis, and receptor-mediated endocytosis).

● *Active Transport* The process by which substances, usually ions, are transported across plasma membranes typically from an area of low concentration to an area of high concentration is called **active transport.** In order to transport a substance against a concentration gradient, the membrane uses energy supplied by ATP. In fact, a typical body cell probably expends up to 40 percent of its ATP for active transport. Although the molecular events involved in active transport are not completely understood, integral proteins in the plasma membrane do assume a role.

As just noted, glucose can be transported across cell membranes via facilitated diffusion from areas of high to low concentration. Glucose can also be moved by the cells lining the digestive tract from the cavity of the tract into the blood, even though blood concentration of glucose is higher. This movement involves active transport. One proposed mechanism is that glucose (or other substance) enters a channel in an integral membrane protein. When the glucose molecule makes contact with an active site in the channel, the energy from ATP induces a change in the membrane protein that expels the glucose on the opposite side of the membrane. Active transport is also an important process in maintaining the concentrations of some ions inside body cells and other ions outside body cells.

● *Endocytosis* Large molecules and particles pass through plasma membranes by a process called **endocytosis,** in which a segment of the plasma membrane surrounds the substance, encloses it, and brings it into the cell. The export of substances from the cell by the reverse process is called **exocytosis,** a very important mechanism for secretory cells. There are three basic kinds of endocytosis: phagocytosis, pinocytosis, and receptor-mediated endocytosis.

In **phagocytosis** (fag'-ō-sī-TŌ-sis), or "cell eating," projections of cytoplasm, called *pseudopodia* (soo'-dō-PŌ-dē-a), engulf large solid particles external to the cell (Figure 2-3a, b). Once the particle is surrounded, the membrane folds inwardly, forming a membrane sac around the particle. This newly formed sac, called a *phagocytic vesicle,* breaks off from the outer cell membrane, and the solid material inside the vesicle is digested. Indigestible particles and cell products are removed from the cell by a reverse phagocytosis. Phagocytosis is important because molecules and particles of material that would normally be restricted from crossing the plasma

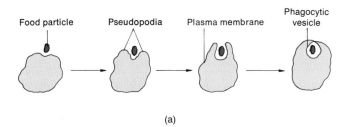

(a)

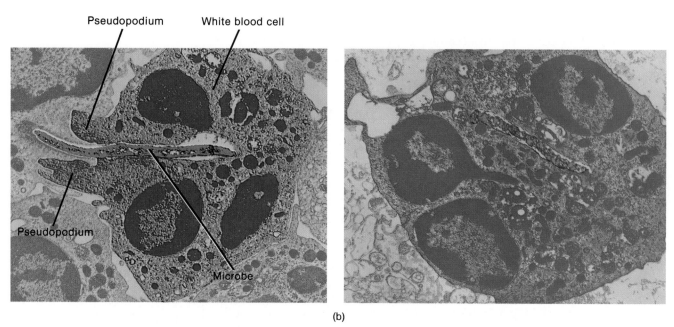

Pseudopodium   White blood cell

Pseudopodium

Microbe

(b)

FIGURE 2-3 Endocytosis. (a) Diagram of phagocytosis. (b) Color-enhanced photomicrographs of phagocytosis. The photomicrograph on the left shows a human white blood cell (neutrophil) engulfing a microbe, and the photomicrograph on the right shows a later stage of phagocytosis in which the engulfed microbe is being destroyed. (Courtesy of Abbott Laboratories.) (c) Two variations of pinocytosis. In the variation on the left, the ingested substance enters a channel formed by the plasma membrane and becomes enclosed in a pinocytic vesicle at the base of the channel. In the variation on the right, the ingested substance becomes enclosed in a pinocytic vesicle that forms at the surface of the cell and detaches. (d) Receptor-mediated endocytosis.

membrane because of their large size can be brought into or removed from the cell. The phagocytic white blood cells of the body constitute a vital defense mechanism. Through phagocytosis, the white blood cells engulf and destroy bacteria and other foreign substances.

In **pinocytosis** (pi'-nō-sī-TŌ-sis), or "cell drinking," the engulfed material consists of an extracellular liquid rather than a solid (Figure 2-3c). Moreover, no cytoplasmic projections are formed. Instead, a minute droplet of liquid is attracted to the surface of the membrane. The membrane folds inwardly, forms a pinocytic vesicle that surrounds the liquid, and detaches from the rest of the intact membrane. Few cells are capable of phagocytosis, but many cells carry on pinocytosis. Examples include cells in the kidneys and urinary bladder.

Receptor-mediated endocytosis is a highly selective process in which cells can take up large molecules or particles. This process is accomplished as follows. The plasma membrane contains protein receptors that have binding sites for extracellular large molecules or particles. Such extracellular materials are known as *ligands* (Figure 2-3d). The binding between receptor and ligand causes

the plasma membrane to invaginate, forming a membrane-bounded *vesicle* around the ligand. As vesicles move inward from the plasma membrane, they fuse with each other and form larger structures called *endosomes*. Each endosome gives rise to an even larger structure called a *CURL* (*compartment of uncoupling of receptor and ligand*). Within the CURL, the ligands separate from the receptors. The ligands are distributed into the vesicular portion of the CURL and the receptors accumulate in the tubular portion of the CURL. Following this segregation, the vesicular portion fuses with a lysosome, where the ligands are broken down by powerful digestive enzymes. The tubular portion recycles receptors to the plasma membrane for reuse.

MODIFIED PLASMA MEMBRANES

Electron-microscope studies have revealed that plasma membranes of certain cells are modified in various ways for specific purposes. For example, the membranes of some cells lining the small intestine have small, cylindrical

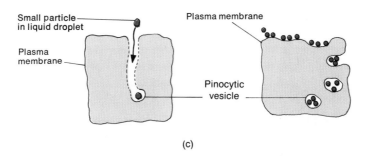

(c)

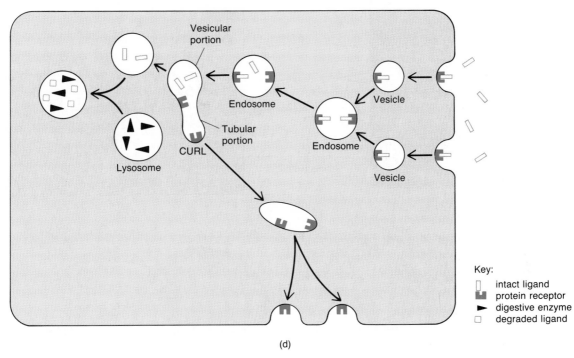

Key:
- ▯ intact ligand
- ▮ protein receptor
- ► digestive enzyme
- ▢ degraded ligand

(d)

projections called **microvilli** (see Figure 2-1). These fingerlike projections enormously increase the absorbing area of the cell surface. A single cell may have as many as 3,000 microvilli, and a 1-sq-mm (0.00155 sq in) area of small intestine may contain as many as 200 million microvilli, which would increase the surface area for absorption by 20 times.

Another membrane modification is found in the rods and cones of the eye. They serve as photoreceptors, or light-receiving cells. The upper portion of each rod contains two-layered, disc-shaped membranes called **sacs** that contain the pigments involved in vision.

Another example of a membrane modification is the **stereocilia.** They are found only in cells lining a duct (ductus epididymis) of the male reproductive system. They appear by light microscopy as long, slender, branching processes of the free surfaces of the lining cells (see Figure 25-7). Electron micrographs show stereocilia to be microvilli.

A final example of a membrane modification is the **myelin sheath** that surrounds portions of certain nerve cells (see Figure 16-2). It is thought that the myelin sheath increases the velocity of impulse conduction, protects the portion of the nerve cell it surrounds, is related to the nutrition of the nerve cell, and is necessary for its repair and regeneration.

CYTOPLASM

The substance inside the cell's plasma membrane and external to the nucleus is called **cytoplasm** (Figure 2-4a, b). It is the matrix, or ground substance, in which various cellular components are found. Physically, cytoplasm may be described as a thick, semitransparent, elastic fluid containing suspended particles and a series of minute tubules and filaments that form a cytoskeleton. Chemically, cytoplasm is 75 to 90 percent water plus solid components. Proteins, carbohydrates, lipids, and inorganic substances compose the bulk of the solid components. The inorganic substances and most carbohydrates are soluble in water and are present as a true solution. Many organic compounds, however, are found as colloids—large molecules that remain suspended in solution. Since the particles of a colloid bear electrical charges that repel each other, they remain suspended and separated from each other.

Functionally, cytoplasm is the substance in which chemical reactions occur. The cytoplasm receives raw materials from the external environment and converts them into usable energy by decomposition reactions. Cytoplasm is also the site where new substances are synthesized for cellular use. It packages chemicals for transport to other parts of the cell or other cells of the body and facilitates the excretion of waste materials.

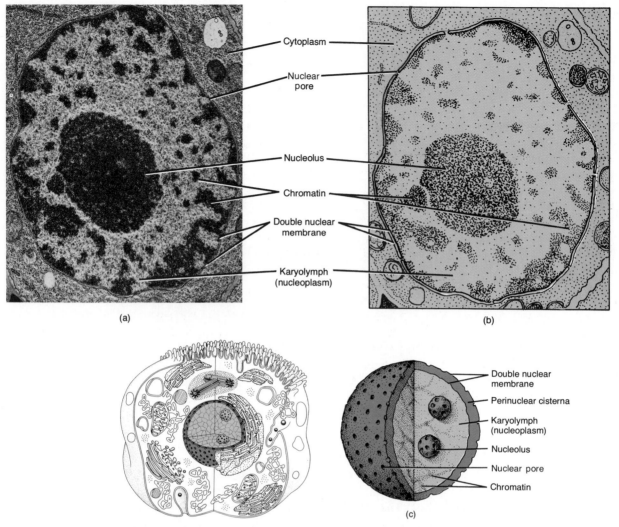

FIGURE 2-4 Cytoplasm and nucleus. (a) Electron micrograph of cytoplasm and the nucleus at a magnification of 31,600×. (Courtesy of Biophoto Associates/Dr. Myron C. Ledbetter, Brookhaven National Laboratory.) (b) Diagram of the electron micrograph. (c) Diagram of a nucleus with two nucleoli.

ORGANELLES

Despite the myriad chemical activities occurring simultaneously in the cell, there is little interference of one reaction with another. The cell has a system of compartmentalization provided by structures called **organelles.** These structures are specialized portions of the cell with characteristic morphology that assume specific roles in growth, maintenance, repair, and control.

NUCLEUS

The **nucleus** is generally a spherical or oval organelle and is the largest structure in the cell (Figure 2-4a–c). It contains hereditary factors of the cell, called genes, which control cellular structure and direct many cellular activities. Mature red blood cells do not have nuclei. These cells carry on only limited types of chemical activity and are not capable of growth or reproduction.

The nucleus is separated from the cytoplasm by a double membrane called the *nuclear membrane,* or *envelope* (Figure 2-4c). Between the two layers of the nuclear membrane is a space called the *perinuclear cisterna.* Each of the nuclear membranes resembles the structure of the plasma membrane. Minute pores in the nuclear membrane allow the nucleus to communicate with a membranous network in the cytoplasm called the endoplasmic reticulum. Substances entering and exiting the nucleus are believed to pass through the tiny pores.

Three prominent structures are visible internal to the nuclear membrane. The first of these is a gel-like fluid that fills the nucleus called *karyolymph* (*nucleoplasm*). One or more spherical bodies called the *nucleoli* are also present. These structures are composed of protein, DNA, and RNA. DNA synthesizes the RNA that is stored in the nucleoli. The RNA, as you will see shortly, assumes a function in protein synthesis. Finally, there is the *genetic material,* consisting principally of DNA. When the cell is not reproducing, the genetic material appears as a threadlike mass called *chromatin*. Prior to cellular reproduction the chromatin shortens and coils into rod-shaped bodies called *chromosomes*.

Studies of chromosomes have provided a great deal of information about their chemistry and detailed structure. Although it has been known for some time that chromosomes consist of DNA and proteins called *histones*, their organization and arrangement in chromosomes have only recently been determined. In specially prepared electron micrographs, it can be seen that chromosomes consist of elementary subunits of structure called **nucleosomes**. Each nucleosome has a diameter of about 100 Å. A nucleosome consists of four pairs of different histone proteins (collectively called an *octamer*) associated with a relatively fixed length of DNA (about 200 nitrogenous base pairs). Present evidence indicates that the DNA is wrapped around the histones in some way. A fifth type of histone maintains adjacent nucleosomes in a helical coil (*solenoid*). Nucleosomes are separated by stretches of DNA only, so that the arrangement resembles beads on a string. The functional significance of nucleosome structure is still speculative. It is believed that histones may facilitate changes in chromosomal structure that expose activated genes (DNA) to perform a specific task in the cell.

RIBOSOMES

Ribosomes are tiny granules, 250 Å at their largest dimension, that are composed of a type of RNA called ribosomal RNA (rRNA) and a number of specific ribosomal proteins. The rRNA is manufactured by DNA in the nucleolus. Ribosomes were so named because of their high content of rRNA. Structurally, a ribosome consists of two subunits, one about half the size of the other. Recently scientists have provided three-dimensional models of the structure of a ribosome (Figure 2-5d). Functionally, ribosomes are the sites of protein synthesis; they receive genetic instructions and translate them into protein. Amino acids are joined one at a time on ribosomes into a protein chain. The completed chain then folds into a protein molecule that can serve as part of the cell's structure or as an enzyme.

Some ribosomes, called *free ribosomes*, are scattered in the cytoplasm; they have no attachments to other parts of the cell. The free ribosomes occur singly or in clusters, and they are primarily concerned with synthesizing proteins for use inside the cell. Other ribosomes are attached to a cellular structure called the endoplasmic reticulum. These ribosomes are concerned with the synthesis of proteins for export from the cell.

ENDOPLASMIC RETICULUM (ER)

Within the cytoplasm, there is a system of pairs of parallel membranes enclosing narrow cavities of varying shapes. This system is known as the **endoplasmic reticulum,** or **ER** (Figure 2-5a–c). The ER, in other words, is a network of channels (*cisternae*) running through the cytoplasm. These channels are continuous with the nuclear membrane.

On the basis of its association with ribosomes, ER is distinguished into two types. **Granular (rough) ER** is studded with ribosomes; **agranular (smooth) ER** is free of ribosomes. Recent studies using radioactive tracers suggest that agranular ER is synthesized from granular ER.

Numerous functions related to homeostasis are attributed to the ER. It contributes to the mechanical support and distribution of the cytoplasm. It also conducts intracellular nerve impulses, such as in muscle cells, where it is called sarcoplasmic reticulum. The ER is involved in the intracellular exchange of materials with the cytoplasm and provides a surface area for chemical reactions. Various products are transported from one portion of the cell to another via the ER, so the ER is considered an intracellular circulatory system. The ER also serves as a storage area for synthesized molecules. And together with a cellular structure called the Golgi complex, the ER assumes a role in the synthesis and packaging of molecules.

GOLGI COMPLEX

Another structure found in the cytoplasm is the **Golgi** (GOL-jē) **complex.** This structure consists of four to eight flattened, baglike channels, stacked upon each other with expanded areas at their ends. Like those of the ER, the stacked elements are called *cisternae,* and the expanded, terminal areas are *vesicles* (Figure 2-6). Generally, the Golgi complex is located near the nucleus.

One function of the Golgi complex is the packaging of secreted proteins. *Secretion* is the production and release from a gland cell of a fluid that usually contains a variety of substances. Proteins synthesized at ribosomes associated with granular ER are transported into the ER cisternae. The proteins then pass in membrane sacs pinched off from the ER into the Golgi complex. As proteins accumulate and concentrate in the cisternae of the Golgi complex, the cisternae expand to form vesicles. After a certain size is reached, the vesicles pinch off from the cisternae. The protein and its associated vesicle are referred to as a *secretory granule*. The secretory granule then moves toward the surface of the cell where the protein is secreted. The contents of the granule are discharged into the extracellular space, and the granule membrane is incorporated into the plasma membrane. Cells of the digestive tract that secrete protein enzymes utilize this mechanism. The secretory granule prevents digestion of the cytoplasm of the cells by the enzymes as it moves toward the cell surface. Other vesicles that pinch off from the Golgi complex are loaded with special digestive enzymes and remain within the cell. They become cellular structures called lysosomes.

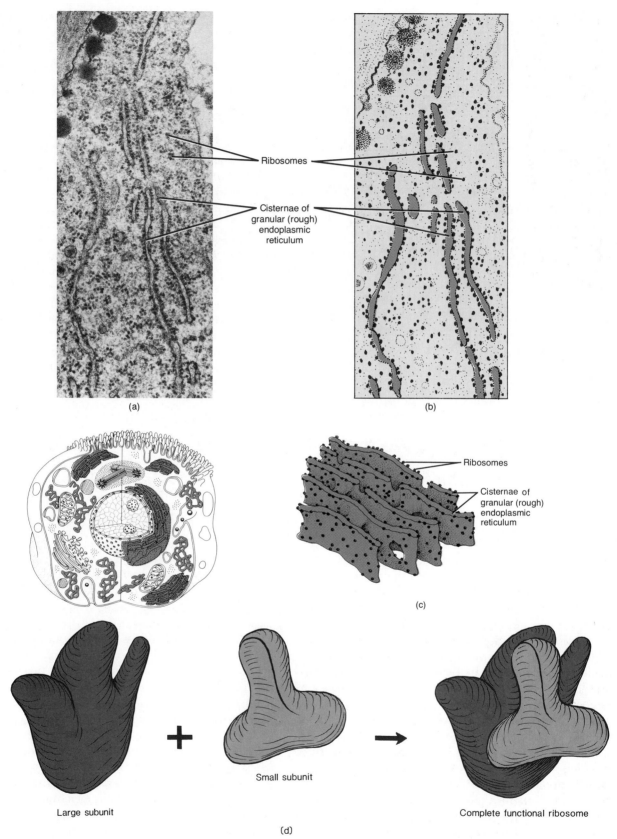

FIGURE 2-5 Endoplasmic reticulum and ribosomes. (a) Electron micrograph of the endoplasmic reticulum and ribosomes at a magnification of 76,000×. (Courtesy of Biophoto Associates/Dr. Myron C. Ledbetter, Brookhaven National Laboratory.) (b) Diagram of the electron micrograph. (c) Diagram of the endoplasmic reticulum and ribosomes. See if you find the agranular (smooth) endoplasmic reticulum in Figure 2-6. (d) Diagram of the three-dimensional structure of ribosomes.

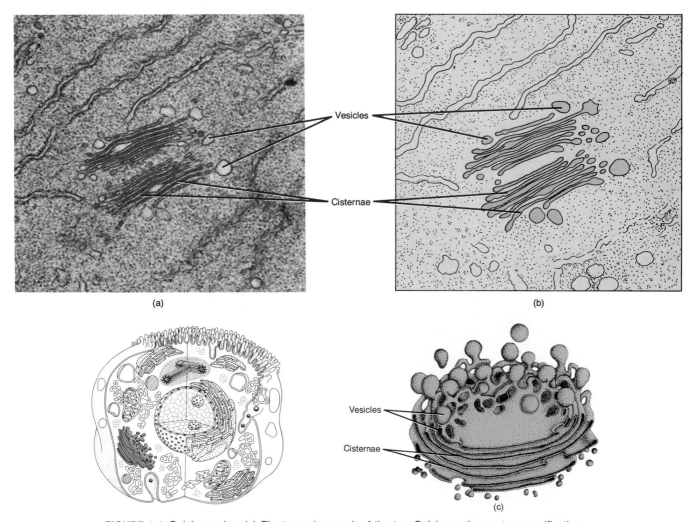

FIGURE 2-6 Golgi complex. (a) Electron micrograph of the two Golgi complexes at a magnification of 78,000×. (Courtesy of Biophoto Associates/Dr. Myron C. Ledbetter, Brookhaven National Laboratory.) (b) Diagram of the electron micrograph. (c) Diagram of a Golgi complex.

Another function of the Golgi complex is associated with lipid secretion. It occurs in essentially the same way as protein secretion, except the lipids are synthesized by the agranular ER. The lipids pass through the ER into the Golgi complex. As in the mechanism just described, the lipids migrate into the cisternae and vesicles and are discharged at the surface of the cell. In the course of moving through the cytoplasm, the vesicle may release lipids into the cytoplasm before being discharged from the cell. These appear in the cytoplasm as lipid droplets. Among the lipids secreted in this manner are steroids.

The Golgi complex also functions in the synthesis of carbohydrates. Recent evidence indicates that carbohydrates synthesized by the Golgi complex are combined with proteins synthesized by ribosomes associated with granular ER to form glycoproteins. As these carbohydrate-protein complexes are assembled, they accumulate in the flattened channels of the Golgi complex. The channels expand and form vesicles. After a critical size is reached, the vesicles pinch off from the channel, migrate

through the cytoplasm, and discharge their contents through the plasma membrane. The Golgi complex is well developed and highly active in secretory cells, such as those found in the pancreas and the salivary glands.

MITOCHONDRIA

Small, spherical, rod-shaped, or filamentous structures called **mitochondria** (mī'-tō-KON-drē-a) appear throughout the cytoplasm. When sectioned and viewed under an electron microscope, each reveals an elaborate internal organization (Figure 2-7). A mitochondrion consists of two membranes, each of which is similar in structure to the plasma membrane. The outer mitochondrial membrane is smooth, but the inner membrane is arranged in a series of folds called *cristae*. The center of the mitochondrion is called the *matrix*.

Because of the nature and arrangement of the cristae, the inner membrane provides an enormous surface area for chemical reactions. Enzymes involved in energy-re-

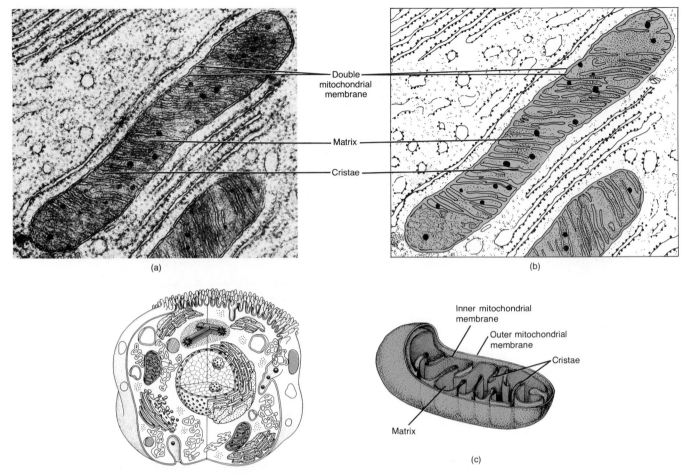

FIGURE 2-7 Mitochondria. (a) Electron micrograph of an entire mitochondrion (above) and a portion of another (below). (Courtesy of Lester V. Bergman & Associates.) (b) Diagram of the electron micrograph. (c) Diagram of a mitochondrion.

leasing reactions that form ATP are located on the cristae. Mitochondria are frequently called the "powerhouses of the cell" because they are the sites for the production of ATP. Active cells, such as muscle and liver cells, have a large number of mitochondria because of their high energy expenditure.

Mitochondria are self-replicative, that is, they can divide to form new ones. The replication process is controlled by DNA that is incorporated into the mitochondrial structure. Self-replication usually occurs in response to increased cellular need for ATP.

LYSOSOMES

When viewed under the electron microscope, **lysosomes** (*lysis* = dissolution; *soma* = body) appear as membrane-enclosed spheres (Figure 2-8). They are formed from Golgi complexes, have a double membrane, and lack detailed structure. They contain powerful digestive enzymes capable of breaking down many kinds of molecules. You will see in Chapter 18 that Tay-Sachs disease results from a deficiency of a lysosomal enzyme. These enzymes are also capable of digesting bacteria that enter the cell. White blood cells, which ingest bacteria by phagocytosis, contain large numbers of lysosomes.

Scientists have wondered why these powerful enzymes do not also destroy their own cells. Perhaps the lysosomal membrane in a healthy cell is impermeable to enzymes, so they cannot move out into the cytoplasm. However, when a cell is injured, the lysosomes release their enzymes. The enzymes then promote reactions that break the cell down into its chemical constituents. This process of self-destruction by cells is known as **autolysis.** The chemical remains are either reused by the body or excreted. Because of this function, lysosomes have been called "suicide packets."

Lysosomes are crucial in the removal of cell parts, whole cells, and even extracellular material, which may be the process underlying bone removal. In bone reshaping, especially during the growth process, special bone-destroying cells called osteoclasts secrete extracellular enzymes that dissolve bone. Bone tissue cultures given excess amounts of vitamin A remove bone apparently through an activation process involving lysosomes.

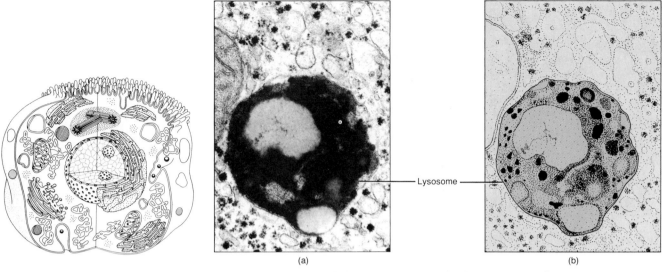

FIGURE 2-8 Lysosome. (a) Electron micrograph of a lysosome at a magnification of 55,000×. (Courtesy of F. Van Hoof, Université Catholique de Louvain.) (b) Diagram of the electron micrograph.

CLINICAL APPLICATION

Animals overfed on vitamin A interestingly develop spontaneous fractures, suggesting greatly increased lysosomal activity. On the other hand, cortisone and hydrocortisone, steroid hormones produced by the adrenal gland, have a stabilizing effect on lysosomal membranes. The steroid hormones are well known for their antiinflammatory properties, which suggests that they reduce destructive cellular activity by lysosomes.

PEROXISOMES

Organelles similar in structure to lysosomes, but smaller, are called **peroxisomes** (pe-ROKS-i-sōms). They are abundant in liver cells and contain several enzymes related to the metabolism of hydrogen peroxide (H_2O_2), a substance that is toxic to body cells. One of the enzymes in peroxisomes, called *catalase,* immediately breaks down H_2O_2 into water and oxygen:

$$\underset{\substack{\text{Hydrogen} \\ \text{peroxide}}}{H_2O_2} \xrightarrow{\text{Catalase}} \underset{\text{Water}}{H_2O} + \underset{\text{Oxygen}}{O_2}$$

MICROFILAMENTS AND MICROTUBULES: THE CYTOSKELETON

In recent years, electron microscopy has revealed that the cytoplasm of cells is more than just a structureless medium in which the various cellular components are suspended. It has been shown that cytoplasm actually has a complex internal structure, consisting of a series of exceedingly small microfilaments and microtubules, together referred to as the **cytoskeleton** (see Figure 2-1).

Microfilaments are rodlike structures ranging from 30 to 120 Å in diameter. They are of variable length and may occur in bundles, randomly scattered throughout the cytoplasm, or arranged in a meshwork, depending on the type of cell in which they are found. Some microfilaments consist of a protein called *actin;* others consist of a protein called *myosin.* In muscle tissue, actin microfilaments (thin myofilaments) and myosin microfilaments (thick myofilaments) are involved in the contraction of muscle cells. This mechanism is described in Chapter 9. In nonmuscle cells, microfilaments help to provide support and shape and assist in the movement of entire cells (phagocytes and cells of developing embryos) and movements within cells (secretion, phagocytosis, pinocytosis).

Microtubles are relatively straight, slender, cylindrical structures that range in diameter from 180 to 300 Å, usually averaging about 240 Å. They consist of a protein called *tubulin.* Microtubules dispersed in the cytoplasm, together with microfilaments, help to provide support and shape for cells. Some evidence suggests that microtubules may form conducting channels through which various substances can move throughout the cytoplasm. This mechanism has been studied extensively in nerve cells. Microtubules also assist in the movement of pseudopodia that are characteristic of phagocytes. As you will see shortly, microtubules form the structure of flagella and cilia (cellular appendages involved in motility), centrioles (organelles that may direct the assembly of microtubules), and the mitotic spindle (structures that are involved in cell division).

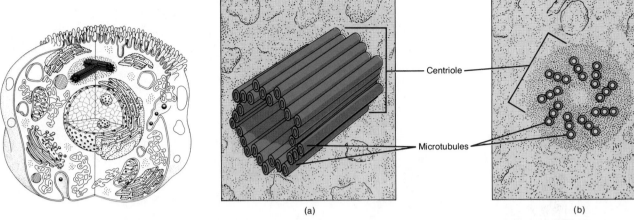

FIGURE 2-9 Centrosome and centrioles. (a) Diagram of a centriole in longitudinal view. (b) Diagram of a centriole in cross section. (c) Electron micrograph of a centriole in cross section at a magnification of 67,000×. (Courtesy of Biophoto Associates.)

CENTROSOME AND CENTRIOLES

A dense area of cytoplasm, generally spherical and located near the nucleus, is called the **centrosome (centrosphere).** Within the centrosome is a pair of cylindrical structures: the **centrioles** (Figure 2-9). Each centriole is composed of a ring of nine evenly spaced bundles. Each bundle, in turn, consists of three microtubules. The two centrioles are situated so that the long axis of one is at right angles to the long axis of the other. Centrioles assume a role in cell reproduction by serving as centers about which microtubules involved in chromosome movement are organized. This role will be described shortly as part of cell division. Certain cells, such as most mature nerve cells, do not have a centrosome and so do not reproduce. This is why they cannot be replaced if they are destroyed. Like mitochondria, centrioles contain DNA that controls their self-replications.

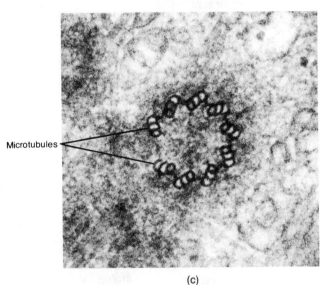

(c)

FLAGELLA AND CILIA

Some body cells possess projections for moving the entire cell or for moving substances along the surface of the cell. These projections contain cytoplasm and are bounded by the plasma membrane. If the projections are few and long in proportion to the size of the cell, they are called **flagella.** The only example of a flagellum in the human body is the tail of a sperm cell, used for locomotion (see Figure 25-6). If the projections are numerous and short, resembling many hairs, they are called **cilia.** In humans, ciliated cells of the respiratory tract move mucus that has trapped foreign particles over the surface of the tissue (see Figure 22-6b). Electron microscopy has revealed no fundamental structural difference between cilia and flagella (Figure 2-10). Both consist of nine pairs of microtubules that form a ring around two single microtubules in the center.

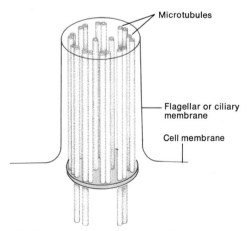

FIGURE 2-10 Structure of a flagellum or cilium.

CELL INCLUSIONS

Cell inclusions are a large and diverse group of chemical substances, some of which have recognizable shapes. These products are principally organic and may appear or disappear at various times in the life of the cell. *Melanin* is a pigment stored in certain cells of the skin, hair, and eyes. It protects the body by screening out harmful ultraviolet rays from the sun. *Glycogen* is a polysaccharide that is stored in the liver, skeletal muscle cells, and the vaginal mucosa. When the body requires quick energy, liver cells can break down the glycogen into glucose and release it. *Lipids,* which are stored in fat cells, may be decomposed for producing energy. A final example of an inclusion is *mucus,* which is produced by cells that line organs. Its function is to provide lubrication and protection.

The major parts of the cell and their functions are summarized in Exhibit 2-1.

EXTRACELLULAR MATERIALS

The substances that lie outside cells are called **extracellular materials.** They include body fluids, which provide a medium for dissolving, mixing, and transporting substances. Among the body fluids are interstitial fluid, the fluid that fills the microscopic spaces (interstitial spaces), and plasma, the liquid portion of blood. Extracellular materials also include secreted inclusions like mucus and special substances that form the matrix in which some cells are embedded.

The matrix materials are produced by certain cells and deposited outside their plasma membranes. The matrix supports the cells, binds them together, and gives strength and elasticity to the tissue. Some matrix materials are *amorphous* (*a* = without; *morpho* = shape); they have no specific shape. These include hyaluronic acid and chondroitin sulfate. **Hyaluronic** (hī'-a-loo-RON-ik) **acid** is a viscous, fluidlike substance that binds cells together, lubricates joints, and maintains the shape of the eyeballs. **Chondroitin** (kon-DROY-tin) **sulfate** is a jelly-like substance that provides support and adhesiveness in cartilage, bone, heart valves, the cornea of the eye, and the umbilical cord.

Other matrix materials are *fibrous,* or threadlike. Fibrous materials provide strength and support for tissues. Among these are **collagenous** (*kolla* = glue) **fibers** consisting of the protein **collagen.** These fibers are found in all types of connective tissue, especially in bones, cartilage, tendons, and ligaments. **Reticular** (*rete* = net) **fibers** consisting of the protein collagen and a coating of glycoprotein, form a network around fat cells, nerve fibers, muscle cells, and blood vessels. They also form the framework,

EXHIBIT 2-1 CELL PARTS AND THEIR FUNCTIONS

PART	FUNCTIONS
PLASMA MEMBRANE	Protects cellular contents; makes contact with other cells; provides receptors for hormones, enzymes, and antibodies; mediates the entrance and exit of materials.
CYTOPLASM	Serves as the ground substance in which chemical reactions occur.
ORGANELLES	
Nucleus	Contains genes and controls cellular activities.
Ribosomes	Sites of protein synthesis.
Endoplasmic reticulum (ER)	Contributes to mechanical support; conducts intracellular nerve impulses in muscle cells; facilitates intracellular exchange of materials with cytoplasm; provides a surface area for chemical reactions; provides a pathway for transporting chemicals; serves as a storage area; together with Golgi complex synthesizes and packages molecules for export.
Golgi complex	Packages synthesized proteins for secretion in conjunction with endoplasmic reticulum; forms lysosomes; secretes lipids; synthesizes carbohydrates; combines carbohydrates with proteins to form glycoproteins for secretion.
Mitochondria	Sites for production of ATP.
Lysosomes	Digest substances and foreign microbes; may be involved in bone removal.
Peroxisomes	Contains several enzymes, such as catalase, related to hydrogen peroxide metabolism.
Microfilaments	Form cytoskeleton with microtubules; involved in muscle cell contraction; provide support and shape; assist in cellular and intracellular movement.
Microtubules	Form cytoskeleton with microfilaments; provide support and shape; form intracellular conducting channels; assist in cellular movement; form the structure of flagella, cilia, centrioles, and spindle fibers.
Centrioles	Help organize mitotic spindle during cell division.
Flagella and cilia	Allow movement of entire cell (flagella) or movement of particles along surface of cell (cilia).
INCLUSIONS	Melanin (pigment in skin, hair, eyes) screens out ultraviolet rays; glycogen (stored glucose) can be decomposed to provide energy; lipids (stored in fat cells) can be decomposed to produce energy; mucus provides lubrication and protection.

or stroma, for many soft organs of the body such as the spleen. **Elastic fibers,** consisting of the protein *elastin,* give elasticity to skin and to tissues forming the walls of blood vessels.

NORMAL CELL DIVISION

Most of the cell activities mentioned thus far maintain the life of the cell on a day-to-day basis. However, cells become damaged, diseased, or worn out and then die. New cells must be produced as replacements and for growth.

Cell division is the process by which cells reproduce themselves. It consists of a nuclear division and a cytoplasmic division. Because nuclear division can be of two types, two kinds of cell division are recognized. The first is the mechanism by which sperm and egg cells are produced, preliminary to the formation of a new organism. The process consists of a nuclear division called **meiosis** plus a cytoplasmic division called **cytokinesis.** It is often called reproductive cell division and is discussed in detail in Chapter 25.

In the second kind of division, often called somatic cell division, a single parent cell duplicates itself. This process consists of a nuclear division called **mitosis** and

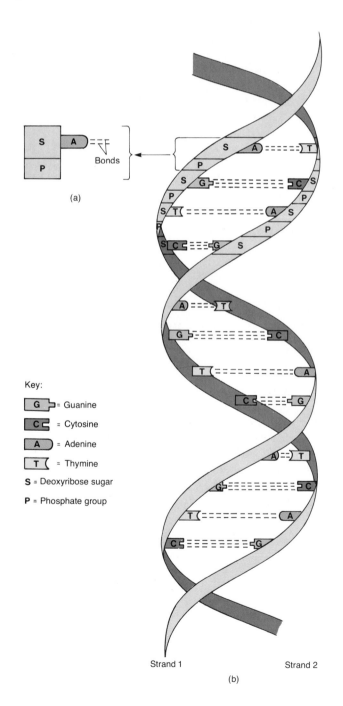

Key:

G = Guanine

C = Cytosine

A = Adenine

T = Thymine

S = Deoxyribose sugar

P = Phosphate group

Strand 1   Strand 2

(b)

(c)

FIGURE 2-11 DNA molecule. (a) Adenine nucleotide. (b) Portion of an assembled DNA molecule. (c) Space-filling (three-dimensional) model to show relative size and location of atoms. (Courtesy of Ealing Corp.)

cytokinesis. The process ensures that each new daughter cell has the same *number* and *kind* of chromosomes as the original parent cell. After the process is complete, the two daughter cells have the same hereditary material and genetic potential as the parent cell. This kind of cell division results in an increase in the number of body cells. In a 24-hour period, the average adult loses trillions of cells from different parts of the body. Obviously, these cells must be replaced. Cells that have a short life span—the cells of the outer layer of skin, the cornea of the eye, the digestive tract—are continually being replaced. Mitosis and cytokinesis are the means by which dead or injured cells are replaced and new cells are added for body growth.

When a cell reproduces, it must replicate its chromosomes so its hereditary traits may be passed on to succeeding generations of cells. A **chromosome** is a highly coiled DNA molecule that is partly covered by protein. The protein causes changes in the length and thickness of the chromosomes. Hereditary information is contained in the DNA portion of the chromosome in units called **genes.** Each human chromosome consists of about 20,000 genes.

Before taking a look at the relationship between chromosomes and cell division, it will first be necessary to examine briefly the structure of DNA, the basic component of chromosomes.

A molecule of DNA is a chain composed of repeating units called *nucleotides.* Each nucleotide of DNA consists of three basic parts (Figure 2-11a): (1) It contains one of four possible *nitrogen bases,* which are ring-shaped structures containing atoms of C, H, O, and N. The nitrogen bases found in DNA are named adenine, thymine, cytosine, and guanine. (2) It contains a sugar called *deoxyribose.* (3) It has a phosphoric acid called the *phosphate group.* The nucleotides are named according to the nitrogen base that is present. Thus, a nucleotide containing thymine is called a *thymine nucleotide,* one containing adenine is called an *adenine nucleotide,* and so on.

The chemical composition of the DNA molecule was known before 1900, but it was not until 1953 that a model of the organization of the chemicals was constructed. This model was proposed by J. D. Watson and F. H. C. Crick on the basis of data from many investigations. Figure 2-11b shows the following structural characteristics of the DNA molecule: (1) The molecule consists of two strands with crossbars. The strands twist about each other in the form of a *double helix* so that the shape resembles a twisted ladder. (2) The uprights of the DNA ladder consist of alternating phosphate groups and the deoxyribose portions of the nucleotides. (3) The rungs of the ladder contain paired nitrogen bases. As shown, adenine always pairs off with thymine, and cytosine always pairs off with guanine.

When a cell is between divisions it is said to be in **interphase,** or the **metabolic phase.** It is during this stage that the replication (synthesis) of chromosomes occurs and the RNA and protein needed to produce structures required for doubling all cellular components are manufactured.

The period of interphase during which chromosomes are replicated is referred to as the **S** (for synthesis) **period.** When DNA replicates, its helical structure partially uncoils. Those portions of DNA that remain coiled stain darker than the uncoiled portions. This unequal distribution of stain causes the DNA to appear as a granular mass called **chromatin** (see Figure 2-12a). During uncoiling, DNA separates at the points where the nitrogen bases are connected. Each exposed nitrogen base then picks up a complementary nitrogen base (with associated sugar and phosphate group) from the cytoplasm of the cell. This uncoiling and complementary base pairing continues until each of the two original DNA strands is matched and joined with two newly formed DNA strands. The original DNA molecule has become two DNA molecules.

The S period is preceded by a **G₁** (for gap or growth) **period,** during which cells are engaged in growth, metabolism, and the production of substances required for division. Following chromosomal replication in the S period, there is another G period called the **G₂ period.** During this gap period there is activity similar to that occurring in the G₁ period. Since the G periods are times during which there are no events related to chromosomal replication, they are thought of as gaps in the divisional cycle (see Figure 2-13).

A microscopic view of a cell during interphase shows a clearly defined nuclear membrane, nucleoli, karyolymph, chromatin, and a pair of centrioles. Once a cell completes its replication of DNA and its production of RNA and proteins during interphase, mitosis begins.

MITOSIS

The succession of events that takes place during mitosis and cytokinesis is plainly visible under a microscope after the cells have been stained in the laboratory.

The process called **mitosis** is the distribution of the two sets of chromosomes into two separate and equal nuclei following the replication of the chromosomes of the parent nucleus. For convenience, biologists divide the process into four stages: prophase, metaphase, anaphase, and telophase. These are arbitrary classifications. Mitosis is actually a continuous process, one stage merging imperceptibly into the next.

Prophase

During **prophase** (Figure 2-12b), the first stage of mitosis, chromatin shortens and coils into chromosomes. The nucleoli become less distinct and the nuclear membrane disappears. Each prophase "chromosome" is actually composed of a pair of structures called **chromatids.** Each

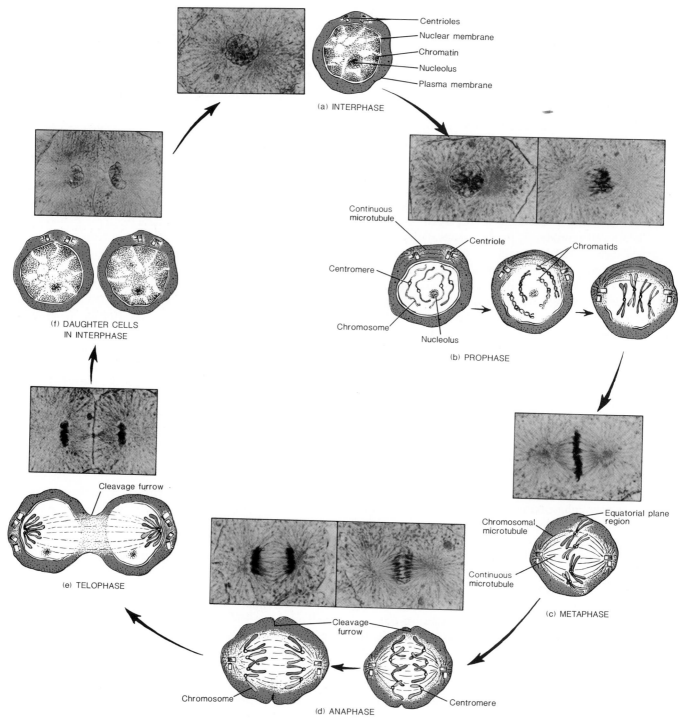

FIGURE 2-12 Cell division: mitosis and cytokinesis. Photomicrographs and diagrammatic representations of the various stages of cell division in whitefish eggs. Read the sequence starting at (a), and move clockwise until you complete the cycle. (Courtesy of Carolina Biological Supply Company.)

chromatid is a complete chromosome consisting of a double-stranded DNA molecule, and each is attached to its chromatid pair by a small spherical body called a **centromere.** During prophase, the chromatid pairs assemble near the equatorial plane region (equator) of the cell.

Also during prophase, the paired centrioles separate, and each pair moves to an opposite pole (end) of the cell. Between the centrioles, a series of microtubules is organized into two groups of fibers. The *continuous (interpolar) microtubules* originate from the vicinity of each pair of centrioles and grow toward each other. Thus, they extend from one pole of the cell to another. As they grow toward each other, the second group of microtubules develops. These are called *chromosomal microtu-*

bules and grow out of the centromeres; they extend from a centromere to a pole of the cell. Together, the continuous and chromosomal microtubules constitute the **mitotic spindle** and, with the centrioles, are referred to as the **mitotic apparatus.**

Metaphase

During **metaphase** (Figure 2-12c), the second stage of mitosis, the centromeres of the chromatid pairs line up on the equatorial plane of the cell. The centromeres of each chromatid pair form a chromosomal microtubule that attaches the centromere to a pole of the cell.

Anaphase

The third stage of mitosis, **anaphase** (Figure 2-12d), is characterized by the division of the centromeres and the movement of complete identical sets of chromatids, now called chromosomes, to opposite poles of the cell. During this movement, the centromeres attached to the chromosomal microtubules seem to drag the trailing parts of the chromosomes toward opposite poles.

The mechanism by which the chromosomes move to opposite poles is not completely understood. Several theories have been presented and the key points of two of them may be summarized as follows. One suggests that the movement of chromosomes is caused by the assembly and disassembly of microtubules. As the chromosomal microtubules are shortened by the removal of tubulin subunits, the chromosomes are pulled toward the poles. Simultaneously, the continuous microtubules are lengthened by the addition of tubulin subunits, further contributing to chromosomal movement.

A second theory is that chromosomal movement and pole separation are caused by mechanochemical interactions between continuous and chromosomal microtubules. This is accomplished by temporary cross bridges between microtubules. As a result of reactions that use energy, the cross bridges cause the continuous and chromosomal microtubules to slide past each other in a manner similar to the closing of an extension ladder.

Telophase

Telophase (Figure 2-12e), the final stage of mitosis, consists of a series of events nearly the reverse of prophase. By this time, two identical sets of chromosomes have reached opposite poles. As telophase progresses, new nuclear membranes begin to enclose the chromosomes, the chromosomes start to assume their chromatin form, nucleoli reappear, and the mitotic spindles disappear. The centrioles also replicate so that each cell has two centriole pairs. The formation of two nuclei identical to those of cells in interphase terminates telophase. A mitotic cycle has thus been completed (Figure 2-12f).

Time Required

The time required for mitosis varies with the kind of cell, its location, and the influence of factors such as temperature. Furthermore, the different stages of mitosis are not equal in duration. Prophase is usually the longest stage, lasting from one to several hours. Metaphase is considerably shorter, ranging from 5 to 15 minutes. Anaphase is the shortest stage, lasting from 2 to 10 minutes. Telophase lasts from 10 to 30 minutes. These lengths of time are only approximate, however.

CYTOKINESIS

Division of the cytoplasm, called **cytokinesis** (sī'-tō-ki-NĒ-sis), often begins during late anaphase and terminates at the same time as telophase. Cytokinesis begins with the formation of a *cleavage furrow* that extends around the cell's equator. The furrow progresses inward, resembling a constricting ring, and cuts completely through the cell to form two separate portions of cytoplasm (Figure 2-12d–f).

If we consider the cell cycle in its entirety, the sequence of events can be summarized as follows: G_1 period ⟶ S period ⟶ G_2 period ⟶ Mitosis ⟶ Cytokinesis (Figure 2-13).

A summary of the events that occur during the interphase and cell division is presented in Exhibit 2-2.

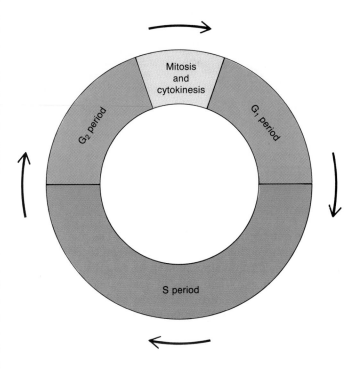

FIGURE 2-13 Various phases and periods in a cell cycle. Relative amounts of time are indicated by the size of the compartments.

EXHIBIT 2-2   SUMMARY OF EVENTS ASSOCIATED WITH INTERPHASE AND CELL DIVISION

PERIOD OR STAGE	ACTIVITY
INTERPHASE	Cell is between divisions.
G₁ period	Cell engages in growth, metabolism, and production of substances required for division; no chromosomal replication.
S period	Chromosomal replication occurs.
G₂ period	Same as for G₁ period.
CELL DIVISION	Single parent cell produces two identical daughter cells.
Prophase	Chromatin shortens and coils into chromosomes (chromatids), nucleoli and nuclear membrane become less distinct, centrioles separate and move to opposite poles of cell, and mitotic spindle forms.
Metaphase	Centromeres of chromatid pairs line up on equatorial plane of cell and form chromosomal microtubules that attach centromeres to poles of cell.
Anaphase	Centromeres divide and identical sets of chromosomes move to opposite poles of cell.
Telophase	Nuclear membrane reappears and encloses chromosomes, chromosomes resume chromatin form, nucleoli reappear, mitotic spindle disappears, and centrioles duplicate.
Cytokinesis	Cleavage furrow forms around equatorial plane of cell, progresses inward, and separates cytoplasm into two separate and equal portions.

ABNORMAL CELL DIVISION: CANCER

DEFINITION

Cancer is not a single disease but many. The human body contains more than a hundred different types of cells, each of which can malfunction in its own distinctive way to cause cancer. When cells in some area of the body duplicate without control, the excess of tissue that develops is called a growth, or **tumor.** The study of tumors is called **oncology** (*onco* = swelling or mass; *logos* = study of), and a physician who specializes in this field is called an **oncologist.** Tumors may be cancerous and sometimes fatal or they may be quite harmless. A cancerous growth is called a **malignant tumor,** or **malignancy.** A noncancerous growth is called a **benign growth.** Benign tumors are composed of cells that do not spread to other

parts of the body, and they may be removed if they interfere with a normal body function or are disfiguring.

The majority of cancer patients do not die from the **primary tumor** that develops. Rather, they die from secondary infections due to lowered resistance following **metastasis** (me-TAS-ta-sis), the spread of the disease to other parts of the body. One of the unique properties of a malignant tumor is its ability to metastasize. Metastatic groups of cells are harder to detect and eliminate than primary tumors.

TYPES

At present, cancers are classified by their microscopic appearance and the body site from which they arise. At least 100 different cancers have been identified in this way. If finer details of appearance are taken into consideration, the number can be increased to 200 or more. The name of the cancer is derived from the type of tissue in which it develops. **Carcinoma** (*carc* = cancer; *oma* = tumor) refers to a malignant tumor consisting of epithelial cells. A tumor that develops from a gland is called an **adenosarcoma** (*adeno* = gland). **Sarcoma** is a general term for any cancer arising from connective tissue. **Osteogenic sarcomas** (*osteo* = bone; *genic* = origin), the most frequent type of childhood cancer, destroy normal bone tissue and eventually spread to other areas of the body. **Myelomas** (*myelos* = marrow) are malignant tumors, occurring in middle-aged and older people, that interfere with the blood-cell-producing function of bone marrow and cause anemia. **Chondrosarcomas** are cancerous growths of cartilage (*chondro* = cartilage).

POSSIBLE CAUSES

What triggers a perfectly normal cell to lose control and become abnormal? Scientists are uncertain. First, there are environmental agents: substances in the air we breathe, the water we drink, the food we eat. A chemical or other environmental agent that produces cancer is called a **carcinogen.** The World Health Organization estimates that carcinogens may be associated with 60 to 90 percent of all human cancer. Examples of carcinogens are the hydrocarbons found in cigarette tar. Ninety percent of all lung cancer patients are smokers. Another environmental factor is radiation. Ultraviolet (UV) light from the sun, for example, may cause genetic mutations in exposed skin cells and lead to cancer, especially among light-skinned people.

Viruses are a second cause of cancer, at least in animals. These agents are tiny packages of nucleic acids, either DNA or RNA, that are capable of infecting cells and converting them to virus-producers. Virologists have linked tumor viruses with cancer in many species of birds and mammals, including primates. Since these experiments have not been performed on humans, there is no absolute proof that viruses cause human cancer. Nevertheless, with over 100 separate viruses identified as carcinogens in many species and tissues of animals, it is also probable that at least some cancers in humans are due to viruses. For example, *human T-cell leukemia-lymphoma virus-1* (*HTLV-1*) is strongly associated with *leukemia,* a malignant disease of blood-forming tissues, and *lymphoma,* a cancer of lymphoid tissue. A variant of HTLV-1, known as HTLV-3, is the causative agent of acquired immune deficiency syndrome (AIDS). The *Epstein-Barr virus* (*EBV*), the causative agent of infectious mononucleosis, has recently been linked as the causative agent of three human cancers—*Burkitt's lymphoma* (a tumor of the jaw in African children and young adults), *nasopharyngeal carcinoma* (common in Chinese males), and *Hodgkin's disease* (a cancer of the lymphatic system). The *hepatitis B virus* (*HBV*) has been associated with cancer of the liver. Also, *type 2 herpes simplex virus,* the causative agent of genital herpes, has been implicated in cancer of the cervix of the uterus, and the *papilloma virus,* a virus that causes warts, has been associated with cancer of the cervix and cancer of the colon.

A great deal of cancer research is now directed toward studying **oncogenes,** genes that have the ability to transform a normal cell into a cancerous cell. Oncogenes are slightly altered versions of their normal genes, referred to as **proto-oncogenes.** It appears that some proto-oncogenes are activated to oncogenes by various types of mutations in which the DNA of the proto-oncogenes is altered. Such mutations are induced by carcinogens. Other proto-oncogenes are activated by viruses.

Currently, scientists are also trying to establish a relationship between stress and cancer. Some believe that stress may play a role not only in the development but also in the metastasis of cancer.

CELLS AND AGING

Aging is a progressive failure of the body's homeostatic adaptive responses. It is a general response that produces observable changes in structure and function and increased vulnerability to environmental stress and disease. Disease and aging probably accelerate each other. The specialized branch of medicine that deals with the medical problems and care of elderly persons is called **geriatrics** (jer'-ē-AT-riks; *geras* = old age; *iatrike* = surgery, medicine).

The obvious characteristics of aging are well known: graying and loss of hair, loss of teeth, wrinkling of skin, decreased muscle mass, and increased fat deposits. The physiological signs of aging are gradual deterioration in function and capacity to respond to environmental stress. Thus basic kidney and digestive metabolic rates decrease, as does the ability to respond effectively to changes in temperature, diet, and oxygen supply in order to maintain a constant internal environment. These manifestations of aging are related to a net decrease in the number of cells in the body (thousands of brain cells are lost each day) and to the dysfunctioning of the cells that remain.

The extracellular components of tissues also change with age. Collagen fibers, responsible for the strength in tendons, increase in number and change in quality with aging. These changes in the collagen of arterial walls are as much responsible for their loss of extensibility as are the deposits associated with atherosclerosis, the deposition of fatty materials in arterial walls. Elastin, another extracellular component, is responsible for the elasticity of blood vessels and skin. It thickens, fragments, and acquires a greater affinity for calcium with age—changes that may also be associated with the development of atherosclerosis.

Several kinds of cells in the body—heart cells, skeletal muscle cells, neurons—are incapable of replacement. Experiments have proved that certain other cell types are limited when it comes to cell division. Cells grown outside the body divided only a certain number of times and then stopped. The number of divisions correlated with the donor's age. The number of divisions also correlated with the normal life span of the different species from which the cells were obtained—strong evidence for the hypothesis that cessation of mitosis is a normal, genetically programmed event. According to this view, an "aging" gene is part of the genetic blueprint at birth and it turns on at a preprogrammed time, slowing down or halting processes vital to life.

Another theory of aging is the free radical theory. Free radicals, or oxygen radicals, are oxygen molecules that bear free electrons and are highly reactive and can easily tie up and weaken proteins. Over time, cells containing free radicals grow more rigid as nutrients are excluded and wastes are kept in. Such effects are exhibited as wrinkled skin, stiff joints, and hardened arteries. Free radicals may also cause damage to DNA. Among the factors that produce free radicals are pollution, radiation, and certain foods we eat. Other substances in the diet, such as vitamin E, vitamin C, beta-carotene, and selenium, are antioxidants and inhibit the formation of free radicals.

Whereas some theories of aging explain the process at the cellular level, others concentrate on regulatory mechanisms operating within the entire organism. For example, one such theory holds that the immune system, which manufactures antibodies against foreign invaders, turns on its own cells. This autoimmune response might be caused by changes in the surfaces of cells, causing antibodies to attack the body's own cells. As surface changes in cells increase, the autoimmune response intensifies, producing the well-known signs of aging. Another organismic theory suggests that aging is programmed in the pituitary gland, a hormone-producing gland attached to the undersurface of the brain. Supposedly, at a set time in life, the gland releases a hormone that triggers age-associated disruptions.

The effects of aging on the various body systems are discussed in their respective chapters.

A reminder: As mentioned in the preface, each chapter in this text that discusses a major system of the body is followed by a glossary of **key medical terms.** Both normal and pathological conditions of the system are included in these glossaries. You should familiarize yourself with the terms since they will play an essential role in your medical vocabulary.

Some of these disorders, as well as disorders discussed in the text, are referred to as local or systemic. A **local disease,** condition, or injury is one that involves one part or a limited area of the body. A **systemic** disease or systemic effects of a condition or injury involve several parts or the entire body.

The science that deals with why, when, and where diseases occur and how they are transmitted in a human community is known as **epidemiology** (ep'-i-dē'-mē-OL-ō-jē; *epidemios* = prevalent; *logos* = study of). The science that deals with the effects and uses of drugs in the treatment of disease is called **pharmacology** (far'-ma-KOL-ō-jē; *pharmakon* = medicine; *logos* = study of).

KEY MEDICAL TERMS ASSOCIATED WITH CELLS

Atrophy (*a* = without; *tropho* = nourish)   A decrease in the size of cells with subsequent decrease in the size of the affected tissue or organ; wasting away.

Biopsy (*bio* = life; *opsis* = vision)   The removal and microscopic examination of tissue from the living body for diagnosis.

Deterioration (*deterior* = worse or poorer)   The process or state of growing worse; disintegration or wearing away.

Dysplasia (*dys* = abnormal; *plassein* = to form)   Alteration in the size, shape, and organization of cells due to chronic irritation or inflammation; may progress to neoplasia or revert to normal if the stress is removed.

Hyperplasia (*hyper* = over; *plas* = grow)   Increase in the number of cells due to an increase in the frequency of cell division.

Hypertrophy   Increase in the size of cells without cell division.

Insidious   Hidden, not apparent, as a disease that does not exhibit distinct symptoms of its arrival.

Metaplasia (*meta* = change)   The transformation of one cell into another.

Metastasis (*stasis* = standing still)   The transfer of disease from one part of the body to another that is not directly connected with it.

Necrosis (*necros* = death; *osis* = condition)   Death of a group of cells.

Neoplasm (*neo* = new)   Any abnormal formation or growth, usually a malignant tumor.

Progeny (*progignere* = to bring forth)   Offspring or descendants.

Senescence   The process of growing old.

STUDY OUTLINE

Generalized Animal Cell (p. 29)

1. A cell is the basic, living, structural and functional unit of the body.
2. A generalized cell is a composite that represents various cells of the body.
3. Cytology is the science concerned with the study of cells.
4. The principal parts of a cell are the plasma (cell) membrane, cytoplasm, organelles, and inclusions. Extracellular materials are manufactured by the cell and deposited outside the plasma membrane.

Plasma (Cell) Membrane (p. 29)

Chemistry and Structure

1. The plasma (cell) membrane surrounds the cell and separates it from other cells and the external environment.
2. It is composed primarily of proteins and phospholipids. According to the fluid mosaic model, the membrane consists of a phospholipid bilayer with integral and peripheral proteins.

Functions

1. Functionally, the plasma membrane facilitates contact with other cells, provides receptors, and mediates the passage of materials.
2. The membrane's selectively permeable nature restricts the passage of certain substances. Substances can pass through the membrane depending on their molecular size, lipid solubility, electrical charges, and the presence of carriers.

Movement of Materials Across Plasma Membranes

1. Passive processes involve the kinetic energy of individual molecules.
2. Diffusion is the net movement of molecules or ions from an area of high concentration to an area of low concentration until an equilibrium is reached.
3. In facilitated diffusion, certain molecules, such as glucose, combine with a carrier to become soluble in the phospholipid portion of the membrane.
4. Osmosis is the movement of water through a selectively permeable membrane from an area of high water concentration to an area of low water concentration.
5. In an isotonic solution, red blood cells maintain their normal shape; in a hypotonic solution, they undergo hemolysis; in a hypertonic solution, they undergo crenation.
6. Filtration is the movement of water and dissolved substances

across a selectively permeable membrane by pressure.
7. Active processes involve the use of ATP by the cell.
8. Active transport is the movement of ions across a cell membrane from lower to higher concentration.
9. Endocytosis is the movement of substances through plasma membranes in which the membrane surrounds the substance, encloses it, and brings it into the cell.
10. Phagocytosis is the ingestion of solid particles by pseudopodia. It is an important process used by white blood cells to destroy bacteria that enter the body.
11. Pinocytosis is the ingestion of a liquid by the plasma membrane. In this process, the liquid becomes surrounded by a vacuole.
12. Receptor-mediated endocytosis is the selective uptake of large molecules and particles by cells.

Modified Plasma Membranes
1. Membranes of certain cells are modified for specific functions.
2. Microvilli are microscopic, fingerlike projections of the plasma membrane that increase the surface area for absorption.
3. Rods of the eye contain sacs of light-sensitive pigments.
4. Stereocilia are long, slender, branching cells that line the ductus epididymis.
5. The myelin sheath of nerve cells protects, aids impulse conduction, and provides nutrition.

Cytoplasm (p. 35)
1. Cytoplasm is the substance inside the cell that contains organelles and inclusions.
2. It is composed mostly of water plus proteins, carbohydrates, lipids, and inorganic substances. The chemicals in cytoplasm are either in solution or in a colloid (suspended) form.
3. Functionally, cytoplasm is the medium in which chemical reactions occur.

Organelles (p. 36)
1. Organelles are specialized portions of the cell with characteristic morphology that carry on specific activities.
2. They assume specific roles in cellular growth, maintenance, repair, and control.

Nucleus
1. Usually the largest organelle, the nucleus controls cellular activities and contains the genetic information.
2. Cells without nuclei, such as mature red blood cells, do not grow or reproduce.
3. The parts of the nucleus include the nuclear membrane, karyolymph, nucleoli, and genetic material (DNA), comprising the chromosomes.
4. Chromosomes consist of DNA and histones and consist of subunits called nucleosomes.

Ribosomes
1. Ribosomes are granular structures consisting of ribosomal RNA and ribosomal proteins.
2. They occur free (singly or in clusters) or in conjunction with endoplasmic reticulum.
3. Functionally, ribosomes are the sites of protein synthesis.

Endoplasmic Reticulum (ER)
1. The ER is a network of parallel membranes continuous with the plasma membrane and nuclear membrane.
2. Granular, or rough, ER has ribosomes attached to it. Agranular, or smooth, ER does not contain ribosomes.
3. The ER provides mechanical support, conducts intracellular nerve impulses in muscle cells, exchanges materials with cytoplasm, transports substances intracellularly, stores synthesized molecules, and helps export chemicals from the cell.

Golgi Complex
1. The Golgi complex consists of four to eight stacked, flattened channels (cisternae) with expanded terminal areas (vesicles).
2. In conjunction with the ER, the Golgi complex secretes proteins and lipids and synthesizes and secretes glycoproteins.
3. It is particularly prominent in secretory cells such as those in the pancreas or salivary glands.

Mitochondria
1. Mitochondria consist of a smooth outer membrane and a folded inner membrane surrounding the interior matrix. The inner folds are called cristae.
2. The mitochondria are called "powerhouses of the cell" because ATP is produced in them.

Lysosomes
1. Lysosomes are spherical structures that contain digestive enzymes. They are formed from Golgi complexes.
2. They are found in large numbers in white blood cells, which carry on phagocytosis.
3. If the cell is injured, lysosomes release enzymes and digest the cell. Thus they are called "suicide packets."
4. Lysosomes may be involved in bone removal.

Peroxisomes
1. Peroxisomes are similar to lysosomes, but smaller.
2. They contain enzymes (e.g., catalase) involved in the metabolism of hydrogen peroxide.

Microfilaments and Microtubules: The Cytoskeleton
1. Together microfilaments and microtubules form the cytoskeleton.
2. Microfilaments are rodlike structures consisting of the protein actin or myosin. They are involved in muscular contraction, support, and movement.
3. Microtubules are cylindrical structures consisting of the protein tubulin. They support, provide movement, and form the structure of flagella, cilia, centrioles, and the mitotic spindle.

Centrosome and Centrioles
1. The dense area of cytoplasm containing the centrioles is called a centrosome.
2. Centrioles are paired cylinders arranged at right angles to one another. They assume an important role in cell reproduction.

Flagella and Cilia
1. These cellular projections have the same basic structure and are used in movement.
2. If projections are few and long, they are called flagella. If they are numerous and hairlike, they are called cilia.
3. The flagellum on a sperm cell moves the entire cell. The cilia on cells of the respiratory tract move foreign matter

trapped in mucus along the cell surfaces toward the throat for elimination.

Cell Inclusions (p. 43)

1. Cell inclusions are chemical substances produced by cells. They are usually organic and may have recognizable shapes.
2. Examples are melanin, glycogen, lipids, and mucus.

Extracellular Materials (p. 43)

1. These are all the substances that lie outside the plasma membrane.
2. They provide support and a medium for the diffusion of nutrients and wastes.
3. Some, like hyaluronic acid and chondroitin sulfate, are amorphous. Others, like collagenous, reticular, and elastic fibers, are fibrous.

Normal Cell Division (p. 44)

1. Cell division is the process by which cells reproduce themselves. It consists of nuclear division and cytoplasmic division (cytokinesis).
2. Cell division that results in the production of sperm and eggs consists of a nuclear division called meiosis and cytokinesis.
3. Cell division that results in an increase in body cells involves a nuclear division called mitosis and cytokinesis.
4. Prior to mitosis and cytokinesis, the DNA molecules, or chromosomes, replicate themselves so the same chromosomal complement can be passed on to future generations of cells.
5. A cell carrying on every life process except division is said to be in interphase, or the metabolic phase.
6. During interphase, the replication of chromosomes occurs during the S period, and the manufacture of RNA and proteins required for producing structures for cell division takes place during the G_1 and G_2 periods.

Mitosis

1. Mitosis is the distribution of two sets of chromosomes into separate and equal nuclei following their replication.
2. It consists of prophase, metaphase, anaphase, and telophase.

Cytokinesis

1. Cytokinesis begins in late anaphase and terminates in telophase.
2. A cleavage furrow forms at the cell's equator and progresses inward, cutting through the cell to form two separate portions of cytoplasm.

Abnormal Cell Division: Cancer (p. 48)

1. Cancerous tumors are referred to as malignant; noncancerous tumors are called benign; the study of tumors is called oncology.
2. The spread of cancer from its primary site is called metastasis.
3. Carcinogens include environmental agents and viruses.

Cells and Aging (p. 49)

1. Aging is a progressive failure of the body's homeostatic adaptive responses.
2. Many theories of aging have been proposed, including a genetically programmed cessation of cell division and excessive immune responses, but none successfully answers all the experimental objections.
3. All of the various body systems exhibit definitive and sometimes extensive changes with aging.

REVIEW QUESTIONS

1. Define a cell. What are the four principal portions of a cell? What is meant by a generalized cell?
2. Discuss the chemistry and structure of the plasma membrane with respect to the fluid mosaic model.
3. Describe the various functions of the plasma membrane. What determines selective permeability?
4. Define and give an example of each of the following: diffusion, facilitated diffusion, osmosis, filtration, active transport, phagocytosis, pinocytosis, and receptor-mediated endocytosis.
5. Describe the structure and function of microvilli, rod sacs, myelin sheaths, and stereocilia as membrane modifications.
6. Discuss the chemical composition and physical nature of cytoplasm. What is its function?
7. What is an organelle? By means of a labeled diagram, indicate the parts of a generalized animal cell.
8. Describe the structure and functions of the nucleus of a cell. What are nucleosomes?
9. Discuss the distribution of the ribosomes. What is their function?
10. Distinguish between granular (rough) and agranular (smooth) endoplasmic reticulum (ER). What are the functions of ER?
11. Describe the structure and functions of the Golgi complex.
12. Why are mitochondria referred to as "powerhouses of the cell"?
13. List and describe the various functions of lysosomes.
14. What is the importance of peroxisomes?
15. Contrast the structure and functions of microfilaments and microtubules.
16. Describe the structure and function of centrioles.
17. How are cilia and flagella distinguished on the basis of structure and function?
18. Define a cell inclusion. Provide examples and indicate their functions.
19. What is an extracellular material? Give examples and the functions of each.
20. Distinguish between the two types of cell division. Why is each important?
21. Define interphase. Explain what happens during the G_1, S, and G_2 periods of interphase.
22. Describe the principal events of each stage of mitosis and of cytokinesis.
23. What is a tumor? Distinguish between malignant and benign tumors.
24. Discuss how certain carcinogens may be related to cancer.
25. What is aging? List some of the characteristics of aging. Briefly describe the various theories regarding aging.
26. Refer to the glossary of key medical terms associated with cells. Be sure that you can define each term.

Tissues

Student Objectives

Define a tissue.

Classify the tissues of the body into four major types and define each type.

Describe the distinguishing characteristics of epithelial tissue.

Contrast the structural and functional differences between covering and lining epithelium and glandular epithelium.

Compare the layering arrangements and cell shapes of covering and lining epithelium.

Explain the various kinds of cell junctions that exist between epithelial cells.

List the structure, location, and function for the following types of epithelium: simple squamous, simple cuboidal, simple columnar (nonciliated and ciliated), stratified squamous, stratified cuboidal, stratified columnar, transitional, and pseudostratified.

Define a gland and distinguish between exocrine and endocrine glands.

Classify exocrine glands according to structural complexity and function and give an example of each.

Identify the distinguishing characteristics of connective tissue.

Contrast the structural and functional differences between embryonic and adult connective tissues.

Describe the ground substance, fibers, and cells that constitute connective tissue.

List the structure, function, and location of loose (areolar) connective tissue, adipose tissue, and dense, elastic, and reticular connective tissue.

List the structure, function, and location of the three types of cartilage.

Describe the structure and functions of osseous tissue (bone) and vascular tissue (blood).

Contrast the three types of muscle tissue with regard to structure and location.

Describe the structural features and functions of nervous tissue.

Define an epithelial membrane.

List the location and function of mucous, serous, synovial, and cutaneous membranes.

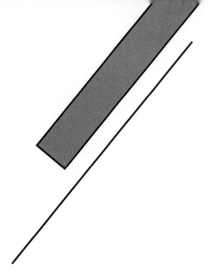

Cells are highly organized units, but they do not function in isolation. They work together in a group of similar cells called a tissue.

TYPES OF TISSUES

A **tissue** is a group of similar cells and their intercellular substance functioning together to perform a specialized activity. The science that deals with the study of tissues is called **histology** (*histo* = tissue; *logos* = study of). Certain tissues function to move body parts. Others move food through body organs. Some tissues protect and support the body. Others produce chemicals such as enzymes and hormones. The various tissues of the body are classified into four principal types according to their function and structure.

1. Epithelial (ep'-i-THĒ-lē-al) **tissue,** which covers body surfaces or tissues, lines body cavities, and forms glands.

2. Connective tissue, which protects and supports the body and its organs and binds organs together.

3. Muscular tissue, which is responsible for movement.

4. Nervous tissue, which initiates and transmits nerve impulses that coordinate body activities.

Epithelial tissue and connective tissue, except for bone and blood, will be discussed in detail in this chapter. The general features of bone tissue and blood will be introduced here, but their detailed discussion occurs later in the book. Similarly, the detailed discussion of muscle tissue and nervous tissue will be postponed.

EPITHELIAL TISSUE

The tissues in this principal type perform many activities in the body, ranging from protection of underlying tissues against microbial invasion, drying out, and harmful environmental factors to secretion. **Epithelial tissue,** or more simply **epithelium,** may be divided into two subtypes: (1) *covering and lining epithelium* and (2) *glandular epithelium.* Covering and lining epithelium forms the outer covering of external body surfaces and the outer covering of some internal organs. It lines the body cavities and the interiors of the respiratory and digestive tracts, blood vessels, and ducts. It makes up, along with nervous tissue, the parts of the sense organs that are sensitive to the stimuli that produce smell and hearing sensations. And it is the tissue from which gametes (sperm and eggs) develop. Glandular epithelium constitutes the secreting portion of glands.

Both types of epithelium consist largely or entirely of closely packed cells with little or no intercellular material between adjacent cells. (Such intercellular material is also called the matrix.) Epithelial cells are arranged in continuous sheets that may be either single or multilayered. Nerves may extend through these sheets, but blood vessels do not. They are *avascular* (*a* = without; *vascular* = blood vessels). The vessels that supply nutrients and remove wastes are located in underlying connective tissue.

Both types of epithelium overlie and adhere firmly to the connective tissue, which holds the epithelium in position and prevents it from being torn. The surface of attachment between the epithelium and the connective tissue is a thin extracellular layer called the **basement membrane.** With few exceptions, epithelial cells secrete a material along their basal surfaces consisting of a special type of collagen and glycoproteins. This structure averages 500 to 800 Å in thickness and is referred to as the **basal lamina.** Frequently, the basal lamina is reinforced by an underlying *reticular lamina* consisting of reticular fibers and glycoproteins. This lamina is produced by cells in the underlying connective tissue. The combination of the basal lamina and reticular lamina constitutes the basement membrane.

Since all epithelium is subjected to a certain degree of wear, tear, and injury, its cells must divide and produce new cells to replace those that are destroyed.

COVERING AND LINING EPITHELIUM

Arrangement of Layers

Covering and lining epithelium is arranged in several different ways related to location and function. If the epithelium is specialized for absorption or filtration and is in an area that has minimal wear and tear, the cells of the tissue are arranged in a single layer. Such an arrangement is called **simple epithelium.** If the epithelium is not specialized for absorption or filtration and is found in an area with a high degree of wear and tear, then the cells are stacked in several layers. This tissue is referred to as **stratified epithelium.** A third, less common, arrangement of epithelium is called **pseudostratified.** Like simple epithelium, pseudostratified epithelium has only one layer of cells. However, some of the cells do not reach the surface—an arrangement that gives the tissue a multilayered, or stratified, appearance. The pseudostratified cells that do reach the surface either secrete mucus or contain cilia that move mucus and foreign particles for eventual elimination from the body.

Cell Shapes

In addition to classifying covering and lining epithelium according to the number of its layers, we may also categorize it by cell shape. The cells may be flat, cubelike, or columnar or may resemble a cross between shapes. **Squamous** (SKWĀ-mus) cells are flattened and scalelike. They are attached to each other like a mosaic. **Cuboidal** cells are usually cube-shaped in cross section. They sometimes appear as hexagons. **Columnar** cells are tall and cylindrical, appearing as rectangles set on end. **Transitional** cells often have a combination of shapes and are found where there is a great degree of distention or expansion in the body. Transitional cells in the bottom layer of an epithelial tissue may range in shape from cuboidal to columnar. In the intermediate layer, they may be cuboidal or polyhedral. In the superficial layer, they may range from cuboidal to squamous, depending on how much they are pulled out of shape during certain body functions.

Cell Junctions

Epithelial cells, unlike connective tissue cells, are tightly joined to form a close functional unit. The points of attachment between adjacent plasma membranes of epithelial cells (and a few other types of cells as well) are called **cell junctions (junctional complexes).** In addition to providing cell-to-cell attachments, cell junctions prevent the movement of materials into certain cells and provide channels for communication between other cells. Four categories of cell junctions are recognized: (1) desmosome (macula adherens), (2) tight junction (zonula occludens), (3) intermediate junction (zonula adherens), and (4) gap junction (nexus). These cell junctions are illustrated in Figure 3-1.

- *Desmosome (Macula adherens)* One of the most frequent types of cell junctions is called a **desmosome** (*desmos* = bond; *soma* = body), or **macula adherens** (*macula* = spot; *adhaereo* = to stick). Desmosomes are numerous in stratified epithelium, especially the epidermis of the skin. They are also found in the epithelial lining of the small intestine. Desmosomes are scattered over adjacent cell surfaces somewhat like spot welds and form firm intercellular attachments between cells. The adjacent plasma membranes at a desmosome appear thickened because there is a dense proteinaceous *plaque* of cytoplasmic material on the inner surfaces of the neighboring membranes. Cytoplasmic filaments called *tonofilaments,* converge in the region of the plaques and make a U-turn back into the cytoplasm. The plasma membranes of the adjacent cells at a desmosome are separated by an intercellular space measuring about 220 to 240 Å. In addition to serving as firm sites of adhesion between adjacent cells, desmosomes also help to support the entire epithelial tissue.

In the basal cells of epithelium, the plasma membranes are in contact with the underlying basement membrane rather than adjacent plasma membranes. Here, there are structures that look like half a desmosome, called **hemidesmosomes** (*hemi* = half). Hemidesmosomes are assumed to provide anchorage for the basal epithelial cell plasma membranes to the basement membrane.

- *Tight Junction (Zonula occludens)* The **tight junction,** or **zonula occludens** (*zonula* = small zone; *occludens* = occlude), is typically located along the lateral surfaces of adjacent plasma membranes, closest to the luminal (free) surface just below microvilli. In a tight junction, the outer layers of adjacent plasma membranes are fused together at various points. The junction completely encircles each cell that is joined and occludes the intercellular spaces. Tight junctions are common in epithelial cells lining the small intestine. Here, they prevent the movement of substances across epithelial cells via the intercellular route. Some substances, however, can selectively enter the small intestinal cells through microvilli by absorption.

- *Intermediate Junction (Zonula adherens)* The **intermediate junction,** or **zonula adherens,** like the tight junction, completely encircles each cell that is joined. The intermediate junction is located just below the tight junction between small intestinal epithelial cells. In an intermediate junction, the adjacent plasma membranes are separated by an intercellular space measuring about 200 Å. The inner surfaces of the neighboring plasma membranes contain a moderately dense area of cytoplasmic

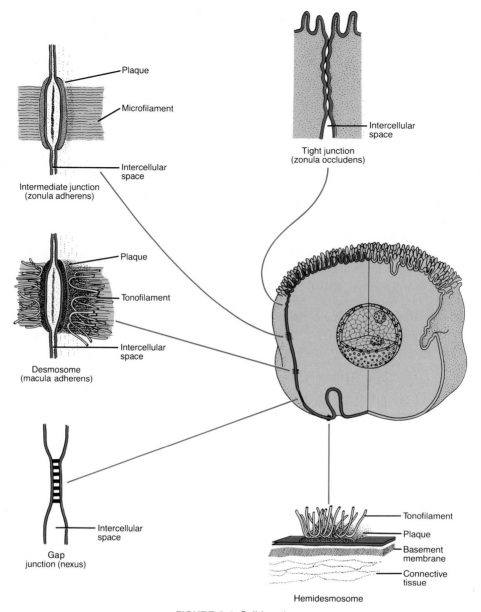

FIGURE 3-1 Cell junctions.

material (not as dense as the plaques in desmosomes) in which *microfilaments* are embedded. These filaments are smaller in diameter than the tonofilaments of desmosomes. Intermediate junctions are believed to form firm connections between adjacent cells.

● *Gap Junction (Nexus)* In a **gap junction,** or **nexus** (*nexus* = bond), the outer layers of adjacent plasma membranes approach each other, leaving a gap of about 20 Å between them. The gap is bridged by structures that link the cells together. These structures may be minute channels formed by membrane proteins that enable adjacent cells to communicate. It is postulated that ions and small molecules pass rapidly through gap junctions from one cell to the next, resulting in spontaneous impulse conduction. Gap junctions are found between smooth

and cardiac muscle cells, as well as between epithelial cells. The presence of gap junctions between muscle cells presumably enables the cells to conduct the impulses rapidly and contract in an orderly and coordinated manner.

Classification

Considering layers and cell type in combination, we may classify covering and lining epithelium as follows:

Simple
    1. Squamous
    2. Cuboidal
    3. Columnar
Statified
    1. Squamous

2. Cuboidal
3. Columnar
4. Transitional
Pseudostratified

Each of the epithelial tissues described in the following sections is illustrated in Exhibit 3-1.

Simple Epithelium

● *Simple Squamous Epithelium* This type of simple epithelium consists of a single layer of flat, scalelike cells. Its surface resembles a tiled floor. The nucleus of each cell is centrally located and oval or spherical. Since simple squamous epithelium has only one layer of cells, it is highly adapted to diffusion, osmosis, and filtration. Thus it lines the air sacs of the lungs, where oxygen is exchanged with carbon dioxide. It is present in the part of the kidney that filters the blood. It also lines the inner surfaces of the membranous labyrinth and tympanic membrane of the ear. Simple squamous epithelium is found in body parts that have little wear and tear.

A tissue similar to simple squamous epithelium is endothelium. **Endothelium** lines the heart, blood vessels, and lymph vessels and forms the walls of capillaries. Another tissue similar to simple squamous epithelium that forms the epithelial layer of serous membranes is **mesothelium.** This tissue lines the thoracic and abdominopelvic cavities and covers the viscera within them.

● *Simple Cuboidal Epithelium* Viewed from above, the cells of simple cuboidal epithelium appear as closely fitted polygons. The cuboidal nature of the cells is obvious only when the tissue is sectioned at right angles. Like simple squamous epithelium, these cells possess a central nucleus. Simple cuboidal epithelium covers the surface of the ovaries, lines the anterior surface of the capsule of the lens of the eye, and forms the pigmented epithelium of the retina of the eye. In the kidneys, where it forms the kidney tubules and contains microvilli, it functions in water reabsorption. It also lines the smaller ducts of some glands and the secreting units of glands, such as the thyroid.

This tissue performs the functions of secretion and absorption. **Secretion,** usually a function of epithelium, is the production and release by cells of a fluid that may contain a variety of substances such as mucus, perspiration, or enzymes. **Absorption** is the intake of fluids or other substances by cells of the skin or mucous membranes.

● *Simple Columnar Epithelium* The surface view of simple columnar epithelium is similar to that of simple cuboidal tissue. When sectioned at right angles to the surface, however, the cells appear somewhat rectangular. The nuclei are located near the bases of the cells.

The luminal surfaces (surfaces adjacent to the lumen, or cavity of a hollow organ, vessel, or duct) of simple columnar epithelial cells are modified in several ways, depending on location and function. Simple columnar epithelium lines the digestive tract from the cardia of the stomach to the anus, the gallbladder, and excretory ducts of many glands. In such sites, the cells protect the underlying tissues. Many of them are also modified to aid in food-related activities. In the small intestine especially, the plasma membranes of the cells are folded into **microvilli** (see Figure 2-1). The microvilli arrangement increases the surface area of the plasma membrane and thereby allows larger amounts of digested nutrients and fluids to be absorbed into the body.

Interspersed among the typical columnar cells of the intestine are other modified columnar cells called **goblet cells.** These cells, which secrete mucus, are so named because the mucus accumulates in the upper half of the cell, causing the area to bulge out. The whole cell resembles a goblet or wine glass. The secreted mucus serves as a lubricant between the food and the walls of the digestive tract.

A third modification of columnar epithelium is found in cells with hairlike processes called **cilia.** In a few portions of the upper respiratory tract, ciliated columnar cells are interspersed with goblet cells. Mucus secreted by the goblet cells forms a film over the respiratory surface. This film traps foreign particles that are inhaled. The cilia wave in unison and move the mucus, with any foreign particles, toward the throat, where it can be swallowed or eliminated. Air is filtered by this process before entering the lungs. Ciliated columnar epithelium is also found in the uterus and uterine (Fallopian) tubes of the female reproductive system, some paranasal sinuses, and the central canal of the spinal cord.

Stratified Epithelium

In contrast to simple epithelium, stratified epithelium consists of at least two layers of cells. Thus it is durable and can protect underlying tissues from the external environment and from wear and tear. Some stratified epithelium cells also produce secretions. The name of the specific kind of stratified epithelium depends on the shape of the surface cells.

● *Stratified Squamous Epithelium* In the more superficial layers of this type of epithelium, the cells are flat, whereas in the deep layers, cells vary in shape from cuboidal to columnar. The basal, or bottom, cells are continually multiplying by cell division. As new cells grow, they compress the cells on the surface and push them outward. The basal cells continually shift upward and outward. As they move farther from the deep layer and their blood supply, they become dehydrated, shrink, and become harder. At the surface, the cells are rubbed off. New cells continually emerge, are sloughed off, and replaced.

EXHIBIT 3-1 EPITHELIAL TISSUES

COVERING AND LINING EPITHELIUM

Simple squamous (250×)
Description: Single layer of flat, scalelike cells; large, centrally located nuclei.
Location: Lines air sacs of lungs, glomerular capsule of kidneys, and inner surface of the membraneous labyrinth and tympanic membrane of the ear. Called endothelium when it lines heart and blood and lymphatic vessels and forms capillaries. Called mesothelium when it lines the ventral body cavity and covers viscera as part of a serous membrane.
Function: Filtration, absorption, and secretion in serous membranes.

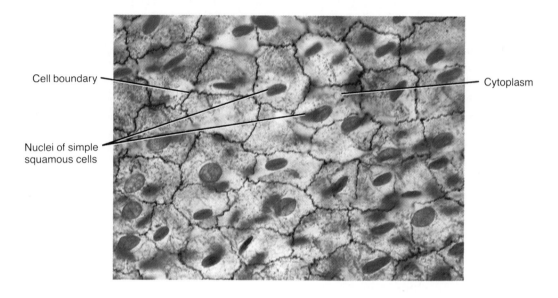

Simple cuboidal (450×)
Description: Single layer of cube-shaped cells; centrally located nuclei.
Location: Covers surface of ovary, lines anterior surface of capsule of the lens of eyes, forms pigmented epithelium of retina of eye, and lines kidney tubules and smaller ducts of many glands.
Function: Secretion and absorption.

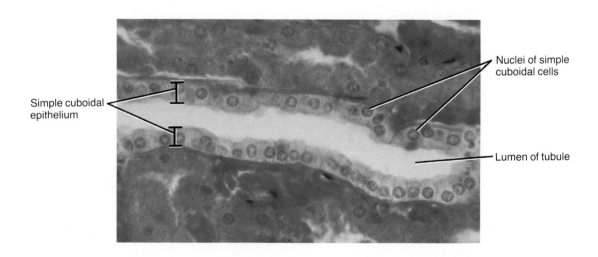

Simple columnar (nonciliated) (600×)

Description: Single layer of nonciliated rectangular cells; contains goblet cells; nuclei at bases of cells.

Location: Lines the digestive tract from the cardia of the stomach to the anus, excretory ducts of many glands, and gallbladder.

Function: Secretion and absorption.

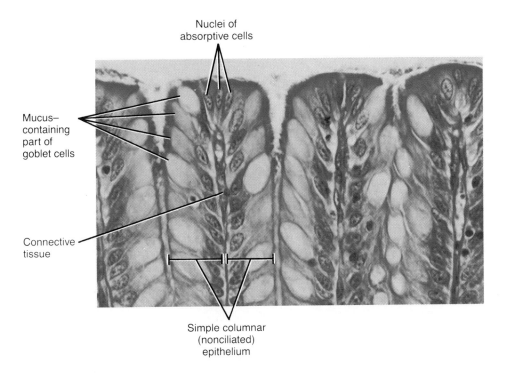

Nuclei of
absorptive cells

Mucus–
containing
part of
goblet cells

Connective
tissue

Simple columnar
(nonciliated)
epithelium

Simple columnar (ciliated) (400×)

Description: Single layer of ciliated columnar cells; contains goblet cells; nuclei at bases of cells.

Location: Lines a few portions of upper respiratory tract, uterine (Fallopian) tubes, uterus, some paranasal sinuses, and central canal of spinal cord.

Function: Moves mucus by ciliary action.

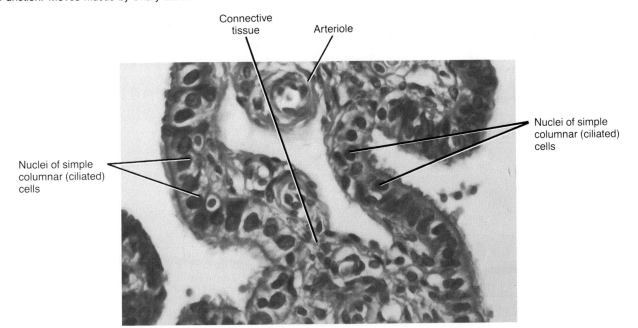

Connective
tissue

Arteriole

Nuclei of simple
columnar (ciliated)
cells

Nuclei of simple
columnar (ciliated)
cells

EXHIBIT 3-1 (*Continued*)

Stratified squamous (185×)
Description: Several layers of cells; cuboidal to columnar shape in deep layers; scalelike shape in superficial layers; basal cells replace surface cells as they are lost.
Location: Nonkeratinized variety lines wet surfaces such as lining of the mouth, tongue, esophagus, part of epiglottis, and vagina.
Function: Protection.

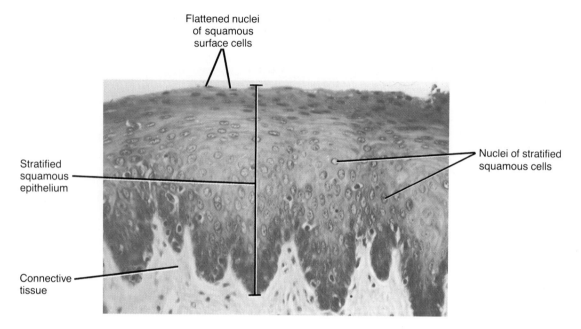

Flattened nuclei
of squamous
surface cells

Stratified
squamous
epithelium

Nuclei of stratified
squamous cells

Connective
tissue

Stratified cuboidal (185×)
Description: Two or more layers of cells in which the surface cells are cube-shaped.
Location: Ducts of adult sweat glands, fornix of conjunctiva of eye, cavernous urethra of male urogenital system, pharynx, and epiglottis.
Function: Protection.

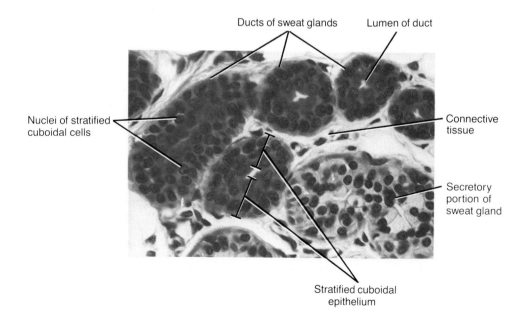

Ducts of sweat glands   Lumen of duct

Nuclei of stratified
cuboidal cells

Connective
tissue

Secretory
portion of
sweat gland

Stratified cuboidal
epithelium

Stratified columnar (600×)
Description: Several layers of polyhedral cells; columnar cells only in superficial layer.
Location: Lines part of male urethra, large excretory ducts of some glands, and small areas in anal mucous membrane.
Function: Protection and secretion.

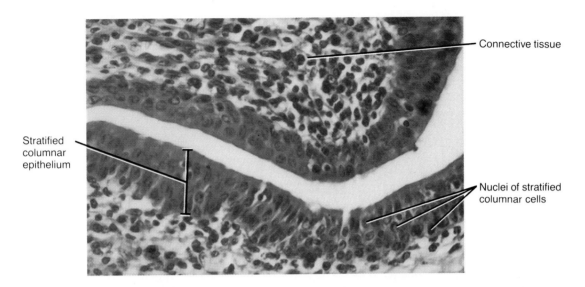

Connective tissue

Stratified columnar epithelium

Nuclei of stratified columnar cells

Stratified transitional (185×)
Description: Resembles nonkeratinized stratified squamous tissue, except that superficial cells are larger and more rounded.
Location: Lines urinary bladder.
Function: Permits distention.

Lumen of urinary bladder

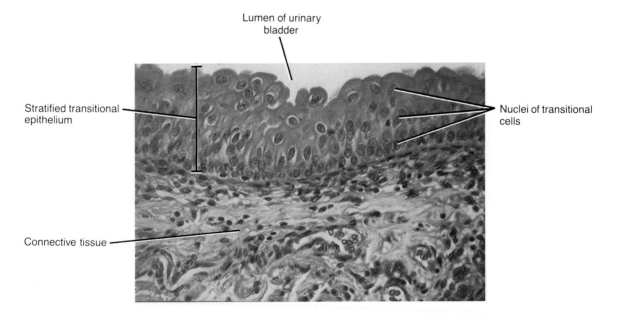

Stratified transitional epithelium

Nuclei of transitional cells

Connective tissue

EXHIBIT 3-1 (*Continued*)

Pseudostratified (500×)

Description: Not a true stratified tissue; nuclei of cells at different levels; all cells attached to basement membrane, but not all reach surface.

Location: Lines larger excretory ducts of many large glands, male urethra, and auditory (Eustachian) tubes; ciliated variety with goblet cells lines most of the upper respiratory tract and some ducts of male reproduction system.

Function: Secretion and movement of mucus by ciliary action.

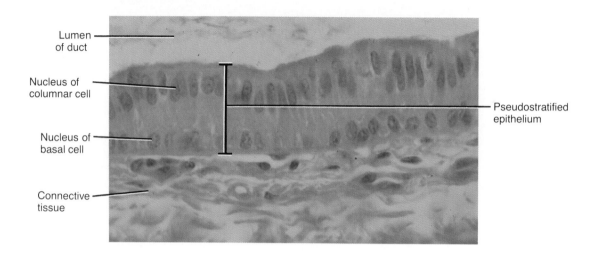

Lumen of duct

Nucleus of columnar cell

Nucleus of basal cell

Connective tissue

Pseudostratified epithelium

GLANDULAR EPITHELIUM

Exocrine gland (300×)

Description: Secretes products into ducts.

Location: Sweat, oil, wax, and mammary glands of the skin; digestive glands such as salivary glands, which secrete into mouth cavity.

Function: Produces mucus, perspiration, oil, wax, milk, or digestive enzymes.

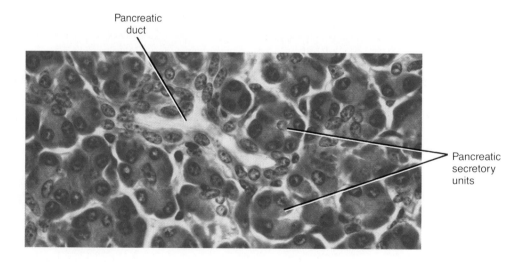

Pancreatic duct

Pancreatic secretory units

Endocrine gland (180×)
Description: Secretes hormones into blood.
Location: Pituitary at base of brain, thyroid and parathyroids near larynx, adrenals above kidneys, pancreas below stomach, ovaries in pelvic cavity and testes in scrotum, pineal at base of brain, and thymus between lungs.
Function: Produces hormones that regulate various body activities.

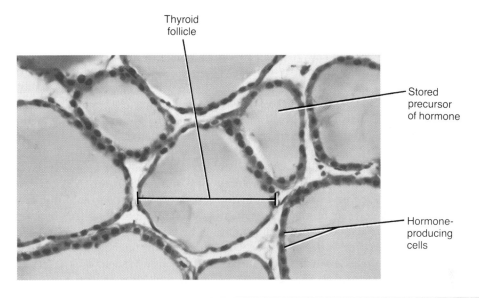

Thyroid follicle

Stored precursor of hormone

Hormone-producing cells

One form of stratified squamous epithelium is called **nonkeratinized stratified squamous epithelium.** This tissue is found on wet surfaces that are subjected to considerable wear and tear and does not perform the function of absorption—the lining of the mouth, the tongue, the esophagus, and the vagina. Another form of stratified squamous epithelium is called **keratinized stratified squamous epithelium.** The surface cells of this type are modified into a tough layer of material containing keratin. **Keratin** is a protein that is waterproof and resistant to friction and helps to resist bacterial invasion. The outer layer of skin consists of keratinized tissue.

● *Stratified Cuboidal Epithelium* This relatively rare type of epithelium is found in the ducts of the sweat glands of adults, fornix of the conjunctiva of the eye, cavernous urethra of the male urogenital system, pharynx, and epiglottis. It sometimes consists of more than two layers of cells. Its function is mainly protective.

● *Stratified Columnar Epithelium* Like stratified cuboidal epithelium, this type of tissue is also uncommon in the body. Usually the basal layer or layers consist of shortened, irregularly polyhedral cells. Only the superficial cells are columnar in form. This kind of epithelium lines part of the male urethra, some larger excretory ducts such as lactiferous (milk) ducts in the mammary glands,

and small areas in the anal mucous membrane. If functions in protection and secretion.

● *Transitional Epithelium* This kind of epithelium is very much like nonkeratinized stratified squamous epithelium. The distinction is that cells of the outer layer in transitional epithelium tend to be large and rounded rather than flat. This feature allows the tissue to be stretched (distended) without the outer cells breaking apart from one another. When stretched, they are drawn out into squamouslike cells. Because of this arrangement, transitional epithelium lines hollow structures that are subjected to expansion from within, such as the urinary bladder. Its function is to help prevent a rupture of the organ.

Pseudostratified Epithelium

The third category of covering and lining epithelium consists of columnar cells and is called pseudostratified epithelium. The nuclei of the cells are at varying depths. Even though all the cells are attached to the basement membrane in a single layer, some do not reach the surface. This feature gives the impression of a multilayered tissue, the reason for the designation *pseudo*stratified epithelium. It lines the larger excretory ducts of many glands, parts of the male urethra, and the auditory (Eustachian) tubes, the tubes that connect the middle ear cavity and upper part of the throat. Pseudostratified epithelium may be ciliated and may contain goblet cells. In this form, it lines most of the upper respiratory tract and certain ducts of the male reproductive system.

GLANDULAR EPITHELIUM

The function of glandular epithelium is secretion, accomplished by glandular cells that lie in clusters deep to the covering and lining epithelium. A **gland** may consist of one cell or a group of highly specialized epithelium cells that secrete substances into ducts, onto a surface, or into the blood. The production of such substances always requires active work by the glandular cells and results in an expenditure of energy.

All glands of the body are classified as exocrine or endocrine according to whether they secrete substances into ducts (or directly onto a free surface) or into the blood. **Exocrine glands** secrete their products into ducts (tubes) that empty at the surface of covering and lining epithelium or directly onto a free surface. The product of an exocrine gland may be released at the skin surface or into the lumen of a hollow organ. The secretions of exocrine glands include mucus, perspiration, oil, wax, and digestive enzymes. Examples of exocrine glands are sweat glands, which eliminate perspiration to cool the skin, sali-

vary glands, which secrete a digestive enzyme, and goblet cells, which produce mucus. **Endocrine glands** are ductless and secrete their products into the blood. The secretions of endocrine glands are always hormones, chemicals that regulate various physiological activities. The pituitary, thyroid, and adrenal glands are endocrine glands.

Structural Classification of Exocrine Glands

Exocrine glands are classified into two structural types: unicellular and multicellular. **Unicellular glands** are single-celled. A good example of a unicellular gland is a goblet cell (Exhibit 3-1). Goblet cells are found in the epithelial lining of the digestive, respiratory, urinary, and reproductive systems. They produce mucus to lubricate the free surfaces of these membranes.

Multicellular glands occur in several different forms (Figure 3-2). If the secretory portions of a gland are tubular, it is referred to as a **tubular gland.** If they are flasklike, it is called an **acinar** (AS-i-nar) **gland.** If the gland contains both tubular and flasklike secretory portions, it is called a **tubuloacinar gland.** Further, if the duct of the gland does not branch, it is referred to as a **simple gland;** if the duct does branch, it is called a **compound gland.** By combining the shape of the secretory portion with the degree of branching of the duct, we arrive at the following structural classification for exocrine glands:

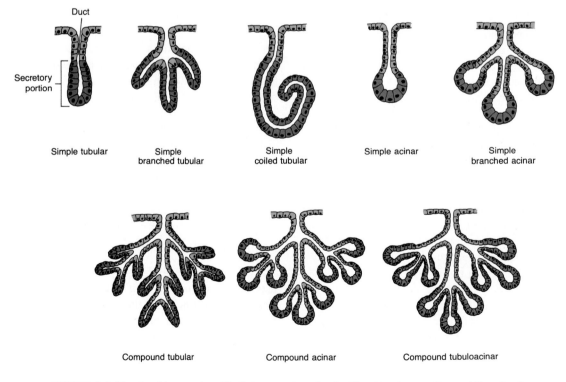

FIGURE 3-2 Structural types of multicellular exocrine glands. The secretory portions of the glands are indicated in purple. The blue areas represent the ducts of the glands.

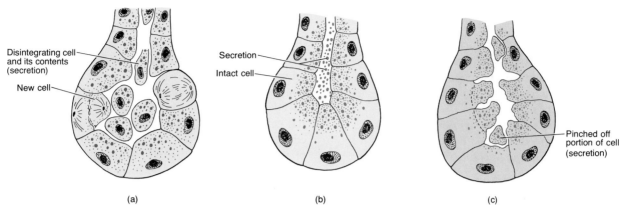

Disintegrating cell
and its contents
(secretion)

New cell

Secretion

Intact cell

Pinched off
portion of cell
(secretion)

(a)   (b)   (c)

FIGURE 3-3 Functional classification of multicellular exocrine glands. (a) Holocrine gland. (b) Merocrine gland. (c) Apocrine gland.

I. **Unicellular.** Single-celled gland that secretes mucus. Example: goblet cell of the digestive and respiratory systems.
II. **Multicellular.** Many-celled glands.
   A. *Simple.* Single, nonbranched duct.
      1. **Tubular.** The secretory portion is straight and tubular. Example: intestinal glands.
      2. **Branched tubular.** The secretory portion is branched and tubular. Examples: gastric and uterine glands.
      3. **Coiled tubular.** The secretory portion is coiled. Example: sudoriferous (sweat) glands.
      4. **Acinar.** The secretory portion is flasklike. Example: seminal vesicle glands.
      5. **Branched acinar.** The secretory portion is branched and flasklike. Example: sebaceous (oil) glands.
   B. *Compound.* Branched duct.
      1. **Tubular.** The secretory portion is tubular. Examples: bulbourethral glands, testes, and liver.
      2. **Acinar.** The secretory portion is flasklike. Examples: salivary glands (sublingual and submandibular).
      3. **Tubuloacinar.** The secretory portion is both tubular and flasklike. Examples: salivary glands (parotid) and pancreas.

Functional Classification of Exocrine Glands

The functional classification of exocrine glands is based on how the gland releases its secretion. The three recognized categories are holocrine, merocrine, and apocrine glands. **Holocrine glands** accumulate a secretory product in their cytoplasm. The cell then dies and is discharged with its contents as the glandular secretion (Figure 3-3a). The discharged cell is replaced by a new cell. One example of a holocrine gland is a sebaceous gland of the skin. **Merocrine glands** simply form the secretory product and discharge it from the cell (Figure 3-3b). Examples of merocrine glands are the salivary glands and pancreas. **Apocrine glands** accumulate their secretory product at the apical (outer) margin of the secreting cell. That portion of the cell pinches off from the rest of the cell to form the secretion (Figure 3-3c). The remaining part of the cell repairs itself and repeats the process. An example of an apocrine gland is the mammary gland.

CONNECTIVE TISSUF

The most abundant tissue in the body is **connective tissue.** This binding and supporting tissue usually is highly vascular and thus has a rich blood supply. An exception is cartilage, which is avascular. The cells are widely scattered rather than closely packed, and there is considerable intercellular material, or matrix. In contrast to epithelium, connective tissues do not occur on free surfaces, such as the surfaces of a body cavity or the external surface of the body. The general functions of connective tissues are protection, support, and the binding together of various organs.

The intercellular substance in a connective tissue largely determines the tissue's qualities. This substance is nonliving and may consist of fluid, semifluid, or mucoid (mucuslike) material. In cartilage, the intercellular material is firm but pliable. In bone, it is considerably harder and not pliable. The cells of connective tissue produce the intercellular substances. The cells may also store fat, ingest bacteria and cell debris, form anticoagulants, or give rise to antibodies that protect against disease.

CLASSIFICATION

Connective tissue may be classified in several ways. We will classify them as follows:

I. **Embryonic connective tissue**
    A. Mesenchyme
    B. Mucous connective tissue
II. **Adult connective tissue**
    A. Connective tissue proper
        1. Loose (areolar) connective tissue
        2. Adipose tissue
        3. Dense (collagenous) connective tissue
        4. Elastic connective tissue
        5. Reticular connective tissue
    B. Cartilage
        1. Hyaline cartilage
        2. Fibrocartilage
        3. Elastic cartilage
    C. Osseous (bone) tissue
    D. Vascular (blood) tissue

Each of the connective tissues described in the following sections is illustrated in Exhibit 3-2.

EMBRYONIC CONNECTIVE TISSUE

Connective tissue that is present primarily in the embryo or fetus is called **embryonic connective tissue.** The term *embryo* refers to a developing human from fertilization through the first two months of pregnancy; a *fetus* refers to a developing human from the third month of pregnancy to birth.

One example of embryonic connective tissue found almost exclusively in the embryo is **mesenchyme (MEZ-en-kīm)**—the tissue from which all other connective tissues eventually arise. Mesenchyme may be observed beneath the skin and along the developing bones of the embryo. Some mesenchymal cells are scattered irregularly throughout adult connective tissue, most frequently around blood vessels. Here mesenchymal cells differentiate into fibroblasts that assist in wound healing.

Another kind of embryonic connective tissue is **mucous connective tissue (Wharton's jelly),** found primarily in the fetus. This tissue is located in the umbilical cord of the fetus, where it supports the wall of the cord.

ADULT CONNECTIVE TISSUE

Adult connective tissue is connective tissue that exists in the newborn and that does not change after birth. It is subdivided into several kinds.

Connective Tissue Proper

Connective tissue that has a more or less fluid intercellular material and a fibroblast as the typical cell is termed **connective tissue proper.** Five examples of such tissues may be distinguished.

● *Loose (Areolar) Connective Tissue* Loose, or areolar (a-RĒ-ō-lar), connective tissue is one of the most widely distributed connective tissues in the body. Structurally, it consists of fibers and several kinds of cells embedded in a semifluid intercellular substance. The term "loose" refers to the loosely woven arrangement of fibers in the intercellular substance. The fibers are neither abundant nor arranged to prevent stretching.

The intercellular substance consists of a viscous material called **hyaluronic acid.** Hyaluronic acid normally facilitates the passage of nutrients from the blood vessels of the connective tissue into adjacent cells and tissues, although the thick consistency of this acid may impede the movement of some drugs. But if an enzyme called **hyaluronidase** is injected into the tissue, hyaluronic acid changes to a watery consistency. This feature is of clinical importance because the reduced viscosity hastens the absorption and diffusion of injected drugs and fluids through the tissue and thus can lessen tension and pain. Some bacteria, white blood cells, and sperm cells produce hyaluronidase.

The three types of fibers embedded between the cells of loose connective tissue are collagenous, elastic, and reticular fibers. **Collagenous (white) fibers** are very tough and resistant to a pulling force, yet are somewhat flexible because they are usually wavy. These fibers often occur in bundles. They are composed of many minute fibers called fibrils lying parallel to one another. The bundle arrangement affords a great deal of strength. Chemically, collagenous fibers consist of the protein collagen. **Elastic (yellow) fibers,** by contrast, are smaller than collagenous fibers and freely branch and rejoin one another. Elastic fibers consist of a protein called elastin. These fibers also provide strength and have great elasticity, up to 50 percent of their length. **Reticular fibers** also consist of collagen, plus some glycoprotein. They are very thin fibers that branch extensively and are not as strong as collagenous fibers. Some authorities believe that reticular fibers are immature collagenous fibers. Like collagenous fibers, reticular fibers provide support and strength and form the *stroma* (framework) of many soft organs.

The cells in loose connective tissue are numerous and varied. Most are **fibroblasts**—large, flat cells with branching processes. If the tissue is injured, the fibroblasts are believed to form collagenous fibers, elastic fibers, and the viscous ground substance. Mature fibroblasts are referred to as **fibrocytes.** The basic distinction between the two is that fibroblasts (or *blasts* of any form of cell) are involved in the formation of immature tissue or repair of mature tissue, and fibrocytes (or *cytes* of any form of cell) are inactive cells that no longer produce fibers or matrix.

Other cells found in loose connective tissue are called **macrophages (MAK-rō-fā-jēz;** *macro* = large; *phagein* = to eat). They are irregular in form with short branching projections and are capable of engulfing bacteria and cellular debris by the process of phagocytosis. Thus they provide a vital defense for the body.

EXHIBIT 3-2   CONNECTIVE TISSUES

EMBRYONIC

Messenchymal (180×)

Description: Consists of highly branched mesenchymal cells embedded in a fluid substance.

Location: Under skin and along developing bones of embryo; some mesenchymal cells found in adult connective tissue, especially along blood vessels.

Function: Forms all other kinds of connective tissue.

Blood vessels

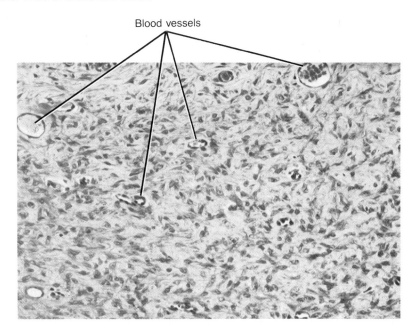

Mucous (320×)

Description: Consists of flattened or spindle-shaped cells embedded in a mucuslike substance containing fine collagenous fibers.

Location: Umbilical cord of fetus.

Function: Support.

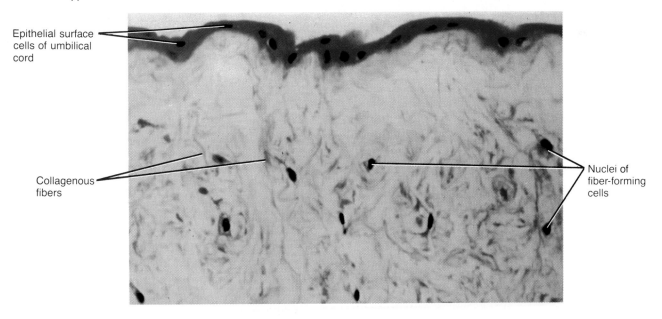

Epithelial surface cells of umbilical cord

Collagenous fibers

Nuclei of fiber-forming cells

EXHIBIT 3-2 (*Continued*)

ADULT

Loose or areolar (180×)

Description: Consists of fibers (collagenous, elastic, and reticular) and several kinds of cells (fibroblasts, macrophages, plasma cells, and mast cells) embedded in a semifluid ground substance.

Location: Subcutaneous layer of skin, mucous membranes, blood vessels, nerves, and body organs.

Function: Strength, elasticity, and support.

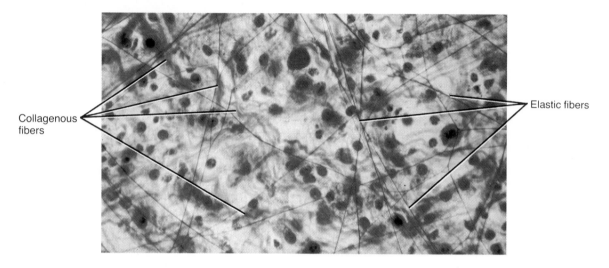

Collagenous fibers

Elastic fibers

Adipose (250×)

Description: Consists of adipocytes, "signet ring"-shaped cells with peripheral nuclei, that are specialized for fat storage.

Location: Subcutaneous layer of skin, around heart and kidneys, marrow of long bones, and padding around joints.

Function: Reduces heat loss through skin, serves as an energy reserve, supports, and protects.

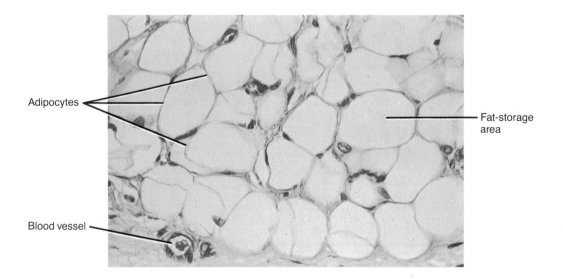

Adipocytes

Fat-storage area

Blood vessel

Dense or collagenous (250×)
Description: Consists of predominately collagenous, or white, fibers arranged in bundles; fibroblasts present in rows between bundles.
Location: Forms tendons, ligaments, aponeuroses, membranes around various organs, and fasciae.
Function: Provides strong attachment between various structures.

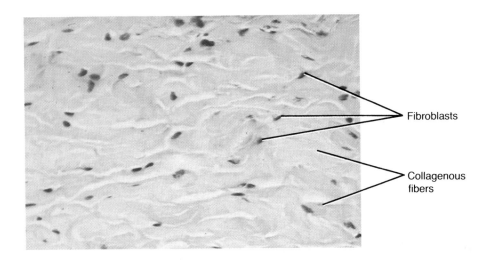

Fibroblasts

Collagenous
fibers

Elastic (180×)
Description: Consists of predominately freely branching elastic, or yellow, fibers; fibroblasts present in spaces between fibers.
Location: Lung tissue, cartilage of larynx, wall of arteries, trachea, bronchial tubes, true vocal cords, and ligamenta flava of vertebrae.
Function: Allows stretching of various organs.

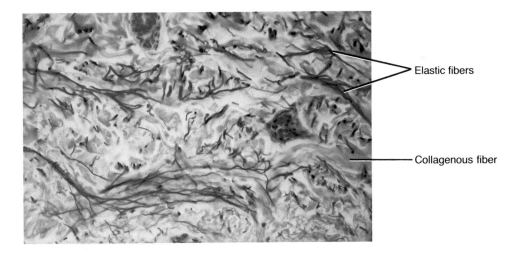

Elastic fibers

Collagenous fiber

EXHIBIT 3-2 (*Continued*)

Reticular (250×)
Description: Consists of a network of interlacing reticular fibers with thin, flat cells wrapped around fibers.
Location: Liver, spleen, and lymph nodes.
Function: Forms stroma of organs; binds together smooth muscle tissue cells.

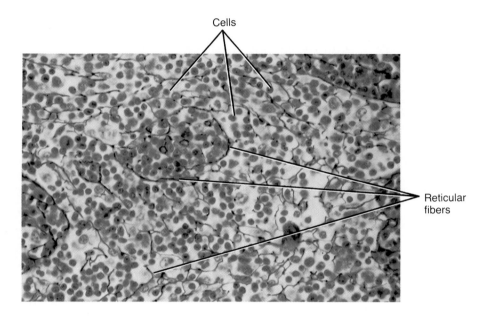

Cells

Reticular fibers

Hyaline cartilage (350×)
Description: Also called gristle; appears as a bluish white, glossy mass; contains numerous chondrocytes; is the most abundant type of cartilage.
Location: Ends of long bones, ends of ribs, nose, parts of larynx, trachea, bronchi, bronchial tubes, and embryonic skeleton.
Function: Provides movement at joints, flexibility, and support.

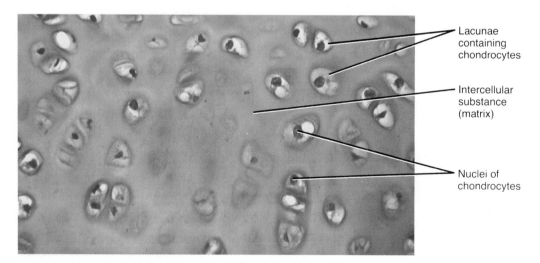

Lacunae containing chondrocytes

Intercellular substance (matrix)

Nuclei of chondrocytes

Fibrocartilage (180×)
Description: Consists of chondrocytes scattered among bundles of collagenous fibers.
Location: Symphysis pubis, intervertebral discs, and menisci of knee.
Function: Support and fusion.

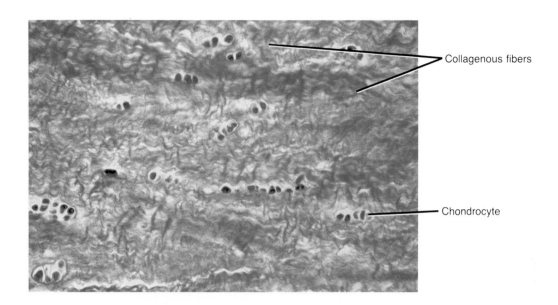

Collagenous fibers

Chondrocyte

Elastic cartilage (350×)
Description: Consists of chondrocytes located in a threadlike network of elastic fibers.
Location: Epiglottis of larynx, external ear, and auditory (Eustachian) tubes.
Function: Gives support and maintains shape.

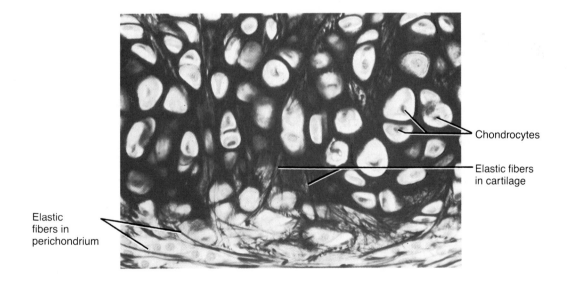

Chondrocytes

Elastic fibers
in cartilage

Elastic
fibers in
perichondrium

EXHIBIT 3-2 (*Continued*)

Osseous (bone) (150×)

Description: Compact bone consists of osteons (Haversian systems) that contain lamellae, lacunae, osteocytes, canaliculi, and central (Haversian) canals.

Location: Both compact and spongy bone comprise the various bones of the body.

Function: Support, protection, movement, storage, blood cell production.

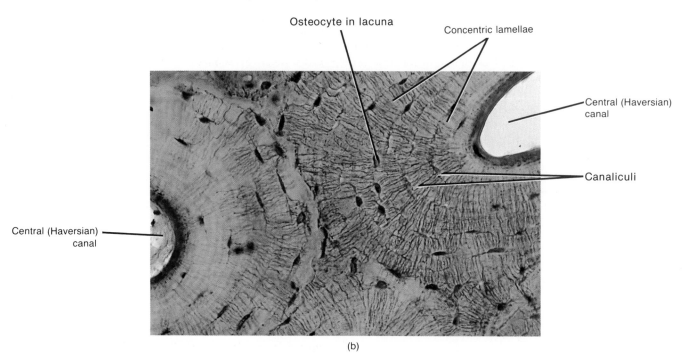

(b)

Vascular (blood) (1800×)

Description: Consists of plasma (intercellular substance) and formed elements (erythrocytes, leucocytes, and thrombocytes).

Location: Within blood vessels (arteries, arterioles, capillaries, venules, and veins).

Functions: Erythrocytes transport oxygen and carbon dioxide, leucocytes carry on phagocytosis and are involved in allergic reactions and immunity, and thrombocytes bring about blood clotting.

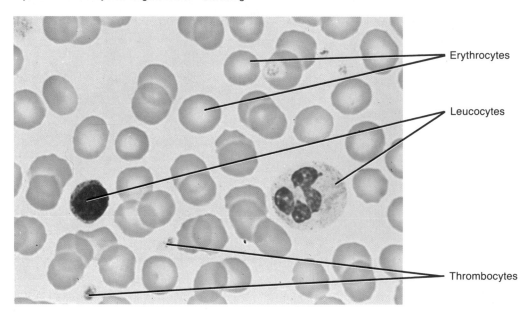

A third kind of cell in loose connective tissue is the **plasma cell.** These cells are small and either round or irregular. Plasma cells develop from a type of white blood cell called a lymphocyte (B cell). They give rise to antibodies and, accordingly, provide a defensive mechanism through immunity. Plasma cells are found in many places of the body, but most are found in connective tissue, especially that of the digestive tract and the mammary glands.

Another cell in loose connective tissue is the **mast cell.** It may develop from another type of white blood cell called a basophil. The mast cell is somewhat larger than a basophil and is found in abundance along blood vessels. It forms heparin, an anticoagulant that prevents blood from clotting in the vessels. Mast cells are also believed to produce histamine and serotonin, chemicals that dilate small blood vessels.

Other cells in losse connective tissue include **melanocytes** (pigment cells), fat cells, and white blood cells.

Loose connective tissue is continuous throughout the body. It is present in all mucous membranes and around all blood vessels and nerves. And it occurs around body organs and in the papillary region of the dermis of the skin. Combined with adipose tissue, it forms the **subcutaneous** (sub'-kyoo-TĀ-nē-us; *sub* = under, *cut* = skin) **layer**—the layer of tissue that attaches the skin to underlying tissues and organs. The subcutaneous layer is also referred to as the **superficial fascia** (FASH-ē-a), or **hypodermis.**

• *Adipose Tissue* Adipose tissue is basically a form of loose connective tissue in which the cells, called **adipocytes,** are specialized for fat storage. Adipocytes are derived from fibroblasts, and the cells have the shape of a "signet ring" because the cytoplasm and nucleus are pushed to the edge of the cell by a large droplet of fat. Adipose tissue is found wherever loose connective tissue is located. Specifically, it is in the subcutaneous layer below the skin, around the kidneys, at the base and on the surface of the heart, in the marrow of long bones, as a padding around joints, and behind the eyeball in the orbit. Adipose tissue is a poor conductor of heat and therefore reduces heat loss through the skin. It is also a major energy reserve and generally supports and protects various organs.

CLINICAL APPLICATION

A surgical procedure, called **suction lipectomy** (*lipo* = fat; *ectomy* = removal of), involves suctioning out small amounts of fat from certain areas of the body. The technique can be used in a number of areas of the body in which fat accumulates, such as the inner and outer thighs, buttocks, area behind the knee, underside of the arms, breasts, and abdomen. However, suction lipectomy is not a treatment for obesity. The ideal candidate for suction lipectomy is an individual under 40 years of age who has good skin elasticity and who is about 10 to 20 pounds overweight.

The procedure begins with a small incision, about 1.27 cm (½ in), into which is inserted a long, hollow cannula (tube). Then, tunnels of fat are sucked out horizontally through the cannula, which is attached to a suction pump. After repeated insertions of the cannula into the fatty area, a Swiss-cheeselike pattern is left in the fat. Finally, the surgeon pushes down on the remaining fat to collapse the gaps and the area is wrapped in bandages to keep the gaps from filling with body fluids.

• *Dense (Collagenous) Connective Tissue* Dense (collagenous) connective tissue is characterized by a close packing of fibers and less intercellular substance than in loose connective tissue. The fibers can be either irregularly or regularly arranged. In areas of the body where tensions are exerted in various directions, the fiber bundles are interwoven and without regular orientation. Such a dense connective tissue is referred to as **irregularly arranged** and occurs in sheets. It forms most fasciae, the reticular region of the dermis of the skin, the periosteum of bone, the perichondrium of cartilage, and the *membrane (fibrous) capsules* around organs, such as the kidneys, liver, testes, and lymph nodes.

In other areas of the body, dense connective tissue is adapted for tension in one direction, and the fibers have an orderly, parallel arrangement. Such a dense connective tissue is known as **regularly arranged.** The most common variety of dense regularly arranged connective tissue has a predominance of collagenous (white) fibers arranged in bundles. Fibroblasts are placed in rows between the bundles. The tissue is silvery white, tough, yet somewhat pliable. Because of its great strength, it is the principal component of *tendons,* which attach muscles to bones; many *ligaments* (collagenous ligaments), which hold bones together at joints; and *aponeuroses* (ap'-ō-noo-RŌ-sēz), which are flat bands connecting one muscle with another or with bone.

CLINICAL APPLICATION

Scientists have developed a carbon fiber implant for **reconstructing severely torn ligaments and tendons.** The implant consists of carbon fibers coated with a plastic called polylactic acid. The coated fibers are sewn in and around torn ligaments and tendons to reinforce them and to provide a scaffolding around which the body's own collagenous fibers grow. Within two weeks, the polylactic acid is absorbed by the body and the carbon fibers eventually fracture. By this time, the fibers are completely clad in collagen produced by fibroblasts.

• *Elastic Connective Tissue* Unlike collagenous connective tissue, elastic connective tissue has a predominance of freely branching elastic fibers. These fibers give the tissue a yellowish color. Fibroblasts are present only in the spaces between the fibers. Elastic connective tissue can be stretched and will snap back into shape. It is a component of the cartilages of the larynx, the walls of elastic arteries, the trachea, the bronchial tubes to the lungs, and the lungs themselves. Elastic connective tissue provides stretch and strength, allowing structures to perform their functions efficiently. Yellow elastic ligaments, as contrasted with collagenous ligaments, are composed mostly of elastic fibers; they form the ligamenta flava of the vertebrae (ligaments between successive vertebrae), the suspensory ligament of the penis, and the true vocal cords.

• *Reticular Connective Tissue* Reticular connective tissue consists of interlacing reticular fibers. It helps to form the stroma of many organs, including the liver, spleen, and lymph nodes. Reticular connective tissue also helps to bind together the cells of smooth muscle tissue. It is especially adapted to providing strength and support.

Cartilage

One type of connective tissue, called cartilage, is capable of enduring considerably more stress than the tissues just discussed. Unlike other connective tissues, cartilage has no blood vessels (except for the perichondrium) or nerves of its own. **Cartilage** consists of a dense network of collagenous fibers and elastic fibers firmly embedded in chondroitin sulfate, a jellylike substance. The cells of mature cartilage, called **chondrocytes** (KON-drō-sīts), occur singly or in groups within spaces called **lacunae** (la-KOO-nē) in the intercellular substance. The surface of cartilage is surrounded by irregularly arranged dense connective tissue called the **perichondrium** (per'-i-KON-drē-um; *peri* = around; *chondro* = cartilage). Three kinds of cartilage are recognized: hyaline cartilage, fibrocartilage, and elastic cartilage (Exhibit 3-2).

• *Hyaline Cartilage* This cartilage, also called gristle, appears as a bluish-white, glossy, homogeneous mass. The collagenous fibers, although present, are not visible with ordinary staining techniques, and the prominent chondrocytes are found in lacunae. Hyaline cartilage is the most abundant kind of cartilage in the body. It is found at joints over the ends of long bones (where it is called *articular cartilage*) and forms the *costal cartilages* at the ventral ends of the ribs. Hyaline cartilage also helps to form the nose, larynx, trachea, bronchi, and bronchial tubes leading to the lungs. Most of the embryonic skeleton consists of hyaline cartilage. It affords flexibility and support.

• *Fibrocartilage* Chondrocytes scattered through many bundles of visible collagenous fibers are found in this type of cartilage. Fibrocartilage is found at the symphysis pubis, the point where the coxal (hip) bones fuse anteriorly at the midline. It is also found in the intervertebral discs between vertebrae and the menisci of the knees. This tissue combines strength and rigidity.

• *Elastic Cartilage* In this tissue, chondrocytes are located in a threadlike network of elastic fibers. Elastic cartilage provides strength and maintains the shape of certain organs—the epiglottis of the larynx, the external part of the ear (the pinna), and the auditory (Eustachian) tubes.

Osseous Tissue (Bone)

Together, cartilage and **osseous** (OS-ē-us) **tissue (bone)** comprise the skeletal system. Like other connective tissues, osseous tissue contains abundant intercellular substance surrounding widely separated cells. Mature bone cells are called **osteocytes.** The intercellular substances consist of mineral salts, primarily calcium phosphate and calcium carbonate, and collagenous fibers. The salts are responsible for the hardness of bone.

Bone tissue is classified as either compact (dense) or spongy (cancellous), depending on how the intercellular substance and cells are organized. At this point, we will discuss compact bone only. The basic unit of compact bone is called an **osteon (Haversian system).** Each osteon consists of **lamellae,** concentric rings of hard, intercellular substance; **lacunae,** small spaces between lamellae that contain osteocytes; **canaliculi,** radiating minute canals that provide numerous routes so that nutrients can reach osteocytes and wastes can be removed from them; and a **central (Haversian) canal** that contains blood vessels and nerves. Notice that whereas osseous tissue is vascular, cartilage is not. Also, the lacunae of osseous tissue are interconnected by canaliculi; those of cartilage are not.

Functionally, the skeletal system supports soft tissues, protects delicate structures, acts with skeletal muscles to produce movement, stores calcium and phosphorus, and houses red marrow, which produces several kinds of blood cells.

The details of compact and spongy bone are discussed in Chapter 5.

Vascular Tissue (Blood)

Vascular tissue (blood) is a liquid connective tissue that consists of an intercellular substance called plasma and formed elements (cells and cell-like structures). **Plasma** is a straw-colored liquid that consists mostly of water plus some dissolved substances (nutrients, enzymes, hormones, respiratory gases, and ions). The formed elements are erythrocytes (red blood cells), leucocytes (white blood cells), and thrombocytes (platelets).

Erythrocytes are produced in red bone marrow and function in transporting oxygen to body cells and carbon dioxide from them. **Leucocytes** are divided into two groups: granulocytes and agranulocytes. Granulocytes have granules in their cytoplasm and lobed or irregular nuclei. They develop from red bone marrow. Examples are *neutrophils,* which are actively phagocytic; *eosinophils,* which are involved in allergic reactions; and *basophils,* which are also involved in allergic reactions. Agranulocytes do not have granules in their cytoplasm and their nuclei are round or only slightly indented. They develop from bone marrow and lymphoid tissue. Examples are *lymphocytes,* which are involved in immunity, and *monocytes,* which are actively phagocytic. **Thrombocytes** are produced in bone marrow and function in blood clotting.

The details of blood are considered in Chapter 12.

MUSCLE TISSUE

Muscle tissue consists of highly specialized cells that are modified for contraction. As a result of this characteristic, muscle tissue provides motion, maintenance of posture, and heat production. On the basis of certain structural and functional characteristics, muscle tissue is classified into three types: skeletal, cardiac, and smooth (Exhibit 3-3).

Skeletal muscle tissue is named for its location—attached to bones. It is also *striated,* that is, the cells contain alternating light and dark bands (striations) that are perpendicular to the long axes of the cells. The striations are visible under a microscope. Skeletal muscle tissue is also *voluntary* because it can be made to contract by conscious control. A single skeletal muscle cell, or fiber, is cylindrical, and the cells are parallel to each other in a tissue. Each muscle fiber contains a plasma membrane, the *sarcolemma,* surrounding the cytoplasm, or *sarcoplasm.* Skeletal muscle fibers are multinucleate (they have more than one nucleus), and the nuclei lie close to the sarcolemma. The contractile elements of skeletal muscle fibers are proteins called *myofibrils.* They contain wide, transverse, dark bands and narrow light ones that give the fibers the striated appearances.

Cardiac muscle tissue forms the bulk of the wall of the heart. Like skeletal muscle tissue, it is striated. However, unlike skeletal muscle tissue, it is usually involuntary; its contraction is usually not under conscious control. Cardiac muscle fibers are roughly quadrangular and branch to form networks throughout the tissue. The fibers usually have only one nucleus that is centrally located. Cardiac muscle fibers are separated from each other by transverse thickenings of the sarcolemma called *intercalated discs.* These are unique to cardiac muscle and serve to strengthen the tissue and aid impulse conduction.

Smooth muscle tissue is located in the walls of hollow internal structures such as blood vessels, the stomach, intestines, and urinary bladder. Smooth muscle fibers are usually involuntary and they are *nonstriated* (*smooth*). Each smooth muscle cell is spindle-shaped with either end tapering to a point and contains a single, centrally located nucleus.

A more detailed discussion of muscle tissue is considered in Chapter 9.

NERVOUS TISSUE

Despite the tremendous complexity of the nervous system, it consists of only two principal kinds of cells: neurons and neuroglia. **Neurons,** or nerve cells, are highly specialized cells that are capable of picking up stimuli; converting stimuli to nerve impulses; and conducting nerve impulses to other neurons, muscle fibers, or glands. Neurons are the structural and functional units of the nervous system. Each consists of three basic portions: cell body and two kinds of processes called dendrites and axons (Exhibit 3-4). The *cell body* (*perikaryon*) contains the nucleus and other organelles. *Dendrites* are highly branched processes of the cell body that conduct nerve impulses toward the cell body. *Axons* are single, long processes of the cell body that conduct nerve impulses away from the cell body.

Neuroglia are cells that protect and support neurons. They are of clinical interest because they are frequently the sites of tumors of the nervous system.

The detailed structure and function of neurons and neuroglia are considered in Chapter 16.

MEMBRANES

The combination of an epithelial layer and an underlying connective tissue layer constitutes an **epithelial membrane.** The principal epithelial membranes of the body are mucous membranes, serous membranes, and the cutaneous membrane, or skin. Another kind of membrane, a **synovial membrane,** does not contain epithelium.

MUCOUS MEMBRANES

A **mucous membrane,** or **mucosa,** lines a body cavity that opens directly to the exterior. Mucous membranes line the entire digestive, respiratory, excretory, and reproductive tracts (see Figure 23-2). The surface tissue of a mucous membrane may vary in type—it is stratified squamous epithelium in the esophagus and simple columnar epithelium in the intestine.

The epithelial layer of a mucous membrane secretes mucus, which prevents the cavities from drying out. It also traps dust in the respiratory passageways and lubri-

EXHIBIT 3-3   MUSCLE TISSUE

Skeletal muscle tissue (800×)
Description: Cylindrical, striated fibers with several peripheral nuclei; voluntary.
Location: Attached to bones.
Function: Motion, posture, heat production.

Cardiac muscle tissue (400×)
Description: Quadrangular, branching, striated fibers with one centrally located nucleus; contains intercalated discs; usually involuntary.
Location: Heart wall.
Function: Motion (contraction of heart).

Capillary   Nucleus of muscle fiber   Striations

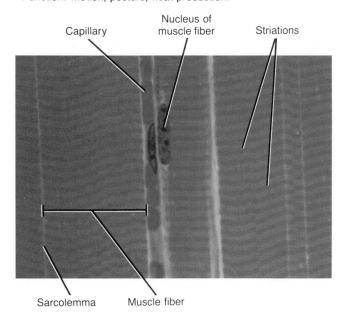

Sarcolemma   Muscle fiber

Intercalated discs   Muscle fibers

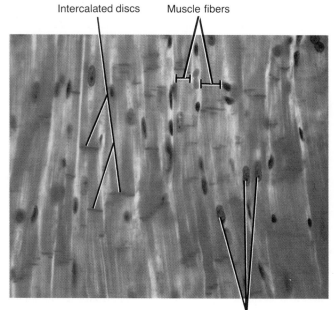

Nuclei of muscle fibers

Smooth muscle tissue (840×)
Description: Spindle-shaped, nonstriated fibers with one centrally located nucleus; usually involuntary.
Location: Walls of hollow internal structures such as blood vessels, stomach, intestines, and urinary bladder.
Function: Motion (constriction of blood vessels, propulsion of foods through gastrointestinal tract, contraction of gallbladder).

Muscle fibers   Nuclei of muscle fibers

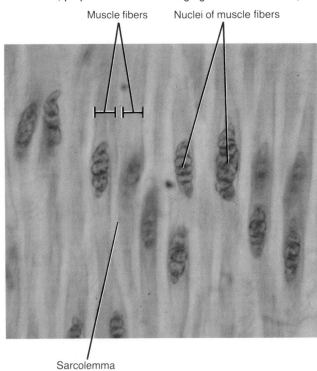

Sarcolemma

EXHIBIT 3-4 NERVOUS TISSUE

Neurons (640×)

Description: Neurons (nerve cells) consist of a cell body and processes extending from the cell body called dendrites or axons.

Location: Nervous system.

Function: Pick up stimuli, convert stimuli to nerve impulses, and conduct nerve impulses to other neurons, muscle fibers, or glands.

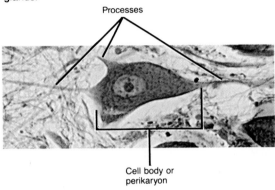

Processes

Cell body or perikaryon

cates food as it moves through the digestive tract. In addition, the epithelial layer is responsible for the secretion of digestive enzymes and the absorption of food.

The connective tissue layer of a mucous membrane is called the *lamina propria.* It binds the epithelium to the underlying structures and allows some flexibility of the membrane. It also holds the blood vessels in place, protects underlying muscles from abrasion or puncture, provides the epithelium covering it with oxygen and nutrients, and removes wastes.

SEROUS MEMBRANES

A **serous membrane,** or **serosa,** lines a body cavity that does not open directly to the exterior, and it covers the organs that lie within the cavity. Serous membranes consist of thin layers of loose connective tissue covered by a layer of mesothelium, and they are in the form of invaginated double-walled sacs. The part attached to the cavity wall is called the *parietal* (pa-RĪ-i-tal) *portion.* The part that covers the organs inside these cavities is the *visceral portion.* The serous membrane lining the thoracic cavity and covering the lungs is called the *pleura* (see Figure 1-7b). The membrane lining the heart cavity and covering the heart is the *pericardium* (*cardio* = heart). The serous membrane lining the abdominal cavity and covering the abdominal organs and some pelvic organs is called the *peritoneum.*

The epithelial layer of a serous membrane secretes a lubricating fluid that allows the organs to glide easily against one another or against the walls of the cavities. The connective tissue layer of the serous membrane consists of a relatively thin layer of loose connective tissue.

CUTANEOUS MEMBRANE

The **cutaneous membrane,** or skin, constitutes an organ of the integumentary system and is discussed in the next chapter.

SYNOVIAL MEMBRANES

Synovial membranes line the cavities of joints (see Figure 8-1). Like serous membranes, they line structures that do not open to the exterior. Unlike mucous, serous, and cutaneous membranes, they do not contain epithelium and are therefore not epithelial membranes. They are composed of loose connective tissue with elastic fibers and varying amounts of fat. Synovial membranes secrete *synovial fluid,* which lubricates the ends of bones as they move at joints and nourishes the articular cartilage covering the bones that form the joints.

STUDY OUTLINE

Types of Tissues (p. 54)

1. A tissue is a group of similar cells and their intercellular substance that are specialized for a particular function.
2. Depending on their function and structure, the various tissues of the body are classified into four principal types: epithelial, connective, muscular, and nervous.

Epithelial Tissue (p. 54)

1. Epithelium has many cells, little intercellular material, and no blood vessels (avascular). It is attached to connective tissue by a basement membrane. It can replace itself.
2. The subtypes of epithelium include covering and lining epithelium and glandular epithelium.
3. Cell junctions found between epithelial cells include desmo-

somes, tight junctions, intermediate junctions, and gap junctions.

Covering and Lining Epithelium

1. Layers are arranged as simple (one layer), stratified (several layers), and pseudostratified (one layer that appears as several); cell shapes include squamous (flat), cuboidal (cubelike), columnar (rectangular), and transitional (variable).
2. Simple squamous epithelium is adapted for diffusion and filtration and is found in lungs and kidneys. Endothelium lines the heart and blood vessels. Mesothelium lines the thoracic and abdominopelvic cavities and covers the organs within them.
3. Simple cuboidal epithelium is adapted for secretion and ab-

sorption. It is found covering ovaries, in kidneys and eyes, and lining some glandular ducts.

4. Nonciliated simple columnar epithelium lines most of the digestive tract. Specialized cells containing microvilli perform absorption. Goblet cells secrete mucus. In a few portions of the respiratory tract, the cells are ciliated to move foreign particles trapped in mucus out of the body.

5. Stratified squamous epithelium is protective. It lines the upper digestive tract and vagina and forms the outer layer of skin.

6. Stratified cuboidal epithelium is found in adult sweat glands, pharynx, epiglottis, and portions of urethra.

7. Stratified columnar epithelium protects and secretes. It is found in the male urethra and large excretory ducts.

8. Transitional epithelium lines the urinary bladder and is capable of stretching.

9. Pseudostratified epithelium has only one layer but gives the appearance of many. It lines larger excretory ducts, parts of urethra, auditory (Eustachian) tubes, and most upper respiratory structures, where it protects and secretes.

Glandular Epithelium

1. A gland is a single cell or a mass of epithelial cells adapted for secretion.

2. Exocrine glands (sweat, oil, and digestive glands) secrete into ducts or directly onto a free surface.

3. Structural classification includes unicellular and multicellular glands; multicellular glands are further classified as tubular, acinar, tubuloacinar, simple, and compound.

4. Functional classification includes holocrine, merocrine, and apocrine glands.

5. Endocrine glands secrete hormones directly into the blood.

Connective Tissue (p. 65)

1. Connective tissue is the most abundant body tissue. It has few cells, an extensive intercellular substance, and a rich blood supply (vascular), except for cartilage. It does not occur on free surfaces.

2. The intercellular substance determines the tissue's qualities.

3. Connective tissue protects, supports, and binds organs together.

4. Connective tissue is classified into two principal types: embryonic and adult.

Embryonic Connective Tissue

1. Mesenchyme forms all other connective tissues.

2. Mucous connective tissue is found in the umbilical cord of the fetus, where it gives support.

Adult Connective Tissue

1. Adult connective tissue is connective tissue that exists in the newborn and does not change after birth. It is subdivided into several kinds: connective tissue proper, cartilage, bone tissue, and vascular tissue.

2. Connective tissue proper has a more or less fluid intercellular material, and a typical cell is the fibroblast. Five examples of such tissues may be distinguished.

3. Loose (areolar) connective tissue is one of the most widely distributed connective tissues in the body. Its intercellular substance (hyaluronic acid) contains fibers (collagenous, elastic, and reticular) and various cells (fibroblasts, macro-

phages, plasma, mast, and melanocytes). Loose connective tissue is found in all mucous membranes, around body organs, and in the subcutaneous layer.

4. Adipose tissue is a form of loose connective tissue in which the cells, called adipocytes, are specialized for fat storage. It is found in the subcutaneous layer and around various organs.

5. Dense (collagenous) connective tissue has a close packing of fibers (regularly or irregularly arranged). It is found as a component of fascia, membranes of organs, tendons, ligaments, and aponeuroses.

6. Elastic connective tissue has a predominance of freely branching elastic fibers that give it a yellow color. It is found in the cartilages of the larynx, elastic arteries, trachea, bronchial tubes, and true vocal cords.

7. Reticular connective tissue consists of interlacing reticular fibers and forms the stroma of the liver, spleen, and lymph nodes.

8. Cartilage has a jellylike matrix containing collagenous and elastic fibers and chondrocytes.

9. Hyaline cartilage is found in the embryonic skeleton, at the ends of bones, in the nose, and in respiratory structures. It is flexible, allows movement, and provides support.

10. Fibrocartilage connects the pelvic bones and the vertebrae. It provides strength.

11. Elastic cartilage maintains the shape of organs such as the larynx, auditory (Eustachian) tubes, and external ear.

12. Osseous tissue (bone) consists of mineral salts and collagenous fibers that contribute to the hardness of bone and cells called osteocytes. It supports, protects, helps provide movement, stores minerals, and houses red marrow.

13. Vascular tissue (blood) consists of plasma and formed elements (erythrocytes, leucocytes, and thrombocytes). Functionally, its cells transport, carry on phagocytosis, participate in allergic reactions, provide immunity, and bring about blood clotting.

Muscle Tissue (p. 75)

1. Muscle tissue is modified for contraction and thus provides motion, maintenance of posture, and heat production.

2. Skeletal muscle tissue is attached to bones, is striated, and is voluntary.

3. Cardiac muscle tissue forms most of the heart wall, is striated, and is usually involuntary.

4. Smooth muscle tissue is found in the walls of hollow internal structures (blood vessels and viscera), is nonstriated, and is usually involuntary.

Nervous Tissue (p. 75)

1. The nervous system is composed of neurons (nerve cells) and neuroglia (protective and supporting cells).

2. Neurons consist of a cell body and two types of processes called dendrites and axons.

3. Neurons are specialized to pick up stimuli, convert stimuli into nerve impulses, and conduct nerve impulses.

Membranes (p. 75)

1. An epithelial membrane is an epithelial layer overlying a connective tissue layer. Examples are mucous, serous, and cutaneous membranes.

2. Mucous membranes line cavities that open to the exterior, such as the digestive tract.
3. Serous membranes (pleura, pericardium, peritoneum) line closed cavities and cover the organs in the cavities. These membranes consist of parietal and visceral portions.
4. The cutaneous membrane is the skin.
5. Synovial membranes line joint cavities and do not contain epithelium.

REVIEW QUESTIONS

1. Define a tissue. What are the four basic kinds of human tissue?
2. Distinguish covering and lining epithelium from glandular epithelium. What characteristics are common to all epithelium?
3. Describe the origin and composition of the basement membrane.
4. Describe the various layering arrangements and cell shapes of epithelium.
5. Define a cell junction. List several functions of cell junctions.
6. Compare desmosomes, hemidesmosomes, tight junctions, intermediate junctions, and gap junctions on the basis of structure, location, and function.
7. How is epithelium classified? List the various types.
8. For each of the following kinds of epithelium, briefly describe the microscopic appearance, location in the body, and functions: simple squamous, simple cuboidal, simple columnar, stratified squamous, stratified cuboidal, stratified columnar, transitional, and pseudostratified.
9. Define the following terms: endothelium, mesothelium, secretion, absorption, goblet cell, and keratin.
10. What is a gland? Distinguish between endocrine and exocrine glands.
11. Describe the classification of exocrine glands according to structure and function and give at least one example of each.
12. Enumerate the ways in which connective tissue differs from epithelium.
13. How are connective tissues classified? List the various types.
14. How are embryonic connective tissue and adult connective tissue distinguished?
15. Describe the following connective tissues with regard to microscopic appearance, location in the body, and function: loose (areolar), adipose, dense (collagenous), elastic, reticular, hyaline cartilage, fibrocartilage, elastic cartilage, osseous tissue (bone), and vascular tissue (blood).
16. Define the following terms: hyaluronic acid, collagenous fiber, elastic fiber, reticular fiber, fibroblast, macrophage, plasma cell, mast cell, melanocyte, adipocyte, chondrocyte, lacuna, osteocyte, lamella, and canaliculi.
17. Describe the histology of muscle tissue. How is it classified? What are its functions?
18. Distinguish between neurons and neuroglia. Describe the structure and function of neurons.
19. Define the following kinds of membranes: mucous, serous, cutaneous, and synovial. Where is each located in the body? What are their functions?
20. Following are some descriptions of various tissues of the body. For each description, name the tissue described.
    a. An epithelium that permits distention (stretching).
    b. A single layer of flat cells concerned with filtration and absorption.
    c. Forms all other kinds of connective tissue.
    d. Specialized for fat storage.
    e. An epithelium with waterproofing qualities.
    f. Forms the framework of many organs.
    g. Produces perspiration, wax, oil, or digestive enzymes.
    h. Cartilage that shapes the external ear.
    i. Contains goblet cells and lines the intestine.
    j. Most widely distributed connective tissue.
    k. Forms tendons, ligaments, and aponeuroses.
    l. Specialized for the secretion of hormones.
    m. Provides support in the umbilical cord.
    n. Lines kidney tubules and is specialized for absorption and secretion.
    o. Permits extensibility of lung tissue.
    p. Stores red marrow, protects, supports.
    q. Nonstriated, usually involuntary muscle tissue.
    r. Consists of granulocytes and agranulocytes.
    s. Composed of a cell body, dendrites, and axons.

4

The Integumentary System

Student Objectives

Define the integumentary system.

Explain how the skin is structurally divided into epidermis and dermis.

List the various layers of the epidermis and describe their functions.

Describe the composition and function of the dermis.

Explain the basis for skin color.

Explain the basic pattern of epidermal ridges and grooves.

Describe the blood supply of the skin.

Describe the development, distribution, structure, color, growth, and replacement of hair.

Compare the structure, distribution, and functions of sebaceous (oil) and sudoriferous (sweat) glands.

List the parts of a nail and describe their composition.

Describe the effects of aging on the integumentary system.

Describe the development of the epidermis, its derivatives, and the dermis.

Describe the causes and effects for the following skin disorders: acne, systemic lupus erythematosus (SLE), psoriasis, decubitus ulcers, sunburn, and skin cancer.

Classify burns into first, second, and third degrees and describe how to estimate the extent of a burn.

Define key medical terms associated with the integumentary system.

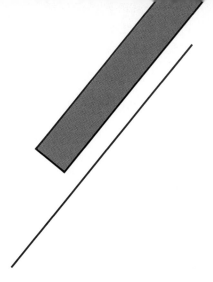

An aggregation of tissues that performs a specific function is an **organ.** The next higher level of organization is a **system**—a group of organs operating together to perform specialized functions. The skin and its derivatives—such as hair, nails, glands, and several specialized receptors—constitute the **integumentary** (in'-teg-yoo-MEN-tar-ē) **system** of the body.

The developmental anatomy of the integumentary system is considered at the end of the chapter.

SKIN

The **skin** is an organ because it consists of tissues structurally joined together to perform specific activities. It is not just a simple, thin covering that keeps the body together and gives it protection. The skin is quite complex in structure and performs several functions essential for survival. **Dermatology** (*dermato* = skin; *logos* = study of) is the medical specialty that deals with the diagnosis and treatment of skin disorders.

The skin is one of the larger organs of the body in terms of surface area. For the average adult, the skin occupies a surface area of approximately 19,355 sq cm (3,000 sq in). It varies in thickness from 0.5 to 3 mm (0.02 to 0.12 in). In general, skin is thicker on the dorsal surface of a part of the body than it is on the ventral surface, but this condition is reversed in the hand and the foot. The skin of the palmar surface of the hand and the plantar surface of the foot is thicker than on their dorsal aspects. Thick skin has a relatively thick epidermis, whereas thin skin has a relatively thin epidermis. On most of the body, the epidermis is from 0.07 to 0.12 mm thick but can be 0.8 mm on the palms and 1.4 mm on the soles.

FUNCTIONS

The skin covers the body and protects the underlying tissues from bacterial invasion, drying out, and harmful light rays. It helps to maintain body temperature; prevents excessive loss of inorganic and organic materials; receives stimuli from the environment; stores chemical compounds; excretes water, salts, and several organic compounds; and synthesizes the hormone, vitamin D. The role of the skin in receiving various stimuli is considered in detail in Chapter 20.

STRUCTURE

Structurally, the skin consists of two principal parts (Figure 4-1). The outer, thinner portion, which is composed of epithelium, is called the **epidermis.** The epidermis is cemented to the inner, thicker, connective tissue part called the **dermis.** Beneath the dermis is a **subcutaneous layer.** This layer, also called the **superficial fascia,** or **hypodermis,** consists of areolar and adipose tissues. Fibers from the dermis extend down into the superficial fascia and anchor the skin to the subcutaneous layer. The superficial fascia, in turn, is firmly attached to underlying tissues and organs.

EPIDERMIS

The **epidermis** is composed of stratified squamous epithelium and contains three distinct types of cells. The most numerous is known as a *keratinocyte,* a cell that undergoes keratinization. This process will be described shortly. The function of these cells is to produce keratin, which helps waterproof and protect the skin and underlying tissues. A second type of cell in the epidermis is called a *melanocyte.* Its role is to produce melanin, one of the pigments responsible for skin color. The third type of cell is called a *Langerhans' cell.* These cells probably arise from bone marrow and invade the epidermis and other stratified squamous epithelial sites of the body where they function in phagocytosis.

The epidermis is organized in four or five cell layers, depending on its location in the body (see Figure 4-1 and Figure 4-2). Where exposure to friction is greatest, such as in the palms and soles, the epidermis has five layers. In all other parts it has four layers. The names

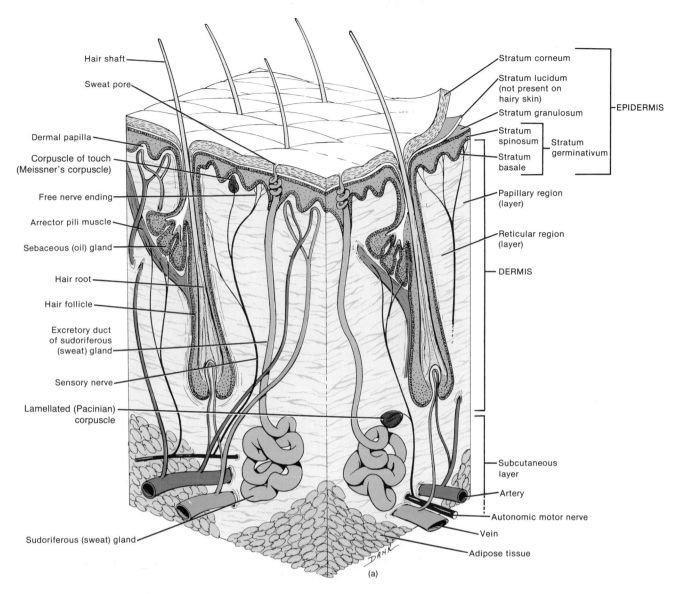

Hair shaft

Sweat pore

Dermal papilla

Corpuscle of touch
(Meissner's corpuscle)

Free nerve ending

Arrector pili muscle

Sebaceous (oil) gland

Hair root

Hair follicle

Excretory duct
of sudoriferous
(sweat) gland

Sensory nerve

Lamellated (Pacinian)
corpuscle

Sudoriferous (sweat) gland

Stratum corneum

Stratum lucidum
(not present on
hairy skin)

Stratum granulosum

Stratum spinosum

Stratum basale

Stratum germinativum

EPIDERMIS

Papillary region
(layer)

Reticular region
(layer)

DERMIS

Subcutaneous
layer

Artery

Autonomic motor nerve

Vein

Adipose tissue

(a)

FIGURE 4-1 Skin. (a) Structure of the skin and underlying
subcutaneous layer. (b) Scanning electron micrograph of the
skin and several hairs at a magnification of 260×. (From
*Tissues and Organs: A Text-Atlas of Scanning Electron Mi-
croscopy,* by Richard K. Kessel and Randy H. Kardon. Copyright © 1979 by Scientific American, Inc.)

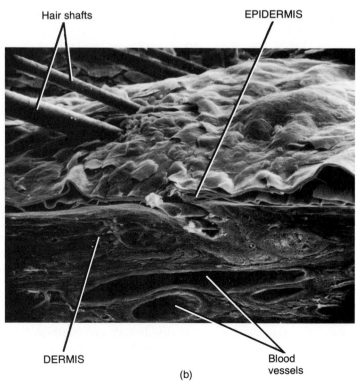

Hair shafts

EPIDERMIS

DERMIS

Blood
vessels

(b)

82

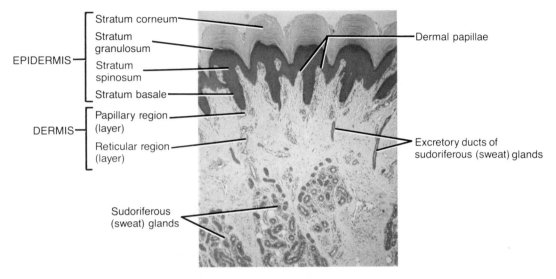

EPIDERMIS
- Stratum corneum
- Stratum granulosum
- Stratum spinosum
- Stratum basale

DERMIS
- Papillary region (layer)
- Reticular region (layer)

Sudoriferous (sweat) glands

Dermal papillae

Excretory ducts of sudoriferous (sweat) glands

FIGURE 4-2 Photomicrograph of thick skin at a magnification of 50×. The stratum lucidum is not shown here, but is illustrated in Figure 4-1a. (© 1983 by Michael H. Ross. Used by permission.)

of the five layers from the deepest to the most superficial are as follows:

1. Stratum basale. This single layer of cuboidal to columnar cells is capable of continued cell division. As these cells multiply, they push up toward the surface and become part of the layers to be described next. Their nuclei degenerate, and the cells die. Eventually, the cells are shed from the top layer of the epidermis.

2. Stratum spinosum. This layer of the epidermis contains eight to ten rows of polyhedral (many-sided) cells that fit closely together. The surfaces of these cells may assume a prickly appearance when prepared for microscope examination (*spinosum* = prickly). The stratum basale and stratum spinosum are sometimes collectively referred to as the **stratum germinativum** (jer'-mi-na-TĒ-vum) to indicate the layers where new cells are germinated. The deeper layers of the epidermis of hairless skin contain nerve endings sensitive to touch called *tactile (Merkel's) discs* (see Figure 20-1).

3. Stratum granulosum. The third layer of the epidermis consists of three to five rows of flattened cells that contain darkly staining granules of a substance called *keratohyalin* (ker'-a-tō-HĪ-a-lin). This compound is involved in the first step of keratin formation. *Keratin* is a waterproofing protein found in the top layer of the epidermis. The nuclei of the cells in the stratum granulosum are in various stages of degeneration. As these nuclei break down, the cells are no longer capable of carrying out vital metabolic reactions and die.

4. Stratum lucidum. This layer is quite pronounced in the thick skin of the palms and soles. It consists of several rows of clear, flat, dead cells that contain droplets of a substance called *eleidin* (el-Ē-i-din). The layer is

so named because eleidin is translucent (*lucidum* = clear). Eleidin is formed from keratohyalin and is eventually transformed to keratin.

5. Stratum corneum. This layer consists of 25 to 30 rows of flat, dead cells completely filled with keratin. These cells are continuously shed and replaced. The stratum corneum serves as an effective barrier against light and heat waves, bacteria, and many chemicals

In the process of keratinization, newly formed cells produced in the basal layers are pushed to more superficial layers. As the cells move toward the surface, the cytoplasm, nucleus, and other organelles are replaced by keratin. Eventually, the keratinized cells are worn off and replaced by underlying cells that, in turn, become keratinized and die. The whole process by which a cell forms in the basal layers, rises to the surface, becomes keratinized, and sloughs off takes about 1 month.

DERMIS

The second principal part of the skin, the **dermis,** is composed of connective tissue containing collagenous and elastic fibers (see Figure 4-1). The dermis is very thick in the palms and soles and very thin in the eyelids, penis, and scrotum. It also tends to be thicker on the dorsal aspects of the body than the ventral and thicker on the lateral aspects of extremities than medial aspects. Numerous blood vessels, nerves, glands, and hair follicles are embedded in the dermis.

The upper region of the dermis, about one-fifth of the thickness of the total layer, is named the **papillary region,** or **layer.** Its surface area is greatly increased by small, fingerlike projections called **dermal papillae** (pa-PIL-ē).

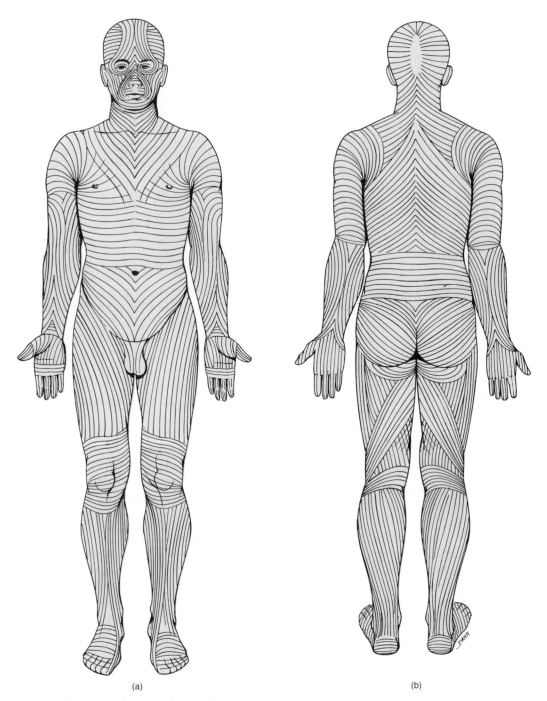

FIGURE 4-3 Lines of cleavage (tension lines) seen in (a) anterior and (b) posterior views.

These structures project into the concavities between ridges in the deep surface of the epidermis, and many contain loops of capillaries. Some dermal papillae contain *touch (Meissner's) corpuscles,* nerve endings sensitive to touch. The papillary region consists of loose connective tissue containing fine elastic fibers.

The remaining portion of the dermis is called the **reticular region,** or **layer.** It consists of dense, irregularly arranged connective tissue containing interlacing bundles of collagenous and coarse elastic fibers. It is named the reticular layer because the bundles of collagenous fibers interlace in a netlike manner. Spaces between the fibers are occupied by a small quantity of adipose tissue, hair follicles, nerves, oil glands, and the ducts of sweat glands. Varying thicknesses of the reticular region are responsible for differences in the thickness of the skin.

The combination of collagenous and elastic fibers in the reticular region provides the skin with strength, extensibility, and elasticity. (Extensibility is the ability to stretch; elasticity is the ability to return to original shape after extension or contraction.) The ability of the skin to stretch can readily be seen during conditions of pregnancy, obesity, and edema. The small tears that occur during extreme stretching are initially red and remain visible afterward as silvery white streaks called *striae* (STRĪ-ē).

The reticular region is attached to underlying organs, such as bone and muscle, by the subcutaneous layer. The subcutaneous layer also contains nerve endings called *lamellated (Pacinian) corpuscles* that are sensitive to pressure (see Figure 20-1).

CLINICAL APPLICATION

The collagenous fibers in the dermis run in all directions, but in particular regions of the body they tend to run more in one direction than another. The predominant direction of the underlying collagenous fibers is indicated in the skin by **lines of cleavage (tension lines).** The lines are especially evident on the palmar surfaces of the fingers where they run parallel to the long axis of the digit. The characteristic lines for each part of the body are shown in Figure 4-3.

Lines of cleavage are of particular interest to a surgeon because an incision running parallel to the collagen fibers will heal with only a fine scar. An incision made across the rows of fibers disrupts the collagen, and the wound tends to gape open and heal in a broad, thick scar.

SKIN COLOR

The color of the skin is due to melanin, a pigment in the epidermis; carotene, a pigment mostly in the dermis; and blood in the capillaries in the dermis. The amount of **melanin** varies the skin color from pale yellow to black. This pigment is found primarily in the basale and spinosum layers. Melanin is synthesized in cells called **melanocytes** (me-LAN-ō-sīts), located either just beneath or between cells of the stratum basale. Since the number of melanocytes is about the same in all races, differences in skin color are due to the amount of pigment the melanocytes produce and disperse. An inherited inability of an individual in any race to produce melanin results in **albinism** (AL-bi-nizm). The pigment is absent in the hair and eyes as well as the skin. An individual affected with albinism is called an **albino** (al-BĪ-no). In some people, melanin tends to form in patches called **freckles.**

Melanocytes synthesize melanin from the amino acid *tyrosine* in the presence of an enzyme called *tyrosinase.* Exposure to ultraviolet radiation increases the enzymatic activity of melanocytes and leads to increased melanin production. The cell bodies of melanocytes send out long processes between epidermal cells (Figure 4-4). On contact with the processes, epidermal cells take up the melanin by phagocytosis. When the skin is again exposed to ultraviolet radiation, both the amount and the darkness of melanin increase, tanning and further protecting the body against radiation. Thus melanin serves a vital protective function. Melanocyte-stimulating hormone (MSH) produced by the anterior pituitary gland causes increased melanin synthesis and dispersion through the epidermis.

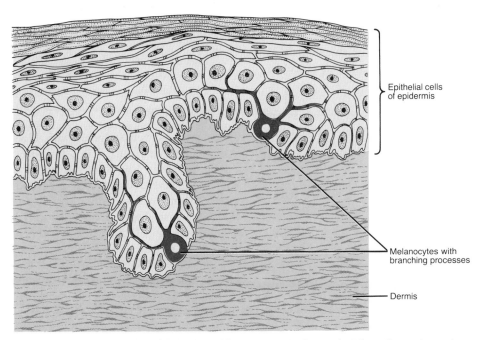

Epithelial cells of epidermis

Melanocytes with branching processes

Dermis

FIGURE 4-4 Melanocytes. Note the branching processes that project from the melanocytes and supply the epithelial cells of the epidermis with melanin.

CLINICAL APPLICATION

Overexposure of the skin to the ultraviolet light of the sun may lead to skin cancer. Among the most malignant and lethal skin cancers is **melanoma** (*melano* = dark-colored; *oma* = tumor), cancer of the melanocytes. Fortunately, most skin cancers are operable.

Another pigment, called **carotene** (KAR-o-tēn), is found in the stratum corneum and fatty areas of the dermis in people of Asian origin. Together, carotene and melanin account for the yellowish hue of their skin.

The pink color of Caucasian skin is due to blood in capillaries in the dermis. The redness of the vessels is not heavily masked by pigment. The epidermis has no blood vessels, a characteristic of all epithelia.

EPIDERMAL RIDGES AND GROOVES

The outer surface of the skin of the palms and fingers and soles and toes is marked by a series of ridges and grooves that appear either as fairly straight lines or as a pattern of loops and whorls, as on the tips of the digits.

The **epidermal ridges** develop during the third and fourth fetal months as the epidermis conforms to the contours of the underlying dermal papillae (Figure 4-5). Since the ducts of sweat glands open on the summits of the epidermal ridges, fingerprints (or footprints) are left when a smooth object is touched. The ridge pattern, which is genetically determined, is unique for each individual. It does not change throughout life, except to enlarge, and thus can serve as the basis for identification through fingerprints or footprints. The function of the ridges is to increase the grip of the hand or foot by increasing friction.

Epidermal grooves on other parts of the skin divide the surface into a number of diamond-shaped areas. Examine the dorsum of the hand as an example. Note that hairs typically emerge at the points of intersection of the grooves. Note also that grooves increase in frequency and depth as regions of free joint movement are approached.

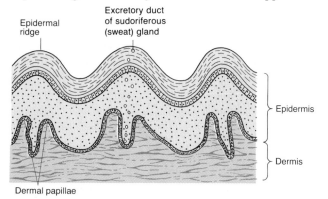

FIGURE 4-5 Relationship between epidermal ridges and dermal papillae.

CLINICAL APPLICATION

Scientists have now made it possible in many cases to improve the appearance of scars from deep acne, chickenpox, burns, cleft lip, and other disorders and to make age-related wrinkles disappear. The procedure that makes this possible is called a **collagen implant.** Collagen is the body's principal structural material. It is found in skin, bone, cartilage, tendons, ligaments, and various viscera and accounts for almost one-third of the total protein of the body.

The collagen implant is prepared from cattle collagen that is suspended in a saline solution containing lidocaine, a local anesthetic. Once injected into the skin, the solution disappears and the collagen, under the influence of body temperature, becomes a stationary fleshlike substance that is incorporated into the skin. The collagen implant becomes colonized by blood vessels and cells and acts as a natural structural framework in the skin.

Another procedure, called **chemical exfoliation (skin peel),** is used for more superficial problems such as frown lines, wrinkles, age spots, and superficial scars. It is performed under sedation or with light anesthesia. In this procedure, the natural oils in the skin are removed with solvents. Then, keratolytic agents (keratin-dissolving chemicals) are mixed with surgical soap or croton oil, applied thickly, and wrapped in bandages for a few days. When the bandages are removed, the keratin-containing layers of the epidermis are peeled away. When the new surface layers grow back, the superficial problems are either greatly reduced or disappear completely. Chemical exfoliation is used only for patients with fair skin since people with darker skin react to the procedure by forming uneven pigmentation.

BLOOD SUPPLY

The dermis is well supplied with blood; the epidermis is avascular (see Figure 4-1a). The arteries supplying the dermis are generally derived from branches of arteries supplying skeletal muscles in a particular region. Some arteries supply the skin directly. One plexus of arteries, the **cutaneous plexus,** is located at the junction of the dermis and subcutaneous layer and sends branches that supply the sebaceous and sudoriferous glands, the deep portions of hair follicles, and adipose tissue. The **papillary plexus,** formed at the level of the papillary layer, sends branches that supply the capillary loops in the dermal papillae, sebaceous glands, and superficial portions of hair follicles. The arterial plexuses are accompanied by venous plexuses that drain blood from the dermis into larger subcutaneous veins.

SKIN AND TEMPERATURE REGULATION

Humans, like other mammals, are **homeotherms,** or warm-blooded organisms. This means that we are able to maintain a remarkably constant body temperature of 37°C (98.6°F) even though the environmental temperature may vary over a broad range.

Suppose you are in an environment where the temperature is 37.8°C (100°F). A sequence of events is set into operation to counteract this above-normal temperature. Sensing devices in the skin called receptors pick up the stimulus—in this case, heat—and activate nerves that send the message to the brain. A temperature-regulating area of the brain then sends nerve impulses to the sudoriferous glands, which produce more perspiration. As the perspiration evaporates from the surface of the skin, the skin is cooled and your body temperature is lowered.

Temperature regulation by perspiration represents only one mechanism by which we maintain normal body temperature. Other mechanisms include adjusting blood flow to the skin, regulating metabolic rate, and regulating skeletal muscle contractions.

EPIDERMAL DERIVATIVES

Structures developed from the embryonic epidermis—hair, glands, nails—perform functions that are necessary and sometimes vital. Hair and nails protect the body. The sweat glands help regulate body temperature.

HAIR

Growth of the epidermis variously distributed over the body are **hairs (pili).** The primary function of hair is protection. Though the protection is limited, hair guards the scalp from injury and the sun's rays. Eyebrows and eyelashes protect the eyes from foreign particles. Hair in the nostrils and external ear canal protects these structures from insects and dust.

Development and Distribution

During the third and fourth months of fetal life, the epidermis develops downgrowths into the dermis called **hair follicles.** Originating from these follicles are small hairs. By the fifth or six month, the follicles produce delicate hairs called **lanugo** (lan-YOO-gō; *lana* = wool) that covers the fetus. The lanugo is usually shed prior to birth, except in the regions of the eyebrows, eyelids, and scalp. Here they persist and become stronger. Several months after birth, these hairs are shed and are replaced by still coarser ones, while the remainder of the body develops a new hair growth called the **vellus** (fleece). At puberty, coarse hairs develop in the axillary and pubic regions of both sexes and on the face and to a lesser extent on

other parts of the body in the male. The coarse hairs that develop at puberty, plus the hairs of the scalp and eyebrows, are referred to as **terminal hairs.**

CLINICAL APPLICATION

Hirsutism (HER-soot-izm) is a term usually applied to an excessive growth of hair in females and children, with a distribution similar to that found in adult males. What happens is that vellus hairs are converted into large terminal hairs as a result of higher than normal levels of androgens. The causes may be related to hypothalamic, adrenal gland, or ovarian disorders, among others.

Hairs are distributed on nearly all parts of the body. It has been estimated that an average adult has about five million hairs, of which about 100,000 are in the scalp. They are absent from the palms of the hands, the soles of the feet, the dorsal surfaces of the distal phalanges, lips, nipples, clitoris, glans penis, inner surface of the prepuce, inner surfaces of the labia majora and minora, and outer surfaces of the labia minora. Straight hairs are oval or cylindrical in cross section, whereas curly hairs are flattened. Straight hairs are stronger than curly ones.

Structure

Each hair consists of a shaft and a root (Figure 4-6). The **shaft** is the superficial portion, most of which projects above the surface of the skin. The shaft of coarse hairs consists of three principal parts. The inner **medulla** is composed of rows of polyhedral cells containing granules of eleidin and air spaces. The medulla is poorly developed or not present at all in fine hairs. The second principal part of the shaft is the middle **cortex.** It forms the major part of the shaft and consists of elongated cells that contain pigment granules in dark hair, but mostly air in white hair. The **cuticle of the hair,** the outermost layer, consists of a single layer of thin, flat, scalelike cells that are the most heavily keratinized. They are arranged like shingles on the side of a house, but the free edges of the cuticle cells point upward rather than downward like shingles. The **root** is the portion below the surface that penetrates into the dermis and even into the subcutaneous layer and, like the shaft, contains a medulla, cortex, and cuticle.

Surrounding the root is the **hair follicle,** which is made up of an external zone of epithelium (the external root sheath) and an internal zone of epithelium (the internal root sheath). The **external root sheath** is a downward continuation of the basale and spinosum layers of the epidermis. Near the surface it contains all the epidermal layers. As it descends, it does not exhibit the superficial

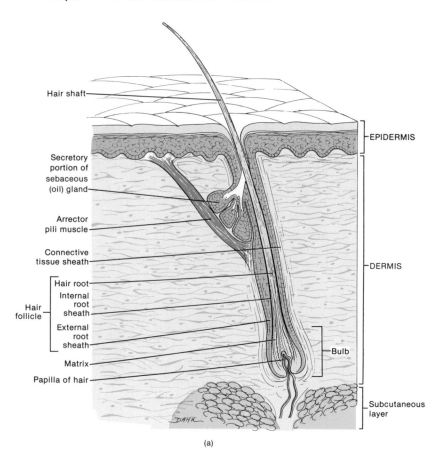

Hair shaft

EPIDERMIS

Secretory portion of sebaceous (oil) gland

Arrector pili muscle

Connective tissue sheath

DERMIS

Hair follicle
Hair root
Internal root sheath
External root sheath

Matrix

Papilla of hair

Bulb

Subcutaneous layer

(a)

DAHK

FIGURE 4-6 Hair. (a) Principal parts of a hair and associated structures.

epidermal layers. At the bottom of the hair follicle, the external root sheath contains only the stratum basale. The **internal root sheath** is formed from proliferating cells of the matrix (described shortly) and takes the form of a cellular tubular sheath deep to the external root sheath. The internal root sheath extends only partway up the follicle and consists of (1) an inner layer, the **cuticle of the internal root sheath,** which is a single layer of flattened cells with atrophied nuclei, (2) a middle **granular (Huxley's) layer,** which is one to three layers of flattened nucleated cells, and (3) an outer **pallid (Henle's) layer,** which is a single layer of cuboidal cells with flattened nuclei.

The base of each follicle is enlarged into an onion-shaped structure, the **bulb.** This structure contains an indentation, the **papilla of the hair,** filled with loose connective tissue. The papilla of the hair contains many blood vessels and provides nourishment for the growing hair. The bulb also contains a region of cells called the **matrix,** a germinal layer. The cells of the matrix produce new hairs by cell division when older hairs are shed. This replacement occurs within the same follicle.

A substance that removes superfluous hair is called a **depilatory.** It dissolves the protein in the hair shaft, turning it into a gelatinous mass that can be wiped away. Since the hair root is not affected, regrowth of the hair occurs. In **electrolysis,** the hair bulb is destroyed by an electric current so that the hair cannot regrow.

Sebaceous glands and a bundle of smooth muscle are also associated with hair. Details of the sebaceous glands will be discussed shortly. The smooth muscle is called **arrector pili;** it extends from the dermis of the skin to the side of the hair follicle (Figure 4-6a). In its normal position hair is arranged at an angle to the surface of the skin. The arrectores pilorum muscles contract under stresses of fright and cold and pull the hairs into a vertical position. This contraction results in "goosebumps" or "gooseflesh" because the skin around the shaft forms slight elevations.

Around each hair follicle are nerve endings, called *root hair plexuses,* that are sensitive to touch (see Figure 20-1). They respond if a hair shaft is moved.

Color

The color of hair is due primarily to melanin. It is formed by melanocytes distributed in the matrix of the bulb of the follicle. There are two basic classes of melanin: eumelanin (brown-black) and pheomelanin (yellow to reddish). One distinguishing characteristic is that pheomelanin has a higher content of sulfur. The many variations in hair color are combinations of different amounts of the pigments. Light-colored hair has a predominance of pheomelanin, as does red hair. Dark-colored hair contains mostly eumelanin. Graying of hair is the loss of pigment believed to be the result of a progressive inability of the

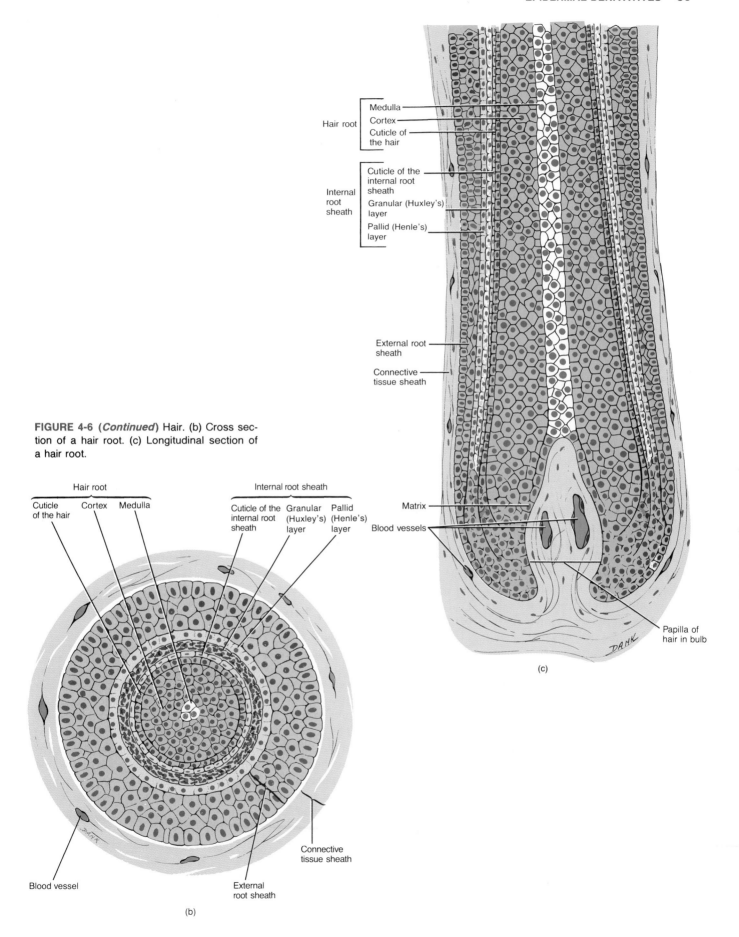

FIGURE 4-6 (*Continued*) Hair. (b) Cross section of a hair root. (c) Longitudinal section of a hair root.

Hair root

Medulla
Cortex
Cuticle of the hair

Internal root sheath

Cuticle of the internal root sheath
Granular (Huxley's) layer
Pallid (Henle's) layer

External root sheath

Connective tissue sheath

Matrix

Blood vessels

Papilla of hair in bulb

(c)

Hair root

Cuticle of the hair   Cortex   Medulla

Internal root sheath

Cuticle of the internal root sheath   Granular (Huxley's) layer   Pallid (Henle's) layer

Connective tissue sheath

Blood vessel

External root sheath

(b)

melanocytes to make tyrosinase, the enzyme necessary for the synthesis of melanin. White hair results from air in the medullary shaft.

Growth and Replacement

Hair replacement occurs according to a cyclic pattern, alternating between growing and resting periods. During the growing phase of the cycle, the cells of the matrix are active. They increase in number by cell division and are pushed upward and eventually die, a situation similar to epidermal growth. The product is a hair, which is essentially dead protein tissue. Hair grows about 1 mm (0.04 in) every 3 days.

During the resting phase, the matrix becomes inactive and undergoes atrophy. At this point the root of the hair detaches from the matrix and the hair slowly moves up the follicle. It may remain there for some time until pulled out, shed, or pushed up by a replacing hair.

Either before or after the hair comes out of the follicle, proliferation of cells by the external root sheath forms a new matrix. The new matrix undergoes cell division, forming a new hair that grows up the follicle and replaces the old hair. You might be interested to know that shaving or cutting the hair has no effect on its growth.

The cycle of hair growth varies in different parts of the body. In the scalp, each hair grows steadily and continuously for 2 to 6 years; growth then stops, and after 3 months, the hair is shed. After another 3 months of a resting phase, a new hair starts to grow from the same follicle. Eyebrows, by contrast, have a growing phase of only about 10 weeks. It is for this reason that these hairs do not grow very long.

Normal hair loss in an adult scalp is about 70 to 100 hairs per day. Both the rate of growth and the replacement cycle may be altered by illness, diet, and other factors. For example, high fever, major illness, major surgery, blood loss, or severe emotional stress may increase the rate of shedding. Rapid weight-loss diets involving severe restriction of calories or protein also increase hair loss. An increase in the rate of shedding can also occur for 3 to 4 months after childbirth. Certain drugs and radiation therapy are also factors in increasing hair loss.

CLINICAL APPLICATION

The age of onset, degree of thinning, and ultimate hair pattern associated with **common baldness** ("male-pattern" baldness) are determined by male hormones, called androgens, and heredity. Androgens are involved in promoting normal sexual development.

GLANDS

Three kinds of glands associated with the skin are sebaceous glands, sudoriferous glands, and ceruminous glands.

Sebaceous (Oil) Glands

Sebaceous (se-BĀ-shus), or **oil, glands,** with few exceptions, are connected to hair follicles (Figure 4-7). The secreting portions of the glands lie in the dermis and those glands associated with hairs open into the necks of hair follicles. Sebaceous glands not associated with hair follicles open directly onto the surface of the skin (lips, glans penis, labia minora, and tarsal glands of the eyelids). Sebaceous glands are simple branched acinar glands. Absent in the palms and soles, they vary in size and shape in other regions of the body. For example, they are small in most areas of the trunk and extremities, but large in the skin of the breasts, face, neck, and upper chest.

The sebaceous glands secrete an oily substance called **sebum** (SĒ-bum), a mixture of fats, cholesterol, proteins, and inorganic salts. Sebum helps keep hair from drying and becoming brittle, forms a protective film that prevents excessive evaporation of water from the skin, and keeps the skin soft and pliable.

CLINICAL APPLICATION

When sebaceous glands of the face become enlarged because of accumulated sebum, **blackheads** develop. Since sebum is nutritive to certain bacteria, **pimples** or **boils** often result. The color of blackheads is due to melanin and oxidized oil, not dirt.

Sudoriferous (Sweat) Glands

Sudoriferous (soo'-dor-IF-er-us; *sudor* = sweat; *ferre* = to bear), or **sweat, glands** are divided into two principal types on the basis of structure and location. **Apocrine sweat glands** are simple, branched tubular glands. Their distribution is limited primarily to the skin of the axilla, pubic region, and pigmented areas (areolae) of the breasts. The secretory portion of apocrine sweat glands is located in the dermis or hypodermis and the excretory duct opens into hair follicles. Apocrine sweat glands begin to function at puberty and produce a more viscous secretion than eccrine sweat glands.

Eccrine sweat glands are much more common than apocrine sweat glands and are simple, coiled tubular glands. They are distributed throughout the skin except for the margins of the lips, nail beds of the fingers and toes, glans penis, glans clitoris, labia minora, and eardrums. Eccrine sweat glands are most numerous in the

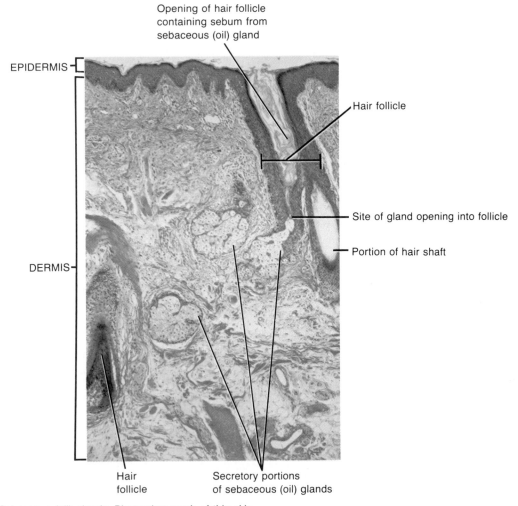

Opening of hair follicle
containing sebum from
sebaceous (oil) gland

EPIDERMIS

Hair follicle

DERMIS

Site of gland opening into follicle

Portion of hair shaft

Hair
follicle

Secretory portions
of sebaceous (oil) glands

FIGURE 4-7 Sebaceous (oil) glands. Photomicrograph of thin skin of the face showing the location of sebaceous glands at a magnification of 60×. (© 1983 by Michael H. Ross. Used by permission.)

skin of the palms and the soles; their density can be as high as 3,000 per square inch in the palms. The secretory portion of eccrine sweat glands is located in the subcutaneous layer, and the excretory duct projects upward through the dermis and epidermis to terminate at a pore at the surface of the epidermis (see Figure 4-1). Eccrine sweat glands function throughout life and produce a secretion that is more watery than that of apocrine sweat glands.

Perspiration, or **sweat,** is the substance produced by sudoriferous glands. It is a mixture of water, salts (mostly NaCl), urea, uric acid, amino acids, ammonia, sugar, lactic acid, and ascorbic acid. Its principal function is to help maintain body temperature. It also helps to eliminate wastes.

Since the mammary glands are actually modified sudoriferous glands, they could be considered here. However, because of their relationship to the reproductive system, they are considered in Chapter 25.

Ceruminous Glands

In certain parts of the skin, sudoriferous glands are modified as **ceruminous** (se-ROO-mi-nus) **glands.** Such modified glands are simple, coiled tubular glands present in the external auditory meatus (canal). The secretory portions of ceruminous glands lie in the submucosa, deep to sebaceous glands, and the excretory ducts open either directly onto the surface of the external auditory meatus or into sebaceous ducts. The combined secretion of the ceruminous and sebaceous glands is called **cerumen** (*cera* = wax).

CLINICAL APPLICATION

Some people produce an abnormal amount of cerumen, or earwax, in the external auditory canal. It then becomes impacted and prevents sound waves from reaching the tympanic membrane. The treatment for **impacted cerumen** is usually periodic ear irrigation or removal of wax with a blunt instrument.

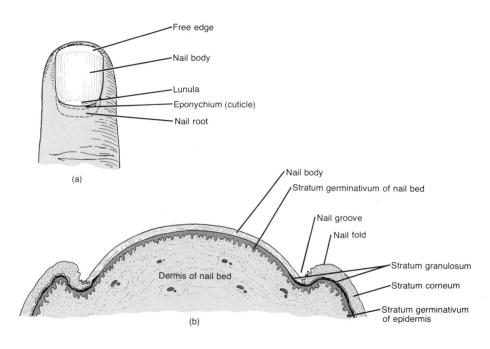

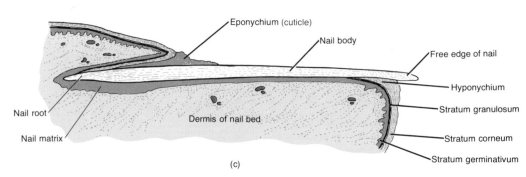

FIGURE 4-8 Structure of nails. (a) Superior view of fingernail. (b) Cross section of a fingernail and nail bed. (c) Sagittal section of a fingernail and nail bed.

NAILS

Hard, keratinized cells of the epidermis are referred to as **nails.** The cells form a clear, solid covering over the dorsal surfaces of the terminal portions of the fingers and toes. Each nail (Figure 4-8) is about 0.5 mm (0.02 in) thick and consists of a **nail body,** a **free edge,** and a **nail root.** The nail body is the portion of the nail that is visible, the free edge is the part that projects beyond the distal end of the digit, and the nail root is the portion that is hidden in the nail groove. Most of the nail body is pink because of the underlying vascular tissue. The whitish semilunar area at the proximal end of the body is called the **lunula** (LOO-nyoo-la). It appears whitish because the vascular tissue underneath does not show through.

The fold of skin that extends around the proximal and lateral borders of the nail is known as the **nail fold,** and the epidermis beneath the nail constitutes the **nail bed.** The furrow between the two is the **nail groove.**

The **eponychium** (ep'-ō-NIK-ē-um), or **cuticle,** is a narrow band of epidermis that extends from the margin of the nail wall (lateral border), adhering to it. It occupies the proximal border of the nail and consists of stratum corneum. The thickened area of stratum corneum below the free edge of the nail is referred to as the **hyponychium** (hī'-pō-NIK-ē-um).

The epithelium of the proximal part of the nail bed is known as the **nail matrix.** Its function is to bring about the growth of nails. Essentially, growth occurs by the transformation of superficial cells of the matrix into nail cells. In the process, the outer, harder layer is pushed forward over the stratum germinativum. The average growth in the length of fingernails is about 1 mm (0.04 in) per week. The growth rate is somewhat slower in toenails. For some reason, the longer the digit, the faster the nail grows. Adding supplements such as gelatin to an otherwise healthy diet has no effect on making nails grow faster or stronger. Functionally, nails help us to grasp and manipulate small objects in various ways.

AGING AND THE INTEGUMENTARY SYSTEM

Although skin is constantly aging, the pronounced effects do not occur until a person reaches the late forties. Collagen fibers stiffen, break apart, and form into a shapeless, matted tangle. Elastic fibers thicken into clumps and fray and, as a result, the skin wrinkles. Fibroblasts, which produce both collagenous and elastic fibers, decrease in number and macrophages become less efficient phagocytes. In addition, there is a loss of subcutaneous fat; atrophy of sebaceous (oil) glands, producing dry and broken skin that is susceptible to infection; a decrease in the number of functioning melanocytes, resulting in gray hair and atypical skin pigmentation; and an increase in the size of some melanocytes that produces pigmented blotching. Aged skin also becomes susceptible to pathological conditions such as senile pruritus (itching), decubitus ulcers (bedsores), and herpes zoster (shingles). Prolonged exposure to the ultraviolet (UV) rays of sunlight accelerates the aging of skin.

DEVELOPMENTAL ANATOMY OF THE INTEGUMENTARY SYSTEM

In this and subsequent chapters the developmental anatomy of the systems of the body will be discussed. Since the principal features of embryonic development are not treated in detail until Chapter 26, it will be necessary to explain a few terms at this point so that you can follow the development of organ systems.

As part of the early development of a fertilized egg, a portion of the developing embryo differentiates into three layers of tissue called primary germ layers. On the basis of position, the primary germ layers are referred to as **ectoderm, mesoderm,** and **endoderm.** They are the embryonic tissues from which all tissues and organs of the body will eventually develop (see Exhibit 26-1).

The *epidermis* is derived from the **ectoderm.** At the beginning of the second month, the ectoderm consists of simple cuboidal epithelium. These cells become flattened and are known as the **periderm.** By the fourth month, all layers of the epidermis are formed and each layer assumes its characteristic structure.

Nails are developed during the third month, Initially, they consist of a thick layer of epithelium called the **primary nail field.** The nail itself is keratinized epithelium and grows forward from its base. It is not until the ninth month that the nails actually reach the tips of the digits.

Hair follicles develop as downgrowths of the stratum basale of the epidermis into the dermis below, between the third and fourth months. The downgrowths soon differentiate into the bulb, papilla of the hair, beginnings of the epithelial portions of sebaceous glands, and other structures associated with hair follicles. By the fifth or sixth month, the follicles produce **lanugo** (delicate fetal hair) first on the head and then on other parts of the body. The lanugo is usually shed prior to birth.

The epithelial (secretory) portions of *sebaceous (oil) glands* develop from the sides of the hair follicles and remain connected to the follicles.

The epithelial portions of *sudoriferous (sweat) glands* are also derived from downgrowths of the stratum basale of the epidermis into the dermis. They appear during the fourth month on the palms and soles and a little later in other regions. The connective tissue and blood vessels associated with the glands develop from **mesoderm.**

The *dermis* is derived from wandering **mesenchymal (mesodermal) cells.** The mesenchyme becomes arranged in a zone beneath the ectoderm and there undergoes changes into the connective tissues that form the dermis.

APPLICATIONS TO HEALTH

ACNE

Acne is an inflammation of sebaceous glands and usually begins at puberty. In fact, a few comedones (blackheads or whiteheads) on the face may be the first signs of approaching puberty. At puberty, the sebaceous glands, under the influence of androgens (male hormones), grow in size and increase production of their complex lipid product, sebum. Although testosterone, a male hormone, appears to be the most potent circulating androgen for sebaceous cell stimulation, adrenal and ovarian androgens can stimulate sebaceous secretions as well.

Acne occurs predominantly in sebaceous follicles. The four basic types of acne lesions in order of increasing severity are comedones, papules, pustules, and cysts. The sebaceous follicles are rapidly colonized by microorganisms such as staphylococcal species and *Propionibacterium acne* bacteria that thrive in the lipid-rich environment of the follicles. When this occurs, the cyst or sac of connective tissue cells can destroy and displace epidermal cells, resulting in permanent scarring. This type of acne, called *cystic acne,* may be treated successfully with a drug called Accutane, a synthetic form of vitamin A. Care must be taken to avoid squeezing, pinching, or scratching the lesions.

SYSTEMIC LUPUS ERYTHEMATOSUS (SLE)

An autoimmune, inflammatory disease that occurs mostly in young women in their reproductive years is called **systemic lupus erythematosus** (er-i'-them-a-TO-sus), **SLE,** or **lupus.** An *autoimmune disease* is one in which the body attacks its own tissues, failing to differentiate between what is foreign and what is not. In SLE, damage to blood vessel walls results in the release of chemicals that mediate the inflammatory response. The blood vessel damage can be associated with virtually every body system.

The cause of SLE is not known. It is not contagious and is thought to be hereditary. There seems to be a strong incidence of other connective tissue disorders—especially rheumatoid arthritis (RA) and rheumatic fever—in relatives of SLE victims. The disease may be triggered by medication, such as penicillin, sulfa, or tetracycline, exposure to excessive sunlight, injury, emotional upset, infection, or other stress. These triggering factors, once recognized, are to be avoided by the patient in the future.

Symptoms include low-grade fever, aches, fatigue, photosensitivity, rapid loss of large amounts of scalp hair, and sometimes an eruption across the bridge of the nose and cheeks called a "butterfly rash." Other skin lesions may occur with blistering and ulceration. The erosive nature of some of the SLE skin lesions was thought to resemble the damage inflicted by the bite of a wolf, thus the term *lupus.* The most serious complications of the disease involve inflammation of the kidneys, liver, spleen, lungs, heart, and the central nervous system.

PSORIASIS

Psoriasis (sō-RĪ-a-sis) is a chronic, occasionally acute, noncontagious, relapsing skin disease borne by 6 to 8 million people in the United States. Psoriasis is characterized by distinct, reddish, slightly raised plaques or papules (small, round skin elevations) covered with scales. Itching is seldom severe, and the lesions heal without scarring. Psoriasis ordinarily involves the scalp, the elbows and knees, the back, and the buttocks. Occasionally the disease is generalized.

In its severe forms, psoriasis is a disabling and disfiguring affliction. It may begin at any age, although it is most severe between ages 10 and 50. It is thought to be hereditary in at least one-third of the patients. Studies have traced the complex cause of psoriasis to an abnormally high rate of mitosis in epidermal cells that may be related to a substance carried in the blood, a defect in the immune system, or a virus. Triggering factors such as trauma, infections, seasonal and hormonal changes, and emotional stress can initiate and intensify the skin eruptions.

Psoriasis is treated by a number of methods, including steroid ointments and creams, natural sunlight, tar preparations, retinoids (chemicals similar to vitamin A), and PUVA, a therapy that combines psoralen (a chemical that increases the skin's reaction to light) and artificial ultraviolet (UV) light.

DECUBITUS ULCERS

Decubitus (dē-KYOO-be-tus) **ulcers,** also known as **bedsores, pressure sores,** or **trophic ulcers,** are caused by a constant deficiency of blood to tissues overlying a bony projection that has been subjected to prolonged pressure against an object such as a bed, cast, or splint. The deficiency results in tissue ulceration. Small breaks in the epidermis become infected, and the sensitive subcutaneous and deeper tissues are damaged. Eventually the tissue is destroyed.

Decubitus ulcers are seen most frequently in patients who are bedridden for long periods of time. The most common areas involved are the skin over the sacrum, heels, ankles, buttocks, and other large bony projections. The chief causes are pressure from infrequent turning of the patient, trauma and maceration of the skin, and malnutrition. Maceration of the skin often follows soaking of bed and clothing by perspiration, urine, or feces.

SUNBURN

Sunburn is injury to the skin as a result of acute, prolonged exposure to the ultraviolet (UV) rays of sunlight. It usually occurs 2 to 8 hours after exposure. Acute redness and pain are maximal in about 12 hours and dissipate approximately 72 to 96 hours later. Sunburn is without doubt one of the most common and painful skin afflictions. The damage to skin cells caused by sunburn is due to inhibition of DNA and RNA synthesis, which leads to cell death. There can also be damage to blood vessels as well as other structures in the dermis. Overexposure over a period of years results in a leathery skin texture, wrinkles, skin folds, sagging skin, warty growths called keratoses, freckling, a yellow discoloration due to abnormal elastic tissue, premature aging of the skin, and skin cancer.

The tendency to sunburn is determined by type of skin. There are basically four skin groups: those who burn and never tan, those who burn but keep some tan, those who burn slightly and develop a good tan, and those who never burn and always tan.

Extreme and severe sunburn may produce *sunstroke*—a disruption of the body's heat-regulating mechanism. It is characterized by high fever and collapse and sometimes by convulsions, coma, and death.

Covering the body remains the best protection against the sun's rays, while a topical sunscreen will help protect uncovered parts.

SKIN CANCER

Excessive sun exposure can result in **skin cancer.** Everyone, regardless of skin pigmentation, is a potential victim of skin cancer if exposure to sunlight is sufficiently intense and continuous. Natural skin pigment can never give complete protection. If you must be in direct sunlight for long periods of time, use a suitable sunscreen. One of the best agents for protection against overexposure to ultraviolet rays of the sun is para-aminobenzoic acid (PABA). The alcohol preparations of PABA are best because the active ingredient binds to the stratum corneum of the skin.

A number of common plants, widely prescribed drugs (tetracyclines, sulfa drugs, and others), constituents of foods (such as riboflavins), and saccharin and cyclamates in dietetic drinks are potential photosensitizers. When activated in the body by light, they may produce substances that can damage the tissues in sensitive individuals. A typical response is the appearance of a rash on parts of the body exposed to the sun. Repeated and persistent exposure could produce permanent changes leading to skin cancer.

BURNS

Tissues may be damaged by thermal (heat), electrical, radioactive, or chemical agents. These agents can destroy the proteins in the exposed cells and cause cell injury or death. Such damage is a **burn.** The injury to tissues directly or indirectly in contact with the damaging agent, such as the skin or the linings of the respiratory and digestive tracts, is the local effect of a burn. Generally, however, the systemic effects of a burn are a greater threat to life than the local effects. The systemic effects of a burn may include (1) a large loss of water, plasma, and plasma proteins, which causes shock; (2) bacterial infection; (3) reduced circulation of blood; and (4) decreased production of urine.

Classification

A burn may extend through the entire thickness of the skin or it may damage or destroy only part of the skin. Clinically, the depth of a burn is determined by its color, the presence or absence of sensation, blister formation, or loss of elasticity. A **first-degree burn** is characterized by mild pain and erythema (redness) and involves only the surface epithelium. Generally, a first-degree burn will heal in about 2 to 3 days and may be accompanied by flaking or peeling. A typical sunburn is an example of a first-degree burn.

A **second-degree burn** involves the deeper layers of the epidermis or the upper levels of the dermis. In a superficial second-degree burn, the deeper layers of the epidermis are injured and there is characteristic erythema,

blister formation, edema, and pain. Blisters beneath or within the epidermis are called *bullae* (BYOOL-ē), meaning bubbles. Such an injury usually heals within 7 to 10 days with only mild scarring. In a deep second-degree burn, there is destruction of the epidermis as well as the upper levels of the dermis. Epidermal derivatives, such as hair follicles, sebaceous glands, and sweat glands, are usually not injured. If there is no infection, deep second-degree burns heal without grafting in about 3 to 4 weeks. Scarring may result.

First- and second-degree burns are collectively referred to as **partial-thickness burns.** A **third-degree burn,** or **full-thickness burn,** involves destruction of the epidermis, dermis, and the epidermal derivatives. Such burns vary in appearance from marble-white to mahogany colored to charred, dry wounds. There is little if any edema and such a burn is usually not painful to the touch due to destruction of nerve endings. Regeneration is slow and much granulation tissue forms before being covered by epithelium. Even if skin grafting is quickly begun, third-degree burns quickly contract and produce scarring.

The severity of a burn is measured in terms of the depth of injury and the amount of surface area affected. Ultrasonography is now being used to help physicians determine burn depth. An ultrasonic wave moves smoothly through burned skin because damaged collagen uncoils and becomes jellylike. On the other hand, an ultrasonic wave is reflected in nonburned skin since collagen here is coiled. A sonar image can thus provide information regarding the depth of a burn. A fairly accurate method for estimating the amount of surface area affected by a burn is to apply the *Lund-Browder method.* This method estimates the extent by measuring the areas affected against the percentage of total surface area for body parts shown in Figure 4-9. For example, if the anterior of the head and neck of an adult is affected, the burn covers 4½ percent of the body surface. Because the proportions of the body change with growth, the percentages vary for different ages. Thus, the extent of burn damage can be made fairly accurately for any age group.

Treatment

A severely burned individual should be moved as quickly as possible to a hospital. Treatment may then include:

1. Cleansing the burn wounds thoroughly.

2. Removing all dead tissue (debridement) so antibacterial agents can directly contact the wound surface and thereby prevent infection.

3. Replacing lost body fluids and electrolytes (ions).

4. Covering wounds with temporary protection as soon as possible.

5. Removing a thin layer of skin from another part of the burn victim's body and transplanting it to the injured area (skin graft).

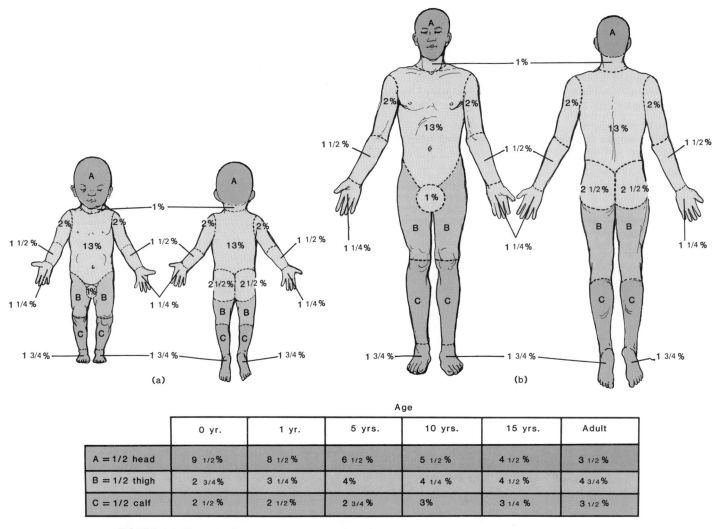

		Age				
	0 yr.	1 yr.	5 yrs.	10 yrs.	15 yrs.	Adult
A = 1/2 head	9 1/2 %	8 1/2 %	6 1/2 %	5 1/2 %	4 1/2 %	3 1/2 %
B = 1/2 thigh	2 3/4 %	3 1/4 %	4%	4 1/4 %	4 1/2 %	4 3/4 %
C = 1/2 calf	2 1/2 %	2 1/2 %	2 3/4 %	3%	3 1/4 %	3 1/2 %

FIGURE 4-9 The Lund-Browder method for estimating the extent of burns. Relative proportions of various body regions in (a) a young child compared to (b) an adult.

KEY MEDICAL TERMS ASSOCIATED WITH THE INTEGUMENTARY SYSTEM

Athlete's foot A superficial fungus infection of the skin of the foot.

Callus (callosity) An area of hardened and thickened skin that is usually seen in palms and soles and is due to pressure and friction.

Carbuncle (*carbunculus* = little coal) A hard, round, deep, painful inflammation of the subcutaneous tissue that causes necrosis (death) and pus formation (abscess).

Cold sore Lesion, usually in oral mucous membrane, caused by type 1 herpes simplex virus (HSV), transmitted by oral or respiratory routes. Triggering factors include ultraviolet (UV) radiation, hormonal changes, and emotional stress. Also called a fever blister.

Comedo (*comedo* = to eat up) A collection of sebaceous material and dead cells in the hair follicle and excretory duct of the sebaceous gland. Usually found over the face, chest, and back, and more commonly during adolescence. Also called blackhead or whitehead.

Cyst (*cyst* = sac containing fluid) A sac with a distinct connective tissue wall, containing a fluid or other material.

Dermabrasion (*derm* = skin) Removal of acne scars, tattoos, or moles by sandpaper or a high-speed brush.

Erythema (*erythema* = redness) Redness of the skin caused by engorgement of capillaries in lower layers of the skin. Erythema occurs with any skin injury, infection, or inflammation.

Furuncle A boil; an abscess resulting from infection of a hair follicle.

Impetigo Superficial skin infection caused by staphylococci or streptococci; most common in children.

Keratosis (*kera* = horn) Formation of a hardened growth of tissue.

Nevus A round, pigmented, flat, or raised skin area that may be present at birth or develop later. Varying in color from yellow-brown to black. Also called a mole or birthmark.

Papule A small, round skin elevation varying in size from a pinpoint to that of a split pea. One example is a pimple.

Pruritus (*pruire* = to itch) Itching, one of the most common dermatological disorders. It may be caused by skin disorders (infections), systemic disorders (cancer, kidney failure), or psychogenic factors (emotional stress).

Pustule A small, round elevation of the skin containing pus.

Subcutaneous (*sub* = under) Beneath the skin; also called **hypodermis.**

Topical Pertaining to a definite area; local. Also, of a medication, applied to the surface rather than ingested or injected.

Wart Mass produced by uncontrolled growth of epithelial skin cells; caused by a virus (papovavirus). Most warts are noncancerous.

STUDY OUTLINE

Skin (p. 81)

1. The skin and its derivatives (hair, glands, and nails) constitute the integumentary system.
2. The skin is one of the larger organs of the body. It performs the functions of protection; maintaining body temperature; preventing excessive loss of inorganic and organic materials; receiving stimuli; storage of chemical compounds; synthesis of vitamin D; and excretion of water, salts, and several organic compounds.
3. The principal parts of the skin are the outer epidermis and inner dermis. The dermis overlies the subcutaneous layer.
4. The epidermal layers, from deepest to most superficial, are the strata basale, spinosum, granulosum, lucidum, and corneum. The basale and spinosum undergo continuous cell division and produce all other layers.
5. The dermis consists of a papillary region and a reticular region. The papillary region is loose connective tissue containing blood vessels, nerves, hair follicles, dermal papillae, and touch (Meissner's) corpuscles. The reticular region is dense, irregularly arranged connective tissue containing adipose tissue, hair follicles, nerves, sebaceous (oil) glands, and ducts of sudoriferous (sweat) glands.
6. Lines of cleavage indicate the direction of collagenous fiber bundles in the dermis and are considered during surgery.
7. The color of skin is due to melanin, carotene, and blood in capillaries in the dermis.
8. Epidermal ridges increase friction for better grasping ability and provide the basis for fingerprints and footprints.
9. The epidermis is avascular; the dermis is abundantly supplied by the cutaneous plexus and papillary plexus.
10. In response to nerve impulses from the brain, sudoriferous (sweat) glands adjust their output of perspiration to help maintain normal body temperature.

Epidermal Derivatives (p. 87)

1. Epidermal derivatives are structures developed from the embryonic epidermis.
2. Among the epidermal derivatives are hair, skin glands (sebaceous, sudoriferous, and ceruminous), and nails.

Hair

1. Hairs are epidermal growths that function in protection.
2. Hair consists of a shaft above the surface, a root that penetrates the dermis and subcutaneous layer, and a hair follicle.
3. Associated with hairs are sebaceous (oil) glands, arrectores pilorum muscles, and root hair plexuses.
4. Hair color is due to combinations of various amounts of the three hair pigments, black melanin, brown melanin, and pheomelanin (yellow). Graying is due to the loss of melanin.
5. New hairs develop from cell division of the matrix in the bulb; hair replacement and growth occurs in a cyclic pattern. "Male-pattern" baldness is caused by androgens and heredity.

Glands

1. Sebaceous (oil) glands are usually connected to hair follicles; they are absent in the palms and soles. Sebaceous glands produce sebum, which moistens hairs and waterproofs the skin. Enlarged sebaceous glands may produce blackheads, pimples, and boils.
2. Sudoriferous (sweat) glands are distinguished into apocrine and eccrine. Apocrine sweat glands are limited in distribution to the skin of the axilla, pubis, and areolae; their ducts open into hair follicles. Eccrine sweat glands have an extensive distribution; their ducts terminate at pores at the surface of the epidermis. Sudoriferous glands produce perspiration, which carries small amounts of wastes to the surface and assists in maintaining body temperature.
3. Ceruminous glands are modified sudoriferous glands that secrete cerumen. They are found in the external auditory meatus.

Nails

1. Nails are hard, keratinized epidermal cells over the dorsal surfaces of the terminal portions of the fingers and toes.
2. The principal parts of a nail are the body, free edge, root, lunula, eponychium, hyponychium, and matrix. Cell division of the matrix cells produces new nails.

Aging and the Integumentary System (p. 93)

1. Most effects of aging occur when an individual reaches the late forties.
2. Among the effects of aging are wrinkling, loss of subcutaneous fat, atrophy of sebaceous glands, and decrease in the number of melanocytes.

Developmental Anatomy of the Integumentary System (p. 93)

1. The epidermis is derived from ectoderm. Hair, nails, and skin glands are epidermal derivatives.
2. The dermis is derived from wandering mesodermal cells.

Applications to Health (p. 93)

1. Acne is an inflammation of sebaceous glands.
2. Systemic lupus erythematosus (SLE) is an autoimmune disease of connective tissue.
3. Psoriasis is a chronic skin disease characterized by reddish, raised plaques or papules.
4. Decubitus ulcers are caused by a chronic deficiency of blood to tissues subjected to prolonged pressure.
5. Sunburn is a skin injury resulting from prolonged exposure to the ultraviolet (UV) rays of sunlight.
6. Skin cancer can be caused by excessive exposure to sunlight.
7. Tissue damage that destroys protein is called a burn. Depending on the depth of damage, skin burns are classified as first-degree and second-degree (partial-thickness) and third-degree (full-thickness). One method employed for determining the extent of a burn is the Lund-Browder method. Burn treatment may include cleansing the wound, removing dead tissue, replacing lost body fluids, covering wounds with temporary protection, and skin grafting.

REVIEW QUESTIONS

1. What is the integumentary system?
2. List the principal functions of the skin.
3. Compare the structure of epidermis and dermis. What is the subcutaneous layer?
4. List and describe the epidermal layers from the deepest outward. What is the importance of each layer?
5. Contrast the structural differences between the papillary and reticular regions of the dermis.
6. What are lines of cleavage? What is their importance during surgery?
7. Explain the factors that produce skin color. What is an albino?
8. Describe how melanin is synthesized and distributed to epidermal cells.
9. How are epidermal ridges formed? Why are they important?
10. List the receptors in the epidermis, dermis, and subcutaneous layer, and indicate the location and role of each.
11. Describe the blood supply of the skin.
12. Describe the structure of a hair. What produces "goose-bumps" or "gooseflesh"?
13. Describe how hair grows and is replaced.
14. Contrast the locations and functions of sebaceous (oil) glands, sudoriferous (sweat) glands, and ceruminous glands. What are the names and chemical components of the secretions of each?
15. Describe the principal parts of a nail.
16. Describe the effects of aging on the integumentary system.
17. Describe the origin of the epidermis, its derivatives, and the dermis.
18. Define each of the following disorders of the integumentary system: acne, systemic lupus erythematosus (SLE), psoriasis, decubitus ulcers, sunburn, and skin cancer.
19. Define a burn. Classify burns according to degree.
20. Explain how the Lund-Browder method is used to estimate the extent of a burn.
20. Refer to the glossary of key medical terms associated with the integumentary system. Be sure that you can define each term.

5

Osseous Tissue

Student Objectives

Identify the components and functions of the skeletal system.

List and describe the gross features of a long bone.

Describe the histological features of compact and spongy bone tissue.

Contrast the steps involved in intramembranous and endochondral ossification.

Identify the zones and growth pattern of the epiphyseal plate.

Describe the processes of bone construction and destruction involved in bone remodeling.

Describe the conditions necessary for normal bone growth and replacement.

Describe the blood and nerve supply of bone tissue.

Explain the effects of aging on the skeletal system.

Describe the development of the skeletal system.

Define rickets and osteomalacia as vitamin deficiency disorders.

Contrast the causes and clinical symptoms associated with osteoporosis, Paget's disease, and osteomyelitis.

Define a fracture and describe several common kinds of fractures.

Describe the sequence of events involved in fracture repair.

Define key medical terms associated with the skeletal system.

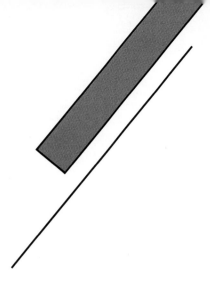

Without the skeletal system we would be unable to perform movements, such as walking or grasping. The slightest jar to the head or chest could damage the brain or heart. It would even be impossible to chew food. The framework of bones and cartilage that protects our organs and allows us to move is called the **skeletal system.** The specialized branch of surgery that deals with the preservation and restoration of the skeletal system, articulations (joints), and associated structures is called **orthopedics** (*ortho* = correct).

The developmental anatomy of the skeletal system is considered at the end of the chapter.

FUNCTIONS

The skeletal system performs several basic functions. First, it **supports** the soft tissues of the body so that the form of the body and an erect posture can be maintained. Second, the system **protects** delicate structures—the brain, spinal cord, lungs, heart, and major blood vessels in the thoracic cavity. Third, the bones serve as levers to which the muscles of the body are attached. When the muscles contract, the bones acting as levers produce **movement.** Fourth, the bones serve as **storage areas** for mineral salts, especially calcium and phosphorus, and fat. Fifth, **blood-cell production** occurs in the red marrow of the bones. This process is referred to as **hematopoiesis** (he'-mat-ō-poy-Ē-sis), or **hemopoiesis.** Red marrow consists of blood cells in immature stages, fat cells, and macrophages. It is responsible for producing red blood cells, some white blood cells, and platelets.

HISTOLOGY

Structurally, the skeletal system consists of two types of connective tissue: cartilage and bone. We described the microscopic structure of cartilage in Chapter 3. Here, our attention will be directed to a detailed discussion of the microscopic structure of bone tissue.

Like other connective tissues, **bone,** or **osseous** (OS-ē-us) **tissue,** contains a great deal of intercellular substance surrounding widely separated cells. Mature bone cells are called **osteocytes.** Unlike other connective tissues, the intercellular substance of bone contains abundant mineral salts, primarily calcium phosphate—$Ca_3(PO_4)_2 \cdot (OH)_2$—and some calcium carbonate—$CaCO_3$. Together these salts are referred to as *hydroxyapatites.* As these salts are deposited in the framework formed by the collagenous fibers of the intercellular substance, the tissue hardens, that is, becomes ossified. The hydroxyapatites compose 67 percent of the weight of bone, and the collagenous fibers make up the remaining 33 percent.

The microscopic structure of bone may be analyzed by first considering the anatomy of a long bone such as the humerus (arm bone) shown in Figure 5-1. A typical long bone consists of the following parts:

1. **Diaphysis** (dī-AF-i-sis). The shaft or long, main portion of the bone.
2. **Epiphyses** (ē-PIF-i-sēz). The extremities or ends of the bone.
3. **Metaphysis** (me-TAF-i-sis). The region in a mature bone where the diaphysis joins the epiphysis. In a growing bone, it is the region (epiphyseal plate) where calcified cartilage is reinforced and then replaced by bone (described later in the chapter).
4. **Articular cartilage.** A thin layer of hyaline cartilage covering the epiphysis where the bone forms a joint with another bone.
5. **Periosteum** (per'-ē-OS-tē-um). A dense, white, fibrous covering around the remaining surface of the bone. The periosteum (*peri* = around; *osteo* = bone) consists of two layers. The outer **fibrous layer** is composed of connective tissue containing blood vessels, lymphatic vessels, and nerves that pass into the bone. The inner **osteogenic** (os'-tē-ō-JEN-ik) **layer** contains elastic fibers, blood vessels, and **osteoblasts**—cells responsible for forming new bone during growth and repair. The word *blast* means a germ, or bud. It denotes an immature cell or tissue that later develops into a specialized form. The

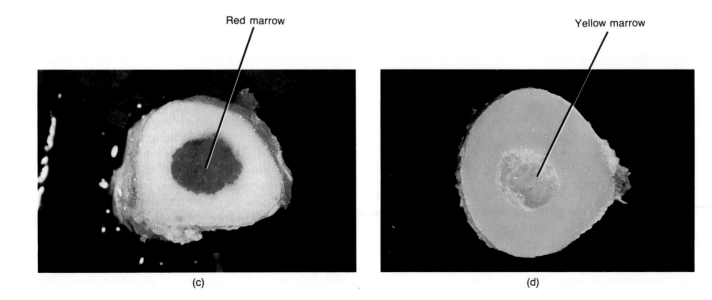

Proximal epiphysis

Articular cartilage

Spongy bone (contains red marrow)

Endosteum

Compact bone

Periosteum

Medullary cavity (contains yellow marrow)

Blood vessel

Nutrient foramen

Diaphysis

Distal epiphysis

(a)

Central (Haversian) canals

Compact bone

Spongy bone trabeculae

Perforating (Volkmann's) canals

(b)

FIGURE 5-1 Osseous tissue. (a) Macroscopic appearance of a long bone that has been partially sectioned. (b) Histological structure of bone. (c) Appearance of red marrow in the medullary cavity of a child's long bone. (d) Appearance of yellow marrow in the medullary cavity of an adult's long bone. (Courtesy of C. Yokochi and J. W. Rohen, *Photographic Anatomy of the Human Body,* 2nd ed., 1979, IGAKU-SHOIN, Ltd., Tokyo, New York.)

Red marrow

Yellow marrow

(c)

(d)

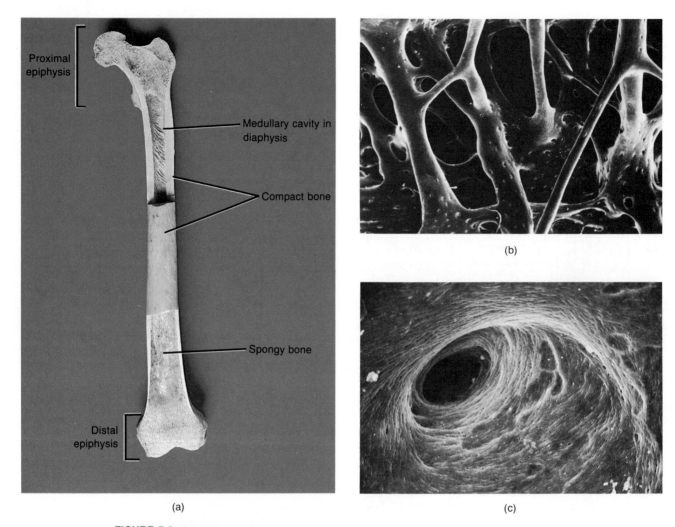

Proximal
epiphysis

Medullary cavity in
diaphysis

Compact bone

Spongy bone

Distal
epiphysis

(a)

(b)

(c)

FIGURE 5-2 Spongy and compact bone. (a) Photograph of a section through the femur to illustrate the positional and structural differences between spongy and compact bone. (Courtesy of Lester V. Bergman & Associates.) (b) Scanning electron micrograph of spongy bone trabeculae at a magnification of 25×. (c) Scanning electron micrograph of a view inside a central (Haversian) canal of compact bone at a magnification of 250×. (Electron micrographs courtesy of Fisher Scientific Company and S.T.E.M. Laboratories, Inc., Copyright 1975.)

periosteum is essential for bone growth, repair, and nutrition. It also serves as a point of attachment for ligaments and tendons.

6. Medullary (MED-yoo-lar'-ē), or **marrow, cavity.** The space within the diaphysis that contains the fatty **yellow marrow** in adults. Yellow marrow consists primarily of fat cells and a few scattered blood cells.

7. Endosteum (en-DOS-tē-um). A layer of osteoblasts that lines the medullary cavity and contains scattered osteoclasts (cells that may assume a role in the removal of bone).

Bone is not a completely solid, homogeneous substance. In fact, all bone has some spaces between its hard components. The spaces provide channels for blood ves-

sels that supply bone cells with nutrients. The spaces also make bones lighter. Depending on the size and distribution of the spaces, the regions of a bone may be categorized as spongy or compact (Figure 5-2).

Spongy, or **cancellous,** bone tissue contains many large spaces filled with red marrow. It makes up most of the bone tissue of short, flat, and irregularly shaped bones and most of the epiphyses of long bones. Spongy bone tissue also provides a storage area for marrow. **Compact,** or **dense,** bone tissue, by contrast, contains few spaces. It is deposited in a layer over the spongy bone tissue. The layer of compact bone is thicker in the diaphysis than the epiphyses. Compact bone tissue provides protection and support and helps the long bones resist the stress of weight placed on them.

COMPACT BONE

We can compare the differences between spongy and compact bone tissue by looking at the highly magnified, transverse sections in Figure 5-3. One main difference is that adult compact bone has a concentric-ring structure, whereas spongy bone does not. Blood vessels and nerves from the periosteum penetrate the compact bone through **perforating (Volkmann's) canals.** The blood vessels of these canals connect with blood vessels and nerves of the medullary cavity and those of the **central (Haversian) canals.** The central canals run longitudinally through the bone. Around them are **concentric lamellae (la-MEL-ē)**— concentric rings of hard, calcified, intercellular substance. Between the lamellae are small spaces called **lacunae (la-KOO-nē),** where osteocytes are found. **Osteocytes** are mature osteoblasts that have lost their ability to produce new bone tissue. Radiating in all directions from the lacunae are minute canals called **canaliculi (kan'-a-LIK-yoo-lī),** which contain slender processes of osteocytes. The canaliculi connect with other lacunae and, eventually, with the central canals. Thus, an intricate network is formed throughout the bone. This branching network of canaliculi provides numerous routes so that nutrients can reach the osteocytes and wastes can be removed. Each central canal, with its surrounding lamellae, lacunae, osteocytes, and canaliculi, is called an **osteon (Haversian system).** Osteons are characteristic of adult bone. The areas between osteons contain **interstitial lamellae.** These also possess lacunae with osteocytes and canaliculi, but their lamellae are usually not connected to the osteons. Interstitial lamellae are fragments of older osteons that have been partially destroyed during bone remodeling.

SPONGY BONE

In contrast to compact bone, spongy bone does not contain true osteons. It consists of an irregular latticework of thin plates of bone called **trabeculae (tra-BEK-yoo-lē).** The spaces between the trabeculae of some bones are filled with red marrow. The cells of red marrow are responsible for producing blood cells. Within the trabeculae lie the small spaces called lacunae, which contain the osteocytes. Blood vessels from the periosteum penetrate through to the spongy bone, and osteocytes in the trabeculae are nourished directly from the blood circulating through the marrow cavities.

Most people think of all bone as a very hard, white material. Yet the bones of an infant are not hard at all, and a child's bones are generally more pliable than those of an adult. The final shape and hardness of adult bones require many years to develop and depend on a complex series of chemical changes. Let us now see how bones are formed and how they grow.

OSSIFICATION

The process by which bone forms in the body is called **ossification (osteogenesis).** The "skeleton" of a human embryo is composed of either fibrous membranes or hyaline cartilage. Both are shaped like bones and provide the medium for ossification. Ossification begins around the sixth or seventh week of embryonic life and continues throughout adulthood. Two kinds of bone formation occur. The first is called intramembranous (in'-tra-MEM-bra-nus) ossification. This term refers to the formation of bone directly on or within the fibrous membranes (*intra* = within; *membranous* = membrane). The second kind, endochondral (en'-dō-KON-dral) ossification, refers to the formation of bone in cartilage (*endo* = within; *chondro* = cartilage). These two kinds of ossification do *not* lead to differences in the structure of mature bones. They simply indicate different methods of bone formation. Both mechanisms involve the replacement of a preexisting connective tissue with bone.

The first stage in the development of bone is the migration of embryonic connective tissue cells (mesenchymal cells) into the area where bone formation is about to begin. These cells increase in number and size. In some skeletal structures, they become chondroblasts; in others, some cells become osteoblasts. The **chondroblasts** will be responsible for cartilage formation. The osteoblasts will form bone tissue by intramembranous or endochondral ossification.

INTRAMEMBRANOUS OSSIFICATION

Of the two types of bone formation, the simpler and more direct is **intramembranous ossification.** Most of the surface skull bones and the clavicles (collar bones) are formed in this way. The essentials of this process are as follows.

Osteoblasts formed from mesenchymal cells cluster in the fibrous membrane. The site of such a cluster is called a *center of ossification.* The osteoblasts then secrete intercellular substances. These substances are partly composed of collagenous fibers that form a framework, or matrix, in which calcium salts are quickly deposited. The deposition of calcium salts is called *calcification.* When a cluster of osteoblasts is completely surrounded by the calcified matrix, it is called a *trabecula (tra-BEK-yoo-la).* As trabeculae form in nearby ossification centers, they fuse into the open latticework characteristic of spongy bone. With the formation of successive layers of bone, some osteoblasts become entrapped in the minute spaces called lacunae. The entrapped osteoblasts lose their ability to form bone and are called osteocytes. The spaces between the trabeculae fill with red marrow. The original connective tissue that surrounds the growing mass of bone then becomes the periosteum. The ossified area has now become true spongy bone. Eventually, the surface layers of the

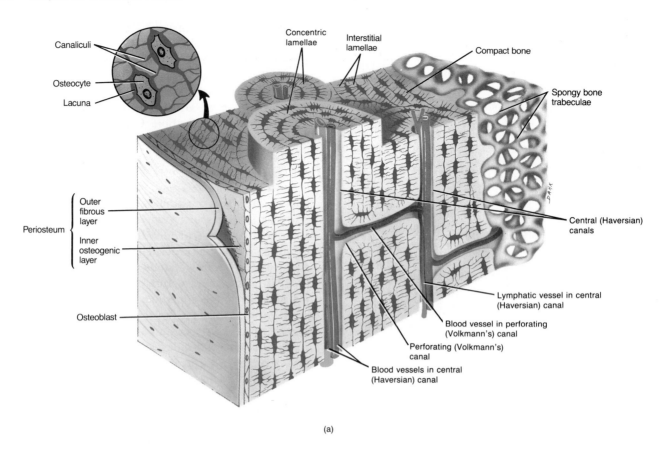

(a)

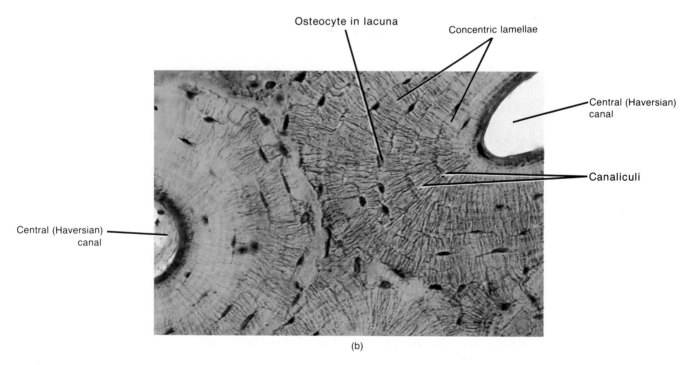

(b)

FIGURE 5-3 Histology of osseous tissue. (a) Enlarged aspect of several osteons (Haversian systems) in compact bone. (b) Photomicrograph of portions of several osteons with interstitial lamellae between at a magnification of 150×. (Courtesy of Carolina Biological Supply Company.)

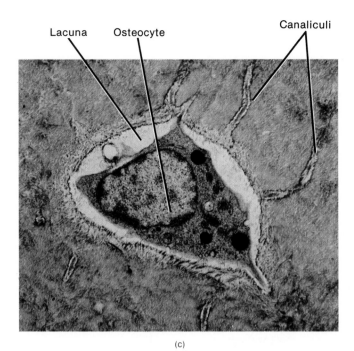

Lacuna   Osteocyte   Canaliculi

(c)

FIGURE 5-3 (*Continued*) Histology of osseous tissue. (c) Electron micrograph of an osteocyte at a magnification of 10,000×. (Courtesy of Biophoto Associates.)

spongy bone will be reconstructed into compact bone. Much of this newly formed bone will be destroyed and reformed so the bone may reach its final adult size and shape.

ENDOCHONDRAL OSSIFICATION

The replacement of cartilage by bone is called **endochondral (intracartilaginous) ossification.** Most bones of the body, including parts of the skull, are formed in this way. Since this type of ossification is best observed in a long bone, we will investigate the tibia, or shin bone (Figure 5-4).

Early in embryonic life, a cartilage model, or template, of the future bone is laid down. This model is covered by a membrane called the *perichondrium* (per-i-KON-dre-um). Midway along the shaft of this model, a blood vessel penetrates the perichondrium, stimulating the cells in the internal layer of the perichondrium to enlarge and become osteoblasts. The cells begin to form a collar of compact bone around the middle of the diaphysis of the cartilage model. Once the perichondrium starts to form bone, it is then called the *periosteum.* Simultaneously with the appearance of the bone collar and the penetration of blood vessels, changes occur in the cartilage in the center of the diaphysis. In this area, called the **primary ossification center,** cartilage cells hypertrophy (increase in size)—probably because they accumulate glycogen for energy and produce enzymes to catalyze future chemical reactions. When the hypertrophied cells burst, there is a change in extracellular pH to a more alkaline pH, caus-

ing the intercellular substance to become *calcified,* that is, minerals are deposited within it. Once the cartilage becomes calcified, nutritive materials required by the cartilage cells can no longer diffuse through the intercellular substance and this may cause the cartilage cells to die. Then the intercellular substance begins to degenerate, leaving large cavities in the cartilage model. The blood vessels grow along the spaces where cartilage cells were previously located and enlarge the cavities further. Gradually, these spaces in the middle of the shaft join with each other, and the marrow cavity is formed.

As these developmental changes are occurring, the osteoblasts of the periosteum deposit successive layers of bone on the outer surface so that the collar thickens, becoming thickest in the diaphysis. The cartilage model continues to grow at its ends, steadily increasing in length. Eventually, blood vessels enter the epiphyses and **secondary ossification centers** appear in the epiphyses and also lay down spongy bone. In the tibia, one secondary ossificatin center develops in the proximal epiphysis soon after birth. The other center develops in the distal epiphysis during the child's second year.

After the two secondary ossification centers have formed, bone tissue completely replaces cartilage, except in two regions. Cartilage continues to cover the articular surfaces of the epiphyses, where it is called **articular cartilage.** It also remains as a plate between the epiphysis and diaphysis, where it is called the **epiphyseal plate** (metaphysis of a growing bone).

The epiphyseal (ep'-i-FIZ-e-al) plate consists of four zones (Figure 5-5). The *zone of reserve cartilage* is adjacent to the epiphysis of the bone. It consists of small chondrocytes that are scattered irregularly throughout the intercellular matrix. The cells of this zone do not function in bone growth. The zone of reserve cartilage functions in anchoring the epiphyseal plate to the bone of the epiphysis. Its blood vessels also provide nutrients for the other zones of the epiphyseal plate.

The second zone, the *zone of proliferating cartilage,* consists of slightly larger chondrocytes that are arranged like stacks of coins. The function of this zone is to make new chondrocytes by cell division to replace those that die at the diaphyseal surface of the epiphyseal plate.

The third zone, the *zone of hypertrophic (hi'-per-TROF-ik) cartilage,* consists of even larger chondrocytes that are also arranged in columns. The cells are in various stages of maturation, with the more mature cells closer to the diaphysis. Lengthwise expansion of the epiphyseal plate is the result of cellular proliferation of the zone of proliferating cartilage and maturation of the cells in the zone of hypertrophic cartilage. Near the diaphyseal end of the bone, the intercellular matrix of the cells of the zone of hypertrophic cartilage becomes calcified and dies.

The fourth zone, the *zone of calcified matrix,* is only a few cells thick and consists mostly of dead cells because

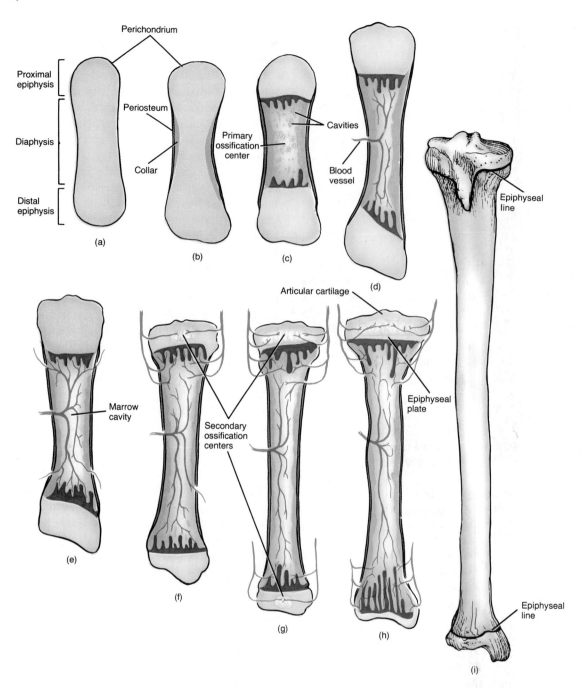

FIGURE 5-4 Endochondral ossification of the tibia. (a) Cartilage model. (b) Collar formation. (c) Development of primary ossification center. (d) Entrance of blood vessels. (e) Marrow-cavity formation. (f) Thickening and lengthening of collar. (g) Formation of secondary ossification centers. (h) Remains of cartilage as the articular cartilage and epiphyseal plate. (i) Formation of the epiphyseal lines.

the intercellular matrix around them has calcified. The calcified matrix is taken up by osteoclasts, and the area is invaded by osteoblasts and capillaries from the bone in the diaphysis. These cells lay down bone on the calcified cartilage that persists. As a result, the diaphyseal border of the epiphyseal plate is firmly cemented to the bone of the diaphysis.

The region between the diaphysis and epiphysis of a bone where the calcified matrix is replaced by bone is

called the *metaphysis* (me-TAF-i-sis). The activity of the epiphyseal plate is the only mechanism by which the diaphysis can increase in length. Unlike cartilage, which can grow by both interstitial and appositional growth, bone can grow only by appositional growth.

The epiphyseal plate allows the diaphysis of the bone to increase in length until early adulthood. As the child grows, cartilage cells are produced by mitosis on the epiphyseal side of the plate. Cartilage cells are then de-

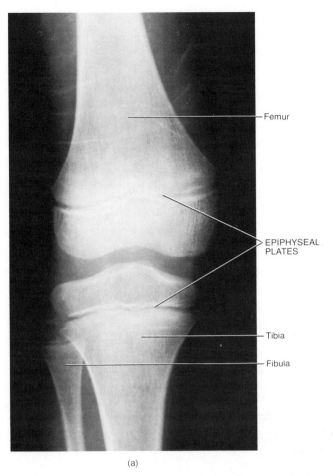

— Femur

— EPIPHYSEAL PLATES

— Tibia

— Fibula

(a)

FIGURE 5-5 Epiphyseal plate. (a) Anteroposterior projection showing epiphyseal plates in the femur and tibia of a 10-year-old child. (Courtesy of Arthur Provost, R.T.) (b) Photomicrograph of epiphysis of a long bone showing the epiphyseal plate at a magnification of 160×. (© 1983 by Michael H. Ross. Used by permission.)

stroyed, and the cartilage is replaced by bone on the diaphyseal side of the plate. In this way, the thickness of the epiphyseal plate remains fairly constant, but the bone on the diaphyseal side increases in length. Growth in diameter occurs along with growth in length. In this process, the bone lining the marrow cavity in destroyed so that the cavity increases in diameter. At the same time, osteoblasts from the periosteum add new osseous tissue around the outer surface of the bone. Initially, diaphyseal and epiphyseal ossification produce only spongy bone. Later, by reconstruction, the outer region of spongy bone is reorganized into compact bone.

CLINICAL APPLICATION

The epiphyseal cartilage cells stop dividing and the cartilage is replaced by bone at about age 18 in females and age 20 in males. The newly formed bony structure is called the **epiphyseal line,** a remnant of the once active epiphyseal plate. With the appearance of the epiphyseal line, appositional bone growth stops. The clavicle is the last bone to stop growing. Ossification of all bones is usually completed by age 25.

BONE REPLACEMENT

Bones undergoing either intramembranous or endochondral ossification are continually remodeled from the time that initial calcification occurs until the final structure appears. **Remodeling** is the replacement of old bone tissue by new bone tissue. Compact bone is formed by the transformation of spongy bone. The diameter of a long bone

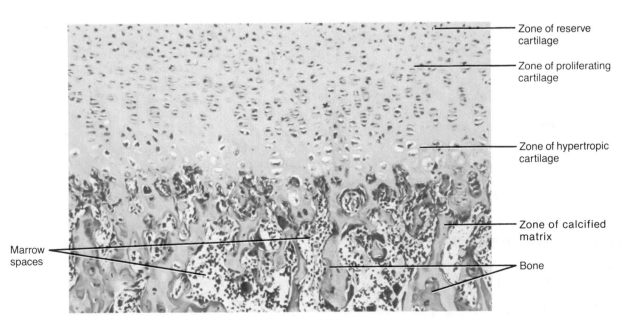

— Zone of reserve cartilage

— Zone of proliferating cartilage

— Zone of hypertropic cartilage

— Zone of calcified matrix

— Bone

Marrow spaces —

(b)

is increased by the destruction of the bone closest to the marrow cavity and the construction of new bone around the outside of the diaphysis. However, even after bones have reached their adult shapes and sizes, old bone is perpetually destroyed and new osseous tissue is formed in its place.

Bone shares with skin the feature of replacing itself throughout adult life. This remodeling takes place at different rates in various body regions. The distal portion of the femur (thighbone) is replaced about every 4 months. By contrast, bone in certain areas of the shaft will not be completely replaced during the individual's life. Remodeling allows worn or injured bone to be removed and replaced with new tissue. It also allows bone to serve as the body's storage area for calcium. Many other tissues in the body need calcium in order to perform their functions. For example, nerve cells need calcium for their activities and muscles needs calcium in order to contract. The nerve and muscle cells take their calcium from the blood. However, the blood itself needs calcium in order to clot. The blood continually trades off calcium with the bones, removing calcium when other tissues are not receiving enough of this element and resupplying the bones with dietary calcium when available to keep them from losing too much bone mass.

The cells believed to be responsible for the resorption (loss of a substance through a physiological or pathological process) of bone tissue are called **osteoclasts** (*clast* = break). In the healthy adult, a delicate homeostasis is maintained between the action of the osteoclasts in removing calcium and the action of the bone-making osteoblasts in depositing calcium. Should too much new tissue be formed, the bones become abnormally thick and heavy. If too much calcium is deposited in the bone, the surplus may form thick bumps, or spurs, on the bone that interfere with movement at joints. A loss of too much tissue or calcium weakens the bones and allows them to break easily or to become very flexible. As you will see later, a greatly accelerated remodeling process results in a condition called Paget's disease.

In the process of resorption, it is believed that osteoclasts send out projections that secrete proteolytic enzymes released from lysosomes and several acids (lactic and citric). The enzymes may function by digesting the collagen and other organic substances, while the acids may cause the bone salts (minerals) to dissolve in solution. It is also presumed that the osteoclastic projections may phagocytose whole fragments of collagen and bone salts.

Normal bone growth in the young and bone replacement in the adult depend on several factors. First, sufficient quantities of calcium and phosphorus, components of the primary salt that makes bone hard, must be included in the diet.

Second, the individual must obtain sufficient amounts of vitamins A, C, and D, substances that are responsible for the proper utilization of calcium and phosphorus by the body. Vitamin D, for example, is required for the absorption of calcium from the digestive tract into the blood and has important effects on bone deposition and resorption.

Third, the body must manufacture the proper amounts of the hormones responsible for bone tissue activity (Chapter 21). Growth hormone (GH), secreted by the pituitary gland, is responsible for the general growth of bones. Too much or too little of these hormones during childhood makes the adult abnormally tall or short. Other hormones specialize in regulating the osteoclasts. Calcitonin (CT), produced by the thyroid gland, inhibits osteoclastic activity, while parathormone (PTH), synthesized by the parathyroid glands, and 1,25-dihydroxyvitamin D (the physiologically active form of vitamin D) increase the number and activity of osteoclasts. And still others, especially the sex hormones, aid osteoblastic activity and thus promote the growth of new bone. The sex hormones act as a double-edged sword. They aid in the growth of new bone, but they also bring about the degeneration of all the cartilage cells in the epiphyseal plates. Because of the sex hormones, the typical adolescent experiences a spurt of growth during puberty, when sex hormone levels start to increase. The individual then quickly completes the growth process as the epiphyseal cartilage disappears. Premature puberty can actually prevent one from reaching an average adult height because of the simultaneous premature degeneration of the plates.

BLOOD AND NERVE SUPPLY

Bone is richly supplied with blood, and blood vessels are especially abundant in portions of bone containing red bone marrow. Blood vessels pass into bones from the periosteum. Although the blood supply to bones varies according to their shape, here we will consider the blood supply to a long bone only.

Because the artery to the diaphysis of a long bone is usually the largest, it is referred to as the **nutrient artery.** The artery first enters the bone during the early development and passes through an opening in the diaphysis, the *nutrient foramen* (see Figure 5-1a), which leads into a *nutrient canal*. On entering the medullary cavity, the nutrient artery divides into a proximal and a distal branch, each of which supplies most of the marrow, inner portion of compact bone of the diaphysis, and metaphysis. As the nutrient artery passes through compact bone on its way to the medullary cavity, it sends branches into the central (Haversian) canals. Branches of the **articular arteries** enter many vascular foramina in the epiphysis and supply the marrow, metaphysis, and bony tissue of the epiphysis. **Periosteal arteries,** accompanied by nerves, enter the diaphysis at numerous points through perforating (Volkmann's) canals (see Figure 5-3a). These blood vessels run in the central canals and supply the outer part of the compact bone of the diaphysis. Veins accompany the several types of arteries; the principal veins leave the bone by numerous vascular foramina at the epiphyses of the bone.

Although the nerve supply to bone is not extensive, some nerves (vasomotor) accompany blood vessels and some (sensory) occur in the periosteum. The nerves of the periosteum are primarily concerned with pain, as might be associated with a fracture or tumor.

AGING AND THE SKELETAL SYSTEM

There are two principal effects of aging on the skeletal system. The first effect is the loss of calcium from bones. This loss usually begins after age 40 in females and continues thereafter until as much as 30 percent of the calcium in bones is lost by age 70. In males, calcium loss typically does not begin until after age 60. The loss of calcium from bones is one of the factors related to a condition called osteoporosis, which will be described shortly.

The second principal effect of aging on the skeletal system is a decrease in the rate of protein formation that results in a decreased ability to produce the organic portion of bone matrix. As a consequence, bone matrix accumulates a lesser proportion of organic matrix and a greater proportion of inorganic matrix. In some elderly individuals, this process can cause their bones to become quite brittle and more susceptible to fracture.

DEVELOPMENTAL ANATOMY OF THE SKELETAL SYSTEM

Bone forms about the sixth or seventh week by either of two processes, **intramembranous ossification** or **endochondrial ossification.** Both processes begin when **mesenchymal (mesodermal) cells** migrate into the area where bone formation will occur. In some skeletal structures, the mesenchymal cells develop into **chondroblasts** that form *cartilage.* In other skeletal structures, the mesenchymal calls develop into **osteoblasts** that form *bone tissue* by intramembranous or endochondral ossification. (The details of ossification have already been discussed earlier in the chapter).

Discussion of the development of the skeletal system provides us with an excellent opportunity to trace the development of the extremities. The *extremities* make their appearance about the fifth week as small elevations at the sides of the trunk called **limb buds** (Figure 5-6). They consist of masses of general **mesoderm** surrounded by **ectoderm.** At this point, a mesenchymal skeleton exists in the limbs; some of the masses of mesoderm surrounding the developing bones will develop into the skeletal muscles of the extremities. By the sixth week, the limb buds develop a constriction around the middle portion. The con-

striction produces distal segments of the upper buds called **hand plates** and distal segments of the lower buds called **foot plates.** These plates represent the beginnings of the *hands* and *feet,* respectively. At this stage of limb development, a cartilaginous skeleton is present. By the seventh week, the *arm, forearm,* and *hand* are evident in the upper limb bud, and the *thigh, leg,* and *foot* appear in the lower limb bud. Endochondral ossification has begun. By the eighth week, the upper limb bud is appropriately called the *upper extremity* as the *shoulder, elbow,* and *wrist* areas become apparent, and the lower limb bud is referred to as the *lower extremity* with the appearance of the *knee* and *ankle* areas.

The **notochord** is a flexible rod of tissue that lies in a position where the future vertebral column will develop (see Figure 9-9a). As the vertebrae develop, the notochord becomes surrounded by the developing vertebral bodies and the notochord eventually disappears, except for remnants that persist as the *nucleus pulposus* of the intervertebral discs.

APPLICATIONS TO HEALTH

Many bone disorders result from deficiencies in vitamins or minerals or from too much or too little of the hormones that regulate bone homeostasis. Infection and tumors are also responsible for certain bone disorders.

VITAMIN DEFICIENCIES

Rickets

A deficiency of vitamin D in children results in a condition called **rickets.** It is characterized by an inability of the body to transport calcium and phosphorus from the digestive tract into the blood for utilization by bones. As a result, epiphyseal cartilage cells cease to degenerate and new cartilage continues to be produced. Epiphyseal cartilage thus becomes wider than normal. At the same time, the soft matrix laid down by the osteoblasts in the diaphysis fails to calcify. As a result, the bones stay soft. When the child walks, the weight of the body causes the bones in the legs to bow. Malformations of the head, chest, and pelvis also occur.

The cure and prevention of rickets consist of adding generous amounts of calcium, phosphorus, and vitamin D to the diet. Exposing the skin to the ultraviolet rays of sunlight also aids the body in manufacturing additional vitamin D.

Osteomalacia

A deficiency of vitamin D in the adult causes the bones to give up excessive amounts of calcium and phosphorus. This loss, called *demineralization,* is especially heavy in

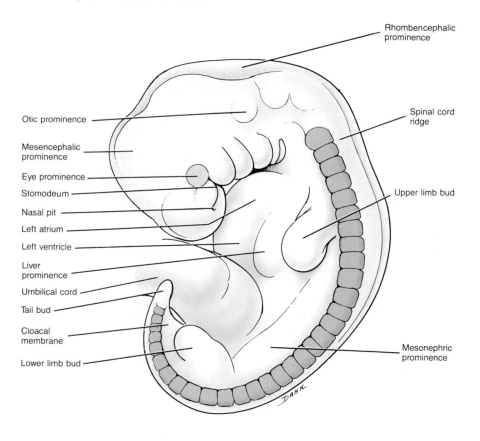

Rhombencephalic
prominence

Otic prominence

Mesencephalic
prominence

Eye prominence

Stomodeum

Nasal pit

Left atrium

Left ventricle

Liver
prominence

Umbilical cord

Tail bud

Cloacal
membrane

Lower limb bud

Spinal cord
ridge

Upper limb bud

Mesonephric
prominence

(a)

Metencephalic area

Myelencephalic area

Mesencephalic
area

Optic cup

Lens
vesicle

Nasal pit

Umbilical
cord

Tail bud

Foot plate

Stomodeum

Heart
prominence

Liver
prominence

Hand plate

(b)

FIGURE 5-6 External features of a developing embryo at various stages of development.
(a) Fifth week. (b) Sixth week. (c) Seventh week. (d) Eighth week.

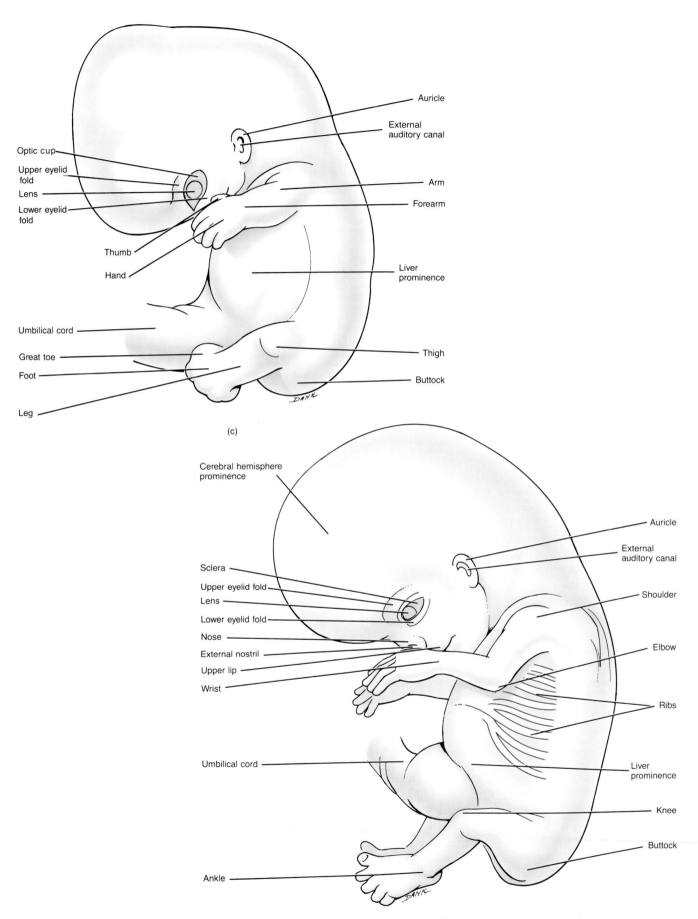

Optic cup

Upper eyelid fold

Lens

Lower eyelid fold

Thumb

Hand

Umbilical cord

Great toe

Foot

Leg

Auricle

External auditory canal

Arm

Forearm

Liver prominence

Thigh

Buttock

(c)

Cerebral hemisphere prominence

Sclera

Upper eyelid fold

Lens

Lower eyelid fold

Nose

External nostril

Upper lip

Wrist

Umbilical cord

Ankle

Auricle

External auditory canal

Shoulder

Elbow

Ribs

Liver prominence

Knee

Buttock

(d)

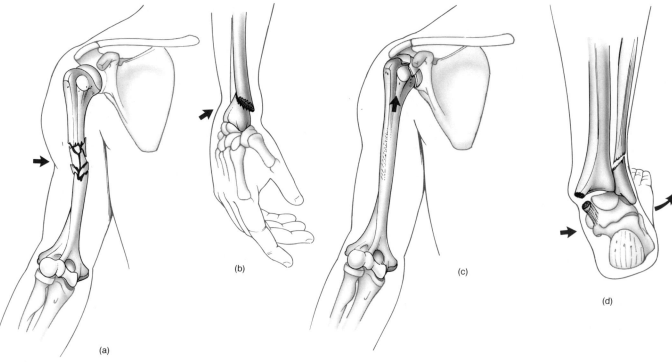

FIGURE 5-7 Types of fractures. (a) Comminuted. (b) Colles'. (c) Impacted. (d) Pott's. (e) Greenstick.

the bones of the pelvis, legs, and spine. Demineralization caused by vitamin D deficiency is called **osteomalacia** (os'-tē-ō-ma-LA-shē-a; *malacia* = softness). After the bones demineralize, the weight of the body produces a bowing of the leg bones, shortening of the backbone, and flattening of the pelvic bones. Osteomalacia mainly affects women who live on poor cereal diets devoid of milk, are seldom exposed to the sun, and have repeated pregnancies that deplete the body of calcium. The condition responds to the same treatment as rickets; if the disease is severe enough and threatens life, large doses of vitamin D are given.

Osteomalacia may also result from failure to absorb fat (steatorrhea) because vitamin D is soluble in fats and calcium combines with fats. As a result, vitamin D and calcium remain with the unabsorbed fat and are lost in the feces.

OSTEOPOROSIS

Osteoporosis (os'-tē-ō-pō-RŌ-sis) is an age-related disorder characterized by decreased bone mass and increased susceptibility to fractures. It primarily affects the middle-aged and elderly people—white women more than men and whites more than blacks. Between puberty and the middle years, the sex hormones maintain osseous tissue by stimulating the osteoblasts to form new bone. Women produce smaller amounts of sex hormones after menopause, and both men and women produce smaller amounts

during old age. As a result, the osteoblasts become less active and there is a decrease in bone mass. Osteoporosis can also occur during nursing and pregnancy and in individuals exposed to prolonged treatment with cortisone. Osteoporosis is responsible for shrinkage of the backbone and height loss, hunched backs, hip fractures, and considerable pain. Osteoporosis affects the entire skeletal system, especially the spine, legs, and feet.

Among the factors implicated in bone loss are a decrease in levels of estrogens, calcium deficiency and malabsorption, vitamin D deficiency, loss of muscle mass, and inactivity. Estrogen replacement therapy (ERT), calcium supplements, and exercise are prescribed to treat osteoporosis.

PAGET'S DISEASE

Paget's disease is characterized by a greatly accelerated remodeling process in which osteoclastic resorption is massive and new bone formation by osteoblasts is extensive. As a result, there is an irregular thickening and softening of the bones and greatly increased vascularity. It rarely occurs in individuals under 50. The cause, or **etiology** (ē'-tē-OL-ō-jē), of the disease is unknown.

OSTEOMYELITIS

The term **osteomyelitis** (os'-tē-ō-mī-i-LĪ-tis) includes all infectious diseases of bone. These diseases may be localized or widespread and may also involve the periosteum, marrow, and cartilage. Various microorganisms may give rise to bone infection, but the most frequent are bacteria known as *Staphylococcus aureus,* commonly called "staph." These bacteria may reach the bone by various means: the bloodstream, an injury such as a fracture, or an infection such as a sinus infection or a tooth abscess. The infection may destroy extensive areas of bone, spead to nearby joints, and, in rare cases, lead to death by producing abscesses. Antibiotics have been effective in treating the disease and in preventing it from spreading through extensive areas of bone.

FRACTURES

In simplest terms, a **fracture** is any break in a bone. Usually, the fracture is restored to normal position by manipulation without surgery. This procedure of setting a fracture is called *closed reduction.* In other cases, the fracture must be exposed by surgery before the break is rejoined. This procedure is known as *open reduction.*

Types

Although fractures may be classified in several different ways, the following scheme is useful (Figure 5-7):

1. Partial (incomplete). A fracture in which the break across the bone is incomplete.

2. Complete. A fracture in which the break across the bone is complete, so that the bone is broken into two pieces.

3. Closed, or **simple.** A fracture in which the bone does not break through the skin.

4. Open, or **compound.** A fracture in which the broken ends of the bone protrude through the skin.

5. Comminuted (KOM-i-nyoo'-ted). A fracture in which the bone is splintered at the site of impact and smaller fragments of bone are found between the two main fragments.

6. Greenstick. A partial fracture in which one side of the bone is broken and the other side bends; occurs only in children.

7. Spiral. A fracture in which the bone is usually twisted apart.

8. Transverse. A fracture at right angles to the long axis of the bone.

9. Impacted. A fracture in which one fragment is firmly driven into the other.

10. Pott's. A fracture of the distal end of the fibula, with serious injury of the distal tibial articulation.

11. Colles' (KOL-ēz). A fracture of the distal end of the radius in which the distal fragment is displaced posteriorly.

12. Displaced. A fracture in which the anatomical alignment of the bone fragments is not preserved.

13. Nondisplaced. A fracture in which the anatomical alignment of the bone fragments is preserved.

14. Stress. A partial fracture resulting from inability to withstand repeated stress due to a change in training, harder surfaces, longer distances, and greater speed. About 25 percent of all stress fractures involve the fibula, typically the distal third.

Fracture Repair

Unlike the skin, which may repair itself within days, or muscle, which may mend in weeks, a bone sometimes requires months to heal. A fractured femur, for example, may take six months to heal. Sufficient calcium to strengthen and harden new bone is deposited only gradually. Bone cells also grow and reproduce slowly. Moreover, the blood supply to bone is decreased, which helps to explain the difficulty in the healing of an infected bone.

The following steps occur in the repair of a fracture:

1. As a result of the fracture, blood vessels crossing the fracture line are broken. These vessels are found in the periosteum, osteons (Haversian systems), and marrow cavity. As blood pours from the torn ends of the vessels, it coagulates and forms a clot in and about the site of the fracture. This clot, called a **fracture hematoma** (hē-ma-TŌ-ma), usually occurs 6 to 8 hours after the injury. Since the circulation of blood ceases when the fracture hematoma forms, bone cells and periosteal cells at the fracture line die. The hematoma serves as a focus for subsequent cellular invasion.

2. A growth of new bone tissue—a **callus**—develops in and around the fractured area. It forms a bridge between separated areas of bone. The part of the callus that forms from the osteogenic cells of the torn periosteum and develops around the outside of the fracture is called an *external callus.* The part of the callus that forms from the osteogenic cells of the endosteum and develops between the two ends of bone fragments and between the two marrow cavities is called the *internal callus.*

Approximately 48 hours after a fracture occurs, the cells that ultimately repair the fracture become actively mitotic. These cells come from the osteogenic layer of the periosteum, the endosteum of the marrow cavity, and the bone marrow. As a result of their accelerated mitotic activity, the cells of the three regions grow toward the fracture. During the first week following the fracture, the cells of the endosteum and bone marrow form new trabeculae in the marrow cavity near the line of fracture. This is the internal callus. During the next few days, osteogenic cells of the periosteum form a collar around each bone fragment. The collar, or external callus, is replaced by trabeculae. The trabeculae of the calli are joined to living and dead portions of the original bone fragments.

3. The final phase of fracture repair is the **remodeling** of the calli. Dead portions of the original fragments are gradually resorbed. Compact bone replaces spongy bone around the periphery of the fracture. In some cases, the healing is so complete that the fracture line is undetectable, even by x ray. However, a thickened area on the surface of the bone usually remains as evidence of the fracture site.

CLINICAL APPLICATION

In the past when a fracture failed to unite, the patient could either wait, hoping that nature and time would solve the problem, or choose surgery. Today there is another alternative. The new procedure, called **pulsating electromagnetic fields (PEMFs)**, involves electrotherapy to stimulate bone repair.

Treatment with PEMFs was first used clincally in 1974. Essentially, the fracture is exposed to a weak electrical current generated from coils that are fastened around the cast. Osteoblasts adjacent to the fracture site become more active metabolically in response to the stimulation of the alteration of electrical charges on the surfaces of their membranes. The increased activity apparently causes an acceleration of calcification, vascularization, and endochondral ossification, resulting in acceleration of fracture repair. It was noted earlier that parathyroid hormone (PTH) increases osteoclastic activity and thus stimulates bone destruction. A recent hypothesis suggests that electricity also heals fractures by keeping PTH from acting on osteoclasts, thus increasing bone formation and repair.

Although the original application of PEMFs was to treat improperly healing fractures, research is now underway to determine if electrical stimulation can be used to regenerate limbs, stop the growth of tumor cells in humans, reduce postoperative swelling, and improve the healing of tendons and ligaments.

KEY MEDICAL TERMS ASSOCIATED WITH OSSEOUS TISSUE

Achondroplasia (*a* = without; *chondro* = cartilage; *plasia* = growth)   Imperfect ossification within cartilage of long bones during fetal life; also called *fetal rickets.*

Craniotomy (*cranium* = skull; *tome* = a cutting)   Any surgery that requires cutting through the bones surrounding the brain.

Necrosis (*necros* = death; *osis* = condition)   Death of tissues or organs; in the case of bone, necrosis results from deprivation of blood supply; could result from fracture, extensive removal of periosteum in surgery, exposure to radioactive substances, and other causes.

Osteitis (*osteo* = bone)   Inflammation or infection of bone.

Osteoarthritis (*arthro* = joint)   A degenerative condition of bone and also the joint.

Osteoblastoma (*oma* = tumor)   A benign tumor of the osteoblasts.

Osteochondroma (*chondro* = cartilage)   A benign tumor of the bone and cartilage.

Osteoma   A benign bone tumor.

Osteosarcoma (*sarcoma* = connective tissue tumor)   A malignant tumor composed of osseous tissue.

Pott's disease   Inflammation of the backbone, caused by the microorganism that produces tuberculosis.

STUDY OUTLINE

Functions (p. 100)
1. The skeletal system consists of all bones attached at joints and cartilage between joints.
2. The functions of the skeletal system include support, protection, leverage, mineral storage, and blood-cell production.

Histology (p. 100)
1. Osseous tissue consists of widely separated cells surrounded by large amounts of intercellular substance. The intercellular substance contains collagenous fibers and abundant hydroxyapatites (mineral salts).
2. Parts of a typical long bone are the diaphysis (shaft), epiphyses (ends), metaphysis, articular cartilage, periosteum, medullary (or marrow) cavity, and endosteum.

3. Compact (dense) bone consists of osteons (Haversian systems) with little space between them. Compact bone lies over spongy bone and composes most of the bone tissue of the diaphysis. Functionally, compact bone protects, supports, and resists stress.
4. Spongy (cancellous) bone consists of trabeculae surrounding many red-marrow-filled spaces. It forms most of the structure of short, flat, and irregular bones and the epiphyses of long bones. Functionally, spongy bone stores marrow and provides some support.

Ossification (p. 103)
1. Bone forms by a process called ossification, or osteogenesis, which begins when mesenchymal cells become transformed into osteoblasts.

2. The process begins during the sixth or seventh week of embryonic life and continues throughout adulthood. The two types of ossification, intramembranous and endochondral, involve the replacement of a preexisting connective tissue with bone.

3. Intramembranous ossification occurs within fibrous membranes of the embryo and the adult.

4. Endochondral ossification occurs within a cartilage model. The primary ossification center of a long bone is in the diaphysis. Cartilage degenerates, leaving cavities that merge to form the marrow cavity. Osteoblasts lay down bone. Next, ossification occurs in the epiphyses, where bone replaces cartilage, except for the epiphyseal plate.

5. The anatomical zones of the epiphyseal plate are the zones of reserve cartilage, proliferating cartilage, hypertrophic cartilage, and calcified matrix.

6. Because of the activity of the epiphyseal plate, the diaphysis of a bone increases in length by appositional growth.

7. In both types of ossification, spongy bone is laid down first. Compact bone is later reconstructed from spongy bone.

Bone Replacement (p. 107)

1. The growth and development of bone depends on a balance between bone formation and resorption.

2. Old bone is constantly destroyed by osteoclasts, while new bone is constructed by osteoblasts. This process is called remodeling.

3. Normal growth depends on calcium, phosphorus, and vitamins A, C, and D and is controlled by hormones that are responsible for bone mineralization and resorption.

Blood and Nerve Supply (p. 108)

1. Long bones are supplied by nutrient, articular, and periosteal arteries; veins accompany the arteries.

2. The nerve supply to bones consists of vasomotor and sensory nerves.

Aging and the Skeletal System (p. 109)

1. The principal effect of aging is a loss of calcium from bones, which may result in osteoporosis.

2. Another principal effect is a decreased production of organic matrix, which makes bones more susceptible to fracture.

Developmental Anatomy of the Skeletal System (p. 109)

1. Bone forms from mesoderm by intramembranous or endochondral ossification.

2. Extremities develop from limb buds, which consist of mesoderm and ectoderm.

Applications to Health (p. 109)

1. Rickets is a vitamin D deficiency in children in which the body does not absorb calcium and phosphorus. The bones soften and bend under the body's weight.

2. Osteomalacia is a vitamin D deficiency in adults caused by demineralization.

3. Osteoporosis is a decrease in the amount and strength of bone tissue due to decreases in hormone output.

4. Paget's disease is the irregular thickening and softening of bones, related to a greatly accelerated remodeling process.

5. Osteomyelitis is a term for the infectious diseases of bones, marrow, and periosteum. It is frequently caused by "staph" bacteria.

6. A fracture is any break in a bone.

7. The types of fractures include: partial, complete, simple, compound, comminuted, greenstick, spiral, transverse, impacted, Pott's, Colles', displaced, nondisplaced, and stress.

8. Fracture repair consists of forming a fracture hematoma, forming a callus, and remodeling.

9. Treatment by pulsating electromagnetic fields (PEMFs) has provided dramatic results in healing fractures that would otherwise not have mended properly. Its application for limb regeneration and stopping the growth of tumor cells is being investigated.

REVIEW QUESTIONS

1. Define the skeletal system. What are its five principal functions?

2. Why is osseous tissue considered a connective tissue? What is its composition?

3. Diagram the parts of a long bone, and list the functions of each part.

4. Distinguish between spongy and compact bone in terms of microscopic appearance, location, and function.

5. Diagram the microscopic appearance of compact bone, and indicate the functions of the various components.

6. Distinguish between the two principal kinds of ossification.

7. Describe the histology of the various zones of the epiphyseal plate. How does the plate grow? What is the significance of the epiphyseal line?

8. How does balance between osteoblast activity and osteoclast activity control the replacement of bone?

9. Describe the blood and nerve supply of a long bone.

10. Explain the effects of aging on the skeletal system.

11. Describe the development of the skeletal system.

12. Distinguish between rickets and osteomalacia. What do the two diseases have in common?

13. What are the principal symptoms of osteoporosis, Paget's disease, and osteomyelitis? What is the etiology of each?

14. What is a fracture? Distinguish several principal kinds. Outline the three basic steps involved in fracture repair.

15. Refer to the glossary of key medical terms associated with the skeletal system. Be sure that you can define each term.

The Skeletal System:
The Axial Skeleton

Student Objectives

Define the four principal types of bones in the skeleton.

Describe the various markings on the surfaces of bones.

List the components of the axial and appendicular skeletons.

Identify the bones of the skull and the major markings associated with each.

Identify the sutures and fontanels of the skull.

Identify the paranasal sinuses of the skull.

Identify the principal foramina of the skull.

Identify the bones of the vertebral column and their principal markings.

List the defining characteristics and curves of each region of the vertebral column.

Identify the bones of the thorax and their principal markings.

Contrast herniated (slipped) disc, curvatures, spina bifida, and fractures of the vertebral column as disorders associated with the skeletal system.

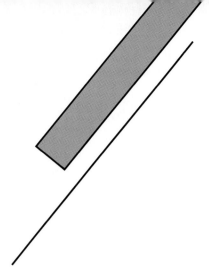

The skeletal system forms the framework of the body. For this reason, a familiarity with the names, shapes, and positions of individual bones will help you understand some of the other organ systems. For example, movements such as throwing a ball, typing, and walking require the coordinated use of bones and muscles. To understand how muscles produce different movements, you need to learn the parts of the bones to which the muscles attach. The respiratory system is also highly dependent on bone structure. The bones in the nasal cavity form a series of passageways that help clean, moisten, and warm inhaled air. Furthermore, the bones of the thorax are specially shaped and positioned so the chest can expand during inhalation. Many bones also serve as landmarks to students of anatomy as well as to surgeons. Blood vessels and nerves often run parallel to bones. These structures can be located more easily if the bone is identified first.

We will study bones by examining the various regions of the body. For instance, we will look at the skull first and see how the bones of the skull relate to each other. Then we will move on to the chest. This regional approach will allow you to see how all the many bones of the body relate to each other.

TYPES OF BONES

Almost all the bones of the body may be classified into four principal types: long, short, flat, and irregular. **Long bones** have greater length than width and consist of a diaphysis and two epiphyses. They are slightly curved for strength. A curved bone is structurally designed to absorb the stress of the body weight at several different points so the stress is evenly distributed. If such bones were straight, the weight of the body would be unevenly distributed and the bone would easily fracture. Examples of long bones include bones of the thighs, legs, toes, arms, forearms, and fingers. Figure 5-1a shows the parts of a long bone.

Short bones are somewhat cube-shaped and nearly equal in length and width. Their texture is spongy except at the surface, where there is a thin layer of compact bone. Examples of short bones are the wrist and ankle bones.

Flat bones are generally thin and composed of two more or less parallel plates of compact bone enclosing a layer of spongy bone. The term *diploe* is applied to the spongy bone of the cranial bones. Flat bones afford considerable protection and provide extensive areas for muscle attachment. Examples of flat bones include the cranial bones (which protect the brain), the sternum and ribs (which protect organs in the thorax), and the scapulas.

Irregular bones have complex shapes and cannot be grouped into any of the three categories just described. They also vary in the amount of spongy and compact bone present. Such bones include the vertebrae and certain facial bones.

There are two additional types of bones that are not included in this classification by shape. **Sutural** (SOO-chur-al), or **Wormian, bones** are small clusters of bones between the joints of certain cranial bones (see Figure 6-2e). Their number varies greatly from person to person. **Sesamoid bones** are small bones in tendons where considerable pressure develops, for instance, in the wrist. These bones, like the sutural bones, are also variable in number. Two sesamoid bones, the patellas, or kneecaps, are present in all individuals.

SURFACE MARKINGS

The surfaces of bones reveal various structural features adapted to specific functions. These features are called **surface markings.** Long bones that bear a great deal of weight have large, rounded ends that can form sturdy joints. Other bones have depressions that receive the rounded ends. Rough areas serve as points of attachment for muscles, tendons, and ligaments. Grooves in the surfaces of bones provide for the passage of blood vessels. Openings occur where blood vessels and nerves pass through the bone. Exhibit 6-1 describes the different markings and their functions.

EXHIBIT 6-1   BONE MARKINGS

MARKING	DESCRIPTION	EXAMPLE
DEPRESSIONS AND OPENINGS		
Fissure (FISH-ur)	A narrow, cleftlike opening between adjacent parts of bones through which blood vessels or nerves pass.	Superior orbital fissure of the sphenoid bone (Figure 6-2).
Foramen (fō-RĀ-men; *foramen* = hole)	An opening through which blood vessels, nerves, or ligaments pass.	Infraorbital foramen of the maxilla (Figure 6-2).
Meatus (mē-Ā-tus; *meatus* = canal)	A tubelike passageway running within a bone.	External auditory meatus of the temporal bone (Figure 6-2).
Paranasal sinus (*sin* = cavity)	An air-filled cavity within a bone connected to the nasal cavity.	Frontal sinus of the frontal bone (Figure 6-8).
Groove or sulcus (*sulcus* = ditchlike groove)	A furrow or depression that accommodates a soft structure such as a blood vessel, nerve, or tendon.	Intertubercular sulcus of the humerus (Figure 7-4).
Fossa (*fossa* = basinlike depression)	A depression in or on a bone.	Mandibular fossa of the temporal bone (Figure 6-4).
PROCESSES **Processes that form joints**	Any prominent projection.	Mastoid process of the temporal bone (Figure 6-2).
Condyle (KON-dīl; *condylus* = knucklelike process)	A large, rounded articular prominence.	Medial condyle of the femur (Figure 7-10).
Head	A rounded, articular projection supported on the constricted portion (neck) of a bone.	Head of the femur (Figure 7-10).
Facet	A smooth, flat surface.	Articular facet for the tubercle of rib on a vertebra (Figure 6-18).
Processes to which tendons, ligaments, and other connective tissues attach		
Tubercle (TOO-ber-kul; *tuber* = knob)	A small, rounded process.	Greater tubercle of the humerus (Figure 7-4).
Tuberosity	A large, rounded, usually roughened process.	Ischial tuberosity of the hipbone (Figure 7-8).
Trochanter (trō-KAN-ter)	A large, blunt projection found only on the femur.	Greater trochanter of the femur (Figure 7-10).
Crest	A prominent border or ridge on a bone.	Iliac crest of the hipbone (Figure 7-7).
Line	A ridge less prominent than a crest.	Linea aspera of the femur (Figure 7-10).
Spinous process, or spine	A sharp, slender process.	Spinous process of a vertebra (Figure 6-12).
Epicondyle (*epi* = above)	A prominence above a condyle.	Medial epicondyle of the femur (Figure 7-10).

DIVISIONS OF THE SKELETAL SYSTEM

The adult human skeleton usually consists of 206 bones grouped in two principal divisions: the **axial skeleton** and the **appendicular skeleton.** The longitudinal *axis,* or center, of the human body is a straight line that runs vertically along the body's center of gravity. This imaginary line runs through the head and down to the space between the feet. The midsagittal plane is drawn through this line. The axial division of the skeleton consists of the bones that lie around the axis: ribs, breastbone, bones of the skull, and backbone.

The appendicular division contains the bones of the free *appendages,* which are the upper and lower extremities, plus the bones called *girdles,* which connect the free appendages to the axial skeleton.

The 80 bones of the axial division and the 126 bones of the appendicular division are typically grouped as shown in Exhibit 6-2.

Now that you understand how the skeleton is organized into axial and appendicular divisions, refer to Figure 6-1 to see how the two divisions are joined to form the skeleton. The bones of the axial skeleton are shown in yellow. Be certain to locate the following regions of the

EXHIBIT 6-2 DIVISIONS OF THE SKELETAL SYSTEM

REGIONS OF THE SKELETON	NUMBER OF BONES
AXIAL SKELETON	
Skull	
Cranium	8
Face	14
Hyoid	1
Auditory ossicles	6
(3 in each ear)*	
Vertebral column	26
Thorax	
Sternum	1
Ribs	24
	80
APPENDICULAR SKELETON	
Pectoral (shoulder) girdles	
Clavicle	2
Scapula	2
Upper extremities	
Humerus	2
Ulna	2
Radius	2
Carpals	16
Metacarpals	10
Phalanges	28
Pelvic (hip) girdle	
Coxal, pelvic, or hip bone	2
Lower extremities	
Femur	2
Fibula	2
Tibia	2
Patella	2
Tarsals	14
Metatarsals	10
Phalanges	28
	126
	Total = 206

* Discussed in Chapter 20.

skeleton: skull, cranium, face, hyoid bone, vertebral column, thorax, shoulder girdle, upper extremity, pelvic girdle, and lower extremity.

SKULL

The **skull,** which contains 22 bones, rests on the superior end of the vertebral column and is composed of two sets of bones: cranial bones and facial bones. The **cranial bones** enclose and protect the brain and the organs of sight, hearing, and balance. The 8 cranial bones are the frontal bone, parietal bones (2), temporal bones (2), occipital bone, sphenoid bone, and ethmoid bone. There are 14 **facial bones:** nasal bones (2), maxillae (2), zygomatic bones (2), mandible, lacrimal bones (2), palatine bones (2), inferior nasal conchae (2), and vomer. Be sure you can locate all the skull bones in the anterior, lateral, median, and posterior views of the skull (Figure 6-2).

SUTURES

A **suture** (SOO-chur), meaning seam or stitch, is an immovable joint found only between skull bones. Very little connective tissue is found between the bones of the suture. Four prominent skull sutures are:

1. Coronal suture between the frontal bone and the two parietal bones.

2. Sagittal suture between the two parietal bones.

3. Lambdoidal (lam-DOY-dal) **suture** between the parietal bones and the occipital bones.

4. Squamosal (skwa-MŌ-sal) **suture** between the parietal bones and the temporal bones.

Refer to Figures 6-2 and 6-3 for the locations of these sutures. Several other sutures are also shown. Their names are descriptive of the bones they connect. For example, the frontonasal suture is between the frontal bone and the nasal bones. These sutures are indicated in Figures 6-2 to 6-5.

FONTANELS

The "skeleton" of a newly formed embryo consists of cartilage or fibrous membrane structures shaped like bones. Gradually the cartilage or fibrous membrane is replaced by bone. At birth, membrane-filled spaces called **fontanels** (fon'-ta-NELS; = little fountains) are found between cranial bones (Figure 6-3). These "soft spots" are areas where the bone-making process is not yet complete. They allow the skull to be compressed during birth. Physicians find the fontanels helpful in determining the position of the infant's head prior to delivery. Although an infant may have many fontanels at birth, the form and location of six are fairly constant.

The **anterior (frontal) fontanel** is located between the angles of the two parietal bones and the two segments of the frontal bone. This fontanel is roughy diamond-shaped, and it is the largest of the six fontanels. It usually closes 18 to 24 months after birth.

The **posterior (occipital;** ok-SIP-i-tal) **fontanel** is situated between the two parietal bones and the occipital bone. This diamond-shaped fontanel is considerably smaller than the anterior fontanel. It generally closes about 2 months after birth.

The **anterolateral (sphenoidal;** sfē-NOY-dal) **fontanels** are paired. One is located on each side of the skull at the junction of the frontal, parietal, temporal, and sphenoid bones. These fontanels are quite small and irregular in shape. They normally close 3 months after birth.

The **posterolateral (mastoid) fontanels** are also paired. One is situated on each side of the skull at the junction of the parietal, occipital, and temporal bones. These fontanels are irregularly shaped. They begin to close 1 or 2 months after birth, but closure is not generally complete until 12 months.

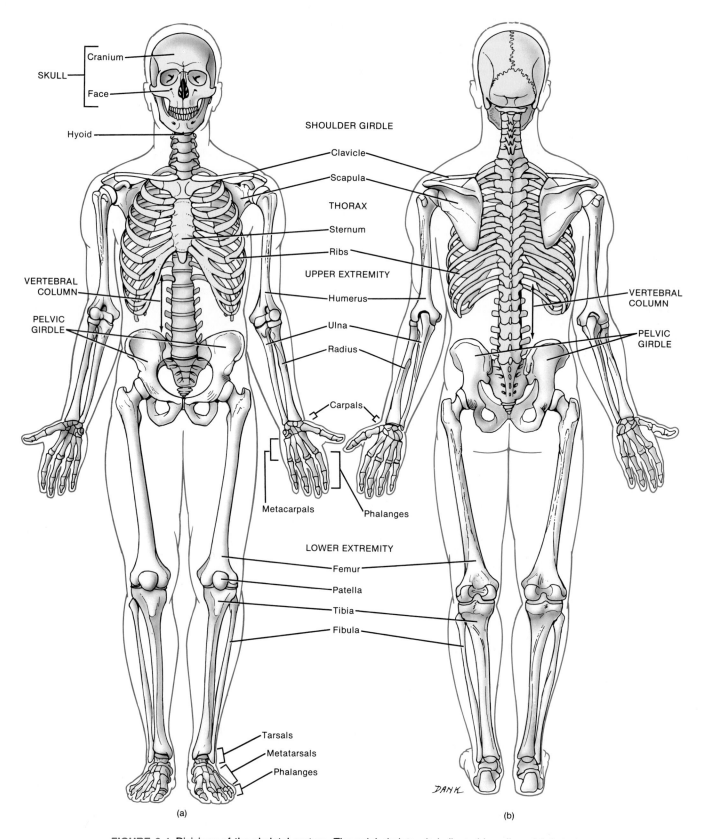

FIGURE 6-1 Divisions of the skeletal system. The axial skeleton is indicated in yellow. (a) Anterior view. (b) Posterior view.

FRONTAL BONE

The **frontal bone** forms the forehead (the anterior part of the cranium), the roofs of the *orbits* (eye sockets), and most of the anterior part of the cranial floor. Soon after birth, the left and right parts of the frontal bone are united by a suture. The suture usually disappears by age 6. If, however, the suture persists throughout life, it is referred to as the **metopic suture.**

If you examine the anterior and lateral views of the skull in Figure 6-2, you will note the **frontal squama** (SKWĀ-ma), or **vertical plate** (*squam* = scale). This scalelike plate, which corresponds to the forehead, gradually slopes down from the coronal suture, then turns abruptly downward. It projects slightly above its lower edge on either side of the midline to form the **frontal eminences.** Inferior to each eminence is a horizontal ridge, the **superciliary arch,** caused by the projection of the frontal sinuses posterior to the eyebrow.

Between the eminences and the arches just superior to the nose is a flattened area, the **glabella.** A thickening of the frontal bone inferior to the superciliary arches is called the **supraorbital margin.** From this margin the frontal bone extends posteriorly to form the roof of the orbit and part of the floor of the cranial cavity.

CLINICAL APPLICATION

Just above the supraorbital margin is a relatively sharp ridge that overlies the frontal sinus. A blow to the ridge frequently lacerates the skin over it, resulting in bleeding. Bruising of the skin over the ridge causes tissue fluid and blood to accumulate in the surrounding connective tissue and gravitate into the upper eyelid. The resulting swelling and discoloration is called a **"black eye."**

Within the supraorbital margin, slightly medial to its midpoint, is a hole called the **supraorbital foramen** (*foramen* = hole). The supraorbital nerve and artery pass through this foramen. The **frontal sinuses** lie deep to the superciliary arches. These mucus-lined cavities act as sound chambers that give the voice resonance.

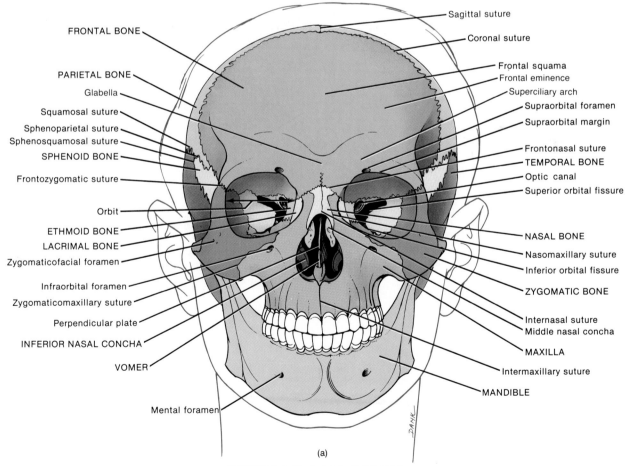

FIGURE 6-2 Skull. (a) Anterior view.

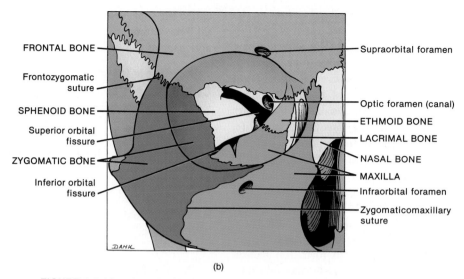

(b)

FIGURE 6-2 (*Continued*) Skull. (b) Detail of the right orbit in anterior view.

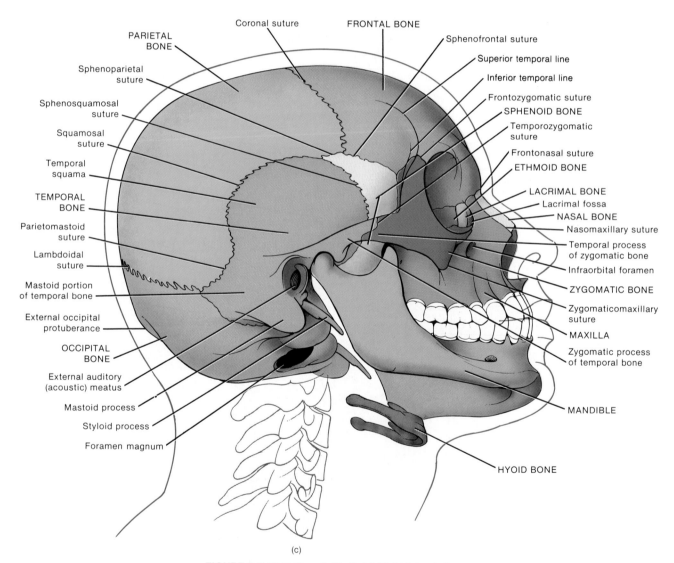

(c)

FIGURE 6-2 (*Continued*) Skull. (c) Right lateral view.

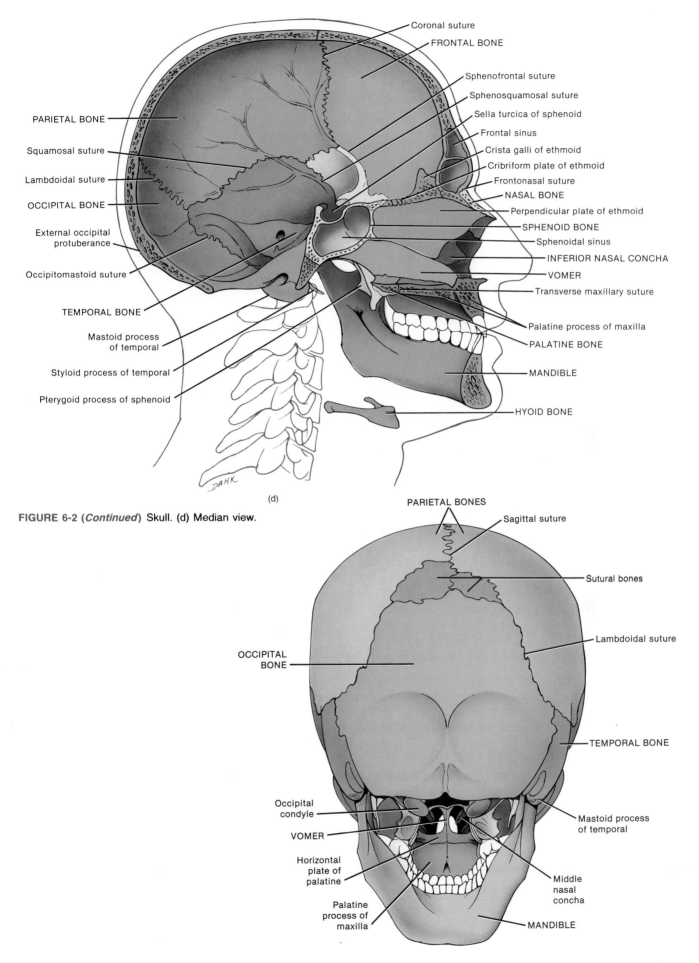

(d)

FIGURE 6-2 (*Continued*) Skull. (d) Median view.

FIGURE 6-2 (*Continued*) Skull. (e) Posterior view.

(e)

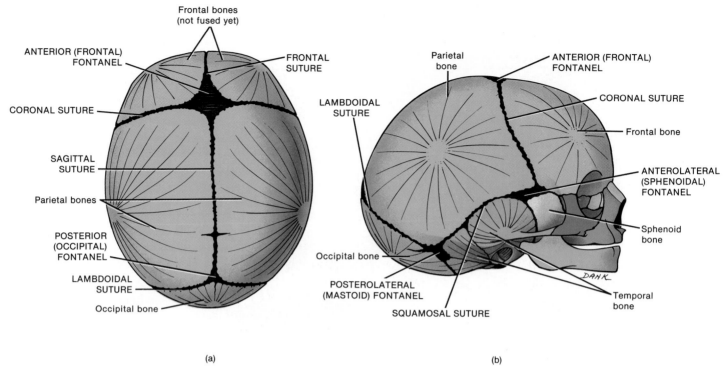

FIGURE 6-3 Fontanels of the skull at birth. (a) Superior view. (b) Right lateral view.

PARIETAL BONES

The two **parietal bones** (pa-RĪ-i-tal; *paries* = wall) form the greater portion of the sides and roof of the cranial cavity.

The external surface contains two slight ridges that may be observed by looking at the lateral view of the skull in Figure 6-2. These are the **superior temporal line** and a less conspicuous **inferior temporal line** below it. The internal surface has many eminences and depressions that accommodate the blood vessels supplying the outer meninx (covering) of the brain called the *dura mater.*

TEMPORAL BONES

The two **temporal bones** form the inferior sides of the cranium and part of the cranial floor. The term *tempora* pertains to the temples.

In the lateral view of the skull in Figure 6-2, notice the **temporal squama**—a thin, large, expanded area that forms the anterior and superior part of the temple. Projecting from the inferior portion of the temporal squama is the **zygomatic process,** which articulates with the temporal process of the zygomatic bone. The zygomatic process of the temporal bone together with the temporal process of the zygomatic bone constitutes the **zygomatic arch.**

At the floor of the cranial cavity, shown in Figure

6-5, is the **petrous portion** of the temporal bone. This portion is triangular and located at the base of the skull between the sphenoid and occipital bones. The petrous portion contains the internal ear, the essential part of the organ of hearing. It also contains the **carotid foramen (canal)** through which the internal carotid artery passes (see Figure 6-4). Posterior to the carotid foramen and anterior to the occipital bone is the **jugular foramen (fossa)** through which the internal jugular vein and the glossopharyngeal (IX) nerve, vagus (X) nerve, and accessory (XI) nerve pass. (As you will see later, the Roman numerals associated with cranial nerves indicate the order in which the nerves arise from the brain, from front to back.)

Between the squamous and petrous portions is a socket called the **mandibular fossa.** Anterior to the mandibular fossa is a rounded eminence, the **articular tubercle.** The mandibular fossa and articular tubercle articulate with the condylar process of the mandible (lower jawbone) to form the temporomandibular (TM) joint. The mandibular fossa and articular tubercle are seen best in Figure 6-4.

In the lateral view of the skull in Figure 6-2, you will see the **mastoid portion** of the temporal bone, located posterior and inferior to the external auditory meatus, or ear canal. In the adult, this portion of the bone contains a number of **mastoid air "cells."** These air spaces are separated from the brain only by thin bony partitions.

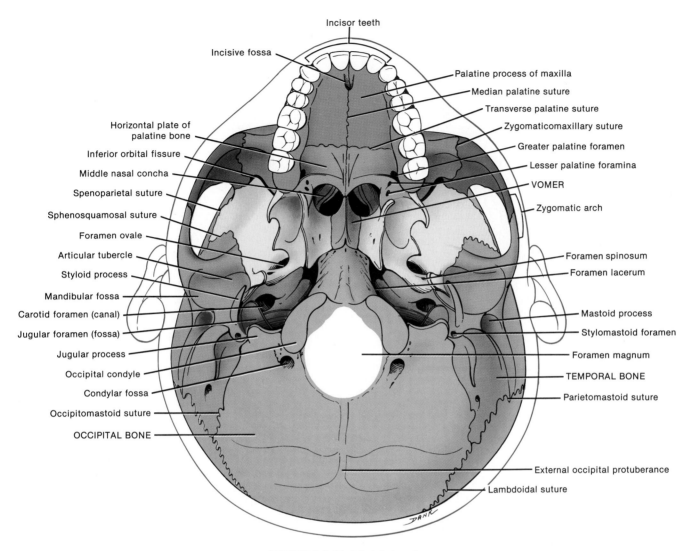

Incisor teeth

Incisive fossa

Palatine process of maxilla

Median palatine suture

Transverse palatine suture

Zygomaticomaxillary suture

Horizontal plate of palatine bone

Greater palatine foramen

Inferior orbital fissure

Lesser palatine foramina

Middle nasal concha

VOMER

Spenoparietal suture

Zygomatic arch

Sphenosquamosal suture

Foramen ovale

Articular tubercle

Foramen spinosum

Styloid process

Foramen lacerum

Mandibular fossa

Carotid foramen (canal)

Mastoid process

Jugular foramen (fossa)

Stylomastoid foramen

Jugular process

Foramen magnum

Occipital condyle

TEMPORAL BONE

Condylar fossa

Parietomastoid suture

Occipitomastoid suture

OCCIPITAL BONE

External occipital protuberance

Lambdoidal suture

FIGURE 6-4 Skull in inferior view.

CLINICAL APPLICATION

> If **mastoiditis,** the inflammation of these bony cells, occurs, the infection may spread to the brain or its outer covering. The mastoid air cells do not drain as do the paranasal sinuses.

The **mastoid process** is a rounded projection of the temporal bone posterior to the external auditory meatus. It serves as a point of attachment for several neck muscles. Near the posterior border of the mastoid process is the **mastoid foramen** through which a vein (emissary) to the transverse sinus and a small branch of the occipital artery to the dura mater pass. The **external auditory meatus** is the canal in the temporal bone that leads to the middle ear. The **internal acoustic meatus** is superior to the jugular forman (see Figure 6-5a). It transmits the facial (VII) and vestibulocochlear (VIII) nerves and the internal auditory artery. The **styloid process** projects downward from the undersurface of the temporal bone and serves as a point of attachment for muscles and ligaments of the tongue and neck. Between the styloid process and the mastoid process is the **stylomastoid foramen,** which transmits the facial (VII) nerve and stylomastoid artery (see Figure 6-4).

OCCIPITAL BONE

The **occipital** (ok-SIP-i-tal) **bone** forms the posterior and prominent portion of the base of the cranium (Figure 6-4).

The **foramen magnum** is a large hole in the inferior part of the bone through which the medulla oblongata and its membranes, the spinal portion of the accessory (XI) nerve, and the vertebral and spinal arteries pass.

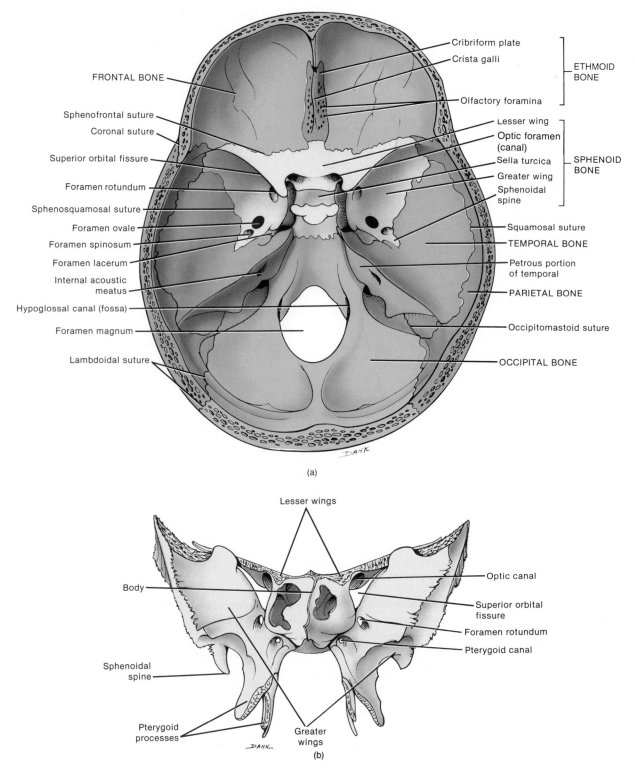

FRONTAL BONE

Sphenofrontal suture
Coronal suture
Superior orbital fissure
Foramen rotundum
Sphenosquamosal suture
Foramen ovale
Foramen spinosum
Foramen lacerum
Internal acoustic meatus
Hypoglossal canal (fossa)
Foramen magnum
Lambdoidal suture

Cribriform plate
Crista galli

ETHMOID BONE

Olfactory foramina

Lesser wing
Optic foramen (canal)
Sella turcica
Greater wing
Sphenoidal spine

SPHENOID BONE

Squamosal suture
TEMPORAL BONE
Petrous portion of temporal
PARIETAL BONE

Occipitomastoid suture

OCCIPITAL BONE

(a)

Lesser wings

Body

Sphenoidal spine

Pterygoid processes

Greater wings

Optic canal

Superior orbital fissure

Foramen rotundum

Pterygoid canal

(b)

FIGURE 6-5 Sphenoid bone. (a) Viewed in the floor of the cranium from above. (b) Anterior view.

The **occipital condyles** are oval processes with convex surfaces, one on either side of the foramen magnum, which articulate with depressions on the first cervical vertebra. Extending laterally from the posterior portion of the condyles are quadrilateral plates of bone called the **jugular processes.** At the base of the condyles is the **hypoglossal canal (fossa)** through which the hypoglossal (XII) nerve passes (see Figure 6-5).

The **external occipital protuberance** is a prominent projection on the posterior surface of the bone just superior to the foramen magnum. You can feel this structure as a definite bump on the back of your head, just above your neck. The protuberance is also visible in Figure 6-2d.

SPHENOID BONE

The **sphenoid** (SFĒ-noyd) **bone** is situated at the middle part of the base of the skull (Figure 6-5). The combining form *spheno* means wedge. This bone is referred to as the keystone of the cranial floor because it articulates with all the other cranial bones. If you view the floor of the cranium from above, you will note that the sphenoid articulates with the temporal bones anteriorly and the occipital bone posteriorly. It lies posterior and slightly superior to the nasal cavities and forms part of the floor and sidewalls of the orbit (eye socket). The shape of the sphenoid is frequently described as a bat with outstretched wings.

The **body** of the sphenoid is the cubelike central portion between the ethmoid and occipital bones. It contains the **sphenoidal sinuses,** which drain into the nasal cavity (see Figure 6-8). On the superior surface of the sphenoid body is a depression called the **sella turcica** (SEL-a TUR-si-ka; = Turk's Saddle). This depression houses the pituitary gland.

The **greater wings** of the sphenoid are lateral projections from the body and form the anterolateral floor of the cranium. The greater wings also form part of the lateral wall of the skull just anterior to the temporal bone. The posterior portion of each greater wing contains a triangular projection, the **sphenoidal spine,** that fits into the angle between the squama and petrous portion of the temporal bone.

The **lesser wings** are anterior and superior to the greater wings. They form part of the floor of the cranium and the posterior part of the orbit. Between the body and lesser wing, you can locate the **optic foramen (canal)** through which the optic (II) nerve and ophthalmic artery pass. Lateral to the body between the greater and lesser wings is a somewhat triangular slit called the **superior orbital fissure.** It is an opening for the oculomotor (III) nerve, trochlear (IV) nerve, ophthalmic branch of the trigeminal (V) nerve, and abducens (VI) nerve. This fissure may also be seen in the anterior view of the skull in Figure 6-2.

On the inferior part of the sphenoid bone you can see the **pterygoid** (TER-i-goyd) **processes.** These structures project inferiorly from the points where the body and greater wings unite. The pterygoid processes form part of the lateral walls of the nasal cavities.

At the base of the lateral pterygoid process in the greater wing is the **foramen ovale** through which the mandibular branch of the trigeminal (V) nerve passes. Another foramen, the **foramen spinosum,** lies at the posterior angle of the sphenoid and transmits the middle meningeal vessels. The **foramen lacerum** is bounded anteriorly by the sphenoid bone, and medially by the sphenoid and occipital bones. Although the foramen is covered in part by a layer of fibrocartilage in living subjects, it transmits the internal carotid artery and the meningeal branch of the ascending pharyngeal artery. A final foramen associated with the sphenoid bone is the **foramen rotundum** through which the maxillary branch of the trigeminal (V) nerve passes. It is located at the junction of the anterior and medial parts of the sphenoid bone.

ETHMOID BONE

The **ethmoid bone** is a light, spongy bone located in the anterior part of the floor of the cranium between the orbits. It is anterior to the sphenoid and posterior to the nasal bones (Figure 6-6). The ethmoid forms part of the anterior portion of the cranial floor, the medial wall of the orbits, the superior portions of the nasal septum, or partition, and most of the sidewalls of the nasal roof. The ethmoid is the principal supporting structure of the nasal cavity.

Its **lateral masses (labyrinths)** compose most of the wall between the nasal cavity and the orbits. They contain several air spaces, or "cells," ranging in number from 3 to 18. It is from these "cells" that the bone derives its name (*ethmos* = sieve). The ethmoid "cells" together form the **ethmoidal sinuses.** The sinuses are shown in Figure 6-8. The **perpendicular plate** forms the superior portion of the nasal septum (see Figure 6-7). The **cribriform (horizontal) plate** lies in the anterior floor of the cranium and forms the roof of the nasal cavity. The cribriform plate contains the **olfactory foramina** through which the olfactory (I) nerves pass. These nerves function in smell. Projecting upward from the horizontal plate is a triangular process called the **crista galli,** which means cock's comb. This structure serves as a point of attachment for the membranes that cover the brain.

The labyrinths contain two thin, scroll-shaped bones on either side of the nasal septum. These are called the **superior nasal concha** (KONG-ka; *concha* = shell) and the **middle nasal concha.** The conchae allow for the efficient circulation and filtration of inhaled air before it passes into the trachea, the bronchi, and the lungs.

CRANIAL FOSSAE

If you examine Figure 6-5a, you can see that the floor of the cranium contains three distinct levels, from anterior to posterior, called **cranial fossae.** The fossae contain depressions for the various brain convolutions, grooves for cranial blood vessels, and numerous foramina. The highest level, the *anterior cranial fossa,* is formed largely by the portion of the frontal bone that constitutes the roof of the orbits and nasal cavity, the crista galli and cribriform plate of the ethmoid bone, and the lesser wings and part of the body of the sphenoid bone. This fossa

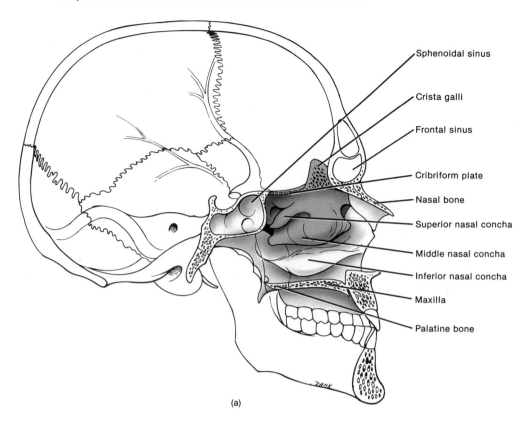

Sphenoidal sinus

Crista galli

Frontal sinus

Cribriform plate

Nasal bone

Superior nasal concha

Middle nasal concha

Inferior nasal concha

Maxilla

Palatine bone

(a)

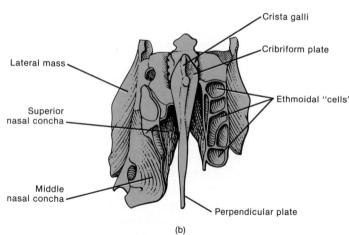

Crista galli

Cribriform plate

Lateral mass

Superior
nasal concha

Ethmoidal "cells"

Middle
nasal concha

Perpendicular plate

(b)

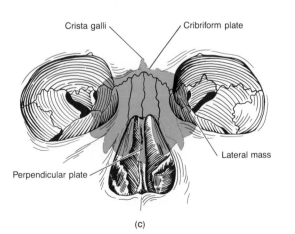

Crista galli

Cribriform plate

Perpendicular plate

Lateral mass

(c)

FIGURE 6-6 Ethmoid bone. (a) Median view showing the ethmoid bone on the inner aspect of the left part of the skull. (b) Anterior view. A frontal section has been made through the left side of the bone to expose the ethmoidal "cells." (c) Highly diagrammatic representation showing the approximate position of the ethmoid bone in the skull in anterior view.

houses the frontal lobes of the cerebral hemispheres. The next fossa is below the level of the anterior cranial fossa and is called the *middle cranial fossa*. It is shaped like a butterfly and has a small median portion and two expanded lateral portions. The median portion is formed by part of the body of the sphenoid bone, and the lateral portions are formed by the greater wings of the sphenoid bone, the temporal squama, and the parietal bone. The middle cranial fossa cradles the temporal lobes of the cerebral hemispheres. The last fossa, at the lowest level, is the *posterior cranial fossa,* the largest of the fossae. It is formed largely by the occipital bone and the petrous and mastoid portions of the temporal bone. It is a very deep fossa that accommodates the cerebellum, pons, and medulla.

NASAL BONES

The paired **nasal bones** are small, oblong bones that meet at the middle and superior part of the face (see Figures 6-2 and 6-7). Their fusion forms the superior part of the bridge of the nose. The inferior portion of the nose, indeed the major portion, consists of cartilage.

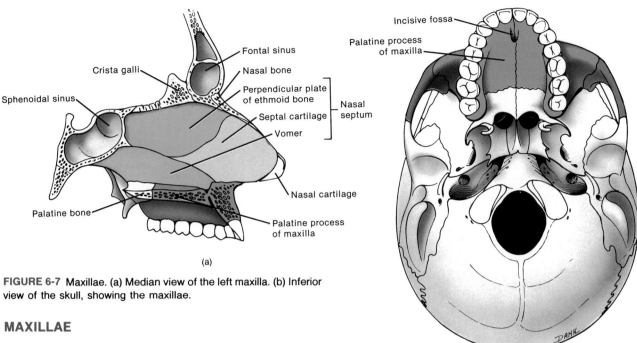

FIGURE 6-7 Maxillae. (a) Median view of the left maxilla. (b) Inferior view of the skull, showing the maxillae.

MAXILLAE

The paired maxillary bones unite to form the upper jaw-bone (Figure 6-7). The **maxillae** (mak-SIL-ē) articulate with every bone of the face except the mandible, or lower jawbone. They form part of the floors of the orbits, part of the roof of the mouth (most of the hard palate), and part of the lateral walls and floor of the nasal cavity.

Each maxillary bone contains a **maxillary sinus** that empties into the nasal cavity (see Figure 6-8). The **alveolar** (al-VĒ-ō-lar) **process** (*alveolus* = hollow) contains the **alveoli** (bony sockets) into which the maxillary (upper) teeth are set. The **palatine process** is a horizontal projection of the maxilla that forms the anterior three-fourths of the hard palate, or anterior portion of the roof of the oral cavity. The two portions of the maxillary bones unite, and the fusion is normally completed before birth.

CLINICAL APPLICATION

If the palatine processes of the maxillary bones do not unite before birth, a condition called **cleft palate** results. The condition may also involve incomplete fusion of the horizontal plates of the palatine bones (see Figure 6-4). Another form of this condition, called **cleft lip,** involves a split in the upper lip. Cleft lip is often associated with cleft palate. Depending on the extent and position of the cleft, speech and swallowing may be affected. A surgical procedure can sometimes improve the condition.

The **infraorbital foramen,** which can be seen in the anterior view of the skull in Figure 6-2, is an opening in the maxilla inferior to the orbit. The infraorbital nerve and artery are transmitted through this opening. Another prominent fossa in the maxilla is the **incisive fossa** just posterior to the incisor teeth. Through it pass branches of the descending palatine vessels and the nasopalatine nerve. A final fossa associated with the maxilla and sphenoid bone is the **inferior orbital fissure.** It is located between the greater wing of the sphenoid and the maxilla (see Figure 6-4). It transmits the maxillary branch of the trigeminal (V) nerve, the infraorbital vessels, and the zygomatic nerve.

PARANASAL SINUSES

Paired cavities called **paranasal sinuses** are located in certain bones near the nasal cavity (Figure 6-8). The paranasal sinuses are lined with mucous membranes that are continuous with the lining of the nasal cavity. Cranial bones containing paranasal sinuses are the frontal, sphenoid, ethmoid, and maxillae. (The paranasal sinuses were described in the discussion of each of these bones.) Besides producing mucus, the paranasal sinuses lighten the skull bones and serve as resonant chambers for sound.

CLINICAL APPLICATION

Secretions produced by the mucous membranes of the paranasal sinuses drain into the nasal cavity. An inflammation of the membranes due to an allergic reaction or infection is called **sinusitis.** If the membranes swell enough to block drainage into the nasal cavity, fluid pressure builds up in the paranasal sinuses and a sinus headache results.

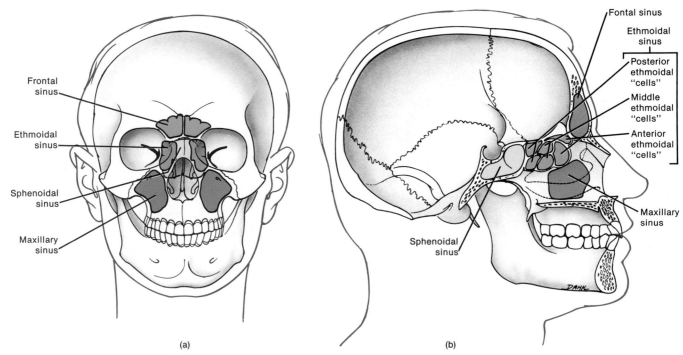

FIGURE 6-8 Paranasal sinuses. (a) Anterior view. (b) Median view. A roentgenogram of the paranasal sinuses is shown in Figure 1-11a.

ZYGOMATIC BONES

The two **zygomatic bones (malars),** commonly referred to as the cheekbones, form the prominences of the cheeks and part of the outer wall and floor of the orbits (Figure 6-2b).

The **temporal process** of the zygomatic bone projects posteriorly and articulates with the zygomatic process of the temporal bone. These two processes form the **zygomatic arch.** A foramen associated with the zygomatic bone is the **zygomaticofacial foramen** near the center of the bone (see Figure 6-2). It transmits the zygomaticofacial nerve and vessels.

MANDIBLE

The **mandible,** or lower jawbone, is the largest, strongest facial bone (Figure 6-9). It is the only movable skull bone.

In the lateral view you can see that the mandible consists of a curved, horizontal portion called the **body** and two perpendicular portions called the **rami.** The **angle** of the mandible is the area where each ramus meets the body. Each ramus has a **condylar (KON-di-lar) process** that articulates with the mandibular fossa and articular tubercle of the temporal bone to form the temporomandibular (TM) joint. It also has a **coronoid (KOR-ō-noyd) process** to which the temporalis muscle attaches. The depression between the coronoid and condylar processes is called the **mandibular notch.** The **alveolar process** is an arch containing the **alveoli** (sockets) for the mandibular (lower) teeth.

CLINICAL APPLICATION

Opening of the mouth very wide, as in a very large yawn, may cause displacement of the condylar process of the mandible from the mandibular fossa of temporal bone, thereby producing a **dislocation of the jaw.** The displacement is usually anterior and bilateral and the patient is unable to close the mouth.

The **mental foramen** (*mentum* = chin) is approximately below the first molar tooth. The mental nerve and vessels pass through this opening. It is through this foramen that dentists sometimes reach the nerve when injecting anesthetics. Another foramen associated with the mandible is the **mandibular foramen** on the medial surface of the ramus, another site frequently used by dentists to inject anesthetics. It transmits the inferior alveolar nerve and vessels. The **mandibular foramen** is the beginning of the **mandibular canal,** which runs forward in the ramus deep to the roots of the teeth. The canal carries branches of the inferior alveolar nerve and vessels to the teeth. Parts of these nerves and vessels emerge through the mental foramen.

LACRIMAL BONES

The paired **lacrimal bones** (LAK-ri-mal; *lacrima* = tear) are thin bones roughly resembling a fingernail in size and shape. They are the smallest bones of the face. These bones are posterior and lateral to the nasal bones in the

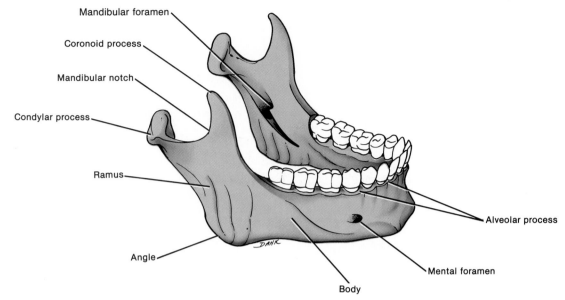

FIGURE 6-9 Mandible in right lateral view.

Labels in figure: Mandibular foramen, Coronoid process, Mandibular notch, Condylar process, Ramus, Angle, Body, Mental foramen, Alveolar process

medial wall of the orbit. They can be seen in the anterior and lateral views of the skull in Figure 6-2.

The lacrimal bones form a part of the medial wall of the orbit. They also contain the **lacrimal fossae** through which the tear ducts pass into the nasal cavity (see Figure 6-2).

PALATINE BONES

The two **palatine** (PAL-a-tīn) **bones** are L-shaped and form the posterior portion of the hard palate, part of the floor and lateral walls of the nasal cavities, and a small portion of the floor of the orbit. The posterior portion of the hard palate, which separates the nasal cavity from the oral cavity, is formed by the **horizontal plates** of the palatine bones. These can be seen in Figure 6-4.

Two foramina associated with the palatine bones are the greater and lesser palatine foramina (Figure 6-4). The **greater palatine foramen,** at the posterior angle of the hard palate, transmits the greater palatine nerve and descending palatine vessels. The **lesser palatine foramina,** usually two or more on each side, are posterior to the greater palatine foramina. They transmit the lesser palatine nerve.

INFERIOR NASAL CONCHAE

Refer to the views of the skull in Figures 6-2a and 6-6a. The two **inferior nasal conchae** (KONG-kē) are scroll-like bones that form a part of the lateral wall of the nasal cavity and project into the nasal cavity inferior to the superior and middle nasal conchae of the ethmoid bone. They serve the same function as the superior and middle nasal conchae, that is, they allow for the circulation and filtration of air before it passes into the lungs. The inferior nasal conchae are separate bones and not part of the ethmoid.

VOMER

The **vomer** (= *plowshare*) is a roughly triangular bone that forms the inferior and posterior part of the nasal septum. It is clearly seen in the anterior view of the skull in Figure 6-2a and the inferior view in Figure 6-4.

The inferior border of the vomer articulates with the cartilage septum that divides the nose into a right and left nostril. Its superior border articulates with the perpendicular plate of the ethmoid bone. Thus, the structures that form the **nasal septum,** or partition, are the perpendicular plate of the ethmoid, the septal cartilage, and the vomer (Figure 6-7a). If the vomer is deviated, that is, pushed to one side, the nasal chambers are of unequal size.

CLINICAL APPLICATION

A **deviated nasal septum** is deflected laterally from the midline of the nose. The deviation usually occurs at the junction of bone with the septal cartilage. If the deviation is severe, it may entirely block the nasal passageway. Even though the blockage may not be complete, infection and inflammation develop and cause nasal congestion, blockage of the paranasal sinus openings, and chronic sinusitis.

EXHIBIT 6-3 SUMMARY OF FORAMINA OF THE SKULL

FORAMEN	LOCATION	STRUCTURES PASSING THROUGH
Carotid (Figure 6-4)	Petrous portion of temporal.	Internal carotid artery.
Greater palatine Figure 6-4)	Posterior angle of hard palate.	Greater palatine nerve and greater palatine vessels.
Hypoglossal (Figure 6-5)	Superior to base of occipital condyles.	Hypoglossal (XII) nerve and branch of ascending pharyngeal artery.
Incisive (Figure 6-7)	Posterior to incisor teeth.	Branches of descending palatine vessels and naso-palatine nerve.
Inferior orbital (Figure 6-4)	Between greater wing of sphenoid and maxilla.	Maxillary branch of trigeminal (V) nerve, zygomatic nerve, and infraorbital vessels.
Infraorbital (Figure 6-2a)	Inferior to orbit in maxilla.	Infraorbital nerve and artery.
Jugular (Figure 6-4)	Posterior to carotid canal between petrous portion of temporal and occipital.	Internal jugular vein, glossopharyngeal (IX) nerve, vagus nerve (X), accessory (XI) nerve, and sigmoid sinus.
Lacrimal (Figure 6-2c)	Lacrimal bone.	Lacrimal (tear) duct.
Lesser palatine (Figure 6-4)	Posterior to greater palatine foramen.	Lesser palatine nerves and artery.
Magnum (Figure 6-4)	Occipital bone.	Medulla oblongata and its membranes, accessory (XI) nerve, and the vertebral and spinal arteries and meninges.
Mandibular	Medial surface of ramus of mandible.	Inferior alveolar nerve and vessels.
Mastoid	Posterior border of mastoid process of temporal bone.	Emissary vein to transverse sinus and branch of occipital artery to dura mater.
Mental (Figure 6-9)	Inferior to second premolar tooth in mandible.	Mental nerve and vessels.
Olfactory (Figure 6-5)	Cribriform plate of ethmoid.	Olfactory (I) nerve.
Optic (Figure 6-5)	Between upper and lower portions of small wing of sphenoid.	Optic (II) nerve and ophthalmic artery.
Ovale (Figure 6-5)	Greater wing of sphenoid.	Mandibular branch of trigeminal (V) nerve.
Rotundum (Figure 6-5)	Junction of anterior and medial parts of sphenoid.	Maxillary branch of trigeminal (V) nerve.
Spinosum (Figure 6-5)	Posterior angle of sphenoid.	Middle meningeal vessels.
Stylomastoid (Figure 6-4)	Between styloid and mastoid processes of temporal.	Facial (VII) nerve and stylomastoid artery.
Superior orbital (Figure 6-5)	Between greater and lesser wings of sphenoid.	Oculomotor (III) nerve, trochlear (IV) nerve, ophthalmic branch of trigeminal (V) nerve, and abducens (VI) nerve.
Supraorbital (Figure 6-2a)	Supraorbital margin of orbit.	Supraorbital nerve and artery.
Zygomaticofacial (Figure 6-2a)	Zygomatic bone.	Zygomaticofacial nerve and vessels.

FORAMINA

The various foramina of the skull were mentioned along with the descriptions of the cranial and facial bones with which they are associated. As preparation for studying other systems of the body, especially the nervous and cardiovascular systems, the foramina and the structures passing through them are listed in Exhibit 6-3. For your convenience and for future reference, the foramina are listed alphabetically.

HYOID BONE

The single **hyoid bone** (*hyoedes* = U-shaped) is a unique component of the axial skeleton because it does not articulate with any other bone. Rather, it is suspended from the styloid process of the temporal bone by ligaments and muscles. The hyoid is located in the neck between the mandible and larynx. It supports the tongue and provides attachment for some of its muscles. Refer to the anterior and lateral views of the skull in Figure 6-2 to see the position of the hyoid bone.

The hyoid consists of a horizontal **body** and paired projections called the **lesser cornua** (*cornu* = horn) and the **greater cornua** (Figure 6-10). Muscles and ligaments attach to these paired projections.

VERTEBRAL COLUMN

DIVISIONS

The **vertebral column (spine),** together with the sternum and ribs, constitutes the skeleton of the **trunk** of the body. The vertebral column is composed of a series of bones called **vertebrae.** In the average adult, the column measures about 71 cm (28 in) in length. In effect, the vertebral column is a strong, flexible rod that moves anteriorly, posteriorly, and laterally. It encloses and protects the spinal cord, supports the head, and serves as a point of attachment for the ribs and the muscles of the back. Between the vertebrae are openings called **intervertebral foramina.** The nerves that connect the spinal cord to various parts of the body pass through these openings.

The adult vertebral column typically contains 26 vertebrae (Figure 6-11a). These are distributed as follows: 7 **cervical vertebrae** (*cervix* = neck) in the neck region; 12 **thoracic vertebrae** posterior to the thoracic cavity; 5 **lumbar vertebrae** (*lumbus* = loin) supporting the lower back; 5 **sacral vertebrae** fused into one bone called the **sacrum;** and usually 4 **coccygeal** (kok-SIJ-ē-al) **vertebrae** fused into one or two bones called the **coccyx** (KOK-six). Prior to the fusion of the sacral and coccygeal vertebrae, the total number of vertebrae is 33.

Between adjacent vertebrae from the axis to the sacrum are fibrocartilaginous **intervertebral discs.** Each disc is composed of an outer fibrous ring consisting of fibrocartilage called the *annulus fibrosus* and an inner soft, pulpy, highly elastic structure called the *nucleus pulposus*. The discs form strong joints, permit various movements of the vertebral column, and absorb vertical shock. Under compression, they flatten, broaden, and bulge from their intervertebral spaces (Figure 6-11c).

CURVES

When viewed from the side, the vertebral column shows four **curves** (Figure 6-11b). From the anterior view, these are alternately convex, meaning they curve toward the viewer, and concave, meaning they curve away from the viewer. The curves of the column, like the curves in a long bone, are important because they increase its strength, help maintain balance in the upright position, absorb shocks from walking, and help protect the column from fracture.

In the fetus, there is only a single anteriorly concave curve. At approximately the third postnatal month, when an infant begins to hold its head erect, the **cervical curve** develops. Later, when the child stands and walks, the **lumbar curve** develops. The cervical and lumbar curves are anteriorly convex. Because they are modifications of the fetal positions, they are called **secondary curves.** The other two curves, the **thoracic curve** and the **sacral curve,** are anteriorly concave. Since they retain the anterior concavity of the fetus, they are referred to as **primary curves.**

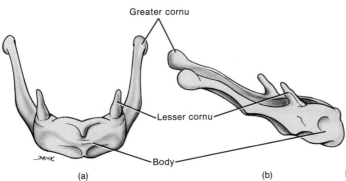

Greater cornu

Lesser cornu

Body

(a)   (b)

FIGURE 6-10 Hyoid bone. (a) Anterior view. (b) Right lateral view.

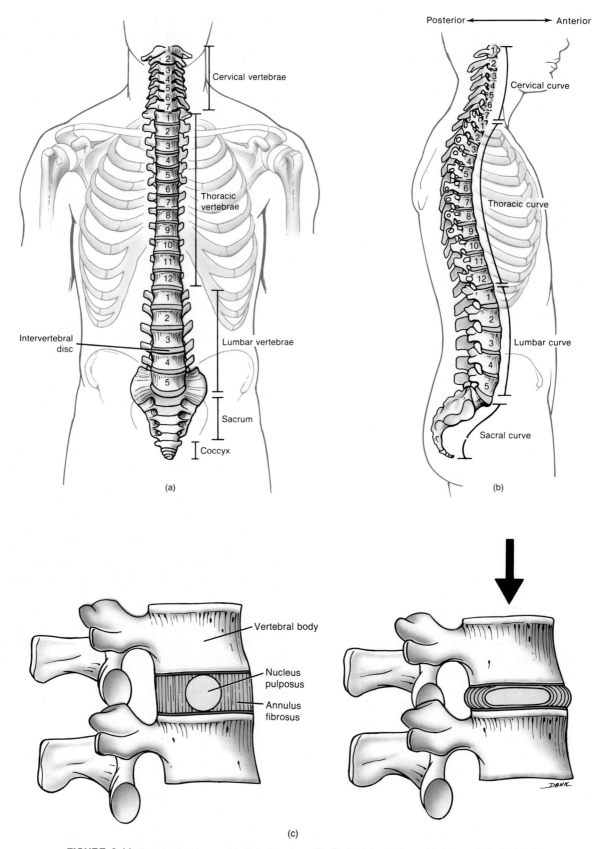

FIGURE 6-11 Vertebral column. (a) Anterior view. (b) Right lateral view. (c) Intervertebral disc in its normal position (left) and under compression (right). The relative size of the disc has been exaggerated for emphasis.

CLINICAL APPLICATION

In recent years, **gravity inversion** has been used by increasing numbers of people to decompress the backbone by using gravity and the body's own weight. This is accomplished by hanging upside down by the ankles in gravity inversion boots suspended from a horizontal bar. Despite the popularity of this method of traction and exercise, there are some potentially harmful effects. For example, it has been found that some healthy young individuals may experience increases in general blood pressure, pulse rate, intraocular pressure, and blood pressure in the eyes. Moreover, gravity inversion is not recommended for people with glaucoma, high blood pressure, heart disease, hiatal hernias, or disorders of the vertebral column.

TYPICAL VERTEBRA

Although there are variations in size, shape, and detail in the vertebrae in different regions of the column, all the vertebrae are basically similar in structure (Figure 6-12). A typical vertebra consists of the following components.

1. The **body** is the thick, disc-shaped anterior portion that is the weight-bearing part of a vertebra. Its superior and inferior surfaces are roughened for the attachment of intervertebral discs. The anterior and lateral surfaces contain nutrient foramina for blood vessels.

2. The **vertebral (neural) arch** extends posteriorly from the body of the vertebra. With the body of the vertebra, it surrounds the spinal cord. It is formed by two short, thick processes, the **pedicles** (PED-i-kuls), which project posteriorly from the body to unite with the laminae. The **laminae** (LAM-i-nē) are the flat parts that join to form the posterior portion of the vertebral arch. The space that lies between the vertebral arch and body contains the spinal cord. This space is known as the **vertebral foramen.** The vertebral foramina of all vertebrae together form the **vertebral (spinal) canal.** The pedicles are notched superiorly and inferiorly in such a way that, when they are arranged in the column, there is an opening between vertebrae on each side of the column. This opening, the **intervertebral foramen,** permits the passage of the spinal nerves.

3. Seven **processes** arise from the vertebral arch. At the point where a lamina and pedicle join, a **transverse process** extends laterally on each side. A single **spinous process (spine)** projects posteriorly and inferiorly from the junction of the laminae. These three processes serve as points of muscular attachment. The remaining four processes form joints with other vertebrae. The two **superior articular processes** of a vertebra articulate with the vertebra immediately superior to them. The two **inferior articular processes** of a vertebra articulate with the vertebra inferior to them. The articulating surfaces of the articular processes are referred to as **facets.**

CLINICAL APPLICATION

In a procedure called **laminectomy,** the spinous processes and laminae of the vertebral arch are removed. The procedure is used to expose a patient's spinal cord and its coverings (meninges) and the nerve roots that originate from the spinal cord. Laminectomy may be used to relieve pressure on neural structures caused by herniated (slipped) discs, tumors, blood clots, or bony fragments.

CERVICAL REGION

When viewed from above, it can be seen that the bodies of **cervical vertebrae** are smaller than those of thoracic vertebrae (Figure 6-13). The arches, however, are larger. The spinous processes of the second through sixth cervical vertebrae are often *bifid,* that is, with a cleft. All cervical vertebrae have three foramina: the vertebral foramen and two transverse foramina. Each cervical transverse process contains a **transverse foramen** through which the vertebral artery and its accompanying vein and nerve fibers pass.

The first two cervical vertebrae differ considerably from the others. The first cervical vertebra (C1), the **atlas,** is named for its support of the head. Essentially, the atlas is a ring of bone with **anterior** and **posterior arches** and large **lateral masses.** It lacks a body and a spinous process. The superior surfaces of the lateral masses, called **superior articular facets,** are concave and articulate with the occipital condyles of the occipital bone. This articulation permits the movement seen when nodding the head. The inferior surfaces of the lateral masses, the **inferior articular facets,** articulate with the second cervical vertebra. The transverse processes and transverse foramina of the atlas are quite large.

The second cervical vertebra (C2), the **axis,** does have a body. A peglike process called the **dens (odontoid process)** projects up through the ring of the atlas. The dens makes a pivot on which the atlas and head rotate. This arrangement permits side-to-side rotation of the head.

CLINICAL APPLICATION

In various instances of trauma, the dens of the axis may be driven into the medulla oblongata of the brain, usually with instantly fatal results. This injury is the usual cause of fatality in **whiplash injuries.**

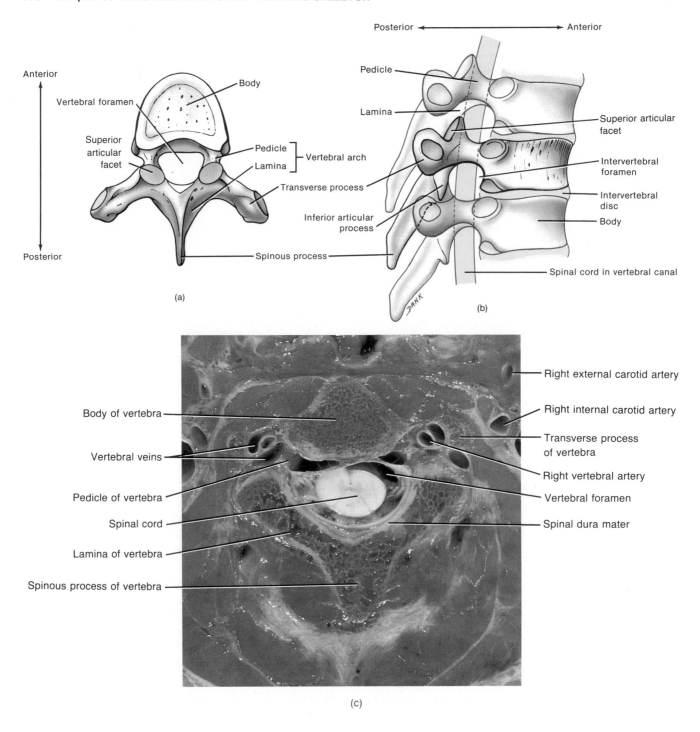

FIGURE 6-12 Typical vertebra. (a) Superior view. (b) Right lateral view. (c) Photograph of a cross section through the upper cervical region of the vertebral column to show the relationship of a vertebra to surrounding structures. (Courtesy of Stephen A. Kieffer and E. Robert Heitzman, *An Atlas of Cross-Sectional Anatomy,* Harper & Row, Publishers, Inc., New York, 1979.)

The third through sixth cervical vertebrae (C3–C6) correspond to the structural pattern of the typical cervical vertebra previously described.

The seventh cervical vertebra (C7), called the **vertebra prominens,** is somewhat different. It is marked by a large, nonbifid spinous process that may be seen and felt at the base of the neck (see Figure 6-11b).

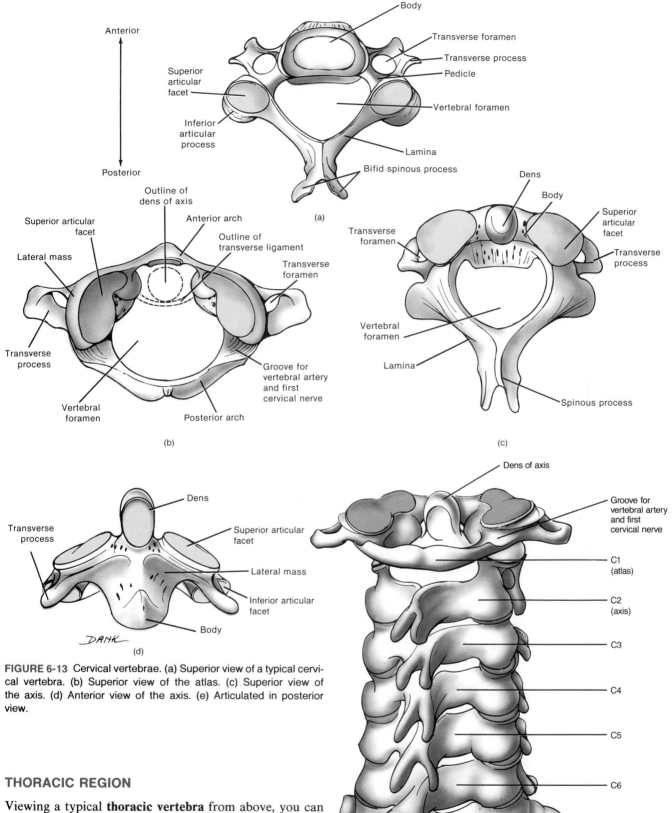

FIGURE 6-13 Cervical vertebrae. (a) Superior view of a typical cervical vertebra. (b) Superior view of the atlas. (c) Superior view of the axis. (d) Anterior view of the axis. (e) Articulated in posterior view.

THORACIC REGION

Viewing a typical **thoracic vertebra** from above, you can see that it is considerably larger and stronger than a vertebra of the cervical region (Figure 6-14). In addition, the spinous process on each vertebra is long, pointed, and directed inferiorly. Thoracic vertebrae also have longer and heavier transverse processes than cervical vertebrae.

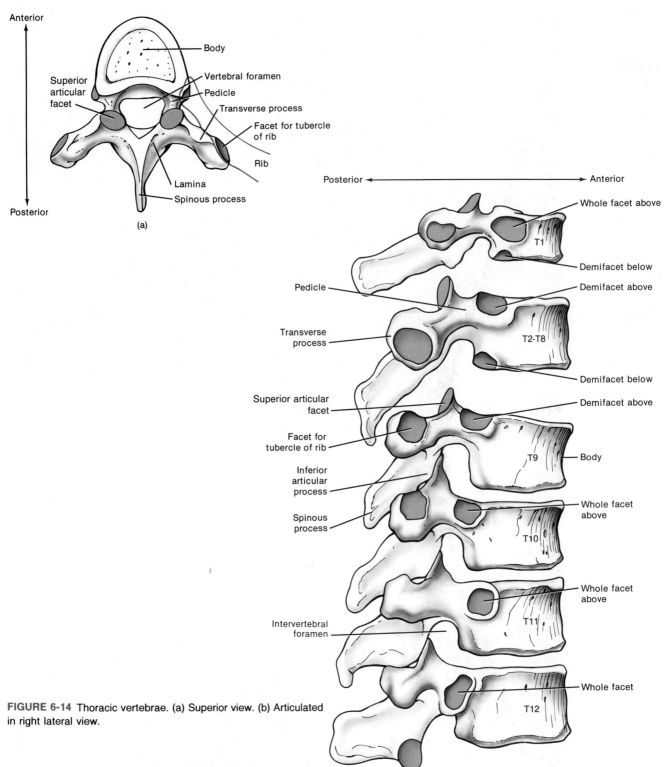

FIGURE 6-14 Thoracic vertebrae. (a) Superior view. (b) Articulated in right lateral view.

Except for the eleventh and twelfth thoracic vertebrae, the transverse processes have **facets** for articulating with the tubercles of the ribs. The bodies of thoracic vertebrae also have whole **facets** or half-facets, called **demifacets,** for articulation with the heads of the ribs. The first thoracic vertebra (T1) has, on either side of its body, a superior whole facet and an inferior demifacet. The superior facet articulates with the first rib, and the inferior demifacet, together with the superior demifacet of the second thora-cic vertebra (T2), forms a facet for articulation with the second rib. The second through eighth thoracic vertebrae (T2–T8) have two demifacets on each side, a larger supe-rior demifacet and a smaller inferior demifacet. When the vertebrae are articulated, they form whole facets for

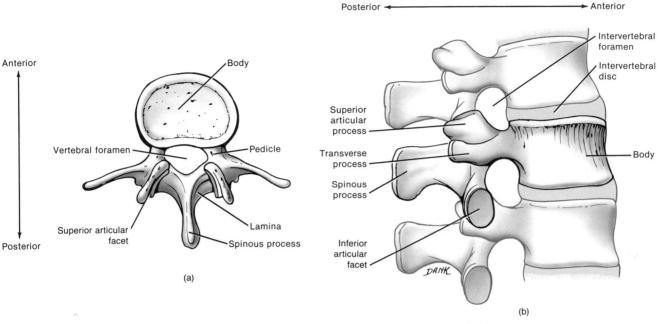

FIGURE 6-15 Lumbar vertebrae. (a) Superior view. (b) Articulated in right lateral view.

the heads of the ribs. The ninth thoracic vertebra (T9) has a single superior demifacet on either side of its body. The tenth through twelfth thoracic vertebrae (T10–T12) have whole facets on either side of their bodies.

LUMBAR REGION

The **lumbar vertebrae** (L1–L5) are the largest and strongest in the column (Figure 6-15). Their various projections are short and thick. The superior articular processes are directed medially instead of superiorly. The inferior articular processes are directed laterally instead of inferiorly. The spinous process is quadrilateral in shape, thick and broad, and projects nearly straight posteriorly. The spinous process is well adapted for the attachment of the large back muscles.

SACRUM AND COCCYX

The **sacrum** is a triangular bone formed by the union of five sacral vertebrae. These are indicated in Figure 6-16 as S1 through S5. The sacrum serves as a strong foundation for the pelvic girdle. It is positioned at the posterior portion of the pelvic cavity between the two hipbones.

The concave anterior side of the sacrum faces the pelvic cavity. It is smooth and contains four **transverse lines** that mark the joining of the sacral vertebral bodies. At the ends of these lines are four pairs of **anterior sacral (pelvic) foramina.**

The convex, posterior surface of the sacrum is irregular. It contains a **median sacral crest,** the fused spinous processes of the upper sacral vertebrae; a **lateral sacral crest,** the transverse processes of the sacral vertebrae; and four pairs of **posterior sacral (dorsal) foramina.** These foramina communicate with the anterior sacral foramina through which nerves and blood vessels pass. The **sacral canal** is a continuation of the vertebral canal. The laminae of the fifth sacral vertebra, and sometimes the fourth, fail to meet. This leaves an inferior entrance to the vertebral canal called the **sacral hiatus** (hi-Ā-tus). On either side of the sacral hiatus are the **sacral cornua,** the inferior articular processes of the fifth sacral vertebra. They are connected by ligaments to the coccygeal cornua of the coccyx.

CLINICAL APPLICATION

Anesthetic agents that act on the sacral and coccygeal nerves are sometimes injected through the sacral hiatus, a procedure called **caudal anesthesia.** Since the sacral hiatus is between the sacral cornua, the cornua are important bony landmarks for locating the hiatus. Anesthetic agents may also be injected through the posterior sacral (dorsal) foramina.

The superior border of the sacrum exhibits an anteriorly projecting border, the **sacral promontory** (PROM-on-tō-rē). It is an obstetrical landmark for measurements of the pelvis. An imaginary line running from the superior surface of the symphysis pubis to the sacral promontory separates the abdominal and pelvic cavities. Laterally,

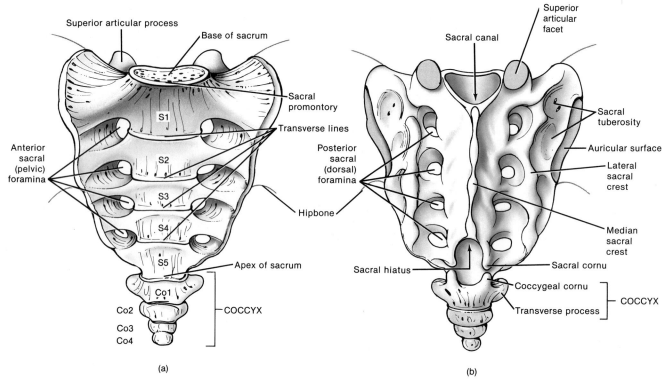

FIGURE 6-16 Sacrum and coccyx. (a) Anterior view. (b) Posterior view.

the sacrum has a large **auricular surface** for articulating with the ilium of the hipbone. Dorsal to the auricular surface is a roughened surface, the **sacral tuberosity,** which contains depressions for the attachment of ligaments. Its **superior articular processes** articulate with the fifth lumbar vertebra.

The **coccyx** is also triangular in shape and is formed by the fusion of the coccygeal vertebrae, usually the last four. These are indicated in Figure 6-16 as Co1 through Co4. The dorsal surface of the body of the coccyx contains two long **coccygeal cornua** that are connected by ligaments to the sacral cornua. The coccygeal cornua are the pedicles and superior articular process of the first coccygeal vertebra. On the lateral surfaces of the body of the coccyx are a series of **transverse processes,** the first pair being the largest. The coccyx articulates superiorly with the sacrum.

THORAX

Anatomically, the term **thorax** refers to the chest. The skeletal portion of the thorax is a bony cage formed by the sternum, costal cartilage, ribs, and the bodies of the thoracic vertebrae (Figure 6-17a).

The thoracic cage is roughly cone-shaped, the narrow portion being superior and the broad portion inferior. It is flattened from front to back. The thoracic cage en-

closes and protects the organs in the thoracic cavity. It also provides support for the bones of the shoulder girdle and upper extremities.

STERNUM

The **sternum,** or breastbone, is a flat, narrow bone measuring about 15 cm (6 in) in length. It is located in the median line of the anterior thoracic wall.

The sternum (see Figure 6-17a, b) consists of three basic portions: the **manubrium** (ma-NOO-brē-um), the triangular, superior portion; the **body,** the middle, largest portion; and the **xiphoid** (ZĪ-foyd) **process,** the inferior, smallest portion. The manubrium has a depression on its superior surface called the **jugular (suprasternal) notch.** On each side of the jugular notch are **clavicular notches** that articulate with the medial ends of the clavicles. The manubrium also articulates with the first and second ribs. The body of the sternum articulates directly or indirectly with the second through tenth ribs. The xiphoid process has no ribs attached to it but provides attachment for some abdominal muscles. The xiphoid process consists of hyaline cartilage during infancy and childhood and does not ossify completely until about age 40. If the hands of a rescuer are mispositioned during cardiopulmonary resuscitation (CPR), there is the danger of fracturing the ossified xiphoid process, separating it from the body, and driving it into the liver.

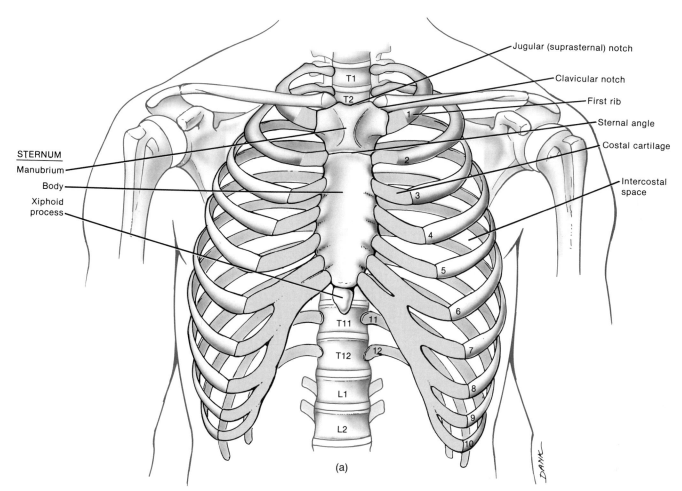

Jugular (suprasternal) notch
Clavicular notch
First rib
Sternal angle
Costal cartilage
Intercostal space

STERNUM
Manubrium
Body
Xiphoid process

T1
T2

T11
T12
L1
L2

11
12

1
2
3
4
5
6
7
8
9
10

(a)

FIGURE 6-17 Skeleton of the thorax. (a) Anterior view. (b) Right lateral view of the sternum.

CLINICAL APPLICATION

Since the sternum possesses red bone marrow throughout life and because it is readily accessible, it is a common site for **marrow biopsy.** Under a local anesthetic, a wide-bore needle is introduced into the marrow cavity of the sternum for aspiration of a sample of red bone marrow. This procedure is called a **sternal puncture.**

The sternum may also be split in the midsagittal plane to allow surgeons access to mediastinal structures such as the thymus gland, heart, and great vessels of the heart.

RIBS

Twelve pairs of **ribs** make up the sides of the thoracic cavity (see Figure 6-17). The ribs increase in length from the first through seventh. Then they decrease in length to the twelfth rib. Each rib articulates posteriorly with its corresponding thoracic vertebra.

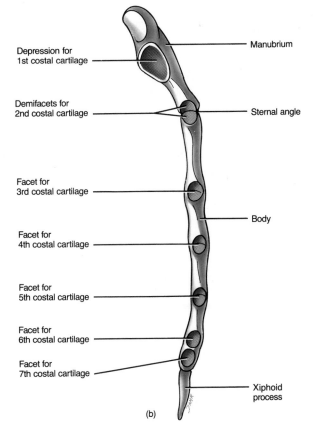

Manubrium
Depression for 1st costal cartilage
Sternal angle
Demifacets for 2nd costal cartilage
Facet for 3rd costal cartilage
Body
Facet for 4th costal cartilage
Facet for 5th costal cartilage
Facet for 6th costal cartilage
Facet for 7th costal cartilage
Xiphoid process

(b)

The first through seventh ribs have a direct anterior attachment to the sternum by a strip of hyaline cartilage, called **costal cartilage** (*costa* = rib). These ribs are called **true (vertebrosternal) ribs.** The remaining five pairs of ribs are referred to as **false ribs** because their costal cartilages do not attach directly to the sternum. The cartilages of the eighth, ninth, and tenth ribs attach to each other and then to the cartilage of the seventh rib. These false ribs are thus called **vertebrochondral ribs.** The eleventh and twelfth false ribs are designated as **floating (vertebral)** ribs because their anterior ends do not attach even indirectly to the sternum. They attach only to the muscles of the body wall.

Although there is some variation in rib structure, we will examine the parts of a typical (third through ninth) rib when viewed from the right side and from behind (Figure 6-18). The **head** of a typical rib is a projection at the posterior end of the rib. It is wedge-shaped and consists of one or two **facets** that articulate with facets on the bodies of adjacent thoracic vertebrae. The facets

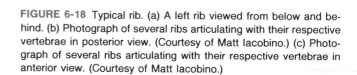

FIGURE 6-18 Typical rib. (a) A left rib viewed from below and behind. (b) Photograph of several ribs articulating with their respective vertebrae in posterior view. (Courtesy of Matt Iacobino.) (c) Photograph of several ribs articulating with their respective vertebrae in anterior view. (Courtesy of Matt Iacobino.)

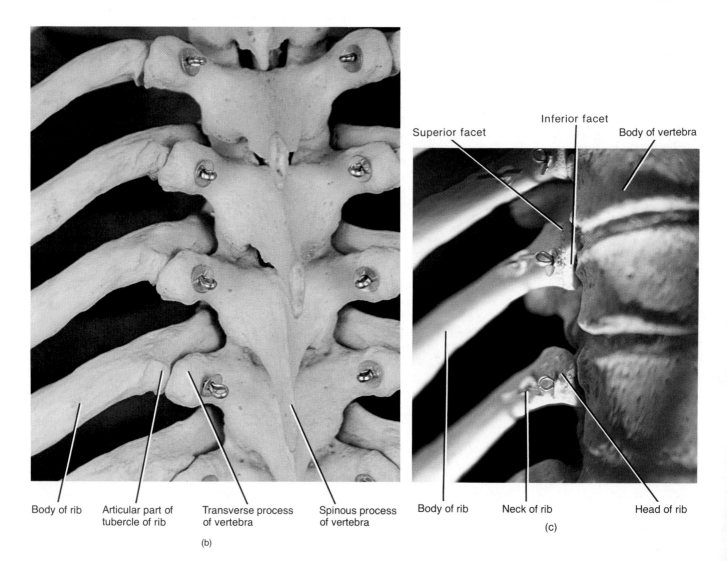

on the head of a rib are separated by a horizontal **interarticular crest.** The inferior facet on the head of a rib is larger than the superior facet. The **neck** is a constricted portion just lateral to the head. A knoblike structure on the posterior surface where the neck joins the body is called a **tubercle.** It consists of a **nonarticular part,** which affords attachment to the ligament of the tubercle, and an **articular part,** which articulates with the facet of a transverse process of the inferior of the two vertebrae to which the head of the rib is connected. The **body (shaft)** is the main part of the rib. A short distance beyond the tubercle, there is an abrupt change in the curvature of the shaft. This point is called the **costal angle** and represents the greatest change in curvature of the rib. The inner surface of the rib has a **costal groove** that protects blood vessels and a small nerve, artery, and vein.

CLINICAL APPLICATION

Rib fractures usually result from direct blows, most commonly from steering wheel impact, falls, and crushing injuries to the chest. Ribs tend to break at the weakest point, that is, the site of greatest curvature just anterior to the costal angle. In children, the ribs are highly elastic and fractures are less frequent than in adults. Since the first two ribs are protected by the clavicle and pectoralis major muscle and the last two ribs are mobile, they are the least commonly injured. The middle ribs are the ones most commonly fractured. In some cases, fractured ribs may cause damage to the heart, great vessels of the heart, lungs, trachea, bronchi, esophagus, spleen, and liver.

The posterior portion of the rib is connected to a thoracic vertebra by its head and articular portion of a tubercle. The facet of the head fits into a facet on the body of a vertebra, and the articular portion of the tubercle articulates with the facet of the transverse process of the vertebra. Each of the second through ninth ribs articulates with the bodies of two adjacent vertebrae. The first, tenth, eleventh, and twelfth ribs articulate with only one vertebra each. On the eleventh and twelfth ribs, there is no articulation between the tubercles and the transverse processes of their corresponding vertebrae.

Spaces between ribs, called **intercostal spaces,** are occupied by intercostal muscles, blood vessels, and nerves.

CLINICAL APPLICATION

Surgical access to the lungs or structures in the mediastinum is commonly undertaken through an intercostal space. Special rib retractors are used to create a wide separation between ribs. The costal cartilages are sufficiently elastic to permit considerable bending.

APPLICATIONS TO HEALTH

HERNIATED (SLIPPED) DISC

In their function as shock absorbers, intervertebral discs are subject to compressional forces. The discs between the fourth and fifth lumbar vertebrae and between the fifth lumbar vertebra and sacrum usually are subject to more forces than other discs. If the anterior and posterior ligaments of the discs become injured or weakened, the pressure developed in the nucleus pulposus may be great enough to rupture the surrounding fibrocartilage. If this occurs, the nucleus pulposus may protrude posteriorly or into one of the adjacent vertebral bodies (herniate). This condition is called a **herniated (slipped) disc.**

Most often the nucleus pulposus slips posteriorly toward the spinal cord and spinal nerves (Figure 6-19). This movement exerts pressure on the spinal nerves, causing considerable, sometimes very acute, pain. If the roots of the sciatic nerve, which passes from the spinal cord to the foot, are pressured, the pain radiates down the back of the thigh, through the calf, and occasionally into the foot. If pressure is exerted on the spinal cord itself, nervous tissue may be destroyed.

Traction, bed rest, and analgesia usually relieve the pain. If such treatment is ineffective, surgical decompression of the spinal nerves by laminectomy or removal of some of the nucleus pulposus may be necessary to relieve pain. Removal may involve conventional surgery or **chemonucleolysis,** the use of a proteolytic enzyme called chymopapain (Chymodiactin). This enzyme is extracted from a papaya plant and is injected into a herniated disc where it dissolves the nucleus pulposus, thus relieving pressure on spinal nerves and pain.

CURVATURES

As a result of various conditions, the normal curves of the vertebral column may become exaggerated or the column may acquire a lateral bend. Such conditions are called **curvatures** of the spine.

Scoliosis (skō'-lē-Ō-sis; *scolio* = bent) is a lateral bending of the vertebral column, usually in the thoracic region. This is the most common of the curvatures. It may be congenital, due to an absence of the lateral half of a vertebra (hemivertebra). It may also be acquired from a persistent severe sciatica. Poliomyelitis may cause scoliosis by the paralysis of muscles on one side of the body, which produces a lateral deviation of the trunk toward the unaffected side. Poor posture is also a contributing factor. Scoliosis can also result from one leg being shorter than the other.

Until recently, a child with progressive scoliosis faced either years of treatment with a brace or major corrective surgery. Several research teams are experimenting with using electrical stimulation of muscles to limit the pro-

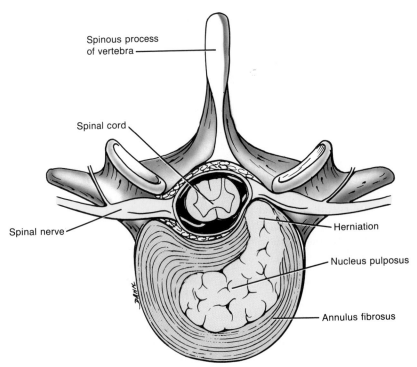

Spinous process
of vertebra

Spinal cord

Spinal nerve

Herniation

Nucleus pulposus

Annulus fibrosus

FIGURE 6-19 Superior view of a herniated (slipped) disc.

gression of scoliosis and even to reduce curvatures in some cases. The skeletal muscles on the convex side of the spinal curve are electrically stimulated.

Kyphosis (kī-FŌ-sis; *kypho* = hunchback) is an exaggeration of the thoracic curve of the vertebral column. In tuberculosis of the spine, vertebral bodies may partially collapse, causing an acute angular bending of the vertebral column. In the elderly, degeneration of the intervertebral discs leads to kyphosis. Kyphosis may also be caused by rickets and poor posture. The term "round-shouldered" is an expression for mild kyphosis. Electrical muscle stimulation is being studied to assess its effects on kyphosis as well as on scoliosis.

Lordosis (lor-DŌ-sis; *lordo* = swayback) is an exaggeration of the lumbar curve of the vertebral column. It may result from increased weight of abdominal contents as in pregnancy or extreme obesity. Other causes include poor posture, rickets, and tuberculosis of the spine.

SPINA BIFIDA

Spina bifida (SPĪ-na BIF-i-da) is a congenital defect of the vertebral column in which laminae fail to unite at the midline. The lumbar vertebrae are involved in about 50 percent of all cases. In less serious cases, the defect is small and the area is covered with skin. Only a dimple or tuft of hair may mark the site. Symptoms are mild, with perhaps intermittent urinary problems. An operation is not ordinarily required. Larger defects in the vertebral arches with protrusion of the membranes around the spi-

nal cord or the spinal cord itself produce serious problems, such as partial or complete paralysis, partial or complete loss of urinary bladder control, and the absence of reflexes.

FRACTURES OF THE VERTEBRAL COLUMN

Fractures of the vertebral column most commonly involve T12, L1, and L2. They usually result from a flexion-compression type of injury such as might be sustained in landing on the feet or buttocks after a fall from a height or having a heavy weight fall on the shoulders. The forceful compression wedges the involved vertebrae. If, in addition to compression, there is forceful forward movement, one vertebra may displace forward on its adjacent vertebra below, with either dislocation or fracture of the articular facets between the two (fracture dislocation) and with rupture of the interspinous ligaments.

The cervical vertebrae may be fractured or, more commonly, dislodged by a fall on the head with acute flexion of the neck, as might happen on diving into shallow water. Dislocation may even result from the sudden forward jerk which may occur when an automobile or airplane crashes ("whiplash"). The relatively horizontal intervertebral facets of the cervical vertebrae allow dislocation to take place without their being fractured, whereas the relatively vertical thoracic and lumbar intervertebral facets nearly always fracture in forward dislocation of the thoracolumbar region. Spinal nerve damage may occur as a result of fractures of the vertebral column.

STUDY OUTLINE

Types of Bones (p. 117)
1. On the basis of shape, bones are classified as long, short, flat, or irregular.
2. Sutural (Wormian) bones are found between the sutures of certain cranial bones. Sesamoid bones develop in tendons or ligaments.

Surface Markings (p. 117)
1. Markings are areas on the surfaces of bones.
2. Each marking is structured for a specific function—joint formation, muscle attachment, or passage of nerves and blood vessels.
3. Terms that describe markings include fissure, foramen, meatus, fossa, process, condyle, head, facet, tuberosity, crest, and spine.

Divisions of the Skeletal System (p. 118)
1. The axial skeleton consists of bones arranged along the longitudinal axis. The parts of the axial skeleton are the skull, hyoid bone, auditory ossicles, vertebral column, sternum, and ribs.
2. The appendicular skeleton consists of the bones of the girdles and the upper and lower extremities. The parts of the appendicular skeleton are the shoulder girdles, the bones of the upper extremities, the pelvic girdle, and the bones of the lower extremities.

Skull (p. 119)
1. The skull consists of the cranium and the face. It is composed of 22 bones.
2. Sutures are immovable joints between bones of the skull. Examples are coronal, sagittal, lambdoidal, and squamosal sutures.
3. Fontanels are membrane-filled spaces between the cranial bones of fetuses and infants. The major fontanels are the anterior, posterior, anterolaterals, and posterolaterals.
4. The 8 cranial bones include the frontal, parietal (2), temporal (2), occipital, sphenoid, and ethmoid.
5. The 14 facial bones are the nasal (2), maxillae (2), zygomatic (2), mandible, lacrimal (2), palatine (2), inferior nasal conchae (2), and vomer.
6. Paranasal sinuses are cavities in bones of the skull that com-

municate with the nasal cavity. They are lined by mucous membranes. The cranial bones containing the paranasal sinuses are the frontal, sphenoid, ethmoid, and maxillae.
7. The foramina of the skull bones provide passages for nerves and blood vessels.

Hyoid Bone (p. 133)
1. The hyoid bone is a U-shaped bone that does not articulate with any other bone.
2. It supports the tongue and provides attachment for some of its muscles.

Vertebral Column (p. 133)
1. The vertebral column, the sternum, and the ribs constitute the skeleton of the trunk.
2. The bones of the adult vertebral column are the cervical vertebrae (7), thoracic vertebrae (12), lumbar vertebrae (5), the sacrum (5, fused), and the coccyx (4, fused).
3. The vertebral column contains primary curves (thoracic and sacral) and secondary curves (cervical and lumbar). These curves give strength, support, and balance.
4. The vertebrae are similar in structure, each consisting of a body, vertebral arch, and seven processes. Vertebrae in the different regions of the column vary in size, shape, and detail.

Thorax (p. 140)
1. The thoracic skeleton consists of the sternum, the ribs and costal cartilages, and the thoracic vertebrae.
2. The thorax protects vital organs in the chest area.

Applications to Health (p. 143)
1. Protrusion of the nucleus pulposus of an intervertebral disc posteriorly or into an adjacent vertebral body is called a herniated (slipped) disc.
2. Exaggeration of a normal curve of the vertebral column is called a curvature. Examples include scoliosis, kyphosis, and lordosis.
3. The imperfect union of the vertebral laminae at the midline, a congenital defect, is referred to as spina bifida.
4. Fractures of the vertebral column most often involve T12, L1, and L2.

REVIEW QUESTIONS

1. What are the four principal types of bones? Give an example of each. Distinguish between a sutural and a sesamoid bone.
2. What are surface markings? Describe and give an example of each.
3. Distinguish between the axial and appendicular skeletons. What subdivisions and bones are contained in each?
4. What are the bones that compose the skull? The cranium? The face?
5. Define a suture. What are the four prominent sutures of the skull? Where are they located?
6. What is a fontanel? Describe the location of the six fairly constant fontanels.
7. What is a paranasal sinus? What cranial bones contain paranasal sinuses?
8. Define the following: mastoiditis, cleft palate, cleft lip, sinusitis, dislocation of the jaw, and deviated nasal septum.
9. What is the hyoid bone? In what respect is it unique? What is its function?
10. What bones form the skeleton of the trunk? Distinguish between the number of nonfused vertebrae found in the adult vertebral column and that of a child.
11. What are the normal curves in the vertebral column? How are primary and secondary curves differentiated? What are the functions of the curves?

12. What are the principal distinguishing characteristics of the bones of the various regions of the vertebral column?
13. What bones form the skeleton of the thorax? What are the functions of the thoracic skeleton?
14. How are ribs classified on the basis of their attachment to the sternum?
15. What is a herniated (slipped) disc? Why does it cause pain?
    How is it treated?
16. What is a curvature? Describe the symptoms and causes of scoliosis, kyphosis, and lordosis.
17. Define spina bifida.
18. What are some common causes of fractures of the vertebral column?

7

The Skeletal System: The Appendicular Skeleton

Student Objectives

Identify the bones of the pectoral (shoulder) girdle and their major markings.

Identify the upper extremity, its component bones, and their markings.

Identify the components of the pelvic (hip) girdle and their principal markings.

Identify the lower extremity, its component bones, and their markings.

Define the structural features and importance of the arches of the foot.

Compare the principal structural differences between female and male skeletons, especially those that pertain to the pelvis.

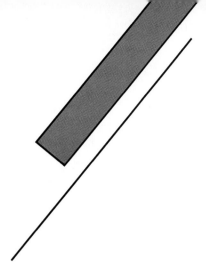

In this chapter we will discuss the bones of the appendicular skeleton, that is, the bones of the pectoral (shoulder) and pelvic (hip) girdles and extremities. The principal structural differences between female and male skeletons are also compared.

PECTORAL (SHOULDER) GIRDLE

The **pectoral** (PEK-tō-ral), or **shoulder, girdles** attach the bones of the upper extremities to the axial skeleton (Figure 7-1). Each of the two pectoral girdles consists of two bones: a clavicle and a scapula. The clavicle is the anterior component and articulates with the sternum. This articulation is referred to as the sternoclavicular joint. The posterior component of the pectoral girdle, the scapula, is positioned freely by complex muscle attachments and articulates with the clavicle and humerus. The pectoral girdles have no articulation with the vertebral column. Although the shoulder joints are not very stable, they are freely movable and thus allow movement in many directions.

CLAVICLE

The **clavicles** (KLAV-i-kuls), or collarbones, are long, slender bones with a double curvature (Figure 7-2). The two bones lie horizontally in the superior and anterior part of the thorax superior to the first rib.

The medial end of the clavicle is known as the **sternal extremity.** It is rounded and articulates with the sternum. The broad, flat, lateral end is referred to as the **acromial** (a-KRŌ-mē-al) **extremity.** It articulates with the acromion of the scapula. This joint is called the **acromioclavicular joint.** (Refer to Figure 7-1 for a view of these articulations.) The **conoid tubercle** on the inferior surface of the lateral end of the bone serves as a point of attachment for a ligament. The **costal tuberosity** on the inferior surface of the medial end also serves as a point of attachment for a ligament.

SCAPULA

The **scapulae** (SCAP-yoo-lē), or shoulder blades, are large, triangular, flat bones situated in the dorsal part of the thorax between the levels of the second and seventh ribs (Figure 7-3). Their medial borders are located about 5 cm (2 in) from the vertebral column.

A sharp ridge, the **spine,** runs diagonally across the dorsal surface of the flattened, triangular **body.** The end of the spine projects as a flattened, expanded process called the **acromion** (a-KRŌ-mē-on). This process articulates with the clavicle. Inferior to the acromion is a depression called the **glenoid cavity.** This cavity articulates with the head of the humerus to form the shoulder joint.

The thin edge of the body near the vertebral column is the **medial (vertebral) border.** The thick edge closer to the arm is the **lateral (axillary) border.** The medial and lateral borders join at the **inferior angle.** The superior edge of the scapular body, called the **superior border,** joins the vertebral border at the **superior angle.**

At the lateral end of the superior border is a projection of the anterior surface called the **coracoid** (KOR-a-koyd) **process** to which muscles attach. Above and below the spine are two fossae: the **supraspinous** (soo'-pra-SPĪ-nus) **fossa** and the **infraspinous fossa,** respectively. Both serve as surfaces of attachment for shoulder muscles. On the ventral (costal) surface is a lightly hollowed-out area called the **subscapular fossa,** also a surface of attachment for shoulder muscles.

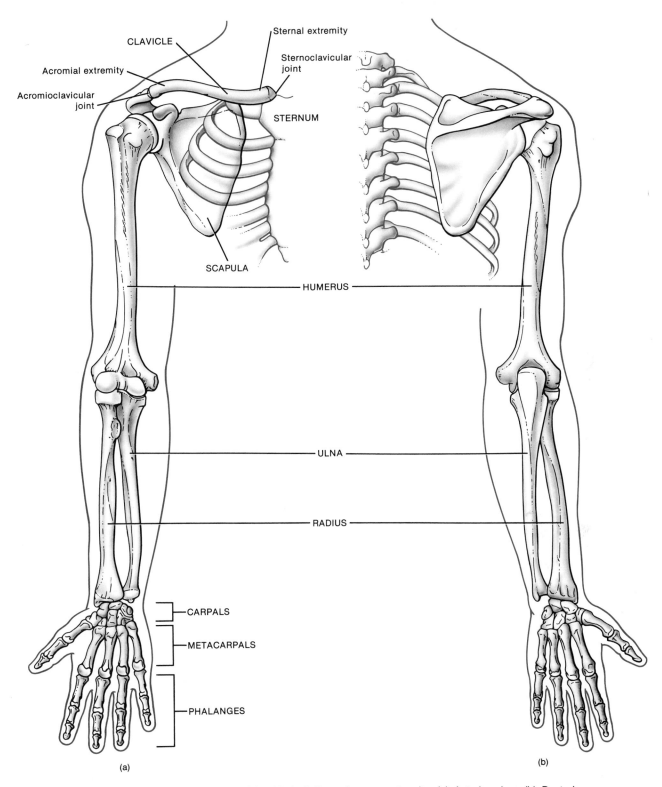

CLAVICLE

Sternal extremity

Sternoclavicular joint

Acromial extremity

Acromioclavicular joint

STERNUM

SCAPULA

HUMERUS

ULNA

RADIUS

CARPALS

METACARPALS

PHALANGES

(a)

(b)

FIGURE 7-1 Right pectoral (shoulder) girdle and upper extremity. (a) Anterior view. (b) Posterior view.

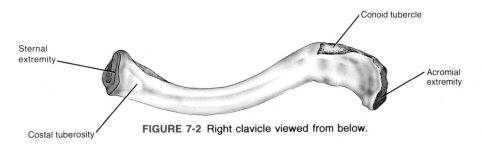

Conoid tubercle

Sternal extremity

Acromial extremity

Costal tuberosity

FIGURE 7-2 Right clavicle viewed from below.

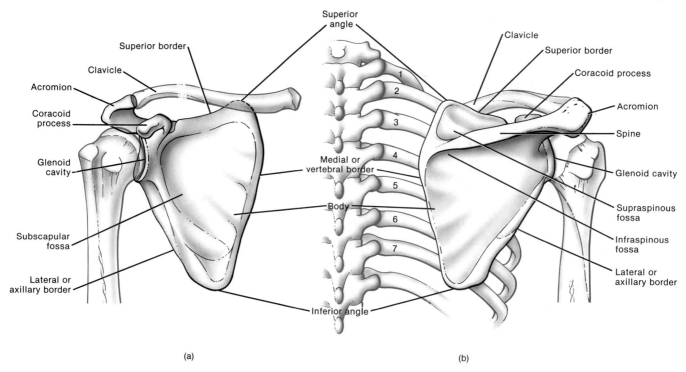

(a)

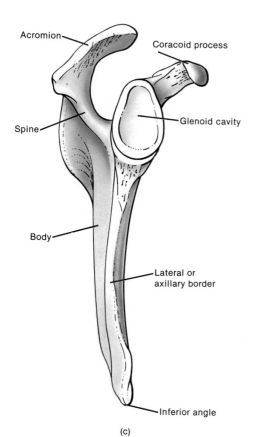

(c)

FIGURE 7-3 Right scapula. (a) Anterior view. (b) Posterior view. (c) Lateral border view.

UPPER EXTREMITY

The **upper extremities** consist of 60 bones. The skeleton of the right upper extremity is shown in Figure 7-1. Each upper extremity includes the humerus in the arm, ulna and radius in the forearm, carpals (wrist bones), metacarpals (palm bones), and phalanges in the fingers of the hand.

HUMERUS

The **humerus** (HYOO-mer-us), or arm bone, is the longest and largest bone of the upper extremity (Figure 7-4). It articulates proximally with the scapula and distally at the elbow with both ulna and radius.

The proximal end of the humerus consists of a **head** that articulates with the glenoid cavity of the scapula. It also has an **anatomical neck,** which is an oblique groove just distal to the head. The **greater tubercle** is a lateral projection distal to the neck. The **lesser tubercle** is an anterior projection. Between these tubercles runs an **intertubercular sulcus (bicipital groove).** The **surgical neck** is a constricted portion just distal to the tubercles. It is so named because of its liability to fracture.

The **body (shaft)** of the humerus is cylindrical at its proximal end. It gradually becomes triangular and is flattened and broad at its distal end. Along the middle portion of the shaft, there is a roughened, V-shaped area called the **deltoid tuberosity.** This area serves as a point of attachment for the deltoid muscle.

The following parts are found at the distal end of the humerus. The **capitulum** (ka-PIT-yoo-lum) is a rounded knob that articulates with the head of the radius. The **radial fossa** is a depression that receives the head of the

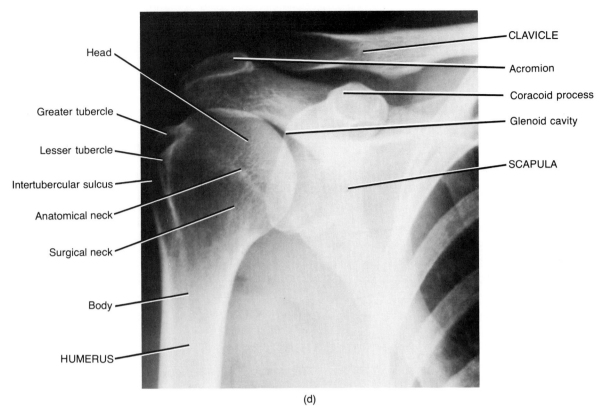

Head

Greater tubercle

Lesser tubercle

Intertubercular sulcus

Anatomical neck

Surgical neck

Body

HUMERUS

CLAVICLE

Acromion

Coracoid process

Glenoid cavity

SCAPULA

(d)

FIGURE 7-3 (*Continued*) Right scapula. (d) Anteroposterior projection of the right scapula and proximal end of the humerus. (Courtesy of Garret J. Pinke, Sr., R.T.)

radius when the forearm is flexed. The **trochlea** (TRŌK-lē-a) is a pulleylike surface that articulates with the ulna. The **coronoid fossa** is an anterior depression that receives part of the ulna when the forearm is flexed. The **olecranon** (ō-LEK-ra-non) **fossa** is a posterior depression that receives the olecranon of the ulna when the forearm is extended. The **medial epicondyle** and **lateral epicondyle** are rough projections on either side of the distal end.

ULNA AND RADIUS

The **ulna** is the medial bone of the forearm (Figure 7-5). In other words, it is located at the little finger side. The proximal end of the ulna presents an **olecranon (olecranon process),** which forms the prominence of the elbow. The **coronoid process** is an anterior projection that, together with the olecranon, receives the trochlea of the humerus. The **trochlear (semilunar) notch** is a curved area between the olecranon and the coronoid process. The trochlea of the humerus fits into this notch. The **radial notch** is a depression located laterally and inferiorly to the trochlear notch. It receives the head of the radius. The distal end of the ulna consists of a **head** that is separated from the wrist by a fibrocartilage disc. A **styloid process** is on the posterior side of the distal end.

The **radius** is the lateral bone of the forearm, that is, it is situated on the thumb side. The proximal end of the radius has a disc-shaped **head** that articulates with

the capitulum of the humerus and radial notch of the ulna. It also has a raised, roughened area on the medial side called the **radial tuberosity.** This is a point of attachment for the biceps muscle. The shaft of the radius widens distally to form a concave inferior surface that articulates with two bones of the wrist called the lunate and scaphoid bones. Also at the distal end is a **styloid process** on the lateral side and a medial, concave **ulnar notch** for articulation with the distal end of the ulna.

CLINICAL APPLICATION

When one falls on the outstretched arm, the radius bears the brunt of forces transmitted through the hand. If a fracture occurs in such a fall, it is usually a transverse break about 3 cm (1 in) from the distal end of the bone. In this type of injury, called a **Colles' fracture,** the hand is displaced backward and upward (see Figure 5-7b).

CARPUS, METACARPUS, AND PHALANGES

The **carpus** (wrist) consists of eight small bones, the **carpals,** united to each other by ligaments (Figure 7-6). The bones are arranged in two transverse rows, with four bones in each row. The proximal row of carpals, from

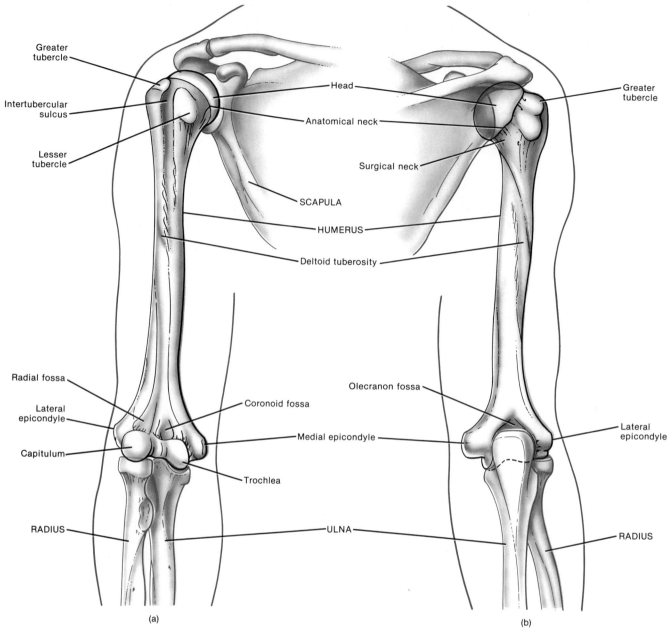

FIGURE 7-4 Right humerus. (a) Anterior view. (b) Posterior view.

the lateral to medial position, consists of the **scaphoid, lunate, triquetral,** and **pisiform.** In about 70 percent of cases involving carpal fractures, only the scaphoid is involved. The distal row of carpals, from the lateral to medial position, consists of the **trapezium, trapezoid, capitate,** and **hamate.**

The five bones of the **metacarpus** constitute the palm of the hand. Each metacarpal bone consists of a proximal **base,** a **shaft,** and a distal **head.** The metacarpal bones are numbered I to V, starting with the lateral bone. The bases articulate with the distal row of carpal bones and with one another. The heads articulate with the proximal phalanges of the fingers. The heads of the metacarpals

are commonly called the "knuckles" and are readily visible when the fist is clenched.

The **phalanges** (fa-LAN-jēz), or bones of the fingers, number 14 in each hand. Each consists of a proximal **base,** a **shaft,** and a distal **head.** There are two phalanges in the first digit, called the thumb, or *pollex,* and three phalanges in each of the remaining four digits. These digits, moving medially from the thumb, are commonly referred to as the index finger, middle finger, ring finger, and little finger. The first row of phalanges, the **proximal row,** articulates with the metacarpal bones and second row of phalanges. The second row of phalanges, the **middle row,** articulates with the proximal row and the third

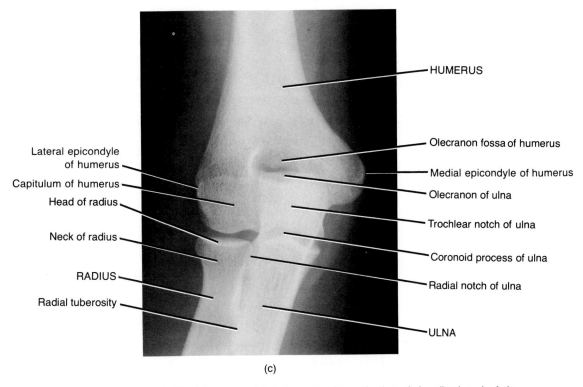

HUMERUS

Olecranon fossa of humerus

Medial epicondyle of humerus

Olecranon of ulna

Trochlear notch of ulna

Coronoid process of ulna

Radial notch of ulna

ULNA

Lateral epicondyle of humerus

Capitulum of humerus

Head of radius

Neck of radius

RADIUS

Radial tuberosity

(c)

FIGURE 7-4 (*Continued*) Right humerus. (c) Anteroposterior projection of the distal end of the humerus and proximal ends of the ulna and radius. (Courtesy of Garret J. Pinke, Sr., R.T.) An anteroposterior projection of the proximal end of the humerus is shown in Figure 7-3d.

row. The third row of phalanges, the **distal row,** articulates with the middle row. A single finger bone is referred to as a **phalanx** (FĀ-lanks). The thumb has no middle phalanx.

PELVIC (HIP) GIRDLE

The **pelvic (hip) girdle** consists of the two **coxal** (KOK-sal) **bones,** commonly called the pelvic, innominate, or hipbones (Figure 7-7). The pelvic girdle provides a strong and stable support for the lower extremities on which the weight of the body is carried. The coxal bones are united to each other anteriorly at the symphysis (SIM-fi-sis) pubis. They unite posteriorly to the sacrum.

Together with the sacrum and coccyx, the two bones of the pelvic girdle form the basinlike structure called the **pelvis.** The pelvis is divided into a greater pelvis and a lesser pelvis by an oblique plane that passes through sacral promontory (posterior), iliopectineal lines (laterally), and symphysis pubis (anteriorly). The circumference of this oblique plane is called the **brim of the pelvis.** The **greater (false) pelvis** represents the expanded portion situated superior to the brim of the pelvis. The greater pelvis consists laterally of the superior portions of the ilia and posteriorly of the superior portion of the sacrum. There is no bony component in the anterior aspect of

the greater pelvis. Rather, the front is formed by the walls of the abdomen.

The **lesser (true) pelvis** is inferior and posterior to the brim of the pelvis. It is formed by the inferior portions of the ilia and sacrum, the coccyx, and the pubes. The lesser pelvis contains a superior opening called the **pelvic inlet** and an inferior opening called the **pelvic outlet.**

CLINICAL APPLICATION

Pelvimetry is the measurement of the size of the inlet and outlet of the birth canal. Measurement of the pelvic cavity is important to the physician, because the fetus must pass through the narrower opening of the lesser pelvis at birth.

Each of the two coxal bones of a newborn consists of three components: a superior **ilium,** an inferior and anterior **pubis,** and an inferior and posterior **ischium** (Figure 7-8). Eventually, the three separate bones fuse into one. The area of fusion is a deep, lateral fossa called the **acetabulum** (as'-e-TAB-yoo-lum). Although the adult coxae are both single bones, it is common to discuss the bones as if they still consisted of three portions.

The ilium is the largest of the three subdivisions of the coxal bone. Its superior border, the **iliac crest,** ends

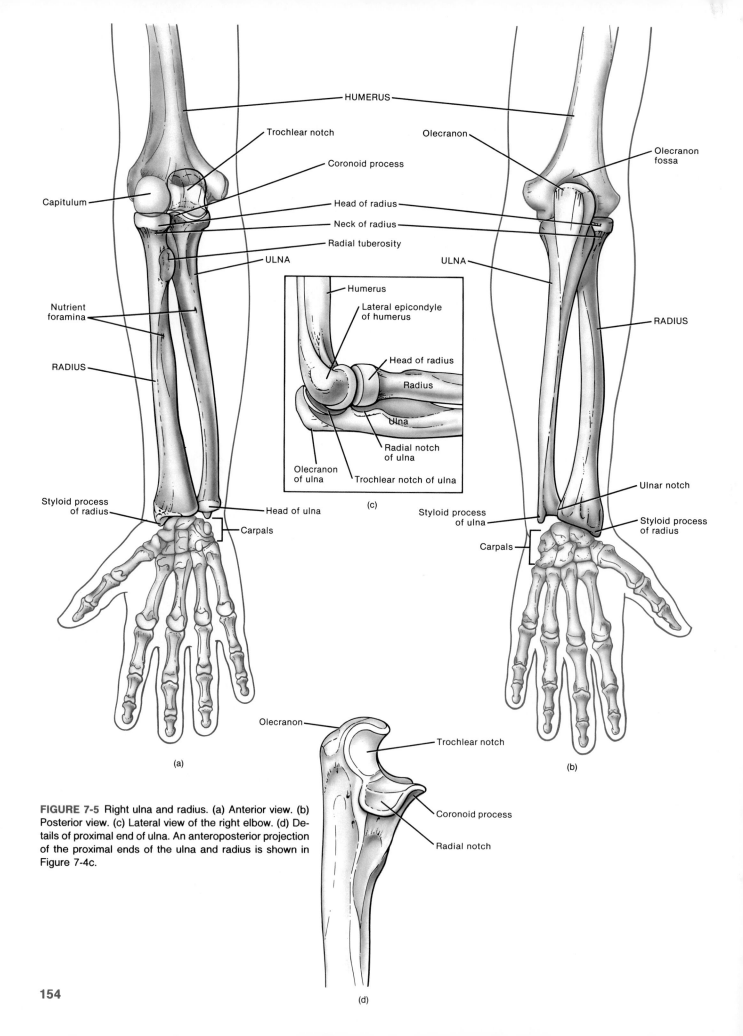

FIGURE 7-5 Right ulna and radius. (a) Anterior view. (b) Posterior view. (c) Lateral view of the right elbow. (d) Details of proximal end of ulna. An anteroposterior projection of the proximal ends of the ulna and radius is shown in Figure 7-4c.

anteriorly in the **anterior superior iliac spine.** The **anterior inferior iliac spine** is located inferior to the anterior superior spine. Posteriorly, the iliac crest ends in the **posterior superior iliac spine.** The **posterior inferior iliac spine** is just inferior. The spines serve as points of attachment for muscles of the abdominal wall. Slightly inferior to the posterior inferior iliac spine is the **greater sciatic** (sī-AT-ik) **notch.** The internal surface of the ilium seen from the medial side is the **iliac fossa.** It is a concavity where the iliacus muscle attaches. Posterior to this fossa are the **iliac tuberosity,** a point of attachment for the sacro-iliac ligament, and the **auricular surface,** which articulates with the sacrum. The other conspicuous markings of the ilium are three arched lines on its gluteal (buttock) surface called the **posterior gluteal line,** the **anterior gluteal line,** and the **inferior gluteal line.** The gluteal muscles attach to the ilium between these lines.

The ischium is the inferior, posterior portion of the coxal bone. It contains a prominent **ischial spine,** a **lesser sciatic notch** below the spine, and an **ischial tuberosity.** The rest of the ischium, the **ramus,** joins with the pubis and together they surround the **obturator** (OB-tyoo-rā'-ter) **foramen.**

The pubis is the anterior and inferior part of the coxal bone. It consists of a **superior ramus,** an **inferior ramus,**

and a **body** that contributes to the formation of the symphysis pubis.

The **symphysis pubis** is the joint between the two coxal bones (see Figure 7-7). It consists of fibrocartilage. The **acetabulum** is the fossa formed by the ilium, ischium, and pubis. It is the socket for the head of the femur. Two-fifths of the acetabulum is formed by the ilium, two-fifths by the ischium, and one-fifth by the pubis. On the inferior portion of the acetabulum is the **acetabular notch.**

LOWER EXTREMITY

The **lower extremities** are composed of 60 bones (Figure 7-9). Each extremity includes the femur in the thigh, patella (kneecap), fibula and tibia in the leg, tarsals (ankle bones), metatarsals, and phalanges in the toes.

FEMUR

The **femur,** or thighbone, is the longest and heaviest bone in the body (Figure 7-10). Its proximal end articulates with the coxal bone. Its distal end articulates with the tibia. The shaft of the femur bows medially so that it approaches the femur of the opposite thigh. As a result

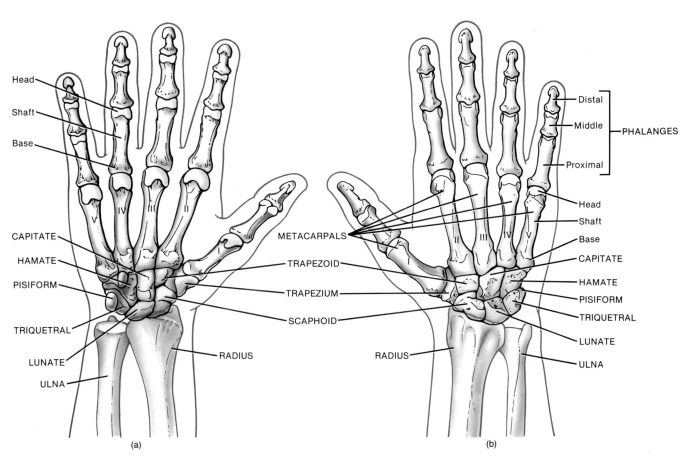

FIGURE 7-6 Right wrist and hand. (a) Anterior view. (b) Posterior view.

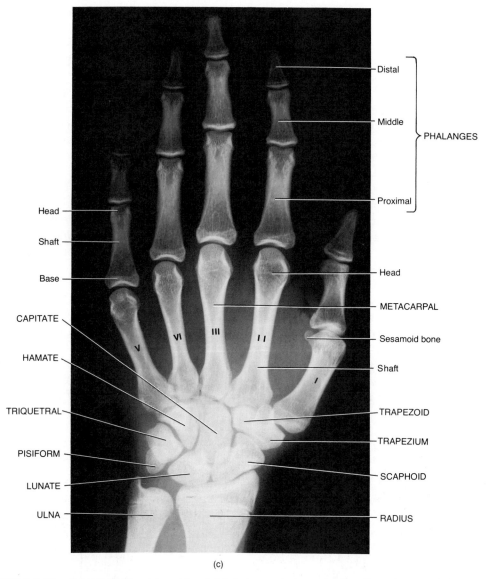

Head

Shaft

Base

CAPITATE

HAMATE

TRIQUETRAL

PISIFORM

LUNATE

ULNA

Distal

Middle

PHALANGES

Proximal

Head

METACARPAL

Sesamoid bone

Shaft

TRAPEZOID

TRAPEZIUM

SCAPHOID

RADIUS

V   VI   III   I I   I

(c)

FIGURE 7-6 (*Continued*) Right wrist and hand. (c) Anteroposterior projection. Note the sesamoid bone. (Courtesy of Daniel Sorrentino, R.T.)

of this convergence, the knee joints are brought nearer to the body's line of gravity. The degree of convergence is greater in the female because the female pelvis is broader.

The proximal end of the femur consists of a rounded **head** that articulates with the acetabulum of the coxal bone. The **neck** of the femur is a constricted region distal to the head. A fairly common fracture in the elderly occurs at the neck of the femur. Apparently the neck becomes so weak that it fails to support the body. The **greater trochanter** (trō-KAN-ter) and **lesser trochanter** are projections that serve as points of attachment for some of the thigh and buttock muscles. Between the trochanters on the anterior surface is a narrow **intertrochan-**

teric line. Between the trochanters on the posterior surface is an **intertrochanteric crest.**

The shaft of the femur contains a rough vertical ridge on its posterior surface called the **linea aspera.** This ridge serves for the attachment of several of the thigh muscles.

The distal end of the femur is expanded and includes the **medial condyle** and **lateral condyle.** These articulate with the tibia. Superior to the condyles are the **medial epicondyle** and **lateral epicondyle.** A depressed area between the condyles on the posterior surface is called the **intercondylar** (in'-ter-KON-di-lar) **fossa.** The **patellar surface** is located between the condyles on the anterior surface.

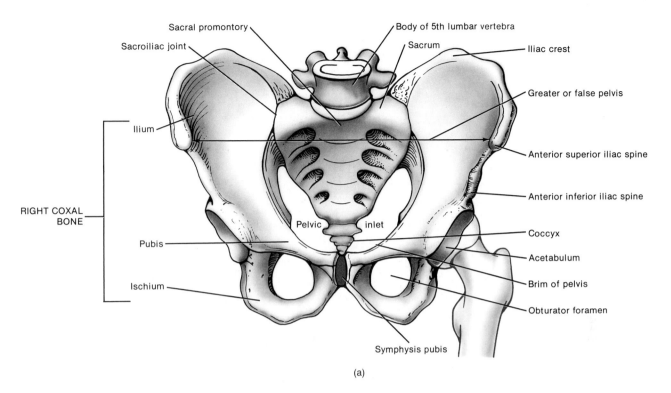

Sacral promontory

Sacroiliac joint

Ilium

RIGHT COXAL
BONE

Pubis

Ischium

Body of 5th lumbar vertebra

Sacrum

Iliac crest

Greater or false pelvis

Anterior superior iliac spine

Anterior inferior iliac spine

Coccyx

Acetabulum

Brim of pelvis

Obturator foramen

Pelvic inlet

Symphysis pubis

(a)

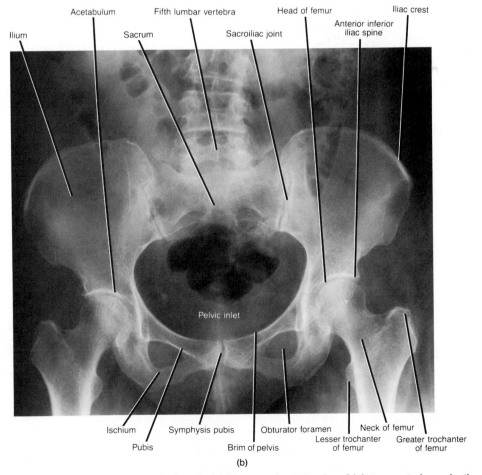

Ilium

Acetabulum

Sacrum

Fifth lumbar vertebra

Sacroiliac joint

Head of femur

Anterior inferior
iliac spine

Iliac crest

Pelvic inlet

Ischium

Pubis

Symphysis pubis

Brim of pelvis

Obturator foramen

Lesser trochanter
of femur

Neck of femur

Greater trochanter
of femur

(b)

FIGURE 7-7 Pelvic (hip) girdle of a female. (a) Diagram of anterior view. (b) Anteroposterior projection. (Courtesy of John C. Bennett, St. Mary's Hospital, San Francisco.)

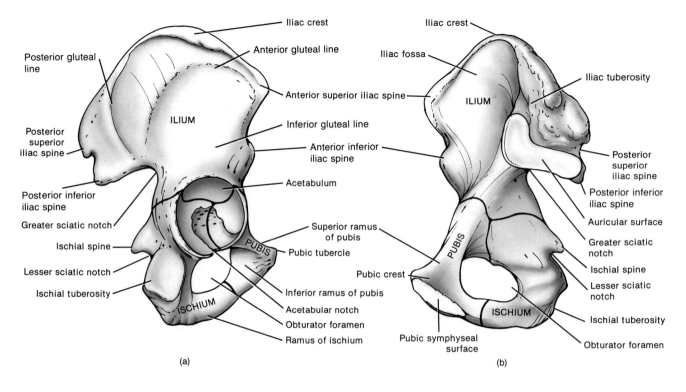

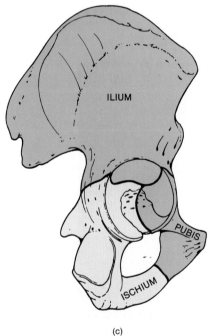

FIGURE 7-8 Right coxal bone. (a) Lateral view. (b) Medial view. The lines of fusion of the ilium, ischium, and pubis are not actually visible in an adult bone. (c) Three divisions of the coxal bone.

PATELLA

The **patella,** or kneecap, is a small, triangular bone anterior to the knee joint (Figure 7-11). It is a sesamoid bone that develops in the tendon of the quadriceps femoris muscle. The broad superior end of the patella is called the **base.** The pointed inferior end is the **apex.** The posterior surface contains two **articular facets,** one for the medial condyle and the other for the lateral condyle of the femur.

TIBIA AND FIBULA

The **tibia,** or shinbone, is the larger, medial bone of the leg (Figure 7-12). It bears the major portion of the weight of the leg. The tibia articulates at its proximal end with the femur and fibula and at its distal end with the fibula of the leg and talus of the ankle.

The proximal end of the tibia is expanded into a **lateral condyle** and a **medial condyle.** These articulate with the condyles of the femur. The inferior surface of the lateral condyle articulates with the head of the fibula. The slightly concave condyles are separated by an upward projection called the **intercondylar eminence.** The **tibial tuberosity** on the anterior surface is a point of attachment for the patellar ligament.

The medial surface of the distal end of the tibia forms the **medial malleolus** (mal-LĒ-ō-lus). This structure articulates with the talus bone of the ankle and forms the prominence that can be felt on the medial surface of your ankle. The **fibular notch** articulates with the fibula.

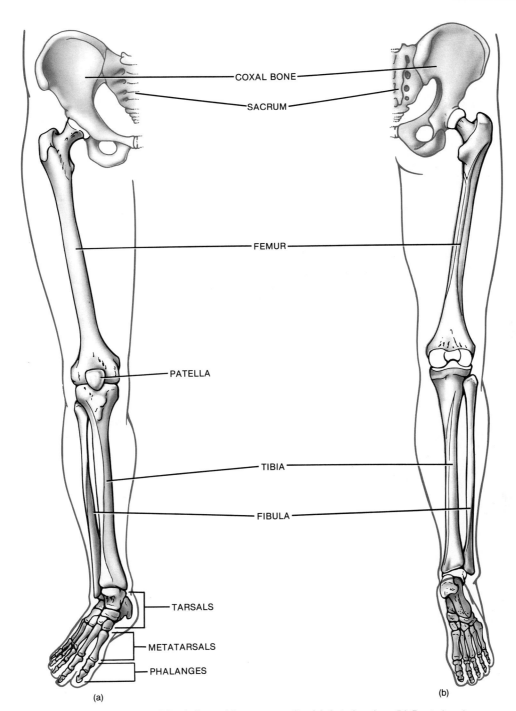

FIGURE 7-9 Right pelvic girdle and lower extremity. (a) Anterior view. (b) Posterior view.

CLINICAL APPLICATION

Shinsplints (tibia stress syndrome) is the name given to soreness or pain along the tibia. It is probably caused by inflammation of the periosteum (periostitis) brought on by the repeated tugging of the muscles and tendons attached to the periosteum. The condition is often caused by walking or running up and down hills or by vigorous activity of the legs following a period of relative inactivity. The condition is commonly treated by rest. Patients who do not respond to rest may be given local injections of cortisonelike steroid drugs or may have to undergo minor surgery to release pressure in the soft tissues around the bone.

The **fibula** is parallel and lateral to the tibia. It is considerably smaller than the tibia. The **head** of the fibula, the proximal end, articulates with the inferior surface of the lateral condyle of the tibia below the level of the

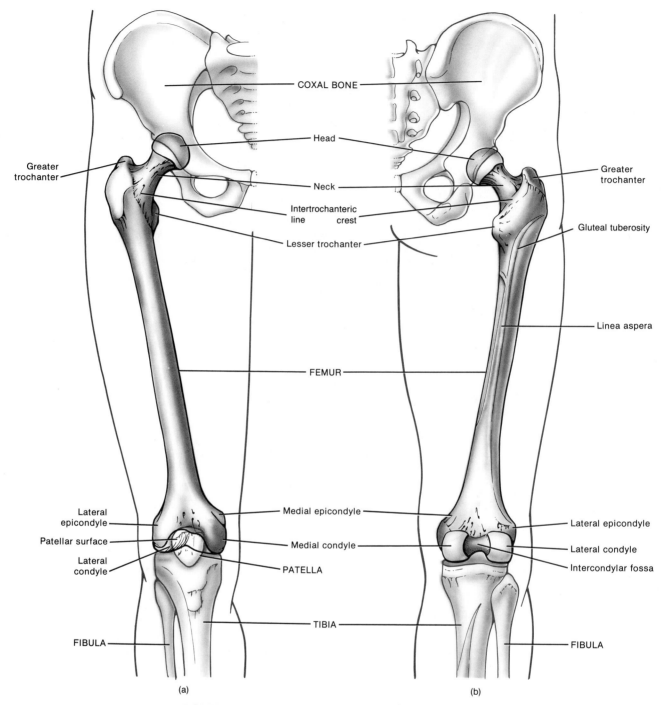

COXAL BONE

Greater
trochanter

Head

Neck

Intertrochanteric
line   crest

Greater
trochanter

Gluteal tuberosity

Lesser trochanter

Linea aspera

FEMUR

Lateral
epicondyle

Medial epicondyle

Lateral epicondyle

Patellar surface

Medial condyle

Lateral condyle

Lateral
condyle

PATELLA

Intercondylar fossa

TIBIA

FIBULA

FIBULA

(a)   (b)

FIGURE 7-10 Right femur. (a) Anterior view. (b) Posterior view.

knee joint. The distal end has a projection called the **lateral malleolus,** which articulates with the talus bone of the ankle. This forms the prominence on the lateral surface of the ankle. The inferior portion of the fibula also articulates with the tibia at the fibular notch. A fracture of the lower end of the fibula with injury to the tibial articulation is called a **Pott's fracture.**

TARSUS, METATARSUS, AND PHALANGES

The **tarsus** is a collective designation for the seven bones of the ankle called **tarsals** (Figure 7-13). The term *tarsos* pertains to a broad, flat surface. The **talus** and **calcaneus** (kal-KĀ-nē-us) are located on the posterior part of the foot. The anterior part contains the **cuboid, navicular,**

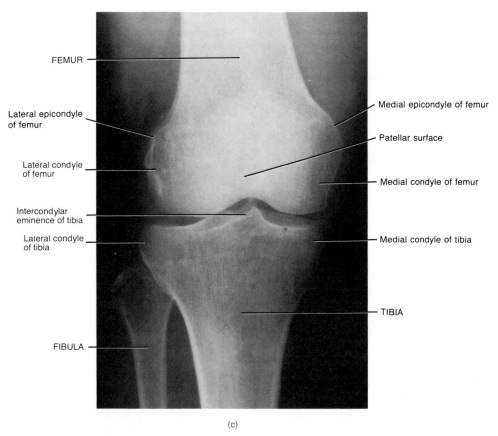

(c)

FIGURE 7-10 (*Continued*) Right femur. (c) Anteroposterior projection of the distal end of the right femur and proximal ends of the right tibia and fibula. (Courtesy of Daniel Sorrentino, R.T.) An anteroposterior projection of the proximal end of the femur is shown in Figure 7-7b.

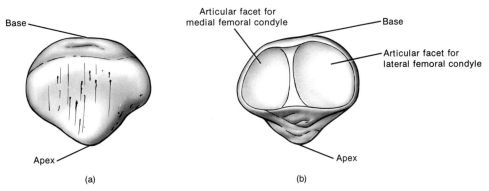

FIGURE 7-11 Right patella. (a) Anterior view. (b) Posterior view.

and three **cuneiform bones** called the **first (medial), second (intermediate),** and **third (lateral)** cuneiform. The talus, the uppermost tarsal bone, is the only bone of the foot that articulates with the fibula and tibia. It is surrounded on one side by the medial malleolus of the tibia and on the other side by the lateral malleolus of the fibula. During walking, the talus initially bears the entire weight of the

extremity. About half the weight is then transmitted to the calcaneus. The remainder is transmitted to the other tarsal bones. The calcaneus, or heel bone, is the largest and strongest tarsal bone.

The **metatarsus** consists of five metatarsal bones numbered I to V from the medial to lateral position. Like the metacarpals of the palm of the hand, each metatarsal

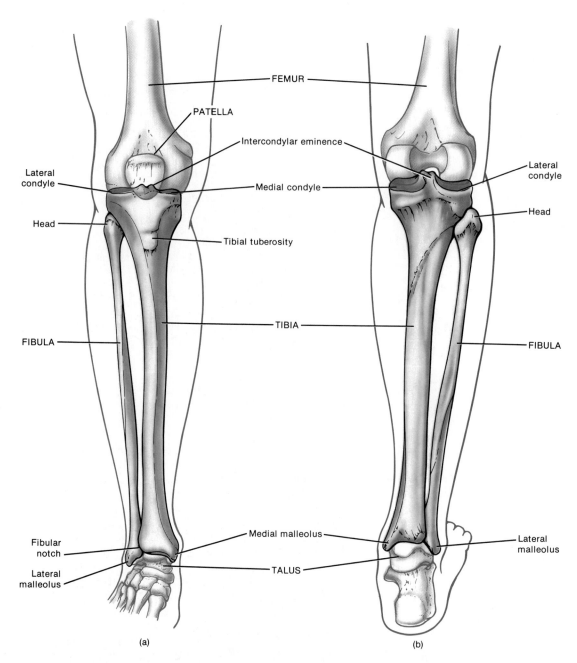

FEMUR

PATELLA

Intercondylar eminence

Lateral condyle

Lateral condyle

Medial condyle

Head

Head

Tibial tuberosity

TIBIA

FIBULA

FIBULA

Fibular notch

Medial malleolus

Lateral malleolus

Lateral malleolus

TALUS

(a)   (b)

FIGURE 7-12 Right tibia and fibula. (a) Anterior view. (b) Posterior view. An anteroposterior projection of the proximal ends of the tibia and fibula is shown in Figure 7-10C.

consists of a proximal **base,** a **shaft,** and a distal **head.** The metatarsals articulate proximally with the first, second, and third cuneiform bones and with the cuboid. Distally, they articulate with the proximal row of phalanges. The first metatarsal is thicker than the others because it bears more weight.

The **phalanges** of the foot resemble those of the hand both in number and arrangement. Each also consists of a proximal **base,** a **shaft,** and a distal **head.** The great (big) toe, or *hallux,* has two large, heavy phalanges called proximal and distal phalanges. The other four toes each have three phalanges—proximal, middle, and distal.

ARCHES OF THE FOOT

The bones of the foot are arranged in two **arches** (Figure 7-14). These arches enable the foot to support the weight of the body and provide leverage while walking. The arches are not rigid. They yield as weight is applied and spring back when the weight is lifted.

The **longitudinal arch** has two parts. Both consist of tarsal and metatarsal bones arranged to form an arch from the anterior to the posterior part of the foot. The **medial,** or inner, part of the longitudinal arch originates at the calcaneus. It rises to the talus and descends through

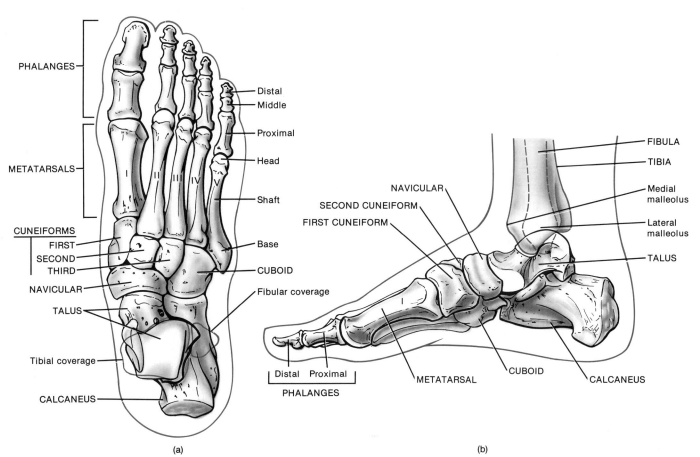

PHALANGES

Distal
Middle

Proximal

Head

METATARSALS

Shaft

CUNEIFORMS
FIRST
SECOND
THIRD

Base

NAVICULAR

CUBOID

TALUS

Fibular coverage

Tibial coverage

CALCANEUS

(a)

FIBULA
TIBIA

NAVICULAR

SECOND CUNEIFORM

FIRST CUNEIFORM

Medial malleolus

Lateral malleolus

TALUS

Distal   Proximal

PHALANGES

METATARSAL

CUBOID

CALCANEUS

(b)

FIGURE 7-13 Right foot. (a) Superior view. (b) Medial view. (c) Plantar view. (Courtesy of John C. Bennett, St. Mary's Hospital, San Francisco.)

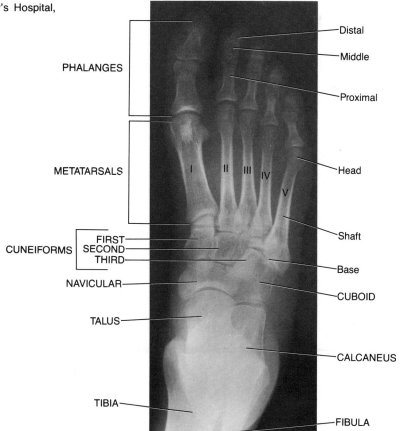

PHALANGES

Distal
Middle

Proximal

METATARSALS

Head

Shaft

CUNEIFORMS
FIRST
SECOND
THIRD

Base

NAVICULAR

CUBOID

TALUS

CALCANEUS

TIBIA

FIBULA

(c)

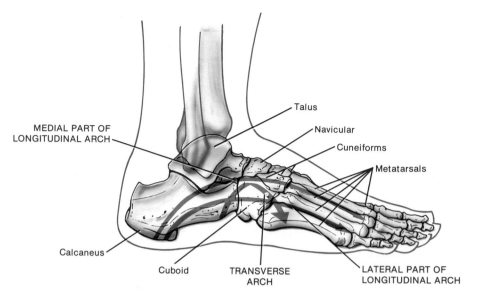

MEDIAL PART OF LONGITUDINAL ARCH

Talus

Navicular

Cuneiforms

Metatarsals

Calcaneus

Cuboid

TRANSVERSE ARCH

LATERAL PART OF LONGITUDINAL ARCH

FIGURE 7-14 Arches of the right foot in lateral view.

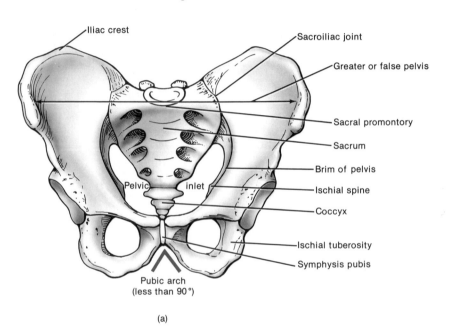

Iliac crest

Sacroiliac joint

Greater or false pelvis

Sacral promontory

Sacrum

Brim of pelvis

Ischial spine

Coccyx

Ischial tuberosity

Symphysis pubis

Pelvic   inlet

Pubic arch (less than 90°)

(a)

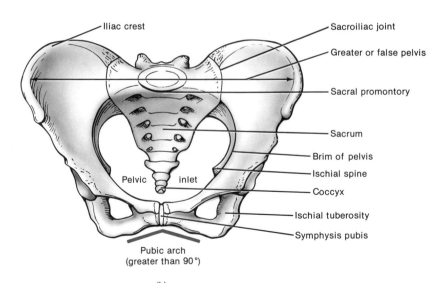

Iliac crest

Sacroiliac joint

Greater or false pelvis

Sacral promontory

Sacrum

Brim of pelvis

Ischial spine

Coccyx

Ischial tuberosity

Symphysis pubis

Pelvic   inlet

Pubic arch (greater than 90°)

(b)

FIGURE 7-15 Pelvis. (a) Male pelvis in anterior view. (b) Female pelvis in anterior view.

EXHIBIT 7-1 COMPARISON OF TYPICAL FEMALE AND MALE PELVIS

POINT OF COMPARISON	FEMALE	MALE
General structure	Light and thin.	Heavy and thick.
Joint surfaces	Small.	Large.
Muscle attachments	Rather indistinct.	Well marked.
Greater pelvis	Shallow.	Deep.
Pelvic inlet	Larger and more oval.	Heart-shaped.
Pelvic outlet	Comparatively large.	Comparatively small.
First piece of sacrum	Superior surface of the body spans about a third the width of sacrum.	Superior surface of the body spans nearly half the width of the sacrum.
Sacrum	Short, wide, flat, curving forward in lower part.	Long, narrow, with smooth concavity.
Auricular surface	Extends only to upper border of third piece of the sacrum.	Extends well down the third piece of the sacrum.
Pubic arch	Greater than a 90° angle.	Less than a 90° angle.
Inferior ramus of pubis	Everted surface not present.	Presents strong everted surface for attachment of the crus of the penis.
Symphysis pubis	Less deep.	More deep.
Ischial spine	Turned inward less.	Turned inward.
Ischial tuberosity	Turned outward.	Turned inward.
Ilium	Less vertical	More vertical.
Iliac fossa	Shallow.	Deep.
Iliac crest	Less curved.	More curved.
Anterior superior iliac spine	Wider apart.	Closer.
Acetabulum	Small.	Large.
Obturator foramen	Oval.	Round.
Greater sciatic notch	Wide.	Narrow.

the navicular, the three cuneiforms, and the three medial metatarsals. The talus is the keystone of this arch. The **lateral,** or outer, part of the longitudinal arch also begins at the calcaneus. It rises at the cuboid and descends to the two lateral metatarsals. The cuboid is the keystone of this arch.

The **transverse arch** is formed by the calcaneus, navicular, cuboid, and the posterior parts of the five metatarsals.

CLINICAL APPLICATION

The bones composing the arches are held in position by ligaments and tendons. If these ligaments and tendons are weakened, the height of the medial longitudinal arch may decrease or "fall." The result is **flatfoot.**

Clawfoot is a condition in which the medial longitudinal arch is abnormally elevated. It is frequently caused by muscle imbalance, such as may result from poliomyelitis.

A **bunion** (hallux valgus, *valgus* = bent outward) is a deformity of the great toe. Although the condition

may be inherited, it is typically caused by wearing tightly fitting shoes. It is characterized by lateral deviation of the great toe, with partial dislocation at its junction with metatarsal I. As a result, the head of metatarsal I is displaced medially. The condition produces inflammation of bursae (fluid-filled sacs at the joint), bone spurs, and calluses.

FEMALE AND MALE SKELETONS

The bones of the male are generally larger and heavier than those of the female. The articular ends are thicker in relation to the shafts. In addition, since certain muscles of the male are larger than those of the female, the points of attachment—tuberosities, lines, ridges—are larger in the male skeleton.

Many significant structural differences between female and male skeletons are noted in the pelvis; most are related to pregnancy and childbirth. The typical differences are listed in Exhibit 7-1 and illustrated in Figure 7-15.

STUDY OUTLINE

Pectoral (Shoulder) Girdle (p. 148)
1. Each pectoral (shoulder) girdle consists of a clavicle and scapula.
2. Each attaches an upper extremity to the trunk.

Upper Extremity (p. 150)
The bones of each upper extremity include the humerus, ulna, radius, carpals, metacarpals, and phalanges.

Pelvic (Hip) Girdle (p. 153)
1. The pelvic girdle consists of two coxal bones or hipbones.
2. It attaches the lower extremities to the trunk at the sacrum.
3. Each coxal bone consists of three fused components—ilium, pubis, and ischium.

Lower Extremity (p. 155)
1. The bones of each lower extremity include the femur, tibia, fibula, tarsals, metatarsals, and phalanges.
2. The bones of the foot are arranged in two arches, the longitudinal arch and the transverse arch, to provide support and leverage.

Female and Male Skeletons (p. 165)
1. The female pelvis is adapted for pregnancy and childbirth. Differences in pelvic structure are listed in Exhibit 7-1.
2. Male bones are generally larger and heavier than female bones and have more prominent markings for muscle attachment.

REVIEW QUESTIONS

1. What is the pectoral (shoulder) girdle? Why is it important?
2. What are the bones of the upper extremity? What is a Colles' fracture?
3. What is the pelvic (hip) girdle? Why is it important?
4. What are the bones of the lower extremity? What is a Pott's fracture?
5. What is pelvimetry? What is its clinical importance?
6. In what ways do the upper extremity and lower extremity differ structurally?
7. Describe the structure of the longitudinal and transverse arches of the foot. What is the function of an arch?
8. How do flatfeet, clawfeet, and bunions arise?
9. What are the principal structural differences between typical male and female skeletons? Use Exhibit 7-1 as a guide in formulating your response.

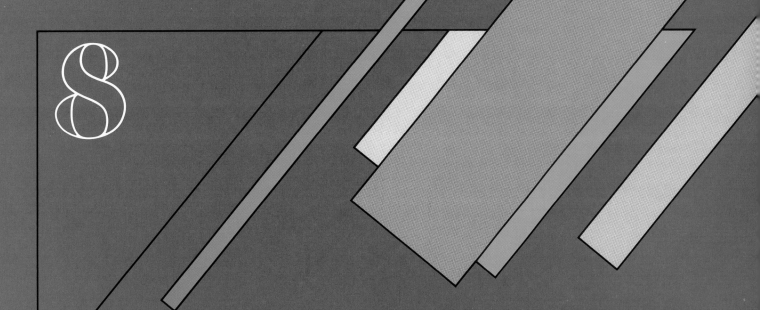

Articulations

Student Objectives

Define an articulation and identify the factors that determine the degree of movement at a joint.

Contrast the structure, kind of movement, and location of fibrous, cartilaginous, and synovial joints.

Discuss and compare the movements possible at various synovial joints.

Describe selected joints of the body with respect to the bones that enter into their formation, structural classification, and anatomical components.

Describe the causes and symptoms of common joint disorders, including rheumatism, rheumatoid arthritis (RA), osteoarthritis, gouty arthritis, bursitis, dislocation, sprain, and tendinitis.

Define key medical terms associated with joints.

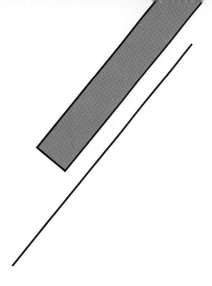

Bones are too rigid to bend without damage. Fortunately, the skeletal system consists of many separate bones, which are held together at joints by flexible connective tissue. All movements that change the positions of the bony parts of the body occur at joints. You can understand the importance of joints if you imagine how a cast over the knee joint prevents flexing the leg or how a splint on a finger limits the ability to manipulate small objects.

An **articulation (joint)** is a point of contact between bones or between cartilage and bones. The scientific study of joints is referred to as **arthrology** (*arthro* = joint; *logos* = study of). The joint's structure determines how it functions. Some joints permit no movement, others permit slight movement, and still others afford considerable movement. In general, the closer the fit at the point of contact, the stronger the joint. At tightly fitted joints, however, movement is restricted. The looser the fit, the greater the movement. Unfortunately, loosely fitted joints are prone to dislocation. Movement at joints is also determined by the flexibility of the connective tissue that binds the bones together and by the position of ligaments, muscles, and tendons.

CLASSIFICATION

FUNCTIONAL

The functional classification of joints takes into account the degree of movement they permit. Functionally, joints are classified as **synarthroses** (sin'-ar-THRŌ-sēz), which are immovable joints; **amphiathroses** (am'-fē-ar-THRŌ-sēz), which are slightly movable joints; and **diarthroses** (dī'-ar-THRŌ-sēz), which are freely movable joints.

STRUCTURAL

The structural classification of joints is based on the presence or absence of a joint cavity (a space between the articulating bones) and the kind of connective tissue that binds the bones together. Structurally, joints are classified as **fibrous,** in which there is no joint cavity and the bones are held together by fibrous connective tissue; **cartilaginous,** in which there is no joint cavity and the bones are held together by cartilage; and **synovial,** in which there is a joint cavity and the bones forming the joint are united by a surrounding articular capsule and frequently by accessory ligaments (described in detail later). We will discuss the joints of the body based on their structural classification, but with reference to their functional classification as well.

FIBROUS JOINTS

Fibrous joints lack a joint cavity, and the articulating bones are held very closely together by fibrous connective tissue. They permit little or no movement. The three types of fibrous joints are (1) sutures, (2) syndesmoses, and (3) gomphoses.

SUTURE

Sutures are found between bones of the skull. In a suture, the bones are united by a thin layer of dense fibrous connective tissue. Based on the form of the margins of the bones, several types of sutures can be distinguished. In a *serrated suture,* the margins of the bones are serrated like the teeth of a saw. An example is the sagittal suture between the two parietal bones (see Figure 6-2a, e). In a *squamous suture,* the margin of one bone overlaps that of the adjacent bone. The squamosal suture between the parietal and temporal bones is an example (see Figure 6-2c). In a *plane suture,* the even, fairly regular margins of the adjacent bones are brought together. An example is the intermaxillary suture between maxillary bones (see Figure 6-2a). Since sutures are immovable, they are functionally classified as synarthroses.

Some sutures, present during growth, are replaced by bone in the adult. In this case they are called **synostoses** (sin'-os-TŌ-sēz), or bony joints—joints in which is a complete fusion of bone across the suture line. An example is the frontal suture between the left and right sides of the frontal bone, which begins to fuse during infancy (see Figure 6-3a). Synostoses are also functionally classified as synarthroses.

SYNDESMOSIS

A **syndesmosis** (sin'-dez-MŌ-sis) is a fibrous joint in which the uniting fibrous connective tissue is present in a much greater amount than in a suture, but the fit between the bones is not quite as tight. The fibrous connective tissue forms an interosseous membrane or ligament. A syndesmosis is slightly movable because the bones are separated more than in a suture and some flexibility is permitted by the interosseous membrane or ligament. Thus, syndesmoses are functionally classified as amphiarthrotic. Examples of syndesmoses include the distal articulation of the tibia and fibula (see Figure 7-12) and the articulations between the shafts of the ulna and radius.

GOMPHOSIS

A **gomphosis** (gom-FŌ-sis) is a type of fibrous joint in which a cone-shaped peg fits into a socket. The intervening substance is the periodontal ligament. A gomphosis is functionally classified as synarthrotic. Examples are the articulations of the roots of the teeth with the alveoli (sockets) of the maxillae and mandible.

CARTILAGINOUS JOINTS

Another joint that has no joint cavity is the **cartilaginous joint.** Here the articulating bones are tightly connected by cartilage. Like fibrous joints, they allow little or no movement. The two types of cartilaginous joints are (1) synchondroses and (2) symphyses.

SYNCHONDROSIS

A **synchondrosis** (sin'-kon-DRŌ-sis) is a cartilaginous joint in which the connecting material is hyaline cartilage. The most common type of synchondrosis is the epiphyseal plate (see Figure 5-7). Such a joint is found between the epiphysis and diaphysis of a growing bone and is immovable. Thus, it is synarthrotic. Since the hyaline cartilage is eventually replaced by bone when growth ceases, the joint is temporary. It is replaced by a synostosis. Another example of a synchondrosis is the joint between the first rib and the sternum. The cartilage in this joint undergoes ossification during adult life.

SYMPHYSIS

A **symphysis** (SIM-fi-sis) is a cartilaginous joint in which the connecting material is a broad, flat disc of fibrocarti-

lage. This joint is found between bodies of vertebrae (see Figure 8-8). A portion of the intervertebral disc is cartilaginous material. The symphysis pubis between the anterior surfaces of the coxal bones is another example (see Figure 7-7). These joints are slightly movable, or amphiarthrotic.

SYNOVIAL JOINTS

STRUCTURE

A joint in which there is a space between articulating bones is called a **synovial** (si-NŌ-ve-al) **joint.** The space, or **joint cavity,** is also called a **synovial cavity** (Figure 8-1). Because of this cavity and because of the arrangement of the articular capsule and accessory ligaments, synovial joints are freely movable. Thus, synovial joints are functionally classified as diarthrotic.

Synovial joints are also characterized by the presence of **articular cartilage.** Articular cartilage covers the surfaces of the articulating bones but does not bind the bones together. The articular cartilage of synovial joints is hyaline cartilage.

Synovial joints are surrounded by a sleevelike **articular capsule** that encloses the joint cavity and unites the articulating bones. The articular capsule is composed of two layers. The outer layer, the *fibrous capsule,* consists of dense connective (collagenous) tissue. It is attached to the periosteum of the articulating bones at a variable distance from the edge of the articular cartilage. The flexibility of the fibrous capsule permits movement at a joint, whereas its great tensile strength resists dislocation. The fibers of some fibrous capsules are arranged in parallel bundles and are therefore highly adapted to resist recurrent strain. Such fibers are called *ligaments* and are given special names. The strength of the ligaments is one of the principal factors in holding bone to bone.

The inner layer of the articular capsule is formed by a *synovial membrane.* The synovial membrane is composed of loose connective tissue with elastic fibers and a variable amount of adipose tissue. It secretes *synovial fluid,* which lubricates the joint and provides nourishment for the articular cartilage. Synovial fluid also contains phagocytic cells that remove microbes and debris resulting from wear and tear in the joint. Synovial fluid consists of hyaluronic acid and an interstitial fluid formed from blood plasma and is similar in appearance and consistency to egg white. When there is no joint movement, the fluid is viscous, but as movement increases, the fluid becomes less viscous. The amount of synovial fluid varies in different joints of the body, ranging from a thin, viscous layer to about 3.5 ml (about 0.1 oz) of free fluid in a large joint such as the knee. The amount present in each joint is sufficient only to form a thin film over the surfaces within an articular capsule.

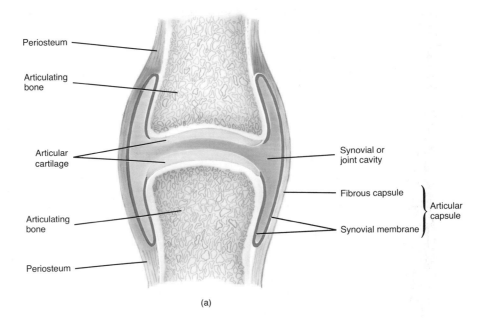

(a)

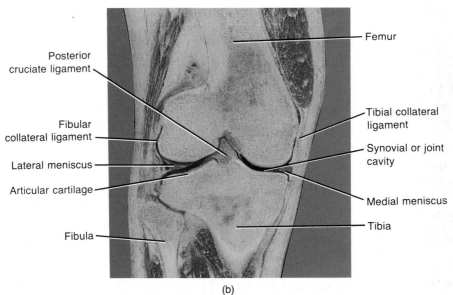

(b)

FIGURE 8-1 Synovial joint. (a) Diagram of a generalized synovial joint in frontal section. (b) Photograph of the internal structure of the right knee joint seen in frontal section. (Courtesy of C. Yokochi and J. W. Rohen, *Photographic Anatomy of the Human Body,* 2nd ed., 1979, IGAKU-SHOIN, Ltd., Tokyo, New York.)

Many synovial joints also contain **accessory ligaments,** which are called extracapsular ligaments and intracapsular ligaments. *Extracapsular ligaments* are outside of the articular capsule. An example is the fibular collateral ligament of the knee joint (see Figure 8-14). *Intracapsular ligaments* occur within the articular capsule but are excluded from the joint cavity by reflections of the synovial membrane. Examples are the cruciate ligaments of the knee joint (see Figure 8-14).

Inside some synovial joints, there are pads of fibrocarti-

lage that lie between the articular surfaces of the bones and are attached by their margins to the fibrous capsule. These pads are called **articular discs (menisci).** The discs usually subdivide the joint cavity into two separate spaces. Articular discs allow two bones of different shapes to fit tightly; they modify the shape of the joint surfaces of the articulating bones. Articular discs also help to maintain the stability of the joint and direct the flow of synovial fluid to areas of greatest friction.

CLINICAL APPLICATION

A tearing of articular discs in the knee, commonly called **torn cartilage,** occurs frequently among athletes. Such damaged cartilage requires surgical removal (meniscectomy) or it will begin to wear and cause arthritis. Until recently, knee joint surgery for torn cartilage necessitated cutting through layers of healthy tissue and removing much, if not all, of the cartilage. This procedure is usually painful and expensive and does not always provide full recovery.

These problems have been overcome by **arthroscopy (arthroscopic surgery).** The arthroscope is a device, several inches long, that resembles a pencil and is designed like a telescope with a series of lenses aligned one above the other. Its optic fibers give off light, like a lamp on a miner's hard hat. It is inserted into the knee joint through an incision as small as a quarter-inch. A second small incision is made to insert a tube through which a salt solution is injected into the joint. Another small incision is used for insertion of an instrument that shaves off and reshapes the damaged cartilage and then sucks out the shaved cartilage along with the salt solution. Some orthopedic surgeons attach a lightweight television camera to the arthroscope so that the image from inside the knee can be projected onto a screen. Since arthroscopy requires only small incision, recovery is usually speedy and there is very little pain or discomfort. Although arthroscopy is used mostly to remove torn cartilage, it can also be used for other types of knee surgery, for surgery on other joints of the body, to obtain tissue samples, diagnose certain pathologies, decide whether surgery is needed, and plan surgical procedures.

The various movements of the body create friction between moving parts. To reduce this friction, saclike structures called **bursae** are situated in the body tissues. These sacs resemble joints in that their walls consist of connective tissue lined by a synovial membrane. They are also filled with a fluid similar to synovial fluid. Bursae are located between the skin and bone in places where skin rubs over bone. They are also found between tendons and bones, muscles and bones, and ligaments and bones. As fluid-filled sacs, they cushion the movement of one part of the body over another. An inflammation of a bursa is called **bursitis.**

The articular surfaces of synovial joints are kept in contact with each other by several factors. One factor is the fit of the articulating bones. This interlocking is very obvious at the hip joint, where the head of the femur articulates with the acetabulum of the coxal bone. Another factor is the strength of the joint ligaments. This is especially important in the hip joint. A third factor is the tension of the muscles around the joint. For exam-

ple, the fibrous capsule of the knee joint is formed principally from tendinous expansions by muscles acting on the joint.

MOVEMENTS

The movements permitted at synovial joints are limited by several factors. One is the **apposition of soft parts.** For example, during bending of the elbow, the anterior surface of the forearm is pressed against the anterior surface of the arm. This apposition limits movement. A second factor is the **tension of ligaments.** The different components of a fibrous capsule are tense only when the joint is in certain positions. Tense ligaments not only restrict the range of movement, but also direct the movement of the articulating bones with respect to each other. In the knee joint, for example, the major ligaments are lax when the knee is bent but tense when the knee is straightened. Also, when the knee is straightened, the surfaces of the articulating bones are in fullest contact with each other. A third factor that restricts movement at a synovial joint is **muscle tension,** which reinforces the restraint placed on a joint by ligaments. A good example of the effect of muscle tension on a joint is seen at the hip joint. When the thigh is raised with the knee straight, the movement is restricted by the tension of the hamstring muscles on the posterior surface of the thigh. But if the knee is bent, the tension on the hamstring muscles is lessened and the thigh can be raised further.

Following is a description of the specific movements that occur at synovial joints.

Gliding

A **gliding movement** is the simplest kind that can occur at a joint. One surface moves back and forth and from side to side over another surface without angular or rotary motion. Some joints that glide are those between the carpals and between the tarsals. The heads and tubercles or ribs glide on the bodies and transverse processes of vertebrae.

Angular

Angular movements increase or decrease the angle between bones. Among the angular movements are flexion, extension, abduction, and adduction (Figure 8-2). **Flexion** usually involves a decrease in the angle between the anterior surfaces of the articulating bones. An exception to this definition is flexion of the knee and the toe joints, in which there is a decrease in the angle between the posterior surfaces of the articulating bones. Examples of flexion include bending the head forward (the joint is between the occipital bone and the atlas), bending the elbow, and bending the knee.

Extension involves an increase in the angle between the anterior surfaces of the articulating bones, with the

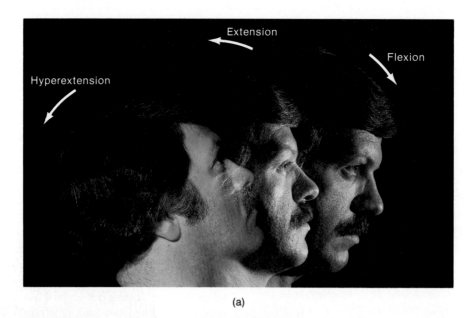

(a)

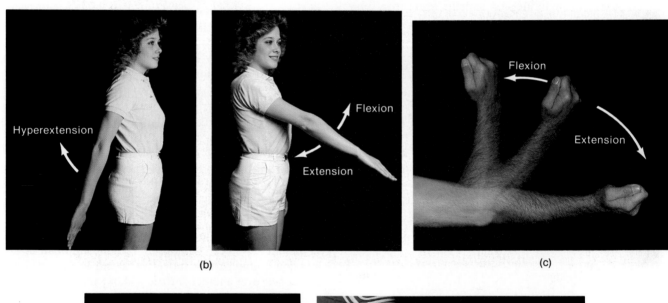

(b)

(c)

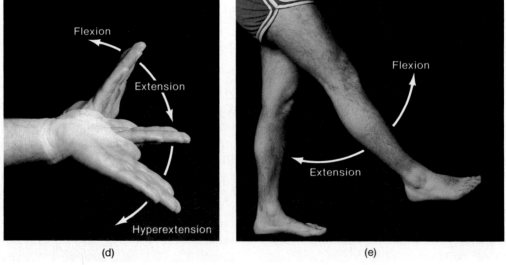

(d)

(e)

FIGURE 8-2 Angular movements at synovial joints. (© 1983 by Gerard J. Tortora. Courtesy of Lynne Tortora and James Borghesi.)

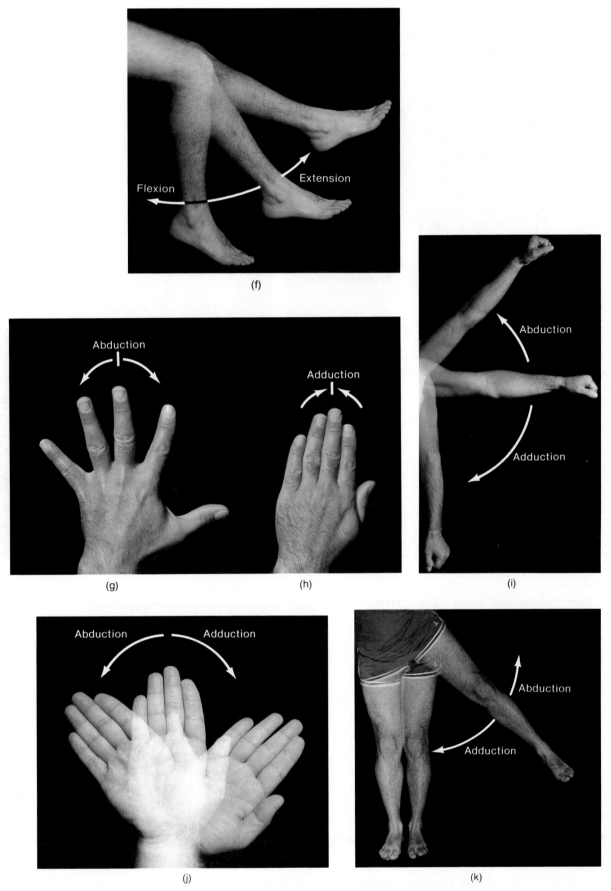

FIGURE 8-2 (*Continued*) Angular movements at synovial joints.

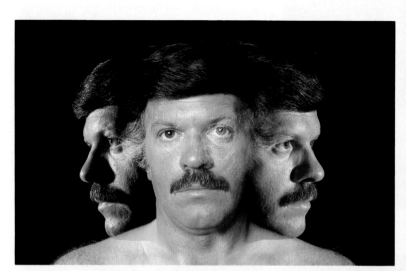

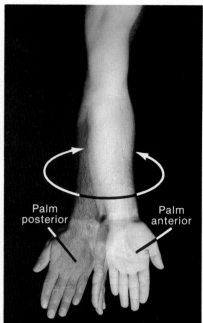

(a)

FIGURE 8-3 Rotation and circumduction. (a) Rotation at the atlantoaxial joint (left) and rotation of the humerus (right). (b) Circumduction of the humerus at the shoulder joint. (© 1983 by Gerard J. Tortora.)

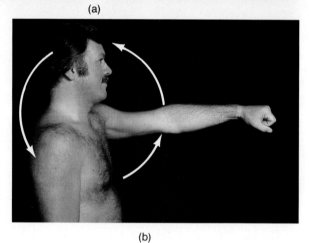

(b)

same exceptions of knee and toe joints. Extension restores a body part to its anatomical position after it has been flexed. Examples of extension are returning the head to the anatomical position after flexion, straightening the arm after flexion, and straightening the leg after flexion. Continuation of extension beyond the anatomical position, as in bending the head backward, is called **hyperextension.**

Abduction usually means movement of a bone *away from* the midline of the body. An example of abduction is moving the arm upward and away from the body until it is held straight out at right angles to the chest. With the fingers and toes, however, the midline of the body is not used as the line of reference. Abduction of the fingers is a movement away from an imaginary line drawn through the middle finger; in other words, it is spreading the fingers. Abduction of the toes is relative to an imaginary line drawn through the second toe.

Adduction is usually movement of a part *toward* the midline of the body. An example of adduction is returning the arm to the side after abduction. As in abduction, adduction of the fingers is relative to the middle finger, and adduction of the toes is relative to the second toe.

Rotation

Rotation is the movement of a bone around its own longitudinal axis. During rotation, no other motion is permitted. In *medial rotation,* the anterior surface of a bone or extremity moves toward the midline. In *lateral rotation,* the anterior surface moves away from the midline. We rotate the atlas around the odontoid process of the axis when we shake the head from side to side. Moving from the shoulder and turning the forearm up, then palm down, and then palm up again is an example of slight lateral and medial rotation of the humerus (Figure 8-3a).

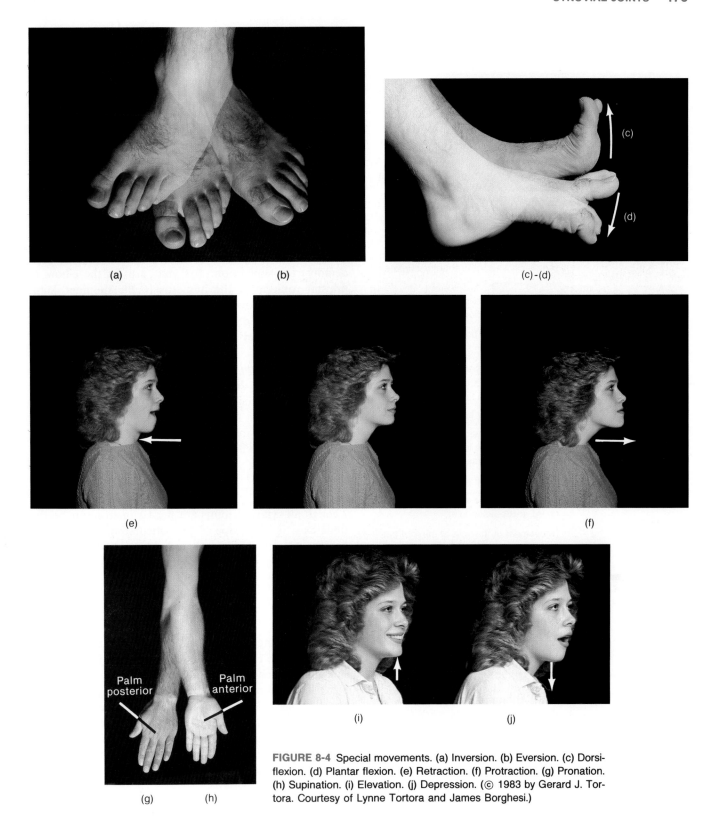

(a)   (b)   (c)-(d)

(e)   (f)

Palm
posterior   Palm
anterior

(g)   (h)   (i)   (j)

FIGURE 8-4 Special movements. (a) Inversion. (b) Eversion. (c) Dorsi-flexion. (d) Plantar flexion. (e) Retraction. (f) Protraction. (g) Pronation. (h) Supination. (i) Elevation. (j) Depression. (© 1983 by Gerard J. Tortora. Courtesy of Lynne Tortora and James Borghesi.)

Circumduction

Circumduction is a movement in which the distal end of a bone moves in a circle while the proximal end remains stable. The bone describes a cone in the air. Circumduction typically involves flexion, abduction, adduction, extension, and rotation. It involves a 360° rotation. An

example is moving the outstretched arm in a circle to wind up to pitch a ball (Figure 8-3b).

Special

Special movements are those found only at the joints indicated in Figure 8-4. **Inversion** is the movement of

EXHIBIT 8-1 SUMMARY OF MOVEMENTS AT SYNOVIAL JOINTS

MOVEMENT	DEFINITION
Gliding	One surface moves back and forth and from side to side over another surface without angular or rotary motion.
Angular	There is an increase or decrease at the angle between bones.
Flexion	Usually involves a decrease in the angle between the anterior surfaces of articulating bones.
Extension	Usually involves an increase in the angle between the anterior surfaces of articulating bones.
Hyperextension	Continuation of extension beyond the anatomical position.
Abduction	Movement of a bone away from the midline.
Adduction	Movement of a bone toward the midline.
Rotation	Movement of a bone around its longitudinal axis; may be medial or lateral.
Circumduction	A movement in which the distal end of a bone moves in a circle while the proximal end remains stable.
Special	Occur at specific joints.
Inversion	Movement of the sole of the foot inward at the ankle joint.
Eversion	Movement of the sole of the foot outward at the ankle joint.
Dorsiflexion	Flexion of the foot at the ankle joint.
Plantar flexion	Extension of the foot at the ankle joint.
Protraction	Movement of the mandible or clavicle forward on a plane parallel to the ground.
Retraction	Movement of a protracted part backward on a plane parallel to the ground.
Supination	Movement of the forearm in which the palm is turned anterior or superior.
Pronation	Movement of the flexed forearm in which the palm is turned posterior or inferior.
Elevation	Movement of a part of the body upward.
Depression	Movement of a part of the body downward.

the sole of the foot inward (medially) at the ankle joint. **Eversion** is the movement of the sole outward (laterally) at the ankle joint. Flexion of the foot at the ankle joint is called **dorsiflexion.** Extension of the foot at the ankle joint is **plantar flexion.**

Protraction is the movement of the mandible or clavicle forward on a plane parallel to the ground. Thrusting the jaw outward is protraction of the mandible. Bringing your arms forward until the elbows touch requires protraction of the clavicle. **Retraction** is the movement of a protracted part of the body backward on a plane parallel to the ground. Pulling the lower jaw back in line with the upper jaw is retraction of the mandible.

Supination is a movement of the forearm in which the palm of the hand is turned anterior or superior. To demonstrate supination, flex your arm at the elbow to prevent rotation of the humerus in the shoulder joint. **Pronation** is a movement of the flexed forearm in which the palm is turned posterior or inferior.

Elevation is an upward movement of a part of the body. You elevate your mandible when you close your mouth. **Depression** is a downward movement of a part of the body. You depress your mandible when you open your mouth. The shoulders can also be elevated and depressed.

A summary of movements that occur at synovial joints is presented in Exhibit 8-1.

TYPES

Though all synovial joints are similar in structure, variations exist in the shape of the articulating surfaces. Accordingly, synovial joints are divided into six subtypes: gliding, hinge, pivot, ellipsoidal, saddle, and ball-and-socket joints.

Gliding

The articulating surfaces of bones in **gliding joints,** or **arthrodia** (ar-THRŌ-dē-a), are usually flat. Only side-to-side and back-and-forth movements are permitted (Figure 8-5a). Twisting and rotation are inhibited at gliding joints generally because ligaments or adjacent bones restrict the range of movement. Since gliding joints do not move around an axis, they are referred to as *nonaxial*. Examples are the joints between carpal bones, tarsal bones, the sternum and clavicle, and the scapula and clavicle.

Hinge

A **hinge,** or **ginglymus** (JIN-gli-mus), **joint** is one in which the convex surface of one bone fits into the concave surface of another bone. Movement is primarily in a single plane, and the joint is therefore known as *monaxial,* or *uniaxial* (Figure 8-5b). The motion is similar to that of a hinged door. Movement is usually flexion and extension. Examples of hinge joints are the elbow, ankle, and interphalan-

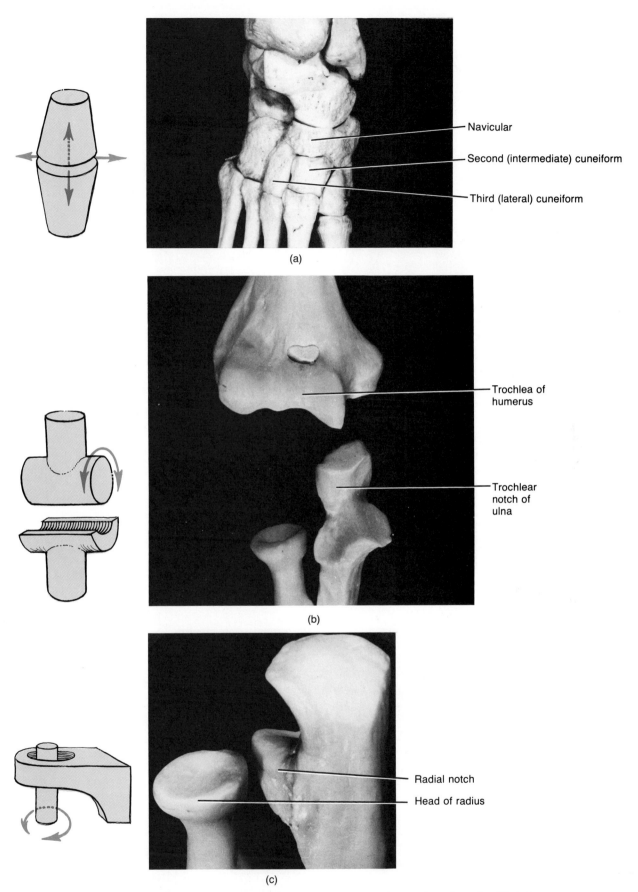

FIGURE 8-5 Subtypes of synovial joints. For each subtype shown, there is a simplified diagram and a photograph of the actual joint. (a) Gliding joint between the navicular and second and third cuneiforms of the tarsus. (b) Hinge joint at the elbow between the trochlea of the humerus and trochlear notch of the ulna. (c) Pivot joint between the head of the radius and the radial notch of the ulna. (© 1985 by Michael H. Ross. Used by permission.)

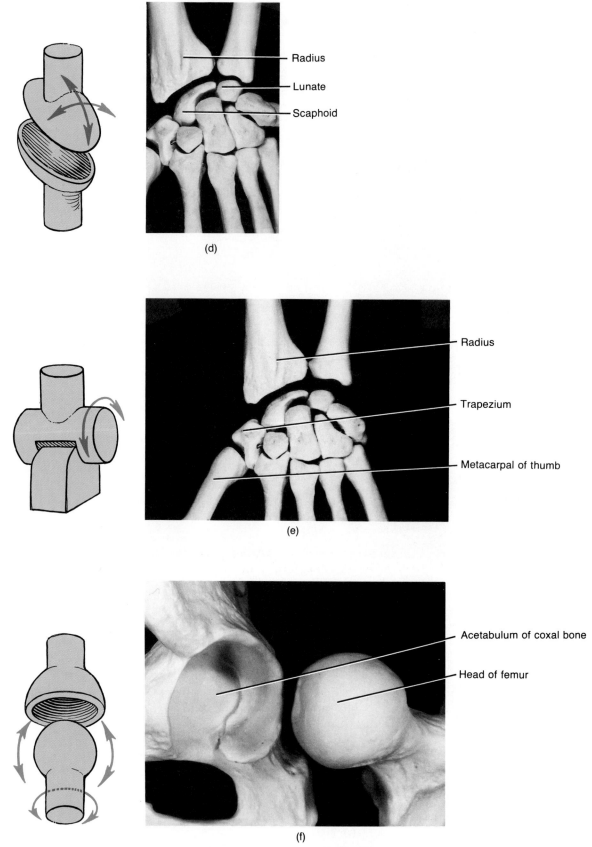

FIGURE 8-5 (*Continued*) Subtypes of synovial joints. (d) Ellipsoidal joint at the wrist between the distal end of the radius and the scaphoid and lunate bones of the carpus. (e) Saddle joint between the trapezium of the carpus and the metacarpal of the thumb. (f) Ball-and-socket joint between the head of the femur and the acetabulum of the coxal bone. (© 1985 by Michael H. Ross. Used by permission.)

geal joints. The movement allowed by a hinge joint is illustrated by flexion and extension at the elbow (see Figure 8-2c).

Pivot

In a **pivot,** or **trochoid** (TRŌ-koyd), **joint,** a rounded, pointed, or conical surface of one bone articulates within a ring formed partly by bone and partly by a ligament. The primary movement permitted is rotation, and the joint is therefore *monaxial* (Figure 8-5c). Examples include the joint between the atlas and axis (atlantoaxial) and between the proximal ends of the radius and ulna. Movement at a pivot joint is illustrated by supination and pronation of the palms and rotation of the head from side to side (see Figure 8-3a).

Ellipsoidal

In an **ellipsoidal,** or **condyloid** (KON-di-loyd), **joint,** an oval-shaped condyle of one bone fits into an elliptical cavity of another bone. Since the joint permits side-to-side and back-and-forth movements, it is *biaxial* (Figure 8-5d). The joint at the wrist between the radius and carpals is ellipsoidal. The movement permitted by such a joint is illustrated when you flex and extend and abduct and adduct the wrist (see Figure 8-2d, k).

Saddle

In a **saddle,** or **sellaris** (sel-A-ris), **joint,** the articular surfaces of both bones are saddle-shaped, that is, concave in one direction and convex in the other. Essentially, the saddle joint is a modified ellipsoidal joint in which the movement is somewhat freer. Movements at a saddle joint are side to side and back and forth. Thus the joint is *biaxial* (Figure 8-5e). The joint between the trapezium and metacarpal of the thumb is an example of a saddle joint.

Ball-and-Socket

A **ball-and-socket,** or **spheroid** (SFĒ-royd), **joint** consists of a ball-like surface of one bone fitted into a cuplike depression of another bone. Such a joint permits *triaxial* movement, or movement in three planes of motion: flexion-extension, abduction-adduction, and rotation (Figure 8-5f). Examples of ball-and-socket joints are the shoulder joint and hip joint. The range of movement at a ball-and-socket joint is illustrated by circumduction of the arm (Figure 8-3b).

SUMMARY OF JOINTS

The summary of joints presented in Exhibit 8-2 is based on the anatomy of the joints. If we rearrange the types of joints into a classification based on movement, we arrive at the following:

Synarthroses: immovable joints
1. **Suture**
2. **Synchondrosis**
3. **Gomphosis**

Amphiarthroses: slightly movable joints
1. **Syndesmosis**
2. **Symphysis**

Diarthroses: freely movable joints
1. **Gliding**
2. **Hinge**
3. **Pivot**
4. **Ellipsoidal**
5. **Saddle**
6. **Ball-and-socket**

SELECTED ARTICULATIONS OF THE BODY

We will now examine in some detail selected articulations of the body. In order to simplify your learning efforts, a series of exhibits has been prepared. Each exhibit considers a specific articulation and contains (1) a definition, that is, a description of the bones that form the joint; (2) the type of joint (its structural classification); and (3) its anatomical components—a description of the major connecting ligaments, articular disc, articular capsule, and other distinguishing features of the joint. Each exhibit also refers you to an illustration of the joint.

EXHIBIT 8-2   JOINTS

TYPE	DESCRIPTION	MOVEMENT	EXAMPLES
FIBROUS	No joint cavity; bones held together by a thin layer of fibrous tissue or dense fibrous tissue.		
Suture	Found only between bones of the skull; articulating bones separated by a thin layer of fibrous tissue.	None (synarthrotic).	Lambdoidal suture between occipital and parietal bones.
Syndesmosis	Articulating bones united by dense fibrous tissue.	Slight (amphiarthrotic).	Distal ends of tibia and fibula.
Gomphosis	Cone-shaped peg fits into a socket; articulating bones separated by periodontal ligament.	None (synarthrotic).	Roots of teeth in alveoli (sockets).
CARTILAGINOUS	No joint cavity; articulating bones united by cartilage.		
Synchondrosis	Connecting material is hyaline cartilage.	None (synarthrotic).	Temporary joint between the diaphysis and epiphyses of a long bone and permanent joint between first rib and sternum.
Symphysis	Connecting material is a broad, flat disc of fibrocartilage.	Slight (amphiarthrotic).	Intervertebral joints and symphysis pubis.
SYNOVIAL	Joint cavity and articular cartilage present; articular capsule composed of an outer fibrous capsule and an inner synovial membrane; may contain accessory ligaments, articular discs (menisci), and bursae.	Freely movable (diarthrotic).	
Gliding	Articulating surfaces usually flat.	Nonaxial.	Intercarpal and intertarsal joints.
Hinge	Spool-like surface fits into a concave surface.	Monaxial (flexion-extension).	Elbow, ankle, and interphalangeal joints.
Pivot	Rounded, pointed, or concave surface fits into a ring formed partly by bone and partly by a ligament.	Monaxial (rotation).	Atlantoaxial and radioulnar joints.
Ellipsoidal	Oval-shaped condyle fits into an elliptical cavity.	Biaxial (flexion-extension, abduction-adduction).	Radiocarpal joint.
Saddle	Articular surfaces concave in one direction and convex in opposite direction.	Biaxial (flexion-extension, abduction-adduction).	Carpometacarpal joint of thumb.
Ball-and-socket	Ball-like surface fits into a cuplike depression.	Triaxial (flexion-extension, abduction-adduction, rotation).	Shoulder and hip joints.

EXHIBIT 8-3   TEMPOROMANDIBULAR (TM) JOINT (Figure 8-6)

DEFINITION	Joint formed by mandibular condyle of the mandible and mandibular fossa and articular tubercle of temporal bone. The temporomandibular (TM) joint is the only movable joint between skull bones; all other skull joints are sutures and therefore immovable.
TYPE OF JOINT	Synovial; combined hinge (ginglymus) and gliding (arthrodial) type.
ANATOMICAL COMPONENTS	1. **Articular disc (meniscus).** Fibrocartilage disc that separates the joint cavity into a superior and inferior compartment, each of which has a synovial membrane. 2. **Articular capsule.*** Thin, fairly loose envelope around the circumference of the joint. 3. **Lateral ligament.** Two short bands on the lateral surface of the articular capsule that extend inferiorly and posteriorly from the lower border and tubercle of the zygomatic arch to the lateral and posterior aspect of the neck of the mandible. It is covered by the parotid gland and helps prevent displacement of the mandible. 4. **Sphenomandibular ligament.** Thin band that extends inferiorly and anteriorly from the spine of the sphenoid bone to the ramus of the mandible. 5. **Stylomandibular ligament.** Thickened band of deep cervical fascia that extends from the styloid process of the temporal bone to the inferior and posterior border of the ramus of the mandible. The ligament separates the parotid from the submandibular gland.

* Also referred to as the **capsular ligament.**

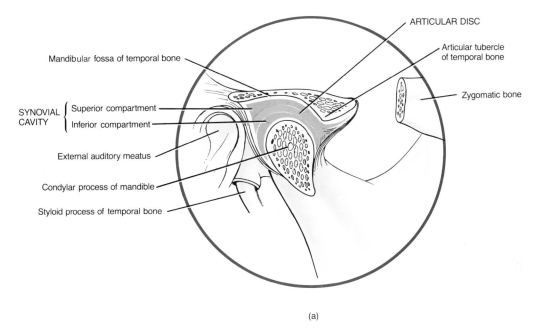

(a)

FIGURE 8-6 Temporomandibular joint. (a) Sagittal section through the temporomandibular joint.

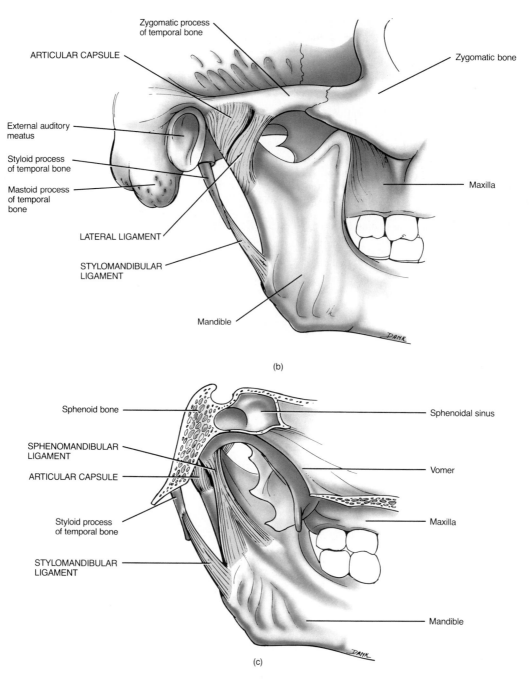

(b)

(c)

FIGURE 8-6 (*Continued*) Temporomandibular joint. (b) Right lateral view. (c) Median view.

EXHIBIT 8-4 ATLANTOOCCIPITAL JOINTS (Figure 8-7)

DEFINITION	Joints formed by the superior articular surfaces of the atlas and the occipital condyles of the occipital bone.
TYPE OF JOINT	Synovial; ellipsoidal (condyloid) type.
ANATOMICAL COMPONENTS	1. **Articular capsules.** Thin, loose envelopes that surround the occipital condyles and attach them to the articular processes of the atlas.
	2. **Anterior atlantooccipital ligament.** Broad, thick ligament that extends from the anterior surface of the foramen magnum of the occipital bone to the anterior arch of the atlas.
	3. **Posterior atlantooccipital ligament.** Broad, thin ligament that extends from the posterior surface of the foramen magnum of the occipital bone to the posterior arch of the atlas. The inferior portion of the ligament is sometimes ossified.
	4. **Lateral atlantooccipital ligaments.** Thickened portions of the articular capsules that also contain fibrous tissue bundles and extend from the jugular process of the occipital bone to the transverse process of the atlas.

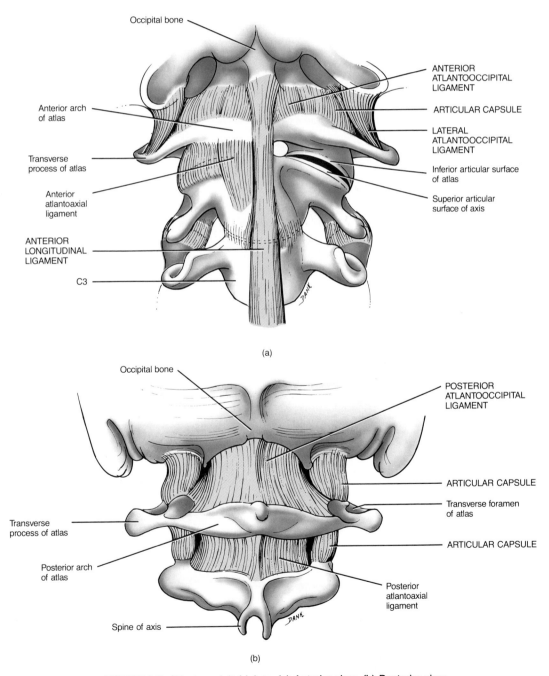

(a)

(b)

FIGURE 8-7 Atlantooccipital joints. (a) Anterior view. (b) Posterior view.

EXHIBIT 8-5 INTERVERTEBRAL JOINTS (Figure 8-8)

DEFINITION

Joints formed between (1) vertebral bodies and (2) vertebral arches.

TYPE OF JOINT

Joints between vertebral bodies—cartilaginous, symphysis type. Joints between vertebral arches—synovial, gliding (arthrodial) type.

ANATOMICAL COMPONENTS OF JOINTS BETWEEN VERTEBRAL BODIES

1. **Anterior longitudinal ligament.** Broad, strong band that extends along the anterior surfaces of the vertebral bodies from the axis to the sacrum. It is firmly attached to the intervertebral discs.
2. **Posterior longitudinal ligament.** Extends along the posterior surfaces of the vertebral bodies within the vertebral canal from the axis to the sacrum. The free surface of the ligament is separated from the spinal dura mater by loose (areolar) tissue.
3. **Intervertebral discs.** Unite with adjacent surfaces of the vertebral bodies from the axis to the sacrum. Each disc is composed of a peripheral **anulus fibrosus** consisting of fibrous tissue and fibrocartilage and a central **nucleus pulposus** composed of a soft, pulpy, highly elastic substance.

ANATOMICAL COMPONENTS OF JOINTS BETWEEN VERTEBRAL ARCHES

1. **Articular capsules.** Thin, loose ligaments attached to the margins of the articular processes of adjacent vertebrae.
2. **Ligamenta flava.** Contain elastic tissue and connect the laminae of adjacent vertebrae from the axis to the first segment of the sacrum.
3. **Supraspinous ligament.** Strong fibrous cord that connects the spinous processes from the seventh cervical vertebra to the sacrum.
4. **Ligamentum nuchae.** Fibrous ligament that represents an enlargement of the supraspinous ligament in the neck. It extends from the external occipital protuberance of the occipital bone to the spinous process of the seventh cervical vertebra.
5. **Interspinous ligaments.** Relatively weak bands that run between adjacent spinous processes.
6. **Intertransverse ligaments.** Bands between transverse processes that are readily apparent in the lumbar region.

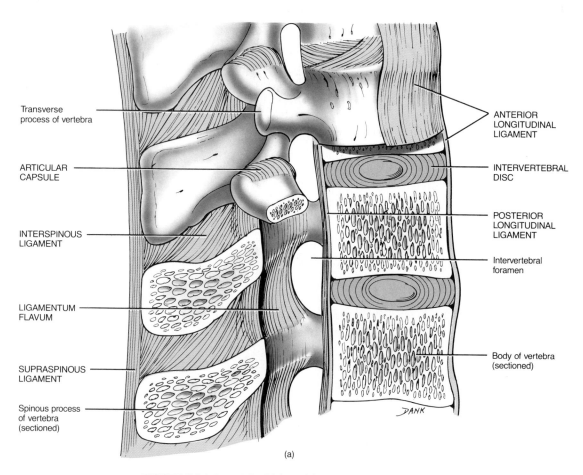

(a)

FIGURE 8-8 Intervertebral joints. (a) Diagram of median view.

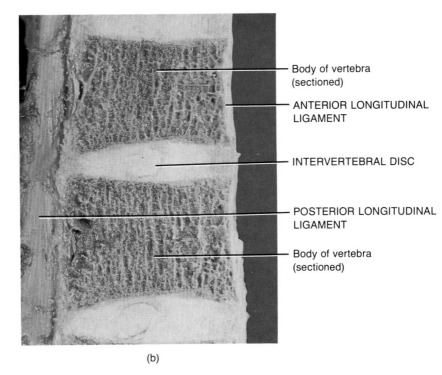

Body of vertebra (sectioned)

ANTERIOR LONGITUDINAL LIGAMENT

INTERVERTEBRAL DISC

POSTERIOR LONGITUDINAL LIGAMENT

Body of vertebra (sectioned)

(b)

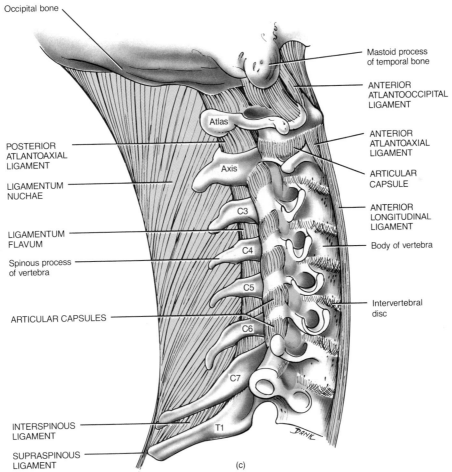

Occipital bone

Mastoid process of temporal bone

ANTERIOR ATLANTOOCCIPITAL LIGAMENT

Atlas

ANTERIOR ATLANTOAXIAL LIGAMENT

POSTERIOR ATLANTOAXIAL LIGAMENT

Axis

ARTICULAR CAPSULE

LIGAMENTUM NUCHAE

ANTERIOR LONGITUDINAL LIGAMENT

C3

LIGAMENTUM FLAVUM

C4

Body of vertebra

Spinous process of vertebra

C5

ARTICULAR CAPSULES

C6

Intervertebral disc

C7

INTERSPINOUS LIGAMENT

T1

SUPRASPINOUS LIGAMENT

(c)

FIGURE 8-8 (Continued) Intervertebral joints. (b) Photograph of an intervertebral disc between two adjacent vertebral bodies. (Courtesy of C. Yokochi and J. W. Rohen, *Photographic Anatomy of the Human Body,* 2nd ed., 1979, IGAKU-SHOIN, Ltd., Tokoyo, New York.) (c) Diagram of right lateral view.

EXHIBIT 8-6   LUMBOSACRAL JOINT (Figure 8-9)

DEFINITION	Joint formed by the body of the fifth lumbar vertebra and the superior surface of the first sacral vertebra of the sacrum.
TYPE OF JOINT	Joint between the bodies of the fifth lumbar vertebra and the first sacral vertebra—cartilaginous joint, symphysis type. Joint between the articular processes—synovial joint, gliding (arthrodial) type.
ANATOMICAL COMPONENTS	1. The lumbosacral joint is similar to the joints between typical vertebrae, united by an **intervertebral disc, anterior** and **posterior longitudinal ligaments, ligamenta flava, interspinous** and **supraspinous ligaments,** and **articular capsules** between articular processes. The lumbosacral joint also contains an iliolumbar ligament.
	2. **Iliolumbar ligament.** Strong ligament that connects the transverse process of the fifth lumbar vertebra with the base of the sacrum and the iliac crest.

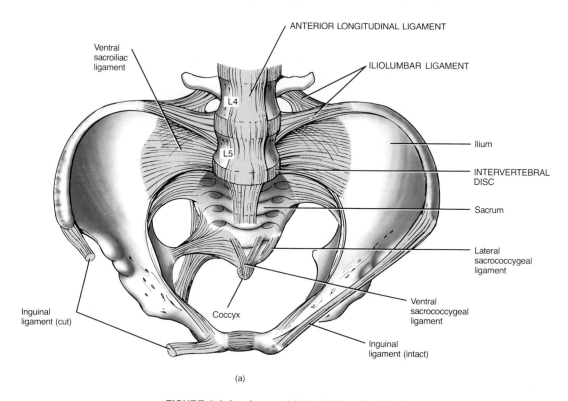

(a)

FIGURE 8-9 Lumbosacral joint. (a) Anterior view.

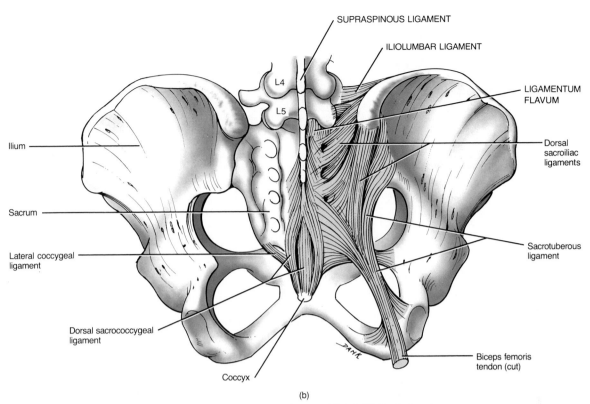

SUPRASPINOUS LIGAMENT

ILIOLUMBAR LIGAMENT

LIGAMENTUM FLAVUM

L4

L5

Ilium

Dorsal sacroiliac ligaments

Sacrum

Lateral coccygeal ligament

Sacrotuberous ligament

Dorsal sacrococcygeal ligament

Coccyx

Biceps femoris tendon (cut)

(b)

FIGURE 8-9 (*Continued*) Lumbosacral joint. (b) Posterior view.

EXHIBIT 8-7   HUMEROSCAPULAR (SHOULDER) JOINT (Figure 8-10)

DEFINITION	Joint formed by the head of the humerus and the glenoid cavity of the scapula.
TYPE OF JOINT	Synovial joint, ball-and-socket (spheroid) type.
ANATOMICAL COMPONENTS	1. **Articular capsule.** Loose sac that completely envelops the joint, extending from the circumference of the glenoid cavity to the anatomical neck of the humerus.

1. **Articular capsule.** Loose sac that completely envelops the joint, extending from the circumference of the glenoid cavity to the anatomical neck of the humerus.
2. **Coracohumeral ligament.** Strong, broad ligament that extends from the coracoid process of the scapula to the greater tubercle of the humerus.
3. **Glenohumeral ligaments.** Three thickenings of the articular capsule over the ventral surface of the joint.
4. **Transverse humeral ligament.** Narrow sheet extending from the greater tubercle to the lesser tubercle of the humerus.
5. **Glenoid labrum.** Narrow rim of fibrocartilage around the edge of the glenoid cavity.
6. Among the bursae associated with the shoulder joint are:
   a. **Subscapular bursa** between the tendon of the subscapularis muscle and the underlying joint capsule.
   b. **Subdeltoid bursa** between the deltoid muscle and joint capsule.
   c. **Subacromial bursa** between the acromion and joint capsule.
   d. **Subcoracoid bursa** either lies between the coracoid process and joint capsule or appears as an extension from the subacromial bursa.

CLINICAL APPLICATION

The strength and stability of the shoulder joint are not provided by the shape of the articulating bones or its ligaments. Instead, four deep muscles of the shoulder and their tendons—subscapularis, supraspinatus, infraspinatus, and teres minor—strengthen and stabilize the shoulder joint (see Figure 10-16). The muscles and their tendons are so arranged as to form a nearly complete encirclement of the joint. This arrangement is referred to as the **rotator (musculotendinous) cuff** and is a common site of injury to baseball pitchers, especially tearing of the supraspinatus muscle.

Dislocations of the shoulder joint are based on the relationship of the displaced head of the humerus to the glenoid cavity. Anterior displacement of the humeral head accounts for almost all shoulder dislocations.

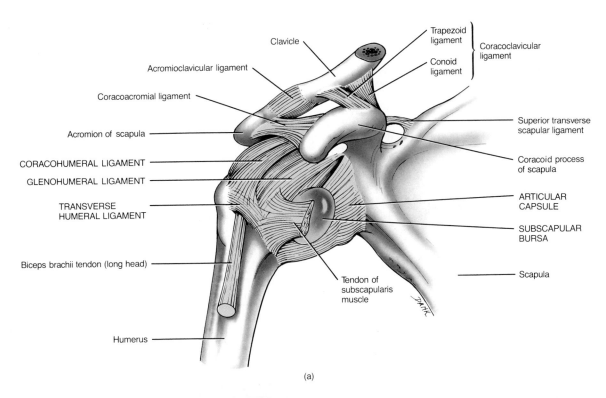

(a)

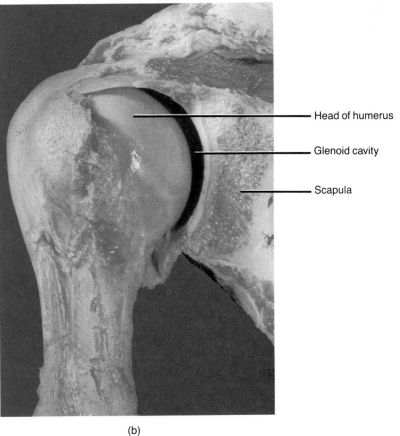

(b)

FIGURE 8-10 Humeroscapular joint. (a) Diagram of anterior view. (b) Photograph of anterior view. (Courtesy of C. Yokochi and J. W. Rohen, *Photographic Anatomy of the Human Body,* 2nd ed., 1979, IGAKU-SHOIN, Ltd., Tokyo, New York.)

EXHIBIT 8-8 ELBOW JOINT (Figure 8-11)

DEFINITION	Joint formed by the trochlea of the humerus, the trochlear notch of the ulna, the capitulum of the humerus, and the head of the radius.
TYPE OF JOINT	Synovial joint, hinge (ginglymus) type.
ANATOMICAL COMPONENTS	1. **Articular capsule.** The anterior part covers the anterior part of the joint from the radial and coronoid fossae of the humerus to the coronoid process of the ulna and the annular ligament of the radius. The posterior part extends from the capitulum, olecranon fossa, and lateral epicondyle of the humerus to the annular ligament of the radius, the olecranon of the ulna, and the ulna posterior to the radial notch. 2. **Ulnar collateral ligament.** Thick, triangular ligament that extends from the medial epicondyle of the humerus to the coronoid process and olecranon of the ulna. 3. **Radial collateral ligament.** Strong, triangular ligament that extends from the lateral epicondyle of the humerus to the annular ligament of the radius and the radial notch of the ulna.

CLINICAL APPLICATION

Tennis elbow most commonly refers to pain at or near the lateral epicondyle of the humerus, usually caused by an improperly executed backhand. In this maneuver, the extensor muscles, especially the extensor carpi radialis brevis muscle, strain or sprain. Also, the extensors rub and roll over the lateral epicondyle and head of the radius, resulting in pain.

Little-league elbow typically develops as a result of a heavy pitching schedule, especially among youngsters. The disorder occurs mostly on the medial side of the elbow, in which there is some degree of involvement of the medial epicondylar epiphysis that may cause it to enlarge, fragment, or separate. Disorders on the lateral side of the elbow are more severe and chronic, and are caused by the repeated jamming of the radial head against the capitulum of the humerus, which results in necrosis of the head of the radius, osteochondritis, and osteoarthritis.

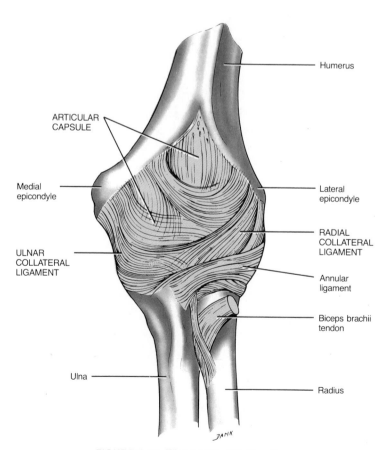

FIGURE 8-11 Elbow joint. Anterior view.

EXHIBIT 8-9 RADIOCARPAL (WRIST) JOINT (Figure 8-12)

DEFINITION	Joint formed by the distal end of the radius, the distal surface of the articular disc separating the carpal and distal radioulnar joint, and the scaphoid, lunate, and triquetral carpal bones.
TYPE OF JOINT	Synovial joint, ellipsoidal (condyloid) type.
ANATOMICAL COMPONENTS	1. **Articular capsule.** Extends from the styloid processes of the ulna and radius to the proximal row of carpal bones.
	2. **Palmar radiocarpal ligament.** Thick, strong ligament that extends from the anterior border of the distal end of the radius to the proximal row of carpals. Some longer fibers extend to the capitate in the distal row.
	3. **Dorsal radiocarpal ligament.** Extends from the posterior border of the distal end of the radius to the proximal row of carpals, especially the triquetral.
	4. **Ulnar collateral ligament.** Extends from the styloid process of the ulna to the triquetral and pisiform bones and the transverse carpal ligament.
	5. **Radial collateral ligament.** Extends from the styloid process of the radius to the scaphoid and pisiform bones.

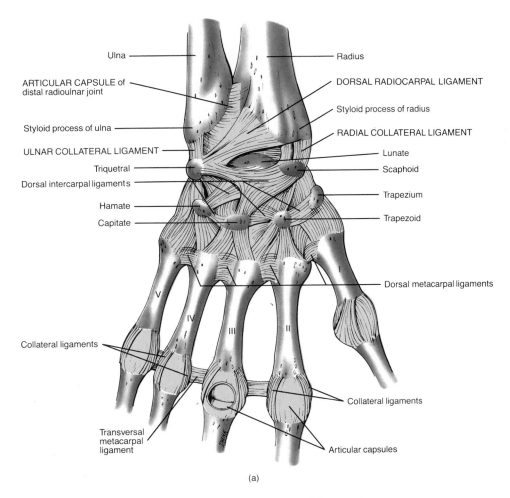

(a)

FIGURE 8-12 Radiocarpal joint. (a) Dorsal view.

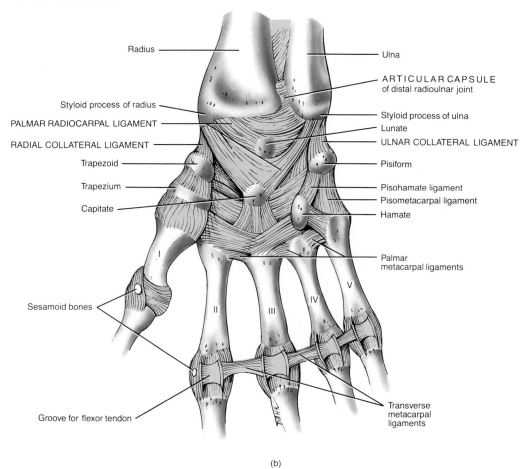

(b)

FIGURE 8-12 (*Continued*) Radiocarpal joint. (b) Palmar view.

EXHIBIT 8-10   COXAL (HIP) JOINT (Figure 8-13)

DEFINITION	Joint formed by the head of the femur and the acetabulum of the coxal bone.
TYPE OF JOINT	Synovial, ball-and-socket (spheroid) type.
ANATOMICAL COMPONENTS	1. **Articular capsule.** Extends from the rim of the acetabulum to the neck of the femur. One of the strongest ligaments of the body, the capsule consists of circular and longitudinal fibers. The circular fibers, called the **zona orbicularis,** form a collar around the neck of the femur. The longitudinal fibers are reinforced by accessory ligaments known as the iliofemoral ligament, pubofemoral ligament, and ischiofemoral ligament.
	2. **Iliofemoral ligament.** Thickened portion of the articular capsule that extends from the anterior inferior iliac spine of the coxal bone to the intertrochanteric line of the femur.
	3. **Pubofemoral ligament.** Thickened portion of the articular capsule that extends from the pubic part of the rim of the acetabulum to the neck of the femur.
	4. **Ischiofemoral ligament.** Thickened portion of the articular capsule that extends from the ischial wall of the acetabulum to the neck of the femur.
	5. **Ligament of the head of the femur (capitate ligament).** Flat, triangular band that extends from the fossa of the acetabulum to the head of the femur.
	6. **Acetabular labrum.** Fibrocartilage rim attached to the margin of the acetabulum.
	7. **Transverse ligament of the acetabulum.** Strong ligament that crosses over the acetabular notch, converting it to a foramen. It supports part of the acetabular labrum and is connected with the ligament of the head of the femur and the articular capsule.

CLINICAL APPLICATION

Dislocation of the hip among adults is quite rare because of (1) the stability of the ball-and-socket joint, (2) the strong, tough, articular capsule, (3) the strength of the intracapsular ligaments, and (4) the extensive musculature over the joint.

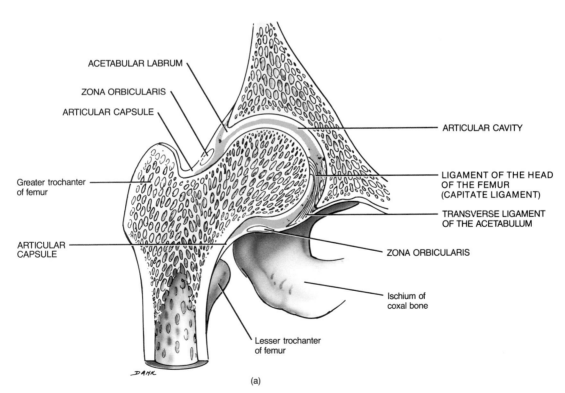

FIGURE 8-13 Coxal (hip) joint. (a) Frontal section.

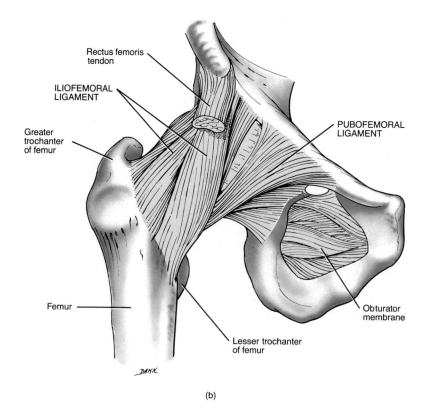

(b)

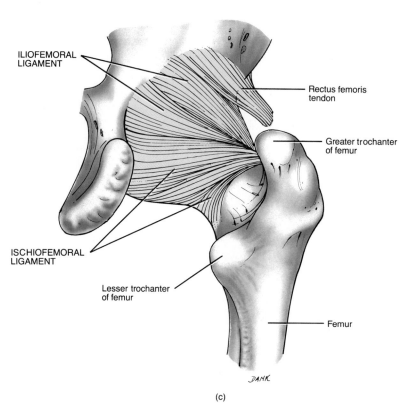

(c)

FIGURE 8-13 (*Continued*) Coxal (hip) joint. (b) Anterior view. (c) Posterior view.

EXHIBIT 8-11   TIBIOFEMORAL (KNEE) JOINT (Figure 8-14)

DEFINITION	The largest joint of the body, actually consisting of three joints: (1) an intermediate patellofemoral joint between the patella and the patellar surface of the femur, (2) a lateral tibiofemoral joint between the lateral condyle of the femur, lateral meniscus, and lateral condyle of the tibia, and (3) a medial tibiofemoral joint between the medial condyle of the femur, medial meniscus, and medial condyle of the tibia.
TYPE OF JOINT	Patellofemoral joint—partly synovial, gliding (arthrodial) type. Lateral and medial tibiofemoral joints—synovial, hinge (ginglymus) type.

ANATOMICAL COMPONENTS

1. **Articular capsule.** No complete, independent capsule unites the bones. The ligamentous sheath surrounding the joint consists mostly of muscle tendons or expansions of them. There are, however, some capsular fibers connecting the articulating bones.
2. **Medial and lateral patellar retinacula.** Fused tendons of insertion of the quadriceps femoris muscle and the fascia lata that strengthen the anterior surface of the joint.
3. **Patellar ligament.** Central portion of the common tendon of insertion of the quadriceps femoris muscle that extends from the patella to the tibial tuberosity. This ligament also strengthens the anterior surface of the joint. The posterior surface of the ligament is separated from the synovial membrane of the joint by an **infrapatellar fat pad.**
4. **Oblique popliteal ligament.** Broad, flat ligament that connects the intercondylar fossa of the femur to the head of the tibia. The tendon of the semimembranous muscle is superficial to the ligament and passes from the medial condyle of the tibia to the lateral condyle of the femur. The ligament and tendon afford strength for the posterior surface of the joint.
5. **Arcuate popliteal ligament.** Extends from the lateral condyle of the femur to the styloid process of the head of the fibula. It strengthens the lower lateral part of the posterior surface of the joint.
6. **Tibial collateral ligament.** Broad, flat ligament on the medial surface of the joint that extends from the medial condyle of the femur to the medial condyle of the tibia. The ligament is crossed by tendons of the sartorius, gracilis, and semitendinosus muscles, all of which strengthen the medial aspect of the joint.
7. **Fibular collateral ligament.** Strong, rounded ligament on the lateral surface of the joint that extends from the lateral condyle of the femur to the lateral side of the head of the fibula. The ligament is covered by the tendon of the biceps femoris muscle. The tendon of the popliteal muscle is deep to the tendon.
8. **Intraarticular ligaments.** Ligaments within the capsule that connect the tibia and femur.
   a. **Anterior cruciate ligament.** Extends posteriorly and laterally from the area anterior to the intercondylar eminence of the tibia to the posterior part of the medial surface of the lateral condyle of the femur. This ligament is stretched or torn in about 70 percent of all serious knee injuries. Both O. J. Simpson and Gale Sayers had their careers ended by torn anterior cruciate ligaments.
   b. **Posterior cruciate ligament.** Extends anteriorly and medially from the posterior intercondylar fossa of the tibia and lateral meniscus to the anterior part of the medial surface of the medial condyle of the femur.
9. **Articular discs.** Fibrocartilage discs between the tibial and femoral condyles. They help to compensate for the incongruence of the articulating bones.
   a. **Medial meniscus.** Semicircular piece of fibrocartilage. Its anterior end is attached to the anterior intercondylar fossa of the tibia, in front of the anterior cruciate ligament. Its posterior end is attached to the posterior intercondylar fossa of the tibia between the attachments of the posterior cruciate ligament and lateral meniscus.
   b. **Lateral meniscus.** Circular piece of fibrocartilage. Its anterior end is attached anterior to the intercondylar eminence of the tibia and lateral and posterior to the anterior cruciate ligament. Its posterior end is attached posterior to the intercondylar eminence of the tibia and anterior to the posterior end of the medial meniscus. The medial and lateral menisci are connected to each other by the **transverse ligament** and to the margins of the head of the tibia by the **coronary ligaments.**
10. The principal **bursae** of the knee include:
    a. **Anterior bursae:** (1) between the patella and skin **(prepatellar bursa),** (2) between upper part of tibia and patellar ligament **(infrapatellar bursa),** (3) between lower part of tibial tuberosity and skin, and (4) between lower part of femur and deep surface of quadriceps femoris muscle **(suprapatellar bursa).**
    b. **Medial bursae:** (1) between medial head of gastrocnemius muscle and the articular capsule, (2) superficial to the tibial collateral ligament between the ligament and tendons of the sartorius, gracilis, and semitendinosus muscles, (3) deep to the tibial collateral ligament between the ligament and the tendon of the semimembranosus muscle, (4) between the tendon of the semimembranosus muscle and the head of the tibia, and (5) between the tendons of the semimembranosus and semitendinosus muscles.
    c. **Lateral bursae:** (1) between the lateral head of the gastrocnemius muscle and articular capsule, (2) between the tendon of the biceps femoris muscle and fibular collateral ligament, (3) between the tendon of the popliteal muscle and fibular collateral ligament, and (4) between the lateral condyle of the femur and the popliteal muscle.

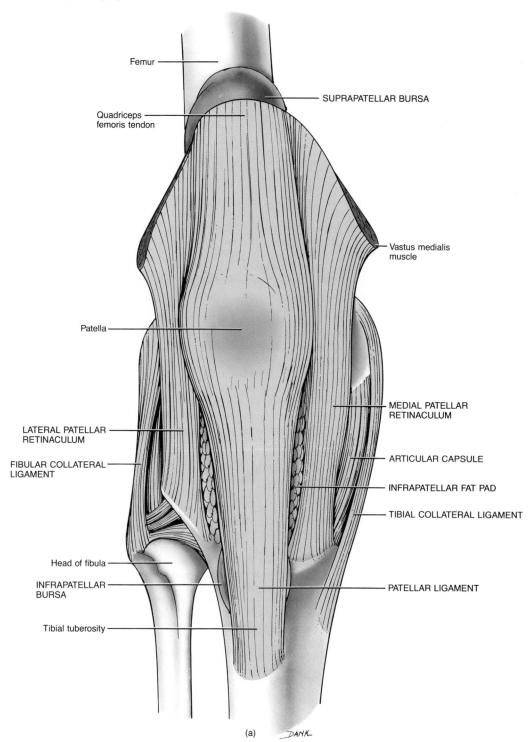

Femur

SUPRAPATELLAR BURSA

Quadriceps
femoris tendon

Vastus medialis
muscle

Patella

MEDIAL PATELLAR
RETINACULUM

LATERAL PATELLAR
RETINACULUM

ARTICULAR CAPSULE

FIBULAR COLLATERAL
LIGAMENT

INFRAPATELLAR FAT PAD

TIBIAL COLLATERAL LIGAMENT

Head of fibula

INFRAPATELLAR
BURSA

PATELLAR LIGAMENT

Tibial tuberosity

(a)   DANK

FIGURE 8-14 Tibiofemoral (knee) joint. (a) Diagram of anterior view.

EXHIBIT 8-11 (*Continued*)

CLINICAL APPLICATION

The most common type of **knee injury** in football is rupture of the tibial collateral ligament, often associated with tearing of the anterior cruciate ligament and medial meniscus (torn cartilage). It is caused by a blow to the lateral side of the knee. When a knee is examined for such an injury, the three C's are kept in mind: collateral ligament, cruciate ligament, and cartilage.

Runner's knee refers to pain around the front of a runner's knee that may result from excessive pronation of the foot or tight hamstrings.

A **swollen knee** may occur immediately or be delayed.

The immediate swelling is due to escape of blood from rupture of the anterior cruciate ligament, torn menisci, fractures, or collateral ligament sprains. Delayed swelling is due to an excessive production of synovial fluid as a result of conditions that irritate the synovial membrane.

A **dislocated knee** refers to the displacement of the tibia relative to the femur. Accordingly, such dislocations are classified as anterior, posterior, medial, lateral, or rotatory. The most common type is anterior dislocation, resulting from hyperextension of the knee. A frequent consequence of a dislocated knee is damage to the popliteal artery.

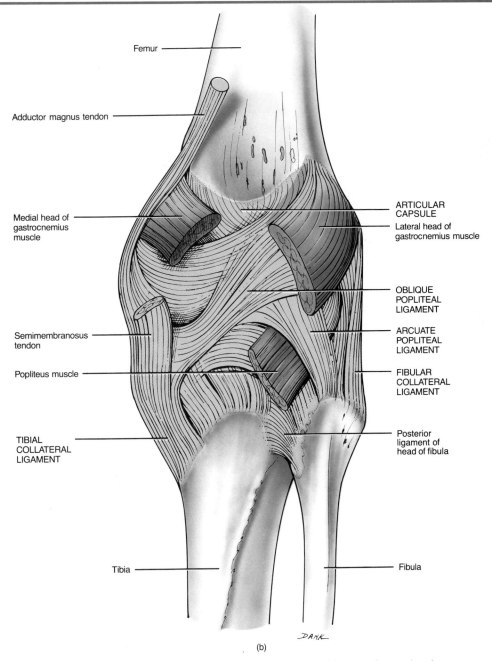

(b)

FIGURE 8-14 (*Continued*) Tibiofemoral (knee) joint. (b) Diagram of posterior view.

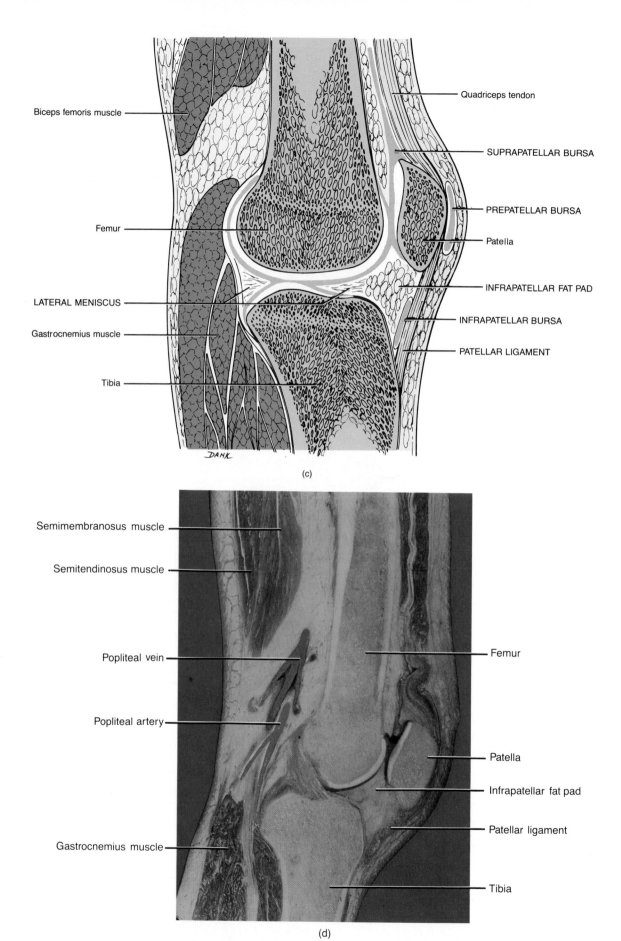

Biceps femoris muscle

Femur

LATERAL MENISCUS

Gastrocnemius muscle

Tibia

Quadriceps tendon

SUPRAPATELLAR BURSA

PREPATELLAR BURSA

Patella

INFRAPATELLAR FAT PAD

INFRAPATELLAR BURSA

PATELLAR LIGAMENT

DAHK

(c)

Semimembranosus muscle

Semitendinosus muscle

Popliteal vein

Popliteal artery

Gastrocnemius muscle

Femur

Patella

Infrapatellar fat pad

Patellar ligament

Tibia

(d)

FIGURE 8-14 (*Continued*) Tibiofemoral (knee) joint. (c) Diagram of sagittal section. (d) Photograph of sagittal section. (Courtesy of C. Yokochi and J. W. Rohen, *Photographic Anatomy of the Human Body,* 2nd ed., 1979, IGAKU-SHOIN, Ltd., Tokyo, New York.)

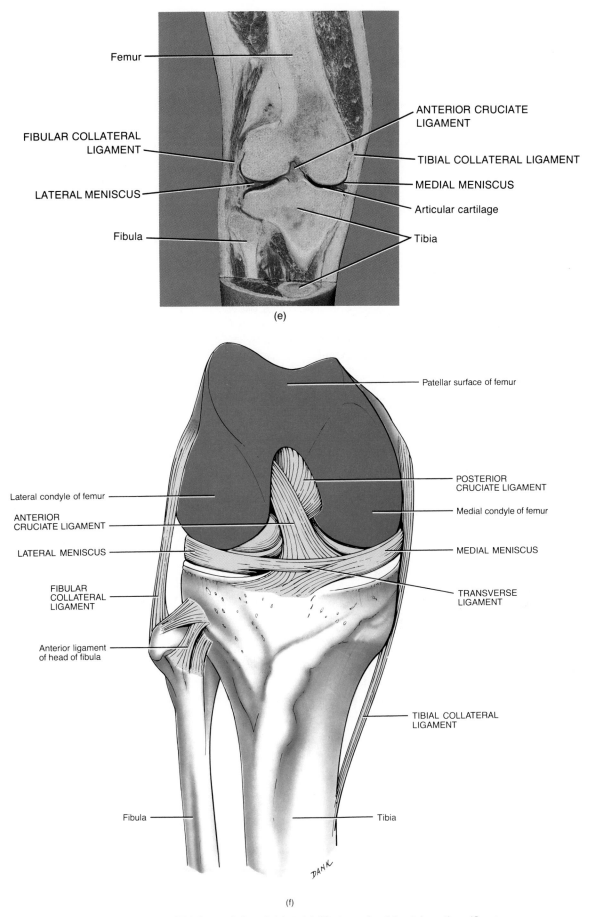

(e)

(f)

FIGURE 8-14 (*Continued*) Tibiofemoral (knee) joint. (e) Photograph of frontal section. (*Courtesy of C. Yokochi and J. W. Rohen, *Photographic Anatomy of the Human Body,* 2nd ed., 1979, IGAKU-SHOIN, Ltd., Tokyo, New York.) (f) Diagram of anterior view (flexed).

EXHIBIT 8-12 TALOCRURAL (ANKLE) JOINTS (Figure 8-15)

DEFINITION	Joints between (1) the distal end of the tibia and its medial malleolus and the talus and (2) the lateral malleolus of the fibula and the talus.
TYPE OF JOINT	Both joints—synovial, hinge (ginglymus) type.
ANATOMICAL COMPONENTS	1. **Articular capsule.** Surrounds the joint and extends from the borders of the tibia and malleoli to the talus.
	2. **Deltoid (medial) ligament.** Strong, triangular ligament that extends from the medial malleolus of the tibia to the navicular, calcaneus, and talus.
	3. **Anterior talofibular ligament.** Extends from the anterior margin of the lateral malleolus of the fibula to the talus.
	4. **Posterior talofibular ligament.** Extends from the posterior margin of the lateral malleolus of the fibula to the talus.
	5. **Calcaneofibular ligament.** Extends from the apex of the lateral malleolus of the fibula to the talus. Together, the anterior talofibular, posterior talofibular, and calcaneofibular ligaments are referred to as the **lateral ligament.**

CLINICAL APPLICATION

The most common **ankle sprain** is a lateral sprain. A person is more likely to turn the foot inward rather than outward due to normal adduction that occurs during running and the fact that the lateral ligament is weaker than the deltoid (medial) ligament. Most lateral ankle sprains involve the anterior talofibular ligament and frequently occur on irregular surfaces.

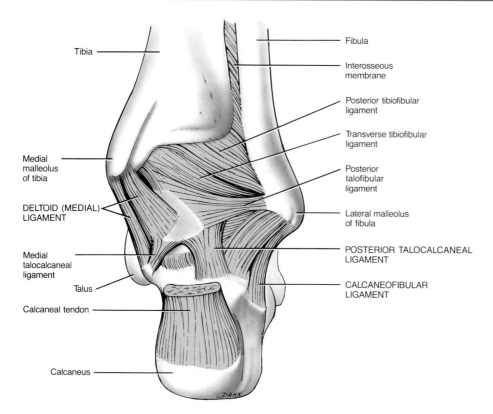

FIGURE 8-15 Talocrural (ankle) joints. Posterior view.

Labels: Tibia, Fibula, Interosseous membrane, Posterior tibiofibular ligament, Transverse tibiofibular ligament, Posterior talofibular ligament, Medial malleolus of tibia, Lateral malleolus of fibula, DELTOID (MEDIAL) LIGAMENT, POSTERIOR TALOCALCANEAL LIGAMENT, Medial talocalcaneal ligament, CALCANEOFIBULAR LIGAMENT, Talus, Calcaneal tendon, Calcaneus

APPLICATIONS TO HEALTH

RHEUMATISM

Rheumatism (*rheumat* = subject to flux) refers to any painful state of the supporting structures of the body—its bones, ligaments, joints, tendons, or muscles. Arthritis is a form of rheumatism in which the joints have become inflamed.

ARTHRITIS

The term **arthritis** refers to at least 25 different diseases, the most common of which are rheumatoid arthritis, osteoarthritis, and gouty arthritis. All these ailments are characterized by inflammation in one or more joints. Inflammation, pain, and stiffness may also be present in adjacent parts of the body, such as the muscles near the joint. Recent evidence suggests that the chronic pain that accompanies various arthritic conditions may be related to the patient's inability to produce endorphins. These chemicals are naturally produced painkillers.

The causes of arthritis are unknown. In some cases, it follows the stress of sprains, infections, and joint injury. Some researchers think that the cause is a bacterium or virus, whereas others suspect an allergy. Some believe the nervous system or hormones are involved, whereas others suspect a metabolic disorder. Still others believe that certain types of prolonged psychological stress, such as inhibited hostility, can upset homeostatic balance and bring on arthritic attacks.

Rheumatoid Arthritis (RA)

Rheumatoid (ROO-ma-toyd) **arthritis (RA)** is the most common inflammatory form of arthritis. It involves inflammation of the joint, swelling, pain, and a loss of function. Usually this form occurs bilaterally—if your left knee is affected, your right knee may also be affected, although usually not to the same degree.

The primary symptom of rheumatoid arthritis is inflammation of the synovial membrane. If it is untreated, the following sequential pathology may occur. The membrane thickens and synovial fluid accumulates. The resulting pressure causes pain and tenderness. The membrane then produces an abnormal tissue called *pannus,* which adheres to the surface of the articular cartilage. The pannus formation sometimes erodes the cartilage completely. When the cartilage is destroyed, fibrous tissue joins the exposed bone ends. The tissue ossifies and fuses the joint so that it is immovable—the ultimate crippling effect of rheumatoid arthritis. Most cases do not progress to this stage, but the range of motion of the joint is greatly inhibited by the severe inflammation and swelling.

Damaged joints may be surgically replaced, either partly or entirely, with artificial joints. The artificial parts are inserted after removal of the diseased portion of the articulating bone and its cartilage. The new metal or plastic joint is fixed in place with a special acrylic cement. When freshly mixed in the operating room, it hardens as strong as bone in minutes. These new parts function nearly as well as a normal joint—and much better than a diseased joint.

Osteoarthritis

A degenerative joint disease far more common than rheumatoid arthritis, and usually less damaging, is **osteoarthritis** (os'-tē-ō-ar-THRĪ-tis). It apparently results from a combination of aging, irritation of the joints, and wear and abrasion.

Degenerative joint disease is a noninflammatory, progressive disorder of movable joints, particularly weight-bearing joints. It is characterized pathologically by the deterioration of articular cartilage and by formation of new bone in the subchondral areas and at the margins of the joint. The cartilage slowly degenerates, and as the bone ends become exposed, small bumps, or *spurs,* of new osseous tissue are deposited on them. These spurs decrease the space of the joint cavity and restrict joint movement. Unlike rheumatoid arthritis, osteoarthritis usually affects only the articular cartilage. The synovial membrane is rarely destroyed, and other tissues are unaffected.

Gouty Arthritis

Uric acid is a waste product produced during the metabolism of nucleic acids. Normally, all the acid is quickly excreted in the urine. In fact, it gives urine its name. The person who suffers from *gout* either produces excessive amounts of uric acid or is not able to excrete normal amounts. The result is a buildup of uric acid in the blood. This excess acid then reacts with sodium to form a salt called sodium urate. Crystals of this salt are deposited in soft tissues. Typical sites are the kidneys and the cartilage of the ears and joints.

In **gouty** (GOW-tē) **arthritis,** sodium urate crystals are deposited in the soft tissues of the joints. The crystals irritate the cartilage, causing inflammation, swelling, and acute pain. Eventually, the crystals destroy all the joint tissues. If the disorder is not treated, the ends of the articulating bones fuse and the joint becomes immovable.

Gouty arthritis occurs primarily in males of any age. It is believed to be the cause of 2 to 5 percent of all chronic joint diseases. Numerous studies indicate that gouty arthritis is sometimes caused by an abnormal gene. As a result of this gene, the body manufactures unusually large amounts of uric acid. Diet and environmental factors such as stress and climate are also suspected causes of gouty arthritis.

Although other forms of arthritis cannot be treated with complete success, the treatment of gouty arthritis with the use of various drugs has been quite effective. A chemical called colchicine has been utilized periodically since the sixth century to relieve the pain, swelling, and tissue destruction that occur during attacks of gouty arthritis. This chemical is derived from the variety of crocus plant from which the spice saffron is obtained. Other drugs, which either inhibit uric acid production or assist in the elimination of excess uric acid by the kidneys, are used to prevent further attacks. The drug allopurinal is used for the treatment of gouty arthritis because it prevents the formation of uric acid without interfering with nucleic acid synthesis.

BURSITIS

An acute chronic inflammation of a bursa is called **bursitis.** The condition may be caused by trauma, by an acute or chronic infection (including syphilis and tuberculosis), or by rheumatoid arthritis. Repeated excessive friction often results in a bursitis with local inflammation and the accumulation of fluid. Bunions are frequently associated with a friction bursitis over the head of the first metatarsal bone. Symptoms include pain, swelling, tenderness, and the limitation of motion involving the inflamed bursa. The prepatellar or subcutaneous infrapatellar bursa may become inflamed in individuals who spend a great deal of time kneeling. This bursitis is usually called **"housemaid's knee" ("carpet layer's knee").**

Aspiration of the knee joint may be necessary to relieve pressure, to evacuate blood, or to obtain a fluid sample for laboratory studies. In this procedure, the needle is inserted somewhat proximal and lateral to the patella through the tendinous part of the vastus lateralis muscle and is directed toward the middle of the joint. Through the same route, the joint cavity may be anesthetized or irrigated, and steroids may be administered in the treatment of knee pathologies.

DISLOCATION

A **dislocation,** or **luxation** (luks-Ā-shun), is the displacement of a bone from a joint with tearing of ligaments, tendons, and articular capsules. A partial or incomplete dislocation is called a **subluxation.** The most common dislocations are those involving a finger, thumb, or shoulder. Those of the mandible, elbow, knee, or hip are less common. Symptoms include loss of motion, temporary paralysis of the involved joint, pain, swelling, and occasionally shock. A dislocation is usually caused by a blow or fall, although unusual physical effort may lead to this condition.

SPRAIN AND STRAIN

A **sprain** is the forcible wrenching or twisting of a joint with partial rupture or other injury to its attachments without luxation. It occurs when the attachments are stressed beyond their normal capacity. There may be damage to the associated blood vessels, muscles, tendons, ligaments, or nerves. A sprain is more serious than a **strain,** which is the overstretching of a muscle. Severe sprains may be so painful that the joint cannot be moved. There is considerable swelling, with reddish to blue discoloration due to hemorrhage from ruptured blood vessels. The ankle joint is most often sprained; the low back area is another frequent location for sprains.

DMSO stands for *dimethyl sulfoxide* (Rimso-50). It is a clear, colorless, and practically odorless liquid that has been used as an industrial solvent since the 1940s. Although the FDA has approved the use of a 50 percent solution of DMSO for the treatment of interstitial cystitis, a specific urinary bladder disorder, a 90 percent solution (approved for veterinary use) and a 99 percent solution (used as an industrial solvent) are also being used as an antiinflammatory, without FDA approval, to treat human disorders such as chronic arthritis, sprains, and strains.

The FDA points to the potential for side effects of DMSO since the drug is rapidly absorbed from the surface of the skin into the bloodstream. Among the side effects are allergic reactions (erythema and itching), headache, nausea, diarrhea, burning on urination, and disturbances in vision. Users also experience an oysterlike taste in their mouths and a garliclike body odor for up to 72 hours after administration. Lens opacities (cloudiness) have also been reported in laboratory animals. Studies of the effectiveness of DMSO for human disorders are now underway.

KEY MEDICAL TERMS ASSOCIATED WITH JOINTS

Ankylosis (*agkyle* = stiff joint; *osis* = condition)   Severe or complete loss of movement at a joint.
Arthralgia (*arth* = joint; *algia* = pain)   Pain in a joint.
Arthrosis   Refers to an articulation; also a disease of a joint.
Bursectomy (*ectomy* = removal of)   Removal of a bursa.

Chondritis (*chondro* = cartilage)   Inflammation of cartilage.
Rheumatology (*rheumat* = subject to flux)   The medical specialty devoted to arthritis.
Synovitis (*synov* = joint)   Inflammation of a synovial membrane in a joint.

STUDY OUTLINE

Classification (p. 168)
1. A joint or articulation is a point of contact between two or more bones.

2. Functional classification of joints is based on the degree of movement permitted. Joints may be synarthroses, amphiarthroses, or diarthroses.

3. Structural classification is based on the presence of a joint cavity and type of connecting tissue. Structurally, joints are classified as fibrous, cartilaginous, or synovial.

Fibrous Joints (p. 168)

1. Bones held by fibrous connective tissue, with no joint cavity, are fibrous joints.
2. These joints include immovable sutures (found in the skull), slightly movable syndesmoses (such as the tibiofibular articulation), and immovable gomphoses (roots of teeth in alveoli of mandible and maxilla).

Cartilaginous Joints (p. 169)

1. Bones held together by cartilage, with no joint cavity, are cartilaginous joints.
2. These joints include immovable synchondroses united by hyaline cartilage (temporary cartilage between diaphysis and epiphyses) and partially movable symphyses united by fibrocartilage (the symphysis pubis).

Synovial Joints (p. 169)

1. Synovial joints contain a joint (synovial) cavity, articular cartilage, and a synovial membrane; some also contain ligaments, articular discs, and bursae.
2. All synovial joints are freely movable.
3. Movements at synovial joints are limited by the apposition of soft parts, tension of ligaments, and muscle tension.
4. Types of movements at synovial joints include gliding movements, angular movements, rotation, circumduction, inversion and eversion, protraction and retraction, supination and pronation, and elevation and depression.
5. Types of synovial joints include gliding joints (wrist bones), hinge joints (elbow), pivot joints (radioulnar), ellipsoidal joints (radiocarpal), saddle joints (carpometacarpal), and ball-and-socket joints (shoulder and hip).
6. A joint may be described according to the number of planes of movement it allows as nonaxial, monaxial, biaxial, or triaxial.

Selected Articulations of the Body (p. 179)

1. Temporomandibular (TM) is between the mandible and temporal bone.
2. Atlantooccipitals are between the atlas and occipital bones.
3. Intervertebrals are between vertebral bodies and between vertebral arches.
4. Lumbosacral is between the fifth lumbar vertebra and the first sacral vertebra of the sacrum.
5. Humeroscapular is between the humerus and scapula.
6. Elbow is between the humerus and ulna and between the humerus and radius.
7. Radiocarpal is between the radius and scaphoid, lunate, and triquetral carpal bones.
8. Coxal is between the femur and coxal bone.
9. Tibiofemoral is between the patella and femur and between the femur and tibia.
10. Talocrurals are between the tibia and talus and between the fibula and talus.

Applications to Health (p. 200)

1. Rheumatism is a painful state of supporting body structures such as bones, ligaments, tendons, joints, and muscles.
2. Arthritis refers to several disorders characterized by inflammation of joints, often accompanied by stiffness of adjacent structures.
3. Rheumatoid arthritis (RA) refers to inflammation of a joint accompanied by pain, swelling, and loss of function.
4. Osteoarthritis is a degenerative joint disease characterized by deterioration of articular cartilage and spur formation.
5. Gouty arthritis is a condition in which sodium urate crystals are deposited in the soft tissues of joints and eventually destroy the tissues.
6. Bursitis is an acute or chronic inflammation of bursae.
7. A dislocation, or luxation, is a displacement of a bone from its joint; a partial dislocation is called subluxation.
8. A sprain is the forcible wrenching or twisting of a joint with partial rupture to its attachments without dislocation, while a strain is the stretching of a muscle.

REVIEW QUESTIONS

1. Define an articulation. What factors determine the degree of movement at joints?
2. Distinguish among the three kinds of joints on the basis of structure and function. List the subtypes. Be sure to include degree of movement and specific examples.
3. Explain the components of a synovial joint. Indicate the relationship of ligaments and tendons to the strength of the joint and restrictions on movement.
4. Explain how the articulating bones in a synovial joint are held together.
5. What is an accessory ligament? Define the two principal types.
6. What is an articular disc? Why are they important?
7. Describe the principle and importance of arthroscopy.
8. What are bursae? What is their function?
9. Define the following principal movements: gliding, angular, rotation, circumduction, and special. Name a joint where each occurs.
10. Have another person assume the anatomical position and execute for you each of the movements at joints discussed in the text. Reverse roles, and see if you can execute the same movements.
11. Contrast nonaxial, monaxial, biaxial, and triaxial planes of movement. Give examples of each, and name a joint at which each occurs.
12. For each joint of the body discussed in Exhibits 8-3 through 8-12, be sure that you can name the bones that form the joint, identify the joint by type, and list the anatomical components of the joint.
13. Distinguish between rheumatoid arthritis (RA), osteoarthritis, and gouty arthritis with respect to causes and symptoms.
14. Define bursitis. How is it caused?
15. Define dislocation. What are the symptoms of dislocation?
16. Distinguish between a sprain and a strain.
17. Refer to the glossary of key medical terms at the end of the chapter and be sure that you can define each term.

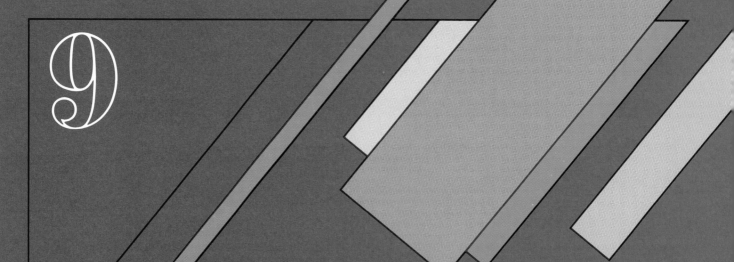

9

Muscle Tissue

Student Objectives

List the characteristics and functions of muscle tissue.

Compare the location, microscopic appearance, nervous control, and functions of the three kinds of muscle tissue.

Define fascia, epimysium, perimysium, endomysium, tendons, and aponeuroses, and list their modes of attachment to muscles.

Describe the relationship of blood vessels and nerves to skeletal muscles.

Identify the histological characteristics of skeletal muscle tissue.

Describe the structure and importance of a neuromuscular junction (motor end plate) and a motor unit.

Compare fast-twitch (white) muscle with slow-twitch (red) muscle.

Contrast cardiac muscle tissue with smooth muscle tissue.

Explain the effects of aging on muscle tissue.

Describe the development of the muscular system.

Define such common muscular disorders as fibrosis, fibrositis, "charleyhorse," muscular dystrophy, and myasthenia gravis.

Compare spasms, cramps, convulsions, and fibrillation.

Define key medical terms associated with the muscular system.

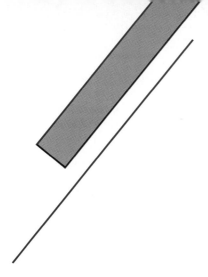

Although bones and joints provide leverage and form the framework of the body, they are not capable of moving the body by themselves. Motion is an essential body function that results from the contraction and relaxation of muscles.

Muscle tissue constitutes about 40 to 50 percent of the total body weight and is composed of highly specialized cells. The scientific study of muscles is known as **myology** (*myo* = muscle; *logos* = study of).

The developmental anatomy of the muscular system is considered at the end of the chapter.

CHARACTERISTICS

Muscle tissue has four principal characteristics that assume key roles in maintaining homeostasis.

1. Excitability is the ability of muscle tissue to receive and respond to stimuli. A stimulus is a change in the internal or external environment strong enough to initiate a nerve impulse (action potential).

2. Contractility is the ability to shorten and thicken, or contract, when a sufficient stimulus is received.

3. Extensibility is the ability of muscle tissue to be stretched. Many skeletal muscles are arranged in opposing pairs. While one is contracting, the other is relaxed and is undergoing extension.

4. Elasticity is the ability of muscle to return to its original shape after contraction or extension.

FUNCTIONS

Through contraction, muscle performs three important functions:

1. Motion.
2. Maintenance of posture.
3. Heat production.

Motion is obvious in movements involving the whole body, such as walking and running, and in localized movements, such as grasping a pencil or nodding the head. All of these movements rely on the integrated functioning of the bones, joints, and muscles attached to the bones. Less noticeable kinds of motion produced by muscles are the beating of the heart, the churning of food in the stomach, the pushing of food through the intestines, the contraction of the gallbladder to release bile, and the contraction of the urinary bladder to expel urine.

In addition to the movement function, muscle tissue also enables the body to maintain posture. The contraction of skeletal muscles holds the body in stationary positions, such as standing and sitting.

The third function of muscle tissue is heat production. Skeletal muscle contractions produce heat and are thereby important in maintaining normal body temperature.

TYPES

Types of muscle tissue are categorized by location, microscopic structure, and nervous control.

Skeletal muscle tissue, which is named for its location, is attached to bones. It is **striated** muscle tissue because striations, or bandlike structures, are visible when the tissue is examined under a microscope. It is a **voluntary** muscle tissue because it can be made to contract by conscious control.

Cardiac muscle tissue forms the bulk of the wall of the heart. It is **striated** and **involuntary,** that is, its contraction is usually not under conscious control.

Smooth muscle tissue is located in the walls of hollow internal structures, such as blood vessels, the stomach, and the intestines. It is referred to as **nonstriated** because it lacks striations. It is involuntary muscle tissue.

Thus all muscle tissues are classified in the following way: (1) skeletal, striated, voluntary muscle, (2) cardiac, striated, involuntary muscle, and (3) smooth, nonstriated, involuntary muscle.

SKELETAL MUSCLE TISSUE

To understand the fundamental mechanisms of muscle movement, you will need some knowledge of its connective tissue components, nerve and blood supply, and histology.

FASCIA

The term **fascia** (FASH-ē-a) is applied to a sheet or broad band of fibrous connective tissue beneath the skin or around muscles and other organs of the body. Fasciae may be divided into three types: superficial, deep, and subserous.

The **superficial fascia (subcutaneous layer)** is immediately deep to the skin. It covers the entire body and varies in thickness in different regions. On the back, or dorsum, of the hand it is quite thin, whereas over the inferior abdominal wall it is thick. The superficial fascia is composed of adipose tissue and loose connective tissue. The outer layer usually contains fat and varies considerably in thickness, while the inner layer is thin and elastic. Between the two layers are found arteries, veins, lymphatics, nerves, the mammary glands, and the facial muscles. Superficial fascia has a number of important functions.

1. It serves as a storehouse for water and particularly for fat. Much of the fat of an overweight person is in the superficial fascia.

2. It forms a layer of insulation protecting the body from loss of heat.

3. It provides mechanical protection from blows.

4. It provides a pathway for nerves and vessels.

The **deep fascia** is by far the most extensive of the three types. It is a dense connective tissue composed of a superficial layer (external investing layer) and a deep layer (internal investing layer). Unlike the superficial fascia, it does not contain fat. The deep fascia lines the body wall and extremities and holds muscles together, separating them into functioning groups. Functionally,

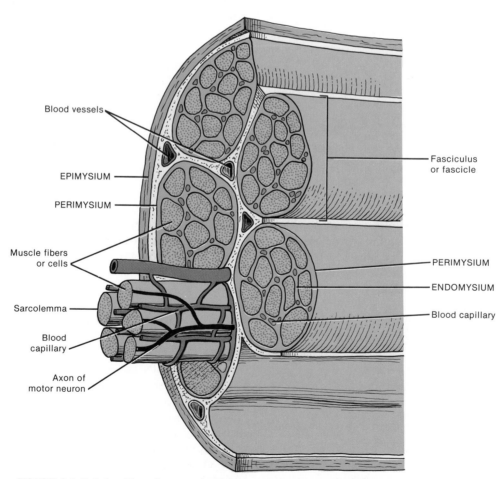

FIGURE 9-1 Relationships of connective tissue to skeletal muscle. Shown is a cross section and longitudinal section of a portion of a skeletal muscle indicating the relative positions of the epimysium, perimysium, and endomysium. Compare this figure with the photomicrograph in Figure 9-3b.

deep fascia allows free movement of muscles, carries nerves and blood vessels, fills spaces between muscles, and sometimes provides the origin for muscles.

The **subserous (visceral) fascia** is located between the internal investing layer of deep fascia and a serous membrane. It is composed of loose connective tissue. It forms the fibrous layer of serous membranes, covers and supports the viscera, and attaches the parietal layer of serous membranes to the internal surface of the body wall.

CLINICAL APPLICATION

Lines of fusion of fascial sheets are essentially avascular and for this reason they are often sites for **surgical incisions.** Surgeons also prefer fascial junctional areas for anchoring of sutures because of the strength of fasciae and the strong union that results from wound healing.

CONNECTIVE TISSUE COMPONENTS

Skeletal muscles are further protected, strengthened, and attached to other structures by several connective tissue components (Figure 9-1). The entire muscle is usually wrapped with a substantial quantity of fibrous connective tissue called the **epimysium** (ep'-i-MĪZ-ē-um). The epimysium is an extension of deep fascia. When the muscle is cut in cross section, invaginations of the epimysium are seen to divide the muscle into bundles of fibers (cells) called **fasciculi** (fa-SIK-yoo-lī), or **fascicles** (FAS-i-kuls). These invaginations of the epimysium are called the **perimysium** (per'-i-MĪZ-ē-um). Perimysium, like epimysium, is an extension of deep fascia. In turn, invaginations of the perimysium, called **endomysium** (en'-dō-MĪZ-ē-um), penetrate into the interior of each fascicle and separate the muscle cells. Endomysium is also an extension of deep fascia.

The epimysium, perimysium, and endomysium are all continuous with the connective tissue that attaches the muscle to another structure, such as bone or other muscle. All three elements may be extended beyond the muscle cells as a **tendon**—a cord of connective tissue that attaches a muscle to the periosteum of a bone. When the connective tissue elements extend as a broad, flat layer, the tendon is called an **aponeurosis.** This structure also attaches to the coverings of a bone or another muscle. When a muscle contracts, the tendon and its corresponding bone or muscle are pulled toward the contracting muscle. In this way skeletal muscles produce movement. Certain tendons, especially those of the wrist and ankle, are enclosed by tubes of fibrous connective tissue called **tendon sheaths.** They are similar in structure to bursae. The inner layer of the tendon sheath, the visceral layer, is applied to the surface of the tendon. The outer layer is known as the parietal layer. Between the layers is a cavity that contains

a film of synovial fluid. Tendon sheaths permit tendons to slide easily and also prevent the tendons from slipping out of place.

CLINICAL APPLICATION

Tendinitis, or **tenosynovitis** (ten'-ō-sin-ō-VĪ-tis), frequently occurs as inflammation involving the tendon sheaths and synovial membrane surrounding certain joints. The wrists, shoulders, elbows (tennis elbow), finger joints (trigger finger), ankles, and associated tendons are most often affected. The affected sheaths may become visibly swollen because of fluid accumulation, or they may remain dry. Local tenderness is variable, and there may be disabling pain with movement of the body part. The condition often follows some form of trauma, strain, or excessive exercise.

NERVE AND BLOOD SUPPLY

Skeletal muscles are well supplied with nerves and blood vessels. This innervation and vascularization is directly related to contraction, the chief characteristic of muscle. For a skeletal muscle cell to contract, it must first be stimulated by an impulse from a nerve cell. Muscle contraction also requires a good deal of energy and therefore large amounts of nutrients and oxygen. Moreover, the waste products of these energy-producing reactions must be eliminated. Thus muscle action depends on the blood supply.

Generally, an artery and one or two veins accompany each nerve that penetrates a skeletal muscle. The larger branches of the blood vessel accompany the nerve branches through the connective tissue of the muscle (Figure 9-2). Microscopic blood vessels called capillaries are arranged in the endomysium. Each muscle cell is thus in close contact with one or more capillaries. Each skeletal muscle cell usually makes contact with a portion of a nerve cell.

HISTOLOGY

When a typical skeletal muscle is teased apart and viewed microscopically, it can be seen to consist of many elongated, cylindrical cells called **muscle fibers,** or **myofibers** (Figure 9-3a, b). These fibers lie parallel to one another and range from 10 to 100 μm in diameter. Some fibers may reach lengths of 30 cm (12 in) or more. Each muscle fiber is enveloped by a plasma membrane called the **sarcolemma** (*sarco* = flesh; *lemma* = sheath). The sarcolemma surrounds a quantity of cytoplasm called **sarcoplasm.** Within the sarcoplasm of a muscle fiber and lying close to the sarcolemma are many nuclei and a number of mitochondria. Skeletal muscle fibers are thus multinucleate. Also within a muscle fiber is the **sarcoplasmic**

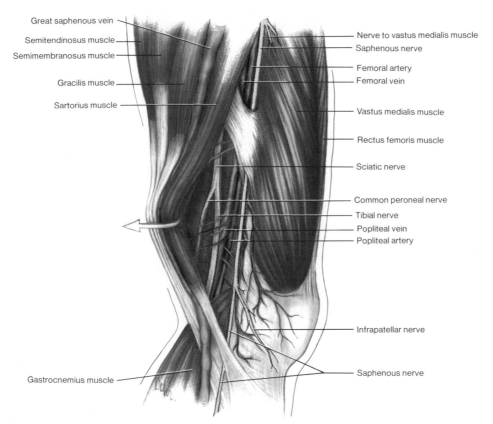

FIGURE 9-2 Relationship of blood vessels and nerves to skeletal muscles as seen in a diagram of the left thigh and knee in medial view.

reticulum (sar'-kō-PLAZ-mik rē-TIK-yoo-lum), a network of membrane-enclosed tubules comparable to smooth endoplasmic reticulum (Figure 9-3c). Running transversely through the fiber and perpendicularly to the sarcoplasmic reticulum are **T tubules (transverse tubules).** The tubules are extensions of the sarcolemma that open to the outside of the fiber. A **triad** consists of a T tubule and the segments of sarcoplasmic reticulum on either side.

A highly magnified view of skeletal muscle fibers reveals threadlike structures, about 1 or 2 μm in diameter, called **myofibrils** (Figure 9-3c, d, e). The myofibrils, ranging in number from several hundred to several thousand, run longitudinally through the muscle fiber and consist of two kinds of even smaller structures called **myofilaments.** The **thin myofilaments** are about 6 nm in diameter. The **thick myofilaments** are about 16 nm in diameter.

The myofilaments of a myofibril do not extend the entire length of a muscle fiber—they are stacked in compartments called **sarcomeres.** Sarcomeres are separated from one another by narrow zones of dense material called **Z lines.** Each sarcomere is about 2.6 μm long. In a relaxed muscle fiber, that is, one that is not contracting, the thin and thick myofilaments overlap and form a dark,

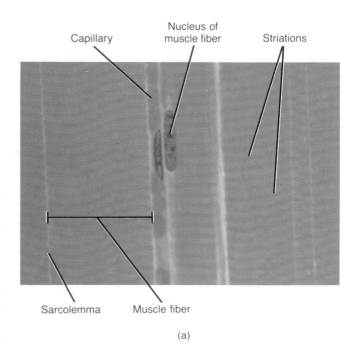

(a)

FIGURE 9-3 Histology of skeletal muscle tissue. (a) Photomicrograph of several muscle fibers in longitudinal section at a magnification of 800×. (© 1983 by Michael H. Ross. Used by permission.)

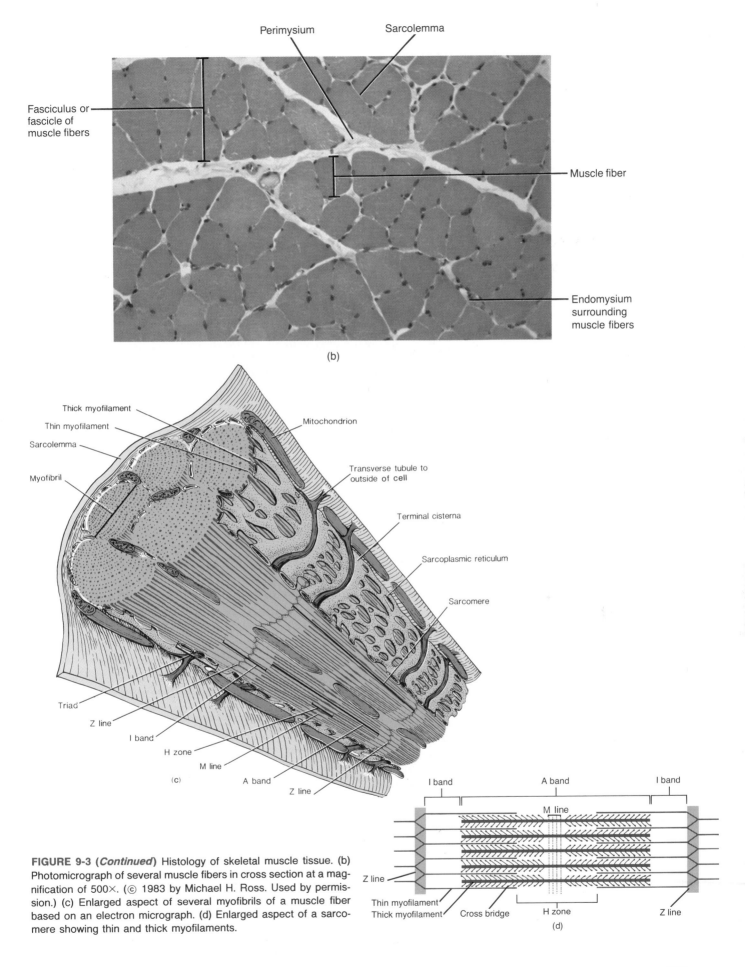

Perimysium

Sarcolemma

Fasciculus or fascicle of muscle fibers

Muscle fiber

Endomysium surrounding muscle fibers

(b)

Thick myofilament

Thin myofilament

Sarcolemma

Myofibril

Mitochondrion

Transverse tubule to outside of cell

Terminal cisterna

Sarcoplasmic reticulum

Sarcomere

Triad

Z line

I band

H zone

M line

A band

Z line

(c)

I band

A band

I band

M line

Z line

Thin myofilament

Thick myofilament

Cross bridge

H zone

Z line

(d)

FIGURE 9-3 (*Continued*) Histology of skeletal muscle tissue. (b) Photomicrograph of several muscle fibers in cross section at a magnification of 500×. (© 1983 by Michael H. Ross. Used by permission.) (c) Enlarged aspect of several myofibrils of a muscle fiber based on an electron micrograph. (d) Enlarged aspect of a sarcomere showing thin and thick myofilaments.

209

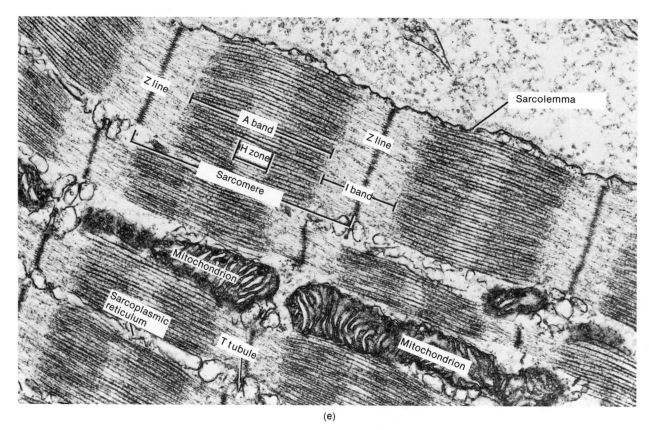

(e)

FIGURE 9-3 (*Continued*) Histology of skeletal muscle tissue. (e) Electron micrograph of several sarcomeres at a magnification of 35,000×. (Courtesy of D. E. Kelly, from *Introduction to the Musculoskeletal System* by Cornelius Rosse and D. Kay Clawson, Harper & Row, Publishers, Inc., New York, 1970.)

dense band called the **anisotropic band,** or **A band** (Figure 9-3d). Each A band is about 1.6 μm long. A light-colored, less dense area called the **isotropic band,** or **I band,** is composed of thin myofilaments only. The I bands are about 1 μm long. This combination of alternating dark and light bands gives the muscle fiber its striated (striped) appearance. A narrow **H zone** contains thick myofilaments only. The H zone is about 0.5 μm long. In the center of the H zone is the **M line,** a series of fine threads that appear to connect the middle parts of the adjacent thick filaments.

The thin myofilaments are composed mostly of the protein **actin.** The actin molecules are arranged in two single strands that entwine helically and give the thin myofilaments their characteristic shape (Figure 9-4a). Each actin molecule contains a *myosin-binding site* that interacts with a cross bridge of a myosin molecule (described shortly). Besides actin, the thin myofilaments contain two other protein molecules, **tropomyosin** and **troponin,** that are involved in the regulation of muscle contractions. Tropomyosin is arranged in strands that are loosely attached to the actin helices. Troponin is located at regular intervals on the surface of tropomyosin. Together, tropomyosin and troponin are referred to as a **tropomyosin-troponin complex.**

The thick myofilaments are composed mostly of the protein **myosin.** A myosin molecule is shaped like a golf club. The tails (handles of the golf club) are arranged parallel to each other forming the shaft of the thick myofilament. The heads of the golf clubs project outward from the shaft and are arranged spirally on the surface of the shaft. The projecting heads are referred to as **cross bridges** and contain an *actin-binding site* and an *ATP-binding site* (Figure 9-4b).

CLINICAL APPLICATION

Steroids are chemical substances derived from cholesterol. Most steroids are hormones. Some steroid hormones are produced by the ovaries (estrogens and progesterone) and testes (testosterone) and help initiate and maintain sexual traits. Other steroid hormones are produced by the adrenal cortex (aldosterone, cortisol, corticosterone, and cortisone) and assume various roles in sodium and water retention, normal metabolism, resisting stress, and antiinflammatory responses.

At the IX Pan American Games in Caracas, Venezuela, held in 1983, worldwide attention was focused on the use of **muscle-building anabolic steroids** by ama-

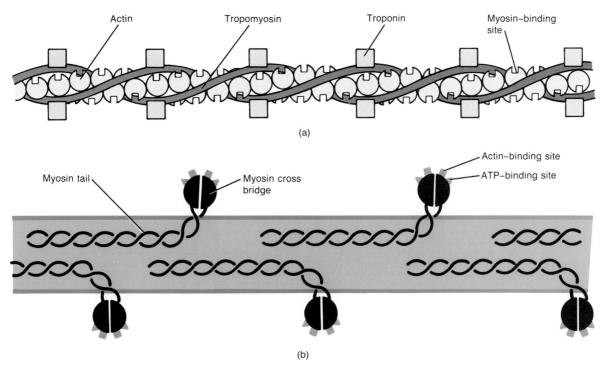

Actin   Tropomyosin   Troponin   Myosin–binding site

(a)

Myosin tail   Myosin cross bridge   Actin–binding site   ATP–binding site

(b)

FIGURE 9-4 Detailed structure of myofilaments. (a) Thin myofilament. (b) Thick myofilament.

teur athletes. These steroids, a variant of the hormone testosterone, were used by the athletes supposedly to build muscle proteins and therefore increase strength and endurance during athletic events. However, physicians point out that use of anabolic steroids may cause a number of side effects, including liver and kidney damage, increased irritability and aggressive behavior, the development of facial hair and deepening of the voice in females, and excessive development of breast glands and diminished hormone secretion and sperm production by the testes in males. Their use may also increase the risk of heart disease.

CONTRACTION

Sliding-Filament Theory

During muscle contraction, the thin myofilaments slide inward toward the H zone. The sarcomere shortens, but the lengths of the thin and thick myofilaments do not change. The myosin cross bridges of the thick myofilaments connect with portions of actin of the thin myofilaments. The myosin cross bridges move like the oars of a boat on the surface of the thin myofilaments, and the thin and thick myofilaments slide past each other. As the thin myofilaments move past the thick myofilaments, the H zone narrows and even disappears when the thin myofilaments meet at the center of the sarcomere (Figure 9-5). In fact, the myosin cross bridges may pull the thin

myofilaments of each sarcomere so far inward that their ends overlap. As the thin myofilaments slide inward, the Z lines are drawn toward each other and the sarcomere is shortened. The sliding of myofilaments and shortening of sarcomeres cause the shortening of the muscle fibers. All these events associated with the movement of myofilaments are known as the **sliding-filament theory** of muscle contraction.

Neuromuscular Junction (Motor End Plate)

For a skeletal muscle fiber to contract, a stimulus must be applied to it. Such a stimulus is normally converted to a nerve impulse and the impulse is transmitted by nerve cells, or **neurons.** A neuron has a threadlike process called a fiber, or axon, that may run 91 cm (3 ft) or more to a muscle. A bundle of such fibers from many different neurons composes a nerve. A neuron that transmits a nerve impulse to muscle tissue is called a **motor neuron.**

On entering a skeletal muscle, the axon of a motor neuron branches into axon terminals (telondendria) that come into close approximation with a portion of the sarcolemma of a muscle fiber. The term **neuromuscular (myoneural) junction,** or **motor end plate,** refers to the axon terminal of a motor neuron together with the portion of the sarcolemma of a muscle fiber in close approximation with the axon terminal (Figure 9-6). Close examination of a neuromuscular junction reveals that the distal ends of the axon terminals are expanded into bulblike structures called **synaptic end bulbs.** The bulbs contain mem-

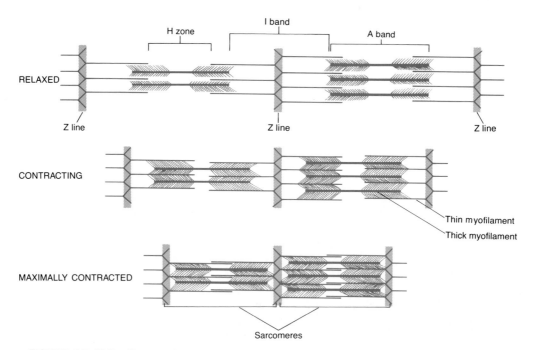

FIGURE 9-5 Sliding-filament theory of muscle contraction. Shown are the positions of the various parts of two sarcomeres in relaxed, contracting, and maximally contracted states. Note the movement of the thin myofilaments and the relative size of the H zone.

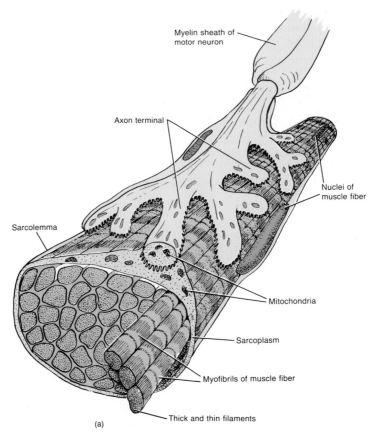

FIGURE 9-6 Neuromuscular junction (motor end plate). (a) Diagram based on a photomicrograph.

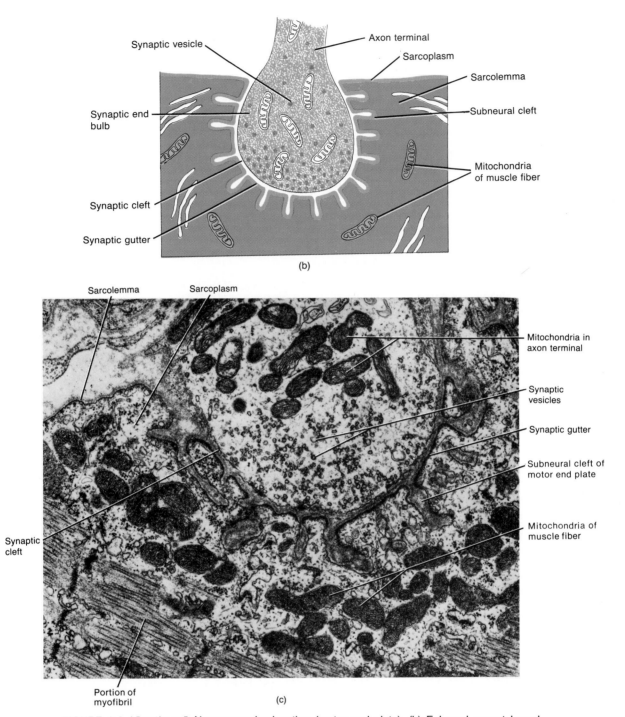

FIGURE 9-6 (*Continued*) Neuromuscular junction (motor end plate). (b) Enlarged aspect based on an electron micrograph. (c) Electron micrograph at a magnification of 30,000×. (Courtesy of Cornelius Rosse and D. Kay Clawson, from *Introduction to the Musculoskeletal System,* Harper & Row, Publishers, Inc., New York, 1970.)

brane-enclosed sacs, the **synaptic vesicles,** that store chemicals called **neurotransmitters.** These chemicals determine whether a nerve impulse is passed on to a muscle, gland, or another nerve cell. The invaginated area of the sarcolemma under the axon terminal is referred to as a **synaptic gutter (trough),** and the space between the axon terminal and sarcolemma is known as a **synaptic cleft.**

There are numerous folds of the sarcolemma along the synaptic gutter called **subneural clefts,** which greatly increase the surface area of the synaptic gutter.

When a nerve impulse reaches an axon terminal, it initiates a sequence that liberates neurotransmitter molecules from synaptic vesicles. The neurotransmitter released at neuromuscular junctions is **acetylcholine** (as'-

ē-til-KŌ-lēn) or **ACh.** It transmits the nerve impulse from the axon terminal of a motor neuron across the synaptic cleft to the sarcolemma of the muscle fiber, thus initiating contraction.

Motor Unit

A motor neuron, together with all the muscle fibers it stimulates, is referred to as a **motor unit.** A single motor neuron may innervate about 150 muscle fibers, depending on the region of the body. This means that stimulation of one neuron will tend to cause the simultaneous contraction of about 150 muscle fibers. In addition, all the muscle fibers of a motor unit that are sufficiently stimulated will contact and relax together. Muscles that control precise movements, such as the extrinsic eye muscles, have fewer than 10 muscle fibers to each motor unit. Muscles of the body that are responsible for gross movements, such as the biceps brachii and gastrocnemius, may have as many as 500 muscle fibers in each motor unit.

EXHIBIT 9-1 SUMMARY OF EVENTS INVOLVED IN CONTRACTION AND RELAXATION OF A SKELETAL MUSCLE FIBER

1. A nerve impulse causes synaptic vesicles in motor axon terminals to release acetylcholine (ACh).
2. Acetylcholine diffuses across the synaptic cleft and initiates an impulse that spreads over the surface of the sarcolemma.
3. The impulse enters the T tubules and sarcoplasmic reticulum and stimulates the sarcoplasmic reticulum to release calcium ions from storage into the sarcoplasm.
4. Calcium ions combine with troponin, causing the tropomyosin-troponin complex to move, thus exposing the myosin-binding sites on actin.
5. A nerve impulse also causes ATP $\longrightarrow$ ADP + P. The released energy activates the myosin cross bridges, which combine with the exposed myosin-binding sites on actin, and causes the myosin cross bridges to move toward the H zone. This movement results in the sliding of the thin myofilaments past the thick myofilaments.
6. The sliding draws the Z lines toward each other, the sarcomere shortens, the muscle fibers contract, and the muscle contracts.
7. Acetylcholine is inactivated by acetylcholinesterase (AChE), thus inhibiting nerve impulse conduction from axon terminals to the sarcolemma.
8. Once the nerve impulse is inhibited, calcium ions are actively transported back into the sarcoplasmic reticulum by a transport protein called calsequestrin, using energy from ATP breakdown.
9. The low calcium concentration in the sarcoplasm permits the tropomyosin-troponin complex to reattach to actin. As a result, myosin-binding sites of actin become covered, myosin cross bridges separate from actin, ADP is resynthesized into ATP (which reattaches to the ATP-binding site of the myosin cross bridge), and the thin myofilaments return to their relaxed position.
10. Sarcomeres return to their resting lengths, muscle fibers relax, and the muscle relaxes.

Mechanism

Rather than discuss the details of the mechanism of muscle contraction, the process is outlined in Exhibit 9-1. By following the step-by-step summary, you should understand the basics of muscle contraction.

CLINICAL APPLICATION

Following death, certain chemical changes occur in muscle tissue that affect the status of the muscles. Due to a lack of ATP, myosin cross bridges remain attached to the actin myofilaments, thus preventing relaxation. The resulting condition, in which muscles are in a state of partial contraction, is called **rigor mortis** (rigidity of death). The time elapsing between death and the onset of rigor mortis varies greatly among individuals. Those who have had long, wasting illnesses undergo rigor mortis more quickly.

All-or-None Principle

According to the **all-or-none principle,** individual muscle fibers of a motor unit will contract to their fullest extent or will not contract at all, provided conditions remain constant. In other words, *muscle fibers* do not partly contract. The principle does not mean that the entire muscle must be either fully relaxed or full contracted because, of the many fibers that comprise the entire muscle, some are contracting and some are relaxing. Thus, the muscle as a whole can have graded contractions. The strength of contraction may be decreased by fatigue, lack of nutrients, or lack of oxygen. The weakest stimulus from a neuron that can initiate a contraction is called a **threshold (liminal) stimulus.** A stimulus of lesser intensity, or one that cannot initiate contraction, is referred to as a **subthreshold (subliminal) stimulus.**

Muscle Tone

A muscle may be in a state of partial contraction even though contraction of the muscle fibers is always complete. At any given time, some cells in a muscle are contracted while others are relaxed. This contraction tightens a muscle, but there may not be enough fibers contracting at the time to produce movement. Asynchronous firing allows the contraction to be sustained for long periods.

A sustained partial contraction of portions of a skeletal muscle in response to activation of stretch receptors results in **muscle tone.** Tone is essential for maintaining posture. For example, when the muscles in the back of the neck are in tonic contraction, they keep the head in the anatomical position and prevent it from slumping forward onto the chest, but they do not apply enough force to pull the head back into hyperextension. The degree of tone in a skeletal muscle is monitored by receptors in the muscle called **muscle spindles.** They provide feedback information on tone to the brain so that adjustments can be made (Chapter 20).

CLINICAL APPLICATION

The term **flaccid** (FLAK-sid) is applied to muscles with less than normal tone. Such a loss of tone may be the result of damage or disease of the nerve that conducts a constant flow of impulses to the muscle. If the muscle does not receive impulses for an extended period of time, it may progress from flaccidity to **atrophy** (AT-rō-fē), which is a state of wasting away. Individual muscle cells decrease in size due to a progressive loss of myofibrils. Muscles may also become flaccid and atrophied if they are not used. Bedridden individuals and people with casts may experience atrophy because the flow of impulses to the inactive muscle is greatly reduced. If the nerve supply to a muscle is cut, it will undergo complete atrophy. In about six months to two years, the muscle will be one-quarter its original size and the muscle fibers will be replaced by fibrous tissue. The transition to fibrous tissue, when complete, cannot be reversed. Battery-operated transcutaneous muscle stimulations (TMSs) are used to maintain the strength of atrophied muscles immobilized by a cast by stimulating the muscles to contract. TMSs are also used to prevent atrophy of muscles whose motor nerves have been temporarily damaged by trauma or stroke. Some athletes even use TMSs to strengthen certain muscles.

Muscular hypertrophy (hī-PER-trō-fē) is the reverse of atrophy. It refers to an increase in the diameters of muscle fibers due to the production of more myofibrils, mitochondria, sarcoplasmic reticulum, nutrients (glycogen and triglycerides), and energy-supplying molecules (ATP and creatine phosphate). Hypertrophic muscles are capable of more forceful contractions. Weak muscular activity does not produce significant hypertrophy. It results from very forceful muscular activity or repetitive muscular activity at moderate levels.

FAST-TWITCH (WHITE) AND SLOW-TWITCH (RED) MUSCLE

The duration of contraction of various muscles of the body varies with the functions of the muscles in maintaining homeostasis. For example, the duration of contraction of eye muscles is less than $\frac{1}{100}$ sec, whereas the duration of contraction of the gastrocnemius muscle on the posterior aspect of the leg is about $\frac{1}{30}$ sec. Eye movements must be extremely rapid in order to fix the eyes on differ-

ent objects very quickly. Movements of the legs during walking or running need not be as rapid as eye movements.

Muscles of the eye contain a predominance of **fast-twitch muscle fibers,** or **type II fibers.** Such fibers have a more extensive sarcoplasmic reticulum for the rapid release and uptake of calcium ions needed for rapid contractions. They also have more glycogen and enzymes that are more active in releasing energy for contraction. Fast-twitch muscle fibers can produce bursts of power at high speeds but tire quickly. Muscles like the gastrocnemius have a predominance of **slow-twitch muscle fibers,** or **type I fibers.** These fibers are smaller and more aerobic, have more blood capillaries and mitochondria, and have a large amount of myoglobin in the sarcoplasm. **Myoglobin** is a substance similar to the hemoglobin in red blood cells that can store oxygen until needed by the mitochondria. Slow-twitch muscle fibers contract more slowly and with less power than fast-twitch muscle fibers, and slow fibers do not tire as quickly as fast fibers. Since slow-twitch muscle has a reddish tint due to myoglobin and the large amount of red blood cells in capillaries, it is also known as **red muscle.** Fast-twitch muscle has little myoglobin and has fewer capillaries and is thus also known as **white muscle.**

An average individual has an equal distribution of fast- and slow-twitch muscle fibers throughout the body. However, the leg muscles of world-class long-distance runners average about 80 percent slow-twitch muscle fibers, whereas the leg muscles of world-class sprinters average about 75 percent fast-twitch muscle fibers. The percentage of slow- and fast-twitch muscle fibers is genetically determined and neither type can be changed by exercise.

CARDIAC MUSCLE TISSUE

The principal constituent of the heart wall is **cardiac muscle tissue.** Although it is striated in appearance like skeletal muscle, it is involuntary. The fibers of cardiac muscle tissue are roughly quadrangular and usually have only a single centrally located nucleus (Figure 9-7). Skeletal muscle fibers contain several nuclei that are peripherally located. The thin sarcolemma of cardiac muscle fibers is similar to that of skeletal muscle, but the sarcoplasm is more abundant and the mitochondria are larger and more numerous. Cardiac muscle fibers have the same arrangement of actin and myosin and the same bands, zones, and lines as skeletal muscle fibers. Myofilaments in cardiac muscle fibers are not arranged in discrete myofibrils as in skeletal muscle. The T tubules of mammalian cardiac muscle are larger than those of skeletal muscle and are located at the Z lines rather than at the A-I band junctions as in skeletal muscle fibers. The sarcoplasmic reticulum of cardiac muscle is less well-developed than that in skeletal muscle.

Intercalated discs   Muscle fibers

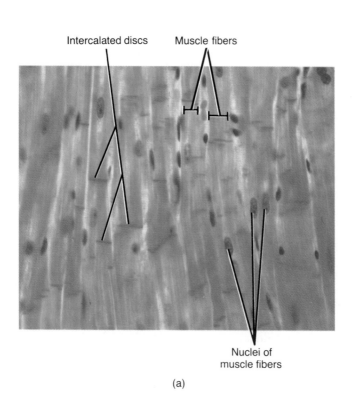

Nuclei of
muscle fibers

(a)

Endomysium surrounding
muscle fibers   Capillaries

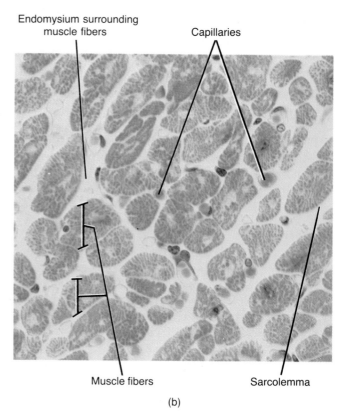

Muscle fibers   Sarcolemma

(b)

FIGURE 9-7 Histology of cardiac muscle tissue. (a) Photomicrograph of several muscle fibers in longitudinal section at a magnification of 400×. (© 1983 by Michael H. Ross. Used by permission.) (b) Photomicrograph of several muscle fibers in cross section at a magnification of 640×. (© 1983 by Michael H. Ross. Used by permission.)

While the groups of skeletal muscle fibers are arranged in a parallel fashion, those of cardiac muscle branch freely with other fibers to form two separate networks. The muscular walls and septum of the upper chambers of the heart (atria) compose one network. The muscular walls and septum of the lower chambers of the heart (ventricles) compose the other network. When a single fiber of either network is stimulated, all the fibers in the network become stimulated as well. Thus, each network contracts as a functional unit. The fibers of each network were once thought to be fused together into a multinucleated mass called a syncytium. But it is now known that each fiber in a network is separated from the next fiber by an irregular transverse thickening of the sarcolemma called an **intercalated** (in-TER-ka-lāt-ed) **disc.** These discs strengthen the cardiac muscle tissue and aid in impulse conduction.

Under normal conditions, cardiac muscle tissue contracts and relaxes rapidly, continuously, and rhythmically about 72 times a minute without stopping while a person is at rest. This is a major physiological difference between cardiac and skeletal muscle tissue. Another difference is the source of stimulation. Skeletal muscle tissue ordinarily contracts only when stimulated by a nerve impulse. In contrast, cardiac muscle tissue can contract without nerve

stimulation. Its source of stimulation is a conducting tissue of specialized muscle within the heart. Nerve stimulation merely causes the conducting tissue to increase or decrease its rate of discharge. Cardiac muscle tissue also has an extra-long refractory period.

SMOOTH MUSCLE TISSUE

Like cardiac muscle tissue, **smooth muscle tissue** is usually involuntary. However, it is nonstriated. A single fiber of smooth muscle tissue is about 5 to 10 μm in diameter and 30 to 200 μm long. It is spindle-shaped, and within the fiber is a single, oval, centrally located nucleus (Figure 9-8). Smooth muscle cells contain actin and myosin filaments, but because the filaments are not as orderly as in skeletal and cardiac muscle tissue, the well-differentiated striations and sarcomeres do not occur.

Two kinds of smooth muscle tissue, visceral and multiunit, are recognized. The more common type is called **visceral muscle tissue.** It is found in wraparound sheets that form part of the walls of the hollow viscera such as the stomach, intestines, uterus, and urinary bladder. The terms *smooth muscle tissue* and *visceral muscle tissue* are sometimes used interchangeably. The fibers in visceral

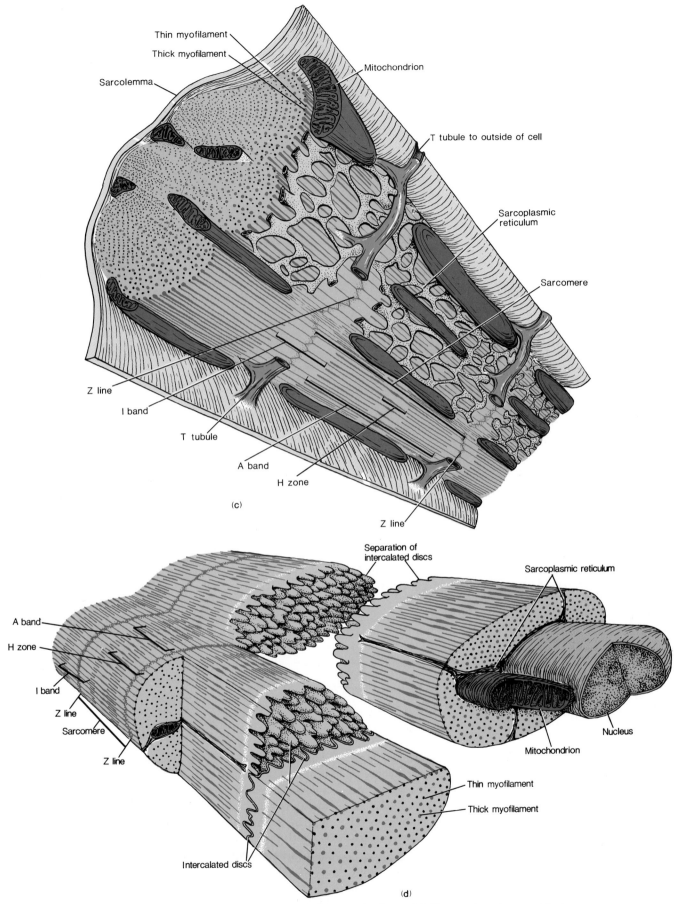

Thin myofilament
Thick myofilament
Sarcolemma
Mitochondrion
T tubule to outside of cell
Sarcoplasmic reticulum
Sarcomere
Z line
I band
T tubule
A band
H zone
Z line

(c)

Separation of intercalated discs
Sarcoplasmic reticulum
A band
H zone
I band
Z line
Sarcomere
Z line
Nucleus
Mitochondrion
Thin myofilament
Thick myofilament
Intercalated discs

(d)

FIGURE 9-7 (Continued) Histology of cardiac muscle tissue. (c) Diagram based on an electron micrograph showing several myofibrils. (d) Diagram based on an electron micrograph showing intercalated discs and related structures.

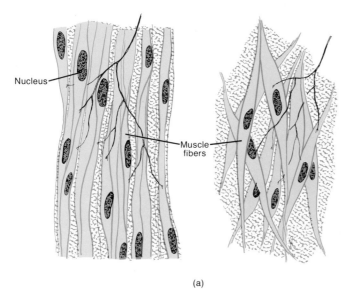

Nucleus

Muscle fibers

(a)

FIGURE 9-8 Histology of smooth muscle tissue. (a) Diagram of visceral smooth muscle tissue (left) and multiunit smooth muscle tissue (right). (b) Photomicrograph of several muscle fibers in longitudinal section at a magnification of 840×. (© 1983 by Michael H. Ross. Used by permission.) (c) Photomicrograph of several muscle fibers in cross section at a magnification of 840×. (© 1983 by Michael H. Ross. Used by permission.)

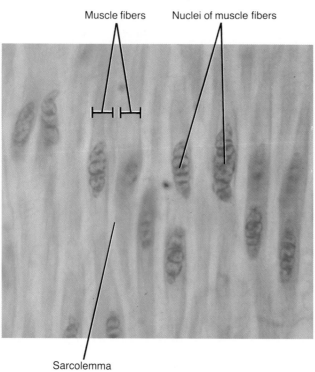

Muscle fibers   Nuclei of muscle fibers

Sarcolemma

(b)

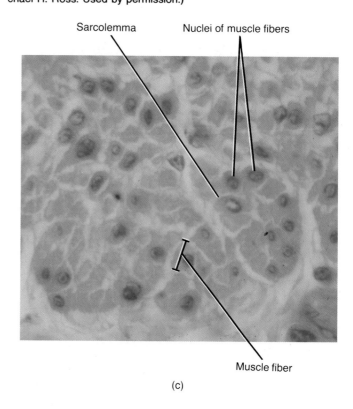

Sarcolemma   Nuclei of muscle fibers

Muscle fiber

(c)

muscle tissue are tightly bound together to form a continuous network. When a neuron stimulates one fiber, the impulse travels over the other fibers so that contraction occurs in a wave over many adjacent fibers. Whereas skeletal muscle cells contract as individual units, visceral muscle cells contract in sequence as the impulse spreads from one cell to another.

The second kind of smooth muscle tissue, **multiunit smooth muscle tissue,** consists of individual fibers each with its own motor-nerve endings. Whereas stimulation of a single visceral muscle fiber causes contraction of many adjacent fibers, stimulation of a single multiunit fiber causes contraction of only that fiber. In this respect, multiunit muscle tissue is like skeletal muscle tissue. Multiunit smooth muscle tissue is found in the walls of blood vessels, in the arrector pili muscles that attach to hair follicles, and in the intrinsic muscles of the eye, such as the iris.

Both kinds of smooth muscle tissue contract and relax more slowly than skeletal muscle tissue. This characteristic is probably due to the arrangement of the thin and thick myofilaments of smooth muscle. Also, both kinds of smooth muscle can maintain a forceful contraction longer than skeletal muscle. Finally, unlike skeletal muscle fibers, smooth muscle fibers can stretch without devel-

**EXHIBIT 9-2   SUMMARY OF THE PRINCIPAL CHARAC-
TERISTICS OF MUSCLE TISSUE**

	LOCATION	MICROSCOPIC APPEARANCE	NERVOUS CONTROL
Skeletal	Attached to bones.	Striated, multinucleated, unbranched fibers.	Voluntary.
Cardiac	Heart.	Striated, uninucleated, branched fibers containing intercalated discs.	Involuntary.
Smooth	Walls of hollow viscera and blood vessels.	Nonstriated, uninucleated, spindle-shaped fibers.	Involuntary.

oping tension. Thus, the smooth muscle in the wall of hollow organs such as the stomach, intestines, and urinary bladder can stretch as the viscera distend, while the pressure within them remains the same.

A summary of the principal characteristics of the three types of muscle tissue is presented in Exhibit 9-2.

AGING AND MUSCLE TISSUE

Beginning at nearly 30 years of age, there is a progressive loss of skeletal muscle mass, and the lost muscle tissue is largely replaced by fat. Accompanying the loss of muscle mass, there is a decrease in maximal strength and a diminishing of muscle reflexes.

DEVELOPMENTAL ANATOMY OF THE MUSCULAR SYSTEM

In this brief discussion of the development of the muscular system, we will concentrate mostly on skeletal muscles. Except for the muscles of the iris of the eyes and the arrector pili muscles attached to hairs, all muscles of the body are derived from **mesoderm.** As the mesoderm develops, a portion of it becomes arranged in dense columns on either side of the developing nervous system. These columns of mesoderm undergo segmentation into a series of blocks of cells called **somites** (Figure 9-9a). The first pair of somites appears on the twentieth day. Eventually, 44 pairs of somites are formed by the thirtieth day.

With the exception of the skeletal muscles of the head and extremities, *skeletal muscles* develop from the **mesoderm of somites.** Since there are very few somites in the head region of the embryo, most of the skeletal muscles there develop from the **general mesoderm** in the head

region. The skeletal muscles of the limbs develop from masses of general mesoderm around developing bones in embryonic limb buds (origins of future extremities; see Figure 5-6).

The cells of a somite are differentiated into three regions: (1) **myotome,** which forms most of the skeletal muscles; (2) **dermatome,** which forms the connective tissues, including the dermis, under the epidermis; and (3) **sclerotome,** which gives rise to the vertebrae (Figure 9-9b).

In the development of a skeletal muscle from a myotome of a somite, certain patterns emerge. For example, a myotome may split longitudinally into two or more portions. This represents the manner in which the trapezius muscle forms. Other myotomes split into two or more layers. Such a pattern represents how the external oblique, internal oblique, and transversus abdominis muscles develop. In other instances, several myotomes fuse to form a single muscle. Such an example is the rectus abdominis muscle. Muscles may also migrate, wholly or in part, from their sites of origin. The latissimus dorsi, for example, originates from the cervical myotomes, but extends all the way down to the thoracic and lumbar vertebrae and hip bone. Another trend that may be observed is a change in the direction of muscle fibers. Initially, muscle fibers in a myotome are parallel to the long axis of the embryo. However, nearly all developed skeletal muscles do not have fibers that are parallel to the long axis. One example is the external oblique. Finally, *fasciae, ligaments,* and *aponeuroses* may form as a result of degeneration of all or parts of myotomes.

Smooth muscle develops from **mesodermal cells** that migrate to and envelop the developing gastrointestinal tract and viscera.

Cardiac muscle develops from **mesodermal cells** that migrate to and envelop the developing heart while it is still in the form of primitive heart tubes (see Figure 13-10).

APPLICATIONS TO HEALTH

Disorders of the muscular system are related to disruptions of homeostasis. The disorders may involve a lack of nutrients, the accumulation of toxic products, disease, injury, disuse, or faulty nervous connections (innervations).

FIBROSIS

The formation of fibrous (containing fibers) connective tissue in locations where it normally does not exist is called **fibrosis.** Skeletal and cardiac muscle fibers cannot undergo mitosis, and dead muscle fibers are normally replaced with fibrous connective tissue. Fibrosis, then, is often a consequence of muscle injury or degeneration.

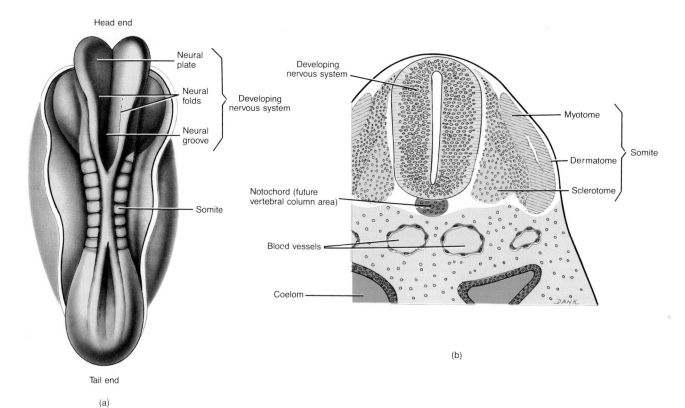

FIGURE 9-9 Development of the muscular system. (a) Dorsal aspect of an embryo indicating the location of somites. (b) Cross section of a somite.

FIBROSITIS

Fibrositis is an inflammation of fibrous tissue. If it occurs in the lumbar region, it is termed **lumbago** (lum-BĀ-gō). Fibrositis is a common condition characterized by pain, stiffness, or soreness of fibrous tissue, especially in the muscle coverings. It is not destructive or progressive. It may persist for years or spontaneously disappear. Attacks of fibrositis may follow an injury, repeated muscular strain, or prolonged muscular tension.

"CHARLEYHORSE"

Fibromyositis refers to a group of symptoms that include pain, tenderness, and stiffness of the joints, muscles, or adjacent structures. These symptoms ordinarily occur in various combinations. When the quadriceps femoris muscle of the thigh is involved, it is usually called **"charleyhorse."** It results from a contusion (bruising) and tearing of muscle fibers that produce a hematoma (collection of blood). "Charleyhorse" is characterized by the sudden onset of pain and is aggravated by motion. In some cases, local muscle spasms are noted. The condition is relieved with heat, massage, and rest and completely disappears, although occasionally it may become chronic or recur at frequent intervals.

MUSCULAR DYSTROPHY

The term **muscular dystrophy** (*dystrophy* = degeneration) applies to a number of inherited myopathies, or muscle-destroying diseases. The disease is characterized by degeneration of the individual muscle cells, which leads to a progressive atrophy of the skeletal muscle. Usually the voluntary skeletal muscles are weakened equally on both sides of the body, whereas the internal muscles, such as the diaphragm, are not affected. Histologically, the changes that occur include the variation in muscle fiber size, degeneration of fibers, and deposition of fat.

The cause of muscular dystrophy has been variously attributed to a genetic defect, faulty metabolism of potassium, protein deficiency, and inability of the body to utilize creatine.

There is no specific drug therapy for this disorder. Treatment involves attempts to prolong ambulation by muscle-strengthening exercises, corrective surgical measures, and appropriate braces. Patients are encouraged to keep physically active.

A simple and reliable screening test for a wide variety of skeletal muscle disorders has been developed. Increased levels of an enzyme called *creatine kinase* (*CK*) in the blood are typical not only of the dystrophies but of other wasting diseases as well. The measurement of CK, the most consistently elevated enzyme in cases of dystrophy, provides a diagnostic test even before clinical symptoms appear. Electromyography (EMG) studies and muscle biopsy are also used to diagnose muscular dystrophy.

MYASTHENIA GRAVIS (MG)

Myasthenia (mī'-as-THĒ-nē-a) **gravis (MG)** is a weakness of the skeletal muscles. It is caused by an abnormality at the neuromuscular junction that prevents the muscle fibers from contracting. Recall that motor neurons stimulate the skeletal muscle fibers to contract by releasing acetylcholine. Myasthenia gravis is an autoimmune disorder caused by antibodies directed against ACh receptors of the muscle fiber sarcolemma. The antibodies bind to the receptors and hinder the attachment of ACh to the receptors (see Chapter 18). As the disease progresses, more neuromuscular junctions become affected. The muscle becomes increasingly weaker and may eventually cease to function altogether.

Myasthenia gravis is more common in females, occurring most frequently between the ages of 20 and 50. The muscles of the face and neck are most apt to be involved. Initial symptoms include a weakness of the eye muscles and difficulty in swallowing. Later, the individual has difficulty chewing and talking. Eventually, the muscles of the limbs may become involved. Death may result from paralysis of the respiratory muscles, but usually the disorder does not progress to this stage.

Anticholinesterase drugs such as neostigmine and pyridostigmine, derivatives of physostigmine, have been the primary treatment for the disease. They act as inhibitors of acetylcholinesterase, thus raising the level of ACh to bind with available receptors. More recently, steroid drugs, such as prednisone, have been used with great success to reduce antibody levels. Immunosuppressant drugs are also used to decrease the production of antibodies that interfere with normal muscle contraction. Another recent treatment involves *plasmapheresis,* a procedure that separates blood cells from the plasma that contains the unwanted antibodies. The blood cells are then mixed with a plasma substitute and pumped back into the individual. In some individuals, surgical removal of the thymus gland is indicated (thymectomy).

ABNORMAL CONTRACTIONS

One kind of abnormal contraction of a muscle is **spasm,** a sudden, involuntary contraction of short duration. A **cramp** is a painful spasmodic contraction of a muscle. It is an involuntary, complete tetanic contraction. **Convulsions** are violent, involuntary tetanic contractions of an entire group of muscles. Convulsions occur when motor neurons are stimulated by fever, poisons, hysteria, or changes in body chemistry due to withdrawal of certain drugs. The stimulated neurons send many bursts of seemingly disordered impulses to the muscle fibers. **Fibrillation** is the uncoordinated contraction of individual muscle fibers preventing the smooth contraction of the muscle. A **tic** is a spasmodic twitching made involuntarily by muscles that are ordinarily under voluntary control. Twitching of the eyelid and face muscles are examples. In general, tics are of psychological origin.

KEY MEDICAL TERMS ASSOCIATED WITH THE MUSCULAR SYSTEM

Electromyography or **EMG** (*electro* = electricity; *myo* = muscle; *graph* = to write)   The recording and study of the electrical changes that occur in muscle tissue.

Gangrene (*gangraena* = an eating sore)   Death of a soft tissue, such as muscle, that results from interruption of its blood supply. It is caused by various species of *Clostridium,* bacteria that live anaerobically in the soil.

Myalgia (*algia* = painful condition)   Pain in or associated with muscles.

Myoma (*oma* = tumor)   A tumor consisting of muscle tissue.

Myomalacia (*malaco* = soft)   Softening of a muscle.

Myopathy (*pathos* = disease)   Any disease of muscle tissue.

Myosclerosis (*scler* = hard)   Hardening of a muscle.

Myositis (*itis* = inflammation of)   Inflammation of muscle fibers (cells).

Myospasm   Spasm of a muscle.

Myotonia (*tonia* = tension)   Increased muscular excitability and contractility with decreased power of relaxation; tonic spasm of the muscle.

Paralysis (*para* = beyond; *lyein* = to loosen)   Loss or impairment of motor (muscular) function resulting from a lesion of nervous or muscular origin.

Trichinosis   A myositis caused by the parasitic worm *Trichinella spiralis,* which may be found in the muscles of humans, rats, and pigs. People contract the disease by eating insufficiently cooked infected pork.

Volkmann's contracture (*contra* = against)   Permanent contraction of a muscle due to replacement of destroyed muscle cells with fibrous tissue that lacks ability to stretch. Destruction of muscle cells may occur from interference with circulation caused by a tight bandage, a piece of elastic, or a cast.

Wryneck or **torticollis** (*tortus* = twisted; *collum* = neck)   Spasmodic contraction of several of the superficial and deep muscles of the neck; produces twisting of the neck and an unnatural position of the head.

STUDY OUTLINE

Characteristics (p. 205)
1. Excitability is the property of receiving and responding to stimuli.
2. Contractility is the ability to shorten and thicken (contract).
3. Extensibility is the ability to be stretched or extended.
4. Elasticity is the ability to return to original shape after contraction or extension.

Functions (p. 205)
1. Through contraction, muscle tissue performs the three important functions of motion, maintenance of posture, and heat production.

Types (p. 205)
1. Skeletal muscle tissue is attached to bones. It is striated and voluntary.
2. Cardiac muscle tissue forms the wall of the heart. It is striated and involuntary.
3. Visceral muscle tissue is located in viscera. It is nonstriated (smooth) and involuntary.

Skeletal Muscle Tissue (p. 206)
1. The term fascia is applied to a sheet or broad band of fibrous connective tissue underneath the skin or around muscles and organs of the body. There are three types of fascia: superficial, deep, and subserous.
2. Connective tissue components are epimysium, covering the entire muscle; perimysium, covering fasciculi; and endomysium, covering fibers.
3. Tendons and aponeuroses are extensions of connective tissue beyond muscle cells that attach the muscle to bone or other muscle.
4. Nerves convey impulses for muscular contraction.
5. Blood provides nutrients and oxygen for contraction.
6. Skeletal muscle consists of fibers covered by a sarcolemma. The fibers contain sarcoplasm, nuclei, sarcoplasmic reticulum, and T tubules.
7. Each fiber contains myofibrils that consist of thin and thick myofilaments. The myofilaments are compartmentalized into sarcomeres.
8. Thin myofilaments are composed of actin, tropomyosin, and troponin; thick myofilaments consist of myosin.

Contraction

Sliding-Filament Theory
1. A nerve impulse travels over the sarcolemma and enters the T tubules and sarcoplasmic reticulum.
2. The nerve impulse leads to the release of calcium ions from the sarcoplasmic reticulum, triggering the contractile process.
3. Actual contraction is brought about when the thin myofilaments of a sarcomere slide toward each other.

Neuromuscular Junction (Motor End Plate)
1. A motor neuron transmits a nerve impulse to a skeletal muscle for contraction.
2. A neuromuscular junction (motor end plate) refers to an axon terminal of a motor neuron and the portion of the muscle fiber sarcolemma in close approximation with it.

Motor Unit
1. A motor neuron and the muscle fibers it stimulates form a motor unit.
2. A single motor unit may innervate up to 500 muscle fibers.

Mechanism
1. When a nerve impulse reaches an axon terminal, the synaptic vesicles of the terminal release acetylcholine (ACh), which transmits the impulse to the muscle fiber sarcolemma and then into the T tubules and sarcoplasmic reticulum.
2. The transmitted impulse releases calcium ions that combine with troponin, causing it to pull on tropomyosin, thus exposing myosin-binding sites on actin.
3. The energy released from the breakdown of ATP causes myosin cross bridges to attach to actin, and their movement results in the sliding of thin myofilaments.

All-or-None Principle
1. Muscle fibers of a motor unit contract to their fullest extent or not at all.
2. The weakest stimulus capable of causing contraction is a liminal, or threshold, stimulus.
3. A stimulus not capable of inducing contraction is a subliminal, or subthreshold, stimulus.

Muscle Tone
1. A sustained partial contraction of portions of a skeletal muscle results in muscle tone.
2. Tone is essential for maintaining posture.
3. Flaccidity is a condition of less than normal tone. Atrophy is a wasting away or decrease in size; hypertrophy is an enlargement or overgrowth.

Fast-Twitch (White) and Slow-Twitch (Red) Muscle
1. The duration of contraction of various muscles of the body varies with the functions of the muscles in maintaining homeostasis.
2. White, or fast-twitch, muscles have an extensive sarcoplasmic reticulum.
3. Red, or slow-twitch, muscles have smaller fibers, more blood capillaries, and a large amount of myoglobin.

Cardiac Muscle Tissue (p. 215)
1. This muscle is found only in the heart. It is striated and involuntary.
2. The cells are quadrangular and usually contain a single centrally placed nucleus.
3. Compared to skeletal muscle tissue, cardiac muscle tissue has more sarcoplasm, more mitochondria, less well-developed sarcoplasmic reticulum, and larger T tubules located at Z lines rather than at A-I band junctions. Myofilaments are not arranged in discrete myofibrils.
4. The fibers branch freely to form two continuous networks, each of which contracts as a functional unit.

5. Intercalated discs provide strength and aid impulse conduction.

Smooth Muscle Tissue (p. 216)
1. Smooth muscle is nonstriated and involuntary.
2. Visceral smooth muscle is found in the walls of viscera. The fibers are arranged in a network.
3. Multiunit smooth muscle is found in blood vessels and the eye. The fibers operate singly rather than as a unit.

Aging and Muscle Tissue (p. 219)
1. At about 30 years of age, there is a progressive loss of skeletal muscle, which is replaced by fat.
2. There is also a decrease in muscle strength and diminished muscle reflexes.

Developmental Anatomy of the Muscular System (p. 219)
1. With few exceptions, muscles develop from mesoderm.

2. Skeletal muscles of the head and extremities develop from general mesoderm; the remainder of the skeletal muscles develop from the mesoderm of somites.

Applications to Health (p. 219)
1. Fibrosis is the formation of fibrous tissue where it normally does not exist; it frequently occurs in damaged muscle tissue.
2. Fibrositis is an inflammation of fibrous tissue. If it occurs in the lumbar region, it is called lumbago.
3. "Charleyhorse" refers to pain, tenderness, and stiffness of joints, muscles, and related structures in the thigh.
4. Muscular dystrophy is a hereditary disease of muscles characterized by degeneration of individual muscle cells.
5. Myasthenia gravis (MG) is a disease characterized by great muscular weakness and fatigability resulting from improper neuromuscular transmission.
6. Abnormal contractions include spasms, cramps, convulsions, fibrillations, and tics.

REVIEW QUESTIONS

1. How is the skeletal system related to the muscular system? What are the three basic functions of the muscular system?
2. What are the four characteristics of muscle tissue?
3. How can the three types of muscle tissue be distinguished?
4. What is fascia? What are the three different types of fascia and where are they found in the body?
5. Define epimysium, perimysium, endomysium, tendon, and aponeurosis. Describe the nerve and blood supply to a skeletal muscle.
6. Discuss the microscopic structure of skeletal muscle tissue.
7. In considering the contraction of skeletal muscle tissue, describe the following: neuromuscular junction (motor end plate), motor unit, role of calcium, sources of energy, and sliding-filament theory.
8. What is the all-or-none principle? Relate it to a threshold and subthreshold stimulus.
9. What is muscle tone? Why is it important? Distinguish between atrophy and hypertrophy.
10. Compare fast-twitch (white) and slow-twitch (red) muscle with respect to structure and function.
11. Compare cardiac and skeletal muscle with regard to microscopic structure, functions, and locations.
12. Compare cardiac and smooth muscle with regard to microscopic structure, functions, and locations.
13. Describe the effects of aging on muscle tissue.
14. Discuss the development of the muscular system.
15. Define fibrosis and fibrositis.
16. What is a "charleyhorse"?
17. What is muscular dystrophy?
18. What is myasthenia gravis (MG)? In this disease, why do the muscles not contract normally?
19. Define each of the following abnormal muscular contractions: spasm, cramp, convulsion, fibrillation, and tic.
20. Refer to the glossary of key medical terms associated with the muscular system. Be sure that you can define each term.

10

The Muscular System

Student Objectives

Describe the relationship between bones and skeletal muscles in producing body movements.

Define a lever and fulcrum and compare the three classes of levers on the basis of placement of the fulcrum, effort, and resistance.

Describe most body movements as activities of groups of muscles by explaining the roles of the prime mover, antagonist, synergist, and fixator.

Identify the various arrangements of muscle fibers in a skeletal muscle and relate the arrangements to the strength of contractions and range of movement.

Define the criteria employed in naming skeletal muscles.

Identify the principal skeletal muscles in different regions of the body by name, origin, insertion, action, and innervation.

Compare the common sites for intramuscular injections.

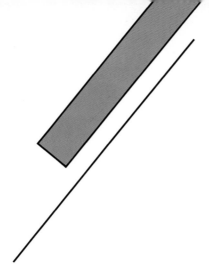

The term **muscle tissue** refers to all the contractile tissues of the body: skeletal, cardiac, and smooth muscle. The **muscular system,** however, refers to the *skeletal* muscle system: the skeletal muscle tissue and connective tissues that make up individual muscle organs, such as the biceps. Cardiac muscle tissue is located in the heart and is therefore considered part of the circulatory system. Smooth muscle tissue of the intestine is part of the digestive system, while smooth muscle tissue of the urinary bladder is part of the urinary system. In this chapter, we discuss only the muscular system. We will see how skeletal muscles produce movement and describe the principal skeletal muscles.

HOW SKELETAL MUSCLES PRODUCE MOVEMENT

ORIGIN AND INSERTION

Skeletal muscles produce movements by exerting force on tendons, which in turn pull on bones. Most muscles cross at least one joint and are attached to the articulating bones that form the joint (Figure 10-1). When such a muscle contracts, it draws one articulating bone toward the other. The two articulating bones usually do not move equally in response to the contraction. One is held nearly in its original position because other muscles contract to pull it in the opposite direction or because its structure makes it less movable. Ordinarily, the attachment of a muscle tendon to the stationary bone is called the **origin.** The attachment of the other muscle tendon to the movable bone is the **insertion.** A good analogy is a spring on a door. The part of the spring attached to the door represents the insertion; the part attached to the frame is the origin. The fleshy portion of the muscle between the tendons of the origin and insertion is called the **belly (gaster).** The origin is usually proximal and the insertion distal, especially in the appendages. In addition, muscles that move a body part generally do not cover the moving part. Figure 10-1a shows that although contraction of the biceps brachii muscle moves the forearm, the belly of the muscle lies over the humerus.

LEVER SYSTEMS AND LEVERAGE

In producing a body movement, bones act as levers and joints function as fulcrums of these levers. A **lever** may be defined as a rigid rod that moves about on some fixed point called a **fulcrum.** A fulcrum may be symbolized as Ⓐ. A lever is acted on at two different points by two different forces: the *resistance* ⓡ and the *effort* (E). The resistance may be regarded as a force to be overcome, whereas the effort is exerted to overcome the resistance. The resistance may be the weight of the body part that is to be moved. The muscular effort (contraction) is applied to the bone at the insertion of the muscle and produces motion. Consider the biceps brachii flexing the forearm at the elbow as a weight is lifted (Figure 10-1b). When the forearm is raised, the elbow is the fulcrum. The weight of the forearm plus the weight in the hand is the resistance. The shortening of the biceps brachii pulling the forearm up is the effort.

Levers are categorized into three types according to the positions of the fulcrum, the effort, and the resistance.

1. In a **first-class lever,** the fulcrum is between the effort and resistance (Figure 10-2a). An example of a first-class lever is a seesaw. There are not many first-class levers in the body. One example is the head resting on the vertebral column. When the head is raised, the facial portion of the skull is the resistance. The joint between the atlas and occipital bone (atlantooccipital joint) is the fulcrum. The contraction of the muscles of the back is the effort.

2. Second-class levers have the fulcrum at one end, the effort at the opposite end, and the resistance in between (Figure 10-2b). They operate like a wheelbarrow. Most authorities agree that there are very few examples of second-class levers in the body. One example is raising

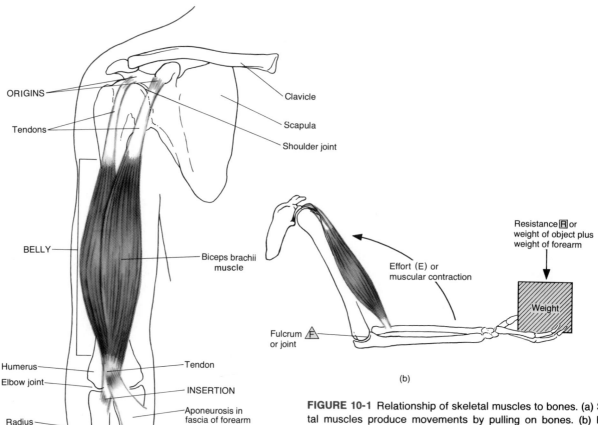

ORIGINS

Tendons

Clavicle

Scapula

Shoulder joint

BELLY

Biceps brachii
muscle

Humerus

Elbow joint

Tendon

INSERTION

Aponeurosis in
fascia of forearm

Radius

Ulna

(a)

Resistance Ⓡ or
weight of object plus
weight of forearm

Effort (E) or
muscular contraction

Weight

Fulcrum Ⓕ
or joint

(b)

FIGURE 10-1 Relationship of skeletal muscles to bones. (a) Skeletal muscles produce movements by pulling on bones. (b) Bones serve as levers and joints act as fulcrums for the levers. Here the lever-fulcrum principle is illustrated by the movement of the forearm lifting a weight. Note where the resistance and effort are applied in this example.

the body on the toes. The body is the resistance, the ball of the foot is the fulcrum, and the contraction of the calf muscles to pull the heel upward is the effort.

3. Third-class levers consist of the fulcrum at one end, the resistance at the opposite end, and the effort between them (Figure 10-2c). They are the most common levers in the body. A common example is flexing the forearm at the elbow. The weight of the forearm is the resistance, the contraction of the biceps brachii is the effort, and the elbow joint is the fulcrum.

Leverage—the mechanical advantage gained by a lever—is largely responsible for a muscle's strength and range of movement. Consider strength first. Suppose we have two muscles of the same strength crossing and acting on a joint. Assume also that one is attached farther from the joint and one is nearer. The muscle attached farther will produce the more powerful movement. Thus, strength of movement depends on the placement of muscle attachments.

In considering the range of movement, again assume that we have two muscles of the same strength crossing and acting on a joint and that one is attached farther from the joint than the other. The muscle inserting closer

to the joint will produce the greater range of movement. Thus range of movement also depends on the placement of muscle attachments. Since strength increases with distance from the joint and range of movement decreases, maximal strength and maximal range are incompatible, and strength and range vary inversely.

ARRANGEMENT OF FASCICULI

Recall from Chapter 9 that skeletal muscle fibers are arranged within the muscle in bundles called fasciculi (fascicles). The muscle fibers are arranged in a parallel fashion within each bundle, but the arrangement of the fasciculi with respect to the tendons may take one of four characteristic patterns.

The first pattern is called **parallel.** The fasciculi are parallel with the longitudinal axis and terminate at either end in flat tendons. The muscle is typically quadrilateral in shape. An example is the stylohyoid muscle (see Figure 10-7). In a modification of the parallel arrangement, called *fusiform,* the fasciculi are nearly parallel with the longitudinal axis and terminate at either end in flat tendons, but the muscle tapers toward the tendons, where the di-

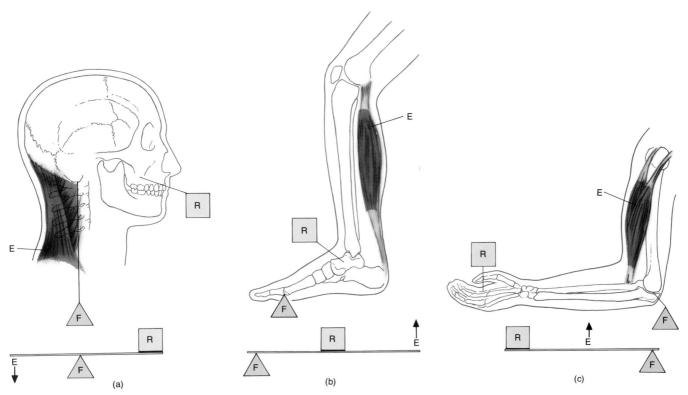

FIGURE 10-2 Classes of levers. Each is defined on the basis of the placement of the fulcrum, effort, and resistance. (a) First-class lever. (b) Second-class lever. (c) Third-class lever.

ameter is less than that of the belly. An example is the biceps brachii muscle (see Figure 10-17).

The second distinct pattern is called **convergent.** A broad origin of fasciculi converges to a narrow, restricted insertion. Such a pattern gives the muscle a triangular shape. An example is the deltoid muscle (see Figure 10-16).

The third distinct pattern is referred to as **pennate.** The fasciculi are short in relation to the entire length of the muscle and the tendon extends nearly the entire length of the muscle. The fasciculi are directed obliquely toward the tendon like the plumes of a feather. If the fasciculi are arranged on only one side of a tendon, as in the extensor digitorum muscle, the muscle is referred to as *unipennate* (see Figure 10-18). If the fasciculi are arranged on both sides of a centrally positioned tendon, as in the rectus femoris muscle, the muscle is referred to as *bipennate* (see Figure 10-21).

The final distinct pattern is referred to as **circular.** The fasciculi are arranged in a circular pattern and enclose an orifice. An example is the orbicularis oris muscle (see Figure 10-4).

Fascicular arrangement is correlated with the power of a muscle and range of movement. When a muscle fiber contracts, it shortens to a length just slightly greater than half of its resting length. Thus, the longer the fibers in a muscle, the greater the range of movement it can produce. By contrast, the strength of a muscle depends on the total number of fibers it contains, since a short fiber can contract as forcefully as a long one. Since a given muscle can either contain a small number of long fibers or a large number of short fibers, fascicular arrangement represents a compromise between power and range of movement. Pennate muscles, for example, have a larger number of fasciculi distributed over their tendons, giving them greater power but a smaller range of movement. Parallel muscles, on the other hand, have comparatively few fasciculi that extend the length of the muscle. Thus, they have a greater range of movement but less power.

GROUP ACTIONS

Most movements are coordinated by several skeletal muscles acting in groups rather than individually. Consider flexing the forearm at the elbow, for example. A muscle that causes a desired action is referred to as the **prime mover (agonist).** In this instance, the biceps brachii is the prime mover (see Figure 10-17). Simultaneously with the contraction of the biceps brachii, another muscle, called the **antagonist,** is relaxing. In this movement, the triceps brachii serves as the antagonist (see Figure 10-17). The antagonist has an effect opposite to that of the prime mover, that is, the antagonist relaxes and yields to the movement of the prime mover. If we consider extending the forearm at the elbow, the triceps brachii is the prime mover and the biceps brachii is the antagonist. If both the prime mover and antagonist contacted simultaneously with equal force, there would not be any movement. Most skeletal muscles are arranged in opposing pairs at various joints, that is, flexors-extensors, abductors-adductors, and so on. You should not assume that the biceps brachii is always the prime mover and the triceps brachii is always the antagonist. For example, when extending the forearm at the elbow, the triceps brachii serves as the prime mover and the biceps brachii functions as the antagonist. Their roles are reversed.

In addition to prime movers and antagonists, most movements also involve muscles called **synergists.** These muscles serve to steady a movement. In this way unwanted movements are prevented and the prime mover functions more efficiently. For example, flex your hand at the wrist and then make a fist. Note how difficult this is to do. Now, extend your hand at the wrist and then make a fist. Note how much easier it is to clench your fist. In this case, the extensor muscles of the wrist act as synergists in cooperation with the flexor muscles of the fingers acting as prime movers. The extensor muscles of the fingers serve as antagonists (see Figure 10-18).

Some muscles in a group also act as **fixators.** Fixators stabilize the origin of the prime mover so that the prime mover can act more efficiently. For example, the scapula is a freely movable bone in the pectoral (shoulder) girdle that serves as an origin for several muscles that move the arm. However, in order for the scapula to serve as a firm origin for muscles that move the arm, it must be held steady. This is accomplished by fixator muscles that hold the scapula firmly against the chest. In abduction of the arm, the deltoid muscle serves as the prime mover, whereas fixators (pectoralis minor, rhomboid, subclavius, and serratus anterior) hold the scapula firmly (see Figure

10-15). These fixators stabilize the origin of the deltoid muscle on the scapula while the insertion of the muscle pulls on the humerus to abduct the arm. Many muscles are, at various times, prime movers, antagonists, synergists, or fixators under different conditions, depending on the movement.

NAMING SKELETAL MUSCLES

The names of most of the nearly 700 skeletal muscles are based on several types of characteristics. Learning the terms used to indicate specific characteristics will help you remember the names of muscles.

1. Muscle names may indicate the **direction of the muscle fibers.** *Rectus* fibers usually run parallel to the midline of the body. *Transverse* fibers run perpendicular to the midline. *Oblique* fibers are diagonal to the midline. Muscles named according to these three directions include the rectus abdominis, transversus abdominis, and external oblique.

2. A muscle may be named according to **location.** The temporalis is near the temporal bone. The tibialis anterior is near the tibia.

3. Size is another criterion. The term *maximus* means largest, *minimus* means smallest, *longus* means long, and *brevis* means short. Examples include the gluteus maximus, gluteus minimus, adductor longus, and peroneus brevis.

4. Some muscles are named for their **number of origins.** The *biceps brachii* has two origins, the *triceps brachii* three, and the *quadriceps femoris* four.

5. Other muscles are named on the basis of **shape.** Common examples include the deltoid (meaning triangular) and trapezius (meaning trapezoid).

EXHIBIT 10-1   PRINCIPAL ACTIONS OF MUSCLES

ACTION	DEFINITION	EXAMPLE
Flexor	Usually decreases the anterior angle at a joint; some decrease the posterior angle.	Flexor carpi radialis.
Extensor	Usually increases the anterior angle at a joint; some increase the posterior angle.	Extensor carpi ulnaris.
Abductor	Moves a bone away from the midline.	Abductor hallucis longus.
Adductor	Moves a bone closer to the midline.	Adductor longus.
Levator	Produces an upward movement.	Levator scapulae.
Depressor	Produces a downward movement.	Depressor labii inferioris.
Supinator	Turns the palm upward or anteriorly.	Supinator.
Pronator	Turns the palm downward or posteriorly.	Pronator teres.
Sphincter	Decreases the size of an opening.	External anal sphincter.
Tensor	Makes a body part more rigid.	Tensor fasciae latae.
Rotator	Moves a bone around its longitudinal axis.	Obturator.

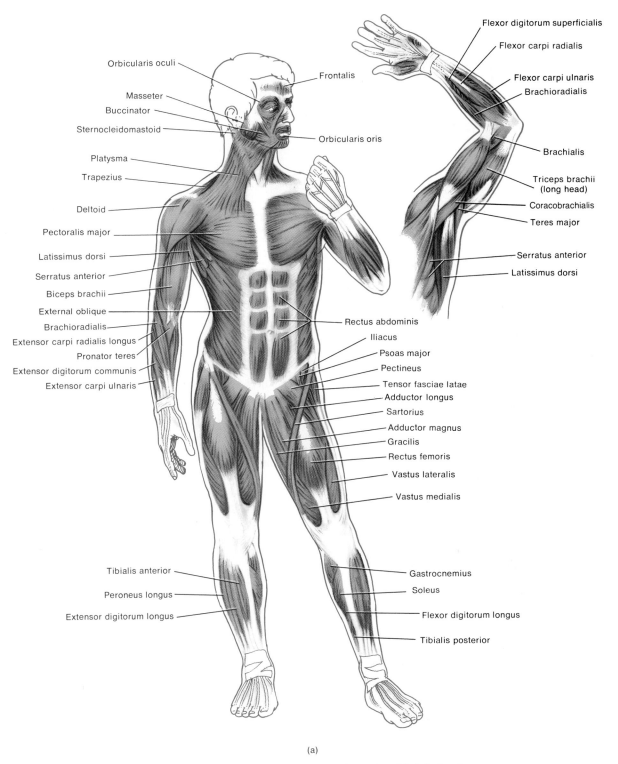

Orbicularis oculi
Frontalis
Masseter
Buccinator
Sternocleidomastoid
Orbicularis oris
Platysma
Trapezius
Deltoid
Pectoralis major
Latissimus dorsi
Serratus anterior
Biceps brachii
External oblique
Brachioradialis
Extensor carpi radialis longus
Pronator teres
Extensor digitorum communis
Extensor carpi ulnaris

Flexor digitorum superficialis
Flexor carpi radialis
Flexor carpi ulnaris
Brachioradialis
Brachialis
Triceps brachii (long head)
Coracobrachialis
Teres major
Serratus anterior
Latissimus dorsi

Rectus abdominis
Iliacus
Psoas major
Pectineus
Tensor fasciae latae
Adductor longus
Sartorius
Adductor magnus
Gracilis
Rectus femoris
Vastus lateralis
Vastus medialis

Tibialis anterior
Peroneus longus
Extensor digitorum longus

Gastrocnemius
Soleus
Flexor digitorum longus
Tibialis posterior

(a)

FIGURE 10-3 Principal superficial skeletal muscles. (a) Anterior view.

6. Muscles may be named after their **origin** and **insertion.** The sternocleidomastoid originates on the sternum and clavicle and inserts at the mastoid process of the temporal bone; the stylohyoideus originates on the styloid process of the temporal bone and inserts at the hyoid bone.

7. Still another criterion used for naming muscles is **action.** Exhibit 10-1 lists the principal actions of muscles, their definitions, and examples of muscles that perform the actions. For convenience, the actions are grouped as antagonistic pairs where possible.

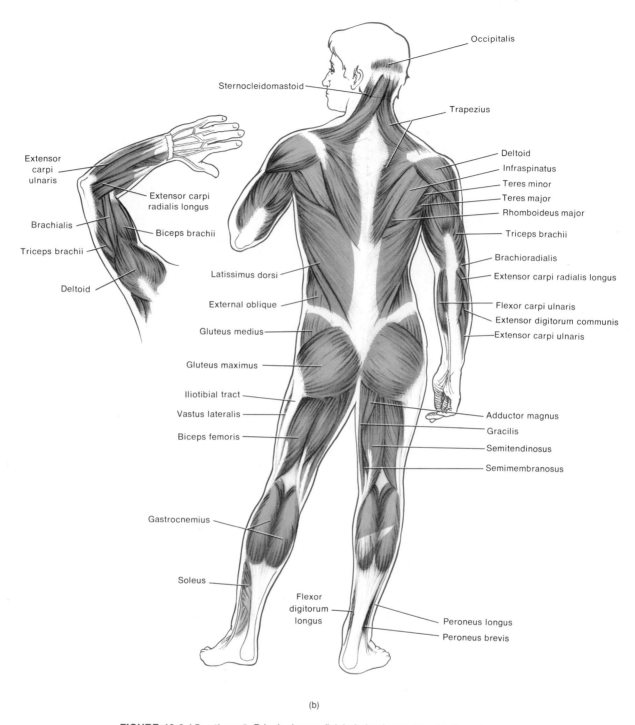

(b)

FIGURE 10-3 (*Continued*) Principal superficial skeletal muscles. (b) Posterior view.

PRINCIPAL SKELETAL MUSCLES

Exhibits 10-2 through 10-21 list the principal muscles of the body with their origins, insertions, actions, and innervations. (By no means have all the muscles of the body been included.) Refer to Chapters 6 and 7 to review bone markings, since they serve as points of origin and insertion for muscles. The muscles are divided into groups according to the part of the body on which they act. If you have mastered the naming of the muscles, their actions will have more meaning. Figure 10-3

shows general anterior and posterior views of the muscular system. Do not try to memorize all these muscles yet. As you study groups of muscles in the following exhibits, refer to Figure 10-3 to see how each group is related to all others.

The figures that accompany the exhibits contain superficial and deep, anterior and posterior, or medial and lateral views to show the muscles' position as clearly as possible. An attempt has been made to show the relationship of the muscles under consideration to other muscles in the area you are studying.

EXHIBIT 10-2 MUSCLES OF FACIAL EXPRESSION (Figure 10-4)

MUSCLE	ORIGIN	INSERTION	ACTION	INNERVATION
Epicranius (*epi* = over; *crani* = skull)	This muscle is divisible into two portions: the frontalis over the frontal bone and the occipitalis over the occipital bone. The two muscles are united by a strong aponeurosis, the galea aponeurotica, which covers the superior and lateral surfaces of the skull.			
Frontalis (*front* = forehead)	Galea aponeurotica.	Skin superior to supraorbital line.	Draws scalp forward, raises eyebrows, and wrinkles skin of forehead horizontally.	Facial (VII) nerve.
Occipitalis (*occipito* = base of skull)	Occipital bone and mastoid process of temporal bone.	Galea aponeurotica.	Draws scalp backward.	Facial (VII) nerve.
Orbicularis oris (*orb* = circular; *or* = mouth)	Muscle fibers surrounding opening of mouth.	Skin at corner of mouth.	Closes lips, compresses lips against teeth, protrudes lips, and shapes lips during speech.	Facial (VII) nerve.
Zygomaticus major (*zygomatic* = cheek bone; *major* = greater)	Zygomatic bone.	Skin at angle of mouth and orbicularis oris.	Draws angle of mouth upward and outward as in smiling or laughing.	Facial (VII) nerve.
Levator labii superioris (*levator* = raises or elevates; *labii* = lip; *superioris* = upper)	Superior to infraorbital foramen of maxilla.	Skin at angle of mouth and orbicularis oris.	Elevates (raises) upper lip.	Facial (VII) nerve.
Depressor labii inferioris (*depressor* = depresses or lowers; *inferioris* = lower)	Mandible.	Skin of lower lip.	Depresses (lowers) lower lip.	Facial (VII) nerve.
Buccinator (*bucc* = cheek)	Alveolar processes of maxilla and mandible and pterygomandibular raphe (fibrous band extending from the pterygoid hamulus to the mandible).	Orbicularis oris.	Major cheek muscle; compresses cheek as in blowing air out of mouth and causes cheeks to cave in, producing the action of sucking.	Facial (VII) nerve.
Mentalis (*mentum* = chin)	Mandible.	Skin of chin.	Elevates and protrudes lower lip and pulls skin of chin up as in pouting.	Facial (VII) nerve.
Platysma (*platy* = flat, broad)	Fascia over deltoid and pectoralis major muscles.	Mandible, muscles around angle of mouth, and skin of lower face.	Draws outer part of lower lip downward and backward as in pouting; depresses mandible.	Facial (VII) nerve.
Risorius (*risor* = laughter)	Fascia over parotid (salivary) gland.	Skin at angle of mouth.	Draws angle of mouth laterally as in tenseness.	Facial (VII) nerve.
Orbicularis oculi (*ocul* = eye)	Medial wall of orbit.	Circular path around orbit.	Closes eye.	Facial (VII) nerve.
Corrugator supercilii (*corrugo* = to wrinkle; *supercilium* = eyebrow)	Medial end of superciliary arch of frontal bone.	Skin of eyebrow.	Draws eyebrow downward as in frowning.	Facial (VII) nerve.
Levator palpebrae superioris (*palpebrae* = eyelids) (see Figure 10-6b)	Roof of orbit (lesser wing of sphenoid bone).	Skin of upper eyelid.	Elevates upper eyelid.	Oculomotor (III) nerve.

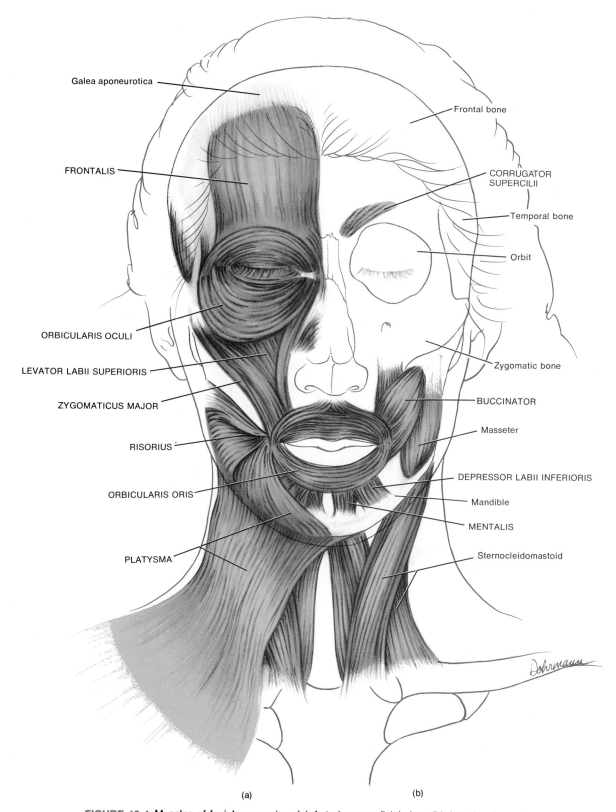

Galea aponeurotica

Frontal bone

FRONTALIS

CORRUGATOR SUPERCILII

Temporal bone

Orbit

ORBICULARIS OCULI

LEVATOR LABII SUPERIORIS

Zygomatic bone

ZYGOMATICUS MAJOR

BUCCINATOR

Masseter

RISORIUS

ORBICULARIS ORIS

DEPRESSOR LABII INFERIORIS

Mandible

MENTALIS

Sternocleidomastoid

PLATYSMA

(a)

(b)

FIGURE 10-4 Muscles of facial expression. (a) Anterior superficial view. (b) Anterior deep view.

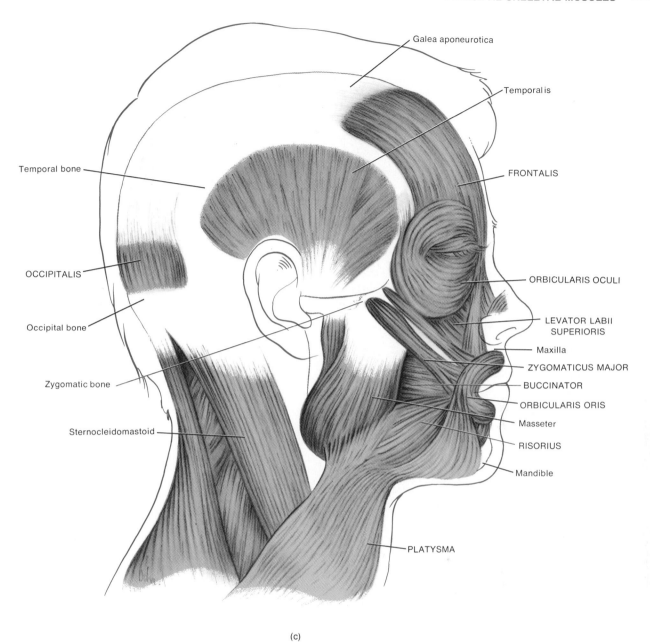

Galea aponeurotica

Temporalis

FRONTALIS

Temporal bone

ORBICULARIS OCULI

OCCIPITALIS

LEVATOR LABII
SUPERIORIS

Occipital bone

Maxilla

Zygomatic bone

ZYGOMATICUS MAJOR

BUCCINATOR

ORBICULARIS ORIS

Sternocleidomastoid

Masseter

RISORIUS

Mandible

PLATYSMA

(c)

FIGURE 10-4 (*Continued*) Muscles of facial expression. (c) Right lateral superficial view.

EXHIBIT 10-3   MUSCLES THAT MOVE THE LOWER JAW (Figure 10-5)

MUSCLE	ORIGIN	INSERTION	ACTION	INNERVATION
Masseter (*maseter* = chewer)	Maxilla and zygomatic arch.	Angle and ramus of mandible.	Elevates mandible as in closing mouth and protracts (protrudes) mandible.	Mandibular branch of trigeminal (V) nerve.
Temporalis (*tempora* = temples)	Temporal bone.	Coronoid process of mandible.	Elevates and retracts mandible.	Temporal nerve from mandibular division of trigeminal (V) nerve.
Medial pterygoid (*medial* = closer to midline; *pterygoid* = like a wing; pterygoid plate of sphenoid bone)	Medial surface of lateral pterygoid plate of sphenoid; maxilla.	Angle and ramus of mandible.	Elevates and protracts mandible and moves mandible from side to side.	Mandibular branch of trigeminal (V) nerve.
Lateral pterygoid (*lateral* = farther from midline)	Greater wing and lateral surface of lateral pterygoid plate of sphenoid.	Condyle of mandible; temporomandibular articulation.	Protracts mandible, opens mouth, and moves mandible from side to side.	Mandibular branch of trigeminal (V) nerve.

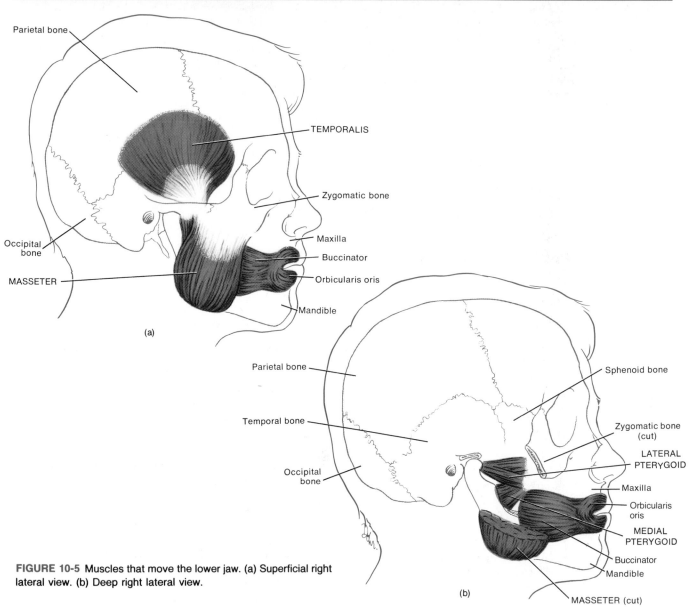

FIGURE 10-5 Muscles that move the lower jaw. (a) Superficial right lateral view. (b) Deep right lateral view.

EXHIBIT 10-4 MUSCLES THAT MOVE THE EYEBALLS—EXTRINSIC MUSCLES* (Figure 10-6)

MUSCLE	ORIGIN	INSERTION	ACTION	INNERVATION
Superior rectus (*superior* = above; *rectus* = in this case, muscle fibers running parallel to long axis of eyeball)	Tendinous ring attached to bony orbit around optic foramen.	Superior and central part of eyeball.	Rolls eyeball upward.	Oculomotor (III) nerve.
Inferior rectus (*inferior* = below)	Same as above.	Inferior and central part of eyeball.	Rolls eyeball downward.	Oculomotor (III) nerve.
Lateral rectus	Same as above.	Lateral side of eyeball.	Rolls eyeball laterally.	Abducens (VI) nerve.
Medial rectus	Same as above.	Medial side of eyeball.	Rolls eyeball medially.	Oculomotor (III) nerve.
Superior oblique (*oblique* = in this case, muscle fibers running diagonally to long axis of eyeball)	Same as above.	Eyeball between superior and lateral recti.	Rotates eyeball on its axis; directs cornea downward and laterally; note that it moves through a ring of fibrocartilaginous tissue called the trochlea (*trochlea* = pulley).	Trochlear (IV) nerve.
Inferior oblique	Maxilla (front of orbital cavity).	Eyeball between inferior and lateral recti.	Rotates eyeball on its axis; directs cornea upward and laterally.	Oculomotor (III) nerve.

* Muscles situated on the outside of the eyeball.

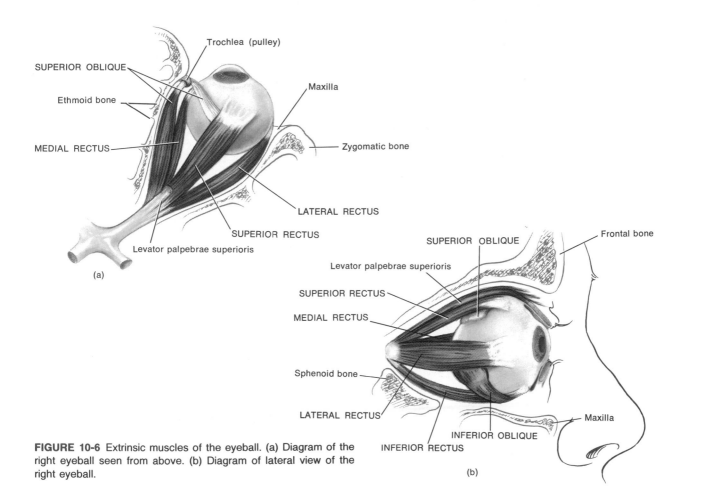

FIGURE 10-6 Extrinsic muscles of the eyeball. (a) Diagram of the right eyeball seen from above. (b) Diagram of lateral view of the right eyeball.

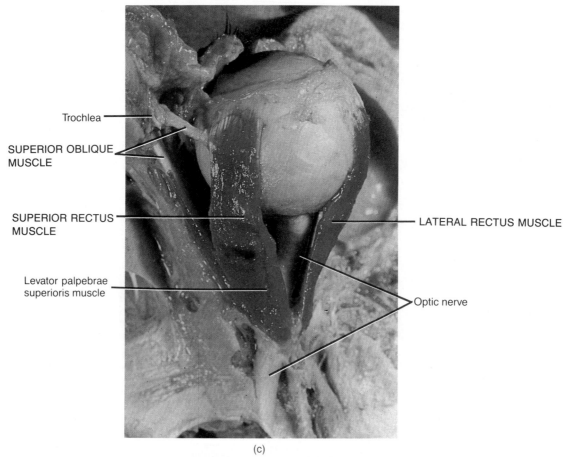

Trochlea

SUPERIOR OBLIQUE MUSCLE

SUPERIOR RECTUS MUSCLE

Levator palpebrae superioris muscle

LATERAL RECTUS MUSCLE

Optic nerve

(c)

FIGURE 10-6 (*Continued*) Extrinsic muscles of the eyeball. (c) Photograph of superior view of the right eyeball. (Courtesy of C. Yokochi and J. W. Rohen, *Photographic Anatomy of the Human Body,* 2nd ed., 1979, IGAKU-SHOIN, Ltd., Tokyo, New York.)

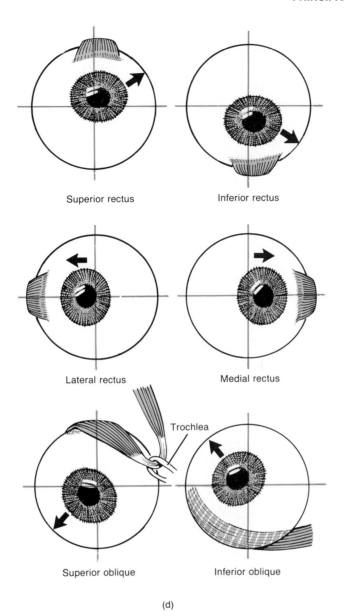

Superior rectus

Inferior rectus

Lateral rectus

Medial rectus

Trochlea

Superior oblique

Inferior oblique

(d)

FIGURE 10-6 (*Continued*) Extrinsic muscles of the eyeball. (d) Movements of the right eyeball in response to contraction of the extrinsic muscles.

EXHIBIT 10-5 MUSCLES THAT MOVE THE TONGUE (Figure 10-7)

MUSCLE	ORIGIN	INSERTION	ACTION	INNERVATION
Genioglossus (*geneion* = chin; *glossus* = tongue)	Mandible.	Undersurface of tongue and hyoid bone.	Depresses tongue and thrusts it forward (protraction).	Hypoglossal (XII) nerve.
Styloglossus (*stylo* = stake or pole: styloid process of temporal bone)	Styloid process of temporal bone.	Side and undersurface of tongue.	Elevates tongue and draws it backward (retraction).	Hypoglossal (XII) nerve.
Palatoglossus* (*palato* = palate)	Anterior surface of soft palate.	Side of tongue.	Elevates tongue and draws soft palate down on tongue.	Hypoglossal (XII) nerve.
Hyoglossus	Body of hyoid bone.	Side of tongue.	Depresses tongue and draws down its sides.	Hypoglossal (XII) nerve.

* Not illustrated.

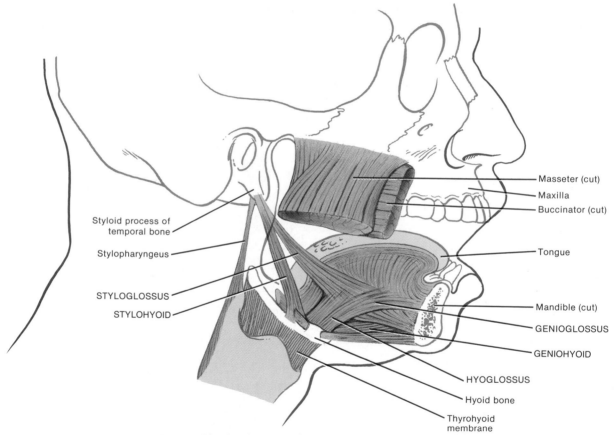

FIGURE 10-7 Muscles that move the tongue viewed from the right side.

EXHIBIT 10-6 MUSCLES OF THE PHARYNX (Figure 10-8)

MUSCLE	ORIGIN	INSERTION	ACTION	INNERVATION
Inferior constrictor (*inferior* = below; *constrictor* = decreases diameter of a lumen)	Cricoid and thyroid cartilages of larynx.	Posterior median raphe of pharynx.	Constricts inferior portion of pharynx to propel a bolus into esophagus.	Pharyngeal plexus.
Middle constrictor	Greater and lesser cornu of hyoid bone and stylohyoid ligament.	Posterior median raphe of pharynx.	Constricts middle portion of pharynx to propel a bolus into esophagus.	Pharyngeal plexus.
Superior constrictor (*superior* = above)	Pterygoid process, pterygomandibular raphe, and mylohyoid line of mandible.	Posterior median raphe of pharynx.	Constricts superior portion of pharynx to propel a bolus into esophagus.	Pharyngeal plexus.
Stylopharyngeus (*stylo* = stake or pole; styloid process of temporal bone; *pharyngo* =pharynx) (see also Figure 10-7)	Medial side of base of styloid process.	Lateral aspects of pharynx and thyroid cartilage.	Elevates larynx and dilates pharynx to help bolus descend.	Glossopharyngeal (IX) nerve.
Salpingopharyngeus (*salping* = pertaining to the auditory or uterine tube)	Inferior portion of auditory tube.	Posterior fibers of palatopharyngeus muscle.	Elevates superior portion of lateral wall of pharynx during swallowing and opens orifice of auditory tube.	Pharyngeal plexus.
Palatopharyngeus (*palato* = palate)	Soft palate.	Posterior border of thyroid cartilage and lateral and posterior wall of pharynx.	Elevates larynx and pharynx and helps close nasopharynx during swallowing.	Pharyngeal plexus.

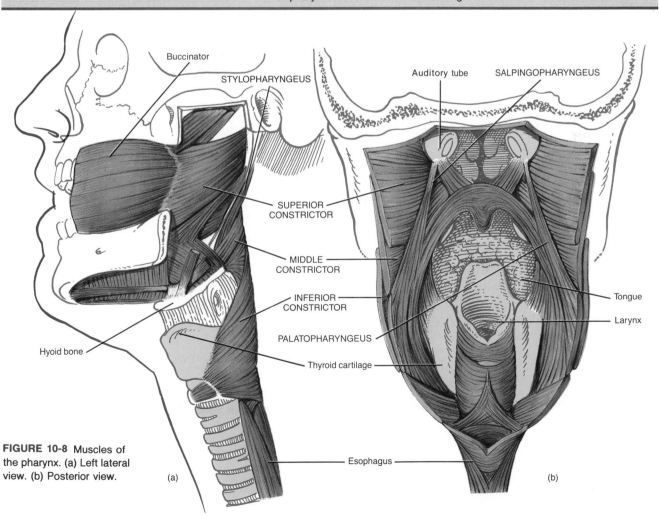

FIGURE 10-8 Muscles of the pharynx. (a) Left lateral view. (b) Posterior view.

EXHIBIT 10-7   MUSCLES OF THE FLOOR OF THE ORAL CAVITY* (Figure 10-9)

MUSCLE	ORIGIN	INSERTION	ACTION	INNERVATION
Digastric (*di* = two; *gaster* = belly)	Anterior belly from inner side of lower border of mandible; posterior belly from mastoid process of temporal bone.	Body of hyoid bone via an intermediate tendon.	Elevates hyoid bone and depresses mandible as in opening the mouth.	Anterior belly from mandibular branch of trigeminal (V) nerve; posterior belly from facial (VII) nerve.
Stylohyoid (*stylo* = stake or pole, styloid process of temporal bone; *hyoedes* = U-shaped, pertaining to hyoid bone; see also Figure 10-7)	Styloid process of temporal bone.	Body of hyoid bone.	Elevates hyoid bone and draws it posteriorly.	Facial (VII) nerve.
Mylohyoid	Inner surface of mandible.	Body of hyoid bone.	Elevates hyoid bone and floor of mouth and depresses mandible.	Mandibular branch of trigeminal (V) nerve.
Geniohyoid (*geneion* = chin) (see Figure 10-7 also)	Inner surface of mandible.	Body of hyoid bone.	Elevates hyoid bone, draws hyoid bone and tongue anteriorly, depresses mandible.	Hypoglossal (XII) nerve.

* These muscles together are referred to as **suprahyoid muscles.**

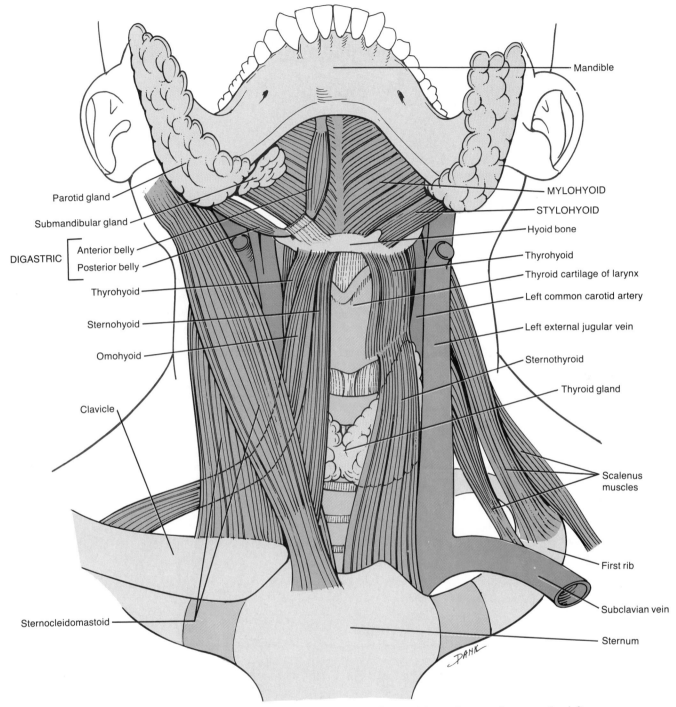

FIGURE 10-9 Muscles of the floor of the oral cavity. Superficial muscles are shown on the left side of the illustration; deep muscles are shown on the right side of the illustration.

EXHIBIT 10-8   MUSCLES OF THE LARYNX (Figure 10-10)

MUSCLE	ORIGIN	INSERTION	ACTION	INNERVATION
EXTRINSIC				
Omohyoid* (*omo* = relationship to the shoulder; *hyoedes* = U-shaped; pertaining to hyoid bone)	Superior border of scapula and superior transverse ligament.	Body of hyoid bone.	Depresses hyoid bone.	Branches of ansa cervicalis nerve (C1–C3).
Sternohyoid* (*sterno* = sternum)	Medial end of clavicle and manubrium of sternum.	Body of hyoid bone.	Depresses hyoid bone.	Branches of ansa cervicalis nerve (C1–C3).
Sternothyroid* (*thyro* = thyroid gland)	Manubrium of sternum.	Thyroid cartilage of larnyx.	Depresses thyroid cartilage.	Branches of ansa cervicalis nerve (C1–C3).
Thyrohyoid*	Thyroid cartilage of larynx.	Greater cornu of hyoid bone.	Elevates thyroid cartilage and depresses hyoid bone.	Cervical nerves C1–C2 and descending hypoglossal (XII) nerve.
Stylopharyngeus	See Exhibit 10-6.			
Palatopharyngeus	See Exhibit 10-6.			
Inferior constrictor	See Exhibit 10-6.			
Middle constrictor	See Exhibit 10-6.			
INTRINSIC				
Cricothyroid (*crico* = cricoid cartilage of larynx)	Anterior and lateral portion of cricoid cartilage of larynx.	Anterior border of inferior cornu of thyroid cartilage of larynx and posterior part of inferior border of lamina of thyroid cartilage.	Produces tension and elongation of vocal folds.	External laryngeal branch of vagus (X) nerve.

* These muscles together are referred to as **infrahyoid,** or **strap, muscles.**

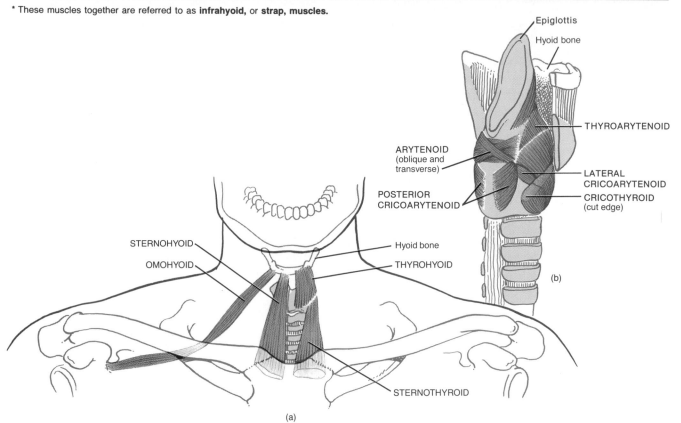

FIGURE 10-10 Muscles of the larynx. (a) Anterior view. (b) Right posterolateral view.

EXHIBIT 10-9 MUSCLES THAT MOVE THE HEAD

MUSCLE	ORIGIN	INSERTION	ACTION	INNERVATION
Sternocleidomastoid (*sternum* = breastbone; *cleido* = clavicle; *mastoid* = mastoid process of temporal bone) (see Figure 10-15)	Sternum and clavicle.	Mastoid process of temporal bone.	Contraction of both muscles flex the cervical part of the vertebral column, draw the head forward, and elevate chin; contraction of one muscle rotates face toward side opposite contracting muscle.	Accessory (XI) nerve; cervical nerves C2–C3.
Semispinalis capitis (*semi* = half; *spine* = spinous process; *caput* = head) (see Figure 10-19	Articular process of seventh cervical vertebra and transverse processes of first six thoracic vertebrae.	Occipital bone.	Both muscles extend head; contraction of one muscle rotates face toward same side as contracting muscle.	Dorsal rami of spinal nerves.
Splenius capitis (*splenion* = bandage) (see Figure 10-19)	Ligamentum nuchae and spines of seventh cervical vertebra and first four thoracic vertebrae.	Occipital bone and mastoid process of temporal bone.	Both muscles extend head; contraction of one rotates it to same side as contracting muscle.	Dorsal rami of middle and lower cervical nerves.
Longissimus capitis (*longissimus* = longest) (see Figure 10-19)	Transverse processes of last four cervical vertebrae.	Mastoid process of temporal bone.	Extends head and rotates face toward side opposite contracting muscle.	Dorsal rami of middle and lower cervical nerves.

EXHIBIT 10-10 MUSCLES THAT ACT ON THE ANTERIOR ABDOMINAL WALL (Figure 10-11)

MUSCLE	ORIGIN	INSERTION	ACTION	INNERVATION
Rectus abdominis (*rectus* = fibers parallel to midline; *abdomino* = belly)	Pubic crest and symphysis pubis.	Cartilage of fifth to seventh ribs and xiphoid process.	Flexes vertebral column.	Branches of thoracic nerves T7–T12.
External oblique (*external* = closer to surface; *oblique* = fibers diagonal to midline)	Lower eight ribs.	Iliac crest and linea alba (midline aponeurosis).	Contraction of both compresses abdomen; contraction of one side alone bends vertebral column laterally.	Branches of thoracic nerves T7–T12 and iliohypogastric nerve.
Internal oblique (*internal* = farther from surface)	Iliac crest, inguinal ligament, and thoracolumbar fascia.	Cartilage of last three or four ribs.	Compresses abdomen; contraction of one side alone bends vertebral column laterally.	Branches of thoracic nerves T8–T12, iliohypogastric, and ilioinguinal nerves.
Transversus abdominis (*transverse* = fibers perpendicular to midline)	Iliac crest, inguinal ligament, lumbar fascia, and cartilages of last six ribs.	Xiphoid process, linea alba, and pubis.	Compresses abdomen.	Branches of thoracic nerves T8–T12, iliohypogastric, and ilioinguinal nerves.

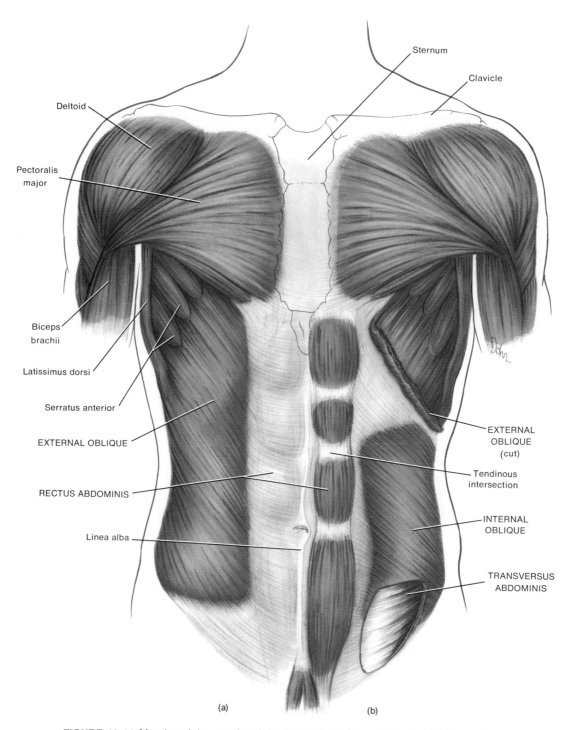

FIGURE 10-11 Muscles of the anterior abdominal wall. (a) Superficial view. (b) Deep view.

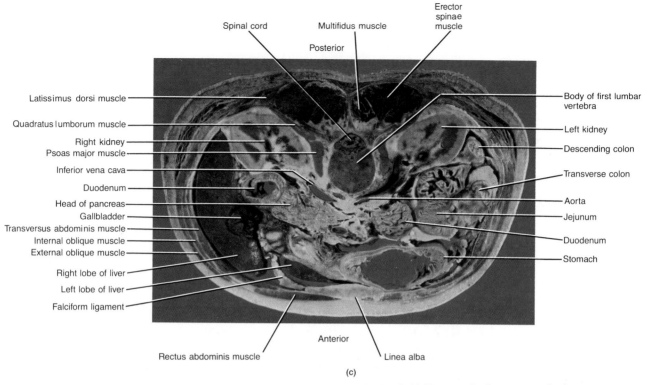

Spinal cord

Posterior

Multifidus muscle

Erector spinae muscle

Latissimus dorsi muscle

Quadratus lumborum muscle

Right kidney

Psoas major muscle

Inferior vena cava

Duodenum

Head of pancreas

Gallbladder

Transversus abdominis muscle

Internal oblique muscle

External oblique muscle

Right lobe of liver

Left lobe of liver

Falciform ligament

Body of first lumbar vertebra

Left kidney

Descending colon

Transverse colon

Aorta

Jejunum

Duodenum

Stomach

Rectus abdominis muscle

Anterior

Linea alba

(c)

FIGURE 10-11 (*Continued*) Muscles of the anterior abdominal wall. (c) Photograph of a cross section through the abdomen showing the musculature and related viscera. (Courtesy of Stephen A. Kieffer and E. Robert Heitzman, *An Atlas of Cross-Sectional Anatomy,* Harper & Row, Publishers, Inc., New York, 1979.) The surface anatomy of the anterior abdominal muscle is illustrated in Figure 11-8.

EXHIBIT 10-11 MUSCLES USED IN BREATHING (Figure 10-12)

MUSCLE	ORIGIN	INSERTION	ACTION	INNERVATION
Diaphragm (*dia* = across; *phragma* = wall)	Xiphoid process, costal cartilages of last six ribs, and lumbar vertebrae.	Central tendon.	Forms floor of thoracic cavity; pulls central tendon downward during inspiration and thus increases vertical length of thorax.	Phrenic nerve.
External intercostals (*external* = closer to surface; *inter* = between; *costa* = rib)	Inferior border of rib above.	Superior border of rib below.	Elevate ribs during inspiration and thus increase lateral and anteroposterior dimensions of thorax.	Intercostal nerves.
Internal Intercostals (*internal* = farther from surface)	Superior border of rib below.	Inferior border of rib above.	Draw adjacent ribs together during forced expiration and thus decrease lateral and anteroposterior dimensions of thorax.	Intercostal nerves.

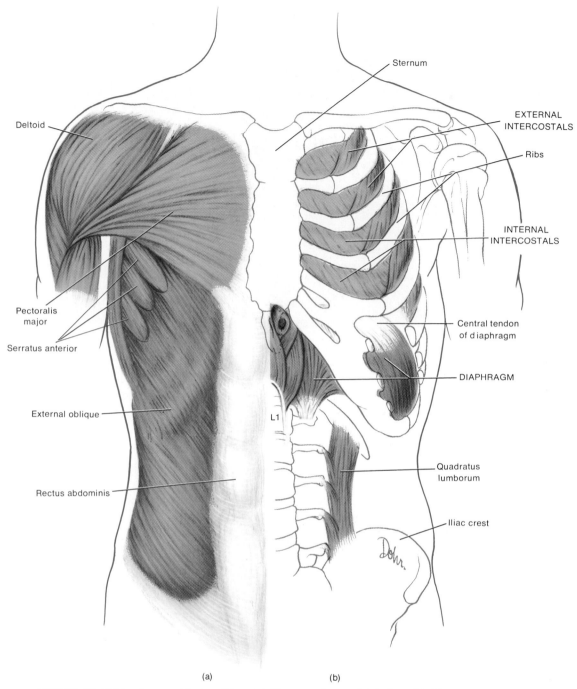

Deltoid

Pectoralis
major

Serratus anterior

External oblique

Rectus abdominis

Sternum

EXTERNAL
INTERCOSTALS

Ribs

INTERNAL
INTERCOSTALS

Central tendon
of diaphragm

DIAPHRAGM

L1

Quadratus
lumborum

Iliac crest

(a)   (b)

FIGURE 10-12 Muscles used in breathing. (a) Diagram of superficial view. (b) Diagram of deep view.

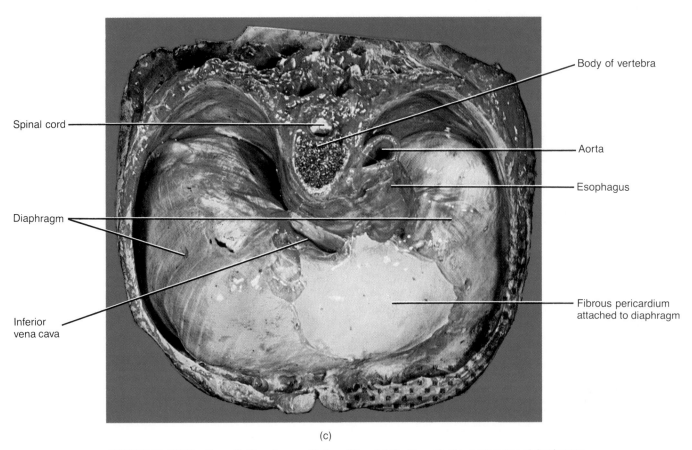

Spinal cord

Diaphragm

Inferior
vena cava

Body of vertebra

Aorta

Esophagus

Fibrous pericardium
attached to diaphragm

(c)

FIGURE 10-12 (*Continued*) Muscles used in breathing. (c) Photograph of superior view of diaphragm.
(Courtesy of C. Yokochi and J. W. Rohen, *Photographic Anatomy of the Human Body,* 1969, IGAKU-
SHOIN, Ltd., Tokyo, New York.)

EXHIBIT 10-12   MUSCLES OF THE PELVIC FLOOR* (Figure 10-13)

MUSCLE	ORIGIN	INSERTION	ACTION	INNERVATION
Levator ani (*levator* = raises; *ani* = anus)	This muscle is divisible into two parts, the pubococcygeus muscle and the iliococcygeus muscle. It forms the funnel-shaped floor of the pelvic cavity and supports the pelvic structures. It contains openings for the anal canal and urethra in both sexes and the vagina in the female.			
Pubococcygeus (*pubo* = pubis; *coccygeus* = coccyx)	Pubis.	Coccyx, urethra, anal canal, and central tendon of perineum.	Supports and slightly raises pelvic floor, resists increased intraabdominal pressure, and draws anus toward pubis and constricts it.	Sacral nerves S3–S4 or S4 and perineal branch of pudendal nerve.
Iliococcygeus (*ilio* = ilium)	Ischial spine.	Coccyx.	Supports and slightly raises pelvic floor, resists increased intraabdominal pressure, and draws anus toward pubis and constricts it.	Sacral nerves S3–S4 or S4 and perineal branch of pudendal nerve.
Coccygeus	Ischial spine.	Lower sacrum and upper coccyx.	Supports and slightly raises pelvic floor, resists intraabdominal pressure, and pulls coccyx forward following defecation or parturition.	Sacral nerve S3 or S4.

* The muscles of the pelvic floor, together with the fasciae covering their internal and external surfaces, are collectively referred to as the **pelvic diaphragm.**

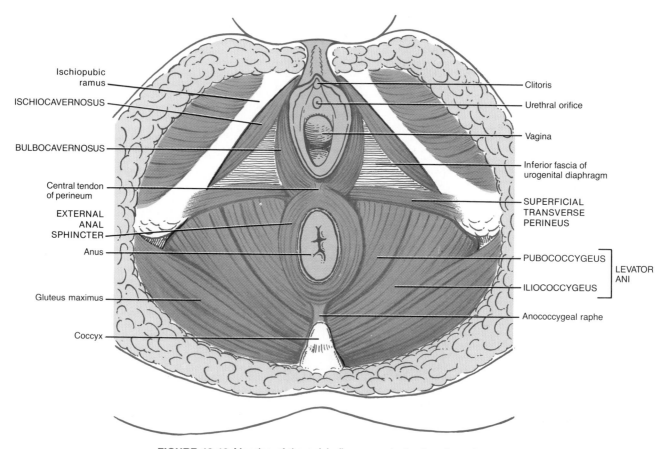

FIGURE 10-13 Muscles of the pelvic floor seen in the female perineum.

EXHIBIT 10-13 MUSCLES OF THE PERINEUM* (Figure 10-14)

MUSCLE	ORIGIN	INSERTION	ACTION	INNERVATION
Superficial transverse perineus (*superficial* = closer to surface; *transverse* = across; *perineus* = perineum)	Ischial tuberosity.	Central tendon of perineum.	Helps to stabilize the central tendon of the perineum.	Perineal branch of pudendal nerve.
Bulbocavernosus (*bulbus* = bulb; *caverna* = hollow place)	Central tendon of perineum.	Inferior fascia of urogenital diaphragm, corpus spongiosum of penis, and deep fascia on dorsum of penis in male; pubic arch and root and dorsum of clitoris in female.	Helps expel last drops of urine during micturition; helps propel semen along urethra, and assists in erection of the penis in male; decreases vaginal orifice and assists in erection of clitoris in female.	Perineal branch of pudendal nerve.
Ischiocavernosus (*ischion* = hip)	Ischial tuberosity and ischial and pubic rami.	Corpus cavernosum of penis in male and clitoris in female.	Maintains erection of penis in male and clitoris in female.	Perineal branch of pudendal nerve.
Deep transverse perineus † (*deep* = farther from surface)	Ischial rami.	Central tendon of perineum.	Helps eject last drops of urine and semen in male and urine in female.	Perineal branch of pudendal nerve.
Urethral sphincter † (*sphincter* = circular muscle that decreases size of an opening; *urethrae* = urethra)	Ischial and pubic rami.	Median raphe in male and vaginal wall in female.	Helps eject last drops of urine and semen in male and urine in female.	Perineal branch of pudendal nerve.
External anal sphincter	Anococcygeal raphe.	Central tendon of perineum.	Keeps anal canal and orifice closed.	Sacral nerve S4 and inferior rectal branch of pudendal nerve.

* The **perineum** is the entire outlet of the pelvis. It is a diamond-shaped area at the lower end of the trunk between the thighs and buttocks. It is bordered anteriorly by the symphysis pubis, laterally by the ischial tuberosities, and posteriorly by the coccyx. A transverse line drawn between the ischial tuberosities divides the perineum into an anterior **urogenital triangle** that contains the external genitals and a posterior **anal triangle** that contains the anus.

† The deep transverse perineus, the urethral sphincter, and a fibrous membrane constitute the **urogenital diaphragm.** It surrounds the urogenital ducts and helps to strengthen the pelvic floor.

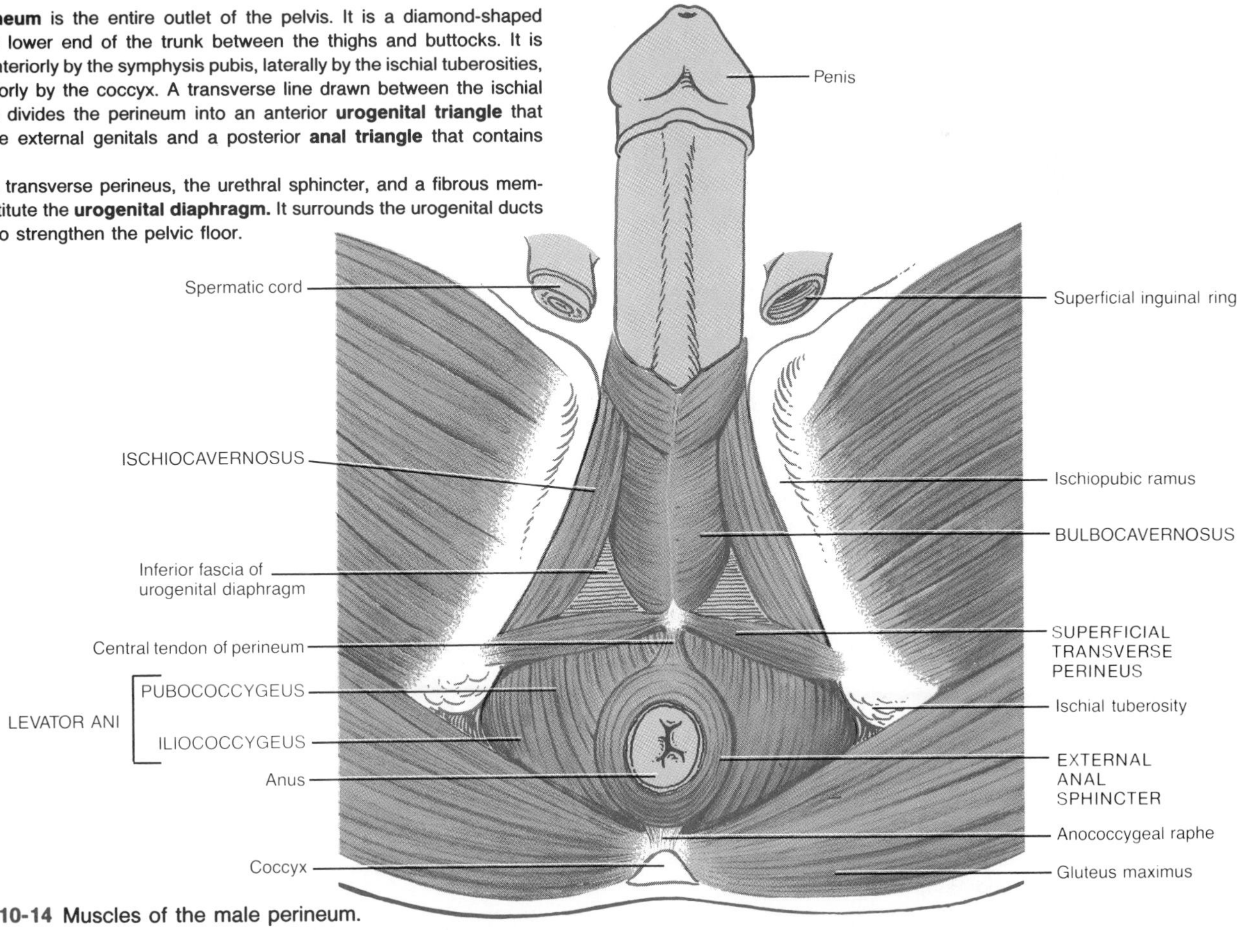

FIGURE 10-14 Muscles of the male perineum.

EXHIBIT 10-14 MUSCLES THAT MOVE THE PECTORAL (SHOULDER) GIRDLE (Figure 10-15)

MUSCLE	ORIGIN	INSERTION	ACTION	INNERVATION
Subclavius (*sub* = under; *clavius* = clavicle)	First rib.	Clavicle.	Depresses clavicle.	Nerve to subclavius.
Pectoralis minor (*pectus* = breast, chest, thorax; *minor* = lesser)	Third through fifth ribs.	Coracoid process of scapula.	Depresses scapula, rotates shoulder joint anteriorly, and elevates third through fifth ribs during forced inspiration when scapula is fixed.	Medial pectoral nerve.
Serratus anterior (*serratus* = saw-toothed; *anterior* = front)	Upper eight or nine ribs.	Vertebral border and inferior angle of scapula.	Rotates scapula laterally and elevates ribs when scapula is fixed.	Long thoracic nerve.
Trapezius (*trapezoides* = trapezoid-shaped)	Occipital bone, ligamentum nuchae, and spines of seventh cervical and all thoracic vertebrae.	Clavicle and acromion and spine of scapula.	Elevates clavicle, adducts scapula, elevates or depresses scapula, and extends head.	Accessory (XI) nerve and cervical nerves C3–C4.
Levator scapulae (*levator* = raises; *scapulae* = scapula)	Upper four or five cervical vertebrae.	Vertebral border of scapula.	Elevates scapula.	Dorsal scapular nerve and cervical nerves C3–C5.
Rhomboideus major (*rhomboides* = rhomboid or diamond-shaped)	Spines of second to fifth thoracic vertebrae.	Vertebral border of scapula.	Adducts scapula and slightly rotates it upward.	Dorsal scapular nerve.
Rhomboideus minor	Spines of seventh cervical and first thoracic vertebrae.	Superior angle of scapula.	Adducts scapula.	Dorsal scapular nerve.

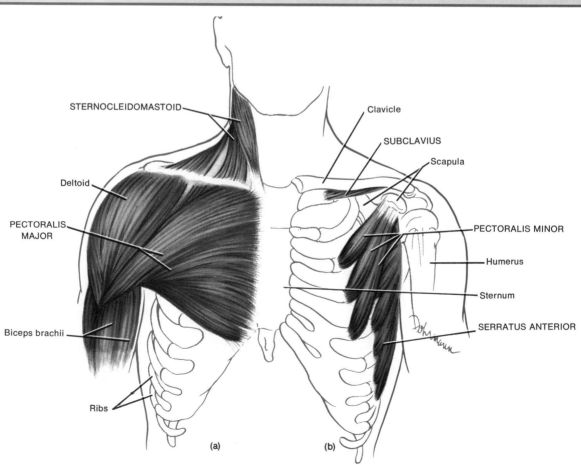

FIGURE 10-15 Muscles that move the pectoral (shoulder) girdle. (a) Anterior superficial view. (b) Anterior deep view.

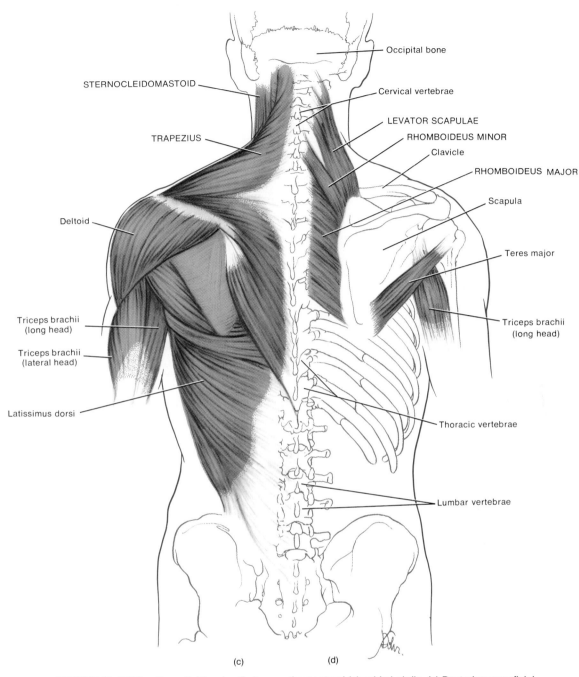

STERNOCLEIDOMASTOID

TRAPEZIUS

Deltoid

Triceps brachii
(long head)

Triceps brachii
(lateral head)

Latissimus dorsi

Occipital bone

Cervical vertebrae

LEVATOR SCAPULAE

RHOMBOIDEUS MINOR

Clavicle

RHOMBOIDEUS MAJOR

Scapula

Teres major

Triceps brachii
(long head)

Thoracic vertebrae

Lumbar vertebrae

(c)   (d)

FIGURE 10-15 (*Continued*) Muscles that move the pectoral (shoulder) girdle. (c) Posterior superficial view. (d) Posterior deep view. The surface anatomy of the muscles that move the pectoral (shoulder) girdle is illustrated in Figures 11-6 and 11-7.

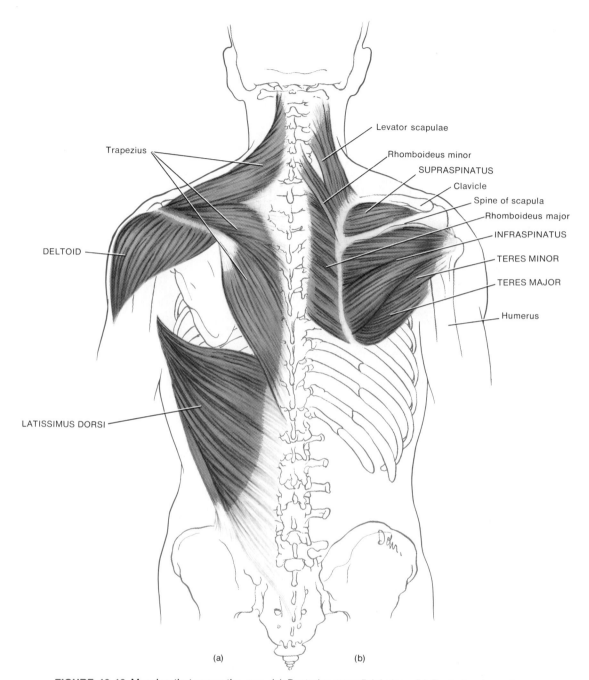

FIGURE 10-16 Muscles that move the arm. (a) Posterior superficial view. (b) Posterior deep view.

EXHIBIT 10-15 MUSCLES THAT MOVE THE ARM (Figure 10-16)

MUSCLE	ORIGIN	INSERTION	ACTION	INNERVATION
Pectoralis major (see Figure 10-15a)	Clavicle, sternum, cartilages of second to sixth ribs.	Greater tubercle of humerus.	Flexes, adducts, and rotates arm medially.	Medial and lateral pectoral nerve.
Latissimus dorsi (*latissimus* = widest; *dorsum* = back)	Spines of lower six thoracic vertebrae, lumbar vertebrae, crests of sacrum and ilium, lower four ribs.	Intertubercular groove of humerus.	Extends, adducts, and rotates arm medially; draws shoulder downward and backward.	Thoracodorsal nerve.
Deltoid (*delta* = triangular)	Clavicle and acromion process and spine of scapula.	Deltoid tuberosity of humerus.	Abducts arm.	Axillary nerve.
Supraspinatus (*supra* = above; *spinatus* = spine of scapula)	Fossa superior to spine of scapula.	Greater tubercle of humerus.	Assists deltoid muscle in abducting arm.	Suprascapular nerve.
Infraspinatus (*infra* = below)	Fossa inferior to spine of scapula.	Greater tubercle of humerus.	Rotates arm laterally.	Suprascapular nerve.
Teres major (*teres* = long and round)	Inferior angle of scapula.	Distal to lesser tubercle of humerus.	Extends arm and draws it down; assists in adduction and medial rotation of arm.	Lower subscapular nerve.
Teres minor	Lateral border of scapula.	Greater tubercle of humerus.	Rotates arm laterally.	Axillary nerve.

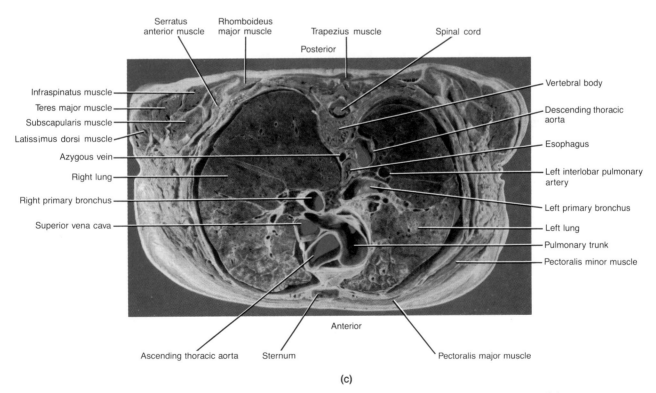

(c)

FIGURE 10-16 (*Continued*) Muscles that move the arm. (c) Photograph of a cross section of the thorax showing the musculature and related viscera. (Courtesy of Stephen A. Kieffer and E. Robert Heitzman, *An Atlas of Cross-Sectional Anatomy,* Harper & Row, Publishers, Inc., New York, 1979.) The surface anatomy of the muscles that move the arm is illustrated in Figures 11-6 and 11-7.

EXHIBIT 10-16   MUSCLES THAT MOVE THE FOREARM (Figure 10-17)

MUSCLE	ORIGIN	INSERTION	ACTION	INNERVATION
Biceps, brachii (*biceps* = two heads of origin; *brachion* = arm)	Long head originates from tubercle above glenoid cavity; short head originates from coracoid process of scapula.	Radial tuberosity and bicipital aponeurosis.	Flexes and supinates forearm.	Musculocutaneous nerve.
Brachialis	Anterior surface of humerus.	Tuberosity and coronoid process of ulna.	Flexes forearm.	Musculocutaneous, radial, and median nerves.
Brachioradialis (*radialis* = radius) (see also Figure 10-18)	Supracondyloid ridge of humerus.	Superior to styloid process of radius.	Flexes forearm.	Radial nerve.
Triceps brachii (*triceps* = three heads of origin)	Long head originates from infraglenoid tuberosity of scapula; lateral head originates from lateral and posterior surface of humerus superior to radial groove; medial head originates from posterior surface of humerus inferior to radial groove.	Olecranon of ulna.	Extends forearm.	Radial nerve.
Coracobrachialis (*coraco* = coracoid process)	Coracoid process of scapula.	Middle of medial surface of shaft of humerus.	Flexes and adducts forearm.	Musculocutaneous nerve.
Anconeus (*anconeal* = pertaining to elbow) (see Figure 10-18)	Lateral epicondyle of humerus.	Olecranon and superior portion of shaft of ulna.	Extends forearm.	Radial nerve.
Supinator (*supination* = turning palm upward or anteriorly)	Lateral epicondyle of humerus and ridge on ulna.	Oblique line of radius.	Supinates forearm.	Deep radial nerve.
Pronator teres (*pronation* = turning palm downward or posteriorly)	Medial epicondyle of humerus and coronoid process of ulna.	Midlateral surface of radius.	Pronates forearm.	Medial nerve.
Pronator quadratus (*quadratus* = squared, four-sided)	Distal portion of shaft of ulna.	Inferior portion of shaft of radius.	Pronates and rotates forearm.	Median nerve.

FIGURE 10-17 Muscles that move the forearm. (a) Anterior view. (b) Posterior view. (c) Photograph of a cross section of the arm showing the musculature and related structures. (Courtesy of Stephen A. Kieffer and E. Robert Heitzman, *An Atlas of Cross-Sectional Anatomy,* Harper & Row, Publishers, Inc., New York, 1979.) The surface anatomy of the muscles that move the forearm is illustrated in Figures 11-10 and 11-11.

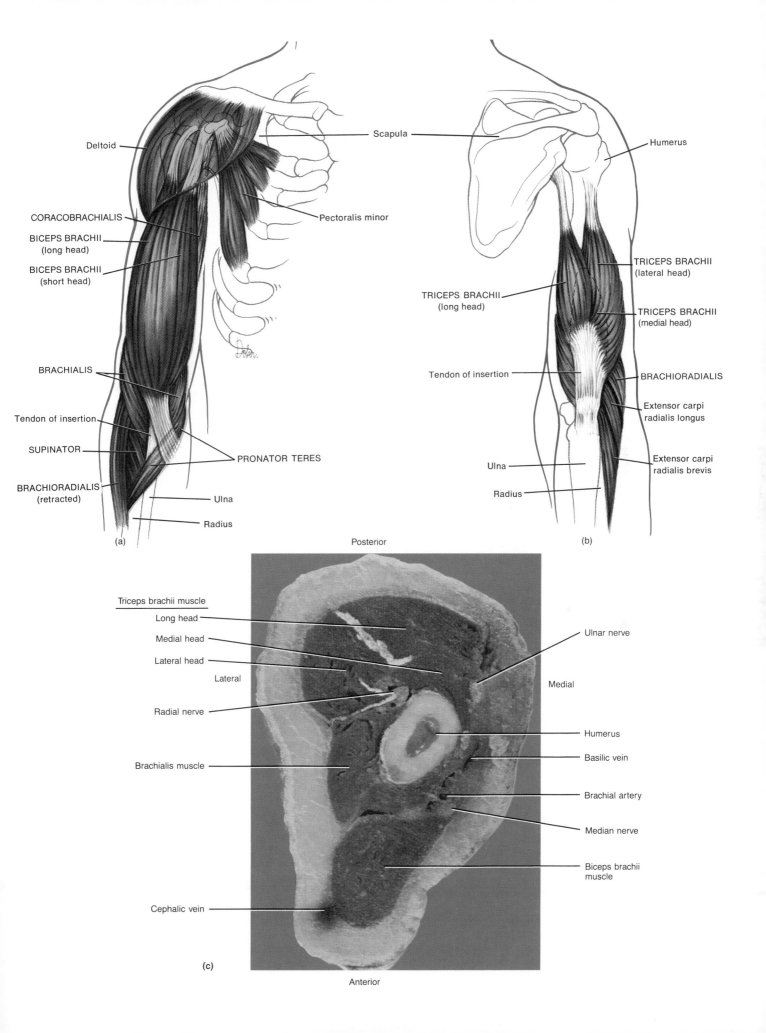

Deltoid

Scapula

Humerus

CORACOBRACHIALIS

Pectoralis minor

BICEPS BRACHII
(long head)

BICEPS BRACHII
(short head)

TRICEPS BRACHII
(lateral head)

TRICEPS BRACHII
(long head)

TRICEPS BRACHII
(medial head)

BRACHIALIS

Tendon of insertion

BRACHIORADIALIS

Tendon of insertion

SUPINATOR

PRONATOR TERES

Extensor carpi
radialis longus

BRACHIORADIALIS
(retracted)

Ulna

Extensor carpi
radialis brevis

Radius

Ulna

Radius

(a)

(b)

Posterior

Triceps brachii muscle

Long head

Ulnar nerve

Medial head

Lateral head

Lateral

Medial

Radial nerve

Humerus

Brachialis muscle

Basilic vein

Brachial artery

Median nerve

Biceps brachii
muscle

Cephalic vein

(c)

Anterior

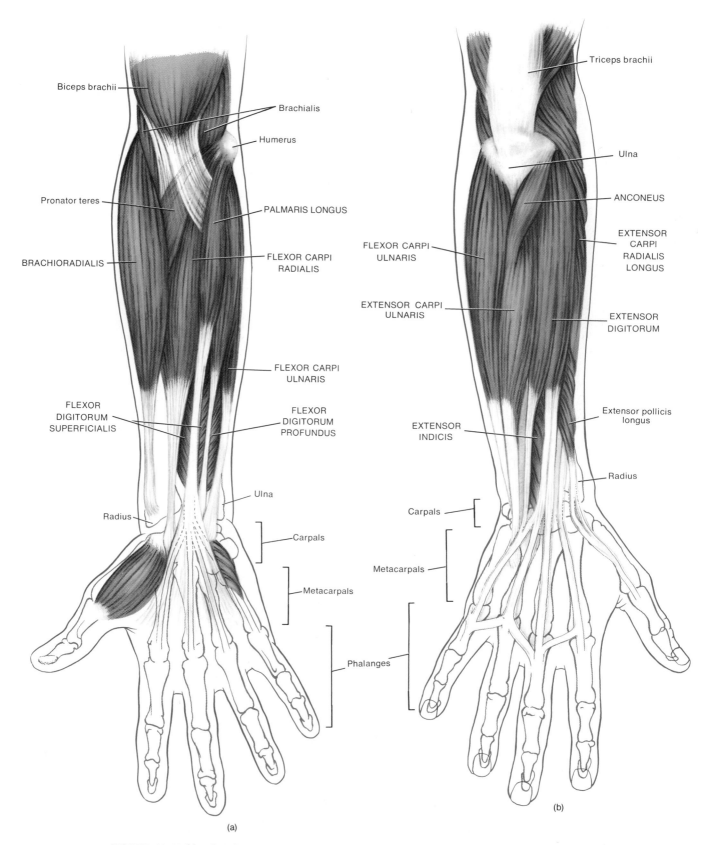

FIGURE 10-18 Muscles that move the wrist and fingers. (a) Anterior view. (b) Posterior view. (c) Photograph of a cross section of the forearm showing the musculature and related structures. (Courtesy of Stephen A. Kieffer and E. Robert Heitzman, *An Atlas of Cross-Sectional Anatomy,* Harper & Row, Publishers, Inc., New York, 1979.) The surface anatomy of the muscles that move the wrist and fingers is illustrated in Figures 11-11 through 11-13.

EXHIBIT 10-17   MUSCLES THAT MOVE THE WRIST AND FINGERS (Figure 10-18)

MUSCLE	ORIGIN	INSERTION	ACTION	INNERVATION
Flexor carpi radialis (*flexor* = decreases angle at joint; *carpus* = wrist; *radialis* = radius)	Medial epicondyle of humerus.	Second and third metacarpals.	Flexes and abducts wrist.	Median nerve.
Flexor carpi ulnaris (*ulnaris* = ulna)	Medial epicondyle of humerus and upper dorsal border of ulna.	Pisiform, hamate, and fifth metacarpal.	Flexes and adducts wrist.	Ulnar nerve.
Palmaris longus (*palma* = palm)	Medial epicondyle of humerus.	Transverse carpal ligament and palmar aponeurosis.	Flexes wrist and tenses palmar aponeurosis.	Median nerve.
Extensor carpi radialis longus (*extensor* = increases angle at joint; *longus* = long)	Lateral epicondyle of humerus.	Second metacarpal.	Extends and abducts wrist.	Radial nerve.
Extensor carpi ulnaris	Lateral epicondyle of humerus and dorsal border of ulna.	Fifth metacarpal.	Extends and adducts wrist.	Deep radial nerve.
Flexor digitorum profundus (*digit* = finger or toe; *profundus* = deep)	Anterior medial surface of body of ulna.	Bases of distal phalanges.	Flexes distal phalanges of each finger.	Median and ulnar nerves.
Flexor digitorum superficialis (*superficialis* = closer to surface)	Medial epicondyle of humerus, coronoid process of ulna, and oblique line of radius.	Middle phalanges.	Flexes middle phalanges of each finger.	Median nerve.
Extensor digitorum	Lateral epicondyle of humerus.	Middle and distal phalanges of each finger.	Extends phalanges.	Deep radial nerve.
Extensor indicis (*indicis* = index)	Dorsal surface of ulna.	Tendon of extensor digitorum of index finger.	Extends index finger.	Deep radial nerve.

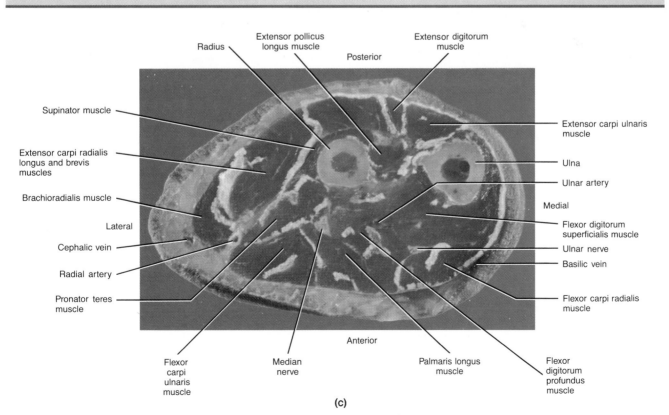

(c)

EXHIBIT 10-18   MUSCLES THAT MOVE THE VERTEBRAL COLUMN (Figure 10-19)

MUSCLE	ORIGIN	INSERTION	ACTION	INNERVATION
Rectus abdominis (see Figure 10-11)	Pubic crest and symphysis pubis.	Cartilages of fifth through seventh ribs.	Flexes vertebral column at lumbar spine and compresses abdomen.	Intercostal (thoracic) nerves T7–T12.
Quadratus lumborum (*quadratus* = squared, four-sided; *lumb* = lumbar region)	Iliac crest.	Twelfth rib and upper four lumbar vertebrae.	Flexes vertebral column laterally.	Intercostal nerve T12 and lumbar nerve L1.
Sacrospinalis (**erector spinae**)	This posterior muscle consists of three groupings: iliocostalis, longissimus, and spinalis. These groups, in turn, consist of a series of overlapping muscles. The iliocostalis group is laterally placed, the longissimus group is intermediate in placement, and the spinalis is medially placed.			
LATERAL				
Iliocostalis lumborum (*ilium* = flank; *costa* = rib)	Iliac crest.	Lower six ribs.	Extends lumbar region of vertebral column.	Dorsal rami of lumbar nerves.
Iliocostalis thoracis (*thorax* = chest)	Lower six ribs.	Upper six ribs.	Maintains erect position of spine.	Dorsal rami of thoracic (intercostal) nerves.
Iliocostalis cervicis (*cervix* = neck)	First six ribs.	Transverse processes of fourth to sixth cervical vertebrae.	Extends cervical region of vertebral column.	Dorsal rami of cervical nerves.
INTERMEDIATE				
Longissimus thoracis	Transverse processes of lumbar vertebrae.	Transverse processes of all thoracic and upper lumbar vertebrae and ninth and tenth ribs.	Extends thoracic region of vertebral column.	Dorsal rami of spinal nerves.
Longissimus cervicis	Transverse processes of fourth and fifth thoracic vertebrae.	Transverse processes of second to sixth cervical vertebrae.	Extends cervical region of vertebral column.	Dorsal rami of spinal nerves.
Longissimus capitis (*caput* = head)	Transverse processes of upper four thoracic vertebrae.	Mastoid process of temporal bone.	Extends head and rotates it to opposite side.	Dorsal rami of middle and lower cervical nerves.
MEDIAL				
Spinalis thoracis	Spines of upper lumbar and lower thoracic vertebrae.	Spines of upper thoracic vertebrae.	Extends vertebral column.	Dorsal rami of spinal nerves.

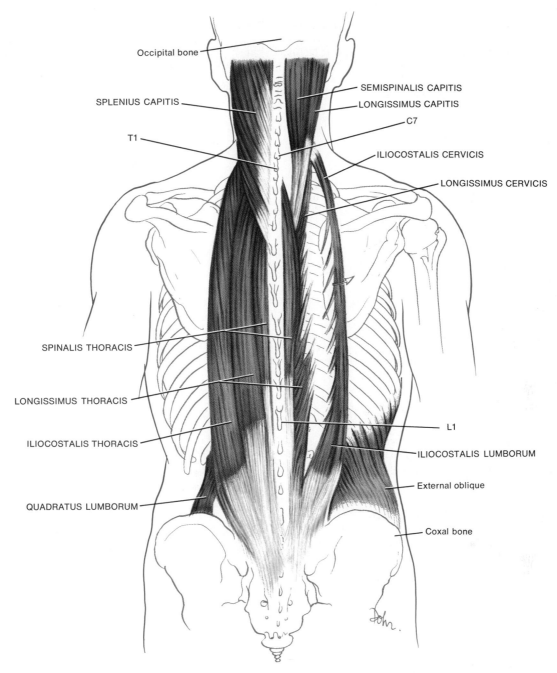

FIGURE 10-19 Muscles that move the vertebral column.

EXHIBIT 10-19 MUSCLES THAT MOVE THE THIGH (Figure 10-20)

MUSCLE	ORIGIN	INSERTION	ACTION	INNERVATION
Psoas major* (*psoa* = muscle of loin)	Transverse processes and bodies of lumbar vertebrae.	Lesser trochanter of femur.	Flexes and rotates thigh laterally; flexes vertebral column.	Lumbar nerves L2–L3.
Iliacus* (*iliac* = ilium)	Iliac fossa.	Tendon of psoas major.	Flexes and rotates thigh laterally; flexes vertebral column slightly.	Femoral nerve.
Gluteus maximus (*glutos* = buttock; *maximus* = largest)	Iliac crest, sacrum, coccyx, and aponeurosis of sacrospinalis.	Iliotibial tract of fascia lata and gluteal tuberosity of femur.	Extends and rotates thigh laterally.	Inferior gluteal nerve.
Gluteus medius (*media* = middle)	Ilium.	Greater trochanter of femur.	Abducts and rotates thigh medially.	Superior gluteal nerve.
Gluteus minimus (*minimus* = smallest)	Ilium.	Greater trochanter of femur.	Abducts and rotates thigh medially.	Superior gluteal nerve.
Tensor fasciae latae (*tensor* = makes tense; *fascia* = band; *latus* = wide)	Iliac crest.	Tibia by way of the iliotibial tract.	Flexes and abducts thigh.	Superior gluteal nerve.
Adductor longus (*adductor* = moves part closer to midline; *longus* = long)	Pubic crest and symphysis pubis.	Linea aspera of femur.	Adducts, rotates, and flexes thigh.	Obturator nerve.
Adductor brevis (*brevis* = short)	Inferior ramus of pubis.	Linea aspera of femur.	Adducts, rotates, and flexes thigh.	Obturator nerve.
Adductor magnus (*magnus* = large)	Inferior ramus of pubis and ischium to ischial tuberosity.	Linea aspera of femur.	Adducts, flexes, and extends thigh (anterior part flexes, posterior part extends).	Obturator and sciatic nerves.
Piriformis (*pirum* = pear; *forma* = shape)	Sacrum.	Greater trochanter of femur.	Rotates thigh laterally and abducts it.	Sacral nerves S2 or S1–S2.
Obturator internus (*obturator* = obturator foramen; *internus* = inside)	Margin of obturator foramen, pubis, and ischium.	Greater trochanter of femur.	Rotates thigh laterally and abducts it.	Nerve to obturator internus.
Pectineus (*pecten* = comb-shaped)	Fascia of pubis.	Pectineal line of femur.	Flexes, adducts, and rotates thigh laterally.	Femoral nerve.

CLINICAL APPLICATION

A common sports injury is a **"pulled groin."** Very simply this means there is a strain, stretching, and probably some tearing away of the tendinous origins of the adductor muscles that move the thigh.

* Together the psoas major and iliacus are sometimes termed the **iliopsoas muscle.**

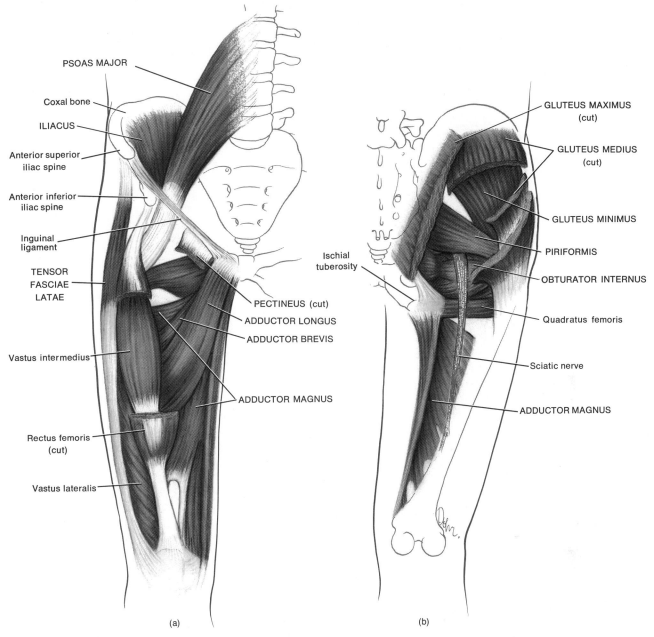

FIGURE 10-20 Muscles that move the thigh. (a) Anterior view. (b) Posterior view. The surface anatomy of the muscles that move the thigh is illustrated in Figure 11-14.

EXHIBIT 10-20   MUSCLES THAT ACT ON THE LEG (Figure 10-21)

MUSCLE	ORIGIN	INSERTION	ACTION	INNERVATION
Quadriceps femoris (*quadriceps* = four heads of origin; *femoris* = femur)	This composite muscle includes four distinct parts, usually described as four separate muscles. The common tendon that includes the patella and attaches to the tibial tuberosity is known as the patellar ligament.			
Rectus femoris (*rectus* = fibers parallel to midline)	Anterior inferior iliac spine.	Upper border of patella.	All four heads extend leg; rectus portion alone also flexes thigh.	Femoral nerve.
Vastus lateralis (*vastus* = large; *lateralis* = lateral)	Greater trochanter and linea aspera of femur.	Upper border and sides of patella; tibial tuberosity through patellar ligament (tendon of quadriceps).		Femoral nerve.
Vastus medialis (*medialis* = medial)	Linea aspera of femur.			Femoral nerve.
Vastus intermedius (*intermedius* = middle)	Anterior and lateral surfaces of body of femur.			Femoral nerve.
Hamstrings	A collective designation for three separate muscles.			
Biceps femoris (*biceps* = two heads of origin)	Long head arises from ischial tuberosity; short head arises from linea aspera of femur.	Head of fibula and lateral condyle of tibia.	Flexes leg and extends thigh.	Tibial nerve from sciatic nerve.
Semitendinosus (*semi* = half; *tendo* = tendon)	Ischial tuberosity.	Proximal part of medial surface of body of tibia.	Flexes leg and extends thigh.	Tibial nerve from sciatic nerve.
Semimembranosus (*membran* = membrane)	Ischial tuberosity.	Medial condyle of tibia.	Flexes leg and extends thigh.	Tibial nerve from sciatic nerve.
Gracilis (*gracilis* = slender)	Symphysis pubis and pubic arch.	Medial surface of body of tibia.	Flexes leg and adducts thigh.	Obturator nerve.
Sartorius (*sartor* = tailor; refers to cross-legged position of tailors)	Anterior superior spine of ilium.	Medial surface of body of tibia.	Flexes leg; flexes thigh and rotates it laterally, thus crossing leg.	Femoral nerve.

CLINICAL APPLICATION

"Pulled hamstrings" are common sports injuries in individuals who run very hard. Sometimes the violent muscular exertion required to perform a feat tears off part of the tendinous origins of the hamstrings from the ischial tuberosity. This is usually accompanied by a contusion (bruising) and tearing of some of the muscle fibers and rupture of blood vessels producing a hematoma (collection of blood).

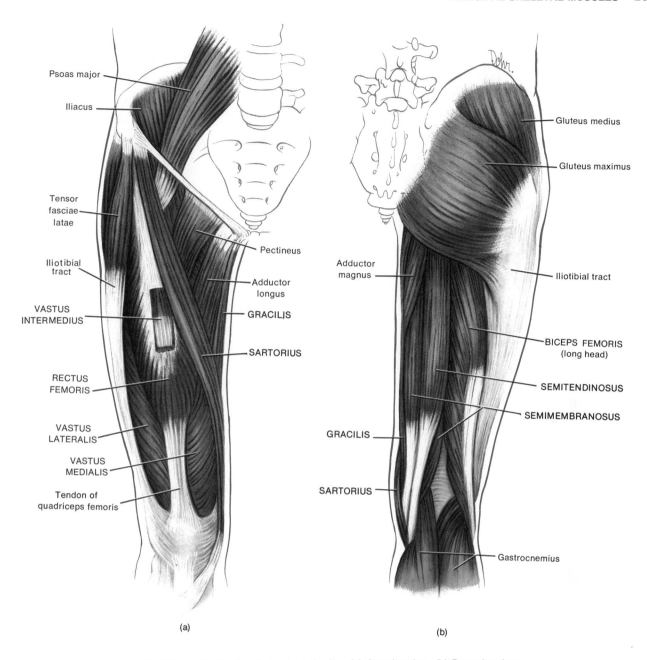

FIGURE 10-21 Muscles that act on the leg. (a) Anterior view. (b) Posterior view.

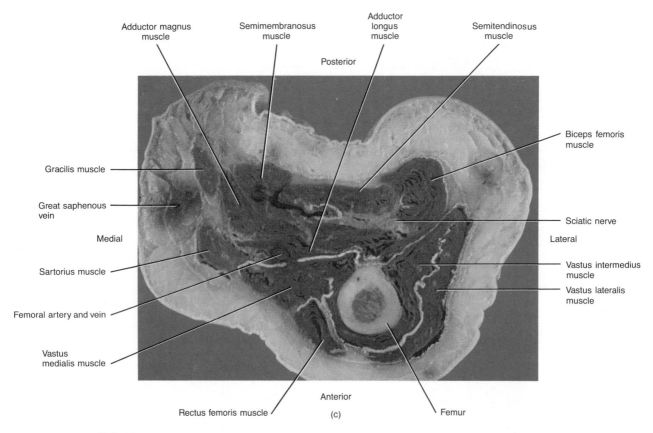

Adductor magnus muscle

Semimembranosus muscle

Adductor longus muscle

Posterior

Semitendinosus muscle

Biceps femoris muscle

Gracilis muscle

Great saphenous vein

Medial

Sartorius muscle

Femoral artery and vein

Vastus medialis muscle

Rectus femoris muscle

Anterior

(c)

Femur

Sciatic nerve

Lateral

Vastus intermedius muscle

Vastus lateralis muscle

FIGURE 10-21 (*Continued*) Muscles that act on the leg. (c) Photograph of a cross section of the thigh showing the musculature and related structures. (Courtesy of Stephen A. Kieffer and E. Robert Heitzman, *An Atlas of Cross-Sectional Anatomy,* Harper & Row, Publishers, Inc., New York, 1979.) The surface anatomy of the muscles that act on the leg is illustrated in Figure 11-14.

EXHIBIT 10-21   MUSCLES THAT MOVE THE FOOT AND TOES (Figure 10-22)

MUSCLE	ORIGIN	INSERTION	ACTION	INNERVATION
Gastrocnemius (*gaster* = belly; *kneme* = leg)	Lateral and medial condyles of femur and capsule of knee.	Calcaneus by way of calcaneal ("Achilles") tendon.	Plantar flexes foot.	Tibial nerve.
Soleus (*soleus* = sole of foot)	Head of fibula and medial border of tibia.	Calcaneus by way of calcaneal ("Achilles") tendon.	Plantar flexes foot.	Tibial nerve.
Peroneus longus (*perone* = fibula; *longus* = long)	Head and body of fibula and lateral condyle of tibia.	First metatarsal and first cuneiform.	Plantar flexes and everts foot.	Superficial peroneal nerve.
Peroneus brevis (*brevis* = short)	Body of fibula.	Fifth metatarsal.	Plantar flexes and everts foot.	Superficial peroneal nerve.
Peroneus tertius (*tertius* = third)	Distal third of fibula.	Fifth metatarsal.	Dorsiflexes and everts foot.	Deep peroneal nerve.
Tibialis anterior (*tibialis* = tibia; *anterior* = front)	Lateral condyle and body of tibia.	First metatarsal and first cuneiform.	Dorsiflexes and inverts foot.	Deep peroneal nerve.
Tibialis posterior (*posterior* = back)	Interosseus membrane between tibia and fibula.	Second, third, and fourth metatarsals; navicular; third cuneiform; and cuboid.	Plantar flexes and inverts foot.	Tibial nerve.
Flexor digitorum longus (*flexor* = decreases angle at joint; *digitorum* = finger or toe)	Tibia.	Distal phalanges of four outer toes.	Flexes toes and plantar flexes and inverts foot.	Tibial nerve.
Extensor digitorum longus (*extensor* = increases angle at joint)	Lateral condyle of tibia and anterior surface of fibula.	Middle and distal phalanges of four outer toes.	Extends toes and dorsiflexes and everts foot.	Deep peroneal nerve.

CLINICAL APPLICATION

The calcaneal tendon is the **strongest tendon of the body,** able to withstand a 1,000-pound force without tearing. Despite this, however, the calcaneal tendon ruptures more frequently than any other tendon in the body.

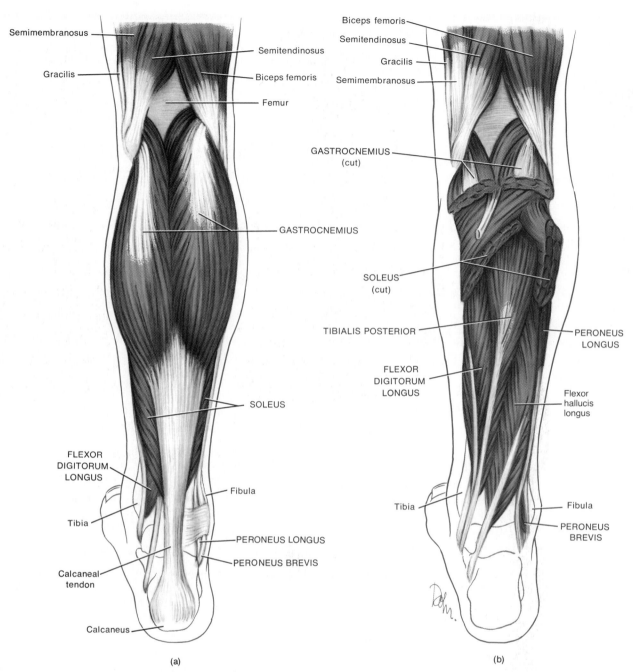

FIGURE 10-22 Muscles that move the foot and toes. (a) Superficial posterior view. (b) Deep posterior view.

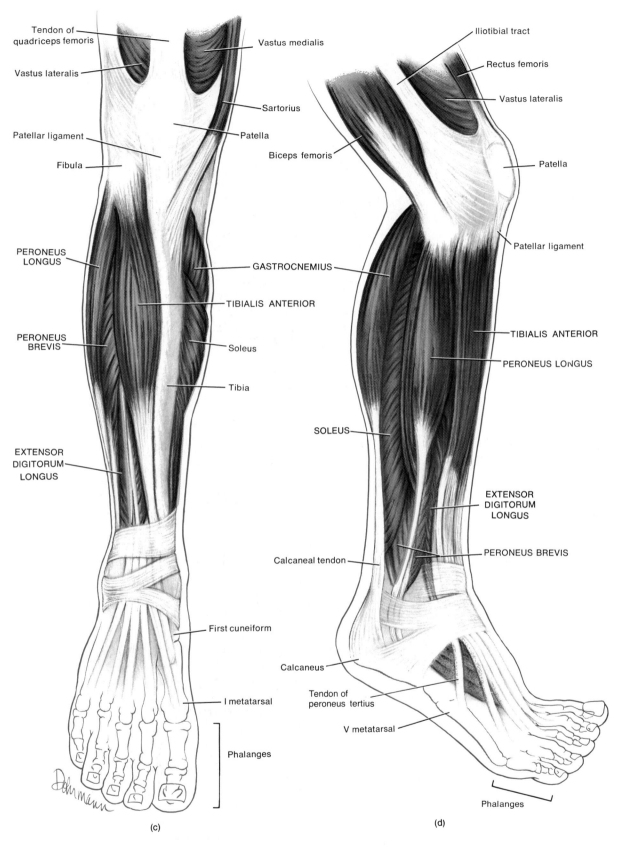

Tendon of quadriceps femoris

Vastus lateralis

Patellar ligament

Fibula

PERONEUS LONGUS

PERONEUS BREVIS

EXTENSOR DIGITORUM LONGUS

Vastus medialis

Sartorius

Patella

GASTROCNEMIUS

TIBIALIS ANTERIOR

Soleus

Tibia

First cuneiform

I metatarsal

Phalanges

(c)

Iliotibial tract

Rectus femoris

Vastus lateralis

Biceps femoris

Patella

Patellar ligament

TIBIALIS ANTERIOR

PERONEUS LONGUS

SOLEUS

EXTENSOR DIGITORUM LONGUS

PERONEUS BREVIS

Calcaneal tendon

Calcaneus

Tendon of peroneus tertius

V metatarsal

Phalanges

(d)

FIGURE 10-22 (*Continued*) Muscles that move the foot and toes. (c) Superficial anterior view. (d) Superficial lateral view.

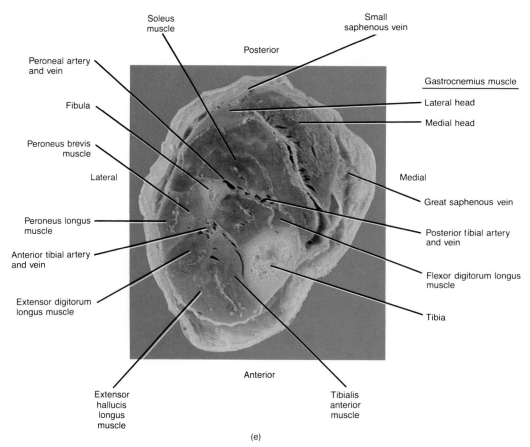

Soleus muscle

Posterior

Small saphenous vein

Peroneal artery and vein

Gastrocnemius muscle

Fibula

Lateral head

Medial head

Peroneus brevis muscle

Lateral

Medial

Great saphenous vein

Peroneus longus muscle

Posterior tibial artery and vein

Anterior tibial artery and vein

Flexor digitorum longus muscle

Extensor digitorum longus muscle

Tibia

Anterior

Extensor hallucis longus muscle

Tibialis anterior muscle

(e)

FIGURE 10-22 (*Continued*) Muscles that move the foot and toes. (e) Photograph of a cross section of the leg showing the musculature and related structures. (Courtesy of Stephen A. Kieffer and E. Robert Heitzman, *An Atlas of Cross-Sectional Anatomy,* Harper & Row, Publishers, Inc., New York, 1979.) The surface anatomy of the muscles that move the foot and toes is illustrated in Figures 11-15 and 11-16.

INTRAMUSCULAR INJECTIONS

An **intramuscular injection** penetrates the skin and subcutaneous tissue to enter the muscle itself. Intramuscular injections are preferred when prompt absorption is desired, when larger doses than can be given cutaneously are indicated, or when the drug is too irritating to give subcutaneously. The common sites for intramuscular injections include the buttock, lateral side of the thigh, and the deltoid region of the arm. Muscles in these areas, especially the gluteal muscles in the buttock, are fairly thick. Because of the large number of muscle fibers and extensive fascia, the drug has a large surface area for absorption. Absorption is further promoted by the extensive blood supply to muscles. Ideally, intramuscular injections should be given deep within the muscle and away from major nerves and blood vessels.

For many intramuscular injections, the preferred site is the gluteus medius muscle of the buttock (Figure 10-23a). The buttock is divided into four quadrants and the upper outer quadrant is used as the injection site. The iliac crest serves as a landmark for this quadrant. The spot for injection is about 5 to 7.5 cm (2 to 3 in) below the iliac crest. The upper outer quadrant is chosen because, in this area, the muscle is quite thick and contains few nerves. Thus there is less chance of injury to the sciatic nerve, which can cause paralysis of the lower extremity. The probability of injecting the drug into a blood vessel is also remote in this area. After the needle is inserted into the muscle, the plunger is pulled up for a few seconds. If the syringe fills with blood, the needle is in a blood vessel and a different injection site on the opposite buttock is chosen.

Injections given in the lateral side of the thigh are inserted into the midportion of the vastus lateralis muscle (Figure 10-23b). This site is determined by using the knee and the greater trochanter of the femur as landmarks. The midportion of the muscle is located by measuring a handbreadth above the knee and a handbreadth below the greater trochanter.

The deltoid injection is given in the midportion of the muscle about two to three fingerbreadths below the acromion of the scapula and lateral to the axilla (Figure 10-23c).

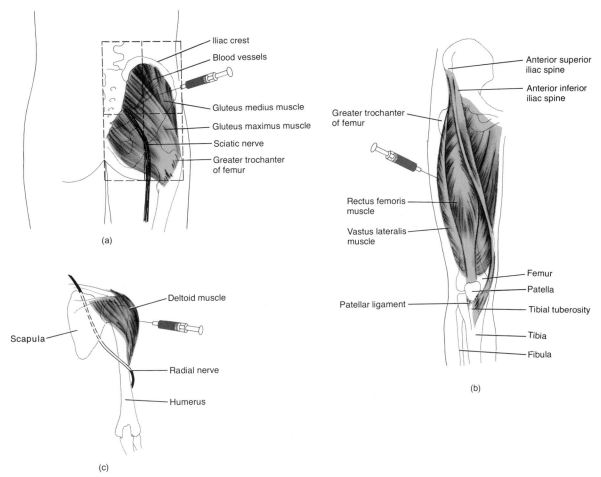

FIGURE 10-23 Intramuscular injections. Shown are the three common sites for intramuscular injections. (a) Buttock. (b) Lateral surface of the thigh. (c) Deltoid region of the arm.

STUDY OUTLINE

How Skeletal Muscles Produce Movement (p. 225)
1. Skeletal muscles produce movement by pulling on bones.
2. The attachment to the stationary bone is the origin. The attachment to the movable bone is the insertion.
3. Bones serve as levers and joints as fulcrums. The lever is acted on by two different forces: resistance and effort.
4. Levers are categorized into three types—first-class, second-class, and third-class—according to the position of the fulcrum, effort, and resistance on the lever.
5. Fascicular arrangements include parallel, convergent, pennate, and circular. Fascicular arrangement is correlated with the power of a muscle and the range of movement.
6. The prime mover produces the desired action. The antagonist produces an opposite action. The synergist assists the prime mover by reducing unnecessary movement. The fixator stabilizes the origin of the prime mover so that it can act more efficiently.

Naming Skeletal Muscles (p. 228)
Skeletal muscles are named on the basis of distinctive criteria: direction of fibers, location, size, number of origins (or heads), shape, origin and insertion, and action.

Principal Skeletal Muscles (p. 230)
The principal skeletal muscles of the body are grouped according to region in Exhibits 10-2 through 10-21.

Intramuscular Injections (p. 269)
1. Advantages of intramuscular injections are prompt absorption, use of larger doses than can be given cutaneously, and minimal irritation.
2. Common sites for intramuscular injections are the buttock, lateral side of the thigh, and the deltoid region of the arm.

REVIEW QUESTIONS

1. Using the terms origin, insertion, and belly in your discussion, describe how skeletal muscles produce body movements by pulling on bones.
2. What is a lever? Fulcrum? Apply these terms to the body, and indicate the nature of the forces that act on levers. Describe the three classes of levers, and provide one example for each in the body.
3. Describe the various arrangements of fasciculi. How is fascicular arrangement correlated with the strength of a muscle and its range of movement?
4. Define the role of the prime mover, antagonist, synergist, and fixator in producing body movements.
5. Select at random several muscles presented in Exhibits 10-2 through 10-21 and see if you can determine the criterion or criteria employed for naming each. In addition, refer to the prefixes, suffixes, roots, and definitions in each exhibit as a guide. Select as many muscles as you wish, as long as you feel you understand the concept involved.
6. What muscles would you use to do the following: (a) frown, (b) pout, (c) show surprise, (d) show your upper teeth, (e) pucker your lips, (f) squint, (g) blow up a balloon, (h) smile?
7. What would happen if you lost tone in the masseter and temporalis muscles?
8. What muscles move the eyeball? In what direction does each muscle move the eyeball?
9. What tongue, facial, and mandibular muscles would you use when chewing a piece of gum?
10. What muscles constrict the pharynx? What muscle dilates the pharynx?
11. What muscles would you use to signify "yes" and "no" by moving your head?
12. What muscles accomplish compression of the anterior abdominal wall?
13. What are the principal muscles involved in breathing? What are their actions?
14. Describe the actions of the muscles of the pelvic floor and perineum.
15. What muscles are used to (a) raise your shoulders, (b) lower your shoulders, (c) join your hands behind your back, (d) joint your hands in front of your chest?
16. What muscles move the arm? In which directions do these movements occur?
17. What muscles move the forearm and what actions are used when striking a match?
18. How many muscles and actions of the wrist and fingers used when writing can you list?
19. Can you perform an exercise that would involve the use of each of the muscles listed in Exhibit 10-18?
20. Review the various movements involved in your favorite kind of dancing. What muscles listed in Exhibit 10-19 would you be using and what actions would you be performing?
21. Determine the muscles and their actions listed in Exhibit 10-20 that you would use in climbing a ladder to a diving board, diving into the water, swimming the length of a pool, and then sitting at pool side.
22. Discuss the muscles that plantar flex, evert, pronate, dorsiflex, and supinate the foot.
23. What are the advantages of intramuscular injections?
24. Describe how you would locate the sites for an intramuscular injection in the buttock, lateral side of the thigh, and deltoid region of the arm.

11

Surface Anatomy

Student Objectives

Define surface anatomy.

Identify the principal regions of the human body.

Describe the surface anatomy features of the head.

Describe the surface anatomy features of the neck.

Describe the surface anatomy features of the trunk.

Describe the surface anatomy features of the upper extremities.

Describe the surface anatomy features of the lower extremities.

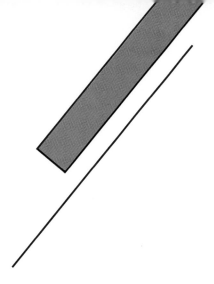

In Chapter 1, several branches of anatomy were defined and their importance to an understanding of the structure of the body was noted. Now that you have a fairly good idea how the body is organized, we can take a closer look at surface anatomy. Very simply, **surface anatomy** is the study of the form and markings of the surface of the body. A knowledge of surface anatomy will help you to identify certain superficial structures through visual inspection or palpation through the skin. **Palpation** means to feel with the hand.

To introduce you to surface anatomy, we will consider some key features of each of the principal regions of the body. These regions are outlined as follows and may be reviewed in Figure 1-2.

Head (cephalic region, or **caput)**
    Cranium (skull, or **brain case)**
    Face (facies)

Neck (collum)
Trunk
    Back (dorsum)
    Chest (thorax)
    Abdomen (venter)
    Pelvis

Upper extremity
    Armpit (axilla)
    Shoulder (acromial, or **omos)**
    Arm (brachium)
    Elbow (cubitus)
    Forearm (antebrachium)
    Wrist (carpus)
    Hand (manus)
        1. Metacarpus (dorsum and palm)
        2. Fingers (phalanges, or digits)

Lower extremity
    Buttocks (gluteal region)
    Thigh (femoral region)
    Knee (genu)
    Leg (crus)
    Ankle (tarsus)

Foot (pes)
    1. Metatarsus (dorsum and sole)
    2. Toes (phalanges, or digits)

HEAD

The **head (cephalic region,** or **caput)** is divided into the cranium and face. The **cranium (skull,** or **brain case)** consists of a *frontal region* (front of skull, or sinciput, that includes the forehead), *parietal region* (crown of skull, or vertex), *temporal region* (side of skull, or tempora), and *occipital region* (base of skull, or occiput).

The **face (facies)** is subdivided into an *orbital* or *ocular region* that includes the eyeballs (bulbi oculorum), eyebrows (supercilia), and eyelids (palpebrae); *nasal region* (nose, or nasus); *infraorbital region* (inferior to orbit); *oral region* (mouth); *mental region* (anterior portion of mandible); *buccal region* (cheek); *parotid-masseteric region* (external to the parotid gland and masseter muscle); *zygomatic region* (inferolateral to orbit); and *auricular region* (ear).

Examine Exhibit 11-1. It contains illustrations of various surface features of the head with an accompanying description of each of the features.

NECK

The **neck (collum)** is divided into an *anterior cervical region,* two *lateral cervical regions,* and a *posterior region,* or *nucha.*

The most prominent structure in the midline of the anterior cervical region is the *thyroid cartilage (Adam's apple)* of the larynx. Just superior to it, the *hyoid bone* can be palpated. Inferior to the thyroid cartilage, the *cricoid cartilage* of the larynx can be felt.

A major portion of the lateral cervical regions is formed by the *sternocleidomastoid muscles.* Each muscle extends from the mastoid process of the temporal bone, felt as a bump behind the auricle of the ear, to the sternum

EXHIBIT 11-1 SURFACE ANATOMY OF THE HEAD

HEAD (Figure 11-1)

1. **Zygomatic region.** Inferolateral to orbit.
2. **Orbital (ocular) region.** Includes eyeball, eyelids, and eyebrows.
3. **Infraorbital region.** Inferior to orbit.
4. **Nasal region.** Nose.
5. **Buccal region.** Cheek.
6. **Oral region.** Mouth.
7. **Mental region.** Anterior portion of mandible.
8. **Parotid-masseteric region.** External to parotid gland and masseter muscle.
9. **Occipital region.** Base of skull (occiput).
10. **Auricular region.** Ear.
11. **Parietal region.** Crown of skull (vertex).
12. **Temporal region.** Side of skull (tempora).
13. **Frontal region.** Front of skull (sinciput).

EYE (Figure 11-2)

1. **Pupil.** Opening of center of iris of eyeball for light transmission.
2. **Iris.** Circular pigmented muscular membrane behind cornea.
3. **Sclera.** "White" of eye; a coat of fibrous tissue that covers entire eyeball except for cornea.
4. **Conjunctiva.** Membrane that covers exposed surface of eyeball and lines eyelids.
5. **Palpebrae (eyelids).** Folds of skin and muscle lined by conjunctiva.

6. **Palpebral fissure.** Space between eyelids when they are open.
7. **Medial commissure.** Site of union of upper and lower eyelids near nose.
8. **Lateral commissure.** Site of union of upper and lower eyelids away from nose.
9. **Lacrimal caruncle.** Fleshy, yellowish projection of medial commissure that contains modified sweat and sebaceous glands.
10. **Eyelashes.** Hairs on margins of eyelids, usually arranged in two or three rows.
11. **Supercilia (eyebrows).** Several rows of hairs superior to upper eyelids.

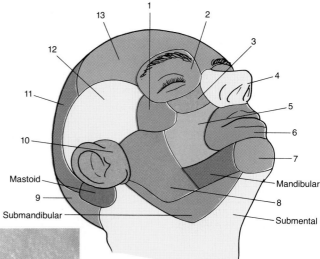

FIGURE 11-1 Regions of the head in anterolateral view.

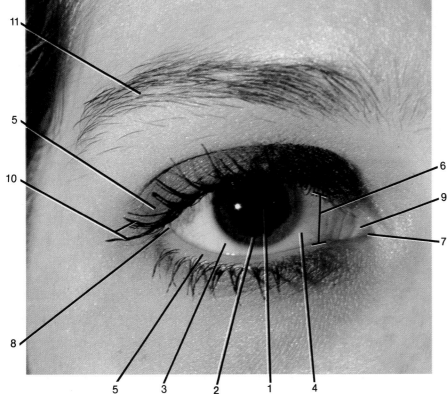

FIGURE 11-2 Surface anatomy of the right eye in anterior view. (© 1982 by Gerard J. Tortora. Courtesy of Lynne Tortora.)

EXHIBIT 11-1 *(Continued)*

EAR (Figure 11-3)

1. **Pinna.** Portion of external ear not contained in head; also called auricle or trumpet.

2. **Tragus.** Cartilaginous projection anterior to external opening to ear.

3. **Antitragus.** Cartilaginous projection opposite tragus.

4. **Concha.** Hollow of pinna.

5. **Helix.** Superior and posterior free margin of pinna.

6. **Antihelix.** Semicircular ridge posterior and superior to concha.

7. **Triangular fossa.** Depression in superior portion of antihelix.

8. **Lobule.** Inferior portion of pinna devoid of cartilage.

9. **External auditory meatus.** Canal extending from external ear to eardrum.

NOSE AND LIPS (Figure 11-4)

1. **Root.** Superior attachment of nose at forehead located between eyes.

2. **Apex.** Tip of nose.

3. **Dorsum nasi.** Rounded anterior border connecting root and apex; in profile, may be straight, convex, concave, or wavy.

4. **Nasofacial angle.** Point at which side of nose blends with tissues of face.

5. **Ala.** Convex flared portion of inferior lateral surface; unites with upper lip.

6. **External naris.** External opening into nose.

7. **Bridge.** Superior portion of dorsum nasi, superficial to nasal bones.

8. **Philtrum.** Vertical groove in medial portion of upper lip.

9. **Lips.** Upper and lower fleshy border of the oral cavity.

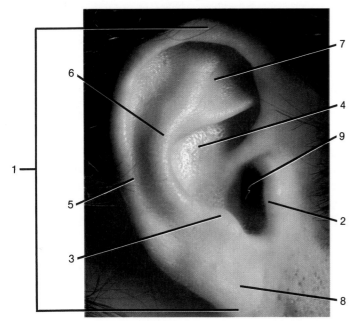

FIGURE 11-3 Surface anatomy of the right ear in lateral view. (© 1982 by Gerard J. Tortora. Courtesy of James Borghesi.)

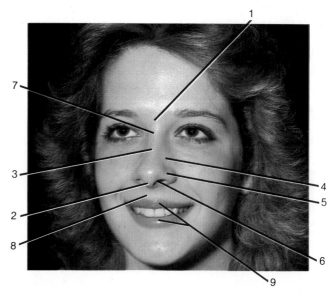

FIGURE 11-4 Surface anatomy of the nose and lips in anterior view. (© 1982 by Gerard J. Tortora. Courtesy of Lynne Tortora.)

and clavicle. Each sternocleidomastoid muscle divides its portion of the neck into an anterior triangle and a posterior (lateral) triangle. The *anterior triangle* is bordered superiorly by the mandible, inferiorly by the sternum, medially by the cervical midline, and laterally by the anterior border of the sternocleidomastoid muscle. The *posterior (lateral) triangle* is bordered inferiorly by the clavicle, anteriorly by the posterior border of the sternocleidomastoid muscle, and posteriorly by the anterior border of the trapezius muscle.

A muscle that extends downward and outward from the base of the skull and occupies a portion of the lateral cervical region is the *trapezius muscle.*

A "stiff neck" is frequently associated with an inflammation of this muscle. A very prominent vein that runs along the lateral surface of the neck is the *external jugular vein.* It is readily seen if you are angry or if your collar is too tight. The locations of a few surface features of the neck are shown in Exhibit 11-2.

EXHIBIT 11-2 SURFACE ANATOMY OF THE NECK (Figure 11-5)

1. **Anterior triangle of neck.** Bordered superiorly by mandible, inferiorly by sternum, medially by cervical midline, and laterally by anterior border of sternocleidomastoid muscle.

2. **Posterior triangle of neck.** Bordered inferiorly by clavicle, anteriorly by posterior border of sternocleidomastoid muscle, and posteriorly by anterior border of trapezius muscle.

3. **Trapezius muscle.** Occupies a portion of lateral surface of neck; helps raise, lower, and draw shoulders backward and extends head.

4. **Sternocleidomastoid muscle.** Forms major portion of lateral surface of neck; flexes head and rotates it to opposite side.

5. **Cricoid cartilage.** Inferior laryngeal cartilage that attaches larynx to trachea; can be palpated by running your fingertip down from your chin over the thyroid cartilage (after you pass the cricoid cartilage, your fingertip sinks in).

6. **Thyroid cartilage (Adam's apple).** Triangular laryngeal cartilage in the midline of the anterior cervical region.

7. **Hyoid bone.** Lies just superior to thyroid cartilage; it is the first resistant structure palpated in the midline below the chin.

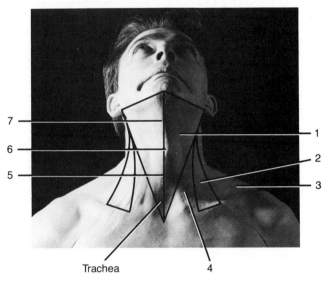

FIGURE 11-5 Surface anatomy of the neck in anterior view. (Courtesy of Carroll H. Weiss, Camera M.D. Studios, Inc.)

TRUNK

The **trunk** is divided into the back, chest, abdomen, and pelvis.

One of the most striking surface features of the **back (dorsum)** is the *vertebral spines,* the posteriorly pointed projections of the vertebrae. A very prominent vertebral spine is the *vertebra prominens* of the seventh cervical vertebra. It is easily seen when the head is flexed. Another easily identifiable surface landmark of the back is the *scapula.* In fact, several parts of the scapula (axillary border, vertebral border, inferior angle, spine, and acromion) may also be seen or palpated. In lean individuals the *ribs* may also be seen. The vertebral border of the scapula crosses ribs 2 to 7. Among the superficial muscles of the back that can be seen are the *latissimus dorsi, sacrospinalis, infraspinatus, trapezius,* and *teres major.* These, as well as the other surface features of the back, are described in Exhibit 11-3.

The **chest (thorax)** presents a number of anatomical landmarks. At its superior region are the *clavicles.* The *sternum* lies in the midline of the chest and is divisible into a superior manubrium, a middle body, and an inferior xiphoid process. Its superior border attaches to the clavicles. Between the medial ends of the clavicles, there is a depression on the superior surface of the sternum called the *jugular (suprasternal) notch.* The trachea can be palpated posterior to the jugular notch. The *sternal angle* is formed by a junction line between the manubrium and body of the sternum and is palpable under the skin. It locates the costal cartilage of the second rib and is the most reliable surface landmark of the chest. At or inferior to the sternal angle and slightly to the right, the trachea bifurcates (divides in two) into the left and right primary bronchi. The inferior portion of the sternum, the *xiphoid process,* may be palpated. Also visible or palpable are the *ribs.* The apex of the lung rises as high as the neck of the first rib, superior to the clavicle. The *costal margins,* the inferior edges of the costal cartilages of ribs 7 through 10, can also be seen or palpated. The margins are very near the iliac crests, sometimes only 1 to 2 cm away. Among the more prominent superficial chest muscles are the *pectoralis major* and *serratus anterior.* These and other surface features of the chest are described in Exhibit 11-3. The surface features of the heart are shown in Figure 13-5.

The **abdomen (venter)** and **pelvis** have already been discussed in terms of their nine regions or four quadrants (see Figures 1-8 and 1-9). External features of the abdomen and pelvis include the *umbilicus, linea alba, external oblique muscle, rectus abdominis muscle, tendinous intersections,* and *symphysis pubis.* The surface features of the abdomen and pelvis are described in Exhibit 11-3.

UPPER EXTREMITY

The **upper extremity** consists of the armpit, shoulder, arm, elbow, forearm, wrist, and hand.

At the **shoulder (acromial** or **omos)** moving laterally along the top of the clavicle, it is possible to palpate a slight elevation at the lateral end of the clavicle, the *acromioclavicular joint.* Less than 2.5 cm (1 in) distal to this joint, one can also feel the *acromion* of the scapula, which forms the tip of the shoulder. The rounded prominence of the shoulder is formed by the *deltoid muscle,* a frequent site for intramuscular injections (Exhibit 11-4).

EXHIBIT 11-3 SURFACE ANATOMY OF THE TRUNK

BACK (Figure 11-6)

1. **Vertebral spines.** Posteriorly pointed projections of vertebrae.

2. **Scapula.** Shoulder blade (leader is on vertebral border).

3. **Latissimus dorsi muscle.** Broad muscle of back that helps draw shoulders backward and downward.

4. **Sacrospinalis (erector spinae) muscle.** Parallel to vertebral column; moves vertebral column in various directions.

5. **Infraspinatus muscle.** Located inferior to spine of scapula; helps laterally rotate humerus.

6. **Trapezius muscle.** Extends from cervical and thoracic vertebrae to spine of scapula and lateral end of clavicle; occupies portion of lateral surface of neck (see Figure 11-5); helps raise, lower, and draw shoulders backward and extend head.

7. **Teres major muscle.** Located inferior to infraspinatus; helps extend, adduct, and medially rotate humerus.

8. **Posterior axillary fold.** Formed by the latissimus dorsi and teres major muscles; can be palpated between the finger and thumb.

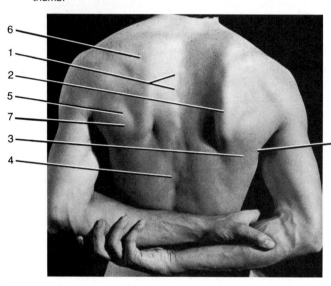

FIGURE 11-6 Surface anatomy of the back in posterior view. (Courtesy of Victor B. Eichler, © 1980.)

CHEST (Figure 11-7)

1. **Jugular (suprasternal) notch of sternum.** Depression on superior border of manubrium of sternum between medial ends of clavicles; trachea can be palpated in the notch.

2. **Clavicle.** Collarbone.

3. **Sternal angle of sternum.** Formed by junction between manubrium and body of sternum.

4. **Nipples.** Superficial to fourth intercostal space or fifth rib about 10 cm (4 in) from the midline in males and most females. The position of the nipples in females is variable depending on the size and pendulousness of the breasts. The right dome of the diaphragm is just inferior to the right nipple, the left dome is about 2 to 3 cm (1 in) inferior to the left nipple, and the central tendon is at the level of the junction of the body of the sternum and xiphoid process.

5. **Serratus anterior muscle.** Inferior and lateral to pectoralis major muscle; helps rotate scapula laterally and elevate ribs; also illustrated in Exhibit 11-3, Abdomen.

Ribs. Form bony cage of thoracic cavity (not illustrated).

Mammary glands. Accessory organs of the female reproductive system located inside the breasts. They overlie the pectoralis major muscle (⅔) and serratus anterior muscle (⅓). After puberty, they enlarge to their hemispherical shape, and in young adult females, they extend from the second through sixth ribs and from the lateral margin of the sternum to the midaxillary line (see Figure 25-21).

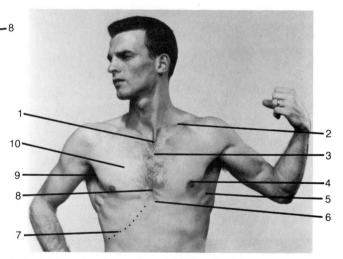

FIGURE 11-7 Surface anatomy of the chest in anterior view. (Courtesy of Vincent P. Destro, Mayo Foundation.)

6. **Xiphoid process of sternum.** Inferior portion of sternum, between the seventh costal cartilages.

7. **Costal margin.** Inferior edges of costal cartilages of ribs 7 through 10.

8. **Body of sternum.** Midportion of sternum.

9. **Anterior axillary fold.** Formed by the lateral border of the pectoralis major muscle; can be palpated between the fingers and thumb.

10. **Pectoralis major muscle.** Principal upper chest muscle; flexes, adducts, and medially rotates humerus.

EXHIBIT 11-3 (*Continued*)

ABDOMEN (Figure 11-8)

1. **Umbilicus.** Also called navel; previous site of attachment of umbilical cord in fetus. It is level with the intervertebral disc between the bodies of L3 and L4. The abdominal aorta bifurcates into the right and left common iliac arteries anterior to the body of vertebra L4. The inferior vena cava lies to the right of the abdominal aorta and is wider; it arises anterior to the body of vertebra L5.

2. **External oblique muscle.** Located inferior to serratus anterior; helps compress abdomen and bend vertebral column laterally.

3. **Rectus abdominis muscle.** Located just lateral to midline of abdomen; helps compress abdomen and flex vertebral column.

4. **Linea alba.** Flat, tendinous raphe forming a furrow along midline between rectus abdominis muscles. The furrow extends from the xiphoid process to the symphysis pubis. It is broad above the umbilicus and narrow below it.

5. **Tendinous intersections.** Fibrous bands that run transversely or obliquely across the rectus abdominis muscles.

6. **McBurney's point.** An important landmark at the junction of the lateral and middle thirds of a line joining the umbilicus and anterior superior iliac spine. An oblique incision through McBur-

ney's point is made for appendectomy. Pressure of the finger on McBurney's point produces tenderness in acute appendicitis.

Symphysis pubis. Anterior joint of hipbones; palpated as a firm resistance in the midline at the inferior portion of the anterior abdominal wall (not illustrated).

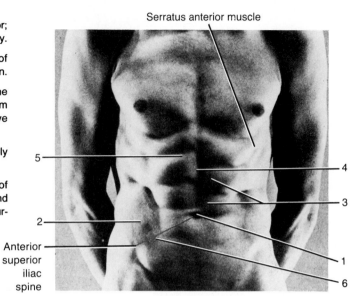

FIGURE 11-8 Surface anatomy of the abdomen in anterior view. (Courtesy of R. D. Lockhart, *Living Anatomy,* Faber & Faber, London, 1974.)

Most of the anterior surface of the **arm (brachium)** is occupied by the *biceps brachii muscle,* whereas most of the posterior surface is occupied by the *triceps brachii muscle.* The *brachial artery,* which provides the main arterial supply to the arm, begins at the lower border of the teres major muscle as the continuation of the axillary artery and terminates at the level of the neck of the radius by dividing into the ulnar and radial arteries. The radial artery is palpable throughout its length (Exhibit 11-4).

At the **elbow (cubitus)** it is possible to locate three bony protuberences. The *medial* and *lateral epicondyles* of the humerus form visible eminences on the dorsum of the elbow. The *olecranon* of the ulna forms the large eminence in the middle of the dorsum of the elbow and lies between and slightly superior to the epicondyles when the forearm is extended. The *ulnar nerve* can be palpated in a groove behind the medial epicondyle. The triangular space of the anterior region of the elbow is the *cubital fossa.* The *median cubital vein* usually crosses the cubital fossa obliquely. This vein is the one frequently selected for removal of blood for diagnosis, transfusions, and intravenous therapy (Exhibit 11-4).

One of the most prominent landmarks of the **forearm (antebrachium)** is the *styloid process* of the ulna. It may be seen as a protuberance on the medial side of the wrist. The ulna is the medial bone of the forearm, and the radius

is the lateral bone of the forearm. On the lateral side of the upper forearm is the *brachioradialis muscle.* Next to it is the *flexor carpi radialis muscle.* On the medial side of the upper forearm is the *flexor carpi ulnaris muscle* (Exhibit 11-4).

At the **wrist (carpus),** several structures may be palpated or seen. On the anterior surface are *bracelet flexure lines* where the skin is firmly attached to the underlying deep fascia. On the anterior surface of the wrist, it is possible to see the *tendon of the palmaris longus muscle* by making a fist. Next to this tendon, as you move toward the thumb, you can feel the *tendon of the flexor carpi radialis muscle.* If you continue toward the thumb, you can palpate the *radial artery* just medial to the styloid process of the radius. This artery is frequently used to take the pulse. The *pisiform bone,* the medial bone of the proximal carpals, can be palpated as a projection distal to the styloid process of the ulna. By bending the thumb backward, two prominent tendons may be located along the posterior surface of the wrist. The one closer to the styloid process of the radius is the *tendon of the extensor pollicus brevis muscle.* The one closer to the styloid process of the ulna is the *tendon of the extensor pollicus longus muscle.* The depression between these two tendons is known as the *"anatomical snuffbox."* By palpating the depression, you can feel the radial artery (Exhibit 11-4).

EXHIBIT 11-4 SURFACE ANATOMY OF THE UPPER EXTREMITY

SHOULDER (Figure 11-9)

1. **Acromion.** Expanded end of spine of scapula; forms tip of shoulder; clearly visible in some individuals and can be palpated about 2.5 cm (1 in) distal to acromioclavicular joint.
2. **Deltoid muscle.** Triangular muscle that forms rounded prominence of shoulder; abducts arm.

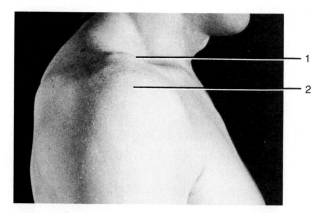

FIGURE 11-9 Surface anatomy of the right shoulder in lateral view. (Courtesy of Victor B. Eichler, © 1980.)

ARM AND ELBOW (Figure 11-10)

1. **Biceps brachii muscle.** Forms bulk of anterior surface of arm; helps flex forearm.
2. **Triceps brachii muscle.** Forms bulk of posterior surface of arm; helps extend forearm.
3. **Medial epicondyle.** Medial projection at distal end of humerus.
4. **Lateral epicondyle.** Lateral projection at distal end of humerus.
5. **Olecranon.** Projection of proximal end of ulna; forms elbow.
6. **Cubital fossa.** Triangular space in anterior region of elbow; contains tendon of biceps brachii muscle, brachial artery and its terminal branches (radial and ulnar arteries), and parts of median and radial nerves.
7. **Median cubital vein** (not illustrated). Crosses cubital fossa obliquely.
8. **Brachial artery** (not illustrated). Passes posterior to coracobrachialis muscle and then medial to biceps brachii muscle. It enters the middle of the cubital fossa and passes under the bicipital aponeurosis, which separates it from the median cubital vein. The artery is frequently used to take blood pressure.
9. **Bicipital aponeurosis.** An aponeurotic band that inserts the biceps brachii muscle into the deep fascia in the medial aspect of the forearm. It can be felt when the muscle contracts.

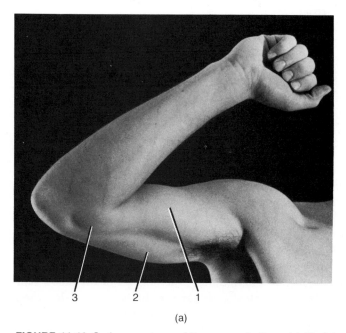

(a)

FIGURE 11-10 Surface anatomy of the arm and elbow. (a) Medial view of right upper extremity. (Courtesy of Victor B. Eichler, © 1980.)

EXHIBIT 11-4 (*Continued*)

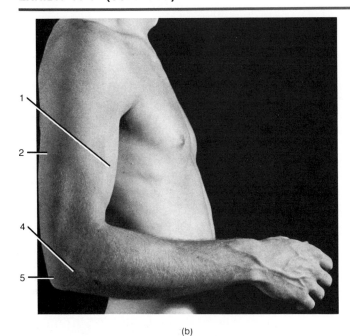

(b)

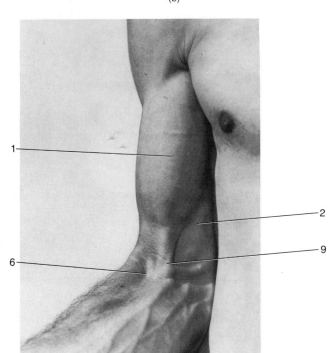

(c)

FIGURE 11-10 (*Continued*) Surface anatomy of the arm and elbow. (b) Lateral view of right upper extremity. (Courtesy of Victor B. Eichler, © 1980.) (c) Anterior view of right arm. (Courtesy of Vincent P. Destro, Mayo Foundation.)

FOREARM (Figure 11-11)

1. **Styloid process of ulna.** Projection of distal end of ulna at medial side of wrist.

2. **Brachioradialis muscle.** Located at superior and lateral aspect of forearm; helps flex forearm.

3. **Flexor carpi radialis muscle.** Located along midportion of forearm; helps flex wrist.

4. **Flexor carpi ulnaris muscle.** Located at medial aspect of forearm; helps flex wrist.

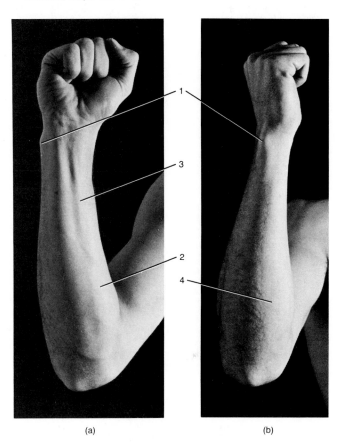

(a)   (b)

FIGURE 11-11 Surface anatomy of the right forearm. (a) Anterior view. (b) Medial view. (Courtesy of Victor B. Eichler, © 1980.)

EXHIBIT 11-4 (*Continued*)

WRIST (Figure 11-12)

1. **Bracelet flexure lines.** Several more or less constant lines on anterior aspect of wrist where skin is firmly attached to underlying deep fascia.

2. **Tendon of palmaris longus muscle.** By making a fist, the tendon can be seen on anterior surface of wrist nearer ulna; muscle helps to flex wrist.

3. **Tendon of flexor carpi radialis muscle.** Tendon on anterior surface of wrist lateral to tendon of palmaris longus.

4. **Radial artery.** Can be palpated just medial to styloid process of ulna; frequently used to take pulse.

5. **Pisiform bone.** Medial bone of proximal carpals; easily palpated as a projection distal to styloid process of ulna.

6. **Tendon of extensor pollicus brevis muscle.** Tendon closer to styloid process of radius along posterior surface of wrist, best seen when thumb is bent backward; muscle extends thumb.

7. **Tendon of extensor pollicus longus muscle.** Tendon closer to styloid process of ulna along posterior surface of wrist, best seen when thumb is bent backward; muscle extends thumb.

8. **"Anatomical snuffbox."** Depression between tendons of extensor pollicus brevis and extensor pollicus longus muscles; radial artery can be palpated in the depression.

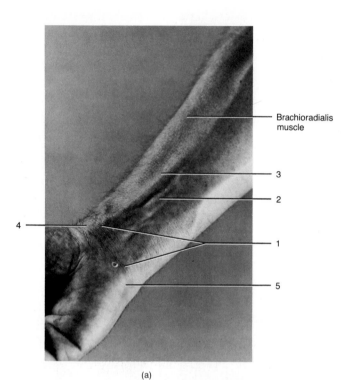

(a)

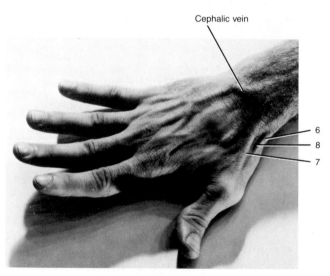

(b)

FIGURE 11-12 Surface anatomy of the right forearm. (a) Anterior view. (b) Dorsum. (Courtesy of Vincent P. Destro, Mayo Foundation.)

EXHIBIT 11-4 (*Continued*)

HAND (Figure 11-13)

1. **"Knuckles."** Commonly refers to dorsal aspects of distal ends of metacarpals II, III, IV, and V; also includes dorsal aspects of metacarpophalangeal and interphalangeal joints.

2. **Thenar eminence.** Lateral rounded contour on palm of hand formed by muscles of thumb.

3. **Hypothenar eminence.** Medial rounded contour on palm of hand formed by muscles of little finger.

4. **Digital flexion creases.** Skin creases on anterior surface of fingers.

5. **Palmar flexion creases.** Skin creases on palm of hand.

6. **Tendon of extensor digiti minimi muscle.** Extensor tendon in line with phalanx V (little finger); muscle extends little finger.

7. **Tendons of extensor digitorum muscle.** Extensor tendons in line with phalanges II, III, and IV; muscle extends fingers and wrist.

8. **Dorsal venous arch.** Superficial veins on dorsum of hand that form cephalic vein.

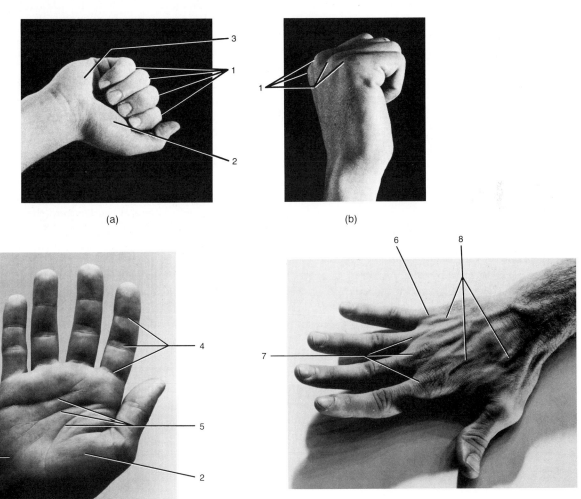

(a)　　　　　(b)

(c)　　　　　(d)

FIGURE 11-13 Surface anatomy of the right hand. (a) Palmar and dorsal surfaces. (b) Medial view. (c) Palmar surface. (d) Dorsum. [(a) and (b) Courtesy of Victor B. Eichler, © 1980. (c) Courtesy of Carroll H. Weiss, Camera M.D. Studios, Inc. (d) Courtesy of Vincent P. Destro, Mayo Foundation.]

On the dorsum of the **hand (manus)** the distal ends of the second through fifth metacarpal bones are commonly referred to as the *"knuckles."* The term also includes the joints between metacarpals and phalanges and the joints between the phalanges of the fingers. The *dorsal venous arch,* the superficial veins on the dorsum of the hand, can be displayed by compressing the blood vessels at the wrist for a few moments as the hand is opened and closed. The *tendon of the extensor digiti minimi muscle* can be seen in line with phalanx V (little finger), and the *tendons of the extensor digitorum muscle* can be seen in line with phalanges II, III, and IV. By examining the palm of the hand, it is possible to see a number of *skin creases,* as well as the location of the joints of the fingers. The skin creases of the palm are collectively called *palmar flexion creases.* The skin creases on the anterior surface of the fingers are called *digital flexion creases.* Also on the palm are the *thenar eminence,* a rounded contour formed by thumb muscles, and the *hypothenar eminence,* a rounded contour formed by little finger muscles (Exhibit 11-4).

LOWER EXTREMITY

The **lower extremity** consists of the buttocks, thigh, knee, leg, ankle, and foot.

The outline of the superior border of the **buttock (gluteal region)** is formed by the *iliac crest.* The *posterior superior iliac spine,* the posterior termination of the iliac crest, lies deep to a dimple (skin depression) about 4 cm (1.5 in) lateral to the midline. Most of the prominence of the buttocks is formed by the *gluteus maximus* and *gluteus medius muscles.* The depression that separates the buttocks is the *gluteal cleft,* whereas the inferior limit of the buttock formed by the inferior margin of the gluteus maximus muscle is called the *gluteal fold.* The bony prominence in each buttock is the *ischial tuberosity* of the hipbone. This structure bears the weight of your body when you are seated. About 20 cm (8 in) below the highest portion of the iliac crest, the *greater trochanter* of the femur can be felt on the lateral side of the thigh. The iliac crest and greater trochanter are useful landmarks when giving an intramuscular injection in the gluteal muscle (Exhibit 11-5).

Among the prominent superficial muscles on the anterior surface of the **thigh (femoral region)** are the *sartorius* and three of the four components of the *quadriceps femoris* (*vastus lateralis, vastus medialis,* and *rectus femoris*). The vastus lateralis is frequently used as an injection site for diabetics when administering insulin. The superficial medial thigh muscles include the *adductor magnus, adductor brevis, adductor longus, gracilis, obturator externus,* and *pectineus.* The superficial posterior thigh muscles are the *hamstrings* (*semitendinosus, semimembranosus,* and *biceps femoris*). See Exhibit 11-5.

On the anterior surface of the **knee (genu)**, the *patella,* or kneecap, is observable. Below it is the *patellar ligament.* On the posterior surface of the knee is a diamond-shaped space, the *popliteal fossa.* Just below the patella, on either side of the patellar ligament, the *medial* and *lateral condyles* of the femur and tibia can be felt (Exhibit 11-5).

The bony prominence inferior to the patella in the middle of the **leg (crus)** is the *tibial tuberosity.* The tibia is the medial bone of the leg, and the fibula is the lateral bone of the leg. Prominent superficial muscles of the leg include the *tibialis anterior, gastrocnemius,* and *soleus* (Exhibit 11-5).

At the **ankle (tarsus)** the *medial malleolus* of the tibia and the *lateral malleolus* of the fibula can be noted as two prominent eminences (Exhibit 11-5).

On the dorsum of the **foot (pes)** you can see the *tendon of the extensor hallucis longus muscle* in line with phalanx V (great toe), the *tendons of the extensor digitorum longus muscle* in line with phalanges II through V, and the *dorsal venous arch,* superficial veins that unite to form the small and great saphenous veins. Arising from the *heel bone* (*calcaneus*) is the *calcaneal (Achilles) tendon* (Exhibit 11-5).

EXHIBIT 11-5 SURFACE ANATOMY OF THE LOWER EXTREMITY

BUTTOCKS AND THIGH (Figure 11-14)

1. **Iliac crest.** Superior margin of ilium of hipbone; forms outline of superior border of buttock; when you rest your hands on your hips, they rest on the iliac crests.

2. **Posterior superior iliac spine.** Posterior termination of iliac crest; lies deep to a dimple (skin depression) about 4 cm (1.5 in) lateral to midline; dimple forms because skin and underlying fascia are attached to bone. The spine marks the inferior limit of cerebrospinal fluid in the subarachnoid space around the spinal cord.

3. **Gluteus maximus muscle.** Forms major portion of prominence of buttock; extends and laterally rotates thigh.

4. **Gluteus medius muscle.** Superolateral to gluteus maximus; abducts and medially rotates thigh; frequent site for intramuscular injections.

5. **Gluteal cleft.** Depression along midline that separates the buttocks.

6. **Gluteal fold.** Inferior limit of buttock formed by inferior margin of gluteus maximus muscle.

7. **Ischial tuberosity.** Bony prominence of ischium of hipbone; bears weight of body when seated.

8. **Greater trochanter.** Projection of proximal end of femur on lateral surface of thigh; can be palpated about 20 cm (8 in) inferior to iliac crest.

9. **Hamstrings.** Superficial posterior thigh muscles that flex leg and extend thigh.

10. **Sartorius muscle.** Superficial anterior thigh muscle that flexes leg and flexes and laterally rotates thigh.

11. **Rectus femoris muscle.** Component of quadriceps femoris group located at midportion of anterior aspect of thigh; extends leg (in conjunction with other components of quadriceps femoris) and flexes thigh (acting alone).

12. **Vastus lateralis muscle.** Component of quadriceps femoris group located at anterolateral aspect of thigh; extends leg.

13. **Vastus medialis muscle.** Component of quadriceps femoris group located at anteromedial aspect of thigh; extends leg.

14. **Adductor magnus muscle.** Located on medial aspect of thigh; adducts, flexes, and extends thigh.

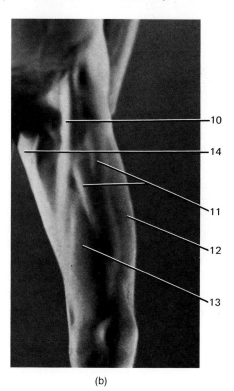

(a) (b)

FIGURE 11-14 Surface anatomy of the buttocks and thigh. (a) Posterior view of buttocks. (b) Anterior view of left thigh. [(a) Courtesy of Lester Bergman & Associates. (b) Courtesy of R. D. Lockhart, *Living Anatomy,* Faber & Faber, London, 1974.]

EXHIBIT 11-5 (*Continued*)

KNEE (Figure 11-15)

1. Patella. Kneecap; located within quadriceps femoris tendon on anterior surface of knee along midline; margins of condyles (described shortly) can be felt on either side of it.

2. Medial condyle of femur. Medial projection of distal end of femur.

3. Medial condyle of tibia. Medial projection of proximal end of tibia.

4. Patellar ligament. Continuation of quadriceps femoris tendon inferior to patella.

5. Popliteal fossa. Diamond-shaped space on posterior aspect of knee visible when knee is flexed; fossa is bordered superolaterally by the biceps femoris muscle, superomedially by the semimembranosus and semitendinosus muscles, and inferolaterally and inferomedially by the lateral and medial heads of the gastrocnemius muscle, respectively.

6. Lateral condyle of tibia. Lateral projection of proximal end of tibia.

7. Lateral condyle of femur. Lateral projection of distal end of femur.

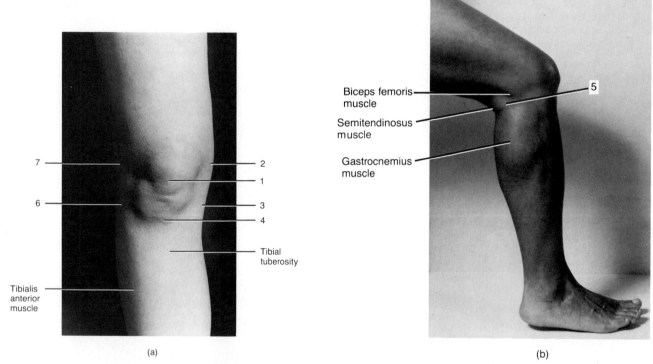

(a)

(b)

FIGURE 11-15 Surface anatomy of the right knee. (a) Anterior view. (b) Posterolateral view. [(a) Courtesy of Carroll H. Weiss, Camera M.D. Studios, Inc. (b) Courtesy of J. J. Martin, Mayo Clinic.]

EXHIBIT 11-5 (*Continued*)

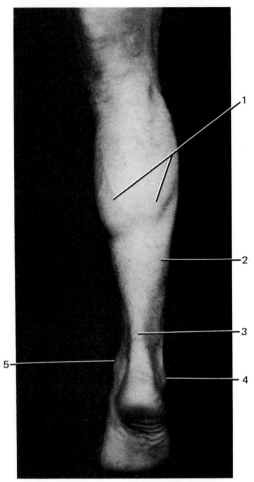

FIGURE 11-16 Surface anatomy of the right leg and ankle in posterior view. (Courtesy of Donald Castellaro and Richard Sollazzo.)

LEG AND ANKLE (Figure 11-16)

Tibial tuberosity. Bony prominence of tibia into which patellar ligament inserts (see Figure 11-15a).

Tibialis anterior muscle. Located at anterior surface of leg along midportion; dorsiflexes and inverts foot (see Figure 11-15a).

1. **Gastrocnemius muscle.** Forms bulk of mid and upper portion of posterior aspect of leg; plantar flexes foot.

2. **Soleus muscle.** Located deep to gastrocnemius muscle; plantar flexes foot.

3. **Calcaneal (Achilles) tendon.** Conspicuous tendon of gastrocnemius and soleus muscles that inserts into calcaneus (heel) bone of foot.

4. **Lateral malleolus of fibula.** Projection of distal end of fibula that forms lateral prominence of ankle.

5. **Medial malleolus of tibia.** Projection of distal end of tibia that forms medial prominence of ankle.

FOOT (Figure 11-17)

1. **Calcaneus.** Heel bone.

2. **Dorsal venous arch.** Superficial veins on dorsum of foot that unite to form small and great saphenous veins.

3. **Tendons of extensor digitorum longus muscle.** Visible in line with phalanges II through V; muscle extends toes and dorsiflexes and everts foot.

4. **Tendon of extensor hallucis longus muscle.** Visible in line with phalanx I (great toe); muscle extends great toe and dorsiflexes ankle.

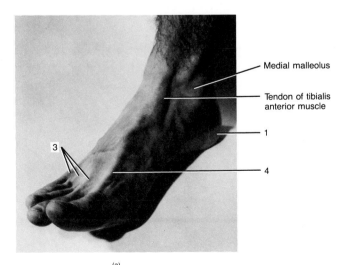

(a)

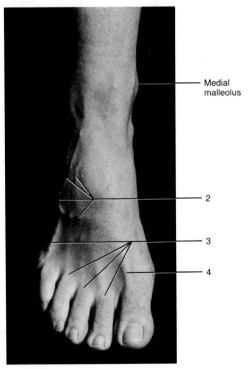

(b)

FIGURE 11-17 Surface anatomy of the right foot. (a) Dorsomedial view. (Courtesy of Vincent P. Destro, Mayo Foundation.) (b) Dorsum. (Courtesy of Carroll H. Weiss, Camera M.D. Studios, Inc.)

STUDY OUTLINE

1. Surface anatomy is the study of the form and markings of the surface of the body.
2. Surface anatomy features may be noted by visual inspection or palpation.
3. The principal regions of the body used to study surface anat-omy are the head, neck, trunk, upper extremity, and lower extremity.
4. A review of surface anatomy is presented in Exhibits 11-1 through 11-5.

REVIEW QUESTIONS

1. Define surface anatomy. What is meant by palpation?
2. List the principal regions of the body and their subdivisions.
3. Using Exhibit 11-1 as an outline, locate as many of the surface features of the head as you can on your partner's body, wall charts, models, photographs, and skeletons.
4. Using Exhibit 11-2 as an outline, locate as many of the surface features of the neck as you can on your partner's body, wall charts, models, photographs, and skeletons.
5. Using Exhibit 11-3 as an outline, locate as many of the surface features of the trunk as you can on your partner's body, wall charts, models, photographs, and skeletons.
6. Using Exhibit 11-4 as an outline, locate as many of the surface features of the upper extremity as you can on your partner's body, wall charts, models, photographs, and skeletons.
7. Using Exhibit 11-5 as an outline, locate as many of the surface features of the lower extremity as you can on your partner's body, wall charts, models, photographs, and skeletons.

12

The Cardiovascular System: Blood

Student Objectives

Describe the principal physical characteristics of blood and its functions in the body.

Compare the origins of the formed elements in blood.

Discuss the structure of erythrocytes and their function in the transport of oxygen and carbon dioxide.

List the structural features and types of leucocytes.

Explain the significance of a differential count.

Discuss the role of leucocytes in phagocytosis and antibody production.

Discuss the structure of thrombocytes and explain their role in blood clotting.

List the components of plasma and explain their importance.

Contrast the causes and clinical symptoms of nutritional, pernicious, hemorrhagic, hemolytic, aplastic, and sickle cell anemia.

Define polycythemia, infectious mononucleosis (IM), and leukemia.

Define key medical terms associated with blood.

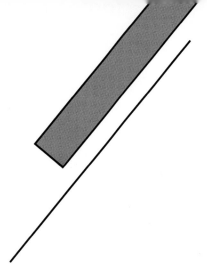

The more specialized a cell becomes, the less capable it is of carrying on an independent existence. For instance, a specialized cell is less capable of protecting itself from extreme temperatures, toxic chemicals, and changes in pH. It may be incapable of devouring whole bits of food. And, if it is firmly implanted in a tissue, it cannot go looking for food or move away from its own wastes. The substance that bathes the cell and carries out these vital functions for it is called **interstitial fluid** (also known as **intercellular**, or **tissue, fluid**).

Interstitial fluid, in turn, must be serviced by blood and lymph. The blood picks up oxygen from the lungs, nutrients from the digestive tract, hormones from the endocrine glands, and enzymes from still other parts of the body. The blood then transports these substances to all the tissues where they diffuse from the capillaries (the smallest blood vessels) into the interstitial fluid. In the interstitial fluid, the substances are passed on to the cells and exchanged for wastes.

Since the blood must service all the tissue of the body, it can be an important medium for the transport of disease-causing organisms. To protect itself from the spread of disease, the body has a lymphatic system—a collection of vessels containing a fluid called lymph. The lymph picks up materials, including wastes, from the interstitial fluid, cleanses them of bacteria, and returns them to the blood. The blood then carries the wastes to the lungs, kidneys, and sweat glands, where they are eliminated from the body. The blood also takes wastes to the liver, where they are detoxified and recycled.

The blood, heart, and blood vessels constitute the **cardiovascular system.** The lymph, lymph vessels, and lymph glands make up the **lymphatic system.** In this chapter we will take a look at the substance known as blood. The branch of science concerned with the study of blood and blood-forming tissues and the disorders associated with them is called **hematology** (*hem* = blood; *logos* = study of).

The developmental anatomy of blood and blood vessels is considered in Chapter 13.

PHYSICAL CHARACTERISTICS

The red body fluid that flows through all the vessels except the lymph vessels is called **blood.** Blood is a viscous fluid, that is, it is thicker and more adhesive than water. Water is considered to have a viscosity of 1.0. The viscosity of blood, by comparison, ranges from 4.5 to 5.5. It flows 4½ to 5½ times more slowly than water, at least in part because of its viscosity. The adhesive quality of blood, or its stickiness, may be felt by touching it. Blood is also slightly heavier than water.

Other physical characteristics of blood include a temperature of about 38°C (100.4°F), a pH range of 7.35 to 7.45 (slightly alkaline), and a 0.85 to 0.90 percent concentration of salt (NaCl).

Blood constitutes about 8 percent of the total body weight. The blood volume of an average-sized man is between 5 and 6 liters (about 5 to 6 qt). An average-sized woman has 4 to 5 liters.

The physical characteristics are summarized in Exhibit 12-1.

FUNCTIONS

Despite its simple appearance, blood is a complex liquid that performs a number of critical functions.

EXHIBIT 12-1   SUMMARY OF PHYSICAL CHARACTERISTICS OF BLOOD

Viscosity	4.5–5.5
Temperature	38°C (100.4°F)
pH	7.35–7.45
NaCl concentration	0.85–0.90 percent
Total body weight	8 percent
Volume	5–6 liters for males and 4–5 liters for females

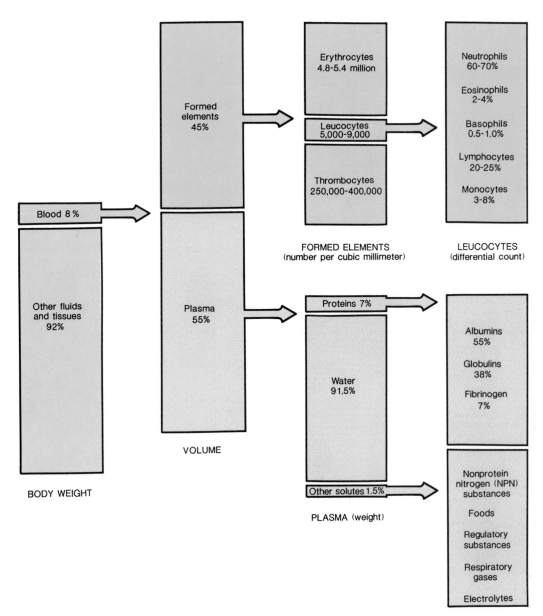

FIGURE 12-1 Components of blood in a normal adult.

1. It transports: oxygen from the lungs to the cells of the body; carbon dioxide from the cells to the lungs; nutrients from the digestive organs to the cells; waste products from the cells to the kidneys, lungs, and sweat glands; hormones from endocrine glands to the cells; enzymes to various cells.

2. It regulates: pH through buffers and amino acids; normal body temperature through the heat-absorbing and coolant properties of its water content; the water content of cells, principally through dissolved sodium ions.

3. It protects against: blood loss through the clotting mechanism; toxins and foreign microbes through special combat-unit cells.

COMPONENTS

Microscopically, blood is composed of two portions: formed elements (cells and cell-like structures) and plasma (liquid containing dissolved substances). The formed elements compose about 45 percent of the volume of blood; plasma constitutes about 55 percent (Figure 12-1).

FORMED ELEMENTS

In clinical practice, the most common classification of the **formed elements** of the blood is the following:

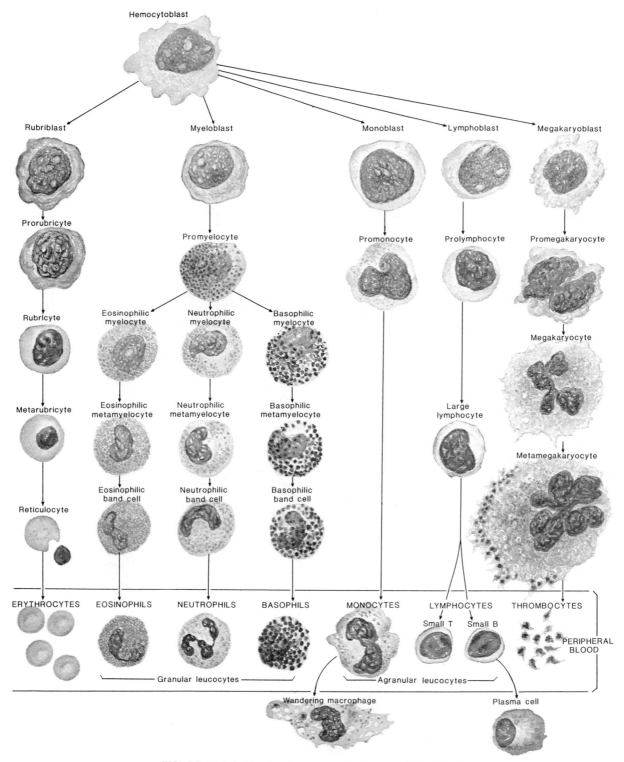

Hemocytoblast

Rubriblast — Myeloblast — Monoblast — Lymphoblast — Megakaryoblast

Prorubricyte — Promyelocyte — Promonocyte — Prolymphocyte — Promegakaryocyte

Rubricyte — Eosinophilic myelocyte — Neutrophilic myelocyte — Basophilic myelocyte — Megakaryocyte

Metarubricyte — Eosinophilic metamyelocyte — Neutrophilic metamyelocyte — Basophilic metamyelocyte — Large lymphocyte — Metamegakaryocyte

Reticulocyte — Eosinophilic band cell — Neutrophilic band cell — Basophilic band cell

ERYTHROCYTES — EOSINOPHILS — NEUTROPHILS — BASOPHILS — MONOCYTES — LYMPHOCYTES (Small T, Small B) — THROMBOCYTES — PERIPHERAL BLOOD

— Granular leucocytes — — Agranular leucocytes —

Wandering macrophage — Plasma cell

FIGURE 12-2 Origin, development, and structure of blood cells.

Erythrocytes (red blood cells)
Leucocytes (white blood cells)
   Granular leucocytes (granulocytes)
      Neutrophils
      Eosinophils
      Basophils
   Agranular leucocytes (agranulocytes)
      Lymphocytes
      Monocytes
Thrombocytes (platelets)

Origin

The process by which blood cells are formed is called **hemopoiesis** (hē-mō-poy-Ē-sis), or **hematopoiesis.** During embryonic and fetal life, there are no clear-cut centers for blood-cell production. The yolk sac, liver, spleen, thymus gland, lymph nodes, and bone marrow all participate at various times in producing the formed elements. In the adult, however, we can pinpoint the production process to the red bone marrow in the sternum, ribs, vertebrae, and pelvis and to lymphoid tissue. Red blood cells, granular leucocytes, and platelets are produced in red bone marrow (myeloid tissue). Agranular leucocytes arise from both myeloid tissue and from lymphoid tissue—spleen, tonsils, lymph nodes. Undifferentiated mesenchymal cells in red bone marrow are transformed into **hemocytoblasts** (hē'-mō-SĪ-tō-blasts), immature cells that are eventually capable of developing into mature blood cells (Figure 12-2). The hemocytoblasts undergo differentiation into five types of cells from which the major types of blood cells develop.

   1. Rubriblasts (proerythroblasts) form mature red blood cells.
   2. Myeloblasts form mature neutrophils, eosinophils, and basophils.
   3. Megakaryoblasts form mature thrombocytes (platelets).
   4. Lymphoblasts form lymphocytes.
   5. Monoblasts form monocytes.

Erythrocytes

● *Structure* Microscopically, **red blood cells (rbc's),** or **erythrocytes** (e-RITH-rō-sīts), appear as biconcave discs when viewed externally in profile. Mature red blood cells average about 6.5 μm in diameter and are quite simple in structure. They lack a nucleus and can neither reproduce nor carry on extensive metabolic activities. Their plasma membrane is selectively permeable and consists of protein (stromatin) and lipids (lecithin and cholesterol). The membrane encloses cytoplasm and a red pigment called *hemoglobin* (*heme* = iron). Hemoglobin, which constitutes about 33 percent of the cell weight, is responsible for the red color of blood. Normal values for hemoglo-

FIGURE 12-3 Diagram of a hemoglobin molecule. The globin (protein) portions of the molecule are indicated in green and blue and the four heme (iron-containing) portions are in the center of each globin molecule.

bin are 14 to 20 gm/100 ml in infants, 12 to 15 gm/100 ml in adult females, and 14 to 16.5 gm/100 ml in adult males.

● *Functions* The hemoglobin in erythrocytes combines with oxygen to form oxyhemoglobin and with carbon dioxide to form carbaminohemoglobin and transports them through the blood vessels. The hemoglobin molecule consists of a protein called globin and a pigment called heme, which contains iron (Figure 12-3). As the erythrocytes pass through the lungs, each of the four iron atoms in the hemoglobin molecules combines with a molecule of oxygen. The oxygen is transported in this state to other tissues of the body. In the tissues, the iron-oxygen reaction reverses, and the oxygen is released to diffuse into the interstitial fluid. On the return trip, the globin portion combines with a molecule of carbon dioxide from the interstitial fluid to form carbaminohemoglobin. This complex is transported to the lungs, where the carbon dioxide is released and then exhaled. Although about 23 percent of carbon dioxide is transported by hemoglobin in this manner, the greater portion, about 70 percent, is transported in blood plasma as the bicarbonate ion, HCO_3^-.

Red blood cells are highly specialized for their transport function. They contain a large number of hemoglobin molecules in order to increase their oxygen-carrying capacity. One estimate is 280 million molecules of hemoglobin per erythrocyte. The biconcave shape of a red blood cell has a much greater surface area than, say, a sphere or a cube. The erythrocyte thus presents the maximum surface area for the diffusion of gas molecules that pass through the membrane to combine with hemoglobin.

CLINICAL APPLICATION

Clinical trials have been performed to test the effectiveness of a **blood substitute** called *Fluosol-DA*. It is a slippery, white liquid that has a very high solubility of oxygen. Since its only function is to transport oxygen, it is actually a hemoglobin substitute. It may eventually turn out to be useful in providing oxygen to tissues that are oxygen-deficient as a result of carbon monoxide poisoning, sickle cell anemia, strokes, heart attacks, and burns. Its use is limited by the fact that it must be kept frozen until just before use and recipients must receive oxygen by mask. Nevertheless, patients would be free of the risks of transfusion reactions and posttransfusion hepatitis.

• *Life Span and Number*   The plasma membrane of a red blood cell becomes fragile and the cell is nonfunctional in about 120 days. The plasma membrane of worn-out red blood cells is removed from circulation by macrophages in the spleen, liver, and bone marrow. The hemoglobin is broken down into hemosiderin, an iron-containing pigment; bilirubin, a noniron-containing pigment; and globin, a protein. The hemosiderin is stored or used in bone marrow to produce new hemoglobin for new red blood cells. Bilirubin is secreted by the liver into bile and globin is metabolized by the liver. A healthy male has about 5.4 million red blood cells per cubic millimeter (mm^3) of blood, and a healthy female has about 4.8 million. The higher value in the male is because of his higher rate of metabolism. To maintain normal quantities of erythrocytes, the body must produce new mature cells at the astonishing rate of 2 million per second. In the adult, production takes place in the red bone marrow in the spongy bone of the cranium, ribs, sternum, bodies of vertebrae, and proximal epiphyses of the humerus and femur.

• *Production*   The process by which erythrocytes are formed is called **erythropoiesis** (e-rith'-rō-poy-Ē-sis). It starts in red bone marrow with the transformation of a hemocytoblast into a rubriblast (see Figure 12-2).

CLINICAL APPLICATION

The rate of erythropoiesis is measured by a procedure called a **reticulocyte** (re-TIK-yoo-lō-sīt) **count.** Some reticulocytes are normally released into the bloodstream before they become mature red blood cells. If the number of reticulocytes in a sample of blood is less than 0.5 percent of the number of mature red blood cells in the sample, erythropoiesis is occurring too slowly. A low reticulocyte count might confirm a diagnosis of nutritional or pernicious anemia. Or it might indicate a kidney disease that prevents the kidney cells from producing erythropoietin. If the reticulocytes number more than 1.5 percent of the mature red blood cells, erythropoiesis is abnormally rapid. Any number of problems may be responsible for a high reticulocyte count: anemia, oxygen deficiency, and uncontrolled red blood cell production caused by a cancer in the bone marrow. A high count may also indicate that treatment for a condition causing anemia has been effective and the bone marrow is making up for lost time.

Another test with important clinical applications is the hematocrit. **Hematocrit** is the percentage of blood that is made up of red blood cells. It is determined by centrifuging blood and noting the ratio of red blood cells to whole blood. The average hematocrit for males is 47 percent. This means that in 100 ml of blood there are 47 ml of cells and 53 ml of plasma. The average hematocrit for females is 42 percent. Anemic blood may have a hematocrit of 15 percent; polycythemic blood (an abnormal increase in the number of functional erythrocytes) may have a hematocrit of 65 percent. Athletes, however, may have higher than normal hematocrits, reflecting constant physical activity rather than a pathological condition.

Leucocytes

• *Structure and Types*   Unlike red blood cells, **leucocytes** (LOO-kō-sīts), or **white blood cells (wbc's),** have nuclei and do not contain hemoglobin (see Figure 12-2). Leucocytes fall into two major groups. The first group is the **granular leucocytes.** They develop from red bone marrow, have granules in their cytoplasm, and possess lobed nuclei. The three kinds of granular leucocytes are **neutrophils,** or **polymorphs** (10 to 12 μm in diameter), **eosinophils** (10 to 12 μm in diameter), and **basophils** (8 to 10 μm in diameter).

The second principal group of leucocytes is the **agranular leucocytes.** They develop from lymphoid and myeloid tissue, no cytoplasmic granules can be seen under a light microscope, and their nuclei are usually spherical. The two kinds of agranular leucocytes are **lymphocytes** (7 to 15 μm in diameter) and **monocytes** (14 to 19 μm in diameter).

• *Functions*   The skin and mucous membranes of the body are continuously exposed to microbes and their toxins. Some of these microbes are capable of invading deeper tissues to cause disease. The general function of leucocytes is to combat these microbes once they enter the body by phagocytosis or antibody production. Neutrophils and monocytes are actively **phagocytotic**—they can ingest bacteria and dispose of dead matter (see Figure 2-3b). Neutrophils (NOO-trō-fils) are the most active leucocytes in response to tissue destruction by bacteria. In addition to carrying on phagocytosis, they release the enzyme lysozyme, which destroys certain bacteria. Apparently monocytes (MON-ō-sīts) take longer to reach the site of infection than do neutrophils, but once they arrive, they do so in larger numbers and destroy more microbes. Monocytes that have migrated to infected tissues are called **wandering macrophages.** They clean up cellular debris following an infection.

A number of different chemicals in inflamed tissue cause phagocytes to migrate toward the tissue. This phenomenon is called **chemotaxis.** Among the substances that provide stimuli for chemotaxis are toxins and degenerative products of damaged tissues.

Most leucocytes possess, to some degree, the ability to crawl through the minute spaces between the cells that form the walls of capillaries and through connective and epithelial tissue. This movement, like that of amoebas, is called **diapedesis** (dī'-a-pe-DĒ-sis). First, part of the cell membrane stretches out like an arm. Then the cytoplasm and nucleus flow into the projection. Finally, the rest of the membrane snaps up into place. Another projection is made, and so on, until the cell has crawled to its destination.

Eosinophils (ē'-ō-SIN-ō-fils) are believed to combat the irritants that cause allergies. Eosinophils leave the capillaries, enter the tissue fluid, and produce antihistamines that destroy antigen-antibody complexes. Eosinophils might also be phagocytic.

Basophils (BĀ-sō-fils) are also believed to be involved in allergic reactions. Basophils leave the capillaries, enter the tissues, and become the mast cells of the tissues, liberating heparin, histamine, and serotonin.

Lymphocytes (LIM-fō-sīts) are involved in the production of antibodies. **Antibodies** (AN-ti-bod'-ēz) are special proteins that inactivate antigens. An **antigen** (AN-ti-jen) is any substance that will stimulate the production of antibodies and is capable of reacting specifically with the antibody. Most antigens are proteins, and most are not synthesized by the body. Many of the proteins that make up the cell structures and enzymes of bacteria are antigens. The toxins released by bacteria are also antigens. The branch of science that deals with the responses of the body when challenged by antigens is referred to as **immunology** (*immunis* = free; im'-yoo-NOL-ō-jē). When antigens enter the body, they react chemically with substances in the lymphocytes and stimulate some lymphocytes, called **B cells,** to become **plasma cells.** The plasma cells then produce antibodies, globulin-type proteins that attach to antigens, much as enzymes attach to substrates.

Like enzymes, a specific antibody will generally attach only to a certain antigen. However, unlike enzymes, which enhance the reactivity of the substrate, antibodies "cover" their antigens so the antigens cannot come in contact with other chemicals in the body. In this way, bacterial poisons can be sealed up and rendered harmless. The bacteria themselves are destroyed by the antibodies. This process is called the **antigen-antibody response.** Phagocytes in tissues destroy the antigen-antibody complexes.

CLINICAL APPLICATION

Multiple myeloma is a malignant disorder of plasma cells in bone marrow. The symptoms are caused by the growing tumor cell mass itself and antibodies produced by the malignant cells. Among the effects are pain and osteoporosis due to multiple osteolytic (bone-dissolving) lesions in bones such as the sternum, ribs, backbone, clavicles, skull, pelvis, and proximal extremities. Other effects include hypercalcemia (excessive blood calcium), anemia, leucopenia, thrombocytopenia, spinal cord compression, enlarged liver, kidney damage, and increased susceptibility to infection.

Other lymphocytes are called **T cells.** One group of T cells, the **killer T cells,** are activated by certain antigens and react by destroying them directly or indirectly by recruiting other lymphocytes and macrophages. T cells are especially effective against bacteria, viruses, fungi, transplanted cells, and cancer cells.

The antigen-antibody response helps us combat infection and gives us immunity to some diseases. It is also responsible for blood types, allergies, and the body's rejection of organs transplanted from an individual with a different genetic makeup.

CLINICAL APPLICATION

An increase in the number of white cells present in the blood indicates a state of inflammation or infection. Because each type of white cell plays a different role, determining which types have increased aids in diagnosis of the condition. A **differential count** is the number of each kind of white cell in 100 white blood cells. A normal differential count falls within the following percentages:

Neutrophils	60–70%
Eosinophils	2–4%
Basophils	0.5–1%
Lymphocytes	20–25%
Monocytes	3–8%
	100%

Particular attention is paid to the neutrophils in a differential count. More often than not, a high neutrophil count indicates damage by invading bacteria. An increase in the number of monocytes generally indicates a chronic infection. Eosinophils and basophils are elevated during allergic reactions. High lymphocyte counts indicate antigen-antibody reactions.

• *Life Span and Number* Foreign bacteria exist everywhere in the environment and have continuous access to the body through the mouth, nose, and pores of the skin. Furthermore, many cells, especially those of epithelial tissue, age and die, and their remains must be disposed of daily. Even when the body is healthy, the leucocytes actively ingest bacteria and debris. However, a leucocyte can phagocytose only a certain number of substances before they interfere with the leucocyte's normal metabolic activities and bring on its death. Consequently, the life span of most leucocytes is very short. In a healthy body, some white blood cells will live only a few days. During a period of infection they may live only a few hours.

Leucocytes are far less numerous than red blood cells, averaging from 5,000 to 9,000 cells per cubic millimeter of blood. Red blood cells, therefore, outnumber white blood cells about 700 to 1. The term **leucocytosis** (loo'-kō-sī-TŌ-sis) refers to an increase in the number of white blood cells. If the increase exceeds 10,000, a pathological condition is usually indicated. An abnormally low level of white blood cells (below 5,000/mm³) is termed **leucopenia** (loo-kō-PĒ-nē-a).

• *Production* Granular leucocytes are produced in red bone marrow (myeloid tissue); agranular leucocytes are produced in both myeloid and lymphoid tissue. The developmental sequences for the five types of leucocytes are shown in Figure 12-2.

CLINICAL APPLICATION

Bone marrow, as indicated earlier, contains hemocytoblasts that differentiate into erythrocytes, leucocytes, and platelets. Among the leucocytes are cells that combat infections and cause tissue rejection. Until recently, the use of **bone marrow transplantation,** the transfer of bone marrow from a donor to a recipient, required that the marrow of the donor had to be obtained from a relative and closely matched to that of the recipient. In cases where the match is not identical, T cells in donor marrow recognize the recipient's tissues as foreign and attack them. The principal tissues rejected in this way are in the skin, liver, and gastrointestinal tract. Several recent advances now permit bone marrow transplantation in which the tissue match between donor and recipient is not even close. In addition to diminishing the effects of the rejection phenomenon from mismatched tissues by using drugs such as cyclosporine to suppress T cells, scientists have now developed ways to remove T cells from donor marrow by using chemicals like lectin and monoclonal antibodies.

Donor marrow is aspirated from the iliac crests of the hip bones. After about 750 ml is removed, it is mixed with heparin (an anticoagulant) and then passed through screens. The suspension of bone marrow cells is treated to remove T cells and given to the recipient just like a blood transfusion. The cells in the suspension pass through the lungs, enter general

EXHIBIT 12-2 SUMMARY OF THE FORMED ELEMENTS IN BLOOD

FORMED ELEMENT	NUMBER	DIAMETER (in μm)	LIFE SPAN	FUNCTION
Erythrocyte (red blood cell)	4.8 million/mm³ in females 5.4 million/mm³ in males	7.5	120 days	Transports oxygen and carbon dioxide.
Leucocyte (white blood cell)	5,000–9,000/mm³		A few hours to a few days	
Granular				
Neutrophil	60–70% of total	10–12		Phagocytosis.
Eosinophil	2–4% of total	10–12		Combats allergens.
Basophil	0.5–1% of total	8–10		Combats allergens.
Agranular				
Lymphocyte	20–25% of total	7–15		Immunity (antigen-antibody reactions).
Monocyte	3–8% of total	14–19		Phagocytosis.
Thrombocyte (platelet)	250,000–400,000/mm³	2–4	5–9 days	Blood clotting.

circulation, and reseed and grow in the marrow cavities of the recipient's bones.

Bone marrow transplantation has been used to treat aplastic anemia, certain types of leukemia, and severe combined immunodeficiency disease (SCID), an inherited deficiency of infection-fighting blood cells. The technique is also being expanded to treat other kinds of leukemia, non-Hodgkin's lymphoma, thalassemia, multiple myeloma, sickle cell anemia, and hemolytic anemia.

Thrombocytes

● *Structure* In addition to the immature cell types that develop into erythrocytes and leucocytes, hemocytoblasts differentiate into still another kind of cell, called a megakaryoblast (see Figure 12-2a). Megakaryoblasts are transformed into megakaryocytes, large cells that shed fragments of cytoplasm. Each fragment becomes enclosed by a piece of the cell membrane and is called a **thrombocyte** (THROM-bō-sīt), or **platelet.** Platelets are disc-shaped cells without a nucleus. They average from 2 to 4 μm in diameter.

● *Function* Platelets prevent fluid loss by initiating a chain of reactions that results in blood clotting.

● *Life Span and Number* Like the other formed elements of the blood, platelets have a short life, probably only 5 to 9 days. Between 250,000 and 400,000 platelets appear in each cubic millimeter of blood.

● *Production* Platelets are produced in red bone marrow according to the developmental sequence shown in Figure 12-2.

A summary of the formed elements in blood is presented in Exhibit 12-2.

CLINICAL APPLICATION

The most commonly ordered hematology test is the **complete blood count (CBC).** It generally includes determination of hemoglobin, hematocrit, red blood cell count, white blood cell count, differential count, and comments about red blood cell, white blood cell, and platelet morphology.

PLASMA

When the formed elements are removed from blood, a straw-colored liquid called **plasma** is left. Exhibit 12-3 outlines the chemical composition of plasma. Note that about 7 percent of the solutes are proteins. Some of these

EXHIBIT 12-3   CHEMICAL COMPOSITION AND DESCRIPTION OF SUBSTANCES IN PLASMA

CONSTITUENT	DESCRIPTION
WATER	Liquid portion of blood; constitutes about 91.5 percent of plasma. Ninety percent of water derived from absorption from digestive tract; 10 percent from cellular respiration. Acts as solvent and suspending medium for solid components of blood and absorbs heat.
SOLUTES Proteins	Constitute about 8.5 percent of plasma.
Albumins	Smallest plasma proteins. Produced by liver and provide blood with viscosity, a factor related to maintenance and regulation of blood pressure. Also exert considerable osmotic pressure to maintain water balance between blood and tissues and regulate blood volume.
Globulins	Protein group to which antibodies belong. Gamma globulins attack measles, hepatitis, and polio viruses and tetanus bacterium.
Fibrinogen	Produced by liver. Plays essential role in clotting.
Nonprotein nitrogen (NPN) substances	Contain nitrogen but are not proteins. Include urea, uric acid, creatine, creatinine, and ammonium salts. Represent breakdown products of protein metabolism and are carried by blood to organs of excretion.
Food substances	Products of digestion passed into blood for distribution to all body cells. Include amino acids (from proteins), glucose (from carbohydrates), fatty acids, glycerides, and glycerol (from fats).
Regulatory substances	Enzymes, produced by body cells, to catalyze chemical reactions. Hormones, produced by endocrine glands, to regulate growth and development in body.
Respiratory gases	Oxygen and carbon dioxide. These gases are more closely associated with hemoglobin of red blood cells than plasma itself.
Electrolytes	Inorganic salts of plasma. Cations include Na^+, K^+, Ca^{2+}, Mg^{2+}; anions include Cl^-, PO_4^{3-}, SO_4^{2-}, HCO_3^-. Help maintain osmotic pressure, normal pH, physiological balance between tissues and blood.

proteins are also found elsewhere in the body, but in blood they are called **plasma proteins. Albumins,** which constitute 55 percent of plasma proteins, are largely responsible for blood's viscosity. The concentration of albumins is about four times higher in plasma than in interstitial fluid. Along with the electrolytes, albumins also help regulate blood volume by preventing the water in the blood from diffusing into the interstitial fluid. Recall that water moves by osmosis from an area of high water (low solute) concentration to an area of low water (high solute) concentration. **Globulins,** which comprise 38 percent of plasma proteins, are antibody proteins released by plasma cells. Gamma globulin is especially well known because it is able to form an antigen-antibody complex with the proteins of the hepatitis and measles viruses and the tetanus bacterium, among others. **Fibrinogen** makes up about 7 percent of plasma proteins and takes part in the blood-clotting mechanism along with the platelets.

CLINICAL APPLICATION

Apheresis (a-FER-e-sis; *aphairesis* = removal) refers to a procedure in which blood is withdrawn from the body, its components are selectively separated, the undesirable component is removed, and the remainder is returned to the body. Blood is withdrawn by a needle or catheter, mixed with an anticoagulant, and pumped through a separator where red blood cells, white blood cells, platelets, and plasma are separated by centrifugal force. The process usually takes 3 to 5 hours and no more than 15 percent of the patient's total blood volume is allowed outside the body.

Based on the component selectively removed, specific names are given to the procedure. Accordingly, *erythropheresis* refers to the removal of red blood cells, *leucopheresis* refers to the removal of white blood cells, *plateletpheresis* refers to the removal of platelets, and *plasmapheresis* refers to the removal of plasma. Apheresis is used in the treatment of certain blood diseases such as aplastic anemia, hemolytic anemia, multiple myeloma, chronic myelogenous leukemia, thrombocytemia, posttransfusion Rh incompatibility, and life-threatening sickle cell crisis. Because of some potentially serious complications of apheresis (hypovolemia, shock, congestive heart failure, pulmonary edema, muscle twitching, thrombosis), the procedure is used discriminately when more conservative therapy is unsuccessful.

APPLICATIONS TO HEALTH

ANEMIA

Anemia is a sign, not a diagnosis. Many kinds of anemia exist, all characterized by insufficient erythrocytes or hemoglobin. These conditions lead to fatigue and intolerance to cold, both of which are related to lack of oxygen needed for energy and heat production, and to paleness, which is due to low hemoglobin content.

Nutritional Anemia

Nutritional anemia arises from an inadequate diet, one that provides insufficient amounts of iron, the necessary amino acids, or vitamin B_{12}.

Pernicious Anemia

Pernicious anemia is an insufficient production of erythrocytes resulting from an inability of the body to produce intrinsic factor.

Hemorrhagic Anemia

An excessive loss of erythrocytes through bleeding is called **hemorrhagic anemia.** Common causes are large wounds, stomach ulcers, and heavy menstrual bleeding. If bleeding is extraordinarily heavy, the anemia is termed acute. Excessive blood loss can be fatal. Slow, prolonged bleeding is apt to produce a chronic anemia; the chief symptom is fatigue.

Hemolytic Anemia

If erythrocyte plasma membranes rupture prematurely, the cells remain as "ghosts" and their hemoglobin pours out into the plasma. A characteristic sign of this condition, called **hemolytic anemia,** is distortions in the shapes of erythrocytes that are progressing toward hemolysis. There may also be a sharp increase in the number of reticulocytes, since the destruction of red blood cells stimulates erythropoiesis.

The premature destruction of red cells may result from inherent defects, such as hemoglobin defects, abnormal red cell enzymes, or defects of the red cell membrane. Agents that may cause hemolytic anemia are parasites, toxins, and antibodies from incompatible blood (Rh⁻ mother and Rh⁺ fetus, for instance). *Erythroblastosis fetalis* (*hemolytic disease of newborn*) is an example of a hemolytic anemia.

The term **thalassemia** (thal'-a-SĒ-mē-a) represents a group of hereditary hemolytic anemias resulting from a defect in the synthesis of hemoglobin, which produces extremely thin and fragile erythrocytes. It occurs primarily in populations from countries bordering the Mediterranean Sea. Treatment generally consists of blood transfusions.

Aplastic Anemia

Destruction or inhibition of the red bone marrow results in **aplastic anemia.** Typically, the marrow is replaced by fatty tissue, fibrous tissue, or tumor cells. Toxins, gamma radiation, and certain medications are causes. Many of the medications inhibit the enzymes involved in hemopoiesis. Bone marrow transplants can now be done with

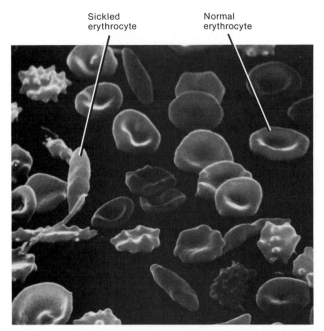

Sickled erythrocyte

Normal erythrocyte

FIGURE 12-4 Scanning electron micrograph of erythrocytes as they appear in sickle cell anemia at a magnification of 5,000×. (Courtesy of Fisher Scientific Company and S.T.E.M. Laboratories, Inc., Copyright 1975.)

a reasonable hope of success in patients with aplastic anemia. They are done early, before the victim is sensitized by transfusions. Immunosuppressive drugs are given for 4 days before the transplant and with decreasing frequency for 100 days afterward.

Sickle Cell Anemia

The erythrocytes of a person with **sickle cell anemia** manufacture an abnormal kind of hemoglobin. When an erythrocyte gives up its oxygen to the interstitial fluid, its hemoglobin tends to lose its integrity in places of low oxygen tension and forms long, stiff, rodlike structures that bend the erythrocyte into a sickle shape (Figure 12-4). The sickled cells rupture easily. Even though erythropoiesis is stimulated by the loss of the cells, it cannot keep pace with the hemolysis. The individual consequently suffers from a hemolytic anemia that reduces the amount of oxygen that can be supplied to the tissues. Prolonged oxygen reduction may eventually cause extensive tissue necrosis, pain, and organ damage. Sickled cells also increase blood viscosity, which leads to occlusion of small blood vessels. Furthermore, because of the shape of the sickled cells, they tend to get stuck in blood vessels and can cut off the blood supply to an organ altogether.

Sickle cell anemia is characterized by several symptoms. In young children, hand-foot syndrome is present, in which there is swelling and pain in the wrists and feet. Older patients experience pain in the back and extremities without swelling and abdominal pain. Complications include neurological disorders (meningitis, seizures,

stroke), impaired pulmonary function, orthopedic abnormalities (femoral head necrosis, osteomyelitis), genitourinary tract disorders (involuntary urination, blood in urine, kidney failure), ocular disturbances (hemorrhage, detached retina, blindness), and obstetric complications (convulsions, coma, infection).

Sickle cell anemia is inherited. The gene responsible for the tendency of the erythrocytes to sickle during hypoxia also seems to prevent erythrocytes from rupturing during a malarial crisis. The gene also alters the permeability of the plasma membranes of sickled cells, causing potassium to leak out. Low levels of potassium kill the malarial parasites that infect sickled cells. Sickle cell genes are found primarily among populations, or descendants of populations, that live in the malaria belt around the world, including parts of Mediterranean Europe and subtropical Africa and Asia. A person with only one of the sickling genes is said to have sickle cell trait. Such an individual has a high resistance to malaria—a factor that may have tremendous survival value—but does not develop the anemia. Only people who inherit a sickling gene from both parents get sickle cell anemia.

Treatment consists of administration of analgesics to relieve pain, antibiotics to counter infections, and transfusion therapy.

POLYCYTHEMIA

The term **polycythemia** (pol'-ē-sī-THĒ-mē-a) refers to an abnormal increase in the number of red blood cells. The hematocrit is an important factor in diagnosing the condition. Increases of 2 to 3 million cells per cubic millimeter above normal are considered to be polycythemic. The blood's viscosity is greatly increased because of the extra red blood cells. The increased viscosity causes a rise in blood pressure and contributes to thrombosis and hemorrhage. The thrombosis results from too many red blood cells piling up as they try to enter smaller vessels. The hemorrhage is due to widespread hyperemia (unusually large amount of blood in an organ part).

INFECTIOUS MONONUCLEOSIS (IM)

Infectious mononucleosis (IM) is a contagious disease primarily affecting lymphoid tissue throughout the body. It is caused by the *Epstein-Barr virus (EBV)*, the same agent that has been linked to Burkett's lymphoma, nasopharyngeal carcinoma, and Hodgkin's disease. It occurs mainly in children and young adults, with the peak incidence at 15 to 20 years of age. The virus most commonly enters the body through intimate oral contact, multiplies in lymphoid tissues, and spreads into the blood where it infects and multiplies in B lymphocytes, the primary host cells. As a result of this infection, the B lymphocytes become enlarged and abnormal in appearance and resemble monocytes, the primary reason for which the disease receives its name mononucleosis. Infectious mononucleosis is characterized by an elevated white blood cell count with an abnormally high percentage of lymphocytes. Symptoms include sore throat, lymphadenopathy (enlarged and tender lymph nodes), fever, brilliant red throat and soft palate, stiff neck, cough, and malaise. The spleen may also enlarge. Secondary complications involving the liver, eyes, heart, kidneys, and nervous system may develop. There is no cure for infectious mononucleosis, and treatment consists of watching for and treating complications. Usually the disease runs its course in a few weeks, and the individual generaly suffers no permanent ill effects.

LEUKEMIA

In its simplest terms, **leukemia** refers to a malignant disease of blood-forming tissues characterized by uncontrolled production and accumulation of leucocytes. Many of the cells fail to reach maturity. The *human T-cell leukemia-lymphoma virus-1* (*HTLV-1*) is strongly associated with leukemia. Clinically, leukemia is classified on the basis of the duration and character of the disease, that is, acute or chronic. It is also classified according to the identity and site of origins of the predominant cell involved, such as myelocytic (myelogenous, myeloblastic, granulocytic), lymphocytic (lymphogenous, lymphatic), and monocytic.

As with most cancers, the symptoms result not so much from the cancer cells themselves as from their interference with normal body processes. The anemia and bleeding problems commonly seen in leukemia result from the crowding out of normal bone marrow cells, preventing normal production of red blood cells and platelets. The most common cause of death from leukemia is internal hemorrhaging, especially cerebral hemorrhage that destroys the vital centers in the brain. Another frequent cause of death is uncontrolled infection due to lack of mature or normal white blood cells. The abnormal accumulation of leucocytes may be reduced by using x rays and antileukemic drugs. Partial or complete remissions may be induced, with some lasting as long as 15 years.

KEY MEDICAL TERMS ASSOCIATED WITH BLOOD

Autologous (*auto* = self) **transfusion**   Donating one's own blood before elective surgery to ensure an abundant supply and reduce transfusion complications.

Citrated whole blood   Whole blood protected from coagulation by CPD (citrate phosphate dextrose) or a similar compound.

Direct (immediate) transfusion (*trans* = through)   Transfer of blood directly from one person to another without exposing the blood to air.

Exchange transfusion   Removing blood from the recipient while simultaneously replacing it with donor blood. This method is used for erythroblastosis fetalis and poisoning.

Gamma globulin (immune serum globulin)   Solution of globulins from nonhuman blood consisting of antibodies that react with specific pathogens, such as measles, epidemic hepatitis, tetanus, and possibly poliomyelitis viruses. It is prepared by injecting the specific virus into animals, removing blood from the animals after antibodies have accumulated, isolating antibodies, and injecting them into a human for short-term immunity.

Hemorrhage (*rrhage* = bursting forth)   Bleeding, either internal (from blood vessels into tissues) or external (from blood vessels directly to the surface of the body).

Indirect (mediate) transfusion   Transfer of blood from a donor to a container and then to the recipient, permitting blood to be stored for an emergency. The blood may be separated into its components so that a patient will receive only a needed part.

Platelet concentrates   A preparation of platelets obtained from freshly drawn whole blood and used for transfusions in platelet-deficiency disorders such as hemophilia.

Reciprocal transfusion   Transfer of blood from a person who has recovered from a contagious infection into the vessels of a patient suffering with the same infection. An equal amount of blood is returned from the patient to the well person. This method allows the patient to receive antibody-bearing lymphocytes from the recovered person.

Septicemia (*sep* = decay; *emia* = condition of blood)   Toxins or disease-causing bacteria in the blood. Also called "blood poisoning."

Thrombocytopenia (*thrombo* = clot; *penia* = poverty)   Very low platelet count that results in a tendency to bleed from capillaries.

Transfusion   Transfer of whole blood, blood components (red blood cells only or plasma only), or bone marrow directly into the bloodstream.

Venesection (*veno* = vein)   Opening of a vein for withdrawal of blood.

Whole blood   Blood containing all formed elements, plasma, and plasma solutes in natural concentration.

STUDY OUTLINE

Physical Characteristics (p. 288)

1. The cardiovascular system consists of blood, the heart, and blood vessels. The lymphatic system consists of lymph, lymph vessels, and lymph glands.

2. Physical characteristics of blood include viscosity, 4.5 to 5.5; temperature, 38°C (100.4°F); pH, 7.35 to 7.45; and salinity,

0.85 to 0.90 NaCl. Blood constitutes about 8 percent of body weight.

Functions (p. 288)

1. Blood transports oxygen, carbon dioxide, nutrients, wastes, hormones, and enzymes.
2. It helps to regulate pH, body temperature, and water content of cells.
3. It prevents blood loss through clotting and combats toxins and microbes through special combat-unit cells.

Components (p. 289)

1. The formed elements in blood include erythrocytes (red blood cells), leucocytes (white blood cells), and thrombocytes (platelets).
2. Blood cells are formed by a process called hemopoiesis.
3. Red bone marrow (myeloid tissue) is responsible for producing red blood cells, granular leucocytes, and platelets; lymphoid tissue and myeloid tissue produce agranular leucocytes.

Erythrocytes

1. Erythrocytes are biconcave discs without nuclei and containing hemoglobin.
2. The function of red blood cells is to transport oxygen and carbon dioxide.
3. Red blood cells live about 120 days. A healthy male has about 5.4 million/mm³ of blood; a healthy female, about 4.8 million/mm³.
4. Erythrocyte formation, called erythropoiesis, occurs in adult red marrow of certain bones.
5. A reticulocyte count is a diagnostic test that indicates the rate of erythropoiesis.
6. A hematocrit measures the percentage of red blood cells in whole blood.

Leucocytes

1. Leucocytes are nucleated cells. Two principal types are granular (neutrophils, eosinophils, basophils) and agranular (lymphocytes and monocytes).
2. The general function of leucocytes is to combat inflammation and infection. Neutrophils and monocytes (wandering macrophages) do so through phagocytosis.
3. Eosinophils and basophils are believed to be involved in combating allergic reactions.
4. Lymphocytes, in response to the presence to foreign substances called antigens, differentiate into tissue plasma cells that produce antibodies. Antibodies attach to the antigens and render them harmless. This antigen-antibody response combats infection and provides immunity.
5. A differential count is a diagnostic test in which white blood cells are enumerated.
6. White blood cells usually live for only a few hours or a few days. Normal blood contains 5,000 to 9,000/mm³.

Thrombocytes

1. Thrombocytes are disc-shaped structures without nuclei.
2. They are formed from megakaryocytes and are involved in clotting.
3. Normal blood contains 250,000 to 400,000/mm³.

Plasma

1. The liquid portion of blood, called plasma, consists of 91.5 percent water and 8.5 percent solutes.
2. Principal solutes include proteins (albumins, globulins, fibrinogen), nonprotein nitrogen (NPN) substances, foods, enzymes and hormones, respiratory gases, and electrolytes.

Applications to Health (p. 296)

1. Anemia is a decreased erythrocyte count or hemoglobin deficiency. Kinds of anemia include nutritional, pernicious, hemorrhagic, hemolytic, aplastic, and sickle cell anemia.
2. Polycythemia is an abnormal increase in the number of erythrocytes.
3. Infectious mononucleosis (IM) is a contagious disease that primarily affects lymphoid tissue. It is characterized by an elevated white blood cell count, with an abnormally high percentage of lymphocytes. The cause is the Epstein-Barr virus (EBV).
4. Leukemia is a malignant disease of blood-forming tissues characterized by the uncontrolled production of white blood cells that interferes with normal clotting and vital body activities.

REVIEW QUESTIONS

1. How are blood, interstitial fluid, and lymph related?
2. List the principal physical characteristics of blood.
3. List the functions of blood and its relationship to other systems of the body.
4. Describe the origin of blood cells.
5. Describe the microscopic appearance of erythrocytes. What is the essential function of erythrocytes?
6. What is a reticulocyte count? What is its diagnostic significance?
7. Define hematocrit. Compare the hematocrits of anemic and polycythemic blood.
8. Describe the classification of leucocytes. What are their functions?
9. What is a differential count? What is its significance?
10. Describe the antigen-antibody response. How is it protective?
11. Describe the structure and function of thrombocytes.
12. Compare erythrocytes, leucocytes, and thrombocytes with respect to: size, number per mm³, and life span.
13. What are the major constituents in plasma? What do they do?
14. Define anemia. Contrast the causes of nutritional, pernicious, hemorrhagic, hemolytic, aplastic, and sickle cell anemias.
15. What is infectious mononucleosis (IM)?
16. What is leukemia? What are the causes of some of its symptoms?
17. Refer to the glossary of key medical terms associated with blood. Be sure that you can define each term.

13

The Cardiovascular System: The Heart

Student Objectives

Describe the location of the heart in the mediastinum and identify the borders of the heart.

Describe the structure of the parietal pericardium (pericardial sac).

Contrast the structure and location of the epicardium, myocardium, and endocardium of the heart wall.

Identify the chambers, great vessels, and valves of the heart.

Describe the surface anatomy features of the heart.

Explain the structural and functional features of the conduction system of the heart.

Describe an electrocardiogram (ECG) and explain its significance.

Describe the principal events of a cardiac cycle.

Contrast the effects of sympathetic and parasympathetic stimulation of the heart.

Discuss the route of blood in coronary (cardiac) circulation.

Describe the structure and function of an artificial heart.

Describe the development of the heart.

List the risk factors involved in heart disease.

Describe how atherosclerosis and coronary artery spasm contribute to coronary artery disease (CAD).

Contrast coarctation of the aorta, patent ductus arteriosus, septal defects, valvular stenosis, and tetralogy of Fallot as congenital heart defects.

Define atrioventricular (AV) block, atrial flutter, atrial fibrillation, and ventricular fibrillation as abnormalities of the conduction system of the heart (arrhythmias).

Define congestive heart failure (CHF) and cor pulmonale (CP).

Define key medical terms associated with the heart.

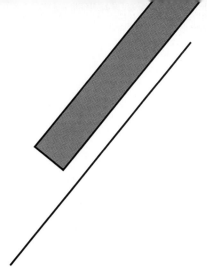

The **heart** is the center of the cardiovascular system. Whereas the term *cardio* refers to the heart, the term *vascular* refers to blood vessels (or an abundant blood supply). The heart is a hollow, muscular organ that weighs about 342 grams (11 oz) and beats over 100,000 times a day to pump 3,784 liters (1,000 gallons) of blood per day through over 60,000 miles of blood vessels. The blood vessels form a network of tubes that carry blood from the heart to the tissues of the body and then return it to the heart. The specific aspects of the heart that we will consider are its location, covering, wall and chambers, great vessels, valves, conduction system, cardiac cycle, surface anatomy, and autonomic control. Several disorders related to the heart will also be considered.

The study of the normal heart and diseases associated with it is known as **cardiology.**

The developmental anatomy of the heart is considered at the end of the chapter.

LOCATION

The heart is situated obliquely between the lungs in the mediastinum. Recall that the mediastinum is the space, actually the mass of tissue, between the pleurae of the lungs that extends from the sternum to the vertebral column. Specifically, the heart is in the middle mediastinum (see Figure 1-7). About two-thirds of its mass lies to the left of the body's midline (Figure 13-1). The heart is shaped like a blunt cone about the size of your closed fist—12 cm (5 in) long, 9 cm (3½ in) wide at its broadest point, and 6 cm (2½ in) thick.

Its pointed end, the *apex,* is formed by the tip of the left ventricle, projects inferiorly, anteriorly, and to the left, and lies superior to the central depression of the diaphragm. Anteriorly, the apex is in the fifth intercostal space—about 7.5 to 8 cm (3 in) from the midline of the body.

The *left border* is formed almost entirely by the left ventricle, although the left atrium forms part of the upper end of the border. The left border of the heart is indicated approximately as a curved line drawn from the left fifth costochondral junction to the left second costochondral junction.

The *superior border,* where the great vessels enter and leave the heart, is formed by both atria. It is represented by a line joining the second left intercostal space to the third right costal cartilage.

The *base* of the heart projects superiorly, posteriorly, and to the right. It is formed by the atria, mostly the left atrium. It lies opposite the fifth to ninth thoracic vertebrae. Anteriorly, it lies just inferior to the second rib.

The *right border* is formed by the right atrium and corresponds to a curved line drawn from the xiphisternal articulation to about the middle of the right third costal cartilage.

The *inferior border* is formed by the right ventricle and slightly by the left ventricle. It is represented by a line passing from the inferior end of the right border through the xiphisternal joint to the apex of the heart.

The *sternocostal (anterior) surface* is formed mainly by the right ventricle and right atrium, whereas the *diaphragmatic (inferior) surface* is formed by the left and right ventricles, mostly the left.

PARIETAL PERICARDIUM (PERICARDIAL SAC)

The heart is enclosed in a loose-fitting serous membrane called the **parietal pericardium,** or **pericardial sac** (Figure 13-1). It consists of two layers: the fibrous layer and the serous layer (Figure 13-2). The *fibrous layer (fibrous pericardium)* is the outer layer and consists of a tough, fibrous connective tissue. It is attached to the large blood vessels entering and leaving the heart, to the diaphragm, and to the inside of the sternal wall of the thorax. It also adheres to the parietal pleurae. The fibrous pericardium prevents overdistension of the heart, surrounds it with

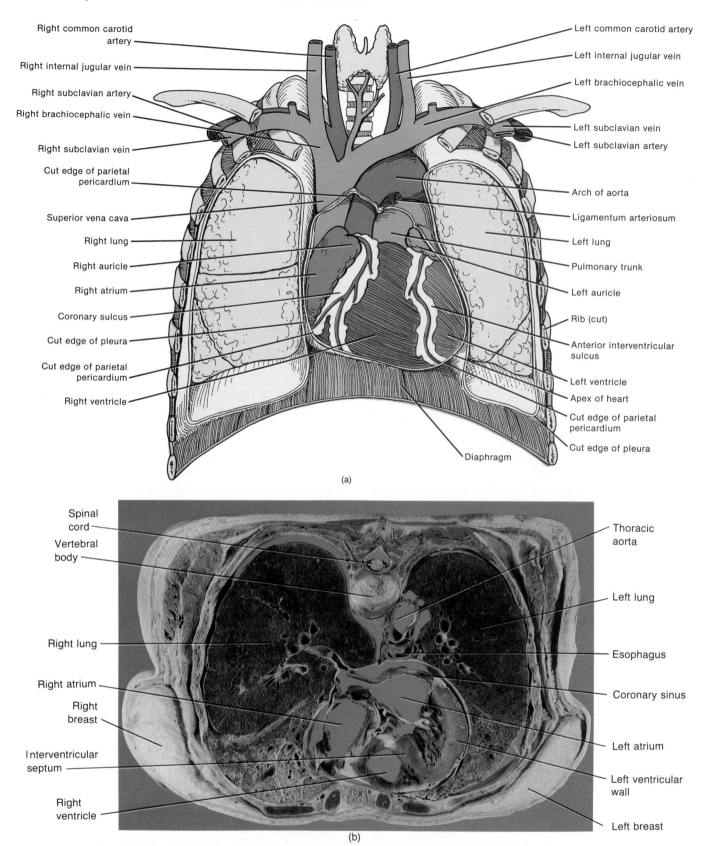

FIGURE 13-1 Position of the heart and associated blood vessels in the thoracic cavity. (a) Diagram. In this and subsequent illustrations, vessels that carry oxygenated blood are colored red; vessels that carry deoxygenated blood are colored blue. (b) Photograph of a cross section through the lower portion of the thoracic cavity. (Courtesy of Stephen A. Kieffer and E. Robert Heitzman, *An Atlas of Cross-Sectional Anatomy,* Harper & Row, Publishers, Inc., New York, 1979.)

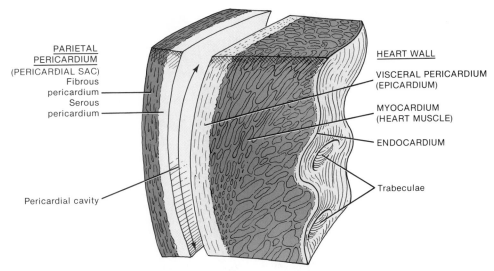

FIGURE 13-2 Structure of the parietal pericardium and heart wall.

a tough protective membrane, and anchors it in the mediastinum.

The inner layer of the parietal pericardium, known as the *serous layer* (*serous pericardium*), is thinner and more delicate. It is continuous with the visceral pericardium (the outer layer of the wall of the heart) at the base of the heart and around the large blood vessels.

An inflammation of the parietal pericardium is known as **pericarditis.** Pericarditis with a buildup of pericardial fluid or extensive bleeding into the pericardial sac, if untreated, is a life-threatening condition. Since the pericardial sac cannot stretch to accommodate the excessive fluid or blood buildup, the heart is subjected to compression. This compression is known as **cardiac tamponade** (tam'-pon-ĀD) and can result in cardiac failure.

HEART WALL

The wall of the heart (Figure 13-2) is divided into three layers: the epicardium (external layer), myocardium (middle layer), and endocardium (inner layer). The **epicardium (visceral pericardium)** is the thin, transparent outer layer of the wall. It is composed of serous tissue and mesothelium. Between the serous pericardium and the epicardium is a potential space called the *pericardial cavity.* The cavity contains a watery fluid, known as *pericardial fluid,* which prevents friction between the membranes as the heart moves.

The **myocardium,** which is cardiac muscle tissue, constitutes the bulk of the heart. Cardiac muscle fibers are involuntary, striated, and branched, and the tissue is arranged in interlacing bundles of fibers. The myocardium is responsible for the contraction of the heart.

The **endocardium** is a thin layer of endothelium overlying a thin layer of connective tissue pierced by tiny blood vessels and bundles of smooth muscle. It lines the inside of the myocardium and covers the valves of the heart and the tendons that hold them open. It is continuous with the endothelial lining of the large blood vessels of the heart.

Inflammation of the epicardium, myocardium, and endocardium is referred to as **epicarditis, myocarditis,** and **endocarditis,** respectively.

CHAMBERS OF THE HEART

The interior of the heart is divided into four **chambers,** or cavities, which receive the circulating blood (Figure 13-3). The two upper chambers are called the right and left **atria.** Each atrium has an appendage called an *auricle* (AWR-i-kul; *auris* = ear), so named because its shape resembles a dog's ear. The auricle increases the atrium's surface area. The lining of the atria is smooth, except for the anterior atrial walls and the lining of the auricles, which contain projecting muscle bundles that are parallel to one another and resemble the teeth of a comb: the *musculi pectinati* (MUS-kyoo-lī pek-ti-NA-tē). These bundles give the lining of the auricles a ridged appearance.

The atria are separated by a partition called the *interatrial septum.* A prominent feature of this septum is an oval depression, the *fossa ovalis,* which corresponds to the site of the foramen ovale, an opening in the interatrial septum of the fetal heart. The fossa ovalis faces the opening of the inferior vena cava and is located in the septal wall of the right atrium.

The two lower chambers are the right and left **ventricles.** They are separated by an *interventricular septum.*

The muscle tissue of the atria and ventricles is separated by connective tissue that also forms the valves. This "cardiac skeleton" effectively divides the myocardium

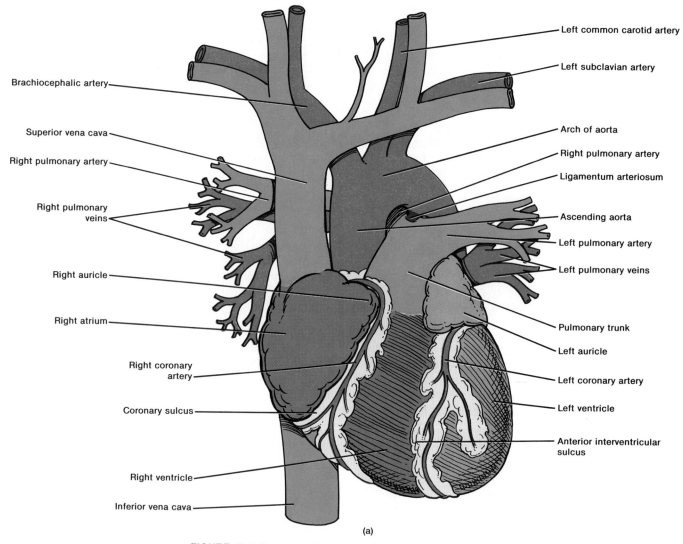

Left common carotid artery

Left subclavian artery

Arch of aorta

Right pulmonary artery

Ligamentum arteriosum

Ascending aorta

Left pulmonary artery

Left pulmonary veins

Pulmonary trunk

Left auricle

Left coronary artery

Left ventricle

Anterior interventricular sulcus

Brachiocephalic artery

Superior vena cava

Right pulmonary artery

Right pulmonary veins

Right auricle

Right atrium

Right coronary artery

Coronary sulcus

Right ventricle

Inferior vena cava

(a)

FIGURE 13-3 Structure of the heart. (a) Anterior external view.

into two separate muscle masses. Externally, a groove known as the *coronary sulcus* (SUL-kus) separates the atria from the ventricles. It encircles the heart and houses the coronary sinus and circumflex branch of the left coronary artery. The *anterior interventricular sulcus* and *posterior interventricular sulcus* separate the right and left ventricles externally. The sulci contain coronary blood vessels and a variable amount of fat (Figure 13-3).

GREAT VESSELS OF THE HEART

The right atrium receives blood from all parts of the body except the lungs. It receives the blood through three veins. The **superior vena cava** brings blood from parts of the body superior to the heart; the **inferior vena cava** brings blood from parts of the body inferior to the heart;

and the **coronary sinus** drains blood from most of the vessels supplying the wall of the heart (Figure 13-3e). The right atrium then delivers the blood into the right ventricle, which pumps it into the **pulmonary trunk.** The pulmonary trunk divides into a **right** and **left pulmonary artery,** each of which carries blood to the lungs. In the lungs, the blood releases its carbon dioxide and takes on oxygen. It returns to the heart via four **pulmonary veins** that empty into the left atrium. The blood then passes into the left ventricle, which pumps the blood into the **ascending aorta.** From here the blood is passed into the **coronary arteries, arch of the aorta, thoracic aorta,** and **abdominal aorta.** These blood vessels transport the blood to all body parts.

During fetal life, there is a temporary blood vessel, called the ductus arteriosus, that connects the pulmonary trunk with the aorta. Its purpose is to redirect blood so

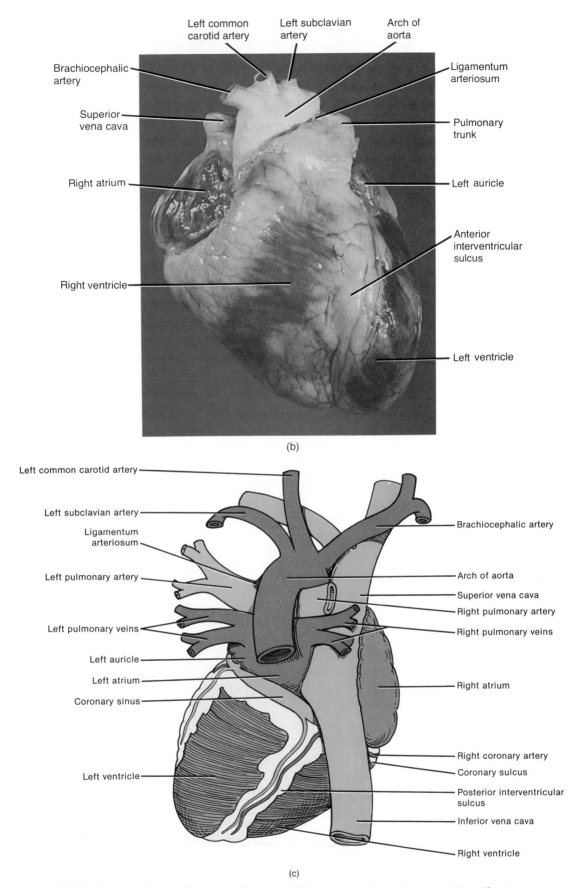

(b)

(c)

FIGURE 13-3 (*Continued*) Structure of the heart. (b) Photograph of anterior external view. (Courtesy of C. Yokochi and J. W. Rohen, *Photographic Anatomy of the Human Body,* 2nd ed., 1979, IGAKU-SHOIN, Ltd., Tokyo, New York.) (c) Diagram of posterior external view.

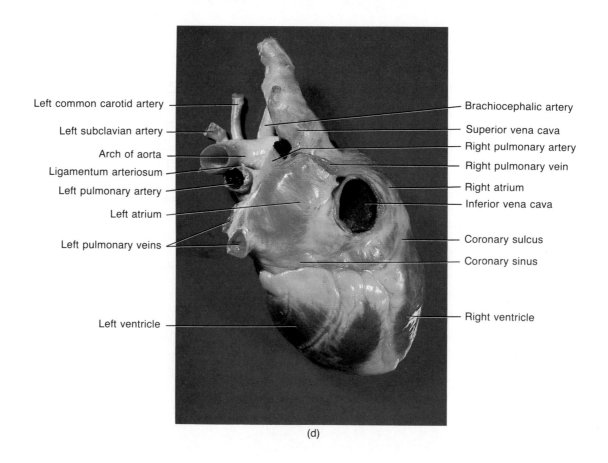

Left common carotid artery

Left subclavian artery

Arch of aorta

Ligamentum arteriosum

Left pulmonary artery

Left atrium

Left pulmonary veins

Left ventricle

Brachiocephalic artery

Superior vena cava

Right pulmonary artery

Right pulmonary vein

Right atrium

Inferior vena cava

Coronary sulcus

Coronary sinus

Right ventricle

(d)

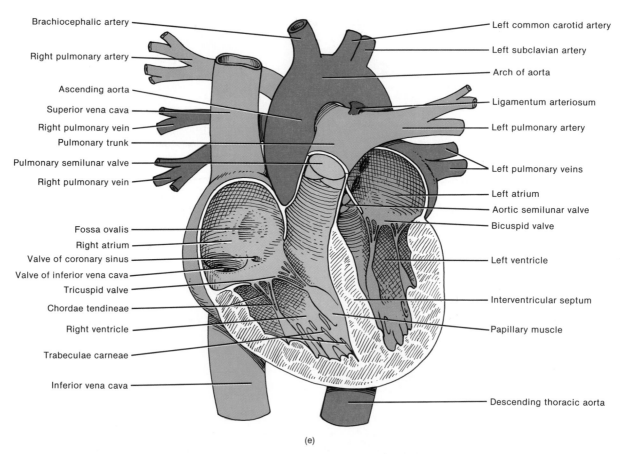

Brachiocephalic artery

Right pulmonary artery

Ascending aorta

Superior vena cava

Right pulmonary vein

Pulmonary trunk

Pulmonary semilunar valve

Right pulmonary vein

Fossa ovalis

Right atrium

Valve of coronary sinus

Valve of inferior vena cava

Tricuspid valve

Chordae tendineae

Right ventricle

Trabeculae carneae

Inferior vena cava

Left common carotid artery

Left subclavian artery

Arch of aorta

Ligamentum arteriosum

Left pulmonary artery

Left pulmonary veins

Left atrium

Aortic semilunar valve

Bicuspid valve

Left ventricle

Interventricular septum

Papillary muscle

Descending thoracic aorta

(e)

FIGURE 13-3 (*Continued*) Structure of the heart. (d) Photograph of posterior external view. (Courtesy of C. Yokochi and J. W. Rohen, *Photographic Anatomy of the Human Body,* 2nd ed., 1979, IGAKU-SHOIN, Ltd., Tokyo, New York.) (e) Diagram of anterior internal view.

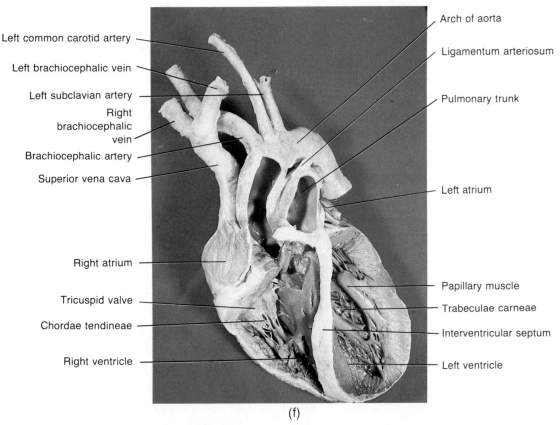

Left common carotid artery

Left brachiocephalic vein

Left subclavian artery

Right brachiocephalic vein

Brachiocephalic artery

Superior vena cava

Right atrium

Tricuspid valve

Chordae tendineae

Right ventricle

Arch of aorta

Ligamentum arteriosum

Pulmonary trunk

Left atrium

Papillary muscle

Trabeculae carneae

Interventricular septum

Left ventricle

(f)

FIGURE 13-3 (*Continued*) Structure of the heart. (f) Photograph of anterior internal view. (Courtesy of C. Yokochi and J. W. Rohen, *Photographic Anatomy of the Human Body,* 2nd ed., 1979, IGAKU-SHOIN, Ltd., Tokyo, New York.)

that it does not enter the fetal lungs, which are nonfunctional. The ductus arteriosus normally closes shortly after birth. Its remnant is known as the **ligamentum arteriosum.**

The sizes of the four chambers vary according to function (Figure 13-3e, f). The right atrium, which must collect blood coming from almost all parts of the body, is slightly larger than the left atrium, which receives blood only from the lungs. The thickness of the chamber wall varies too. The atria are thin-walled because they need only enough cardiac muscle tissue to deliver the blood into the ventricles with the aid of a reduced pressure created by the expanding ventricles. The right ventricle has a thicker layer of myocardium than the atria, since it must send blood to the lungs and around back to the left atrium. The left ventricle has the thickest wall, since it must pump blood at high pressure through literally thousands of miles of vessels in the head, trunk, and extremities.

VALVES OF THE HEART

As each chamber of the heart contracts, it pushes a portion of blood into a ventricle or out of the heart through an artery. In order to keep the blood from flowing backward, the heart has structures composed of collagen called **valves.**

ATRIOVENTRICULAR (AV) VALVES

Atrioventricular (AV) valves lie between the atria and ventricles (Figure 13-3e, f). The right atrioventricular valve between the right atrium and right ventricle is also called the **tricuspid valve** because it consists of three flaps, or cusps. These flaps are fibrous tissues that grow out of the walls of the heart and are covered with endocardium. The pointed ends of the cusps project into the ventricle. Cords called *chordae tendineae* (KOR-dē TEN-di-nē) connect the pointed ends to small conical projections—the *papillary muscles* (muscular columns)—located on the inner surface of the ventricles. The irregular surface of ridges and folds of the myocardium in the ventricles is known as the *trabeculae carneae* (tra-BEK-yoo-lē KAR-nē). The chordae tendineae and their papillary muscles keep the flaps pointing in the direction of the blood flow. In order for blood to pass from the right atrium to the right ventricle, the tricuspid valve opens down as the right atrium contracts, the papillary muscles relax, and the chordae tendineae slacken (Figure 13-4a). When the right ventricle pumps blood out of the heart into the pulmonary trunk, any blood driven back toward the atrium is pushed between the flaps and the ventricle wall (Figure 13-4b). This action drives the cusps upward until their edges meet and close the opening. At the same time, contraction of the papillary muscles and tightening

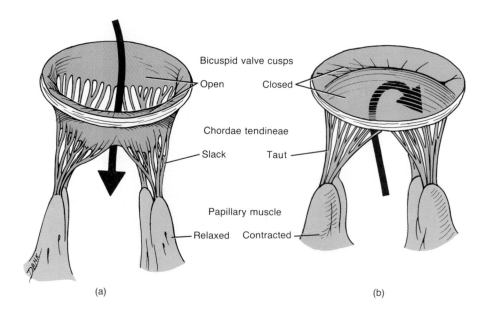

(a)   (b)

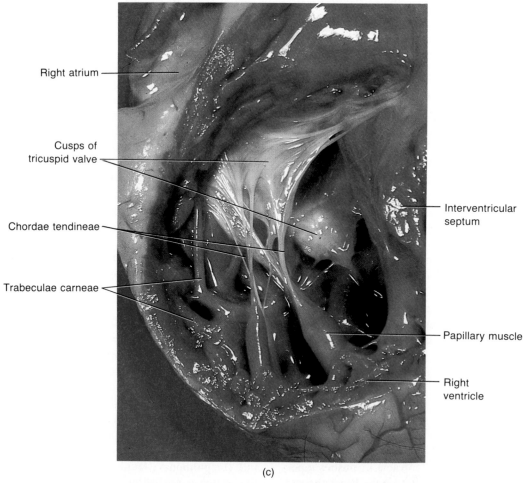

(c)

FIGURE 13-4 Function of atrioventricular (AV) valves. (a) Bicuspid valve open. (b) Bicuspid valve closed. The tricuspid valve operates in a similar manner. (c) Photograph of the tricuspid valve. (Courtesy of C. Yokochi and J. W. Rohen, *Photographic Anatomy of the Human Body,* 2nd ed., 1979, IGAKU-SHOIN, Ltd., Tokyo, New York.)

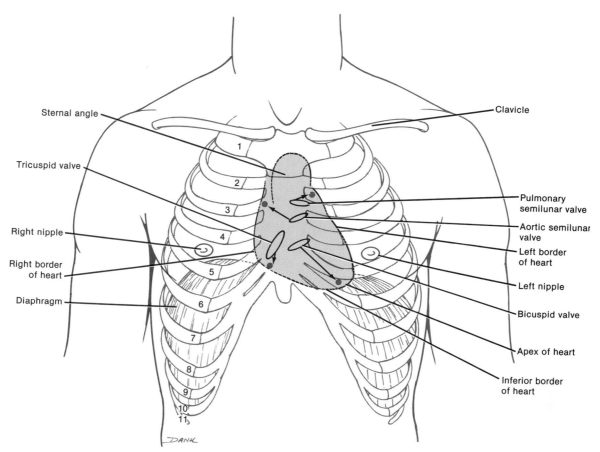

FIGURE 13-5 Surface projection of the heart. The red circles indicate where heart sounds caused by the respective valves are best heard.

of the chordae tendineae help prevent the valve from swinging upward into the atrium.

The atrioventricular valve between the left atrium and left ventricle is called the **bicuspid (mitral) valve.** It is also known as the left atrioventricular (AV) valve. It has two cusps that work in the same way as the cusps of the tricuspid valve. Its cusps are also attached by way of the chordae tendineae to papillary muscles.

SEMILUNAR VALVES

Both arteries that leave the heart have a valve that prevents blood from flowing back into the heart. These are the **semilunar valves** (see Figure 13-3e, f). The **pulmonary semilunar valve** lies in the opening where the pulmonary trunk leaves the right ventricle. The **aortic semilunar valve** is situated at the opening between the left ventricle and the aorta.

Both valves consist of three semilunar (half-moon or crescent-shaped) cusps. Each cusp is attached by its convex margin to the artery wall. The free borders of the cusps curve outward and project into the opening inside the blood vessel. Like the atrioventricular valves, the semilunar valves permit blood to flow in only one direction— in this case, from the ventricles into the arteries.

The valves of the heart may be identified by surface

projection (Figure 13-5). The pulmonary and aortic semilunar valves are represented on the surface by a line about 2.5 cm (1 in) in length. The pulmonary semilunar valve lies horizontally behind the inner end of the left third costal cartilage and the adjoining part of the sternum. The aortic semilunar valve is placed obliquely behind the left side of the sternum, opposite the third intercostal space. The tricuspid valve lies behind the sternum, extending from the midline at the level of the fourth costal cartilage down toward the right sixth chondrosternal junction. The bicuspid lies behind the left side of the sternum obliquely at the level of the fourth costal cartilage. It is represented by a line about 3 cm in length.

Although heart sounds are produced in part by the closure of valves, they are not necessarily heard best over these valves. Each sound tends to be clearest in a slightly different location closest to the surface of the body (Figure 13-5).

CLINICAL APPLICATION

Heart sounds provide valuable information about the valves. Unusual sounds are called **murmurs.** Some murmurs are caused by the noise made by a little blood bubbling back up into an atrium because of improper closure of an atrioventricular valve.

CONDUCTION SYSTEM

The heart is innervated by the autonomic nervous system (Chapter 19), but the autonomic neurons only increase or decrease the time it takes to complete a cardiac cycle (heartbeat), that is, they do not initiate contraction. The chamber walls can go on contracting and relaxing, contracting and relaxing, without any direct stimulus from the nervous system. This action is possible because the heart has an intrinsic regulating system called the **conduction system.** The conduction system is composed of specialized muscle tissue that generates and distributes the electrical impulses that stimulate the cardiac muscle fibers to contract. These tissues are the sinoatrial (sinuatrial) node, the atrioventricular (AV) node, the atrioventricular (AV) bundle (of His), the bundle branches, and the conduction myofibers (Purkinje fibers). The cells of the conduction system develop during embryological life from certain cardiac muscle cells. These cells lose their ability to contract and become specialized for impulse transmission.

A **node** of the conducting system is a compact mass of conducting cells. The **sinoatrial (sinuatrial) node,** also known as the **SA node,** or **pacemaker,** is located in the right atrial wall inferior to the opening of the superior vena cava (Figure 13-6a). The SA node initiates each cardiac cycle and thereby sets the basic pace for the heart rate—hence its common name, pacemaker. The rate set by the SA node may be altered by nervous impulses from the autonomic nervous system or by certain blood-borne chemicals such as thyroid hormones and epinephrine.

CLINICAL APPLICATION

Sinus node dysfunction refers to a disorder of impulse formation within the SA node and impulse conduction out of the node. The condition is characterized by dizziness, lightheadedness, and/or syncope (fainting). Less commonly, mental changes, stroke, congestive heart failure (CHF), and angina (chest pain) may occur. Sinus node dysfunction may be caused by certain heart diseases, autonomic nervous system dysfunction, and certain drugs.

Once an action potential is initiated by the SA node, the impulse spreads out over both atria, causing them to contract and at the same time depolarizing the **atrioventricular (AV) node.** Because of its location near the inferior portion of the interatrial septum, the AV node is one of the last portions of the atria to be depolarized.

From the AV node, a tract of conducting fibers called the **atrioventricular (AV) bundle (bundle of His)** runs through the cardiac skeleton to the top of the interventricular septum. It then continues down both sides of the septum as the **right** and **left bundle branches.** The atrioventricular bundle distributes the action potential over the medial surfaces of the ventricles. Actual contraction of the ventricles is stimulated by **conduction myofibers (Purkinje fibers)** that emerge from the bundle branches and pass into the cells of the myocardium.

ELECTROCARDIOGRAM

In a normal heartbeat, the two atria contract simultaneously while the two ventricles relax. Then, when the two ventricles contract, the two atria relax. The term **systole** (SIS-tō-lē) refers to the phase of contraction; **diastole** (dī-AS-tō-lē) is the phase of relaxation. A **cardiac cycle,** or complete heartbeat, consists of the systole and diastole of both atria plus the systole and diastole of both ventricles.

Impulse transmission through the conduction system generates electrical currents that can be detected on the body's surface. A recording of the electrical changes that accompany the cardiac cycle is called an **electrocardiogram (ECG, or EKG).** The instrument used to record the changes is an *electrocardiograph.*

Each portion of the cardiac cycle produces a different electrical impulse. These impulses are transmitted from the electrodes to a recording pen that graphs the impulses as a series of up-and-down waves called *deflection waves.* In a typical record (Figure 13-6b), three clearly recognizable waves accompany each cardiac cycle. The first, called the **P wave,** is a small upward wave. It indicates atrial depolarization—the spread of an impulse from the SA node through the muscle of the two atria. A fraction of a second after the P wave begins, the atria contract. The second wave, called the **QRS wave (complex),** begins as a downward deflection, continues as a large, upright, triangular wave, and ends as a downward wave at its base. This deflection represents ventricular depolarization, that is, the spread of the electrical impulse through the ventricles. The third recognizable deflection is a dome-shaped **T wave.** This wave indicates ventricular repolarization. There is no deflection to show atrial repolarization because the stronger QRS wave masks this event.

The ECG is useful in diagnosing abnormal cardiac rhythms and conduction patterns and following the course of recovery from a heart attack. It can also detect the presence of fetal life or determine the presence of several fetuses.

CARDIAC CYCLE

Two phenomena control the movement of blood through the heart as part of its cardiac cycle (heartbeat): the opening and closing of the valves and the contraction and relaxation of the myocardium. Both these activities occur

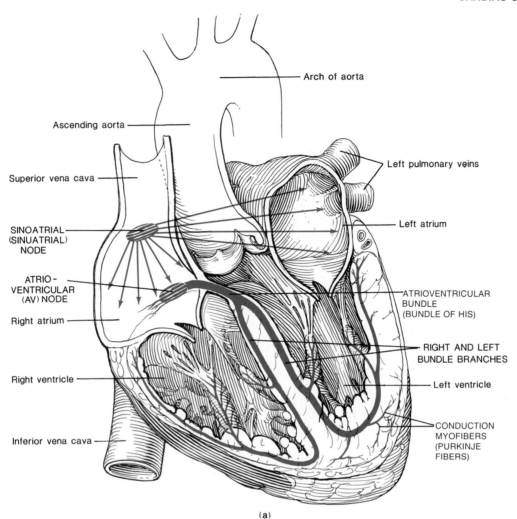

(a)

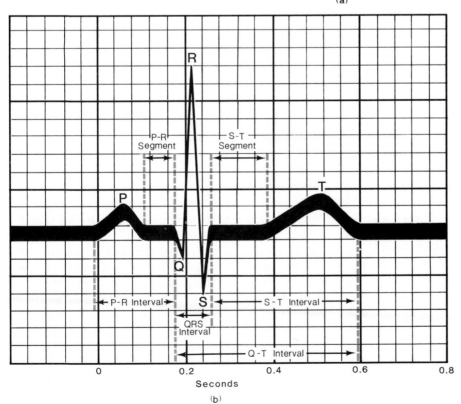

0   0.2   0.4   0.6   0.8

Seconds

(b)

FIGURE 13-6 Conduction system of the heart. (a) Location of the nodes and bundles of the conduction system. (b) Normal electrocardiogram of a single heartbeat, enlarged for emphasis.

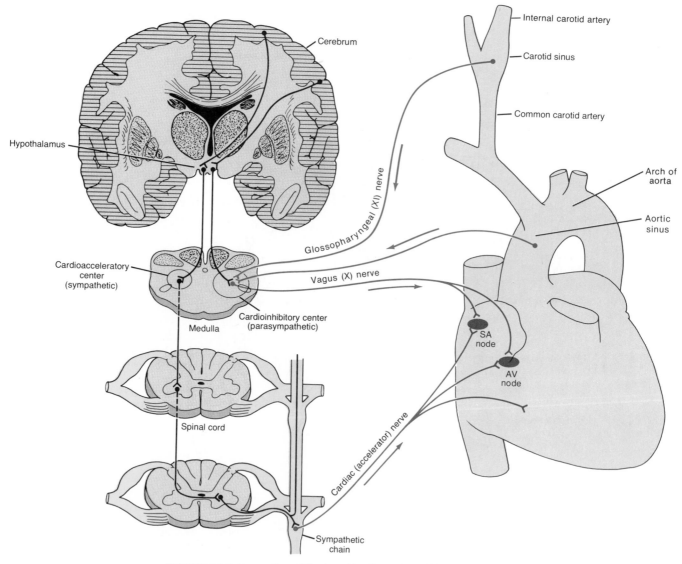

FIGURE 13-7 Innervation of the heart by the autonomic nervous system.

without direct stimulation from the nervous system. The valves are controlled by pressure changes in each heart chamber. The contraction of the cardiac muscle is stimulated by its conduction system.

If we assume that the average heart beats 75 times per minute, then each cardiac cycle requires about 0.8 sec. During the first 0.1 sec, the atria contract and the ventricles relax. The atrioventricular valves are open, and the semilunar valves are closed. For the next 0.3 sec, the atria are relaxing and the ventricles are contracting. During the first part of this period, all valves are closed; during the second part, the semilunar valves are open. The last 0.4 sec of the cycle is the relaxation, or quiescent, period and all chambers are in diastole. In a complete cycle, then, the atria are in systole 0.1 sec and in diastole 0.7 sec; the ventricles are in systole 0.3 sec and in diastole 0.5 sec. For the first part of the quiescent period, all valves are closed; during the latter part, the atrioventricu-

lar valves open and blood starts draining into the ventricles. When the heart beats faster than normal, the quiescent period is shortened accordingly.

The sound of the heartbeat comes primarily from turbulence in blood flow created by the closure of the valves, not from the contraction of the heart muscle. The first sound, which can be described as a **lubb** ($\overline{oo}$) sound, is a long, booming sound. The lubb is the sound created by the closure of the atrioventricular valves soon after ventricular systole begins. The second sound, which is heard as a short, sharp sound, can be described as a **dupp** (ŭ) sound. Dupp is the sound created as the semilunar valves close toward the end of ventricular systole. A pause between the second sound and the first sound of the next cycle is about two times longer than the pause between the first and second sound of each cycle. Thus the cardiac cycle can be heard as a lubb, dupp, pause; lubb, dupp, pause; lubb, dupp, pause.

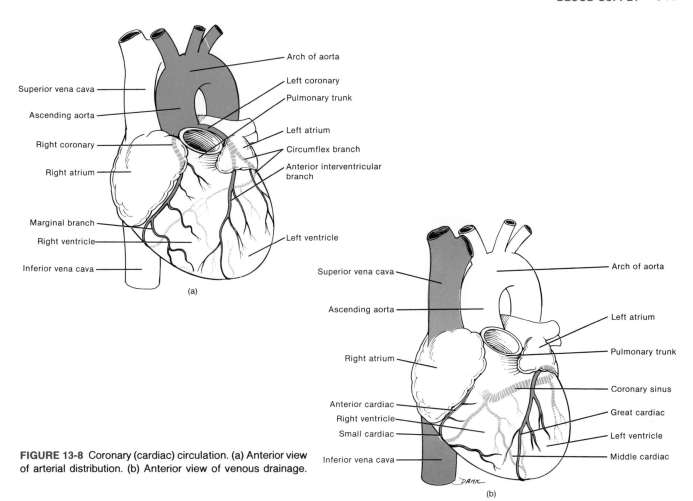

Left coronary

Superior vena cava

Ascending aorta

Right coronary

Right atrium

Marginal branch

Right ventricle

Inferior vena cava

Arch of aorta

Pulmonary trunk

Left atrium

Circumflex branch

Anterior interventricular branch

Left ventricle

(a)

Superior vena cava

Ascending aorta

Right atrium

Anterior cardiac

Right ventricle

Small cardiac

Inferior vena cava

Arch of aorta

Left atrium

Pulmonary trunk

Coronary sinus

Great cardiac

Left ventricle

Middle cardiac

(b)

FIGURE 13-8 Coronary (cardiac) circulation. (a) Anterior view of arterial distribution. (b) Anterior view of venous drainage.

AUTONOMIC CONTROL

The pacemaker initiates contraction and, left to itself, would set an unvarying heart rate. However, the body's need for blood supply varies under different conditions. There are several regulatory mechanisms stimulated by factors such as chemicals present in the body, temperature, emotional state, and age. The most important control of heart rate and strength of contraction is the effect of the autonomic nervous system.

Within the medulla of the brain is a group of neurons called the **cardioacceleratory center (CAC).** Arising from this center are sympathetic fibers that travel down a tract in the spinal cord and then pass outward in the *cardiac (accelerator) nerves* and innervate the SA node, AV node, and portions of the myocardium (Figure 13-7). When the cardioacceleratory center is stimulated, nerve impulses travel along the sympathetic fibers and cause them to release norepinephrine, which increases the rate of heartbeat and the strength of contraction.

The medulla also contains a group of neurons that form the **cardioinhibitory center (CIC).** Arising from this center are parasympathetic fibers that reach the heart via the *vagus (X) nerve.* These fibers innervate the SA node and AV node. When this center is stimulated, nerve impulses transmitted along the parasympathetic fibers cause the release of acetylcholine, which decreases the rate of heartbeat and strength of contraction.

The autonomic control of the heart is therefore the result of opposing sympathetic (stimulatory) and parasympathetic (inhibitory) influences. Sensory impulses from receptors in different parts of the cardiovascular system act on the centers so that a balance between stimulation and inhibition is maintained. Nerve cells capable of responding to changes in blood pressure are called **pressoreceptors (baroreceptors).** These receptors affect the rate of heartbeat.

Heart rate may also be influenced by certain chemicals (epinephrine, sodium, potassium), temperature, emotions, sex, and age.

BLOOD SUPPLY

The wall of the heart, like any other tissue, including large blood vessels, has its own blood vessels. Nutrients could not possibly diffuse through all the layers of cells that make up the heart tissue. The flow of blood through the numerous vessels that pierce the myocardium is called the **coronary (cardiac) circulation** (Figure 13-8). The term

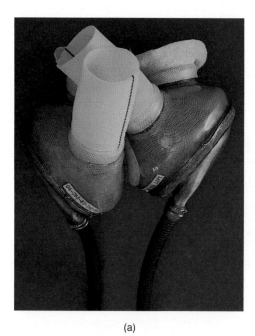

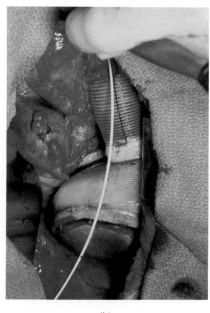

(a)   (b)

FIGURE 13-9 Artificial heart. (a) Photograph of an artificial heart prior to implantation. (b) Photograph of an implanted artificial heart. (Courtesy Nelson, University of Utah.)

coronary refers to the fact that the blood vessels of the heart resemble a crown.

The vessels that serve the myocardium include the *left coronary artery,* which originates as a branch of the ascending aorta. This artery runs under the left atrium and divides into the anterior interventricular and circumflex branches. The *anterior interventricular branch* follows the anterior interventricular sulcus and supplies oxygenated blood to the walls of both ventricles. The *circumflex branch* distributes oxygenated blood to the walls of the left ventricle and left atrium.

The *right coronary artery* also originates as a branch of the ascending aorta. It runs under the right atrium and divides into the posterior interventricular and marginal branches. The *posterior interventricular branch* follows the posterior interventricular sulcus and supplies the walls of the two ventricles with oxygenated blood. The *marginal branch* transports oxygenated blood to the myocardium of the right ventricle and right atrium. The left ventricle receives the most abundant blood supply because of the enormous work it must do.

As blood passes through the atrial system of the heart, it delivers oxygen and nutrients and collects carbon dioxide and wastes. Most of the deoxygenated blood, which carries the carbon dioxide and wastes, is collected by a large vein, the *coronary sinus,* which empties into the right atrium. A vascular sinus is a vein with a thin wall that has no smooth muscle to alter its diameter. The principal tributaries of the coronary sinus are the *great cardiac vein,* which drains the anterior aspect of the heart, and the *middle cardiac vein,* which drains the posterior aspect of the heart.

CLINICAL APPLICATION

Most heart problems result from faulty coronary circulation. If a reduced oxygen supply weakens the cells but does not actually kill them, the condition is called **ischemia** (is-KĒ-mē-a). **Angina pectoris** (an-JĪ-na, or AN-ji-na, PEK-tō-ris), meaning "chest pain," is ischemia of the myocardium. (Pain impulses originating from most visceral muscles are referred to an area on the surface of the body.) The symptoms of angina pectoris include chest pain, accompanied by constriction or tightness, labored breathing, and a sensation of foreboding. Sometimes weakness, dizziness, and perspiration occur.

A much more serious problem is **myocardial infarction** (in-FARK-shun), commonly called a heart attack. **Infarction** means the death of an area of tissue because of an interrupted blood supply. Myocardial infarction may result from a thrombus or embolus in one of the coronary arteries. The tissue distal to the obstruction dies and is replaced by noncontractile scar tissue. Thus, the heart muscle loses at least some of its strength. The aftereffects depend partly on the size and location of the infarcted, or dead, area.

ARTIFICIAL HEART

On December 2, 1982, Dr. Barney B. Clark, a 61-year-old retired dentist, made medical history by becoming the first human to receive an **artificial heart.** The 7½-

hour operation was performed by Dr. William DeVries, a surgeon at the University of Utah Medical Center.

The artificial heart, known as Jarvik-7 after its inventory, Dr. Robert Jarvik, consists of an aluminum base and a pair of rigid plastic chambers that serve as ventricles (Figure 13-9). Each chamber contains a flexible diaphragm driven by compressed air that forces blood past mechanical valves into the major blood vessels. The surgery essentially consisted of removing the ventricles from Dr. Clark's heart but leaving the atria intact. Then Dacron connectors were sutured onto the atria, aorta, and pulmonary trunk. The artificial heart was then snapped into position via the connectors.

The external power source for the artificial heart is a 375-pound unit that includes two compressors and a back-up compressor that force compressed air through 6-foot-long air tubes into the artificial heart. These tubes enter through the abdomen. The unit also contains a 3-hour supply of pressurized air should there be a power failure, a dehumidifier that removes moisture from the compressed air, and mechanisms that control air pressure and heart rate. The unit is always within 6 feet of the patient, the length of the air tubes. The artificial heart is designed to pump sufficient blood to sustain only moderate activity.

On the 112th day after the implantation, Dr. Clark died from massive circulatory collapse, shock, and failure of all of his natural organs. This was probably due to complications of antibiotic therapy to treat pneumonia and the damage that his body suffered just prior to implantation of the artificial heart. While in his body, the artificial heart beat almost 13 million times for 2,688 hours. Except for a defective bicuspid valve that had to be replaced, the heart showed no evidence of clots, infection, or physical deterioration. Dr. Clark's courage and determination made him one of the outstanding pioneers in the advancement of medical research; his efforts have helped scientists to develop even more successful artificial hearts.

On November 25, 1984, William Schroeder became the second person to receive a Jarvick-7 permanent artificial heart. The 6½-hour operation took place at Humana Heart Institute in Louisville, Kentucky, and was also performed by Dr. William DeVries. The Jarvik-7 implanted in Mr. Schroeder was a modified version of the first implant. In the modified version, the previously faulty, two-piece valves were replaced with stronger, one-piece titanium valves, and the heart could be hooked up for up to 3 hours a day to a portable power pack.

DEVELOPMENTAL ANATOMY OF THE HEART

The *heart,* a derivative of **mesoderm,** begins to develop before the end of the third week. It begins its development in the ventral region of the embryo beneath the foregut (see Figure 23-22). The first step is the formation of a pair of tubes, the **endothelial (endocardial) tubes,** from mesodermal cells (Figure 13-10). These tubes then unite to form a common tube, the **primitive heart tube.** Next, the primitive heart tube undergoes a series of constrictions so that five regions develop: **truncus arteriosus, bulbus cordis, ventricle, atrium,** and **sinus venosus.** Since the bulbus cordis and ventricle subsequently grow more rapidly than the others, the heart assumes a U-shape and later an S-shape. The flexures of the heart reorient the regions so that the atrium and sinus venosus eventually come to lie superior to the bulbus cordis, ventricle, and truncus anteriosus.

At about the seventh week a partition, the **atrial septum,** forms in the atrial region, dividing it into a *right* and *left atrium.* The opening in the partition is the **foramen ovale,** which normally closes at birth and later forms a depression called the *fossa ovalis.* A **ventricular septum** also develops and partitions the ventricular region into a *right* and *left ventricle.* The bulbus cordis and truncus arteriosus divide into two vessels, the *aorta* (arising from the left ventricle) and the *pulmonary trunk* (arising from the right ventricle). Recall that the ductus arteriosus is a temporary vessel between the aorta and pulmonary trunk until birth. The great veins of the heart, *superior vena cava* and *inferior vena cava,* develop from the venous end of the primitive heart tube.

APPLICATIONS TO HEALTH

RISK FACTORS IN HEART DISEASE

It is estimated that one in every five persons who reaches age 60 will have a myocardial infarction (heart attack). One in every four persons between 30 and 60 has the potential to be stricken. Heart disease is epidemic in this country, despite the fact that some of the causes can be foreseen and prevented. The results of research indicate that people who develop combinations of certain risk factors eventually have heart attacks. Risk factors are characteristics, symptoms, or signs present in a person free of disease that are statistically associated with an excessive rate of development of a disease. Among the risk factors in heart disease are:

1. High blood cholesterol level.
2. High blood pressure.
3. Cigarette smoking.
4. Obesity.
5. Lack of regular exercise.
6. Diabetes mellitus.
7. Genetic predisposition (family history of heart disease at an early age).

The first five risk factors all contribute to increasing the heart's work load. High blood cholesterol is discussed shortly and hypertension is discussed in the next chapter.

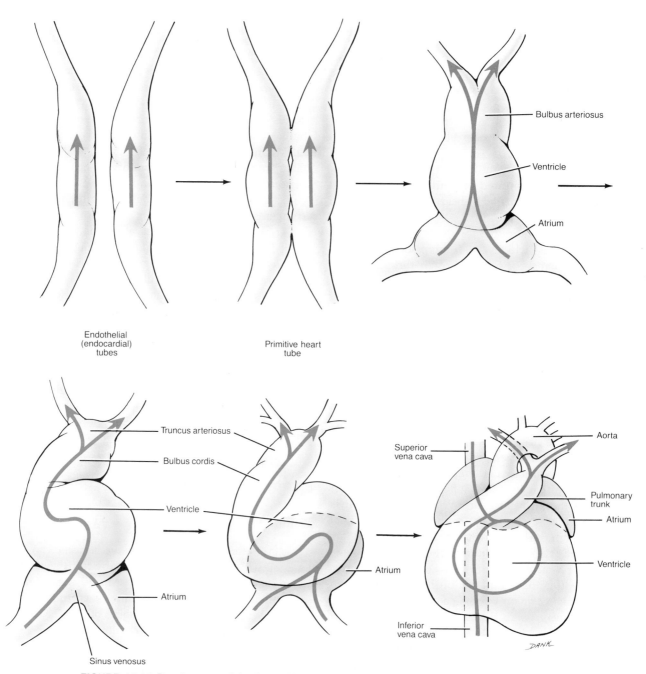

Endothelial (endocardial) tubes

Primitive heart tube

Bulbus arteriosus

Ventricle

Atrium

Truncus arteriosus

Bulbus cordis

Ventricle

Atrium

Sinus venosus

Superior vena cava

Inferior vena cava

Aorta

Pulmonary trunk

Atrium

Ventricle

Atrium

FIGURE 13-10 Development of the heart. The red arrows show the direction of the flow of blood from the venous to the arterial end.

Cigarette smoking, through the effects of nicotine, stimulates the adrenal gland to oversecrete aldosterone, epinephrine, and norepinephrine—powerful vasoconstrictors. Overweight people develop miles of extra capillaries to nourish fat tissue. The heart has to work harder to pump the blood through more vessels. Without exercise, venous return gets less help from contracting skeletal muscles. In addition, regular exercise strengthens the smooth muscle of blood vessels and enables them to assist general circulation. Exercise also increases cardiac efficiency and output. In diabetes mellitus, fat metabolism dominates

glucose metabolism. As a result, cholesterol levels get progressively higher and result in plaque formation, a situation that may lead to high blood pressure. High blood pressure drives fat into the vessel wall, encouraging atherosclerosis.

CORONARY ARTERY DISEASE (CAD)

Coronary artery disease (CAD) is a condition in which the heart muscle receives an inadequate amount of blood because of an interruption of its blood supply. It is pres-

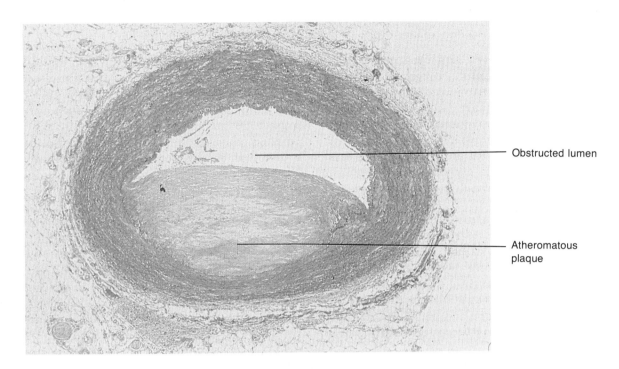

Obstructed lumen

Atheromatous plaque

FIGURE 13-11 Photomicrograph of a cross section of an artery partially obstructed by an atherosclerotic plaque. (Courtesy of Lester V. Bergman & Associates.)

ently the leading cause of death in the United States. Depending on the degree of interruption, symptoms can range from a mild chest pain to a full-scale heart attack. Generally, symptoms manifest themselves when there is about a 75 percent narrowing of coronary artery lumina. The underlying causes of CAD are many and varied. Two of the principal ones are atherosclerosis and coronary artery spasm. Another is a thrombus or an embolus in a coronary artery.

Atherosclerosis

Atherosclerosis (ath'-er-ō-skle-RŌ-sis) is a process in which fatty substances, especially cholesterol and triglycerides (ingested fats), are deposited in the walls of medium-sized and large arteries in response to certain stimuli. It is believed that the first event in atherosclerosis is damage to the endothelial lining of the artery. Contributing factors include high blood pressure (hypertension), carbon monoxide in cigarettes, and a high cholesterol level. Following endothelial damage, platelets aggregate in the area of injury, adhere to the injured surface, and begin releasing their clot-forming chemicals. One of the chemicals increases permeability of the endothelial lining of the tunica interna (inner coat) and modifies the cholesterol so that it is taken up by smooth muscle cells in the tunica media (middle coat). The more cholesterol there is in the blood, the faster the process occurs. Platelets also release a chemical that stimulates the smooth muscle cells to proliferate. The result of enlarged, prolifer-

ating muscle cells is the development of an **atherosclerotic plaque** which deforms the arterial wall (Figure 13-11).

The plaque looks like a pearly gray or yellow mound of tissue. As it grows, it may obstruct blood flow in the affected artery and damage the tissues the artery supplies. An additional danger is that the plaque may provide a roughened surface for clot formation, and a thrombus may form. If the clot breaks off and forms an embolus, it may obstruct small arteries, capillaries, and veins quite a distance from the site of formation.

Coronary Artery Spasm

Atherosclerosis results in a fixed obstruction to blood flow. Recently, it was learned that obstruction can also be caused by **coronary artery spasm,** a condition in which the smooth muscle of a coronary artery undergoes a sudden contraction, resulting in vasoconstriction. Coronary artery spasm typically occurs in individuals with atherosclerosis and may result in chest pain during rest (variant angina), chest pain during exertion (typical angina), heart attacks, and sudden death. Although the causes of coronary artery spasm are unknown, several factors are receiving attention. These include smoking, stress, and a vasoconstrictor chemical released by platelets. There is considerable interest in aspirin and other drugs that might inhibit the vasoconstrictor chemical. Coronary artery spasm is treated with long-acting nitrates (nitroglycerin) and calcium-blocking agents.

CONGENITAL DEFECTS

One baby in every 100 is born with a malformed heart. Such an unborn defect is called a **congenital defect.** In some, holes exist where partitions should be solid, blood vessels that should close remain open, and blood vessels are misconnected. In other congenital defects, the chambers of the heart are too narrow, a major artery is almost pinched shut, and heart valves are malformed and leak. Many defects are not serious and may go unnoticed for a lifetime. Others heal themselves. But some are life-threatening and must be mended by surgical techniques ranging from simple suturing to replacement of malfunctioning parts with synthetic materials.

One congenital defect that can occur is **coarctation** (kō'-ark-TĀ-shun) **of the aorta.** In this condition, a segment of the aorta is too narrow. As a result, the flow of oxygenated blood to the body is reduced, the left ventricle is forced to pump harder, and high blood pressure develops. The condition may be corrected by insertion of a synthetic graft.

Another common congenital defect is **patent ductus arteriosus.** The ductus arteriosus between the aorta and the pulmonary trunk normally closes shortly after birth in response to increased oxygenation of blood flowing through the ductus arteriosus. Only a few hours after birth, the muscular wall of the ductus arteriosus constricts and, within a week, constriction is sufficient to stop all blood flow. During the second month, the ductus arteriosus becomes occluded by the formation of fibrous tissue in its lumen. In some babies, the ductus arteriosus remains open. As a result, aortic blood flows into the lower-pressure pulmonary trunk, thus increasing the pulmonary trunk blood pressure and overworking both ventricles and the heart. Treatment consists of a surgical procedure in which the ductus arteriosus is ligated or cut so that the cut ends are tied off.

A **septal defect** is an opening in the septum that separates the interior of the heart into a left and right side. **Interatrial septal defect** is failure of the fetal foramen ovale between the two atria to close after birth. Because pressure in the right atrium is low, interatrial septal defect generally allows a good deal of blood to flow from the left atrium to the right without going through systemic circulation. This defect, an example of left-to-right shunt, overloads the pulmonary circulation, produces fatigue, and increases respiratory infections. If it occurs early in life, it inhibits growth because the systemic circulation may be deprived of a considerable portion of the blood destined for the organs and tissues of the body. **Interventricular septal defect** is caused by an incomplete closure of the interventricular septum. Due to the higher pressure in the left ventricle, blood is shunted directly from the left ventricle into the right ventricle. This is another example of a left-to-right shunt. Septal openings can now be sewn shut or covered with synthetic patches.

Valvular stenosis is a narrowing, or *stenosis,* of one of the valves regulating blood flow in the heart. Stenosis may occur in the valve itself, most commonly in the mitral valve from rheumatic heart disease or in the aortic valve from sclerosis or rheumatic fever. Or it may occur near a valve. All stenoses are serious because they place a severe work load on the heart by making it work harder to push the blood through the abnormally narrow valve openings. As a result of mitral stenosis, blood pressure is increased. Angina pectoris and heart failure may accompany this disorder. Most stenosed valves are totally replaced with artificial valves.

Tetralogy of Fallot (tet-RAL-ō-jē fal-Ō) is a combination of four defects: an interventricular septal defect, an aorta that emerges from both ventricles instead of from the left ventricle only, a stenosed pulmonary semilunar valve, and an enlarged right ventricle. The condition is an example of a right-to-left shunt, in which blood is shunted from the right ventricle to the left ventricle without going through pulmonary circulation. Because there is stenosis of the pulmonary semilunar valve, the increased right ventricular pressure forces deoxygenated blood from the right ventricle to enter the left ventricle through the interventricular septum. As a result, deoxygenated blood gets mixed with the oxygenated blood that is pumped into systemic circulation. Also, because the aorta emerges from the right ventricle and the pulmonary artery is stenosed, very little blood ever gets to the lungs, and pulmonary circulation is bypassed almost completely. The insufficient amount of oxygenated blood in systemic circulation results in *cyanosis* (sī-a-NŌ-sis), a blue or dark purple discoloration of skin. For this reason the tetralogy of Fallot is one of the conditions that causes a "blue baby." Lung disorders and suffocation also result in cyanosis.

Today it is possible to correct cases of tetralogy of Fallot when the patient is of proper age and condition. Open-heart operations are performed in which the narrowed pulmonary valve is cut open and the interventricular septal defect is sealed with a Dacron patch.

ARRHYTHMIAS

An **arrhythmia** (a-RITH-mē-a) is a general term that refers to an abnormality or irregularity in the heart rhythm. More physicians are using the term **dysrhythmia** since it implies an abnormal rhythm, whereas arrhythmia implies no rhythm. An arrhythmia results when there is a disturbance in the conduction system of the heart, either due to faulty production of electrical impulses or faulty conduction of impulses as they pass through the system.

There are many different types of arrhythmias that can occur, some normal and some quite serious. Arrhythmias may be caused by factors such as caffeine, nicotine, alcohol, anxiety, certain drugs, hyperthyroidism, potassium deficiency, and certain heart diseases. Serious arrhythmias can result in cardiac arrest if the heart cannot supply its own oxygen demands, as well as those of the rest of the body. They can be controlled, and the normal heart rhythm can be reestablished, if they are detected and treated early enough.

Heart Block

One serious arrhythmia is called a **heart block.** Perhaps the most common blockage is in the atrioventricular (AV) node, which conducts impulses from the atria to the ventricles. This disturbance is called *atrioventricular (AV) block.* It usually indicates a myocardial infarction, atherosclerosis, rheumatic heart disease, diphtheria, or syphilis. With complete AV block, patients may have vertigo, unconsciousness, or convulsions. These symptoms result from a decreased cardiac output with diminished cerebral blood flow and cerebral hypoxia or lack of sufficient oxygen.

Among the causes of AV block are excessive stimulation by the vagus (X) nerves (which depresses conductivity of the junctional fibers), destruction of the AV bundle as a result of coronary infarct, atherosclerosis, myocarditis, or depression caused by various drugs.

Flutter and Fibrillation

Two other abnormal rhythms are **flutter** and **fibrillation.** In **atrial flutter** the artial rhythm averages between 240 and 360 beats per minute. The condition is essentially rapid atrial contractions accompanied by a second-degree AV block. It generally indicates severe damage to heart muscle. Atrial flutter usually becomes fibrillation after a few hours, days, or weeks. **Atrial fibrillation** is asynchronous contraction of the atrial muscles that causes the atria to contract irregularly and still faster. Atrial flutter and fibrillation occur in myocardial infarction, acute and chronic rheumatic heart disease, and hyperthyroidism. Atrial fibrillation results in complete uncoordination of atrial contraction so that atrial pumping ceases altogether. When the muscle fibrillates, the muscle fibers of the atrium quiver individually instead of contracting together. The quivering cancels out the pumping of the atrium. In a strong heart, atrial fibrillation reduces the pumping effectiveness of the heart by only 25 to 30 percent.

Ventricular fibrillation is another abnormality that almost always indicates imminent cardiac arrest and death unless corrected quickly. It is characterized by asynchronous, haphazard, ventricular muscle contractions. The rate may be rapid or slow. The impulse travels to the different parts of the ventricles at different rates. Thus, part of the ventricle may be contracting while other parts are still unstimulated. Ventricular contraction becomes ineffective and circulatory failure and death occur. Ventricular fibrillation may be caused by coronary occlusion. It sometimes occurs during surgical procedures on the heart or pericardium. It may be the cause of death in electrocution.

Premature Contraction (Extrasystole)

Another form of arrhythmia arises when a small region of the heart becomes more excitable than normal, causing an occasional abnormal impulse to be generated between normal impulses. The region from which the abnormal impulse is generated is called an *ectopic focus.* As a wave of depolarization spreads outward from the ectopic focus, it causes a **premature contraction (extrasystole).** The contraction occurs early in diastole before the SA node is normally scheduled to discharge its impulses. A person with such a ventricular premature contraction might feel a thump in the chest. Ectopic foci may be caused by emotional stress, excessive intake of stimulants such as caffeine or nicotine, lack of sleep, and local ischemic areas of cardiac muscle.

CONGESTIVE HEART FAILURE (CHF)

Congestive heart failure (CHF) may be defined as a chronic or acute state that results when the heart is not capable of supplying the oxygen demands of the body. Symptoms and signs of CHF are produced by diminished blood flow to the various tissues of the body and by accumulation of excess blood in the various organs because the heart is unable to pump out the blood returned to it by the great veins. Both diminished blood flow and accumulation of excess blood in the organs occur together, but certain symptoms result from congestion, while others are produced by poor tissue nutrition. CHF is usually caused by coronary artery disease in which atherosclerosis leads to myocardial ischemia.

COR PULMONALE (CP)

Cor pulmonale (kor pul-mōn-ALE; *cor* = heart; *pulmon* = lung), or **CP**, refers to right ventricular hypertrophy from disorders that bring about hypertension (high blood pressure) in the pulmonary circulation. Certain disorders such as those involving the respiratory center in the brain, lung diseases, diseases involving airways in the lungs, deformities of the thoracic cage, and neuromuscular disorders can result in hypoxia in the lungs. As a result, vasoconstriction of pulmonary arteries and arterioles occurs and blood flow to the lungs is decreased. Also, any conditions such as tumors, aneurysms, and diseased blood vessels in the lungs can result in decreased blood flow. The reduction in blood flow and increased resistance in the lungs brings about pulmonary hypertension. Ultimately, this causes increased pressure in the pulmonary arteries and pulmonary trunk, resulting in right ventricular hypertrophy.

Individuals with CP may exhibit fatigue, weakness, dyspnea, RUQ pain due to an enlarged liver, warm and cyanotic extremities, cyanosis of the face, distended neck veins, chest pain, and edema.

KEY MEDICAL TERMS ASSOCIATED WITH THE HEART

Aortic insufficiency   An improper closure of the aortic semilunar valve that permits a backflow of blood.

Cardiac arrest   Complete stoppage of the heartbeat.

Cardiomegaly (*mega* = large)   Heart enlargement. Long-distance runners and weight lifters frequently develop heart enlargement as a natural adaptation to increased work load produced by regular exercise.

Commissurotomy   An operation that is performed to widen the opening in a heart valve that has become narrowed by scar tissue.

Compensation   A change in the circulatory system made to compensate for some abnormality; an adjustment of the size of the heart or rate of heartbeat made to counterbalance a defect in structure or function; often used specifically to describe the maintenance of adequate circulation in spite of the presence of heart disease.

Compliance   The passive or diastolic stiffness properties of the left ventricle. A hypertrophied or fibrosed heart with a stiff wall, for example, has decreased compliance. Also, the stiffness properties of the lung.

Constrictive pericarditis   A shrinking and thickening of the pericardium that prevents the heart muscle from expanding and contracting normally.

Incompetent valve   Any valve that does not close properly, thus permitting a backflow of blood; also called **valvular insufficiency.**

Palpitation   A fluttering of the heart or abnormal rate of rhythm of the heart.

Pancarditis (*pan* = all)   Inflammation of the whole heart including inner layer (endocardium), heart muscle (myocardium), and outer sac (pericardium).

Paroxysmal tachycardia   A period of rapid heartbeats that begins and ends suddenly.

Percussion   Tapping a part of the body as an aid in diagnosing the condition of parts of the body by the sound obtained.

Stokes-Adams syndrome   Sudden attacks of unconsciousness, sometimes with convulsions, that may accompany heart block.

Syncope   A temporary cessation of consciousness; a fainting spell. One cause might be insufficient blood supply to the brain.

STUDY OUTLINE

Location (p. 301)

1. The heart is situated obliquely between the lungs in the mediastinum.
2. About two-thirds of its mass is to the left of the midline.

Parietal Pericardium (Pericardial Sac) (p. 301)

1. The parietal pericardium, consisting of an outer fibrous layer and an inner serous layer, encloses the heart.
2. Between the serous pericardium and the epicardium is the pericardial cavity, a space filled with pericardial fluid that prevents friction between the two membranes.

Wall; Chambers; Vessels; and Valves (p. 303)

1. The wall of the heart has three layers: the epicardium, myocardium, and endocardium.
2. The chambers include two upper atria and two lower ventricles.
3. The blood flows through the heart from the superior and inferior venae cavae and the coronary sinus to the right atrium, through the tricuspid valve to the right ventricle, through the pulmonary trunk to the lungs, through the pulmonary veins into the left atrium, through the bicuspid valve to the left ventricle, and out through the aorta.
4. Valves prevent backflow of blood in the heart.
5. Atrioventricular (AV) valves, between the atria and their ventricles, are the tricuspid valve on the right side of the heart and the bicuspid (mitral) valve on the left.
6. The chordae tendineae and their muscles keep the flaps of the valves pointing in the direction of blood flow.
7. The two arteries that leave the heart both have a semilunar valve.

Conduction System (p. 310)

1. The conduction system consists of tissue specialized for impulse conduction.
2. Components of this system are the sinoatrial (SA) node (pacemaker), atrioventricular (AV) node, atrioventricular (AV) bundle (bundle of His), bundle branches, and conduction myofibers (Purkinje fibers).

Electrocardiogram (p. 310)

1. A cardiac cycle consists of the systole (contraction) and diastole (relaxation) of both atria plus the systole and diastole of both ventricles followed by a short pause.
2. The record of electrical changes during each cardiac cycle is referred to as an electrocardiogram (ECG).
3. A normal ECG consists of a P wave (spread of impulse from SA node over atria), QRS wave (spread of impulse through ventricles), and T wave (ventricular repolarization).
4. The ECG is invaluable in diagnosing abnormal cardiac rhythms and conduction patterns, detecting the presence of fetal life, determining the presence of several fetuses, and following the course of recovery from a heart attack.

Cardiac Cycle (p. 310)

1. Blood flow through the heart is controlled by valves and myocardial action.
2. An average cardiac cycle lasts 0.8 sec.
3. Heart sounds are caused by the action of the valves.

Autonomic Control (p. 313)

1. Heart rate and strength of contraction may be increased by sympathetic stimulation from the cardioaccelerary cen-

ter (CAC) in the medulla and decreased by parasympathetic stimulation from the cardioinhibitory center (CIC) in the medulla.

2. Pressoreceptors are nerve cells that respond to changes in blood pressure. They act on the cardiac centers in the medulla.

3. Other influences on heart rate include chemicals (epinephrine, sodium, potassium), temperature, emotion, sex, and age.

Blood Supply (p. 313)

1. The coronary (cardiac) circulation takes oxygenated blood through the arterial system of the myocardium.

2. Deoxygenated blood returns to the right atrium via the coronary sinus.

3. Complications of this system are angina pectoris and myocardial infarction.

Artificial Heart (p. 314)

1. The artificial heart consists of an aluminum base and rigid plastic chambers that serve as ventricles.

2. When it is implanted, the recipient's atria are left intact.

Developmental Anatomy (p. 315)

1. The heart develops from mesoderm.

2. The endothelial tubes develop into the four-chambered heart and great vessels of the heart.

Applications to Health (p. 315)

1. Risk factors in heart disease include high blood cholesterol, high blood pressure, cigarette smoking, obesity, lack of regular exercise, diabetes mellitus, and genetic predisposition.

2. Coronary artery disease (CAD) refers to a condition in which the myocardium receives inadequate blood due to atherosclerosis, coronary artery spasm, or thrombi or emboli; it may be treated with drugs or surgically.

3. Congenital heart defects include coarctation of the aorta, patent ductus arteriosus, septal defects, valvular stenosis, and tetralogy of Fallot.

4. Arrhythmias result in heart block, flutter, and fibrillation.

5. Congestive heart failure (CHF) results when the heart cannot supply the oxygen demands of the body.

6. Cor pulmonale (CP) refers to right ventricular hypertrophy secondary to pulmonary hypertension.

REVIEW QUESTIONS

1. Describe the location of the heart in the mediastinum and identify the various borders and surfaces of the heart.

2. Distinguish the subdivisions of the parietal pericardium. What is the purpose of this structure?

3. Compare the three layers of the heart wall according to composition, location, and function.

4. Define atria and ventricles. What vessels enter or exit the atria and ventricles?

5. Describe the principal valves in the heart and how they operate.

6. Describe how the valves of the heart may be identified by surface projection.

7. Describe the structure and function of the heart's conducting system.

8. Define and label the deflection waves of a normal electrocardiogram. Explain why the ECG is an important diagnostic tool.

9. Define a cardiac cycle and describe its timing. What causes heart sounds?

10. Distinguish between the cardioacceleratory center (CAC) and cardioinhibitory center (CIC) with respect to the regulation of heart rate.

11. Describe the route of blood in coronary (cardiac) circulation. Distinguish between angina pectoris and myocardial infarction.

12. Explain how an artificial heart is implanted and how it operates.

13. Describe how the heart develops.

14. Describe the risk factors involved in heart disease.

15. What is coronary artery disease (CAD)? What factors contribute to it?

16. Define each of the following congenital heart defects: coarctation of the aorta, patent ductus arteriosus, septal defects, valvular stenosis, and tetralogy of Fallot.

17. Define an arrhythmia. Give several examples.

18. Describe the symptoms of congestive heart failure (CHF) and cor pulmonale (CP).

19. Refer to the glossary of key medical terms associated with the heart. Be sure that you can define each term.

14

The Cardiovascular System: Blood Vessels

Student Objectives

Contrast the structure and function of arteries, arterioles, capillaries, venules, and veins.

Define a blood reservoir and explain its importance.

Identify the principal arteries and veins of systemic circulation.

Trace the route of blood involved in hepatic portal circulation and explain its importance.

Identify the major blood vessels of pulmonary circulation.

Contrast fetal and adult circulation.

Describe the importance of cerebral circulation.

Explain the fate of fetal circulation structures once postnatal circulation is established.

Describe the effects of aging on the cardiovascular system.

Describe the development of blood vessels and blood.

List the causes and symptoms of aneurysms and hypertension.

Define medical terminology associated with blood vessels.

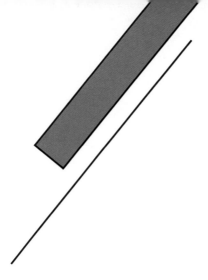

Blood vessels form an extensive network of tubes that carry blood away from the heart, transport it to the tissues of the body, and then return it to the heart. **Arteries** are the vessels that carry blood from the heart to the tissues. Large, elastic arteries leave the heart and divide into medium-sized, muscular vessels that head toward the various regions of the body. The medium-sized arteries divide into small arteries, which, in turn, divide into still smaller arteries called **arterioles.** As the arterioles enter a tissue, they branch into countless microscopic vessels called **capillaries.** Through the walls of the capillaries, substances are exchanged between the blood and body tissues. Before leaving the tissue, groups of capillaries reunite to form small veins called **venules.** These, in turn, merge to form progressively larger tubes called veins. **Veins,** then, are blood vessels that convey blood from the tissues back to the heart. Since blood vessels require oxygen and nutrients just like other tissues of the body, they also have blood vessels in their own walls called **vasa vasorum.**

The developmental anatomy of blood vessels and blood is considered later in the chapter.

ARTERIES

Arteries have walls constructed of three coats, or tunics, and a hollow core, called a *lumen,* through which the blood flows (Figures 14-1 and 14-4). The inner coat of an arterial wall is the **tunica interna (intima).** It is composed of a lining of endothelium (simple squamous epithelium) that is in contact with the blood and a layer of elastic tissue called the internal elastic membrane. The middle coat, or **tunica media,** is usually the thickest layer. It consists of elastic fibers and smooth muscle. The outer coat, the **tunica externa (adventitia),** is composed principally of elastic and collagenous fibers. An external elastic membrane may separate the tunica media from the tunica externa.

As a result of the structure of the middle coat especially, arteries have two major properties: elasticity and contractility. When the ventricles of the heart contract and eject blood into the large arteries, the arteries expand to accommodate the extra blood. Then, as the ventricles relax, the elastic recoil of the arteries forces the blood onward. The contractility of an artery comes from its smooth muscle. The smooth muscle is arranged longitudinally and in rings around the lumen somewhat like a doughnut and is innervated by sympathetic branches of the autonomic nervous system. When there is sympathetic stimulation, the smooth muscle contracts, squeezes the wall around the lumen, and narrows the vessel. Such a decrease in the size of the lumen is called **vasoconstriction.** Conversely, when sympathetic stimulation is removed, the smooth muscle fibers relax and the size of the arterial lumen increases. This increase is called **vasodilation** and is often due to the inhibition of vasoconstriction.

The contractility of arteries also serves a function in stopping bleeding. This function is called vascular spasm, one of the three mechanisms involved in hemostasis. The blood flowing through an artery is under a great deal of pressure. Thus great quantities of blood can be quickly lost from a broken artery. When an artery is cut, its wall constricts so that blood does not escape quite so rapidly. However, there is a limit to how much vasoconstriction can help.

ELASTIC (CONDUCTING) ARTERIES

Large arteries are referred to as **elastic (conducting) arteries.** They include the aorta and brachiocephalic, common carotid, subclavian, vertebral, and common iliac arteries. The wall of elastic arteries is relatively thin in proportion to their diameter, and their tunica media contains more elastic fibers and less smooth muscle. As the heart alternately contracts and relaxes, the rate of blood flow tends to be intermittent. When the heart contracts and forces blood into the aorta, the wall of the elastic arteries stretches. During relaxation of the heart, the wall of the elastic arteries recoils, moving the blood forward in a more continuous flow. Elastic arteries are called conducting arteries because they conduct blood from the heart to medium-sized arteries.

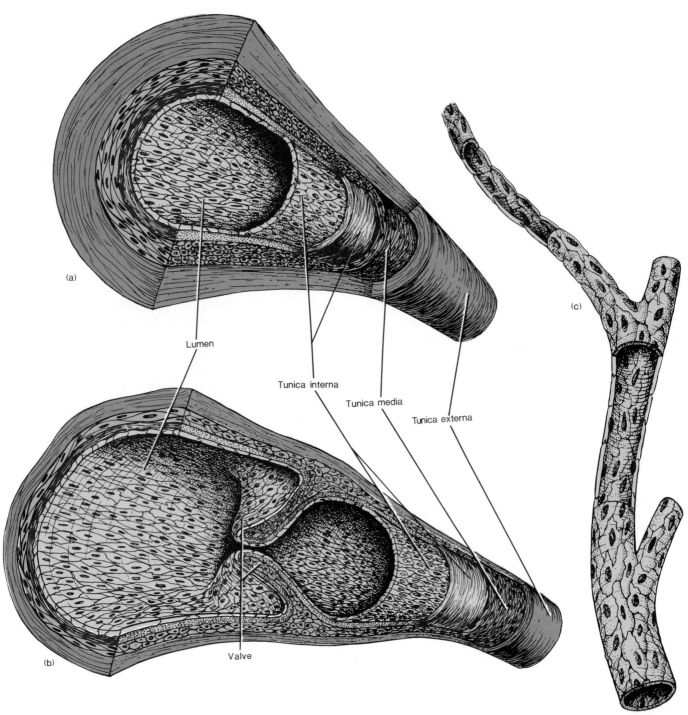

FIGURE 14-1 Comparative structure of (a) an artery, (b) a vein, and (c) a capillary. The relative size of the capillary is enlarged for emphasis.

MUSCULAR (DISTRIBUTING) ARTERIES

Medium-sized arteries are called **muscular (distributing) arteries.** They include the axillary, brachial, radial, intercostal, splenic, mesenteric, femoral, popliteal, and tibial arteries. Their tunica media contains more smooth muscle than elastic fibers, and they are capable of vasoconstriction and vasodilation to adjust the volume of blood to suit the needs of the structure supplied. The wall of muscular arteries is relative thick, mainly due to the large amounts of smooth muscle. Muscular arteries are called distributing arteries because they distribute blood to various parts of the body.

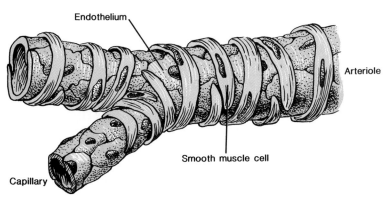

Endothelium

Arteriole

Smooth muscle cell

Capillary

FIGURE 14-2 Structure of an arteriole.

ANASTOMOSES

Most parts of the body receive branches from more than one artery. In such areas the distal ends of the vessels unite. The junction of two or more vessels supplying the same body region is called an **anastomosis** (a-nas-tō-MŌ-sis). Anastomoses may also occur between the origins of veins and between arterioles and venules. Anastomoses between arteries provide alternate routes by which blood can reach a tissue or organ. Thus if a vessel is occluded by disease, injury, or surgery, circulation to a part of the body is not necessarily stopped. The alternate route of blood to a body part through an anastomosis is known as **collateral circulation.** An alternate blood route may also be supplied by nonanastomosing vessels that supply the same region of the body.

Arteries that do not anastomose are known as **end arteries.** Occlusion of an end artery interrupts the blood supply to a whole segment of an organ, producing necrosis (death) of that segment.

CLINICAL APPLICATION

A procedure that provides physicians with a much clearer look at diseased arteries is called **digital subtraction angiography (DSA).** The procedure employs a computer technique that compares an x-ray image of the same region of the body before and after a contrast substance has been introduced intravenously. Any tissue or blood vessels that show up in the first image can be subtracted (erased) from the second image, leaving an unobstructed view of an artery. DSA figuratively lifts an artery out of the body so that it can be studied in isolation. DSA helps in diagnosing lesions in the carotid arteries leading to the brain, a potential cause of strokes, and in evaluating patients before surgery and after coronary bypass surgery and certain transplant operations.

ARTERIOLES

An **arteriole** is a very small, almost microscopic artery that delivers blood to capillaries. Arterioles closer to the arteries from which they branch have a tunica interna like that of arteries, a tunica media composed of smooth muscle and very few elastic fibers, and a tunica externa composed mostly of elastic and collagenous fibers (see Figure 14-4c). As arterioles get smaller in size, the tunics change character so that arterioles closest to capillaries consist of little more than a layer of endothelium surrounded by a few scattered smooth muscle cells (Figure 14-2).

Arterioles play a key role in regulating blood flow from arteries into capillaries. The smooth muscle of arterioles, like that of arteries, is subject to vasoconstriction and vasodilation. During vasoconstriction, blood flow into capillaries is restricted; during vasodilation, the flow is significantly increased. The relationship of arterioles to blood flow will be considered shortly.

CAPILLARIES

Capillaries are microscopic vessels that usually connect arterioles and venules (see Figure 14-1). They are found near almost every cell in the body. The distribution of capillaries in the body varies with the activity of the tissue. For example, in places where activity is higher, such as the liver, kidneys, lungs, muscles, and nervous system, there are rich capillary supplies. In areas where activity is lower, such as tendons and ligaments, the capillary supply is not as extensive. The epidermis, cornea of the eye, and cartilage are devoid of capillaries.

The primary function of capillaries is to permit the exchange of nutrients and wastes between the blood and tissue cells. The structure of the capillaries is admirably suited to this purpose. Capillary walls are composed of only a single layer of cells (endothelium). They have no tunica media or tunica externa. Thus a substance in the blood must pass through the plasma membrane of just one cell to reach tissue cells. This vital exchange of materials occurs only through capillary walls—the thick walls of arteries and veins present too great a barrier.

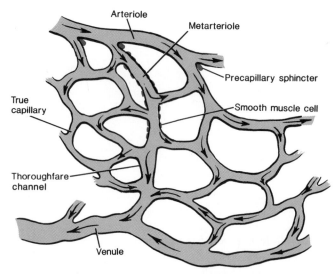

FIGURE 14-3 Details of a capillary network.

Although capillaries pass directly from arterioles to venules in some places in the body, in other places they form extensive branching networks. These networks increase the surface area for diffusion and thereby allow a rapid exchange of large quantities of materials. In most tissues, blood normally flows through only a small portion of the capillary network when metabolic needs are low. But, when a tissue becomes active, the entire capillary network fills with blood.

The flow of blood through capillaries is regulated by vessels with smooth muscle in their walls. A **metarteriole** (*met* = beyond) is a vessel that emerges from an arteriole, traverses the capillary network, and empties into a venule (Figure 14-3). The proximal portions of metarterioles are surrounded by scattered smooth muscle cells whose contraction and relaxation help to regulate the amount and force of blood. The distal portion of a metarteriole has no smooth muscle cells and is called a *thoroughfare channel.* It serves as a low-resistance channel that increases blood flow. **True capillaries** emerge from arterioles or metarterioles and are not on the direct flow route from arteriole to venule. At their sites of origin, there is a ring of multiunit smooth muscle called a **precapillary sphincter** that controls the flow of blood entering a true capillary.

Autoregulation refers to a local, automatic adjustment of blood flow in a given region of the body in response to the particular needs of the tissue. In most body tissues, oxygen is the principal stimulus for autoregulation. A suggested mechanism for autoregulation is as follows. In response to low oxygen supplies, the cells in the immediate area produce and release *vasodilator substances.* Such substances are thought to include potassium ions, hydrogen ions, carbon dioxide, lactic acid, and adenosine. Once released, the vasodilator substances produce a local dila-

tion of arterioles and relaxation of precapillary sphincters. The result is an increased flow of blood into the tissue, which restores oxygen levels to normal. The autoregulation mechanism is important in meeting the nutritional demands of active tissues, such as muscle tissue, where the demand might increase as much as tenfold.

Some capillaries of the body, such as those found in muscle tissue and other locations, are referred to as **continuous capillaries.** These capillaries are so named because the cytoplasm of the endothelial cells is continuous when viewed in cross section through a microscope; the cytoplasm appears as an uninterrupted ring, except for the endothelial junction. Other capillaries of the body are referred to as **fenestrated capillaries.** They differ from continuous capillaries in that their endothelial cells have numerous fenestrae (pores) where the cytoplasm is absent. The fenestrae range from 700 to 1000 Å in diameter and are closed by a thin diaphragm, except in the capillaries in the kidneys where they are assumed to be open. Fenestrated capillaries are also found in the villi of the small intestine, choroid plexuses of the ventricles in the brain, ciliary processes of the eyes, and endocrine glands.

Microscopic blood vessels in certain parts of the body, such as the liver, are termed **sinusoids.** They are wider than capillaries and more tortuous. Also, instead of the usual endothelial lining, sinusoids are lined largely by phagocytic cells. In the liver, such cells are called *stellate reticuloendothelial (Kupffer's) cells.* Like capillaries, sinusoids convey blood from arterioles to venules. Other regions containing sinusoids include the spleen, adenohypophysis, parathyroid glands, and adrenal cortex.

VENULES

When several capillaries unite, they form small veins called **venules** (see Figure 14-4c). Venules collect blood from capillaries and drain it into veins. The venules closest to the capillaries consist of a tunica interna of endothelium and a tunica externa of connective tissue. As the venules approach the veins, they also contain the tunica media characteristic of veins.

VEINS

Veins are composed of essentially the same three coats as arteries, but they have considerably less elastic tissue and smooth muscle and contain more white fibrous tissue (Figures 14-1 and 14-4). However, they are still distensible enough to adapt to variations in the volume and pressure of blood passing through them.

By the time the blood leaves the capillaries and moves into the veins, it has lost a great deal of pressure. The difference in pressure can be observed in the blood flow

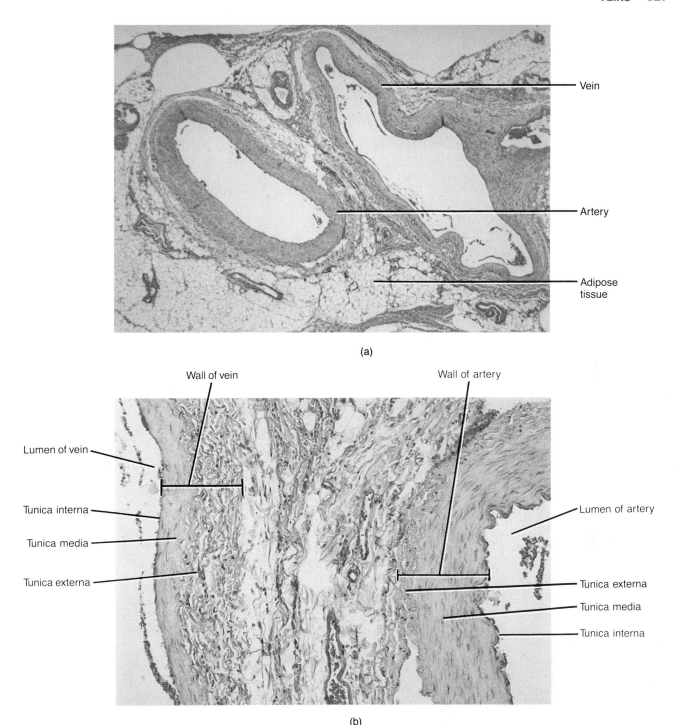

(a)

(b)

FIGURE 14-4 Histology of blood vessels. (a) Photomicrograph of an artery and its accompanying vein at a magnification of 50×. (b) Photomicrograph of a portion of the wall of an artery and its accompanying vein at a magnification of 250×.

from a cut vessel; blood leaves a cut vein in an even flow rather than in the rapid spurts characteristic of arteries. Most of the structural differences between arteries and veins reflect this pressure difference. For example, veins do not need walls as strong as those of arteries. The low pressure in veins, however, has its disadvantages.

When you stand, the pressure pushing blood up the veins in your lower extremities is barely enough to balance the force of gravity pushing it back down. For this reason, many veins, especially those in the limbs, contain valves that prevent backflow (Figure 14-5). Normal valves ensure the flow of blood toward the heart.

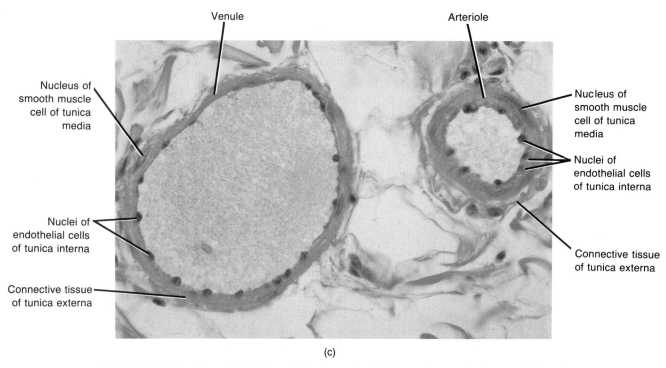

(c)

FIGURE 14-4 (*Continued*) Histology of blood vessels. (c) Photomicrograph of an arteriole and a venule at a magnification of 800×. (© 1983 by Michael H. Ross. Used by permission.)

CLINICAL APPLICATION

In people with weak venous valves, gravity forces large quantities of blood back down into distal parts of the vein. This pressure overloads the vein and pushes its wall outward. After repeated overloading, the walls lose their elasticity and become stretched and flabby. Such dilated and tortuous veins caused by incompetent valves are called **varicose veins.** They may be due to heredity, mechanical factors (prolonged standing and pregnancy), or aging. Because a varicosed wall is not able to exert a firm resistance against the blood, blood tends to accumulate in the pouched-out area of the vein, causing it to swell and forcing fluid into the surrounding tissue. Veins close to the surface of the legs are highly susceptible to varicosities. Veins that lie deeper are not as vulnerable because surrounding skeletal muscles prevent their walls from overstretching.

Varicosities are also common in the veins that lie in the wall of the anal canal. These varicosities are called **hemorrhoids** (HEM-o-royds). Hemorrhoids may be caused by constipation. Repeated straining during defecation forces blood down into the superior hemorrhoidal plexus, increasing pressure in these veins. Constipation is related to low-fiber diets, especially in North America. One hypothesis links increased intra-abdominal pressure, caused by straining during evacuation of firm feces, directly to hemorrhoids or varicose

veins. The most widely used treatment for hemorrhoids is *rubber band ligation,* in which a rubber band is tied around the hemorrhoid, cutting off its blood supply. In a few days, the hemorrhoid dries up and falls off. An even newer treatment is called *infrared photo-coagulation,* in which a very high-energy light beam coagulates the hemorrhoid.

A **vascular (venous) sinus** is a vein with a thin endothelial wall that has no smooth muscle to alter its diameter. Surrounding tissue replaces the tunica media and tunica externa to provide support. Intracranial vascular sinuses, which are supported by the dura mater, return cerebrospinal fluid and deoxygenated blood from the brain to the heart. Another example of a vascular sinus is the coronary sinus of the heart.

BLOOD RESERVOIRS

The volume of blood in various parts of the cardiovascular system varies considerably. Veins, venules, and venous sinuses contain about 59 percent of the blood in the system, arteries about 13 percent, pulmonary vessels about 12 percent, the heart about 9 percent, and arterioles and capillaries about 7 percent. Since systemic veins contain so much of the blood, they are referred to as **blood reservoirs.** They serve as storage depots for blood, which can

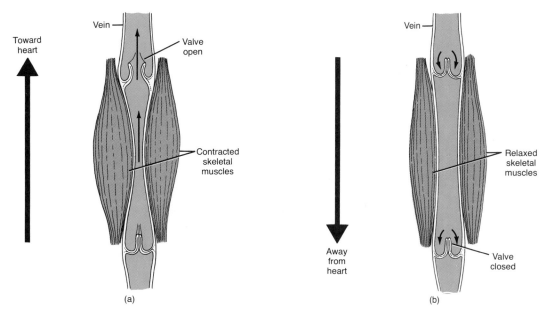

Vein

Toward heart

Valve open

Contracted skeletal muscles

(a)

Vein

Relaxed skeletal muscles

Away from heart

Valve closed

(b)

FIGURE 14-5 Role of skeletal muscle contractions and venous valves in returning blood to the heart. (a) When skeletal muscles contract, the valves open, and blood is forced toward the heart. (b) When skeletal muscles relax, the valves close to prevent the backflowing of blood from the heart.

be moved quickly to other parts of the body if the need arises. When there is increased muscular activity, the vasomotor center sends increasing sympathetic impulses to veins that serve as blood reservoirs. The result is vasoconstriction, which permits the distribution of blood from venous reservoirs to skeletal muscles, where it is needed most. A similar mechanism operates in cases of hemorrhage when blood volume and pressure decrease. Vasoconstriction of veins in venous reservoirs helps to compensate for the blood loss. Among the principal blood reservoirs are the veins of the abdominal organs (especially the liver and spleen) and the veins of the skin.

CIRCULATORY ROUTES

The arteries, arterioles, capillaries, venules, and veins are organized into definite routes that circulate the blood throughout the body. We can now look at the basic routes the blood takes as it is transported through its vessels.

Figure 14-6 shows a number of basic **circulatory routes** through which the blood travels. **Systemic circulation** includes all the oxygenated blood that leaves the left ventricle through the aorta and the deoxygenated blood that returns to the right atrium after traveling to all the organs, including the nutrient arteries to the lungs. Two of the many subdivisions of the systemic circulation are the **coronary (cardiac) circulation** (see Figure 13-8), which supplies the myocardium of the heart, and the **hepatic portal circulation,** which runs from the digestive tract to the liver. Blood leaving the aorta and traveling through the systemic arteries is a bright red color. As it moves through

the capillaries, it loses its oxygen and takes on carbon dioxide, which gives the blood in the systemic veins its dark red color.

When blood returns to the heart from the systemic route, it goes out of the right ventricle through the **pulmonary circulation** to the lungs. In the lungs, it loses its carbon dioxide and takes on oxygen. It is now bright red again. It returns to the left atrium of the heart and reenters the systemic circulation.

Another major route—**fetal circulation**—exists only in the fetus and contains special structures that allow the developing fetus to exchange materials with its mother.

Cerebral circulation (cerebral arterial circle, or circle of Willis) is discussed in Exhibit 14-3.

SYSTEMIC CIRCULATION

The flow of blood from the left ventricle to all parts of the body and back to the right atrium is called the **systemic circulation.** The purpose of systemic circulation is to carry oxygen and nutrients to body tissues and to remove carbon dioxide and other wastes from the tissues. All systemic arteries branch from the *aorta,* which arises from the left ventricle of the heart.

As the aorta emerges from the left ventricle, it passes upward and deep to the pulmonary artery. At this point, it is called the *ascending aorta.* The ascending aorta gives off two coronary branches to the heart muscle. Then it turns to the left, forming the *arch of the aorta* before descending to the level of the fourth thoracic vertebra as the *descending aorta.* The descending aorta lies close

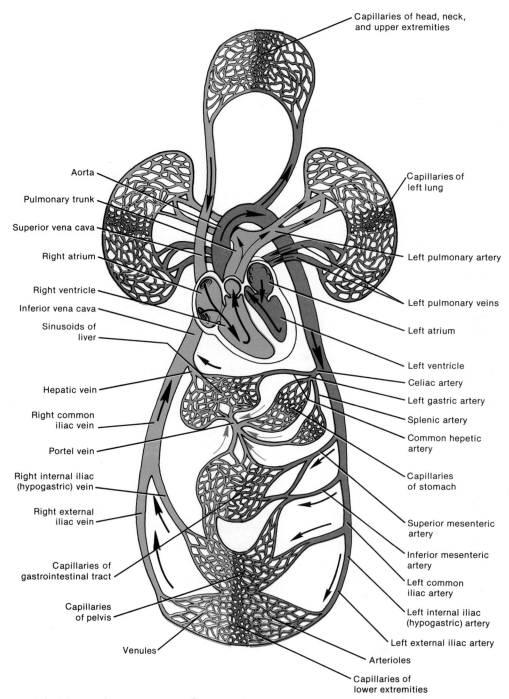

FIGURE 14-6 Circulatory routes. Systemic circulation is indicated by heavy black arrows, pulmonary circulation by thin black arrows, and hepatic portal circulation by thin colored arrows. Refer to Figure 13-8 for the details of coronary (cardiac) circulation and Figure 14-19 for the details of fetal circulation.

to the vertebral bodies, passes through the diaphragm, and terminates at the level of the fourth lumbar vertebra by dividing into two *common iliac arteries,* which carry blood to the lower extremities. The section of the descending aorta between the arch of aorta and the diaphragm is referred to as the *thoracic aorta.* The section between the diaphragm and the common iliac arteries is termed the *abdominal aorta.* Each section of the aorta gives off arteries that continue to branch into distributing arteries

leading to organs and finally into the arterioles and capillaries that pierce the tissues.

Blood is returned to the heart through the systemic veins. All the veins of the systemic circulation flow into either the *superior* or *inferior vena cava* or the *coronary sinus.* They in turn empty into the right atrium. The principal arteries and veins of systemic circulation are described and illustrated in Exhibits 14-1 to 14-12 and Figures 14-7 to 14-16.

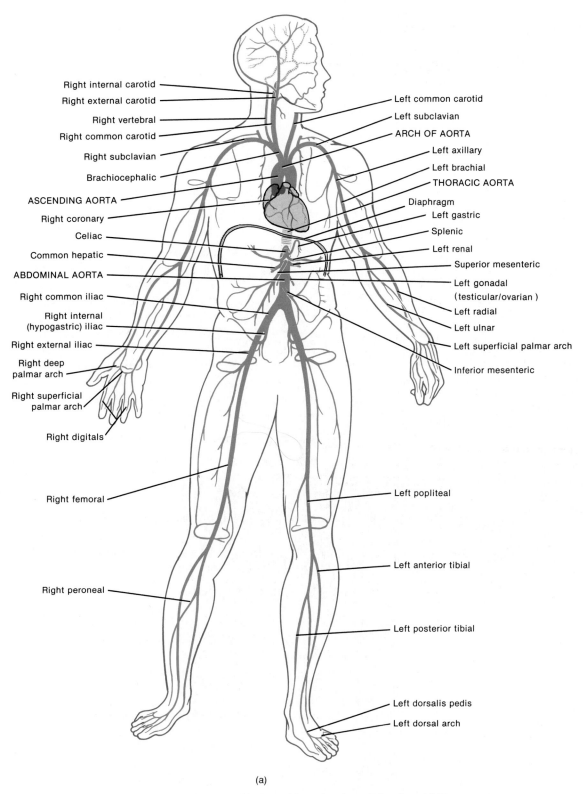

Right internal carotid
Right external carotid
Right vertebral
Right common carotid
Right subclavian
Brachiocephalic
ASCENDING AORTA
Right coronary
Celiac
Common hepatic
ABDOMINAL AORTA
Right common iliac
Right internal
(hypogastric) iliac
Right external iliac
Right deep
palmar arch
Right superficial
palmar arch
Right digitals
Right femoral
Right peroneal

Left common carotid
Left subclavian
ARCH OF AORTA
Left axillary
Left brachial
THORACIC AORTA
Diaphragm
Left gastric
Splenic
Left renal
Superior mesenteric
Left gonadal
(testicular/ovarian)
Left radial
Left ulnar
Left superficial palmar arch
Inferior mesenteric

Left popliteal

Left anterior tibial

Left posterior tibial

Left dorsalis pedis
Left dorsal arch

(a)

FIGURE 14-7 Aorta and its principal branches in anterior view. (a) Diagram.

EXHIBIT 14-1   AORTA AND ITS BRANCHES (Figure 14-7)

DIVISION OF AORTA	ARTERIAL BRANCH	REGION SUPPLIED
Ascending aorta	Right and left coronary	Heart.
Arch of aorta	Brachiocephalic ⟨ Right common carotid / Right subclavian	Right side of head and neck. / Right upper extremity.
	Left common carotid	Left side of head and neck.
	Left subclavian	Left upper extremity.
Thoracic aorta	Intercostals	Intercostal and chest muscles and pleurae.
	Superior phrenics	Posterior and superior surfaces of diaphragm.
	Bronchials	Bronchi of lungs.
	Esophageals	Esophagus.
Abdominal aorta	Inferior phrenics	Inferior surface of diaphragm.
	Celiac ⟨ Common hepatic	Liver.
	←Left gastric	Stomach and esophagus.
	Splenic	Spleen, pancreas, and stomach.
	Superior mesenteric	Small intestine, cecum, and ascending and transverse colons.
	Suprarenals	Adrenal (suprarenal) glands.
	Renals	Kidneys.
	Gonadals ⟨ Testicular	Testes.
	Ovarians	Ovaries.
	Inferior mesenteric	Transverse, descending, and sigmoid colons and rectum.
	Common iliacs ⟨ External iliacs	Lower extremities.
	Internal iliacs (hypogastrics)	Uterus, prostate, muscles of buttocks, and urinary bladder.

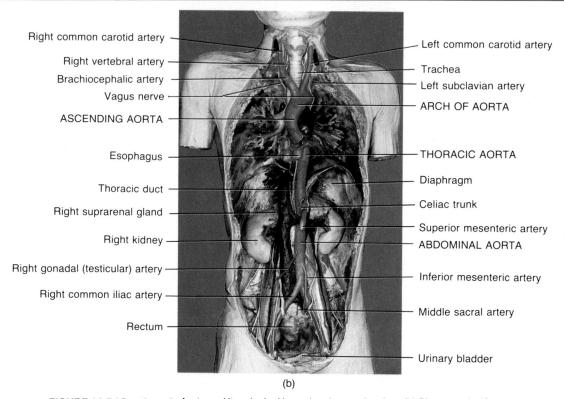

(b)

FIGURE 14-7 (*Continued*) Aorta and its principal branches in anterior view. (b) Photograph. (Courtesy of C. Yokochi and J. W. Rohen, *Photographic Anatomy of the Human Body*, 2nd ed., 1979, IGAKU-SHOIN, Ltd., Tokyo, New York.)

EXHIBIT 14-2 ASCENDING AORTA (Figure 14-8)

BRANCH	DESCRIPTION AND REGION SUPPLIED
Coronary arteries	Right and left branches arise from ascending aorta just superior to aortic semilunar valve. They form crown around heart, giving off branches to atrial and ventricular myocardium.

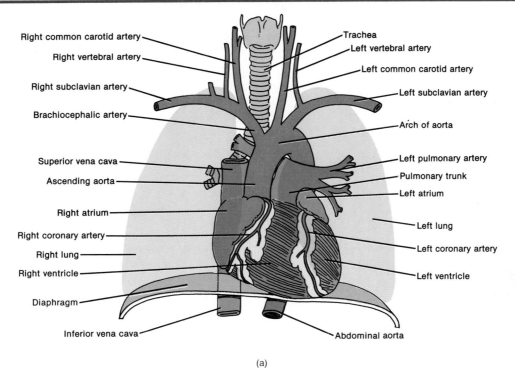

Right common carotid artery
Right vertebral artery
Right subclavian artery
Brachiocephalic artery
Superior vena cava
Ascending aorta
Right atrium
Right coronary artery
Right lung
Right ventricle
Diaphragm
Inferior vena cava

Trachea
Left vertebral artery
Left common carotid artery
Left subclavian artery
Arch of aorta
Left pulmonary artery
Pulmonary trunk
Left atrium
Left lung
Left coronary artery
Left ventricle
Abdominal aorta

(a)

FIGURE 14-8 Ascending aorta and its branches in anterior view.
(a) Diagram. (b) Scheme of distribution.

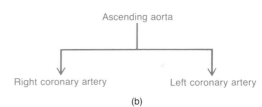

Ascending aorta

Right coronary artery   Left coronary artery

(b)

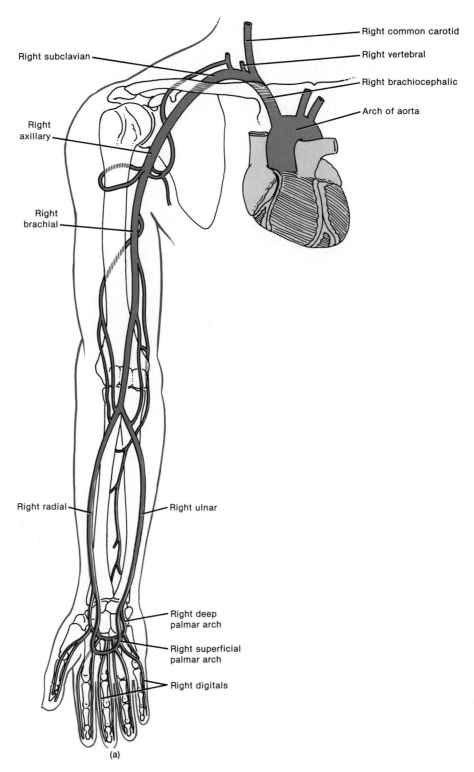

Right common carotid

Right subclavian

Right vertebral

Right brachiocephalic

Right axillary

Arch of aorta

Right brachial

Right radial

Right ulnar

Right deep palmar arch

Right superficial palmar arch

Right digitals

(a)

FIGURE 14-9 Arch of the aorta and its branches. (a) Diagram of anterior view of the arteries of the right upper extremity.

EXHIBIT 14-3 ARCH OF AORTA (Figure 14-9)

BRANCH	DESCRIPTION AND REGION SUPPLIED
Brachiocephalic	**Brachiocephalic artery** is first branch off arch of aorta. It divides to form right subclavian artery and right common carotid artery. **Right subclavian artery** extends from brachiocephalic to first rib and then passes into armpit (axilla) and supplies arm, forearm, and hand. Continuation of right subclavian into axilla is called **axillary artery.*** From here, it continues into arm as **brachial artery.** At bend of elbow, brachial artery divides into medial **ulnar** and lateral **radial arteries.** These vessels pass down to palm, one on each side of forearm. In palm, branches of two arteries anastomose to form two palmar arches—**superficial palmar arch** and **deep palmar arch.** From these arches arise **digital arteries,** which supply fingers and thumb. Before passing into axilla, right subclavian gives off major branch to brain called **vertebral artery.** Right vertebral artery passes through foramina of transverse processes of cervical vertebrae and enters skull through foramen magnum to reach undersurface of brain. Here it unites with left vertebral artery to form **basilar artery.** **Right common carotid artery** passes upward in neck. At upper level of larynx, it divides into **right external** and **right internal carotid arteries.** External carotid supplies right side of thyroid gland, tongue, throat, face, ear, scalp, and dura mater. Internal carotid supplies brain, right eye, and right sides of forehead and nose. Anastomoses of left and right internal carotids along with basilar artery form arterial circle at base of brain called **cerebral arterial circle (circle of Willis).** From this anastomosis arise arteries supplying brain. Essentially, cerebral arterial circle is formed by union of **anterior cerebral arteries** (branches of internal carotids) and **posterior cerebral arteries** (branches of basilar artery). Posterior cerebral arteries are connected with internal carotids by **posterior communicating arteries.** Anterior cerebral arteries are connected by **anterior communicating arteries.** Cerebral arterial circle equalizes blood pressure to brain and provides alternate routes for blood to brain should arteries become damaged.
Left common carotid	**Left common carotid** is second branch off arch of aorta (see Figure 14-8). Corresponding to right common carotid, it divides into basically same branches with same names, except that arteries are now labeled "left" instead of "right."
Left subclavian	**Left subclavian artery** is third branch off arch of aorta (see Figure 14-8). It distributes blood to left vertebral artery and vessels of left upper extremity. Arteries branching from left subclavian are named like those of right subclavian.

* The right subclavian artery is a good example of the practice of giving the same vessel different names as it passes through different regions.

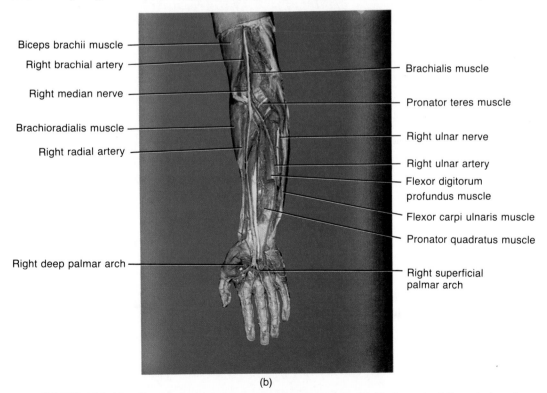

(b)

FIGURE 14-9 (*Continued*) Arch of the aorta and its branches. (b) Photograph of the arteries of the right upper extremity. (Courtesy of C. Yokochi and J. W. Rohen, *Photographic Anatomy of the Human Body,* 2nd ed., 1979, IGAKU-SHOIN, Ltd., Tokyo, New York.)

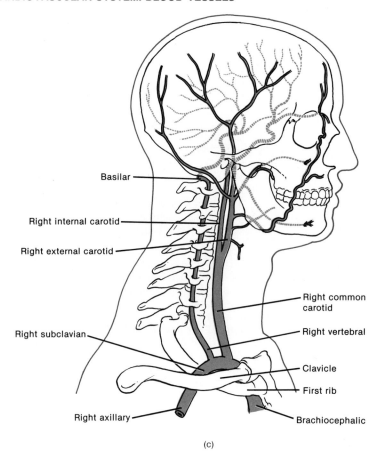

(c)

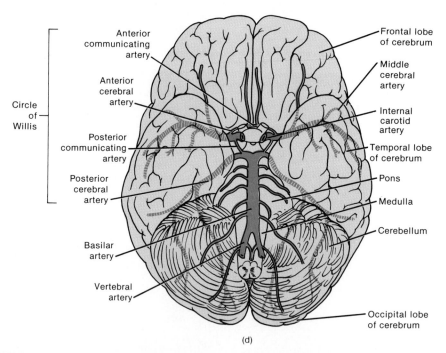

(d)

FIGURE 14-9 (*Continued*) Arch of the aorta and its branches. (c) Right lateral view of the arteries of the neck and head. (d) Arteries of the base of the brain. Note the arteries that comprise the cerebral arterial circle.

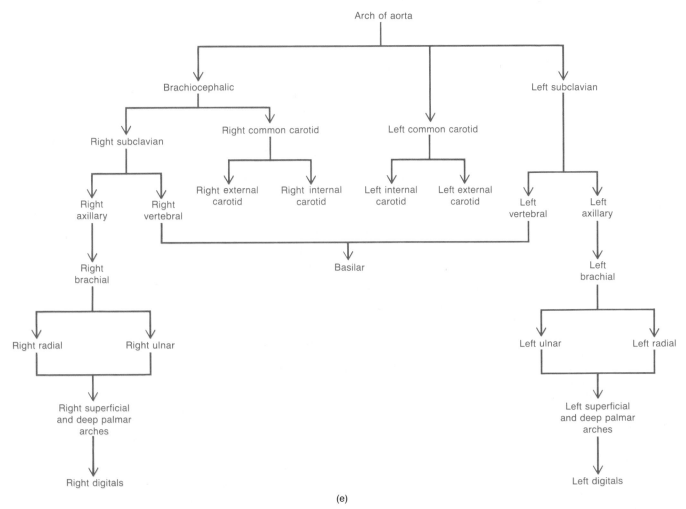

(e)

FIGURE 14-9 (*Continued*) Arch of the aorta and its branches. (e) Scheme of distribution.

EXHIBIT 14-4   THORACIC AORTA (Figure 14-10)

BRANCH	DESCRIPTION AND REGION SUPPLIED
	Thoracic aorta runs from fourth to twelfth thoracic vertebra. Along its course, it sends off numerous small arteries to viscera and skeletal muscles of the chest. Branches of an artery that supply viscera are called **visceral branches.** Those that supply body wall structures are **parietal branches.**
VISCERAL	
Pericardial	Several minute **pericardial arteries** supply blood to dorsal aspect of pericardium.
Bronchial	One **right** and two **left bronchial arteries** supply the bronchial tubes, areolar tissue of the lungs, bronchial lymph nodes, and esophagus.
Esophageal	Four or five **esophageal arteries** supply the esophagus.
Mediastinal	Numerous small **mediastinal arteries** supply blood to structures in the posterior mediastinum.
PARIETAL	
Posterior intercostal	Nine pairs of **posterior intercostal arteries** supply the intercostal, pectoral, and abdominal muscles; overlying subcutaneous tissue and skin; mammary glands; and vertebral canal and its contents.
Subcostal	The **left** and **right subcostal arteries** have a distribution similar to that of the posterior intercostals.
Superior phrenic	Small **superior phrenic arteries** supply the posterior surface of the diaphragm.

EXHIBIT 14-5   ABDOMINAL AORTA (Figure 14-10)

BRANCH	DESCRIPTION AND REGION SUPPLIED
VISCERAL	
Celiac	**Celiac artery (trunk)** is first visceral aortic branch below diaphragm. It has three branches: (1) **common hepatic artery,** which supplies tissues of liver, (2) **left gastric artery,** which supplies stomach, and (3) **splenic artery,** which supplies spleen, pancreas, and stomach.
Superior mesenteric	**Superior mesenteric artery** distributes blood to small intestine and part of large intestine.
Suprarenals	Right and left **suprarenal arteries** supply blood to adrenal (suprarenal) glands.
Renals	Right and left **renal arteries** carry blood to kidneys.
Gonadals (testiculars or ovarians)	Right and left **testicular arteries** extend into scrotum and terminate in testes; right and left **ovarian arteries** are distributed to ovaries.
Inferior mesenteric	**Inferior mesenteric artery** supplies major part of large intestine and rectum.
PARIETAL	
Inferior phrenics	**Inferior phrenic arteries** are distributed to undersurface of diaphragm.
Lumbars	**Lumbar arteries** supply spinal cord and its meninges and muscles and skin of lumbar region of back.
Middle sacral	**Middle sacral artery** supplies sacrum, coccyx, gluteus maximus muscles, and rectum.

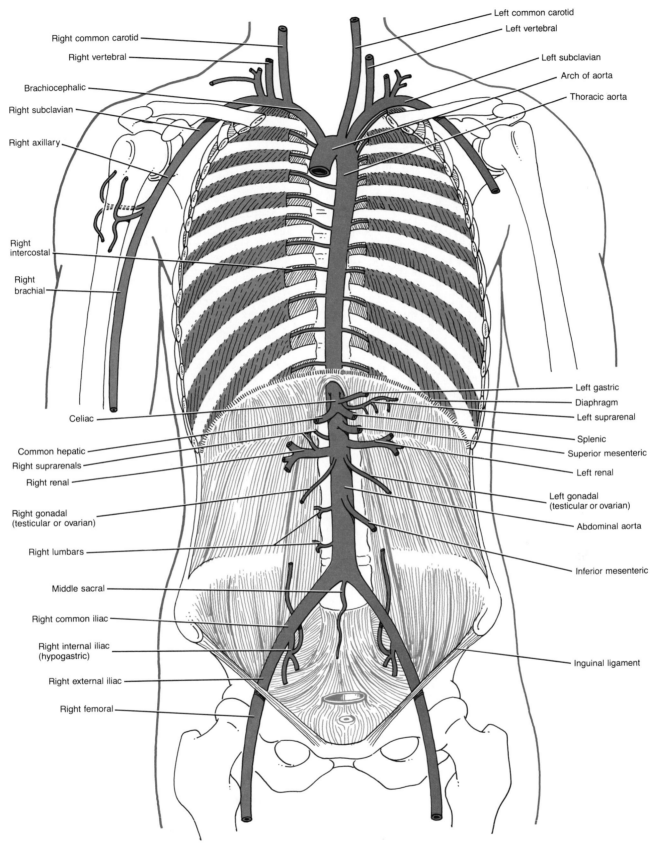

FIGURE 14-10 Thoracic and abdominal aorta and their principal branches in anterior view.

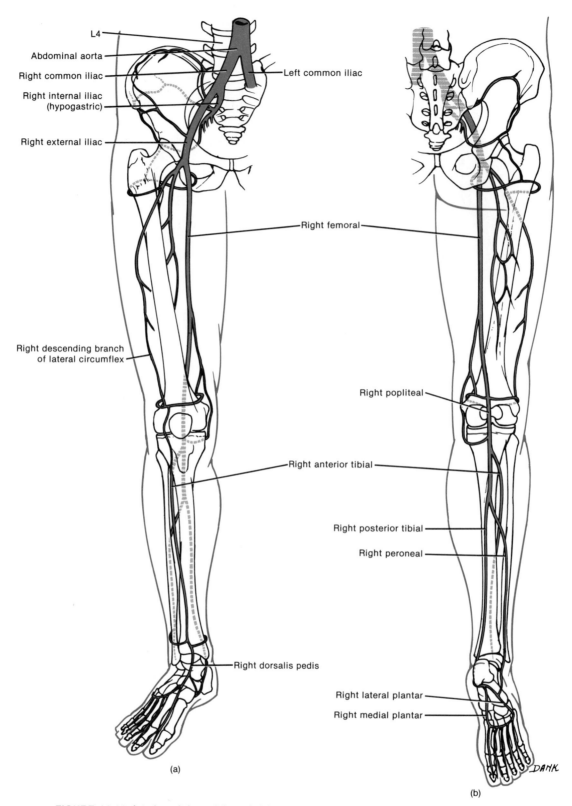

L4

Abdominal aorta

Right common iliac

Right internal iliac
(hypogastric)

Right external iliac

Left common iliac

Right femoral

Right descending branch
of lateral circumflex

Right popliteal

Right anterior tibial

Right posterior tibial

Right peroneal

Right dorsalis pedis

Right lateral plantar

Right medial plantar

(a)

(b)

DANK

FIGURE 14-11 Arteries of the pelvis and right lower extremity. (a) Anterior view. (b) Posterior view.

EXHIBIT 14-6   ARTERIES OF PELVIS AND LOWER EXTREMITIES (Figure 14-11)

BRANCH	DESCRIPTION AND REGION SUPPLIED
Common iliacs	At about level of fourth lumbar vertebra, abdominal aorta divides into right and left **common iliac arteries.** Each passes downward about 5 cm (2 in) and gives rise to two branches: internal iliac and external iliac.
Internal iliacs	**Internal iliac (hypogastric) arteries** form branches that supply psoas major, quadratus lumborum, and medial side of each thigh, urinary bladder, rectum, prostate gland, ductus deferens, uterus, and vagina.
External iliacs	**External iliac arteries** diverge through pelvis, enter thighs, and here become right and left **femoral arteries.** Both femorals send branches back up to genitals and wall of abdomen. Other branches run to muscles of thigh. Femoral continues down medial and posterior side of thigh at back of knee joint, where it becomes **popliteal artery.** Between knee and ankle, popliteal runs down back of leg and is called **posterior tibial artery.** Below knee, **peroneal artery** branches off posterior tibial to supply structures on medial side of fibula and calcaneus. In calf, **anterior tibial artery** branches off popliteal and runs along front of leg. At ankle, it becomes **dorsalis pedis artery.** At ankle, posterior tibial divides into **medial** and **lateral plantar arteries.** These arteries anastomose with dorsalis pedis and supply blood to foot.

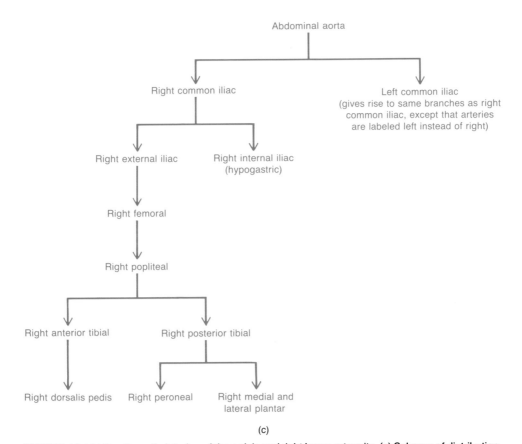

(c)

FIGURE 14-11 (*Continued*) Arteries of the pelvis and right lower extremity. (c) Scheme of distribution.

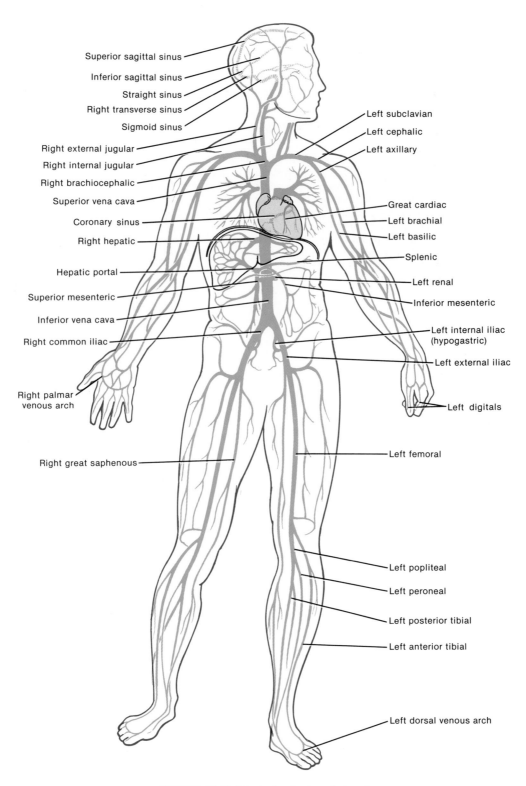

Superior sagittal sinus
Inferior sagittal sinus
Straight sinus
Right transverse sinus
Sigmoid sinus
Right external jugular
Right internal jugular
Right brachiocephalic
Superior vena cava
Coronary sinus
Right hepatic
Hepatic portal
Superior mesenteric
Inferior vena cava
Right common iliac
Right palmar venous arch
Right great saphenous

Left subclavian
Left cephalic
Left axillary
Great cardiac
Left brachial
Left basilic
Splenic
Left renal
Inferior mesenteric
Left internal iliac (hypogastric)
Left external iliac
Left digitals
Left femoral
Left popliteal
Left peroneal
Left posterior tibial
Left anterior tibial
Left dorsal venous arch

FIGURE 14-12 Principal veins in anterior view.

EXHIBIT 14-7   VEINS OF SYSTEMIC CIRCULATION (Figure 14-12)

VEIN	DESCRIPTION AND REGION DRAINED
	All systemic and coronary (cardiac) veins return blood to the right atrium of the heart through one of three large vessels. Return flow in coronary circulation is taken up by **cardiac veins,** which empty into the large vein of the heart, the **coronary sinus.** From here, the blood empties into the right atrium of the heart (see Figure 13-8). Return flow in systemic circulation empties into the superior vena cava or inferior vena cava.
Superior vena cava	Veins that empty into the **superior vena cava** are veins of the head and neck, upper extremities, and some from the thorax.
Inferior vena cava	Veins that empty into the **inferior vena cava** are some from the thorax and veins of the abdomen, pelvis, and lower extremities.

CLINICAL APPLICATION

The inferior vena cava is commonly compressed during the later stages of pregnancy due to the enlargement of the uterus. This produces edema of the ankles and feet and temporary varicose veins.

EXHIBIT 14-8   VEINS OF HEAD AND NECK (Figure 14-13)

VEIN	DESCRIPTION AND REGION DRAINED
Internal jugulars	Right and left **internal jugular veins** receive blood from face and neck. They arise as continuation of **sigmoid sinuses** at base of skull. Intracranial vascular sinuses are located between layers of dura mater and receive blood from brain. Other sinuses that drain into internal jugular include **superior sagittal sinus, inferior sagittal sinus, straight sinus,** and **transverse (lateral) sinuses.** Internal jugulars descend on either side of neck and pass behind clavicles, where they join with right and left subclavian veins. Unions of internal jugulars and subclavians form right and left brachiocephalic veins. From here blood flows into superior vena cava.
External jugulars	Right and left **external jugular veins** run down neck along outside of internal jugulars. They drain blood from parotid (salivary) glands, facial muscles, scalp, and other superficial structures into subclavian veins.

CLINICAL APPLICATION

In cases of heart failure, the venous pressure in the right atrium may rise. In such instances, the pressure in the column of blood in the external jugular vein rises so that, even with the patient at rest and sitting in a chair, the external jugular vein will be visibly distended. Temporary distention of the vein is often seen in healthy adults when the intrathoracic pressure is raised in singing, coughing, and physical exertion.

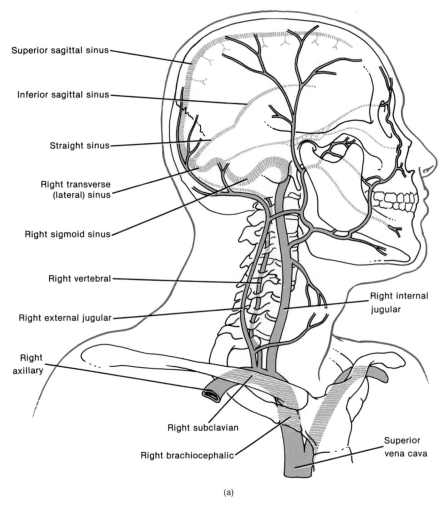

Superior sagittal sinus

Inferior sagittal sinus

Straight sinus

Right transverse (lateral) sinus

Right sigmoid sinus

Right vertebral

Right external jugular

Right axillary

Right internal jugular

Right subclavian

Right brachiocephalic

Superior vena cava

(a)

FIGURE 14-13 Veins of the head and neck. (a) Right lateral view. (b) Scheme of drainage.

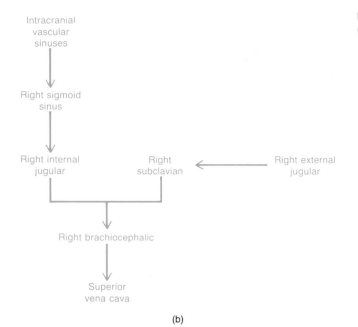

Intracranial vascular sinuses

Right sigmoid sinus

Right internal jugular

Right subclavian

Right external jugular

Right brachiocephalic

Superior vena cava

(b)

EXHIBIT 14-9 VEINS OF UPPER EXTREMITIES (Figure 14-14)

VEIN	DESCRIPTION AND REGION DRAINED
	Blood from each upper extremity is returned to the heart by deep and superficial veins. Both sets of veins contain valves. **Deep veins** are located deep in the body. They usually accompany arteries, and many have the same names as corresponding arteries. **Superficial veins** are located just below the skin and are often visible. They anastomose extensively with each other and deep veins.
SUPERFICIAL **Cephalics**	**Cephalic vein** of each upper extremity begins in the medial part of **dorsal venous arch** and winds upward around radial border of forearm. In front of elbow, it is connected to basilic vein by the **median cubital vein.** Just below elbow, cephalic vein unites with **accessory cephalic vein** to form cephalic vein of upper extremity. Ultimately, cephalic vein empties into axillary vein.
Basilics	**Basilic vein** of each upper extremity originates in the ulnar part of **dorsal venous arch.** It extends along posterior surface of ulna to point below elbow where it receives **median cubital vein.** If a vein must be punctured for an injection, transfusion, or removal of a blood sample, median cubitals are preferred. The median cubital vein joins the basilic vein to form the axillary vein.
Median antebrachials	**Median antebrachial veins** drain **palmar venous arch,** ascend on ulnar side of anterior forearm, and end in median cubital veins.
DEEP **Radials**	**Radial veins** receive **dorsal metacarpal veins.**
Ulnars	**Ulnar veins** receive tributaries from **palmar venous arch.** Radial and ulnar veins unite in bend of elbow to form brachial vein.
Brachials	Located on either side of brachial artery, **brachial veins** join into axillary veins.
Axillaries	**Axillary veins** are a continuation of brachials and basilics. Axillaries end at first rib, where they become subclavians.
Subclavians	Right and left **subclavian veins** unite with internal jugulars to form brachiocephalic veins. Thoracic duct of lymphatic system delivers lymph into left subclavian vein at junction with internal jugular. Right lymphatic duct delivers lymph into right subclavian vein at corresponding junction.

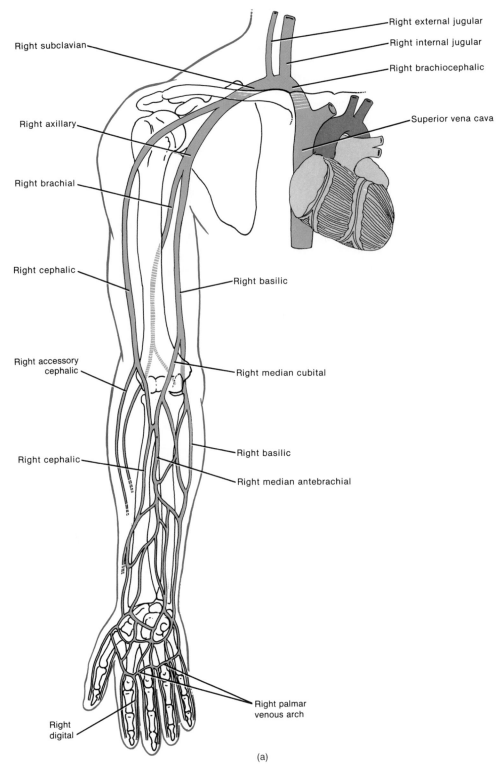

FIGURE 14-14 Veins of the right upper extremity in anterior view. (a) Diagram.

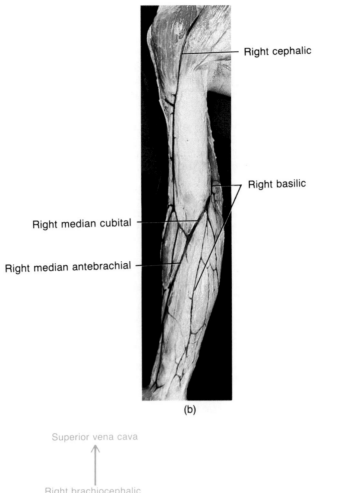

(b)

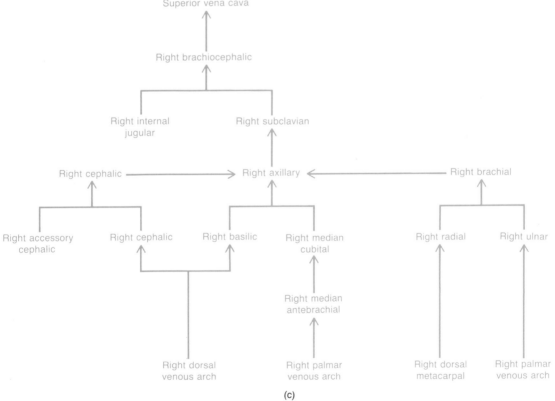

(c)

FIGURE 14-14 (*Continued*) Veins of the right upper extremity in anterior view. (b) Photograph. (Courtesy of C. Yokochi and J. W. Rohen, *Photographic Anatomy of the Human Body*, 2nd ed., 1979, IGAKU-SHOIN, Ltd., Tokyo, New York.) (c) Scheme of drainage.

EXHIBIT 14-10   VEINS OF THORAX (Figure 14-15)

VEIN	DESCRIPTION AND REGION DRAINED
Brachiocephalic	Right and left **brachiocephalic veins,** formed by union of subclavians and internal jugulars, drain blood from head, neck, upper extremities, mammary glands, and upper thorax. Brachiocephalics unite to form superior vena cava.
Azygos veins	**Azygos veins,** besides collecting blood from thorax, may serve as bypass for inferior vena cava that drains blood from lower body. Several small veins directly link azygos veins with inferior vena cava. And large veins that drain lower extremities and abdomen dump blood into azygos. If inferior vena cava or hepatic portal vein becomes obstructed, azygos veins can return blood from lower body to superior vena cava.
Azygos	**Azygos vein** lies in front of vertebral column, slightly right of midline. It begins as continuation of right ascending lumbar vein. It connects with inferior vena cava, right common iliac, and lumbar veins. Azygos receives blood from **right intercostal veins** that drain chest muscles; from hemiazygos and accessory hemiazygos veins; from several **esophageal, mediastinal,** and **pericardial veins;** and from right **bronchial vein.** Vein ascends to fourth thoracic vertebra, arches over right lung, and empties into superior vena cava.
Hemiazygos	**Hemiazygos vein** is in front of vertebral column and slightly left of midline. It begins as continuation of left ascending lumbar vein. It receives blood from lower four or five **intercostal veins** and some **esophageal** and **mediastinal veins.** At level of ninth thoracic vertebra, it joins azygos vein.
Accessory hemiazygos	**Accessory hemiazygos vein** is also in front and to left of vertebral column. It receives blood from three or four **intercostal veins** and left **bronchial vein.** It joins azygos at level of eighth thoracic vertebra.

EXHIBIT 14-11   VEINS OF ABDOMEN AND PELVIS (Figure 14-15)

VEIN	DESCRIPTION AND REGION DRAINED
Inferior vena cava	**Inferior vena cava** is the largest vein of the body. It is formed by union of two common iliac veins that drain lower extremities and abdomen. Inferior vena cava extends upward through abdomen and thorax to right atrium. Numerous small veins enter the inferior vena cava. Most carry return flow from branches of abdominal aorta, and names correspond to names of arteries.
Common iliacs	**Common iliac veins** are formed by union of internal (hypogastric) and external iliac veins and represent distal continuation of inferior vena cava at its bifurcation.
Internal iliacs	Tributaries of **internal iliac (hypogastric) veins** basically correspond to branches of internal iliac arteries. Internal iliacs drain gluteal muscles, medial side of thigh, urinary bladder, rectum, prostate gland, ductus deferens, uterus, and vagina.
External iliacs	**External iliac veins** are continuation of femoral veins and receive blood from lower extremities and inferior part of anterior abdominal wall.
Renals	**Renal veins** drain kidneys.
Gonadals (testiculars or ovarians)	**Testicular veins** drain testes (left testicular vein empties into left renal vein); **ovarian veins** drain ovaries (left ovarian vein empties into left renal vein).
Suprarenals	**Suprarenal veins** drain adrenal (suprarenal) glands (left suprarenal vein empties into left renal vein).
Inferior phrenics	**Inferior phrenic veins** drain diaphragm (left inferior phrenic vein sends tributary to left renal vein).
Hepatics	**Hepatic veins** drain liver.
Lumbars	A series of parallel **lumbar veins** drain blood from both sides of posterior abdominal wall. Lumbars connect at right angles with right and left **ascending lumbar veins,** which form origin of corresponding azygos or hemiazygos vein. Lumbars drain blood into ascending lumbars and then run to inferior vena cava, where they release remainder of flow.

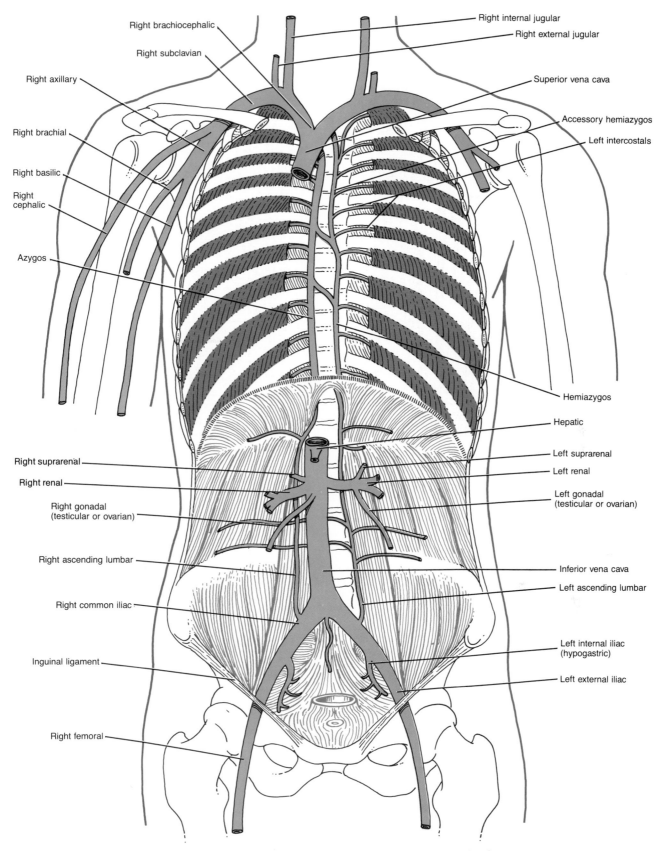

FIGURE 14-15 Veins of the thorax, abdomen, and pelvis in anterior view.

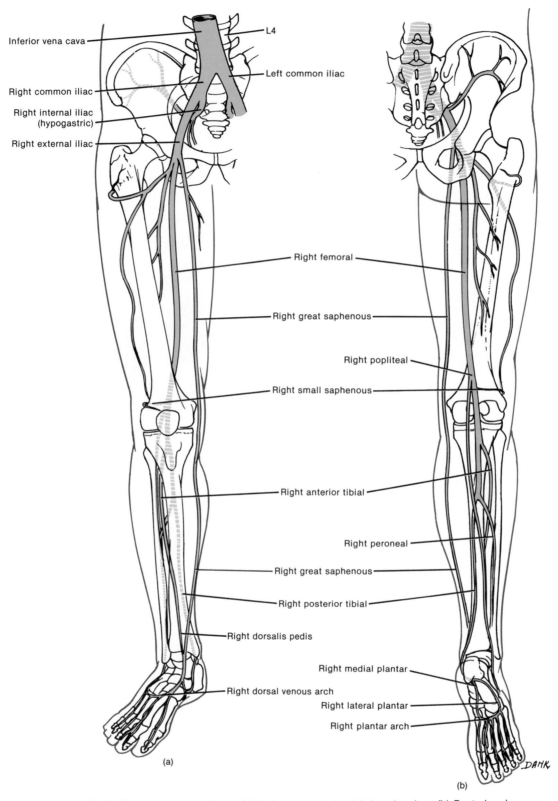

Inferior vena cava

L4

Left common iliac

Right common iliac

Right internal iliac
(hypogastric)

Right external iliac

Right femoral

Right great saphenous

Right popliteal

Right small saphenous

Right anterior tibial

Right peroneal

Right great saphenous

Right posterior tibial

Right dorsalis pedis

Right medial plantar

Right dorsal venous arch

Right lateral plantar

Right plantar arch

(a)

(b)

DANK

FIGURE 14-16 Veins of the pelvis and right lower extremity. (a) Anterior view. (b) Posterior view.

EXHIBIT 14-12 VEINS OF LOWER EXTREMITIES (Figure 14-16)

VEIN	DESCRIPTION AND REGION DRAINED
	Blood from each lower extremity is returned by superficial set and deep set of veins.
SUPERFICIAL VEINS **Great saphenous**	**Great saphenous vein,** longest vein in body, begins at medial end of **dorsal venous arch** of foot. It passes in front of medial malleolus and then upward along medial aspect of leg and thigh. It receives tributaries from superficial tissues and connects with deep veins as well. It empties into femoral vein in groin.

CLINICAL APPLICATION

The great saphenous vein is very constant in its position anterior to the medial malleolus. It is frequently used for prolonged administration of intravenous fluids, which is particularly important in very young babies and in patients of any age who are in shock and whose veins are collapsed. It and the small saphenous vein are subject to varicosity.

VEIN	DESCRIPTION AND REGION DRAINED
Small saphenous	**Small saphenous vein** begins at lateral end of dorsal venous arch of foot. It passes behind lateral malleolus and ascends under skin of back of leg. It receives blood from foot and posterior portion of leg. It empties into popliteal vein behind knee.
DEEP VEINS **Posterior tibial**	**Posterior tibial vein** is formed by union of **medial** and **lateral plantar veins** behind medial malleolus. It ascends deep in muscle at back of leg, receives blood from **peroneal vein,** and unites with anterior tibial vein just below knee.
Anterior tibial	**Anterior tibial vein** is upward continuation of **dorsalis pedis veins** in foot. It runs between tibia and fibula and unites with posterior tibial to form popliteal vein.
Popliteal	**Popliteal vein,** just behind knee, receives blood from anterior and posterior tibials and small saphenous vein.
Femoral	**Femoral vein** is upward continuation of popliteal just above knee. Femorals run up posterior of thighs and drain deep structures of thighs. After receiving great saphenous veins in groin, they continue as right and left external iliac veins.

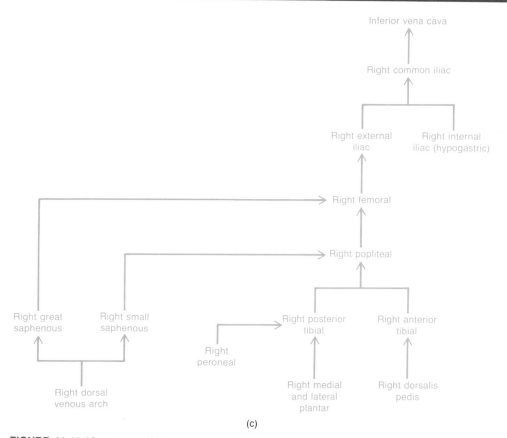

(c)

FIGURE 14-16 (*Continued*) Veins of the pelvis and right lower extremity. (c) Scheme of drainage.

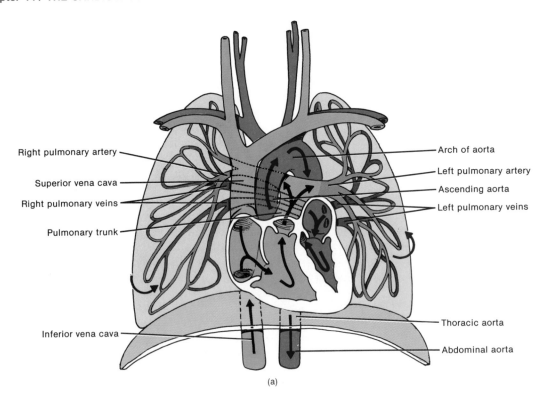

Right pulmonary artery

Superior vena cava

Right pulmonary veins

Pulmonary trunk

Inferior vena cava

Arch of aorta

Left pulmonary artery

Ascending aorta

Left pulmonary veins

Thoracic aorta

Abdominal aorta

(a)

PULMONARY CIRCULATION

The flow of deoxygenated blood from the right ventricle to the lungs and the return of oxygenated blood from the lungs to the left atrium is called **pulmonary circulation** (Figure 14-17). The *pulmonary trunk* emerges from the right ventricle and passes upward, backward, and to the left. It then divides into two branches. The *right pulmonary artery* runs to the right lung; the *left pulmonary artery* goes to the left lung. On entering the lungs, the branches divide and subdivide until ultimately they form capillaries around the alveoli in the lungs. Carbon dioxide is passed from the blood into the alveoli to be breathed out of the lungs. Oxygen breathed in by the lungs is passed from the alveoli into the blood. The capillaries unite, venules and veins are formed, and eventually, two *pulmonary veins* exit from each lung and transport the oxygenated blood to the left atrium. The pulmonary veins are the only postnatal veins that carry oxygenated blood. Contractions of the left ventricle then send the blood into the systemic circulation.

CLINICAL APPLICATION

Blood flow can be measured to diagnose certain circulatory disorders. **Circulation time** is the time required for blood to pass from the right atrium, through pulmonary circulation, back to the left ventricle, through systemic circulation down to the foot, and back again to the right atrium. Such a trip usually takes about 1 minute.

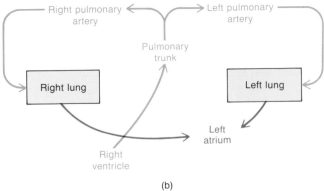

Right pulmonary artery

Left pulmonary artery

Pulmonary trunk

Right lung

Left lung

Left atrium

Right ventricle

(b)

FIGURE 14-17 Pulmonary circulation. (a) Diagram (b) Scheme of circulation.

HEPATIC PORTAL CIRCULATION

Blood enters the liver from two sources. The hepatic artery delivers oxygenated blood from the systemic circulation; the hepatic portal vein delivers deoxygenated blood from the digestive organs. The term **hepatic portal circulation** refers to this flow of venous blood from the digestive organs to the liver before returning to the heart (Figure 14-18). Hepatic portal blood is rich with substances absorbed from the digestive tract. The liver monitors these substances before they pass into the general circulation. For example, the liver stores nutrients such as glucose. It modifies other digested substances so they may be used by cells. It detoxifies harmful substances that have been absorbed by the digestive tract and destroys bacteria by phagocytosis.

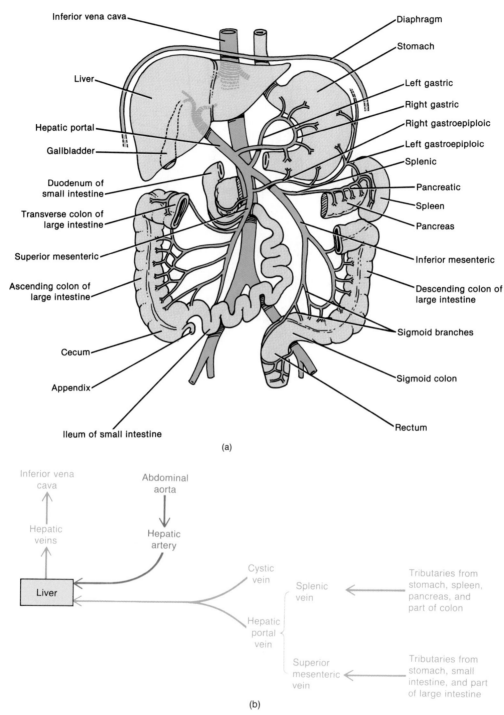

FIGURE 14-18 Hepatic portal circulation. (a) Diagram. (b) Scheme of blood flow through the liver, including arterial circulation. Deoxygenated blood is indicated in blue; oxygenated blood in red.

The hepatic portal system includes veins that drain blood from the pancreas, spleen, stomach, intestines, and gallbladder and transport it to the portal vein of the liver. The *hepatic portal vein* is formed by the union of the superior mesenteric and splenic veins. The *superior mesenteric vein* drains blood from the small intestine and portions of the large intestine and stomach. The *splenic vein* drains the spleen and receives tributaries from the stomach, pancreas, and portions of the colon. The tributaries from the stomach are the *gastric, pyloric,* and *gastroepiploic veins.* The *pancreatic veins* come from the pancreas, and the *inferior mesenteric veins* come from the portions of the colon. Before the hepatic portal vein enters the liver, it receives the *cystic vein* from the gallbladder and other veins. Ultimately, blood leaves the liver through the *hepatic veins,* which enter the inferior vena cava.

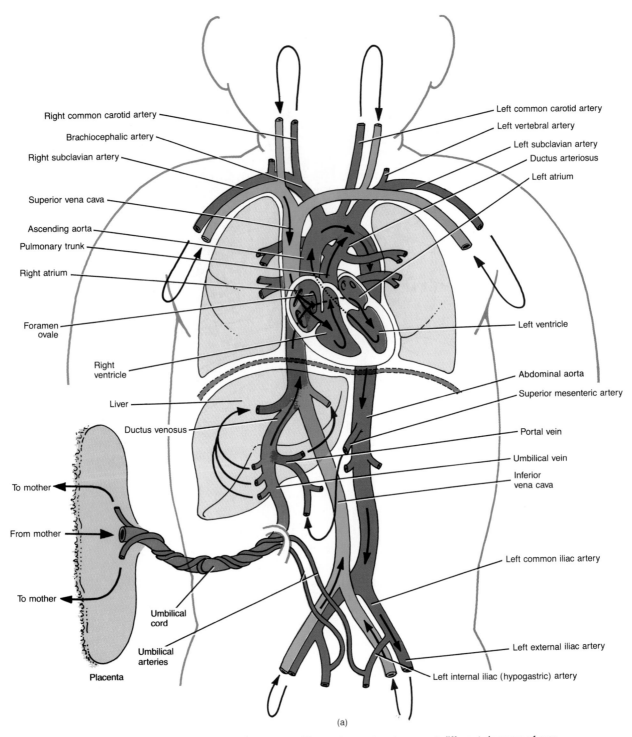

Right common carotid artery

Brachiocephalic artery

Right subclavian artery

Superior vena cava

Ascending aorta

Pulmonary trunk

Right atrium

Foramen ovale

Right ventricle

Liver

Ductus venosus

To mother

From mother

To mother

Umbilical cord

Umbilical arteries

Placenta

Left common carotid artery

Left vertebral artery

Left subclavian artery

Ductus arteriosus

Left atrium

Left ventricle

Abdominal aorta

Superior mesenteric artery

Portal vein

Umbilical vein

Inferior vena cava

Left common iliac artery

Left external iliac artery

Left internal iliac (hypogastric) artery

(a)

FIGURE 14-19 Fetal circulation. (a) Diagram. The various colors represent different degrees of oxygenation of blood ranging from greatest (red) to intermediate (two shades of purple) to least (blue).

FETAL CIRCULATION

The circulatory system of a fetus, called **fetal circulation,** differs from an adult's because the lungs, kidneys, and digestive tract of a fetus are nonfunctional. The fetus derives its oxygen and nutrients from the maternal blood and eliminates its carbon dioxide and wastes into the maternal blood. (Figure 14-19).

The exchange of materials between fetal and maternal circulation occurs through a structure called the *placenta* (pla-SEN-ta). It is attached to the umbilicus of the fetus by the umbilical (um-BIL-i-kal) cord, and it communi-

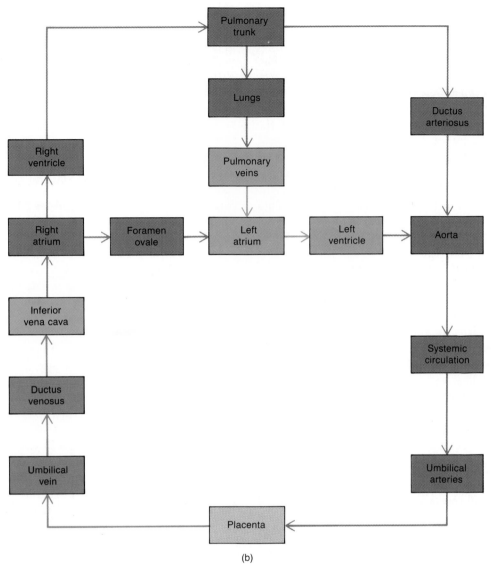

FIGURE 14-19 (Continued) Fetal circulation. (b) Scheme of circulation.

cates with the mother through countless small blood vessels that emerge from the uterine wall. The umbilical cord contains blood vessels that branch into capillaries in the placenta. Wastes from the fetal blood diffuse out of the capillaries, into spaces containing maternal blood (intervillous spaces) in the placenta, and finally into the mother's uterine blood vessels. Nutrients travel the opposite route—from the maternal blood vessels to the intervillous spaces to the fetal capillaries. Normally there is no mixing of maternal and fetal blood since all exchanges occur through capillaries.

Blood passes from the fetus to the placenta via two *umbilical arteries.* These branches of the internal iliac (hypogastric) arteries are included in the umbilical cord. At the placenta, the blood picks up oxygen and nutrients and eliminates carbon dioxide and wastes. The oxygenated blood returns from the placenta to the fetus via a single

umbilical vein. This vein ascends to the liver of the fetus, where it divides into two branches. Some blood flows through the branch that joins the hepatic portal vein and enters the liver. Although the fetal liver manufactures red blood cells, it does not function in digestion. Therefore, most of the blood flows into the second branch, the *ductus venosus* (DUK-tus ve-NŌ-sus). The ductus venosus eventually passes its blood to the inferior vena cava, bypassing the liver.

In general, circulation through other portions of the fetus is not unlike postnatal circulation. Deoxygenated blood returning from the lower regions is mingled with oxygenated blood from the ductus venosus in the inferior vena cava. This mixed blood then enters the right atrium. The circulation of blood through the upper portion of the fetus is also similar to postnatal flow. Deoxygenated blood returning from the upper regions of the fetus is collected by the superior vena cava, and it also passes into the right atrium.

Most of the blood does not pass through the right ventricle to the lungs, as it does in postnatal circulation, since the fetal lungs do not operate. In the fetus, an opening called the *foramen ovale* (fō-RĀ-men ō-VAL-ē) exists in the septum between the right and left atria. A valve in the inferior vena cava directs about a third of the blood through the foramen ovale so that it may be sent directly into the systemic circulation. The blood that does descend into the right ventricle is pumped into the pulmonary trunk, but little of this blood actually reaches the lungs. Most blood in the pulmonary trunk is sent through the *ductus arteriosus* (ar-tē-rē-Ō-sus). This small vessel connecting the pulmonary trunk with the aorta enables blood in excess of nutrient requirements to bypass the fetal lungs. The blood in the aorta is carried to all parts of the fetus through its systemic branches. When the common iliac arteries branch into the external and internal iliacs (hypogastric), part of the blood flows into the internal iliacs (hypogastrics). It then goes to the umbilical arteries and back to the placenta for another exchange of materials. The only vessel that carries fully oxygenated blood is the umbilical vein.

At birth, when lung, renal, digestive, and liver functions are established, the special structures of fetal circulation are no longer needed and the following changes occur.

1. The umbilical arteries atrophy to become the *lateral umbilical ligaments.*
2. The umbilical vein atrophies and becomes the *round ligament* of the liver.
3. The placenta is delivered by the mother as the *"afterbirth."*
4. The ductus venosus becomes the *ligamentum venosum,* a fibrous cord in the liver.
5. The foramen ovale normally closes shortly after birth to become the *fossa ovalis,* a depression in the interatrial septum.
6. The ductus arteriosus closes, atrophies, and becomes the *ligamentum arteriosum.*

Anatomical defects resulting from failure of these changes to occur are described in Chapter 13.

AGING AND THE CARDIOVASCULAR SYSTEM

General changes associated with aging and the cardiovascular system include loss of extensibility of the aorta, reduction in cardiac muscle cell size, progressive loss of cardiac muscular strength, and a reduced output of blood by the heart. There is also an increase in blood pressure. Coronary artery disease (CAD) increases and represents the major cause of heart disease and death in older Americans. Congestive heart failure occurs and is viewed as a set of symptoms associated with the impaired pumping performance of the heart. Changes in blood vessels such as hardening of the arteries and cholesterol deposits in arteries that serve brain tissue can reduce its nourishment and result in the malfunction or death of brain cells.

DEVELOPMENTAL ANATOMY OF BLOOD AND BLOOD VESSELS

Since the human egg and yolk sac have little yolk to nourish the developing embryo, blood and blood vessel formation starts as early as 15 to 16 days. The development begins in the **mesoderm** of the yolk sac, chorion, and body stalk.

Blood vessels develop from isolated masses and cords of mesenchyme in the mesoderm called **blood islands** (Figure 14-20). Spaces soon appear in the islands and become the lumens of the blood vessels. Some of the mesenchymal cells immediately around the spaces give rise to the *endothelial lining of the blood vessels.* Mesenchyme around the endothelium forms the *tunics* (intima, media, externa) of the larger blood vessels. Growth and fusion of blood islands form an extensive network of blood vessels throughout the embryo.

Blood plasma and *blood cells* are produced by the endothelial cells and appear in the blood vessels of the yolk sac and allantois quite early. Blood formation in the embryo itself begins at about the second month in the liver, spleen, bone marrow, and lymph nodes.

APPLICATIONS TO HEALTH

ANEURYSM

An **aneurysm** (AN-yoo-rizm) is a thin, weakened section of the wall of an artery or a vein that bulges outward forming a balloonlike sac of the blood vessel. Some aneurysms also result in permanent dilation of the blood vessel. Common causes of aneurysms include atherosclerosis, syphilis, congenital blood vessel defects, and trauma. If an aneurysm goes untreated, it grows larger and larger until the blood vessel wall becomes so thin that it may burst, causing massive hemorrhage with shock, severe pain, stroke, or death depending on which vessel is involved. Even an unruptured aneurysm can lead to damage by interrupting blood flow or putting pressure on adjacent blood vessels, organs, or bones.

Aneurysms commonly occur in blood vessels of the cerebral arterial circle. Here, they can cause strokes and mental impairment. The aorta has a higher incidence of

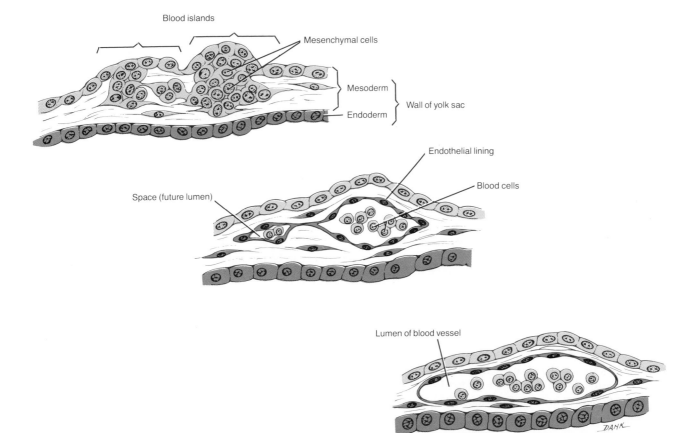

FIGURE 14-20 Development of blood vessels and blood cells from blood islands.

aneurysms than any other artery, mostly inferior to the renal arteries, probably because of its curved shape, large size, and high pressure. In the thorax, aneurysms usually occur in the ascending or thoracic aorta but seldom in the arch of the aorta. Symptoms of thoracic aortic aneurysm depend on pressure exerted on adjacent structures. For example, pressure on the inferior (recurrent) laryngeal nerve causes hoarseness and a brassy cough. Pressure on the esophagus may cause dysphagia (difficulty in swallowing). Dyspnea (difficulty in breathing) may follow pressure on the trachea, the root of the lung, or the phrenic nerve.

Surgical repair of an aneurysm consists of temporarily clamping the damaged artery above and below the aneurysm and then opening the aneurysm. A graft, usually of Dacron, is then sutured to healthy segments of the artery to reestablish normal blood flow.

HYPERTENSION

Hypertension, or high blood pressure, is the most common disease affecting the heart and blood vessels. Statistics indicate that hypertension afflicts one out of every five American adults. Although there is some disagree-

ment as to what defines hypertension, a strong consensus has emerged suggesting that a blood pressure of 120/80 is normal and desirable in a healthy adult. A reading of 140/90 is generally regarded as the threshold of hypertension, while higher values, especially those over 160/95, are classified as dangerously hypertensive.

Primary hypertension (*essential hypertension*) is a persistently elevated blood pressure that cannot be attributed to any particular organic cause. Approximately 85 percent of all hypertension cases fit this definition. The other 15 percent are *secondary hypertension.* Secondary hypertension has an identifiable underlying cause such as atherosclerosis, kidney disease, and adrenal hypersecretion. Atherosclerosis increases blood pressure by reducing the elasticity of the arterial wall and narrowing the lumen through which the blood can flow. Kidney disease and obstruction of blood flow may cause the kidneys to release renin into the blood. This enzyme catalyzes the formation of angiotensin II from a plasma protein. Angiotensin II is a powerful blood-vessel constrictor—and the most potent agent known for raising blood pressure. It also stimulates aldosterone release. Aldosteronism, the hypersecretion of aldosterone, may also cause an increase in blood pressure. Aldosterone is the adrenal cortex hormone that promotes the retention of salt and water by the kidneys. It thus tends to increase plasma volume. Pheochromocytoma (fē-ō-krō'-mō-sī-TŌ-ma) is a tumor of the adrenal medulla. It produces and releases into the blood large quantities of norepinephrine and epinephrine. These hormones also raise blood pressure. Epinephrine causes an increase in heart rate and norepinephrine causes vasoconstriction.

High blood pressure is of considerable concern because of the harm it can do to the heart, brain, and kidneys if it remains uncontrolled. The heart is most commonly affected by high blood pressure. When pressure is high, the heart uses more energy in pumping. Because of the increased effort, the heart muscle thickens and the heart becomes enlarged. The heart also needs more oxygen. If it cannot meet the demands put on it, angina pectoris or even myocardial infarction may develop. Hypertension is also a factor in the development of atherosclerosis. Continued high blood pressure may produce a cerebral vascular accident (CVA), or stroke. In this case, severe strain has been imposed on the cerebral arteries that supply the brain. These arteries are usually less protected by surrounding tissues than are the major arteries in other parts of the body. These weakened cerebral arteries may finally rupture, and a brain hemorrhage follows.

The kidneys are also prime targets of hypertension. The principal site of damage is in the arterioles that supply them. The continual high blood pressure pushing against the walls of the arterioles causes them to thicken, thus narrowing the lumen. The blood supply to the kidneys is thereby gradually reduced. In response, the kidneys may secrete renin, which raises the blood pressure even higher and complicates the problem. The reduced blood flow to the kidney cells may eventually lead to the death of the cells.

Medical science cannot cure hypertension. However, almost all cases of hypertension, whether mild or severe, can be controlled in a number of ways. The overweight person with hypertension will usually be placed on a reducing diet, because blood pressure often falls with weight loss. Treatment often involves the restriction of sodium intake. Sodium restriction curbs fluid retention by the body and also tends to reduce blood volume. Recent evidence suggests that less fat and more potassium and calcium may also lead to a reduction in blood pressure. Since nicotine is a vasoconstrictor, it elevates blood pressure. Stopping smoking may help to decrease blood pressure. As indicated earlier, exercise can help to reduce hypertension. In recent years, some physicians have advocated relaxation techniques (yoga, meditation, and biofeedback) to treat hypertension. For those individuals who require medication, a number of drugs are available. Many people can be treated with diuretics, which eliminate large amounts of water and sodium, thus decreasing blood volume and reducing blood pressure. Vasodilators are often used in combination with diuretics. They relax the smooth muscle in arterial walls, causing vasodilation and thus lowering blood pressure. Beta blockers are also used to lower blood pressure, often in combination with diuretics.

DEEP VENOUS THROMBOSIS (DVT)

Venous thrombosis, the presence of a thrombus in a vein, typically occurs in deep veins of the lower extremities. In such cases, the condition is referred to as **deep venous thrombosis (DVT).** The two most serious complications of DVT are pulmonary embolism, in which the thrombus dislodges and finds its way into the pulmonary arterial blood flow, and postphlebitic syndrome, which consists of edema, pain, and skin changes due to destruction of venous valves. Treatment consists of intravenous heparin therapy and elevation of the extremity, fibrinolytic therapy (streptokinase or urokinase), and, in rare cases, thrombectomy.

KEY MEDICAL TERMS ASSOCIATED WITH BLOOD VESSELS

Angiocardiography (*angio* = vessel; *cardio* = heart; *graph* = writing) X-ray examination of the heart and great blood vessels after injection of a radiopaque dye into the blood stream.

Aortography X-ray examination of the aorta and its main branches after injection of a radiopaque dye.

Claudication Pain and lameness or limping caused by defective circulation of the blood in the vessels of the limbs.

Coronary endarterectomy The removal of the obstructing area within the lumen of the vessel.

Cyanosis (*cyano* = blue) Slightly bluish, dark purple skin coloration due to oxygen deficiency in systemic blood.

Hypercholesteremia (*hyper* = over; *heme* = blood) An excess of cholesterol in the blood.

Hypotension (*hypo* = below; *tension* = pressure) Low blood pressure; most commonly used to describe an acute drop in blood pressure, as occurs in circulatory shock.

Normotensive Characterized by normal blood pressure.

Occlusion The closure or obstruction of the lumen of a structure such as a blood vessel.

Phlebitis (*phleb* = vein) Inflammation of a vein often in a leg.

Shunt A passage between two blood vessels or between the two sides of the heart.

Thrombectomy (*thrombo* = clot) An operation to remove a blood clot from a blood vessel.

Thrombophlebitis Inflammation of a vein with clot formation.

STUDY OUTLINE

Arteries (p. 323)
1. Arteries carry blood away from the heart. Their wall consists of a tunica interna, tunica media (which maintains elasticity and contractility), and tunica externa.
2. Large arteries are referred to as elastic (conducting) arteries and medium-sized arteries are called muscular (distributing) arteries.
3. Many arteries anastomose—the distal ends of two or more vessels unite. An alternate blood route from an anastomosis is called collateral circulation. Arteries that do not anastomose are called end arteries.

Arterioles (p. 325)
1. Arterioles are small arteries that deliver blood to capillaries.
2. Through constriction and dilation they assume a key role in regulating blood flow from arteries into capillaries.

Capillaries (p. 325)
1. Capillaries are microscopic blood vessels through which materials are exchanged between blood and tissue cells; some capillaries are continuous, others are fenestrated.
2. Capillaries branch to form an extensive capillary network throughout the tissue. This network increases the surface area, allowing a rapid exchange of large quantities of materials.
3. Precapillary sphincters regulate blood flow through capillaries.
4. In response to low levels of oxygen, cells produce vasodilator substances that cause dilation of arterioles and relaxation of precapillary sphincters, a phenomenon called autoregulation.
5. Microscopic blood vessels in the liver are called sinusoids.

Venules (p. 326)
1. Venules are small vessels that continue from capillaries and merge to form veins.
2. They drain blood from capillaries into veins.

Veins (p. 326)
1. Veins consist of the same three tunics as arteries but have less elastic tissue and smooth muscle.
2. They contain valves to prevent backflow of blood.
3. Weak valves can lead to varicose veins or hemorrhoids.
4. Vascular (venous) sinuses are veins with very thin walls.

Blood Reservoirs (p. 328)
1. Systemic veins are collectively called blood reservoirs.
2. They store blood which through vasoconstriction can move to other parts of the body if the need arises.
3. The principal reservoirs are the veins of the abdominal organs (liver and spleen) and skin.

Circulatory Routes (p. 329)
1. The largest circulatory route is the systemic circulation.
2. Two of the many subdivisions of the systemic circulation are coronary (cardiac) circulation and hepatic portal circulation.

3. Other routes include the cerebral, fetal, and hepatic portal circulation.

Systemic Circulation
1. The systemic circulation takes oxygenated blood from the left ventricle through the aorta to all parts of the body including lung tissue.
2. The aorta is divided into the ascending aorta, the arch of the aorta, and the descending aorta. Each section gives off arteries that branch to supply the whole body.
3. Blood is returned to the heart through the systemic veins. All the veins of the systemic circulation flow into either the superior or inferior vena cava or the coronary sinus. They in turn empty into the right atrium.

Pulmonary Circulation
1. The pulmonary circulation takes deoxygenated blood from the right ventricle to the lungs and returns oxygenated blood from the lungs to the left atrium.
2. It allows blood to be oxygenated for systemic circulation.

Hepatic Portal Circulation
1. The hepatic portal circulation collects blood from the veins of the pancreas, spleen, stomach, intestines, and gallbladder and directs it into the hepatic portal vein of the liver.
2. This circulation enables the liver to utilize nutrients and detoxify harmful substances in the blood.

Fetal Circulation
1. The fetal circulation involves the exchange of materials between fetus and mother.
2. The fetus derives its oxygen and nutrients and eliminates its carbon dioxide and wastes through the maternal blood supply by means of a structure called the placenta.
3. At birth, when lung, digestive, and liver functions are established, the special structures of fetal circulation are no longer needed.

Aging and the Cardiovascular System (p. 356)
1. General changes include loss of elasticity of blood vessels, reduction in cardiac muscle size, and reduced cardiac output.
2. The incidence of coronary artery disease (CAD), congestive heart failure, and atherosclerosis increases with age.

Developmental Anatomy of Blood and Blood Vessels (p. 356)
1. Blood vessels develop from isolated masses of mesenchyme in mesoderm called blood islands.
2. Blood is produced by the endothelium of blood vessels.

Applications to Health (p. 356)
1. An aneurysm is a thin, weakened section of the wall of an artery or vein that bulges outward forming a balloonlike sac.
2. Hypertension, or high blood pressure, is classified as primary and secondary.
3. Deep venous thrombosis (DVT) refers to a blood clot in a deep vein, especially in the lower extremities.

REVIEW QUESTIONS

1. Describe the structural and functional differences among arteries, arterioles, capillaries, venules, and veins.
2. Discuss the importance of the elasticity and contractility of arteries.
3. Distinguish between elastic and muscular arteries in terms of location, histology, and function.
4. What is an anastomosis? What is collateral circulation?
5. Describe how capillaries are structurally adapted for exchanging materials with body cells. Distinguish between true, continuous, and fenestrated capillaries.
6. What are blood reservoirs? Why are they important?
7. What is meant by a circulatory route? Define systemic circulation.
8. Diagram the major divisions of the aorta, their principal arterial branches, and the regions supplied.
9. Trace a drop of blood from the arch of the aorta through its systemic circulatory route to the tip of the big toe on your left foot and back to the heart again. Remember that the major branches of the arch are the brachiocephalic artery, left common carotid artery, and left subclavian artery. Be sure to indicate which veins return the blood to the heart.
10. What is the cerebral arterial circle (circle of Willis)? Why is it important?
11. What major organs are supplied by branches of the thoracic aorta? How is blood returned from these organs to the heart?
12. What organs are supplied by the celiac, superior mesenteric, renal, inferior mesenteric, inferior phrenic, and middle sacral arteries? How is blood returned to the heart?
13. Trace a drop of blood from the brachiocephalic artery into the digits of the right upper extremity and back again to the right atrium.
14. What are the three major groups of systemic veins?
15. Define pulmonary circulation. Prepare a diagram to indicate the route. What is the purpose of the route?
16. What is hepatic portal circulation? Describe the route by means of a diagram. Why is this route significant?
17. Discuss in detail the anatomy and physiology of fetal circulation. Be sure to indicate the function of the umbilical arteries, umbilical vein, ductus venosus, foramen ovale, and ductus arteriosus.
18. Describe the effects of aging on the cardiovascular system.
19. Describe the development of blood vessels and blood.
20. What is an aneurysm? Why is an aneurysm a serious problem?
21. Compare the causes of primary and secondary hypertension. How does hypertension affect the body? How is hypertension treated?
22. Refer to the glossary of key medical terms associated with blood vessels. Be sure that you can define each term.

15

The Lymphatic System

Student Objectives

Describe the components of the lymphatic system and list its functions.

Describe the structure and origin of lymphatics and contrast them with veins.

Describe the histological aspects of lymph nodes and explain their functions.

Trace the general plan of lymph circulation from lymphatics into the thoracic duct or right lymphatic duct.

Discuss how edema develops.

Describe the principal lymph nodes of the head and neck, extremities, and trunk, their location, and the areas they drain.

Contrast the locations and functions of the tonsils, spleen, and thymus gland as lymphatic organs.

Describe the development of the lymphatic system.

Describe the clinical symptoms of the following disorders: hypersensitivity (allergy), tissue rejection, autoimmune diseases, and acquired immune deficiency syndrome (AIDS).

Define key medical terms associated with the lymphatic system.

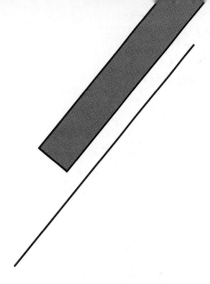

Lymph, lymph vessels, a series of small masses of lymphoid tissue called lymph nodes, and three organs—the tonsils, thymus, and spleen—make up the **lymphatic (lim-FAT-ik) system.** The primary function of the lymphatic system is to drain from the tissue spaces protein-containing fluid that escapes from the blood capillaries. Such proteins cannot be directly reabsorbed. Other functions of the lymphatic system are to transport fats from the digestive tract to the blood, to produce lymphocytes, and to develop antibodies.

The developmental anatomy of the lymphatic system is considered later in the chapter.

LYMPHATIC VESSELS

Lymphatic vessels originate as microscopic vessels in spaces between cells called **lymph capillaries** (Figure 15-1a). Lymph capillaries may occur singly or in extensive plexuses. They originate throughout the body, but not in avascular tissue, the central nervous system, splenic pulp, and bone marrow. They are slightly larger and more permeable than blood capillaries.

Lymph capillaries also differ from blood capillaries in that they end blindly; blood capillaries have an arterial and a venous end. In addition, lymph capillaries are structurally adapted to ensure the return of proteins to the circulation when they leak out of blood capillaries. Close examination of lymph capillaries reveals that the endothelial cells making up the capillary wall overlap each other, forming pores (Figure 15-1b). This overlapping arrangement permits fluid to flow easily into the capillary but prevents the flow of fluid out of the capillary, much like a one-way valve would operate. Note also that the outer surfaces of the endothelial cells of the capillary wall are attached to the surrounding tissue by structures called *anchoring filaments.* During edema, there is an excessive accumulation of fluid in the tissue, causing tissue swelling. This swelling produces a pull on the anchoring filaments, opening the pores even more so that more fluid can flow into the lymph capillary.

Just as blood capillaries converge to form venules and veins, lymph capillaries unite to form larger and larger lymph vessels called **lymphatics** (Figure 15-2). Lymphatics resemble veins in structure, but have thinner walls and more valves and contain lymph nodes at various intervals. Lymphatics of the skin travel in loose subcutaneous tissue and generally follow veins. Lymphatics of the viscera generally follow arteries, forming plexuses around them. Ultimately, lymphatics converge into two main channels—the thoracic duct and the right lymphatic duct. These will be described shortly.

CLINICAL APPLICATION

Lymphangiography (lim-fan'-jē-OG-ra-fē) is the x-ray examination of lymphatic vessels and lymph organs after they are filled with a radiopaque substance. Such an x ray is called a **lymphangiogram** (lim-FAN-jē-ō-gram). Lymphangiograms are useful in detecting edema and carcinomas and in locating lymph nodes for surgical or radiotherapeutic treatment.

STRUCTURE OF LYMPH NODES

The oval or bean-shaped structures located along the length of lymphatics are called **lymph nodes.** They are scattered throughout the body, usually in groups, and range from 1 to 25 mm (0.04 to 1 in) in length. A lymph node contains a slight depression on one side called a *hilus* (HĪ-lus), or *hilum,* where blood vessels and efferent lymphatic vessels leave the node (Figure 15-3). Each node is covered by a *capsule* of fibrous connective tissue that extends into the node. The capsular extensions are called *trabeculae* (tra-BEK-yoo-lē). The capsule, trabeculae, and hilus constitute the stroma (framework) of a lymph node. The parenchyma of a lymph node is specialized into two regions. The outer *cortex* contains densely packed lymphocytes arranged in masses called *lymph nodules.* The nodules often contain lighter-staining central areas, the

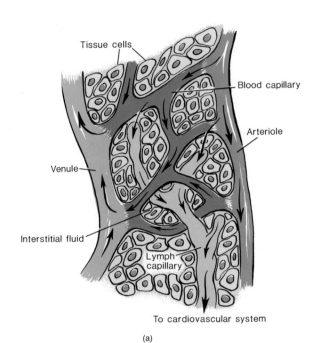

Tissue cells

Blood capillary

Arteriole

Venule

Interstitial fluid

Lymph capillary

To cardiovascular system

(a)

germinal centers, where lymphocytes are produced. The inner region of a lymph node is called the *medulla.* In the medulla, the lymphocytes are arranged in strands called *medullary cords.*

The circulation of lymph through a node involves afferent (to convey toward a center) lymphatic vessels, sinuses in the node, and efferent (to convey away from a center) lymphatic vessels. *Afferent lymphatic vessels* enter the convex surface of the node at several points. They contain valves that open toward the node so that the lymph is directed inward. Once inside the node, the lymph enters the sinuses, which are a series of irregular channels. Lymph from the afferent lymphatic vessels enters the *cortical sinuses* under the capsule. From here it circulates to the *medullary sinuses* between the medullary cords. From these sinuses the lymph usually circulates into a single *efferent lymphatic vessel.* This vessel is located at the hilus of the lymph node. The efferent vessel is wider than the afferent vessels and contains valves that open away from the node to convey the lymph out of the node.

FIGURE 15-1 Lymph capillaries. (a) Relationship of lymph capillaries to tissue cells and blood capillaries. (b) Details of a lymph capillary.

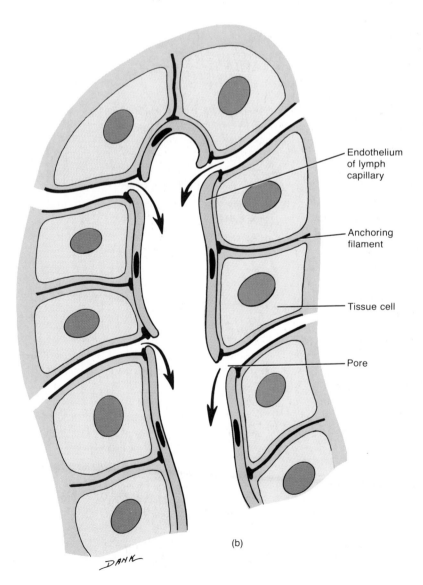

Endothelium of lymph capillary

Anchoring filament

Tissue cell

Pore

(b)

DANK

SYSTEMIC CIRCULATION   PULMONARY CIRCULATION

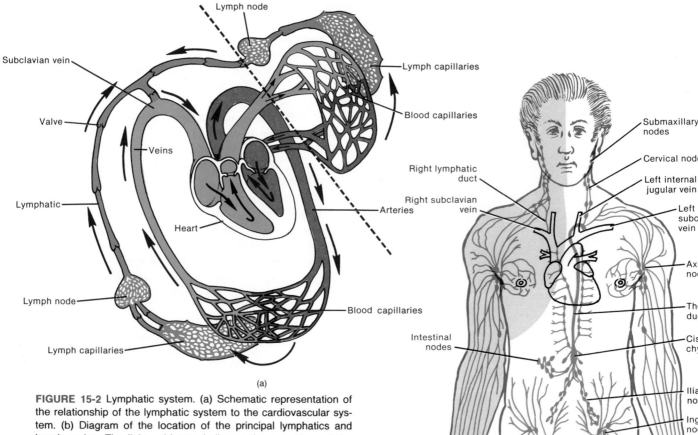

(a)

FIGURE 15-2 Lymphatic system. (a) Schematic representation of the relationship of the lymphatic system to the cardiovascular system. (b) Diagram of the location of the principal lymphatics and lymph nodes. The light gold area indicates those portions of the body drained by the right lymphatic duct. All other areas of the body are drained by the thoracic duct.

As the lymph circulates through the nodes, it is processed by macrophages. The macrophages are fixed phagocytic cells of the reticuloendothelial system that line the sinuses. They remove bacteria, foreign material, and cell debris from the lymph. Lymph nodes also give rise to lymphocytes and plasma cells. The plasma cells produce antibodies.

Histological features of lymphatics and lymph nodes are shown in Figure 15-4.

CLINICAL APPLICATION

At times the number of microbes entering the nodes in the lymph is so great that the macrophages and lymphocytes cannot remove or detoxify them. The nodes become infected. **Infected lymph nodes** become enlarged and tender. Knowledge of the regions drained by the nodes may be helpful in diagnosing the site of an infection.

Knowledge of the location of the lymph nodes and the direction of lymph flow is also important in the diagnosis and prognosis of the spread of cancer by

(b)

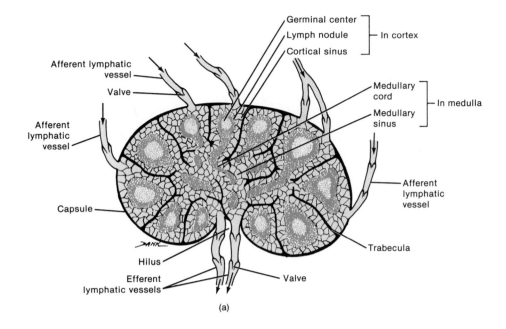

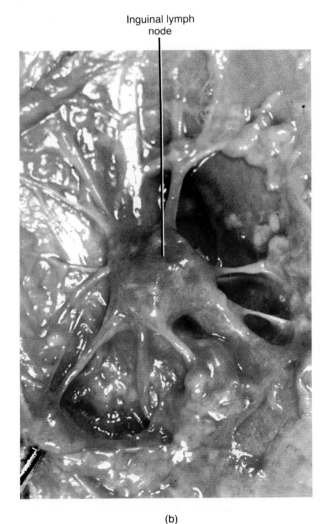

Inguinal lymph node

(b)

FIGURE 15-3 Structure of a lymph node. (a) Diagram showing the path taken by circulating lymph. (b) Photograph of an inguinal lymph node. (Courtesy of C. Yokochi and J. W. Rohen, *Photographic Anatomy of the Human Body,* 2nd ed., 1979, IGAKU-SHOIN, Ltd., Tokyo, New York.)

metastasis. Cancer cells usually spread by way of the lymphatic system and produce aggregates of tumor cells where they lodge. Such secondary tumor sites are predictable by the direction of lymph flow from the organ primarily involved.

LYMPH CIRCULATION

ROUTE

When plasma is filtered by blood capillaries, it passes into the interstitial spaces; it is then known as interstitial fluid. Fluid movement between blood capillaries and body cells depends on hydrostatic and osmotic pressures. When this fluid passes from interstitial spaces into lymph capillaries, it is called lymph. Lymph from lymph capillary plexuses is then passed to lymphatics that run toward lymph nodes. At the nodes, afferent vessels penetrate the capsules at numerous points and the lymph passes through the sinuses of the nodes. Efferent vessels from the nodes either run with afferent vessels into another node of the same group or pass on to another group of nodes. From the most proximal group of each chain of nodes, the efferent vessels unite to form **lymph trunks.** The principal trunks are the *lumbar, intestinal, bronchomediastinal, subclavian,* and *jugular trunks* (Figure 15-5).

Thoracic (Left Lymphatic) Duct

The principal trunks pass their lymph into two main channels, the thoracic duct and the right lymphatic duct. The **thoracic (left lymphatic) duct** is about 38 to 45 cm (15 to 18 in) in length and begins as a dilation in front of

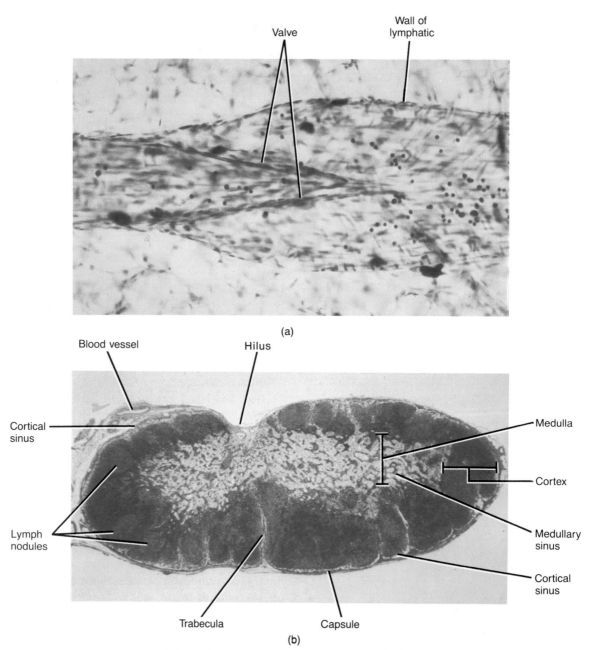

FIGURE 15-4 Histology of lymphatics and lymph nodes. (a) Photomicrograph of a lymphatic at a magnification of 180×. (b) Photomicrograph of a lymph node at a magnification of 25×. (Photomicrographs © 1983 by Michael H. Ross. Used by permission.)

the second lumbar vertebra called the *cisterna chyli* (sis-TER-na KĪ-lē). The thoracic duct is the main collecting duct of the lymphatic system and receives lymph from the left side of the head, neck, and chest, the left upper extremity, and the entire body below the ribs.

The cisterna chyli receives lymph from the right and left lumbar trunks and from the intestinal trunk. The lumbar trunks drain lymph from the lower extremities, wall and viscera of the pelvis, kidneys, suprarenals, and the deep lymphatics from most of the abdominal wall. The intestinal trunk drains lymph from the stomach, in-

testines, pancreas, spleen, and visceral surface of the liver. In the neck, the thoracic duct also receives lymph from the left jugular, left subclavian, and left bronchomediastinal trunks. The left jugular trunk drains lymph from the left side of the head and neck; the left subclavian trunk drains lymph from the upper left extremity; and the left bronchomediastinal trunk drains lymph from the left side of the deeper parts of the anterior thoracic wall, upper part of the anterior abdominal wall, anterior part of the diaphragm, left lung, and left side of the heart.

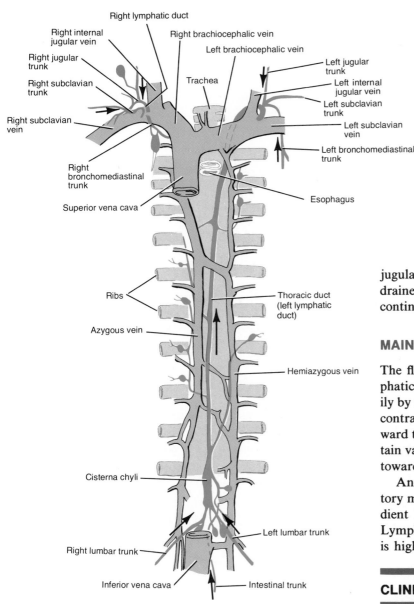

FIGURE 15-5 Relationship of lymph trunks to the thoracic duct and right lymphatic duct. See also Figure 15-2c.

Right Lymphatic Duct

The **right lymphatic duct** is about 1.25 cm (0.5 in) long and drains lymph from the upper right side of the body (see Figure 15-2b). The right lumphatic duct collects lymph from its trunks as follows (Figure 15-5). It receives lymph from the right jugular trunk, which drains the right side of the head and neck, from the right subclavian trunk, which drains the right upper extremity, and from the right bronchomediastinal trunk, which drains the right side of the thorax, right lung, right side of the heart, and part of the convex surface of the liver.

Ultimately, the thoracic duct empties all of its lymph into the junction of the left internal jugular vein and left subclavian vein, and the right lymphatic duct empties all of its lymph into the junction of the right internal

jugular vein and right subclavian vein. Thus, lymph is drained back into the blood and the cycle repeats itself continuously.

MAINTENANCE

The flow of lymph from tissue spaces to the large lymphatic ducts to the subclavian veins is maintained primarily by the milking action of muscle tissue. Skeletal muscle contractions compress lymph vessels and force lymph toward the subclavian veins. Lymph vessels, like veins, contain valves, and the valves ensure the movement of lymph toward the subclavian veins (see Figure 15-4a).

Another factor that maintains lymph flow is respiratory movements. These movements create a pressure gradient between the two ends of the lymphatic system. Lymph flows from the tissue spaces, where the pressure is higher, toward the thoracic region, where it is lower.

CLINICAL APPLICATION

Edema, an excessive accumulation of interstitial fluid in tissue spaces, may be caused by an obstruction, such as an infected node or a blockage of vessels, in the pathway between the lymphatic capillaries and the subclavian veins. Another cause is excessive lymph formation and increased permeability of blood capillary walls. A rise in capillary blood pressure, in which interstitial fluid is formed faster than it is passed into lymphatics, also may result in edema.

PRINCIPAL GROUPS OF LYMPH NODES

Lymph nodes usually appear in groups and typically are arranged in two sets: *superficial* and *deep.*

Exhibits 15-1 through 15-5 list the principal groups of lymph nodes of the body by region and the general areas of the body they drain.

EXHIBIT 15-1   PRINCIPAL LYMPH NODES OF THE HEAD AND NECK (Figure 15-6)

LYMPH NODES	LOCATION AND AREAS DRAINED
LYMPH NODES OF THE HEAD	
Occipital	Near trapezius and semispinalis capitis muscles. They drain the occipital portion of scalp and upper neck.
Retroauricular	Behind ear. They drain skin of ear and posterior parietal region of scalp.
Preauricular	Anterior to tragus. They drain pinna and temporal region of the scalp.
Parotid	Embedded in and below parotid gland. They drain root of nose, eyelids, anterior temporal region, external auditory meatus, tympanic cavity, nasopharynx, and posterior portions of nasal cavity.
Facial	Consist of three groups: infraorbital, buccal, and mandibular.
Infraorbital	Below the orbit. They drain eyelids and conjunctiva.
Buccal	At angle of mouth. They drain the skin and mucous membrane of nose and cheek.
Mandibular	Over mandible. They drain the skin and mucous membrane of nose and cheek.
LYMPH NODES OF THE NECK	
Submandibular	Along inferior border of mandible. They drain chin, lips, nose, nasal cavity, cheeks, gums, lower surface of palate, and anterior portion of tongue.
Submental	Between digastric muscles. They drain chin, lower lip, cheeks, tip of tongue, and floor of mouth.
Superficial cervical	Along external jugular vein. They drain lower part of ear and parotid region.
Deep cervical	Largest group of nodes in neck, consisting of numerous large nodes forming a chain extending from base of skull to root of neck. They are arbitrarily divided into superior deep cervical nodes and inferior deep cervical nodes.
Superior deep cervical	Under sternocleidomastoid muscle. They drain posterior head and neck, pinna, tongue, larynx, esophagus, thyroid gland, nasopharynx, nasal cavity, and palate.
Inferior deep cervical	Near subclavian vein. They drain posterior scalp and neck, superficial pectoral region, and part of arm.

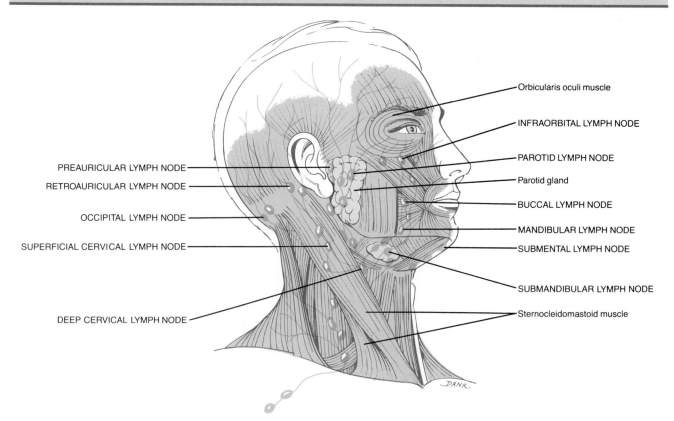

FIGURE 15-6 Principal lymph nodes of the head and neck in lateral view.

EXHIBIT 15-2   PRINCIPAL LYMPH NODES OF THE UPPER EXTREMITIES (Figure 15-7)

LYMPH NODES	LOCATION AND AREAS DRAINED
Supratrochlear	Above medial epicondyle of humerus. They drain medial fingers, palm, and forearm.
Deltopectoral	Below clavicle. They drain lymphatic vessels on radial side of upper extremity.
Axillary	Most deep lymph nodes of the upper extremities are in the axilla and are called the axillary nodes. They are large in size and may be grouped as follows.
Lateral	Medial and posterior aspects of axillary artery. They drain most of whole upper extremity. Since infection or malignancy of upper extremity may cause tenderness and swelling in axilla, the axillary nodes, especially the lateral group, are clinically important since they filter lymph from much of upper extremity.
Pectoral (anterior)	Along inferior border of the pectoralis minor muscle. They drain skin and muscles of anterior and lateral thoracic walls and central and lateral portions of mammary gland.
Subscapular (posterior)	Along subscapular artery. They drain skin and muscles of posterior part of neck and thoracic wall.
Central (intermediate)	Base of axilla embedded in adipose tissue. They drain lateral, pectoral (anterior), and subscapular (posterior) nodes.
Subclavicular (medial)	Posterior and superior to pectoralis minor muscle. They drain deltopectoral nodes.

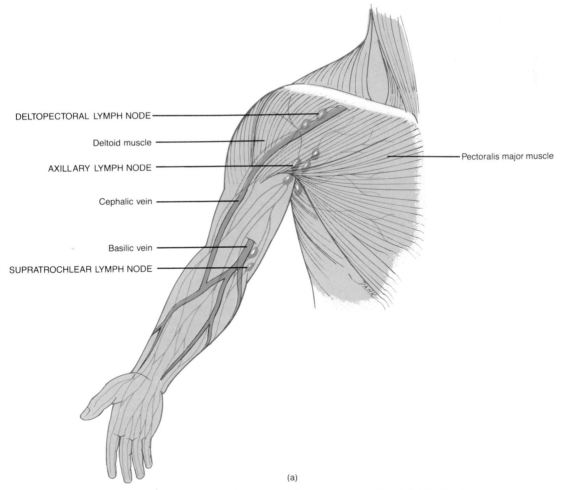

(a)

FIGURE 15-7 Principal lymph nodes of the upper extremities. (a) Anterior view.

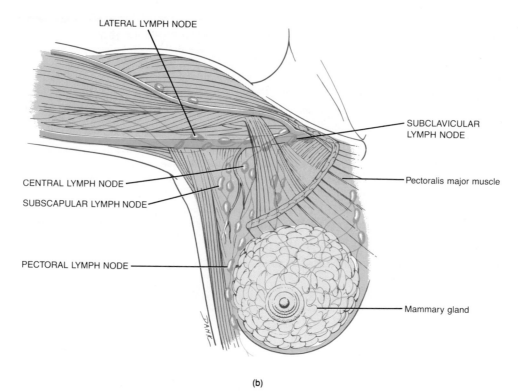

LATERAL LYMPH NODE

SUBCLAVICULAR LYMPH NODE

CENTRAL LYMPH NODE

SUBSCAPULAR LYMPH NODE

Pectoralis major muscle

PECTORAL LYMPH NODE

Mammary gland

(b)

FIGURE 15-7 (*Continued*) Principal lymph nodes of the upper extremities. (b) Anterior view.

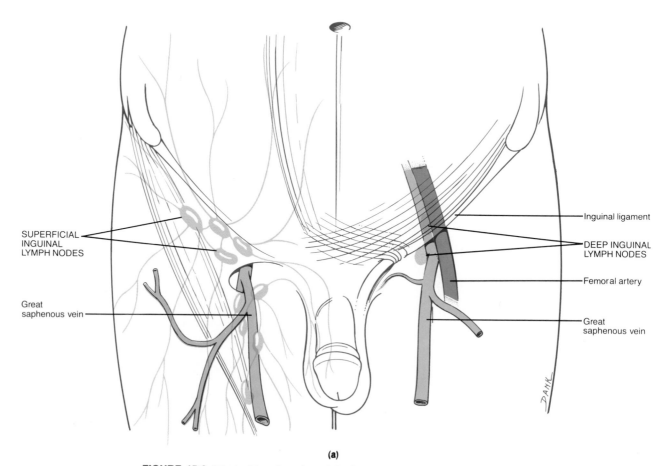

SUPERFICIAL INGUINAL LYMPH NODES

Inguinal ligament

DEEP INGUINAL LYMPH NODES

Femoral artery

Great saphenous vein

Great saphenous vein

(a)

FIGURE 15-8 Principal lymph nodes of the lower extremities. (a) Anterior view.

EXHIBIT 15-3 PRINCIPAL LYMPH NODES OF THE LOWER EXTREMITIES (Figure 15-8)

LYMPH NODES	LOCATION AND AREAS DRAINED
Popliteal	In adipose tissue in popliteal fossa. They drain knee.
Superficial inguinal	Parallel to saphenous vein. They drain anterior and lateral abdominal wall to level of umbilicus, gluteal region, external genitals, perineal region, and entire superficial lymphatics of lower extremity.
Deep inguinal	Medial to femoral vein. They drain deep lymphatics of lower extremity, penis, and clitoris.

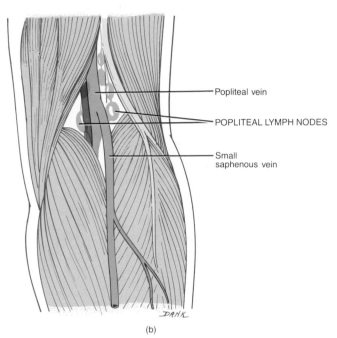

Popliteal vein

POPLITEAL LYMPH NODES

Small
saphenous vein

(b)

FIGURE 15-8 (*Continued*) Principal lymph nodes of the lower extremities. (b) Posterior view.

EXHIBIT 15-4 PRINCIPAL LYMPH NODES OF THE ABDOMEN AND PELVIS (Figure 15-9)

LYMPH NODES	LOCATION AND AREAS DRAINED
	Lymph nodes of the abdomen and pelvis are divided into *parietal lymph nodes* that are retroperitoneal (behind the parietal peritoneum) and in close association with larger blood vessels and *visceral lymph nodes* found in association with visceral arteries.
PARIETAL **External iliac**	Arranged about external iliac vessels. They drain the deep lymphatics of abdominal wall below umbilicus, adductor region of thigh, urinary bladder, prostate gland, ductus deferens, seminal vesicles, prostatic and membranous urethra, uterine tubes, uterus, and vagina.
Common iliac	Arranged along course of common iliac vessels. They drain pelvic viscera.
Internal iliac	Near internal iliac artery. They drain pelvic viscera, perineum, gluteal region, and posterior surface of thigh.
Sacral	In hollow of sacrum. They drain rectum, prostate gland, and posterior pelvic wall.
Lumbar	From aortic bifurcation to diaphragm; arranged around aorta and designated as **right lateral aortic nodes, left lateral aortic nodes, preaortic nodes,** and **retroaortic nodes.** They drain the efferents from testes, ovaries, uterine tubes, uterus, kidneys, suprarenal glands, abdominal surface of diaphragm, and lateral abdominal wall.

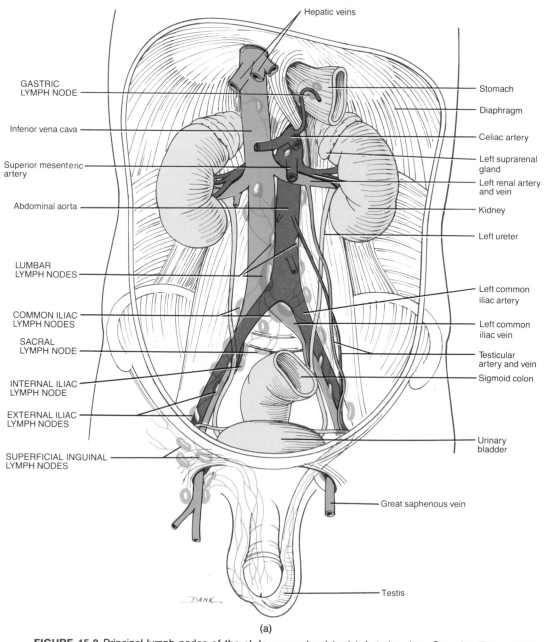

(a)

FIGURE 15-9 Principal lymph nodes of the abdomen and pelvis. (a) Anterior view. See also Figure 15-2c.

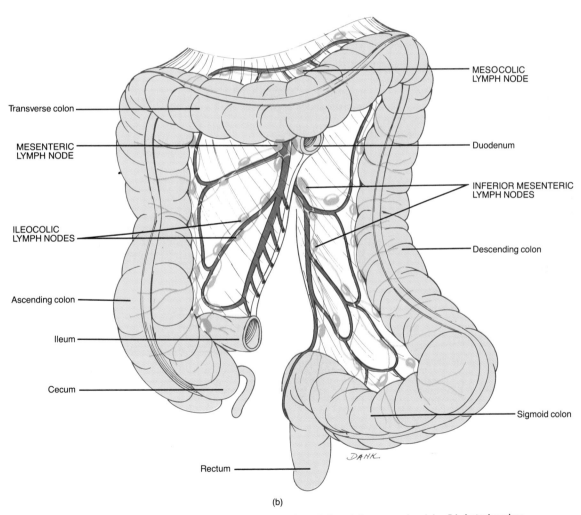

(b)

FIGURE 15-9 (*Continued*) Principal lymph nodes of the abdomen and pelvis. (b) Anterior view.

EXHIBIT 15-5 PRINCIPAL LYMPH NODES OF THE THORAX (Figure 15-10)

LYMPH NODES	LOCATION AND AREAS DRAINED
	Lymph nodes of the thorax are divided into *parietal lymph nodes,* which drain the wall of the thorax, and *visceral lymph nodes,* which drain the viscera.
PARIETAL	
Sternal	Alongside internal thoracic artery. They drain central and lateral parts of mammary gland, deeper structures of anterior abdominal wall above umbilicus, diaphragmatic surface of liver, and deeper parts of anterior portion of thoracic wall.
Intercostal	Near heads of ribs at posterior parts of intercostal spaces. They drain posterolateral aspect of thoracic wall.
Phrenic (diaphragmatic)	Located on thoracic aspect of the diaphragm and divisible into three sets called anterior phrenic, middle phrenic, and posterior phrenic.
Anterior phrenic	Behind base of xiphoid process. They drain convex surface of liver, diaphragm, and anterior abdominal wall.
Middle phrenic	Close to phrenic nerves where they pierce diaphragm. They drain middle part of diaphragm and convex surface of liver.
Posterior phrenic	Back of diaphragm near aorta. They drain posterior part of diaphragm.
VISCERAL	
Anterior mediastinal	Anterior part of superior mediastinum anterior to arch of aorta. They drain thymus gland and pericardium.
Posterior mediastinal	Posterior to pericardium. They drain esophagus, posterior aspect of the pericardium, diaphragm, and convex surface of liver.
Tracheobronchial	The tracheobronchial nodes drain lungs, bronchi, thoracic part of trachea, and heart and are divisible into four groups:
Tracheal	On either side of trachea.
Bronchial	At inferior part of trachea and in angle between two bronchi.
Bronchopulmonary	In the hilus of each lung.
Pulmonary	Within lungs on larger bronchial branches.

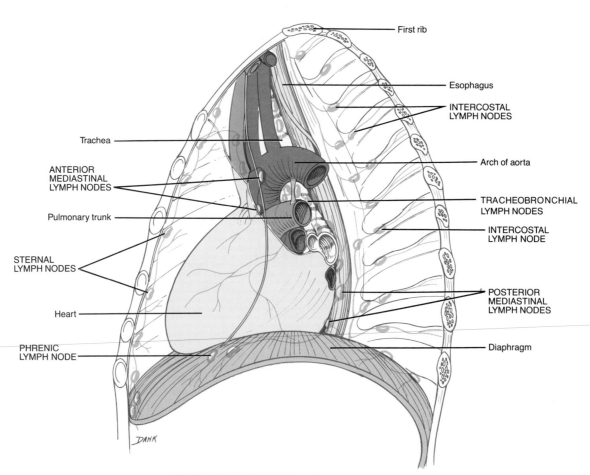

FIGURE 15-10 Principal lymph nodes of the thorax.

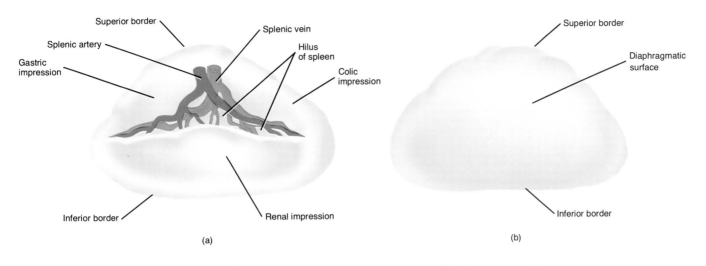

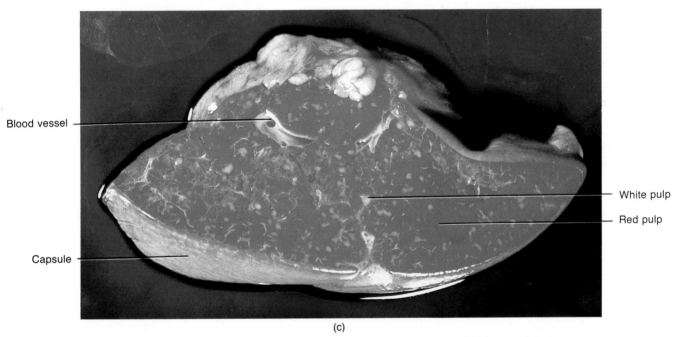

FIGURE 15-11 Gross structure of the spleen. (a) Diagram of visceral surface. (b) Diagram of diaphragmatic surface. (c) Photograph of a section through the spleen showing white and red pulp. (Courtesy of C. Yokochi and J. W. Rohen, *Photographic Anatomy of the Human Body,* 2nd ed., 1979, IGAKU-SHOIN, Ltd., Tokyo, New York.)

LYMPHATIC ORGANS

TONSILS

Tonsils are basically masses of lymphoid tissue embedded in mucous membrane. The **pharyngeal** (fa-RIN-jē-al) **tonsil,** or **adenoid,** is embedded in the posterior wall of the nasopharynx (see Figure 22-3a). The **palatine** (PAL-a-tīn) **tonsils** are situated in the tonsillar fossae between the pharyngopalatine and glossopalatine arches (see Figure 23-4). These are the ones commonly removed by a tonsillectomy. The **lingual** (LIN-gwal) **tonsil** is located at the base of the tongue and may also have to be removed by a tonsillectomy (see Figure 23-5a).

SPLEEN

The oval **spleen** is the largest mass of lymphatic tissue in the body, measuring about 12 cm (5 in) in length. It is situated in the left hypochondriac region between the fundus of the stomach and diaphragm (see Figure 1-8d). Its *visceral surface* (Figure 15-11a) contains the contours of the organs adjacent to it—the gastric impression (stomach), renal impression (left kidney), and colic impression (left flexure of colon). The *diaphragmatic* (dī-a-fra-MAT-ik) *surface* (Figure 15-11b) is smooth and convex and conforms to the concave surface of the diaphragm to which it is adjacent.

The spleen is surrounded by a capsule of fibroelastic tissue and scattered smooth muscle (Figure 15-11c). The

Capsule　　Trabecular vein　　White pulp　　Red pulp

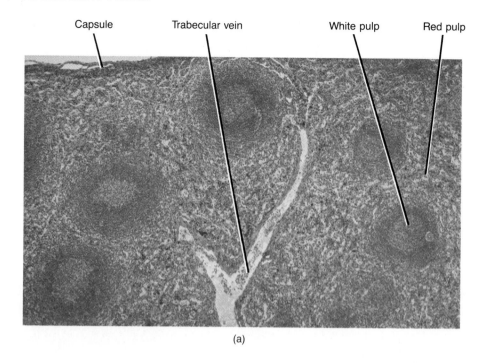

(a)

Red pulp　　　　　　White pulp

Central
artery

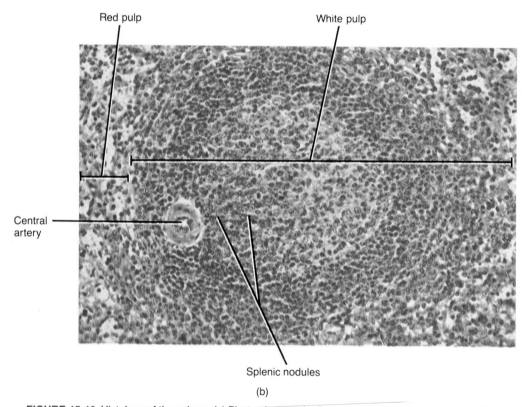

Splenic nodules

(b)

FIGURE 15-12 Histology of the spleen. (a) Photomicrograph of a portion of the spleen at a magnification of 60×. (© 1983 by Michael H. Ross. Used by permission.) (b) Photomicrograph of an enlarged aspect of white pulp at a magnification of 150×.

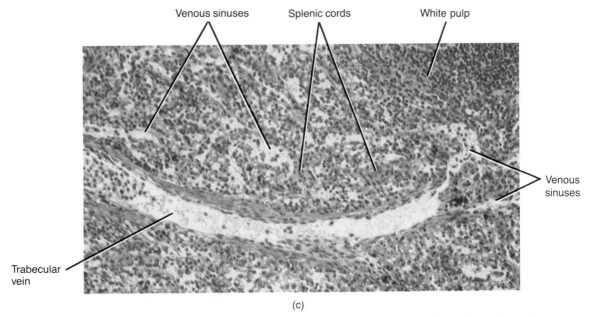

Venous sinuses   Splenic cords   White pulp

Venous sinuses

Trabecular vein

(c)

FIGURE 15-12 (*Continued*) Histology of the spleen. (c) Photomicrograph of an enlarged aspect of red pulp at a magnification of 150×. (Photomicrographs © 1983 by Michael H. Ross. Used by permission.)

capsule, in turn, is covered by a serous membrane, the peritoneum. Like lymph nodes, the spleen contains trabeculae and a hilus. The capsule, trabeculae, and hilus constitute the stroma of the spleen.

The parenchyma of the spleen consists of two different kinds of tissue called white pulp and red pulp (Figure 15-12). *White pulp* is essentially lymphoid tissue arranged around arteries. The clusters of lymphocytes surrounding the arteries at intervals and the expansions are referred to as *splenic nodules* (*Malpighian corpuscles*). The *red pulp* consists of *venous sinuses* filled with blood and cords of splenic tissue called *splenic* (*Billroth's*) *cords.* Veins are closely associated with the red pulp.

The splenic artery and vein and the efferent lymphatics pass through the hilus. Since the spleen has no afferent lymphatic vessels or lymph sinuses, it does not filter lymph. The spleen phagocytoses bacteria and worn-out red blood cells and platelets. It also produces lymphocytes and antibody-producing plasma cells. In addition, the spleen stores and releases blood in case of demand, such as during hemorrhage. The release seems to be purely sympathetic. The sympathetic impulses cause the smooth muscle of the spleen to contract.

About 10 percent of the population has *accessory spleens*. They are most commonly found near the hilus of the primary spleen or embedded in the tail of the pancreas. In general, accessory spleens are about 1 cm (0.5 in) in diameter.

CLINICAL APPLICATION

Although the spleen is protected from traumatic injuries, it is the most frequently damaged organ in cases of abdominal trauma, particularly those involving severe blows over the lower left chest or upper abdomen that fracture the protecting ribs. Such a crushing injury may **rupture the spleen,** which causes severe intraperitoneal hemorrhage and shock. Prompt removal of the spleen, called a **splenectomy,** is needed to prevent the patient from bleeding to death. The functions of the spleen are then assumed by other reticuloendothelial organs.

THYMUS GLAND

The **thymus gland** is a bilobed mass of lymphatic tissue in the upper thoracic cavity. It is found along the trachea behind the sternum (see Figure 21-13). The thymus is relatively large in children. It reaches its maximum size at puberty and then undergoes involution. Eventually, most of it is replaced by fat and connective tissue. Its role in immunity is to help produce cells that destroy invading microbes directly or cells that manufacture antibodies against invading microbes.

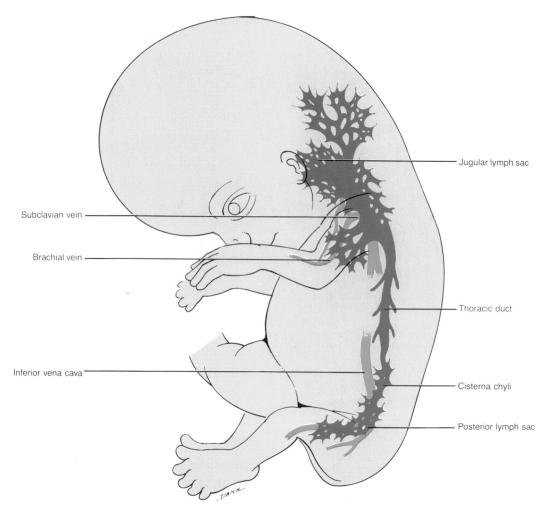

Subclavian vein

Brachial vein

Inferior vena cava

Jugular lymph sac

Thoracic duct

Cisterna chyli

Posterior lymph sac

FIGURE 15-13 Development of the lymphatic system.

DEVELOPMENTAL ANATOMY OF THE LYMPHATIC SYSTEM

The lymphatic system begins its development by the end of the fifth week. *Lymphatic vessels* develop from **lymph sacs** that arise from developing veins. Thus, the lymphatic system is also derived from **mesoderm.**

The first lymph sacs to appear are the paired **jugular lymph sacs** at the junction of the internal jugular and subclavian veins (Figure 15-13). From the jugular lymph sacs, capillary plexuses spread to the *thorax, upper extremities, neck,* and *head.* Some of the plexuses enlarge and form lymphatics in their respective regions. Each jugular lymph sac retains at least one connection with its jugular vein, the left one developing into the superior portion of the thoracic duct (left lymphatic duct). The next lymph sac to appear is the unpaired **retroperitoneal lymph sac** at the root of the mesentery of the intestine. It develops from the primitive vena cava and mesonephric (primitive kidney) veins. Capillary plexuses and lymphatics spread from the retroperitoneal lymph sac to the *ab-*

dominal viscera and *diaphragm.* The sac establishes connections with the cisterna chyli but loses its connections with neighboring veins.

At about the time the retroperitoneal lymph sac is developing, another lymph sac, the **cisterna chyli,** develops below the diaphragm on the posterior abdominal wall. It gives rise to the inferior portion of the *thoracic duct* and the *cisterna chyli* of the thoracic duct. Like the retroperitoneal lymph sac, the cisterna chyli also loses its connections with surrounding veins.

The last of the lymph sacs, the paired **posterior lymph sacs,** develop from the iliac veins at their union with the posterior cardinal veins. The posterior lymph sacs produce capillary plexuses and lymphatics of the *abdominal wall, pelvic region,* and *lower extremity.* The posterior lymph sacs join the cisterna chyli and lose their connections with adjacent veins.

With the exception of the anterior part of the sac from which the cisterna chyli develops, all lymph sacs become invaded by **mesenchymal cells** and are converted into groups of *lymph nodes.*

APPLICATIONS TO HEALTH

The antigen-antibody response is essential to survival and typically provides a state of immunity. Under certain circumstances, however, it may create problems. Three such problems are hypersensitivity (allergy), tissue rejection, and autoimmunity.

HYPERSENSITIVITY (ALLERGY)

A person who is overly reactive to an antigen is said to be **hypersensitive,** or **allergic.** Whenever an allergic reaction occurs, there is tissue injury. The antigens that induce an allergic reaction are called **allergens.** Almost any substance can be an allergen for some individual. Common allergens include certain foods (milk, eggs), antibiotics such as penicillin, cosmetics, chemicals in plants such as poison ivy, pollens, dust, molds, and even microbes.

One type of hypersensitivity is called **anaphylaxis** (an'-a-fi-LAK-sis), the interaction of humoral antibodies (IgE) with mast cells and basophils. Anaphylaxis literally means "against protection." Some anaphylactic reactions are localized and produce *allergic rhinitis* (hay fever), *bronchial asthma, atopic dermatitis* (eczema), and *urticaria* (hives). A severe anaphylactic reaction, called *acute anaphylaxis* (*anaphylactic shock*), may produce life-threatening systemic effects such as circulatory shock and asphyxia.

TISSUE REJECTION

Transplantation involves the replacement of an injured or diseased tissue or organ. Usually, the body recognizes the proteins in the transplanted tissue or organ as foreign and produces antibodies against them. This phenomenon is known as **tissue rejection.** Rejection can be somewhat reduced by matching donor and recipient and by administering drugs that inhibit the body's ability to form antibodies.

Until recently, **immunosuppressive drugs** suppressed not only the recipient's immune rejection of the donor organ, but the immune response to all antigens as well. This caused patients to become very susceptible to infectious diseases. A new drug called cyclosporin A, derived from a fungus, has largely overcome this problem. It is a selective immunosuppressive drug. It inhibits T cells, which are responsible for tissue rejection, but has only a minimal effect on B cells. Thus, rejection is avoided and resistance against disease is still maintained.

AUTOIMMUNE DISEASES

Under normal conditions, the body's immune mechanism is able to recognize its own tissues and chemicals. It normally does not produce T cells or B cells against its own substances. Such recognition of self is called **immunologic tolerance.** Although the mechanism of tolerance is not completely understood, it is believed that suppressor T cells may inhibit the differentiation of B cells into antibody-producing plasma cells or inhibit helper T cells, which cooperate with B cells to amplify antibody production.

At times, however, immunologic tolerance breaks down and the body has difficulty in discriminating between its own antigens and foreign antigens. This loss of immunologic tolerance leads to an **autoimmune disease (autoimmunity).** Such diseases are immunologic responses mediated by antibodies against a person's own tissue antigens. Among human autoimmune diseases are rheumatoid arthritis (RA), systemic lupus erythematosus (SLE), thyroiditis, rheumatic fever, glomerulonephritis, encephalomyelitis, hemolytic and pernicious anemias, myasthenia gravis, and multiple sclerosis (MS).

ACQUIRED IMMUNE DEFICIENCY SYNDROME (AIDS)

Never before has science been confronted with an epidemic in which the primary disease only lowers the victim's immunity and then a second unrelated disease produces the symptoms that may result in death. This primary disease, first recognized in 1981, is called **acquired immune deficiency syndrome (AIDS)** and has a mortality rate of nearly 40 percent. One of the principal problems associated with AIDS is that its victims have too few T cells. In addition, the ratio of helper T cells to suppressor T cells, normally 2:1, is reversed. Symptoms of AIDS may develop for months or years and include malaise; a low-grade fever or night sweats; coughing; shortness of breath; sore throat; extreme fatigue; muscle aches; unexplained weight loss; enlarged lymph nodes in the neck, axilla, and groin; and blue-violet or brownish spots on the skin, usually the lower extremities.

The two diseases that most often kill AIDS victims are Kaposi's sarcoma and *Pneumocystis carinii* pneumonia. Kaposi's sarcoma is a deadly form of skin cancer prevalent in equatorial Africa but previously almost unknown in the United States. *Pneumocystis carinii* pneumonia is a rare form of pneumonia caused by the protozoan *Pneumocystis carinii.* AIDS victims are also subject to a form of herpes that attacks the central nervous system and a bacterial infection that usually causes tuberculosis in chickens and pigs.

Among the high-risk groups who are susceptible to AIDS are homosexual and bisexual males and intravenous drug users. Other risk groups include hemophiliacs and Haitians. It is also found among infants and other patients who have received blood transfusions. The cause of AIDS is a virus called *human T-cell leukemia virus-3* (*HTLV-3*). A variant of this virus, known as HTLV-1, causes a rare type of leukemia in humans and attacks and transforms T cells. The pattern of AIDS transmission closely resembles the occurrence of hepatitis B (serum hepatitis), a liver disease that commonly strikes homosexual drug addicts using contaminated needles, and sometimes patients getting blood transfusions. Like hepatitis B, AIDS appears to be transmitted by intimate direct contact involving mucosal surfaces or by parenteral (injection) routes. Airborne transmission and spread through casual contact seem unlikely.

KEY MEDICAL TERMS ASSOCIATED WITH THE LYMPHATIC SYSTEM

Adenitis (*adeno* = gland; *itis* = inflammation of) Enlarged, tender, and inflamed lymph nodes resulting from an infection.

Elephantiasis Great enlargement of a limb (especially lower limbs) and scrotum resulting from obstruction of lymph glands or vessels by a parasitic worm.

Hypersplenism (*hyper* = over) Abnormal splenic activity involving highly increased blood cell destruction.

Lymphadenectomy (*ectomy* = removal) Removal of a lymph node.

Lymphadenopathy (*patho* = disease) Enlarged, sometimes tender lymph glands.

Lymphangioma (*angio* = vessel; *oma* = tumor) A benign tumor of the lymph vessels.

Lymphangitis Inflammation of the lymphatic vessels.

Lymphedema (*edema* = swelling) Accumulation of lymph fluid producing subcutaneous tissue swelling.

Lymphoma Any tumor composed of lymph tissue. Malignancy of reticuloendothelial cells of lymph nodes is called Hodgkin's disease.

Lymphostasis (*stasis* = halt) A lymph flow stoppage.

Splenomegaly (*mega* = large) Enlarged spleen.

STUDY OUTLINE

Lymphatic Vessels (p. 362)
1. The lymphatic system consists of lymph, lymphatic vessels, lymph nodes, and lymph organs.
2. Lymphatic vessels begin as blind-ended lymph capillaries in tissue spaces between cells.
3. Lymph capillaries merge to form larger vessels, called lymphatics, which ultimately converge into the thoracic duct or right lymphatic duct.
4. Lymphatics have thinner walls and more valves than veins.

Structure of Lymph Nodes (p. 362)
1. Lymph nodes are oval structures located along lymphatics.
2. Lymph enters nodes through afferent lymphatic vessels and exits through efferent lymphatic vessels.
3. Lymph passing through the nodes is processed by macrophages.

Lymph Circulation (p. 365)
1. The passage of lymph is from interstitial fluid, to lymph capillaries, to lymphatics, to lymph trunks, to the thoracic duct or right lymphatic trunk, to the subclavian veins.
2. Lymph flows as a result of skeletal muscle contractions and respiratory movements. It is also aided by valves in the lymphatics.

Principal Groups of Lymph Nodes (p. 367)
1. Lymph nodes are scattered throughout the body as superficial and deep groups.
2. The principal groups of lymph nodes are found in the head and neck, upper extremities, lower extremities, abdomen and pelvis, and thorax.

Lymphatic Organs (p. 375)
1. Tonsils are masses of lymphoid tissue embedded in mucous membranes. They include the pharyngeal, palatine, and lingual tonsils.
2. The spleen functions as a lymphatic organ in phagocytosis of bacteria and worn-out cells and production of lymphocytes and plasma cells. It also acts as a reservoir for blood.
3. The thymus gland functions in immunity by processing T cells and stimulating B cells to develop into antibody-producing plasma cells.

Developmental Anatomy (p. 378)
1. Lymphatic vessels develop from lymph sacs, which develop from veins. Thus they are derived from mesoderm.
2. Lymph nodes develop from lymph sacs that become invaded by mesenchymal cells.

Applications to Health (p. 379)
1. Hypersensitivity is overreactivity to an antigen. Localized anaphylactic reactions include hay fever, asthma, eczema, and hives; acute anaphylaxis is a severe reaction with systemic effects.
2. Tissue rejection of a transplanted tissue or organ involves antibody production against the proteins (antigens) in the transplant. It may be overcome with immunosuppressive drugs.
3. Autoimmune diseases result when the body does not recognize "self" antigens and produces antibodies against them. Several human autoimmune diseases are rheumatoid arthritis (RA), systemic lupus erythematosus (SLE), rheumatic fever, hemolytic and pernicious anemias, myasthenia gravis, and multiple sclerosis (MS).
4. Acquired immune deficiency syndrome (AIDS) lowers the body's immunity by decreasing the number of T cells and reversing the ratio of helper T cells to suppressor T cells. AIDS victims frequently develop Karposi's sarcoma and *Pneumocystis carinii* pneumonia.

REVIEW QUESTIONS

1. Identify the components and functions of the lymphatic system.
2. How do lymphatic vessels originate? Compare veins and lymphatics with regard to structure.
3. What is a lymphangiogram? What is its diagnostic value?
4. Describe the structure of a lymph node. What functions do lymph nodes serve?
5. Construct a diagram to indicate the route of lymph circulation.
6. List and explain the various factors involved in the maintenance of lymph circulation.
7. Define edema. What are some of its causes?
8. For each of the following regions of the body, list the major lymph nodes and the areas of the body they drain: head, neck, upper extremities, lower extremities, abdomen and pelvis, and thorax.
9. Identify the tonsils by location.
10. Describe the location, gross anatomy, histology, and functions of the spleen. What is a splenectomy?
11. Describe the role of the thymus gland in immunity.
12. Describe how the lymphatic system develops.
13. Define hypersensitivity.
14. Why does tissue rejection occur? How is this problem overcome?
15. Define an autoimmune disease. Give several examples.
16. Describe the symptoms of acquired immune deficiency syndrome (AIDS). What are the complications of AIDS?
17. Refer to the glossary of key medical terms associated with the lymphatic system. Be sure that you define each term.

16

Nervous Tissue

Student Objectives

Classify the organs of the nervous system into central and peripheral divisions.

Contrast the histological characteristics and functions of neuroglia and neurons.

Classify neurons by structure and function.

Define a nerve impulse.

Explain how a nerve impulse is conducted across a synapse and a neuromuscular junction.

Explain the organization of neurons in the nervous system.

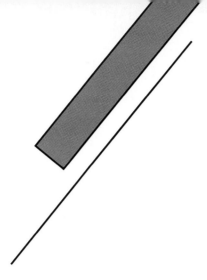

The **nervous system** is the body's control center and communications network. In humans, the nervous system serves three broad functions. First, it senses changes within the body and in the outside environment. This is its sensory function. Second, it interprets the changes. This is its integrative function. Third, it responds to the interpretation by initiating action in the form of muscular contractions or glandular secretions. This is its motor function.

Through sensation, integration, and response, the nervous system represents the body's most rapid means of maintaining homeostasis. Its split-second reactions, carried out by nerve impulses, can normally make the adjustments necessary to keep the body functioning efficiently. As you will see later, the nervous system shares the maintenance of homeostasis with the endocrine system. Although the adjustments made by hormones secreted by endocrine glands are slower than those made by nerve impulses, they are no less effective.

The branch of medical science that deals with the normal functioning and disorders of the nervous system is called **neurology** (*neuro* = nerve, or nervous system; *logos* = study of).

The developmental anatomy of the nervous system is considered in Chapter 18.

ORGANIZATION

The nervous system may be divided into two principal divisions, the central nervous system and the peripheral nervous system, and several subdivisions (Figure 16-1).

The **central nervous system (CNS)** is the control center for the entire system and consists of the brain and spinal cord. All body sensations must be relayed from receptors to the central nervous system if they are to be interpreted and acted on. All the nerve impulses that stimulate muscles to contract and glands to secrete must also pass from the central nervous system.

The various nerve processes that connect the brain and spinal cord with receptors, muscles, and glands constitute the **peripheral** (pe-RIF-er-al) **nervous system (PNS).** The peripheral nervous system may be divided into an afferent system and an efferent system. The **afferent** (AF-er-ent) **system** consists of nerve cells that convey information from receptors in the periphery of the body to the central nervous system. These nerve cells, called *afferent* (*sensory*) *neurons,* are the first cells to pick up incoming information. The **efferent** (EF-er-ent) **system** consists of nerve cells that convey information from the central nervous system to muscles and glands. These nerve cells are called *efferent* (*motor*) *neurons.*

The efferent system is subdivided into a somatic nervous system and an autonomic nervous system. The **somatic nervous system** (*soma* = body), or **SNS,** consists of efferent neurons that conduct impulses from the central nervous system to skeletal muscle tissue. Since the somatic nervous system produces movement only in skeletal muscle tissue, it is under conscious control and therefore voluntary. The **autonomic nervous system** (*auto* = self; *nomos* = law), or **ANS,** by contrast, contains efferent neurons that convey impulses from the central nervous system to smooth muscle tissue, cardiac muscle tissue, and glands. Since it produces reponses only in involuntary muscles and glands, it is usually considered to be involuntary.

With few exceptions, the viscera receive nerve fibers from the two divisions of the autonomic nervous system: the **sympathetic division** and the **parasympathetic division.** In general, the fibers of one division stimulate or increase an organ's activity, while the fibers from the other inhibit or decrease activity (see Chapter 19).

HISTOLOGY

Despite the organizational complexity of the nervous system, it consists of only two principal kinds of cells: neurons and neuroglia. Neurons make up the nervous tissue that forms the structural and functional portion of the system. They are highly specialized for impulse conduction and for all special functions attributed to the nervous

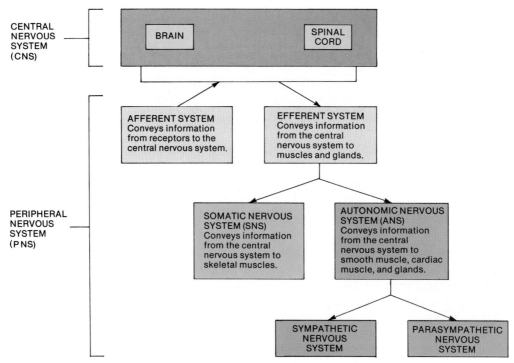

FIGURE 16-1 Organization of the nervous system.

system: thinking, controlling muscle activity, regulating glands. Neuroglia serve as a special supporting and protective component of the nervous system.

NEUROGLIA

The cells of the nervous system that perform the functions of support and protection are called **neuroglia** (noo-ROG-lē-a; *neuro* = nerve; *glia* = glue), or **glial cells** (Figure 16-2). Neuroglia are generally smaller than neurons and outnumber them by 5 to 10 times. Many of the neuroglial cells form a supporting network by twining around nerve cells or lining certain structures in the brain and spinal cord. Others bind nervous tissue to supporting structures and attach the neurons to their blood vessels. A few types of neuroglial cells also serve specialized protective functions. For example, many nerve fibers are coated with a phospholipid sheath, called a myelin sheath, produced by one type of neuroglia. Certain small glial cells are phagocytic; they protect the central nervous system from disease by engulfing invading microbes and clearing away debris. Neuroglia are of clinical interest because they are a common source of tumors (gliomas) of the nervous system. It is estimated that gliomas account for 40 to 45 percent of intracranial tumors. Unfortunately, gliomas are very invasive. Exhibit 16-1 lists the neuroglial cells and summarizes their functions.

NEURONS

Nerve cells, called **neurons,** are responsible for conducting impulses from one part of the body to another. They

are the structural and functional units of the nervous system.

Structure

A neuron consists of three distinct portions: (1) cell body, (2) dendrites, and (3) axon (Figure 16-3a). The **cell body, soma,** or **perikaryon** (per'-i-KAR-ē-on), contains a well-defined nucleus and nucleolus surrounded by a granular cytoplasm. Within the cytoplasm are typical organelles such as lysosomes, mitochondria, and Golgi complexes. Many neurons also contain cytoplasmic inclusions such as *lipofuscin* pigment that occurs as clumps of yellowish-brown granules. Lipofuscin may be a by-product of lysosomal activity. Although its significance is unknown, lipofuscin is related to aging; the amount of pigment increases with age. Also located in the cytoplasm are structures characteristic of neurons: chromatophilic substance and neurofibrils. The *chromatophilic substance* (*Nissl bodies*) is an orderly arrangement of granular (rough) endoplasmic reticulum whose function is protein synthesis. Newly synthesized proteins pass from the perikaryon into the neuronal processes, mainly the axon, at the rate of about 1 mm (0.04 in)/day. These proteins replace those lost during metabolism and are used for growth of neurons and regeneration of peripheral nerve fibers. *Neurofibrils* are long, thin fibrils composed of microtubules. They may assume a function in support and the transportation of nutrients. Mature neurons do not contain a mitotic apparatus. The significance of this absence is noted shortly.

The cytoplasmic processes of neurons generally depend

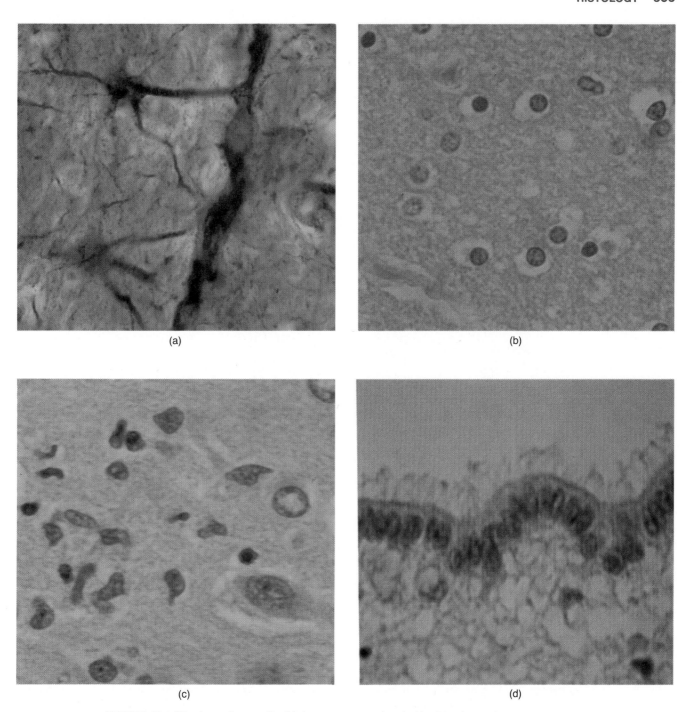

(a)

(b)

(c)

(d)

FIGURE 16-2 Histology of neuroglia. (a) Astrocytes associated with a blood vessel at a magnification of 1000×. (Courtesy of Reschme, Peter Arnold, Inc.) (b) Oligodendrocytes (cells in clear areas) at a magnification of 400×. (c) Microglia at a magnification of 400×. (d) Ependyma lining the fourth ventricle at a magnification of 400×. Note the cilia. (b–d © 1984, Rotker, Taurus.)

on the direction in which they conduct impulses. There are two kinds: dendrites and axons. **Dendrites** (*dendro* = tree) are highly branched, thick extensions of the cytoplasm of the cell body. They typically contain chromatophilic substance, mitochondria, and other cytoplasmic organelles. A neuron usually has several main dendrites. Their function is to conduct an impulse toward the cell body.

The second type of cytoplasmic process, called an **axon (axis cylinder),** is a single, highly specialized, long, thin process that conducts impulses away from the cell body to another neuron or tissue. It usually originates from the cell body as a small conical elevation called the *axon hillock.* An axon contains mitochondria and neurofibrils but no chromatophilic substance; thus it does not carry on protein synthesis. Its cytoplasm, called *axoplasm,* is

EXHIBIT 16-1   NEUROGLIA OF CENTRAL NERVOUS SYSTEM

TYPE	DESCRIPTION	FUNCTION
Astrocytes (*astro* = star; *cyte* = cell)	Star-shaped cells with numerous processes. *Protoplasmic astrocytes* are found in the gray matter of the CNS, and *fibrous astrocytes* are found in the white matter of the CNS.	Twine around nerve cells to form supporting network in brain and spinal cord; attach neurons to their blood vessels.
Oligodendrocytes (*oligo* = few; *dendro* = tree)	Resemble astrocytes in some ways, but processes are fewer and shorter.	Give support by forming semirigid connective tissue rows between neurons in brain and spinal cord; produce a phospholipid myelin sheath around axons of neurons of central nervous system.
Microglia (*micro* = small; *glia* = glue)	Small cells with few processes; derived from monocytes; normally stationary, but may migrate to site of injury; also called brain macrophages.	Engulf and destroy microbes and cellular debris; may migrate to area of injured nervous tissue and function as small macrophages.
Ependyma (ependymocytes) (*ependyma* = upper garment)	Epithelial cells arranged in a single layer and ranging in shape from squamous to columnar; many are ciliated.	Form a continuous epithelial lining for the ventricles of the brain (spaces that form and circulate cerebrospinal fluid) and the central canal of the spinal cord; probably assist in the circulation of cerebrospinal fluid.

surrounded by a plasma membrane known as the *axolemma* (*lemma* = sheath, or husk). Axons vary in length from a few millimeters (1 mm = 0.04 in) in the brain to a meter (3.28 ft) or more between the spinal cord and toes. Along the course of an axon, there may be side branches called *axon collaterals*. The axon and its collaterals terminate by branching into many fine filaments called *axon terminals*, or *telodendria*. The distal ends of axon terminals are expanded into bulblike structures called *synaptic end bulbs*, which are important in nerve impulse conduction. They contain membrane-enclosed sacs called *synaptic vesicles* that store chemicals (neurotransmitters) which determine whether impulse conduction occurs or not.

The cell body of a neuron is essential for the synthesis of many substances that sustain the life of the nerve cell. Neurons have two types of intracellular systems for transporting synthesized materials from the cell body. The slower one, called **axoplasmic flow,** conveys axoplasm in one direction only—from the cell body toward axon terminals. This mechanism may occur by protoplasmic streaming and supplies new axoplasm for developing or regenerating axons and renews axoplasm in growing and mature axons. The faster type of intracellular transport is called **axonal transport.** It conveys materials in both directions—away from the cell body and toward the cell body—possibly along tracks formed by microtubules and filaments. Axonal transport moves various organelles and materials that form the membranes of the axolemma, synaptic end bulbs, and synaptic vesicles. Materials returning to the cell body are degraded or recycled.

CLINICAL APPLICATION

The route taken by materials back to the cell body by axonal transport is the route by which the **herpes virus** and **rabies virus** make their way back to nerve cell bodies where they multiply and cause their damage. The toxin produced by the **tetanus bacterium** uses the same route to reach the central nervous system. In fact, the time delay between the release of the toxin and the first appearance of symptoms is in part due to the time required for movement of the toxin by axonal transport.

The term **nerve fiber** is applied to an axon and its sheaths. Figure 16-3b shows two sections of a nerve fiber of the peripheral nervous system. Many axons, especially

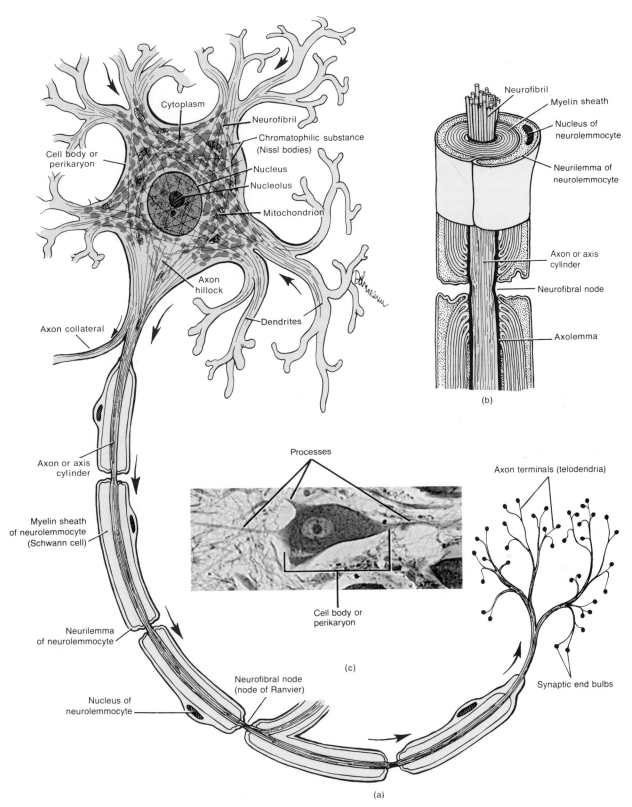

Cytoplasm

Neurofibril

Chromatophilic substance
(Nissl bodies)

Cell body or
perikaryon

Nucleus

Nucleolus

Mitochondrion

Axon
hillock

Dendrites

Axon collateral

Axon or axis
cylinder

Myelin sheath
of neurolemmocyte
(Schwann cell)

Neurilemma
of neurolemmocyte

Nucleus of
neurolemmocyte

Neurofibral node
(node of Ranvier)

Neurofibril

Myelin sheath

Nucleus of
neurolemmocyte

Neurilemma of
neurolemmocyte

Axon or axis
cylinder

Neurofibral node

Axolemma

(b)

Processes

Cell body or
perikaryon

(c)

Axon terminals (telodendria)

Synaptic end bulbs

(a)

FIGURE 16-3 Structure of a typical neuron as exemplified by an efferent (motor) neuron. (a) An entire efferent (motor) neuron. Arrows indicate the direction in which nerve impulses travel. (b) Sections through a myelinated fiber. (c) Photomicrograph of an efferent neuron at a magnification of 640×. (Courtesy of Biophoto Associates.)

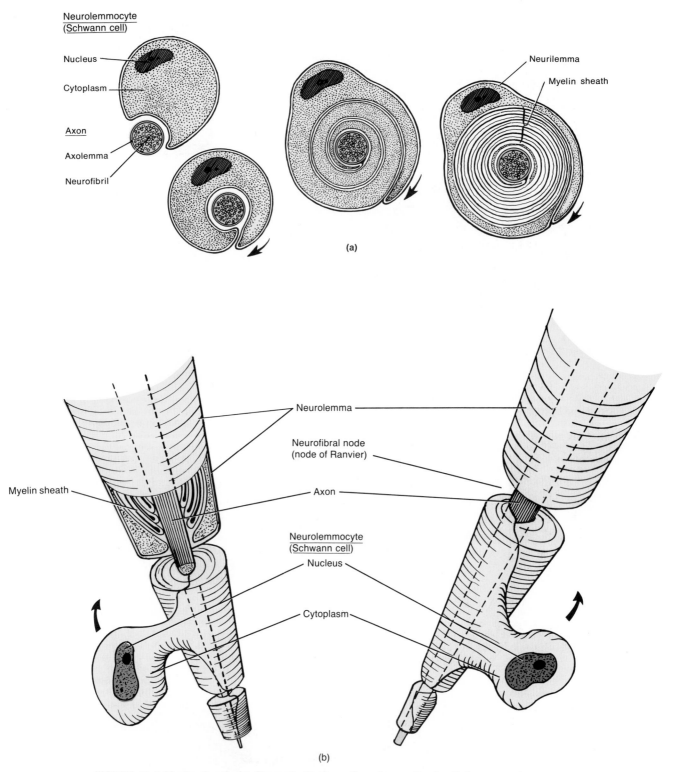

FIGURE 16-4 Myelin sheath. (a) Stages in the formation of a myelin sheath by a neurolemmocyte (Schwann cell). (b) Relationship of neurolemmocytes to axons.

large, peripheral axons, are surrounded by a multilayered white, phospholipid, segmented covering called the **myelin sheath.** Axons containing such a covering are *myelinated,* while those without it are *unmyelinated.* The function of the myelin sheath is to increase the speed of nerve impulse conduction and to insulate and maintain the axon. Myelin is responsible for the color of the white matter in the nerves, brain, and spinal cord.

The myelin sheath of axons of the peripheral nervous system is produced by flattened cells, called **neurolemmo-**

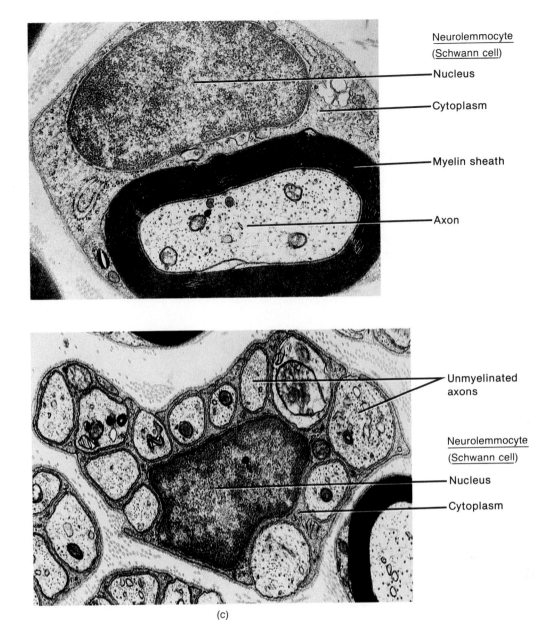

Neurolemmocyte
(Schwann cell)

Nucleus

Cytoplasm

Myelin sheath

Axon

Unmyelinated
axons

Neurolemmocyte
(Schwann cell)

Nucleus

Cytoplasm

(c)

FIGURE 16-4 (*Continued*) Myelin sheath. (c) Electron micrographs of a myelinated axon (above) and several unmyelinated axons (below). (© Biology Media, Schultz, 1980, Photo Researchers.)

cytes (Schwann cells), located along the axons. In the formation of a sheath, a developing neurolemmocyte encircles the axon until its ends meet and overlap (Figure 16-4). The cell then winds around the axon many times and, as it does so, the cytoplasm and nucleus are pushed to the outside layer. The inner portion, consisting of several layers of neurolemmocyte membrane, is the myelin sheath. The peripheral nucleated cytoplasmic layer of the neurolemmocyte (the outer layer that encloses the sheath) is called the **neurilemma.**

The neurilemma is found only around fibers of the peripheral nervous system. Its function is to assist in the regeneration of injured axons. Between the segments of the myelin sheath are unmyelinated gaps called **neurofibral nodes (nodes of Ranvier).** Unmyelinated fibers are also enclosed by neurolemmocytes but without multiple wrappings.

Nerve fibers of the central nervous system may also be myelinated or unmyelinated. Myelination of central nervous system axons is accomplished by oligodendrocytes in somewhat the same manner that neurolemmocytes myelinate peripheral nervous system axons (see Ex-

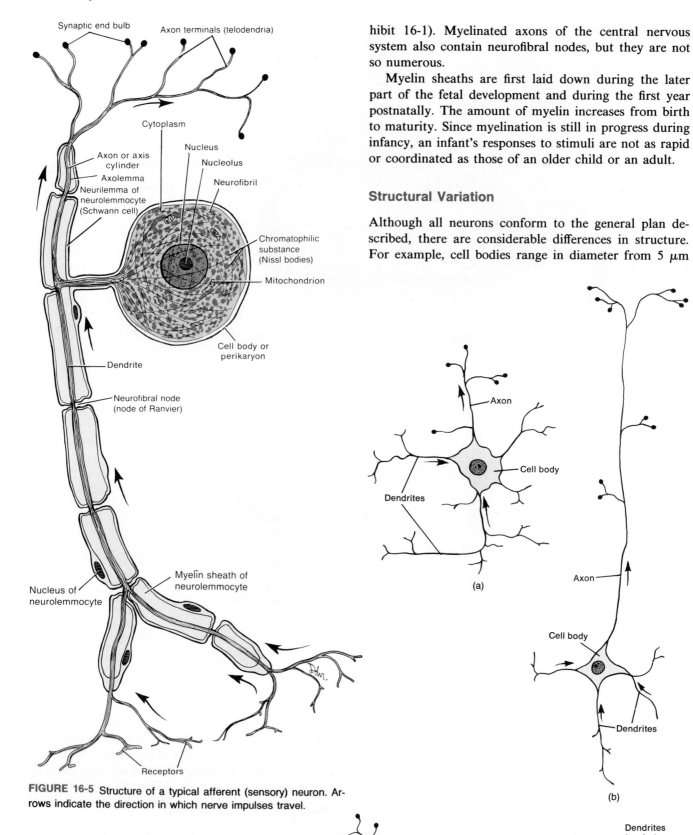

FIGURE 16-5 Structure of a typical afferent (sensory) neuron. Arrows indicate the direction in which nerve impulses travel.

FIGURE 16-6 Representative association (connecting) neurons. (a) Stellate cell. (b) Cell of Martinotti. (c) Horizontal cell of Cajal.

hibit 16-1). Myelinated axons of the central nervous system also contain neurofibral nodes, but they are not so numerous.

Myelin sheaths are first laid down during the later part of the fetal development and during the first year postnatally. The amount of myelin increases from birth to maturity. Since myelination is still in progress during infancy, an infant's responses to stimuli are not as rapid or coordinated as those of an older child or an adult.

Structural Variation

Although all neurons conform to the general plan described, there are considerable differences in structure. For example, cell bodies range in diameter from 5 μm for the smallest cells to 135 μm for large motor neurons. The pattern of dendritic branching is varied and distinctive for neurons in different parts of the body. The axons of very small neurons are only a fraction of a millimeter in length and lack a myelin sheath, while axons of large neurons are over a meter long and are usually enclosed in a myelin sheath.

Classification

The different neurons in the body may be classified by structure and function.

The structural classification is based on the number of processes extending from the cell body. **Multipolar neurons** have several dendrites and one axon (see Figure 16-3a). Most neurons in the brain and spinal cord are of this type. **Bipolar neurons** have one dendrite and one axon and are found in the retina of the eye, the inner ear, and the olfactory area. **Unipolar neurons** have only one process extending from the cell body. The single pro-

cess divides into a central branch, which functions as an axon, and a peripheral branch, which functions as a dendrite (see Figure 16-5). Unipolar neurons originate in the embryo as bipolar neurons. During development, the axon and dendrite fuse into a single process and become a functional axon. Unipolar neurons are found in posterior (sensory) root ganglia of spinal nerves.

The functional classification of neurons is based on the direction in which they transmit impulses. **Sensory neurons,** called **afferent neurons,** transmit impulses from receptors in the skin, sense organs, and viscera to the brain and spinal cord. They are usually unipolar (Figure 16-5). **Motor neurons,** called **efferent neurons,** convey impulses from the brain and spinal cord to effectors, which may be either muscles or glands (see Figure 16-3a). Other neurons, called **association (connecting,** or **internuncial) neurons,** carry impulses from sensory neurons to motor neurons and are located in the brain and spinal cord (Figure 16-6). Examples of association neurons shown are a *stellate cell,* a *cell of Martinotti* (mar'-ti-NOT-ē),

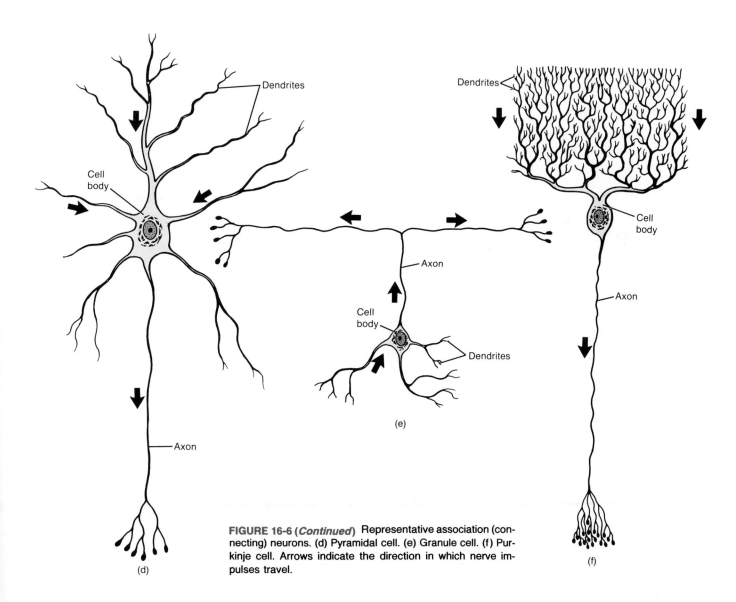

FIGURE 16-6 (*Continued***)** Representative association (connecting) neurons. (d) Pyramidal cell. (e) Granule cell. (f) Purkinje cell. Arrows indicate the direction in which nerve impulses travel.

a *horizontal cell of Cajal* (kā-HAL), and a *pyramidal* (pi-RAM-i-dal) *cell.* All are found in the cerebral cortex, the outer layer of the cerebrum. The *granule cell* and *Purkinje* (pur-KIN-jē) *cell* are association neurons in the cortex of the cerebellum. Most neurons in the body, perhaps 90 percent, are association neurons.

The processes of afferent and efferent neurons are arranged into bundles called *nerves.* Since nerves lie outside the central nervous system, they belong to the peripheral nervous system. The functional components of nerves are the nerve fibers, which may be grouped according to the following scheme.

1. General somatic afferent fibers conduct impulses from the skin, skeletal muscles, and joints to the central nervous system.

2. General somatic efferent fibers conduct impulses from the central nervous system to skeletal muscles. Impulses over these fibers cause the contraction of skeletal muscles.

3. General visceral afferent fibers convey impulses from the viscera and blood vessels to the central nervous system.

4. General visceral efferent fibers belong to the autonomic nervous system and are also called **autonomic fibers.** They convey impulses from the central nervous system to cause contractions of smooth and cardiac muscles and secretion by glands.

In addition to being grouped as nerves, neural tissue is also organized into other structures such as ganglia, tracts, nuclei, and horns. These are described in Chapter 17.

REGENERATION

Two striking features of nervous tissue are its limited ability to regenerate and its highly developed ability to produce and transmit electrical messages called nerve impulses.

Unlike the cells of epithelial tissue, neurons have only limited powers for **regeneration,** that is, a natural ability to renew themselves. Around six months of age, the cell bodies of most developing nerve cells lose their mitotic apparatus (centrioles and mitotic spindles) and their ability to reproduce. Thus, when a neuron is damaged or destroyed, it cannot be replaced by the daughter cells of other neurons. A neuron destroyed is permanently lost and only some types of damage may be repaired.

Damage to some types of myelinated axons often can be repaired if the cell body remains intact and if the cell that performs the myelination remains active. Axons in the peripheral nervous system are myelinated by neurolemmocytes. They proliferate following axonal damage and their neurilemmas form a tube that assists in regeneration (Chapter 17). Axons in the brain and spinal cord

(central nervous system) are myelinated by oligodendroglial cells. These cells do not form neurilemmas to assist in regeneration and do not survive following axonal damage. An added complication in the central nervous system is that following axonal damage, the affected region is rapidly converted into a special form of scar tissue by astroglial proliferation. The scar tissue forms a barrier to regeneration. Thus, an injury to the brain or spinal cord is also permanent because axonal regeneration is blocked by rapid scar tissue formation. An injury to a nerve in the arm (peripheral nervous system) may repair itself before scar tissue forms and so some nerve function may be restored.

NERVE IMPULSE

Although it is beyond the scope of this text to describe the details of a nerve impulse, certain concepts must be understood to know how the nervous system works. Very simply, a **nerve impulse** is a wave of negativity that self-propagates along the surface of the membrane of a neuron. Among other things, a nerve impulse depends on the movement of sodium, potassium, and other ions between interstitial fluid and the inside of a neuron. For a nerve impulse to begin, a stimulus of adequate strength must be applied to the neuron. A **stimulus** is a change in the environment of sufficient strength to initiate an impulse. The ability of a neuron to respond to a stimulus and convert it into an impulse is known as **excitability.**

The nerve impulse is the most rapid way that the body can respond to environmental changes. It provides the quickest means for achieving homeostasis. The speed of a nerve impulse is determined by the size, type, and physiological condition of the nerve fiber. For example, myelinated fibers with the largest diameters can transmit impulses at speeds up to about 100 m (328 ft)/sec. Unmyelinated fibers with the smallest diameters can transmit impulses at the rate of about 0.5 m (1.5 ft)/sec.

In addition to excitability, neurons are also characterized by **conductivity,** the ability to transmit an impulse to another neuron or another tissue, such as a muscle or a gland. The junction between two neurons is called a **synapse** (*synapse* = to join). The synapse is essential for homeostasis because of its ability to transmit certain impulses and inhibit others. Much of an organism's ability to learn will probably be explained in terms of synapses. Moreover, most diseases of the brain and many psychiatric disorders result from a disruption of synaptic communication. And synapses are the sites of action for most drugs that affect the brain, including therapeutic and addictive substances. Figure 16-7 shows that within a synapse is a minute gap, about 200 Å across, called the **synaptic cleft.** A **presynaptic neuron** is a neuron located before a synapse. A **postsynaptic neuron** is located after a synapse.

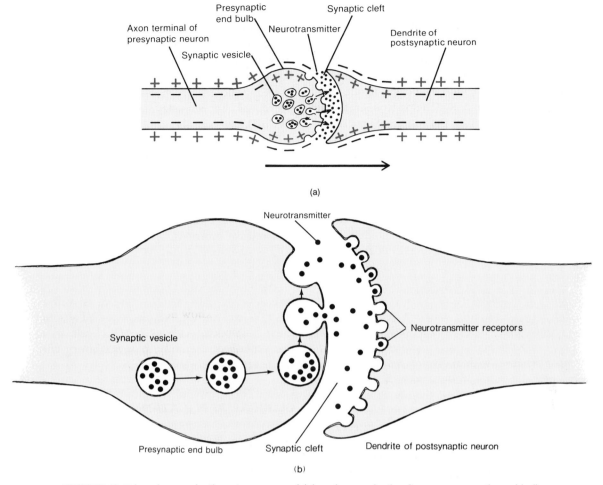

(a)

(b)

FIGURE 16-7 Impulse conduction at synapses. (a) Impulse conduction from a presynaptic end bulb across a synapse to a postsynaptic dendrite. The negative charges on the outside of the membrane indicate impulse conduction; the positive charges on the outside of the membrane indicate that no impulse is conducted. The arrow indicates the direction of impulse conduction. (b) Details of impulse conduction in which synaptic vesicles fuse with the presynaptic membrane and discharge a neurotransmitter into the synaptic cleft.

Impulses are conducted from a neuron to a muscle cell across an area of contact called a **neuromuscular (myoneural) junction,** or **motor end plate,** the details of which were described in Chapter 9. The area of contact between a neuron and glandular cells is known as a **neuroglandular junction.** Together, neuromuscular and neuroglandular junctions are known as **neuroeffector junctions.**

The axon terminals of axons terminate in expanded bulblike structures referred to as **synaptic end bulbs.** The synaptic end bulbs of a presynaptic neuron commonly synapse with the dendrites, cell body, or axon hillock of a postsynaptic neuron. Accordingly, synapses may be classified as *axodendritic, axosomatic,* and *axoaxonic.* The synaptic end bulbs from a single presynaptic neuron may synapse with several postsynaptic neurons. Such an arrangement, called **divergence,** permits a single presynaptic neuron to influence several postsynaptic neurons or several muscle or gland cells at the same time (see Figure

16-9a). In another arrangement, called **convergence,** the synaptic end bulbs of several presynaptic neurons synapse with a single postsynaptic neuron (see Figure 16-9b). This arrangement permits stimulation or inhibition of the postsynaptic neuron.

At a synapse there is only **one-way impulse conduction**—from a presynaptic axon to a postsynaptic dendrite, cell body, or axon hillock. Impulses must move forward over their pathways. They cannot back up into another presynaptic neuron. Such a mechanism is crucial in preventing impulse conduction along improper pathways.

Whether an impulse is conducted across a synapse, neuromuscular junction, or neuroglandular junction depends on the presence of chemicals called **neurotransmitters.** These chemicals are made by the neuron, usually from amino acids. Following its production and transportation to the synaptic end bulbs, the neurotransmitter is stored in the bulbs in small membrane-enclosed sacs called

synaptic vesicles (Figure 16-7). Each of the thousands of synaptic vesicles present may contain between 10,000 and 100,000 neurotransmitter molecules.

When a nerve impulse arrives at a synaptic end bulb of a presynaptic neuron, it is believed that a small amount of calcium ions leaks into the bulb, attracts synaptic vesicles to the plasma membrane, and helps to liberate the neurotransmitter molecules from the vesicles. Some scientists believe that the synaptic vesicles fuse with the plasma membrane of the presynaptic neuron, form openings, and release the neurotransmitter through the openings into the synaptic cleft. Other scientists think that the neurotransmitter leaves through small channels to enter the synaptic cleft. In either case, the neurotransmitter enters the synaptic cleft and, depending on the chemical nature of the neurotransmitter and the interaction of the neurotransmitter with receptors of the postsynaptic plasma membrane, several things can happen.

An **excitatory transmitter-receptor interaction** is one that generates a nerve impulse across a synapse, while an **inhibitory transmitter-receptor interaction** is one that inhibits a nerve impulse across a synapse. A single postsynaptic neuron synapses with many presynaptic neurons. Some presynaptic end bulbs produce excitation and some produce inhibition. The sum of all the effects, excitatory and inhibitory, determines the effect on the postsynaptic neuron. Thus the postsynaptic neuron is an **integrator.** It receives signals, integrates them, and then responds accordingly. The postsynaptic neuron may respond in the following ways:

1. If the excitatory effect is greater than the inhibitory effect, but equal to or less than the threshold (minimal) level of stimulation, the result is facilitation, that is, near excitation so that subsequent stimuli can generate an impulse.

2. If the excitatory effect is greater than the inhibitory effect, but equal to or higher than the threshold level of stimulation, the result is generation of an impulse.

3. If the inhibitory effect is greater than the excitatory effect, the result is inhibition of the impulse.

Perhaps the best studied neurotransmitter substance is **acetylcholine** (as'-ē-til-KŌ-lēn), or **ACh.** It is a neurotransmitter released by many neurons outside the brain and spinal cord and by some neurons inside the brain and cord. This neurotransmitter was discussed in Chapter 9 in relation to impulse conduction from a motor neuron to a muscle fiber across the neuromuscular junction (motor end plate). At neuromuscular junctions, ACh binds to receptor sites on the muscle fiber membrane and increases the membrane's permeability to Na^+ ions. The acetylcholine receptor is an integral protein in the plasma membrane of the muscle fiber. When ACh molecules bind to the ACh receptor, a change occurs in the receptor, causing its channel to open. As a result, there is an inward movement of Na^+ ions. In the absence of ACh, the recep-

tor channel remains closed. The inward movement of Na^+ ions leads to a sequence of events that generates a nerve impulse, causing the muscle fiber to contract. As long as ACh is present in the synaptic cleft, it can stimulate a muscle fiber almost indefinitely. The transmission of a continuous succession of impulses by ACh is normally prevented by an enzyme called **acetylcholinesterase** (AChE), or simply **cholinesterase** (kō'-lin-ES-ter-ās). AChE is found on the surfaces of the subneural clefts of the membrane of muscle fibers. Within $\frac{1}{500}$ sec, AChE inactivates ACh. This action permits the membrane of the muscle fiber to discontinue impulse transmission immediately so that another impulse may be generated. When the next impulse comes through, the synaptic vesicles release more ACh, the impulse is conducted, and AChE again inactivates ACh. This cycle is repeated over and over again.

ACh is released at some neuromuscular junctions that have cardiac and smooth muscle, as well as those that have skeletal muscle. It is also released at some neuroglandular junctions. Although ACh leads to excitation in many parts of the body, it is inhibitory with respect to the heart (vagus nerve).

CLINICAL APPLICATION

There are many ways that **synaptic conduction can be altered** by disease, drugs, and pressure. In Chapter 9 it was noted that *myasthenia gravis* results from antibodies directed against acetylcholine receptors on muscle fiber membranes at neuromuscular junctions, causing dysfunctions in muscular contractions. *Alkalosis,* an increase in pH above 7.45 up to about 8.0, results in increased excitability of neurons that can cause cerebral convulsions, while *acidosis,* a decrease in pH below 7.35 down to about 6.80, results in a depression of neuronal activity and can produce a comatose state.

Curare, the deadly South American Indian arrowhead poison, also competes for acetylcholine receptor sites. Since the muscles of respiration depend on neuromuscular transmission to initiate their contraction, death results from asphyxiation. *Neostigmine* and *physostigmine* are anticholinesterase agents that combine with acetylcholinesterase to inactivate it for several hours. As a result, acetylcholine accumulates and produces repetitive stimulation of muscles. This causes muscular spasm, and death due to laryngeal spasm can ensue. *Diisopropyl fluorophosphate* is a very powerful nerve gas that inactivates acetylcholinesterase for up to several weeks, making it a particularly lethal drug. The *botulism toxin* inhibits the release of acetylcholine, thus inhibiting muscle contraction. The toxin is exceedingly potent in even small amounts (less than 0.0001 mg) and is the substance involved in one type of food

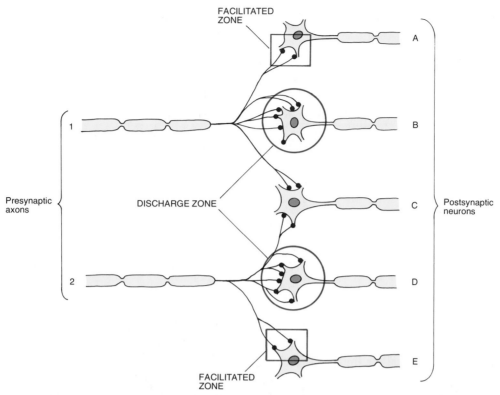

FIGURE 16-8 Relative positions of discharge and facilitated zones in a very simplified version of a neuronal pool.

poisoning. *Hypnotics, tranquilizers,* and *anesthetics* depress synaptic conduction, while *caffeine* and *benzedrine* lead to facilitation (near excitation).

Pressure has an effect on impulse conduction also. If excessive or prolonged pressure is applied to a nerve, impulse conduction is interrupted and part of the body may "go to sleep," producing a tingling sensation. This sensation is caused by an accumulation of waste products and a depressed circulation of blood.

ORGANIZATION OF NEURONS

The central nervous system contains millions of neurons. Their arrangement is not haphazard. They are organized into definite patterns called **neuronal pools.** Each pool differs from all others and has its own role in regulating homeostasis.

A neuronal pool may contain thousands or even millions of neurons. To illustrate the composition of a neuronal pool, a simplified version is given in Figure 16-8. This example contains only five postsynaptic neurons and two incoming presynaptic neurons. The postsynaptic neurons are subject to stimulation by the incoming presynaptic neurons. The postsynaptic neurons in the pool may be stimulated by one or several presynaptic knobs. More-

over, the incoming presynaptic knobs may produce facilitation (near excitation), excitation, or inhibition. A principal feature of a neuronal pool is the location of the presynaptic neurons in relation to the postsynaptic neurons. Compare the location of presynaptic axon 1 with that of postsynaptic neuron B. Since they are aligned, more presynaptic end bulbs of axon 1 synapse with postsynaptic neuron B than with postsynaptic neurons A and C. We say that postsynaptic neuron B is in the center of the field of presynaptic axon 1. Consequently, postsynaptic neuron B usually receives sufficient presynaptic end bulbs from axon 1 to generate an impulse. This region where the neuron in the pool fires is called the **discharge zone.**

Now look outside the field of presynaptic axon 1. Note its relation to postsynaptic neuron A. Here postsynaptic neuron A in the pool is receiving few presynaptic end bulbs from the axon supplying postsynaptic neuron B. Thus there are insufficient presynaptic end bulbs to fire postsynaptic neuron A but enough to cause facilitation. We therefore call this area the **facilitated zone.** Presynaptic axon 1, when stimulated, will cause excitation of postsynaptic neuron B and facilitation of postsynaptic neuron A.

Neuronal pools in the central nervous system are arranged in patterns over which the impulses are conducted. These are termed **circuits. Simple series circuits** are arranged so that a presynaptic neuron stimulates a single

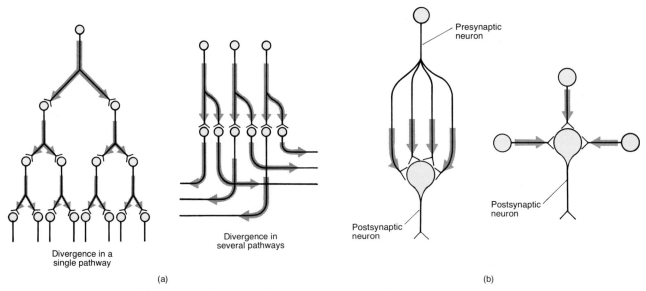

FIGURE 16-9 Diverging and converging circuits. (a) Diverging. (b) Converging.

neuron in a pool. The single neuron then stimulates another, and so on. In other words, the impulse is relayed
from one neuron to another in succession as a new impulse
is generated at each synapse.

Most circuits, however, are more complex. In a **diverging circuit,** the impulse from a single presynaptic neuron
causes the stimulation of increasing numbers of cells along
the circuit (Figure 16-9a). An example of such a circuit
is a single motor neuron in the brain stimulating numerous
other motor neurons in the spinal cord that, in turn,
leave the spinal cord where each stimulates many skeletal
muscle fibers. Thus a single impulse may result in the
contraction of several skeletal muscle fibers. In another
kind of diverging circuit, impulses from one pathway are
relayed to other pathways so the same information travels
in various directions at the same time. This circuit is
common along sensory pathways of the nervous system.

Another kind of circuit is called a **converging circuit**
(Figure 16-9b). In one pattern of convergence, the postsynaptic neuron receives impulses from several fibers of
the same source. Here there is the possibility of strong
excitation or inhibition. In a second pattern, the post-

synaptic neuron receives impulses from several different
sources. Here there is a possibility of reacting the same
way to different stimuli. Suppose your reaction to vomit
is distinctly unpleasant. The smell of vomit (one kind
of stimulus), the sight of vomit (another kind), or just
reading about vomit (still another kind) might all have
the same effect on you—an unpleasant one.

Some circuits in your body are constructed so that
once the presynaptic cell is stimulated, it will cause the
postsynaptic cell to transmit a series of impulses. One
such circuit is called a **reverberating (oscillatory) circuit**
(Figure 16-10a). In this pattern, the incoming impulse
stimulates the first neuron, which stimulates the second,
which stimulates the third, and so on. Branches from
the second and third neurons synapse with the first, however, sending the impulse back through the circuit again
and again. A central feature of the reverberating circuit
is that once fired, the output signal may last from a few
seconds to many hours. The duration depends on the
number and arrangement of neurons in the circuit.
Among the body responses thought to be the result of
output signals from reverberating circuits are the rate

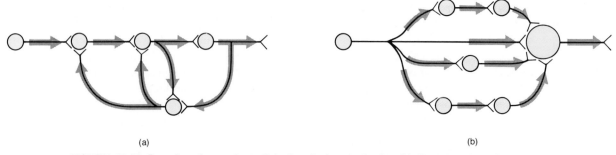

FIGURE 16-10 Reverberating and parallel after-discharge circuits. (a) Reverberating. (b) Parallel
after-discharge.

of breathing, coordinated muscular activities, waking up, and sleeping (when reverberation stops). Some scientists think reverberating circuits are related to short-term memory.

A final circuit worth consideration is the **parallel after-discharge circuit** (Figure 16-10b). Like a reverberating circuit, a parallel after-discharge circuit is constructed so the postsynaptic cell transmits a series of impulses. In a parallel after-discharge circuit, a single presynaptic cell stimulates a group of neurons, each of which synapses

with a common postsynaptic cell. The advantage of this circuit is that the postsynaptic neuron can send out a stream of impulses in succession as they are received. The impulses leave the postsynaptic neuron once every $\frac{1}{2,000}$ second. This circuit has no feedback system. Once all the neurons in the circuit have transmitted their impulses to the postsynaptic neuron, the circuit is broken. It is thought that the parallel after-discharge circuit is employed for precise activities like mathematical calculations.

STUDY OUTLINE

Organization (p. 383)

1. The nervous system controls and integrates all body activities by sensing changes, interpreting them, and reacting to them.
2. The central nervous system consists of the brain and spinal cord.
3. The peripheral nervous system is classified into an afferent system and an efferent system.
4. The efferent system is subdivided into a somatic nervous system and an autonomic nervous system.
5. The somatic nervous system consists of efferent neurons that conduct impulses from the central nervous system to skeletal muscle tissue.
6. The autonomic nervous system contains efferent neurons that convey impulses from the central nervous system to smooth muscle tissue, cardiac muscle tissue, and glands.

Histology (p. 383)
Neuroglia
1. Neuroglia are specialized tissue cells that support neurons, attach neurons to blood vessels, produce the myelin sheath, and carry out phagocytosis.
2. Neuroglial cells include astrocytes, oligodendrocytes, microglia, and ependyma.

Neurons
1. Neurons, or nerve cells, consist of a perikaryon, or cell body, dendrites that pick up stimuli and convey impulses to the cell body, and usually a single axon. The axon transmits impulses from the neuron to the dendrites or cell body of another neuron or to an effector organ of the body.
2. On the basis of structure, neurons are multipolar, bipolar, and unipolar.
3. On the basis of function, sensory (afferent) neurons transmit impulses to the central nervous system; association neurons transmit impulses to other neurons, including motor neurons; and motor (efferent) neurons transmit impulses to effectors.

Regeneration (p. 392)
1. Around the time of birth, the cell body loses its mitotic apparatus and is no longer able to divide.
2. Nerve fibers (axis cylinders) that have a neurilemma are capable of regeneration.

Nerve Impulse (p. 392)
1. A nerve impulse is a wave of negativity that travels along the surface of the membrane of a neuron.
2. The ability of a neuron to respond to a stimulus and convert it into a nerve impulse is called excitability.
3. The speed of a nerve impulse is determined by the size, type, and physiological condition of the nerve fiber.
4. Conductivity is the ability of a neuron to transmit a nerve impulse to another neuron or another tissue.
5. The junction between neurons is called a synapse. Impulse conduction across a synapse requires the release of neurotransmitters by presynaptic neurons.
6. Impulse conduction at a synapse is one-way conduction.
7. An excitatory transmitter-receptor interaction is one that can generate a nerve impulse across the synapse.
8. An inhibitory transmitter-receptor interaction is one that can inhibit an impulse across a synapse.
9. It is thought that the neurotransmitter that causes excitation in a major portion of the central nervous system is acetylcholine (ACh). An enzyme called acetylcholinesterase (AChE) inactivates acetylcholine.

Organization of Neurons (p. 395)
1. Neurons in the central nervous system are organized into definite patterns called neuronal pools. Each pool differs from all others and has its own role in the functioning nervous system.
2. Neuronal pools are organized into circuits. These include simple series, diverging, converging, reverberating, and parallel after-discharge circuits.

REVIEW QUESTIONS

1. Describe the three basic functions of the nervous system. Distinguish between the central and peripheral nervous system and define the subdivisions of the peripheral nervous system.

2. Relate the terms *voluntary* and *involuntary* to the nervous system.
3. What are neuroglia? List the principal types and their functions.

4. Define a neuron. Diagram and label a neuron. Next to each part list its function.
5. Distinguish between axoplasmic flow and axonal transport. Why is axonal transport clinically important?
6. What is a myelin sheath? What is its function? How is the myelin sheath formed?
7. Define the neurilemma. Why is it important?
8. Discuss the structural classification of neurons. Give an example of each.
9. What are the structural differences between a typical afferent and a typical efferent neuron? Give several examples of association neurons.
10. Describe the functional classification of neurons.
11. Distinguish among the following kinds of fibers: general somatic afferent, general somatic efferent, general visceral afferent, and general visceral efferent.
12. What determines neuron regeneration?
13. Define a nerve impulse.
14. Compare excitability and conductivity.
15. What factors determine the rate of impulse conduction?
16. What is a synapse? How are synapses classified?
17. What is a neurotransmitter? Describe the action of acetylcholine.
18. What events are involved in the conduction of a nerve impulse across a synapse?
19. Distinguish between excitatory and inhibitory transmission.
20. Why does one-way impulse conduction occur?
21. Why is the postsynaptic neuron called an integrator?
22. How do disease, drugs, and pressure affect synaptic transmission?
23. Distinguish between the discharge zone and facilitated zone and neuronal pool.
24. What is a neuron circuit? Distinguish among simple series, diverging, converging, reverberating, and parallel after-discharge circuits.

17

The Spinal Cord and the Spinal Nerves

Student Objectives

Define white matter, gray matter, nerve, ganglion, tract, nucleus, and horn.

Describe the gross anatomical features of the spinal cord.

Explain how the spinal cord is protected.

Describe the structure and location of the spinal meninges.

Discuss the location, general technique, purpose, and significance of a spinal puncture.

Describe the structure of the spinal cord in cross section.

Explain the functions of the spinal cord as a conduction pathway and a reflex center.

List the location, origin, termination, and function of the principal ascending and descending tracts of the spinal cord.

Describe the components of a reflex arc.

Describe the composition and coverings of a spinal nerve.

Name the 31 pairs of spinal nerves.

Explain how a spinal nerve branches on leaving the intervertebral foramen.

Explain the composition and distribution of the cervical, brachial, lumbar, sacral, and coccygeal plexuses.

Define an intercostal nerve.

Define a dermatome and state its clinical importance.

Describe spinal cord injury and list the immediate and long-range effects.

Identify the effects of peripheral nerve damage and conditions necessary for its regeneration.

Explain the causes and symptoms of sciatica, neuritis, and shingles.

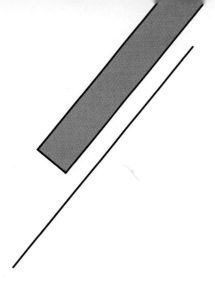

In this chapter, our main concern will be to study the structure and function of the spinal cord and the nerves that originate from it. Keep in mind, however, that the spinal cord is continuous with the brain, and together they constitute the central nervous system.

GROUPING OF NEURAL TISSUE

The term **white matter** refers to aggregations of myelinated axons from many neurons supported by neuroglia. The lipid substance myelin has a whitish color that gives white matter its name. The **gray matter** of the nervous system contains either nerve cell bodies and dendrites or bundles of unmyelinated axons and neuroglia. The absence of myelin in these areas accounts for their gray color.

A **nerve** is a bundle of fibers located outside the central nervous system. Since the dendrites of somatic afferent neurons and axons of somatic efferent neurons of the peripheral nervous system are myelinated, most nerves are white matter. Nerve cell bodies that lie outside the central nervous system are generally grouped with other nerve cell bodies to form **ganglia** (GANG-lē-a; *ganglion* = knot). Ganglia, since they are made up principally of nerve cell bodies, are masses of gray matter.

A **tract** is a bundle of fibers in the central nervous system. Tracts may run long distances up and down the spinal cord. Tracts also exist in the brain and connect parts of the brain with each other and with the spinal cord. The chief spinal tracts that conduct impulses up the cord are concerned with sensory impulses and are called *ascending tracts.* Spinal tracts that carry impulses down the cord are motor tracts and are called *descending tracts.* The major tracts consist of myelinated fibers and are therefore white matter.

A **nucleus** is a mass of nerve cell bodies and dendrites in the central nervous system, all of which have similar functions. It forms gray matter. **Horns,** or **columns,** are the chief areas of gray matter in the spinal cord. The term *horn* describes the two-dimensional appearance of the organization of gray matter in the spinal cord as seen in cross section. The term *column* describes the three-dimensional appearance of the gray matter in longitudinal columns. Since the white matter of the spinal cord is also arranged in columns, we will refer to the gray matter as being arranged in horns.

SPINAL CORD

GENERAL FEATURES

The **spinal cord** is a cylindrical structure that is slightly flattened anteriorly and posteriorly. It begins as a continuation of the medulla oblongata, the inferior part of the brain stem, and extends from the foramen magnum of the occipital bone to the level of the second lumbar vertebra (Figure 17-1). The length of the adult spinal cord ranges from 42 to 45 cm (16 to 18 in). The diameter of the cord varies at different levels.

When the cord is viewed externally, two conspicuous enlargements can be seen. The superior enlargement, the **cervical enlargement,** extends from the fourth cervical to the first thoracic vertebra. Nerves that supply the upper extremities arise from the cervical enlargement. The inferior enlargement, called the **lumbar enlargement,** extends from the ninth to the twelfth thoracic vertebra. Nerves that supply the lower extremities arise from the lumbar enlargement.

Below the lumbar enlargement, the spinal cord tapers to a conical portion known as the **conus medullaris** (KŌ-nus med-yoo-LAR-is). The conus medullaris ends at the level of the intervertebral disc between the first and second lumbar vertebra. Arising from the conus medullaris is the **filum terminale** (FĪ-lum ter-mi-NAL-ē), a nonnervous fibrous tissue of the spinal cord that extends inferiorly to attach to the coccyx. The filum terminale consists mostly of pia mater, the innermost of three membranes that cover and protect the spinal cord and brain. Some nerves that arise from the lower portion of the cord do not leave the vertebral column immediately. They angle

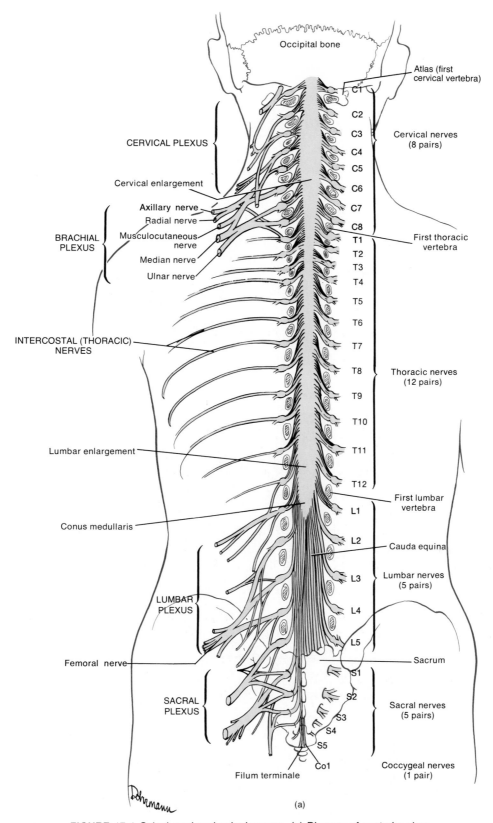

Occipital bone

Atlas (first cervical vertebra)

C1

CERVICAL PLEXUS

C2

C3

C4

C5

Cervical nerves (8 pairs)

Cervical enlargement

C6

C7

Axillary nerve

Radial nerve

C8

BRACHIAL PLEXUS

Musculocutaneous nerve

T1

First thoracic vertebra

Median nerve

T2

Ulnar nerve

T3

T4

T5

T6

INTERCOSTAL (THORACIC) NERVES

T7

T8

Thoracic nerves (12 pairs)

T9

T10

Lumbar enlargement

T11

T12

First lumbar vertebra

Conus medullaris

L1

L2

Cauda equina

L3

Lumbar nerves (5 pairs)

L4

LUMBAR PLEXUS

L5

Femoral nerve

Sacrum

S1

SACRAL PLEXUS

S2

S3

Sacral nerves (5 pairs)

S4

S5

Co1

Filum terminale

Coccygeal nerves (1 pair)

(a)

FIGURE 17-1 Spinal cord and spinal nerves. (a) Diagram of posterior view.

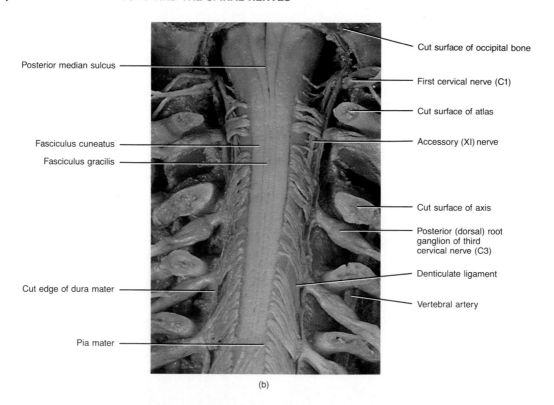

Posterior median sulcus

Fasciculus cuneatus

Fasciculus gracilis

Cut edge of dura mater

Pia mater

Cut surface of occipital bone

First cervical nerve (C1)

Cut surface of atlas

Accessory (XI) nerve

Cut surface of axis

Posterior (dorsal) root ganglion of third cervical nerve (C3)

Denticulate ligament

Vertebral artery

(b)

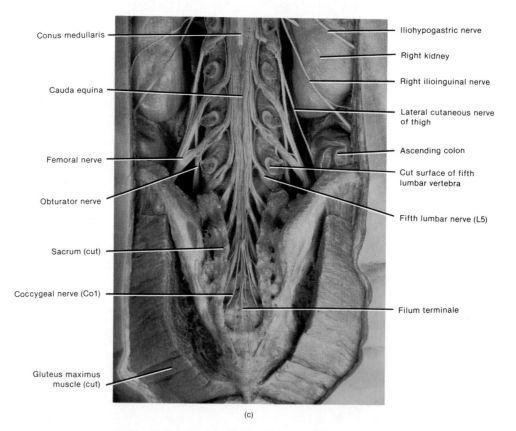

Conus medullaris

Cauda equina

Femoral nerve

Obturator nerve

Sacrum (cut)

Coccygeal nerve (Co1)

Gluteus maximus muscle (cut)

Iliohypogastric nerve

Right kidney

Right ilioinguinal nerve

Lateral cutaneous nerve of thigh

Ascending colon

Cut surface of fifth lumbar vertebra

Fifth lumbar nerve (L5)

Filum terminale

(c)

FIGURE 17-1 (*Continued*) Spinal cord and spinal nerves. (b) Photograph of the posterior aspect of the inferior portion of the medulla and the superior six segments of the cervical portion of the spinal cord. (c) Photograph of the posterior aspect of the conus medullaris and cauda equina. (Courtesy of N. Gluhbegovic and T. H. Williams, *The Human Brain: A Photographic Guide,* Harper & Row, New York, 1980.)

inferiorly in the vertebral canal like wisps of coarse hair flowing from the end of the cord. They are appropriately named the **cauda equina** (KAW-da ē-KWĪ-na), meaning horse's tail.

The spinal cord is a series of 31 segments, each giving rise to a pair of spinal nerves. **Spinal segment** refers to a region of the spinal cord from which a pair of spinal nerves arises. The cord is divided into right and left sides by two grooves (see Figure 17-3). The **anterior median fissure** is a deep, wide groove on the anterior (ventral) surface, and the **posterior median sulcus** is a shallower, narrow groove on the posterior (dorsal) surface.

PROTECTION AND COVERINGS

Vertebral Canal

The spinal cord is located in the vertebral canal of the vertebral column. The canal is formed by the vertebral foramina of all the vertebrae arranged on top of each other. Since the wall of the vertebral canal is essentially a ring of bone surrounding the spinal cord, the cord is well protected. A certain degree of protection is also provided by the meninges, cerebrospinal fluid, and the vertebral ligaments.

Meninges

The **meninges** (me-NIN-jēz) are coverings that run continuously around the spinal cord and brain. Those associated specifically with the cord are known as **spinal meninges** (Figure 17-2). The outer spinal meninx is called the **dura mater** (DYOO-ra MĀ-ter), meaning tough mother. It forms a tube from the level of the second sacral vertebra, where it is fused with the filum terminale, to the foramen magnum, where it is continuous with the dura mater of the brain. It is composed of dense, fibrous connective tissue. Between the dura mater and the wall of the vertebral canal is the *epidural space,* which is filled with fat, connective tissue, and blood vessels. It serves as padding around the cord. The epidural space inferior to the second lumbar vertebra is the site for the injection of anesthetics, such as a saddle block for childbirth.

The middle spinal meninx is called the **arachnoid** (a-RAK-noyd), or spider layer. It is a delicate connective tissue membrane that forms a tube inside the dura mater. It is also continuous with the arachnoid of the brain. Between the dura mater and the arachnoid is the *subdural space,* which contains serous fluid.

The inner meninx is known as the **pia mater** (PĒ-a MĀ-ter), or delicate mother. It is a transparent fibrous membrane that forms a tube around and adheres to the surface of the spinal cord and brain. It contains numerous blood vessels. Between the arachnoid and the pia mater is the *subarachnoid space,* where the cerebrospinal fluid circulates.

All three spinal meninges cover the spinal nerves up to the point of exit from the spinal column through the intervertebral foramina. The spinal cord is suspended in the middle of its dural sheath by membranous extensions of the pia mater. These extensions, called the *denticulate* (den-TIK-yoo-lāt) *ligaments,* are attached laterally to the dura mater along the length of the cord between the ventral and dorsal nerve roots on either side. The ligaments protect the spinal cord against shock and sudden displacement.

CLINICAL APPLICATION

Inflammation of the meninges is known as **meningitis.** If only the dura mater becomes inflamed, the condition is *pachymeningitis.* Inflammation of the arachnoid and pia mater is *leptomeningitis,* the most common form of meningitis.

Cerebrospinal fluid is removed from the subarachnoid space in the inferior lumbar region of the spinal cord by a **spinal (lumbar) puncture,** or **tap.** The procedure is normally performed between the third and fourth or fourth and fifth lumbar vertebrae. The spinous process of the fourth lumbar vertebra is easily located. A line drawn across the highest points of the iliac crests will pass through the spinous process. A lumbar puncture is below the spinal cord and thus poses little danger to it. If the patient lies on one side, drawing the knees and chest together, the vertebrae separate slightly so that a needle can be conveniently inserted. In its course the needle pierces the skin, superficial fascia, supraspinous ligament, interspinous ligament, epidural space, and arachnoid, to enter the subarachnoid space. Lumbar punctures are used to withdraw fluid for diagnostic purposes and to introduce antibiotics (as in the case of meningitis) and radiopaque dyes.

STRUCTURE IN CROSS SECTION

The spinal cord consists of both gray and white matter. Figure 17-3 shows that the gray matter forms an H-shaped area within the white matter. The gray matter consists primarily of nerve cell bodies and unmyelinated axons and dendrites of association and motor neurons. The white matter consists of bundles of myelinated axons of motor and sensory neurons.

The cross bar of the H is formed by the **gray commissure** (KOM-mi-shur). In the center of the gray commissure is a small space called the **central canal.** This canal runs the length of the spinal cord and is continuous with the fourth ventricle of the medulla. It contains cerebrospinal fluid. Anterior to the gray commissure is the **anterior (ventral) white commissure,** which connects the white matter of the right and left sides of the spinal cord.

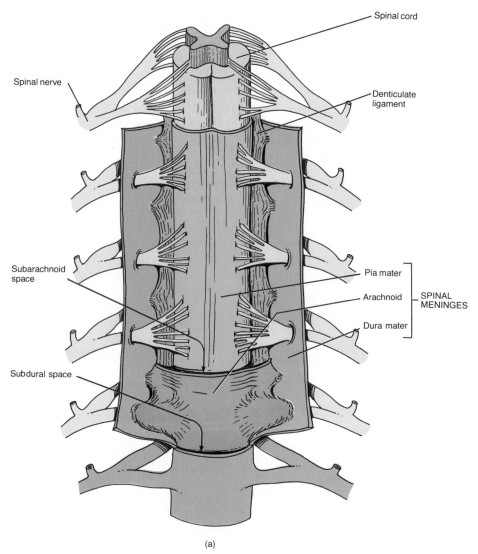

Spinal cord

Spinal nerve

Denticulate ligament

Subarachnoid space

Pia mater

Arachnoid — SPINAL MENINGES

Dura mater

Subdural space

(a)

FIGURE 17-2 Spinal meninges. (a) Location of the spinal meninges as seen in sections of the spinal cord.

The upright portions of the H are further subdivided into regions. Those closer to the front of the cord are called **anterior (ventral) gray horns.** They represent the motor part of the gray matter. The regions closer to the back of the cord are referred to as **posterior (dorsal) gray horns.** They represent the sensory part of the gray matter. The regions between the anterior and posterior gray horns are intermediate **lateral gray horns.** The lateral gray horns are present in the thoracic, upper lumbar, and sacral segments of the cord.

The gray matter of the cord also contains several nuclei that serve as relay stations for impulses and origins for certain nerves. Nuclei are clusters of nerve cell bodies and dendrites in the spinal cord and brain.

The white matter, like the gray matter, is organized into regions. The anterior and posterior gray horns divide the white matter on each side into three broad areas: **anterior (ventral) white columns, posterior (dorsal) white columns,** and **lateral white columns.** Each column (or *funiculus*) in turn consists of distinct bundles of myelinated fibers that run the length of the cord. These bundles are called **tracts,** or **fasciculi** (fa-SIK-yoo-lī). The longer **ascending tracts** consist of sensory axons that conduct impulses that enter the spinal cord upward to the brain. The longer **descending tracts** consist of motor axons that conduct impulses from the brain downward into the spinal cord, where they synapse with other neurons whose axons pass out to muscles and glands. Thus the ascending tracts are sensory tracts and the descending tracts are motor tracts. Still other short tracts contain ascending or descending axons that convey impulses from one level of the cord to another.

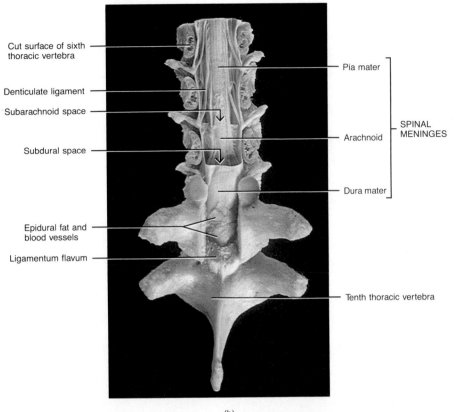

(b)

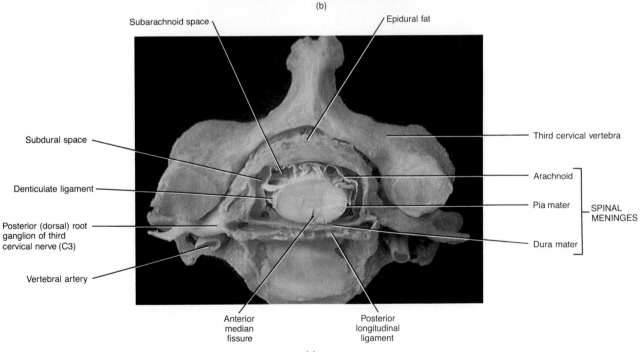

(c)

FIGURE 17-2 (*Continued*) Spinal meninges. (b) Photograph of the posterior aspect of the spinal cord. (c) Photograph of a cross section through the spinal cord between the second and third thoracic vertebrae. (Courtesy of N. Gluhbegovic and T. H. Williams, *The Human Brain: A Photographic Guide,* Harper & Row, New York, 1980.)

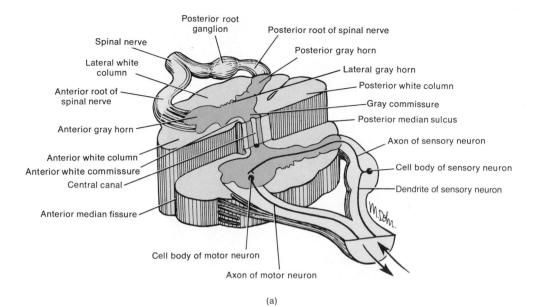

(a)

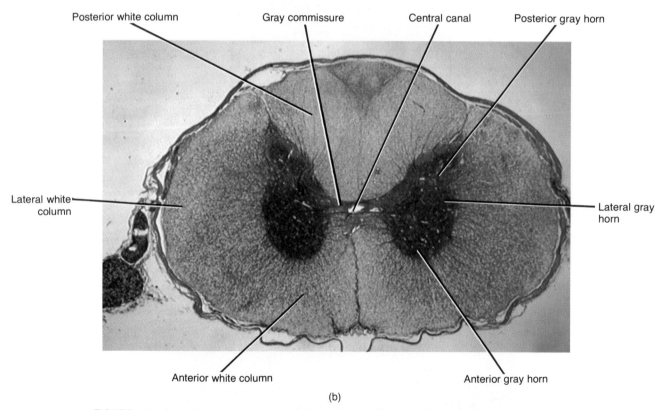

(b)

FIGURE 17-3 Spinal Cord. (a) The organization of gray and white matter in the spinal cord as seen in cross section. The front of the figure has been sectioned at a lower level than the back so that you can see what is inside the posterior root ganglion, posterior root of the spinal nerve, anterior root of the spinal nerve, and the spinal nerve. (b) Photograph of the spinal cord at the seventh cervical segment. (Courtesy of Victor B. Eichler, Wichita State University.)

FUNCTIONS

A major function of the spinal cord is to convey sensory impulses from the periphery to the brain and to conduct motor impulses from the brain to the periphery. This function is accomplished within spinal tracts. A second principal function is to provide a means of integrating reflexes. Both functions are essential to maintaining homeostasis.

Spinal Tracts

The vital function of conveying sensory and motor information to and from the brain is carried out by the ascending and descending tracts of the cord. The names of the tracts indicate the white column (funiculus) in which the tract travels, where the cell bodies of the tract originate, and where the axons of the tract terminate. Since the origin and termination are specified, the direction of impulse conduction is also indicated by the name. For example, the anterior spinothalamic tract is located in the *anterior* white column, it originates in the *spinal cord,* and it terminates in the *thalamus* (a region of the brain). It is an ascending (sensory) tract, since it conveys impulses from the cord upward to the brain.

The principal ascending and descending tracts are listed in Exhibit 17-1 and shown in Figure 17-4.

Reflex Center

The second prinicpal function of the spinal cord is to provide reflexes. Spinal nerves are the paths of communication between the spinal cord tracts and the periphery. Figure 17-3a reveals that each pair of spinal nerves is connected to a segment of the cord by two points of attachment called roots. The **posterior,** or **dorsal (sensory), root** contains sensory nerve fibers only and conducts impulses from the periphery to the spinal cord. These fibers extend into the posterior (dorsal) gray horn. Each dorsal root also has a swelling, the **posterior,** or **dorsal (sensory), root ganglion,** which contains the cell bodies of the sensory neurons from the periphery. The other point of attachment of a spinal nerve to the cord is the **anterior,** or **ventral (motor), root.** It contains motor nerve fibers only and conducts impulses from the spinal cord to the periphery.

The cell bodies of the motor neurons are located in the gray matter of the cord. If the motor impulse supplies a skeletal muscle, the cell bodies are located in the anterior (ventral) gray horn. If, however, the impulse supplies smooth muscle, cardiac muscle, or a gland through the autonomic nervous system, the cell bodies are located in the lateral gray horn.

The path an impulse follows from its origin in the dendrites or cell body of a neuron in one part of the body to its termination elsewhere in the body is called a **conduction pathway.** All conduction pathways consist of circuits of neurons. One pathway is known as a **reflex arc,** the functional unit of the nervous system. A reflex arc contains two or more neurons over which impulses are conducted from a receptor to the brain or spinal cord and then to an effector. The basic components of a reflex arc are as follows (Figure 17-5):

1. Receptor. The distal end of a dendrite or a sensory structure associated with the distal end of a dendrite. Its role in the reflex arc is to respond to a change in the internal or external environment by initiating a nerve impulse in a sensory neuron.

2. Sensory neuron. Passes the impulse from the receptor to its axonal termination in the central nervous system.

3. Center. A region, usually in the central nervous system, where an incoming sensory impulse generates an outgoing motor impulse. In the center, the impulse may be inhibited, transmitted, or rerouted. In the center of some reflex arcs, the sensory neuron directly generates the impulse in the motor neuron. The center may also contain an association neuron between the sensory neuron and the motor neuron leading to a muscle or a gland.

4. Motor neuron. Transmits the impulse generated by the sensory or association neuron in the center to the organ of the body that will respond.

5. Effector. The organ of the body, either muscle or gland, that responds to the motor impulse. This response is called a **reflex action,** or **reflex.**

Reflexes are fast responses to changes in the internal or external environment that allow the body to maintain homeostasis. Reflexes are associated not only with skeletal muscle contraction, but also with body functions such as heart rate, respiration, digestion, urination, and defecation. Reflexes carried out by the spinal cord alone are called **spinal reflexes.** Reflexes that result in the contraction of skeletal muscles are known as **somatic reflexes.** Those that cause the contraction of smooth or cardiac muscle or secretion by glands are **visceral (autonomic) reflexes.**

SPINAL NERVES

NAMES

The 31 pairs of spinal nerves are named and numbered according to the region and level of the spinal cord from which they emerge (see Figure 17-1). The first cervical pair emerges between the atlas and the occipital bone. All other spinal nerves leave the vertebral column from the intervertebral foramina between adjoining vertebrae. There are 8 pairs of cervical nerves, 12 pairs of thoracic nerves, 5 pairs of lumbar nerves, 5 pairs of sacral nerves, and 1 pair of coccygeal nerves.

EXHIBIT 17-1 SELECTED ASCENDING AND DESCENDING TRACTS OF SPINAL CORD

TRACT	LOCATION (WHITE COLUMN)	ORIGIN	TERMINATION	FUNCTION
ASCENDING TRACTS				
Anterior (ventral) spinothalamic	Anterior (ventral) column.	Posterior (dorsal) gray horn on one side of cord, but crosses to opposite side of brain.	Thalamus; impulses eventually conveyed to cerebral cortex.	Conveys sensations for touch and pressure from one side of body to opposite side of thalamus. Eventually sensations reach cerebral cortex.
Lateral spinothalamic	Lateral column.	Posterior (dorsal) gray horn on one side of cord, but crosses to opposite side of brain.	Thalamus; impulses eventually conveyed to cerebral cortex.	Conveys sensations for pain and temperature from one side of body to opposite side of thalamus. Eventually sensations reach cerebral cortex.
Posterior column (Fasciculus gracilis and fasciculus cuneatus)	Posterior (dorsal) column.	Axons of afferent neurons from periphery that enter posterior (dorsal) column on one side of cord and rise to same side of brain.	Nucleus gracilis and nucleus cuneatus of medulla; impulses eventually conveyed to cerebral cortex.	Convey sensations from one side of body to same side of medulla for touch; two-point discrimination (ability to distinguish that two points on skin are touched even though close together); proprioception (awareness of precise position of body parts and their direction of movement); stereognosis (ability to recognize size, shape, and texture of object); weight discrimination (ability to assess weight of an object); and vibration. Eventually sensations may reach cerebral cortex.
Posterior (dorsal) spinocerebellar	Posterior (dorsal) portion of lateral column.	Posterior (dorsal) gray horn on one side of cord and rises to same side of brain.	Cerebellum.	Conveys sensations from one side of body to same side of cerebellum for subconscious proprioception.
Anterior (ventral) spinocerebellar	Anterior (ventral) portion of lateral column.	Posterior (dorsal) gray horn on one side of cord; contains both crossed and uncrossed fibers.	Cerebellum.	Conveys sensations from both sides of body to cerebellum for subconscious proprioception.

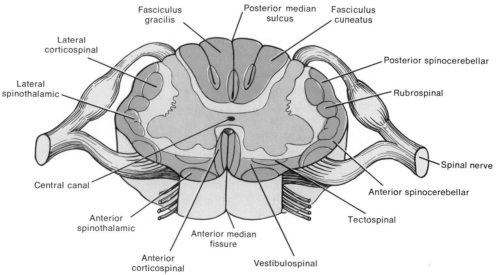

FIGURE 17-4 Selected tracts of the spinal cord. Ascending (sensory) tracts are indicated in pink; descending (motor) tracts are shown in blue.

EXHIBIT 17-1 (*Continued*)

TRACT	LOCATION (WHITE COLUMN)	ORIGIN	TERMINATION	FUNCTION
DESCENDING TRACTS				
Lateral corticospinal	Lateral column.	Cerebral cortex on one side of brain, but crosses in base of medulla to opposite side of cord.	Anterior (ventral) gray horn.	Conveys motor impulses from one side of cortex to anterior gray horn of opposite side. Eventually impulses reach skeletal muscles on opposite side of body that coordinate precise, discrete movements.
Anterior (ventral) corticospinal	Anterior (ventral) column.	Cerebral cortex on one side of brain, uncrossed in medulla, but crosses to opposite side of cord.	Anterior (ventral) gray horn.	Conveys motor impulses from one side of cortex to anterior gray horn of opposite side. Eventually impulses reach skeletal muscles on opposite side of body that coordinate precise, discrete movements.
Rubrospinal	Lateral column.	Midbrain (red nucleus) on one side of brain, but crosses to opposite side of cord.	Anterior (ventral) gray horn.	Conveys motor impulses from one side of midbrain to skeletal muscles on opposite side of body that are concerned with muscle tone and posture.
Tectospinal	Anterior (ventral) column.	Midbrain on one side of brain, but crosses to opposite side of cord.	Anterior (ventral) gray horn.	Conveys motor impulses from one side of midbrain to skeletal muscles on opposite side of body that control movements of head in response to auditory, visual, and cutaneous stimuli.
Vestibulospinal	Anterior (ventral) column.	Medulla on one side of brain and descends to same side of cord.	Anterior (ventral) gray horn.	Conveys motor impulses from one side of medulla to skeletal muscles on same side of body that regulate body tone in response to movements of head (equilibrium).

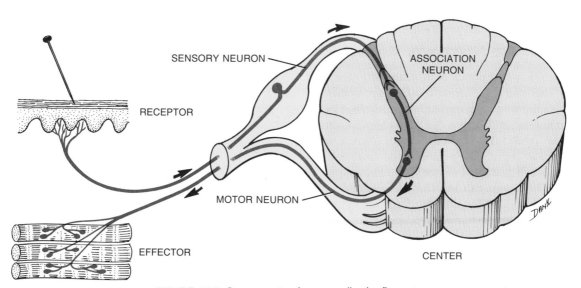

FIGURE 17-5 Components of a generalized reflex arc.

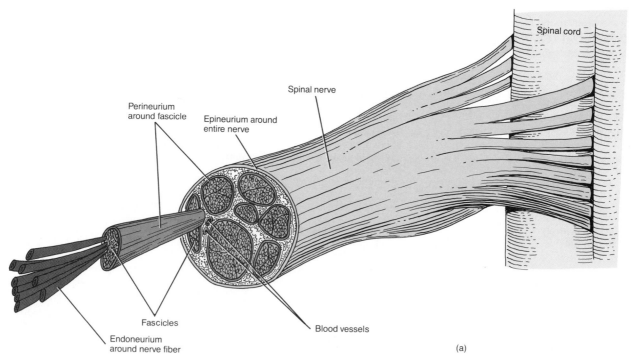

(a)

FIGURE 17-6 Coverings of a spinal nerve. (a) Diagram. (b) Scanning electron micrograph at a magnification of 900×. (Courtesy of Richard G. Kessel and Randy H. Kardon, from *Tissues and Organs: A Text-Atlas of Scanning Electron Microscopy.* Copyright © 1979 by Scientific American, Inc.)

During fetal life, the spinal cord and vertebral column grow at different rates, the cord growing more slowly. Thus not all the spinal cord segments are in line with their corresponding vertebrae. Remember that the spinal cord terminates near the level of the first or second lumbar vertebra. Thus the lower lumbar, sacral, and coccygeal nerves must descend more and more to reach their foramina before emerging from the vertebral column. This arrangement constitutes the cauda equina.

COMPOSITION AND COVERINGS

A **spinal nerve** has two points of attachment to the cord: a posterior root and an anterior root. The posterior and anterior roots unite to form a spinal nerve at the intervertebral foramen. Since the posterior root contains sensory fibers and the anterior root contains motor fibers, a spinal nerve is a *mixed nerve*. The posterior (dorsal) root ganglion contains cell bodies of sensory neurons.

In Figure 17-6, you can see that the nerve contains many fibers surrounded by different coverings. The individual fibers, whether myelinated or unmyelinated, are wrapped in a connective tissue called the **endoneurium** (en'-dō-NYOO-rē-um). Groups of fibers with their endoneurium are arranged in bundles called fascicles, and each

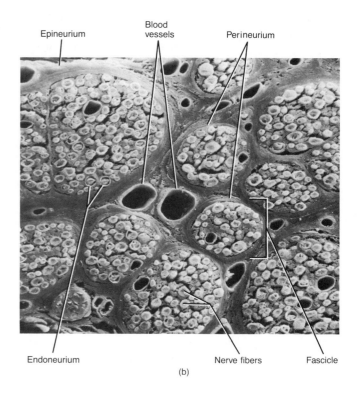

(b)

bundle is wrapped in connective tissue called the **perineurium** (per'-i-NYOO-rē-um). The outermost covering around the entire nerve is the **epineurium** (ep'-i-NYOO-rē-um). The spinal meninges fuse with the epineurium as the nerve exists from the vertebral canal.

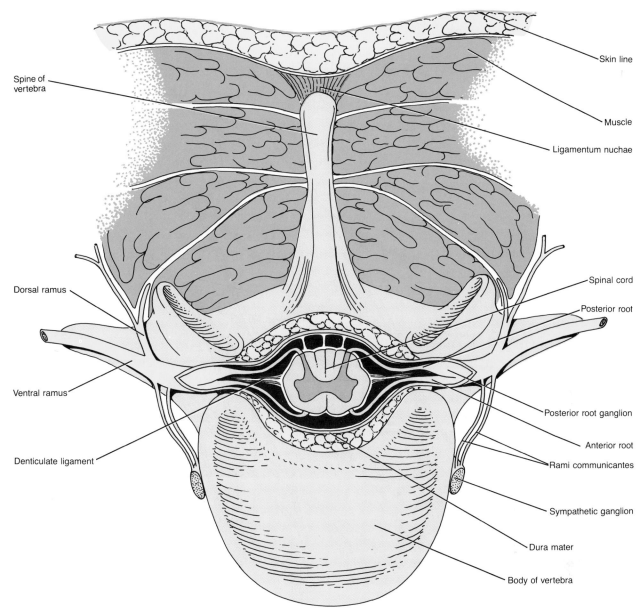

Spine of
vertebra

Skin line

Muscle

Ligamentum nuchae

Spinal cord

Posterior root

Dorsal ramus

Ventral ramus

Posterior root ganglion

Anterior root

Rami communicantes

Denticulate ligament

Sympathetic ganglion

Dura mater

Body of vertebra

FIGURE 17-7 Branches of a typical spinal nerve.

BRANCHES

Shortly after a spinal nerve leaves its intervertebral foramen, it divides into several branches (Figure 17-7). These
branches are known as **rami** (RĀ-mē). The **dorsal ramus**
(RĀ-mus) innervates the deep muscles and skin of the
dorsal surface of the back. The **ventral ramus** of a spinal
nerve innervates the superficial back muscles and all the
structures of the extremities and the lateral and ventral
trunks. In addition to dorsal and ventral rami, spinal
nerves also give off a **meningeal branch.** This branch reenters the spinal canal through the intervertebral foramen
and supplies the vertebrae, vertebral ligaments, blood vessels of the spinal cord, and the meninges. Other branches

of a spinal nerve are the **rami communicantes** (ko-myoo-
nē-KAN-tēz), components of the autonomic nervous system whose structure and function are discussed in Chapter
19.

PLEXUSES

The ventral rami of spinal nerves, except for thoracic
nerves T2 to T11, do not go directly to the structures
of the body they supply. Instead, they form networks
by joining with adjacent nerves on either side of the body.
Such a network is called a **plexus** (*plexus* = braid). The
principal plexuses are the cervical plexus, the brachial
plexus, the lumbar plexus, and the sacral plexus. Emerg-

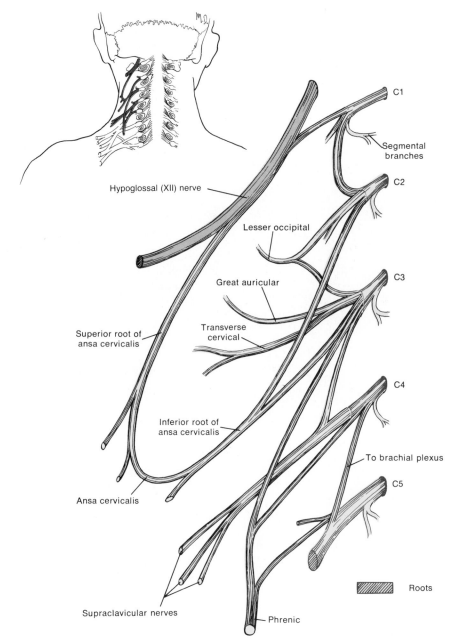

FIGURE 17-8 Cervical plexus. Consult Exhibit 17-2 so that you can determine the distribution of each of the nerves of the plexus.

ing from the plexuses are nerves bearing names that are often descriptive of the general regions they supply or the course they take. Each of the nerves, in turn, may have several branches named for the specific structures they innervate.

Cervical Plexus

The **cervical plexus** is formed by the ventral rami of the first four cervical nerves (C1 to C4) with contributions from C5. There is one on each side of the neck alongside the first four cervical vertebrae (Figure 17-8). The *roots*

of the plexus indicated in the diagram are the ventral rami. The cervical plexus supplies the skin and muscles of the head, neck, and upper part of the shoulders. Branches of the cervical plexus also connect with cranial nerves XI (accessory) and XII (hypoglossal). The phrenic nerves are a major pair of nerves arising from the cervical plexuses that supply the motor fibers to the diaphragm. Damage to the spinal cord above the origin of the phrenic nerves results in paralysis of the diaphragm, since the phrenic nerves no longer send impulses to the diaphragm. Contractions of the diaphragm are essential for normal breathing.

EXHIBIT 17-2   CERVICAL PLEXUS

NERVE	ORIGIN	DISTRIBUTION
SUPERFICIAL OR CUTANEOUS BRANCHES		
Lesser occipital	C2–C3.	Skin of scalp behind and above ear.
Greater auricular	C2–C3.	Skin in front, below, and over ear and over parotid glands.
Transverse cervical	C2–C3.	Skin over anterior aspect of neck.
Supraclavicular	C3–C4.	Skin over upper portion of chest and shoulder.
DEEP OR LARGELY MOTOR BRANCHES		
Ansa cervicalis		This nerve is divided into a superior root and an inferior root.
Superior root	C1–C2.	Infrahyoid, thyrohyoid, and geniohyoid muscles of neck.
Inferior root	C3–C4.	Omohyoid, sternohyoid, and sternothyroid muscles of neck.
Phrenic	C3–C5.	Diaphragm between thorax and abdomen.
Segmental branches	C1–C5.	Prevertebral (deep) muscles of neck, levator scapulae, and middle scalene muscles.

Exhibit 17-2 summarizes the nerves and distributions of the cervical plexus. The relationship of the cervical plexus to the other plexuses is shown in Figure 17-1a.

Brachial Plexus

The **brachial plexus** is formed by the ventral rami of spinal nerves C5 to C8 and T1 with contributions from C4 and T2. On either side of the last four cervical and first thoracic vertebrae, the brachial plexus extends downward and laterally, passes over the first rib behind the clavicle, and then enters the axilla (Figure 17-9). The brachial plexus constitutes the entire nerve supply for the upper extremities and shoulder region.

The *roots* of the brachial plexus, like those of the cervical plexus, are the ventral rami of the spinal nerves. The roots of C5 and C6 unite to form the *superior trunk,* C7 becomes the *middle trunk,* and C8 and T1 form the *inferior trunk.* Each trunk, in turn, divides into an *anterior division* and a *posterior division.* The divisions then unite to form cords. The *posterior cord* is formed by the union of the posterior divisions of the superior, middle, and inferior trunks. The *medial cord* is formed as a continuation of the anterior division of the inferior trunk. The *lateral cord* is formed by the union of the anterior divisions of the superior and middle trunk. The peripheral nerves arise from the cords. Thus the brachial plexus begins as roots that unite to form trunks, the trunks branch into divisions, the divisions form cords, and the cords give rise to the peripheral nerves.

Three important nerves arising from the brachial plexus are the radial, median, and ulnar. The radial nerve supplies the muscles on the posterior aspect of the arm and forearm. The median nerve supplies most of the muscles of the anterior forearm and some of the muscles in the palm. The ulnar nerve supplies the anteromedial muscles of the forearm and most of the muscles of the palm.

CLINICAL APPLICATION

Prolonged use of a crutch that presses into the axilla may result in injury to a portion of the brachial plexus. The usual **crutch palsy** involves the posterior cord of the brachial plexus or, more frequently, just the radial nerve, which, in general, supplies extensors.

Radial nerve damage is indicated by wrist drop: inability to extend the hand at the wrist. Care must be taken not to injure the radial nerve when deltoid intramuscular injections are given. **Median nerve damage** is indicated by numbness, tingling, and pain in the palm and fingers; weak thumb movements; and inability to pronate the forearm and flex the wrist properly. Compression of the median nerve inside the carpal tunnel, formed anteriorly by the transverse carpal ligament and posteriorly by the carpal bones, is known as **carpal tunnel syndrome.** It may be caused by any condition that aggravates compression of the contents of the carpal tunnel, such as trauma, edema, and flexion of the wrist. **Ulnar nerve damage** is indicated by an inability to flex and adduct the wrist and difficulty in spreading the fingers.

A summary of the nerves and distributions of the brachial plexus is given in Exhibit 17-3. The relationship of the brachial plexus to the other plexuses is shown in Figure 17-1a.

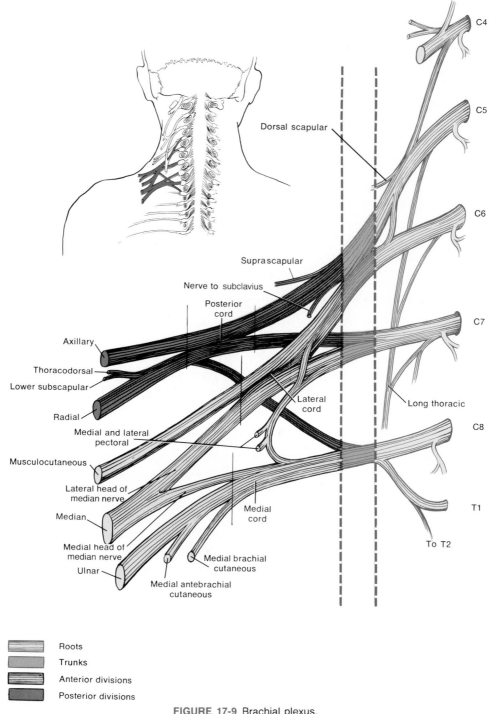

C4

C5

Dorsal scapular

Suprascapular

C6

Nerve to subclavius

Posterior
cord

Axillary

Thoracodorsal

Lower subscapular

Lateral
cord

Long thoracic

C7

Radial

Medial and lateral
pectoral

Musculocutaneous

Medial
cord

C8

Lateral head of
median nerve

Median

T1

Medial head of
median nerve

Ulnar

Medial brachial
cutaneous

To T2

Medial antebrachial
cutaneous

Roots

Trunks

Anterior divisions

Posterior divisions

FIGURE 17-9 Brachial plexus.

Lumbar Plexus

The **lumbar plexus** is formed by the ventral rami of spinal nerves L1 to L4. It differs from the brachial plexus in that there is no intricate interlacing of fibers. It also consists of *roots* and an *anterior* and *posterior division*. On either side of the first four lumbar vertebrae, the lumbar plexus passes obliquely outward behind the psoas major muscle (posterior division) and anterior to the quadratus lumborum muscle (anterior division) and then gives rise to its peripheral nerves (Figure 17-10). The lumbar plexus supplies the anterolateral abdominal wall, external genitals, and part of the lower extremity. The largest nerve arising from the lumbar plexus is the femoral nerve.

EXHIBIT 17-3   BRACHIAL PLEXUS

NERVE	ORIGIN	DISTRIBUTION
ROOT NERVES		
Dorsal scapular	C5.	Levator scapulae, rhomboideus major, and rhomboideus minor muscles.
Long thoracic	C5–C7.	Serratus anterior muscle.
TRUNK NERVES		
Nerve to subclavius	C5–C6.	Subclavius muscle.
Suprascapular	C5–C6.	Supraspinatus and infraspinatus muscles.
LATERAL CORD NERVES		
Musculocutaneous	C5–C7.	Coracobrachialis, biceps brachii, and brachialis muscles.
Median (lateral head)	C5–C7.	See distribution for median (medial head).
Lateral pectoral	C5–C7.	Pectoralis major muscle.
POSTERIOR CORD NERVES		
Upper subscapular	C5–C6.	Subscapularis muscle.
Thoracodorsal	C6–C8.	Latissimus dorsi muscle.
Lower subscapular	C5–C6.	Subscapularis and teres major muscles.
Axillary (circumflex)	C5–C6.	Deltoid and teres minor muscles; skin over deltoid and upper posterior aspect of arm.
Radial	C5–C8 and T1.	Extensor muscles of arm and forearm (triceps brachii, brachioradialis, extensor carpi radialis longus, extensor digitorum, extensor carpi ulnaris, extensor indicis); skin of posterior arm and forearm, lateral two-thirds of dorsum of hand, and fingers over proximal and middle phalanges.
MEDIAL CORD NERVES		
Medial pectoral	C8–T1.	Pectoralis major and pectoralis minor muscles.
Medial brachial cutaneous	C8–T1.	Skin of medial and posterior aspects of lower third of arm.
Medial antebrachial cutaneous	C8–T1.	Skin of medial and posterior aspects of forearm.
Median (medial head)	C5–C8 and T1.	Medial and lateral heads of median nerve form median nerve. Distributed to flexors of forearm (pronator teres, flexor carpi radialis, flexor digitorum superficialis), except flexor carpi ulnaris; skin of lateral two-thirds of palm of hand and fingers.
Ulnar	C8–T1.	Flexor carpi ulnaris and flexor digitorum profundus muscles; skin of medial side of hand, little finger, and medial half of ring finger.
OTHER CUTANEOUS DISTRIBUTIONS		
Intercostobrachial	Second intercostal nerve.	Skin over medial side of arm.
Upper lateral brachial cutaneous	Axillary.	Skin over deltoid muscle and down to elbow.
Posterior brachial cutaneous	Radial.	Skin over posterior aspect of arm.
Lower lateral brachial cutaneous	Radial.	Skin over lateral aspect of elbow.
Lateral antebrachial cutaneous	Musculocutaneous.	Skin over lateral aspect of forearm.
Posterior antebrachial cutaneous	Radial.	Skin over posterior aspect of forearm.

CLINICAL APPLICATION

Injury to the femoral nerve is indicated by an inability to extend the leg and by loss of sensation in the skin over the anteromedial aspect of the thigh.

A summary of the nerves and distributions of the lumbar plexus is presented in Exhibit 17-4. The relationship of the lumbar plexus to the other plexuses is shown in Figure 17-1a.

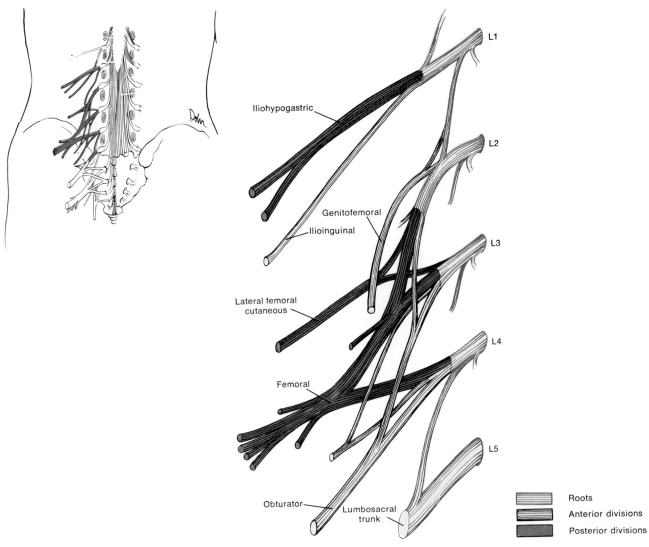

FIGURE 17-10 Lumbar plexus. Consult Exhibit 17-4 so that you can determine the distribution of each of the nerves of the plexus.

EXHIBIT 17-4 LUMBAR PLEXUS

NERVE	ORIGIN	DISTRIBUTION
Iliohypogastric	T12–L1.	Muscles of anterolateral abdominal wall (external oblique, internal oblique, transversus abdominis); skin of lower abdomen and buttock.
Ilioinguinal	L1.	Muscles of anterolateral abdominal wall as indicated above; skin of upper medial aspect of thigh, root of penis and scrotum in male, and labia majora and mons pubis in female.
Genitofemoral	L1–L2.	Cremaster muscle; skin over middle anterior surface of thigh, scrotum in male, and labia majora in female.
Lateral femoral cutaneous	L2–L3.	Skin over lateral, anterior, and posterior aspects of thigh.
Femoral	L2–L4.	Flexor muscles of thigh (iliacus, psoas major, pectineus, rectus femoris, sartorius); extensor muscles of leg (rectus femoris, vastus lateralis, vastus medialis, vastus intermedius); skin on front and over medial aspect of thigh and medial side of leg and foot.
Obturator	L2–L4.	Adductor muscles of leg (obturator externus, pectineus, adductor longus, adductor brevis, adductor magnus, gracilis); skin over medial aspect of thigh.

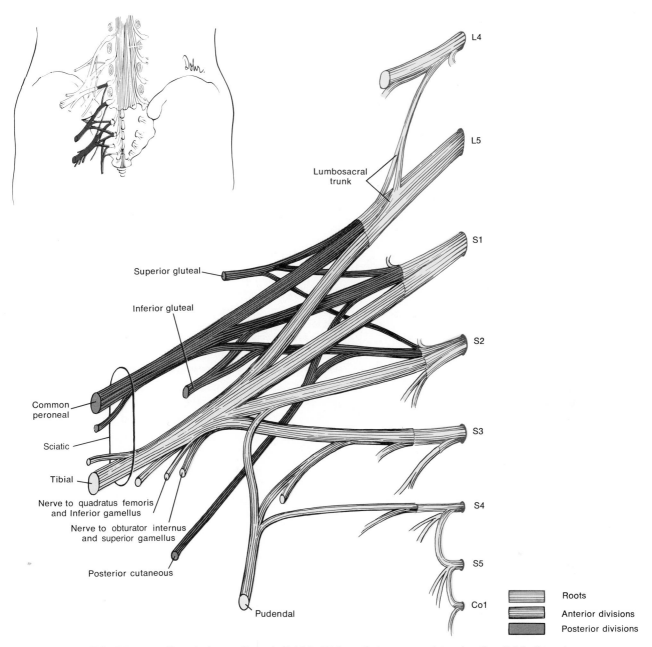

FIGURE 17-11 Sacral plexus. Consult Exhibit 17-5 so that you can determine the distribution of each of the nerves of the plexus.

Sacral Plexus

The **sacral plexus** is formed by the ventral rami of spinal nerves L4 to L5 and S1 to S4. It is situated largely in front of the sacrum (Figure 17-11). Like the lumbar plexus, it contains *roots* and an *anterior* and *posterior division.* The sacral plexus supplies the buttocks, perineum, and lower extremities. The largest nerve arising from the sacral plexus—and, in fact, the largest nerve in the body—is the sciatic nerve. The sciatic nerve supplies the entire musculature of the leg and foot.

CLINICAL APPLICATION

Damage to the sciatic nerve and its branches results in foot drop, an inability to dorsiflex the foot. This nerve may be injured because of a slipped disc, dislocated hip, pressure from the uterus during pregnancy, or an improperly given gluteal intramuscular injection.

A summary of the nerves and distributions of the sacral plexus is given in Exhibit 17-5. The relationship of the

EXHIBIT 17-5 SACRAL PLEXUS

NERVE	ORIGIN	DISTRIBUTION
Superior gluteal	L4–L5 and S1.	Gluteus minimus and gluteus medius muscles and tensor fasciae latae.
Inferior gluteal	L5–S2.	Gluteus maximus muscle.
Nerve to piriformis	S1–S2.	Piriformis muscle.
Nerve to quadratus femoris	L4–L5 and S1.	Inferior gemellus and quadratus femoris muscles.
Nerve to obturator internus	L5–S2.	Superior gemellus and obturator internus muscles.
Perforating cutaneous	S2–S3.	Skin over lower medial aspect of buttock.
Posterior cutaneous	S1–S3.	Skin over anal region, lower lateral aspect of buttock, upper posterior aspect of thigh, upper part of calf, scrotum in male, and labia majora in female.
Sciatic	L4–S3.	Actually two nerves: tibial and common peroneal, bound together by common sheath of connective tissue. It splits into its two divisions, usually at knee. (See below for distributions.) As sciatic nerve descends through thigh, it sends branches to hamstring muscles (biceps femoris, semitendinosus, semimembranosus) and adductor magnus.
Tibial (medial popliteal)	L4–S3.	Gastrocnemius, plantaris, soleus, popliteus, tibialis posterior, flexor digitorum, and hallucis muscles. Branches of tibial nerve in foot are medial plantar nerve and lateral plantar nerve.
Medial plantar		Abductor hallucis, flexor digitorum brevis, and flexor hallucis muscles; skin over medial two-thirds of plantar surface of foot.
Lateral plantar		Remaining muscles of foot not supplied by medial plantar nerve; skin over lateral third of plantar surface of foot.
Common peroneal (lateral popliteal)	L4–S2.	Divides into a superficial peroneal and a deep peroneal branch.
Superficial peroneal		Peroneus longus and peroneus brevis muscles; skin over distal third of anterior aspect of leg and dorsum of foot.
Deep peroneal		Tibialis anterior, extensor hallucis longus, peroneus tertius, and extensor digitorum brevis muscles; skin over great and second toes.
Pudendal	S2–S4.	Muscles of perineum; skin of penis and scrotum in male and clitoris, labia majora, labia minora, and lower vagina in female.

sacral plexus to the other plexuses is shown in Figure 17-1a.

INTERCOSTAL (THORACIC) NERVES

Spinal nerves T2 to T11 do not enter into the formation of plexuses. Nerves T2 to T11 are known as **intercostal (thoracic) nerves** and are distributed directly to the structures they supply in intercostal spaces (see Figure 17-1a). After leaving its intervertebral foramen, the ventral ramus of nerve T2 supplies the intercostal muscles of the second intercostal space and the skin of the axilla and posteromedial aspect of the arm. Nerves T3 and T6 pass in the costal grooves of the ribs and are distributed to the intercostal muscles and skin of the anterior and lateral chest wall. Nerves T7 to T11 supply the intercostal muscles and the abdominal muscles and overlying skin. The dorsal rami of the intercostal nerves supply the deep back muscles and the skin of the dorsal aspect of the thorax.

DERMATOMES

The skin over the entire body is supplied segmentally by spinal nerves. This means that the spinal nerves innervate specific, constant segments of the skin. All spinal nerves except C1 supply branches to the skin. The skin segment supplied by the dorsal root of a spinal nerve is a **dermatome** (Figure 17-12).

In the neck and trunk, the dermatomes form consecutive bands of skin. In the trunk, there is an overlap of adjacent dermatome nerve supply. Thus there is little loss of sensation if only a single nerve supply to a dermatome is interrupted. Most of the skin of the face and scalp is supplied by the trigeminal (V) cranial nerve.

Since physicians know which spinal nerves are associated with each dermatome, it is possible to determine which segment of the spinal cord or spinal nerve is malfunctioning. If a dermatome is stimulated and the sensation is not perceived, it can be assumed that the nerves supplying the dermatome are involved.

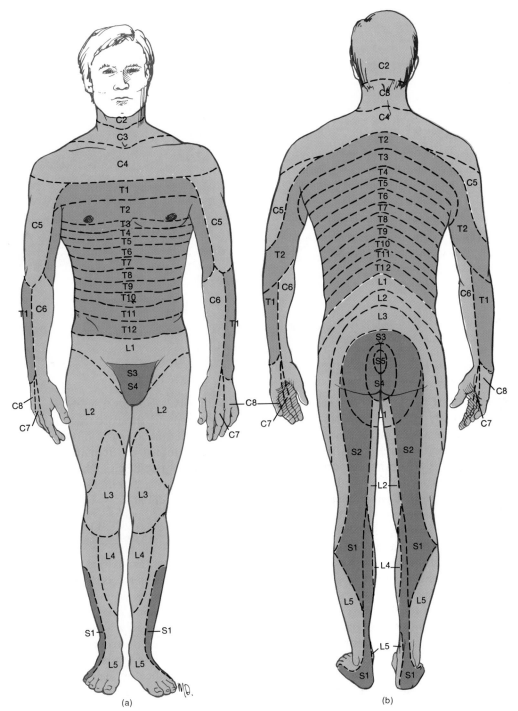

FIGURE 17-12 Distribution of spinal nerves to dermatomes. The lines are not perfectly aligned so that there is often considerable overlap. (a) Anterior view. (b) Posterior view.

APPLICATIONS TO HEALTH

SPINAL CORD INJURY

The spinal cord may be damaged by expanding tumors, blood clots, degenerative and demyelinating disorders, fracture or dislocation of the vertebrae enclosing it, penetrating wounds caused by projectile metal fragments, or other traumatic events such as automobile accidents. Depending on the location and extent of the injury, paralysis may occur. The various types of paralysis may be classified as follows: **monoplegia** (*mono* = one; *plege* = stroke), paralysis of one extremity only; **diplegia** (*di* = two), paralysis of both upper extremities or both lower extremities; **paraplegia** (*para* = beyond), paralysis of both lower extremities; **hemiplegia** (*hemi* = half), paralysis of the upper extremity, trunk, and lower extremity on one side of the body; and **quadriplegia** (*quad* = four), paralysis of the two upper and two lower extremities.

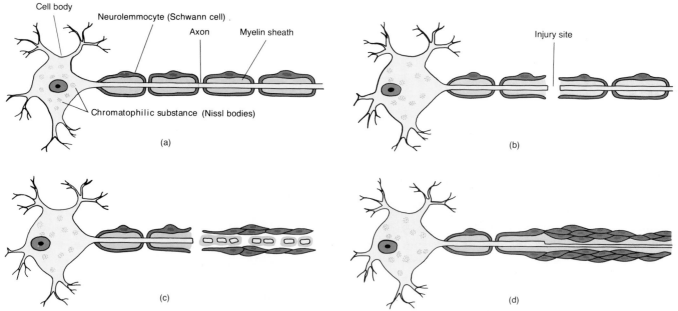

FIGURE 17-13 Peripheral nerve damage and repair. (a) Normal neuron. (b) Chromatolysis. (c) Wallerian degeneration. (d) Regeneration.

Complete transection of the spinal cord means that the cord is cut transversely and severed from one side to the other. It results in a loss of all sensations and voluntary movement below the level of the transection. If the upper cervical cord is transected, quadriplegia results; if the transection is between the cervical and lumbar enlargements, paraplegia results. **Hemisection** of the spinal cord refers to a partial transection. It is characterized, below the hemisection, by a loss of proprioception, tactile discrimination, and feeling of vibration on the same side as the injury; paralysis on the same side; and loss of feelings of pain and temperature on the opposite side. If the hemisection is of the upper cervical cord, hemiplegia results; if the hemisection is of the thoracic cord, paralysis of one lower extremity results (monoplegia).

Following transection, there is an intitial period of **spinal shock** that lasts from a few days to several weeks. During this period, all reflex activity is abolished, a condition called *areflexia* (a'-rē-FLEK-sē-a). In time, however, there is a return of reflex activity. The first reflex to return is the knee jerk. Its reappearance may take several days. Next the flexion reflexes return, over a period of up to several months. Then the crossed extensor reflexes return. Visceral reflexes such as erection and ejaculation are also affected by transection. Moreover, urinary bladder and bowel functions are no longer under voluntary control.

Until recently, severe damage resulting from transection was thought to be irreversible. However, a team of researchers has developed a technique for regenerating severed spinal cords in animals. The technique, called *delayed nerve grafting,* involves cutting away the crushed or injured section of the spinal cord and bridging the gap with nerve segments from the arm or leg. The original severed axons in the cord can then grow through the bridge. If the animal research is successful, the procedure may be attempted on humans. Delayed nerve grafting offers hope for paraplegics who have lost voluntary movements of the lower extremities, as well as urinary, bowel, and sexual functions.

PERIPHERAL NERVE DAMAGE AND REPAIR

As we have seen, axons that have a neurilemma can be repaired as long as the cell body is intact, fibers are in association with neurolemmocytes (Schwann cells), and scar tissue formation does not occur too rapidly. Most nerves that lie outside the brain and spinal cord consist of axons that are covered with a neurilemma. A person who injures a nerve in the upper extremity, for example, has a good chance of regaining nerve function. Axons in the brain and spinal cord do not have a neurilemma. Injury there is permanent.

When there is damage to an axon (or to dendrites of somatic afferent neurons), there are usually changes in the cell body and always changes in the portions of the nerve processes distal to the site of damage. The changes associated with the cell body are referred to as the **axon reaction,** or **retrograde degeneration.** Those associated with the distal portion of the cut fiber are called **Wallerian** (wal-LE-rē-an) **degeneration.** The axon reaction occurs in essentially the same way, whether the damaged fiber is in the central or peripheral nervous system. The Wallerian reaction, however, depends on whether the fiber is central or peripheral.

Axon Reaction

When there is damage to an axon of a central or peripheral neuron, certain structural changes occur in the cell body. One of the most significant features of the axon reaction occurs 24 to 48 hours after damage. The chromatophilic substance (Nissl bodies), arranged in an orderly fashion in an uninjured cell body, breaks down into finely granular masses. This alteration is called *chromatolysis* (krō'-ma-TOL-i-sis; *chromo* = color, *lysis* = dissolution). It begins between the axon hillock and nucleus, but spreads throughout the cell body. As a result of chromatolysis, the cell body swells and the swelling reaches its maximum between 10 and 20 days after injury (Figure 17-13b). Chromatolysis results in a loss of ribosomes by the rough endoplasmic reticulum and an increase in the number of free ribosomes. Another sign of the axon reaction is the off-center position of the nucleus in the cell body. This change makes it possible to identify the cell bodies of damaged fibers through a microscope.

Wallerian Degeneration

The part of the axon distal to the damage becomes slightly swollen and then breaks up into fragments by the third to fifth day. The myelin sheath around the axon also undergoes degeneration (Figure 17-13c). Degeneration of the distal portion of the axon and myelin sheath is called Wallerian degeneration. Following degeneration, there is phagocytosis of the remains by macrophages.

Even though there is degeneration of the axon and myelin sheath, the neurilemma of the neurolemmocytes remains. The neurolemmocytes on either side of the site of injury multiply by mitosis and grow toward each other until they form a tube across the injured area. The tube provides a means for new axons to grow from the proximal area across the injured area into the distal area previously occupied by the original nerve fiber (Figure 17-13d). The growth of new axons will not occur if the gap at the site of injury is too large or if the gap becomes filled with dense collagenous fibers.

Regeneration

Following chromatolysis, there are signs of recovery in the cell body. There is an acceleration of RNA and protein synthesis, which favors regeneration of the axon. Recovery often takes several months and involves the restoration of normal levels of RNA and proteins and the chromatophilic substance to its usual, uninjured patterns.

Accelerated protein synthesis is required for repair of the damaged axon. The proteins synthesized in the cell body pass into the axon by axoplasmic flow at about the rate of 1 mm (0.04 in)/day. The proteins assist in regenerating the damaged axon. During the first few days following damage, regenerating axons begin to invade the tube formed by the neurolemmocytes. Axons from the proximal area grow at the rate of about 1.5 mm (0.06 in)/day across the area of damage, find their way into the distal neurilemmal tubes, and grow toward the distally located receptors and effectors. Thus sensory and motor connections are reestablished. In time, a new myelin sheath is also produced by the neurolemmocytes. However, function is never completely restored after a nerve is severed.

NEURITIS

Neuritis is inflammation of a single nerve, two or more nerves in separate areas, or many nerves simultaneously. It may result from irritation to the nerve produced by direct blows, bone fractures, contusions, or penetrating injuries. Additional causes include vitamin deficiency (usually thiamine) and poisons such as carbon monoxide, carbon tetrachloride, heavy metals, and some drugs.

SCIATICA

Sciatica (sī-AT-i-ka) is a type of neuritis characterized by severe pain along the path of the sciatic nerve or its branches. The term is commonly applied to a number of disorders affecting this nerve. Because of its length and size, the sciatic nerve is exposed to many kinds of injury. Inflammation of or injury to the nerve causes pain that passes from the back or thigh down its length into the leg, foot, and toes.

Probably the most common cause of sciatica is a herniated (slipped) disc. Other causes include irritation from osteoarthritis, back injuries, or pressure on the nerve from certain types of exertion. Sciatica may be associated with diabetes mellitus, gout, or vitamin deficiencies. Other cases are idiopathic (unknown).

SHINGLES

Shingles is an acute infection of the peripheral nervous system and is frequently a relapse of the same case of chickenpox first acquired during childhood. Shingles is caused by a virus called herpes zoster (HER-pēz ZOS-ter), the chickenpox virus. Following recovery from chickenpox, the virus retreats to dorsal root ganglia where it resides. If the virus becomes activated, the immune system usually prevents it from spreading. However, from time to time the activated virus overcomes the immune system. In this case, the virus leaves the ganglion and travels down sensory neurons, causing the pain, and invades the skin where the neurons end, causing a characteristic line of skin blisters and discoloration of the skin. The line of blisters has a shape corresponding to the distribution of a particular nerve. The intercostal nerves and thoracic spinal nerves in the waist area are most commonly affected.

STUDY OUTLINE

Grouping of Neural Tissue (p. 400)

1. White matter is an aggregation of myelinated axons and associated neuroglia.
2. Gray matter is a collection of nerve cell bodies and dendrites or unmyelinated axons along with associated neuroglia.
3. A nerve is a bundle of nerve fibers outside the central nervous system.
4. A ganglion is a collection of cell bodies outside the central nervous system.
5. A tract is a bundle of fibers of similar function in the central nervous system.
6. A nucleus is a mass of nerve cell bodies and dendrites in the gray matter of the brain and spinal cord.
7. A horn or column is an area of gray matter in the spinal cord.

Spinal Cord (p. 400)

General Features

1. The spinal cord begins as a continuation of the medulla oblongata and terminates at about the second lumbar vertebra.
2. It contains cervical lumbar enlargements that serve as points of origin for nerves to the extremities.
3. The tapered portion of the spinal cord is the conus medullaris, from which arise the filum terminale and cauda equina.
4. The spinal cord is partially divided into right and left sides by the anterior median fissure and posterior median sulcus.
5. The gray matter in the spinal cord is divided into horns and the white matter into funiculi, or columns.
6. In the center of the spinal cord is the central canal, which runs the length of the spinal cord and contains cerebrospinal fluid.
7. There are ascending (sensory) tracts and descending (motor) tracts.

Protection and Coverings

1. The spinal cord is protected by the vertebral canal, meninges, cerebrospinal fluid, and vertebral ligaments.
2. The meninges are three coverings that run continuously around the spinal cord and brain: dura mater, arachnoid, and pia mater.
3. Removal of cerebrospinal fluid from the subarachnoid space or ventricle is called a spinal (lumbar) puncture. The procedure is used to diagnose pathologies and to introduce antibiotics.

Structure in Cross Section

1. Parts of the spinal cord observed in cross section are the gray commissure; central canal; anterior, posterior, and lateral gray horns; anterior, posterior, and lateral white columns; and ascending and descending tracts.
2. The spinal cord conveys sensory and motor information by way of the ascending and descending tracts, respectively.

Functions

1. A major function of the spinal cord is to convey sensory impulses from the periphery to the brain and to conduct motor impulses from the brain to the periphery.

2. Another function is to serve as a reflex center. The posterior root, posterior root ganglion, and anterior root are involved in conveying an impulse.
3. A reflex arc is the shortest route that can be taken by an impulse from a receptor to an effector. Its basic components are a receptor, a sensory neuron, a center, a motor neuron, and an effector.
4. A reflex is a quick, involuntary response to a stimulus that passes along a reflex arc. Reflexes represent the body's principal mechanisms for responding to changes in the internal and external environment.

Spinal Nerves (p. 407)

Names

1. The 31 pairs of spinal nerves are named and numbered according to the region and level of the spinal cord from which they emerge.
2. There are 8 pairs of cervical, 12 pairs of thoracic, 5 pairs of lumbar, 5 pairs of sacral, and 1 pair of coccygeal.

Composition and Coverings

1. Spinal nerves are attached to the spinal cord by means of a posterior root and an anterior root. All spinal nerves are mixed.
2. Spinal nerves are covered by endoneurium, perineurium, and epineurium.

Branches

1. After a spinal nerve leaves its intervertebral foramen, it divides into several rami (branches).
2. Branches of a spinal nerve include the dorsal ramus, ventral ramus, meningeal branch, and rami communicantes.

Plexuses

1. The ventral rami of spinal nerves, except for T2 to T11, form networks of nerves called plexuses.
2. Emerging from the plexuses are nerves bearing names that are often descriptive of the general regions they supply or the course they take.
3. The principal plexuses are called the cervical, brachial, lumbar, and sacral, plexuses.
4. Nerves that do not form plexuses are called intercostal nerves (T2 to T11). They are distributed directly to the structures they supply in intercostal spaces.

Dermatomes

1. All spinal nerves except C1 innervate specific, constant segments of the skin. The skin segments are called dermatomes.
2. Knowledge of dermatomes helps a physician determine which segment of the spinal cord or spinal nerve is malfunctioning.

Applications to Health (p. 419)

1. Complete or partial severing of the spinal cord is called transection. It may result in quadriplegia or paraplegia. Partial transection is followed by a period of loss of reflex activity called areflexia.

2. Following peripheral nerve damage, repair is accomplished by an axon reaction, Wallerian degeneration, and regeneration.

3. Inflammation of nerves is known as neuritis.
4. Neuritis of the sciatic nerve and its branches is called sciatica.
5. Shingles is acute infection of peripheral nerves.

REVIEW QUESTIONS

1. Define the following terms: white matter, gray matter, nerve, ganglion, tract, nucleus, and horn.
2. Describe the location of the spinal cord. What are the cervical and lumbar enlargements?
3. Define conus medullaris, filum terminale, and cauda equina. What is a spinal segment? How is the spinal cord partially divided into a right and left side?
4. Explain the location and composition of the spinal meninges. Describe the location of the epidural, subdural, and subarachnoid spaces. What are the denticulate ligaments?
5. Based on your knowledge of the structure of the spinal cord in cross section, define the following: gray commissure, central canal, anterior gray horn, lateral gray horn, posterior gray horn, anterior white column, lateral white column, posterior white column, ascending tract, and descending tract.
6. Describe the function of the spinal cord as a conduction pathway.
7. Using Exhibit 17-1 as a guide, be sure that you can list the location, origin, termination, and function of the principal ascending and descending tracts.
8. Describe how the spinal cord serves as a reflex center.
9. What is a reflex arc? List and define the components of a reflex arc.
10. Define a spinal nerve. Why are all spinal nerves classified as mixed nerves?
11. Describe how a spinal nerve is attached to the spinal cord.
12. Explain how a spinal nerve is enveloped by its several different coverings.
13. Describe the branches and innervations of a typical spinal nerve.
14. What is a plexus? Explain the location, origin, nerves, and distributions of the following plexuses: cervical, brachial, lumbar, and sacral.
15. What are intercostal nerves?
16. Define a dermatome. Why is a knowledge of dermatomes important?
17. What is transection? Distinguish between quadriplegia and paraplegia. What is meant by spinal shock?
18. Outline the principal events that occur as part of the axon reaction, Wallerian degeneration, and regeneration following peripheral nerve damage.
19. Distinguish between neuritis and sciatica.
20. What is shingles?

18

The Brain and the Cranial Nerves

Student Objectives

Identify the principal parts of the brain.

Describe the location of the cranial meninges.

Explain the formation and circulation of cerebrospinal fluid.

Describe the blood supply to the brain and the concept of the blood–brain barrier (BBB).

Compare the components of the brain stem with regard to structure and function.

Identify the structure and functions of the diencephalon.

Describe the lobes, tracts, and basal ganglia of the cerebrum.

Describe the structure and functions of the limbic system.

Compare the motor, sensory, and association areas of the cerebrum.

Describe the principle of the electroencephalograph and its significance in the diagnosis of certain disorders.

Explain brain lateralization and the split-brain concept.

Describe the anatomical characteristics and functions of the cerebellum.

Define a cranial nerve.

Identify the 12 pairs of cranial nerves by name, number, type, location, and function.

Describe the effects of aging on the nervous system.

Describe the development of the nervous system.

List the clinical symptoms of these disorders of the nervous system: brain tumors, poliomyelitis, amyotrophic lateral sclerosis (ALS), cerebral palsy, Parkinson's disease, epilepsy, multiple sclerosis (MS), cerebrovascular accidents (CVAs), dyslexia, Tay-Sachs disease, headache, trigeminal neuralgia, rabies, Reye's syndrome (RS), and Alzheimer's disease.

Define key medical terms associated with the central nervous system.

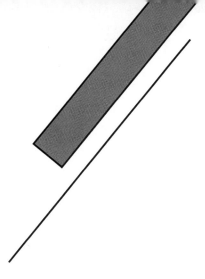

Now we will consider the principal parts of the brain, how the brain is protected, how it is related to the spinal cord, and how it is related to the 12 pairs of cranial nerves.

BRAIN

PRINCIPAL PARTS

To understand the terminology used for the principal parts of the brain, it will first be necessary to describe briefly the embryological development of the brain. The developmental anatomy of the brain is explained in more detail at the end of the chapter.

At the end of the fourth week of the embryonic period, the brain develops from three rudimentary regions of the embryo called primary brain vesicles. These are the **prosencephalon** (forebrain), **mesencephalon** (midbrain), and **rhombencephalon** (hindbrain). During the fifth week of development, the prosencephalon develops into two secondary brain vesicles called the **diencephalon** and **telencephalon;** the rhombencephalon also develops into two secondary brain vesicles called the **myelencephalon** and **metencephalon.** Since the mesencephalon remains unchanged, there are five secondary brain vesicles. Ultimately, the diencephalon develops into the thalamus and hypothalamus of the brain, the telencephalon forms the cerebrum, the mesencephalon becomes the midbrain, the myelencephalon develops into the medulla oblongata, and the metencephalon becomes the pons and cerebellum.

These relationships are summarized in Exhibit 18-1. The **brain** of an average adult is one of the largest organs of the body, weighing about 1,300 g (3 lb). Figure 18-1 shows that the brain is mushroom shaped. It is divided into four principal parts: brain stem, diencephalon, cerebrum, and cerebellum. In some cases, embryological names are retained. The **brain stem,** the stalk of the mushroom, consists of the medulla oblongata, pons, and midbrain, or mesencephalon (mes-en-SEF-a-lon). The lower end of the brain stem is a continuation of the spinal cord. Above the brain stem is the **diencephalon** (dī-en-SEF-a-lon), consisting primarily of the thalamus and hypothalamus. The **cerebrum** spreads over the diencephalon. The cerebrum constitutes about seven-eighths of the total weight of the brain and occupies most of the cranium. Inferior to the cerebrum and posterior to the brain stem is the **cerebellum.**

PROTECTION AND COVERINGS

The brain is protected by the cranial bones (see Figure 6-2). Like the spinal cord, the brain is also protected by meninges. The **cranial meninges** surround the brain, are continuous with the spinal meninges, and have the same basic structure and bear the same names as the spinal meninges: the outermost **dura mater,** middle **arachnoid,** and innermost **pia mater** (Figure 18-2a).

EXHIBIT 18-1 DEVELOPMENT OF THE BRAIN FROM BRAIN VESICLES

PRIMARY BRAIN VESICLE	SECONDARY BRAIN VESICLE	PART OF BRAIN FORMED
Prosencephalon	Diencephalon	Thalamus and hypothalamus
	Telencephalon	Cerebrum
Mesencephalon	Mesencephalon	Midbrain
Rhombencephalon	Myelencephalon	Medulla oblongata
	Metencephalon	Pons and cerebellum

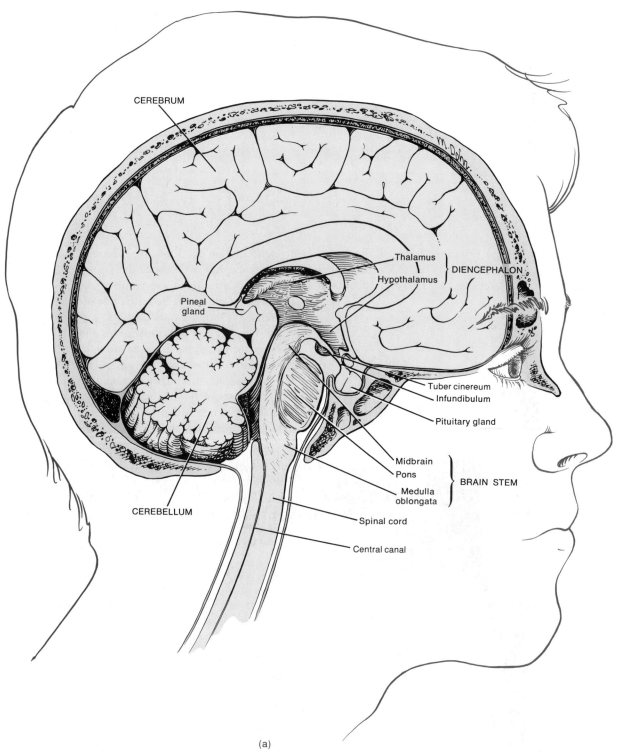

CEREBRUM

Thalamus

Hypothalamus

DIENCEPHALON

Pineal
gland

Tuber cinereum

Infundibulum

Pituitary gland

Midbrain

Pons

Medulla
oblongata

BRAIN STEM

CEREBELLUM

Spinal cord

Central canal

(a)

FIGURE 18-1 Brain. (a) Principal parts of the medial aspect of the brain seen in sagittal section. The infundibulum and pituitary gland are discussed in conjunction with the endocrine system in Chapter 21.

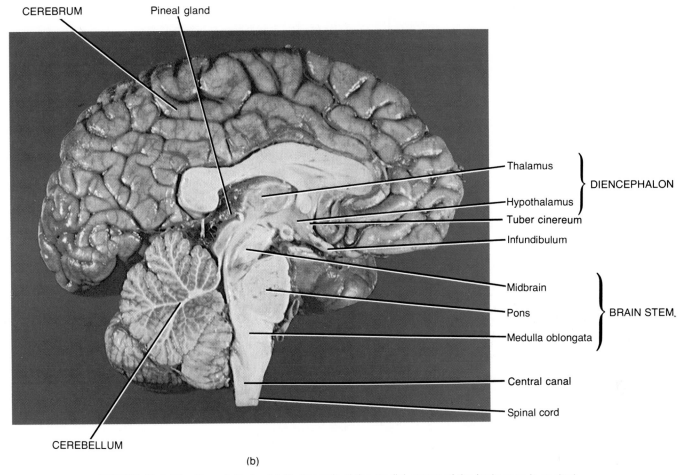

CEREBRUM

Pineal gland

Thalamus ⎫
Hypothalamus ⎬ DIENCEPHALON
Tuber cinereum ⎭
Infundibulum

Midbrain ⎫
Pons ⎬ BRAIN STEM.
Medulla oblongata ⎭

Central canal

Spinal cord

CEREBELLUM

(b)

FIGURE 18-1 (*Continued*) Brain. (b) Photograph of the medial aspect of the brain seen in sagittal section. (Courtesy of C. Yokochi and J. W. Rohen, *Photographic Anatomy of the Human Body,* 2nd ed., 1979, IGAKU-SHOIN, Ltd., Tokyo, New York.)

The cranial dura mater consists of two layers. The thicker, outer layer (periosteal layer) tightly adheres to the cranial bones and serves as periosteum. The thinner, inner layer (meningeal layer) includes a mesothelial layer on its smooth surface. The spinal dura mater corresponds to the meningeal layer of the cranial dura mater.

CLINICAL APPLICATION

The middle meningeal artery, running on the inner surface of the temporal bone between the dura mater and the skull, is closely adherent to the bones of the skull. It and the meningeal veins may be ruptured by a blow on the temple, especially if the bone is fractured. This rupture produces an **extradural hemorrhage,** which results in gradually increasing cranial pressure, drowsiness, unconsciousness, and death unless there is surgical intervention.

CEREBROSPINAL FLUID (CSF)

The brain, as well as the rest of the central nervous system, is further protected against injury by **cerebrospinal fluid (CSF).** This fluid circulates through the subarachnoid space around the brain and spinal cord and through the ventricles of the brain. The subarachnoid space is the area between the arachnoid and pia mater.

The **ventricles** (VEN-tri-kuls) are cavities in the brain that communicate with each other, with the central canal of the spinal cord, and with the subarachnoid space (Figure 18-2a, b). Each of the two **lateral ventricles** is located in a hemisphere (side) of the cerebrum under the corpus callosum. The **third ventricle** is a slit between and inferior to the right and left halves of the thalamus and between the lateral ventricles. Each lateral ventricle communicates with the third ventricle by a narrow, oval opening, the **interventricular foramen.** The **fourth ventricle** lies between the inferior brain stem and the cerebellum. It communicates with the third ventricle via the **cerebral aque-**

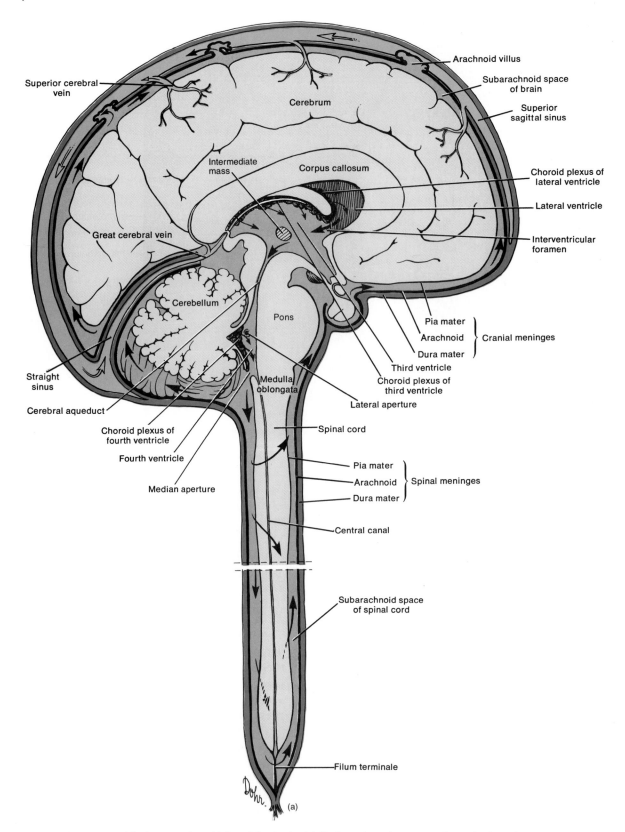

FIGURE 18-2 Meninges and ventricles of the brain. (a) Brain and meninges seen in sagittal section. Arrows indicate the direction of flow of cerebrospinal fluid.

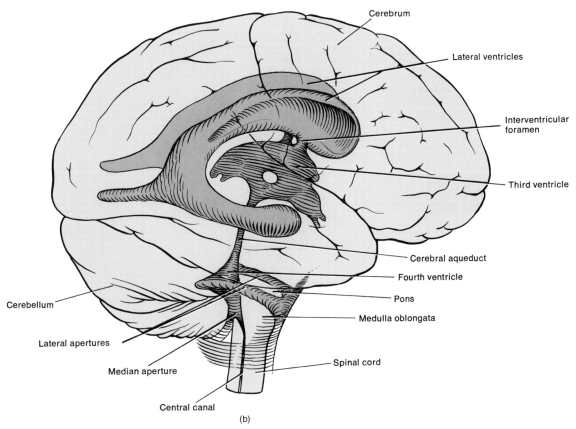

FIGURE 18-2 (*Continued*) Meninges and ventricles of the brain. (b) Diagrammatic lateral projection of the ventricles.

duct, which passes through the midbrain. The roof of the fourth ventricle has three openings: a **median aperture** and two **lateral apertures.** Through these openings, the fourth ventricle also communicates with the subarachnoid space of the brain and cord.

The entire central nervous system contains between 80 and 150 ml (3 to 5 oz) of cerebrospinal fluid. It is a clear, colorless fluid of watery consistency. Chemically, it contains proteins, glucose, urea, and salts. It also contains some lymphocytes. The fluid serves as a shock absorber for the central nervous system. It also circulates nutritive substances filtered from the blood.

Cerebrospinal fluid is formed primarily by filtration and secretion from networks of capillaries, called **choroid** (KŌ-royd; *chorion* = delicate) **plexuses,** located in the ventricles (Figure 18-2a). Various components of the choroid plexuses form a **blood–cerebrospinal fluid barrier** that permits certain substances to enter the fluid but inhibits others. Such a barrier protects the brain and spinal cord from harmful substances. The fluid formed in the choroid plexuses of the lateral ventricles circulates through the interventricular foramina to the third ventricle, where more fluid is added by the choroid plexus of the third ventricle. It then flows through the cerebral aqueduct into the fourth ventricle. Here there are contri-

butions from the choroid plexus of the fourth ventricle. The fluid then circulates through the apertures of the fourth ventricle into the subarachnoid space around the back of the brain. It also passes downward to the subarachnoid space around the posterior surface of the spinal cord, up the anterior surface of the spinal cord, and around the anterior part of the brain. From here it is gradually reabsorbed into veins. Some cerebrospinal fluid may be formed by ependymal (neuroglial) cells lining the central canal of the spinal cord. This small quantity of fluid ascends to reach the fourth ventricle. Most of the fluid is absorbed into the superior sagittal sinus. The absorption actually occurs through **arachnoid villi**–fingerlike projections of the arachnoid that push into the dural venous sinuses, especially the superior sagittal sinus. Normally, cerebrospinal fluid is absorbed as rapidly as it is formed.

The formation, circulation, and absorption of cerebrospinal fluid are summarized in Figure 18-3.

CLINICAL APPLICATION

If an obstruction, such as a tumor, arises in the brain and interferes with the drainage of fluid from the ven-

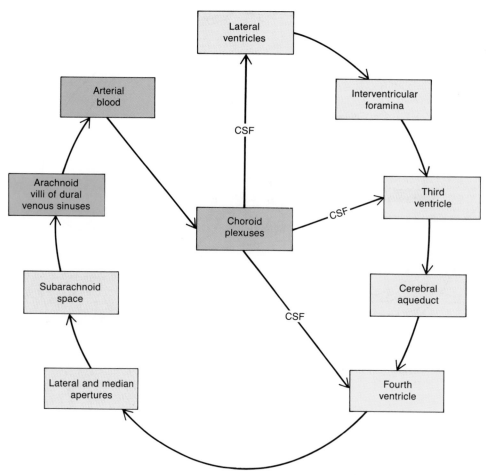

FIGURE 18-3 Summary of the formation, circulation, and absorption of cerebrospinal fluid.

tricles into the subarachnoid space, large amounts of fluid accumulate in the ventricles. Fluid pressure inside the brain increases, and, if the fontanels have not yet closed, the head bulges to relieve the pressure. This condition is called **internal hydrocephalus** (*hydro* = water; *enkephalos* = brain). If an obstruction interferes with drainage somewhere in the subarachnoid space and cerebrospinal fluid accumulates inside the space, the condition is termed **external hydrocephalus.**

BLOOD SUPPLY

The brain is well supplied with oxygen and nutrients by blood vessels that form the cerebral arterial circle (circle of Willis). (Cerebral circulation is outlined in Exhibit 14-3 and Figure 14-9.)

Although the brain composes only about 2 percent of the body weight, it utilizes about 20 percent of the oxygen used by the entire body. The brain is one of the most metabolically active organs of the body and the amount of oxygen used varies with the degree of mental activity. If the blood flow to the brain is interrupted even

briefly, unconsciousness may result. A 1- or 2-minute interruption may weaken the brain cells by starving them of oxygen. If the cells are totally deprived of oxygen for 4 minutes, many are permanently injured. Lysosomes of brain cells are sensitive to decreased oxygen concentration. If the condition persists long enough, lysosomes break open and the released enzymes bring about self-destruction of brain cells. Occasionally during childbirth the oxygen supply from the mother's blood is interrupted before the baby leaves the birth canal and can breathe. Often such babies are stillborn or suffer permanent brain damage that may result in mental retardation, epilepsy, and paralysis.

Blood supplying the brain also contains glucose, the principal source of energy for brain cells. Because carbohydrate storage in the brain is limited, the supply of glucose must be continuous. If blood entering the brain has a low glucose level, mental confusion, dizziness, convulsions, and loss of consciousness may occur.

Both carbon dioxide and oxygen have potent effects on cerebral blood flow. Carbon dioxide increases cerebral blood flow by combining with water to form carbonic acid (H_2CO_3), which breaks down into hydrogen ions

(H^+) and bicarbonate ions (HCO_3^-). The H^+ ions then cause vasodilation of cerebral vessels and increased blood flow. Dilation is almost directly proportional to an increase in H^+ ion concentration. A decrease in oxygen in the blood also causes vasodilation and increased cerebral blood flow.

Glucose, oxygen, and certain ions pass rapidly from the circulating blood into brain cells. Other substances, such as creatinine, urea, chloride, insulin, and sucrose, enter quite slowly. Still other substances—proteins and most antibiotics—do not pass at all from the blood into brain cells. The differential rates of passage of certain materials from the blood into most parts of the brain are based upon a concept called the **blood–brain barrier (BBB).** The barrier is either absent or less selective in the hypothalamus and roof of the fourth ventricle. Electron micrograph studies of the capillaries of the brain reveal that they differ structurally from other capillaries. Brain capillaries are constructed of more densely packed cells and are surrounded by large numbers of neuroglial cells and a continuous basement membrane. These features form a barrier to the passage of certain materials. Substances that cross the barrier are either very small molecules or require the assistance of a carrier molecule to cross by active transport. The blood–brain barrier functions as a selective barrier to protect brain cells from harmful substances. An injury to the brain due to trauma, inflammation, or toxins causes a breakdown of the blood–brain barrier, permitting the passage of normally restricted substances into brain tissue.

CLINICAL APPLICATION

Various **drugs** differ with respect to their passage through the blood–brain barrier. The antibiotics *chloramphenicol* (Chloromycetin), *tetracycline* (Achromycin V), and *sulfonamides* (Sonilyn) cross easily. *Penicillin* (Bicillin) crosses in only very small amounts.

Thiopental sodium (Pentothal), a general anesthetic, rapidly crosses the blood–brain barrier following intravenous injection. *Atropine* (Atropisol), an antispasmodic, also quickly enters the brain. *Anisotropine methylbromide* (Valpin), another antispasmodic, and *phenylbutazone* (Azolid), an antiinflammatory drug, cannot cross the blood–brain barrier.

BRAIN STEM

Medulla Oblongata

The **medulla oblongata** (me-DULL-la ob'-long-GA-ta), or simply **medulla,** is a continuation of the upper portion of the spinal cord and forms the inferior part of the brain stem (Figure 18-4). Its position in relation to the other parts of the brain may be noted in Figure 18-1. It lies just superior to the level of the foramen magnum and extends upward to the inferior portion of the pons. The medulla measures 3 cm (about 1 in) in length.

The medulla contains all ascending and descending tracts that communicate between the spinal cord and various parts of the brain. These tracts constitute the white matter of the medulla. Some tracts cross as they pass through the medulla. Let us see how this crossing occurs and what it means.

On the ventral side of the medulla are two roughly triangular structures called **pyramids** (Figures 18-4 and 18-5). The pyramids are composed of the largest motor tracts that pass from the outer region of the cerebrum (cerebral cortex) to the spinal cord. Just above the junction of the medulla with the spinal cord, most of the fibers in the left pyramid cross to the right side, and most of the fibers in the right pyramid cross to the left. This crossing is called the **decussation** (dē'-ku-SĀ-shun) **of pyramids.** The adaptive value, if any, of this phenomenon is unknown. The principal motor fibers that undergo decussation belong to the lateral corticospinal tracts. These tracts originate in the cerebral cortex and pass inferiorly to the medulla. The fibers cross in the pyramids and descend in the lateral columns of the spinal cord, terminating in the anterior gray horns. Here synapses occur with motor neurons that terminate in skeletal muscles. As a result of the crossing, fibers that originate in the left cerebral cortex activate muscles on the right side of the body, and fibers that originate in the right cerebral cortex activate muscles on the left side. Decussation explains why motor areas of one side of the cerebral cortex control muscular movements on the opposite side of the body.

The dorsal side of the medulla contains two pairs of prominent nuclei: the right and left **nucleus gracilis** (gras-I-lis; *gracilis* = slender) and **nucleus cuneatus** (kyoo-nē-Ā-tus; *cuneus* = wedge). These nuclei receive sensory fibers from ascending tracts (right and left fasciculus gracilis and fasciculus cuneatus) of the spinal cord and relay the sensory information to the opposite side of the medulla. This information is conveyed to the thalamus and then to the sensory areas of the cerebral cortex. Nearly all sensory impulses received on one side of the body cross in the medulla or spinal cord and are perceived in the opposite side of the cerebral cortex.

In addition to its function as a conduction pathway for motor and sensory impulses between the brain and spinal cord, the medulla also contains an area of dispersed gray matter containing some white fibers. This region is called the **reticular formation.** Actually, portions of the reticular formation are also located in the spinal cord, pons, midbrain, and diencephalon. The reticular formation functions in consciousness and arousal.

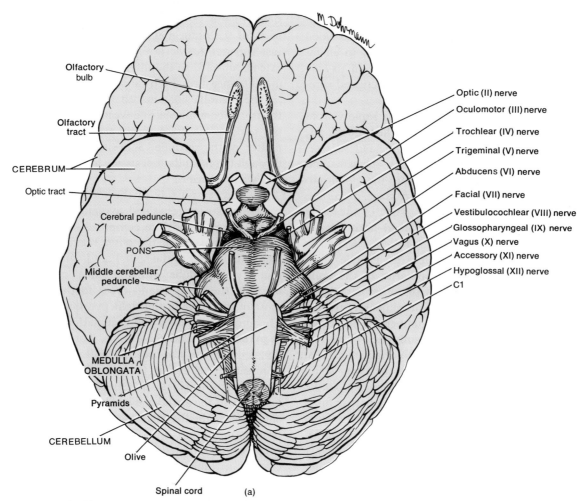

M. Dohrmann

Olfactory bulb

Olfactory tract

CEREBRUM

Optic tract

Cerebral peduncle

PONS

Middle cerebellar peduncle

MEDULLA OBLONGATA

Pyramids

CEREBELLUM

Olive

Spinal cord

Optic (II) nerve
Oculomotor (III) nerve
Trochlear (IV) nerve
Trigeminal (V) nerve
Abducens (VI) nerve
Facial (VII) nerve
Vestibulocochlear (VIII) nerve
Glossopharyngeal (IX) nerve
Vagus (X) nerve
Accessory (XI) nerve
Hypoglossal (XII) nerve
C1

(a)

FIGURE 18-4 Brain stem. (a) Diagram of the ventral surface of the brain showing the structure of the brain stem in relation to the cranial nerves and associated structures. (b) Photograph of the ventral surface of the brain showing the brain stem in relation to associated structures. (Courtesy of Martin Rotker, Taurus Photos). (c) Photograph of the ventral surface of the brain stem. (Courtesy of N. Gluhbegovic and T. H. Williams, *The Human Brain: A Photographic Guide,* Harper & Row, New York, 1980.)

CLINICAL APPLICATION

The most common knockout blow is one that makes contact with the mandible. Such a blow twists and distorts the brain stem and overwhelms the reticular activating system (RAS) by sending a sudden volley of nerve impulses to the brain, resulting in **unconsciousness.**

Within the medulla are three vital reflex centers of the reticular system. The **cardiac center** regulates heartbeat and force of contraction, the **medullary rhythmicity area** adjusts the basic rhythm of breathing, and the **vasomotor (vasoconstrictor) center** regulates the diameter of blood vessels. Other centers in the medulla are considered nonvital and coordinate swallowing, vomiting, coughing, sneezing, and hiccuping.

CLINICAL APPLICATION

Neurosurgeons are now using an ultrasonic device called a **cavitron ultrasonic surgical aspirator (CUSA)** that can shatter certain kinds of brain tumors. High-frequency sound waves stimulate the slender tip of the instrument to vibrate 23,000 times per second. This vibratory action disintegrates the tumors. The instrument also delivers an irrigating saline solution that aspirates the fragmented particles. One major advantage of CUSA is that it decreases the probability of damaging either adjacent normal tissues such as the brain stem, which could cause abnormal heart rhythms, or large blood vessels, which could cause hemorrhage.

The medulla also contains the nuclei of origin for several pairs of cranial nerves (Figures 18-4a and 18-5). These

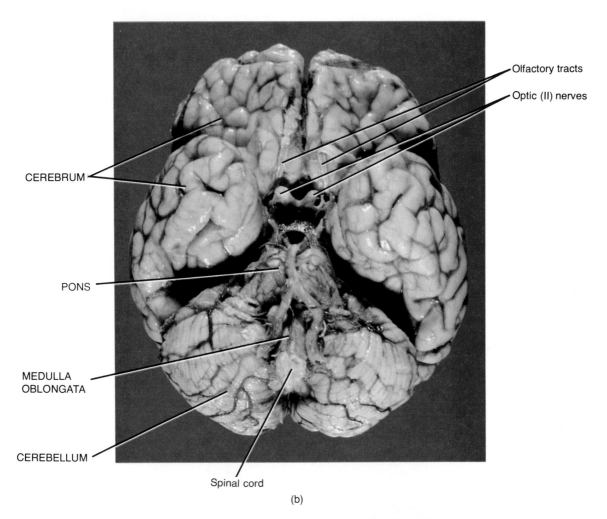

CEREBRUM

Olfactory tracts

Optic (II) nerves

PONS

MEDULLA
OBLONGATA

CEREBELLUM

Spinal cord

(b)

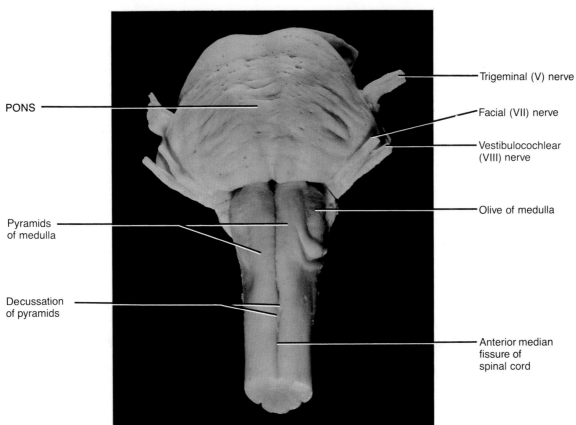

PONS

Trigeminal (V) nerve

Facial (VII) nerve

Vestibulocochlear
(VIII) nerve

Olive of medulla

Pyramids
of medulla

Decussation
of pyramids

Anterior median
fissure of
spinal cord

(c)

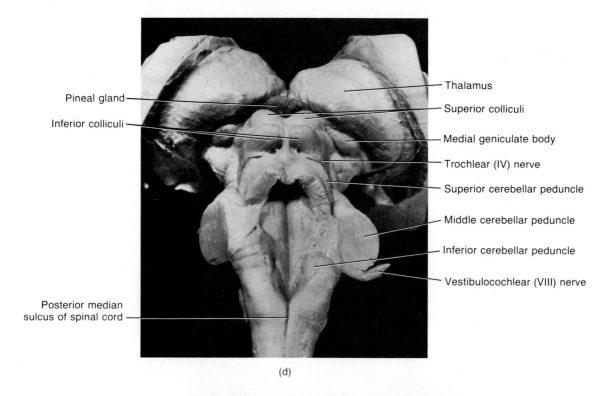

Pineal gland

Inferior colliculi

Posterior median
sulcus of spinal cord

Thalamus

Superior colliculi

Medial geniculate body

Trochlear (IV) nerve

Superior cerebellar peduncle

Middle cerebellar peduncle

Inferior cerebellar peduncle

Vestibulocochlear (VIII) nerve

(d)

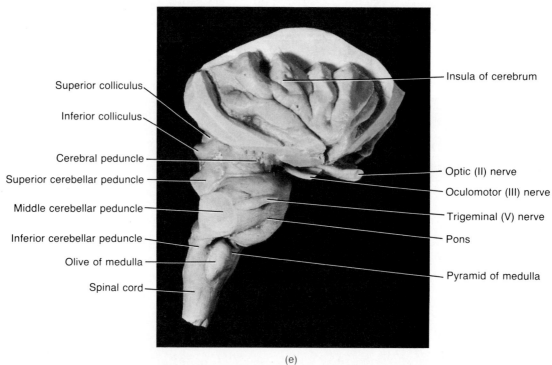

Superior colliculus

Inferior colliculus

Cerebral peduncle

Superior cerebellar peduncle

Middle cerebellar peduncle

Inferior cerebellar peduncle

Olive of medulla

Spinal cord

Insula of cerebrum

Optic (II) nerve

Oculomotor (III) nerve

Trigeminal (V) nerve

Pons

Pyramid of medulla

(e)

FIGURE 18-4 (*Continued*) Brain stem. (d) Photograph of the posterior surface of the brain stem. (Courtesy of N. Gluhbegovic and T. H. Williams, *The Human Brain: A Photographic Guide,* Harper & Row, New York, 1980.) (e) Photograph of the brain stem in right lateral view. (Courtesy of N. Gluhbegovic and T. H. Williams, *The Human Brain: A Photographic Guide,* Harper & Row, New York, 1980.)

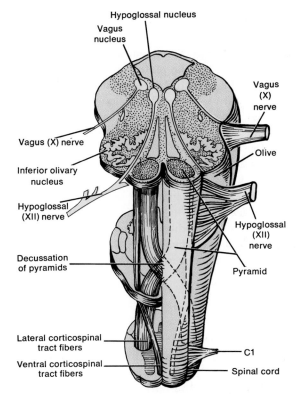

Hypoglossal nucleus
Vagus nucleus
Vagus (X) nerve
Vagus (X) nerve
Inferior olivary nucleus
Hypoglossal (XII) nerve
Olive
Hypoglossal (XII) nerve
Decussation of pyramids
Pyramid
Lateral corticospinal tract fibers
Ventral corticospinal tract fibers
C1
Spinal cord

FIGURE 18-5 Details of the medulla showing the decussation of pyramids.

are the cochlear and vestibular branches of the vestibulocochlear (VIII) nerves, which are concerned with hearing and equilibrium (there is also a nucleus for the vestibular branches in the pons); the glossopharyngeal (IX) nerves, which relay impulses related to swallowing, salivation, and taste; the vagus (X) nerves, which relay impulses to and from many thoracic and abdominal viscera; the cranial portions of the accessory (XI) nerves, which convey impulses related to head and shoulder movements (a part of this nerve also arises from the first five segments of the spinal cord); and the hypoglossal (XII) nerves, which convey impulses that involve tongue movements.

On each lateral surface of the medulla is an oval projection called the **olive** (Figure 18-4c, e). The olive contains an inferior olivary nucleus and two accessory olivary nuclei. The nuclei are connected to the cerebellum by fibers.

Also associated with the medulla is the greater part of the **vestibular nuclear complex.** This nuclear group consists of the *lateral, medial,* and *inferior vestibular nuclei* in the medulla and the *superior vestibular nucleus* in the pons. As you will see later (Chapter 20), the vestibular nuclei assume an important role in helping the body maintain its sense of equilibrium.

In view of the many vital activities controlled by the medulla, it is not surprising that a hard blow to the base of the skull can be fatal. Nonfatal medullary injury may be indicated by cranial nerve malfunctions on the same

side of the body as the area of medullary injury, paralysis and loss of sensation on the opposite side of the body, and irregularities in respiratory control.

Pons

The relationship of the **pons** to other parts of the brain can be seen in Figures 18-1 and 18-4. The pons, which means bridge, lies directly above the medulla and anterior to the cerebellum. It measures about 2.5 cm (1 in) in length. Like the medulla, the pons consists of white fibers scattered throughout with nuclei. As the name implies, the pons is a bridge connecting the spinal cord with the brain and parts of the brain with each other. These connections are provided by fibers that run in two principal directions. The transverse fibers connect with the cerebellum through the *middle cerebellar peduncles.* The longitudinal fibers of the pons belong to the motor and sensory tracts that connect the spinal cord or medulla with the upper parts of the brain stem.

The nuclei for certain paired cranial nerves are also contained in the pons (Figure 18-4a). These include the trigeminal (V) nerves, which relay impulses for chewing and for sensations of the head and face; the abducens (VI) nerves, which regulate certain eyeball movements; the facial (VII) nerves, which conduct impulses related to taste, salivation, and facial expression; and the vestibular branches of the vestibulocochlear (VIII) nerves, which are concerned with equilibrium.

Other important nuclei in the reticular formation of the pons are the **pneumotaxic** (noo-mō-TAK-sik) **area** and the **apneustic** (ap-NOO-stik) **area.** Together with the medullary rhythmicity area in the medulla, they help control respiration.

Midbrain

The **midbrain,** or **mesencephalon** (*meso* = middle; *enkephalos* = brain), extends from the pons to the lower portion of the diencephalon (Figures 18-1 and 18-4). It is about 2.5 cm (1 in) in length. The cerebral aqueduct passes through the midbrain and connects the third ventricle above with the fourth ventricle below.

The ventral portion of the midbrain contains a pair of fiber bundles referred to as **cerebral peduncles** (pe-DUNG-kulz). The cerebral peduncles contain many motor fibers that convey impulses from the cerebral cortex to the pons and spinal cord. They also contain sensory fibers that pass from the spinal cord to the thalamus. The cerebral peduncles constitute the main connection for tracts between upper parts of the brain and lower parts of the brain and the spinal cord.

The dorsal portion of the midbrain is called the **tectum** (*tectum* = roof) and contains four rounded eminences: the **corpora quadrigemina** (KOR-po-ra kwad-ri-JEM-in-a). Two of the eminences are known as the **superior col-**

liculi (ko-LIK-yoo-lī). These serve as reflex centers for movements of the eyeballs and head in response to visual and other stimuli. The other two eminences, the **inferior colliculi,** serve as reflex centers for movements of the head and trunk in response to auditory stimuli. The midbrain also contains the **substantia nigra** (sub-STAN-shē-a NĪ-gra), a large, heavily pigmented nucleus near the cerebral peduncles.

A major nucleus in the reticular formation of the midbrain is the **red nucleus.** Fibers from the cerebellum and cerebral cortex terminate in the red nucleus. The red nucleus is also the origin of cell bodies of the descending rubrospinal tract. Other nuclei in the midbrain are associated with cranial nerves (see Figure 18-4a). These include the oculomotor (III) nerves, which mediate some movements of the eyeballs and changes in pupil size and lens shape, and the trochlear (IV) nerves, which conduct impulses that move the eyeballs.

A structure called the **medial lemniscus** (*lemniskos* = ribbon or band) is common to the medulla, pons, and midbrain. The medial lemniscus is a band of white fibers containing axons that convey impulses for fine touch, proprioception, and vibrations from the medulla to the thalamus.

DIENCEPHALON

The **diencephalon** (*dia* = through; *enkephalos* = brain) consists principally of the thalamus and hypothalamus. The relationship of these structures to the rest of the brain is shown in Figure 18-1.

Thalamus

The **thalamus** (THAL-a-mus; *thalamos* = inner chamber) is an oval structure above the midbrain that measures about 3 cm (1 in) in length and constitutes four-fifths of the diencephalon. It consists of two oval masses of mostly gray matter organized into nuclei that form the lateral walls of the third ventricle (Figure 18-6). The masses are joined by a bridge of gray matter called the **intermediate mass.** Each mass is deeply embedded in a cerebral hemisphere and is bounded laterally by the **internal capsule.**

Although the thalamic masses are primarily gray matter, some portions are white matter. Among the white matter portions are the **stratum zonale,** which covers the dorsal surface; the **external medullary lamina,** covering the lateral surface; and the **internal medullary lamina,** which divides the gray matter masses into an anterior nuclear group, a medial nuclear group, and a lateral nuclear group.

Within each group are nuclei that assume various roles. Some nuclei in the thalamus serve as relay stations for all sensory impulses, except smell, to the cerebral cortex. These include the **medial geniculate** (je-NIK-yoo-lāt) **nu-**

clei (hearing), the **lateral geniculate nuclei** (vision), and the **ventral posterior nuclei** (general sensations and taste). Other nuclei are centers for synapses in the somatic motor system. These include the **ventral lateral nuclei** (voluntary motor actions) and **ventral anterior nuclei** (voluntary motor actions and arousal). The thalamus is the principal relay station for sensory impulses that reach the cerebral cortex from the spinal cord, brain stem, cerebellum, and parts of the cerebrum.

The thalamus also functions as an interpretation center for some sensory impulses, such as pain, temperature, crude touch, and pressure. The thalamus also contains a **reticular nucleus** in its reticular formation, which in some way seems to modify neuronal activity in the thalamus, and an **anterior nucleus** in the floor of the lateral ventricle, which is concerned with certain emotions and memory.

Hypothalamus

The **hypothalamus** (*hypo* = under) is a small portion of the diencephalon. Its relationship to other parts of the brain is shown in Figures 18-1 and 18-6a. The hypothalamus forms the floor and part of the lateral walls of the third ventricle. It is partially protected by the sella turcica of the sphenoid bone.

Despite its small size, nuclei in the hypothalamus control many body activities, most of them related to homeostasis. Although differentiation of the hypothalamic nuclei is far from precise, it is possible to identify certain nuclei. Some of these nuclei are more readily identified in lower animals and are more distinct in fetuses than adults. Also, within a given nucleus there may be several kinds of cells that can be differentiated histologically. The localization of function, with a few exceptions, is not specific to the individual nuclei; certain functions tend to overlap nuclear boundaries. For this reason, functions are attributed to regions rather than specific nuclei. Since the hypothalamic nuclei are useful landmarks in understanding the subsequent discussion of functions, a three-dimensional view of the hypothalamic nuclei is shown in Figure 18-7.

The chief functions of the hypothalamus are as follows:

1. It controls and integrates the autonomic nervous system, which stimulates smooth muscle, regulates the rate of contraction of cardiac muscle, and controls the secretions of many glands. This is accomplished by axons of neurons whose dendrites and cell bodies are in hypothalamic nuclei. The axons form tracts from the hypothalamus to sympathetic and parasympathetic nuclei in the brain stem and spinal cord. Through the autonomic nervous system, the hypothalamus is the main regulator of visceral activities. It regulates heart rate, movement of food through the digestive tract, and contraction of the urinary bladder.

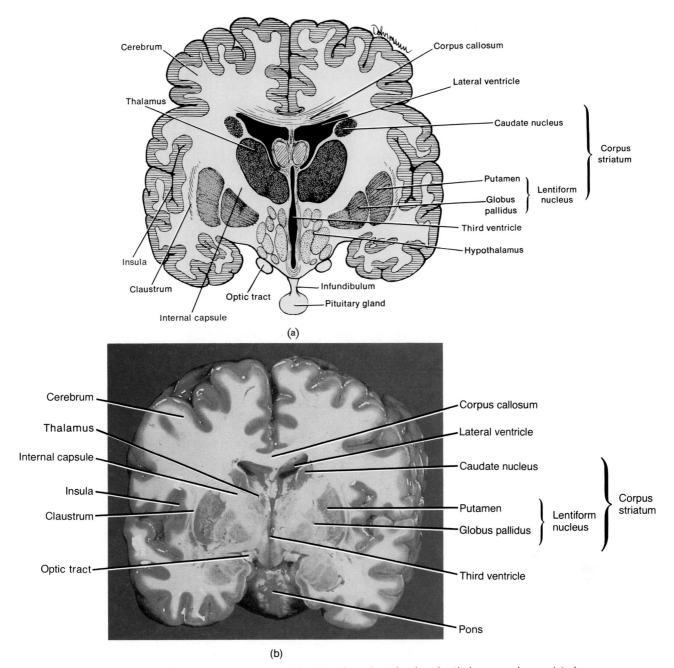

FIGURE 18-6 Thalamus. (a) Diagram of a frontal section showing the thalamus and associated structures. (b) Photograph of a frontal section of the cerebrum showing the thalamus and associated structures. (Courtesy of C. Yokochi and J. W. Rohen, *Photographic Anatomy of the Human Body*, 2nd ed., 1979, IGAKU-SHOIN, Ltd., Tokyo, New York.)

2. It is involved in the reception of sensory impulses from the viscera.

3. It is the principal intermediary between the nervous system and the endocrine system—the two major control systems of the body. The hypothalamus lies just above the pituitary, the main endocrine gland. When the hypothalamus detects certain changes in the body, it releases chemicals called regulating factors that stimulate or inhibit the anterior pituitary gland. The anterior pituitary then releases or holds back hormones that regulate

various physiological activities of the body. The hypothalamus also produces two hormones, antidiuretic hormone (ADH) and oxytocin (OT), which are transported to and stored in the posterior pituitary gland. The hormones are released from storage when needed by the body.

4. It is the center for the mind-over-body phenomenon. When the cerebral cortex interprets strong emotions, it often sends impulses along the tracts that connect the cortex with the hypothalamus. The hypothalamus then directs impulses via the autonomic nervous system and

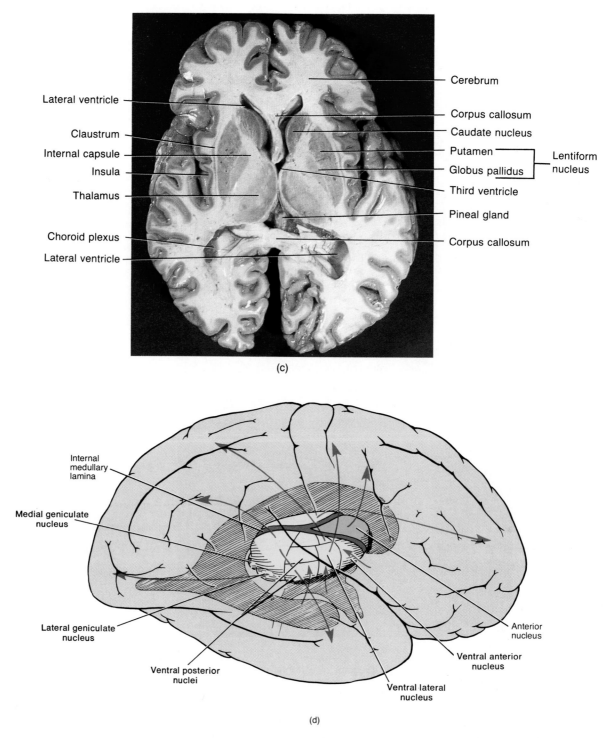

(c)

(d)

FIGURE 18-6 (*Continued*) Thalamus. (c) Photograph of a horizontal section of the cerebrum showing the thalamus and associated structures. (Courtesy of C. Yokochi and J. W. Rohen, *Photographic Anatomy of the Human Body,* 2nd ed., 1979, IGAKU-SHOIN, Ltd., Tokyo, New York.) (d) Diagram of a right lateral view of the thalamic nuclei.

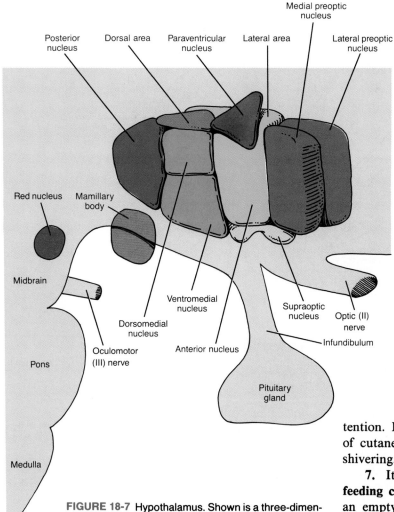

Posterior nucleus
Dorsal area
Paraventricular nucleus
Medial preoptic nucleus
Lateral area
Lateral preoptic nucleus
Red nucleus
Mamillary body
Midbrain
Ventromedial nucleus
Supraoptic nucleus
Optic (II) nerve
Dorsomedial nucleus
Oculomotor (III) nerve
Anterior nucleus
Infundibulum
Pons
Pituitary gland
Medulla

FIGURE 18-7 Hypothalamus. Shown is a three-dimensional view of the hypothalamic nuclei.

also releases chemicals that stimulate the anterior pituitary gland. The result can be a wide range of changes in body activities. For instance, when you panic, impulses leave the hypothalamus to stimulate your heart to beat faster. Likewise, continued psychological stress can produce long-term abnormalities in body function that result in serious illness. These so-called psychosomatic disorders are definitely real.

5. It is associated with feelings of rage and aggression.

6. It controls normal body temperature. Certain cells of the hypothalamus serve as a thermostat. If blood flowing through the hypothalamus is above normal temperature, the hypothalamus directs impulses along the autonomic nervous system to stimulate activities that promote heat loss. Heat can be lost through relaxation of the smooth muscle in the blood vessels, causing vasodilation of cutaneous vessels and increased heat loss from the skin. Heat loss also occurs by sweating. Conversely, if the temperature of the blood is below normal, the hypothalamus generates impulses that promote heat re-

tention. Heat can be retained through the constriction of cutaneous blood vessels, cessation of sweating, and shivering.

7. It regulates food intake through two centers. The **feeding center** is stimulated by hunger sensations from an empty stomach. When sufficient food has been ingested, the **satiety** (sa-TĪ-e-tē) **center** is stimulated and sends out impulses that inhibit the feeding center.

8. It contains a **thirst center.** Certain cells in the hypothalamus are stimulated when the extracellular fluid volume is reduced. The stimulated cells produce the sensation of thirst.

9. It is one of the centers that maintains the waking state and sleep patterns.

10. It exhibits properties of a self-sustained oscillator and, as such, acts as a pacemaker to drive many biological rhythms.

CEREBRUM

Supported on the brain stem and forming the bulk of the brain is the **cerebrum** (see Figure 18-1). The surface of the cerebrum is composed of gray matter 2 to 4 mm (0.08 to 0.16 in) thick and is referred to as the **cerebral cortex** (*cortex* = rind, or bark). The cortex, containing billions of cells, consists of six layers of nerve cell bodies in most areas. Beneath the cortex lies the cerebral white matter.

During embryonic development, when there is a rapid increase in brain size, the gray matter of the cortex en-

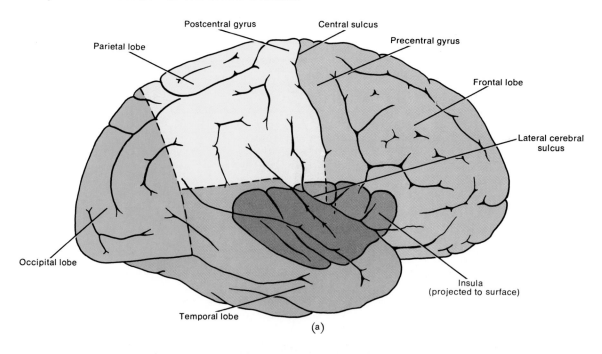

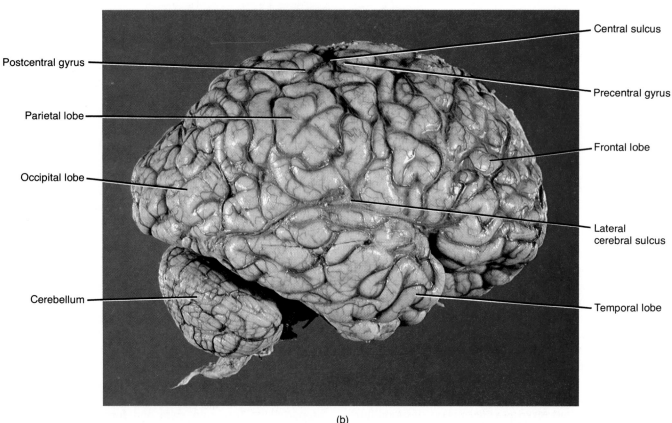

FIGURE 18-8 Lobes and fissures of the cerebrum. (a) Diagram of right lateral view. Since the insula cannot be seen externally, it has been projected to the surface. It can be seen in Figure 18-6a, b and Figure 18-8c. (b) Photograph of the right lateral view. (Courtesy of Martin Rotker, Taurus Photos.)

larges out of proportion to the underlying white matter. As a result, the cortical region rolls and folds upon itself. The folds are called **gyri** (JĪ-rī), or **convolutions** (Figure 18-8). The deep grooves between folds are referred to as **fissures;** the shallow grooves between folds are **sulci**

(SUL-sī). The most prominent fissure, the **longitudinal fissure,** nearly separates the cerebrum into right and left halves, or **hemispheres** (Figure 18-8d–f). The hemispheres, however, are connected internally by a large bundle of transverse fibers composed of white matter called

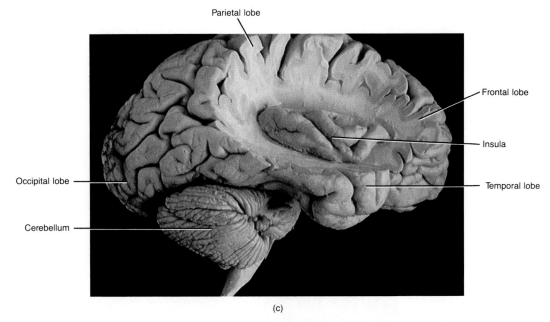

Parietal lobe

Frontal lobe

Insula

Occipital lobe

Temporal lobe

Cerebellum

(c)

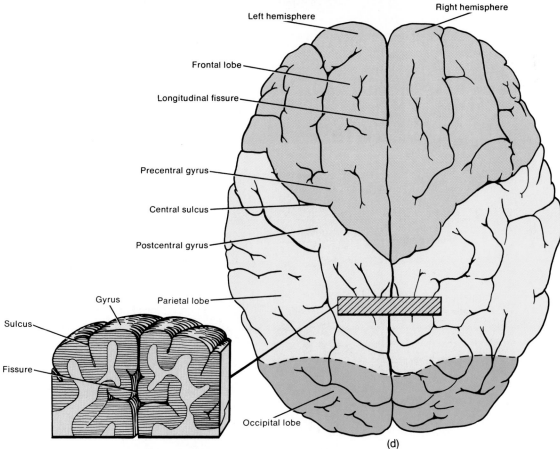

Left hemisphere

Right hemisphere

Frontal lobe

Longitudinal fissure

Precentral gyrus

Central sulcus

Postcentral gyrus

Gyrus

Sulcus

Parietal lobe

Fissure

Occipital lobe

(d)

FIGURE 18-8 (*Continued*) Lobes and fissures of the cerebrum. (c) Photograph of right lateral view showing the insula after removal of a portion of the cerebrum. (Courtesy of N. Gluhbegovic and T. H. Williams, *The Human Brain: A Photographic Guide,* Harper & Row, New York, 1980.) (d) Diagram of superior view. The insert to the left indicates the relative differences among a gyrus, sulcus, and fissure.

the **corpus callosum** (kal-LŌ-sum; *corpus* = body; *callosus* = hard). Between the hemispheres is an extension of the cranial dura mater called the **falx (FALKS) cerebri (cerebral fold).** It encloses the superior and inferior sagittal sinuses.

Lobes

Each cerebral hemisphere is further subdivided into four lobes by deep sulci or fissures (Figure 18-8a–f). The **central sulcus** separates the **frontal lobe** from the **parietal**

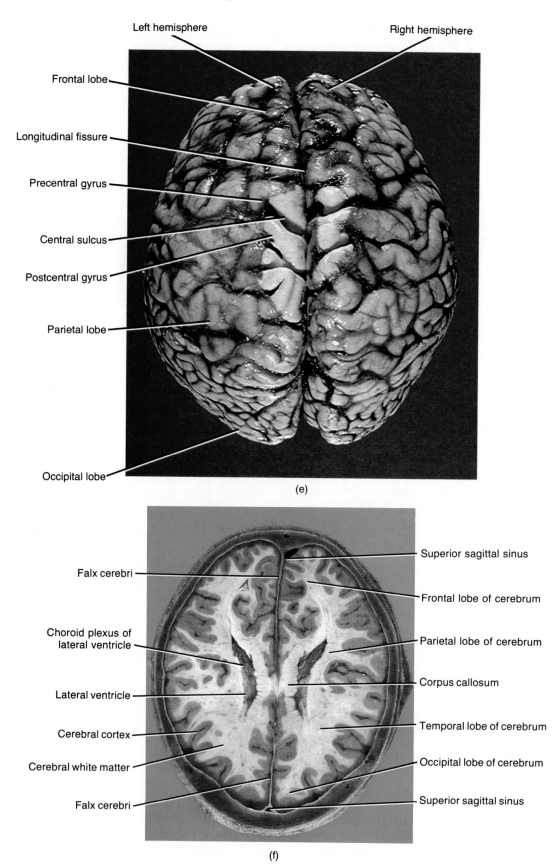

FIGURE 18-8 (*Continued*) Lobes and fissures of the cerebrum. (e) Photograph of superior view. (Courtesy of Martin Rotker, Taurus Photos.) (f) Photograph of a cross section through the cerebrum. (Courtesy of Stephen A. Kieffer and E. Robert Heitzman, *An Atlas of Cross-Sectional Anatomy*, Harper & Row, Publishers, Inc., New York, 1979.)

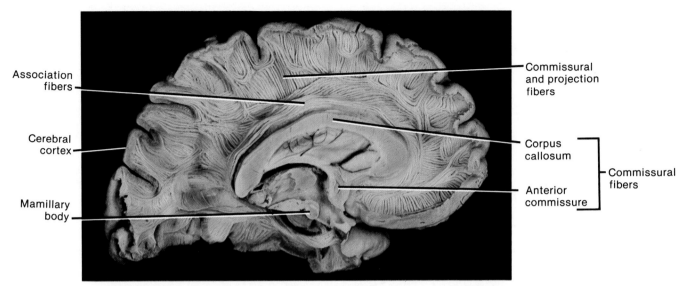

FIGURE 18-9 White matter tracts of the left cerebral hemisphere seen in sagittal section. (Courtesy of N. Gluhbegovic and T. H. Williams, *The Human Brain: A Photographic Guide,* Harper & Row, New York, 1980.)

lobe. A major gyrus, the **precentral gyrus,** is located immediately anterior to the central sulcus. The gyrus is a landmark for the primary motor area of the cerebral cortex. Another major gyrus, the **postcentral gyrus,** is located immediately posterior to the central sulcus. This gyrus is a landmark for the general sensory area of the cerebral cortex. The **lateral cerebral sulcus** separates the **frontal lobe** from the **temporal lobe.** The **parietooccipital sulcus** separates the **parietal lobe** from the **occipital lobe.** Another prominent fissure, the **transverse fissure,** separates the cerebrum from the cerebellum. The frontal lobe, parietal lobe, temporal lobe, and occipital lobe are named after the bones that cover them. A fifth part of the cerebrum, the **insula,** lies deep within the lateral cerebral fissure, under the parietal, frontal, and temporal lobes. It cannot be seen in an external view of the brain (Figure 18-8a–c).

As you will see later, the olfactory (I) and optic (II) nerves are associated with specific lobes of the cerebrum.

White Matter

The white matter underlying the cortex consists of myelinated axons running in three principal directions (Figure 18-9).

1. Association fibers connect and transmit impulses between gyri in the same hemisphere.

2. Commissural fibers transmit impulses from the gyri in one cerebral hemisphere to the corresponding gyri in the opposite cerebral hemisphere. Three important groups of commissural fibers are the *corpus callosum, anterior commissure,* and *posterior commissure.*

3. Projection fibers form ascending and descending tracts that transmit impulses from the cerebrum to other parts of the brain and spinal cord.

Basal Ganglia (Cerebral Nuclei)

The **basal ganglia (cerebral nuclei)** are paired masses of gray matter in each cerebral hemisphere (Figures 18-6 and 18-10). The largest of the basal ganglia of each hemisphere is the **corpus striatum** (strī-Ā-tum; *corpus* = body; *striatus* = striped). It consists of the **caudate** (*cauda* = tail) **nucleus** and the **lentiform** (*lenticula* = shaped like a lentil or lens) **nucleus.** The lentiform nucleus, in turn, is subdivided into a lateral portion called the **putamen** (pu-TĀ-men; *putamen* = shell) and a medial portion called the **globus pallidus** (*globus* = ball; *pallid* = pale).

The portion of the *internal capsule* passing between the lentiform nucleus and the caudate nucleus and between the lentiform nucleus and thalamus is sometimes considered part of the corpus striatum. The internal capsule is made up of a group of sensory and motor white matter tracts that connect the cerebral cortex with the brain stem and spinal cord.

Other structures frequently considered part of the basal ganglia are the claustrum and amygdaloid nucleus. The **claustrum** (KLAWS-trum) is a thin sheet of gray matter lateral to the putamen. The **amygdaloid** (a-MIG-da-loyd; *amygda* = almond) **nucleus** is located at the tail end of the caudate nucleus. Some authorities also consider the **substantia nigra,** the **subthalamic nucleus,** and the **red nucleus** to be part of the basal ganglia. The substantia nigra is a large nucleus in the midbrain whose axons

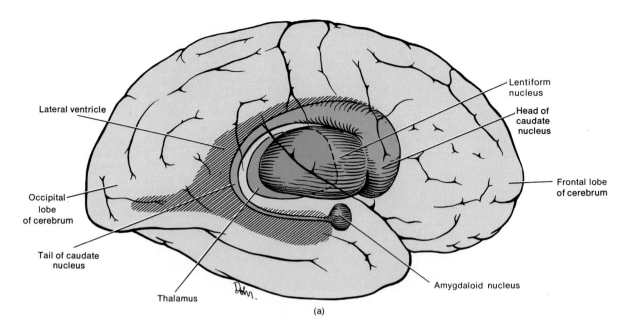

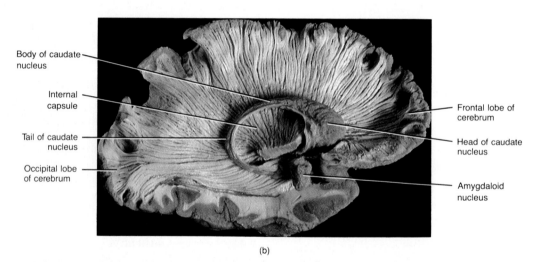

FIGURE 18-10 Basal ganglia. (a) In this diagram of the right lateral view of the cerebrum, the basal ganglia have been projected to the surface. Refer to Figure 18-6a, b to note the positions of the basal ganglia in the frontal section of the cerebrum. (b) Photograph of the medial surface of the left cerebral hemisphere showing portions of the basal ganglia. (Courtesy of N. Gluhbegovic and T. H. Williams, *The Human Brain: A Photographic Guide,* Harper & Row, New York, 1980.)

terminate in the caudate nucleus and putamen. The subthalamic nucleus lies against the internal capsule. Its major connection is with the globus pallidus.

The basal ganglia are interconnected by many fibers. They are also connected to the cerebral cortex, thalamus, and hypothalamus. The caudate nucleus and the putamen control large subconscious movements of the skeletal muscles, such as swinging the arms while walking. Such gross movements are also consciously controlled by the cerebral cortex. The globus pallidus is concerned with the regulation of muscle tone required for specific body movements.

CLINICAL APPLICATION

Damage to the basal ganglia results in abnormal body movements, such as uncontrollable shaking, called **tremor,** and **involuntary movements of skeletal muscles.** Moreover, destruction of a substantial portion of the caudate nucleus almost totally **paralyzes** the side of the body opposite to the damage. The caudate nucleus is an area often affected by a stroke.

A lesion in the subthalamic nucleus results in a

motor disturbance on the opposite side of the body called **hemiballismus** (*hemi* = half; *ballismos* = jumping), which is characterized by involuntary movements occurring suddenly with great force and rapidity. The movements are purposeless and generally of the withdrawal type, although they may be jerky. The spontaneous movements affect the proximal portions of the extremities most severely, especially the arms.

Limbic System

Certain components of the cerebral hemispheres and diencephalon constitute the **limbic** (*limbus* = border) **system.** Among its components are the following regions of gray matter:

1. Limbic lobe. Formed by two gyri of the cerebral hemisphere: the cingulate gyrus and the hippocampal gyrus.

2. Hippocampus. An extension of the hippocampal gyrus that extends into the floor of the lateral ventricle.

3. Amygdaloid nucleus. Located at the tail end of the caudate nucleus.

4. Mammillary bodies of the hypothalamus. Two round masses close to the midline near the cerebral peduncles.

5. Anterior nucleus of the thalamus. Located in the floor of the lateral ventricle.

The limbic system is a wishbone-shaped group of structures that encircles the brain stem and functions in the emotional aspects of behavior related to survival. The hippocampus, together with portions of the cerebrum, also functions in memory. Although behavior is a function of the entire nervous system, the limbic system controls most of its involuntary aspects. Experiments on the limbic system of monkeys and other animals indicate that the amygdaloid nucleus assumes a major role in controlling the overall pattern of behavior.

Other experiments have shown that the limbic system is associated with pleasure and pain. When certain areas of the limbic system of the hypothalamus, thalamus, and midbrain are stimulated, the reactions of experimental animals indicate they are experiencing intense punishment. When other areas are stimulated, the animals' reactions indicate they are experiencing extreme pleasure. In still other studies, stimulation of the perifornical nuclei of the hypothalamus results in a behavioral pattern called *rage.* The animal assumes a defensive posture—extending its claws, raising its tail, hissing, spitting, growling, and opening its eyes wide. Stimulating other areas of the limbic system results in an opposite behavioral pattern: docility, tameness, and affection. Because the limbic system assumes a primary function in emotions such as pain, pleasure, anger, rage, fear, sorrow, sexual feelings, docility,

and affection, it is sometimes called the "visceral" or "emotional" brain.

CLINICAL APPLICATION

Brain injuries are commonly associated with head injuries and result from displacement and distortion of neuronal tissue at the moment of impact. The various degrees of brain injury are described by the following terms:

1. **Concussion.** An abrupt but temporary loss of consciousness following a blow to the head or a sudden stopping of a moving head. A concussion produces no visible bruising of the brain.

2. **Contusion.** A visible bruising of the brain due to trauma and blood leaking from microscopic vessels. The pia mater is stripped from the brain over the injured area and may be torn, allowing blood to enter the subarachnoid space. A contusion usually results in an extended loss of consciousness, ranging from several minutes to many hours.

3. **Laceration.** Tearing of the brain, usually from a skull fracture or gunshot wound. A laceration results in rupture of large blood vessels with bleeding into the brain and subarachnoid space. Consequences include cerebral hematoma, edema, and increased intracranial pressure.

Functional Areas of Cerebral Cortex

The functions of the cerebrum are numerous and complex. In a general way, the cerebral cortex is divided into sensory, motor, and association areas. The **sensory areas** interpret sensory impulses, the **motor areas** control muscular movement, and the **association areas** are concerned with emotional and intellectual processes.

● *Sensory Areas* The **general sensory area,** or **somesthetic** (sō-mes-THET-ik; *soma* = body; *aisthesis* = perception) **area** is located directly posterior to the central sulcus of the cerebrum in the postcentral gyrus of the parietal lobe. It extends from the longitudinal fissure on the top of the cerebrum to the lateral cerebral sulcus. In Figure 18-11 the general sensory area is designated by the areas numbered 1, 2, and 3.*

The general sensory area receives sensations from cutaneous, muscular, and visceral receptors in various parts

---

* These numbers, as well as most of the others shown, are based on K. Brodmann's cytoarchitectural map of the cerebral cortex. His map, first published in 1909, attempts to correlate structure and function.

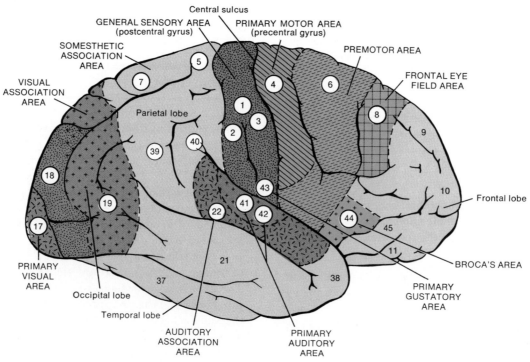

FIGURE 18-11 Functional areas of the cerebrum. This lateral view indicates the sensory and motor areas of the right hemisphere. Although Broca's area is in the left hemisphere of most people, it is shown here to indicate its location.

of the body. Each point of the area receives sensations from specific parts of the body, and essentially the entire body is spatially represented in it. The size of the portion of the sensory area receiving stimuli from body parts is not dependent on the size of the part, but on the number of receptors the part contains. For example, a larger portion of the sensory area receives impulses from the lips than from the thorax (see Figure 20-4). The major function of the general sensory area is to localize exactly the points of the body where the sensations originate. The thalamus is capable of localizing sensations in a general way, that is, it receives sensations from large areas of the body, but cannot distinguish between specific areas of stimulation. This ability is reserved to the general sensory area of the cortex.

Posterior to the general sensory area is the **somesthetic association area.** It corresponds to the areas numbered 5 and 7 in Figure 18-11. The somesthetic association area receives input from the thalamus, other lower portions of the brain, and the general sensory area. Its role is to integrate and interpret sensations. This area permits you to determine the exact shape and texture of an object without looking at it, to determine the orientation of one object to another as they are felt, and to sense the relationship of one body part to another. Another role of the somesthetic association area is the storage of memories of past sensory experiences. Thus you can compare sensations with previous experiences.

Other sensory areas of the cortex include:

1. Primary visual area (area 17). Located on the medial surface of the occipital lobe and occasionally extends around to the lateral surface. It receives sensory impulses from the eyes and interprets shape and color.

2. Visual association area (areas 18 and 19). Located in the occipital lobe. It receives sensory signals from the primary visual area and the thalamus. It relates present to past visual experiences with recognition and evaluation of what is seen.

3. Primary auditory area (areas 41 and 42). Located in the superior part of the temporal lobe near the lateral cerebral sulcus. It interprets the basic characteristics of sound such as pitch and rhythm.

4. Auditory association area (area 22). Inferior to the primary auditory area in the temporal cortex. It determines if a sound is speech, music, or noise. It also interprets the meaning of speech by translating words into thoughts.

5. Primary gustatory area (area 43). Located at the base of the postcentral gyrus above the lateral cerebral sulcus in the parietal cortex. It interprets sensations related to taste.

6. Primary olfactory area. Located in the temporal lobe on the medial aspect. It interprets sensations related to smell.

7. Gnostic (NOS-tik; *gnosis* = knowledge) **area** (areas

5, 7, 39, and 40). This *common integrative area* is located among the somesthetic, visual, and auditory association areas. The gnostic area receives impulses from these areas, as well as from the taste and smell areas, the thalamus, and lower portions of the brain stem. It integrates sensory interpretations from the association areas and impulses from other areas so that a common thought can be formed from the various sensory inputs. It then transmits signals to other parts of the brain to cause the appropriate response to the sensory signal.

CLINICAL APPLICATION

Within the past few years more precise localization of radioisotopes in the body has been made possible by new computer imaging techniques similar to those used in CT scanning. The result is a sophisticated version of radioisotope scanning called **positron emission tomography (PET)**.

The principle behind PET scanning is as follows. Short-lived radioisotopes such as ^{11}C, ^{13}N, or ^{15}O are produced and incorporated into a solution that can be injected into the body. As the radioisotope circulates through the body, it emits positively charged electrons called *positrons*. Positrons collide with negatively charged electrons in body tissues, causing their annihilation and the release of gamma rays. The gamma rays travel in opposite directions and are detected and recorded by PET receptors. A computer then takes the information and constructs a colored **PET scan** that shows where the radioisotopes are being used in the body.

Although PET may not become a clinical tool for several years, it has already provided information that cannot be obtained by any other technique. Using PET, physicians can study the effects of drugs in body organs, measure blood flow through organs such as the brain and heart, identify the extent of damage as a result of strokes or heart attacks, and detect cancers and measure the effects of treatment. PET studies with schizophrenic and manic-depressive patients have revealed that schizophrenics tend to use less glucose in certain brain areas, whereas manic-depressive patients use more glucose during manic phases. PET technology is also now being used to study the chemical changes that occur during epileptic seizures and in persons suffering from senile dementia. It is hoped that PET could eventually become a routine part of psychiatric examinations, helping to diagnose patients who show symptoms of more than one kind of mental illness. As impressive as PET is in the diagnosis of disease, it is also being used by scientists to probe the healthy brain. For example, by detecting and recording changes in glucose metabolism (consumption), it is pos-

sible to identify which specific areas of the brain are involved in specific sensory and motor activities. The PET scans in Figure 18-12 indicate brain activity in response to certain visual and auditory stimulation.

• *Motor Areas*   The **primary motor area** (area 4) is located in the precentral gyrus of the frontal lobe (see Figure 18-11). Like the general sensory area, the primary motor area consists of regions that control specific muscles or groups of muscles (see Figure 20-6). Stimulation of a specific point of the primary motor area results in a muscular contraction, usually on the opposite side of the body.

The **premotor area** (area 6) is anterior to the primary motor area. It is concerned with learned motor activities of a complex and sequential nature. It generates impulses that cause a specific group of muscles to contract in a specific sequence, for example, writing. Thus the premotor area controls skilled movements.

The **frontal eye field area** (area 8) in the frontal cortex is sometimes included in the premotor area. This area controls voluntary scanning movements of the eyes—searching for a word in a dictionary, for instance.

The **language areas** are also significant parts of the motor cortex. The translation of speech or written words into thoughts involves sensory areas—primary auditory, auditory association, primary visual, visual association, and gnostic—as we just described. The translation of thoughts into speech involves the **motor speech area** (area 44), or **Broca's (BRŌ-kaz) area,** located in the frontal lobe just superior to the lateral cerebral sulcus. From this area, a sequence of signals is sent to the premotor regions that control the muscles of the larynx, pharynx, and mouth. The impulses from the premotor area to the muscles result in specific, coordinated contractions that enable you to speak. Simultaneously, impulses are sent from Broca's area to the primary motor area. From here, impulses reach your breathing muscles to regulate the proper flow of air past the vocal cords. The coordinated contractions of your speech and breathing muscles enable you to translate your thoughts into speech.

CLINICAL APPLICATION

Broca's area and other language areas are located in the left cerebral hemisphere of most individuals regardless of whether they are left-handed or right-handed. Injury to the sensory or motor speech areas results in **aphasia** (a-FĀ-zē-a; *a* = without; *phasis* = speech), an inability to speak; **agraphia** (*a* = without; *graph* = write), an inability to write; **word deafness,** an inability to understand spoken words; or **word blindness,** an inability to understand written words.

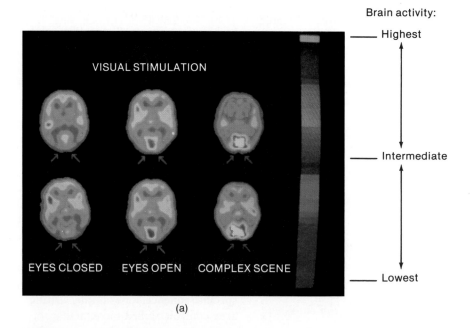

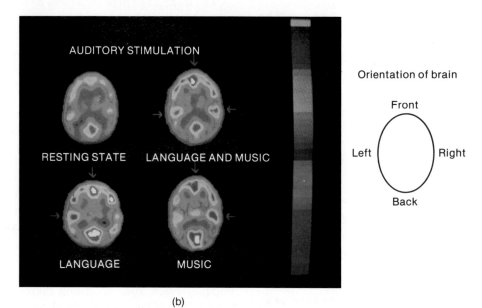

FIGURE 18-12 Positron emission tomography (PET) scans showing brain activity in response to visual and auditory stimulation. (a) Response of the visual areas of the cerebral cortex (areas between arrows) when the eyes are closed, open to white light, and open to a complex scene. (b) Response of auditory areas of the cerebral cortex (areas at horizontal arrows) and the frontal lobe (areas at vertical arrow) in response to language and music, language alone, and music alone. Note that when the auditory stimulation consists of both language and music, the auditory cortex on both sides of the brain is active; when the auditory stimulation is language only, there is a predominant left-sided activation of the auditory cortex; and when the auditory stimulation is music only, there is a predominant right-sided activation of the auditory cortex. Frontal lobe activation probably results from a high level of mentation and the promise that all subjects participating in the experiment would be paid in proportion to how much they could remember about the auditory stimulation for a subsequent test. (Courtesy of Dr. Michael E. Phelps, Division of Biophysics, Department of Radiological Sciences, UCLA School of Medicine, Los Angeles, California.)

● *Association Areas* The **association areas** of the cerebrum are made up of association tracts that connect motor and sensory areas (see Figure 18-9). The association region of the cortex occupies the greater portion of the lateral surfaces of the occipital, parietal, and temporal lobes and the frontal lobes anterior to the motor areas. The association areas are concerned with memory, emotions, reasoning, will, judgment, personality traits, and intelligence.

Brain Waves

Brain cells can generate electrical activity as a result of literally millions of action potentials of individual neurons. These electrical potentials are called **brain waves** and indicate activity of the cerebral cortex. Brain waves pass through the skull easily and can be detected by sensors called electrodes. A record of such waves is called an **electroencephalogram (EEG).** An EEG is obtained by placing electrodes on the head and amplifying the waves with an electroencephalograph. Distinct EEG patterns appear in certain abnormalities. In fact, the EEG is used clinically in the diagnosis of epilepsy, infectious diseases, tumors, trauma, and hematomas. Electroencephalograms also furnish information regarding sleep and wakefulness.

CLINICAL APPLICATION

In cases of doubt, extinction of an EEG, called a flat EEG, is increasingly being taken as one criterion of **brain death.** Other criteria include unconsciousness, no spontaneous breathing, the absence of a light response and dilation of the pupils, the absence of reflexes, unresponsiveness, and lack of normal muscle tone or strength.

BRAIN LATERALIZATION (SPLIT-BRAIN CONCEPT)

Gross examination of the brain would suggest that it is bilaterally symmetric. However, detailed examination by the use of CT scans reveals certain anatomical differences between the two hemispheres. For example, in left-handed people the parietal and occipital lobes of the right hemisphere are usually narrower than the corresponding lobes of the left hemisphere. In addition, the frontal lobe of the left hemisphere of such individuals is typically narrower than that of the right hemisphere.

In addition to the structural differences between both sides of the brain, there are also several important functional differences. It has been shown that the left hemisphere is more important for right-hand control, spoken and written language, numerical and scientific skills, and reasoning in most people. Conversely, it has been shown that the right hemisphere is more important for left-hand control; musical and artistic awareness; space and pattern perception; insight; imagination; and generating mental images of sight, sound, touch, taste, and smell in order to compare relationships.

CEREBELLUM

The **cerebellum** is the second-largest portion of the brain (almost one-eighth of the brain's mass) and occupies the inferior and posterior aspects of the cranial cavity. Specifically, it is posterior to the medulla and pons and below the occipital lobes of the cerebrum. It is separated from the cerebrum by the **transverse fissure** and by an extension of the cranial dura mater called the **tentorium** (*tentorium* = tent) **cerebelli.** The tentorium cerebelli partially encloses the transverse sinuses and supports the occipital lobes of the cerebral hemispheres.

The cerebellum is shaped somewhat like a butterfly. The central constricted area is the **vermis,** which means worm shaped, and the lateral "wings," or lobes, are referred to as **hemispheres** (Figure 18-13). Each hemisphere consists of lobes that are separated by deep and distinct fissures. The **anterior lobe** and **posterior lobe** are concerned with subconscious movements of skeletal muscles. The **flocculonodular lobe** is concerned with the sense of equilibrium (see Chapter 19). Between the hemispheres is another extension of the cranial dura mater: the **falx cerebelli.** It passes only a short distance between the cerebellar hemispheres and contains the occipital sinus.

The surface of the cerebellum, called the **cortex,** consists of gray matter in a series of slender, parallel ridges called **folia.** They are less prominent than the gyri of the cerebral cortex. Beneath the gray matter are **white matter tracts (arbor vitae)** that resemble branches of a tree. Deep within the white matter are masses of gray matter, the **cerebellar nuclei.**

The cerebellum is attached to the brain stem by three paired bundles of fibers called **cerebellar peduncles** (see Figure 18-4d, e). **Inferior cerebellar peduncles** connect the cerebellum with the medulla at the base of the brain stem and with the spinal cord. **Middle cerebellar peduncles** connect the cerebellum with the pons. **Superior cerebellar peduncles** connect the cerebellum with the midbrain.

The cerebellum is a motor area of the brain concerned with certain subconscious movements in the skeletal muscles. These movements are required for coordination, maintenance of posture, and balance. The cerebellar peduncles are the fiber tracts that allow the cerebellum to perform its functions.

Let us now see how the cerebellum produces coordinated movement. Motor areas of the cerebral cortex voluntarily initiate muscle contraction. Once the movement has begun, the sensory areas of the cortex receive impulses

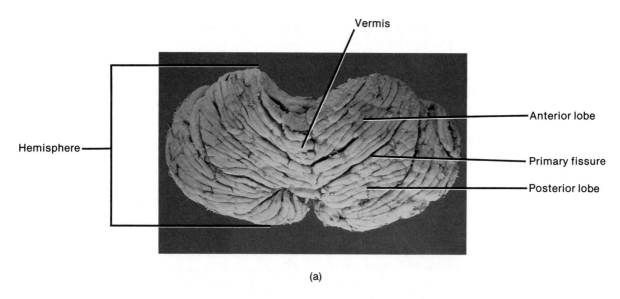

(a)

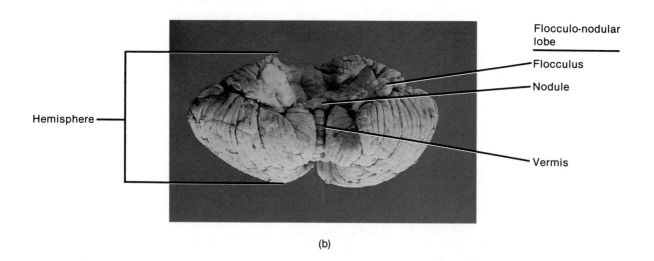

(b)

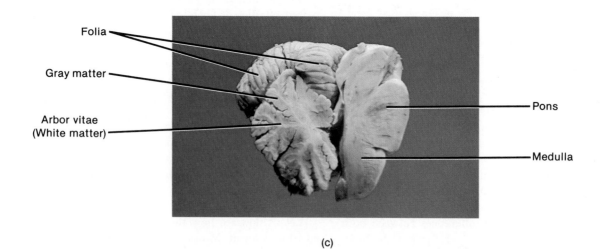

(c)

FIGURE 18-13 Cerebellum. (a) Superior view. (b) Posteroinferior view. (c) Viewed in sagittal section. (Courtesy of Lester V. Bergman & Associates.)

from nerves in the joints. The impulses provide information about the extent of muscle contraction and the amount of joint movement. The term *proprioception* is applied to this sense of the position of one body part relative to another. The cerebral cortex uses the proprioceptive sensations to determine which muscles are required to contract next and with what strength they are to contract in order to continue moving in the desired direction. Then a pattern of impulses is generated by the cerebral cortex along tracts to the pons and midbrain, which relay the impulses over the middle and superior cerebellar peduncles to the cerebellum. The cerebellum then generates subconscious motor impulses along the inferior cerebellar peduncles to the medulla and spinal cord. The impulses pass downward along the spinal cord and out the nerves that stimulate the prime movers and synergists to contract and that inhibit the contraction of the antagonists. The result is a smooth, coordinated movement. A well-functioning cerebellum is essential for delicate movements such as playing the piano.

The cerebellum also transmits impulses that control postural muscles, that is, it maintains normal muscle tone. The cerebellum also maintains body equilibrium. The inner ear contains structures that sense balance. Information such as whether the body is leaning to the left or right is transmitted from the inner ear to the cerebellum. The cerebellum then discharges impulses that cause the contraction of the muscles necessary for maintaining equilibrium.

There is some evidence that the cerebellum may play a role in a person's emotional development, modulating sensations of anger and pleasure.

CLINICAL APPLICATION

Damage to the cerebellum through trauma or disease is characterized by certain symptoms involving skeletal muscles on the same side of the body as the damage. The effects are ipsilateral because of a double crossing of tracts within the cerebellum. There may be lack of muscle coordination, called *ataxia* (*a* = without; *taxis* = order). Blindfolded people with ataxia cannot touch the tip of their nose with a finger because they cannot coordinate movement with their sense of where a body part is located. Another sign of ataxia is a change in the speech pattern due to a lack of coordination of speech muscles. Cerebellar damage may also result in *disturbances of gait,* in which the subject staggers or cannot coordinate normal walking movements, and *severe dizziness.*

A summary of the functions of the various parts of the brain is presented in Exhibit 18-2.

CRANIAL NERVES

Of the 12 pairs of **cranial nerves,** 10 originate from the brain stem, but all leave the skull through foramina of the skull (see Figure 18-4a). The cranial nerves are designated with Roman numerals and with names. The Roman numerals indicate the order in which the nerves arise from the brain (anterior to posterior). The names indicate the distribution or function of the nerves. Cranial nerves that contain both sensory and motor fibers are termed *mixed nerves.* Other cranial nerves contain sensory fibers only. The cell bodies of sensory fibers are located in ganglia outside the brain. The cell bodies of motor fibers lie in nuclei within the brain.

Some motor fibers control subconscious movements, yet the somatic nervous system has been defined as a *conscious* system. The reason for this apparent contradiction is that some fibers of the autonomic nervous system leave the brain bundled together with somatic fibers of the cranial nerves, as is the case for spinal nerves. Therefore, subconscious functions transmitted by the autonomic fibers are described along with the conscious functions of the somatic fibers of the cranial nerves.

Although the cranial nerves are mentioned singly in the following description of their type, location, and function, remember that they are paired structures.

OLFACTORY (I)

The **olfactory (I) nerve** is entirely sensory and conveys impulses related to smell. It arises as bipolar neurons from the olfactory mucosa of the nasal cavity (see Figure 20-8). The dendrites and cell bodies of these neurons are generally limited to the mucosa covering the superior nasal conchae and the adjacent nasal septum. Axons from the neurons pass through the cribriform plate of the ethmoid bone and synapse with other olfactory neurons in the *olfactory bulb,* an extension of the brain lying above the cribriform plate. The axons of these neurons make up the *olfactory tract.* The fibers from the tract terminate in the primary olfactory area in the cerebral cortex.

OPTIC (II)

The **optic (II) nerve** is entirely sensory and conveys impulses related to vision. Impulses initiated by rods and cones of the retina are relayed by bipolar neurons to ganglion cells (see Figure 20-11). Axons of the ganglion cells, the optic nerve fibers, exit the optic foramen, after which the two optic nerves unite to form the *optic chiasma* (kī-AZ-ma). Within the chiasma, fibers from the medial half of each retina cross to the opposite side; those from the lateral half remain on the same side. From the chiasma, the regrouped fibers pass posteriorly to the *optic tracts.* From the optic tracts, the majority of fibers terminate in a nucleus (lateral geniculate) in the thalamus. They

EXHIBIT 18-2 SUMMARY OF FUNCTIONS OF PRINCIPAL PARTS OF BRAIN

PART	FUNCTION
Brain stem	
Medulla	Relays motor and sensory impulses between other parts of the brain and the spinal cord.
	Reticular formation (also in pons, midbrain, and diencephalon) functions in consciousness and arousal.
	Vital reflex centers regulate heartbeat, breathing (together with pons), and blood vessel diameter.
	Nonvital reflex centers coordinate swallowing, vomiting, coughing, sneezing, and hiccuping.
	Contains nuclei of origin for cranial nerves VIII, IX, X, XI, and XII.
	Vestibular nuclear complex helps maintain equilibrium.
Pons	Relays impulses between parts of the brain and the spinal cord and within parts of the brain.
	Contains nuclei of origin for cranial nerves V, VI, VII, and VIII.
	Pneumotaxic area and apneustic area, together with the medulla, help control breathing.
Midbrain	Relays motor impulses from the cerebral cortex to the pons and spinal cord and relays sensory impulses from the spinal cord to the thalamus.
	Superior colliculi coordinate movements of the eyeballs in response to visual and other stimuli, and the inferior colliculi coordinate movements of the head and trunk in response to auditory stimuli.
	Contains nuclei of origin for cranial nerves III and IV.
Diencephalon	
Thalamus	Several nuclei serve as relay stations for all sensory impulses, except smell, to the cerebral cortex.
	Relays motor impulses from the cerebral cortex to the spinal cord.
	Interprets pain, temperature, crude touch, and pressure sensations.
	Anterior nucleus functions in emotions and memory.
Hypothalamus	Controls and integrates the autonomic nervous system.
	Receives sensory impulses from viscera.
	Articulates with the pituitary gland.
	Center for mind-over-body phenomena.
	Functions in rage and aggression.
	Controls normal body temperature, food intake, and thirst.
	Helps maintain the waking state and sleep.
	Functions as a self-sustained oscillator that drives many biological rhythms.
Cerebrum	Sensory areas interpret sensory impulses, motor areas control muscular movement, and association areas function in emotional and intellectual processes.
	Basal ganglia control gross muscle movements and regulate muscle tone.
	Limbic system functions in emotional aspects of behavior related to survival.
Cerebellum	Controls subconscious skeletal muscle contractions required for coordination, posture, and balance.
	Assumes a role in emotional development, modulating sensations of anger and pleasure.

then synapse with neurons that pass to the visual areas of the cerebral cortex (see Figure 20-12). Some fibers from the optic chiasma terminate in the superior colliculi of the midbrain. They synapse with neurons whose fibers terminate in the nuclei that convey impulses to the oculomotor (III), trochlear (IV), and abducens (VI) nerves—nerves that control the extrinsic (external) and intrinsic (internal) eye muscles. Through this relay, there are widespread motor responses to light stimuli.

OCULOMOTOR (III)

The **oculomotor (III) nerve** is a mixed cranial nerve. It originates from neurons in a nucleus in the ventral portion of the midbrain (Figure 18-14a). It runs forward, divides into superior and inferior divisions, both of which pass through the superior orbital fissure into the orbit. The superior branch is distributed to the superior rectus (an extrinsic eyeball muscle) and the levator palpebrae superioris (the muscle of the upper eyelid). The inferior branch is distributed to the medial rectus, inferior rectus, and inferior oblique muscles—all extrinsic eyeball muscles. These distributions to the levator palpebrae superioris and extrinsic eyeball muscles constitute the motor portion of the oculomotor nerve. Through these distributions, impulses are sent that control movements of the eyeball and upper eyelid.

The inferior branch of the oculomotor nerve also provides parasympathetic innervation to the *ciliary ganglion,* a relay center of the autonomic nervous system that connects a nucleus in the midbrain with the intrinsic eyeball muscles. These intrinsic muscles include the ciliary muscle of the eyeball and the sphincter muscle of the iris. Through the ciliary ganglion, the oculomotor nerve controls the smooth muscle (ciliary muscle) responsible for accommodation of the lens for near vision and the smooth muscle (sphincter muscle of iris) responsible for constriction of the pupil.

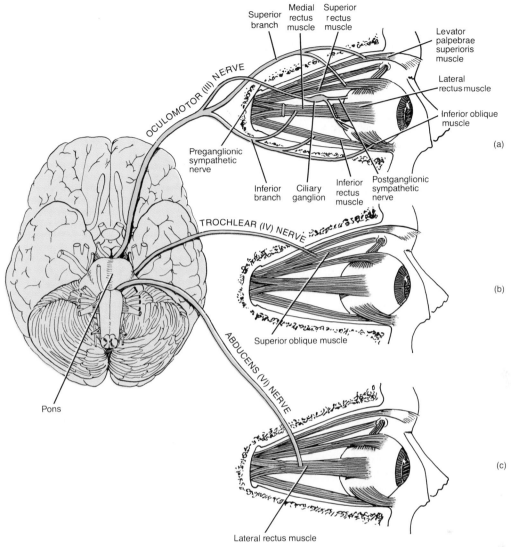

FIGURE 18-14 (a) Oculomotor (III) nerve. (b) Trochlear (IV) nerve. (c) Abducens (VI) nerve.

The sensory portion of the oculomotor nerve consists of afferent fibers from proprioceptors in the eyeball muscles supplied by the nerve to the midbrain. These fibers convey impulses related to muscle sense (proprioception).

Although the oculomotor (III), trochlear (IV), accessory (XI), and hypoglossal (XII) cranial nerves are referred to as mixed nerves because they contain fibers from proprioceptors, they are primarily motor, serving to stimulate skeletal muscle contractions.

TROCHLEAR (IV)

The **trochlear** (TROK-lē-ar) **(IV) nerve** is a mixed cranial nerve. It is the smallest of the 12 cranial nerves. The motor portion originates in a nucleus in the midbrain, and axons from the nucleus pass through the superior orbital fissure of the orbit (Figure 18-14b). The motor fibers innervate the superior oblique muscle of the eyeball,

another extrinsic eyeball muscle. The trochlear nerve controls movement of the eyeball.

The sensory portion of the trochlear nerve consists of afferent fibers that run from proprioceptors in the superior oblique muscle to the nucleus of the nerve in the midbrain. The sensory portion is responsible for muscle sense.

TRIGEMINAL (V)

The **trigeminal (V) nerve** is a mixed cranial nerve and the largest of the cranial nerves. As indicated by its name, the trigeminal nerve has three branches: ophthalmic (of-THAL-mik), maxillary, and mandibular (Figure 18-15). The trigeminal nerve contains two roots on the ventrolateral surface of the pons. The large sensory root has a swelling called the *semilunar* (*Gasserian*) *ganglion* located in a fossa on the inner surface of the petrous portion

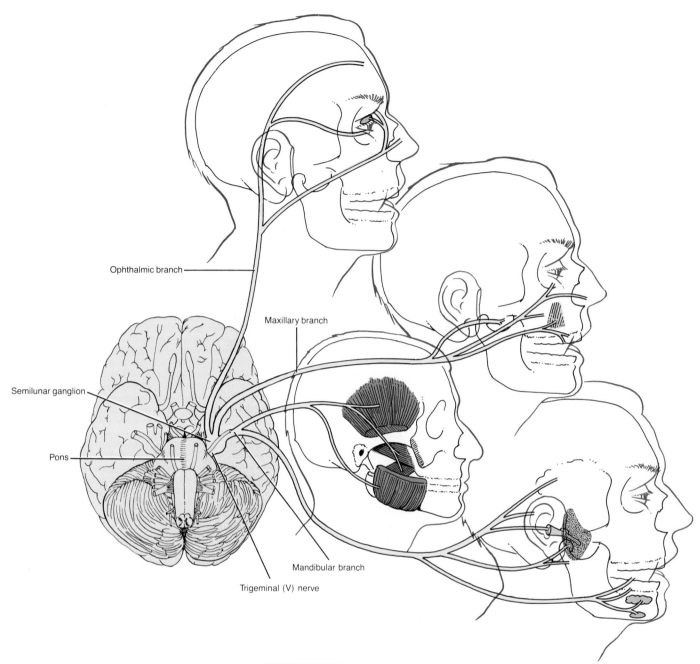

FIGURE 18-15 Trigeminal (V) nerve.

of the temporal bone. From this ganglion, the **ophthalmic branch** enters the orbit via the superior orbital fissure, the **maxillary branch** enters the foramen rotundum, and the **mandibular branch** exits through the foramen ovale. The smaller motor root originates in a nucleus in the pons. The motor fibers join the mandibular branch and supply the muscles of mastication. These motor fibers, which control chewing movements, constitute the motor portion of the trigeminal nerve.

The sensory portion of the trigeminal nerve delivers impulses related to touch, pain, and temperature and consists of the ophthalmic, maxillary, and mandibular branches. The ophthalmic branch receives sensory fibers from the skin over the upper eyelid, eyeball, lacrimal glands, upper part of the nasal cavity, side of the nose, forehead, and anterior half of the scalp. The maxillary branch receives sensory fibers from the mucosa of the nose, palate, parts of the pharynx, upper teeth, upper lip, cheek, and lower eyelid. The mandibular branch transmits sensory fibers from the anterior two-thirds of the tongue (not taste), lower teeth, skin over the mandible and side of the head in front of the ear, and mucosa of the floor of the mouth. Sensory fibers from the three branches of the trigeminal nerve enter the semilunar gan-

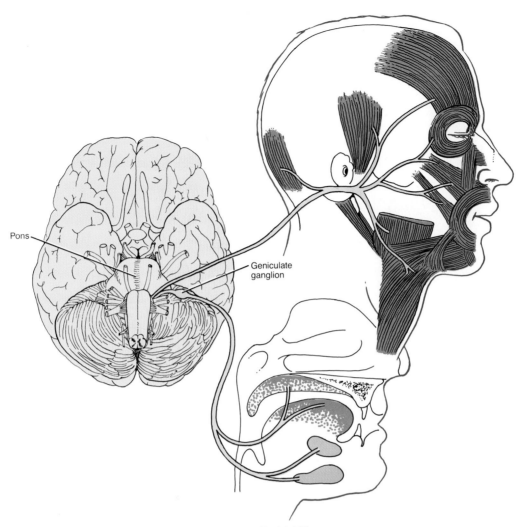

Pons

Geniculate
ganglion

FIGURE 18-16 Facial (VII) nerve.

glion and terminate in a nucleus in the pons. There are also sensory fibers from proprioceptors in the muscles of mastication.

CLINICAL APPLICATION

The inferior alveolar nerve, a branch of the mandibular nerve, supplies all the teeth in one half of the mandible and is frequently **anesthetized in dental procedures.** The same procedure will anesthetize the lower lip because the mental nerve is a branch of the inferior alveolar nerve. Since the lingual nerve runs very close to the inferior alveolar near the mental foramen, it too is often anesthetized at the same time. For anesthesia to the upper teeth, the superior alveolar nerve endings, which are branches of the maxillary branch, are blocked by inserting the needle beneath the mucous membrane; the anesthetic solution is then infiltrated slowly throughout the area of the roots of the teeth to be treated.

ABDUCENS (VI)

The **abducens** (ab-DOO-sens) **(VI) nerve** is a mixed cranial nerve that originates from a nucleus in the pons (see Figure 18-14c). The motor fibers extend from the nucleus to the lateral rectus muscle of the eyeball, an extrinsic eyeball muscle. Impulses over the fibers bring about movement of the eyeball. The sensory fibers run from proprioceptors in the lateral rectus muscle to the pons and mediate muscle sense. The abducens nerve reaches the lateral rectus muscle through the superior orbital fissure of the orbit.

FACIAL (VII)

The **facial (VII) nerve** is a mixed cranial nerve. Its motor fibers originate from a nucleus in the pons, enter the petrous portion of the temporal bone, and are distributed to facial, scalp, and neck muscles (Figure 18-16). Impulses along these fibers cause contraction of the muscles of facial expression. Some motor fibers are also distributed

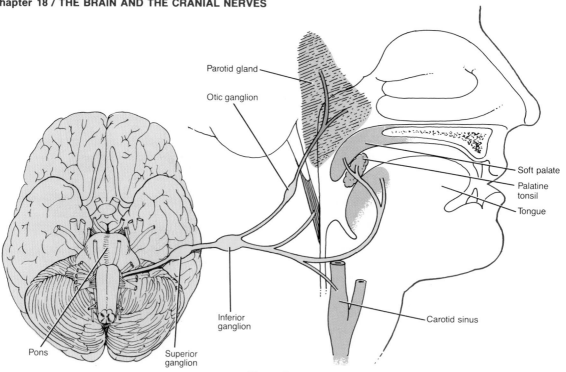

FIGURE 18-17 Glossopharyngeal (IX) nerve.

via parasympathetics to the lacrimal, sublingual, submandibular, nasal, and palatine glands.

The sensory fibers extend from the taste buds of the anterior two-thirds of the tongue to the *geniculate ganglion,* a swelling of the facial nerve. From here, the fibers pass to a nucleus in the pons, which sends fibers to the thalamus for relay to the gustatory area of the cerebral cortex. The sensory portion of the facial nerve also conveys deep general sensations from the face. There are also sensory fibers from proprioceptors in the muscles of the face and scalp.

VESTIBULOCOCHLEAR (VIII)

The **vestibulocochlear** (ves-tib'-yoo-lō-KŌK-lē-ar) **(VIII) nerve,** formerly known as the **acoustic nerve,** is another sensory cranial nerve. It consists of two branches: the cochlear (auditory) branch and the vestibular branch (see Figure 20-17). The **cochlear branch,** which conveys impulses associated with hearing, arises in the spiral organ (organ of Corti) in the cochlea of the internal ear. The cell bodies of the cochlear branch are located in the *spiral ganglion* of the cochlea. From here the axons pass through a nucleus in the medulla and terminate in the medial geniculate nucleus in the thalamus. Ultimately, the fibers synapse with neurons that relay the impulses to the auditory areas of the cerebral cortex.

The **vestibular branch** arises in the semicircular canals, the saccule, and the utricle of the inner ear. Fibers from the semicircular canals, saccule, and utricle extend to the *vestibular ganglion,* where the cell bodies are contained. The cell bodies of the fibers synapse in the ganglion with fibers that extend to a nucleus in the medulla and

pons and terminate in the thalamus. Some fibers also enter the cerebellum. The vestibular branch transmits impulses related to equilibrium.

GLOSSOPHARYNGEAL (IX)

The **glossopharyngeal** (glos'-ō-fa-RIN-jē-al) **(IX) nerve** is a mixed cranial nerve. Its motor fibers originate in a nucleus in the medulla (Figure 18-17). The nerve exits the skull through the jugular foramen. The motor fibers are distributed to the swallowing muscles of the pharynx and the parotid gland (parasympathetic) to mediate swallowing movements and the secretion of saliva.

The sensory fibers of the glossopharyngeal nerve supply the pharynx and taste buds of the posterior third of the tongue. Some sensory fibers also originate from receptors in the carotid sinus, which assumes a major role in blood pressure regulation. The sensory fibers terminate in a nucleus in the thalamus. There are also sensory fibers from proprioceptors in the muscles innervated by this nerve.

VAGUS (X)

The **vagus (X) nerve** is a mixed cranial nerve that is widely distributed from the head and neck into the thorax and abdomen (Figure 18-18). Its motor fibers originate in a nucleus of the medulla and terminate in the muscles of the pharynx, larynx, respiratory passageways, lungs, heart, esophagus, stomach, small intestine, most of the large intestine, and gallbladder. Impulses along the motor fibers generate visceral, cardiac, and skeletal muscle movement. Parasympathetic fibers innervate involuntary muscles and glands of the gastrointestinal (GI) tract.

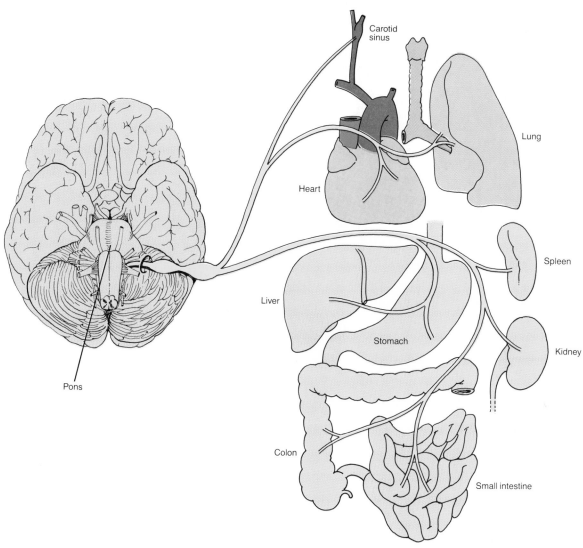

FIGURE 18-18 Vagus (X) nerve.

Sensory fibers of the vagus nerve supply essentially the same structures as the motor fibers. They convey impulses for various sensations from the larynx, the viscera, and the ear. The fibers terminate in the medulla and pons. There are also sensory fibers from proprioceptors in the muscles supplied by this nerve.

ACCESSORY (XI)

The **accessory (XI) nerve** (formerly the **spinal accessory nerve**) is a mixed cranial nerve. It differs from all other cranial nerves in that it originates from both the brain stem and the spinal cord (Figure 18-19). The **bulbar (medullary) portion** originates from nuclei in the medulla, passes through the jugular foramen, and supplies the voluntary muscles of the pharynx, larynx, and soft palate that are used in swallowing. The **spinal portion** originates in the anterior gray horn of the first five segments of the cervical portion of the spinal cord. The fibers from the segments join, enter the foramen magnum, and exit

through the jugular foramen along with the bulbar portion.

The spinal portion conveys motor impulses to the sternocleidomastoid and trapezius muscles to coordinate head movements. The sensory fibers originate from proprioceptors in the muscles supplied by its motor neurons and terminate in upper cervical posterior root ganglia.

HYPOGLOSSAL (XII)

The **hypoglossal (XII) nerve** is a mixed cranial nerve. The motor fibers originate in a nucleus in the medulla, pass through the hypoglossal canal, and supply the muscles of the tongue (Figure 18-20). These fibers conduct impulses related to speech and swallowing.

The sensory portion of the hypoglossal nerve consists of fibers originating from proprioceptors in the tongue muscles and terminating in the medulla. The sensory fibers conduct impulses for muscle sense.

A summary of cranial nerves and clinical applications related to dysfunction is presented in Exhibit 18-3.

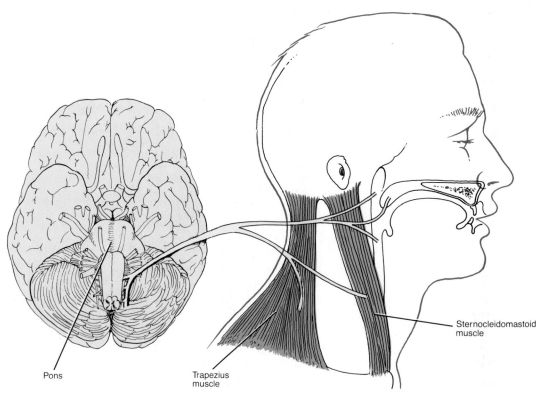

Sternocleidomastoid
muscle

Pons

Trapezius
muscle

FIGURE 18-19 Accessory (XI) nerve.

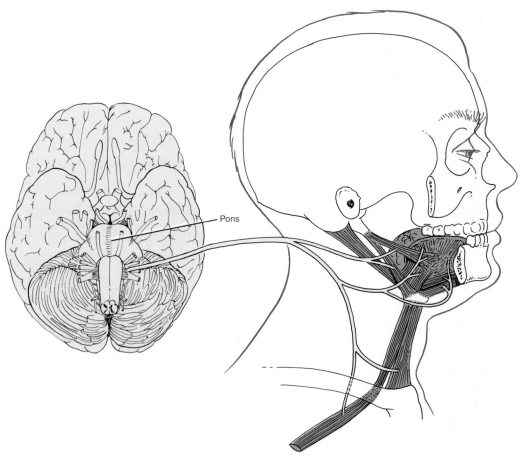

Pons

FIGURE 18-20 Hypoglossal (XII) nerve.

EXHIBIT 18-3 SUMMARY OF CRANIAL NERVES

NERVE	LOCATION	FUNCTION AND CLINICAL APPLICATION
Olfactory (I)	Arises in olfactory mucosa, passes through olfactory bulb and olfactory tract, and terminates in primary olfactory areas of cerebral cortex.	Smell. Loss of the sense of smell, called *anosmia,* may result from head injuries in which the cribriform plate of the ethmoid bone is fractured and from lesions along the olfactory pathway.
Optic (II)	Arises in retina of the eye, forms optic chiasma, passes through optic tracts and lateral geniculate nucleus in thalamus, and terminates in visual areas of cerebral cortex.	Vision. Fractures in the orbit, lesions along the visual pathway, and diseases of the nervous system may result in visual field defects and loss of visual acuity. A defect of vision is called *anopsia.*
Oculomotor (III)	Motor portion: originates in midbrain and is distributed to levator palpebrae superioris of upper eyelid and four extrinsic eyeball muscles (superior rectus, medial rectus, inferior rectus, and inferior oblique); parasympathetic innervation to ciliary muscle of eyeball and sphincter muscle of iris. Sensory portion: consists of afferent fibers from proprioceptors in eyeball muscles and terminates in midbrain.	Motor: movement of eyelid and eyeball, accommodation of lens for near vision, and constriction of pupil. Sensory: muscle sense (proprioception). A lesion in the nerve causes *strabismus* (squinting) *ptosis* (drooping) of the upper eyelid, pupil dilation, the movement of the eyeball downward and outward on the damaged side, a loss of accommodation for near vision, and double vision (*diplopia*).
Trochlear (IV)	Motor portion: originates in midbrain and is distributed to superior oblique muscle, an extrinsic eyeball muscle. Sensory portion: consists of afferent fibers from proprioceptors in superior oblique muscles and terminates in midbrain.	Motor: movement of eyeball. Sensory: muscle sense (proprioception). In trochlear nerve paralysis, the head is tilted to the affected side and diplopia and strabismus occur.
Trigeminal (V)	Motor portion: originates in pons and terminates in muscles of mastication. Sensory portion: consists of three branches: *ophthalmic*—contains sensory fibers from skin over upper eyelid, eyeball, lacrimal glands, nasal cavity, side of nose, forehead, and anterior half of scalp; *maxillary*—contains sensory fibers from mucosa of nose, palate, parts of pharynx, upper teeth, upper lip, cheek, and lower eyelid; *mandibular*—contains sensory fibers from anterior two-thirds of tongue, lower teeth, skin over mandible, and side of head in front of ear. The three branches terminate in pons. Sensory portion also consists of afferent fibers from proprioceptors in muscles of mastication.	Motor: chewing. Sensory: conveys sensations for touch, pain, and temperature from structures supplied; muscle sense (proprioception). Injury results in paralysis of the muscles of mastication and a loss of sensation of touch and temperature. *Neuralgia* (pain) of one or more branches of trigeminal nerve is called *trigeminal neuralgia* (*tic douloureux*).
Abducens (VI)	Motor portion: originates in pons and is distributed to lateral rectus muscle, an extrinsic eyeball muscle. Sensory portion: consists of afferent fibers from proprioceptors in lateral rectus muscle and terminates in pons.	Motor: movement of eyeball. Sensory: muscle sense (proprioception). With damage to this nerve, the affected eyeball cannot move laterally beyond the midpoint and the eye is usually directed medially.
Facial (VII)	Motor portion: originates in pons and is distributed to facial, scalp, and neck muscles; parasympathetic distribution to lacrimal, sublingual, submandibular, nasal, and palatine glands. Sensory portion: arises from taste buds on anterior two-thirds of tongue and passes through geniculate ganglion, a nucleus in pons that sends fibers to thalamus for relay to gustatory areas of cerebral cortex. Also consists of afferent fibers from proprioceptors in muscles of face and scalp.	Motor: facial expression and secretion of saliva and tears. Sensory: taste; muscle sense (proprioception). Injury produces paralysis of the facial muscles, called *Bell's palsy,* loss of taste, and the eyes remain open, even during sleep.

EXHIBIT 18-3 (*Continued*)

NERVE	LOCATION	FUNCTION AND CLINICAL APPLICATION
Vestibulocochlear (VIII)	Cochlear branch: arises in spiral organ (organ of Corti), forms spiral ganglion, passes through a nucleus in the medulla, and terminates in thalamus. Fibers synapse with neurons that relay impulses to auditory areas of cerebral cortex. Vestibular branch: arises in semicircular canals, saccule, and utricle and forms vestibular ganglion; fibers pass through medulla and pons and terminate in thalamus.	Cochlear branch: conveys impulses associated with hearing. Vestibular branch: conveys impulses associated with equilibrium. Injury to the cochlear branch may cause *tinnitus* (ringing) or deafness. Injury to the vestibular branch may cause *vertigo* (a subjective feeling of rotation), *ataxia*, and *nystagmus* (involuntary rapid movement of the eyeball).
Glossopharyngeal (IX)	Motor portion: originates in medulla and is distributed to swallowing muscles of pharynx; parasympathetic distribution to parotid gland. Sensory portion: arises from taste buds on posterior one-third of tongue and from carotid sinus and terminates in thalamus. Also consists of afferent fibers from proprioceptors in swallowing muscles supplied.	Motor: swallowing movements and secretion of saliva. Sensory: taste and regulation of blood pressure; muscle sense (proprioception). Injury results in pain during swallowing, reduced secretion of saliva, loss of sensation in the throat, and loss of taste.
Vagus (X)	Motor portion: originates in medulla and terminates in muscles of pharynx, larynx, respiratory passageways, lungs, esophagus, heart, stomach, small intestine, most of large intestine, and gallbladder; parasympathetic fibers innervate involuntary muscles and glands of the gastrointestinal (GI) tract. Sensory portion: arises from essentially same structures supplied by motor fibers and terminates in medulla and pons. Also consists of afferent fibers from proprioceptors in muscles supplied.	Motor: visceral muscle movement and swallowing movements. Sensory: sensations from organs supplied; muscle sense (proprioception). Severing of both nerves in the upper body interferes with swallowing, paralyzes vocal cords, and interrupts sensations from many organs. Injury to both nerves in the abdominal area has little effect, since the abdominal organs are also supplied by autonomic fibers from the spinal cord.
Accessory (XI)	Motor portion: consists of a bulbar portion and a spinal portion. Bulbar portion originates from medulla and supplies voluntary muscles of pharynx, larynx, and soft palate. Spinal portion originates from anterior gray horn of first five cervical segments of spinal cord and supplies sternocleidomastoid and trapezius muscles. Sensory portion: consists of afferent fibers from proprioceptors in muscles supplied.	Motor: bulbar portion mediates swallowing movements; spinal portion mediates movements of head. Sensory: muscle sense (proprioception). If damaged, the sternocleidomastoid and trapezius muscles become paralyzed, with resulting inability to turn the head or raise the shoulders.
Hypoglossal (XII)	Motor portion: originates in medulla and supplies muscles of tongue. Sensory portion: consists of fibers from proprioceptors in tongue muscles that terminate in medulla.	Motor: movement of tongue during speech and swallowing. Sensory: muscle sense (proprioception). Injury results in difficulty in chewing, speaking, and swallowing. The tongue, when protruded, curls toward the affected side and the affected side becomes atrophied, shrunken, and deeply furrowed.

AGING AND THE NERVOUS SYSTEM

One of the effects of aging on the nervous system is that nerve cells are lost. Associated with this decline, there is a decreased capacity for sending nerve impulses to and from the brain. Conduction velocity decreases, voluntary motor movements slow down, and the reflex time for skeletal muscles is decreased. Parkinson's disease is the most common movement disorder involving the central nervous system. Degenerative changes and disease states involving the sense organs can alter vision, hearing, taste, smell, and touch. The disorders that represent the most common visual problems and may be responsible for serious loss of vision are presbyopia (inability to focus on nearby objects), cataracts (cloudiness of the lens), and glaucoma (excessive fluid pressure in the eyeball). Impaired hearing associated with aging, known as presbycusis, is usually the result of changes in important structures of the inner ear.

DEVELOPMENTAL ANATOMY OF THE NERVOUS SYSTEM

The development of the nervous system occurs early in the third week of development and begins with a thickening of the **ectoderm** called the **neural plate** (Figure 18-21a–c). The plate folds inward and forms a longitudinal groove, the **neural groove.** The raised edges of the neural plate are called **neural folds.** As development continues, the neural folds increase in height, meet, and form a tube, the **neural tube.**

The cells of the wall that encloses the neural tube differentiate into three kinds. The outer or **marginal layer** develops into the *white matter* of the nervous system; the middle or **mantle layer** develops into the *gray matter* of the system; and the inner or **ependymal layer** eventually forms the *lining of the ventricles* of the central nervous system.

The **neural crest** is a mass of tissue between the neural tube and the ectoderm (Figure 18-21). It becomes differentiated and eventually forms the *posterior (dorsal) root ganglia of spinal nerves, spinal nerves, ganglia of cranial nerves, cranial nerves, ganglia of the autonomic nervous system,* and the *adrenal medulla.*

When the neural tube is formed from the neural plate, the anterior portion of the neural tube develops into three enlarged areas called vesicles: (1) **forebrain vesicle (prosencephalon),** (2) **midbrain vesicle (mesencephalon),** and (3) **hindbrain vesicle (rhombencephalon)** (Figure 18-22). The vesicles are fluid-filled enlargements that develop by the fourth week of gestation. Since they are the first vesicles to form, they are called **primary vesicles.** As development progresses, the vesicular region undergoes several flexures (bends), resulting in subdivision of the three primary vesicles so that by the fifth week of development the embryonic brain consists of five **secondary vesicles.** The forebrain vesicle (prosencephalon) divides into an anterior **telencephalon** and a posterior **diencephalon;** the midbrain vesicle (mesencephalon) remains unchanged; the hindbrain vesicle (rhombencephalon) divides into an anterior **metencephalon** and a posterior **myelencephalon.**

Ultimately, the telencephalon develops into the *cerebral hemispheres* and *basal ganglia;* the diencephalon develops into the *thalamus, hypothalamus,* and *pineal gland;*

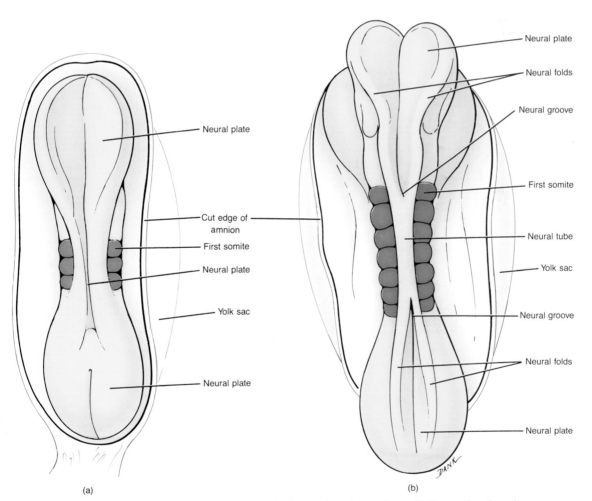

(a)   (b)

FIGURE 18-21 Origin of the nervous system. (a) Dorsal view of an embryo with three pairs of somites showing the neural plate. (b) Dorsal view of an embryo with seven pairs of somites in which the neural folds have united just medial to the somites forming the early neural tube.

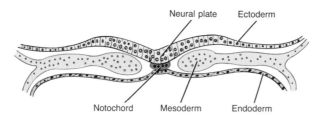

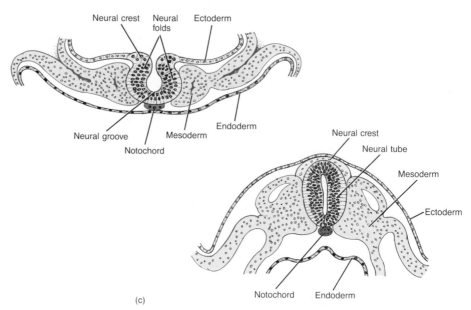

(c)

FIGURE 18-21 (*Continued*) Origin of the nervous system. (c) Cross section through the embryo showing the formation of the neural tube.

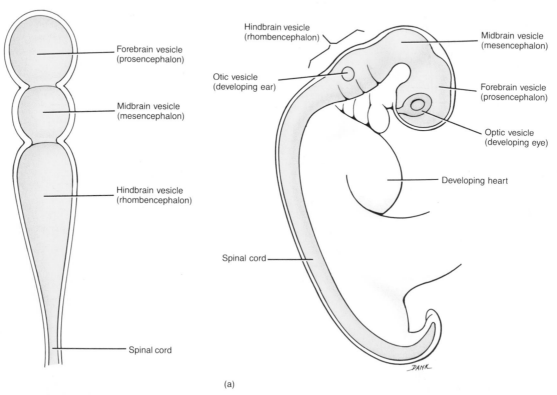

(a)

FIGURE 18-22 Development of the brain and spinal cord. (a) Primary vesicles of the neural tube seen in frontal section (left) and right lateral view (right) at about 3 to 4 weeks.

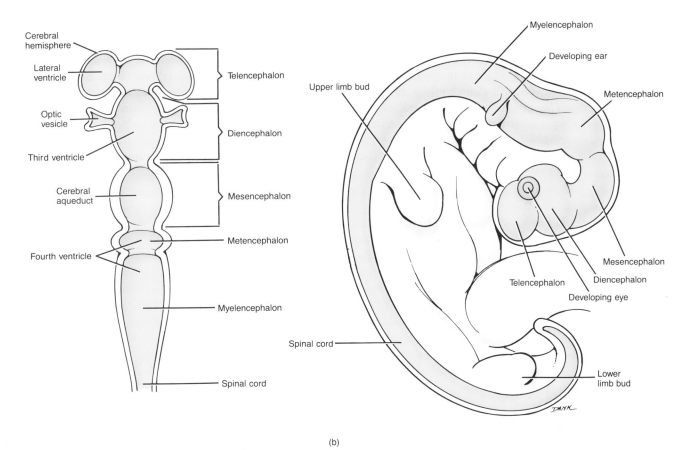

(b)

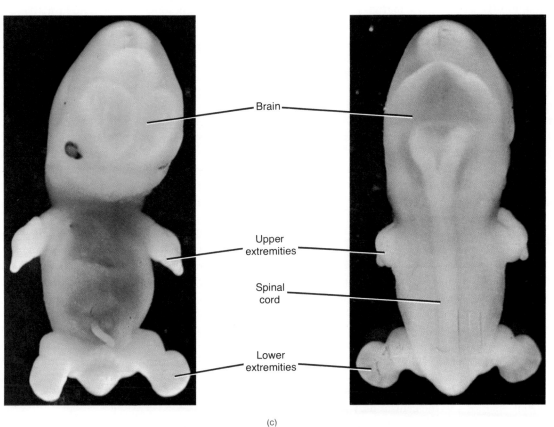

(c)

FIGURE 18-22 (*Continued*) Development of the brain and spinal cord. (b) Secondary vesicles seen in frontal section (left) and right lateral view (right) at about 5 weeks. (c) Photograph of a 40-day embryo seen in anterior view (left) and posterior view (right). Note the brain, spinal cord, and upper and lower extremities. (Courtesy of Roberts Rugh and Landrum B. Shettles, M.D., with Richard N. Einhorn, from *From Conception to Birth: The Drama of Life's Beginnings,* Harper & Row, Publishers, Inc., New York, 1971.)

the midbrain vesicle (mesencephalon) develops into the *midbrain;* the metencephalon develops into the *pons* and *cerebellum;* and the myelencephalon develops into the *medulla oblongata.* The cavities within the vesicles develop into the *ventricles* of the brain, whereas the fluid within them is *cerebrospinal fluid.* The area of the neural tube posterior to the myelencephalon gives rise to the *spinal cord.*

APPLICATIONS TO HEALTH

Many disorders can affect the central nervous system. Some are caused by viruses or bacteria. Others are caused by damage to the nervous system during birth. The origins of many conditions, however, are unknown. Here we discuss the origins and symptoms of some common central nervous system disorders.

BRAIN TUMORS

A **brain tumor** refers to any benign or malignant growth within the cranium. Tumors arising from neuroglial cells in the cerebrum, brain stem, and cerebellum are called *intraaxial neoplasms* and are characterized by infiltration and expansion, mostly within white matter. Tumors arising from supporting or neighboring structures such as cranial nerve coverings, meninges, and the pituitary gland are referred to as *extraaxial neoplasms.* Intracranial pressure from a growing tumor or edema associated with the tumor produce the characteristic signs and symptoms of a brain tumor. Among these are headache, altered consciousness, edema of the optic disc of the eyeball, and vomiting. Brain tumors can also result in seizures, visual problems, cranial nerve abnormalities, hormonal syndromes, personality changes, dementia, and sensory or motor deficits. Treatment of brain tumors involves surgery, radiation therapy, and chemotherapy.

POLIOMYELITIS

Poliomyelitis, also known as **infantile paralysis** or simply **polio,** is a viral infection that is most common during childhood. The causative agent is a virus called poliovirus. The onset of the disease is marked by fever, severe headache, a stiff neck and back, deep muscle pain and weakness, and loss of certain somatic reflexes. The virus can be ingested in drinking water contaminated with feces containing the virus. It may affect almost all parts of the body. In its most serious form, called **bulbar polio,** the virus spreads via blood to the central nervous system, where it destroys the motor nerve cell bodies, specifically those in the anterior horns of the spinal cord and in the nuclei of the cranial nerves. Injury to the spinal gray matter is the basis for the name of this disease (*polio* = gray matter; *myel* = spinal cord). Destruction of the ante-

rior horns produces paralysis. The first sign of bulbar polio is difficulty in swallowing, breathing, and speaking. Poliomyelitis can cause death from respiratory or heart failure if the virus invades the brain cells of the vital medullary centers. The incidence of polio in the United States has decreased markedly since the availability of polio vaccines (Salk vaccine and more recently Sabin vaccine).

AMYOTROPHIC LATERAL SCLEROSIS (ALS)

Amyotrophic lateral sclerosis (ALS) is a progressive neuromuscular disease that typically strikes patients between 40 and 70 years of age. Also known as Lou Gehrig's disease, it is characterized by degeneration of motor cells in the spinal cord that leads to muscular weakness. Early symptoms may include weakness and wasting of muscles of the hands and feet, followed by twitches and cramps in the muscles. As the disease progresses, use of the muscles is severely impaired. Muscles of the palate, pharynx, tongue, neck, and shoulders are commonly affected in advanced stages of the disease, leading to difficulties in swallowing and speech. ALS does not affect the individual's mental abilities.

The average course of the disease is between 3 and 4 years, although some people remain active for 15 years or more with short periods of apparent arrest. At the present time the cause of ALS is unknown and there is no cure.

CEREBRAL PALSY

The term **cerebral palsy** refers to a group of motor disorders caused by damage to the motor areas of the brain during fetal life, birth, or infancy. One cause is infection of the mother with German measles during the first three months of pregnancy. During early pregnancy, certain cells in the fetus are dividing and differentiating in order to lay down the basic structures of the brain. These cells can be abnormally changed by toxin from the measles virus. Radiation during fetal life, temporary oxygen starvation during birth, and hydrocephalus during infancy may also damage brain cells. About 70 percent of cerebral palsy victims appear to be mentally retarded. The apparent mental slowness, however, is often due to the person's inability to speak or hear well. Such individuals are often more mentally acute than they appear. Cerebral palsy is not a progressive disease; it does not worsen as time elapses. Once the damage is done, however, it is irreversible.

PARKINSON'S DISEASE

Parkinson's disease, or **parkinsonism,** is a progressive disorder of the central nervous system that typically affects its victims around age 60. The cause is unknown. The disease is related to pathological changes in the substantia nigra and basal ganglia. The substantia nigra contains cell bodies of neurons that produce the neurotransmitter dopamine (DA) in their axon terminals. The axon terminals release DA in the basal ganglia of the cerebrum. Recall that basal ganglia regulate subconscious contractions of skeletal muscles that aid activities also consciously controlled by the motor areas of the cerebral cortex— swinging the arms when walking, for example. In Parkinson's disease there is a degeneration of DA-producing neurons in the substantia nigra, and the severe reduction of DA in the basal ganglia brings about most of the symptoms of Parkinson's disease.

Diminished levels of DA cause unnecessary skeletal muscle movements that often interfere with voluntary movement. For instance, the muscles of the upper extremity may alternately contract and relax, causing the hand to shake. This shaking is called *tremor*, the most common symptom of Parkinson's disease. The tremor may spread to the ipsilateral lower extremity and then to the contralateral extremities. Motor performance is also impaired by *bradykinesia* (*brady* = slow; *kinesis* = motion), in which activities such as shaving, cutting food, and buttoning a shirt take longer and become increasingly more difficult. Muscular movements are performed not only slowly but with decreasing range of motion (*hypokinesia*). For example, as handwriting continues, letters get smaller, become poorly formed, and eventually become illegible. Some muscles may contract continuously, causing *rigidity* of the involved body part. Rigidity of the facial muscles gives the face a masklike appearance. The expression is characterized by a wide-eyed, unblinking stare and a slightly open mouth with uncontrolled drooling. Decreased DA production also results in impaired walking in which steps become shorter and shuffling and arm swings diminish. There are also changes that lead to stooped posture and loss of postural reflexes, autonomic dysfunction (constipation, retention), sensory complaints (pain, numbness, tingling), and sustained muscle spasms. Vision, hearing, and intelligence are unaffected by the disorder, indicating that Parkinson's disease does not attack the cerebral cortex.

Treatment of the symptoms of Parkinson's disease is directed toward increasing levels of DA. Although people with Parkinson's disease do not manufacture enough DA, injections of it are useless; the blood–brain barrier stops it. However, symptoms are somewhat relieved by a drug developed in the 1960s called levodopa. Administered by itself, levodopa may elevate brain levels of DA, causing undesirable side effects such as low blood pressure, nausea, mental changes, and liver dysfunction. Levodopa has been combined with carbidopa, which inhibits the formation of DA outside the brain. The combined drugs diminish the undesirable side effects. Currently, methods are being devised to help drugs pass through the blood–brain barrier by combining them with fat-soluble chemicals.

Once the complex passes through the barrier, the fat-soluble chemical breaks down and is excreted, leaving the drug in the brain where it exerts its effect.

MULTIPLE SCLEROSIS (MS)

Multiple sclerosis (MS) is the progressive destruction of the myelin sheaths of neurons in the central nervous system accompanied by disappearance of oligodendrocytes and the proliferation of astrocytes. The sheaths deteriorate to *scleroses,* which are hardened scars or plaques, in multiple regions. The destruction of myelin sheaths interferes with the transmission of impulses from one neuron to another, literally short-circuiting conduction pathways. Usually the first symptoms occur between the ages of 20 and 40. Early symptoms are generally produced by the formation of a few plaques and are, consequently, mild. Plaque formation in the cerebellum may produce lack of coordination in one hand. The patient's handwriting becomes strained and irregular. A short-circuiting of pathways in the corticospinal tract may partially paralyze the leg muscles so that the patient drags a foot when walking. Other early symptoms include double vision and urinary tract infections. Following a period of remission during which the symptoms temporarily disappear, a new series of plaques develop and the victim suffers a second attack. One attack follows another over the years. Each time the plaques form, some neurons are damaged by the hardening of their sheaths, while others are uninjured by their plaques. The result is a progressive loss of function interspersed with remission periods during which the undamaged neurons regain their ability to transmit impulses.

The symptoms of MS depend on the areas of the central nervous system most heavily laden with plaques. Sclerosis of the white matter of the spinal cord is common. As the sheaths of the neurons in the corticospinal tract deteriorate, the patient loses the ability to contract skeletal muscles. Damage to the ascending tracts produces numbness and short-circuits impulses related to position of body parts and flexion of joints. Damage to either set of tracts also destroys spinal cord reflexes.

As the disease progresses, most voluntary motor control is eventually lost and the patient becomes bedridden. Death occurs anywhere from 7 to 30 years after the first symptoms appear. The usual cause of death is a severe infection resulting from the loss of motor activity. Without the constricting action of the urinary bladder wall, for example, the bladder never totally empties and stagnant urine provides an environment for bacterial growth. Bladder infection may then spread to the kidney, damaging kidney cells.

Although the etiology of MS is unclear, there is increasing evidence that it might result from a viral infection that precipitates an autoimmune response. Viruses may

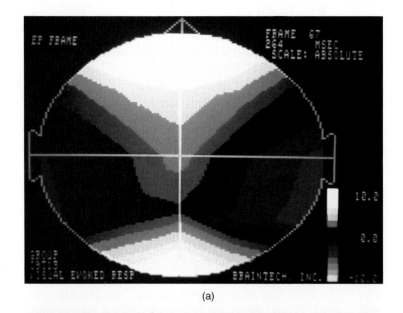

(a)

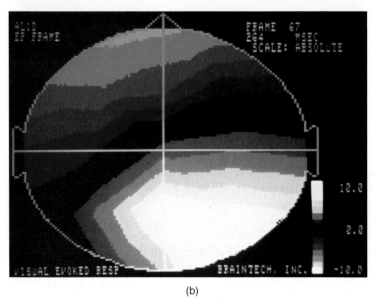

(b)

FIGURE 18-23 Brain electrical activity mapping (BEAM). (a) Image of a visual response (flash stimulus) produced by a normal child. Note the symmetry of the response in the anterior and posterior portions of the brain. (b) Image of a visual response (flash stimulus) produced by a dyslexic child. Note the asymmetry of the response. (Courtesy of Frank H. Duffy, M.D., and Gloria B. McAnulty, Ph.D., The Children's Hospital Medical Center, Boston, Massachusetts.)

trigger the destruction of oligodendrocytes by antibodies and killer cells of the body's immune system. Like other demyelinating diseases, MS is incurable. However, in view of the evidence that it might be an autoimmune disease, immunosuppressive therapy is widely used (glucocorticoids such as prednisone). A recent treatment consists of administering cyclophosphamide (a powerful anticancer drug that suppresses the immune system) in combination with ACTH (a pituitary gland hormone that stimulates secretion of hormones containing cortisone) to patients with active MS. Treatment is also directed at management of complications such as spasticity, facial neuralgia and twitching, urinary bladder problems, and constipation. Electrical stimulation of the spinal cord can also improve function in certain patients. Improvement has also been shown in patients who are administered pure oxygen while in a pressure chamber (hyperbaric oxy-

gen). This supports the idea that the destruction of myelin occurs preferentially in parts of the brain that are relatively low in oxygen.

EPILEPSY

Epilepsy is the second most common neurological disorder after stroke. It is characterized by short, recurrent, periodic attacks of motor, sensory, or psychological malfunction. The attacks, called *epileptic seizures,* are initiated by abnormal and irregular discharges of electricity from millions of neurons in the brain. The discharges stimulate many of the neurons to send impulses over their conduction pathways. As a result, a person undergoing an attack may contract skeletal muscles involuntarily. Lights, noise, or smells may be sensed when the eyes, ears, and nose actually have not been stimulated. The electrical discharges may also inhibit certain brain centers. For instance, the waking center in the brain may be depressed so that the person loses consciousness.

The causes of epilepsy are varied. Many conditions can cause nerve cells to produce periodic bursts of impulses. These causes include head injuries, tumors and abscesses of the brain, and childhood infections, such as mumps, whooping cough, and measles. Epilepsy may also be *idiopathic,* that is, have no demonstrable cause. It should be noted that epilepsy almost never affects intelligence. If frequent severe seizures are allowed to occur over a long period of time, however, some cerebral damage may occasionally result. Damage can be prevented by controlling the seizures with drug therapy.

Epileptic seizures can be eliminated or alleviated by drugs that make neurons more difficult to stimulate. Many of these drugs change the permeability of the neuron cell membrane so that it does not depolarize as easily. One such drug is valproic acid, which increases the quantity of the inhibitory neurotransmitter GABA.

CEREBROVASCULAR ACCIDENTS (CVAs)

The most common brain disorder is a **cerebrovascular accident (CVA),** also called a **stroke,** or **cerebral apoplexy.** A CVA is characterized by a relatively abrupt onset of persisting neurological symptoms due to the destruction of brain tissue (infarction), resulting from disorders in the vessels that supply the brain. Common causes of CVAs are intracerebral hemorrhage from aneurysms, embolism, and atherosclerosis of the cerebral arteries. An *intracerebral hemorrhage* is a rupture of a vessel in the pia mater or brain. Blood seeps into the brain and damages neurons by increasing intracranial fluid pressure. An *embolus* is a blood clot, air bubble, or bit of foreign material, most often debris from an inflammation, that becomes lodged in an artery and blocks circulation. *Atherosclerosis* is the formation of plaques in the artery walls. The plaques may slow down circulation by constricting the vessel. Both emboli and atherosclerosis cause brain damage by reducing the supply of oxygen and glucose needed by brain cells.

DYSLEXIA

Dyslexia (dis-LEK-sē-a; *dys* = difficulty; *lexis* = words) is unrelated to basic intellectual capacity, but it causes a mysterious difficulty in handling words and symbols. Apparently some peculiarity in the brain's organizational pattern distorts the ability to read, write, and count. Letters in words seem transposed, reversed, or upside-down—*dog* becomes *god; b* changes identity with *d;* a sign saying "OIL" inverts into "710." Many dyslexics cannot orient themselves in the three dimensions of space and may show bodily awkwardness.

The exact cause of dyslexia is unknown, since it is unaccompanied by outward scars of detectable neurologi-

cal damage and its symptoms vary from victim to victim. It occurs three times as often among boys as among girls. It has been variously attributed to defective vision, brain damage, lead in the air, physical trauma, or oxygen deprivation during birth. A recent theory holds that it might be related to dysfunction of the vestibular apparatus and semicircular canals of the ear (see Chapter 20).

CLINICAL APPLICATION

Now available to physicians is a new technique called **brain electrical activity mapping (BEAM).** It is a non-invasive procedure that measures and displays the electrical activity of the brain on a color television screen and compares the image produced with a normal image (Figure 18-23). BEAM is used primarily to diagnose dyslexia, but it may have other diagnostic applications for learning disabilities, schizophrenia, depression, dementia, epilepsy, and early tumor occurrence and recurrence.

TAY-SACHS DISEASE

Tay-Sachs disease is a central nervous system affliction that brings death before age 5. The Tay-Sachs gene is carried mostly by individuals descended from the Ashkenazi Jews of Eastern Europe. Approximately 1 in 3,600 of their offspring will be afflicted with Tay-Sachs disease. The disease involves the neuronal degeneration of the central nervous system because of excessive amounts of a lipid called ganglioside in the nerve cells of the brain. The substance accumulates because of a deficient lysosomal enzyme. The afflicted child develops normally until the age of 4 to 8 months. Then the symptoms follow a course of progressive degeneration: paralysis, blindness, inability to eat, decubitus, and death from infection. There is no known cure.

HEADACHE

One of the most common human afflictions is **headache,** or **cephalgia** (*enkephalos* = brain; *algia* = painful condition). Based on origin, two general types are distinguished: intracranial and extracranial. Serious headaches of intracranial origin are caused by brain tumors, blood vessel abnormalities, inflammation of the brain or meninges, decrease in oxygen supply to the brain, and damage to brain cells. Extracranial headaches are related to infections of the eyes, ears, nose, and sinuses and are commonly felt as headaches because of the location of these structures. The sinuses are pain sensitive in only one or two areas and do not usually cause pain except in acute sinusitis or with some obvious chronic sinus problem such as a tumor or cyst formation or pressure on the conchae. Headaches that accompany sore throats and influenza may arise when the infection produces chemical changes in the blood that irritate pain sensors in the brain. Other extracranial causes include oral complications, occular disease such as glaucoma, and orthopedic disorders such as cervical spine disease.

Most headaches require no special treatment. Analgesic and tranquilizing compounds are generally effective for tension headaches, but not for migraine headaches. Drugs that constrict the blood vessels can be helpful for migraine. Biofeedback training and dietary changes may also be of some help. Taking a careful and complete history is an important procedure in determining the cause or causes of a patient's headaches.

TRIGEMINAL NEURALGIA (TIC DOULOUREUX)

As noted earlier, pain arising from irritation of the trigeminal (V) nerve is known as **trigeminal neuralgia,** or **tic douloureux** (doo-loo-ROO). The disorder is characterized by brief but extreme pain in the face and forehead on the affected side. The characteristic pain consists of red-hot, needlelike jabs lasting a few seconds and building to a searing pain like a hot poker being dragged or stuck into the face, lasting 10 to 15 seconds. Many patients describe sensitive regions around the mouth and nose that can cause an attack when touched. Eating, drinking, washing the face, and exposure to cold may also bring on an attack.

Treatment may be palliative (relieving symptoms without curing the disease) or surgical. One technique involves alcohol injections directly into the semilunar (Gasserian) ganglion, which controls the trigeminal (V) nerve. This method is superior to open surgery because it is safer. Moreover, it can bring lasting pain relief with preservation of touch sensation in the face and release of the patient 24 hours after treatment. Acupuncture can also provide relief in some patients.

RABIES

Rabies is an acute infection that may result in fatal encephalitis, an inflammation of the brain. The disease is caused by a virus called rabiesvirus, usually transmitted by the bite of an animal that harbors the virus in its saliva. As the virus multiplies in skeletal muscle and connective tissues, it produces spasms of the muscles of the mouth and pharynx when swallowing liquids. The mere sight or thought of water can trigger the spasms. Encephalitis occurs when the virus migrates along peripheral nerves by axonal transport.

Until recently, antirabies treatment was the Pasteur treatment, which consists of 14 to 21 injections of vaccine under the skin of the abdomen over a period of 2 to 3 weeks. This treatment sometimes causes an allergic reaction since the vaccine is prepared from rabbit brain. Today, antirabies treatment consists of administration of a vaccine prepared from human serum called *rabies immune globulin* (*RIG*) followed by vaccination with *hu-*

man diploid cell vaccine (*HDCV*). HDCV is prepared from human cell culture and is administered in five intramuscular (deltoid) injections over a 4-week period, with a sixth dose about 2 months later. Allergic reactions with the new vaccines are minimal.

REYE'S SYNDROME (RS)

Reye's syndrome (RS), first described in 1963 by the Australian pathologist R. Douglas Reye, seems to occur following a viral infection, particularly chickenpox or influenza. Aspirin at normal doses is believed to be a risk factor in the development of RS. The majority of persons affected are children or teenagers. The disease is characterized by vomiting and brain dysfunction (disorientation, lethargy, and personality changes) and may progress to coma. Also, the liver becomes infiltrated with small lipid droplets and loses some of its ability to detoxify ammonia. The disease runs its course in just a few days.

Brain dysfunction and death are typically caused by swelling of brain cells. The pressure not only kills the cells directly but also results in hypoxia that kills them indirectly. The mortality rate is about 40 percent. Swelling may result in irreversible brain damage, including mental retardation, in children who survive. Therapy is directed at controlling the swelling.

ALZHEIMER'S DISEASE

Alzheimer's (ALTZ-hī-merz) **disease** is a disabling neurological disorder that afflicts more than 5 percent of all persons over 65 years of age. It is characterized by the dysfunction and death of specific cerebral neurons. The disorder typically includes the development of widespread intellectual impairment, including memory loss, shortened attention span, decreased ability to perform mathematical calculations, and loss of orientation in time and space. In addition, personality changes such as irritability and moodiness often accompany the intellectual deficits. Some elderly people also experience delirium, the abrupt development of dramatic changes involving confusion, hallucinations, and sudden fluctuations in alertness.

Two outstanding hallmarks of Alzheimer's disease have been known for years. They are cores of abnormal protein (senile plaques) between nerve cells and twisted fibers (neurofibrillary tangles) in the cell bodies of neurons. Both are prominent in the cerebral cortex and hippocampus, areas of the brain involved in memory and learning. There is evidence that a deficiency in acetylcholine (ACh) might contribute to memory and learning deficits. It has recently been learned that there is a relationship between excessive amounts of aluminum and Alzheimer's disease and that administration of physostigmine (a substance derived from dried Calabar beans) and lecithin (phospholipid) can induce increased levels of acetylcholine and improve memory.

KEY MEDICAL TERMS ASSOCIATED WITH THE CENTRAL NERVOUS SYSTEM

Agnosia (*a* = without; *gnosis* = knowledge) Inability to recognize the significance of sensory stimuli such as auditory, visual, olfactory, gustatory, and tactile.

Analgesia (*an* = without; *algia* = painful condition) Insensibility to pain.

Anesthesia (*esthesia* = feeling) Loss of feeling.

Apraxia (*pratto* = to do) Inability to carry out purposeful movements in the absence of paralysis.

Coma Abnormally deep unconsciousness with an absence of voluntary response to stimuli and with varying degrees of reflex activity. It may be due to illness or to an injury.

Huntington's disease A rare hereditary disease characterized by involuntary jerky movements, depression, and mental deterioration that terminates in dementia and death. Also called **Huntington's chorea** (*choreia* = dance).

Lethargy A condition of functional torpor or sluggishness.

Nerve block Loss of sensation in a region, such as in local dental anesthesia.

Neuralgia (*neur* = nerve) Attacks of pain along the entire course or branch of a peripheral sensory nerve.

Paralysis Diminished or total loss of motor function resulting from damage to nervous tissue or a muscle.

Spastic (*spas* = draw or pull) Resembling sapsms or convulsions.

Stupor Condition of unconsciousness, torpor, or lethargy with suppression of sense or feeling.

Torpor Abnormal inactivity or lack of response to normal stimuli.

Viral encephalitis An acute inflammation of the brain caused by a direct attack by various viruses or by an allergic reaction to any of the many viruses that are normally harmless to the central nervous system. If the virus affects the spinal cord as well, it is called **encephalomyelitis**.

STUDY OUTLINE

Brain (p. 425)
Principal Parts; Protection and Coverings
1. During embryological development, brain vesicles are formed and serve as forerunners of various parts of the brain.
2. The diencephalon develops into the thalamus and hypothalamus, the telencephalon forms the cerebrum, the mesencephalon develops into the midbrain, the myelencephalon forms the medulla, and the metencephalon develops into the pons and cerebellum.
3. The principal parts of the brain are the brain stem (medulla oblongata, pons, and midbrain), diencephalon (thalamus and hypothalamus), cerebrum, and cerebellum.
4. The brain is protected by the cranial bones, cranial meninges, and cerebrospinal fluid.

Cerebrospinal Fluid
1. Cerebrospinal fluid is formed in the choroid plexuses and circulates through the subarachnoid space, ventricles, and central canal. Most of the fluid is absorbed by the arachnoid villi of the superior sagittal sinus.
2. Cerebrospinal fluid protects by serving as a shock absorber. It also circulates nutritive substances from the blood.
3. The accumulation of cerebrospinal fluid in the head is called hydrocephalus. If the fluid accumulates in the ventricles, it is called internal hydrocephalus. If it accumulates in the subarachnoid space, it is called external hydrocephalus.

Blood Supply
1. The blood supply to the brain is via the cerebral arterial circle (circle of Willis).
2. Any interruption of the oxygen supply to the brain can result in weakening, permanent damage, or death of brain cells. Interruption of the mother's blood supply to a child during childbirth before it can breathe may result in paralysis, mental retardation, epilepsy, or death.
3. Glucose deficiency may produce dizziness, convulsions, and unconsciousness.

4. The blood–brain barrier (BBB) is a concept that explains the differential rates of passage of certain material from the blood into the brain.

Brain Stem
1. The medulla oblongata is continuous with the upper part of the spinal cord. It contains nuclei that are reflex centers for regulation of heart rate, respiratory rate, vasoconstriction, swallowing, coughing, vomiting, sneezing, and hiccuping. It also contains the nuclei of origin for cranial nerves VIII (cochlear and vestibular branches) through XII.
2. The pons is superior to the medulla. It connects the spinal cord with the brain and links parts of the brain with one another. It relays impulses related to voluntary skeletal movements from the cerebral cortex to the cerebellum. It contains the nuclei for cranial nerves V through VII and the vestibular branch of VIII. The reticular formation of the pons contains the pneumotaxic center, which helps control respiration.
3. The midbrain connects the pons and diencephalon. It conveys motor impulses from the cerebrum to the cerebellum and cord, sensory impulses from cord to thalamus, and regulates auditory and visual reflexes. It also contains the nuclei of origin for cranial nerves III and IV.

Diencephalon
1. The diencephalon consists of the thalamus and hypothalamus.
2. The thalamus is superior to the midbrain and contains nuclei that serve as relay stations for all sensory impulses, except smell, to the cerebral cortex. It also registers conscious recognition of pain and temperature and some awareness of crude touch and pressure.
3. The hypothalamus is inferior to the thalamus. It controls and integrates the autonomic nervous system, receives sensory impulses from viscera, connects the nervous and endocrine systems, coordinates mind-over-body phenomena, functions in rage and aggression, controls body temperature,

regulates food and fluid intake, maintains the waking state and sleep patterns, and acts as a self-sustained oscillator that drives biological rhythms.

Cerebrum

1. The cerebrum is the largest part of the brain. Its cortex contains convolutions, fissures, and sulci.
2. The cerebral lobes are named frontal, parietal, temporal, and occipital.
3. The white matter is under the cortex and consists of myelinated axons running in three principal directions.
4. The basal ganglia are paired masses of gray matter in the cerebral hemispheres. They help to control muscular movements.
5. The limbic system is found in the cerebral hemispheres and diencephalon. It functions in emotional aspects of behavior and memory.
6. The motor areas of the cerebral cortex are the regions that govern muscular movement. The sensory areas are concerned with the interpretation of sensory impulses. The association areas are concerned with emotional and intellectual processes.
7. Positron emission tomography (PET) helps identify which parts of the brain are involved in specific sensory and motor activities.
8. Brain waves generated by the cerebral cortex are recorded as an EEG. They may be used to diagnose epilepsy, infections, and tumors.

Brain Lateralization (Split-Brain Concept)

1. Recent research indicates that the two hemispheres of the brain are not bilaterally symmetrical, either anatomically or functionally.
2. The left hemisphere is more important for right-handed control, spoken and written language, numerical and scientific skills, and reasoning.
3. The right hemisphere is more important for left-handed control, musical and artistic awareness, space and pattern perception, insight, imagination, and generating mental images of sight, sound, touch, taste, and smell.

Cerebellum

1. The cerebellum occupies the inferior and posterior aspects of the cranial cavity. It consists of two hemispheres and a central, constricted vermis.
2. It is attached to the brain stem by three pairs of cerebellar peduncles.
3. The cerebellum functions in the coordination of skeletal muscles and the maintenance of normal muscle tone and body equilibrium.

Cranial Nerves (p. 451)

1. Twelve pairs of cranial nerves originate from the brain.
2. They are named primarily on the basis of distribution or function and numbered by order of origin from the undersurface of the brain from anterior to posterior. (See Exhibit 18-3 for summary of cranial nerves.)

Aging and the Nervous System (p. 460)

1. Age-related effects involve loss of neurons and decreased capacity for sending nerve impulses.
2. Degenerative changes also affect the sense organs.

Developmental Anatomy of the Nervous System (p. 461)

1. The development of the nervous system begins with a thickening of ectoderm called the neural plate.
2. The parts of the brain develop from primary and secondary vesicles.

Applications to Health (p. 464)

1. Brain tumors are neoplasms (intraaxial or extraaxial) within the cranium.
2. Poliomyelitis is a viral infection that results in paralysis.
3. Amyotrophic lateral sclerosis (ALS) is a progressive neuromuscular disease that affects motor cells of the spinal cord.
4. Cerebral palsy refers to a group of motor disorders caused by damage to motor centers of the cerebral cortex, cerebellum, or basal ganglia during fetal development, childbirth, or early infancy.
5. Parkinson's disease is a progressive degeneration of the dopamine (DA)-producing neurons in the substantia nigra resulting in insufficient DA in the basal ganglia.
6. Multiple sclerosis (MS) is the destruction of myelin sheaths of the neurons of the central nervous system. Impulse transmission is interrupted.
7. Epilepsy results from irregular electrical discharges of brain cells and may be diagnosed by an EEG. Depending on the form of the disease, the victim experiences degrees of motor, sensory, or psychological malfunction.
8. Cerebrovascular accidents (CVAs), also called strokes, involve brain tissue destruction due to hemorrhage, thrombosis, or atherosclerosis.
9. Dyslexia involves an inability of an individual to comprehend written language. It may be diagnosed using brain electrical activity mapping (BEAM).
10. Tay-Sachs disease is an inherited disorder that involves neurological degeneration of the CNS because of excessive amounts of ganglioside.
11. Headaches are of two types: intracranial and extracranial.
12. Irritation of the trigeminal (V) nerve is known as trigeminal neuralgia.
13. Rabies is an acute viral infection that produces muscle spasms and encephalitis.
14. Reye's syndrome (RS) is characterized by vomiting, brain dysfunction, and liver damage.
15. Alzheimer's disease is a disorder of the elderly that involves widespread intellectual impairment, personality changes, and sometimes delirium.

REVIEW QUESTIONS

1. Identify the four principal parts of the brain and the components of each, where applicable. What is the origin of each of the parts?
2. Describe the structure and location of the cranial meninges. What is an extradural hemorrhage?
3. Where is cerebrospinal fluid (CSF) formed? Describe its

circulation. Where is CSF absorbed?

4. What is the blood–brain barrier (BBB)? What is its role?

5. Describe the location and structure of the medulla. Define decussation of pyramids. Why is it important? List the principal functions of the medulla.

6. Describe the location and structure of the pons. What are its functions?

7. Describe the location and structure of the midbrain. What are some of its functions?

8. Describe the location and structure of the thalamus. List some of its functions.

9. Where is the hypothalamus located? Explain some of its major functions.

10. Where is the cerebrum located? Describe the cortex, convolutions, fissures, and sulci of the cerebrum.

11. List and locate the lobes of the cerebrum. How are they separated from one another? What is the insula?

12. Describe the organization of cerebral white matter. Be sure to indicate the function of each group of fibers.

13. What are basal ganglia? Name the important basal ganglia and list the function of each.

14. Define the limbic system. Explain several of its functions.

15. What is meant by a sensory area of the cerebral cortex? List, locate, and give the function of each sensory area.

16. Describe the principle of positron emission tomography (PET). What are its clinical applications?

17. What is meant by a motor area of the cerebral cortex? List, locate, and give the function of each motor area.

18. What conditions may result from damage to sensory or motor speech areas?

19. What is an association area of the cerebral cortex? What are its functions?

20. Define an electroencephalogram (EEG). What is the diagnostic value of an EEG?

21. Describe brain lateralization (split-brain concept).

22. Describe the location of the cerebellum. List the principal parts of the cerebellum.

23. What are cerebellar peduncles? List and explain the function of each.

24. Explain the functions of the cerebellum. What is ataxia?

25. Define a cranial nerve. How are cranial nerves named and numbered? Distinguish between a mixed and a sensory cranial nerve.

26. For each of the 12 pairs of cranial nerves, list (a) its name, number, and type; (b) its location; and (c) its function. In addition, list the effects of damage, where applicable.

27. Describe the effects of aging on the nervous system.

28. Describe how the nervous system develops.

29. Define each of the following: brain tumors, poliomyelitis, amyotrophic lateral sclerosis (ALS), cerebral palsy, Parkinson's disease, multiple sclerosis (MS), epilepsy, cerebrovascular accidents (CVAs), dyslexia, Tay-Sachs disease, headache, trigeminal neuralgia, rabies, Reye's syndrome (RS), and Alzheimer's disease.

30. What is brain electrical activity mapping (BEAM)? What are its clinical applications?

31. Refer to the glossary of key medical terms associated with the nervous system. Be sure that you can define each term.

19

The Autonomic Nervous System

Student Objectives

Compare the structural and functional differences between the somatic efferent and autonomic portions of the nervous system.

Identify the structural features of the autonomic nervous system.

Compare the sympathetic and parasympathetic divisions of the autonomic nervous system in terms of structure, physiology, and chemical transmitters released.

Describe a visceral autonomic reflex and its components.

Explain the role of the hypothalamus and its relationship to the sympathetic and parasympathetic division.

Explain the relationship between biofeedback and the autonomic nervous system.

Describe the relationship between meditation and the autonomic nervous system.

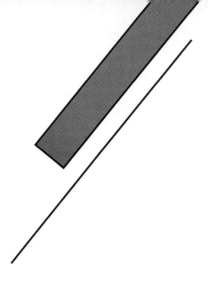

The portion of the nervous system that regulates the activities of smooth muscle, cardiac muscle, and glands is the **autonomic nervous system (ANS).** Structurally, the system consists of visceral efferent neurons organized into nerves, ganglia, and plexuses. Functionally, it usually operates without conscious control. The system was originally named *autonomic* because physiologists thought it functioned with no control from the central nervous system, that it was autonomous, or self-governing. It is now known that the autonomic system is neither structurally nor functionally independent of the central nervous system. It is regulated by centers in the brain, in particular by the cerebral cortex, hypothalamus, and medulla oblongata. However, the old terminology has been retained, and since the autonomic nervous system does differ from the somatic nervous system in some ways, the two are separated for convenience of study.

SOMATIC EFFERENT AND AUTONOMIC NERVOUS SYSTEMS

Whereas the somatic efferent nervous system produces conscious movement in skeletal muscles, the autonomic nervous system (visceral efferent nervous system) regulates visceral activities, and it generally does so involuntarily and automatically. Examples of visceral activities regulated by the autonomic nervous system are changes in the size of the pupil, accommodation for near vision, dilation and constriction of blood vessels, adjustment of the rate and force of the heartbeat, movements of the gastrointestinal tract, and secretion by most glands. These activities usually lie beyond conscious control. They are automatic.

The autonomic nervous system is entirely motor. All its axons are efferent fibers, which transmit impulses from the central nervous system to visceral effectors. Autonomic fibers are called **visceral efferent fibers. Visceral effectors** include cardiac muscle, smooth muscle, and glandular epithelium. This does not mean there are no afferent (sensory) impulses from visceral effectors, however. Impulses that give rise to visceral sensations pass over visceral afferent neurons that have cell bodies located in the posterior (dorsal) root ganglia of spinal nerves. Some functions of these afferent neurons were described with the cranial and spinal nerves. The hypothalamus, which largely controls the autonomic nervous system, also receives impulses from the visceral sensory fibers.

In the motor portion of the neural pathway of the somatic efferent system, an efferent neuron runs from the central nervous system and synapses directly on a skeletal muscle. In the neural pathway of the autonomic nervous system, there are two efferent neurons and a ganglion between them. The first neuron runs from the central nervous system to a ganglion, where it synapses with the second efferent neuron. It is this neuron that ultimately synapses on a visceral effector. Also, whereas fibers of somatic efferent neurons release acetylcholine (ACh) as their neurotransmitter, fibers of autonomic efferent neurons release either ACh or norepinephrine (NE).

The autonomic nervous system consists of two principal divisions: the **sympathetic** and the **parasympathetic.** Many organs innervated by the autonomic nervous system receive visceral efferent neurons from both components of the autonomic system—one set from the sympathetic division, another from the parasympathetic division. In general, impulses transmitted by the fibers of one division stimulate the organ to start or increase activity, whereas impulses from the other division decrease the organ's activity. Organs that receive impulses from both sympathetic and parasympathetic fibers are said to have *dual innervation.* Thus autonomic innervation may be excitatory or inhibitory. In the somatic efferent nervous system, only one kind of motor neuron innervates an organ, which is always a skeletal muscle. Moreover, innervation is always excitatory. When a somatic neuron stimulates a skeletal muscle, the muscle becomes active. When the neuron ceases to stimulate the muscle, contraction stops altogether.

A summary of the principal differences between the somatic efferent and autonomic nervous systems is presented in Exhibit 19-1.

EXHIBIT 19-1 COMPARISON OF SOMATIC EFFERENT AND AUTONOMIC NERVOUS SYSTEMS

	SOMATIC EFFERENT	AUTONOMIC
Effectors	Skeletal muscles.	Cardiac muscle, smooth muscle, glandular epithelium.
Type of control	Voluntary.	Involuntary.
Neural pathway	One efferent neuron extends from CNS and synapses directly on a skeletal muscle.	One efferent neuron extends from the CNS and synapses with another efferent neuron in a ganglion; the second neuron synapses on a visceral effector.
Action on effector	Always excitatory.	May be excitatory or inhibitory, depending on whether stimulation is sympathetic or parasympathetic.
Neurotransmitters	Acetylcholine (ACh).	Acetylcholine (ACh) or norepinephrine (NE).

STRUCTURE OF THE AUTONOMIC NERVOUS SYSTEM

VISCERAL EFFERENT PATHWAYS

Autonomic visceral efferent pathways always consist of two neurons. One extends from the central nervous system to a ganglion. The other extends directly from the ganglion to the effector (muscle or gland).

The first of the visceral efferent neurons in an autonomic pathway is called a **preganglionic neuron** (Figure 19-1). Its cell body is in the brain or spinal cord. Its myelinated axon, called a **preganglionic fiber,** passes out of the central nervous system as part of a cranial or spinal nerve. At some point, the fiber separates from the nerve and courses to an autonomic ganglion, where it synapses with the dendrites or cell body of the postganglionic neuron, the second neuron in the visceral efferent pathway.

The **postganglionic neuron** lies entirely outside the central nervous system. Its cell body and dendrites (if it has dendrites) are located in the autonomic ganglion, where the synapse with the preganglionic fibers occurs. The axon of a postganglionic neuron, called a **postganglionic fiber,** is unmyelinated and terminates in a visceral effector.

Thus preganglionic neurons convey efferent impulses from the central nervous system to autonomic ganglia. Postganglionic neurons relay the impulses from autonomic ganglia to visceral effectors.

Preganglionic Neurons

In the sympathetic division, the preganglionic neurons have their cell bodies in the lateral gray horns of the 12 thoracic segments and first 2 lumbar segments of the spinal cord (Figure 19-2). It is for this reason that the sympathetic division is also called the **thoracolumbar** (thō'-ra-kō-LUM-bar) **division** and the fibers of the sympathetic preganglionic neurons are known as the **thoracolumbar outflow.**

The cell bodies of the preganglionic neurons of the parasympathetic division are located in the nuclei of cranial nerves III, VII, IX, and X in the brain stem and in the lateral gray horns of the second through fourth sacral segments of the spinal cord. Hence the parasympathetic division is also known as the **craniosacral division,** and the fibers of the parasympathetic preganglionic neurons are referred to as the **craniosacral outflow.**

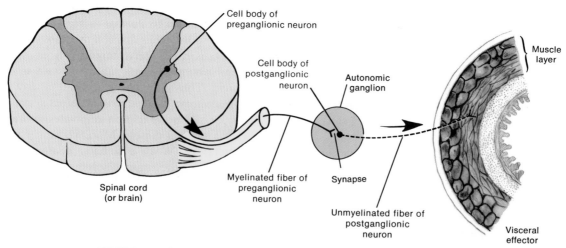

FIGURE 19-1 Relationship between preganglionic and postganglionic neurons.

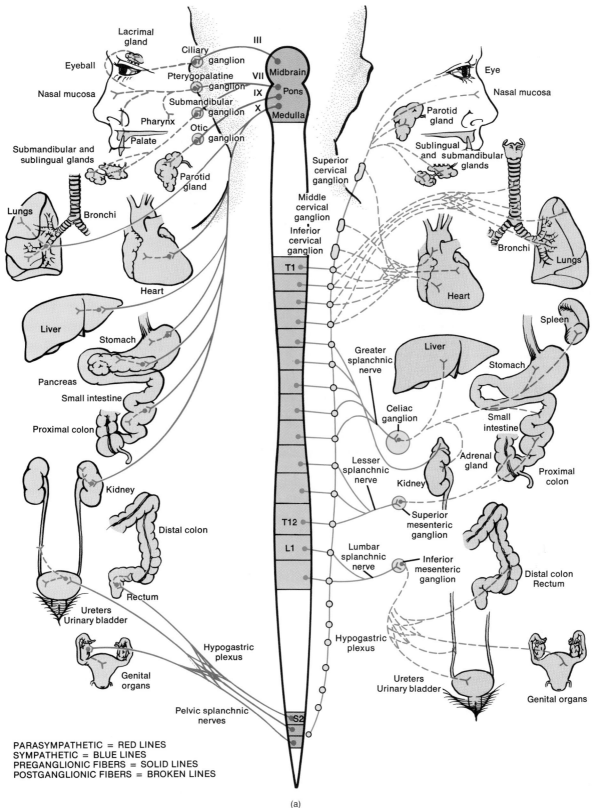

Lacrimal gland

Ciliary ganglion

III

Eyeball

Pterygopalatine ganglion

VII

Midbrain

Eye

Nasal mucosa

IX

Pons

Nasal mucosa

Submandibular ganglion

X

Medulla

Parotid gland

Pharynx

Otic ganglion

Palate

Sublingual and submandibular glands

Submandibular and sublingual glands

Superior cervical ganglion

Parotid gland

Lungs

Bronchi

Middle cervical ganglion

Inferior cervical ganglion

Bronchi

Lungs

Heart

T1

Heart

Liver

Spleen

Stomach

Greater splanchnic nerve

Liver

Stomach

Pancreas

Celiac ganglion

Small intestine

Small intestine

Proximal colon

Lesser splanchnic nerve

Adrenal gland

Proximal colon

Kidney

Kidney

Distal colon

T12

Superior mesenteric ganglion

L1

Lumbar splanchnic nerve

Inferior mesenteric ganglion

Distal colon Rectum

Rectum

Ureters Urinary bladder

Ureters Urinary bladder

Hypogastric plexus

Hypogastric plexus

Genital organs

Pelvic splanchnic nerves

S2

Genital organs

PARASYMPATHETIC = RED LINES
SYMPATHETIC = BLUE LINES
PREGANGLIONIC FIBERS = SOLID LINES
POSTGANGLIONIC FIBERS = BROKEN LINES

(a)

FIGURE 19-2 Structure of the autonomic nervous system. (a) Diagram. Although the parasympathetic division is shown only on the left side of the figure and the sympathetic division is shown only on the right side, keep in mind that each division is actually on both sides of the body (bilateral symmetry).

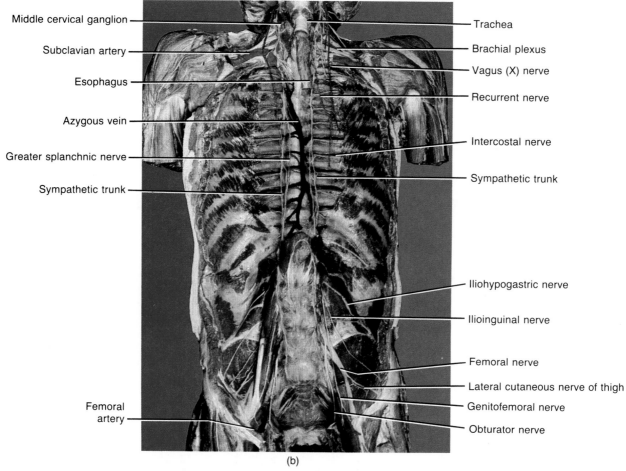

Middle cervical ganglion

Subclavian artery

Esophagus

Azygous vein

Greater splanchnic nerve

Sympathetic trunk

Femoral artery

Trachea

Brachial plexus

Vagus (X) nerve

Recurrent nerve

Intercostal nerve

Sympathetic trunk

Iliohypogastric nerve

Ilioinguinal nerve

Femoral nerve

Lateral cutaneous nerve of thigh

Genitofemoral nerve

Obturator nerve

(b)

FIGURE 19-2 (*Continued*) Structure of the autonomic nervous system. (b) Photograph. (Courtesy of C. Yokochi and J. W. Rohen, *Photographic Anatomy of the Human Body,* 2nd ed., 1979, IGAKU-SHOIN, Ltd., Tokyo, New York.)

CLINICAL APPLICATION

Raynaud's (rā-NŌZ) **disease** is a disorder characterized by spasms of arteries, especially those in the fingers and toes due to overactivity of the sympathetic nervous system. As a result of the spasmodic contractions, the tissues of the digits receive an inadequate blood supply and the digits exhibit pallor (paleness) or cyanosis and severe pain. The disease is typically bilateral and is provoked by exposure to cold. In extreme cases gangrene may occur. Cutting the preganglionic sympathetic fibers supplying the area abolishes symptoms for patients with severe progression of the disease.

Autonomic Ganglia

Autonomic pathways include **autonomic ganglia,** where synapses between visceral efferent neurons occur. Autonomic ganglia differ from posterior root ganglia. The lat-

ter contain cell bodies of sensory neurons and no synapses occur in them.

The autonomic ganglia may be divided into three general groups. The **sympathetic trunk (vertebral chain) ganglia** are a series of ganglia that lie in a vertical row on either side of the vertebral column, extending from the base of the skull to the coccyx (Figure 19-3). They are also known as **paravertebral (lateral) ganglia.** They receive preganglionic fibers only from the thoracolumbar (sympathetic) division (Figure 19-2).

The second kind of autonomic ganglion also belongs to the sympathetic division. It is called a **prevertebral (collateral) ganglion** (Figure 19-3). The ganglia of this group lie anterior to the spinal column and close to the large abdominal arteries from which their names are derived. Examples of prevertebral ganglia so named are the celiac ganglion, on either side of the celiac artery just below the diaphragm; the superior mesenteric ganglion, near the beginning of the superior mesenteric artery in the upper abdomen; and the inferior mesenteric ganglion, located near the beginning of the inferior mesenteric ar-

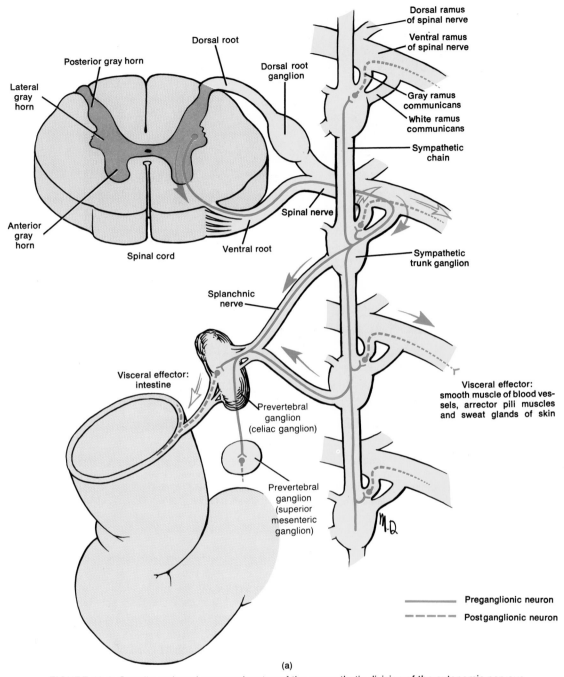

Dorsal ramus of spinal nerve

Ventral ramus of spinal nerve

Gray ramus communicans

White ramus communicans

Sympathetic chain

Sympathetic trunk ganglion

Visceral effector: smooth muscle of blood vessels, arrector pili muscles and sweat glands of skin

Posterior gray horn

Lateral gray horn

Dorsal root

Dorsal root ganglion

Anterior gray horn

Spinal nerve

Spinal cord

Ventral root

Splanchnic nerve

Visceral effector: intestine

Prevertebral ganglion (celiac ganglion)

Prevertebral ganglion (superior mesenteric ganglion)

———— Preganglionic neuron

– – – – – Postganglionic neuron

(a)

FIGURE 19-3 Ganglia and rami communicantes of the sympathetic division of the autonomic nervous system. (a) Diagram.

tery in the middle of the abdomen (Figure 19-2). Prevertebral ganglia receive preganglionic fibers from the thoracolumbar (sympathetic) division.

The third kind of autonomic ganglion belongs to the parasympathetic division and is called a **terminal (intramural) ganglion.** The ganglia of this group are located at the end of a visceral efferent pathway very close to visceral effectors or within the walls of visceral effectors. Terminal ganglia receive preganglionic fibers from the

craniosacral (parasympathetic) division. The preganglionic fibers do not pass through sympathetic trunk ganglia (Figure 19-2).

In addition to autonomic ganglia, the autonomic nervous system contains **autonomic plexuses.** Slender nerve fibers from ganglia containing postganglionic nerve cell bodies arranged in a branching network constitute an autonomic plexus.

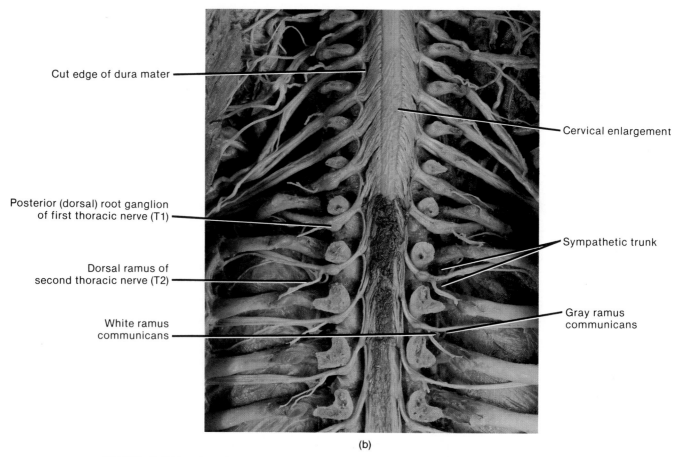

Cut edge of dura mater

Cervical enlargement

Posterior (dorsal) root ganglion
of first thoracic nerve (T1)

Sympathetic trunk

Dorsal ramus of
second thoracic nerve (T2)

Gray ramus
communicans

White ramus
communicans

(b)

FIGURE 19-3 (*Continued*) Ganglia and rami communicantes of the sympathetic division of the autonomic nervous system. (b) Photograph of the posterior aspect of the lower cervical and upper thoracic segments of the spinal cord showing the trunks of the brachial plexus. (Courtesy of N. Gluhbegovic and T. H. Williams, *The Human Brain: A Photographic Guide,* Harper & Row, New York, 1980.)

Postganglionic Neurons

Axons from preganglionic neurons of the sympathetic division pass to ganglia of the sympathetic trunk. They can either synapse in the sympathetic chain ganglia with postganglionic sympathetics or they can continue, without synapsing, through the chain ganglia to end at a prevertebral ganglion where synapses with the postganglionic sympathetics can take place. Each sympathetic preganglionic fiber synapses with several postganglionic fibers in the ganglion, and the postganglionic fibers pass to several visceral effectors. After exiting their ganglia, the postsynaptic fibers innervate their visceral effectors.

Axons from preganglionic neurons of the parasympathetic division pass to terminal ganglia near or within a visceral effector. In the ganglion, the presynaptic neuron usually synapses with only four or five postsynaptic neurons to a single visceral effector. After exiting their ganglia, the postsynaptic fibers supply their visceral effectors.

With this background in mind, we can now examine some specific structural features of the sympathetic and parasympathetic divisions of the autonomic nervous system.

SYMPATHETIC DIVISION

The preganglionic fibers of the sympathetic division have their cell bodies located in the lateral gray horn of the spinal cord in the thoracic and first two lumbar segments (Figure 19-2). The preganglionic fibers are myelinated and leave the spinal cord through the ventral root of a spinal nerve along with the somatic efferent fibers at the same segmental levels. After exiting through the intervertebral foramina, the preganglionic sympathetic fibers enter a white ramus to pass to the nearest sympathetic trunk ganglion on the same side. Collectively, the white rami are called the **white rami communicantes** (kō-myoo-ne-KAN-tēz). Their name indicates that they contain myelinated fibers. Only thoracic and upper lumbar nerves have white rami communicantes. The white rami communicantes connect the ventral ramus of the spinal nerve with the ganglia of the sympathetic trunk.

The paired sympathetic trunks are situated anterolaterally to the spinal cord, one on either side. Each consists of a series of ganglia arranged more or less segmentally. The divisions of the sympathetic trunk are named on the basis of location. Typically, there are 22 ganglia in each chain: 3 cervical, 11 thoracic, 4 lumbar, and 4 sacral. Although the trunk extends downward from the neck, thorax, and abdomen to the coccyx, it receives preganglionic fibers only from the thoracic and lumbar segments of the spinal cord (Figure 19-2).

The cervical portion of each sympathetic trunk is located in the neck anterior to the prevertebral muscles. It is subdivided into a superior, middle, and inferior ganglion (Figure 19-2). The *superior cervical ganglion* is posterior to the internal carotid artery and anterior to the transverse processes of the second cervical vertebra. Postganglionic fibers leaving the ganglion serve the head, where they are distributed to the sweat glands, the smooth muscle of the eye and blood vessels of the face, the nasal mucosa, and the submandibular, sublingual, and parotid salivary glands. Gray rami communicantes from the ganglion also pass to the upper two to four cervical spinal nerves. The *middle cervical ganglion* is situated near the sixth cervical vertebra at the level of the cricoid cartilage. Postganglionic fibers from it innervate the heart. The *inferior cervical ganglion* is located near the first rib, anterior to the transverse processes of the seventh cervical vertebra. Its postganglionic fibers also supply the heart.

The thoracic portion of each sympathetic trunk usually consists of 11 segmentally arranged ganglia, lying ventral to the necks of the corresponding ribs. This portion of the sympathetic trunk receives most of the sympathetic preganglionic fibers. Postganglionic fibers from the thoracic sympathetic trunk innervate the heart, lungs, bronchi, and other thoracic viscera.

The lumbar portion of each sympathetic trunk is found on either side of the corresponding lumbar vertebrae. The sacral portion of the sympathetic trunk lies in the pelvic cavity on the medial side of the sacral foramina. Postganglionic fibers from the lumbar and sacral sympathetic chain ganglia are distributed with the respective spinal nerves via gray rami, or they may join the hypogastric plexus via direct visceral branches.

When a preganglionic fiber of a white ramus communicans enters the sympathetic trunk, it may terminate (synapse) in several ways. Some fibers synapse in the first ganglion at the level of entry. Others pass up or down the sympathetic trunk for a variable distance to form the fibers on which the ganglia are strung. These fibers, known as **sympathetic chains** (Figure 19-3), may not synapse until they reach a ganglion in the cervical or sacral area. Most rejoin the spinal nerves before supplying peripheral visceral effectors such as sweat glands and the smooth muscle in blood vessels and around hair follicles in the extremities. The **gray ramus communicans** (kō-MYOO-ne-kanz) is the structure containing the postganglionic fibers that connect the ganglion of the sympathetic trunk to the spinal nerve (Figure 19-3). The fibers are unmyelinated. All spinal nerves have gray rami communicantes. Gray rami communicantes outnumber the white rami, since there is a gray ramus leading to each of the 31 pairs of spinal nerves.

In most cases, a sympathetic preganglionic fiber terminates by synapsing with a large number, usually 20 or more, of postganglionic cell bodies in a ganglion. Often the postganglionic fibers then terminate in widely separated organs of the body. Thus an impulse that starts in a single preganglionic neuron may reach several visceral effectors. For this reason, most sympathetic responses have widespread effects on the body.

Some preganglionic fibers pass through the sympathetic trunk without terminating in the trunk. Beyond the trunk, they form nerves known as **splanchnic** (SPLANK-nik) **nerves** (Figure 19-2). After passing through the trunk of ganglia, the splanchnic nerves terminate in the *celiac* (SĒ-lē-ak), or *solar, plexus.* In the plexus, the preganglionic fibers synapse in ganglia with postganglionic cell bodies. These ganglia are prevertebral ganglia. The greater splanchnic nerve passes to the celiac ganglion of the celiac plexus. From here, postganglionic fibers are distributed to the stomach, spleen, liver, kidney, and small intestine. The lesser splanchnic nerve passes through the celiac plexus to the superior mesenteric ganglion of the superior mesenteric plexus. Postganglionic fibers from this ganglion innervate the small intestine and colon. The lowest splanchnic nerve, not always present, enters the renal plexus. Postganglionics supply the renal artery and ureter. The lumbar splanchnic nerve enters the inferior mesenteric plexus. In the plexus, the preganglionic fibers synapse with postganglionic fibers in the inferior mesenteric ganglion. These fibers pass through the hypogastric plexus and supply the distal colon and rectum, urinary bladder, and genital organs. As noted earlier, the postganglionic fibers leaving the prevertebral ganglia follow the course of various arteries to abdominal and pelvic visceral effectors.

PARASYMPATHETIC DIVISION

The preganglionic cell bodies of the parasympathetic division are found in nuclei in the brain stem and the lateral gray horn of the second through fourth sacral segments of the spinal cord (Figure 19-2). Their fibers emerge as part of a cranial nerve or as part of the ventral root of a spinal nerve. The **cranial parasympathetic outflow** consists of preganglionic fibers that leave the brain stem by way of the oculomotor (III) nerves, facial (VII) nerves, glossopharyngeal (IX) nerves, and vagus (X) nerves. The **sacral parasympathetic outflow** consists of preganglionic fibers that leave the ventral roots of the second through fourth sacral nerves. The preganglionic fibers of both the cranial and sacral outflows end in terminal ganglia, where they synapse with postganglionic neurons. We will first look at the cranial outflow.

The cranial outflow has five components: four pairs of ganglia and the plexuses associated with the vagus nerve. The four pairs of cranial parasympathetic ganglia innervate structures in the head and are located close to the organs they innervate. The *ciliary ganglion* is near the back of an orbit lateral to each optic nerve. Preganglionic fibers pass with the oculomotor (III) nerve to the ciliary ganglion. Postganglionic fibers from the ganglion innervate smooth muscle cells in the eyeball. Each *pterygopalatine* (ter'-i-gō-PAL-a-tin) *ganglion* is situated lateral to a sphenopalatine foramen. It receives preganglionic fibers from the facial (VII) nerve and transmits postganglionic fibers to the nasal mucosa, palate, pharynx, and lacrimal gland. Each *submandibular ganglion* is found near the duct of a submandibular salivary gland. It receives preganglionic fibers from the facial (VII) nerve and transmits postganglionic fibers that innervate the submandibular and sublingual salivary glands. The *otic ganglia* are situated just below each foramen ovale. The otic ganglion receives preganglionic fibers from the glossopharyngeal (IX) nerve and transmits postganglionic fibers that innervate the parotid salivary gland. Ganglia associated with cranial outflow are classified as terminal ganglia. Since the terminal ganglia are close to their visceral effectors, postganglionic parasympathetic fibers are short. Postganglionic sympathetic fibers are relatively long.

The last component of the cranial outflow consists of the preganglionic fibers that leave the brain via the vagus (X) nerves. This component has the most extensive distribution of the parasympathetic fibers. It provides about 80 percent of the craniosacral outflow. Each vagus (X) nerve enters into the formation of several plexuses in the thorax and abdomen. As it passes through the thorax, it sends fibers to the *superficial cardiac plexus* in the arch of the aorta and the *deep cardiac plexus* anterior to the branching of the trachea. These plexuses contain terminal ganglia, and the postganglionic parasympathetic fibers emerging from them supply the heart. Also in the thorax is the *pulmonary plexus,* in front of and behind the roots of the lungs and within the lungs themselves. It receives preganglionic fibers from the vagus and transmits postganglionic parasympathetic fibers to the lungs and bronchi. Other plexuses associated with the vagus (X) nerve are described in later chapters in conjunction with the appropriate thoracic, abdominal, and pelvic viscera. Postganglionic fibers from these plexuses innervate viscera such as the liver, pancreas, stomach, kidneys, small intestine, and part of the colon.

The sacral parasympathetic outflow consists of preganglionic fibers from the ventral roots of the second through fourth sacral nerves. Collectively, they form the *pelvic splanchnic nerves.* They pass into the hypogastric plexus. From ganglia in the plexus, parasympathetic postganglionic fibers are distributed to the colon, ureters, urinary bladder, and reproductive organs.

The salient structural features of the sympathetic and parasympathetic divisions are compared in Exhibit 19-2.

PHYSIOLOGY OF THE AUTONOMIC NERVOUS SYSTEM

NEUROTRANSMITTERS

Autonomic fibers, like other axons of the nervous system, release neurotransmitters at synapses as well as at points of contact with visceral effectors. These latter points are called **neuroeffector junctions.** Neuroeffector junctions may be either neuromuscular or neuroglandular junctions. On the basis of the neurotransmitter produced, autonomic fibers may be classified as either cholinergic or adrenergic (Figure 19-4).

Cholinergic (kō'-lin-ER-jik) **fibers** release **acetylcholine (ACh)** and include the following: (1) all sympathetic and parasympathetic preganglionic axons, (2) all parasympathetic postganglionic axons, and (3) some sympathetic postganglionic axons. The cholinergic sympathetic postganglionic axons include those to sweat glands and blood vessels in skeletal muscles, to the skin, and to the external genitalia. Since acetylcholine is quickly inactivated by the enzyme **acetylcholinesterase (AChE),** the effects of cholinergic fibers are short-lived and local.

Adrenergic (ad'-ren-ER-jik) **fibers** produce **norepinephrine (NE).** Most sympathetic postganglionic axons are adrenergic. Since norepinephrine is inactivated much more slowly by **catechol-*o*-methyltransferase (COMT)** or **monoamine oxidase (MAO)** than acetylcholine is by acetylcholinesterase, and since norepinephrine may enter the blood stream, the effects of sympathetic stimulation are longer lasting and more widespread than parasympathetic stimulation. The action of NE produced by sympathetic postganglionic axons is augmented by both NE and epinephrine secreted by the adrenal medulla. Since both substances are secreted into the blood, their effects are more

---

EXHIBIT 19-2 STRUCTURAL FEATURES OF SYMPATHETIC AND PARASYMPATHETIC DIVISIONS

SYMPATHETIC	PARASYMPATHETIC
Forms thoracolumbar outflow.	Forms craniosacral outflow.
Contains sympathetic trunk and prevertebral ganglia.	Contains terminal ganglia.
Ganglia are close to the CNS and distant from visceral effectors.	Ganglia are near or within visceral effectors.
Each preganglionic fiber synapses with many postganglionic neurons that pass to many visceral effectors.	Each preganglionic fiber usually synapses with four or five postganglionic neurons that pass to a single visceral effector.
Distributed throughout the body, including the skin.	Distribution limited primarily to head and viscera of thorax, abdomen, and pelvis.

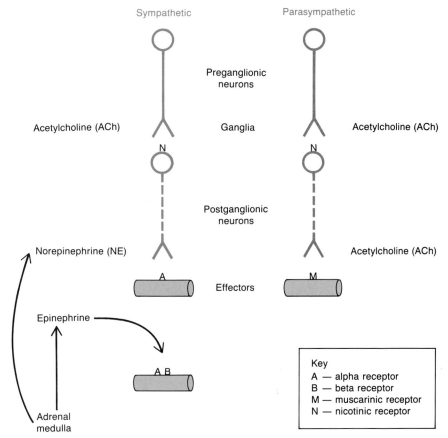

FIGURE 19-4 Neurotransmitters and receptors associated with the autonomic nervous system.

sustained than that of NE released by axons. NE and epinephrine released by the adrenal medulla are ultimately destroyed by enzymes in the liver once they have acted on their visceral effectors.

RECEPTORS

The actual effect produced by ACh is determined by the type of receptor with which it interacts (Figure 19-4). The two types of receptors are known as nicotinic receptors and muscarinic receptors. **Nicotinic receptors** are found on both sympathetic and parasympathetic postganglionic neurons. These receptors are so named because the actions of ACh on such receptors are similar to those produced by nicotine.

Muscarinic receptors are found on effectors innervated by parasympathetic postganglionic axons. These receptors are so named because the actions of ACh on such receptors are similar to those produced by muscarine, a toxin produced by a mushroom.

The effects of NE and epinephrine, like those of ACh, are also determined by the type of receptor with which they interact. Such receptors are found on visceral effectors innervated by most sympathetic postganglionic axons and are referred to as **alpha receptors** and **beta receptors.** Although cells of most effectors contain either alpha or beta receptors, some effector cells contain both. NE, in general, stimulates alpha receptors to a greater extent than beta receptors, and epinephrine, in general, stimulates both alpha and beta receptors.

ACTIVITIES

Most visceral effectors have dual innervation, that is, they receive fibers from both the sympathetic and the parasympathetic divisions. In these cases, impulses from one division stimulate the organ's activities, whereas impulses from the other division inhibit the organ's activities. The stimulating division may be either the sympathetic or the parasympathetic, depending on the organ. For example, sympathetic impulses increase heart activity, whereas parasympathetic impulses decrease it. On the other hand, parasympathetic impulses increase digestive activities, whereas sympathetic impulses inhibit them. The actions of the two systems are carefully integrated to help maintain homeostasis. A summary of the activities of the autonomic nervous system is presented in Exhibit 19-3.

The parasympathetic division is primarily concerned with activities that restore and conserve body energy. It is a **rest-repose system.** Under normal body conditions, for instance, parasympathetic impulses to the digestive glands and the smooth muscle of the digestive system

EXHIBIT 19-3 ACTIVITIES OF AUTONOMIC NERVOUS SYSTEM

VISCERAL EFFECTOR	EFFECT OF SYMPATHETIC STIMULATION	EFFECT OF PARASYMPATHETIC STIMULATION
Eye		
Iris	Contraction of dilator muscle that results in dilation of pupil.	Contraction of sphincter muscle that results in constriction of pupil.
Ciliary muscle	No innervation.	Contraction that results in lens accommodation for near vision.
Glands		
Sweat	Stimulates secretion.	No innervation.
Lacrimal (tear)	Vasoconstriction, which inhibits secretion.	Normal or excessive secretion.
Salivary	Vasoconstriction, which decreases secretion.	Stimulates secretion and vasodilation.
Gastric	Vasoconstriction, which inhibits secretion.	Stimulates secretion.
Intestinal	Vasoconstriction, which inhibits secretion.	Stimulates secretion.
Adrenal medulla	Promotes epinephrine and norepinephrine secretion.	No innervation.
Adrenal cortex	Promotes glucocorticoid secretion.	No innervation.
Lungs (bronchial tubes)	Dilation.	Constriction.
Heart	Increases rate and strength of contraction; dilates coronary vessels that supply blood to heart muscle cells.	Decreases rate and strength of contraction; constricts coronary vessels.
Blood vessels		
Skin	Constriction.	No innervation for most.
Skeletal muscle	Dilation.	No innervation.
Visceral organs (except heart and lungs)	Constriction.	No innervation for most.
Liver	Promotes glycogenolysis; decreases bile secretion.	Promotes glycogenesis; increases bile secretion.
Gallbladder	Relaxation.	Contraction.
Stomach	Decreases motility.	Increases motility.
Intestines	Decreases motility.	Increases motility.
Gastrointestinal sphincters	Increases tone.	Relaxation.
Kidney	Constriction of blood vessels that results in decreased urine volume.	No innervation.
Pancreas	Inhibits secretion.	Promotes secretion.
Spleen	Contraction and discharge of stored blood into general circulation.	No innervation.
Urinary bladder	Relaxation of muscular wall; increases tone in internal sphincter.	Contraction of muscular wall; relaxation of internal sphincter.
Arrector pili of hair follicles	Contraction that results in erection of hairs.	No innervation.
Uterus	Inhibits contraction if nonpregnant; stimulates contraction if pregnant.	Minimal effect.
Sex organs	In male, vasoconstriction of ductus deferens, seminal vesicle, prostate; results in ejaculation. In female, reverse uterine peristalsis.	Vasodilation and erection in both sexes; secretion in female.

dominate over sympathetic impulses. Thus energy-supplying food can be digested and absorbed by the body.

The sympathetic division, by contrast, is primarily concerned with processes involving the expenditure of energy. When the body is in homeostasis, the main function of the sympathetic division is to counteract the parasympathetic effects just enough to carry out normal processes requiring energy. During extreme stress, however, the sympathetic dominates the parasympathetic. When people are confronted with a stress condition, for example, their bodies become alert and they sometimes perform feats of unusual strength. Fear stimulates the sympathetic division.

Activation of the sympathetic division sets into operation a series of physiological responses collectively called the **fight-or-flight response.** It produces the following effects:

1. The pupils of the eyes dilate.
2. The heart rate increases.
3. The blood vessels of the skin and viscera constrict.
4. The remainder of the blood vessels dilate. This reaction causes a rise in blood pressure and a faster flow of blood into the dilated blood vessels of skeletal muscles, cardiac muscle, lungs, and brain—organs involved in fighting off danger.
5. Rapid breathing occurs as the bronchioles dilate to allow faster movement of air in and out of the lungs.
6. Blood sugar level rises as liver glycogen is converted to glucose to supply the body's additional energy needs.
7. The medulla of the adrenal gland is stimulated to produce epinephrine and norepinephrine, hormones that intensify and prolong the sympathetic effects noted above.
8. Processes that are not essential for meeting the stress situation are inhibited. For example, muscular movements of the gastrointestinal tract and digestive secretions are slowed down or even stopped.

CLINICAL APPLICATION

If the sympathetic trunk is cut on one side, the sympathetic supply to that side of the head is removed and the result is **Horner's syndrome,** in which the patient exhibits (on the affected side): ptosis (drooping of the upper eyelid), slight elevation of the lower eyelid, narrowing of the palpebral fissure, enophthalmos (the eye appears sunken), miosis (constricted pupil), anhydrosis (lack of sweating), and flushing of the skin.

VISCERAL AUTONOMIC REFLEXES

A **visceral autonomic reflex** adjusts the activity of a visceral effector. In other words, it results in the contraction of smooth or cardiac muscle or secretion by a gland. Such reflexes assume a key role in activities such as regulating heart action, blood pressure, respiration, digestion, defecation, and urinary bladder functions.

A visceral autonomic reflex arc consists of the following components:

1. **Receptor.** The receptor is the distal end of an afferent neuron in an exteroceptor or enteroceptor.
2. **Afferent neuron.** This neuron, either a somatic afferent or visceral afferent neuron, conducts the sensory impulse to the spinal cord or brain.
3. **Association neurons.** These neurons are found in the central nervous system.
4. **Visceral efferent preganglionic neuron.** In the thoracic and abdominal regions, this neuron is in the lateral

gray horn of the spinal cord. The axon passes through the ventral root of the spinal nerve, the spinal nerve, and the white ramus communicans. It then enters a sympathetic trunk or prevertebral ganglion, where it synapses with a postganglionic neuron. In the cranial and sacral regions, the visceral efferent preganglionic axon leaves the central nervous system and passes to a terminal ganglion, where it synapses with a postganglionic neuron. The role of the visceral efferent preganglionic neuron is to convey a motor impulse from the brain or spinal cord to an autonomic ganglion.

5. **Visceral efferent postganglionic neuron.** This neuron conducts a motor impulse from a visceral efferent preganglionic neuron to the visceral effector.
6. **Visceral effector.** A visceral effector is smooth muscle, cardiac muscle, or a gland.

The basic difference between a somatic reflex arc and a visceral autonomic reflex arc is that in a somatic reflex arc, only one efferent neuron is involved. In a visceral autonomic reflex arc, two efferent neurons are involved.

Visceral sensations do not always reach the cerebral cortex. Most remain at subconscious levels. Under normal conditions, you are not aware of muscular contractions of the digestive organs, heartbeat, changes in the diameter of blood vessels, and pupil dilation and constriction. Your body adjusts such visceral activities by visceral reflex arcs whose centers are in the spinal cord or lower regions of the brain. Among such centers are the cardiac, respiratory, vasomotor, swallowing, and vomiting centers in the medulla and the temperature control center in the hypothalamus. Stimuli delivered by somatic or visceral afferent neurons synapse in these centers, and the returning motor impulses conducted by visceral efferent neurons bring about an adjustment in the visceral effector without conscious recognition. The impulses are interpreted and acted on subconsciously. Some visceral sensations do give rise to conscious recognition: hunger, nausea, and fullness of the urinary bladder and rectum.

CONTROL BY HIGHER CENTERS

The autonomic nervous system is not a separate nervous system. Although little is known about the specific centers in the brain that regulate specific autonomic functions, it is known that axons from many parts of the central nervous system are connected to both the sympathetic and the parasympathetic divisions of the autonomic nervous system and thus exert considerable control over it. Autonomic centers in the cerebral cortex are connected to autonomic centers of the thalamus, for example. These, in turn, are connected to the hypothalamus. In this hierarchy of command, the thalamus sorts incoming impulses before they reach the cerebral cortex. The cerebral cortex then turns over control and integration of visceral activities to the hypothalamus. It is at the level of the hypothalamus that the major control and integration of the autonomic nervous system is exerted.

The hypothalamus receives input from areas of the nervous system concerned with emotions, visceral functions, olfaction, gustation, as well as changes in temperature, osmolarity, and levels of various substances in blood. Anatomically, the hypothalamus is connected to both the sympathetic and the parasympathetic divisions of the autonomic nervous system by axons of neurons whose dendrites and cell bodies are in various hypothalamic nuclei. The axons form tracts from the hypothalamus to sympathetic and parasympathetic nuclei in the brain stem and spinal cord through relays in the reticular formation. The posterior and lateral portions of the hypothalamus appear to control the sympathetic division. When these areas are stimulated, there is an increase in visceral activities—an increase in heart rate, a rise in blood pressure due to vasoconstriction of blood vessels, an increase in the rate and depth of respiration, dilation of the pupils, and inhibition of the digestive tract. On the other hand, the anterior and medial portions of the hypothalamus seem to control the parasympathetic division. Stimulation of these areas results in a decrease in heart rate, lowering of blood pressure, constriction of the pupils, and increased motility of the digestive tract.

Control of the autonomic nervous system by the cerebral cortex occurs primarily during emotional stress. In extreme anxiety, which can result from either conscious or subconscious stimulation in the cerebral cortex, the cortex can stimulate the hypothalamus. This, in turn, stimulates the cardiac and vasomotor centers of the medulla, which increases heart rate and blood pressure. If the cortex is stimulated by an extremely unpleasant sight, the stimulation causes vasodilation of blood vessels, a lowering of blood pressure, and fainting.

Evidence of even more direct control of visceral responses is provided by data gathered from studies of biofeedback and meditation.

BIOFEEDBACK

In the simplest terms, **biofeedback** is a process in which people get constant signals, or feedback, about visceral body functions such as blood pressure, heart rate, and muscle tension. By using special monitoring devices, they can control these visceral functions consciously.

Suppose you are connected to a monitor that informs you of your heartbeat by means of lights. A red light indicates a fast heart beat, an amber light a normal rate, and a green light a slow rate. When you see the red light flash, you know your heart is beating too fast. You have been informed of a visceral response. This is the biofeedback. According to some researchers, you can be taught to slow down the heart rate by thinking of some-

thing pleasant and thus relaxing the body. The green light flashes and your reward is a slower heart rate. In similar experiments, some individuals have learned to control heart rhythm.

Researchers estimate that between 6 and 8 percent of the American population suffers from migraine headaches. Moreover, no effective treatment has been developed that does not have significant side effects and serious risks. One approach to alleviating migraine headaches is biofeedback.* The first clue that biofeedback could be used in this way came when a patient in the voluntary control laboratory of the Menninger Foundation demonstrated that with a 5°C (10°F) rise in hand skin temperature (as a result of vasodilation and increased blood flow) she could spontaneously recover from a migraine headache.

In a study conducted at the Menninger Foundation, subjects suffering from migraine headaches received instructions in the use of a monitor that registers the skin temperature of the right index finger. Subjects were also given a typewritten sheet containing two sets of phrases. The first set was designed to help them relax the entire body. The second set was designed to bring about an increased flow of blood in the hands. The subjects practiced raising their skin temperature at home for 5 to 15 minutes a day. When skin temperature increased, the monitor emitted a high-pitched sound. In time, the monitor was abandoned.

Once the subjects learned how to vasodilate their blood vessels, the migraine headaches lessened. Since migraine headaches are believed to involve a distension of blood vessels in the head, the shunting of blood from head to hands relieved the distension and thus the pain.

Other experiments have shown that biofeedback can be applied to childbirth. Women were given monitors hooked up to their fingers and arms to measure electrical conductivity of the skin and skeletal muscle tension. Both conductivity and tension increase with nervousness and make labor difficult. Muscle tension was recorded as a sirenlike sound that became louder with nervousness. Skin conductivity was recorded as a crackling noise that also increased with nervousness. The monitors kept the women informed of their nervousness. This was the biofeedback. Having pleasant thoughts reduced the sound levels. The reward was less nervousness. The results of the study indicate that the women needed less medication during labor and labor time itself was shortened.

There is no way to determine where biofeedback will lead. Perhaps the outstanding contribution of biofeedback research has been to demonstrate that the autonomic nervous system is not autonomous. Visceral responses can

---

* Much of the following discussion of the use of biofeedback for the treatment of migraine headaches is based on information provided by Dr. Joseph D. Sargent of the Menninger Foundation, Topeka, Kansas.

be controlled. Current therapeutic applications of biofeedback include treatment of asthma, Raynaud's disease, hypertension, gastrointestinal disorders, fecal incontinence, anxiety, pain, and neuromuscular rehabilitation.

MEDITATION

Yoga, which literally means union, is defined as a higher consciousness achieved through a fully rested and relaxed body and a fully awake and relaxed mind. One widely practiced technique for achieving higher consciousness is called transcendental meditation (TM). One sits in a comfortable position with the eyes closed and concentrates on a suitable sound or thought.

Research indicates that transcendental meditation can alter physiological responses. Oxygen consumption decreases drastically along with carbon dioxide elimination.

Subjects have experienced a reduction in metabolic rate and blood pressure. Researchers have also observed a decrease in heart rate, an increase in the intensity of alpha brain waves, a sharp decrease in the amount of lactic acid in the blood, and an increase in the skin's electrical resistance. These last four responses are characteristic of a highly relaxed state of mind. Alpha waves are found in the EEGs of almost all individuals in a resting, but awake, state; they disappear during sleep.

These responses have been called an integrated response—essentially, a hypometabolic state due to inactivation of the sympathetic division of the autonomic nervous system. The response is the exact opposite of the fight-or-flight response, which is a hyperactive state of the sympathetic division. The existence of the integrated response suggests that the central nervous sysem does exert some control over the autonomic nervous system.

STUDY OUTLINE

Somatic Efferent and Autonomic Nervous Systems (p. 473)
1. The somatic efferent nervous system produces conscious movement in skeletal muscles.
2. The autonomic nervous system, or visceral efferent nervous system, regulates visceral activities, that is, activities of smooth muscle, cardiac muscle, and glands, and it usually operates without conscious control.
3. It is regulated by centers in the brain, in particular by the cerebral cortex, the hypothalamus, and the medulla oblongata.
4. A single somatic efferent neuron synapses on skeletal muscles; in the autonomic nervous system, there are two efferent neurons—one from the CNS to a ganglion and one from a ganglion to a visceral effector.
5. Somatic efferent neurons release acetylcholine (ACh) and autonomic efferent neurons release either acetylcholine (ACh) or norepinephrine (NE).

Structure of the Autonomic Nervous System (p. 474)
1. The autonomic nervous system consists of visceral efferent neurons organized into nerves, ganglia, and plexuses.
2. It is entirely motor. All autonomic axons are efferent fibers.
3. Efferent neurons are preganglionic (with myelinated axons) and postganglionic (with unmyelinated axons).
4. The autonomic system consists of two principal divisions: sympathetic (thoracolumbar) and parasympathetic (craniosacral).
5. Autonomic ganglia are classified as sympathetic trunk ganglia (on sides of spinal column), prevertebral ganglia (anterior to spinal column), and terminal ganglia (near or inside visceral effectors).

Physiology of the Autonomic Nervous System (p. 480)
1. Autonomic fibers release neurotransmitters at synapses. On the basis of the neurotransmitter produced, these fibers may be classified as cholinergic or adrenergic.
2. Cholinergic fibers release acetylcholine (ACh). Adrenergic fibers produce norepinephrine (NE).
3. Acetylcholine (ACh) interacts with nicotinic receptors on postganglionic neurons and muscarinic receptors on certain visceral effectors.
4. Norepinephrine (NE) generally interacts with alpha receptors on visceral effectors, and epinephrine generally interacts with alpha and beta receptors on visceral effectors.
5. Sympathetic responses are widespread and, in general, concerned with energy expenditure. Parasympathetic responses are restricted and are typically concerned with energy restoration and conservation.

Visceral Autonomic Reflexes (p. 483)
1. A visceral autonomic reflex adjusts the activity of a visceral effector.
2. A visceral autonomic reflex arc consists of a receptor, afferent neuron, association neuron, visceral efferent preganglionic neuron, visceral efferent postganglionic neuron, and visceral effector.

Control by Higher Centers (p. 483)
1. The hypothalamus controls and integrates the autonomic nervous system. It is connected to both the sympathetic and the parasympathetic divisions.
2. Biofeedback is a process in which people learn to monitor visceral functions and to control them consciously. It has been used to control heart rate, alleviate migraine headaches, and make childbirth easier.
3. Yoga is a higher consciousness that is achieved through a fully rested and relaxed body and a fully awake and relaxed mind.
4. Transcendental meditation (TM) produces the following physiological responses: decreased oxygen consumption and carbon dioxide elimination, reduced metabolic rate, decrease in heart rate, increase in the intensity of alpha brain waves, a sharp decrease in the amount of lactic acid in the blood, and an increase in the skin's electrical resistance.

REVIEW QUESTIONS

1. What are the principal components of the autonomic nervous system? What is its general function? Why is it called involuntary?
2. What are the principal differences between the voluntary nervous system and the autonomic nervous system?
3. Relate the role of visceral efferent fibers and visceral effectors to the autonomic nervous system.
4. Distinguish between preganglionic neurons and postganglionic neurons with respect to location and function.
5. What is an autonomic ganglion? Describe the location and function of the three types of autonomic ganglia. Define white and gray rami communicantes.
6. On what basis are the sympathetic and parasympathetic divisions of the autonomic nervous system differentiated anatomically and functionally?
7. Discuss the distinction between cholinergic and adrenergic fibers of the autonomic nervous system.
8. How is acetylcholine (ACh) related to nicotinic and muscarinic receptors?
9. How are alpha and beta receptors related to norepinephrine (NE) and epinephrine?
10. Give examples of the antagonistic effects of the sympathetic and parasympathetic divisions of the autonomic nervous system.
11. Summarize the principal functional differences between the voluntary nervous system and the autonomic nervous system.
12. Give the *sympathetic response* in a fear situation for each of the following body parts: hair follicles, sweat glands, iris of eye, lungs, spleen, adrenal glands, kidneys, urinary bladder, intestines, gallbladder, liver, heart, blood vessels of the viscera, skeletal muscles, and skin.
13. Define a visceral autonomic reflex and give three examples.
14. Describe a complete visceral autonomic reflex in proper sequence.
15. Describe how the hypothalamus controls and integrates the autonomic nervous system.
16. Define biofeedback. Explain how it could be useful.
17. What is transcendental meditation (TM)? How is the integrated response related to the autonomic nervous system?

20

Sensory Structures

Student Objectives

Define a sensation and list the four prerequisites necessary for its transmission.

Define projection, adaptation, afterimage, and modality as characteristics of sensations.

Classify receptors on the basis of location, stimulus detected, and simplicity or complexity.

List the location and function of the receptors for tactile sensations (touch, pressure, vibration), thermoreceptive sensations (heat and cold), and pain.

Distinguish between somatic, visceral, referred, and phantom pain and describe the various methods used to relieve pain.

Identify the proprioceptive receptors and indicate their functions.

Describe the composition of the somatosensory cortex.

Discuss the origin, neuronal components, and destination of the posterior column and spinothalamic pathways.

Describe the composition of the motor cortex.

Compare the course of the pyramidal and extrapyramidal motor pathways.

Locate the receptors for olfaction and describe the neural pathway for smell.

Identify the gustatory receptors and describe the neural pathway for taste.

Describe the structure of the accessory structures of the eye.

List the structural divisions of the eye.

Identify the afferent pathway of light impulses to the brain.

Define the anatomical subdivisions of the ear.

List the principal events in hearing.

Identify the receptor organs for static and dynamic equilibrium.

Describe the development of the eye and ear.

Contrast the causes and symptoms of cataracts, glaucoma, conjunctivitis, trachoma, deafness, labyrinthine disease, Ménière's syndrome, otitis media, and motion sickness.

Define key medical terms associated with sensory structures.

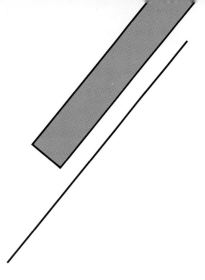

SENSATIONS

Your ability to sense stimuli is vital to your survival. If pain could not be sensed, burns would be common. An inflamed appendix or stomach ulcer would progress unnoticed. A lack of sight would increase the risk of injury from unseen obstacles, a loss of smell would allow a harmful gas to be inhaled, a loss of hearing would prevent recognition of automobile horns, and a lack of taste would allow toxic substances to be ingested. In short, if you could not "sense" your environment and make the necessary homeostatic adjustments, you could not survive on your own.

DEFINITION

In its broadest context, **sensation** refers to a state of awareness of external or internal conditions of the body. **Perception** refers to the conscious registration of a sensory stimulus. For a sensation to occur, four prerequisites must be fulfilled.

1. A **stimulus,** or change in the environment, capable of initiating a response by the nervous system must be present.

2. A **receptor** or **sense organ** must pick up the stimulus and convert it to a nerve impulse. A sense receptor or sense organ may be viewed as specialized nervous tissue that is extremely sensitive to internal or external conditions.

3. The impulse must be **conducted** along a neural pathway from the receptor or sense organ to the brain.

4. A region of the brain must **translate** the impulse into a sensation.

Receptors are capable of converting a specific stimulus into a nerve impulse. The stimulus may be light, heat, pressure, mechanical, energy, or chemical energy. Each stimulus is capable of causing the membrane of the receptor to depolarize. This depolarization is called a **generator (receptor) potential.**

The generator potential is a graded response; within limits, the magnitude increases with stimulus strength and frequency. When the generator potential reaches the threshold level, it initiates an action potential (nerve impulse). Once initiated, the action potential is propagated along the nerve fiber. Whereas a generator potential is a local, graded response, an action potential obeys the all-or-none principle. The function of a generator potential is to initiate an action potential by transducing (converting) a stimulus into a nerve impulse.

A receptor may be quite simple. It may consist of the dendrites of a single neuron in the skin that are sensitive to pain stimuli. Or it may be contained in a complex organ such as the eye. Regardless of complexity, all sense receptors contain the dendrites of sensory neurons. The dendrites occur either alone or in close association with specialized cells of other tissues.

Receptors are at the same time very excitable and very specialized. Except for pain receptors, each has a low threshold of response to its specific stimulus and a high threshold to all others.

Once a stimulus is received by a receptor and converted into an impulse, the impulse is conducted along an afferent pathway that enters either the spinal cord or the brain. Many sensory impulses are conducted to the sensory areas of the cerebral cortex. It is in this region that stimuli produce conscious sensations. Sensory impulses that terminate in the spinal cord or brain stem can initiate motor activities but typically do not produce conscious sensations. The thalamus detects pain sensations but cannot distinguish the intensity or location from which they arise. This is a function of the cerebrum.

CHARACTERISTICS

Most conscious sensations or perceptions occur in the cortical regions of the brain. In other words, you see, hear, and feel in the brain. You seem to see with your eyes, hear with your ears, and feel pain in an injured part of your body only because the cortex interprets the sensation as coming from the stimulated sense receptor. The term **projection** describes this process by which the brain refers sensations to their point of stimulation.

A second characteristic of many sensations is **adapta-**

tion. The perception of a sensation may disappear even though a stimulus is still being applied. When you get into a tub of hot water, you might feel a burning sensation. But soon the sensation decreases to one of comfortable warmth, even though the stimulus (hot water) is still present. Other examples of adaptation include placing a ring on your finger, putting on your shoes or hat, and sitting on a chair. Initially, you are conscious of the sensations involved, but they are lost soon thereafter.

Sensations may also be characterized by **afterimages,** that is, some sensations persist even though the stimulus has been removed. This phenomenon is the reverse of adaptation. One common example of afterimage occurs when you look at a bright light and then look away or close your eyes. You still see the light for several seconds or minutes afterward.

Another characteristic of sensations is **modality:** the specific sensation felt. The sensation may be one of pain, pressure, touch, body position, equilibrium, hearing, vision, smell, or taste. In other words, the distinct property by which one sensation may be distinguished from another is its modality.

CLASSIFICATION OF RECEPTORS

Location

One convenient method of classifying receptors is by their location. **Exteroceptors** (eks'-ter-ō-SEP-tors) provide information about the external environment. They are sensitive to stimuli outside the body and transmit sensations of hearing, sight, smell, taste, touch, pressure, temperature, and pain. Exteroceptors are located near the surface of the body.

Visceroceptors (vis'-er-ō-SEP-tors), or **enteroceptors,** provide information about the internal environment. These sensations arise from within the body and may be felt as pain, pressure, fatigue, hunger, thirst, and nausea. Visceroceptors are located in blood vessels and viscera.

Proprioceptors (prō'-prē-ō-SEP-tors) provide information about body position and movement. Such sensations give us information about muscle tension, the position and tension of our joints, and equilibrium. These receptors are located in muscles, tendons, joints, and the internal ear.

Stimulus Detected

Another method of classifying receptors is by the type of stimuli they detect. **Mechanoreceptors** detect mechanical deformation of the receptor itself or in adjacent cells. Stimuli so detected include those related to touch, pressure, vibration, proprioception, hearing, equilibrium, and blood pressure. **Thermoreceptors** detect changes in temperature. **Nociceptors** detect pain, usually as a result of physical or chemical damage to tissues. **Electromagnetic**

(photo) receptors detect light on the retina of the eye. **Chemoreceptors** detect taste in the mouth, smell in the nose, and chemicals in body fluids, such as oxygen, carbon dioxide, water, and glucose.

Simplicity or Complexity

As will be described shortly, receptors may also be classified according to the simplicity or complexity of their structure and the neural pathway involved. **Simple receptors** and neural pathways are associated with **general senses.** The receptors for general sensations are numerous and widespread. Examples include cutaneous sensations such as touch, pressure, vibration, heat, cold, and pain. **Complex receptors** and neural pathways are associated with **special senses.** The receptors for each special sense are found in only one or two specific areas of the body. Among the special senses are smell, taste, sight, and hearing.

GENERAL SENSES

CUTANEOUS SENSATIONS

Cutaneous sensations include tactile sensations (touch, pressure, vibration), thermoreceptive sensations (cold and heat), and pain. The receptors for these sensations are in the skin, connective tissue, and the ends of the gastrointestinal tract.

The cutaneous receptors are distributed over the body surface in such a way that certain parts of the body are densely populated with receptors and other parts contain only a few. Areas of the body that have few cutaneous receptors are insensitive; those containing many are very sensitive. The following order for these receptors, from greatest sensitivity to least, has been established: tip of tongue, tip of finger, side of nose, back of hand, and back of neck.

Cutaneous receptors have simple structures. They consist of the dendrites of sensory neurons that may or may not be enclosed in a capsule of epithelial or connective tissue. Impulses generated by cutaneous receptors pass along somatic afferent neurons in spinal and cranial nerves, through the thalamus, to the general sensory area of the parietal lobe of the cortex.

Tactile Sensations

Even though the **tactile sensations** are divided into separate sensations of touch, pressure, and vibration, they are all detected by the same types of receptors, that is, receptors subject to deformation.

• *Touch*   **Touch sensations** generally result from stimulation of tactile receptors in the skin or tissues immediately beneath the skin. The term *discriminative touch* re-

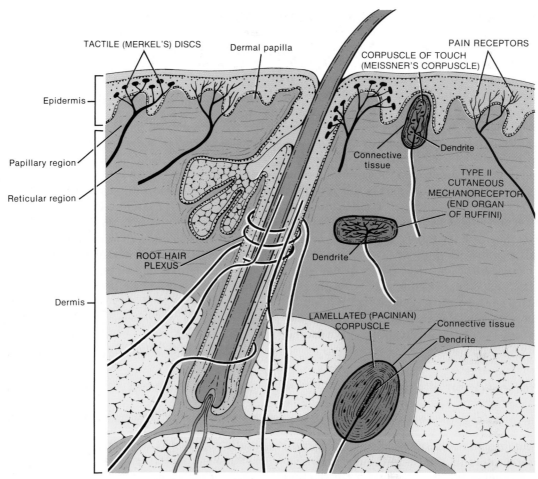

FIGURE 20-1 Structure and location of cutaneous receptors.

fers to the ability to recognize exactly what point of the body is touched. *Light touch* refers to the ability to perceive that something has touched the skin, although its exact location, shape, size, or texture cannot be determined.

Tactile receptors for touch include root hair plexuses, free nerve endings, tactile discs, corpuscles of touch, and type II cutaneous mechanoreceptors (Figure 20-1). **Root hair plexuses** are dendrites arranged in networks around the roots of hairs. They are not surrounded by supportive or protective structures. If a hair shaft is moved, the dendrites are stimulated. Root hair plexuses detect movements mainly on the surface of the body.

Other receptors that are not surrounded by supportive or protective structures are called **free (naked) nerve endings.** Free nerve endings are found everywhere in the skin and many other tissues.

Tactile, or **Merkel's** (MER-kelz), **discs** are receptors for touch that consist of disclike formations of dendrites attached to deeper layers of epidermal cells. They are distributed in many of the same locations as corpuscles of touch.

Corpuscles of touch, or **Meissner's** (MĪS-nerz) **corpus-**

cles, are egg-shaped receptors containing a mass of dendrites enclosed by connective tissue. They are located in the dermal papillae of the skin and are most numerous in the fingertips, palms of the hands, and soles of the feet. They are also abundant in the eyelids, tip of the tongue, lips, nipples, clitoris, and tip of penis.

Type II cutaneous mechanoreceptors, or **end organs of Ruffini,** are embedded deeply in the dermis and in deeper tissues of the body. They detect heavy and continuous touch sensations.

● *Pressure*   **Pressure sensations** generally result from stimulation of tactile receptors in deeper tissues and are longer lasting and have less variation in intensity than touch sensations. Moreover, pressure is felt over a larger area than touch.

Pressure receptors are free nerve endings, type II cutaneous mechanoreceptors, and lamellated corpuscles. **Lamellated,** or **Pacinian** (pa-SIN-ē-an), **corpuscles** (Figure 20-1) are oval structures composed of a capsule resembling an onion and consist of connective tissue layers enclosing dendrites. Lamellated corpuscles are located in the subcutaneous tissue under the skin, the deep subcutaneous tissues that lie under mucous membranes, in serous membranes, around joints and tendons, in the perimysium of muscles, in the mammary glands, in the external genitalia of both sexes, and in certain viscera.

● *Vibration* **Vibration sensations** result from rapidly repetitive sensory signals from tactile receptors.

The receptors for vibration sensations are corpuscles of touch and lamellated corpuscles. Whereas corpuscles of touch detect low-frequency vibration, lamellated corpuscles detect higher-frequency vibration.

Thermoreceptive Sensations

The **thermoreceptive sensations** are heat and cold. The **thermoreceptive receptors** are not known, but might be free nerve endings.

Pain Sensations

The receptors for **pain,** called **nociceptors,** are simply the branching ends of the dendrites of certain sensory neurons (Figure 20-1). Pain receptors are found in practically every tissue of the body. They may respond to any type of stimulus. When stimuli for other sensations, such as touch, pressure, heat, and cold, reach a certain threshold, they stimulate the sensation of pain as well. Excessive stimulation of a sense organ causes pain. Additional stimuli for pain receptors include excessive distension or dilation of a structure, prolonged muscular contractions, muscle spasms, inadequate blood flow to an organ, or the presence of certain chemical substances. Pain receptors, because of their sensitivity to all stimuli, perform a protective function by identifying changes that may endanger the body. Pain receptors adapt only slightly or not at all. Adaptation is the decrease or disappearance of the perception of a sensation even though the stimulus is still present. If there were adaptation to pain, it would cease to be sensed and irreparable damage could result.

Sensory impulses for pain are conducted to the central nervous system along spinal and cranial nerves. The lateral spinothalamic tracts of the spinal cord relay impulses to the thalamus. From here the impulses may be relayed to the postcentral gyrus of the parietal lobe. Recognition of the kind and intensity of most pain is ultimately localized in the cerebral cortex. Some awareness of pain occurs at subcortical levels.

Pain may be divided into two types: somatic and visceral. **Somatic pain** arises from stimulation of receptors in the skin, in which case it is called *superficial somatic pain,* or from stimulation of receptors in skeletal muscles, joints, tendons, and fascia, then called *deep somatic pain.* **Visceral pain** results from stimulation of receptors in the viscera.

The ability of the cerebral cortex to locate the origin of pain is related to past experience. In most instances of somatic pain and in some instances of visceral pain, the cortex accurately projects the pain back to the stimulated area. If you burn your finger, you feel the pain in your finger. If the lining of your pleural cavity is inflamed, you experience pain there. In most instances of visceral pain, however, the sensation is not projected back to the point of stimulation. Rather, the pain may be felt in or just under the skin that overlies the stimulated organ. The pain may also be felt in a surface area far from the stimulated organ. This phenomenon is called **referred pain.** In general, the area to which the pain is referred and the visceral organ involved receive their innervation from the same segment of the spinal cord. Consider the following example. Afferent fibers from the heart as well as from the skin over the heart and along the medial aspect of the left upper extremity enter spinal cord segments T1 to T4. Thus the pain of a heart attack is typically felt in the skin over the heart and along the left arm. Figure 20-2 illustrates cutaneous regions to which visceral pain may be referred.

CLINICAL APPLICATION

A kind of pain frequently experienced by patients who have had a limb amputated is called **phantom pain.** They still experience pain or other sensations in the extremity as if the limb were still there. An explanation for this phenomenon is that the remaining proximal portions of the sensory nerves that previously received impulses from the limb are being stimulated by the trauma of the amputation. Stimuli from these nerves are interpreted by the brain as coming from the nonexistent (phantom) limb.

Pain sensations may be controlled by interrupting the pain impulse between the receptors and the interpretation centers of the brain. This may be done chemically, surgically, or by other means. Most pain sensations respond to pain-reducing drugs, which, in general, act to inhibit nerve impulse conduction at synapses.

Occasionally, however, pain may be controlled only by surgery. The purpose of surgical treatment is to interrupt the pain impulse somewhere between the receptors and the interpretation centers of the brain by severing the sensory nerve, its spinal root, or certain tracts in the spinal cord or brain. *Sympathectomy* is excision of portions of the neural tissue from the autonomic nervous system; *cordotomy* is severing a spinal cord tract, usually the lateral spinothalamic; *rhizotomy* is the cutting of sensory nerve roots; *prefrontal lobotomy* is the destruction of the tracts that connect the thalamus with the prefrontal and frontal lobes of the cerebral cortex. In each instance, the pathway for pain is severed so that pain impulses are no longer conducted to the cortex.

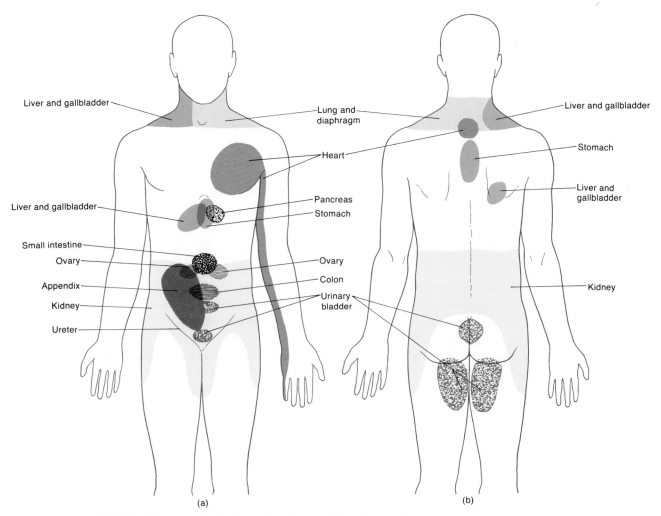

FIGURE 20-2 Referred pain. The colored parts of the diagrams indicate cutaneous areas to which visceral pain is referred. (a) Anterior view. (b) Posterior view.

Another method of inhibiting pain impulses is **acu-puncture** (*acus* = needle; *pungere* = sting). Needles are inserted through selected areas of the skin and then twirled by the acupuncturist or by a mechanical device. After 20 to 30 minutes, pain is deadened for 6 to 8 hours. The location of needle insertion depends on the part of the body the acupuncturist wishes to anesthetize. To pull a tooth, a needle is inserted in the web between thumb and index finger. For a tonsillectomy, a needle is inserted approximately 5 cm (2 in) above the wrist. For removal of a lung, a needle is placed in the forearm midway between wrist and elbow.

There is no satisfactory explanation of how acupuncture works. According to the "gate control" theory, twirling the acupuncture needle stimulates two sets of nerves that eventually enter the spinal cord and synapse with the same association neurons. One nerve contains nerve fibers of small diameter (C fibers) for pain, and the other nerve contains nerve fibers of larger diameter (A fibers) for touch. The impulse passing along the touch-conduct-

ing neurons is faster than that passing along the pain-conducting neurons—fibers with large diameters conduct impulses faster than those with small diameters. Because the touch impulse reaches the posterior gray horn of the cord first, it has priority over the pain impulse. It thus "closes the gate" to the brain before the pain impulse reaches the cord. Since the pain impulse does not pass to the brain, no pain is felt.

PROPRIOCEPTIVE SENSATIONS

An awareness of the activities of muscles, tendons, and joints and equilibrium is provided by the **proprioceptive,** or **kinesthetic** (kin'-es-THET-ik), **sense.** It informs us of the degree to which muscles are contracted, the amount of tension created in the tendons, the change of position of a joint, and the orientation of the head relative to the ground and in response to movements (equilibrium). The proprioceptive sense enables us to recognize the location and rate of movement of one body part in relation

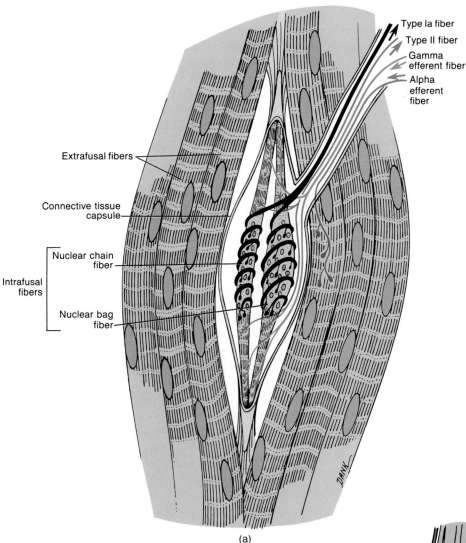

Type Ia fiber
Type II fiber
Gamma efferent fiber
Alpha efferent fiber

Extrafusal fibers

Connective tissue capsule

Nuclear chain fiber

Intrafusal fibers

Nuclear bag fiber

(a)

FIGURE 20-3 Proprioceptive receptors. (a) Muscle spindle. (b) Tendon organ.

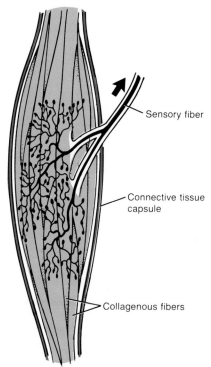

Sensory fiber

Connective tissue capsule

Collagenous fibers

(b)

to others. It also allows us to estimate weight and determine the muscular work necessary to perform a task. With the proprioceptive sense, we can judge the position and movements of our limbs without using our eyes when we walk, type, or dress in the dark.

Receptors

Proprioceptive receptors are located in skeletal muscles, tendons in and around synovial joints, and the internal ear.

● *Muscle Spindles*   **Muscle spindles** are delicate proprioceptive receptors interspersed among skeletal muscle fibers and oriented parallel to the fibers (Figure 20-3a).

The ends of the spindles are anchored to the endomysium and perimysium. Muscle spindles consist of 3 to 10 specialized muscle cells called *intrafusal fibers,* which are partially enclosed in a connective tissue capsule that is filled with lymph. The spindles are surrounded by skeletal muscle fibers of the muscle called *extrafusal fibers.* The central region of each intrafusal fiber has few or no actin and myosin filaments and an accumulation of nuclei. In some intrafusal fibers the nuclei bunch at the center (*nuclear bag fibers*); in others, the nuclei form a chain at the center (*nuclear chain fibers*). The central region of the intrafusal fibers cannot contract and represents the sensory receptor area for a spindle.

Although the central receptor area cannot contract because it lacks myofilaments, it does contain two types of sensory fibers. A large sensory fiber, called a *type Ia fiber,* innervates the exact center of the intrafusal fibers. The branches of the Ia fiber, called *primary (annulospiral) endings,* wrap around the center of the intrafusal fibers. When the central part of the spindle is stretched, the primary endings are stimulated and send impulses to the spinal cord at exceedingly great velocities. The central receptor area is also innervated by two sensory fibers called *type II fibers.* Their branches, known as *secondary (flower spray) endings,* are located on either side of the primary ending. Secondary endings are also stimulated when the central part of the spindle is stretched and they too send impulses to the spinal cord.

The ends of the intrafusal fibers contain actin and myosin myofilaments and represent the contractile portions of the fibers. The ends of the fibers contract when stimulated by *gamma efferent neurons.* These neurons are small motor neurons located in the anterior gray horn of the spinal cord. The fibers of the gamma efferent neurons terminate as motor end plates on the ends of the intrafusal fibers. Extrafusal fibers are innervated by large motor neurons called *alpha efferent neurons.* These neurons are also located in the anterior gray horn of the spinal cord, near gamma efferent neurons.

Muscle spindles are stimulated in response to both sudden and maintained stretches on the central areas of the intrafusal fibers. The muscle spindles monitor changes in the length of a skeletal muscle by responding to the rate and degree of change in length. This information is relayed to the central nervous system to assist in the coordination and efficiency of muscle contraction. The neural pathway involved in transmitting the information to the central nervous system, processing by the central nervous system, and eliciting a motor response is shown in Figure 20-5.

● *Tendon Organs* **Tendon organs (Golgi tendon organs)** are proprioceptive receptors found at the junction of a tendon with a muscle. They help protect tendons and their associated muscles from damage resulting from ex-

cessive tension. Each consists of a thin capsule of connective tissue that encloses a few collagenous fibers (Figure 20-3b). The capsule is penetrated by one or more sensory neurons whose terminal branches entwine among and around the collagenous fibers. When tension is applied to a tendon, tendon organs are stimulated and the information is relayed to the central nervous system. The neural pathway involved in transmitting the information to the central nervous system, processing by the central nervous system, and eliciting a motor response is shown in Figure 20-5.

● *Joint Kinesthetic Receptors* There are several types of **joint kinesthetic receptors** within and around the articular capsules of synovial joints. Encapsulated receptors, similar to type II cutaneous mechanoreceptors (end organs of Ruffini) are present in the capsules of joints and respond to pressure. Small lamellated (Pacinian) corpuscles in the connective tissue outside articular capsules are receptors that respond to acceleration and deceleration. Articular ligaments contain receptors similar to tendon organs that mediate reflex inhibition of the adjacent muscles when excessive strain is placed on the joint.

● *Maculae and Cristae* The proprioceptors in the internal ear are the macula of the saccule and the utricle and cristae in the semicircular ducts. Their function in equilibrium is discussed later in the chapter.

Proprioceptors adapt only slightly. This feature is advantageous since the brain must be apprised of the status of different parts of the body at all times so that adjustments can be made to ensure coordination.

The afferent pathway for muscle sense consists of impulses generated by proprioceptors via cranial and spinal nerves to the central nervous system. Impulses for conscious proprioception pass along ascending tracts in the cord, where they are relayed to the thalamus and cerebral cortex. The sensation is registered in the general sensory area in the parietal lobe of the cerebral cortex posterior to the central sulcus. Proprioceptive impulses that have resulted in reflex action pass to the cerebellum along spinocerebellar tracts.

LEVELS OF SENSATION

As we have said, a receptor converts a stimulus into a nerve impulse, and only after that impulse has been conducted to a region of the spinal cord or brain can it be translated into a sensation. The nature of the sensation and the type of reaction generated vary with the level of the central nervous system at which the sensation is translated.

Sensory fibers terminating in the spinal cord can generate spinal reflexes without immediate action by the brain. Sensory fibers terminating in the lower brain stem bring about far more complex motor reactions than simple spinal reflexes. When sensory impulses reach the lower brain stem, they cause subconscious motor reactions. Sensory impulses that reach the thalamus can be localized crudely in the body. At the thalamic levels sensations are sorted by modality, that is, identified as the *specific* sensation of touch, pressure, pain, position, hearing, vision, smell, or taste. When sensory information reaches the cerebral cortex, we experience precise localization. It is at this level that memories of previous sensory information are stored and the perception of sensation occurs on the basis of past experience.

SENSORY PATHWAYS

SOMATOSENSORY CORTEX

Sensory information from receptors on one side of the body crosses over to the opposite side in the spinal cord or brain stem and then to the **somatosensory cortex (general sensory,** or **somesthetic, area)** of the cerebral cortex where conscious sensations are produced (see Figure 18-11). Areas of the somatosensory cortex have been mapped out which represent the termination of sensory information from all parts of the body. Figure 20-4 shows the location and areas of representation of the somatosensory cortex of the right cerebral hemisphere. The left cerebral hemisphere has a duplicate somatosensory cortex.

Note that some parts of the body are represented by large areas in the somatosensory cortex. These include the lips, face, and thumb. Other parts of the body, such as the trunk and lower extremities, are represented by relatively small areas. The relative sizes of the somatosensory cortex are directly proportional to the number of specialized sensory receptors in each respective part of the body. Thus, there are numerous receptors in the skin of the lips but relatively few in the skin of the trunk. Essentially, the size of the area for a particular part of the body is determined by the functional importance of the part and its need for sensitivity.

Let us now examine how sensory information is transmitted from receptors to the central nervous system. You will find it helpful to review the principal ascending and descending tracts of the spinal cord (see Exhibit 17-1 and Figure 17-4). Sensory information transmitted from the spinal cord to the brain is conducted along two general pathways: the posterior column pathway and the spinothalamic pathway.

POSTERIOR COLUMN PATHWAY

In the **posterior column pathway (fasciculus gracilis** and **fasciculus cuneatus)** to the cerebral cortex there are three separate sensory neurons (Figure 20-5). The **first-order**

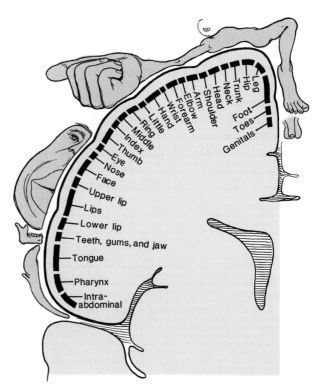

FIGURE 20-4 Somatosensory cortex of the right cerebral hemisphere.

neuron connects the receptor with the spinal cord and medulla on the same side of the body. The cell body of the first-order neuron is in the posterior root ganglion of a spinal or cranial nerve. The first-order neuron synapses with a **second-order neuron.** The second-order neuron passes from the medulla upward to the thalamus. The cell body of the second-order neuron is located in the nucleus cuneatus or gracilis of the medulla. Before passing into the thalamus, the second-order neuron crosses to the opposite side of the medulla and enters the medial lemniscus, a projection tract that terminates at the thalamus. In the thalamus, the second-order neuron synapses with a **third-order neuron.** The third-order neuron terminates in the somesthetic sensory area of the cerebral cortex.

The posterior column pathway conducts impulses related to proprioception, discriminative touch, two-point discrimination, and vibrations.

SPINOTHALAMIC PATHWAY

The **spinothalamic pathway** is also composed of three orders of sensory neurons. The first-order neuron connects a receptor of the neck, trunk, and extremities with the spinal cord. The cell body of the first-order neuron is in the posterior root ganglion also. The first-order neuron synapses with the second-order neuron, which has its cell body in the posterior gray horn of the spinal cord. The fiber of the second-order neuron crosses to the opposite side of the spinal cord and passes upward to the brain stem in the **lateral spinothalamic tract** (pain and temperature) or **ventral spinothalamic tract** (light touch and pressure). The fibers from the second-order neuron terminate in the thalamus. There the second-order neuron synapses with a third-order neuron. The third-order neuron terminates in the somesthetic sensory area of the cerebral cortex. The spinothalamic pathway conveys sensory impulses for pain and temperature as well as crude touch and pressure.

CEREBELLAR TRACTS

Both the **posterior spinocerebellar tract** and **anterior spinocerebellar tract** are concerned with conveying impulses concerned with subconscious proprioception and thus assume a role in posture and muscle tone (see Exhibit 17-1).

MOTOR PATHWAYS

After receiving and interpreting sensations, the central nervous system generates impulses to direct responses to that sensory input. We have already considered somatic reflex arcs and visceral efferent pathways. Now we will look at the transmission of motor impulses that result in the movement of skeletal muscles.

The principal parts of the brain concerned with skeletal muscle control are the cerebral motor cortex, basal ganglia, reticular formation, and cerebellum. The motor cortex assumes the major role for controlling precise, discrete muscular movements. The basal ganglia largely integrate semivoluntary movements like walking, swimming, and laughing. The cerebellum, although not a control center, assists the motor cortex and basal ganglia by making body movements smooth and coordinated. Voluntary motor impulses are conveyed from the brain through the spinal cord by way of two major pathways: the pyramidal pathways and the extrapyramidal pathways.

MOTOR CORTEX

Just as the somatosensory cortex has been mapped to indicate the termination of sensory information from all parts of the body, the **motor cortex** has been mapped to indicate which groups of muscles are controlled by its specific areas (Figure 20-6). The right cerebral hemisphere has a duplicate of the motor cortex in the left cerebral hemisphere. Note that the different muscle groups are not represented equally in the motor cortex. In general, the degree of representation is proportional to the preciseness of movement required of a particular

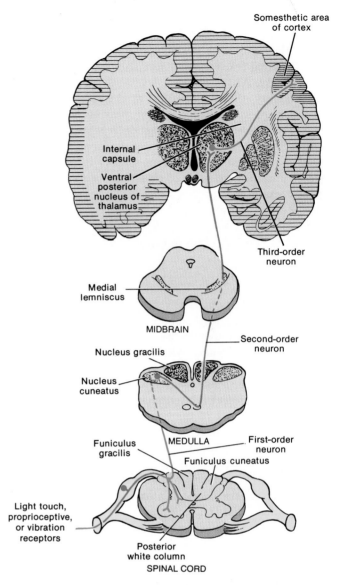

FIGURE 20-5 Sensory pathway for light touch, proprioception, and vibration—the posterior column pathway.

part of the body. For example, the thumb, fingers, lips, tongue, and vocal cords have large representations. The trunk has a relatively small representation. By comparing Figures 20-4 and 20-6, you will see that somatosensory and motor representations are not identical for the same part of the body.

PYRAMIDAL PATHWAYS

Voluntary motor impulses are conveyed from the motor areas of the brain to somatic efferent neurons leading to skeletal muscles via the **pyramidal** (pi-RAM-i-dal) **pathways** (Figure 20-7). Most pyramidal fibers originate from cell bodies in the precentral gyrus. They descend through the internal capsule of the cerebrum and cross

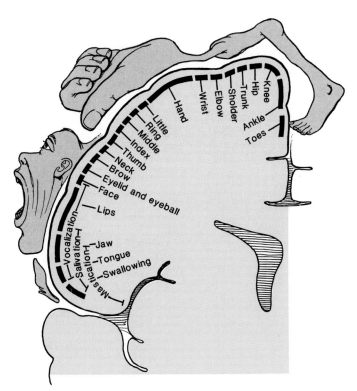

FIGURE 20-6 Motor cortex of the right cerebral hemisphere.

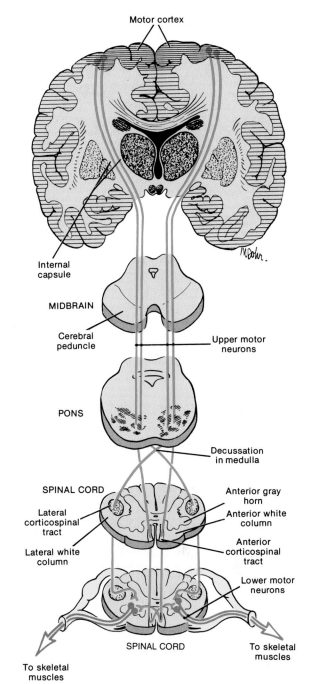

FIGURE 20-7 Pyramidal pathways.

to the opposite side of the brain. They terminate in nuclei of cranial nerves that innervate voluntary muscles or in the anterior gray horn of the spinal cord. A short connecting neuron probably completes the connection of the pyramidal fibers with the motor neurons that activate voluntary muscles.

The pathways over which the impulses travel from the motor cortex to skeletal muscles have two components: **upper motor neurons (pyramidal fibers)** in the brain and **lower motor neurons (peripheral fibers)** in the spinal cord.

The various tracts of the pyramidal system convey impulses from the cortex that result in precise muscular movements. These tracts include the **lateral corticospinal, anterior corticospinal,** and **corticobulbar** (see Exhibit 17-1).

EXTRAPYRAMIDAL PATHWAYS

The **extrapyramidal pathways** include all descending tracts other than the pyramidal tracts. Generally, these include tracts that begin in the basal ganglia and reticular formation. The main extrapyramidal tracts are the **rubrospinal** (muscle tone and posture), **tectospinal** (head movements in response to visual stimuli), and **vestibulospinal** (muscle tone in response to head movements that affect equilibrium); see Exhibit 17-1.

Only one motor neuron carries the impulse from the cerebral cortex to the cranial nerve nuclei or spinal cord: an **upper motor neuron.** Only one motor neuron in the pathway actually terminates in a skeletal muscle: the **lower motor neuron.** This neuron, a somatic efferent neuron, always extends from the central nervous system to the skeletal muscle. Since it is the final transmitting neuron in the pathway, it is also called the **final common pathways.**

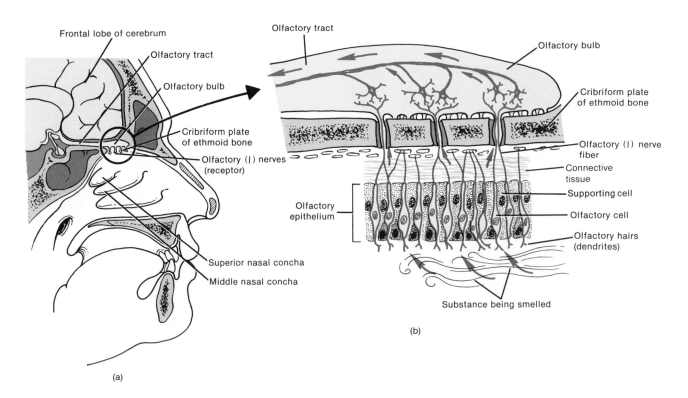

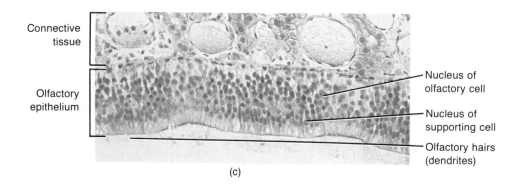

FIGURE 20-8 Olfactory receptors. (a) Location of receptors in nasal cavity. (b) Enlarged aspect of olfactory receptors. (c) Photomicrograph of the olfactory mucosa at a magnification of 300×. (© 1985 by Michael H. Ross. Used by permission.)

CLINICAL APPLICATION

The lower motor and upper motor neurons are important clinically. If the lower motor neuron is damaged or diseased, there is neither voluntary nor reflex action of the muscle it innervates, and the muscle remains in a relaxed state, a condition called **flaccid paralysis.** Injury or disease of upper motor neurons in a motor pathway is characterized by varying degrees of continued contraction of the muscle, a condition called **spasticity,** and exaggerated reflexes. Another characteristic is the **Babinski sign,** which is dorsiflexion of the great toe accompanied by fanning of the lateral toes in response to stroking the plantar surface along the outer border of the foot. As we have seen, this response is normal only in infants; the normal adult response is plantar flexion of the toes.

SPECIAL SENSES

The special senses—smell, taste, sight, hearing, and equilibrium—have receptor organs that are structurally more complex than receptors for general sensations. The sense of smell is the least specialized, as opposed to the sense of sight, which is the most specialized. Like the general senses, however, the special senses allow us to detect changes in our environment.

OLFACTORY SENSATIONS

Structure of Receptors

The receptors for the **olfactory** (ol-FAK-tō-rē) **sense** are located in the nasal epithelium in the superior portion of the nasal cavity on either side of the nasal septum (Figure 20-8). The nasal epithelium consists of two principal kinds of cells: supporting and olfactory. The **supporting cells** are columnar epithelial cells of the mucous membrane lining the nose. Olfactory glands in the mucosa keep the mucous membrane moist. The **olfactory cells** are bipolar neurons whose cell bodies lie between the supporting cells. The distal (free) end of each olfactory cell contains six to eight dendrites, called **olfactory hairs.** The hairs are believed to react to odors in the air and then to stimulate the olfactory cells, thus initiating the olfactory pathway.

Olfactory Pathway

The unmyelinated axons of the olfactory cells unite to form the **olfactory (I) nerves,** which pass through foramina in the cribriform plate of the ethmoid bone (see Figure 18-4a). The olfactory (I) nerves terminate in paired masses of gray matter called the **olfactory bulbs.** The olfactory bulbs lie beneath the frontal lobes of the cerebrum on either side of the crista galli of the ethmoid bone. The first synapse of the olfactory neural pathway occurs in the olfactory bulbs between the axons of the olfactory (I) nerves and the dendrites of neurons inside the olfactory bulbs. Axons of these neurons run posteriorly to form the **olfactory tract.** From here, impulses are conveyed to the primary olfactory area of the cerebral cortex. In the cerebral cortex, the impulses are interpreted as odor and give rise to the sensation of smell. Impulses related to olfaction do not pass through the thalamus.

Both the supporting cells of the nasal epithelium and tear glands are innervated by branches of the trigeminal (V) nerve. The nerve receives stimuli of pain, cold, heat, tickling, and pressure. Olfactory stimuli such as pepper, ammonia, and chloroform are irritating and may cause tearing because they stimulate the lacrimal and nasal mucosal receptors of the trigeminal (V) nerve as well as the olfactory neurons.

GUSTATORY SENSATIONS

Structure of Receptors

The receptors for **gustatory** (GUS-ta-tō'-rē) **sensations,** or sensations of taste, are located in the taste buds (Figure 20-9). Taste buds are most numerous on the tongue, but they are also found on the soft palate and in the throat. The **taste buds** are oval bodies consisting of two kinds of cells. The **supporting cells** are a specialized epithelium that forms a capsule. Inside each capsule are 4 to 20 **gustatory cells.** Each gustatory cell contains a hairlike process (**gustatory hair**) that projects to the external surface through an opening in the taste bud called the **taste pore.** Gustatory cells make contact with taste stimuli through the taste pore.

Taste buds are found in some connective tissue elevations on the tongue called **papillae** (see Figure 23-5a). The papillae give the upper surface of the tongue its rough appearance. **Circumvallate** (ser-kum-VAL-āt), or **vallate, papillae,** the largest type, are circular and form an inverted V-shaped row at the posterior portion of the tongue. **Fungiform** (FUN-ji-form; meaning mushroom-shaped) **papillae** are knoblike elevations found primarily on the tip and sides of the tongue. All circumvallate and most fungiform papillae contain taste buds. **Filiform** (FIL-i-form) **papillae** are pointed threadlike structures that cover the anterior two-thirds of the tongue.

Despite the many substances we seem to taste, there are basically only four primary taste sensations: sour, salt, bitter, and sweet. All other "tastes," such as chocolate, pepper, and coffee, are combinations of these four modified by accompanying olfactory sensations.

CLINICAL APPLICATION

Persons with colds or allergies sometimes complain that they cannot taste their food. Although their taste sensations may be operating normally, their olfactory sensations are not. This illustrates that much of **what we think of as taste is actually smell.** Odors from foods pass upward into the nasopharynx and stimulate the olfactory system. In fact, a given concentration of a substance will stimulate the olfactory system thousands of times more than it stimulates the gustatory system.

Each of the four primary tastes is caused by a different response to different chemicals. Certain regions of the tongue react more strongly than others to certain taste sensations. Although the tip of the tongue reacts to all four primary taste sensations, it is highly sensitive to sweet and salty substances. The posterior portion of the tongue

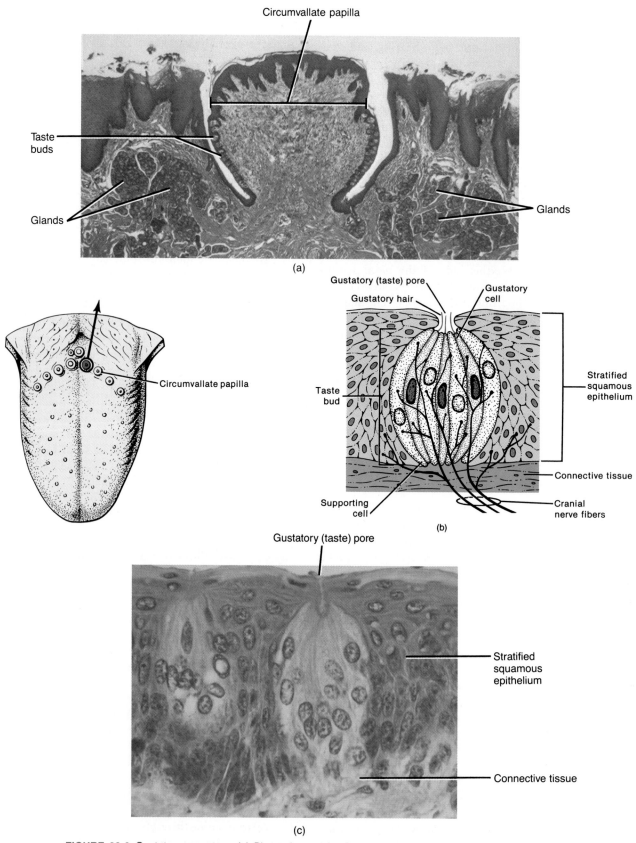

FIGURE 20-9 Gustatory receptors. (a) Photomicrograph of a circumvallate papilla containing taste buds at a magnification of 50×. (© 1983 by Michael H. Ross. Used by permission.) (b) Diagram of the structure of a taste bud. (c) Photomicrograph of a taste bud at a magnification of 600×. (© 1983 by Michael H. Ross. Used by permission.)

is highly sensitive to bitter substances. The lateral edges of the tongue are more sensitive to sour substances (see Figure 23-5a).

Gustatory Pathway

The cranial nerves that supply afferent fibers to taste buds are the facial (VII) nerve and trigeminal (V) nerve, which supply the anterior two-thirds of the tongue; the glossopharyngeal (IX), which supplies the posterior one-third of the tongue; and the vagus (X), which supplies the epiglottis. Taste impulses are conveyed from the gustatory cells in taste buds along the nerves to the medulla and then to the thalamus. They terminate in the primary gustatory area in the parietal lobe of the cerebral cortex.

VISUAL SENSATIONS

The structures related to vision are the eyeball, the optic (II) nerve, the brain, and a number of accessory structures. The study of the structure, function, and diseases of the eye is known as **ophthalmology** (of'-thal-MOL-ō-jē; *ophthalmo* = eye; *logos* = study of). A physician who specializes in the diagnosis and treatment of eye disorders is known as an **ophthalmologist,** whereas an **optometrist** is a person who is licensed to examine and test the eyes and treat visual defects by prescribing corrective lenses.

The developmental anatomy of the eye is considered later in the chapter.

Accessory Structures of Eye

Among the **accessory structures** are the eyebrows, eyelids, eyelashes, and the lacrimal apparatus (Figure 20-10). The **eyebrows** form a transverse arch at the junction of the upper eyelid and forehead. Structurally, they resemble the hairy scalp. The skin of the eyebrows is richly supplied with sebaceous glands. The hairs are generally coarse and directed laterally. Deep to the skin of the eyebrows are the fibers of the orbicularis oculi muscles. The eyebrows help to protect the eyeballs from foreign objects, perspiration, and the direct rays of the sun.

The upper and lower **eyelids,** or **palpebrae** (PAL-pe-brē), have several important roles. They shade the eyes during sleep, protect the eyes from excessive light and foreign objects, and spread lubricating secretions over the eyeballs. The upper eyelid is more movable than the lower and contains in its superior region a special levator muscle known as the **levator palpebrae superioris.** The space between the upper and lower eyelids that exposes the eyeball is called the **palpebral fissure.** Its angles are known as the **lateral commissure** (KOM-i-shūr), which is narrower and closer to the temporal bone, and the **medial commissure,** which is broader and nearer the nasal

bone. In the medial commissure there is a small reddish elevation, the **lacrimal caruncle** (KAR-ung-kul), containing sebaceous and sudoriferous glands. A whitish material secreted by the caruncle collects in the medial commissure.

From superficial to deep, each eyelid consists of epidermis, dermis, subcutaneous areolar connective tissue, fibers of the orbicularis oculi muscle, a tarsal plate, tarsal glands, ciliary glands, and a conjunctiva. The **tarsal plate** is a thick fold of connective tissue that forms much of the inner wall of each eyelid and gives form and support to the eyelids. Embedded in grooves on the deep surface of each tarsal plate is a row of elongated glands known as **tarsal,** or **Meibomian** (mī-BŌ-mē-an), **glands.** These are modified sebaceous glands, and their oily secretion helps keep the eyelids from adhering to each other. Infection of the tarsal glands produces a tumor or cyst on the eyelid called a **chalazion** (ka-LĀ-zē-on). The **conjunctiva** (kon'-junk-TĪ-va) is a thin mucous membrane. It is called the **palpebral conjunctiva** when it lines the inner aspect of the eyelids. It is called the **bulbar (ocular) conjunctiva** when it is reflected from the eyelids onto the anterior surface of the eyeball to the periphery of the cornea. When the blood vessels of the bulbar conjunctiva are dilated and congested due to local irritation or infection, the person has bloodshot eyes.

Projecting from the border of each eyelid, anterior to the tarsal glands, is a row of short, thick hairs, the **eyelashes.** In the upper lid, they are long and turn upward; in the lower lid, they are short and turn downward. Sebaceous glands at the base of the hair follicles of the eyelashes called **sebaceous ciliary glands,** or **glands of Zeis** (ZĪS), pour a lubricating fluid into the follicles. Infection of these glands and eyelash follicles is called a **sty.**

The **lacrimal apparatus** is a term used for a group of structures that manufactures and drains away tears. These structures are the lacrimal glands, the excretory lacrimal ducts, the lacrimal canals, the lacrimal sacs, and the nasolacrimal ducts. A **lacrimal gland** is a compound tubuloacinar gland located at the superior lateral portion of each orbit. Each is about the size and shape of an almond. Leading from the lacrimal glands are 6 to 12 **excretory lacrimal ducts** that empty lacrimal fluid, or tears, onto the surface of the conjunctiva of the upper lid. From here the lacrimal fluid passes medially and enters two small openings called **puncta lacrimalia** that appear as two small pores, one in each papilla of the eyelid, at the medial commissure of the eye. The lacrimal secretion then passes into two ducts, the **lacrimal canals,** and is next conveyed into the lacrimal sac. The lacrimal canals are located in the lacrimal grooves of the lacrimal bones. The **lacrimal sac** is the superior expanded portion of the **nasolacrimal duct,** a canal that transports the lacrimal secretion into the inferior meatus of the nose.

The **lacrimal secretion** is a watery solution containing

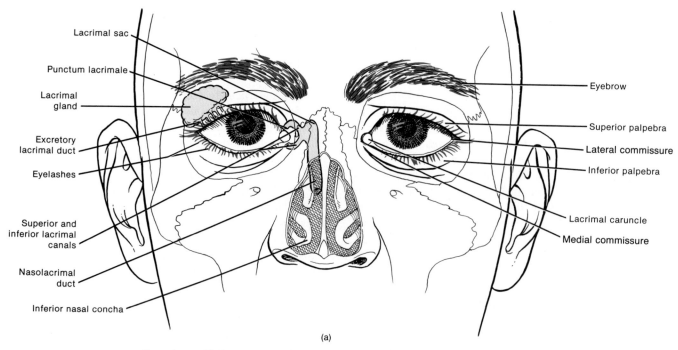

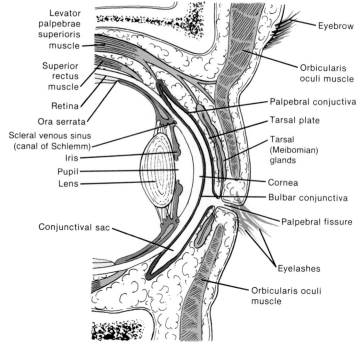

FIGURE 20-10 Accessory structures of the eye. (a) Anterior view. (b) Sagittal section of the eyelids and anterior portion of the eyeball.

CLINICAL APPLICATION

Normally, the lacrimal secretion is carried away by evaporation or by passing into the lacrimal canals and then into the nasal cavities as fast as it is produced. If, however, an irritating substance makes contact with the conjunctiva, the lacrimal glands are stimulated to oversecrete. Tears then accumulate more rapidly than they can be carried away. This is a protective mechanism since the tears dilute and wash away the irritating substance. **"Watery" eyes** also occur when an inflammation of the nasal mucosa, such as a cold, obstructs the nasolacrimal ducts so that drainage of tears is blocked.

Structure of Eyeball

The adult **eyeball** measures about 2.5 cm (1 in) in diameter. Of its total surface area, only the anterior one-sixth is exposed. The remainder is recessed and protected by the orbit into which its fits. Anatomically, the eyeball can be divided into three layers: fibrous tunic, vascular tunic, and retina or nervous tunic (Figure 20-11).

● *Fibrous Tunic*   The **fibrous tunic** is the outer coat of the eyeball. It can be divided into two regions: the poste-

salts, some mucus, and a bactericidal enzyme called **lysozyme.** It cleans, lubricates, and moistens the eyeball. After being secreted by the lacrimal glands, it is spread over the surface of the eyeball by the blinking of the eyelids. Usually, 1 ml per day is produced.

The surface anatomy of the accessory structures of the eye and the eyeball is shown in Figure 11-2.

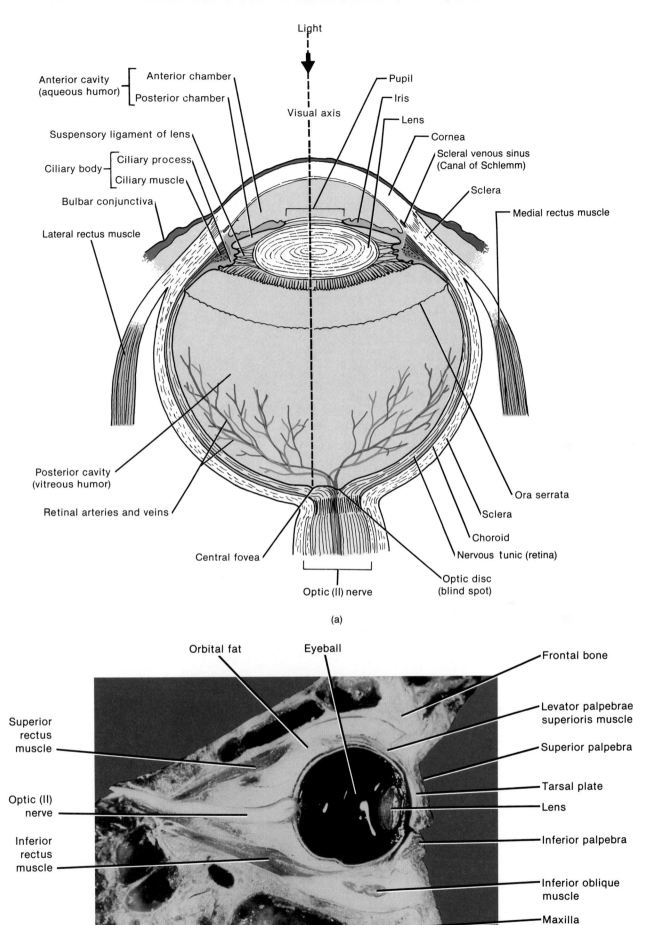

FIGURE 20-11 Structure of the eyeball. (a) Diagram of gross structure in transverse section. (b) Photograph of midsagittal section. (Courtesy of C. Yokochi and J. W. Rohen, *Photographic Anatomy of the Human Body,* 2nd ed., 1979, IGAKU-SHOIN, Ltd., Tokyo, New York.)

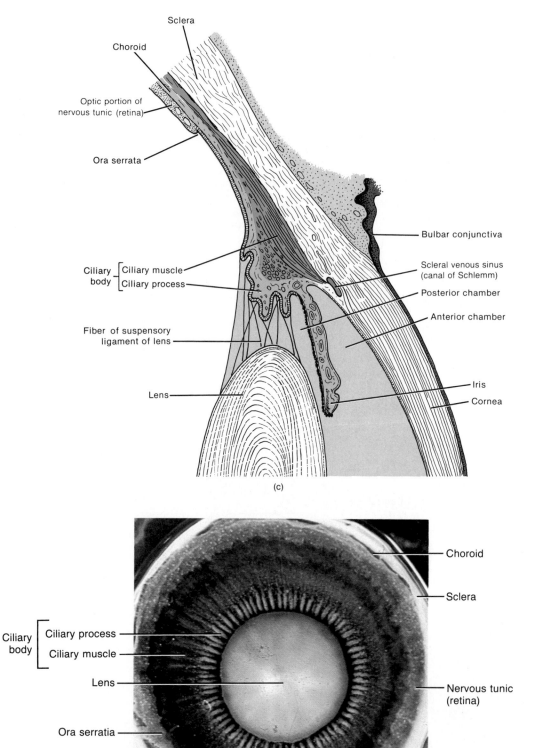

(c)

(d)

FIGURE 20-11 (*Continued*) Structure of the eyeball. (c) Section through the anterior portion of the eyeball at the sclerocorneal junction. (d) Photograph of the anterior portion of the eyeball viewed from behind. (Courtesy of C. Yokochi and J. W. Rohen, *Photographic Anatomy of the Human Body*, 2nd ed., 1979, IGAKU-SHOIN, Ltd., Tokyo, New York.)

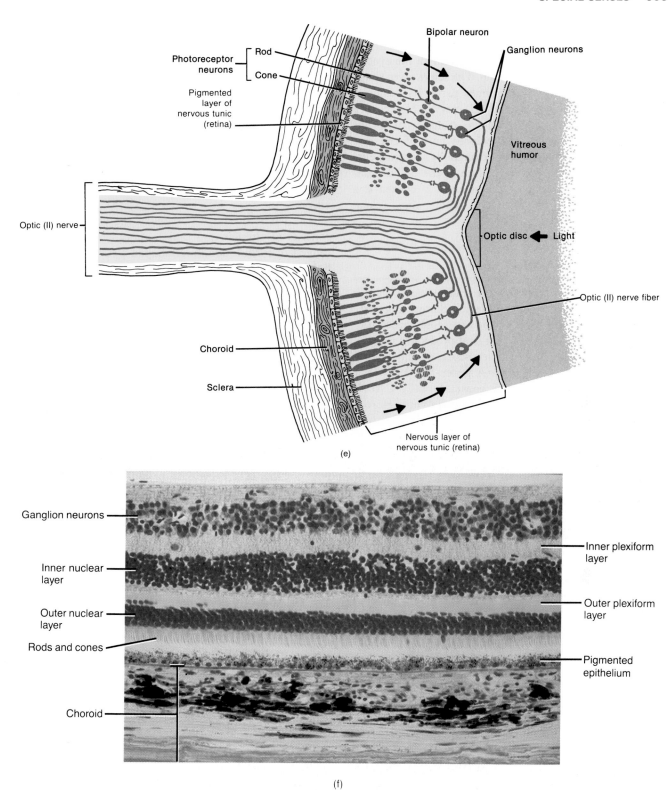

(e)

(f)

FIGURE 20-11 (*Continued*) Structure of the eyeball. (e) Diagram of the microscopic structure of the retina, exaggerated for emphasis. (f) Photomicrograph of a portion of the nervous tunic (retina) at a magnification of 300×. The outer nuclear layer contains nuclei of photoreceptor neurons; the outer plexiform layer is where photoreceptor and bipolar neurons synapse; the inner nuclear layer contains the nuclei of bipolar neurons; and the inner plexiform layer is where bipolar and ganglion neurons synapse. (© 1983 by Michael H. Ross. Used by permission.)

rior portion is the sclera and the anterior portion is the cornea. The **sclera** (SKLE-ra), the "white of the eye," is a white coat of fibrous tissue that covers all the eyeball except the anterior colored portion. The sclera gives shape to the eyeball and protects its inner parts. Its posterior surface is pierced by the optic (II) nerve. The **cornea** (KOR-nē-a) is a nonvascular, nervous, transparent fibrous coat that covers the iris, the colored part of the eye. The cornea's outer surface is covered by an epithelial layer continuous with the epithelium of the bulbar conjunctiva. At the junction of the sclera and cornea is a venous sinus known as the **scleral venous sinus (canal of Schlemm).**

CLINICAL APPLICATION

Each year, about 10,000 **corneal transplants (keratoplasty)** are performed in the United States. Corneal transplants are considered to be the most successful type of transplantation. The surgical procedure is performed under an operating microscope at magnifications of between 6× and 40×. A general or local anesthesia is used. The defective cornea, usually 7 to 8 mm (about 0.3 in) in diameter, is excised with microcorneal scissors and replaced with a donor cornea of similar diameter. The transplanted cornea is sewn into place with nylon sutures. Patients usually remain hospitalized for 3 to 5 days and are later fitted with hard or soft contact lenses or eyeglasses.

In the process of normal vision, light rays pass through the cornea, as well as other structures, to form a clear image on the nervous tunic (retina). In myopia (nearsightedness), the cornea is too curved and images of distant objects fall short of the nervous tunic, causing blurred vision. Many people with myopia may be helped by a surgical procedure called **radial keratotomy.** In the procedure, several microscopic incisions, like the spikes of a wheel, are made in the cornea. This flattens the cornea and improves vision without using corrective lenses.

• *Vascular Tunic* The **vascular tunic** is the middle layer of the eyeball and is composed of three portions: the choroid, the ciliary body, and the iris. Collectively, these three structures are called the **uvea** (YOO-vē-a). The **choroid** (KŌ-royd), the posterior portion of the vascular tunic, is a thin, dark brown membrane that lines most of the internal surface of the sclera. It contains numerous blood vessels and a large amount of pigment. The choroid absorbs light rays so they are not reflected back out of the eyeball. Through its blood supply, it nourishes the retina. Like the sclera, it is pierced by the optic (II) nerve at the back of the eyeball.

In the anterior portion of the vascular tunic, the choroid becomes the **ciliary** (SIL-ē-ar'-ē) **body.** It is the thickest portion of the vascular tunic. It extends from the **ora serrata** (ō-ra ser-RĀ-ta) of the retina (inner tunic) to a point just behind the sclerocorneal junction. The ora serrata is simply the jagged margin of the retina. The ciliary body consists of the ciliary processes and ciliary muscle. The **ciliary processes** consist of protrusions or folds on the internal surface of the ciliary body that secrete aqueous humor. The **ciliary muscle** is a smooth muscle that alters the shape of the lens for near or far vision.

The **iris** is the third portion of the vascular tunic. It consists of circular and radial smooth muscle fibers arranged to form a doughnut-shaped structure. The black hole in the center of the iris is the **pupil,** the area through which light enters the eyeball. The iris is suspended between the cornea and the lens and is attached at its outer margin to the ciliary process. A principal function of the iris is to regulate the amount of light entering the eyeball. When the eye is stimulated by bright light, the circular muscles of the iris contract and decrease the size of the pupil. When the eye must adjust to dim light, the radial muscles of the iris relax and increase the pupil's size.

• *Nervous Tunic (Retina)* The third and inner coat of the eye, the **nervous tunic (retina),** lies only in the posterior portion of the eye. Its primary function is image formation. The retina is the only place in the body where blood vessels can be seen directly. This is done by use of a special reflecting light called an ophthalmoscope. It is possible for an ophthalmologist to examine the retina and detect vascular changes associated with hypertension, atherosclerosis, and diabetes. The retina consists of an inner nervous tissue layer and an outer pigmented layer. The retina covers the choroid. At the edge of the ciliary body, it ends in a scalloped border called the ora serrata, where the nervous layer or visual portion of the retina ends. The pigmented layer extends anteriorly over the back of the ciliary body and the iris as the nonvisual portion of the retina.

CLINICAL APPLICATION

Detachment of the retina may occur in trauma, such as a blow to the head. The tear in the retina causes blindness in the corresponding field of vision. The actual detachment occurs between the sensory part of the retina and the underlying pigmented layer. Fluid accumulates between these layers, forcing the thin, pliable retina to billow out toward the vitreous humor. The retina may be reattached by photocoagulation using a laser beam, cryosurgery, or scleral resection.

The nervous layer contains three zones of neurons. These three zones, named in the order in which they conduct impulses, are the **photoreceptor neurons,** the **bipolar neurons,** and the **ganglion neurons.** The dendrites of the photoreceptor neurons are called rods and cones because of their shapes. They are visual receptors highly specialized for stimulation by light rays. Functionally, rods and cones develop generator potentials. **Rods** are specialized for vision in dim light. They also allow us to discriminate between different shades of dark and light and permit us to see shapes and movement. **Cones** are specialized for color vision and sharpness of vision (*visual acuity*). They are stimulated only by bright light, which is why we cannot see color by moonlight. It is estimated that there are 7 million cones and somewhere between 10 and 20 times as many rods. Cones are most densely concentrated in the **central fovea,** a small depression in the center of the macula lutea. The **macula lutea** (MAK-yoo-la LOO-tē-a), or yellow spot, is the exact center of the posterior portion of the retina, corresponding to the visual axis of the eye. The fovea is the area of sharpest vision because of the high concentration of cones. Rods are absent from the fovea and macula and increase in density toward the periphery of the retina.

CLINICAL APPLICATION

In a disease called **senile macular degeneration (SMD),** new blood vessels grow over the macula lutea. The effect ranges from distorted vision to blindness. SMD accounts for nearly all new cases of blindness in people over 65. Its cause is unknown. Laser beam treatment has been used effectively in arresting blood vessel proliferation and restoring normal vision in some cases. Treatment, however, must be obtained soon after the onset of symptoms. To determine if SMD exists, a simple test can be performed without any assistance or special instruments. Stare, one eye at a time (covering the other with your hand), at any long straight line, such as a door frame. If the line appears bent or twisted, or if a black spot appears, to either eye, a physician should be informed immediately.

When information has passed through the photoreceptor neurons, it is conducted across synapses to the bipolar neurons in the intermediate zone of the nervous layer of the retina. From here it is passed to the ganglion neurons. These cells transmit their signals through optic (II) nerve fibers to the brain in the form of nerve impulses.

The axons of the ganglion neurons extend posteriorly to a small area of the retina called the **optic disc (blind spot).** This region contains openings through which the axons of the ganglion neurons exit as the optic (II) nerve.

Since it contains no rods or cones, an image striking it cannot be perceived. Thus it is called the blind spot.

• *Lens* In addition to the fibrous tunic, vascular tunic, and retina, the eyeball itself contains the lens, just behind the pupil and iris. The **lens** is constructed of numerous layers of protein fibers arranged like the layers of an onion. Normally, the lens is perfectly transparent. It is enclosed by a clear connective tissue capsule and held in position by the **suspensory ligament.** A loss of transparency of the lens is known as a **cataract.**

• *Interior* The interior of the eyeball is a large cavity divided into two smaller ones by the lens. The **anterior cavity,** the division anterior to the lens, is further divided into the **anterior chamber,** which lies behind the cornea and in front of the iris, and the **posterior chamber,** which lies behind the iris and in front of the suspensory ligament and lens. The anterior cavity is filled with a watery fluid, similar to cerebrospinal fluid, called the **aqueous humor.** The fluid is believed to be secreted into the posterior chamber by choroid plexuses of the ciliary processes of the ciliary bodies behind the iris. Once the fluid is formed, it permeates the posterior chamber and then passes forward between the iris and the lens, through the pupil, into the anterior chamber. From the anterior chamber, the aqueous humor, which is continually produced, is drained off into the scleral venous sinus and then into the blood. The anterior chamber thus serves a function similar to the subarachnoid space around the brain and spinal cord. The scleral venous sinus is analogous to a venous sinus of the dura mater. The pressure in the eye, called **intraocular pressure,** is produced mainly by the aqueous humor. The intraocular pressure keeps the retina smoothly applied to the choroid so the retina will form clear images. Normal intraocular pressure (about 24 mm Hg) is maintained by drainage of the aqueous humor through the scleral venous sinus. Excessive intraocular pressure, called **glaucoma** (glow-KŌ-ma), results in degeneration of the retina and blindness. Besides maintaining intraocular pressure, the aqueous humor is also the principal link between the cardiovascular system and the lens and cornea. Neither the lens nor the cornea has blood vessels.

The second, and larger, cavity of the eyeball is the **posterior cavity.** It lies between the lens and the retina and contains a jellylike substance called the **vitreous humor.** This substance contributes to intraocular pressure, helps prevent the eyeball from collapsing, and holds the retina flush against the internal portions of the eyeball. The vitreous humor, unlike the aqueous humor, does not undergo constant replacement. It is formed during embryonic life and is not replaced thereafter.

A summary of structures associated with the eyeball is presented in Exhibit 20-1.

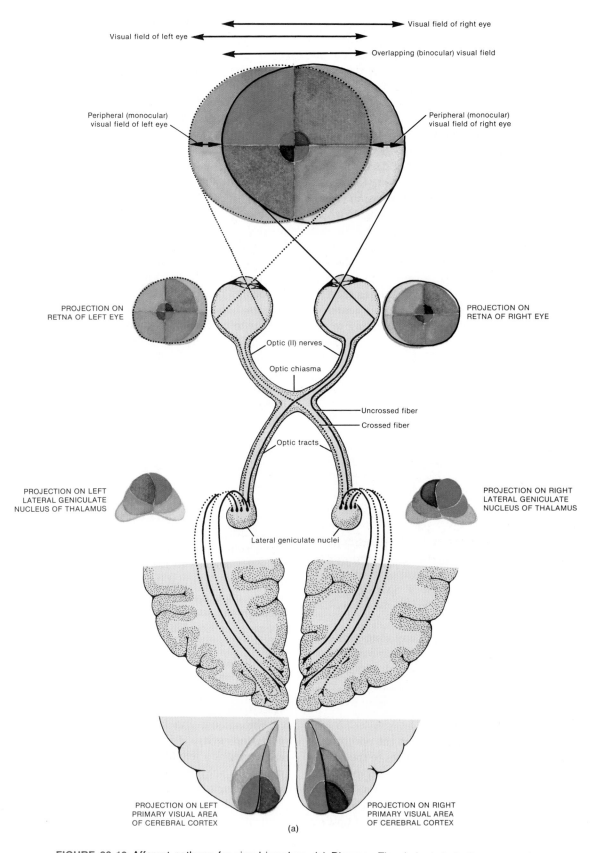

FIGURE 20-12 Afferent pathway for visual impulses. (a) Diagram. The dark circle in the center of the visual fields is the macula lutea. The center of the macula lutea is the central fovea, the area of sharpest vision.

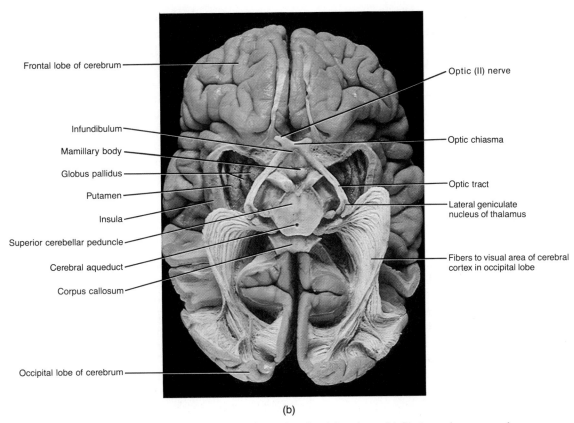

Frontal lobe of cerebrum

Infundibulum

Mamillary body

Globus pallidus

Putamen

Insula

Superior cerebellar peduncle

Cerebral aqueduct

Corpus callosum

Occipital lobe of cerebrum

Optic (II) nerve

Optic chiasma

Optic tract

Lateral geniculate nucleus of thalamus

Fibers to visual area of cerebral cortex in occipital lobe

(b)

FIGURE 20-12 (*Continued*) Afferent pathway for visual impulses. (b) Photograph as seen from the ventral aspect of the brain. (Courtesy of N. Gluhbegovic and T. H. Williams, *The Human Brain: A Photographic Guide,* Harper & Row, New York, 1980.)

EXHIBIT 20-1 SUMMARY OF STRUCTURES ASSOCIATED WITH THE EYEBALL

STRUCTURE	FUNCTION
Fibrous tunic	
Sclera	Provides shape and protects inner parts.
Cornea	Admits and refracts light.
Vascular tunic	
Choroid	Provides blood supply and absorbs light.
Ciliary body	Secretes aqueous humor and alters shape of lens for near or far vision (accommodation).
Iris	Regulates amount of light that enters eyeball.
Nervous tunic (retina)	Receives light, converts light into impulses, and transmits impulses to the optic (II) nerve.
Lens	Refracts light.
Anterior cavity	Contains aqueous humor that helps maintain shape of eyeball, keeps retina applied to choroid, and refracts light.
Posterior cavity	Contains vitreous humor that helps maintain shape of eyeball, keeps retina applied to choroid, and refracts light.

Visual Pathway

From the rods and cones, information is transmitted through bipolar neurons to ganglion cells. Ganglion cells initiate nerve impulses. The cell bodies of the ganglion cells lie in the retina, and their axons leave the eyeball via the **optic (II) nerve** (Figure 20-12). The axons pass through the **optic chiasma** (kī-AZ-ma), a crossing point of the optic (II) nerves. Some fibers cross to the opposite side. Others remain uncrossed. On passing through the optic chiasma, the fibers, now part of the **optic tract,** enter the brain and terminate in the lateral geniculate nucleus of the thalamus. Here the fibers synapse with third-order neurons whose axons pass to the visual areas located in the occipital lobes of the cerebral cortex.

Analysis of the afferent pathway to the brain reveals that the visual field of each eye is divided into two regions: the **nasal (medial) half** and the **temporal (lateral) half.** For each eye, light rays from an object in the nasal half of the visual field fall on the temporal half of the retina. Light rays from an object in the temporal half of the visual field fall on the nasal half of the retina (Figure 20-12). Also, light rays from objects at the top of the visual field of each eye fall on the inferior portion of the retina, and light rays from objects at the bottom of the visual field fall on the superior portion of the retina. In the optic chiasma, nerve fibers from the nasal halves of the retinas cross and continue on to the lateral geniculate nuclei of the thalamus; nerve fibers from the temporal halves of the retinas do not cross but continue directly on to the lateral geniculate nuclei. As a result, the primary visual area of the cerebral cortex of the right occipital lobe interprets visual sensations from the left side of an object via impulses from the temporal half of the retina of the right eye and the nasal half of the retina of the left eye. The primary visual area of the cerebral cortex of the left occipital lobe interprets visual sensations from the right side of an object via impulses from the nasal half of the right eye and the temporal half of the left eye.

CLINICAL APPLICATION

A **scotoma** (skō-TŌ-ma), or **blind spot,** in the field of vision, other than the normal blind spot (optic disc), may indicate a brain tumor along one of the afferent pathways. For instance, a symptom of a tumor in the right optic tract might be an inability to see the left side of a normal field of vision without moving the eyeball.

AUDITORY SENSATIONS AND EQUILIBRIUM

In addition to containing receptors for sound waves, the ear also contains receptors for equilibrium. Anatomically, the ear is divided into three principal regions: the external (outer) ear, the middle ear, and the internal (inner) ear.

The developmental anatomy of the ear is considered later in the chapter.

External (Outer) Ear

The **external (outer) ear** is structurally designed to collect sound waves and direct them inward (Figure 20-13a). It consists of the pinna, the external auditory canal, and the tympanic membrane, also called the eardrum.

The **pinna (auricle)** is a trumpet-shaped flap of elastic cartilage covered by thick skin. The rim of the pinna is called the **helix;** the inferior portion is the **lobule.** The pinna is attached to the head by ligaments and muscles. Additional surface anatomy features of the external ear are shown in Figure 11-3.

The **external auditory canal (meatus)** is a tube about 2.5 cm (1 in) in length that lies in the external auditory meatus of the temporal bone. It leads from the pinna to the eardrum and directs sound waves to the tympanic membrane. The walls of the canal consist of bone lined with cartilage that is continuous with the cartilage of the pinna. The cartilage in the external auditory canal is covered with thin, highly sensitive skin. Near the exterior opening, the canal contains a few hairs and specialized

sebaceous glands called **ceruminous** (se-ROO-mi-nus) **glands,** which secrete **cerumen** (earwax). The combination of hairs and cerumen (se-ROO-min) helps to prevent foreign objects from entering the ear. Some people produce an abnormal amount of cerumen, or earwax, in the external auditory canal. It becomes impacted and prevents sound waves from reaching the tympanic membrane. The treatment for **impacted cerumen** is usually periodic ear irrigation or removal of wax with a blunt instrument.

The **tympanic** (tim-PAN-ik) **membrane (eardrum)** is a thin, semitransparent partition of fibrous connective tissue between the external auditory canal and the middle ear. Its external surface is concave and covered with skin. Its internal surface is convex and covered with a mucous membrane.

CLINICAL APPLICATION

Rupture of the tympanic membrane may result from foreign bodies, pressure, or infection. Severe bleeding and escape of cerebrospinal fluid through a ruptured tympanic membrane may occur following a severe blow to the head and indicate skull fracture.

Middle Ear

The **middle ear,** also called the **tympanic cavity,** is a small, epithelial-lined, air-filled cavity hollowed out of the temporal bone (Figure 20-13a–d). The cavity is separated from the external ear by the eardrum and from the internal ear by a thin bony partition that contains two small openings: the oval window and the round window.

The posterior wall of the cavity communicates with the mastoid air cells of the temporal bone through a chamber called the **tympanic antrum.** This anatomical fact explains why a middle ear infection may spread to the temporal bone, causing mastoiditis, or even to the brain.

The anterior wall of the cavity contains an opening that leads directly into the **auditory (Eustachian) tube.** The auditory tube connects the middle ear with the nasopharynx of the throat. Through this passageway, infections may travel from the throat and nose to the ear. The lumen of the nasopharyngeal end of the auditory tube is normally closed during rest. It is opened by contraction of the tensor palati muscle, as occurs during yawning or swallowing. The function of the tube is to equalize air pressure on both sides of the tympanic membrane. Abrupt changes in external or internal air pressure as might occur during air travel could cause the eardrum to rupture. During swallowing and yawning, the tube opens to allow atmospheric air to enter or leave the middle ear until the internal pressure equals the external pressure. If the pressure is not relieved, intense pain, hearing impairment, tinnitus, and vertigo could develop. Any sudden

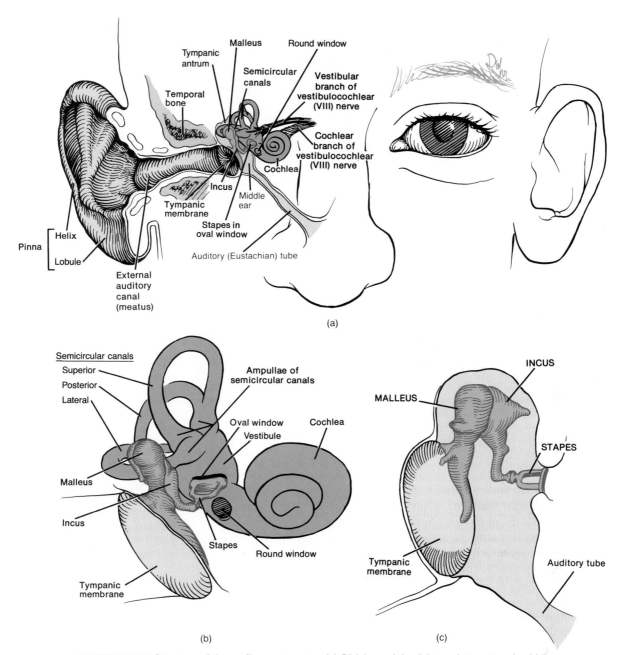

FIGURE 20-13 Structure of the auditory apparatus. (a) Divisions of the right ear into external, middle, and internal portions seen in a frontal section through the right side of the skull. (b) Details of the middle ear and bony labyrinth of the internal ear. (c) Diagram of auditory ossicles of the middle ear.

pressure changes against the eardrum may be equalized by deliberately swallowing or pinching the nose closed, closing the mouth, and gently forcing air from the lungs into the nasopharynx.

Extending across the middle ear are three exceedingly small bones called **auditory ossicles** (OS-si-kuls). The bones, named for their shape, are the malleus, incus, and stapes, commonly called the hammer, anvil, and stirrup, respectively. They are connected by synovial joints. The "handle" of the **malleus** is attached to the internal surface of the tympanic membrane. Its head articulates with the

body of the incus. The **incus** is the intermediate bone in the series and articulates with the head of the stapes. The base, or footplate, of the **stapes** fits into a small opening in the thin bony partition between the middle and inner ear. The opening is called the **fenestra vestibuli** (fe-NES-tra ves-TIB-yoo-lī), or **oval window.** Directly below the oval window is another opening, the **fenestra cochlea** (fe-NES-tra KŌK-lē-a), or **round window.** This opening is enclosed by a membrane called the secondary tympanic membrane. The auditory ossicles are attached to the middle ear by means of ligaments.

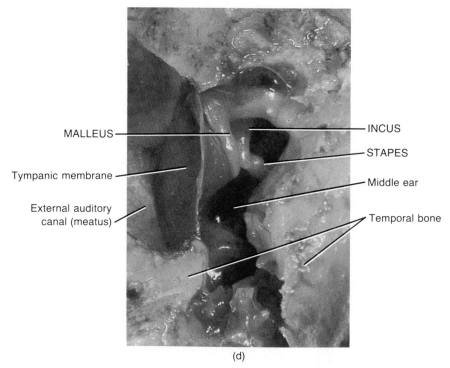

(d)

FIGURE 20-13 (*Continued*) Structure of the auditory apparatus. (d) Photograph of auditory ossicles of the middle ear. (Courtesy of C. Yokochi and J. W. Rohen, *Photographic Anatomy of the Human Body,* 2nd ed., 1979, IGAKU-SHOIN, Ltd., Tokyo, New York.)

Three ligaments are associated with the malleus (Figure 20-14). The *anterior ligament of the malleus* is attached by one end to the anterior process of the malleus and by the other to the anterior wall of the tympanic cavity. The *superior ligament of the malleus* extends from the head of the malleus to the roof of the tympanic cavity. The *lateral ligament of the malleus* extends from the neck of the malleus to the lateral wall of the tympanic cavity. Two ligaments are associated with the incus. The *posterior ligament of the incus* extends from the short crus of the incus to the posterior wall of the tympanic cavity, and the *superior ligament of the incus* extends from the body of the incus to the roof of the tympanic cavity. A single ligament, the *annular ligament of the base of the stapes,* is associated with the stapes. It extends from the base of the stapes to the fenestra vestibuli.

In addition to the ligaments, there are also two muscles attached to the ossicles. The *tensor tympani muscle* originates from the cartilaginous portion of the auditory tube, the greater wing of the sphenoid bone, as well as from the osseous canal in which it is contained. The muscle inserts into the manubrium of the malleus. It is supplied by the mandibular nerve via fibers that pass through the otic ganglion. The action of the muscle is to draw the malleus medially, thus increasing tension on the tympanic membrane and reducing the amplitude of oscillations. This action tends to prevent damage to the inner ear when exposed to loud sounds.

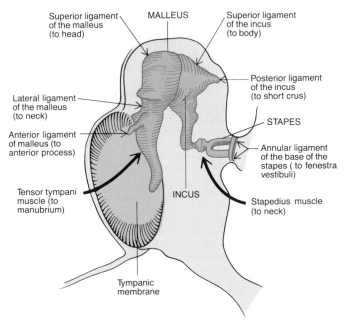

FIGURE 20-14 Areas of attachment of the ligaments and muscles of the auditory ossicles.

The *stapedius muscle* is the smallest of all skeletal muscles. It originates from a projection (pyramid) in the posterior wall of the tympanic cavity and inserts into the neck of the stapes. The muscle is innervated by the facial (VII) nerve. The action of the muscle is to draw

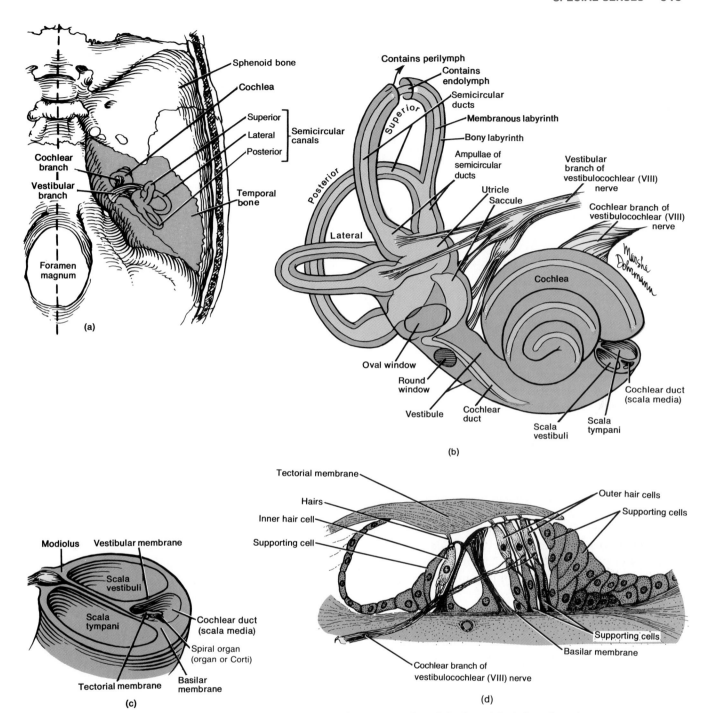

FIGURE 20-15 Details of the internal ear. (a) Relative position of the bony labyrinth projected to the inner surface of the floor of the skull. (b) The outer, blue area belongs to the bony labyrinth. The inner, rust-colored area belongs to the membranous labyrinth. (c) Cross section through the cochlea. (d) Enlargement of the spiral organ (organ of Corti).

the stapes posteriorly, thereby tightening the annular ligament of the base of the stapes and reducing the oscillatory range of the stapes (prevents excessive movement). Like the tensor tympani muscle, the stapedius muscle has a protective function in that it dampens (checks) large vibrations that result from loud noises. It is for this reason that paralysis of the stapedius muscle is associated with *hyperacusia* (acuteness of hearing).

Internal (Inner) Ear

The **internal (inner) ear** is also called the **labyrinth** (LAB-i-rinth) because of its complicated series of canals (Figure 20-15). Structurally, it consists of two main divisions: a body labyrinth and a membranous labyrinth that fits in the bony labyrinth. The **bony labyrinth** is a series of cavities in the petrous portion of the temporal bone. It can

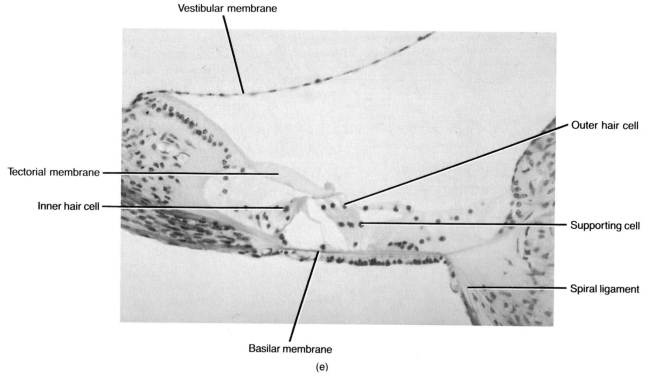

Vestibular membrane

Tectorial membrane

Inner hair cell

Outer hair cell

Supporting cell

Spiral ligament

Basilar membrane

(e)

FIGURE 20-15 (*Continued*) Details of the internal ear. (e) Photomicrograph of the spiral organ (organ of Corti) at a magnification of 300×. (© 1983 by Michael H. Ross. Used by permission.)

be divided into three areas named on the basis of shape: the vestibule, cochlea, and semicircular canals. The bony labyrinth is lined with periosteum and contains a fluid called **perilymph.** This fluid surrounds the **membranous labyrinth,** a series of sacs and tubes lying inside and having the same general form as the bony labyrinth. The membranous labyrinth is lined with epithelium and contains a fluid called **endolymph.**

The **vestibule** constitutes the oval central portion of the bony labyrinth. The membranous labyrinth in the vestibule consists of two sacs called the **utricle** and **saccule.** These sacs are connected to each other by a small duct.

Projecting upward and posteriorly from the vestibule are the three bony **semicircular canals.** Each is arranged at approximately right angles to the other two. On the basis of their positions, they are called superior, posterior, and lateral canals. One end of each canal enlarges into a swelling called the **ampulla** (am-POOL-la). The portions of the membranous labyrinth that lie inside the bony semicircular canals are called the **semicircular ducts (membranous semicircular canals).** These structures are almost identical in shape to the semicircular canals and communicate with the utricle of the vestibule.

Lying in front of the vestibule is the **cochlea** (KŌK-lē-a), so designated because of its resemblance to a snail's shell. The cochlea consists of a bony spiral canal that makes about 2¾ turns around a central bony core called

the **modiolus.** A cross section through the cochlea shows the canal is divided into three separate channels by partitions that together have the shape of the letter Y. The stem of the Y is a bony shelf that protrudes into the canal; the wings of the Y are composed mainly of membranous labyrinth. The channel above the bony partition is the **scala vestibuli;** the channel below is the **scala tympani.** The cochlea adjoins the wall of the vestibule, into which the scala vestibuli opens. The scala tympani terminates at the round window. The perilymph of the vestibule is continuous with that of the scala vestibuli. The third channel (between the wings of the Y) is the membranous labyrinth: the **cochlear duct (scala media).** The cochlear duct is separated from the scala vestibuli by the **vestibular membrane.** It is separated from the scala tympani by the **basilar membrane.**

Resting on the basilar membrane is the **spiral organ (organ of Corti),** the organ of hearing. The spiral organ is a series of epithelial cells on the inner surface of the basilar membrane. It consists of a number of supporting cells and hair cells, which are the receptors for auditory sensations. The inner hair cells are medially placed in a single row and extend the entire length of the cochlea. The outer hair cells are arranged in several rows throughout the cochlea. The hair cells have long hairlike processes at their free ends that extend into the endolymph of the cochlear duct. The basal ends of the hair cells are in contact with fibers of the cochlear branch of the vestibulo-

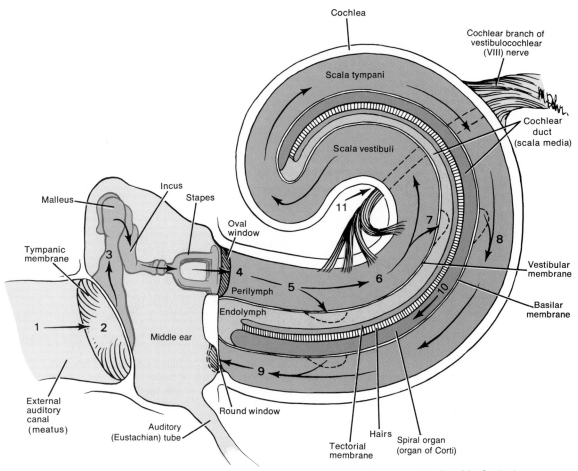

FIGURE 20-16 Physiology of hearing. The numbers correspond to the events listed in the text.

cochlear (VIII) nerve. Projecting over and in contact with the hair cells of the spiral organ is the **tectorial membrane**, a delicate and flexible gelatinous membrane.

Hair cells of the spiral organ are easily damaged by **exposure to high-intensity noises** such as those produced by engines of jet planes, revved-up motorcycles, and loud music. As a result of the noises, hair cells become arranged in disorganized patterns or they and their supporting cells degenerate.

Mechanism of Hearing

Sound waves result from the alternate compression and decompression of air molecules. They originate from a vibrating object, much the same way that waves travel over the surface of water. The events involved in the physiology of hearing sound waves are as follows (Figure 20-16):

1. Sound waves that reach the ear are directed by the pinna into the external auditory canal.

2. When the waves strike the tympanic membrane, the alternate compression and decompression of the air

causes the membrane to vibrate. The distance the membrane moves is always very small and is relative to the force and velocity with which air molecules strike it. It vibrates slowly in response to low-frequency sounds and rapidly in response to high-frequency sounds.

3. The central area of the tympanic membrane is connected to the malleus, which also starts to vibrate. The vibration is then picked up by the incus, which transmits the vibration to the stapes.

4. As the stapes moves back and forth, it pushes the oval window in and out. If sound waves passed directly to the oval window without passing through the tympanic membrane and auditory bones, hearing would be inadequate. A minimal amount of sound energy is required to transmit sound waves through the perilymph of the cochlea. Since the tympanic membrane has a surface area about 22 times larger than that of the oval window, it can collect about 22 times more sound energy. This energy is sufficient to transmit sound waves through the perilymph.

5. The movement of the oval window sets up waves in the perilymph.

6. As the window bulges inward, it pushes the peri-

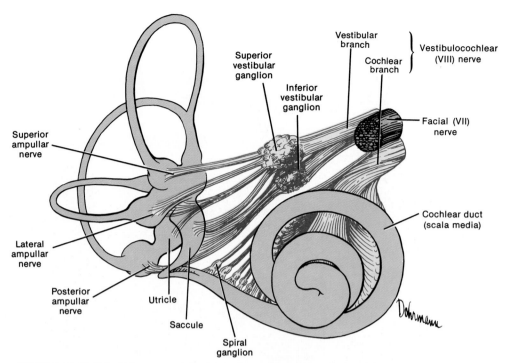

FIGURE 20-17 Principal nerves of the internal ear. Shown here is the right membranous labyrinth with the bony labyrinth removed. Nerve branches from each of the ampullae of the semicircular ducts and from the utricle and saccule form the vestibular branch of the vestibulocochlear (VIII) nerve. The vestibular ganglia contain the cell bodies of the vestibular branch. The cochlear branch of the vestibulocochlear (VIII) nerve arises in the spiral organ (organ of Corti). The spiral ganglia contain the cell bodies of the cochlear branch.

lymph of the scala vestibuli and waves are propagated through the scala vestibuli into the perilymph of the scala tympani.

7. This pressure pushes the vestibular membrane inward and increases the pressure of the endolymph inside the cochlear duct.

8. The basilar membrane gives under the pressure and bulges out into the scala tympani. Due to differences in structure of various areas of the basilar membrane, high-frequency resonance of the membrane occurs near its base and low-frequency resonance of the membrane occurs near its apex.

9. The sudden pressure in the scala tympani pushes the perilymph toward the round window, causing it to bulge back into the middle ear. Conversely, as the sound waves subside, the stapes moves backward and the procedure is reversed. That is, the fluid moves in the opposite direction along the same pathway and the basilar membrane bulges into the cochlear duct.

10. When the basilar membrane vibrates, the hair cells of the spiral organ are moved against the tectorial membrane. In some unknown manner, the movement of the hairs develops generator potentials that ultimately lead to the generation of nerve impulses. Differences in pitch are related to sound waves of various frequencies that cause specific regions of the basilar membrane to vibrate more intensely than other regions. In addition, loudness

is determined by the intensity of sound waves. High-intensity sound waves cause greater vibration of the basilar membrane.

11. The impulses are then passed on to the cochlear branch of the vestibulocochlear (VIII) nerve (Figure 20-17) and cochlear nuclei in the medulla. Here most impulses cross to the opposite side and then travel to the inferior colliculus of the midbrain, medial geniculate body of the thalamus, and finally to the auditory area of the temporal lobe of the cerebral cortex.

CLINICAL APPLICATION

Artificial ears are cochlear implants that translate sounds into electronic signals that can be interpreted by the brain. They take the place of hair cells of the spiral organ, which normally convert sound waves into electrical signals carried to the brain. The implants are used for individuals with cochlear disease, people who make up the majority of those who have no hearing in either ear.

The device consists of electrodes implanted in the cochlea that are connected to a plug attached to the outside of the head. A microprocessor pack is worn on the person's belt and connected by a cord to the plug. Sound waves enter a tiny microphone attached

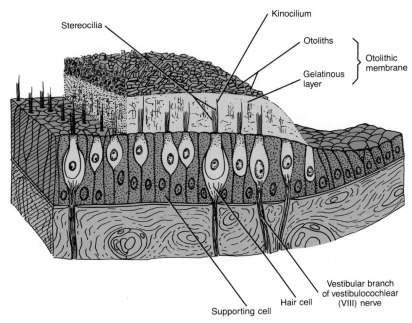

Stereocilia · Kinocilium · Otoliths · Gelatinous layer · Otolithic membrane · Vestibular branch of vestibulocochlear (VIII) nerve · Hair cell · Supporting cell

FIGURE 20-18 Structure of the macula.

to the ear and travel down the cord into the pack where they are converted into electrical signals. The signals then travel to the electrodes in the cochlea where they stimulate nerve endings and transmit the signals to the brain.

Mechanism of Equilibrium

There are two kinds of **equilibrium.** One, called **static equilibrium,** refers to the orientation of the body (mainly the head) relative to the ground (gravity). The second kind, **dynamic equilibrium,** is the maintenance of body position (mainly the head) in response to sudden movements such as rotation, acceleration, and deceleration. The receptor organs for equilibrium are the maculae of the saccule and the utricle and cristae in the semicircular ducts.

• *Static Equilibrium* The walls of both the utricle and saccule contain a small, flat, plaquelike region called a **macula** (Figure 20-18). The maculae are the receptors that are concerned with static equilibrium. They provide sensory information regarding the orientation of the head in space and are essential for maintaining posture.

Microscopically, the two maculae resemble the spiral organ. They consist of differentiated neuroepithelial cells that are innervated by the vestibular branch of the vestibulocochlear (VIII) nerve. The maculae are anatomically located in planes perpendicular to one another and possess two kinds of cells: **hair (receptor) cells** and **supporting cells.** Two shapes of hair cells have been identified: one

is more flask-shaped and the other is more cylindrical. Both show long extensions of the cell membrane consisting of many *stereocilia* (they are actually microvilli) and one *kinocilium* (a conventional cilium) anchored firmly to its basal body and extending beyond the longest microvilli. Some of the microvilli reach lengths of over 100 μm, and some receptor cells have over 80 such projections.

The columnar supporting cells of the maculae are scattered between the hair cells. Floating directly over the hair cells is a thick, gelatinous glycoprotein layer, probably secreted by the supporting cells, called the **otolithic membrane.** A layer of calcium carbonate crystals, called **otoliths** (*oto* = ear; *lithos* = stone), extends over the entire surface of the otolithic membrane. The specific gravity of these otoliths is about 3, which makes them much denser than the endolymph fluid that fills the rest of the utricle.

The otolithic membrane sits on top of the macula like a discus on a greased cookie sheet. If you tilt your head forward, the otolithic membrane (the discus in our analogy) slides downhill over the hair cells in the direction determined by the tilt of your head. In the same manner, if you are sitting upright in a car that suddenly jerks forward, the otolithic membrane, due to its inertia, slides backward and stimulates the hair cells. As the otoliths move, they pull on the gelatinous layer, which pulls on the stereocilia and makes them bend. The movement of the stereocilia initiates a nerve impulse that is then transmitted to the vestibular branch of the vestibulocochlear (VIII) nerve (see Figure 20-17).

Most of the vestibular branch fibers enter the brain stem and terminate in the vestibular nuclear complex in

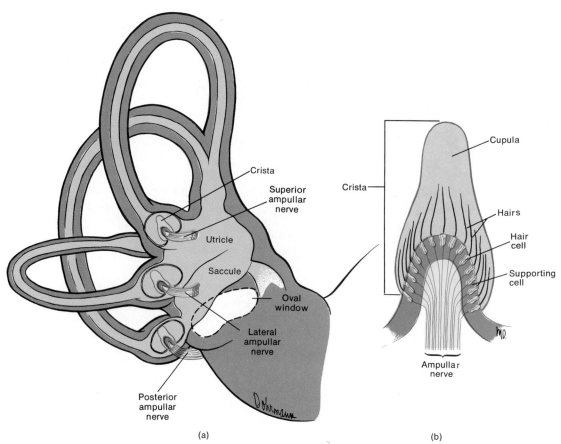

FIGURE 20-19 Semicircular ducts and dynamic equilibrium. (a) Position of the cristae relative to the membranous ampullae. (b) Enlarged aspect of a crista.

the medulla. The remaining fibers enter the flocculonodular lobe of the cerebellum through the inferior cerebellar peduncle. Bidirectional pathways connect the vestibular nuclei and cerebellum. Fibers from all the vestibular nuclei form the medial longitudinal fasciculus, which extends from the brain stem into the cervical portion of the spinal cord. The fasciculus sends impulses to the nuclei of cranial nerves that control eye movements (oculomotor (III), trochlear (IV), and abducens (VI)) and to the accessory (XI) nerve nucleus that helps control movements of the head and neck. In addition, fibers from the lateral vestibular nucleus form the vestibulospinal tract, which conveys impulses to skeletal muscles that regulate body tone in response to head movements. Various pathways between the vestibular nuclei, cerebellum, and cerebrum enable the cerebellum to assume a key role in helping the body to maintain static equilibrium. The cerebellum continuously receives updated sensory information from the utricle and saccule concerning static equilibrium. Using this information, the cerebellum monitors and makes corrective adjustments in the motor activities that originate in the cerebral cortex. Essentially, the cerebellum sends continuous impulses to the motor areas of the cerebrum, in response to input from the utricle and saccule,

causing the motor system to increase or decrease its impulses to specific skeletal muscles to maintain static equilibrium.

● *Dynamic Equilibrium* The cristae in the three semicircular ducts maintain dynamic equilibrium (Figure 20-19). The ducts are positioned at right angles to one another in three planes: frontal (the superior duct), sagittal (the posterior duct), and lateral (the lateral duct). This positioning permits detection of an imbalance in three planes. In the ampulla, the dilated portion of each duct, there is a small elevation called the **crista.** Each crista is composed of a group of **hair (receptor) cells** and **supporting cells** covered by a mass of gelatinous material called the **cupula.** When the head moves, the endolymph in the semicircular ducts flows over the hairs and bends them. The movement of the hairs stimulates sensory neurons, and the impulses pass over the vestibular branch of the vestibulocochlear (VIII) nerve. The impulses follow the same pathways as those involved in static equilibrium and are eventually sent to the muscles that must contract to maintain body balance in the new position.

A summary of the structures of the ear related to hearing and equilibrium is presented in Exhibit 20-2.

EXHIBIT 20-2 SUMMARY OF STRUCTURES OF THE EAR RELATED TO HEARING AND EQUILIBRIUM

STRUCTURE	FUNCTION
External (outer) ear	
Pinna (auricle)	Collects sound waves.
External auditory canal	Directs sound waves to tympanic membrane.
Tympanic membrane	Sound waves cause it to vibrate which, in turn, causes the malleus to vibrate.
Middle ear	
Auditory (Eustachian) tube	Equalizes pressure on both sides of the tympanic membrane.
Auditory ossicles	Transmit sound waves from tympanic membrane to oval window.
Internal (inner) ear	
Utricle	Contains macula, receptor for static equilibrium.
Saccule	Contains macula, receptor for static equilibrium.
Semicircular ducts	Contain cristae, receptors for dynamic equilibrium.
Cochlea	Contains a series of fluids, channels, and membranes that transmit sound waves to the spiral organ (organ of Corti), the organ of hearing; the spiral organ generates nerve impulses and transmits them to the cochlear branch of the vestibulocochlear (VIII) nerve.

DEVELOPMENTAL ANATOMY OF THE EYE AND EAR

The first appearance of the eyes, at about the fourth week of development, is a broad, shallow depression on the medial surface of the diencephalon called the **optic groove** (Figure 20-20a). As development continues, lateral outgrowths from either side of the diencephalon arise. Each outgrowth is called an **optic vesicle** (see Figure 18-22b). The distal part of each vesicle expands to form a hollow bulb, while the proximal part remains narrow and is referred to as the **optic stalk** (Figure 20-20b). Surface **ectoderm** adjacent to the optic vesicle thickens and is known as the **lens placode.**

As each optic vesicle enlarges, it folds in on itself and becomes a double-layered **optic cup.** The outer, thinner layer becomes the *pigmented layer of the retina;* the inner, thicker layer becomes the *nervous layer of the retina.* Along with the invagination of each optic cup, the lens placode invaginates to form the hollow **lens vesicle.** Each lens vesicle separates from the surface ectoderm and becomes surrounded by its optic cup. The lens vesicle will eventually develop into the *lens* of the eye, and the overlying **ectoderm** will develop into the *cornea* of the eye.

Fibers from the nervous layer of the retina pass through the optic stalk on their way to the brain and retinal blood vessels infiltrate the stalk. As a result, the *optic nerve* is formed. The *fibrous tunic* and *vascular tunic* are formed from local **mesenchymal (mesodermal) cells.**

The first rudiment of the *internal ear* is a thickened region of surface **ectoderm** on either side of the head near the hindbrain vesicle. This region, which develops shortly after the first appearance of the optic groove, is called the **otic placode** (Figure 20-20a, c). As the placode develops, it invaginates to form the **otic pit.** As invagination continues, the otic pit separates from the surface ectoderm, forming a closed sac, referred to as the **otic vesicle.** The otic vesicle forms the various parts of the *membranous labyrinth.*

Lateral pouches, called **pharyngeal pouches,** develop from the sides of the pharynx. Over these pouches, the surface **ectoderm** indents forming what are called **branchial grooves.** The *middle ear cavity* and *auditory (Eustachian) tube* develop from the first pharyngeal pouch. The *auditory ossicles* (malleus, incus, and stapes) develop in the middle ear cavity from **mesenchymal (mesodermal) cells.** The *external ear* and *external auditory canal* develop from the first branchial groove.

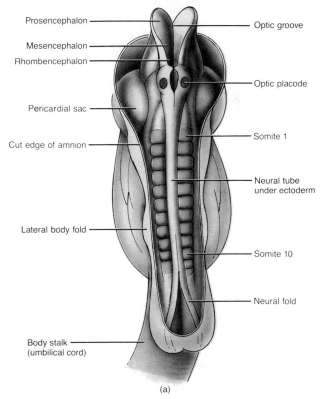

(a)

FIGURE 20-20 Development of the eye and ear. (a) Dorsal aspect of an embryo with 10 pairs of somites showing the postion of the optic groove and otic placode.

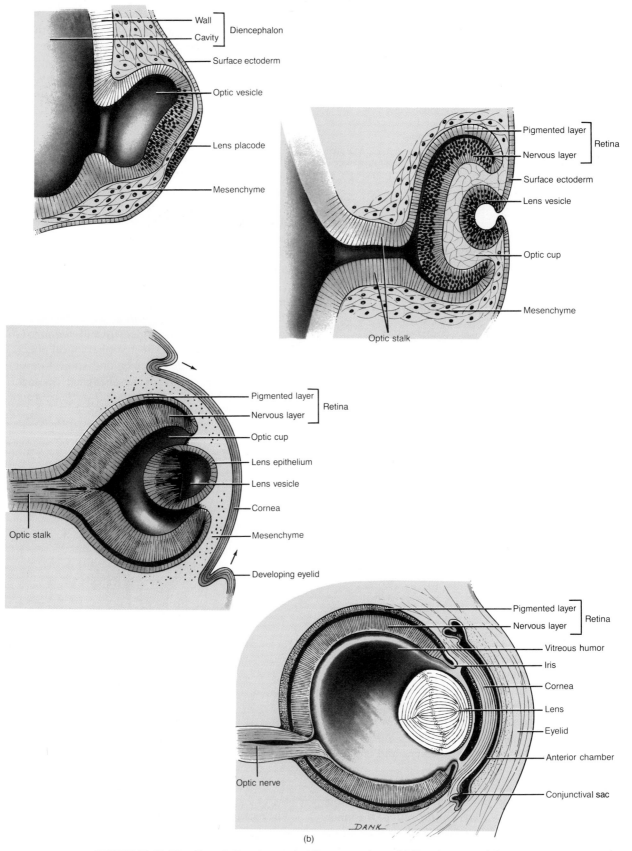

FIGURE 20-20 (*Continued*) Development of the eye and ear. **(b)** Development of the eye.

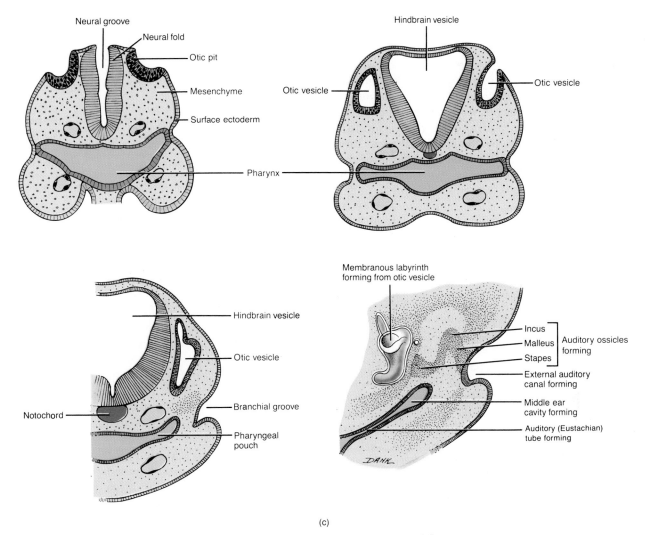

(c)

FIGURE 20-20 (*Continued*) Development of the eye and ear. (c) Development of the ear.

APPLICATIONS TO HEALTH

The special sense organs can be altered or damaged by numerous disorders. The causes of disorder can range from congenital origins to the effects of old age. In this section we discuss a few common disorders of the eyes and ears.

CATARACT

The most prevalent disorder resulting in blindness is cataract formation. A **cataract** is a clouding of the lens or its capsule so that it becomes opaque or milk white. The two basic processes involved in cataract formation are the breakdown of the normal lens protein and an influx of water into the lens. As a result, the light from an object, which normally passes directly through the lens to produce a sharp image, produces only a degraded image. If the cataract is severe enough, no image at all is produced.

Cataracts can occur at any age, but we will discuss the type that develops with old age. As a person gets older, the cells in the lenses may degenerate and be replaced with nontransparent fibrous protein, or the lens may start to manufacture nontransparent protein. The cataract formation results in a progressive, painless loss of vision. The degree of loss depends on the location and extent of the opacity. If vision loss is gradual, frequent changes in glasses may help maintain useful vision for a while. However, when the changes become so extensive that light rays are blocked out altogether, surgery is indicated to remove the opaque lens. An artificial lens is substituted by means of eyeglasses or by surgical implantation of a plastic lens inside the eyeball.

GLAUCOMA

The second most common cause of blindness, especially in the elderly, is **glaucoma.** This disorder is characterized by an abnormally high pressure due to a buildup of aqueous humor inside the eyeball. The aqueous humor does not return into the bloodstream through the scleral venous sinus (canal of Schlemm) as quickly as it is formed. The fluid accumulates and, by compressing the lens into the vitreous humor, puts pressure on the neurons of the retina. If the pressure continues over a long period of time, glaucoma can progress from mild visual impairment to a point where neurons of the retina are destroyed, resulting in degeneration of the optic disc, visual field defects, and blindness. It can affect a person of any age, but 95 percent of the victims are over 40. Glaucoma affects the eyesight of more than 1 million people in the United States.

Treatment is through drugs or surgery. Drug treatment involves administration of topical drugs (epinephrine, pilocarpine, and timolol) that decrease aqueous humor production by inhibiting formation by the choroid plexuses of the ciliary processes and increase aqueous humor outflow by causing sphincter muscles of the iris and ciliary body to contract. These contractions prevent the obstruction of the anterior chamber and the scleral venous sinus, resulting in improved fluid outflow from the eye. Topical therapy may be followed by use of oral drugs, such as acetazolamide, which also decrease production of aqueous humor.

CONJUNCTIVITIS (PINKEYE)

Many different eye inflammations exist, but the most common type is **conjunctivitis (pinkeye),** an inflammation of the conjunctiva, the membrane that lines the insides of the eyelids and covers the cornea. Conjunctivitis can be caused by microorganisms, most often the pneumococci or staphylococci bacteria. In such cases, the inflammation is very contagious. This epidemic type, common in children, is rarely serious. Conjunctivitis may also be caused by a number of irritants, in which case the inflammation is not contagious. Irritants include dust, smoke, wind, pollutants in the air, and excessive glare. The condition may be acute or chronic.

TRACHOMA

A serious form of chronic contagious conjunctivitis is known as **trachoma** (tra-KŌ-ma). It is caused by an organism, called the TRIC agent, that has characteristics of both viruses and bacteria. Trachoma is characterized by many granulations or fleshy projections on the eyelids. If untreated, these projections can irritate and inflame the cornea and reduce vision. The disease produces an excessive growth of subconjunctival tissue and the invasion of blood vessels into the upper half of the front of the cornea. The disease progresses until it covers the entire cornea, bringing about a loss of vision because of corneal opacity.

Antibiotics, such as tetracycline and the sulfa drugs, kill the organisms that cause trachoma and have reduced the seriousness of this infection.

DEAFNESS

Deafness is the lack of the sense of hearing or significant hearing loss. It is generally divided into two principal types: nerve deafness and conduction deafness. *Nerve deafness* is caused by impairment of the cochlea or cochlear branch of the vestibulocochlear (VIII) nerve. *Conduction deafness* is caused by impairment of the middle ear mechanisms for transmitting sounds into the cochlea. Among the factors that contribute to deafness are atherosclerosis, which reduces blood supply to the ears; repeated exposure to loud noise, which destroys hair cells of the spiral organ; certain drugs such as streptomycin; disease; impacted cerumen; injury to the tympanic membrane; and aging, which results in thickening of the tympanic membrane, stiffening of the auditory ossicles, and decreased numbers of hair cells due to diminished cell division.

Some older patients with tinnitus (ringing in the ears) and hearing loss due to degeneration of the inner ear have fully recovered following megadoses of zinc (600 mg daily). However, doses larger than 220 mg per day should not be taken for longer than necessary, due to the potential for liver damage.

LABYRINTHINE DISEASE

Labyrinthine (lab'-i-RIN-thēn) **disease** refers to a malfunction of the inner ear that is characterized by deafness, tinnitus (ringing in the ears), vertigo (hallucination of movement), nausea, and vomiting. There may also be a blurring of vision, nystagmus (rapid, involuntary movement of the eyeballs), and a tendency to fall in a certain direction.

Among the causes of labyrinthine disease are: (1) infection of the middle ear, (2) trauma from brain concussion producing hemorrhage or splitting of the labyrinth, (3) cardiovascular diseases such as arteriosclerosis and blood vessel disturbances, (4) congenital malformation of the labyrinth, (5) excessive formation of endolymph, (6) allergy, (7) blood abnormalities, and (8) aging.

In severe cases of labyrinthine disease, surgical treatment is considered. Treatment with ultrasound, which consists of beaming ultrasound waves into the labyrinth, destroys the semicircular ducts (equilibrium) while preserving the cochlear duct (hearing). This treatment has produced excellent results. Another successful procedure involves electrical coagulation of the labyrinth, which changes the fluid to a jelly or a solid.

MÉNIÈRE'S SYNDROME

Ménière's (men-YAIRZ) **syndrome** is one type of labyrinthine disorder that is characterized by fluctuating hearing loss, attacks of vertigo, and roaring tinnitus. Etiology of Ménière's syndrome is unknown. It is now thought that there is either an overproduction or underabsorption of endolymph in the cochlear duct. The hearing loss is caused by distortions in the basilar membrane of the cochlea. The classic type of Ménière's syndrome involves both the semicircular canals and the cochlea. The natural course of Ménière's syndrome may stretch out over a period of years, with the end result being almost total destruction of the patient's hearing.

The treatment for Ménière's syndrome may be either medical or surgical. Medical treatment is aimed at reducing the excess amount of endolymph to lessen the amount of distention and distortion of the basilar membrane. To accomplish this, the overall extracellular volume is diminished by means of salt restriction, the administration of diuretics, or surgical procedures that involve shunts. A popular treatment for the acute cases is the inhalation of 5 percent CO_2 with 95 percent O_2 for 10 minutes, four times a day. High levels of CO_2 in the cerebral circulation are powerful dilators of the cerebral vessels. For the 20 to 30 percent of Ménière's patients who do not respond to medical treatment or shunt procedures, some type of vestibular surgery is indicated, for relief from vertigo. Here, there are two basic procedures available to the otologic surgeon: those that destroy residual hearing and those that preserve it. As a rule, labyrinthectomy is reserved for advanced cases in which the hearing is poor.

OTITIS MEDIA

Otitis media is an acute infection of the middle ear cavity with a reddening and outward bulging of the eardrum, which may rupture. Second only to the common cold as a disease of childhood, otitis media is the most frequent diagnosis made by physicians who care for children. Infants and young children are at highest risk for otitis media; the peak prevalence is between 6 and 24 months of age.

Abnormal function of the auditory tube appears to be the most important mechanism in the pathogenesis of middle ear disease. This allows bacteria from the nasopharynx, which are a primary cause of middle ear infection, to enter the middle ear. Otitis media can be acute or chronic, and the primary etiology of both forms is bacterial infection. *Streptococcus pneumoniae* is the most common agent found in all age groups, followed by *Hemophilus influenzae.* Hearing loss is by far the most prevalent complication of otitis media.

Antimicrobial therapy has practically eliminated such complications of middle ear infections as mastoiditis and brain abscess, but perforation of the tympanic membrane and labyrinthitis still occur.

MOTION SICKNESS

Motion sickness is a functional disorder brought on by repetitive angular, linear, or vertical motion and characterized by various symptoms, primarily nausea and vomiting. Warning symptoms include yawning, salivation, pallor, hyperventilation, profuse cold sweating, and prolonged drowsiness. Specific kinds of motion sickness are seasickness, airsickness, carsickness, and space-, train-, or swing sickness.

Etiology is excessive stimulation of the vestibular apparatus by motion. The pathways taken by the afferent impulses from the labyrinth to the vomiting center in the medulla have not been defined. Visual stimuli and emotional factors like fear and anxiety can also contribute to motion sickness. Ideally, treatment with drugs of susceptible individuals should be instituted prior to their entering into conditions that would produce motion sickness, since prevention is more successful than treatment of symptoms once they have developed.

KEY MEDICAL TERMS ASSOCIATED WITH SENSORY STRUCTURES

Achromatopsia (*a* = without; *chrom* = color) Complete color blindness.

Ametropia (*ametro* = disproportionate; *ops* = eye) Refractive defect of the eye resulting in an inability to focus images properly on the retina.

Anopsia (*opsia* = vision) A defect of vision.

Astereognosis (*stereos* = solid; *gnosis* = knowledge) Loss of ability to recognize objects or to appreciate their form by touching them.

Blepharitis (*blepharo* = eyelid; *itis* = inflammation of) An inflammation of the eyelid.

Dyskinesia (*dys* = difficult; *kinesis* = movement) Abnormality of motor function characterized by involuntary, purposeless movements.

Eustachitis An inflammation or infection of the auditory (Eustachian) tube.

Keratitis (*kerato* = cornea) An inflammation or infection of the cornea.

Kinesthesis (*aisthesis* = sensation)   The sense of perception of movement.

Labyrinthitis   An inflammation of the labyrinth (inner ear).

Myringitis (*myringa* = eardrum)   An inflammation of the eardrum; also called tympanitis.

Nystagmus   A constant, rapid, involuntary movement of the eyeballs, possibly caused by a disease of the central nervous system.

Otalgia (*oto* = ear; *algia* = pain)   Earache.

Otosclerosis (*sclerosis* = hardening)   Pathological process in which new bone is deposited around the oval window. The result may be immobilization of the stapes, leading to deafness.

Presbyopia (*presby* = old)   Inability to focus on nearby ob-

jects due to loss of elasticity of the crystalline lens. The loss is usually caused by aging.

Ptosis (*ptosis* = fall)   Falling or drooping of the eyelid. (This term is also used for the slipping of any organ below its normal position.)

Retinoblastoma (*blast* = bud; *oma* = tumor)   A common tumor arising from immature retinal cells and accounting for 2 percent of childhood malignancies.

Strabismus   An eye muscle disorder, commonly called "crossed eyes," in which the eyeballs do not move in unison. It may be caused by lack of coordination of the extrinsic eye muscles.

Tinnitus   A ringing in the ears.

STUDY OUTLINE

Sensations (p. 488)

Definition

1. Sensation is a state of awareness of external and internal conditions of the body.
2. The prerequisites for sensation are reception of a stimulus, conversion of the stimulus into a nerve impulse by a receptor, conduction of the impulse to the brain, and translation of the impulse into a sensation by a region of the brain.

Characteristics

1. Projection occurs when the brain refers a sensation to the point of stimulation.
2. Adaptation is the loss of sensation even though the stimulus is still applied.
3. An afterimage is the persistence of a sensation even though the stimulus is removed.
4. Modality is the property by which one sensation is distinguished from another.

Classification of Receptors

1. According to location, receptors are classified as exteroceptors, visceroceptors, and proprioceptors.
2. On the basis of type of stimulus detected, receptors are classified as mechanoreceptors, thermoreceptors, nociceptors, electromagnetic receptors, and chemoreceptors.
3. In terms of simplicity or complexity, simple receptors are associated with general senses and complex receptors are associated with special senses.

General Senses (p. 489)

Cutaneous Sensations

1. Cutaneous sensations include tactile sensations (touch, pressure, vibration), thermoreceptive sensations (heat and cold), and pain. Receptors for these sensations are located in the skin, connective tissues, and the ends of the gastrointestinal tract.
2. Receptors for touch are root hair plexuses, free nerve endings, tactile (Merkel's) discs, corpuscles of touch (Meissner's corpuscles), and type II cutaneous mechanoreceptors (end or-

gans of Ruffini). Receptors for pressure are free nerve endings, type II cutaneous mechanoreceptors, and lamellated (Pacinian) corpuscles. Receptors for vibration are corpuscles of touch and lamellated corpuscles.
3. Pain receptors are located in nearly every body tissue.
4. Two kinds of pain recognized in the parietal lobe of the cortex are somatic and visceral.
5. Referred pain is felt in the skin near or away from the organ sending pain impulses.
6. Phantom pain is the sensation of pain in a limb that has been amputated.
7. Pain impulses may be inhibited by drugs, surgery, and acupuncture.

Proprioceptive Sensations

1. Receptors located in skeletal muscles, tendons, in and around joints, and the internal ear convey impulses related to muscle tone, movement of body parts, and body position.
2. The receptors include muscle spindles, tendon organs (Golgi tendon organs), joint kinesthetic receptors, and the maculae and cristae.

Levels of Sensation

1. Sensory fibers terminating in the lower brain stem bring about far more complex motor reactions than simple spinal reflexes.
2. When sensory impulses reach the lower brain stem, they cause subconscious motor reactions.
3. Sensory impulses that reach the thalamus can be localized crudely in the body.
4. When sensory impulses reach the cerebral cortex, we experience precise localization.

Sensory Pathways (p. 495)

1. Sensory information from all parts of the body terminates in a specific area of the somatosensory cortex.
2. In the posterior column pathway and the spinothalamic pathway there are first-order, second-order, and third-order neurons.
3. The neural pathway for pain and temperature is the lateral spinothalamic pathway.
4. The neural pathway for crude touch and pressure is the

anterior spinothalamic pathway.

5. The neural pathway for light touch, proprioception, and vibration is the posterior column pathway.

6. The pathways to the cerebellum are the anterior and posterior spinocerebellar tracts.

Motor Pathways (p. 496)

1. The muscles of all parts of the body are controlled by a specific area of the motor cortex.

2. Voluntary motor impulses are conveyed from the brain through the spinal cord along the pyramidal pathways and the extrapyramidal pathways.

3. Pyramidal pathways include the lateral corticospinal, anterior corticospinal, and corticobulbar tracts.

4. Major extrapyramidal tracts are the rubrospinal, tectospinal, and vestibulospinal tracts.

Special Senses (p. 499)

Olfactory Sensations

1. The receptors for olfaction are in the nasal epithelium.

2. Olfactory cells convey impulses to olfactory (I) nerves, olfactory bulbs, olfactory tracts, and cerebral cortex.

Gustatory Sensation

1. The receptors for gustation are located in taste buds.

2. The four primary tastes are salt, sweet, sour, and bitter.

3. Gustatory cells convey impulses to cranial nerves V, VII, IX, and X, the medulla, thalamus, and cerebral cortex.

Visual Sensations

1. Accessory structures of the eyes include the eyebrows, eyelids, eyelashes, and the lacrimal apparatus.

2. The eye is constructed of three coats: (a) fibrous tunic (sclera and cornea), (b) vascular tunic (choroid, ciliary body, and iris), and (c) nervous tunic (retina), which contains rods and cones.

3. The anterior cavity contains aqueous humor; the posterior cavity contains vitreous humor.

4. Rods and cones develop generator potentials and ganglion cells initiate nerve impulses.

5. Impulses from ganglion cells are conveyed through the retina to the optic (II) nerve, the optic chiasma, the optic tract, the thalamus, and the cortex.

Auditory Sensations and Equilibrium

1. The ear consists of three anatomical subdivisions: (a) the external or outer ear (pinna, external auditory canal, and tympanic membrane), (b) the middle ear (auditory tube, ossicles, oval window, and round window), and (c) the internal or inner ear (bony labyrinth and membranous labyrinth). The internal ear contains the spiral organ, the organ of hearing.

2. Sound waves enter the external auditory canal, strike the tympanic membrane, pass through the ossicles, strike the oval window, set up waves in the perilymph, strike the vestibular membrane and scala tympani, increase pressure in the endolymph, strike the basilar membrane, and stimulate hairs on the spiral organ. A sound impulse is then initiated.

3. Static equilibrium is the orientation of the body relative to the pull of gravity. The maculae of the utricle and saccule are the sense organs of static equilibrium.

4. Dynamic equilibrium is the maintenance of body position in response to movement. The cristae in the semicircular ducts are the sense organs of dynamic equilibrium.

Developmental Anatomy of the Eye and Ear (p. 519)

1. The eyes first appear as a depression of the diencephalon called the optic groove.

2. The internal ears develop from the otic placode, the middle ears develop from the first pharyngeal pouch, and the external ears develop from the first branchial groove.

Applications to Health (p. 521)

1. Cataract is the loss of transparency of the lens or capsule.

2. Glaucoma is abnormally high intraocular pressure, which destroys neurons of the retina.

3. Conjunctivitis is an inflammation of the conjunctiva.

4. Trachoma is a chronic, contagious inflammation of the conjunctiva.

5. Deafness is the lack of the sense of hearing or significant hearing loss.

6. Labyrinthine disease is basically a malfunction of the inner ear that has a variety of causes.

7. Ménière's syndrome is the malfunction of the inner ear that may cause deafness and loss of equilibrium.

8. Otitis media is an acute infection of the middle ear cavity.

9. Motion sickness is a functional disorder precipitated by repetitive angular, linear, or vertical motion.

REVIEW QUESTIONS

1. Define a sensation and a sense receptor. What prerequisites are necessary for the perception of a sensation?

2. Describe the following characteristics of a sensation: projection, adaptation, afterimage, modality.

3. Classify receptors on the basis of location, stimulus detected, and simplicity or complexity.

4. Distinguish between a general sense and a special sense.

5. What is a cutaneous sensation? Distinguish tactile, thermoreceptive, and pain sensations.

6. For each of the following cutaneous sensations, describe the receptor involved in terms of structure, function, and location: touch, pressure, vibration, and pain.

7. Why are pain receptors important? Differentiate somatic pain, visceral pain, referred pain, and phantom pain.

8. What is the proprioceptive sense? Where are the receptors for this sense located?

9. Describe how various parts of the body are represented in the somatosensory cortex.

10. Distinguish between the posterior column and spinothalamic pathways.

11. Describe how various parts of the body are represented in the motor cortex.

12. Which pathways control voluntary motor impulses from the brain through the spinal cord? How are they distinguished?
13. Identify the receptors for olfaction. Describe the neural pathway for olfaction.
14. Identify the receptors for taste. Describe the neural pathway for taste.
15. How are papillae related to taste buds? Describe the structure and location of the papillae.
16. Describe the structure and importance of the following accessory structures of the eye: eyelids, eyelashes, and eyebrows.
17. What is the function of the lacrimal apparatus? Explain how it operates.
18. By means of a labeled diagram, indicate the principal anatomical structures of the eye.
19. Describe the location and contents of the chambers of the eye. What is intraocular pressure? How is the scleral venous sinus (canal of Schlemm) related to this pressure?
20. Describe the path of a visual impulse from the optic (II) nerve to the brain.
21. Define visual field. Relate the visual field to image formation on the retina.
22. Diagram the principal parts of the outer, middle, and inner ear. Describe the function of each part labeled.
23. Explain the events involved in the transmission of sound from the pinna to the spiral organ.
24. What is the afferent pathway for sound impulses from the cochlear branch of the vestibulocochlear (VIII) nerve to the brain?
25. Compare the function of the maculae in the saccule and utricle in maintaining static equilibrium with the role of the cristae in the semicircular ducts in maintaining dynamic equilibrium.
26. Describe the path of an impulse that results in static and dynamic equilibrium.
27. Describe the development of the eye and ear.
28. Define each of the following: cataract, glaucoma, conjunctivitis, trachoma, deafness, labyrinthine disease, Ménière's syndrome, otitis media, and motion sickness.
29. Refer to the key medical terms associated with the sensory structures. Be sure that you can define each term listed.

21

The Endocrine System

Student Objectives

Define an endocrine gland and an exocrine gland and list the endocrine glands of the body.

Describe the structural and functional division of the pituitary gland into the adenohypophysis and the neurohypophysis.

Describe the location, histology, and blood and nerve supply of the pituitary gland.

Discuss the symptoms of pituitary dwarfism, giantism, acromegaly, and diabetes insipidus as pituitary gland disorders.

Describe the location, histology, and blood and nerve supply of the thyroid gland.

Discuss the symptoms of cretinism, myxedema, exophthalmic goiter, and simple goiter as thyroid gland disorders.

Describe the location, histology, and blood and nerve supply of the parathyroid glands.

Discuss the symptoms of tetany and osteitis fibrosa cystica as parathyroid gland disorders.

Describe the location, histology, and blood and nerve supply of the adrenal (suprarenal) glands.

Explain the subdivisions of the adrenal (suprarenal) glands into cortical and medullary portions.

Discuss the symptoms of aldosteronism, Addison's disease, Cushing's syndrome, and adre-

nogenital syndrome as adrenal cortical disorders.

Discuss the symptoms of pheochromocytomas as an adrenal medullary disorder.

Describe the location, histology, and blood and nerve supply of the pancreas.

Discuss the symptoms of diabetes mellitus and hyperinsulinism as endocrine disorders of the pancreas.

Explain why the ovaries and testes are classified as endocrine glands.

Describe the location, histology, and blood and nerve supply of the pineal gland.

Describe the location, histology, and blood and nerve supply of the thymus gland.

Explain why portions of the gastrointestinal tract, placenta, kidneys, and skin are considered endocrine structures.

Describe the effects of aging on the endocrine system.

Describe the development of the endocrine system.

Define key medical terms associated with the endocrine system.

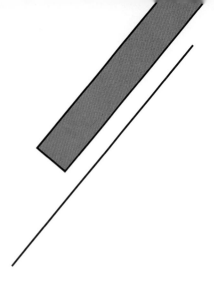

The nervous system controls homeostasis through electrical impulses delivered over neurons. The body's other control system, the **endocrine system,** affects bodily activities by releasing chemical messengers, called hormones, into the bloodstream. Whereas the nervous system causes muscles to contract and glands to secrete, the endocrine system brings about changes in the metabolic activities of body tissues. The release of nerve impulses is much more rapid than that of hormones. Also, the effects of nervous system stimulation are generally brief compared to the effects of endocrine stimulation.

Obviously, the body could not function if the two great control systems were to pull in opposite directions. The nervous and endocrine systems coordinate their activities like an interlocking supersystem. Certain parts of the nervous system stimulate or inhibit the release of hormones, and hormones, in turn, are quite capable of stimulating or inhibiting the flow of nerve impulses.

Although the effects of hormones are many and varied, their actions can be categorized into four broad areas:

1. They help to control the internal environment by regulating its chemical composition and volume.

2. They respond to marked changes in environmental conditions to help the body cope with emergency demands such as infection, trauma, emotional stress, dehydration, starvation, hemorrhage, and temperature extremes.

3. They assume a role in the smooth, sequential integration of growth and development.

4. They contribute to the basic processes of reproduction, including gamete production, fertilization, nourishment of the embryo and fetus, delivery, and nourishment of the newborn.

The developmental anatomy of the endocrine system is considered later in the chapter.

The science concerned with the structure and functions of the endocrine glands and the diagnosis and treatment of disorders of the endocrine system is called **endocrinology** (*endo* = within; *krinein* = to separate; *logos* = study of).

ENDOCRINE GLANDS

The body contains two kinds of glands: exocrine and endocrine. **Exocrine glands** secrete their products onto a free surface or into ducts. The ducts carry the secretions into body cavities, the lumens of various organs, or to the body's surface. Exocrine glands include sweat, sebaceous, mucous, and digestive glands. **Endocrine glands** by contrast, secrete their products (hormones) into the extracellular space around the secretory cells, rather than into ducts. The secretion then passes into capillaries to be transported in the blood. The endocrine glands make up the **endocrine system.** The endocrine glands of the body include the pituitary (hypophysis), thyroid, parathyroids, adrenals (suprarenals), pancreas, ovaries, testes, pineal (epiphysis cerebri), thymus, kidneys, stomach, small intestine, and placenta. (As you will see, some of these glands are mixed, that is, both endocrine and exocrine. (The pancreas is an example). The location of many organs of the endocrine system is illustrated in Figure 21-1.

The secretions of endocrine glands are called **hormones** (*hormone* = set in motion). The one thing all hormones have in common is the function of maintaining homeostasis by changing the physiological activities of cells. The general effects of most hormones are fairly well known. However, the manner in which specific hormones affect different cells of the body is just starting to become clear. The cells that respond to the effects of a hormone are called **target cells.** It appears as though hormones exert their effects on target cells in two quite different ways—one involves interaction with plasma membrane receptors and the other involves activation of genes.

The amount of hormone released by an endocrine gland or tissue is determined by the body's need for the hormone at any given time. Secretion is normally regulated so that there is no overproduction or underproduction of a particular hormone. This regulation is one of the very important ways that the body attempts to maintain homeostasis. Unfortunately, there are times when the regulating mechanism does not operate properly and

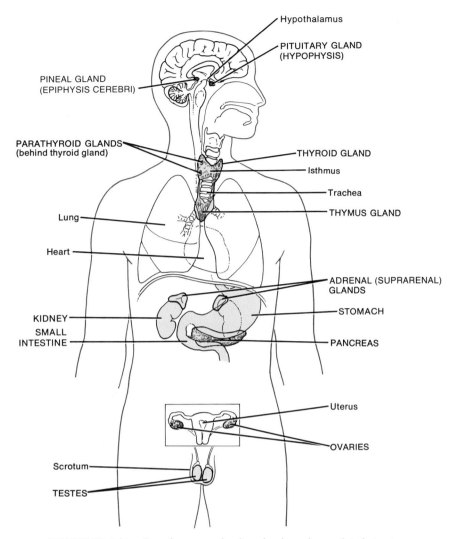

FIGURE 21-1 Location of many endocrine glands and associated structures.

hormonal levels are excessive or deficient. When this happens, disorders result.

The typical way in which hormonal secretions are regulated is by **negative feedback control.** As applied to hormones, information regarding the hormone level or its effect is fed back to the gland, which then responds accordingly. In one type of negative feedback system, the regulation of hormonal secretion does not involve direct participation by the nervous system. For example, blood calcium level is controlled by parathyroid hormone (PTH), produced by the parathyroid glands. If, for some reason, blood calcium level is low, this serves as a stimulus for the parathyroids to release more PTH. PTH then exerts its effects in various parts of the body until the blood calcium level is raised to normal. A high blood calcium level serves as a stimulus for the parathyroids to cease their production of PTH. In the absence of the hormone, other mechanisms take over until blood calcium level is lowered to normal. Note that in negative feedback control the body's response (increased or decreased cal-

cium level) is opposite (negative) to the stimulus (low or high calcium level). Other hormones that are regulated without direct involvement of the nervous system include calcitonin (CT), produced by the thyroid gland; insulin, by the pancreas; and aldosterone, by the adrenal cortex.

In other negative feedback systems, the hormone is released as a direct result of nerve impulses that stimulate the endocrine gland. Epinephrine and norepinephrine (NE) are released from the adrenal medulla in response to sympathetic nerve impulses. Antidiuretic hormone (ADH) is released from the posterior pituitary in response to nerve impulses from the hypothalamus.

One of the few exceptions to the rule of negative feedback control is oxytocin (OT). The regulating system for the release of OT from the pituitary in response to nerve impulses from the hypothalamus is a positive feedback cycle, that is, the output intensifies the input.

There are also negative feedback systems that involve the nervous system through chemical secretions from the hypothalamus, called **regulating factors.** Some regulating

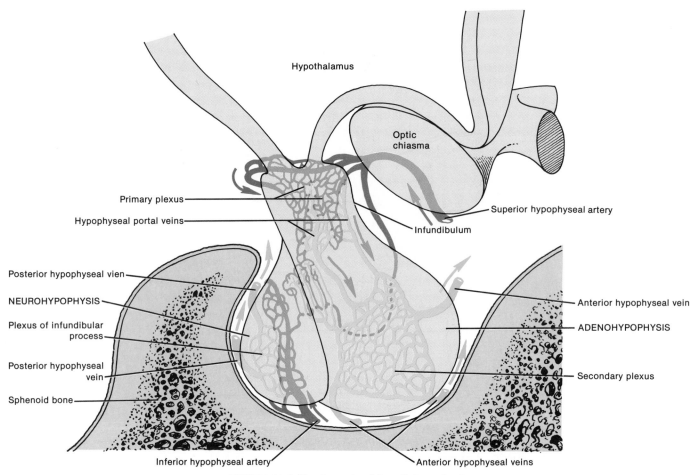

FIGURE 21-2 Blood supply of the pituitary gland.

factors, called **releasing factors,** stimulate the release of the hormone into the blood, so that it can exert its influence. Other regulating factors, called **inhibiting factors,** prevent the release of the hormone.

PITUITARY (HYPOPHYSIS)

The hormones of the **pituitary gland,** also called the **hypophysis** (hī-POF-i-sis), regulate so many body activities that the pituitary has been nicknamed the "master gland." It is a round structure and surprisingly small, measuring about 1.3 cm (0.5 in) in diameter. The pituitary lies in the sella turcica of the sphenoid bone. Posterior to the optic chiasma is a grayish protuberance, the **tuber cinereum,** which is part of the hypothalamus. The **median eminence** is a raised portion of the tuber cinereum to which is attached the **infundibulum,** a stalklike structure that attaches the pituitary gland to the hypothalamus (see Figure 18-1a).

The pituitary is divided structurally and functionally into an anterior lobe and a posterior lobe. Both are connected to the hypothalamus. The **anterior lobe** constitutes about 75 percent of the total weight of the gland.

It is derived from the ectoderm from an embryological invagination of pharyngeal epithelium. Accordingly, the anterior lobe contains many glandular epithelial cells and forms the glandular part of the pituitary. A system of blood vessels connects the anterior lobe with the hypothalamus.

The **posterior lobe** is also derived from the ectoderm, but from an outgrowth of the hypothalamus of the brain. Accordingly, the posterior lobe contains axonic ends of neurons whose cell bodies are located in the hypothalamus. The nerve fibers that terminate in the posterior lobe are supported by cells called pituicytes. Other nerve fibers connect the posterior lobe directly with the hypothalamus.

Between the lobes is a small, relatively avascular zone, the **pars intermedia.** Although much larger and more clearly defined in structure and function in some lower animals, its role in humans is obscure.

ADENOHYPOPHYSIS

The anterior lobe of the pituitary is also called the **adenohypophysis** (ad'-i-nō-hī-POF-i-sis). It releases hormones that regulate a whole range of bodily activities from growth to reproduction. The release of these hormones

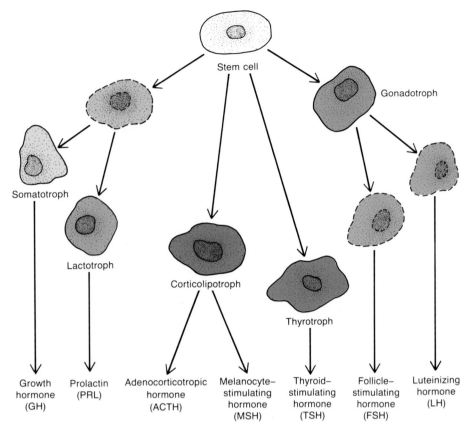

Growth hormone (GH)   Prolactin (PRL)   Adenocorticotropic hormone (ACTH)   Melanocyte–stimulating hormone (MSH)   Thyroid–stimulating hormone (TSH)   Follicle–stimulating hormone (FSH)   Luteinizing hormone (LH)

FIGURE 21-3 Cells of the adenohypophysis as revealed by special stains. Most cells that produce growth hormone (somatotrophs) and prolactin (lactotrophs) are separate cells. However, some normal and tumor cells are single cells that produce both growth hormone and prolactin. The corticolipotroph cell produces adrenocorticotropic hormone and melanocyte-stimulating hormone. Thyrotroph cells synthesize thyroid-stimulating hormone. Most gonadotroph cells produce both follicle-stimulating hormone and luteinizing hormone. However, a few separate cells may exist, some producing follicle-stimulating hormone and some producing luteinizing hormone. (Adapted from a slide provided by Calvin Ezrin, M.D., Clinical Professor of Medicine, U.C.L.A., and Adjunct Professor of Pathology, University of Toronto.)

is either stimulated or inhibited by chemical secretions from the hypothalamus called **regulating factors.** Regulating factors will be considered along with each of the anterior pituitary hormones. These factors constitute an important link between the nervous system and the endocrine system.

The hypothalamic regulating factors are delivered to the adenohypophysis in the following way. The blood supply to the adenohypophysis and infundibulum is derived principally from several *superior hypophyseal* (hī-po-FIZ-ē-al) *arteries.* These arteries are branches of the internal carotid and posterior communicating arteries (Figure 21-2). The superior hypophyseal arteries form a network or plexus of capillaries, the *primary plexus,* in the infundibulum near the inferior portion of the hypothalamus. Regulating factors from the hypothalamus diffuse into this plexus. This plexus drains into the *hypophyseal portal veins,* that pass down the infundibulum. At the inferior portion of the infundibulum, the veins form

a *secondary plexus* in the adenohypophysis. From this plexus, hormones of the adenohypophysis pass into the anterior hypophyseal veins for distribution to tissue cells. Such a delivery system permits regulating factors to act quickly on the adenohypophysis without first circulating through the heart. The short route prevents dilution or destruction of the regulating factors.

When the adenohypophysis receives proper stimulation from the hypothalamus via regulating factors, its glandular cells secrete any one of seven hormones. Recently, special staining techniques have established the division of glandular cells into five principal types (Figure 21-3):

1. Somatotroph cells produce **growth hormone (GH),** which controls general body growth.

2. Lactotroph cells synthesize **prolactin (PRL),** which initiates milk production by the mammary glands.

3. Corticolipotroph cells synthesize **adrenocortico-**

tropic hormone (ACTH), which stimulates the adrenal cortex to secrete its hormones, and **melanocyte-stimulating hormone (MSH)**, which is related to skin pigmentation.

4. Thyrotroph cells manufacture **thyroid-stimulating hormone (TSH)**, which controls the thyroid gland.

5. Gonadotroph cells produce **follicle-stimulating hormone (FSH)**, which stimulates the production of eggs and sperm in the ovaries and testes, respectively, and **luteinizing hormone (LH)**, which stimulates other sexual and reproductive activities.

Except for the growth hormone (GH), melanocyte-stimulating hormone (MSH), and prolactin (PRL), all the secretions are referred to as **tropic hormones** (*trop* = turn on), which means that they stimulate other endocrine glands. Follicle-stimulating hormone (FSH) and luteinizing hormone (LH) are also called **gonadotropic** (gō-nad-ō-TRŌ-pik) **hormones** because they regulate the functions of the gonads. The gonads (ovaries and testes) are the endocrine glands that produce sex steroid hormones.

CLINICAL APPLICATION

Disorders of the endocrine system, in general, involve **hyposecretion** (underproduction) of hormones or **hypersecretion** (overproduction).

Among the clinically interesting disorders related to the adenohypophysis are those involving GH. If GH is hyposecreted during the growth years, bone growth is slow and the epiphyseal plates close before normal height is reached. This condition is called **pituitary dwarfism**. Other organs of the body also fail to grow, and the pituitary dwarf is childlike in many physical respects. Treatment requires administration of GH during childhood before the epiphyseal plates close.

Hypersecretion of GH during childhood results in **giantism**, an abnormal increase in the length of long bones. Hypersecretion during adulthood is called **acromegaly** (ak'-rō-MEG-a-lē). Acromegaly cannot produce further lengthening of the long bones because the epiphyseal plates are already closed. Instead, the bones of the hands, feet, cheeks, and jaws thicken. Other tissues also grow. The eyelids, lips, tongue, and nose enlarge and the skin thickens and furrows, especially on the forehead and soles of the feet.

NEUROHYPOPHYSIS

In a strict sense, the posterior lobe, or **neurohypophysis,** is not an endocrine gland, since it does not synthesize hormones. The posterior lobe consists of cells called **pitui-**

cytes (pi-TOO-i-sītz), which are similar in appearance to the neuroglia of the nervous system. It also contains axon terminations of secretory neurons of the hypothalamus (Figure 21-4). Such neurons are called **neurosecretory cells.** The cell bodies of the neurons originate in nuclei in the hypothalamus. The fibers project from the hypothalamus, form the **hypothalamic-hypophyseal tract,** and terminate on blood capillaries in the neurohypophysis. The cell bodies of the neurosecretory cells produce two hormones: **oxytocin (OT)** and **antidiuretic hormone (ADH).** OT is produced primarily in the paraventricular nucleus and ADH is synthesized primarily in the supraoptic nucleus.

OT stimulates contraction of the smooth muscle of the pregnant uterus during labor and stimulates contractile cells around the ducts of the mammary glands to eject milk. ADH prevents excessive urine production by bringing about water reabsorption and secondarily causes blood pressure to rise by bringing about constriction of arterioles.

Following their production, the hormones are transported in the neuron fibers by a carrier protein called *neurophysin* into the neurohypophysis and stored in the axon terminals. Later, when the hypothalamus is properly stimulated, it sends impulses over the neurosecretory cells. The impulses cause the release of the hormones from the axon terminals into the blood.

The blood supply to the neurohypophysis is from the *inferior hypophyseal arteries,* derived from the internal carotid arteries. In the neurohypophysis, the inferior hypophyseal arteries form a plexus of capillaries called the *plexus of the infundibular process.* From this plexus hormones stored in the neurohypophysis pass into the *posterior hypophyseal veins* for distribution to tissue cells.

CLINICAL APPLICATION

The principal abnormality associated with dysfunction of the neurohypophysis is **diabetes insipidus** (in-SIP-i-dus). Diabetes means overflow and insipidus means tasteless. This disorder should not be confused with diabetes mellitus (*meli* = honey), a disorder of the pancreas characterized by sugar in the urine. Diabetes insipidus is the result of a hyposecretion of ADH, usually caused by damage to the neurohypophysis or the supraoptic nucleus in the hypothalamus. Symptoms include excretion of large amounts of urine and subsequent thirst. Diabetes insipidus is treated by administering ADH.

A summary of pituitary gland hormones, actions, and disorders is presented in Exhibit 21-1.

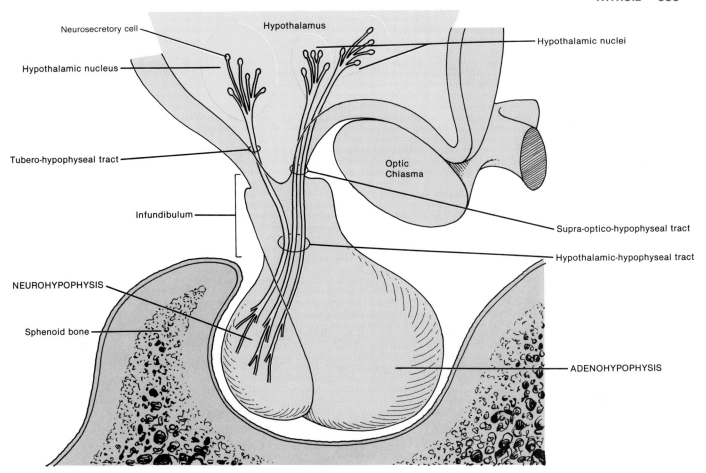

Neurosecretory cell

Hypothalamic nucleus

Tubero-hypophyseal tract

Infundibulum

NEUROHYPOPHYSIS

Sphenoid bone

Hypothalamus

Hypothalamic nuclei

Optic Chiasma

Supra-optico-hypophyseal tract

Hypothalamic-hypophyseal tract

ADENOHYPOPHYSIS

FIGURE 21-4 Hypothalamic-hypophyseal tract.

THYROID

The **thyroid gland** is located just below the larynx. The right and left **lateral lobes** lie one on either side of the trachea. The lobes are connected by a mass of tissue called an **isthmus** (IS-mus) that lies in front of the trachea, just below the cricoid cartilage (Figure 21-5). The **pyramidal lobe,** when present, extends upward from the isthmus. The gland weighs about 25 g (almost 1 oz) and has a rich blood supply, receiving about 80 to 120 ml of blood per minute.

Histologically, the thyroid gland is composed of spherical sacs called **thyroid follicles** (Figure 21-6). The walls of each follicle consist of cells that reach the surface of the lumen of the follicle (**follicular cells)** and cells that do not reach the lumen (**parafollicular,** or **C, cells).** When the cells are inactive, they tend to be cuboidal, but when actively secreting hormones, they become more columnar. The follicular cells manufacture **thyroxine** (thī-ROK-sēn), or T_4, since it contains four atoms of iodine, and **triiodothyronine** (trī-ī'-od-ō-THĪ-rō-nēn), or T_3, since it contains three atoms of iodine. Together these hormones are referred to as the **thyroid hormones.** Thyroxine is normally secreted in greater quantity than triiodothyro-

nine, but triiodothyronine is 3 to 4 times more potent. Moreover, in peripheral tissues, especially the liver and lungs, about one-third of triiodothyronine is converted to thyroxine. Both hormones are similar functionally. They control metabolism, regulate growth and development, and increase reactivity of the nervous system. The parafollicular cells produce **calcitonin** (kal-si-TŌ-nin), or **CT.** This hormone decreases blood levels of calcium and phosphate by inhibiting bone breakdown (osteoclastic activity) and accelerating calcium absorption by bones.

One of the thyroid gland's unique features is its ability to store hormones and release them in a steady flow over a long period of time. In the follicle cells, iodide is oxidized to iodine. Through a series of enzymatically controlled reactions, the iodine combines with the amino acid tyrosine to form the thyroid hormones. This combination occurs within a large glycoprotein molecule, called **thyroglobulin (TGB),** which is secreted by the follicle cells into the follicle.

The main blood supply is from the superior thyroid artery, a branch of the external carotid artery, and the inferior thyroid artery, a branch of the subclavian artery. The thyroid is drained by the superior and middle thyroid veins, which pass into the internal jugular veins, and the

EXHIBIT 21-1 SUMMARY OF PITUITARY GLAND HORMONES, ACTIONS, AND DISORDERS

HORMONE	PRINCIPAL ACTIONS	DISORDERS
Adenohypophyseal hormones		
Growth hormone (GH)	Growth of body cells; protein anabolism.	Hyposecretion of GH during the growth years results in pituitary dwarfism.
Thyroid-stimulating hormone (TSH)	Controls secretion of hormones by thyroid gland.	Hypersecretion of GH during the growth years results in giantism, and hypersecretion of GH during adulthood results in acromegaly.
Adrenocorticotropic hormone (ACTH)	Controls secretion of some hormones by adrenal cortex.	
Follicle-stimulating hormone (FSH)	In female, initiates development of ova and induces ovarian secretion of estrogens. In male, stimulates testes to produce sperm.	
Luteinizing hormone (LH)	In female, together with estrogens, stimulates ovulation and formation of progesterone-producing corpus luteum, prepares uterus for implantation, and readies mammary glands to secrete milk. In male, stimulates interstitial cells in testes to develop and produce testosterone.	
Prolactin (PRL)	Together with other hormones, initiates and maintains milk secretion by the mammary glands.	
Melanocyte-stimulating hormone (MSH)	Stimulates dispersion of melanin granules in melanocytes.	
Neurohypophyseal hormones		
Oxytocin (OT)	Stimulates contraction of smooth muscle cells of pregnant uterus during labor and stimulates contraction of contractile cells of mammary glands for milk ejection.	
Antidiuretic hormone (ADH)	Principal effect is to decrease urine volume; also to raise blood pressure by constricting arteries during severe hemorrhage.	Hyposecretion of ADH results in diabetes insipidus.

inferior thyroid veins, which join the brachiocephalic veins or internal jugular veins.

The nerve supply of the thyroid consists of postganglionic fibers from the superior and middle cervical sympathetic ganglia. Preganglionic fibers from the ganglia are derived from the second through seventh thoracic segments of the spinal cord.

CLINICAL APPLICATION

Hyposecretion of thyroid hormones during the growth years results in **cretinism** (KRĒ-tin-izm). Two outstanding clinical symptoms of the cretin are dwarfism and mental retardation. The first is caused by failure of the skeleton to grow and mature. The second is caused by failure of the brain to develop fully. Recall that one function of thyroid hormones is to control tissue growth and development. Cretins also exhibit

retarded sexual development and a yellowish skin color. Flat pads of fat develop, giving the cretin a characteristic round face and thick nose; a large, thick, protruding tongue; and protruding abdomen. Because the energy-producing metabolic reactions are slow, the cretin has a low body temperature and general lethargy. Carbohydrates are stored rather than utilized. Heart rate is also slow. If the condition is diagnosed early, the symptoms can be eliminated by administering thyroid hormones.

Hypothyroidism during the adult years produces **myxedema** (mix-e-DĒ-ma). A hallmark of this disorder is an edema that causes the facial tissues to swell and look puffy. Like the cretin, the person with myxedema suffers from slow heart rate, low body temperature, hoarseness, dry skin, muscular weakness, general lethargy, constipation, cold intolerance, and a tendency to gain weight easily. The long-term effect of a slow

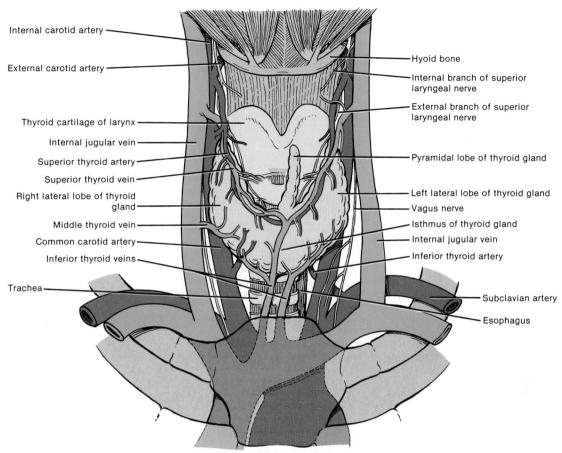

Internal carotid artery

External carotid artery

Thyroid cartilage of larynx

Internal jugular vein

Superior thyroid artery

Superior thyroid vein

Right lateral lobe of thyroid gland

Middle thyroid vein

Common carotid artery

Inferior thyroid veins

Trachea

Hyoid bone

Internal branch of superior laryngeal nerve

External branch of superior laryngeal nerve

Pyramidal lobe of thyroid gland

Left lateral lobe of thyroid gland

Vagus nerve

Isthmus of thyroid gland

Internal jugular vein

Inferior thyroid artery

Subclavian artery

Esophagus

FIGURE 21-5 Location and blood supply of the thyroid gland in anterior view. The pyramidal lobe of the thyroid is not constant and, when present, may be attached to the hyoid bone by a muscle.

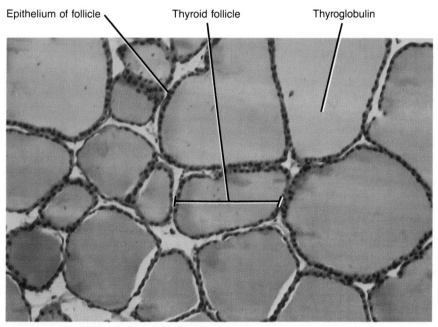

Epithelium of follicle

Thyroid follicle

Thyroglobulin

FIGURE 21-6 Histology of the thyroid gland. Photomicrograph at a magnification of 230×. (© 1983 by Michael H. Ross. Used by permission.)

EXHIBIT 21-2   SUMMARY OF THYROID GLAND HORMONES, ACTIONS, AND DISORDERS

HORMONE	PRINCIPAL ACTIONS	DISORDERS
Thyroid hormones		
Thyroxine (T₄)	Regulates metabolism, growth and development, and activity of nervous system.	Hyposecretion of thyroid hormones during the growth years results in cretinism. Hypothyroidism during adulthood produces myxedema. Hypersecretion of thyroid hormones produces exophthalmic goiter.
Triiodothyronine (T₃)	Same as above.	
Calcitonin (CT)	Lowers blood levels of calcium by accelerating calcium absorption by bones.	

heart rate may overwork the heart muscle, causing the heart to enlarge. Because the brain has already reached maturity, the person with myxedema does not experience mental retardation. However, in moderately severe cases, nerve reactivity may be dulled so that the person lacks mental alertness. Myxedema occurs eight times more frequently in females than in males. Its symptoms are abolished by the administration of thyroxine.

Hypersecretion of thyroid hormones gives rise to **exophthalmic** (ek'-sof-THAL-mik) **goiter** (GOY-ter). This disease, like myxedema, is also more frequent in females. One of its primary symptoms is an enlarged thyroid, called a **goiter,** which may be two to three times its original size. Two other symptoms are an edema behind the eye, which causes the eye to protrude **(exophthalmos),** and an abnormally high metabolic rate. The high metabolic rate produces a range of effects that are generally opposite to those of myxedema—increased pulse, high body temperature, and moist, flushed skin. The person loses weight and is usually full of "nervous" energy. The thyroid hormones also increase the responsiveness of the nervous system, causing the person to become irritable and exhibit tremors of the extended fingers. Hyperthyroidism is usually treated by administering drugs that suppress thyroid hormone synthesis or by surgically removing part of the gland.

Goiter is a symptom of many thyroid disorders. It may also occur if the gland does not receive enough iodine to produce sufficient thyroxine for the body's needs. The follicular cells then enlarge in a futile attempt to produce more thyroid hormones and they secrete large quantities of thyroglobulin (TGB). This condition is called **simple goiter.** Simple goiter is most often caused by a lower than average amount of iodine in the diet. It may also develop if iodine intake is not increased during certain conditions that put a high demand on the body for thyroxine, such as frequent exposure to cold and high fat and protein diets.

A summary of thyroid gland hormones, actions, and disorders is presented in Exhibit 21-2.

PARATHYROIDS

Typically embedded on the posterior surfaces of the lateral lobes of the thyroid are small, round masses of tissue called the **parathyroid glands.** Usually, two parathyroids, superior and inferior, are attached to each lateral thyroid lobe (Figure 21-7). They measure about 3 to 6 mm (0.1 to 0.3 in) in length, 2 to 5 mm (0.07 to 0.2 in) in width, and 0.5 to 2 mm (0.02 to 0.07 in) in thickness.

Histologically, the parathyroids contain two kinds of epithelial cells (Figure 21-8). The more numerous cells, called **principal (chief) cells,** are believed to be the major synthesizer of **parathyroid hormone (PTH).** Some researchers believe that the other kind of cell, called an **oxyphil cell,** synthesizes a reserve capacity of hormone. Functionally, PTH increases blood calcium level and decreases blood phosphate level by increasing the rate of calcium absorption from the gastrointestinal tract into the blood; increases the number and activity of osteoclasts; increases calcium absorption by the kidneys; increases phosphate excretion by the kidneys; and activates vitamin D.

The parathyroids are abundantly supplied with blood from branches of the superior and inferior thyroid arteries. Blood is drained by the superior, middle, and inferior thyroid veins. The nerve supply of the parathyroids is derived from the cervical sympathetic ganglia and the pharyngeal branch of the vagus.

CLINICAL APPLICATION

A normal amount of calcium in the extracellular fluid is necessary to maintain the resting state of neurons. A deficiency of calcium caused by hypoparathyroidism (abnormally diminished function of the parathyroid glands) causes neurons to depolarize without the usual

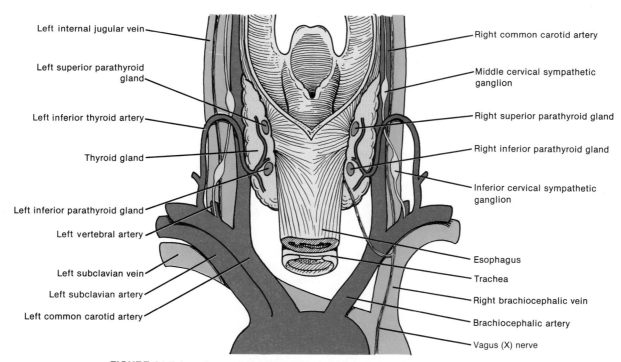

Left internal jugular vein

Left superior parathyroid gland

Left inferior thyroid artery

Thyroid gland

Left inferior parathyroid gland

Left vertebral artery

Left subclavian vein

Left subclavian artery

Left common carotid artery

Right common carotid artery

Middle cervical sympathetic ganglion

Right superior parathyroid gland

Right inferior parathyroid gland

Inferior cervical sympathetic ganglion

Esophagus

Trachea

Right brachiocephalic vein

Brachiocephalic artery

Vagus (X) nerve

FIGURE 21-7 Location and blood supply of the parathyroid glands in posterior view.

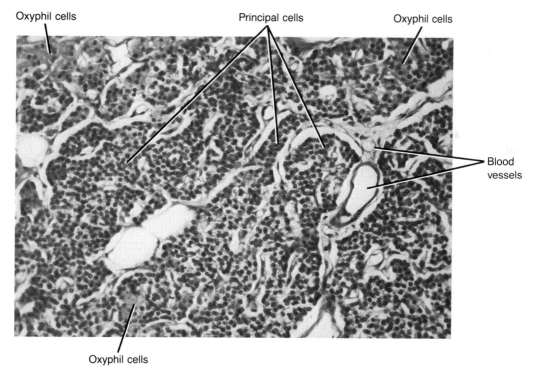

Oxyphil cells

Principal cells

Oxyphil cells

Blood vessels

Oxyphil cells

FIGURE 21-8 Histology of the parathyroid glands. Photomicrograph at a magnification of 180×. (© 1983 by Michael H. Ross. Used by permission.)

EXHIBIT 21-3 SUMMARY OF PARATHYROID GLAND HORMONE, ACTIONS, AND DISORDERS

HORMONE	PRINCIPAL ACTIONS	DISORDERS
Parathyroid hormone (PTH)	Increases blood calcium level and decreases blood phosphate level by increasing rate of calcium absorption from gastrointestinal tract into blood; increases number and activity of osteoclasts; increases calcium absorption by kidneys; increases phosphate excretion by kidneys; and activates vitamin D.	Hypoparathyroidism results in tetany. Hyperparathyroidism produces osteitis fibrosa cystica.

stimulus. As a result, nerve impulses increase and result in muscle twitches, spasms, and convulsions. This condition is called **tetany.** The effects of hypocalcemic tetany are observed in the *Trousseau* (troo-SŌ) and *Chvostek* (VOS-tek) *signs.* The Trousseau sign is observed when the binding of a blood pressure cuff around the arm produces contraction of the fingers and inability to open the hand. The Chvostek sign is a contracture of the facial muscles elicited by tapping the facial nerves at the angle of the jaw. Hypoparathyroidism results from surgical removal of the parathyroids or from parathyroid damage caused by parathyroid disease, infection, hemorrhage, or mechanical injury. Hyperparathyroidism (abnormally excessive function of the parathyroid glands) causes demineralization of bone. This condition is called **osteitis fibrosa cystica** because the areas of destroyed bone tissue are replaced by cavities that fill with fibrous tissue. The bones thus become deformed and are highly susceptible to fracture. Hyperparathyroidism is usually caused by a tumor in the parathyroids.

A summary of parathyroid hormones, actions, and disorders is presented in Exhibit 21-3.

ADRENALS (SUPRARENALS)

The body has two **adrenal (suprarenal) glands,** one of which is located superior to each kidney (Figure 21-9). Each adrenal gland is structurally and functionally differentiated into two sections: the outer **adrenal cortex,** which makes up the bulk of the gland, and the inner **adrenal medulla** (Figure 21-10). Whereas the adrenal cortex is derived from mesoderm, the adrenal medulla is derived from the ectoderm. Covering the gland are an inner, thick layer of fatty connective tissue and an outer, thin fibrous capsule.

The average dimensions of the adult adrenal gland are about 50 mm (2 in) in length, 30 mm (1.1 in) in width, and 10 mm (0.4 in) in thickness. The adrenals, like the thyroid, are among the most vascular organs of the body.

ADRENAL CORTEX

Histologically, the cortex is subdivided into three zones (Figure 21-10). Each zone has a different cellular arrangement and secretes different groups of steroid hormones. The outer zone, directly underneath the connective tissue capsule, is referred to as the **zona glomerulosa.** It comprises about 15 percent of the total cortical volume. Its cells are arranged in arched loops or round balls. Its primary secretions are a group of hormones called **mineralocorticoids** (min'-er-al-ō-KOR-ti-koyds). The principal mineralocorticoid is **aldosterone** (al-dō-STER-ōn), which causes the kidneys to reabsorb sodium and water and to increase potassium excretion.

CLINICAL APPLICATION

Hypersecretion of the mineralocorticoid aldosterone results in **aldosteronism,** characterized by a decrease in the body's potassium concentration. If potassium depletion is great, neurons cannot depolarize and muscular paralysis results. Hypersecretion also brings about excessive retention of sodium and water. The water increases the volume of the blood and causes high blood pressure. It also increases the volume of the interstitial fluid, producing edema.

The middle zone, or **zona fasciculata,** is the widest of the three zones and consists of cells arranged in long, straight cords. The zona fasciculata secretes mainly **glucocorticoids** (gloo'-kō-KOR-ti-koyds). These include **cortisol (hydrocortisone), corticosterone,** and **cortisone.** Together the hormones work with other hormones to promote normal metabolism, help resist stress, and decrease edema caused by the inflammatory response.

CLINICAL APPLICATION

Hyposecretion of glucocorticoids results in the condition called **Addison's disease (primary adrenal insufficiency).** Clinical symptoms include hypoglycemia, which leads to muscular weakness, mental lethargy,

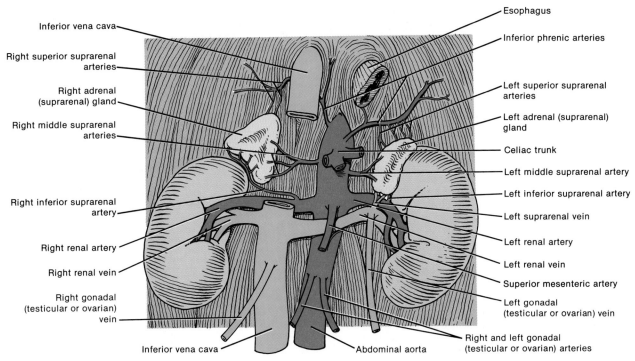

FIGURE 21-9 Location and blood supply of the adrenal (suprarenal) glands.

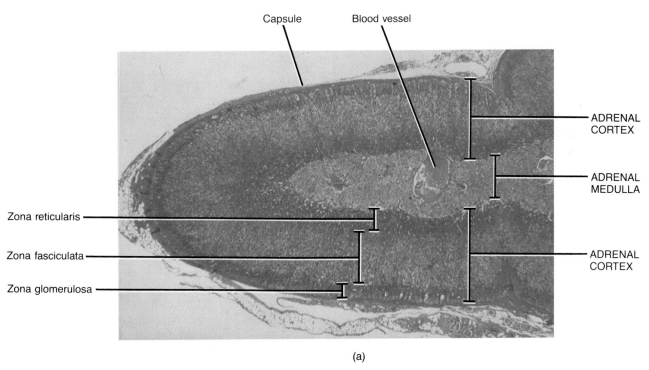

(a)

FIGURE 21-10 Histology of the adrenal (suprarenal) glands. (a) Photomicrograph of a section of the adrenal gland showing its subdivisions and zones at a magnification of 25×.

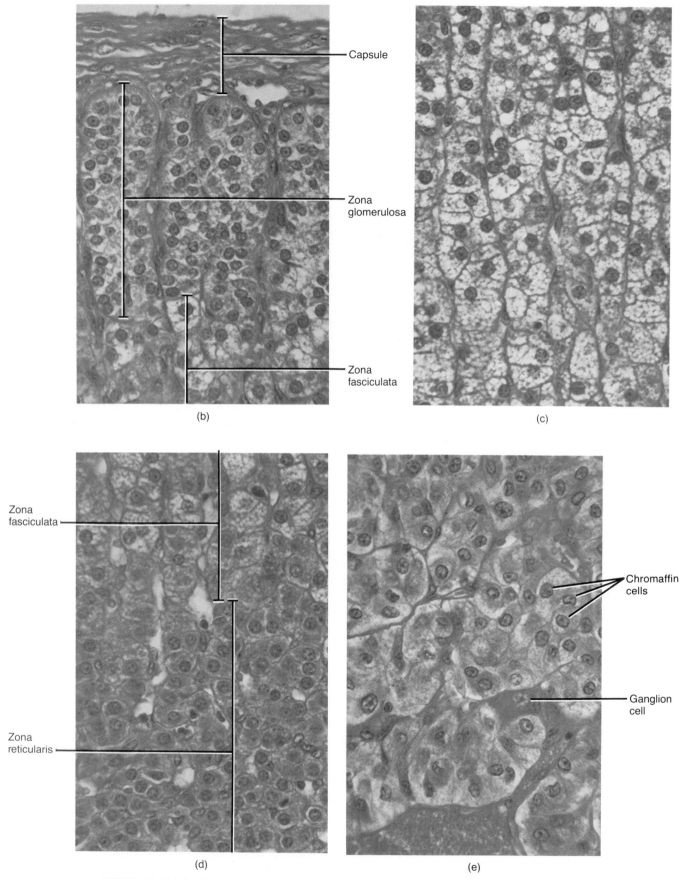

Capsule

Zona
glomerulosa

Zona
fasciculata

(b)

(c)

Zona
fasciculata

Zona
reticularis

(d)

Chromaffin
cells

Ganglion
cell

(e)

FIGURE 21-10 (*Continued*) Histology of the adrenal (suprarenal) glands. (b) Capsule, zona glomerulosa, and part of the zona fasciculata at a magnification of 400×. (c) Zona fasciculata at a magnification
of 400×. (d) Zona fasciculata and zona reticularis at a magnification of 400×. (e) Medulla at a
magnification of 400×. (© 1983 by Michael H. Ross. Used by permission.)

and weight loss. Increased potassium and decreased sodium lead to low blood pressure and dehydration.

Cushing's syndrome is a hypersecretion of glucocorticoids, especially cortisol and cortisone. The condition is characterized by the redistribution of fat. The result is spindly legs accompanied by a "moon face," "buffalo hump" on the back, and pendulous abdomen. Facial skin is flushed, and the skin covering the abdomen develops stretch marks. The individual also bruises easily, and wound healing is poor.

The inner zone, the **zona reticularis,** contains cords of cells that branch freely. This zone synthesizes minute amounts of hormones, mainly **gonadocorticoids** (gō-na-dō-KŌR-ti-koyds), or **sex hormones.** These are estrogens and androgens. Estrogens are several closely related female sex hormones that are also produced by the ovaries and placenta. Androgens are male sex hormones. An important androgen, called testosterone, is produced by the testes. The concentration of sex hormones secreted by normal adult adrenals is usually so low that their effects are insignificant. Their functions will be discussed shortly.

CLINICAL APPLICATION

The **adrenogenital syndrome** usually refers to a group of enzyme deficiencies that block the synthesis of glucocorticoids. In an attempt to compensate, the anterior pituitary secretes more ACTH. As a result, excess androgenic (male) hormones are produced, causing *virilism,* or masculinization. For instance, the female develops extremely virile characteristics such as growth of a beard, development of a much deeper voice, occasionally development of baldness, development of a masculine distribution of hair on the body and on the pubis, growth of the clitoris that resembles a penis, and deposition of proteins in the skin and muscles producing typical masculine characteristics. Such virilism may also result from tumors of the adrenal gland called *virilizing adenomas* (*aden* = gland; *oma* = tumor).

In the prepubertal male, the syndrome causes the same characteristics as in the female, plus rapid development of the male sexual organs and creation of male sexual desires. In the adult male, the virilizing characteristics of the adrenogenital syndrome are usually completely obscured by the normal virilizing characteristics of the testosterone secreted by the testes. As a result, it is often difficult to make a diagnosis of adrenogenital syndrome in the male adult. However, an occasional adrenal tumor secretes sufficient quantities of feminizing hormones that the male patient develops **gynecomastia** (*gyneca* = woman, *mast* = breast), which means excessive growth of the male mammary glands. Such a tumor is called a *feminizing adenoma.*

ADRENAL MEDULLA

The adrenal medulla consists of hormone-producing cells, called **chromaffin** (krō-MAF-in) **cells,** which surround large blood-containing sinuses. Chromaffin cells develop from the same source as the postganglionic cells of the sympathetic division of the nervous system. They are directly innervated by preganglionic cells of the sympathetic division of the autonomic nervous system and may be regarded as postganglionic cells that are specialized to secrete. In all other visceral effectors, preganglionic sympathetic fibers first synapse with postganglionic neurons before innervating the effector. In the adrenal medulla, however, the preganglionic fibers pass directly into the chromaffin cells of the gland. The secretion of hormones from the chromaffin cells is directly controlled by the autonomic nervous system, and innervation by the preganglionic fibers allows the gland to respond rapidly to a stimulus.

The two principal hormones synthesized by the adrenal medulla are **epinephrine** and **norepinephrine (NE),** also called adrenaline and noradrenaline, respectively. Epinephrine constitutes about 80 percent of the total secretion of the gland and is more potent in its action than norepinephrine. Both hormones are **sympathomimetic** (sim'-pa-thō-mi-MET-ik), that is, they produce effects that mimic those brought about by the sympathetic division of the autonomic nervous system. To a large extent, they are responsible for the fight-or-flight response. Like the glucocorticoids of the adrenal cortices, these hormones help the body resist stress. However, unlike the cortical hormones, the medullary hormones are not essential for life.

The main arteries that supply the adrenal glands are the several superior suprarenal arteries arising from the inferior phrenic artery, the middle suprarenal artery from the aorta, and the inferior suprarenal arteries from the renal arteries. The suprarenal vein of the right adrenal gland drains into the inferior vena cava, whereas the suprarenal vein of the left adrenal gland empties into the left renal vein.

The principal nerve supply to the adrenal glands is from preganglionic fibers from the splanchnic nerves and from the celiac and associated sympathetic plexuses. These myelinated fibers end on the secretory cells of the gland found in a region of the medulla.

CLINICAL APPLICATION

Tumors of the chromaffin cells of the adrenal medulla, called **pheochromocytomas** (fē-ō-krō'-mō-sī-TŌ-mas), cause hypersecretion of the medullary hormones. The oversecretion causes high blood pressure, high levels of sugar in the blood and urine, an elevated basal metabolic rate, nervousness, profuse sweating, weight loss,

EXHIBIT 21-4   SUMMARY OF HORMONES PRODUCED BY THE ADRENAL GLANDS, ACTIONS, AND DISORDERS

HORMONE	PRINCIPAL ACTIONS	DISORDERS
Adrenal cortical hormones		
Mineralocorticoids (mainly aldosterone)	Increase blood levels of sodium and water and decrease blood levels of potassium.	Hypersecretion of aldosterone results in aldosteronism.
Glucocorticoids (mainly cortisol)	Help promote normal metabolism, resistance to stress, and counter inflammatory response.	Hyposecretion of glucocorticoids produces Addison's disease. Hypersecretion results in Cushing's syndrome.
Gonadocorticoids	Concentrations secreted by adults are so low that their effects are insignificant.	Androgenital syndrome inhibits synthesis of glucocorticoids that results in excess production of ACTH and androgens, causing virilism. The release of sufficient feminizing hormones in males causes gynecomastia.
Adrenal medullary hormones		
Epinephrine	Sympathomimetic, that is, produces effects that mimic those of ANS during stress.	Hypersecretion of medullary hormones results in a prolonged fight-or-flight response.
Norepinephrine (NE)	Same as above.	

tachycardia, fever, chest tightness, nausea, vomiting, and anxiety. Since the medullary hormones create the same effects as sympathetic nervous stimulation, hypersecretion puts the individual into a prolonged version of the fight-or-flight response. This condition ultimately wears out the body, and the individual eventually suffers from general weakness. The vast majority of pheochromocytoma patients are generally cured after surgery.

A summary of adrenal gland hormones, actions, and disorders is presented in Exhibit 21-4.

PANCREAS

The **pancreas** can be classified as both an endocrine and an exocrine gland. We shall treat its endocrine functions at this point; its exocrine functions are discussed in the chapter on the digestive system (Chapter 22). The pancreas is a flattened organ located posterior and slightly inferior to the stomach (Figure 21-11). The adult pancreas consists of a head, body, and tail. Its average length is about 12.5 cm (6 in), and its average weight is about 85 g (3 oz).

The endocrine portion of the pancreas consists of clusters of cells called **pancreatic islets**, or **islets of Langerhans** (LAHNG-er-hanz) (Figure 21-12). Three kinds of cells are found in these clusters: (1) **alpha cells,** which secrete the hormone **glucagon;** (2) **beta cells,** which secrete the hormone **insulin,** and (3) **delta cells,** which secrete **growth hormone-inhibiting factor (GHIF),** or **somatostatin,** a hormone that inhibits the secretion of insulin. The

islets are surrounded by blood capillaries and by cells (acini) that form the exocrine part of the gland.

Glucagon (GLOO-ka-gon) increases blood sugar level by accelerating the conversion of glycogen in the liver into glucose (glycogenolysis) and the conversion in the liver of other nutrients, such as amino acids, glycerol, and lactic acid, into glucose (gluconeogenesis). The liver then releases the glucose into the blood, and the blood sugar level rises. Insulin decreases blood sugar level in several ways. It accelerates the transport of glucose from the blood into cells, especially skeletal muscle cells. It also accelerates the conversion of glucose into glycogen (glycogenesis). Insulin also decreases glycogenolysis and gluconeogenesis, stimulates the conversion of glucose or other nutrients into fatty acids (lipogenesis), and helps stimulate protein synthesis.

The arterial supply of the pancreas is from the superior and inferior pancreaticoduodenal arteries and from the splenic and superior mesenteric arteries. The nerves to the pancreas are branches of the celiac plexus. The glandular portion of the pancreas is innervated by the craniosacral division of the autonomic nervous system, whereas the blood vessels of the pancreas are innervated by the thoracolumbar division of the autonomic nervous system.

CLINICAL APPLICATION

An absolute or relative deficiency of insulin results in a number of clinical symptoms referred to as **diabetes mellitus** (MEL-it-us). Research has led many to conclude that diabetes mellitus is not a single hereditary disease, but rather a heterogeneous group of hereditary diseases, all of which ultimately lead to an eleva-

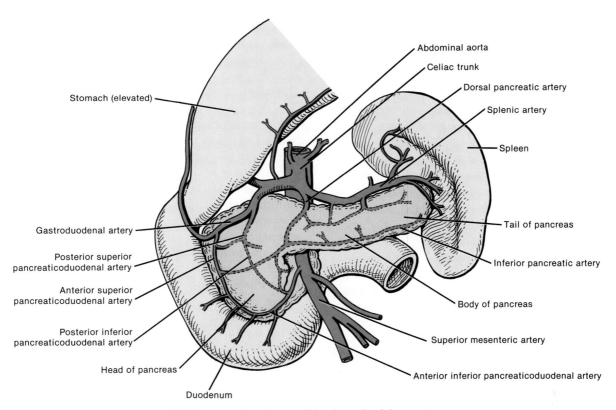

Abdominal aorta

Celiac trunk

Dorsal pancreatic artery

Splenic artery

Spleen

Stomach (elevated)

Tail of pancreas

Inferior pancreatic artery

Body of pancreas

Gastroduodenal artery

Posterior superior
pancreaticoduodenal artery

Anterior superior
pancreaticoduodenal artery

Posterior inferior
pancreaticoduodenal artery

Superior mesenteric artery

Head of pancreas

Anterior inferior pancreaticoduodenal artery

Duodenum

FIGURE 21-11 Location and blood supply of the pancreas.

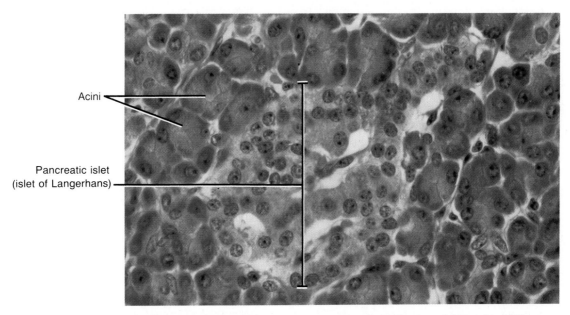

Acini

Pancreatic islet
(islet of Langerhans)

FIGURE 21-12 Histology of the pancreas. Photomicrograph at a magnification of 600×. (© 1983 by Michael H. Ross. Used by permission.)

tion of glucose in the blood (hyperglycemia) and excretion of glucose in the urine as hyperglycemia increases. Diabetes mellitus is also characterized by the three "polys": an inability to reabsorb water, resulting in increased urine production (*polyuria*); excessive thirst (*polydipsia*); and excessive eating (*polyphagia*).

Two major types of diabetes have been distinguished: the maturity-onset type and the juvenile-onset type. *Maturity-onset diabetes* is much more common, representing more than 90 percent of all cases. It most often occurs in people who are over 40 and overweight. Clinical symptoms are mild, and the high glucose levels in the blood can usually be controlled by diet alone. Many maturity-onset diabetics have a sufficiency or even a surplus of insulin in the blood. For these individuals, diabetes arises not from a shortage of insulin, but probably from defects in the molecular machinery that mediates the action of insulin on its target cells. Maturity-onset diabetes is therefore called *noninsulin-dependent diabetes.*

Juvenile-onset diabetes develops in people younger than age 20. Its onset is more abrupt and it is generally more severe than maturity-onset diabetes. It is usually preceded by a virus infection like measles or mumps. The disease is characterized by a marked decline in the number of beta cells in the pancreas leading to insufficient production of insulin and an elevation of glucose in the blood. It is known as *insulin-dependent diabetes.* The deficiency of insulin accelerates the breakdown of the body's reserve of fat resulting in the production of organic acids called ketones. This causes a form of acidosis called *ketosis,* which lowers the pH of the blood and can result in death. The catabolism of stored fats and proteins also causes weight loss. As lipids are transported by the blood from storage depots to hungry cells, lipid particles are deposited on the walls of blood vessels. The deposition leads to atherosclerosis and a multitude of cardiovascular problems.

Some cases of juvenile-onset diabetes may be explained primarily on a genetic basis and others on an environmental basis. Still other cases appear to rise from a complex interaction between genetic background and environment.

Hyperinsulinism is much rarer than hyposecretion and is generally the result of a malignant tumor in an islet. The principal symptom is a decreased blood glucose level, which stimulates the secretion of epinephrine, glucagon, and GH. As a consequence, anxiety, sweating, tremor, increased heart rate, and weakness occur. Moreover, brain cells do not have enough glucose to function efficiently. This condition leads to mental disorientation, convulsions, unconsciousness, shock, and eventual death as the vital centers in the medulla are affected.

A summary of pancreatic hormones, actions, and disorders is presented in Exhibit 21-5.

OVARIES AND TESTES

The female gonads, called the **ovaries,** are paired oval bodies located in the pelvic cavity. The ovaries produce female sex hormones called **estrogens** and **progesterone.** These hormones are responsible for the development and maintenance of the female sexual characteristics. Along with the gonadotropic hormones of the pituitary, the sex hormones also regulate the menstrual cycle, maintain pregnancy, and prepare the mammary glands for lactation. The ovaries (and placenta) also produce a hormone called **relaxin,** which relaxes the symphysis pubis and helps dilate the uterine cervix toward the end of pregnancy.

The male has two oval glands, called **testes,** that lie in the scrotum. The testes produce **testosterone,** the primary male sex hormone, that stimulates the development and maintenance of the male sexual characteristics. The testes also produce the hormone **inhibin** that inhibits secretion of FSH to control sperm production. The detailed structure of the ovaries and testes will be discussed in Chapter 24.

A summary of hormones produced by the ovaries and testes and their actions is presented in Exhibit 21-6.

PINEAL (EPIPHYSIS CEREBRI)

The endocrine gland attached to the roof of the third ventricle is known as the **pineal gland** (because of its resemblance to a pine cone), or **epiphysis cerebri** (see Figure 21-1). The gland is about 5 to 8 mm (0.2 to 0.3 in) long and 5 mm wide. It weighs about 0.2 g. It is covered by a capsule formed by the pia mater and consists of masses of **neuroglial cells** and parenchymal secretory cells called **pinealocytes.** Around the cells are scattered preganglionic sympathetic fibers. The pineal gland starts to calcify at about the time of puberty. Such calcium deposits are referred to as **brain sand.** Contrary to a once widely held belief, there is no evidence that the pineal atrophies with age and that the presence of brain sand is an indication of atrophy. In fact, the presence of brain sand may indicate increased secretory activity.

Although many anatomical facts concerning the pineal gland have been known for years, its physiology is still somewhat obscure. One hormone secreted by the pineal gland is **melatonin,** which appears to inhibit reproductive activities by inhibiting gonadotropic hormones. Some evidence also exists that the pineal secretes a second hormone called **adrenoglomerulotropin** (a-drē'-nō-glō-mer'-yoo-lō-TRŌ-pin). This hormone may stimulate the adrenal cortex to secrete aldosterone. Other substances found in the

EXHIBIT 21-5 SUMMARY OF HORMONES PRODUCED BY THE PANCREAS, ACTIONS, AND DISORDERS

HORMONE	PRINCIPAL ACTIONS	DISORDERS
Glucagon	Raises blood sugar level by accelerating conversion of glucogen into glucose in liver (glycogenolysis) and conversion of other nutrients into glucose in the liver (gluconeogenesis) and releasing glucose into blood.	
Insulin	Lowers blood sugar level by accelerating transport of glucose into cells, converting glucose into glycogen (glycogenesis), and decreasing glycogenolysis and gluconeogenesis; also increases lipogenesis and stimulates protein synthesis.	An absolute or relative deficiency of insulin produces diabetes mellitus. Hypersecretion of insulin results in hyperinsulinism.

EXHIBIT 21-6 SUMMARY OF HORMONES OF THE OVARIES AND TESTES AND THEIR ACTIONS

HORMONE	PRINCIPAL ACTIONS
Ovarian hormones	
Estrogens and progesterone	Development and maintenance of female sexual characteristics. Together with gonadotropic hormones of the adenohypophysis, they also regulate the menstrual cycle, maintain pregnancy, and prepare the mammary glands for lactation.
Relaxin	Relaxes symphysis pubis and helps dilate uterine cervix near the end of pregnancy.
Testicular hormones	
Testosterone	Development and maintenance of male sexual characteristics.
Inhibin	Inhibits secretion of FSH to control sperm production.

EXHIBIT 21-7 SUMMARY OF HORMONES OF THE PINEAL GLAND AND THEIR ACTIONS

HORMONE	PRINCIPAL ACTIONS
Melatonin	May inhibit reproductive activities by inhibiting gonadotropic hormones.
Adrenoglomerulotropin	May stimulate the adrenal cortex to secrete aldosterone.

pineal gland include norepinephrine (NE), serotonin, histamine, GnRF, and GABA.

The posterior cerebral artery supplies the pineal with blood, and the great cerebral vein drains it.

A summary of hormones produced by the pineal gland and their actions is presented in Exhibit 21-7.

THYMUS

Usually a bilobed lymphatic organ, the **thymus gland** is located in the superior mediastinum, posterior to the sternum and between the lungs (Figure 21-13). The two **thymic lobes** are held in close proximity by an enveloping layer of connective tissue. Each lobe is enclosed by a fibrous connective tissue **capsule.** The capsule gives off extensions into the lobes called **trabeculae** (tra-BEK-yoo-

lē), which divide the lobes into **lobules** (Figure 21-14a). Each lobule consists of a deeply staining peripheral **cortex** and a lighter staining central **medulla.** The cortex is composed almost entirely of small, medium, and large tightly packed lymphocytes held in place by reticular tissue fibers (Figure 21-14b). Since the reticular tissue differs in origin and structure from that usually found in other lymphatic organs, it is referred to as **epithelioreticular** supporting tissue. The medulla consists mostly of epithelial cells and more widely scattered lymphocytes, and its reticulum is more cellular than fibrous. In addition, the medulla contains characteristic **thymic (Hassall's) corpuscles,** concentric layers of epithelial cells. Their significance is unknown.

The thymus gland is conspicuous in the infant, and it reaches its maximum size of about 40 grams during puberty. After puberty, most of the thymic tissue is replaced by fat and connective tissue. By the time the person reaches maturity, the gland has atrophied substantially.

The function of the thymus gland is related to immunity. Lymphoid tissue of the body consists primarily of lymphocytes that may be distinguished into two kinds: B cells and T cells. Both are derived originally in the embryo from lymphocytic stem cells in bone marrow. Before migrating to their positions in lymphoid tissue, the descendants of the stem cells follow two distinct pathways. About half of them migrate to the thymus gland,

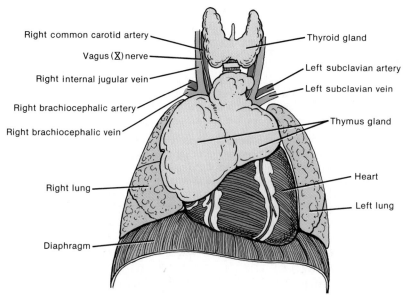

Right common carotid artery

Vagus (X) nerve

Right internal jugular vein

Right brachiocephalic artery

Right brachiocephalic vein

Right lung

Diaphragm

Thyroid gland

Left subclavian artery

Left subclavian vein

Thymus gland

Heart

Left lung

FIGURE 21-13 Location of the thymus gland in a young child.

where they are processed to become thymus-dependent lymphocytes, or **T cells.** The thymus gland confers on some of them the ability to destroy antigens (foreign microbes and substances) directly. The remaining stem cells are processed in some as yet undetermined area of the body, possibly bone marrow, the fetal liver, and spleen, or gut-associated lymphoid tissue, and are known as **B cells.** These cells, under the influence of hormones produced by the thymus gland, called **thymosin, thymic humoral factor (THF), thymic factor (TF),** and **thymopoietin,** differentiate into plasma cells. Plasma cells, in turn, produce antibodies against antigens. Thymosin is also secreted by macrophages.

The arterial supply of the thymus is derived mainly from the internal thoracic and inferior thyroid vessels. The veins that drain the thymus are the internal thoracic, brachiocephalic, and thyroid veins. Postganglionic sympathetic and parasympathetic fibers supply the gland.

A summary of hormones produced by the thymus gland and their actions is presented in Exhibit 21-8.

OTHER ENDOCRINE TISSUES

Before leaving our discussion of hormones, it should be noted that body tissues other than endocrine glands also secrete hormones. The gastrointestinal tract synthesizes several hormones that regulate digestion in the stomach and small intestine. Among these hormones are **stomach gastrin, enteric gastrin, secretin, cholecystokinin (CCK), enterocrinin,** and **gastric inhibitory peptide (GIP).**

The placenta produces **human chorionic gonadotropin (HCG), estrogens, progesterone, relaxin,** and **human cho-**

rionic somatomammotropin (HCS), all of which are related to pregnancy.

When the kidneys (and liver to a lesser extent) become hypoxic, it is believed that they release an enzyme called **renal erythropoietic factor.** It is secreted into the blood where it acts on a plasma protein to bring about the production of a hormone called **erythropoietin** (ē-rith'-rō-POY-ē-tin), which stimulates red blood cell production. The kidneys also help bring about the activation of the hormone **vitamin D.**

Finally, the skin produces vitamin D in the presence of sunlight.

AGING AND THE ENDOCRINE SYSTEM

The endocrine system exhibits a variety of changes, and many researchers look to this system with the hope of finding the key to the aging process. Disorders of the endocrine system are not frequent, and when they do occur, most often they are related to pathologic changes rather than age. Diabetes and thyroid disorders are the most important endocrine problems that have a significant effect on health and function.

EXHIBIT 21-8 SUMMARY OF HORMONES PRODUCED BY THE THYMUS GLAND AND THEIR ACTIONS

HORMONE	PRINCIPAL ACTIONS
Thymosin, thymic humoral factor (THF), thymic factor (TF), and thymopoietin	Induce B cells to differentiate into antibody-producing plasma cells.

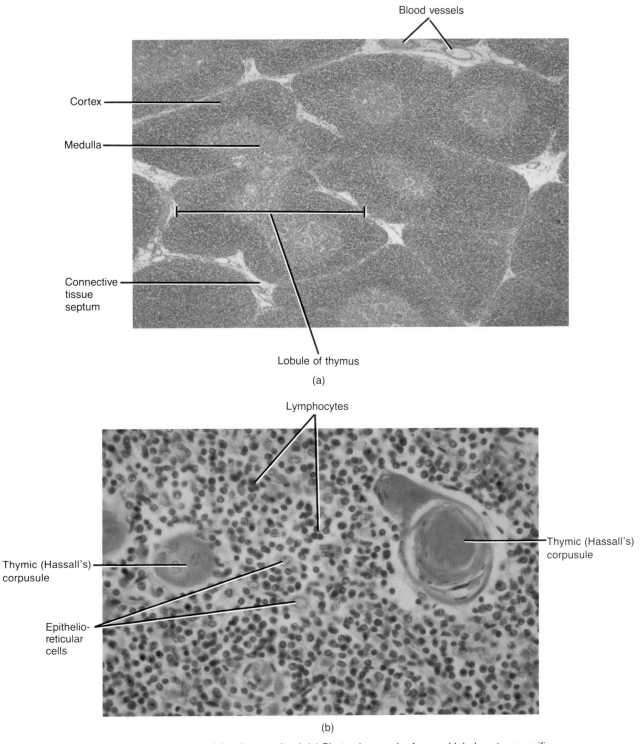

Blood vessels

Cortex

Medulla

Connective
tissue
septum

Lobule of thymus

(a)

Lymphocytes

Thymic (Hassall's)
corpusule

Thymic (Hassall's)
corpusule

Epithelio-
reticular
cells

(b)

FIGURE 21-14 Histology of the thymus gland. (a) Photomicrograph of several lobules at a magnification of 40×. (© 1983 by Michael H. Ross. Used by permission.) (b) Photomicrograph of an enlarged aspect of thymic (Hassall's) corpuscles at a magnification of 600×. (© 1983 by Michael H. Ross. Used by permission.)

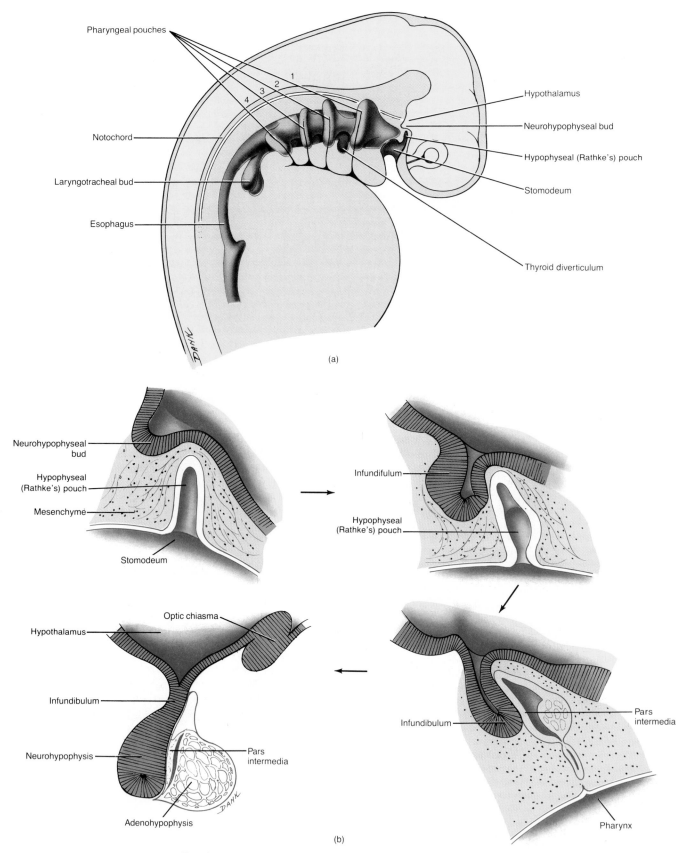

(a)

(b)

FIGURE 21-15 Development of the endocrine system. (a) Position of the neurohypophyseal bud, hypophyseal (Rathke's) pouch, and thyroid diverticulum. (b) Development of the pituitary gland.

DEVELOPMENTAL ANATOMY OF THE ENDOCRINE SYSTEM

The development of the endocrine system is not as localized as the development of other systems. The endocrine organs develop in widely separated parts of the embryo.

The *pituitary gland* (*hypophysis*) has a double origin from different regions of the **ectoderm**. The *neurohypophysis* (posterior lobe) is derived from an outgrowth of ectoderm called the **neurohypophyseal bud,** located on the floor of the hypothalamus (Figure 21-15a,b). The *infundibulum,* also an outgrowth of the neurohypophyseal bud, connects the neurohypophysis to the hypothalamus. The *adenohypophysis* (anterior lobe) is derived from an outgrowth of **ectoderm** from the roof of the stomodeum (mouth) called the **hypophyseal (Rathke's) pouch.** The pouch grows toward the neurohypophyseal bud, and the pouch loses its connection with the roof of the mouth.

The *thyroid gland* develops as a midventral outgrowth of **endoderm,** called the **thyroid diverticulum,** from the floor of the pharynx at the level of the second pair of pharyngeal pouches (Figure 21-15b). The outgrowth projects inferiorly and differentiates into the right and left lateral lobes and the isthmus of the gland.

The *parathyroid glands* develop as outgrowths from the third and fourth **pharyngeal pouches** (Figure 21-15a).

The adrenal cortex and adrenal medulla have completely different embryological origins. The *adrenal cortex* is derived from intermediate **mesoderm** from the same region that produces the gonads (see Figure 25-23). The *adrenal medulla* is **ectodermal** in origin and is derived from the **neural crest,** which also produces sympathetic ganglia and other structures of the nervous system (see Figure 18-21c).

The *pancreas* develops from **endoderm** from dorsal and ventral outgrowths of the part of the **foregut** that later becomes the duodenum (see Figure 23-22). The two outgrowths eventually fuse to form the pancreas. The origin of the ovaries and testes is discussed in the section on the reproductive system.

The *pineal gland* arises from **ectoderm** of the **diencephalon** (see Figure 18-22b), as an outgrowth between the thalamus and colliculi.

The *thymus gland* arises from **endoderm** from the third **pharyngeal pouches** (Figure 21-15a).

KEY MEDICAL TERMS ASSOCIATED WITH THE ENDOCRINE SYSTEM

Hyperplasia (*hyper* = over; *plas* = grow, form) Excessive development of tissue.

Hypoplasia (*hypo* = under) Defective development of tissue.

Neuroblastoma (*neuro* = nerve) Malignant tumor arising from the adrenal medulla associated with metastases to bones.

Thyroid storm An aggravation of all symptoms of hyperthyroidism, resulting from trauma, surgery, and unusual emotional stress or labor.

STUDY OUTLINE

Endocrine Glands (p. 528)

1. Both the endocrine and nervous systems assume a role in maintaining homeostasis.
2. Hormones help regulate the internal environment, respond to stress, help regulate growth and development, and contribute to reproductive processes.
3. Exocrine glands (sweat, sebaceous, digestive) secrete their products through ducts into body cavities or onto body surfaces.
4. Endocrine glands secrete hormones into the blood.

Pituitary (Hypophysis) (p. 530)

1. The pituitary is located in the sella turcica and is differentiated into the adenohypophysis (the anterior lobe and glandular portion) and the neurohypophysis (the posterior lobe and nervous portion).
2. Hormones of the adenohypophysis are released or inhibited by regulating factors produced by the hypothalamus.
3. The blood supply to the adenohypophysis is from the superior hypophyseal arteries.
4. Histologically, the adenohypophysis consists of somatotroph cells that produce growth hormone (GH), which regulates growth; lactotroph cells that produce prolactin (PRL), which helps initiate milk secretion; thyrotroph cells that secrete thyroid-stimulating hormone (TSH), which regulates thyroid gland activities; gonadotroph cells that synthesize follicle-stimulating hormone (FSH), which regulates the activities of the ovaries and testes, and luteinizing hormone (LH), which regulates female and male reproductive activities; and corticolipotroph cells that secrete adrenocorticotropic hormone (ACTH), which regulates the activities of the adrenal cortex, and melanocyte-stimulating hormone (MSH), which increases skin pigmentation.
5. Disorders associated with improper levels of GH are pituitary dwarfism, giantism, and acromegaly.
6. The neural connection between the hypothalamus and neurohypophysis is via the hypothalamic-hypophyseal tract.
7. Hormones made by the hypothalamus and stored in the neurohypophysis are oxytocin, or OT (stimulates contraction of uterus and ejection of milk), and antidiuretic hormone,

or ADH (stimulates water reabsorption by the kidneys and arteriole constriction).

8. A disorder associated with dysfunction of the neurohypophysis is diabetes insipidus.

Thyroid (p. 533)

1. The thyroid gland is located below the larynx.
2. Histologically, the thyroid consists of thyroid follicles composed of follicular cells, which secrete the thyroid hormones thyroxine (T_4) and triiodothyronine (T_3), and parafollicular cells, which secrete calcitonin (CT).
3. Thyroid hormones are synthesized from iodine and tyrosine within thyroglobulin (TGB).
4. Thyroid hormones regulate the rate of metabolism, growth and development, and the reactivity of the nervous system.
5. Cretinism, myxedema, exophthalmic goiter, and simple goiter are disorders associated with dysfunction of the thyroid gland.
6. Calcitonin (CT) lowers the blood level of calcium.

Parathyroids (p. 536)

1. The parathyroids are embedded on the posterior surfaces of the lateral lobes of the thyroid.
2. Histologically, the parathyroids consist of principal and oxyphil cells.
3. Parathyroid hormone (PTH) regulates the homeostasis of calcium and phosphate by increasing blood calcium level and decreasing blood phosphate level.
4. Tetany and osteitis fibrosa cystica are disorders associated with the parathyroid glands.

Adrenals (Suprarenals) (p. 538)

1. The adrenal glands are located superior to the kidneys. They consist of an outer cortex and inner medulla.
2. Histologically, the cortex is divided into a zona glomerulosa, zona fasciculata, and zona reticularis; the medulla consists of chromaffin cells.
3. Cortical secretions are mineralocorticoids, glucocorticoids, and gonadocorticoids.
4. Mineralocorticoids (e.g., aldosterone) increase sodium and water reabsorption and decrease potassium reabsorption.
5. A dysfunction related to aldosterone secretion is aldosteronism.
6. Glucocorticoids (e.g., cortisol) promote normal metabolism, help resist stress, and serve as antiinflammatories.
7. Disorders associated with glucocorticoid secretion are Addison's disease and Cushing's syndrome.
8. Gonadocorticoids secreted by the adrenal medulla have minimal effects. Excessive production results in adrenogenital syndrome.
9. Medullary secretions are epinephrine and norepinephrine (NE), which produce effects similar to sympathetic responses. They are released under stress.
10. Tumors of medullary chromaffin cells are called pheochromocytomas.

Pancreas (p. 542)

1. The pancreas is posterior and slightly inferior to the stomach.
2. Histologically, it consists of pancreatic islets, or islets of Langerhans (endocrine cells), and acini (enzyme-producing cells).

Three types of cells in the endocrine portion are alpha cells, beta cells, and delta cells.

3. Alpha cells secrete glucagon, beta cells secrete insulin, and delta cells secrete growth hormone-inhibiting factor (GHIF), or somatostatin.
4. Glucagon increases blood sugar level and insulin decreases blood sugar level.
5. Disorders associated with insulin production are diabetes mellitus and hyperinsulinism.

Ovaries and Testes (p. 544)

1. Ovaries are located in the pelvic cavity and produce sex hormones related to development and maintenance of female sexual characteristics, menstrual cycle, pregnancy, and lactation.
2. Testes lie inside the scrotum and produce sex hormones related to the development and maintenance of male sexual characteristics.

Pineal (Epiphysis Cerebri) (p. 544)

1. The pineal is attached to the roof of the third ventricle.
2. Histologically, it consists of secretory parenchymal cells called pinealocytes, neuroglial cells, and scattered preganglionic sympathetic fibers. Calcified deposits are referred to as brain sand.
3. It secretes melatonin (possibly regulates reproductive activities by inhibiting gonadotropic hormones) and adrenoglomerulotropin (may stimulate adrenal cortex to secrete aldosterone).

Thymus (p. 545)

1. The thymus is a bilobed lymphatic gland located in the superior mediastinum posterior to the sternum and between the lungs.
2. Histologically, it consists primarily of various sizes of lymphocytes.
3. The gland is necessary for the maturation of the thymus-dependent lymphocytes (T cells) of the immune system; its hormones (thymosin, thymic humoral factor, thymic factor, and thymopoietin) may cause B cells to differentiate into antibody-producing plasma cells.

Other Endocrine Tissues (p. 546)

1. The gastrointestinal tract synthesizes stomach and intestinal gastrin, secretin, cholecystokinin (CCK), enterocrinin, and gastric inhibitory peptide (GIP).
2. The placenta produces human chorionic gonadotropin (HCG), estrogens, progesterone, relaxin, and human chorionic somatomammotropin (HCS).
3. The kidneys release an enzyme that produces erythropoietin.
4. The skin synthesizes vitamin D.

Aging and the Endocrine System (p. 546)

1. Most endocrine disorders are related to pathologies rather than age.
2. Diabetes and thyroid disorders are among the more important endocrine disorders.

Developmental Anatomy of the Endocrine System (p. 549)

1. The adenohypophysis arises from the hypophyseal (Rathke's)

pouch; the neurohypophysis develops from the hypophyseal bud.

2. The thyroid gland develops from the thyroid diverticulum.

3. The parathyroid glands and thymus gland develop from pharyngeal pouches.

4. The adrenal cortex arises from mesoderm; the adrenal medulla develops from ectoderm (neural crest).

5. The pancreas develops from the foregut and the pineal gland develops from the diencephalon.

REVIEW QUESTIONS

1. Distinguish between an endocrine gland and an exocrine gland.

2. What is a hormone? Distinguish between tropic and gonadotropic hormones.

3. Describe the histology of the adenohypophysis. Why does the anterior lobe of the gland have such an abundant blood supply?

4. What hormones are produced by the adenohypophysis? What are their functions?

5. Describe the clinical symptoms of pituitary dwarfism, giantism, and acromegaly.

6. Discuss the histology of the neurohypophysis and the functions of its hormones.

7. Describe the structure and importance of the hypothalamic-hypophyseal tract.

8. What are the clinical symptoms of diabetes insipidus?

9. Describe the location and histology of the thyroid gland.

10. Discuss the physiological effects of the thyroid hormones.

11. Discuss the clinical symptoms of cretinism, myxedema, exophthalmic goiter, and simple goiter.

12. Describe the function of calcitonin.

13. Where are the parathyroids located? What is their histology?

14. What are the functions of the parathyroid hormone?

15. Discuss the clinical symptoms of tetany and osteitis fibrosa cystica.

16. Compare the adrenal cortex and adrenal medulla with regard to location and histology.

17. Describe the hormones produced by the adrenal cortex in terms of type and function.

18. Describe the clinical symptoms of aldosteronism, Addison's disease, Cushing's syndrome, and adrenogenital syndrome.

19. What relationship does the adrenal medulla have to the autonomic nervous system? What is the action of adrenal medullary hormones?

20. What is a pheochromocytoma?

21. Describe the location of the pancreas and the histology of the pancreatic islets (islets of Langerhans).

22. What are the actions of glucagon and insulin?

23. Describe the clinical symptoms of diabetes mellitus and hyperinsulinism.

24. Where is the pineal gland located? What are its assumed functions?

25. Describe the location and histology of the thymus gland. What is its proposed function?

26. Describe the effects of aging on the endocrine system.

27. Describe the development of the endocrine system.

28. Refer to the glossary of key medical terms associated with the endocrine system. Be sure that you can define each term.

22

The Respiratory System

Student Objectives

Identify the organs of the respiratory system.

Compare the structure of the external and internal nose.

Differentiate the three anatomical subdivisions of the pharynx and describe their roles in respiration.

Identify the anatomical features of the larynx related to respiration and voice production.

Describe the location and structure of the tubes that form the bronchial tree.

Contrast tracheostomy and intubation as alternative methods for clearing air passageways.

Identify the coverings of the lungs and the gross anatomical features of the lungs.

Describe the structure of a bronchopulmonary segment and a lobule of the lung.

Explain the structure of the alveolar-capillary membrane and its function in the diffusion of respiratory gasses.

Explain how the respiratory center functions in establishing the basic rhythm of respiration.

Define coughing, sneezing, sighing, yawning, sobbing, crying, laughing, and hiccuping as modified respiratory movements.

Describe the effects of aging on the respiratory system.

Describe the development of the respiratory system.

Define bronchogenic carcinoma, bronchial asthma, bronchitis, emphysema, pneumonia, tuberculosis, respiratory distress syndrome (RDS) of the newborn, respiratory failure, sudden infant death syndrome (SIDS), coryza (common cold), influenza (flu), pulmonary embolism (PE), pulmonary edema, and carbon monoxide poisoning as disorders of the respiratory system.

Define key medical terms associated with the respiratory system.

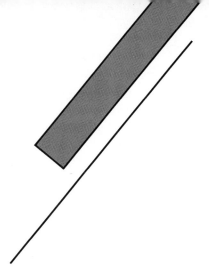

Cells need a continuous supply of oxygen to carry out the activities that are vital to their survival. Many of these activities release quantities of carbon dioxide. Since an excessive amount of carbon dioxide produces acid conditions that are poisonous to cells, the gas must be eliminated quickly and efficiently. The two systems that supply oxygen and eliminate carbon dioxide are the cardiovascular system and the respiratory system. The **respiratory system** consists of organs that exchange gases between the atmosphere and blood. These organs are the nose, pharynx, larynx, trachea, bronchi, and lungs (Figure 22-1). The cardiovascular system transports the gases in the blood between the lungs and the cells. The term *upper respiratory system* refers to the nose and throat and associated structures. The *lower respiratory system* refers to the remainder of the system.

The overall exchange of gases between the atmosphere, the blood, and the cells is **respiration.** Three basic processes are involved. The first process, **pulmonary ventilation,** or breathing, is the inspiration (inflow) and expiration (outflow) of air between the atmosphere and the lungs. The second and third processes involve the exchange of gases within the body. **External respiration** is the exchange of gases between the lungs and blood. **Internal respiration** is the exchange of gases between the blood and the cells.

The respiratory and cardiovascular systems participate equally in respiration. Failure of either system has the same effect on the body: disruption of homeostasis and rapid death of cells from oxygen starvation.

The developmental anatomy of the respiratory system is considered later in the chapter.

ORGANS

NOSE

The **nose** has an external portion and an internal portion inside the skull (Figure 22-2). The external portion consists of a supporting framework of bone and cartilage covered with skin and lined with mucous membrane. The bridge of the nose is formed by the nasal bones, which hold it in a fixed position. Because it has a framework of pliable cartilage, the rest of the external nose is quite flexible. On the undersurface of the external nose are two openings called the **nostrils,** or **external nares** (NA-rez; sing., naris). The surface anatomy of the nose may be examined by referring to Figure 11-4.

The internal portion of the nose is a large cavity in the skull that lies inferior to the cranium and superior to the mouth. Anteriorly, the internal nose merges with the external nose, and posteriorly it communicates with the throat (pharynx) through two openings called the **internal nares (choanae).** Four paranasal sinuses (frontal, sphenoidal, maxillary, and ethmoidal) and the nasolacrimal ducts also open into the internal nose. The lateral walls of the internal nose are formed by the ethmoid, maxillae, and inferior nasal conchae bones. The ethmoid also forms the roof. The floor is formed by the palatine bones and the palatine process of the maxilla, which together comprise the hard palate.

The inside of both the external and internal nose consists of a **nasal cavity,** divided into right and left sides by a vertical partition called the **nasal septum.** The anterior portion of the septum is made primarily of cartilage. The remainder is formed by the vomer and the perpendicular plate of the ethmoid (see Figure 6-7a). The anterior portion of the nasal cavity, just inside the nostrils, is called the **vestibule.** It is surrounded by cartilage. The upper nasal cavity is surrounded by bone.

CLINICAL APPLICATION

Nasal polyps are protruding growths of the mucous membrane that usually hang down from the posterior wall of the nasal septum. The polyps appear as bluish-white tumors and may fill the nasopharynx as they become larger. Nasal polyps usually undergo atrophy if untreated, but they are easily removed by a physician with a nasal snare and cautery.

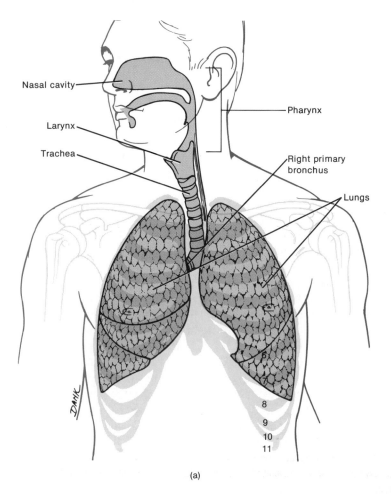

(a)

FIGURE 22-1 Organs of the respiratory system in relation to surrounding structures. (a) Diagram. (b) Photograph. (Courtesy of C. Yokochi and J. W. Rohen, *Photographic Anatomy of the Human Body,* 2nd ed., 1979, IGAKU-SHOIN, Ltd., Tokyo, New York.)

The interior structures of the nose are specialized for three functions: incoming air is warmed, moistened, and filtered; olfactory stimuli are received; and large hollow resonating chambers are provided for speech sounds.

When air enters the nostrils, it passes first through the vestibule. The vestibule is lined by skin containing coarse hairs that filter out large dust particles. The air then passes into the upper nasal cavity. Three shelves formed by projections of the superior, middle, and inferior nasal conchae extend out of the lateral wall of the cavity. The conchae, almost reaching the septum, subdivide each side of the nasal cavity into a series of groovelike passageways—the **superior, middle,** and **inferior meatuses.** Mucous membrane lines the cavity and its shelves. The olfactory receptors lie in the membrane lining the area superior to the superior nasal conchae, which is also called the **olfactory region.** Below the olfactory region, the membrane contains pseudostratified ciliated columnar cells with many goblet cells and capillaries. As the air whirls around the conchae and meatuses, it is warmed by the capillaries. Mucus secreted by the goblet cells moistens the air and traps dust particles. Drainage from the lacrimal ducts and perhaps secretions from the paranasal sinuses also help moisten the air. The cilia move the mucus-dust packages along the pharynx so they can be eliminated from the body.

CLINICAL APPLICATION

Nosebleed, or **epistaxis** (ep'-i-STAK-sis), is common because of the exposure of the nose to trauma and the extensive blood supply of the nose. Bleeding, either arterial or venous, usually occurs on the anterior part

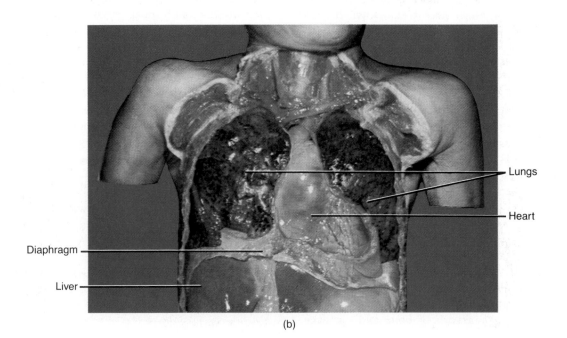

(b)

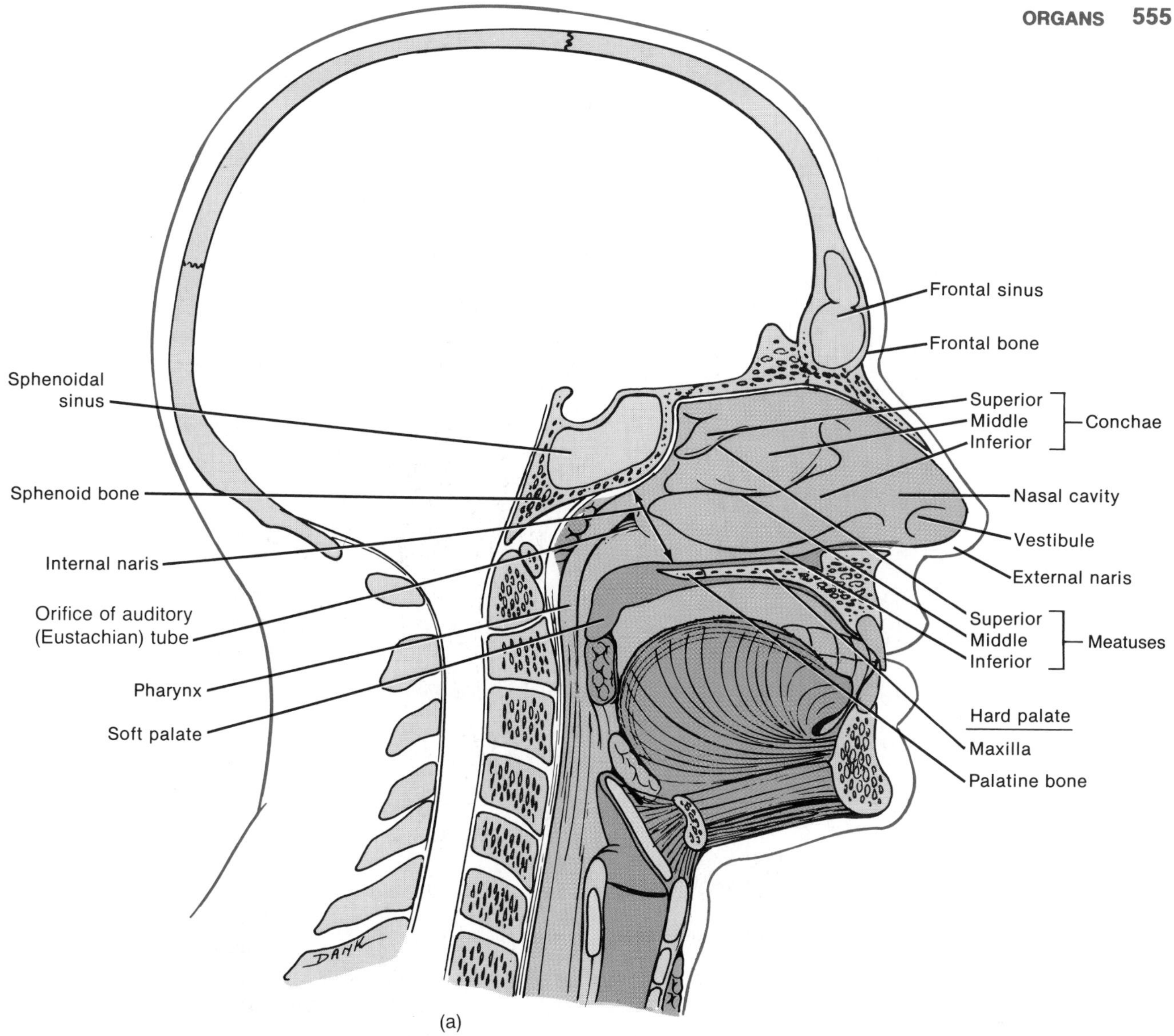

Sphenoidal sinus

Sphenoid bone

Internal naris

Orifice of auditory (Eustachian) tube

Pharynx

Soft palate

Frontal sinus

Frontal bone

Superior
Middle ⎤ Conchae
Inferior

Nasal cavity

Vestibule

External naris

Superior
Middle ⎤ Meatuses
Inferior

Hard palate

Maxilla

Palatine bone

(a)

FIGURE 22-2 Nose. (a) Diagram of the left side seen in sagittal section with the nasal septum removed.

of the septum and can be arrested by firm packing of the external nares. If the point of bleeding is in the posterior region, plugging of both the external and internal nares may be necessary. In extreme emergency, the external carotid artery may have to be ligated (tied) in order to control the hemorrhage.

The arterial supply to the nasal cavity is principally from the sphenopalatine branch of the maxillary artery. The remainder is supplied by the ophthalmic artery. The veins of the nasal cavity drain into the sphenopalatine vein, the facial vein, and the ophthalmic vein.

The nerve supply of the nasal cavity consists of olfactory cells in the olfactory epithelium associated with the olfactory nerve (see Figure 20-8) and the nerves of general sensation. These nerves are branches of the ophthalmic division of the trigeminal (V) nerve and the maxillary division of the trigeminal (V) nerve.

PHARYNX

The **pharynx** (FAR-inks), or throat, is a somewhat funnel-shaped tube about 13 cm (5 in) long that starts at the internal nares and extends partway down the neck (Figure 22-3). It lies just posterior to the nasal cavity and oral cavity and just anterior to the cervical vertebrae. Its wall is composed of skeletal muscles and lined with mucous membrane. The functions of the pharynx are to serve as a passageway for air and food and to provide a resonating chamber for speech sounds. The branch of medicine that deals with the diagnosis and treatment of diseases of the ears, nose, and throat is called **otorhinolaryngology** (ō'-tō-rī'-nō-lar'-in-GOL-ō-jē; *otic* = ear; *rhino* = nose).

The uppermost portion of the pharynx, called the **nasopharynx,** lies posterior to the internal nasal cavity and extends to the plane of the soft palate. There are four openings in its wall: two internal nares and two openings that lead into the auditory (Eustachian) tubes. The posterior wall also contains the pharyngeal tonsil, or adenoid.

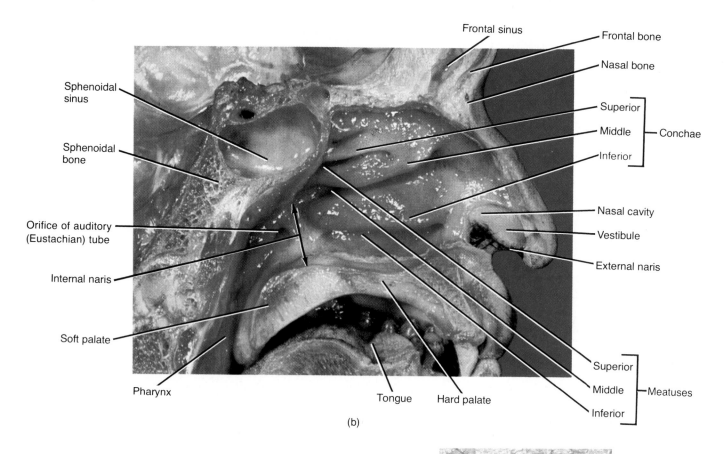

(b)

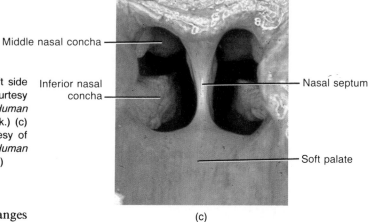

(c)

FIGURE 22-2 (*Continued*) Nose. (b) Photograph of the left side seen in sagittal section with the nasal septum removed. (Courtesy of C. Yokochi and J. W. Rohen, *Photographic Anatomy of the Human Body,* 2nd ed., 1979, IGAKU-SHOIN, Ltd., Tokyo, New York.) (c) Photograph of a posterior view of the nasal cavity. (Courtesy of C. Yokochi and J. W. Rohen, *Photographic Anatomy of the Human Body,* 2nd ed., 1979, IGAKU-SHOIN, Ltd., Tokyo, New York.)

Through the internal nares the nasopharynx exchanges air with the nasal cavities and receives the packages of dust-laden mucus. It is lined with pseudostratified ciliated epithelium, and the cilia move the mucus down toward the mouth. The nasopharynx also exchanges small amounts of air with the auditory tubes so that the air pressure inside the middle ear equals the pressure of the atmospheric air flowing through the nose and pharynx.

The middle portion of the pharynx, the **oropharynx,** lies posterior to the oral cavity and extends from the soft palate inferior to the level of the hyoid bone. It has only one opening, the *fauces* (FAW-sēz), or opening from the mouth. It is lined by stratified squamous epithelium. This portion of the pharynx is both respiratory and digestive in function, since it is a common passageway for

both air and food. Two pairs of tonsils, the palatine and lingual tonsils, are found in the oropharynx. The lingual tonsils lie at the base of the tongue (see also Figure 23-5a).

The lowest portion of the pharynx, the **laryngopharynx** (la-rin'-gō-FAR-inks), extends downward from the hyoid bone and becomes continuous with the esophagus (food tube) posteriorly and the larynx (voice box) anteriorly. Like the oropharynx, the laryngopharynx is a respiratory and a digestive pathway and is lined by stratified squamous epithelium.

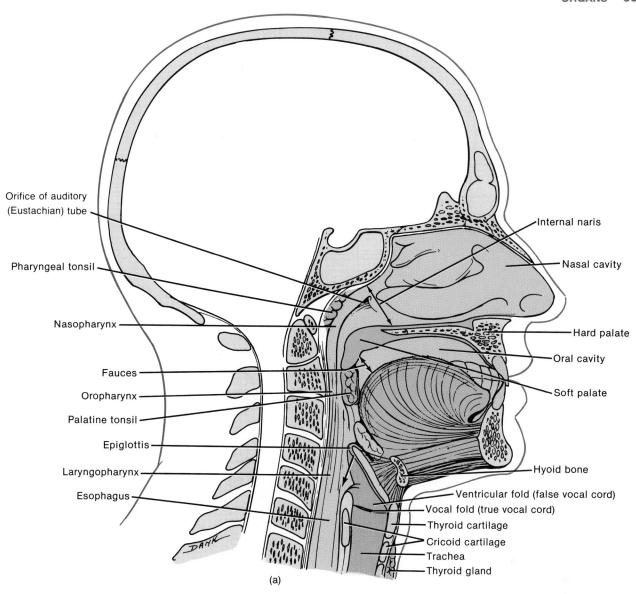

Orifice of auditory
(Eustachian) tube

Pharyngeal tonsil

Nasopharynx

Fauces

Oropharynx

Palatine tonsil

Epiglottis

Laryngopharynx

Esophagus

Internal naris

Nasal cavity

Hard palate

Oral cavity

Soft palate

Hyoid bone

Ventricular fold (false vocal cord)

Vocal fold (true vocal cord)

Thyroid cartilage

Cricoid cartilage

Trachea

Thyroid gland

(a)

FIGURE 22-3 Head and neck seen in sagittal section. (a) Diagram.

CLINICAL APPLICATION

Snoring, obstructive breathing during sleep, occurs from time to time in about 45 percent of adults and habitually in about 25 percent. The noise of snoring comes from vibrations of soft tissue between the internal nares and larynx—soft palate uvula, tonsils and their arches, base of the tongue, and pharyngeal muscles. Three circumstances, alone or in combination, contribute to snoring: (1) nasal airway impairment due to nasal injuries, deformities, polyps, infection, tumors, or deviated nasal septum; (2) pharyngeal airway impairment due to enlarged tonsils, cysts, or tumors; and (3) decreased tone of the muscles of the soft palate, tongue, or pharynx.

The arteries of the pharynx are the ascending pharyngeal, the ascending palatine branch of the facial, the descending palatine and pharyngeal branches of the maxillary, and the muscular branches of the superior thyroid artery. The veins of the pharynx drain into the pterygoid plexus and the internal jugular vein.

Most of the muscles of the pharynx (see Figure 10-8) are innervated by the pharyngeal plexus. This plexus is formed by the pharyngeal branches of the glossopharyngeal and vagal nerves and the superior cervical sympathetic ganglion.

LARYNX

The **larynx,** or voice box, is a short passageway that connects the pharynx with the trachea. It lies in the midline

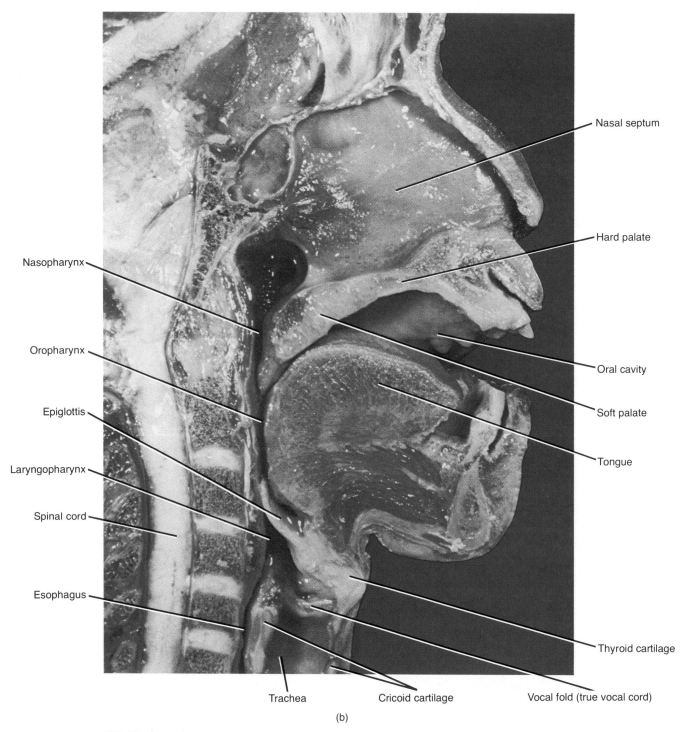

Nasal septum

Hard palate

Nasopharynx

Oropharynx

Oral cavity

Epiglottis

Soft palate

Laryngopharynx

Spinal cord

Tongue

Esophagus

Thyroid cartilage

Trachea   Cricoid cartilage   Vocal fold (true vocal cord)

(b)

FIGURE 22-3 (*Continued*) Head and neck seen in sagittal section. (b) Photograph. (Courtesy of C. Yokochi and J. W. Rohen, *Photographic Anatomy of the Human Body,* 2nd ed., 1979, IGAKU-SHOIN, Ltd., Tokyo, New York.)

of the neck anterior to the fourth through sixth cervical vertebrae.

The wall of the larynx is supported by nine pieces of cartilage (Figure 22-4). Three are single and three are paired. The three single pieces are the thyroid cartilage, epiglottic cartilage (epiglottis), and cricoid cartilage. Of the paired cartilages, the arytenoid cartilages are the most

important. The paired corniculate and cuneiform cartilages are of lesser significance.

The **thyroid cartilage (Adam's apple)** consists of two fused plates that form the anterior wall of the larynx and give it its triangular shape. It is larger in males than in females.

The **epiglottis** is a large, leaf-shaped piece of cartilage

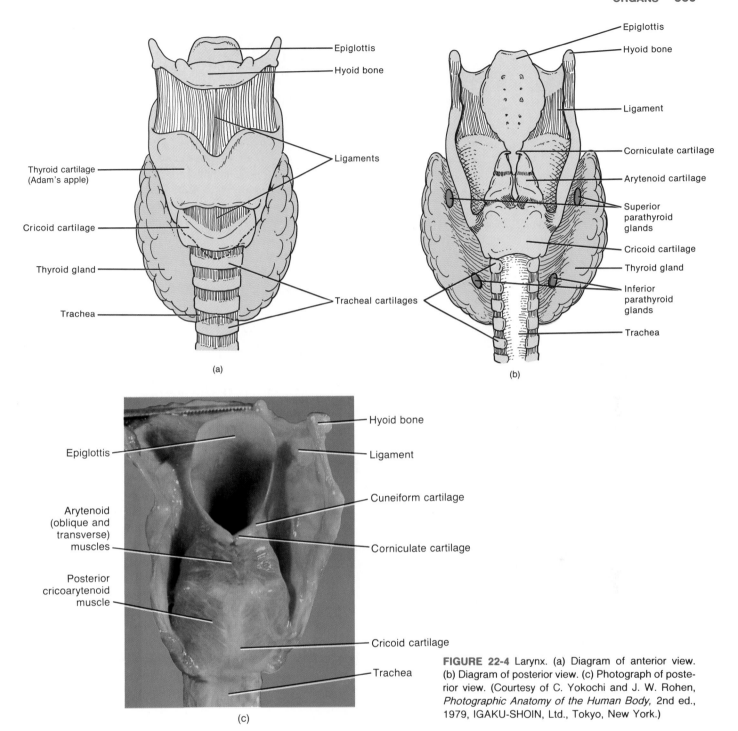

FIGURE 22-4 Larynx. (a) Diagram of anterior view. (b) Diagram of posterior view. (c) Photograph of posterior view. (Courtesy of C. Yokochi and J. W. Rohen, *Photographic Anatomy of the Human Body,* 2nd ed., 1979, IGAKU-SHOIN, Ltd., Tokyo, New York.)

lying on top of the larynx (see also Figure 22-3). The "stem" of the epiglottis is attached to the thyroid cartilage, but the "leaf" portion is unattached and free to move up and down like a trapdoor. During swallowing, there is elevation of the larynx. This causes the free edge of the epiglottis to form a lid over the glottis and closure of the glottis. The **glottis** is the space between the vocal folds (true vocal cords) in the larynx. In this way, the larynx is closed off and liquids and foods are routed into the esophagus and kept out of the trachea. If anything

but air passes into the larynx, a cough reflex attempts to expel the material.

The **cricoid** (KRĪ-koyd) **cartilage** is a ring of cartilage forming the inferior wall of the larynx. It is attached to the first ring of cartilage of the trachea.

The paired **arytenoid** (ar'-i-TĒ-noyd) **cartilages** are pyramidal in shape and located at the superior border of the cricoid cartilage. They attach to the vocal folds and pharyngeal muscles and by their action can move the vocal cords.

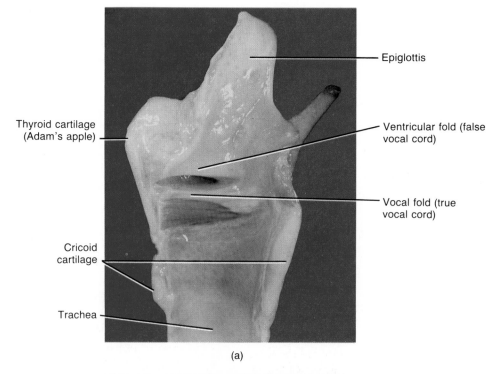

Epiglottis

Thyroid cartilage
(Adam's apple)

Ventricular fold (false
vocal cord)

Vocal fold (true
vocal cord)

Cricoid
cartilage

Trachea

(a)

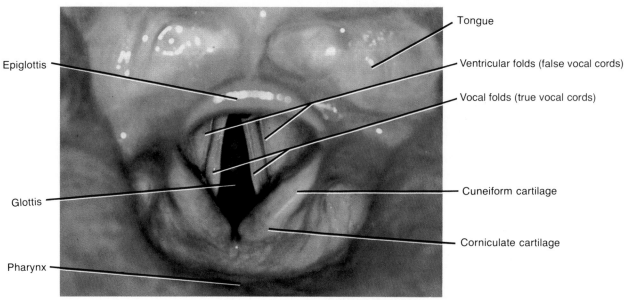

Tongue

Epiglottis

Ventricular folds (false vocal cords)

Vocal folds (true vocal cords)

Glottis

Cuneiform cartilage

Corniculate cartilage

Pharynx

(b)

FIGURE 22-5 Interior view of larynx. (a) Photograph of sagittal section. (Courtesy of C. Yokochi and J. W. Rohen, *Photographic Anatomy of the Human Body,* 2nd ed., 1979, IGAKU-SHOIN, Ltd., Tokyo, New York.) (b) Photograph of the true vocal cords, false vocal cords, and open glottis from above. (Courtesy of C. Yokochi and J. W. Rohen, *Photographic Anatomy of the Human Body,* 2nd ed., 1979, IGAKU-SHOIN, Ltd., Tokyo, New York.)

The paired **corniculate** (kor-NIK-yoo-lāt) **cartilages** are cone-shaped. One is located at the apex of each arytenoid cartilage. The paired **cuneiform** (kyoo-NĒ-i-form) **cartilages** are rod-shaped cartilages that connect the epiglottis to the arytenoid cartilages.

The epithelium lining the larynx below the vocal folds

is pseudostratified. It consists of ciliated columnar cells, goblet cells, and basal cells, and it helps to trap dust not removed in the upper passages.

The mucous membrane of the larynx is arranged into two pairs of folds—an upper pair called the **ventricular folds (false vocal cords)** and a lower pair called simply the **vocal folds (true vocal cords)** (Figure 22-5). When the ventricular folds are brought together they function in holding the breath against pressure in the thoracic cavity such as might occur when a person exerts a strain while lifting a heavy weight. The mucous membrane of the vocal folds is lined by nonkeratinized stratified squamous epithelium. Under the membrane lie bands of elastic ligaments stretched between pieces of rigid cartilage like the strings on a guitar. Skeletal muscles of the larynx, called intrinsic muscles, are attached internally to the pieces of rigid cartilage and to the vocal folds themselves (see Figure 10-10). When the muscles contract, they pull the strings of elastic ligaments tight and stretch the vocal folds out into the air passageways so that the glottis is narrowed. If air is directed against the vocal folds, they vibrate and set up sound waves in the column of air in the pharynx, nose, and mouth. The greater the pressure of air, the louder the sound.

Pitch is controlled by the tension on the vocal folds. If they are pulled taut by the muscles, they vibrate more rapidly and a higher pitch results. Lower sounds are produced by decreasing the muscular tension on the vocal folds. Vocal folds are usually thicker and longer in males than in females, and therefore they vibrate more slowly. Thus men generally have a lower range of pitch than women.

Sound originates from the vibration of the vocal folds, but other structures are necessary for converting the sound into recognizable speech. The pharynx, mouth, nasal cavity, and paranasal sinuses all act as resonating chambers that give the voice its human and individual quality. By constricting and relaxing the muscles in the wall of the pharynx, we produce the vowel sounds. Muscles of the face, tongue, and lips help us to enunciate words.

CLINICAL APPLICATION

Laryngitis is an inflammation of the larynx that is most often caused by a respiratory infection or irritants such as cigarette smoke. Inflammation of the vocal folds themselves causes hoarseness or loss of voice by interfering with the contraction of the folds or by causing them to swell to the point where they cannot vibrate freely. Many long-term smokers acquire a permanent hoarseness from the damage done by chronic inflammation.

The arteries of the larynx are the superior laryngeal, inferior laryngeal, and cricothyroid. The superior and inferior laryngeal and cricothyroid veins accompany the arteries. The superior cricoid vein and cricothyroid vein empty into the superior thyroid vein, and the inferior vein empties into the inferior thyroid vein.

The nerves of the larynx are the superior and recurrent laryngeal branches of the vagus.

TRACHEA

The **trachea** (TRĀ-kē-a), or windpipe, is a tubular passageway for air about 12 cm (4½ in) in length and 2.5 cm (1 in) in diameter. It is located anterior to the esophagus and extends from the larynx to the fifth thoracic vertebra, where it divides into right and left primary bronchi (see Figure 22-7a).

The wall of the trachea consists of a mucosa, submucosa, cartilaginous layer, and adventitia. The tracheal epithelium of the mucosa is pseudostratified. It consists of ciliated columnar cells that reach the luminal surface, goblet cells, and basal cells that do not reach the luminal surface (Figure 22-6). The epithelium provides the same protection against dust as the membrane lining the larynx. Seromucous glands and their ducts are present in the submucosa. The cartilaginous layer consists of 16 to 20 horizontal incomplete rings of hyaline cartilage that look like a series of letter Cs stacked one on top of another. The open parts of the Cs face the esophagus and permit it to expand into the trachea during swallowing. Transverse smooth muscle fibers, called the *trachealis muscle,* attach the open ends of the cartilage rings. The open ends of the rings of cartilage are also attached by elastic connective tissue. The solid parts of the Cs provide a rigid support so the tracheal wall does not collapse inward and obstruct the air passageway.

CLINICAL APPLICATION

At the point where the trachea bifurcates into right and left primary bronchi, there is an internal ridge called the **carina** (ka-RĪ-na). It is formed by a posterior and somewhat inferior projection of the last tracheal cartilage. The mucous membrane of the carina is one of the most sensitive areas of the respiratory system and is associated with the cough reflex. Widening and distortion of the carina, which can be seen in an examination by bronchoscopy, is a serious prognostic sign, since it usually indicates a carcinoma of the lymph nodes around the bifurcation of the trachea. **Bronchoscopy** is the visual examination of the bronchi through a **bronchoscope,** an illuminated, tubular instrument that can be passed through the trachea into the bronchi.

Occasionally the respiratory passageways are unable to protect themselves from obstruction. The rings of cartilage may be accidentally crushed; the mucous membrane may become inflamed and swell so much that it closes off the air space; inflamed membranes secrete a great deal of mucus that may clog the lower respiratory passageways; a large object may be breathed in (aspirated) while the glottis is open; or an aspirated foreign object may cause spasm of the laryngeal muscles. The passageways must be cleared quickly. If the obstruction is above the level of the

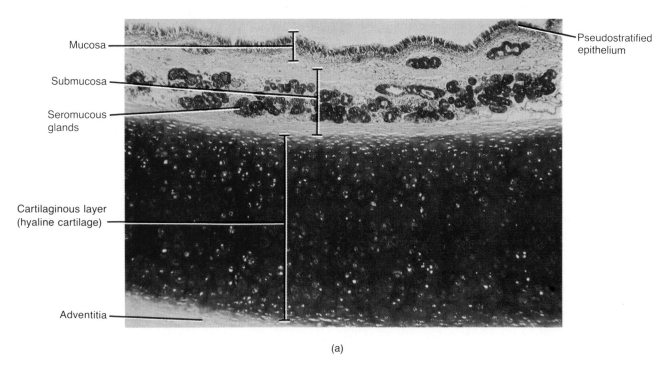

Mucosa

Submucosa

Seromucous glands

Cartilaginous layer (hyaline cartilage)

Adventitia

Pseudostratified epithelium

(a)

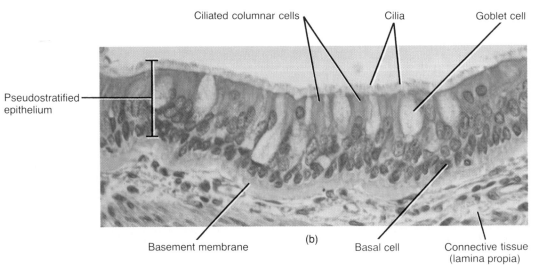

Ciliated columnar cells

Cilia

Goblet cell

Pseudostratified epithelium

Basement membrane

(b)

Basal cell

Connective tissue (lamina propia)

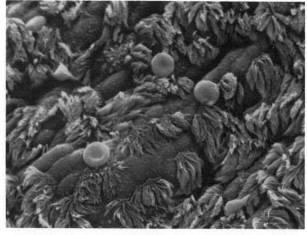

(c)

FIGURE 22-6 Histology of the trachea. (a) Photomicrograph of a portion of the tracheal wall at a magnification of 80×. (© 1983 by Michael H. Ross. Used by permission.) (b) Photomicrograph of an enlarged aspect of the tracheal epithelium at a magnification of 600×. (© 1983 by Michael H. Ross. Used by permission.) (c) Scanning electron micrograph of the tracheal epithelium at a magnification of 200×. (Courtesy of Fisher Scientific Company and S.T.E.M. Laboratories, Inc., Copyright, 1975.)

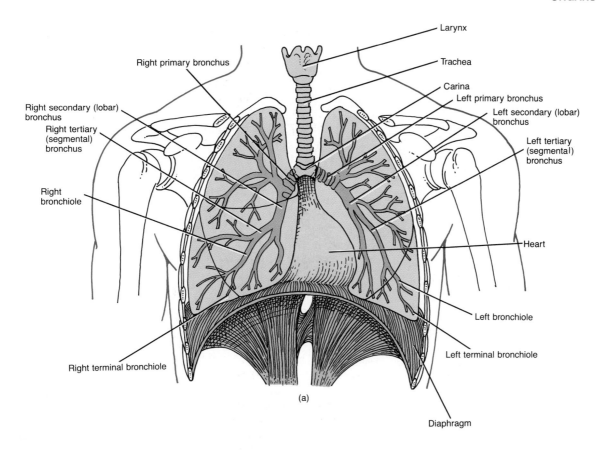

Labels: Larynx, Trachea, Carina, Left primary bronchus, Left secondary (lobar) bronchus, Left tertiary (segmental) bronchus, Heart, Left bronchiole, Left terminal bronchiole, Diaphragm, Right primary bronchus, Right secondary (lobar) bronchus, Right tertiary (segmental) bronchus, Right bronchiole, Right terminal bronchiole

(a)

chest, a **tracheostomy** (trā-kē-OS-tō-mē) may be performed. A midline skin incision is made in the neck from just above the cricoid cartilage to the jugular notch of the sternum. Next, an incision is made in the trachea below the obstructed area. The patient breathes through a metal or plastic tracheal tube inserted through the incision. Another method is **intubation.** A tube is inserted into the mouth or nose and passed down through the larynx and trachea. The firm wall of the tube pushes back any flexible obstruction, and the inside of the tube provides a passageway for air. If mucus is clogging the trachea, it can be suctioned out through the tube.

The arteries of the trachea are branches of the inferior thyroid, internal thoracic, and bronchial arteries. The veins of the trachea terminate in the inferior thyroid veins.

The smooth muscle and glands of the trachea are innervated parasympathetically via the vagus nerve directly and by its recurrent laryngeal branches. Sympathetic innervation is through branches from the sympathetic trunk and its ganglia.

BRONCHI

The trachea terminates in the chest by dividing at the sternal angle into a **right primary bronchus** (BRON-kus), which goes to the right lung, and a **left primary bronchus,**

FIGURE 22-7 Air passageways to the lungs. (a) Diagram of the bronchial tree in relation to the lungs.

which goes to the left lung (Figure 22-7a–b). The right primary bronchus is more vertical, shorter, and wider than the left. As a result, foreign objects in the air passageways are more likely to enter it than the left and frequently lodge in it. Like the trachea, the primary bronchi (BRON-kē) contain incomplete rings of cartilage and are lined by a pseudostratified ciliated epithelium.

On entering the lungs, the primary bronchi divide to form smaller bronchi—the **secondary (lobar) bronchi,** one for each lobe of the lung (the right lung has three lobes; the left lung has two). The secondary bronchi continue to branch, forming still smaller bronchi, called **tertiary (segmental) bronchi** which divide into **bronchioles.** Bronchioles, in turn, branch into even smaller tubes called **terminal bronchioles.** This continuous branching from the trachea resembles a tree trunk with its branches and is commonly referred to as the **bronchial tree.**

As the branching becomes more extensive in the bronchial tree, several structural changes may be noted. First, rings of cartilage are replaced by plates of cartilage that finally disappear in the bronchioles. Second, as the cartilage decreases, the amount of smooth muscle increases. Third, the epithelium changes from pseudostratified ciliated to simple cuboidal in the terminal bronchioles.

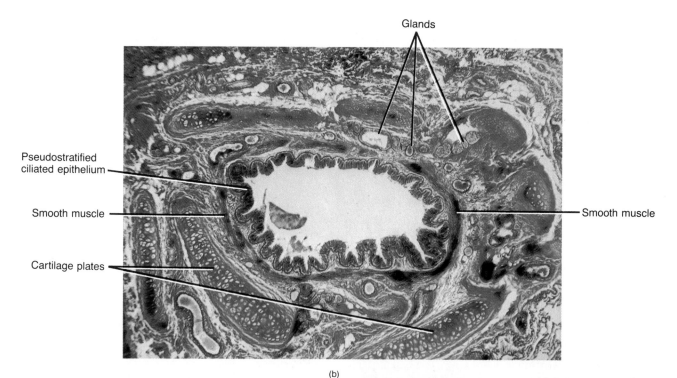

(b)

FIGURE 22-7 (*Continued*) Air passageways to the lungs. (b) Photomicrograph of a cross section of a primary bronchus at a magnification of 100×. (© 1983 by Michael H. Ross. Used by permission. (c) Anteroposterior bronchogram. (Courtesy of Lester W. Paul and John H. Juhl, *The Essentials of Roentgen Interpretation,* 3rd ed., Harper & Row, Publishers, Inc., New York, 1972.)

CLINICAL APPLICATION

The fact that the walls of the bronchioles contain a great deal of smooth muscle but no cartilage is clinically significant. During an **asthma attack** the muscles go into spasm. Because there is no supporting cartilage, the spasms can close off the air passageways.

Bronchography (brong-KOG-ra-fē) is a technique for examining the bronchial tree. An intratracheal catheter is passed into the mouth or nose, through the glottis, and into the trachea. Then, an opaque contrast medium, usually containing iodine, is introduced by means of gravity into the trachea and distributed through the bronchial branches. Roentgenograms of the chest in various positions are taken and the developed film, a **bronchogram** (BRONG-kō-gram), provides a picture of the tree (Figure 22-7c).

The blood supply to the bronchi is via the left bronchial and right bronchial arteries. The veins that drain the bronchi are the right bronchial vein, which enters the azygos vein, and the left bronchial vein, which empties into the hemiazygos vein or the left superior intercostal vein.

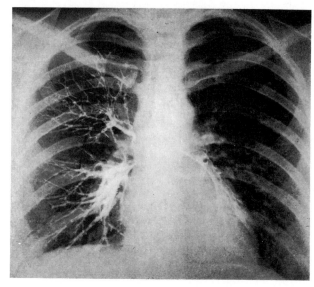

(c)

LUNGS

The **lungs** are paired, cone-shaped organs lying in the thoracic cavity. They are separated from each other by the heart and other structures in the mediastinum (see Figure 13-1). Two layers of serous membrane, collectively called the **pleural membrane,** enclose and protect each lung. The outer layer is attached to the wall of the thoracic cavity and is called the **parietal pleura.** The inner layer, the **visceral pleura,** covers the lungs themselves. Between the visceral and parietal pleura is a small potential space, the **pleural cavity,** which contains a lubricating fluid secreted by the membranes (see Figure 1-7b). This fluid prevents friction between the membranes and allows them to move easily on one another during breathing.

CLINICAL APPLICATION

In certain conditions, the pleural cavity may fill with air (**pneumothorax**), blood (**hemothorax**), or pus. Air in the pleural cavity, most commonly introduced in a surgical opening of the chest, may cause the lung to collapse. Fluid can be drained from the pleural cavity by inserting a needle, usually posteriorly through the seventh intercostal space. The needle is passed along the superior border of the lower rib to avoid damage to the intercostal nerves and blood vessels. Below the seventh intercostal space there is danger of penetrating the diaphragm.

Inflammation of the pleural membrane, or **pleurisy,** causes friction during breathing that can be quite painful when the swollen membranes rub against each other.

Gross Anatomy

The lungs extend from the diaphragm to a point about 1.5 to 2.5 cm (¾ to 1 in) superior to the clavicles and lie against the ribs anteriorly and posteriorly. The broad inferior portion of the lung, the **base,** is concave and fits over the convex area of the diaphragm (Figure 22-8). The narrow superior portion of the lung is termed the **apex (cupula).** The surface of the lung lying against the ribs, the **costal surface,** is rounded to match the curvature of the ribs. The **mediastinal (medial) surface** of each lung contains a vertical slit, the **hilus,** through which bronchi, pulmonary vessels, and nerves enter and exit. The blood vessels, bronchi, and nerves are held together by the pleura and connective tissue, and they constitute the **root** of the lung. Medially, the left lung also contains a concavity, the **cardiac notch,** in which the heart lies.

The right lung is thicker and broader than the left. It is also somewhat shorter than the left because the diaphragm is higher on the right side to accommodate the liver that lies below it.

Lobes and Fissures

Each lung is divided into lobes by one or more fissures. Both lungs have an **oblique fissure,** which extends downward and forward. The right lung also has a **horizontal fissure.** The oblique fissure in the left lung separates the **superior lobe** from the **inferior lobe.** The upper part of the oblique fissure of the right lung separates the superior

lobe from the inferior lobe, whereas the lower part of the oblique fissure separates the inferior lobe from the **middle lobe.** The horizontal fissure of the right lung subdivides the superior lobe, thus forming a middle lobe.

Each lobe receives its own secondary (lobar) bronchus. Thus the right primary bronchus gives rise to three secondary (lobar) bronchi called the **superior, middle,** and **inferior secondary (lobar) bronchi.** The left primary bronchus gives rise to a **superior** and an **inferior secondary (lobar) bronchus.** Within the substance of the lung, the secondary bronchi give rise to the **tertiary (segmental) bronchi,** which are constant in both origin and distribution. The segment of lung tissue that each supplies is called a **bronchopulmonary segment** (Figure 22-9). Bronchial and pulmonary disorders, such as tumors or abscesses, may be localized in a bronchopulmonary segment and may be surgically removed without seriously disrupting surrounding lung tissue.

Lobules

Each bronchopulmonary segment of the lungs is broken up into many small compartments called **lobules** (Figure 22-10a). Each lobule is wrapped in elastic connective tissue and contains a lymphatic vessel, an arteriole, a venule, and a branch from a terminal bronchiole. Terminal bronchioles subdivide into microscopic branches called **respiratory bronchioles** (Figure 22-10a, b). As the respiratory bronchioles become more distal, the epithelial lining changes from cuboidal to squamous. Respiratory bronchioles, in turn, subdivide into several (2 to 11) **alveolar ducts (atria).**

Around the circumference of the alveolar ducts are numerous alveoli and alveolar sacs. An **alveolus** (al-VĒ-ō-lus) is a cup-shaped outpouching lined by epithelium and supported by a thin elastic basement membrane. **Alveolar sacs** are two or more alveoli that share a common opening (Figure 22-10a–c). The alveolar walls consist of two principal types of epithelial cells: *squamous pulmonary epithelial cells* and *spetal cells.* The squamous pulmonary epithelial cells are the larger of the two types of cells and form a continuous lining of the alveolar wall, except for occasional septal cells. Septal cells are much smaller, cuboidal in shape, and are dispersed among the squamous pulmonary epithelial cells. Septal cells produce a phospholipid substance called *surfactant* (sur-FAK-tant), which lowers surface tension. Also found within the alveolar wall are free *alveolar macrophages* (*dust cells*). They are highly phagocytic and serve to remove dust particles or other debris that gain entrance to alveolar spaces. Deep to the layer of squamous pulmonary epithelial cells is an elastic basement membrane. Over the alveoli, the arteriole and venule disperse into a capillary network. The blood capillaries consist of a single layer of endothelial cells and basement membrane.

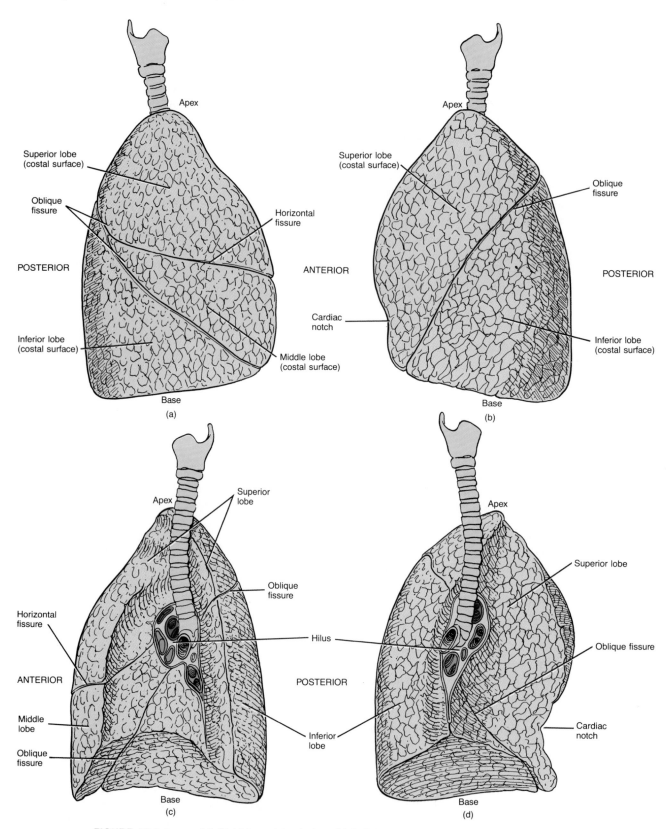

FIGURE 22-8 Lungs. (a) Right lung, lateral view. (b) Left lung, lateral view. (c) Right lung, medial view. (d) Left lung, medial view.

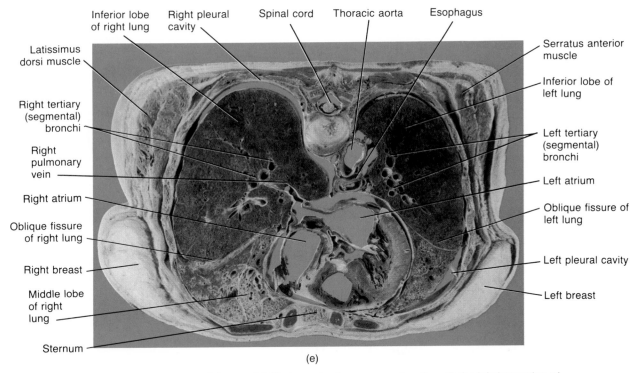

Inferior lobe of right lung

Right pleural cavity

Spinal cord

Thoracic aorta

Esophagus

Serratus anterior muscle

Inferior lobe of left lung

Latissimus dorsi muscle

Right tertiary (segmental) bronchi

Right pulmonary vein

Right atrium

Oblique fissure of right lung

Right breast

Middle lobe of right lung

Sternum

Left tertiary (segmental) bronchi

Left atrium

Oblique fissure of left lung

Left pleural cavity

Left breast

(e)

FIGURE 22-8 (*Continued*) Lungs. (e) Photograph of a cross section through the inferior portion of the thoracic cavity. (Courtesy of Stephen A. Kieffer and E. Robert Heitzman, *An Atlas of Cross-Sectional Anatomy,* Harper & Row, Publishers, Inc., New York, 1979.)

CLINICAL APPLICATION

Many respiratory disorders are treated by means of **nebulization** (neb'-yoo-li-ZĀ-shun). This procedure is the administering of medication in the form of droplets that are suspended in air to selected areas of the respiratory tract. The patient inhales the medication as a fine mist. The number of droplets suspended in the mist and the area of the respiratory tract that the medication will reach both depend on droplet size. Smaller droplets (approximately 2 μm in diameter) can be suspended in greater numbers than can large droplets and will reach the alveolar ducts and sacs. Larger droplets (approximately 7 to 16 μm in diameter) will be deposited mostly in the bronchi and bronchioles. Droplets of 40 μm and larger will be deposited in the upper respiratory tract—the mouth, pharynx, trachea, and main bronchi. Nebulization therapy can be used with many different types of drugs, such as chemicals that relax the smooth muscle of the respiratory passageways, chemicals that reduce the thickness of mucus, and antibiotics.

Alveolar-Capillary (Respiratory) Membrane

The exhange of respiratory gases between the lungs and blood takes place by diffusion across the alveoli and capillary walls. This membrane, through which the respiratory

gases move, is collectively known as the **alveolar-capillary (respiratory) membrane** (Figure 22-10e). It consists of:

1. A layer of squamous pulmonary epithelial cells with septal cells and free alveolar macrophages that constitute the alveolar (epithelial) wall.

2. An epithelial basement membrane underneath the alveolar wall.

3. A capillary basement membrane that is often fused to the epithelial basement membrane.

4. The endothelial cells of the capillary.

Despite the large number of layers, the alveolar-capillary membrane averages only 0.5 μm in thickness. This is of considerable importance to the efficient diffusion of respiratory gases. Moreover, it has been estimated that the lungs contain 300 million alveoli, providing an immense surface area of 70 m² (753 ft²) for the exchange of gases.

Blood and Nerve Supply

The arterial supply of the lungs is derived from the pulmonary trunk. It divides into a left pulmonary artery, which enters the left lung, and a right pulmonary artery, which enters the right lung. The venous return of the oxygenated blood is by way of the pulmonary veins, typically two in number on each side—the right and left superior and inferior pulmonary veins. All four veins drain into the left atrium (see Figure 14-17).

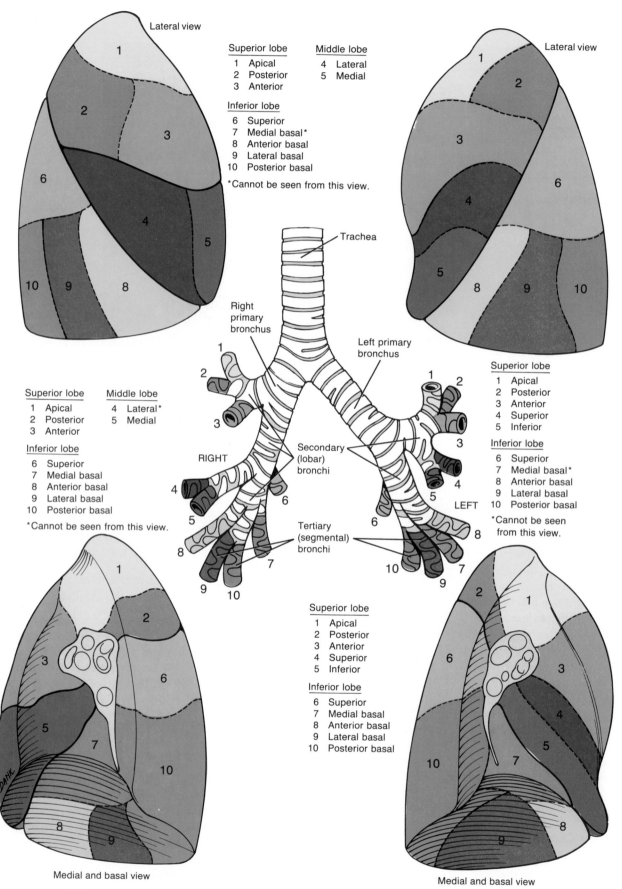

Lateral view

Superior lobe
1 Apical
2 Posterior
3 Anterior

Middle lobe
4 Lateral
5 Medial

Inferior lobe
6 Superior
7 Medial basal*
8 Anterior basal
9 Lateral basal
10 Posterior basal

*Cannot be seen from this view.

Lateral view

Trachea

Right primary bronchus

Left primary bronchus

Superior lobe
1 Apical
2 Posterior
3 Anterior

Middle lobe
4 Lateral*
5 Medial

Inferior lobe
6 Superior
7 Medial basal
8 Anterior basal
9 Lateral basal
10 Posterior basal

*Cannot be seen from this view.

Secondary (lobar) bronchi

RIGHT

Tertiary (segmental) bronchi

LEFT

Superior lobe
1 Apical
2 Posterior
3 Anterior
4 Superior
5 Inferior

Inferior lobe
6 Superior
7 Medial basal*
8 Anterior basal
9 Lateral basal
10 Posterior basal

*Cannot be seen from this view.

Superior lobe
1 Apical
2 Posterior
3 Anterior
4 Superior
5 Inferior

Inferior lobe
6 Superior
7 Medial basal
8 Anterior basal
9 Lateral basal
10 Posterior basal

Medial and basal view

Medial and basal view

FIGURE 22-9 Bronchopulmonary segments of the lungs. The bronchial branches are shown in the center of the figure. The bronchopulmonary segments are numbered and named for convenience.

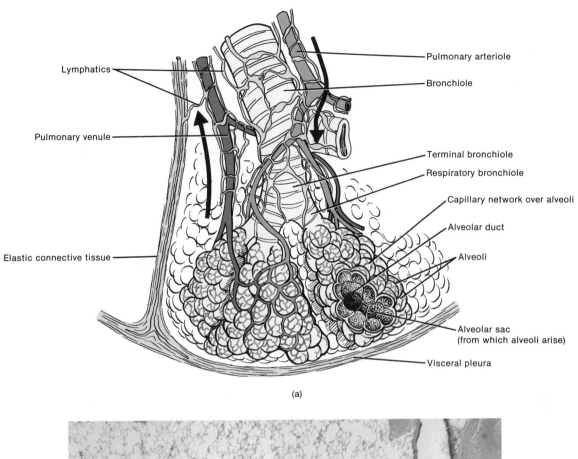

Pulmonary arteriole

Bronchiole

Lymphatics

Pulmonary venule

Terminal bronchiole

Respiratory bronchiole

Capillary network over alveoli

Alveolar duct

Alveoli

Elastic connective tissue

Alveolar sac
(from which alveoli arise)

Visceral pleura

(a)

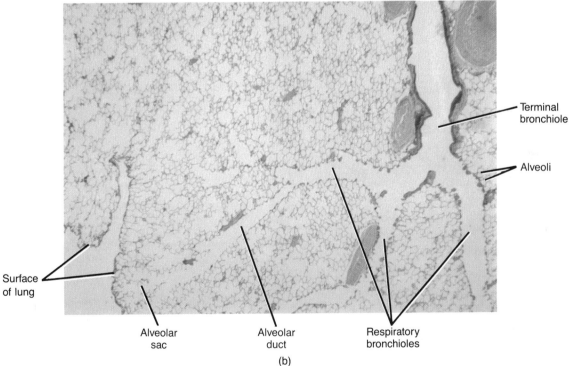

Terminal
bronchiole

Alveoli

Surface
of lung

Alveolar
sac

Alveolar
duct

Respiratory
bronchioles

(b)

FIGURE 22-10 Histology of the lungs. (a) Diagram of a lobule of the lung. (b) Photomicrograph of a terminal bronchiole, respiratory bronchioles, alveolar duct, alveolar sacs, and alveoli at a magnification of 50×. (© 1983 by Michael H. Ross. Used by permission.)

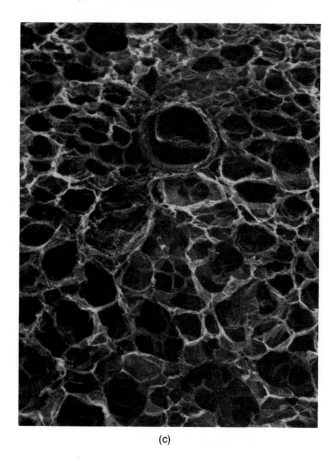

(c)

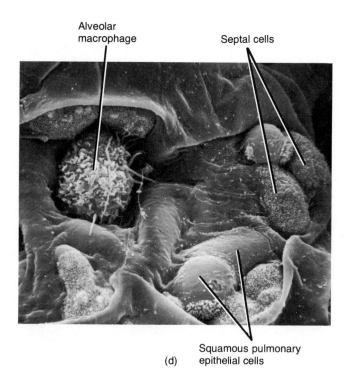

Alveolar
macrophage

Septal cells

Squamous pulmonary
epithelial cells

(d)

FIGURE 22-10 (*Continued*) Histology of the lungs. (c) Scanning electron micrograph of alveolar sacs at a magnification of 100×. (Courtesy of Fisher Scientific Company and S.T.E.M. Laboratories, Inc., Copyright, 1975.) (d) Scanning electron micrograph of an alveolus showing squamous pulmonary alveolar cells and an alveolar macrophage at a magnification of 3,430×. (Courtesy of Richard K. Kessel and Randy H. Kardon, *Tissues and Organs: A Text-Atlas of Scanning Electron Microscopy*, Copyright © 1979 by Scientific American, Inc.) (e) Diagram of the structure of the alveolar-capillary (respiratory) membrane.

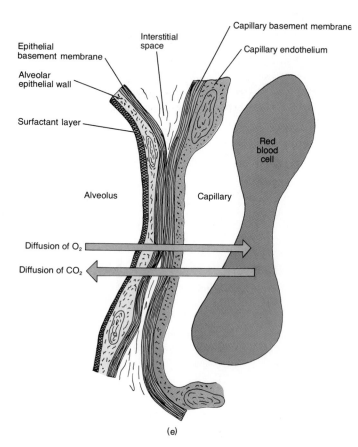

Epithelial
basement membrane

Alveolar
epithelial wall

Surfactant layer

Interstitial
space

Capillary basement membrane

Capillary endothelium

Red
blood
cell

Alveolus

Capillary

Diffusion of O_2

Diffusion of CO_2

(e)

The nerve supply of the lungs is derived from the autonomic nervous system. Parasympathetic fibers are derived from the second, third, and fourth thoracic ganglia.

NERVOUS CONTROL OF RESPIRATION

The basic rhythm of respiration is controlled by portions of the nervous system in the medulla and pons. In response to the demands of the body, this rhythm can be modified.

Inspiration and expiration depend on a pressure differential. If the volume of the lungs is increased, the intrapleural (intrathoracic) pressure—that is, the pressure in the lungs—decreases, and air from the atmosphere enters the lungs. If the volume decreases, intrapleural pressure increases, and air is expelled. The volume of the lungs is controlled by the action of the respiratory muscles—the diaphragm and external intercostal muscles. Their

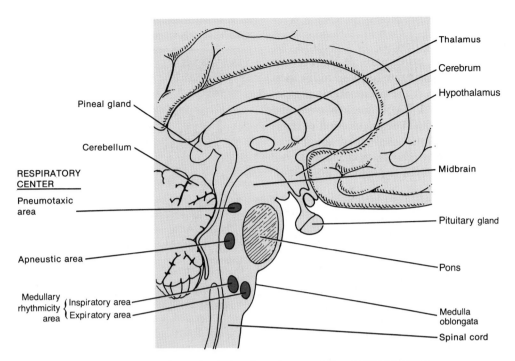

FIGURE 22-11 Approximate location of areas of the respiratory center.

contraction pulls the diaphragm down and the ribs out, thus increasing the size of the thoracic cavity. Their relaxation decreases the size of the cavity.

The term applied to normal quiet breathing is **eupnea** (yoop-NĒ-a; *eu* = normal). Eupnea involves shallow, deep, or combined shallow and deep breathing. Shallow (chest) breathing is called **costal breathing.** It consists of an upward and outward movement of the chest as a result of contraction of the external intercostal muscles. Deep (abdominal) breathing is called **diaphragmatic breathing.** It consists of the outward movement of the abdomen as a result of the contraction and descent of the diaphragm.

The respiratory muscles contract and relax as a result of nerve impulses transmitted to them from centers in the brain. The area from which nerve impulses are sent to respiratory muscles is located bilaterally in the reticular formation of the brain stem; it is referred to as the **respiratory center.** The respiratory center consists of a widely dispersed group of neurons and is functionally divided into three areas: (1) the medullary rhythmicity area, in the medulla, (2) the pneumotaxic (noo-mō-TAK-sik) area, in the pons, and (3) the apneustic (ap-NOO-stik) area, in the pons (Figure 22-11).

MEDULLARY RHYTHMICITY AREA

The function of the **medullary rhythmicity area** is to control the basic rhythm of respiration. In the normal resting state, inspiration usually lasts for about 2 seconds and expiration for about 3 seconds. This is the basic rhythm of respiration. Within the medullary rhythmicity area are

found both inspiratory and expiratory neurons that comprise inspiratory and expiratory areas, respectively. Let us first consider the proposed role of the inspiratory neurons in respiration.

The basic rhythm of respiration is determined by nerve impulses generated in the inspiratory area. At the beginning of expiration, the inspiratory area in inactive, but, after 3 seconds, it suddenly and automatically becomes active. This activity seems to result from an intrinsic excitability of the inspiratory neurons. In fact, when all incoming nerve connections to the inspiratory area are cut or blocked, the area still rhythmically discharges impulses that result in inspiration. Nerve impulses from the active inspiratory area last for about 2 seconds and travel to the muscles of inspiration. The impulses reach the diaphragm by the phrenic nerves and the external intercostal muscles by the intercostal nerves. When the impulses reach the inspiratory muscles, the muscles contract and inspiration occurs. At the end of 2 seconds, the inspiratory muscles become inactive again, and the cycle repeats itself over and over.

It is believed that the expiratory neurons remain inactive during most normal quiet respiration. During quiet respiration, inspiration is accomplished by active contraction of the inspiratory muscles and expiration results from passive recoil of the lungs and thoracic wall as the inspiratory muscles relax. However, during high levels of ventilation, it is believed that impulses from the inspiratory area activate the expiratory area. Impulses discharged from the expiratory area cause contraction of the internal intercostals and abdominal muscles that decrease the size of the thoracic cavity.

PNEUMOTAXIC AREA

Although the medullary rhythmicity area controls the basic rhythm of respiration, other parts of the nervous system help coordinate the transition between inspiration and expiration. One of these is the **pneumotaxic area** in the upper pons (Figure 22-11). It continuously transmits impulses to the inspiratory area. The major effect of these impulses is to help turn off the inspiratory area before the lungs become too full of air. In other words, the impulses limit inspiration and thus facilitate expiration.

APNEUSTIC AREA

Another part of the nervous system that coordinates the transition between inspiration and expiration is the **apneustic area** in the lower pons (Figure 22-11). It sends impulses to the inspiratory area that activate it and prolong inspiration, thus inhibiting expiration. This phenomenon occurs when the pneumotaxic area is inactive. When the pneumotaxic area is active, it overrides the apneustic area.

CORTICAL INFLUENCES

The respiratory center has connections with the cerebral cortex, which means we can voluntarily alter our pattern of breathing. We can even refuse to breathe at all for a short time. Voluntary control is protective because it enables us to prevent water or irritating gases from entering the lungs. The ability to stop breathing is limited by the buildup of CO_2 in the blood, however. When the pCO_2 increases to a certain level, the inspiratory area is stimulated, impulses are sent to inspiratory muscles, and breathing resumes whether or not the person wishes. It is impossible for people to kill themselves by holding their breath.

MODIFIED RESPIRATORY MOVEMENTS

Respirations also provide humans with methods for expressing emotions such as laughing, yawning, sighing, and sobbing. Moreover, respiratory air can be used to expel foreign matter from the upper air passages through actions such as sneezing and coughing. Some of the modified respiratory movements that express emotion or clear the air passageways are listed in Exhibit 22-1. All these movements are reflexes, but some of them also can be initiated voluntarily.

AGING AND THE RESPIRATORY SYSTEM

The cardiovascular and respiratory systems operate as a unit, and damage or disease in one of these organ systems is often secondarily reflected in the other. In pulmo-

EXHIBIT 22-1   MODIFIED RESPIRATORY MOVEMENTS

MOVEMENT	COMMENT
Coughing	A long-drawn and deep inspiration followed by a complete closure of the glottis, which results in a strong expiration that suddenly pushes the glottis open and sends a blast of air through the upper respiratory passages. Stimulus for this reflex act may be a foreign body lodged in the larynx, trachea, or epiglottis.
Sneezing	Spasmodic contraction of muscles of expiration, which forcefully expels air through the nose and mouth. Stimulus may be an irritation of the nasal mucosa.
Sighing	A long-drawn and deep inspiration immediately followed by a shorter but forceful expiration.
Yawning	A deep inspiration through the widely opened mouth producing an exaggerated depression of the lower jaw. It may be stimulated by drowsiness, fatigue, or low levels of oxygen.
Sobbing	A series of convulsive inspirations followed by a single prolonged expiration. The glottis closes earlier than normal after each inspiration so only a little air enters the lungs with each inspiration.
Crying	An inspiration followed by many short convulsive expirations, during which the glottis remains open and the vocal cords vibrate; accompanied by characteristic facial expressions.
Laughing	The same basic movements as crying, but the rhythm of the movements and the facial expressions usually differ from those of crying. Laughing and crying are sometimes indistinguishable.
Hiccuping	Spasmodic contraction of the diaphragm followed by a spasmodic closure of the glottis to produce a sharp inspiratory sound. Stimulus is usually irritation of the sensory nerve endings of the digestive tract.

nary heart disease, the right side of the heart enlarges in response to certain lung diseases. This condition is known as right ventricular hypertrophy, or cor pulmonale. The airways and tissues of the respiratory tract, including the air sacs, become less elastic and more rigid.

In addition to the lungs becoming less elastic, the chest wall also becomes more rigid. As a result, there is a decrease in pulmonary lung capacity. In fact, vital capacity (the amount of air moved by maximal inspiration followed by maximal expiration) can decrease as much as 35 percent by age 70. Also, there is a decrease in blood levels of oxygen, making breathing more difficult while lying down; decreased activity of alveolar macrophages; and diminished ciliary action of the epithelium lining the

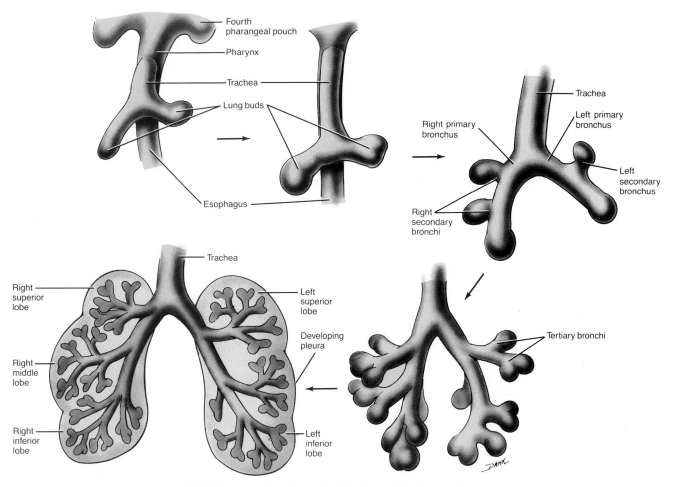

FIGURE 22-12 Development of the bronchial tubes and lungs.

respiratory tract. Due to all of these age-related factors, elderly people are more susceptible to pneumonia, bronchitis, emphysema, and other pulmonary disorders.

DEVELOPMENTAL ANATOMY OF THE RESPIRATORY SYSTEM

The development of the mouth and pharynx is considered in the exhibit on the digestive system. Here, we will consider the remainder of the respiratory system.

At about four weeks, the respiratory system begins as an outgrowth of the **endoderm** of the foregut (precursor of some digestive organs) just behind the pharynx. This outgrowth is called the **laryngotracheal bud** (see Figure 21-15a). As the bud grows, it elongates and differentiates into the future *larynx*. Its proximal end maintains a slit-like opening into the pharynx called the *glottis*. The middle portion of the bud gives rise to the *trachea*. The distal portion divides into two **lung buds**, which grow into the *bronchi* and *lungs* (Figure 22-12).

As the lung buds develop, they branch and rebranch and give rise to all the *bronchial tubes*. After the sixth month, the closed terminal portions of the tubes dilate

and become the *alveoli* of the lungs. The smooth muscle, cartilage, and connective tissues of the bronchial tubes and the pleural sacs of the lungs are contributed by **mesenchymal (mesodermal) cells.**

APPLICATIONS TO HEALTH

BRONCHOGENIC CARCINOMA (LUNG CANCER)

As part of ordinary breathing, many irritating substances are inhaled. Almost all pollutants, including inhaled smoke, have an irritating effect on the bronchial tubes and lungs and may be regarded as stresses or irritating stimuli. The effects of irritation on the bronchial epithelium are important clinically because a common lung cancer, **bronchogenic carcinoma,** starts in the walls of the bronchi.

Microscopic examination of the pseudostratified epithelium of a bronchial tube reveals three kinds of cells, all of which are in contact with the basement membrane. The columnar cells contain cilia and extend upward from the basement membrane and reach the luminal surface. At intervals between the ciliated columnar cells are the mucus-secreting goblet cells. The basal cells do not reach the luminal surface. The basal cells divide continuously, replacing the columnar cells as they wear down and are sloughed off.

The constant irritation by inhaled smoke and pollutants causes an enlargement of the goblet cells of the bronchial epithelium. They respond by secreting excessive mucus. The basal cells also respond to the stress by undergoing cell division so fast that they push into the area occupied by the goblet and columnar cells. As many as 20 rows of basal cells may be produced. Many researchers believe that if the stress is removed at this point, the epithelium can return to normal.

If the stress persists, more and more mucus is secreted and the cilia become less effective. As a result, mucus is not carried toward the throat but remains trapped in the bronchial tubes. The individual then develops a "smoker's cough." Moreover, the constant irritation from the pollutant slowly destroys the alveoli, which are replaced with thick, inelastic connective tissue. Mucus that has accumulated becomes trapped in the air sacs. Millions of sacs rupture, reducing the diffusion surface for the exchange of oxygen and carbon dioxide. The individual has now developed emphysema. If the stress is removed at this point, there is little chance for improvement. Alveolar tissue that has been destroyed cannot be repaired. But removal of the stress can stop further destruction of lung tissue.

If the stress continues, the emphysema gets progressively worse, and the basal cells of the bronchial tubes continue to divide and break through the basement membrane. At this point the stage is set for bronchogenic carcinoma. Columnar and goblet cells disappear and may be replaced with squamous cancer cells. If this happens, the malignant growth spreads throughout the lung and may block a bronchial tube. If the obstruction occurs in a large bronchial tube, very little oxygen enters the lung, and disease-producing bacteria thrive on the mucoid secretions. In the end, the patient may develop emphysema, carcinoma, and a host of infectious diseases. Treatment involves surgical removal of the diseased lung. However, metastasis (spreading) of the growth through the lymphatic or blood system may result in new growths in other parts of the body such as the brain and liver.

Other factors may be associated with lung cancer. For instance, breast, stomach, and prostate malignancies can metastasize to the lungs. People who apparently have not been exposed to pollutants do occasionally develop bronchogenic carcinoma. However, the occurrence of bronchogenic carcinoma is probably over 20 times higher in heavy cigarette smokers than it is in nonsmokers.

BRONCHIAL ASTHMA

Bronchial asthma is a reaction, usually allergic, characterized by attacks of wheezing and difficult breathing. At-

tacks are brought on by spasms of the smooth muscles that lie in the walls of the smaller bronchi and bronchioles, causing the passageways to close partially. The patient has trouble exhaling, and the alveoli may remain inflated during expiration. Usually the mucous membranes that line the respiratory passageways become irritated and secrete excessive amounts of mucus that may clog the bronchi and bronchioles and worsen the attack. About three out of four asthma victims are allergic to edible or airborne substances as common as wheat or dust. Others are sensitive to the proteins of harmless bacteria that inhabit the paranasal sinuses, nose, and throat. Asthma might also have a psychosomatic origin.

BRONCHITIS

Bronchitis is inflammation of the bronchi characterized by hypertrophy and hyperplasia of seromucous glands and goblet cells lining the bronchial airways. The typical symptom is a productive cough in which a thick greenish-yellow sputum is raised. This secretion signifies the presence of the underlying infection that is causing excessive secretion of mucus. Cigarette smoking remains the most important cause of chronic bronchitis, that is, bronchitis that lasts for at least three months of the year for two successive years. Other factors that influence development of the disease are family history of bronchitis or other lung disease, air pollution, carbon monoxide, respiratory infections, and deficient antibodies, especially IgA antibodies.

EMPHYSEMA

In **emphysema** (em'-fi-SĒ-ma), the alveolar walls lose their elasticity and remain filled with air during expiration. The name means "blown up," or "full of air." Reduced forced expiratory volume is the first symptom. Later, alveoli in other areas of the lungs are damaged. Many alveoli may merge to form larger air sacs with a reduced overall volume. The lungs become permanently inflated because they have lost elasticity. To adjust to the increased lung size, the size of the chest cage increases. The patient has to work voluntarily to exhale. Oxygen diffusion does not occur as easily across the damaged alveolar-capillary membrane, blood O_2 is somewhat lowered, and any mild exercise that raises the oxygen requirements of the cells leaves the patient breathless. As the disease progresses, the alveoli are replaced with thick fibrous connective tissue. Even carbon dioxide does not diffuse easily through this fibrous tissue. If the blood cannot buffer all the hydrogen ions that accumulate, the blood pH drops or unusually high amounts of carbon dioxide may dissolve in the plasma. High carbon dioxide levels produce acid conditions that are toxic to brain cells. Consequently, the inspiratory area becomes less active and the respiration rate slows down, further aggravating the problem. The compressed and damaged capillaries around the deteriorating alveoli may no longer be able to receive blood. As a result, resistance to blood flow increases in the pulmonary trunk and the right ventricle overworks as it attempts to force blood through the remaining capillaries.

Emphysema is generally caused by a long-term irritation. Air pollution, occupational exposure to industrial dust, and cigarette smoke are the most common irritants. Cigarette smoke not only deactivates a protein that is seemingly crucial in preventing emphysema but also prevents the repair of affected lung tissue. Chronic bronchial asthma also may produce alveolar damage. Cases of emphysema are becoming more and more frequent in the United States. The irony is that the disease can be prevented and the progressive deterioration can be stopped by eliminating the harmful stimuli.

Diseases such as bronchial asthma, bronchitis, and emphysema have in common some degree of obstruction of the air passageways. The term **chronic obstructive pulmonary disease (COPD)** is used to refer to these disorders.

PNEUMONIA

The term **pneumonia** means an acute infection or inflammation of the alveoli. In this disease, the alveolar sacs fill up with fluid and dead white blood cells, reducing the amount of air space in the lungs. (Remember that one of the cardinal signs of inflammation is edema.) Oxygen has difficulty diffusing through the inflamed alveoli, and the blood O_2 may be drastically reduced. Blood CO_2 usually remains normal because carbon dioxide diffuses through the alveoli more easily than oxygen does. If all the alveoli of a lobe are inflamed, the pneumonia is called *lobar pneumonia.* If only parts of the lobe are involved, it is called *lobular (segmental) pneumonia.* If both the alveoli and the bronchial tubes are included, it is called *bronchopneumonia.*

The most common cause of pneumonia is the pneumococcus bacterium, but other bacteria or a fungus may be the source of trouble. Viral pneumonia is caused by several viruses, including the influenza virus.

TUBERCULOSIS

The bacterium called *Mycobacterium tuberculosis* produces an inflammation called **tuberculosis.** This disease still ranks as the number-one killer in the communicable disease category. Tuberculosis most often affects the lungs and the pleurae. The bacteria destroy parts of the lung tissue, and the tissue is replaced by fibrous connective tissue. Because the connective tissue is inelastic and thick, the affected areas of the lungs do not snap back during expiration and larger amounts of air are retained. Gases no longer diffuse easily through the fibrous tissue.

Tuberculosis bacteria are spread by inhalation. Although they can withstand exposure to many disinfec-

tants, they die quickly in sunlight. Thus, tuberculosis is sometimes associated with crowded, poorly lit housing. Many drugs are successful in treating tuberculosis. Rest, sunlight, and good diet are vital parts of treatment.

RESPIRATORY DISTRESS SYNDROME (RDS) OF THE NEWBORN

Respiratory distress syndrome (RDS) of the newborn, also called glassy-lung disease or hyaline membrane disease (HMD), is responsible for approximately 20,000 newborn infant deaths per year. Before birth, the respiratory passages are filled with fluid. Part of this fluid is amniotic fluid inhaled during respiratory movements in utero. The remainder is produced by the submucosal glands and the goblet cells of the respiratory epithelium.

At birth, this fluid-filled passageway must become an air-filled passageway, and the collapsed primitive alveoli (terminal sacs) must expand and function in gaseous exchange. The success of this transition depends largely on surfactant, the phospholipid produced by alveolar cells that lowers surface tension in the fluid layer lining the primitive alveoli once air enters the lungs. Surfactant is present in the fetus's lungs as early as 23 weeks, and by 28 to 32 weeks the amount present is sufficient to prevent alveolar collapse during breathing. Surfactant is produced continuously by septal alveolar cells. The presence of surfactant can be detected by amniocentesis.

Although in a normal, full-term infant the second and subsequent breaths require less respiratory effort than the first, breathing is not completely normal until about 40 minutes after birth. The entire lung is not inflated fully with the first one or two breaths. In fact, for the first 7 to 10 days, small areas of the lungs may remain uninflated.

In the newborn whose lungs are deficient in surfactant, the effort required for the first breath is essentially the same as that required in normal newborns. However, the surface tension of the alveolar fluid is 7 to 14 times higher than the surface tension of alveolar fluid with a monomolecular layer of surfactant. Consequently, during expiration after the first inspiration, the surface tension of the alveoli increases as the alveoli deflate. The alveoli collapse almost to their original uninflated state.

Idiopathic RDS of the newborn usually appears within a few hours after birth. (An idiopathic condition is one that occurs spontaneously in an individual from an unknown or obscure cause.) Affected infants show difficult and labored breathing with withdrawal of the intercostal and subcostal spaces. Death may occur soon after onset of respiratory difficulty or may be delayed for a few days, although many infants survive. At autopsy, the lungs are underinflated and areas of atelectasis (collapse of lung) are prominent. If the infant survives for at least a few hours after developing respiratory distress, the alveoli are often filled with a fluid of high-protein content that resembles a hyaline (or glassy) membrane. RDS of the newborn occurs frequently in premature infants and also in infants of diabetic mothers, particularly if the diabetes is untreated or poorly controlled.

A treatment called PEEP (positive end expiratory pressure) is used to treat RDS of the newborn. It consists of passing a tube through the air passages to the top of the lungs to provide needed oxygen-rich air at continuous pressures of up to 14 mm Hg. Continuous pressure keeps the baby's alveoli open and available for gas exchange. The addition of human surfactant, obtained from the amniotic fluid of infants born by caesarian section, to the oxygen-rich air counters the effects of RDS of the newborn more effectively and lessens lung damage.

RESPIRATORY FAILURE

Respiratory failure refers to a condition in which the respiratory system cannot supply sufficient oxygen to maintain metabolism or cannot eliminate enough carbon dioxide to prevent respiratory acidosis. Specifically, respiratory failure occurs when the pO_2 of arterial blood is less than 50 mm Hg or when the pCO_2 of arterial blood is greater than 50 mm Hg. Respiratory failure always causes dysfunction in other organs as well.

Among the causes of respiratory failure are lung disorders (chronic obstructive pulmonary disease, pneumonia, and adult respiratory distress syndrome); mechanical disorders that disturb the chest wall or neuromusculature (anesthetic blocking agents, cervical spinal cord injuries, myasthenia gravis, and amyotrophic lateral sclerosis); depression of the respiratory center by drugs, strokes, or trauma; and carbon monoxide poisoning.

Symptoms of respiratory failure include disorientation, malaise, headache, muscular weakness, difficulty with sleep, palpitations, breathlessness, coughing, tachycardia, dysrhythmia, cyanosis, hypertension, edema, stupor, and coma. Treatment consists of providing adequate oxygenation and reversing the respiratory acidosis.

SUDDEN INFANT DEATH SYNDROME (SIDS)

Sudden infant death syndrome (SIDS), also called crib death or cot death, kills approximately 10,000 American babies every year. It kills more infants between the ages of 1 week and 12 months than any other disease. SIDS occurs without warning. Although about half of its victims had an upper respiratory infection within two weeks of death, the babies tended to be remarkably healthy. Two other findings are important: the apparently *silent* nature of death and the *disarray* frequently found at the scene.

Given this death scene, it is not surprising that for centuries "crib death" was mistakenly thought to be due to suffocation. Most medical people, however, believe that a healthy child *cannot* smother in its bedclothes. The diaper of a SIDS victim is usually full of stool and urine, and the urinary bladder is almost always empty. External examination reveals only a blood-tinged froth that often exudes from the nostrils. These signs all point to a sudden lethal episode associated with tonic motor activity, not to a gradually deepening coma leading to death.

The similarity of findings in SIDS cases strengthens the theory that there is one underlying cause. The seasonal distribution of SIDS incidences, being lowest during the summer months and highest in the late fall and winter, strongly suggests an infectious agent, viruses being the most likely possibility. Most cases occur at a time in life when gamma globulin levels are low and the infant is at a critical period of susceptibility.

At least thirteen different hypotheses have been proposed as explanations for SIDS. Although the precise pathogenic mechanisms have yet to be clarified, several valuable observations have emerged as a result of investigations during the past decade. Most, if not all, such cases appear to share a common mechanism ("final common pathway"). Thus, SIDS is probably a single disease entity. According to one study, a likely cause of death is laryngospasm, possibly triggered by a previous mild, nonspecific virus infection of the upper respiratory tract. Another possible factor includes periods of prolonged apnea as a result of malfunction of the respiratory center. Hypoxia may also be involved. Other suspected causes are an allergic reaction or bacterial poisoning.

CORYZA (COMMON COLD) AND INFLUENZA (FLU)

Several viruses are responsible for **coryza (common cold).** Typical symptoms include sneezing, excessive nasal secretion, and congestion. The uncomplicated common cold is not usually accompanied by a fever. Complications include sinusitis, ear infections, and laryngitis. Since common colds are caused by viruses, antibiotics are of no use in treatment. Although over-the-counter drugs may lessen the severity of certain symptoms of the common cold, they do not lessen recovery time.

Influenza (flu) is also caused by a virus. Its symptoms include chills, fever (usually higher than 101°F), headache, and muscular aches. Coldlike symptoms appear as the fever subsides. Diarrhea is not a symptom of influenza.

PULMONARY EMBOLISM (PE)

Pulmonary embolism (PE) refers to the presence of a blood clot or other foreign substance in a pulmonary arterial vessel that obstructs circulation to lung tissue. An embolus may arise in a distant site, such as a leg vein, and pass through the right side of the heart to a lung via a pulmonary artery. The immediate effect of a PE is complete or partial obstruction of the pulmonary arterial blood flow to the lung resulting in dysfunction of the affected lung tissue. A large embolus can produce death in a few minutes. PE rarely develops in patients without one or more risk factors. Among these are immobilization in bed for more than three days; heart disease (especially with congestive heart failure); trauma, such as fractures; malignant disease; obesity; pregnancy; polycythemia; and the use of oral contraceptives.

Among the clinical findings associated with PE are sudden and unexplained dyspnea (labored breathing), chest pain, apprehension, rapid breathing, rales, cough, deep venous thrombosis, pulmonary infarction as evidenced by hemoptysis (expectoration of blood or blood-stained sputum), and syncope.

Treatment of PE consists of intravenous (IV) injections of heparin (an anticoagulant); administration of oral anticoagulants, such as warfarin; and administration of clot-dissolving enzymes, such as streptokinase and urokinase. In some cases surgery is indicated to reverse the lethal effect of massive PE by embolectomy.

PULMONARY EDEMA

Pulmonary edema refers to an abnormal accumulation of interstitial fluid in the interstitial spaces and alveoli of the lungs. The edema may arise from increased pulmonary capillary permeability (pulmonary origin) or increased pulmonary capillary pressure (cardiac origin). The most common symptom is dyspnea. Others include wheezing, tachypnea (rapid respirations), restlessness, feeling of suffocation, cyanosis, pallor (paleness), and diaphoresis (excessive perspiration). Treatment consists of administering oxygen, drugs that dilate the bronchioles and lower blood pressure, and drugs that correct acid-base imbalance; suctioning of airways; and mechanical ventilation.

CARBON MONOXIDE POISONING

Carbon monoxide (CO) is a colorless and odorless gas found in exhaust fumes from automobiles and in tobacco smoke. It is a by-product of burning carbon-containing materials such as coal and wood. One of its interesting features is that it combines with hemoglobin very much as oxygen does, except that the combination of carbon monoxide and hemoglobin is over 210 times as tenacious as the combination of oxygen and hemoglobin. It combines easily and the combination does not break down readily. In concentrations as small as 0.1 percent, carbon monoxide will combine with half the hemoglobin molecules. Thus, the oxygen-carrying capacity of the blood is reduced by half. Increased levels of carbon monoxide lead to hypoxia, and the result is **carbon monoxide poisoning.**

The condition may be treated by administering pure oxygen, which slowly replaces the carbon monoxide combined with the hemoglobin.

KEY MEDICAL TERMS ASSOCIATED WITH THE RESPIRATORY SYSTEM

Apnea (*a* = without; *pnoia* = breath) Absence of ventilatory movements.

Asphyxia (*sphyxis* = pulse) Oxygen starvation due to low atmospheric oxygen or interference with pulmonary ventilation, external respiration, or internal respiration.

Aspiration (*spirare* = breathe) The act of breathing or drawing in.

Atelectasis (*ateles* = incomplete; *ektasis* = dilation) A collapsed lung or portion of a lung.

Bronchiectasis A chronic dilation of the bronchi or bronchioles.

Cheyne-Stokes respiration Irregular breathing beginning with shallow breaths that increase in depth and rapidity, then decrease and cease altogether for 15 to 20 seconds. The cycle repeats itself again and again. Cheyne-Stokes is normal in infants. It is also often seen just before death from pulmonary, cerebral, cardiac, and kidney disease and is referred to as the "death rattle."

Diphtheria (*diphthera* = membrane) An acute bacterial infection that causes the mucous membranes of the oropharynx, nasopharynx, and larynx to enlarge and become leathery. Enlarged membranes may obstruct airways and cause death from asphyxiation.

Dyspnea (*dys* = painful, difficult; *pnoia* = breath) Labored breathing (short-windedness).

Hypoxia (*hypo* = below, under) Reduction in oxygen supply to cells.

Orthopnea (*ortho* = straight) Inability to breathe in a horizontal position.

Pneumonectomy (*pneumo* = lung; *tome* = cutting) Surgical removal of a lung.

Pneumothorax Air in pleural space causing collapse of the lung. Most common cause is surgical opening of chest during heart surgery, making intrathoracic pressure equal atmospheric pressure.

Rales Sounds sometimes heard in the lungs that resemble bubbling or rattling. May be caused by air or an abnormal secretion in the lungs.

Respirator A device for providing artifical respiration or to assist in pulmonary ventilation. Also refers to an apparatus used to qualify air breathed through it.

Rhinitis (*rhino* = nose) Chronic or acute inflammation of the mucous membrane of the nose.

STUDY OUTLINE

Organs (p. 553)

1. Respiratory organs include the nose, pharynx, larynx, trachea, bronchi, and lungs.
2. They act with the cardiovascular system to supply oxygen and remove carbon dioxide from the blood.

Nose

1. The external portion is made of cartilage and skin and is lined with mucous membrane. Openings to the exterior are the external nares.
2. The internal portion communicates with the nasopharynx through the internal nares and the paranasal sinuses.
3. The nasal cavity is divided by a septum. The anterior portion of the cavity is called the vestibule.
4. The nose is adapted for warming, moistening, and filtering air, for olfaction, and for speech.

Pharynx

1. The pharynx, or throat, is a muscular tube lined by a mucous membrane.
2. The anatomic regions are nasopharynx, oropharynx, and laryngopharynx.
3. The nasopharynx functions in respiration. The oropharynx and laryngopharynx function both in digestion and in respiration.

Larynx

1. The larynx is a passageway that connects the pharynx with the trachea.
2. It contains the thyroid cartilage (Adam's apple); the epiglottis, which prevents food from entering the larynx; the cricoid cartilage, which connects the larynx and trachea; and the paired arytenoid, corniculate, and cuneiform cartilages.
3. The larynx contains vocal folds which produce sound. Taut folds produce high pitches, and relaxed ones produce low pitches.

Trachea

1. The trachea extends from the larynx to the primary bronchi.
2. It is composed of smooth muscle and C-shaped rings of cartilage and is lined with pseudostratified epithelium.
3. Two methods of removing obstructions from the respiratory passageways are tracheostomy and intubation.

Bronchi

1. The bronchial tree consists of the trachea, primary bronchi, secondary bronchi, tertiary bronchi, bronchioles, and terminal bronchioles. Walls of bronchi contain rings of cartilage; walls of bronchioles do not.
2. A bronchogram is a roentgenogram of the tree after introduction of an opaque contrast medium usually containing iodine.

Lungs

1. Lungs are paired organs in the thoracic cavity. They are enclosed by the pleural membrane. The parietal pleura is the outer layer; the visceral pleura is the inner layer.
2. The right lung has three lobes separated by two fissures; the left lung has two lobes separated by one fissure and a depression, the cardiac notch.
3. The secondary bronchi give rise to branches called segmental bronchi, which supply segments of lung tissue called bronchopulmonary segments.
4. Each bronchopulmonary segment consists of lobules, which contain lymphatics, arterioles, venules, terminal bronchioles, respiratory bronchioles, alveolar ducts, alveolar sacs, and alveoli.
5. Gas exchange occurs across the alveolar-capillary (respiratory) membranes.

Nervous Control of Respiration (p. 570)

1. The respiratory center consists of a medullary rhythmicity area (inspiratory and expiratory area), pneumotaxic area, and apneustic area.
2. The inspiratory area has an intrinsic excitability that sets the basic rhythm of respiration.
3. The pneumotaxic and apneustic areas coordinate the transition between inspiration and expiration.
4. Respirations may be modified by a number of factors, both in the brain and outside.

Modified Respiratory Movements (p. 572)

1. Modified respiratory movements are used to express emotions and to clear air passageways.
2. Coughing, sneezing, sighing, yawning, sobbing, crying, laughing, and hiccuping are types of modified respiratory movements.

Aging and the Respiratory System (p. 572)

1. Aging results in decreased vital capacity, decreased blood level of O_2, and diminished alveolar macrophage activity.
2. Elderly people are more susceptible to pneumonia, emphysema, bronchitis, and other pulmonary disorders.

Developmental Anatomy of the Respiratory System (p. 573)

1. The respiratory system begins as an outgrowth of endoderm called the laryngotracheal bud.
2. Smooth muscle, cartilage, and connective tissue of the bronchial tubes and pleural sacs develop from mesoderm.

Applications to Health (p. 573)

1. In bronchogenic carcinoma, bronchial epithelial cells are replaced by cancer cells after constant irritation has disrupted the normal growth, division, and function of the epithelial cells.
2. Bronchial asthma occurs when spasms of smooth muscle in bronchial tubes result in partial closure of air passageways, inflammation, inflated alveoli, and excess mucus production.
3. Bronchitis is an inflammation of the bronchial tubes.
4. Emphysema is characterized by deterioration of alveoli leading to loss of their elasticity. Symptoms are reduced expiratory volume, inflated lungs, and enlarged chest.
5. Pneumonia is an acute inflammation or infection of alveoli.
6. Tuberculosis is an inflammation of pleura and lungs produced by the organism *Mycobacterium tuberculosis.*
7. Respiratory distress syndrome (RDS) of the newborn is

an infant disorder in which surfactant is lacking and alveolar ducts and alveoli have a glassy appearance.

8. Respiratory failure means the respiratory system cannot supply sufficient oxygen or eliminate sufficient carbon dioxide.

9. Sudden infant death syndrome (SIDS) has recently been linked to laryngospasm, possibly triggered by a viral infection of the upper respiratory tract.

10. Coryza (common cold) is caused by viruses and is usually not accompanied by a fever, whereas influenza (flu) is usually accompanied by a fever greater than 101°F.

11. Pulmonary embolism (PE) refers to a blood clot or other foreign substance in a pulmonary arterial vessel that obstructs blood flow to lung tissue.

12. Pulmonary edema is an accumulation of interstitial fluid in interstitial spaces and alveoli of the lungs.

13. Carbon monoxide poisoning occurs when CO combines with hemoglobin instead of oxygen. The result is hypoxia.

REVIEW QUESTIONS

1. What organs make up the respiratory system? What function do the respiratory and cardiovascular systems have in common?

2. What are the basic differences among pulmonary ventilation, external respiration, and internal respiration?

3. Describe the structure of the external and internal nose and describe their functions in filtering, warming, and moistening air.

4. What is the pharynx? Differentiate the three regions of the pharynx and indicate their roles in respiration.

5. Describe the structure of the larynx and explain how it functions in respiration and voice production.

6. Describe the location and structure of the trachea. What is tracheostomy? Intubation?

7. What is the bronchial tree? Describe its structure. What is a bronchogram?

8. Where are the lungs located? Distinguish the parietal pleura from the visceral pleura. What is pleurisy?

9. Define each of the following parts of a lung: base, apex, costal surface, medial surface, hilus, root, cardiac notch, and lobe.

10. What is a lobule of the lung? Describe its composition and function in respiration.

11. What is a bronchopulmonary segment?

12. Describe the histology of the alveolar-capillary (respiratory) membrane.

13. How does the medullary rhythmicity area function in controlling respiration? How are the apneustic area and pneumotaxic area related to the control of respiration?

14. Define the various kinds of modified respiratory movements.

15. Describe the effects of aging on the respiratory system.

16. Describe the development of the respiratory system.

17. For each of the following disorders, list the principal clinical symptoms: bronchogenic carcinoma, bronchial asthma, bronchitis, emphysema, pneumonia, tuberculosis, respiratory distress syndrome (RDS) of the newborn, respiratory failure, sudden infant death syndrome (SIDS), coryza (common cold), influenza (flu), pulmonary embolism (PE), pulmonary edema, and carbon monoxide poisoning.

18. Refer to the key medical terms associated with the respiratory system. Be sure that you can define each term.

23

The Digestive System

Student Objectives

Define digestion and distinguish between the chemical and mechanical phases.

Identify the organs of the gastrointestinal (GI) tract and the accessory organs of digestion.

Describe the structure of the wall of the gastrointestinal (GI) tract.

Define the mesentery, mesocolon, falciform ligament, lesser omentum, and greater omentum as extensions of the peritoneum.

Describe the structure of the mouth and its role in mechanical digestion.

Identify the location and histology of the salivary glands.

Identify the parts of a typical tooth and compare deciduous and permanent dentitions.

Describe the anatomy and histology of the esophagus.

Describe the anatomy and histology of the stomach and explain how its structural features are related to mechanical and chemical digestion.

Describe the structure and histology of the pancreas.

Define the structure and histology of the liver.

Describe the structure and histology of the gallbladder.

Describe those structural features of the small intestine that adapt it for digestion and absorption.

Describe the mechanical movements of the small intestine.

Describe the anatomy and histology of the large intestine.

Explain the mechanical movements of the large intestine.

Explain how the large intestine is adapted for absorption, feces formation, and defecation.

Describe the effects of aging on the digestive system.

Describe the development of the digestive system.

Describe the clinical symptoms of the following disorders: dental caries, periodontal disease, peritonitis, peptic ulcers, appendicitis, GI tumors, diverticulitis, cirrhosis, hepatitis, gallstones, anorexia nervosa, and bulimia.

Define key medical terms associated with the digestive system.

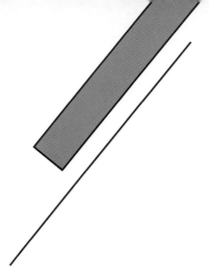

We all know that food is vital to life. Food is required for the chemical reactions that occur in every cell—both those that synthesize new enzymes, cell structures, bone, and all the other components of the body and those that release the energy needed for the building processes. However, the vast majority of foods we eat are simply too large to pass through the plasma membranes of the cells. The breaking down of food molecules for use by body cells is called **digestion,** and the organs that collectively perform this function comprise the **digestive system.**

The medical specialty that deals with the structure, function, diagnosis, and treatment of diseases of the stomach and intestines is called **gastroenterology** (gas'-trō-en'-ter-OL-ō-jē; *gastro* = stomach; *enteron* = intestines).

The developmental anatomy of the digestive system is considered later in the chapter.

DIGESTIVE PROCESSES

The digestive system prepares food for consumption by the cells through five basic activities.

    1. Ingestion. Taking food into the body (eating).
    2. Peristalsis. The movement of food along the digestive tract.
    3. Digestion. The breakdown of food by both chemical and mechanical processes.
    4. Absorption. The passage of digested food from the digestive tract into the cardiovascular and lymphatic systems for distribution to cells.
    5. Defecation. The elimination of indigestible substances from the body.

Chemical digestion is a series of catabolic reactions that break down the large carbohydrate, lipid, and protein molecules we eat into molecules usable by body cells. These products of digestion are small enough to pass through the walls of the digestive organs, into the blood and lymph capillaries, and eventually into the body's cells. **Mechanical digestion** consists of various movements that

aid chemical digestion. Food is prepared by the teeth before it can be swallowed. Then the smooth muscles of the stomach and small intestine churn the food so it is thoroughly mixed with the enzymes that catalyze the reactions.

ORGANIZATION

The organs of digestion are traditionally divided into two main groups. First is the **gastrointestinal (GI) tract,** or **alimentary canal,** a continuous tube running through the ventral body cavity and extending from the mouth to the anus (Figure 23-1). The relationship of the digestive organs to the nine regions of the abdominopelvic cavity may be reviewed in Figure 1-8b. The length of a tract taken from a cadaver is about 9 m (30 ft). In a living person it is somewhat shorter because the muscles in its wall are in a state of tone. Organs composing the gastrointestinal tract include the mouth, pharynx, esophagus, stomach, small intestine, and large intestine. The GI tract contains the food from the time it is eaten until it is digested and prepared for elimination. Muscular contractions in the walls of the GI tract break down the food physically by churning it. Secretions produced by cells along the tract break down the food chemically.

The second group of organs composing the digestive system consists of the **accessory structures**—the teeth, tongue, salivary glands, liver, gallbladder, and pancreas. Teeth protrude into the GI tract and aid in the physical breakdown of food. The other accessory structures, except for the tongue, lie totally outside the tract and produce or store secretions that aid in the chemical breakdown of food. These secretions are released into the tract through ducts.

GENERAL HISTOLOGY

The wall of the GI tract, especially from the esophagus to the anal canal, has the same basic arrangement of

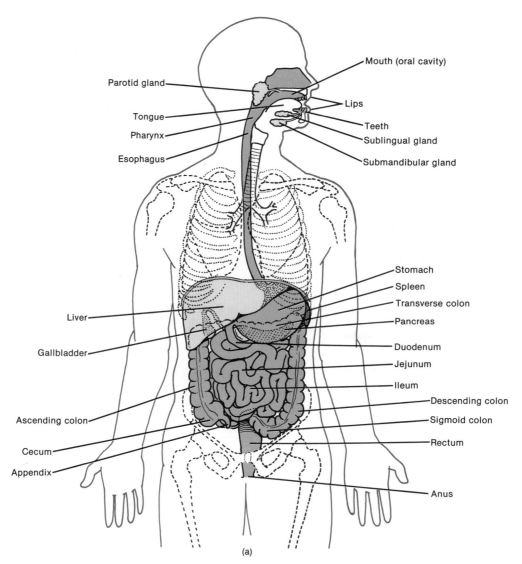

FIGURE 23-1 Organs of the digestive system and related structures. (a) Diagram.

tissues. The four coats (tunics) of the tract from the inside out are the mucosa, submucosa, muscularis, and serosa, or adventitia (Figure 23-2).

Mucosa

The **mucosa,** or inner lining of the tract, is a mucous membrane attached to a thin layer of visceral muscle. Two layers compose the membrane: a *lining epithelium,* which is in direct contact with the contents of the GI tract, and an underlying layer of loose connective tissue called the *lamina propria.* Under the lamina propria is visceral muscle called the *muscularis mucosae.*

The epithelial layer is composed of nonkeratinized cells that are stratified in the mouth and esophagus but are simple throughout the rest of the tract. The functions of the stratified epithelium are protection and secretion. The functions of the simple epithelium are secretion and

absorption. However, the lack of keratin allows some absorption to occur in all parts of the tract.

The lamina propria is made of loose connective tissue containing many blood and lymph vessels and scattered lymph nodules, masses of lymphatic tissue that are not encapsulated. This layer supports the epithelium, binds it to the muscularis mucosae, and provides it with a blood and lymph supply. The blood and lymph vessels are the avenues by which nutrients in the tract reach the other tissues of the body. The lymph tissue also protects against disease. Remember that the GI tract is in contact with the outside environment and contains food that often carries harmful bacteria. Unlike the skin, the mucous membrane of the tract is not protected from bacterial entry by keratin. The lamina propria also contains glandular epithelium that secretes products necessary for chemical digestion.

The muscularis mucosae contains smooth muscle fibers

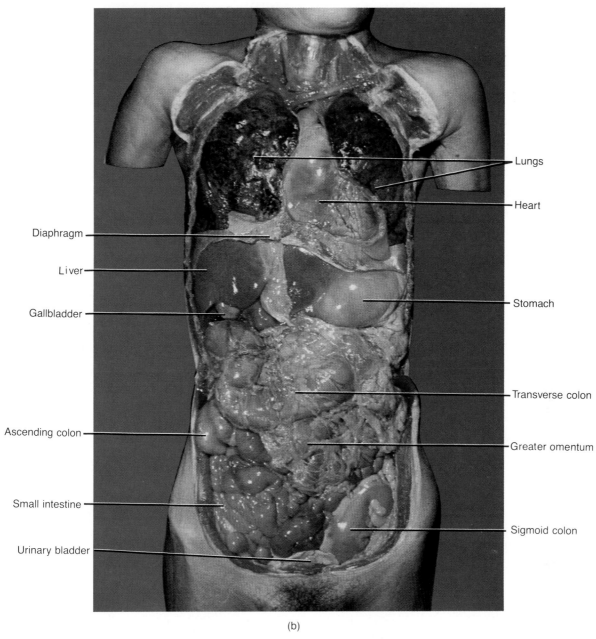

Diaphragm

Liver

Gallbladder

Ascending colon

Small intestine

Urinary bladder

Lungs

Heart

Stomach

Transverse colon

Greater omentum

Sigmoid colon

(b)

FIGURE 23-1 (*Continued*) Organs of the digestive system and related structures. (b) Photograph. (Courtesy of C. Yokochi and J. W. Rohen, *Photographic Anatomy of the Human Body,* 2nd ed., 1979, IGAKU-SHOIN, Ltd., Tokyo, New York.)

that throw the mucous membrane of the intestine into small folds that increase the digestive and absorptive area. With one exception, which will be described later, the other three coats of the intestine contain no glandular epithelium.

Submucosa

The **submucosa** consists of loose connective tissue that binds the mucosa to the third tunic, the muscularis. It is highly vascular and contains a portion of the *submucous plexus* (*plexus of Meissner*), which is part of the auto-

nomic nerve supply to the muscularis mucosae. This plexus is important in controlling secretions by the GI tract.

Muscularis

The **muscularis** of the mouth, pharynx, and esophagus consists in part of skeletal muscle that produces voluntary swallowing. Throughout the rest of the tract, the muscularis consists of smooth muscle that is generally found in two sheets: an inner ring of circular fibers and an outer sheet of longitudinal fibers. Contractions of the smooth

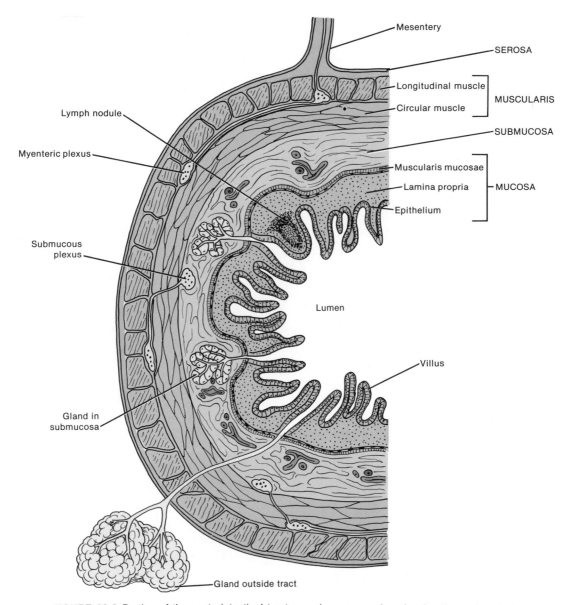

FIGURE 23-2 Portion of the gastrointestinal tract seen in cross section showing the various layers and related structures.

muscles help to break down food physically, mix it with digestive secretions, and propel it through the tract. The muscularis also contains the major nerve supply to the alimentary tract—the *myenteric plexus* (*plexus of Auerbach*), which consists of fibers from both autonomic divisions. This plexus mostly controls GI motility.

Serosa

The **serosa** is the outermost layer of most portions of the canal. It is a serous membrane composed of connective tissue and epithelium. This layer is also called the **visceral peritoneum** and forms a portion of the peritoneum, which we will describe in detail.

PERITONEUM

The **peritoneum** (per'-i-tō-NĒ-um) is the largest serous membrane of the body. Serous membranes are also associated with the heart (pericardium) and lungs (pleurae). Serous membranes consist of a layer of simple squamous epithelium (called mesothelium) and an underlying supporting layer of connective tissue. The **parietal peritoneum** lines the wall of the abdominal cavity. The **visceral peritoneum** covers some of the organs and constitutes their serosa. The potential space between the parietal and visceral portions of the peritoneum is called the **peritoneal cavity** and contains serous fluid. In certain diseases, the peritoneal cavity may become distended by several liters

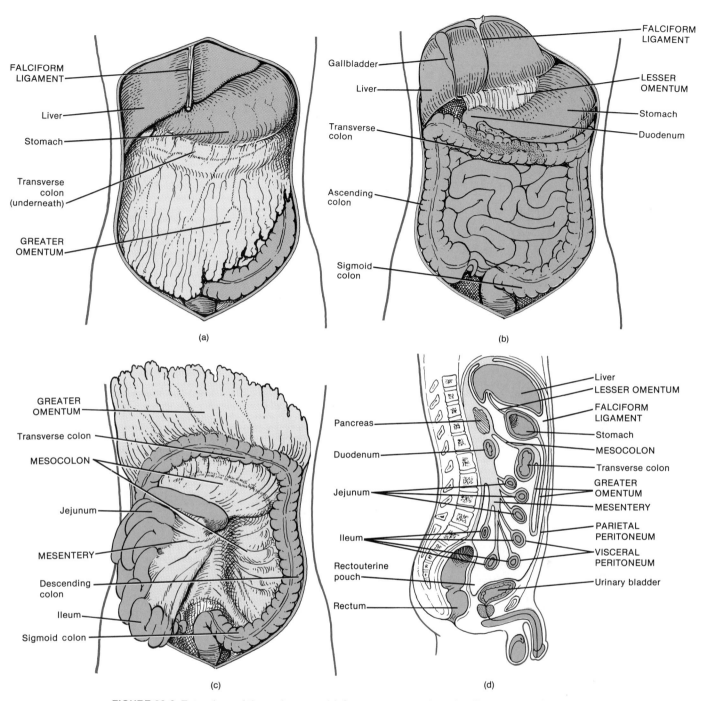

FIGURE 23-3 Extensions of the peritoneum. (a) Greater omentum (see also Figure 23-1b). (b) Lesser omentum. The liver and gallbladder have been lifted. (c) Mesentery. The greater omentum has been lifted. (d) Sagittal section through the abdomen and pelvis, indicating the relationship of the peritoneal extensions to each other.

of fluid so that it forms an actual space. Such an accumulation of serous fluid is called **ascites** (a-SĪ-tēz). As you will see later, some organs lie on the posterior abdominal wall and are covered by peritoneum on their anterior surfaces only. Such organs, including the kidneys and pancreas, are said to be **retroperitoneal.**

Unlike the pericardium and pleurae, the peritoneum contains large folds that weave in between the viscera.

The folds bind the organs to each other and to the walls of the cavity and contain the blood and lymph vessels and the nerves that supply the abdominal organs. One extension of the peritoneum is called the **mesentery** (MEZ-en-ter'-ē). It is an outward fold of the serous coat of the small intestine (Figure 23-3). The tip of the fold is attached to the posterior abdominal wall. The mesentery binds the small intestine to the wall. A similar fold

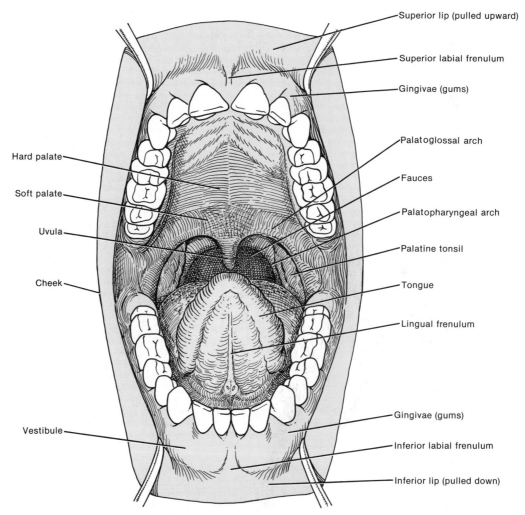

Superior lip (pulled upward)

Superior labial frenulum

Gingivae (gums)

Palatoglossal arch

Fauces

Palatopharyngeal arch

Palatine tonsil

Tongue

Lingual frenulum

Gingivae (gums)

Inferior labial frenulum

Inferior lip (pulled down)

Hard palate

Soft palate

Uvula

Cheek

Vestibule

FIGURE 23-4 Mouth (oral cavity).

of parietal peritoneum, called the **mesocolon** (mez'-ō-KŌ-lon), binds the large intestine to the posterior body wall. It also carries blood vessels and lymphatics to the intestines.

Other important peritoneal folds are the falciform ligament, the lesser omentum, and the greater omentum. The **falciform** (FAL-si-form) **ligament** attaches the liver to the anterior abdominal wall and diaphragm. The **lesser omentum** (ō-MENT-um) arises as two folds in the serosa of the stomach and duodenum suspending the stomach and duodenum from the liver. The **greater omentum** is a four-layered fold in the serosa of the stomach that hangs down like an apron over the front of the intestines. It then passes up to part of the large intestine (the transverse colon), wraps itself around it, and finally attaches to the parietal peritoneum of the posterior wall of the abdominal cavity. Because the greater omentum contains large quantities of adipose tissue, it commonly is called the "fatty apron." The greater omentum contains numerous lymph nodes. If an infection occurs in the intestine, plasma cells formed in the lymph nodes combat the infection and help prevent it from spreading to the peritoneum.

MOUTH (ORAL CAVITY)

The **mouth,** also referred to as the **oral,** or **buccal** (BUK-al), **cavity,** is formed by the cheeks, hard and soft palates, and tongue (Figure 23-4). Forming the lateral walls of the oral cavity are the **cheeks**—muscular structures covered on the outside by skin and lined by nonkeratinized stratified squamous epithelium. The anterior portions of the cheeks terminate in the superior and inferior lips.

The **lips (labia)** are fleshy folds surrounding the orifice of the mouth. They are covered on the outside by skin and on the inside by a mucous membrane. The transition zone where the two kinds of covering tissue meet is called the **vermilion** (ver-MIL-yon). This portion of the lips is nonkeratinized, and the color of the blood in the underlying blood vessels is visible through the transparent surface layer of the vermilion. The inner surface of each lip is attached to its corresponding gum by a midline fold of mucous membrane called the **labial frenulum** (LĀ-bē-al FREN-yoo-lum).

The orbicularis oris muscle and connective tissue lie between the external integumentary covering and the in-

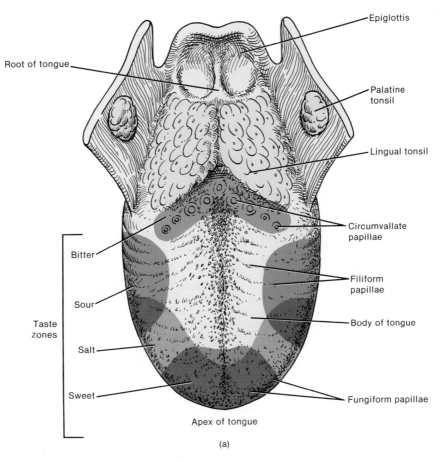

FIGURE 23-5 Tongue. (a) Locations of the papillae and the four taste zones.

ternal mucosal lining. During chewing, the cheeks and lips help to keep food between the upper and lower teeth. They also assist in speech.

The **vestibule** of the oral cavity is bounded externally by the cheeks and lips and internally by the gums and teeth. The **oral cavity proper** extends from the vestibule to the **fauces** (FAW-sēs), the opening between the oral cavity and the pharynx or throat.

The **hard palate,** the anterior portion of the roof of the mouth, is formed by the maxillae and palatine bones, is covered by mucous membrane, and forms a bony partition between the oral and nasal cavities. The **soft palate** forms the posterior portion of the roof of the mouth. It is an arch-shaped muscular partition between the oropharynx and nasopharynx and is lined by mucous membrane.

Hanging from the free border of the soft palate is a conical muscular process called the **uvula** (OO-vyoo-la). On either side of the base of the uvula are two muscular folds that run down the lateral side of the soft palate. Anteriorly, the **palatoglossal arch (anterior pillar)** extends inferiorly, laterally, and anteriorly to the side of the base of the tongue. Posteriorly, the **palatopharyngeal** (PAL-a-tō-fa-rin'-jē-al) **arch (posterior pillar)** projects inferiorly, laterally, and posteriorly to the side of the pharynx.

The palatine tonsils are situated between the arches, and the lingual tonsil is situated at the base of the tongue. At the posterior border of the soft palate, the mouth opens into the oropharynx through the fauces.

TONGUE

The **tongue,** together with its associated muscles, forms the floor of the oral cavity. It is an accessory structure of the digestive system composed of skeletal muscle covered with mucous membrane (Figure 23-5). The tongue is divided into symmetrical lateral halves by a median septum that extends throughout its entire length and is attached inferiorly to the hyoid bone. Each half of the tongue consists of an identical complement of extrinsic and intrinsic muscles.

The **extrinsic muscles** of the tongue originate outside the tongue and insert into it. They include the hyoglossus, chondroglossus, genioglossus, styloglossus, and palatoglossus (see Figure 10-7). The extrinsic muscles move the tongue from side to side and in and out. These movements maneuver food for chewing, shape the food into a rounded mass, called a **bolus,** and force the food to the back of the mouth for swallowing. They also form

Fungiform
papilla

Filiform
papillae

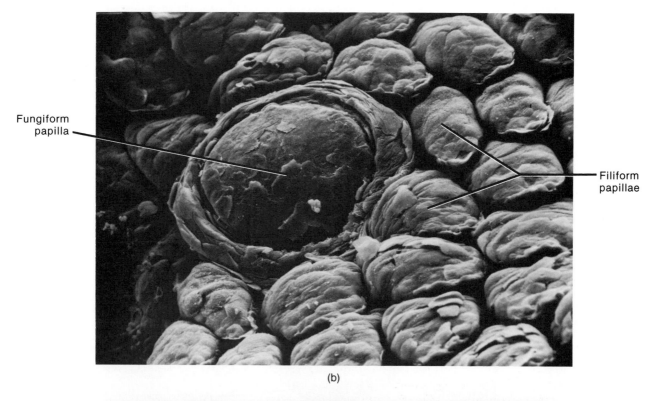

(b)

Gustatory
(taste)
pore

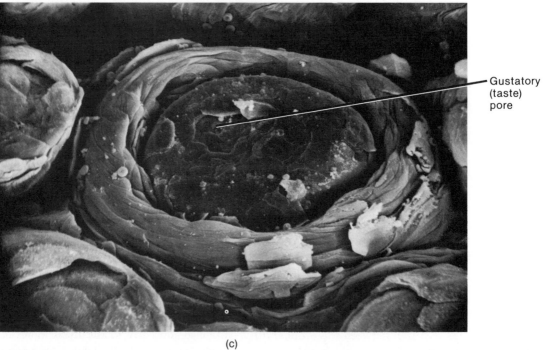

(c)

FIGURE 23-5 (*Continued*) Tongue. (b) Scanning electron micrograph of a fungiform papilla surrounded by filiform papillae at a magnification of 245×. (c) Scanning electron micrograph of a circumvallate papilla at a magnification of 1,020×. (Courtesy of Richard K. Kessel and Randy H. Kardon, *Tissues and Organs: A Text-Atlas of Scanning Electron Microscopy.* Copyright © 1979 by Scientific American, Inc.)

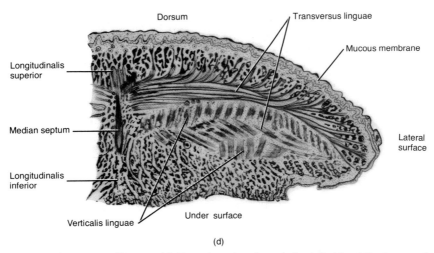

FIGURE 23-5 (*Continued*) Tongue. (d) Frontal section through the left side of the tongue showing its intrinsic muscles.

the floor of the mouth and hold the tongue in position. The **intrinsic muscles** originate and insert within the tongue and alter the shape and size of the tongue for speech and swallowing. The intrinsic muscles include the longitudinalis superior, longitudinalis inferior, transversus linguae, and verticalis linguae (Figure 23-5d). The **lingual frenulum,** a fold of mucous membrane in the midline of the undersurface of the tongue, aids in limiting the movement of the tongue posteriorly.

CLINICAL APPLICATION

If the lingual frenulum is too short, tongue movements are restricted, speech is faulty, and the person is said to be "tongue-tied." This congenital problem is referred to as **ankyloglossia** (ang-ki-lō-GLOSS-ē-a). It can be corrected by cutting the lingual frenulum.

The upper surface and sides of the tongue are covered with **papillae** (pa-PIL-ē), projections of the lamina propria covered with epithelium (Figure 23-5a–c). **Filiform papillae** are conical projections distributed in parallel rows over the anterior two-thirds of the tongue. They are whitish and contain no taste buds. **Fungiform papillae** are mushroomlike elevations distributed among the filiform papillae and more numerous near the tip of the tongue. They appear as red dots on the surface of the tongue, and most of them contain taste buds. **Circumvallate papillae,** 10 to 12 in number, are arranged in the form of an inverted V on the posterior surface of the tongue, and all of them contain taste buds. Note the taste zones of the tongue in Figure 23-5a.

The tongue has a rich blood supply. The lingual artery is the only blood supply, and its veins eventually drain into the internal jugular vein. Most of the muscles of the tongue are supplied by the hypoglossal (XII) nerves. Taste sensations from the tongue are conveyed by the facial (VII) and glossopharyngeal (IX) nerves.

SALIVARY GLANDS

Saliva is a fluid that is continuously secreted by glands in or near the mouth. Ordinarily, just enough saliva is secreted to keep the mucous membranes of the mouth moist, but when food enters the mouth, secretion increases so the saliva can lubricate, dissolve, and begin the chemical breakdown of the food. The mucous membrane lining the mouth contains many small glands, the **buccal glands,** that secrete small amounts of saliva. However, the major portion of saliva is secreted by the **salivary glands,** accessory structures that lie outside the mouth and pour their contents into ducts that empty into the oral cavity. There are three pairs of salivary glands: parotid, submandibular (submaxillary), and sublingual glands (Figure 23-6a).

The **parotid glands** are located under and in front of the ears between the skin and the masseter muscle. They are compound tubuloacinar glands. Each secretes into the oral cavity vestibule via a duct, called the parotid (Stensen's) duct, that pierces the buccinator muscle to open into the vestibule opposite the upper second molar tooth. The **submandibular glands,** which are compound acinar glands, are found beneath the base of the tongue in the posterior part of the floor of the mouth (Figure 23-6b). Their ducts, the submandibular (Wharton's) ducts, run superficially under the mucosa on either side of the midline of the floor of the mouth and enter the oral cavity proper just behind the central incisors. The **sublingual glands,** also compound acinar glands, are anterior to the submandibular glands, and their ducts, the lesser sublingual (Rivinus's) ducts, open into the floor of the mouth in the oral cavity proper.

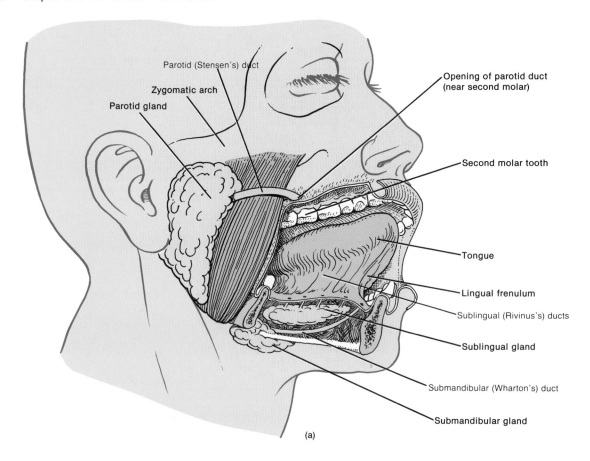

Parotid (Stensen's) duct

Zygomatic arch

Parotid gland

Opening of parotid duct (near second molar)

Second molar tooth

Tongue

Lingual frenulum

Sublingual (Rivinus's) ducts

Sublingual gland

Submandibular (Wharton's) duct

Submandibular gland

(a)

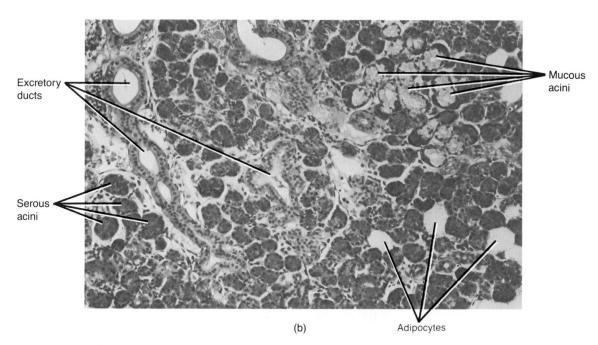

Excretory ducts

Serous acini

Mucous acini

Adipocytes

(b)

FIGURE 23-6 Salivary glands. (a) Location of the salivary glands. (b) Photomicrograph of the submandibular gland showing mostly serous acini and a few mucous acini at a magnification of 120×. The parotids consist of all serous acini and the sublinguals consist of mostly mucous acini and a few serous acini. (© 1983 by Michael H. Ross. Used by permission.)

CLINICAL APPLICATION

Although any of the salivary glands may become infected as a result of a nasopharyngeal infection, the parotids are typically the target of the mumps virus (myxovirus). **Mumps** is an inflammation and enlargement of the parotid glands accompanied by moderate fever, malaise, and extreme pain in the throat, especially when swallowing sour foods or acid juices. Swelling occurs on one or both sides of the face, just anterior to the ramus of the mandible. In about 20 to 35 percent of males past puberty, the testes may also become inflamed, and although it rarely occurs, sterility is a possible consequence. In some people, aseptic meningitis, pancreatitis, and hearing loss may also occur as complications.

The parotid gland receives its blood supply from branches of the external carotid artery and is drained by vessels that are tributaries of the external jugular vein. The submandibular gland is supplied by branches of the facial artery and drained by tributaries of the facial vein. The sublingual gland is supplied by the sublingual branch of the lingual artery and the submental branch of the facial artery and is drained by tributaries of the sublingual and submental veins.

The salivary glands receive both sympathetic and parasympathetic innervation. The sympathetic fibers form plexuses on the blood vessels that supply the glands and serve as vasoconstrictors. The parotid gland receives sympathetic fibers from the plexus on the external carotid artery, whereas the submandibular and sublingual glands receive sympathetic fibers that contribute to the sympathetic plexus and accompany the facial artery to the glands. The parasympathetic fibers of the glands consist of secretomotor fibers to the glands.

The fluids secreted by the buccal glands and the three pairs of salivary glands constitute **saliva.** Amounts of saliva secreted daily vary considerably but range from 1,000 to 1,500 ml. Chemically, saliva is 99.5 percent water and 0.5 percent solutes. Among the solutes are salts—chlorides, bicarbonates, and phosphates of sodium and potassium. Some dissolved gases and various organic substances, including urea and uric acid, serum albumin and globulin, mucin, the bacteriolytic enzyme lysozyme, and the digestive enzyme salivary amylase, are also present. *Salivary amylase* initiates the breakdown of starch.

Each saliva-producing gland supplies different ingredients to saliva. The parotids contain cells that secrete a watery serous liquid that includes the enzyme salivary amylase. The submandibular glands contain cells similar to those found in the parotids plus some mucous cells. Therefore, they secrete a fluid that is thickened with mucus but still contains quite a bit of enzyme. The sublingual glands contain mostly mucous cells, so they secrete a much thicker fluid that contributes only a small amount of enzyme to the saliva.

Saliva continues to be secreted heavily some time after food is swallowed. This flow of saliva washes out the mouth and dilutes and buffers the chemical remnants of irritating substances.

TEETH

The **teeth (dentes)** are accessory structures of the digestive system located in sockets of the alveolar processes of the mandible and maxillae. The alveolar processes are covered by the **gingivae** (jin-JI-vē), or gums, which extend slightly into each socket forming the gingival sulcus (Figure 23-7). The sockets are lined by the **periodontal ligament,** which consists of bundles of collagenous fibers and is attached to the socket walls and the cemental surface of the roots. Thus it anchors the teeth in position and also acts as a shock absorber to dissipate the forces of chewing.

A typical tooth consists of three principal portions. The **crown** is the portion above the level of the gums. The **root** consists of one to three projections embedded in the socket. The **neck** is the constricted junction line of the crown and the root.

Teeth are composed primarily of **dentin,** a bonelike substance that gives the tooth its basic shape and rigidity. The dentin encloses a cavity. The enlarged part of the cavity, the **pulp cavity,** lies in the crown and is filled with **pulp,** a connective tissue containing blood vessels, nerves, and lymphatics. Narrow extensions of the pulp cavity run through the root of the tooth and are called **root canals.** Each root canal has an opening at its base, the **apical foramen.** Through the foramen enter blood vessels bearing nourishment, lymphatics affording protection, and nerves providing sensation. The dentin of the crown is covered by **enamel** that consists primarily of calcium phosphate and calcium carbonate. Enamel is the hardest substance in the body and protects the tooth from the wear of chewing. It is also a barrier against acids that easily dissolve the dentin. The dentin of the root is covered by **cementum,** another bonelike substance, which attaches the root to the periodontal ligament.

The branch of dentistry that is concerned with the prevention, diagnosis, and treatment of diseases that affect the pulp, root, periodontal ligament, and alveolar bone is known as **endodontics** (en'-dō-DON-tiks; *endo* = within; *odous* = tooth).

Dental Terminology

Because of the curvature of the dental arches, it is necessary to use terms other than anterior, posterior, medial, and lateral in describing the surfaces of the teeth. Accordingly, the following directional terms are used. **Labial** refers to the surface of a tooth in contact with or directed toward the lips. **Buccal** refers to the surface in contact

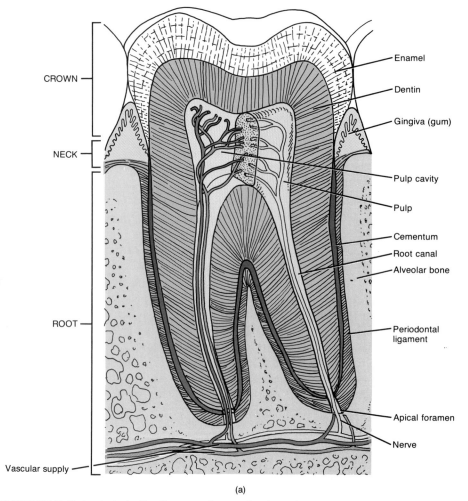

CROWN

NECK

ROOT

Vascular supply

Enamel

Dentin

Gingiva (gum)

Pulp cavity

Pulp

Cementum

Root canal

Alveolar bone

Periodontal ligament

Apical foramen

Nerve

(a)

FIGURE 23-7 Parts of a typical tooth as seen in a section through a molar. (a) Diagram. (b) Photograph. (Courtesy of C. Yokochi and J. W. Rohen, *Photographic Anatomy of the Human Body*, 2nd ed., 1979, IGAKU-SHOIN, Ltd., Tokyo, New York.)

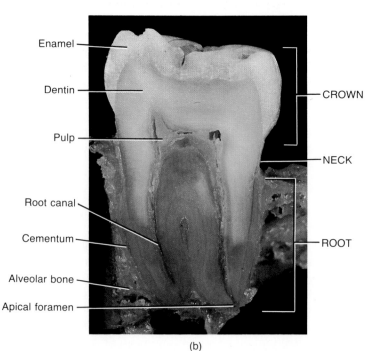

Enamel

Dentin

Pulp

Root canal

Cementum

Alveolar bone

Apical foramen

CROWN

NECK

ROOT

(b)

with or directed toward the cheeks. **Lingual** is restricted to the teeth of the lower jaw and refers to the surface directed toward the tongue. **Palatal,** on the other hand, is restricted to the teeth of the upper jaw and refers to the surface directed toward the palate. The term **mesial** designates the anterior or medial side of the tooth relative to its position in the dental arch. **Distal** refers to the posterior or lateral side of the tooth relative to its position in the dental arch. Essentially, mesial and distal refer to the sides of adjacent teeth that are in contact with each other. Finally, **occlusal** refers to the biting surface of a tooth.

Dentitions

Everyone has two **dentitions,** or sets of teeth. The first of these—the **deciduous teeth, milk teeth,** or **baby teeth**—begin to erupt at about 6 months of age, and one pair appears at about each month thereafter until all 20 are present. Figure 23-8a illustrates the deciduous teeth. The

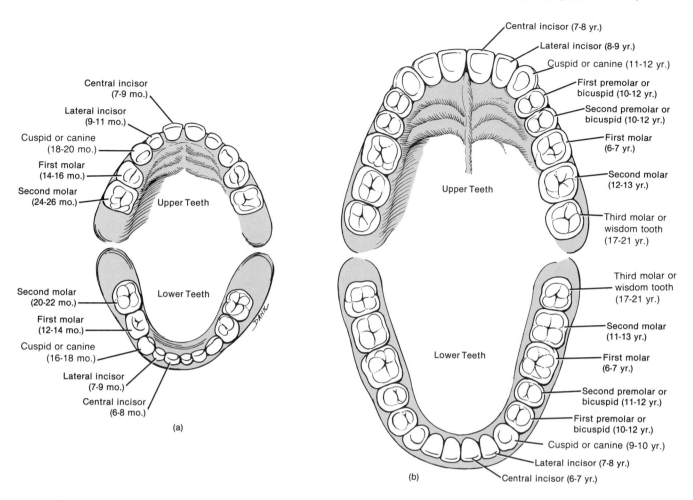

FIGURE 23-8 Dentitions and times of eruptions. The times of eruptions are indicated in parentheses. (a) Deciduous dentition. (b) Permanent dentition.

incisors, which are closest to the midline, are chisel-shaped and adapted for cutting into food. They are referred to as **central** or **lateral incisors** on the basis of their position. Next to the incisors, moving posteriorly, are the **cuspids (canines),** which have a pointed surface called a cusp. Cuspids are used to tear and shred food. The incisors and cuspids have only one root apiece. Behind them lie the **first** and **second molars,** which have four cusps. Upper molars have three roots; lower molars have two roots. The molars crush and grind food.

All the deciduous teeth are lost—generally between 6 and 12 years of age—and are replaced by the **permanent dentition** (Figure 23-8b). The permanent dentition contains 32 teeth that appear between the age of 6 and adulthood. It resembles the deciduous dentition with the following exceptions. The deciduous molars are replaced with the **first** and **second premolars (bicuspids),** which have two cusps and one root (upper first bicuspids have two roots) and are used for crushing and grinding. The permanent molars erupt into the mouth behind the bicuspids. They do not replace any deciduous teeth and erupt

as the jaw grows to accommodate them—the **first molars** at age 6, the **second molars** at age 12, the **third molars (wisdom teeth)** after age 18. The human jaw has become smaller through time and often does not afford enough room behind the second molars for the eruption of the third molars. In this case, the third molars remain embedded in the alveolar bone and are said to be "impacted." Most often they cause pressure and pain and must be surgically removed. In some individuals, third molars may be dwarfed in size or may not develop at all.

Blood and Nerve Supply

The arteries that supply blood to the teeth are distributed to the pulp cavity and surrounding periodontal membrane. The upper incisors and cuspids are supplied by anterior superior alveolar branches of the maxillary artery; the upper premolars and molars are supplied by the posterior superior alveolar branches of the maxillary artery; the lower incisors and cuspids are supplied by the incisive branches of the inferior alveolar artery; and

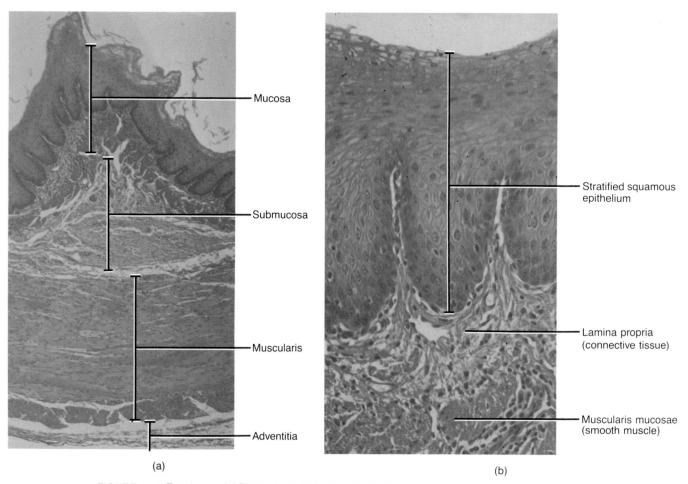

(a)

(b)

FIGURE 23-9 Esophagus. (a) Photomicrograph of a portion of the wall of the esophagus at a magnification of 60×. (b) Photomicrograph of an enlarged aspect of the mucosa of the esophagus at a magnification of 180×. (© 1983 by Michael H. Ross. Used by permission.)

the lower premolars and molars are supplied by the dental branches of the inferior alveolar artery.

The teeth receive sensory fibers from branches of the maxillary and mandibular divisions of the trigeminal (V) nerve—the upper teeth from branches of the maxillary division and the lower teeth from branches of the mandibular division.

Through chewing, or **mastication,** the teeth grind food and mix it with saliva. As a result, the food is reduced to a soft, flexible mass called a **bolus** that is easily swallowed.

PHARYNX

Swallowing, or **deglutition** (dē-gloo-TISH-un), is a mechanism that moves food from the mouth to the stomach. It is facilitated by saliva and mucus and involves the mouth, pharynx, and esophagus. Swallowing is conveniently divided into three stages: (1) the voluntary stage, in which the bolus is moved into the oropharynx, (2)

the pharyngeal stage, the involuntary passage of the bolus through the pharynx into the esophagus, and (3) the esophageal stage, the involuntary passage of the bolus through the esophagus into the stomach.

ESOPHAGUS

The **esophagus** (e-SOF-a-gus), the third organ involved in deglutition, is a muscular, collapsible tube that lies behind the trachea. It is about 23 to 25 cm (10 in) long and begins at the end of the laryngopharynx, passes through the mediastinum anterior to the vertebral column, pierces the diaphragm through an opening called the *esophageal hiatus,* and terminates in the superior portion of the stomach.

HISTOLOGY

The *mucosa* of the esophagus consists of nonkeratinized stratified squamous epithelium, lamina propria, and a

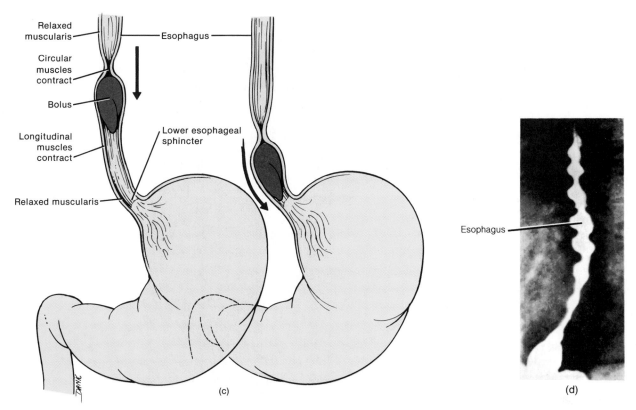

(c)

(d)

FIGURE 23-9 (Continued) Esophagus. (c) Diagram of peristalsis. (d) Anteroposterior projection of peristalsis made during fluoroscopic examination while a patient was swallowing barium. (Courtesy of Lester W. Paul and John H. Juhl, *The Essentials of Roentgen Interpretation,* 3rd ed., Harper & Row, Publishers, Inc., New York, 1972.)

muscularis mucosae (Figure 23-9a, b). The *submucosa* contains connective tissue and blood vessels. The *muscularis* of the upper third is striated, the middle third is striated and smooth, and the lower third is smooth. The outer layer is known as the *adventitia* (ad-ven-TISH-a) and not the serosa because the loose connective tissue of the layer is not covered by epithelium (mesothelium) and because the connective tissue merges with the connective tissue of surrounding structures.

ACTIVITIES

The esophagus does not produce digestive enzymes and does not carry on absorption. It secretes mucus and transports food to the stomach. The passage of food from the laryngopharynx into the esophagus is regulated by a sphincter (thick circle of muscle around an opening) at the entrance to the esophagus called the **upper esophageal** (e-sof'-a-JĒ-al) **sphincter.** It consists of the cricopharyngeus muscle attached to the cricoid cartilage. The elevation of the larynx during the pharyngeal stage of swallowing causes the sphincter to relax and the bolus enters the esophagus. The sphincter also relaxes during expiration.

During the esophageal phase of swallowing, food is pushed through the esophagus by involuntary muscular movements called **peristalsis** (per'-i-STAL-sis) (Figure 23-9c, d). Peristalsis is a function of the muscularis and is controlled by the medulla. In the section of the esophagus lying just above and around the top of the bolus, the circular muscle fibers contract. The contraction constricts the esophageal wall and squeezes the bolus downward. Meanwhile, longitudinal fibers lying around the bottom of and just below the bolus also contract. Contraction of the longitudinal fibers shortens this lower section, pushing its wall outward so it can receive the bolus. The contractions are repeated in a wave that moves down the esophagus, pushing the food toward the stomach. Passage of the bolus is further facilitated by glands secreting mucus. The passage of solid or semisolid food from the mouth to the stomach takes 4 to 8 seconds. Very soft foods and liquids pass through in about 1 second.

Just above the level of the diaphragm, the esophagus is slightly narrowed. This narrowing has been attributed to a physiological sphincter in the inferior part of the esophagus known as the **lower esophageal (gastroesophageal) sphincter.** The lower esophageal sphincter relaxes during swallowing and thus aids the passage of the bolus from the esophagus into the stomach.

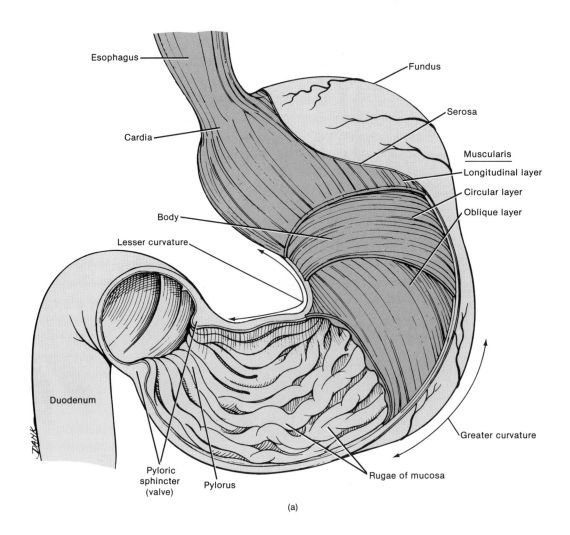

Esophagus

Cardia

Body

Lesser curvature

Duodenum

Pyloric
sphincter
(valve)

Pylorus

Fundus

Serosa

Muscularis
Longitudinal layer
Circular layer
Oblique layer

Greater curvature

Rugae of mucosa

(a)

FIGURE 23-10 External and internal anatomy of the stomach. (a) Diagram in anterior view.

BLOOD AND NERVE SUPPLY

The arteries of the esophagus are derived from the arteries along its length: inferior thyroid, thoracic aorta, intercostal arteries, phrenic, and left gastric arteries. It is drained by the adjacent veins. Innervation of the esophagus is by recurrent laryngeal nerves, the cervical sympathetic chain, and vagi.

STOMACH

The **stomach** is a J-shaped enlargement of the GI tract directly under the diaphragm in the epigastric, umbilical, and left hypochondriac regions of the abdomen (see Figure 1-8b). The superior portion of the stomach is a continuation of the esophagus. The inferior portion empties into the duodenum, the first part of the small intestine. Within each individual, the position and size of the stomach vary continually. For instance, the diaphragm pushes the stomach downward with each inspiration and pulls it upward

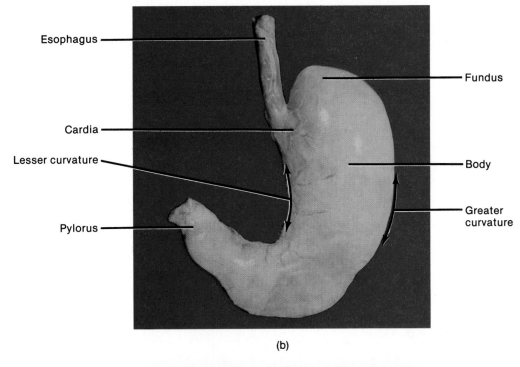

(b)

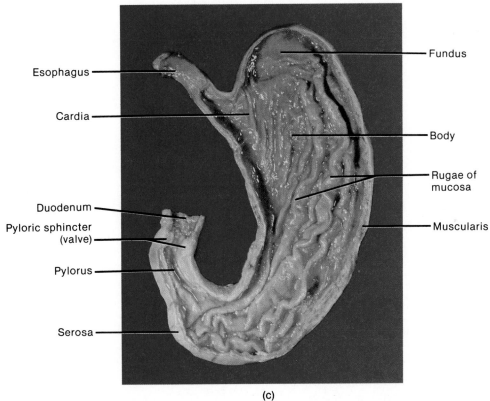

(c)

with each expiration. Empty, it is about the size of a large sausage, but it can stretch to accommodate large amounts of food.

ANATOMY

The stomach is divided into four areas: cardia, fundus, body, and pylorus (Figure 23-10). The **cardia** surrounds the lower esophageal sphincter. The rounded portion

FIGURE 23-10 (*Continued*) External and internal anatomy of the stomach. (b) Photograph of the external surface in anterior view. (c) Photograph of the internal surface in anterior view showing rugae. (Photographs courtesy of C. Yokochi and J. W. Rohen, *Photographic Anatomy of the Human Body,* 2nd ed., 1979, IGAKU-SHOIN, Ltd., Tokyo, New York.)

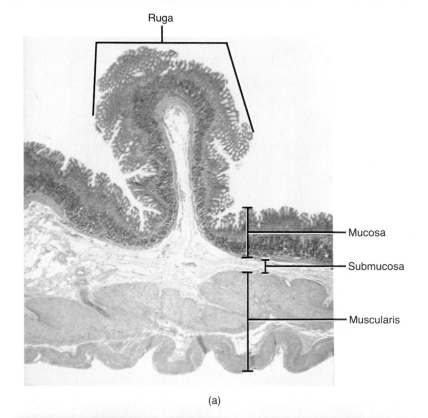

(a)

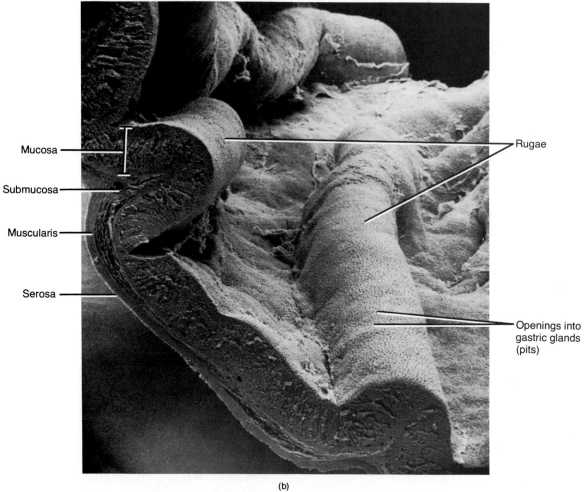

(b)

FIGURE 23-11 Histology of the stomach. (a) Photomicrograph of a portion of the fundic wall of the stomach at a magnification of 18×. (© 1983 by Michael H. Ross. Used by permission.) (b) Scanning electron micrograph of the stomach wall at a magnification of 55×. (Courtesy of Richard K. Kessel and Randy H. Kardon, *Tissues and Organs: A Text-Atlas of Scanning Electron Microscopy.* Copyright © 1979 by Scientific American, Inc.)

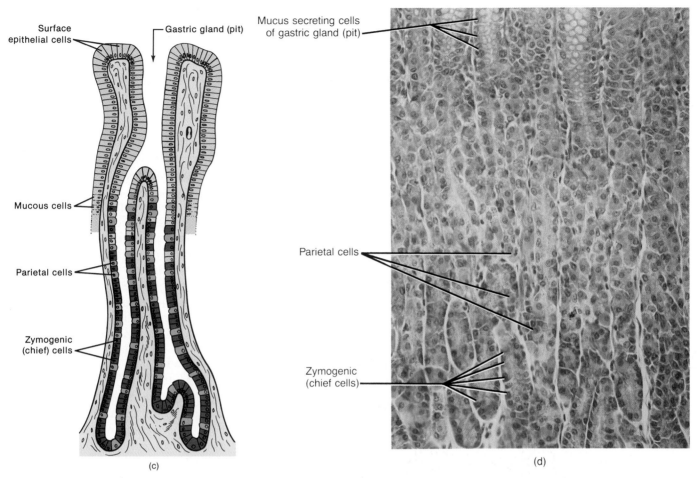

FIGURE 23-11 (*Continued*) Histology of the stomach. (c) Diagram of gastric glands from the fundic wall of the stomach. (d) Photomicrograph of an enlarged aspect of the mucosa of the fundic wall of the stomach at a magnification of 180×. (© 1983 by Michael H. Ross. Used by permission.)

above and to the left of the cardia is the **fundus.** Below the fundus is the large central portion of the stomach, called the **body.** The narrow, inferior region is the **pylorus (antrum).** The concave medial border of the stomach is called the **lesser curvature,** and the convex lateral border is the **greater curvature.** The pylorus communicates with the duodenum of the small intestine via a sphincter called the **pyloric sphincter (valve).**

CLINICAL APPLICATION

Two abnormalities of the pyloric sphincter can occur in infants. **Pylorospasm** is characterized by failure of the muscle fibers encircling the opening to relax normally. It can be caused by hypertrophy or continuous spasm of the sphincter and usually occurs between the second and twelfth weeks of life. Ingested food does not pass easily from the stomach to the small intestine, the stomach becomes overly full, and the infant vomits frequently to relieve the pressure. Pylorospasm is treated by adrenergic drugs that relax the

muscle fibers of the sphincter. **Pyloric stenosis** is a narrowing of the pyloric sphincter caused by a tumor-like mass that apparently is formed by enlargement of the circular muscle fibers. It must be surgically corrected.

HISTOLOGY

The stomach wall is composed of the same four basic layers as the rest of the alimentary canal, with certain modifications. When the stomach is empty, the *mucosa* lies in large folds, called **rugae** (ROO-jē), that can be seen with the naked eye. Microscopic inspection of the mucosa reveals a layer of simple columnar epithelium containing many narrow openings that extend down into the lamina propria (Figure 23-11a–d). These pits—**gastric glands**—are lined with several kinds of secreting cells: zymogenic, parietal, mucous, and enteroendocrine. The **zymogenic (chief) cells** secrete the principal gastric enzyme precursor, pepsinogen. Hydrochloric acid, involved in the conversion of pepsinogen to the active enzyme pepsin, and intrinsic factor, involved in the absorption of vitamin B$_{12}$ for red blood cell production, are produced by the **parietal cells.** Inability to produce intrinsic factor can result in pernicious anemia. The **mucous cells** secrete mucus. The **enteroendocrine cells** secrete stomach gastrin, a hormone that stimulates secretion of hydrochloric acid and pepsinogen, contracts the lower esophageal sphincter, mildly increases motility of the GI tract, and relaxes the pyloric sphincter. Secretions of the gastric glands are together called **gastric juice.**

CLINICAL APPLICATION

The general term **endoscopy** refers to the visual inspection of any cavity of the body using an endoscope, an illuminated tube with lenses. Endoscopes can be used to visualize the entire gastrointestinal tract. In fact, endoscopes can be fitted with special devices that also remove foreign objects from the esophagus and stomach, dilate strictures in the esophagus, remove small gallstones, temporarily stop bleeding, biopsy lesions, and remove polyps from the colon.

Endoscopic examination of the stomach is called **gastroscopy.** During the procedure, the patient is anesthesized and the endoscope is passed into the stomach, which is then inflated with air. The gastric mucosa may be examined for any abnormalities and, if necessary, a gastric mucosal biopsy may be performed.

The *submucosa* of the stomach is composed of loose areolar connective tissue, which connects the mucosa to the muscularis.

The *muscularis,* unlike that in other areas of the alimentary canal, has three layers of smooth muscle: an outer longitudinal layer, a middle circular layer, and an inner oblique layer. This arrangement of fibers allows the stomach to contract in a variety of ways to churn food, break it into small particles, mix it with gastric juice, and pass it to the duodenum.

The *serosa* covering the stomach is part of the visceral peritoneum. At the lesser curvature the two layers of the visceral peritoneum come together and extend upward to the liver as the lesser omentum. At the greater curvature, the visceral peritoneum continues downward as the greater omentum hanging over the intestines.

ACTIVITIES

Several minutes after food enters the stomach, gentle, rippling, peristaltic movements called **mixing waves** pass over the stomach every 15 to 25 seconds. These waves macerate food, mix it with the secretions of the gastric glands, and reduce it to a thin liquid called **chyme** (kīm). Few mixing waves are observed in the fundus, which is primarily a storage area. Foods may remain in the fundus for an hour or more without becoming mixed with gastric juice. During this time, salivary digestion continues.

As digestion proceeds in the stomach, more vigorous mixing waves begin at the body of the stomach and intensify as they reach the pylorus. The pyloric sphincter normally remains almost, but not completely, closed. As food reaches the pylorus, each mixing wave forces a small amount of the gastric contents into the duodenum through the pyloric sphincter. Most of the food is forced back into the body of the stomach where it is subjected to further mixing. The next wave pushes it forward again and forces a little more into the duodenum. The forward and backward movement of the gastric contents are responsible for almost all of the mixing in the stomach.

The principal chemical activity of the stomach is to begin the digestion of proteins. In the adult, digestion is achieved primarily through the enzyme **pepsin.** Another enzyme of the stomach, **gastric lipase,** splits the butterfat molecules found in milk. This enzyme has a limited role in the adult stomach. Adults rely almost exclusively on an enzyme found in the small intestine to digest fats. The infant stomach also secretes **rennin,** which is important in the digestion of milk. Rennin and calcium act on the casein of milk to produce a curd. The coagulation prevents too rapid a passage of milk from the stomach. Rennin is absent in the gastric secretions of adults.

The stomach empties all its contents into the duodenum 2 to 6 hours after ingestion. Food rich in carbohydrate leaves the stomach in a few hours. Protein foods are somewhat slower, and emptying is slowest after a meal containing large amounts of fat.

CLINICAL APPLICATION

Excessive gastric emptying in the wrong direction sometimes occurs. **Vomiting** is the forcible expulsion of the contents of the upper GI tract (stomach and sometimes duodenum) through the mouth. The strongest stimuli for vomiting are irritation and distension of the stomach. Other stimuli include unpleasant sights and dizziness. Impulses are transmitted to the vomiting center in the medulla, and returning impulses to the upper GI tract organs, diaphragm, and abdominal muscles bring about the vomiting act. The steps typically involved in the act are: (1) a deep breath inspiration, (2) opening of the upper esophageal sphincter, (3) closure of the glottis, (4) elevation of the soft palate to close off the internal nares, (5) contraction of the diaphragm and abdominal muscles, and (6) opening of the lower esophageal sphincter. Basically, vomiting involves squeezing the stomach between the diaphragm and abdominal muscles and expulsion of the contents through open esophageal sphincters.

Prolonged vomiting, especially in infants and elderly people, can be serious because the loss of gastric juice and fluids can lead to disturbances in fluid and acid–base balance.

The stomach wall is impermeable to the passage of most materials into the blood, so most substances are not absorbed until they reach the small intestine. However, the stomach does participate in the absorption of some water, electrolytes, certain drugs, and alcohol.

BLOOD AND NERVE SUPPLY

The arterial supply of the stomach is derived from the celiac artery. The right and left gastric arteries form an anastomosing arch along the lesser curvature, and the right and left gastroepiploic arteries form a similar arch on the greater curvature. Short gastric arteries supply the fundus. The veins of the same name accompany the arteries and drain, directly or indirectly, into the hepatic portal vein.

The vagi convey parasympathetic fibers to the stomach. These fibers form synapses within the submucous plexus in the submucosa. The sympathetic nerves arise from the celiac ganglia.

PANCREAS

The next organ of the GI tract involved in the breakdown of food is the small intestine. Chemical digestion in the small intestine depends not only on its own secretions, but on activities of three accessory structures of digestion outside the alimentary canal: the pancreas, liver, and gallbladder.

ANATOMY

The **pancreas** is a soft, oblong tubuloacinar gland about 12.5 cm (6 in) long and 2.5 cm (1 in) thick. It lies posterior to the greater curvature of the stomach and is connected by a duct (sometimes two) to the duodenum (Figure 23-12). The pancreas is divided into a head, body, and tail. The **head** is the expanded portion near the C-shaped curve of the duodenum. Moving superiorly and to the left of the head are the centrally located **body** and the terminal tapering **tail.**

The pancreas is linked to the small intestine by a series of ducts. The products of its secreting cells are dumped into small ducts attached to the cells and eventually leave the pancreas through a large main tube called the **pancreatic duct (duct of Wirsung).** In most people the pancreatic duct unites with the common bile duct from the liver and gallbladder and enters the duodenum in a common duct, called the **hepatopancreatic ampulla (ampulla of Vater).** The ampulla opens on an elevation of the duodenal mucosa known as the **duodenal papilla,** about 10 cm (4 in) below the pylorus of the stomach. An **accessory duct (duct of Santorini)** may also lead from the pancreas and empty into the duodenum about 2.5 cm (1 in) above the ampulla.

HISTOLOGY

The pancreas is made up of small clusters of glandular epithelial cells. About 1 percent of the cells, the **pancreatic islets (islets of Langerhans),** form the endocrine portion of the pancreas and consist of alpha, beta, and delta cells that secrete hormones (glucagon, insulin, and somatostatin, respectively). The functions of these hormones may be reviewed in Chapter 21. The remaining 99 percent of the cells, called **acini** (AS-i-nē), are the exocrine portions of the organ (see Figure 21-12). Secreting cells of the acini release a mixture of digestive enzymes called **pancreatic juice.**

ACTIVITIES

The pancreas produces about 1,200 to 1,500 ml (1.2 to 1.5 qt) of pancreatic juice each day. It is a clear, colorless liquid consisting of mostly water, some salts, sodium bicarbonate, and enzymes. The sodium bicarbonate gives pancreatic juice a slightly alkaline pH (7.1 to 8.2) that stops the action of pepsin from the stomach and creates the proper environment for the enzymes in the small intestine. The enzymes in pancreatic juice include a carbohydrate-digesting enzyme called *pancreatic amylase;* several protein-digesting enzymes called *trypsin* (TRIP-sin), *chymotrypsin* (kī'-mō-TRIP-sin), and *carboxypolypeptidase* (kar-bok'-sē-polē'-PEP-ti-dās); the principal fat-digesting enzyme in the adult body called *pancreatic lipase;* and nucleic acid-digesting enzymes called *ribonuclease* and *deoxyribonuclease.*

BLOOD AND NERVE SUPPLY

The arterial supply of the pancreas is from the splenic artery through its pancreatic branches from the hepatic artery and from the superior mesenteric artery by way of the inferior and superior pancreaticoduodenal arteries. The veins, in general, correspond to the arteries. Venous blood reaches the hepatic portal vein by means of the splenic and superior mesenteric veins.

The nerves to the pancreas are branches of the celiac plexus that accompany the arteries entering the gland. The glandular innervation is from the parasympathetic division of the autonomic nevous system, and the innervation of the blood vessels is from the sympathetic division of the autonomic nervous system.

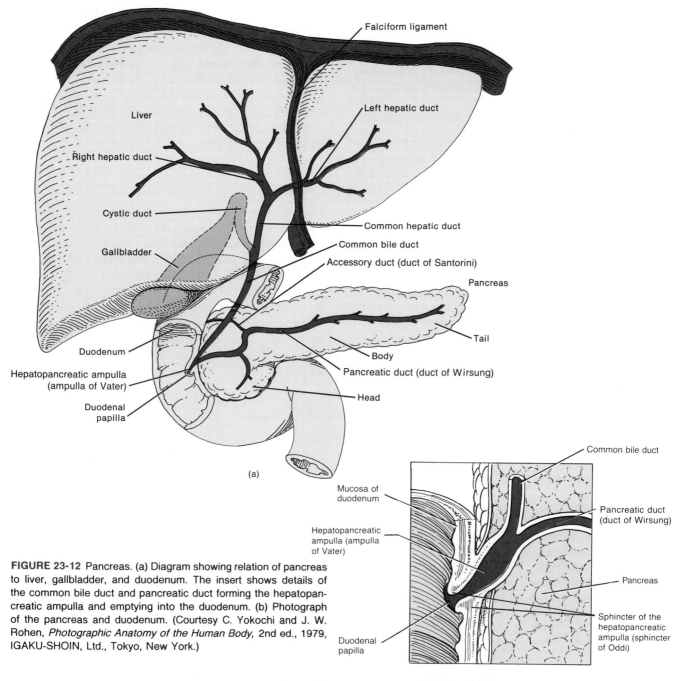

Falciform ligament

Liver

Left hepatic duct

Right hepatic duct

Cystic duct

Common hepatic duct

Gallbladder

Common bile duct

Accessory duct (duct of Santorini)

Pancreas

Duodenum

Tail

Body

Hepatopancreatic ampulla (ampulla of Vater)

Pancreatic duct (duct of Wirsung)

Duodenal papilla

Head

(a)

Common bile duct

Mucosa of duodenum

Pancreatic duct (duct of Wirsung)

Hepatopancreatic ampulla (ampulla of Vater)

Pancreas

Sphincter of the hepatopancreatic ampulla (sphincter of Oddi)

Duodenal papilla

FIGURE 23-12 Pancreas. (a) Diagram showing relation of pancreas to liver, gallbladder, and duodenum. The insert shows details of the common bile duct and pancreatic duct forming the hepatopancreatic ampulla and emptying into the duodenum. (b) Photograph of the pancreas and duodenum. (Courtesy C. Yokochi and J. W. Rohen, *Photographic Anatomy of the Human Body,* 2nd ed., 1979, IGAKU-SHOIN, Ltd., Tokyo, New York.)

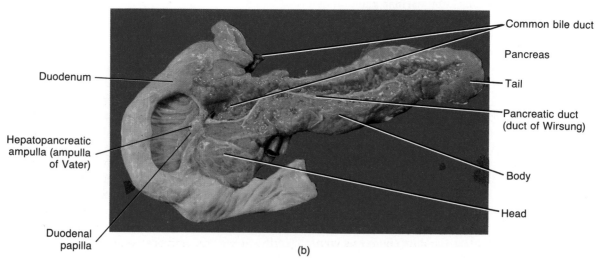

Common bile duct

Pancreas

Duodenum

Tail

Pancreatic duct (duct of Wirsung)

Hepatopancreatic ampulla (ampulla of Vater)

Body

Head

Duodenal papilla

(b)

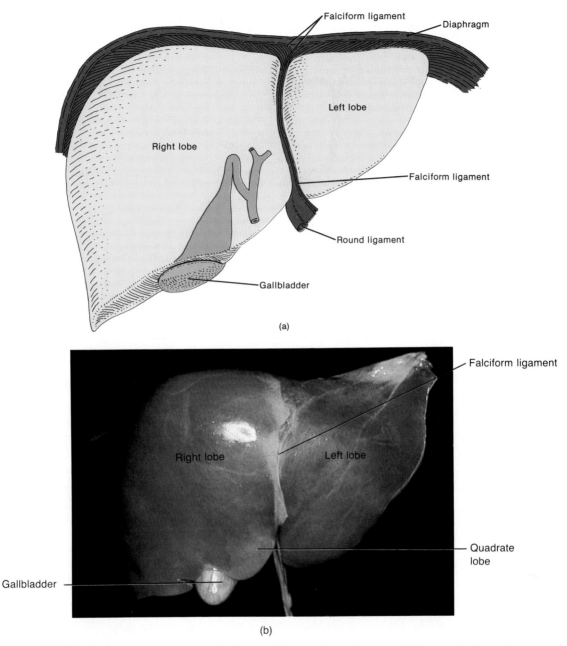

FIGURE 23-13 External anatomy of the liver. (a) Diagram of anterior view. (b) Photograph of anterior view. (© 1985 by Michael H. Ross. Used by permission.)

LIVER

The **liver** weighs about 1.4 kg (about 3 lb) in the average adult. It is located under the diaphragm and occupies most of the right hypochondrium and part of the epigastrium of the abdomen (see Figure 1-8c).

ANATOMY

The liver is almost completely covered by peritoneum and completely covered by a dense connective tissue layer that lies beneath the peritoneum. It is divided into two principal lobes—the **right lobe** and the **left lobe**—separated by the **falciform ligament** (Figure 23-13). Associated with the right lobe are the inferior **quadrate lobe** and posterior **caudate lobe.** The falciform ligament is a reflection of the parietal peritoneum, which extends from the undersurface of the diaphragm to the superior surface of the liver, between the two principal lobes of the liver. In the free border of the falciform ligament is the **ligamentum teres (round ligament).** It extends from the liver to the umbilicus. The ligamentum teres is a fibrous cord derived from the umbilical vein of the fetus.

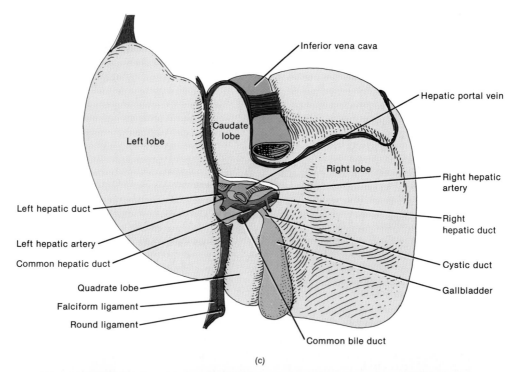

(c)

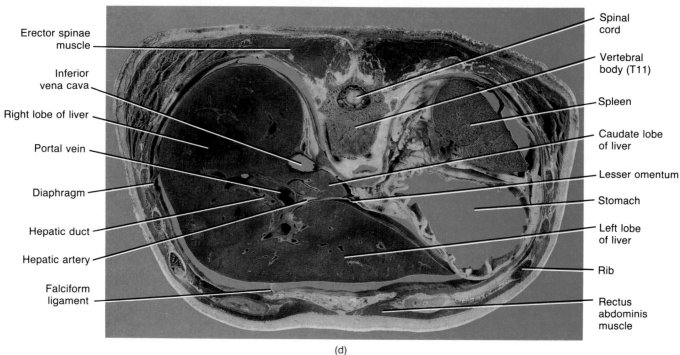

(d)

FIGURE 23-13 (Continued) External anatomy of the liver. (c) Diagram of posteroinferior view. (d) Photograph of a cross section through the abdomen. (Courtesy of Stephen A. Kieffer and E. Robert Heitzman, *An Atlas of Cross-Sectional Anatomy,* Harper & Row, Publishers, Inc., New York, 1979.)

The liver, like the pancreas is connected to the small intestine by a series of ducts. Bile, one of the liver's products, enters **bile capillaries,** or **canaliculi** (kan'-a-LIK-yoo-lī), that empty into small ducts. These small ducts eventually merge to form the larger **right** and **left hepatic ducts,** which unite to leave the liver as the **common hepatic duct** (Figures 23-12 and 23-13c). Further on, the common hepatic duct joins the **cystic duct** from the gallbladder. The two tubes become the **common bile duct.** The common bile duct and pancreatic duct enter the duodenum in a common duct called the **hepatopancreatic ampulla.**

Each day the hepatic cells secrete 800 to 1,000 ml

(about 1 qt) of bile, a yellow, brownish, or olive-green liquid. It has a pH of 7.6 to 8.6. Bile consists mostly of water and bile salts, cholesterol, a phospholipid called lecithin, bile pigments, and several ions.

Bile is partially an excretory product and partially a digestive secretion. Bile salts assume a role in **emulsification,** the breakdown of fat globules into a suspension of fat droplets about 1 μm in diameter, and absorption of fats following their digestion. Cholesterol is made soluble in bile by bile salts and lecithin. The principal bile pigment is **bilirubin.** When red blood cells are broken down, iron, globin, and bilirubin are released. The iron and globin are recycled, but some of the bilirubin is excreted into the bile ducts. Bilirubin eventually is broken down in the intestine, and one of its breakdown products (urobilinogen) give feces their color.

CLINICAL APPLICATION

If sufficient bile salts or lecithin are present in bile or if there is excessive cholesterol, the cholesterol precipitates out of solution and crystalizes to form **gallstones (biliary calculi).** The problems associated with gallstone formation and the diagnosis and treatment of gallstones are discussed in detail at the end of the chapter.

If the liver is unable to remove bilirubin from the blood because of increased destruction of red blood cells or obstruction of bile ducts, large amounts of bilirubin circulate through the bloodstream and collect in other tissues, giving the skin and eyes a yellow color. This condition is called **jaundice.** If the jaundice is due to damaged red blood cells, it is called **hemolytic jaundice;** if it is due to obstruction in the biliary system, it is known as **obstructive jaundice.**

HISTOLOGY

The lobes of the liver are made up of numerous functional units called **lobules,** which may be seen under a microscope (Figure 23-14). A lobule consists of cords of **hepatocytes (liver cells)** arranged in a radial pattern around a **central vein.** Between the cords are endothelial-lined spaces called **sinusoids,** through which blood passes. The sinusoids are also partly lined with phagocytic cells, termed **stellate reticuloendothelial (Kupffer's) cells,** that destroy worn-out white and red blood cells and bacteria.

CLINICAL APPLICATION

Liver tissue for diagnostic purposes may be obtained by **liver biopsy.** In the procedure, the needle is inserted through the seventh, eighth, or ninth intercostal space while the patient is holding his or her breath in full expiration. This lessens the possibility of damage to the lung and contamination of the pleural cavity.

The liver receives a double supply of blood. From the hepatic artery it obtains oxygenated blood, and from the hepatic portal vein it receives deoxygenated blood containing newly absorbed nutrients (see Figures 14-18 and 23-13c). The liver is a common site of metastatic cancer from the gastrointestinal tract or from any region drained by the hepatic portal vein. Branches of both the hepatic artery and the hepatic portal vein carry the blood into the sinusoids of the lobules, where oxygen, most of the nutrients, and certain poisons are extracted by the hepatocytes. Nutrients are stored or used to make new materials. The poisons are stored or detoxified. Products manufactured by the hepatic cells and nutrients needed by other cells are secreted back into the blood. The blood then drains into the central vein and eventually passes into a hepatic vein. Unlike the other products of the liver, bile normally is not secreted into the bloodstream.

The nerve supply to the liver consists of vagal preganglionic parasympathetic fibers and postganglionic sympathetic fibers from the celiac ganglia.

ACTIVITIES

The liver performs many vital functions. Among these are the following:

1. The liver manufactures bile salts, which are used in the small intestine for the emulsification and absorption of fats.

2. The liver, together with mast cells, manufactures the anticoagulant heparin and most of the other plasma proteins, such as prothrombin, fibrinogen, and albumin.

3. The stellate reticuloendothelial cells of the liver phagocytose worn-out red and white blood cells and some bacteria.

4. Liver cells contain enzymes that either break down poisons or transform them into less harmful compounds. When amino acids are burned for energy, for example, they leave behind toxic nitrogenous wastes (such as ammonia) that are converted to urea by the liver cells. Moderate amounts of urea are harmless to the body and are easily excreted by the kidneys and sweat glands.

5. Newly absorbed nutrients are collected in the liver. Depending on the body's needs, it can change any excess monosaccharides into glycogen or fat, both of which can be stored, or it can transform glycogen, fat, and protein into glucose.

6. The liver stores glycogen, copper, iron, and vitamins A, B$_{12}$, D, E, and K. It also stores some poisons that cannot be broken down and excreted. (High levels of DDT are found in the livers of animals, including humans, who eat sprayed fruits and vegetables.)

7. The liver and kidneys participate in the activation of vitamin D.

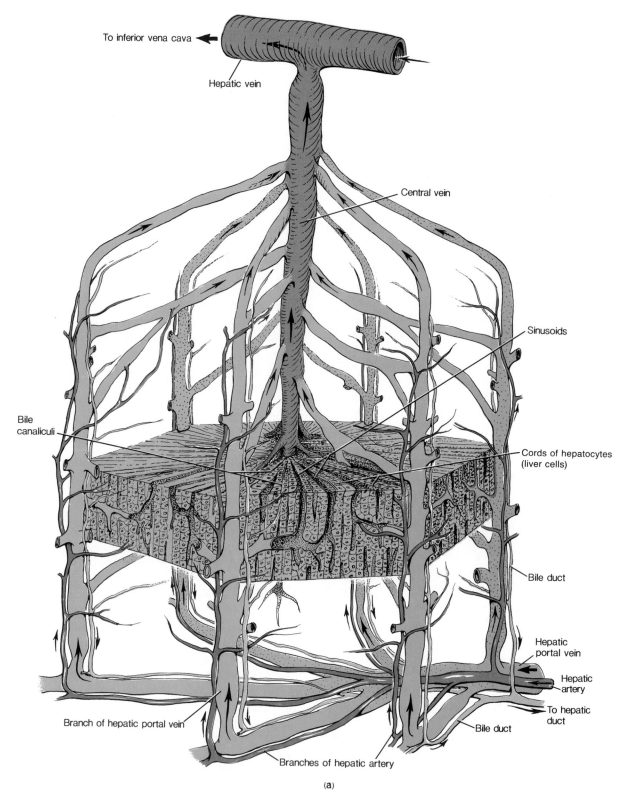

To inferior vena cava

Hepatic vein

Central vein

Sinusoids

Bile
canaliculi

Cords of hepatocytes
(liver cells)

Bile duct

Hepatic
portal vein

Hepatic
artery

To hepatic
duct

Branch of hepatic portal vein

Bile duct

Branches of hepatic artery

(a)

FIGURE 23-14 Histology of the liver. (a) Diagram of the microscopic appearance of a liver lobule.

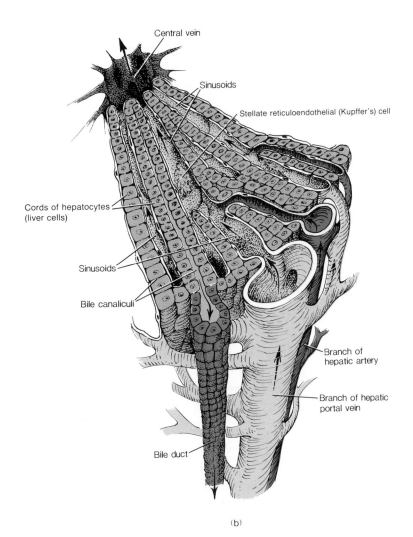

Central vein

Sinusoids

Stellate reticuloendothelial (Kupffer's) cell

Cords of hepatocytes
(liver cells)

Sinusoids

Bile canaliculi

Branch of
hepatic artery

Branch of hepatic
portal vein

Bile duct

(b)

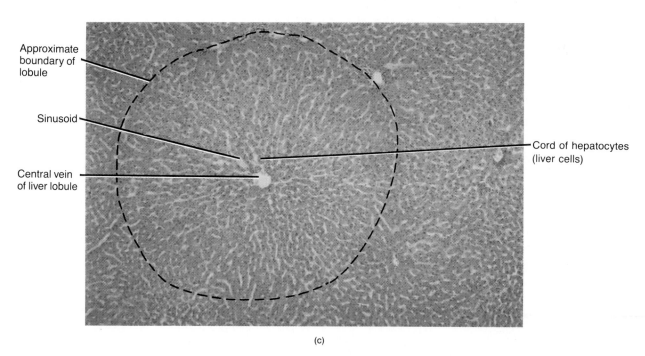

Approximate
boundary of
lobule

Sinusoid

Central vein
of liver lobule

Cord of hepatocytes
(liver cells)

(c)

FIGURE 23-14 (*Continued*) Histology of the liver. (b) Enlarged aspect showing details. (c) Photomicrograph of a liver lobule at a magnification of 65×. (© 1983 by Michael H. Ross. Used by permission.)

Rugae

Mucosa

Muscularis

Blood
vessels

Adventitia

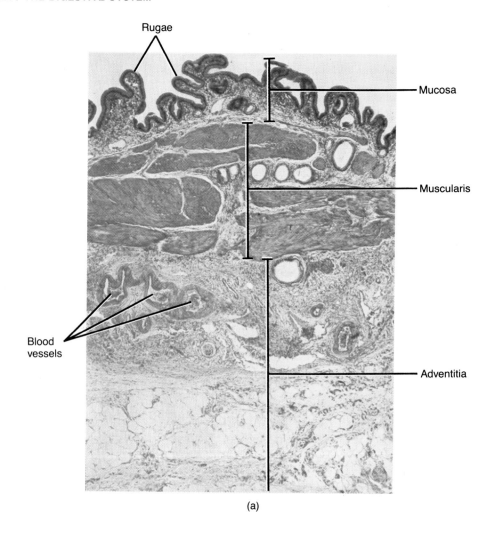

(a)

Lamina propria   Blood vessel   Simple columnar
epithelium

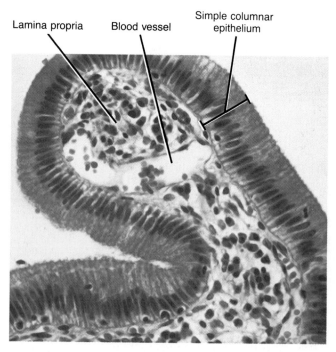

(b)

FIGURE 23-15 Gallbladder. (a) Photomicrograph of a portion of the wall of the gallbladder at a magnification of 40×. (b) Photomicrograph of an enlarged aspect of the mucosa of the gallbladder at a magnification of 600×. (© 1983 by Michael H. Ross. Used by permission.)

GALLBLADDER

The **gallbladder** is a pear-shaped sac about 7 to 10 cm (3 to 4 in) long. It is located in a fossa on the visceral surface of the liver (see Figures 23-12a and 23-13). The inner wall of the gallbladder consists of a mucous membrane arranged in rugae resembling those of the stomach (Figure 23-15). The middle, muscular coat of the wall consists of smooth muscle fibers. Contraction of these fibers by hormonal stimulation ejects the contents of the gallbladder into the cystic duct. The outer coat is the visceral peritoneum.

The gallbladder is supplied by the cystic artery, which arises from the right hepatic artery. The cystic veins drain the gallbladder. The nerves to the gallbladder include branches from the celiac plexus, the vagus, and the right phrenic nerve.

The function of the gallbladder is to store and concentrate bile (up to 10-fold) until it is needed in the small intestine. In the concentration process, water and many ions are absorbed by the gallbladder mucosa. Bile from the liver enters the small intestines through the common bile duct. When the small intestine is empty, a valve around the hepatopancreatic ampulla (ampulla of Vater) called the **sphincter of the hepatopancreatic ampulla (sphincter of Oddi),** closes, and the backed-up bile overflows into the cystic duct to the gallbladder for storage.

SMALL INTESTINE

The major portions of digestion and absorption occur in a long tube called the **small intestine.** The small intestine begins at the pyloric sphincter of the stomach, coils through the central and lower part of the abdominal cavity, and eventually opens into the large intestine. It averages 2.5 cm (1 in) in diameter and about 6.35 m (21 ft) in length.

ANATOMY

The small intestine is divided into three segments (see Figure 23-1). The **duodenum** (doo'-ō-DĒ-num), the shortest part, originates at the pyloric sphincter of the stomach and extends about 25 cm (10 in) until it merges with the jejunum. The **jejunum** (jē-JOO-num) is about 2.5 m (8 ft) long and extends to the ileum. The final portion of the small intestine, the **ileum** (IL-ē-um), measures about 3.6 m (12 ft) and joins the large intestine at the **ileocecal** (il'-ē-ō-SĒ-kal) **valve.**

HISTOLOGY

The wall of the small intestine is composed of the same four tunics that make up most of the GI tract. However, both the mucosa and the submucosa are modified to allow the small intestine to complete the processes of digestion and absorption (Figure 23-16).

The mucosa contains many pits lined with glandular epithelium. These pits—the **intestinal glands**—secrete the intestinal digestive enzymes, collectively called **intestinal juice.** The submucosa of the duodenum contains **duodenal (Brunner's) glands,** which secrete an alkaline mucus to protect the wall of the small intestine from the action of the enzymes and to aid in neutralizing acid in the chyme. Some of the epithelial cells in the mucosa and submucosa have been transformed to goblet cells, which secrete additional mucus.

Intestinal juice is a clear yellow fluid secreted in amounts of about 2 to 3 liters (about 2 to 3 qt) a day. It has a pH of 7.6, which is slightly alkaline, and contains water, mucus, and several enzymes. The intestinal enzymes include three carbohydrate-digesting enzymes called *maltase, sucrase,* and *lactase;* several protein-digesting enzymes called *peptidases;* and two nucleic acid-digesting enzymes, *ribonuclease* and *deoxyribonuclease.* Much of the digestion of foods by enzymes produced by the small intestine actually occurs in epithelial cells lining the small intestine rather than in a fluid outside the cells in the lumen of the tube.

Since almost all the absorption of nutrients occurs in the small intestine, its structure is specially adapted for this function. Its length alone provides a large surface area for absorption and that area is further increased

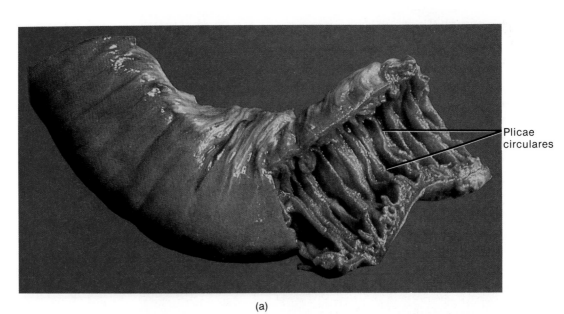

(a)

FIGURE 23-16 Small intestine. Shown are various structures that adapt the small intestine for digestion and absorption. (a) Photograph of a section of the jejunum cut open to expose the plicae circulares. (Courtesy of C. Yokochi and J. W. Rohen, *Photographic Anatomy of the Human Body,* 2nd ed., 1979, IGAKU-SHOIN, Ltd., Tokyo, New York.)

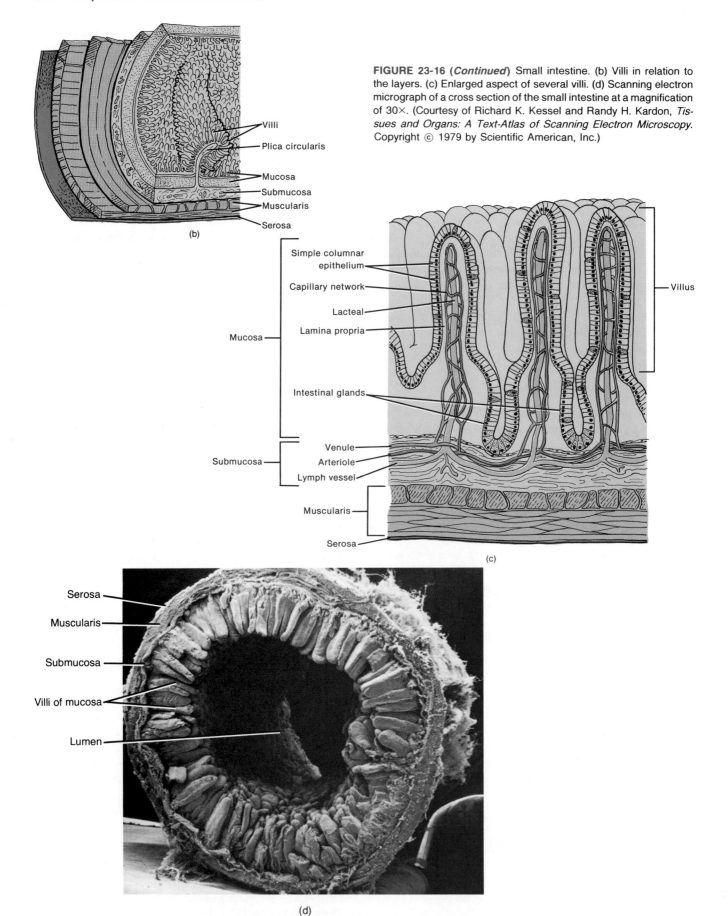

FIGURE 23-16 (*Continued*) Small intestine. (b) Villi in relation to the layers. (c) Enlarged aspect of several villi. (d) Scanning electron micrograph of a cross section of the small intestine at a magnification of 30×. (Courtesy of Richard K. Kessel and Randy H. Kardon, *Tissues and Organs: A Text-Atlas of Scanning Electron Microscopy.* Copyright © 1979 by Scientific American, Inc.)

by modifications in the structure of its wall. The epithelium covering and lining the mucosa consists of simple columnar epithelium. These epithelial cells, except those transformed into goblet cells, contain **microvilli,** fingerlike projections of the plasma membrane. Larger amounts of digested nutrients diffuse into the intestinal wall because the microvilli increase the surface area of the plasma membrane. They also increase the surface area for digestion.

The mucosa lies in a series of **villi,** projections 0.5 to 1 mm high, giving the intestinal mucosa its velvety appearance to the unaided eye. The enormous number of villi (4 to 5 million) vastly increases the surface area of the epithelium available for absorption and digestion. Each villus has a core of lamina propria, the connective tissue layer of the mucosa. Embedded in this connective tissue are an arteriole, a venule, a capillary network, and a **lacteal** (LAK-tē-al) or lymphatic vessel. Nutrients that diffuse through the epithelial cells that cover the villus are able to pass through the capillary walls and the lacteal and enter the cardiovascular and lymphatic systems.

In addition to the microvilli and villi, a third set of projections called **plicae circulares** (PLĪ-kē SER-kyoo-lar-es), or **circular folds,** further increases the surface area for absorption and digestion. The plicae are permanent ridges, about 10 mm (0.394 in) high, in the mucosa. Some of the folds extend all the way around the intestine, and others extend only partway around. The folds begin near the proximal portion of the duodenum and terminate at about the midportion of the ileum. The plicae circulares enhance absorption by causing the chyme to spiral, rather than moving in a straight line, as it passes through the small intestine.

The muscularis of the small intestine consists of two layers of smooth muscle. The outer, thinner layer contains longitudinally arranged fibers. The inner, thicker layer contains circularly arranged fibers. Except for a major portion of the duodenum, the serosa (or visceral peritoneum) completely covers the small intestine. Additional histological aspects of the small intestine are shown in Figure 23-17.

There is an abundance of lymphatic tissue in the form of lymph nodules, masses of lymphatic tissue not covered by a capsule wall. **Solitary lymph nodules** are most numerous in the lower part of the ileum. Groups of lymph nodules, referred to as **aggregated lymphatic follicles (Peyer's patches),** are numerous in the ileum.

ACTIVITIES

The movements of the small intestine are arbitrarily divided into two types: segmentation and peristalsis. **Segmentation** is the major movement of the small intestine. It is strictly a localized contraction in areas containing food. It mixes chyme with the digestive juices and brings the particles of food into contact with the mucosa for

absorption. It does not push the intestinal contents along the tract. Segmentation starts with the contractions of circular muscle fibers in a portion of the small intestine, an action that constricts the intestine into segments. Next, muscle fibers that encircle the middle of each segment also contract, dividing each segment again. Finally, the fibers that contracted first relax, and each small segment unites with an adjoining small segment so that large segments are reformed. This sequence of events is repeated 12 to 16 times a minute, sloshing the chyme back and forth. Segmentation depends mainly on intestinal distension, which initiates impulses to the central nervous system. Returning parasympathetic impulses increase motility. Sympathetic impulses decrease intestinal motility.

Peristalsis propels the chyme onward through the intestinal tract. Peristaltic contractions in the small intestine are normally very weak compared to those in the esophagus or stomach. Chyme moves through the small intestine at a rate of about 1 cm/min. Thus, chyme remains in the small intestine for 3 to 5 hours. Peristalsis, like segmentation, is initiated by distension and controlled by the autonomic nervous system.

Chyme entering the small intestine contains partially digested carbohydrates, partially digested proteins, and essentially undigested lipids. The completion of the digestion of carbohydrates into monosaccharides, proteins into amino acids, and lipids into fatty acids, glycerol, and glycerides is a collective effort of pancreatic juice, bile, and intestinal juice in the small intestine.

All the chemical and mechanical phases of digestion from the mouth down through the small intestine are directed toward changing food into forms that can pass through the epithelial cells lining the mucosa into the underlying blood and lymph vessels. Passage of these digested nutrients from the alimentary canal into the blood or lymph is called **absorption.**

About 90 percent of all absorption of nutrients takes place throughout the length of the small intestine. The other 10 percent occurs in the stomach and large intestine. Any undigested or unabsorbed material left in the small intestine is passed on to the large intestine. Absorption of materials in the small intestine occurs specifically through the villi and depends on diffusion, facilitated diffusion, osmosis, and active transport.

BLOOD AND NERVE SUPPLY

The arterial blood supply of the small intestine is from the superior mesenteric artery and the gastroduodenal artery, coming from the hepatic artery of the celiac trunk. Blood is returned by way of the superior mesenteric vein, which, with the splenic vein, forms the hepatic portal vein.

The nerves to the small intestine are supplied by the superior mesenteric plexus. The branches of the plexus contain postganglionic sympathetic fibers, preganglionic

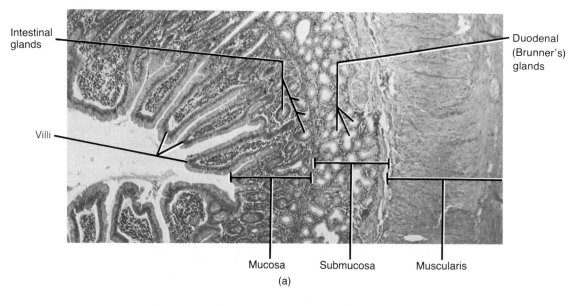

Intestinal glands

Villi

Duodenal (Brunner's) glands

Mucosa   Submucosa   Muscularis

(a)

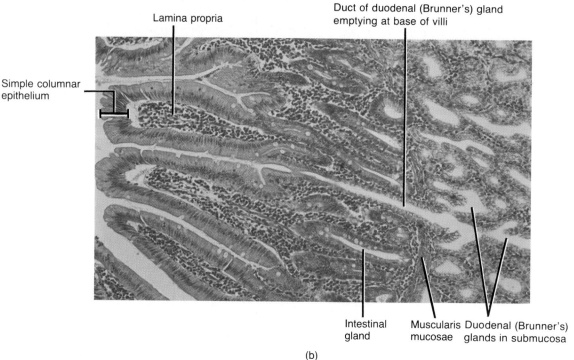

Lamina propria

Duct of duodenal (Brunner's) gland emptying at base of villi

Simple columnar epithelium

Intestinal gland

Muscularis mucosae

Duodenal (Brunner's) glands in submucosa

(b)

FIGURE 23-17 Histology of the small intestine. (a) Photomicrograph of a portion of the wall of the duodenum at a magnification of 40×. (b) Photomicrograph of an enlarged aspect of the duodenal mucosa at a magnification of 165×. (© 1983 by Michael H. Ross. Used by permission.)

parasympathetic fibers, and afferent fibers. The afferent fibers are both vagal and of spinal nerves. In the wall of the small intestine are two autonomic plexuses: the myenteric plexus between the muscular layers and the submucous plexus in the submucosa. The nerve fibers are derived chiefly from the sympathetic division of the autonomic nervous system and partly from the vagus.

LARGE INTESTINE

The overall functions of the large intestine are the completion of absorption, the manufacture of certain vitamins, the formation of feces, and the expulsion of feces from the body.

ANATOMY

The **large intestine** is about 1.5 m (5 ft) in length and averages 6.5 cm (2.5 in) in diameter. It extends from the ileum to the anus and is attached to the posterior abdominal wall by its **mesocolon** of visceral peritoneum.

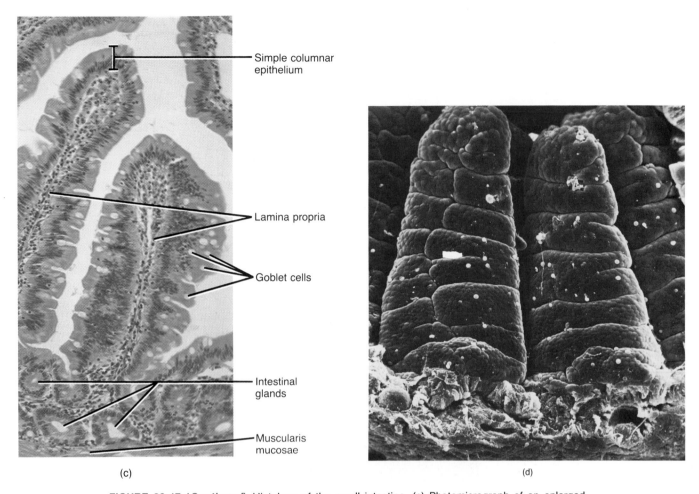

Simple columnar epithelium

Lamina propria

Goblet cells

Intestinal glands

Muscularis mucosae

(c)

(d)

FIGURE 23-17 (*Continued*) Histology of the small intestine. (c) Photomicrograph of an enlarged aspect of two villi from the ileum at a magnification of 120×. (© 1983 by Michael H. Ross. Used by permission.) (d) Scanning electron micrograph of several villi at a magnification of 120×. (Courtesy of Biophoto Associates.)

Structurally, the large intestine is divided into four principal regions: cecum, colon, rectum, and anal canal (Figure 23-18).

The opening from the ileum into the large intestine is guarded by a fold of mucous membrane called the **ileocecal sphincter (valve).** This structure allows materials from the small intestine to pass into the large intestine. Hanging below the ileocecal valve is the **cecum,** a blind pouch about 6 cm (2.5 in) long. Attached to the cecum is a twisted, coiled tube, measuring about 8 cm (3 in) in length, called the **vermiform appendix** (*vermis* = worm). The visceral peritoneum of the appendix, called the **mesoappendix,** attaches the appendix to the inferior part of the ileum and adjacent part of the posterior abdominal wall.

The open end of the cecum merges with a long tube called the **colon.** The colon is divided into ascending, transverse, descending, and sigmoid portions. The **ascending colon** ascends on the right side of the abdomen, reaches the undersurface of the liver, and turns abruptly to the left. Here it forms the **right colic (hepatic) flexure.**

The colon continues across the abdomen to the left side as the **transverse colon.** It curves beneath the lower end of the spleen on the left side as the **left colic (splenic) flexure** and passes downward to the level of the iliac crest as the **descending colon.** The **sigmoid colon** begins at the left iliac crest, projects inward to the midline, and terminates as the rectum at about the level of the third sacral vertebra.

The **rectum,** the last 20 cm (8 in) of GI tract, lies anterior to the sacrum and coccyx. The terminal 2 to 3 cm (1 in) of the rectum is called the **anal canal** (Figure 23-19). The mucous membrane of the anal canal is arranged in longitudinal folds called **anal columns** that contain a network of arteries and veins. The opening of the anal canal to the exterior is called the **anus.** It is guarded by an internal sphincter of smooth muscle and an external sphincter of skeletal muscle. Normally the anus is closed except during the elimination of the wastes of digestion.

The medical specialty that deals with the diagnosis and treatment of disorders of the rectum and anus is called **proctology** (*proct* = rectum; *logos* = study of).

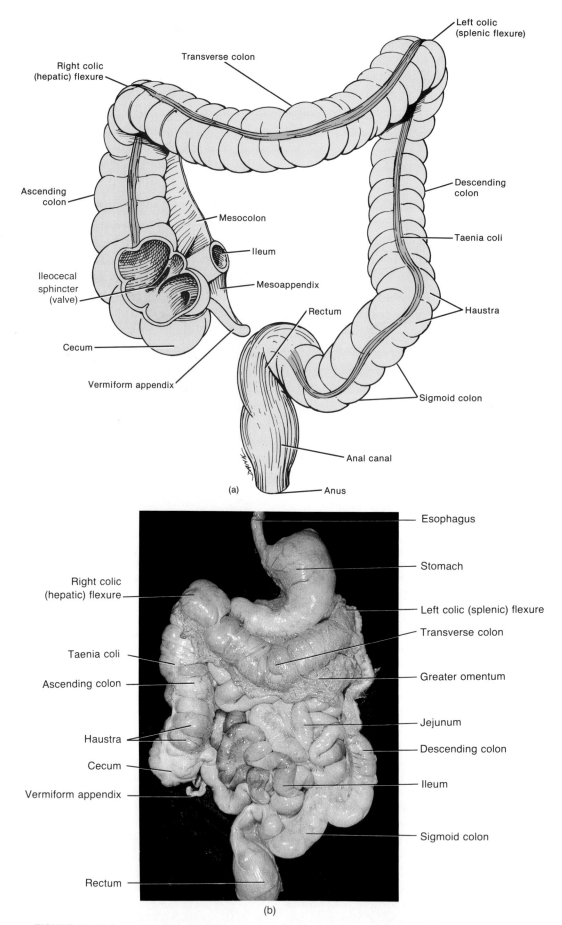

FIGURE 23-18 Large intestine. (a) Diagram of anatomy of the large intestine. (b) Photograph of anatomy of the large intestine in relation to associated structures. (Courtesy of C. Yokochi and J. W. Rohen, *Photographic Anatomy of the Human Body,* 2nd ed., 1979, IGAKU-SHOIN, Ltd., Tokyo, New York.)

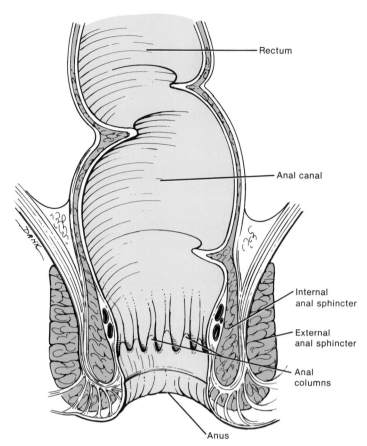

FIGURE 23-19 Anal canal seen in longitudinal section.

Rectum

Anal canal

Internal
anal sphincter

External
anal sphincter

Anal
columns

Anus

CLINICAL APPLICATION

Inflammation and enlargement of the rectal veins (varicose veins) due to weakening of the valves is known as **hemorrhoids (piles).** Initially contained within the anus (first degree), they gradually enlarge until they prolapse or extend outward on defecation (second degree) and finally remain prolapsed through the anal orifice (third degree).

HISTOLOGY

The wall of the large intestine differs from that of the small intestine in several respects (Figure 23-20). No villi or permanent circular folds are found in the mucosa, which does, however, contain simple columnar epithelium with numerous goblet cells. These cells secrete mucus that lubricates the colonic contents as they pass through the colon. Solitary lymph nodules also are found in the mucosa. The submucosa of the large intestine is similar to that found in the rest of the alimentary canal. The muscularis consists of an external layer of longitudinal muscles and an internal layer of circular muscles. Unlike other parts of the digestive tract, the longitudinal muscles do not form a continuous sheet around the wall, but are broken up into three flat bands called **taeniae coli** (TĒ-

ni-ē KŌ-lī). Each band runs the length of most of the large intestine. Tonic contractions of the bands gather the colon into a series of pouches called **haustra** (HAWS-tra), which give the colon its puckered appearance (Figure 23-21). The serosa of the large intestine is part of the visceral peritoneum. Small pouches of visceral peritoneum filled with fat are attached to taeniae coli and are called **epiploic appendages.**

ACTIVITIES

The passage of chyme from the ileum into the cecum is regulated by the action of the ileocecal sphincter. The sphincter normally remains mildly contracted so that the passage of chyme into the cecum is usually a slow process. Movements of the colon begin when substances enter through the ileocecal valve. Since chyme moves through the small intestine at a fairly constant rate, the time required for a meal to pass into the colon is determined by gastric evacuation time. As food passes through the ileocecal sphincter, it fills the cecum and accumulates in the ascending colon.

One movement characteristic of the large intestine is **haustral churning.** In this process, the haustra remain relaxed and distended while they fill up. When the distension reaches a certain point, the walls contract and squeeze the contents into the next haustrum. **Peristalsis** also occurs, although at a slower rate than in other portions of the tract (3 to 12 contractions per minute). A final type of movement is **mass peristalsis,** a strong peristaltic wave that begins in about the middle of the transverse colon and drives the colonic contents into the rectum. Food in the stomach initiates this reflex action in the colon. Thus mass peristalsis usually takes place three or four times a day, during a meal or immediately after.

The last stage of digestion occurs through bacterial, not enzymatic, action. Mucus is secreted by the glands of the large intestine, but no enzymes are secreted. Chyme is prepared for elimination by the action of bacteria. These bacteria ferment any remaining carbohydrates and release hydrogen, carbon dioxide, and methane gas. These gases contribute to flatus (gas) in the colon. They also convert remaining proteins to amino acids and break down the amino acids into simpler substances: indole, skatole, hydrogen sulfide, and fatty acids. Some of the indole and skatole is carried off in the feces and contributes to their odor. The rest are absorbed and transported to the liver, where they are converted to less toxic compounds and excreted in the urine. Bacteria also decompose bilirubin to simpler pigments (urobilinogen), which give feces their brown color. Several vitamins needed for normal metabolism, including some B vitamins and vitamin K, are synthesized by bacterial action and absorbed.

By the time the chyme has remained in the large intestine 3 to 10 hours, it has become solid or semisolid as a result of absorption and is now known as **feces.** Chemi-

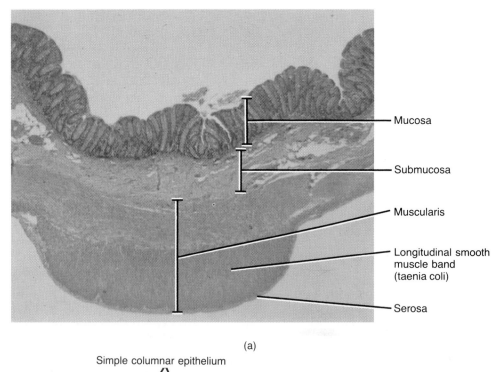

(a)

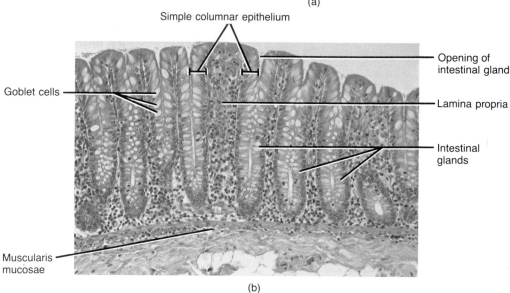

(b)

FIGURE 23-20 Histology of the large intestine. (a) Photomicrograph of a portion of the wall of the large intestine at a magnification of 25×. (b) Photomicrograph of an enlarged aspect of the mucosa of the large intestine at a magnification of 140×. (© 1983 by Michael H. Ross. Used by permission.)

cally, feces consist of water, inorganic salts, sloughed off epithelial cells from the mucosa of the alimentary canal, bacteria, products of bacterial decomposition, and undigested parts of food not attacked by bacteria.

Although most water absorption occurs in the small intestine, the large intestine absorbs enough to make it an important organ in maintaining the body's water balance. Of the 0.5 to 1.0 liter that enters the large intestine, all but about 100 ml is absorbed. The absorption is greatest in the cecum and ascending colon. The large intestine also absorbs electrolytes, including sodium and chloride.

Mass peristaltic movements push fecal material into the rectum. The resulting distension of the rectal wall stimulates pressure-sensitive receptors, initiating a reflex for **defecation,** which is emptying of the rectum. Contraction of the longitudinal rectal muscles shortens the rectum, thereby increasing the pressure inside it. The pressure along with voluntary contractions of the diaphragm and abdominal muscles forces the sphincters open, and the feces are expelled through the anus. Voluntary contractions of the diaphragm and abdominal muscles aid defecation by increasing the pressure inside the abdomen,

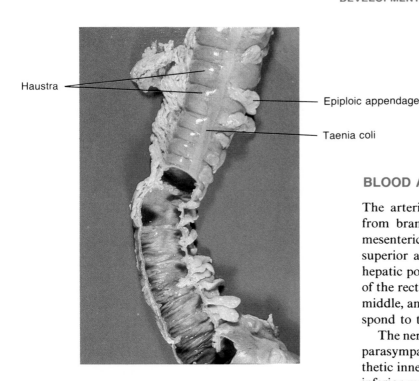

Haustra

Epiploic appendage

Taenia coli

FIGURE 23-21 Photograph of a portion of the large intestine showing taenia coli, haustra, and epiploic appendages. (Courtesy of C. Yokochi and J. W. Rohen, *Photographic Anatomy of the Human Body,* 2nd ed., 1979, IGAKU-SHOIN, Ltd., Tokyo, New York.)

BLOOD AND NERVE SUPPLY

The arterial supply of the cecum and colon is derived from branches of the superior mesenteric and inferior mesenteric arteries. The venous return is by way of the superior and inferior mesenteric veins ultimately to the hepatic portal vein and into the liver. The arterial supply of the rectum and anal canal is derived from the superior, middle, and inferior rectal arteries. The rectal veins correspond to the rectal arteries.

The nerves to the large intestine consist of sympathetic, parasympathetic, and afferent components. The sympathetic innervation is derived from the celiac, superior and inferior mesenteric, and internal iliac plexuses. The fibers reach the plexuses by way of the thoracic and lumbar splanchnic nerves. The parasympathetic innervation is derived from the vagus and pelvic splanchnic nerves.

AGING AND THE DIGESTIVE SYSTEM

Overall general changes associated with aging of the digestive system include decreasing secretory mechanisms, decreasing motility (muscular movement) of the digestive organs, loss of strength and tone of the muscular tissue and its supporting structures, changes in neurosensory feedback on enzyme and hormone release, and the diminished response to pain and internal sensations. Specific changes include reduced sensitivity to mouth irritations and sores, loss of taste, pyorrhea, difficulty in swallowing, hiatal hernia, cancer of the esophagus, gastritis, peptic ulcer, and gastric cancer. Changes in the small intestine include duodenal ulcers, appendicitis, malabsorption, and maldigestion. Other pathologies that increase in incidence are gallbladder problems, jaundice, cirrhosis, and acute pancreatitis. Large-intestinal changes such as constipation, cancer of the colon or rectum, hemorrhoids, and diverticular disease of the colon also occur.

which pushes the walls of the sigmoid colon and rectum inward. If defecation does not occur, the feces remain in the rectum until the next wave of mass peristalsis again stimulates the pressure-sensitive receptors, creating the desire to defecate.

In infants, the defecation reflex causes automatic emptying of the rectum without the voluntary control of the external anal sphincter. In certain instances of spinal cord injury, the reflex is abolished and defecation requires supportive measures, such as cathartics (laxatives).

CLINICAL APPLICATION

Diarrhea refers to frequent defecation of liquid feces caused by increased motility of the intestines. Since chyme passes too quickly through the small intestine and feces passes too quickly through the large intestine, there is not enough time for absorption. Like vomiting, diarrhea can result in dehydration and electrolyte imbalances. Diarrhea may be caused by stress and microbes that irritate the gastrointestinal mucosa.

Constipation refers to infrequent or difficult defecation. It is caused by decreased motility of the intestines in which feces remain in the colon for prolonged periods of time. As it does so, there is considerable absorption and feces become dry and hard. Constipation may be caused by improper bowel habits, spasms of the colon, insufficient bulk in the diet, lack of exercise, and emotions. Usual treatment for constipation is a mild cathartic (laxative) that induces defecation.

DEVELOPMENTAL ANATOMY OF THE DIGESTIVE SYSTEM

About the fourteenth day after fertilization, the cells of the endoderm form a cavity referred to as the **primitive gut** (Figure 23-22). Soon after the mesoderm forms and splits into two layers (somatic and splanchnic), the

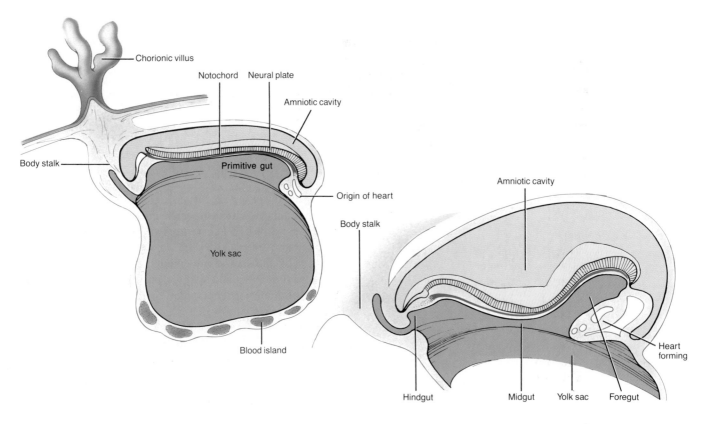

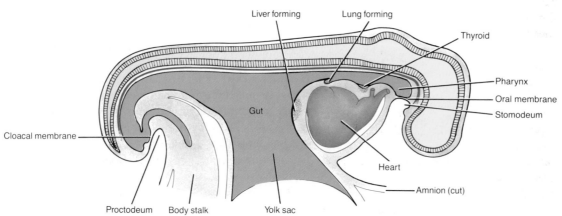

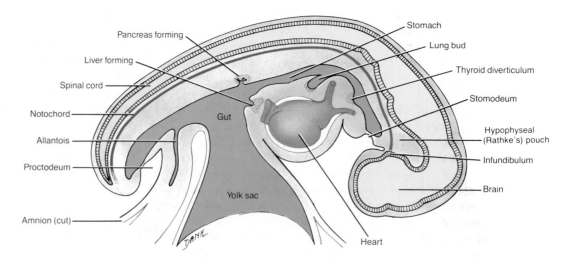

FIGURE 23-22 Development of the digestive system.

splanchnic mesoderm associates with the endoderm of the primitive gut. Thus, the primitive gut has a double-layered wall. The **endodermal layer** gives rise to the *epithelial lining* and *glands* of most of the gastrointestinal tract, and the **mesodermal layer** produces its *smooth muscle* and *connective tissue.*

The primitive gut elongates, and about the latter part of the third week, it differentiates into an anterior **foregut,** a central **midgut,** and a posterior **hindgut.** Until the fifth week of development, the midgut opens into the yolk sac. After that time, the yolk sac constricts, detaches from the midgut, and the midgut seals. In the region of the foregut, a depression consisting of **ectoderm,** the **stomodeum,** appears. This develops into the *oral cavity.* The **oral membrane** that separates the foregut from the stomodeum ruptures during the fourth week of development, so that the foregut is continuous with the outside of the embryo through the oral cavity. Another depression consisting of **ectoderm,** the **proctodeum,** forms in the hindgut and goes on to develop into the *anus.* The **cloacal membrane,** which separates the hindgut from the proctodeum, ruptures, so that the hindgut is continuous with the outside of the embryo through the anus. Thus, the digestive tract forms a continuous tube from the mouth to the anus.

The foregut develops into the *pharynx, esophagus, stomach,* and a *portion of the duodenum.* The midgut is transformed into the *remainder of the duodenum,* the *jejunum,* the *ileum,* and *portions of the large intestine* (cecum, appendix, ascending colon, and most of the transverse colon). The hindgut develops into the *remainder of the large intestine,* except for a portion of the anal canal that is derived from the proctodeum.

As development progresses, the endoderm at various places along the foregut develops into hollow buds that grow into the mesoderm. These buds will develop into the *salivary glands, liver, gallbladder,* and *pancreas.* Each of the glands retains a connection with the gastrointestinal tract via a duct.

APPLICATIONS TO HEALTH

DENTAL CARIES

Dental caries, or tooth decay, involve a gradual demineralization (softening) of the enamel and dentin. If this condition remains untreated, various microorganisms may invade the pulp, causing inflammation and infection with subsequent death (necrosis) of the dental pulp and abscess of the alveolar bone surrounding the root's apex. Such teeth are treated by root canal therapy.

The process of dental caries is initiated when bacteria act on sugars, giving off acids that demineralize the enamel. Microbes that digest sugar into lactic acid are common in the mouth cavity. One that seems to be cariogenic (caries causing) is the bacterium *Streptococcus mu-*

tans. **Dextran,** a sticky polysaccharide produced from sucrose, forms a capsule around the bacteria causing them to stick to the teeth. Masses of bacterial cells, dextran, and other debris adhering to teeth are collectively called **dental plaque.** Saliva cannot reach the tooth surface to buffer the acid because the plaque covers the teeth. Brushing the teeth immediately after eating removes the plaque from flat surfaces before the bacteria have a chance to go to work. Dentists also suggest that the plaque between the teeth be removed every 24 hours with dental floss or by flushing with a water irrigation device.

Preventive measures other than brushing, flossing, and irrigation include prenatal diet supplements (chiefly vitamin D, calcium, and phosphorus), fluoride treatments to protect against acids during the period when teeth are being calcified, and dental sealing. In this last procedure, pits and fissures that serve as reservoirs for dental plaque are sealed by the application of a permanent, durable plastic sealant. The sealant is applied to the prepared biting surfaces of the molar teeth. Dental sealants are used primarily for children and adolescents, although some adults could benefit from them.

PERIODONTAL DISEASE

Periodontal disease is a collective term for a variety of conditions characterized by inflammation and degeneration of the gingivae, alveolar bone, periodontal ligament, and cementum. One such condition is called **pyorrhea.** The initial symptoms are enlargement and inflammation of the soft tissue and bleeding gums. Without treatment, the soft tissue may deteriorate and the alveolar bone may be resorbed, causing loosening of the teeth and recession of the gums.

Periodontal diseases are frequently caused by poor oral hygiene; local irritants, such as bacteria, impacted food, and cigarette smoke; or by a poor "bite." The latter may put a strain on the tissues supporting the teeth. Periodontal diseases may also be caused by allergies, vitamin deficiencies (especially vitamin C), and a number of systemic disorders, especially those that affect bone, connective tissue, or circulation.

PERITONITIS

Peritonitis is an acute inflammation of the serous membrane lining the abdominal cavity and covering the abdominal viscera. One possible cause is contamination of the peritoneum by pathogenic bacteria from the external environment. This contamination could result from accidental or surgical wounds in the abdominal wall or from perforation or rupture of organs with consequent exposure to the outside environment. Another possible cause is perforation of the walls of organs that contain bacteria or chemicals beneficial to the organ but toxic to the peritoneum. For example, the large intestine contains colonies of bacteria that live on undigested nutrients and break them down so they can be eliminated. But if the bacteria enter the peritoneal cavity, they attack the cells of the peritoneum for food and produce acute infection. Moreover, the peritoneum has no natural barriers that keep it from being irritated or digested by chemical substances such as bile and digestive enzymes.

Although it contains a great deal of lymphatic tissue and can combat infection fairly well, the peritoneum is in contact with most of the abdominal organs. If infection gets out of hand, it may destroy vital organs and bring on death. For these reasons, perforation of the alimentary canal from an ulcer is considered serious. A surgeon planning to do extensive surgery on the colon may give the patient high doses of antibiotics for several days prior to surgery to kill intestinal bacteria and reduce the risk of peritoneal contamination.

PEPTIC ULCERS

An **ulcer** is a craterlike lesion in a membrane. Ulcers that develop in areas of the alimentary canal exposed to acid gastric juice are called **peptic ulcers.** Peptic ulcers occasionally develop in the lower end of the esophagus, but most occur on the lesser curvature of the stomach, where they are called **gastric ulcers,** or in the first part of the duodenum, where they are called **duodenal ulcers.**

Hypersecretion of acid gastric juice seems to be the immediate cause of duodenal ulcers. In gastric ulcer patients, because the stomach wall is highly adapted to resist gastric juice through the secretion of mucus, the cause may be hyposecretion of mucus. Hypersecretion of pepsin also may contribute to ulcer formation.

Among the factors believed to stimulate an increase in acid secretion are emotions, certain foods or medications (alcohol, coffee, aspirin), and overstimulation of the vagus (X) nerve. Normally, the mucous membrane lining the stomach and duodenal walls resists the secretions of hydrochloric acid and pepsin. In some people, however, this resistance breaks down and an ulcer develops.

The danger inherent in ulcers is the erosion of the muscular portion of the wall of the stomach or duodenum. This erosion may damage blood vessels and produce fatal hemorrhaging. If an ulcer erodes all the way through the wall, the condition is called *perforation.* Perforation allows bacteria and partially digested food to pass into the peritoneal cavity, producing peritonitis.

APPENDICITIS

Appendicitis is an inflammation of the vermiform appendix. It is preceded by obstruction of the lumen of the appendix by fecal material, inflammation, a foreign body, carcinoma of the cecum, stenosis, or kinking of the organ. The infection that follows as a result of obstruction causes inflammation of all layers of the appendix. This may result in edema, ischemia, gangrene, and perforation. Rupture of the appendix results in peritonitis. Loops of the intestines, the omentum, and the parietal peritoneum may become adherent and form an abscess, either at the site of the appendix or elsewhere in the abdominal cavity.

Typically, appendicitis begins with referred pain in the umbilical region of the abdomen followed by anorexia (lack or loss of appetite for food), nausea, and vomiting. After several hours, the pain localizes in the right lower quadrant and is continuous, dull or severe, and intensified by coughing, sneezing, or body movements.

Early appendectomy (removal of the appendix) is recommended in all suspected cases because it is safer to operate than to risk gangrene, rupture, and peritonitis. Appendectomy may be performed through a muscle-splitting incision in the right lower quadrant in which the cecum is brought into the incision. The base of the appendix is tied, the appendix is excised, and the stump is usually cauterized and then invaginated into the cecum.

TUMORS

Both benign and malignant **tumors** can occur in all parts of the gastrointestinal tract. The benign growths are much more common, but malignant tumors are responsible for 30 percent of all deaths from cancer in the United States. Carcinoma of the colon and rectum is one of the most common malignant diseases, ranking second to that of the lungs in males and lungs and breasts in females. Over 50 percent of colorectal cancers occur in the rectum and sigmoid colon.

Malignant tumors of the rectum may be detected by *digital rectal examination.* The easiest method for detecting cancer of the colon is *fecal occult blood testing,* in which a sample of stool is tested for the presence of occult (hidden) blood, a sign that a malignant tumor might be present. Direct visualization of the rectum and sigmoid colon for the detection of carcinomas is possible by using a *flexible fiberoptic sigmoidoscope.* By using a *flexible fiberoptic endoscope* a physician can visualize the entire gastrointestinal tract. The endoscopic examination of the colon is known as *colonoscopy.* Both the sigmoidoscope and endoscope can also be used to magnify, photograph, biopsy, and remove polyps.

Another test in a routine examination for intestinal disorders is the filling of the gastrointestinal tract with barium, which is either swallowed or given in an enema. Barium, a mineral, shows up on x rays the same way that calcium appears in bones. Tumors as well as ulcers can be diagnosed this way. The only definitive treatment of gastrointestinal carcinomas, if they cannot be removed using the endoscope, is surgery.

DIVERTICULITIS

Diverticula are saclike outpouchings of the wall of the colon when the muscularis becomes weak. The development of diverticula is called **diverticulosis.** Many people who develop diverticulosis are asymptomatic and experience no complications. About 15 percent of people with diverticulosis will eventually develop an inflammation within diverticula, a condition known as **diverticulitis.**

Research done by placing inflated balloons in the colon and measuring the pressure created by segmentation indicates that diverticula form because of lack of sufficient bulk in the colon during segmentation. The powerful contractions, working against insufficient bulk, create a pressure so high that it causes the colonic walls to blow out.

The increase in diverticular disease, from a rarity at the turn of the century to a disorder of an estimated 25 to 33 percent of middle-aged and older Americans today, has been attributed to a shift to a low-residue diet. A recent test result reported that 62 to 70 patients treated for diverticular disease with a high-residue diet containing unprocessed bran showed marked relief of symptoms.

Treatment consists of bed rest, cleansing enemas, and drugs to reduce infection. In severe cases, portions of the affected colon may require surgical removal and temporary colostomy.

CIRRHOSIS

Cirrhosis refers to a distorted or scarred liver as a result of chronic inflammation. The parenchymal (functional) liver cells are replaced by fibrous or adipose connective tissue, a process called stromal repair. The liver has a high capacity for parenchymal regeneration, and stromal repair occurs whenever a parenchymal cell is killed or cells are damaged continuously for a long time. The symptoms of cirrhosis include jaundice, edema in the legs, uncontrolled bleeding, and increased sensitivity to drugs. Cirrhosis may be caused by hepatitis (inflammation of the liver), certain chemicals that destroy liver cells, parasites that infect the liver, and alcoholism.

HEPATITIS

Hepatitis refers to inflammation of the liver and can be caused by viruses, drugs, and chemicals, including alcohol. Clinically several types are recognized.

Hepatitis A (infectious hepatitis) is caused by hepatitis A virus and is spread by fecal contamination of food, clothing, toys, eating utensils, and so forth (fecal-oral route). It is generally a mild disease of children and young adults characterized by anorexia, malaise, nausea, diarrhea, fever, and chills. Eventually jaundice appears. It does not cause lasting liver damage. Most people recover in 4 to 6 weeks.

Hepatitis B (serum hepatitis) is caused by hepatitis B virus and is spread primarily by contaminated syringes and transfusion equipment. It can also be spread by any

secretion of fluid by the body (tears, saliva, semen). Hepatitis B can produce chronic liver inflammation and can persist for years or even a lifetime. Persons who harbor the active hepatitis B virus are at risk for cirrhosis and also become carriers.

Non-A, non-B (NANB) hepatitis is a form of hepatitis that cannot be traced to either hepatitis A or hepatitis B viruses. It is clinically similar to hepatitis B and is often spread by blood transfusions. It is believed to account for considerably more posttransfusion hepatitis than that related to hepatitis B.

GALLSTONES

Gallstones (biliary calculi) are a major health problem in the United States today, affecting more than 15 million people. **Cholecystectomy** (removal of the gallbladder) is the most frequently performed major operation. The ailment is four times more common in women than in men, and the number of persons suffering with gallstones is highest in the 55- to 65-year-old age group. The cholesterol in bile may crystallize at any point between bile canaliculi, where it is first apparent, and the hepatopancreatic ampulla, where the bile enters the duodenum. The fusion of single crystals is the beginning of 95 percent of all gallstones.

Cholesterol gallstones can cause obstruction to the outflow of bile from the gallbladder and irritate the mucosal surface. These two factors, combined with the presence of bacteria, may lead to precipitation of other substances on the pure cholesterol core of the gallstone. Following their formation, gallstones gradually grow in size and number and may cause minimal, intermittent, or complete obstruction to the flow of bile from the gallbladder into the duct system. If obstruction of the outlet occurs and the gallbladder cannot empty as it normally does after eating, the pressure within it increases and the individual may have intense pain or discomfort **(biliary colic).** The pain may radiate to the right shoulder or to the lower back. Bacterial content in the bile is great. With obstruction and stasis, bacteria increase in number. The products of bacterial action may produce a number of additional symptoms, including fever. The longer the calculi are in the gallbladder, the greater the incidence of calculi in the duct system. Small calculi may pass from the gallbladder to the cystic duct and into the common bile duct. Calculi may also ascend into the intrahepatic duct system and obstruct segments of the liver. Complete obstruction of the flow of bile into the duodenum may result in death.

Treatment of gallstones consists of using gallstone-dissolving drugs or surgery. Chenodeoxycholic acid (CDCA) is a drug that is used to dissolve gallstones composed of cholesterol if they are not too large or too heavily coated with calcium. For many people, however, removal of the gallbladder and its contents (cholecystectomy) is the definitive treatment for gallstones.

ANOREXIA NERVOSA

Anorexia nervosa is a disorder characterized by loss of appetite and bizarre patterns of eating. The subconsciously self-imposed starvation appears to be a response to emotional conflicts about self-identification and acceptance of a normal adult sex role. The disorder is found predominantly in young, single females. The physical consequence of the disorder is severe and progressive starvation. Amenorrhea (absence of menstruation) and a lowered basal metabolic rate reflect the depressant effects of the starvation. Individuals may become emaciated and may ultimately die of starvation or one of its complications. Also associated with the disorder are depression, brain abnormalities coupled with impaired mental performance, and a tendency to frequently void large amounts of urine. Treatment consists of psychotherapy and dietary regulation.

BULIMIA

A disorder that typically affects single, middle-class, young, white females is known as **bulimia** (*bous* = ox; *limos* = hunger), or **binge-purge syndrome.** It is characterized by uncontrollable overeating followed by forced vomiting or overdoses of laxatives. This binge-purge cycle occurs in response to fears of being overweight, stress, depression, and physiological disorders such as hypothalamic tumors.

Bulimia can upset the body's electrolyte balance and increase susceptibility to flu, salivary gland infections, dry skin, acne, muscle spasms, loss of hair, kidney and liver diseases, tooth decay, ulcers, hernias, constipation, and hormone imbalances. Treatment of bulimia combines nutrition counseling, psychotherapy, and medical treatment.

KEY MEDICAL TERMS ASSOCIATED WITH THE DIGESTIVE SYSTEM

Botulism (*botulus* = sausage) A type of food poisoning caused by a toxin produced by *Clostridium botulinum.* The bacterium is ingested in improperly cooked or preserved foods. The toxin inhibits nerve impulse conduction at synapses by inhibiting the release of acetylcholine. Symptoms include paralysis, nausea, vomiting, blurred or double vision, difficulty in speech, difficulty in swallowing, dryness of the mouth, and general weakness.

Cholecystitis (*chole* = bile; *kystis* = bladder; *itis* = inflammation of) Inflammation of the gallbladder that often leads to infection. Some cases are caused by obstruction of the cystic duct with bile stones. Stagnating bile salts irritate the mucosa. Dead mucosal cells provide a medium for the growth of bacteria.

Cholelithiasis (*lithos* = stone) The presence of gallstones.

Colitis Inflammation of the colon and rectum. Inflammation of the mucosa reduces absorption of water and salts, producing watery, bloody feces and, in severe cases, dehydration and salt depletion. Irritated muscularis spasms produce cramps.

Colostomy (*stomoun* = provide an opening) An incision of the colon to create an artificial opening or "stoma" to the exterior. This opening serves as a substitute anus through which feces are eliminated. A temporary colostomy may be done to allow a badly inflamed colon to rest and heal. If the rectum is removed for malignancy, the colostomy provides a permanent outlet for feces.

Dysphagia (*dys* = abnormal; *phagein* = to eat) Difficulty in swallowing that may be caused by inflammation, paralysis, obstruction, or trauma.

Enteritis (*enteron* = intestine) An inflammation of the intestine, particularly the small intestine.

Flatus Excessive amounts of air (gas) in the stomach or intestine, usually expelled through the anus. If the gas is expelled through the mouth, it is called **eructation,** or **belching** (burping). Flatus may result from gas released during the breakdown of foods in the stomach or from swallowing air or gas-containing substances such as carbonated drinks.

Gastrectomy (*gastro* = stomach; *tome* = excision) Removal of a portion of or the entire stomach.

Hernia Protrusion of an organ or part of an organ through a membrane or cavity wall, usually the abdominal cavity. *Diaphragmatic* (*hiatal*) *hernia* is the protrusion of the lower esophagus, stomach, or intestine into the thoracic cavity through the opening in the diaphragm (esophageal hiatus) that allows passage of the esophagus. *Umbilical hernia* is the protrusion of abdominal organs through the navel area of the abdominal wall. *Inguinal hernia* is the protrusion of the hernial sac containing the intestine into the inguinal opening. It may extend into the scrotal compartment, causing strangulation of the herniated part.

Inflammatory bowel disease (IBD) Disorder that exists in two forms: (1) Crohn's disease (inflamed intestine) and (2) ulcerative colitis (inflammation of the colon consisting of ulcerations and usually accompanied by rectal bleeding).

Irritable bowel syndrome Disease of the entire gastrointestinal tract characterized by abnormal muscular contractions, especially a spastic colon, excessive mucus in stools, and alternating diarrhea and constipation.

Nausea (*nausia* = seasickness) Discomfort preceding vomiting. Possibly, it is caused by distension or irritation of the gastrointestinal tract, most commonly the stomach.

Pancreatitis Inflammation of the pancreas. The pancreas secretes active trypsin instead of trypsinogen, and the trypsin digests the pancreatic cells and blood vessels.

STUDY OUTLINE

Digestive Processes (p. 581)

1. Food is prepared for use by cells by five basic activities: ingestion, peristalsis, mechanical and chemical digestion, absorption, and defecation.
2. Chemical digestion is a series of catabolic reactions that break down the large carbohydrate, lipid, and protein molecules of food into molecules that are usable by body cells.
3. Mechanical digestion consists of movements that aid chemical digestion.
4. Absorption is the passage of end products of digestion from the digestive tract into blood or lymph for distribution to cells.

Organization (p. 581)

1. The organs of digestion are usually divided into two main groups: those composing the gastrointestinal (GI) tract, or alimentary canal, and accessory structures.
2. The GI tract is a continuous tube running through the ventral body cavity from the mouth to the anus.
3. The accessory structures include the teeth, tongue, salivary glands, liver, gallbladder, and pancreas.
4. The basic arrangement of tissues in the alimentary canal from the inside outward is the mucosa, submucosa, muscularis, and serosa (peritoneum).
5. Extensions of the peritoneum include the mesentery, mesocolon, falciform ligament, lesser omentum, and greater omentum.

Mouth (Oral Cavity) (p. 586)

1. The mouth is formed by the cheeks, palates, lips, and tongue, which aid mechanical digestion.
2. The vestibule is the space between the cheeks and lips and teeth and gums.
3. The oral cavity proper extends from the vestibule to the fauces.

Tongue

1. The tongue, together with its associated muscles, forms the floor of the oral cavity. It is composed of skeletal muscle covered with mucous membrane.
2. The upper surface and sides of the tongue are covered with papillae. Some papillae contain taste buds.

Salivary Glands

1. The major portion of saliva is secreted by the salivary glands, which lie outside the mouth and pour their contents into ducts that empty into the oral cavity.
2. There are three pairs of salivary glands: the parotid, submandibular (submaxillary), and sublingual glands.
3. Saliva lubricates food and starts the chemical digestion of starch.

Teeth

1. The teeth, or dentes, project into the mouth and are adapted for mechanical digestion.
2. A typical tooth consists of three principal portions: crown, root, and cervix.

3. Teeth are composed primarily of dentin covered by enamel, the hardest substance in the body.
4. There are two dentitions—deciduous and permanent.
5. Through mastication food is mixed with saliva and shaped into a bolus.

Pharynx (p. 594)

1. Deglutition or swallowing moves a bolus from the mouth to the stomach.
2. Swallowing consists of a voluntary stage, pharyngeal stage (involuntary) and esophageal stage (involuntary).

Esophagus (p. 594)

1. The esophagus is a collapsible, muscular tube that connects the pharynx to the stomach.
2. It passes a bolus into the stomach by peristalis.
3. It contains an upper and lower esophageal sphincter.

Stomach (p. 596)

1. The stomach begins at the bottom of the esophagus and ends at the pyloric sphincter.
2. The anatomic subdivisions of the stomach include the cardia, fundus, body, and pylorus.
3. Adaptations of the stomach for digestion include rugae; glands that produce mucus, hydrochloric acid, a protein-digesting enzyme, intrinsic factor, and stomach gastrin; and a three-layered muscularis for efficient mechanical movement.
4. Mechanical digestion consists of mixing waves.
5. Chemical digestion consists primarily of starting the digestion of proteins by pepsin.
6. Among the substances absorbed by the stomach are some water, certain electrolytes and drugs, and alcohol.

Pancreas (p. 601)

1. The pancreas is divisible into a head, body, and tail and is connected to the duodenum via the pancreatic duct (duct of Wirsung) and accessory duct (duct of Santorini).
2. Pancreatic islets (islets of Langerhans) secrete hormones and acini secrete pancreatic juice.
3. Pancreatic juice contains enzymes that digest starch (pancreatic amylase), proteins (trypsin, chymotrypsin, and carboxypeptidase), fats (pancreatic lipase), and nucleotides (nucleases).

Liver (p. 603)

1. The liver is divisible into left and right lobes; associated with the right lobe are the caudate and quadrate lobes.
2. The lobes of the liver are made up of lobules that contain hepatocytes (liver cells), sinusoids, stellate reticuloendothelial (Kupffer) cells, and a central vein.
3. Hepatocytes of the liver produce bile that is transported by a duct system to the gallbladder for storage.
4. Bile's contribution to digestion is the emulsification of neutral fats.
5. The liver also functions in producing heparin and plasma proteins, phagocytosis, detoxification, conversion of nutrients, and storage of minerals, vitamins, and glycogen.

Gallbladder (p. 608)

1. The gallbladder is a sac located in a fossa on the visceral surface of the liver.
2. The gallbladder stores and concentrates bile.

Small Intestine (p. 609)

1. The small intestine extends from the pyloric sphincter to the ileocecal sphincter.
2. It is divided into the duodenum, jejunum, and ileum.
3. It is highly adapted for digestion and absorption. Its glands produce enzymes and mucus, and the microvilli, villi, and plicae circulares of its wall provide a large surface area for digestion and absorption.
4. Intestinal enzymes break down carbohydrates (maltase, sucrase, lactase), proteins (aminopeptidase and dipeptidase), and nucleotides (nucleases).
5. Mechanical digestion in the small intestine involves segmentation and peristalsis.
6. Absorption is the passage of the end products of digestion from the alimentary canal into the blood or lymph

Large Intestine (p. 612)

1. The large intestine extends from the ileocecal sphincter to the anus.
2. Its subdivisions include the cecum, colon, rectum, and anal canal.
3. The mucosa contains numerous goblet cells and the muscularis consists of taeniae coli.
4. Mechanical movements of the large intestine include haustral churning, peristalsis, and mass peristalsis.
5. The last stages of chemical digestion occur in the large intestine through bacterial, rather than enzymatic, action. Substances are further broken down and some vitamins are synthesized.
6. The large intestine absorbs water, electrolytes, and vitamins.
7. Feces consists of water, inorganic salts, epithelial cells, bacteria, and undigested foods.
8. The elimination of feces from the large intestine is called defecation.
9. Defecation is a reflex action aided by voluntary contractions of the diaphragm and abdominal muscles.

Aging and the Digestive System (p. 617)

1. General changes include decreased secretory mechanisms, decreased motility, and loss of tone.
2. Specific changes include loss of taste, pyorrhea, hernias, ulcers, constipation, hemorrhoids, and diverticular diseases.

Developmental Anatomy of the Digestive System (p. 617)

1. The endoderm of the primitive gut forms the epithelium and glands of most of the gastrointestinal tract.
2. The mesoderm of the primitive gut forms the smooth muscle and connective tissue of the gastrointestinal tract.

Applications to Health (p. 619)

1. Dental caries are started by acid-producing bacteria that reside in dental plaque.
2. Periodontal diseases are characterized by inflammation and degeneration of gingivae, alveolar bone, periodontal membrane, and cementum.
3. Peritonitis is inflammation of the peritoneum.
4. Peptic ulcers are craterlike lesions that develop in the mucous membrane of the alimentary canal in areas exposed to gastric juice.
5. Appendicitis is an inflammation of the vermiform appendix resulting from obstruction of the lumen of the appendix by inflammation, a foreign body, carcinoma of the cecum, stenosis, or kinking of the organ.
6. Tumors of the gastrointestinal tract may be detected by sigmoidoscopy, colonoscopy, and barium x ray.
7. Diverticulitis is the inflammation of diverticula in the colon.
8. Cirrhosis is a condition in which parenchymal cells of the liver damaged by chronic inflammation are replaced by fibrous or adipose connective tissue.
9. Hepatitis is an inflammation of the liver. Types include hepatitis A; hepatitis B; and non-A, non-B (NANB) hepatitis.
10. The fusion of individual crystals of cholesterol is the beginning of 95 percent of all gallstones. Gallstones can cause obstruction to the outflow of bile in any portion of the duct system.
11. Anorexia nervosa is a disorder characterized by loss of appetite.
12. Bulimia is a binge-purge syndrome of behavior in which uncontrollable overeating is followed by forced vomiting or overdoses of laxatives.

REVIEW QUESTIONS

1. Define digestion. Distinguish between chemical and mechanical digestion.
2. Identify the organs of the gastrointestinal (GI) tract in sequence. How does the gastrointestinal (GI) tract differ from the accessory structures of digestion?
3. Describe the structure of each of the four coats of the gastrointestinal (GI) tract.
4. What is the peritoneum? Describe the location and function of the mesentery, mesocolon, falciform ligament, lesser omentum, and greater omentum.
5. What structures form the oral cavity?
6. Make a simple diagram of the tongue. Indicate the location of the papillae and the four taste zones.
7. Describe the location of the salivary glands and their ducts.
8. How are salivary glands distinguished histologically?
9. What are the principal portions of a typical tooth? What are the functions of each part?
10. Contrast the functions of incisors, cuspids, permolars, and molars.
11. Define deglutition. List the sequence of events involved in

passing a bolus from the mouth to the stomach.

12. Describe the location and histology of the esophagus. What is its role in digestion?

13. Describe the location of the stomach. List and briefly explain the anatomical features of the stomach.

14. What is the importance of rugae, zymogenic cells, parietal cells, mucous cells, and enteroendocrine cells in the stomach?

15. Describe mechanical digestion in the stomach. What is the role of pepsin?

16. Where is the pancreas located? Describe the duct system connecting the pancreas to the duodenum.

17. What are pancreatic acini? Contrast their functions with those of the pancreatic islets.

18. Where is the liver located? What are its principal functions? Describe the anatomy of the liver. Draw a labeled diagram of a liver lobule.

19. Once bile has been formed by the liver, how is it collected and transported to the gallbladder for storage? What is the function of bile?

20. Where is the gallbladder located? How is it connected to the duodenum?

21. What are the subdivisions of the small intestine? How are the mucosa and submucosa of the small intestine adapted for digestion and absorption?

22. Describe the movements in the small intestine.

23. Define absorption.

24. What are the principal subdivisions of the large intestine?

25. How does the muscularis of the large intestine differ from that of the rest of the digestive tract? What are haustra?

26. Describe the mechanical movements that occur in the large intestine.

27. Explain the activities of the large intestine that change its contents into feces. Define defecation. How does it occur?

28. Describe the effects of aging on the digestive system.

29. Describe the development of the digestive system.

30. Describe the causes (where known) and clinical symptoms for: dental caries, periodontal disease, peritonitis, peptic ulcers, appendicitis, tumors, diverticulitis, cirrhosis, hepatitis, gallstones, anorexia nervosa, and bulimia.

31. Refer to the glossary of key medical terms associated with the digestive system. Be sure that you can define each term.

24

The Urinary System

Student Objectives

Identify the external and internal gross anatomical features of the kidneys.

Define the structural features of a nephron.

Describe the blood and nerve supply to the kidneys.

Discuss the principle of hemodialysis.

Describe the location, structure, and physiology of the ureters.

Describe the location, structure, histology, and function of the urinary bladder.

Describe the location, structure, and physiology of the urethra.

Describe the effects of aging on the urinary system.

Describe the development of the urinary system.

Discuss the causes of renal calculi (kidney stones), gout, glomerulonephritis, pyelitis, pyelonephritis, cystitis, nephrosis, and polycystic disease.

Define key medical terms associated with the urinary system.

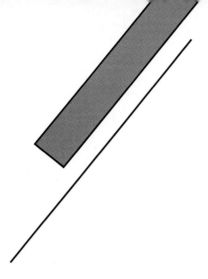

The metabolism of nutrients results in the production of wastes by body cells, including carbon dioxide and excess water and heat. Protein catabolism produces toxic nitrogenous wastes such as ammonia and urea. In addition, many of the essential ions such as sodium, chloride, sulfate, phosphate, and hydrogen tend to accumulate in excess of the body's needs. All the toxic materials and the excess essential materials must be eliminated.

The primary function of the **urinary system** is to help keep the body in homeostasis by controlling the composition and volume of blood. It does so by removing and restoring selected amounts of water and solutes. Two kidneys, two ureters, one urinary bladder, and a single urethra make up the system (Figure 24-1). The kidneys regulate the composition and volume of the blood and remove wastes from the blood in the form of urine. They excrete selected amounts of various wastes, assume a role in erythropoiesis by forming renal erythropoietic factor, help control blood pH, help regulate blood pressure by secreting renin (which activates the renin-angiotensin pathway), and participate in the activation of vitamin D. Urine is excreted from each kidney through its ureter and is stored in the urinary bladder until it is expelled from the body through the urethra. Other systems that aid in waste elimination are the respiratory, integumentary, and digestive systems (Exhibit 24-1).

The specialized branch of medicine that deals with structure, function, and diseases of the male and female urinary systems and the male reproductive system is known as **urology** (yoo-ROL-ō-jē; *uro* = urine or urinary tract; *logos* = study of).

The developmental anatomy of the urinary system is considered later in the chapter.

KIDNEYS

The paired **kidneys** are reddish organs that resemble kidney beans in shape. They are found just above the waist between the parietal peritoneum and the posterior wall of the abdomen. Since they are external to the peritoneal lining of the abdominal cavity, their placement is described as *retroperitoneal* (re'-trō-per-i-tō-NĒ-al). Other retroperitoneal structures include the ureters and adrenal (suprarenal) glands. Relative to the vertebral column, the kidneys are located between the levels of the last thoracic and third lumbar vertebrae and are partially protected by the eleventh and twelfth pairs of ribs. The right kidney is slightly lower than the left because of the large area occupied by the liver.

EXTERNAL ANATOMY

The average adult kidney measures about 10 to 12 cm (4 to 5 in) long, 5.0 to 7.5 cm (2 to 3 in) wide, and 2.5 cm (1 in) thick. Its concave medial border faces the vertebral column. Near the center of the concave border is a notch called the **hilus,** through which the ureter leaves the kidney. Blood and lymph vessels and nerves also enter and exit the kidney through the hilus (Figure 24-2). The hilus is the entrance to a cavity in the kidney called the **renal sinus.**

EXHIBIT 24-1 EXCRETORY ORGANS AND PRODUCTS ELIMINATED

EXCRETORY ORGANS	PRODUCTS ELIMINATED	
	PRIMARY	SECONDARY
Kidneys	Water, nitrogenous wastes from portein catabolism, and inorganic salts.	Heat and carbon dioxide.
Lungs	Carbon dioxide.	Heat and water.
Skin (sudoriferous glands)	Heat.	Carbon dioxide, water, salts, and urea.
Gastrointestinal (GI) tract	Solid wastes and secretions.	Carbon dioxide, water, salts, and heat.

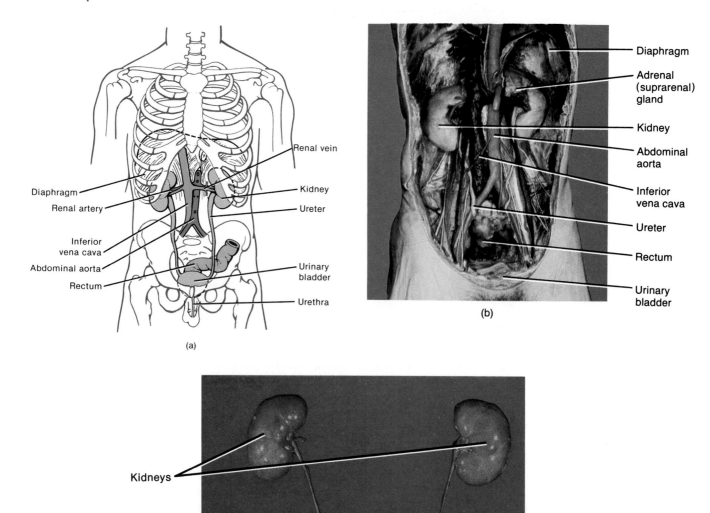

(a)

(b)

(c)

FIGURE 24-1 Organs of the male urinary system. (a) Diagram of urinary organs in relation to surrounding structures. (b) Photograph of urinary organs in relation to surrounding structures. (Courtesy of C. Yokochi and J. W. Rohen, *Photographic Anatomy of the Human Body,* 2nd ed., 1979, IGAKU-SHOIN, Ltd., Tokyo, New York.) (c) Photograph of isolated urinary and some reproductive system organs. (Courtesy of C. Yokochi and J. W. Rohen, *Photographic Anatomy of the Human Body,* 2nd ed., 1979, IGAKU-SHOIN, Ltd., Tokyo, New York.)

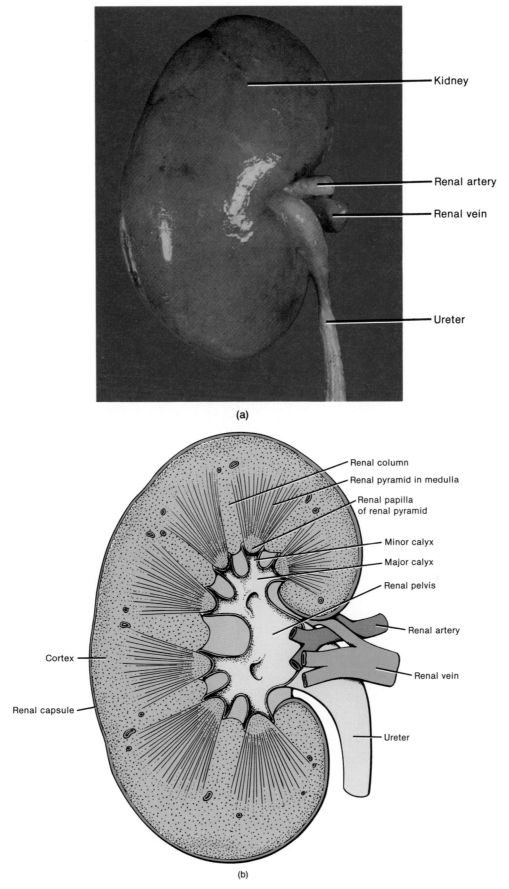

(a)

(b)

FIGURE 24-2 Kidney. (a) Photograph of the external view of the right kidney. (Courtesy of C. Yokochi and J. W. Rohen, *Photographic Anatomy of the Human Body,* 2nd ed., 1979, IGAKU-SHOIN, Ltd., Tokyo, New York.) (b) Diagram of a coronal section of the right kidney illustrating the internal anatomy.

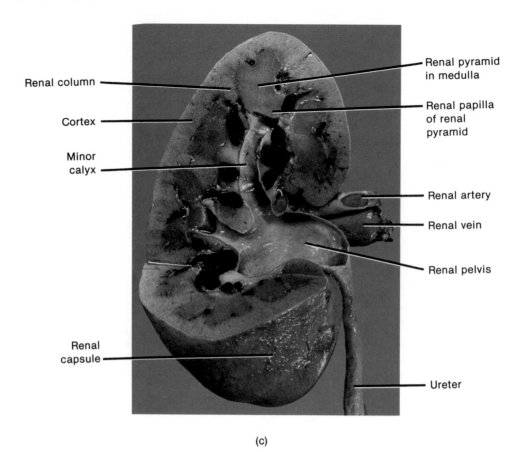

(c)

(d)

FIGURE 24-2 (*Continued*) Kidney. (c) Photograph of a coronal and a cross section of the right kidney illustrating the internal anatomy. (Courtesy of C. Yokochi and J. W. Rohen, *Photographic Anatomy of the Human Body,* 2nd ed., 1979, IGAKU-SHOIN, Ltd., Tokyo, New York.) (d) Photograph of a cross section through the abdomen. (Courtesy of Stephen A. Kieffer and E. Robert Heitzman, *An Atlas of Cross-Sectional Anatomy,* Harper & Row, Publishers, Inc., New York, 1979.)

Three layers of tissue surround each kidney. The innermost layer, the **renal capsule,** is a smooth, transparent, fibrous membrane that can easily be stripped off the kidney and is continuous with the outer coat of the ureter at the hilus. It serves as a barrier against trauma and the spread of infection to the kidney. The second layer, the **adipose capsule,** is a mass of fatty tissue surrounding the renal capsule. It also protects the kidney from trauma and holds it firmly in place within the abdominal cavity. The outermost layer, the **renal fascia,** is a thin layer of fibrous connective tissue that anchors the kidney to its surrounding structures and to the abdominal wall.

CLINICAL APPLICATION

Floating kidney, or **ptosis** (TŌ-sis), occurs when the kidney is no longer held in place securely by the adjacent organs or its covering of fat and slips from its normal position. Individuals, especially thin people, in whom either the adipose capsule or renal fascia is deficient, may develop ptosis. It is dangerous because it may cause kinking of the ureter with reflux of urine and retrograde pressure. Pain occurs if the ureter is twisted. Also, if the kidneys drop below the rib cage, they become susceptible to blows and penetrating injuries.

INTERNAL ANATOMY

A coronal (frontal) section through a kidney reveals an outer, reddish area called the **cortex** and an inner, reddish-brown region called the **medulla** (Figure 24-2b, c). Within the medulla are 8 to 18 striated, triangular structures termed **renal (medullary) pyramids.** The striated appearance is due to the presence of straight tubules and blood vessels. The bases of the pyramids face the cortical area, and their apices, called **renal papillae,** are directed toward the center of the kidney. The cortex is the smooth-textured area extending from the renal capsule to the bases of the pyramids and into the spaces between them. It is thus divided into an outer cortical zone and an inner juxtamedullary zone. The cortical substance between the renal pyramids forms the **renal columns.**

Together the cortex and renal pyramids constitute the parenchyma of the kidney. Structurally, the parenchyma of each kidney consists of approximately 1 million microscopic units called nephrons, collecting ducts, and their associated vascular supply. Nephrons are the functional units of the kidney. They help regulate blood composition and form urine.

In the renal sinus of the kidney is a large cavity called the **renal pelvis.** The edge of the pelvis contains cuplike extensions called **major** and **minor calyces** (KĀ-li-sēz). There are 2 or 3 major calyces and 8 to 18 minor calyces.

Each minor calyx collects urine from collecting ducts of the pyramids. From the major calyces, the urine drains into the pelvis and out through the ureter.

NEPHRON

The functional unit of the kidney is the **nephron** (NEF-ron) (Figure 24-3). Essentially, a nephron is a **renal tubule** and its vascular component. It begins as a double-walled cup, called the **glomerular (Bowman's) capsule,** lying in the cortex of the kidney. The outer wall, or *parietal layer,* is composed of simple squamous epithelium (Figure 24-4). It is separated from the inner wall, known as the *visceral layer,* by the *capsular space.* The visceral layer consists of epithelial cells called podocytes. It surrounds a capillary network called the **glomerulus** (glō-MER-yoo-lus). Collectively, the glomerular capsule and its enclosed glomerulus constitute a renal corpuscle (KŌR-pus-sul).

The visceral layer of the glomerular capsule and the endothelium of the glomerulus form an **endothelial-capsular membrane.** This membrane consists of the following parts in the order in which substances filtered by the kidney must pass through.

1. Endothelium of the glomerulus. This single layer of endothelial cells has completely open pores (fenestrated) averaging 500 to 1,000 Å in diameter.

2. Basement membrane of the glomerulus. This extracellular membrane lies beneath the endothelium and contains no pores. It consists of fibrils in a glycoprotein matrix. It serves as the dializing membrane.

3. Epithelium of the visceral layer of the glomerular capsule. These epithelial cells, because of their peculiar shape, are called **podocytes.** The podocytes contain foot-like structures called **pedicles** (PED-i-sels). The pedicles are arranged parallel to the circumference of the glomerulus and cover the basement membrane except for spaces between them called **filtration slits (slit pores).**

The endothelial-capsular membrane filters water and solutes in the blood. Large molecules, such as proteins, and the formed elements in blood do not normally pass through it. The water and solutes that are filtered out of the blood pass into the capsular space between the visceral and parietal layers of the glomerular capsule and then into the renal tubule.

The glomerular capsule opens into the first section of the renal tubule, called the **proximal convoluted tubule,** which also lies in the cortex. Convoluted means the tubule is coiled rather than straight; proximal signifies that the glomerular capsule is the origin of the tubule. The wall of the proximal convoluted tubule consists of cuboidal epithelium with microvilli. These cytoplasmic extensions, like those of the small intestine, increase the surface area for reabsorption and secretion.

Nephrons are frequently classified into two kinds. A **cortical nephron** usually has its glomerulus in the outer

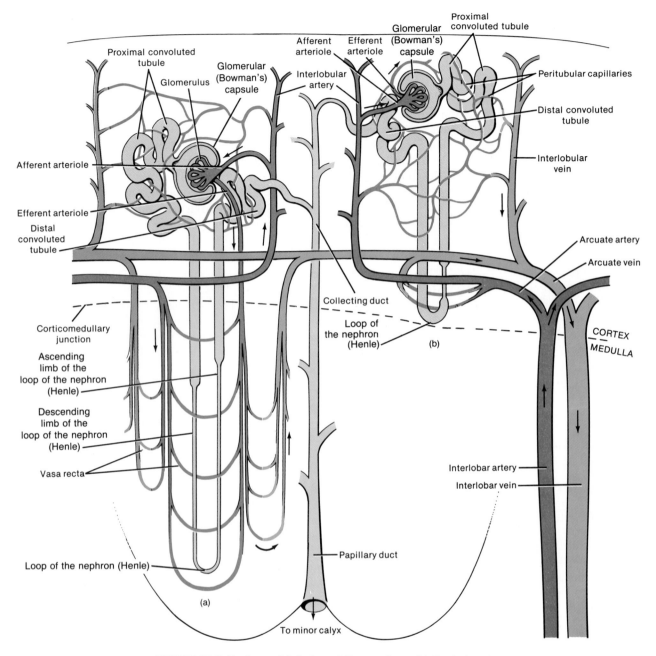

FIGURE 24-3 Nephrons. (a) Juxtamedullary nephron. (b) Cortical nephron.

cortical zone, and the remainder of the nephron rarely penetrates the medulla. A **juxtamedullary nephron** usually has its glomerulus close to the corticomedullary junction, and other parts of the nephron penetrate deeply into the medulla (see Figure 24-3).

In the juxtamedullary nephron, the renal tubule straightens, becomes thinner, and dips into the medulla, where it is called the **descending (thin) limb of the loop of the nephron.** This section consists of squamous epithelium. The tubule then increases in diameter as it bends into a U-shaped structure called the **loop of the nephron (loop of Henle).** It then ascends toward the cortex as the **ascending (thick) limb of the loop of the nephron,** which consists of cuboidal and low columnar epithelium.

In the cortex, the tubule again becomes convoluted. Because of its distance from the point of origin at the glomerular capsule, this section is referred to as the **distal convoluted tubule.** The cells of the distal tubule, like those of the proximal tubule, are cuboidal. Unlike the cells of the proximal tubule, however, the cells of the distal tubule have few microvilli. In a cortical nephron, the proximal section runs into the distal tubule without the intervening limbs and the loop of the nephron. In both types of nephron, the distal tubule terminates by merging with a straight **collecting duct.**

In the medulla, the collecting ducts receive the distal tubules of several nephrons, pass through the renal pyramids, and open at the renal papillae into the minor calyces

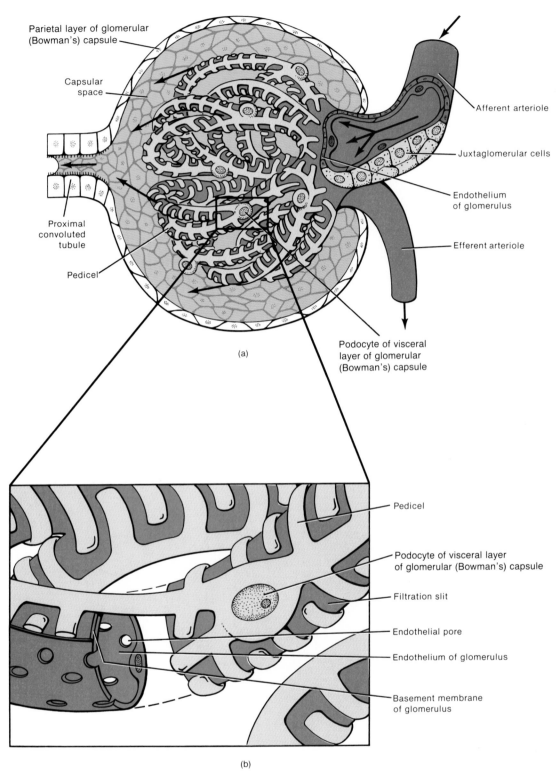

Parietal layer of glomerular (Bowman's) capsule

Capsular space

Proximal convoluted tubule

Pedicel

Afferent arteriole

Juxtaglomerular cells

Endothelium of glomerulus

Efferent arteriole

Podocyte of visceral layer of glomerular (Bowman's) capsule

(a)

Pedicel

Podocyte of visceral layer of glomerular (Bowman's) capsule

Filtration slit

Endothelial pore

Endothelium of glomerulus

Basement membrane of glomerulus

(b)

FIGURE 24-4 Endothelial-capsular membrane. (a) Parts of a renal corpuscle. (b) Enlarged aspect of a portion of the endothelial-capsular membrane.

through a number of large **papillary ducts.** On the average, there are 30 papillary ducts per renal papilla. Cells of the collecting ducts are cuboidal; those of the papillary ducts are columnar.

The histology of a nephron and glomerulus is shown in Figure 24-5.

BLOOD AND NERVE SUPPLY

The nephrons are largely responsible for removing wastes from the blood and regulating its fluid and electrolyte content. Thus they are abundantly supplied with blood vessels. The right and left **renal arteries** transport about

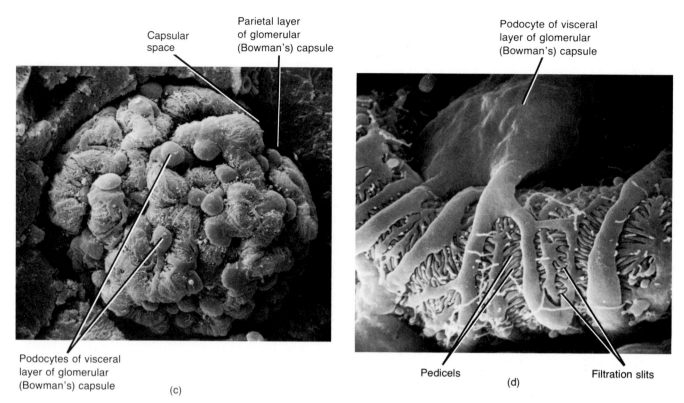

FIGURE 24-4 (*Continued*) Endothelial-capsular membrane. (c) Scanning electron micrograph of a renal corpuscle at a magnification of 1,230×. (Courtesy of Richard K. Kessel and Randy H. Kardon, *Tissues and Organs: A Text-Atlas of Scanning Electron Microscopy.* Copyright © 1979 by Scientific American, Inc.) (d) Scanning electron micrograph of a podocyte at a magnification of 7,800×. (Courtesy of Richard K. Kessel and Randy H. Kardon, *Tissues and Organs: A Text-Atlas of Scanning Electron Microscopy.* Copyright © 1979 by Scientific American, Inc.)

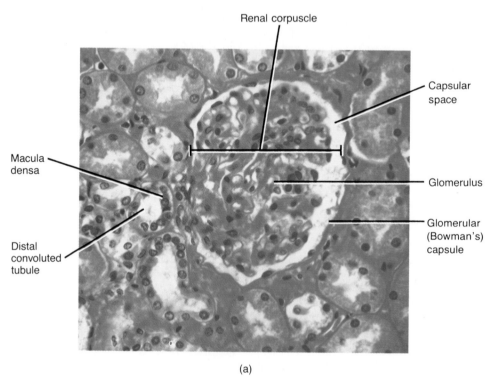

FIGURE 24-5 Histology of a nephron. (a) Photomicrograph of the cortex of the kidney showing a renal corpuscle and surrounding renal tubules at a magnification of 400×. (© 1983 by Michael H. Ross. Used by permission.)

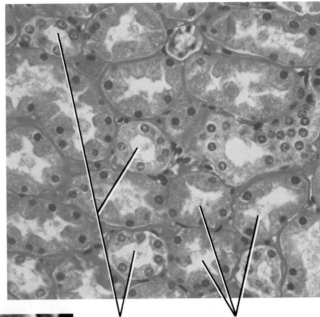

FIGURE 24-5 (*Continued*) Histology of a nephron. (b) Photomicrograph of the cortex of the kidney showing renal tubules at a magnification of 400×. (© 1983 by Michael H. Ross. Used by permission.) (c) Scanning electron micrograph of several renal tubules at a magnification of 500×. (Courtesy of Fisher Scientific Company and S.T.E.M. Laboratories, Inc., Copyright, 1975.)

Distal convoluted tubules

Proximal convoluted tubules

(b)

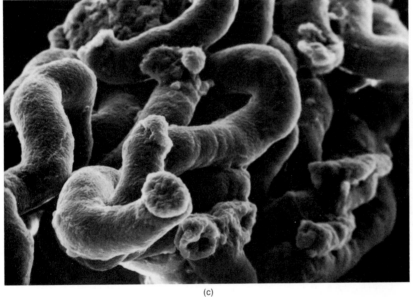

(c)

one-fourth the total cardiac output to the kidneys (Figure 24-6). Approximately 1,200 ml passes through the kidneys every minute.

Before or immediately after entering through the hilus, the renal artery divides into several branches that enter the parenchyma and pass as the **interlobar arteries** between the renal pyramids in the renal columns. At the base of the pyramids, the interlobar arteries arch between the medulla and cortex; here they are known as the **arcuate arteries.** Divisions of the arcuate arteries produce a series of **interlobular arteries,** which enter the cortex and divide into **afferent arterioles** (see Figure 24-3).

One afferent arteriole is distributed to each glomerular capsule, where the arteriole divides into the tangled capillary network termed the **glomerulus.** The glomerular capillaries then reunite to form an **efferent arteriole,** which leads away from the capsule and is smaller in diameter than the afferent arteriole. This variation in diameter helps raise the glomerular pressure. The afferent-efferent arteriole situation is unique because blood usually flows out of capillaries into venules and not into other arterioles.

Each efferent arteriole of a cortical nephron divides to form a network of capillaries, called the **peritubular capillaries,** around the convoluted tubules. The efferent arteriole of a juxtamedullary nephron also forms peritubular capillaries. In addition, it forms long loops of thin-walled vessels called **vasa recta** that dip down alongside the loop of the nephron into the medullary region of the papilla.

The peritubular capillaries eventually reunite to form **interlobular veins.** The blood then drains through the **arcuate veins** to the **interlobar veins** running between

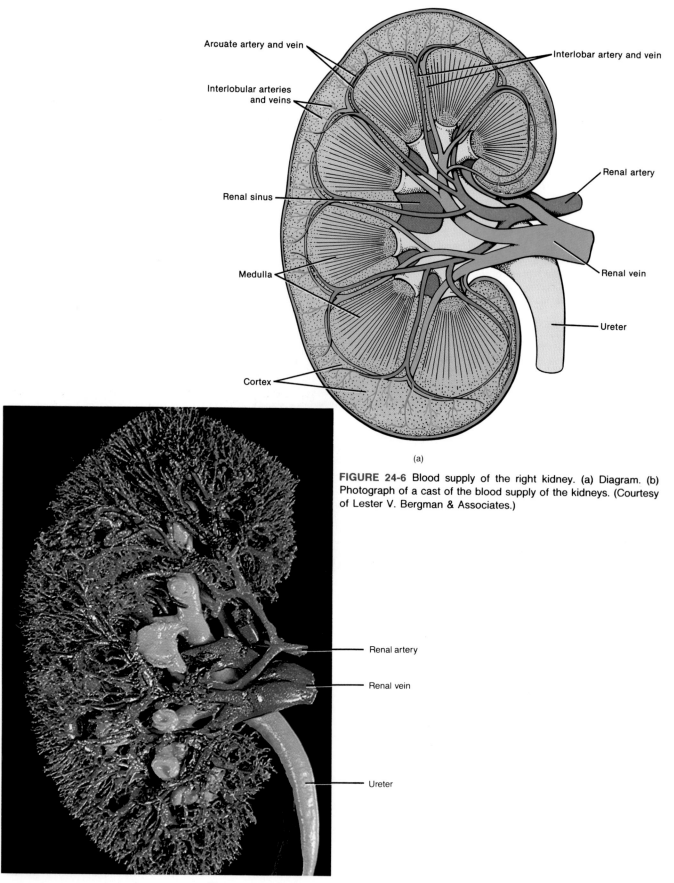

Arcuate artery and vein

Interlobar artery and vein

Interlobular arteries and veins

Renal sinus

Renal artery

Medulla

Renal vein

Ureter

Cortex

(a)

FIGURE 24-6 Blood supply of the right kidney. (a) Diagram. (b) Photograph of a cast of the blood supply of the kidneys. (Courtesy of Lester V. Bergman & Associates.)

Renal artery

Renal vein

Ureter

(b)

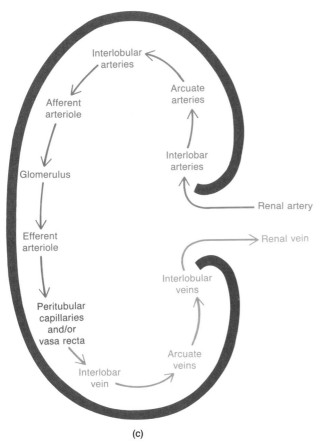

(c)

FIGURE 24-6 (*Continued*) Blood supply of the right kidney. (c) Scheme of circulation.

the pyramids and leaves the kidney through a single **renal vein** that exits at the hilus. The vasa recta pass blood into the interlobular veins. From here, it goes to the arcuate veins, the interlobar veins, and then into the renal vein.

The nerve supply to the kidneys is derived from the **renal plexus** of the autonomic system. Nerves from the plexus accompany the renal arteries and their branches and are distributed to the vessels. Because the nerves are vasomotor, they regulate the circulation of blood in the kidney by regulating the diameters of the arterioles.

JUXTAGLOMERULAR APPARATUS

As the afferent arteriole approaches the renal corpuscle, the smooth muscle cells of the tunica media become modified. Their nuclei become rounded (instead of elongated), and their cytoplasm contains granules (instead of myofibrils). Such modified cells are called **juxtaglomerular cells.** The cells of the distal convoluted tubule adjacent to the afferent arteriole become considerably narrower. Collectively, these cells are known as the **macula densa.** Together with the modified cells of the afferent arteriole they constitute the **juxtaglomerular apparatus** (Figure 24-7). The juxtaglomerular apparatus helps regulate renal blood pressure.

PHYSIOLOGY

The major work of the urinary system is done by the nephrons. The other parts of the system are primarily passageways and storage areas. Nephrons carry out three important functions. They control blood concentration and volume by removing selected amounts of water and solutes. They help regulate blood pH. And they remove toxic wastes from the blood. As the nephrons go about these activities, they remove many materials from the blood, return the ones that the body requires, and eliminate the remainder. The eliminated materials are collectively called **urine.** The entire volume of blood in the body is filtered by the kidneys approximately 60 times a day.

HEMODIALYSIS THERAPY

If the kidneys are so impaired by disease or injury that they are unable to excrete nitrogenous wastes and regulate pH and electrolyte concentration of the plasma, the blood must be filtered by an artificial device. Such filtering of the blood is called **hemodialysis.** *Dialysis* means using a semipermeable membrane to separate large nondiffusible particles from smaller diffusible ones. One of the best-known devices for accomplishing dialysis is the kidney machine (Figure 24-8). A tube connects it with the patient's radial artery. The blood is pumped from the artery through the tubes to one side of a semipermeable dialyzing membrane made of cellulose acetate. The other side of the membrane is continually washed with an artificial solution called the dialysate. The blood that passes through the artificial kidney is treated with an anticoagulant. Only about 500 ml of the patient's blood is in the machine at a time. This volume is easily compensated for by vasoconstriction and increased cardiac output.

All substances (including wastes) in the blood except protein molecules and blood cells can diffuse back and forth across the semipermeable membrane. The electrolyte level of the plasma is controlled by keeping the dialysate electrolytes at the same concentration found in normal plasma. Any excess plasma electrolytes move down the concentration gradient and into the dialysate. If the plasma electrolyte level is normal, it is in equilibrium with the dialysate and no electrolytes are gained or lost. Since the dialysate contains no wastes, substances such as urea move down the concentration gradient and into the dialysate. Thus wastes are removed and normal electrolyte balance is maintained.

A great advantage of the kidney machine is that nutrition can be bolstered by placing large quantities of glucose in the dialyzing solution. While the blood gives up its wastes, the glucose diffuses into the blood. Thus the kidney machine beautifully accomplishes the principal function of the fundamental unit of the kidney—the nephron.

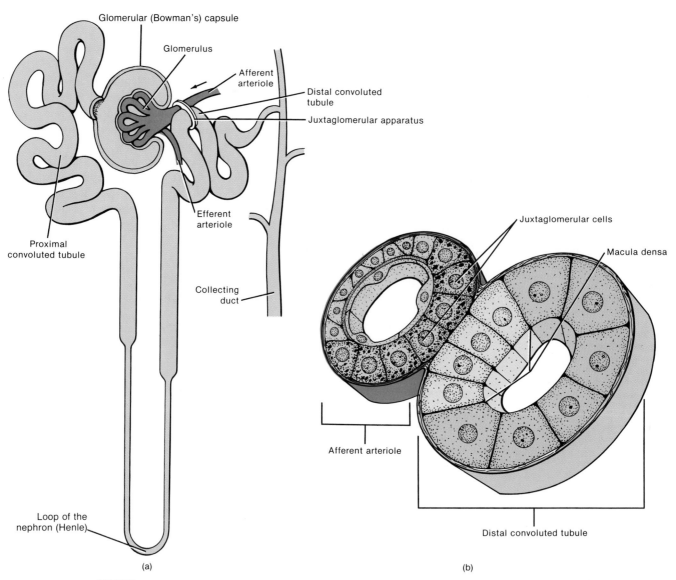

(a)

(b)

FIGURE 24-7 Juxtaglomerular apparatus. (a) External view. (b) Cells of the juxtaglomerular apparatus seen in cross section.

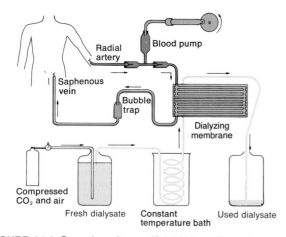

FIGURE 24-8 Operation of an artificial kidney. The blood route is indicated in red and blue. The route of the dialysate is indicated in gold.

There are obvious drawbacks to the artificial kidney, however. Anticoagulants must be added to the blood during dialysis. A large amount of the patient's blood must flow through this apparatus to make the treatment effective, and so the slow rate at which the blood can be processed makes the treatment time-consuming. To date, no artificial kidney has been implanted permanently.

A recent development, called **continuous ambulatory peritoneal dialysis (CAPD),** promises to make hemodialysis more convenient and less time-consuming for many patients. CAPD uses the peritoneum instead of cellulose acetate as the dialyzing membrane. Since the peritoneum is a semipermeable membrane, it permits rapid bidirectional transfer of substances. A catheter is placed in the patient's peritoneal cavity and connected to a supply of dialysate. Gravity feeds the solution into the abdominal cavity from its plastic container. When the process is complete, the dialysate is returned from the abdominal cavity to the plastic container and then discarded.

URETERS

Once urine is formed by the nephrons and collecting ducts, it drains through papillary ducts into the calyces surrounding the renal papillae. The minor calyces join to become the major calyces that unite to become the renal pelvis. From the pelvis, the urine drains into the ureters and is carried by peristalsis to the urinary bladder. From the urinary bladder, the urine is discharged from the body through the single urethra.

STRUCTURE

The body has two **ureters** (yoo-RĒ-ters)—one for each kidney. Each ureter is an extension of the pelvis of the kidney and extends 25 to 30 cm (10 to 12 in) to the urinary bladder (see Figure 24-1). As the ureters descend, their thick walls increase in diameter, but at their widest point they measure less than 1.7 cm (½ in) in diameter. Like the kidneys, the ureters are retroperitoneal in placement. The ureters enter the urinary bladder at the superior lateral angle of its base.

Although there are no anatomical valves at the openings of the ureters into the urinary bladder, there is a functional one that is quite effective. Since the ureters pass under the urinary bladder for several centimeters, pressure in the urinary bladder compresses the ureters and prevents backflow of urine when pressure builds up in the urinary bladder during urination. When this physiological valve is not operating, it is possible for cystitis (urinary bladder inflammation) to develop into kidney infection.

HISTOLOGY

Three coats of tissue form the wall of the ureters (Figure 24-9). The inner coat, or mucosa, is mucous membrane with transitional epithelium. The solute concentration and pH of urine differ drastically from the internal environment of cells that form the wall of the ureters. Mucus secreted by the mucosa prevents the cells from coming in contact with urine. Throughout most of the length of the ureters, the second or middle coat, the muscularis, is composed of inner longitudinal and outer circular layers of smooth muscle. The muscularis of the proximal third of the ureters also contains a layer of outer longitudinal muscle. Peristalsis is the major function of the muscularis. The third, or external, coat of the ureters is a fibrous coat. Extensions of the fibrous coat anchor the ureters in place.

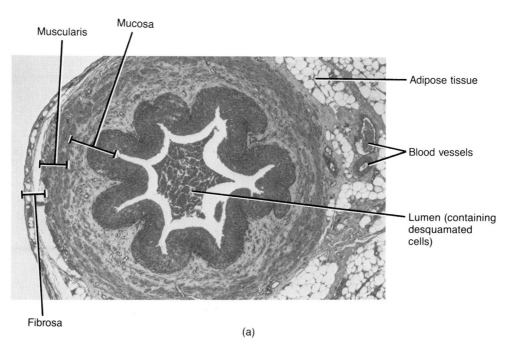

(a)

FIGURE 24-9 Histology of the ureter. (a) Photomicrograph of the ureter seen in cross section at a magnification of 50×.

Transitional epithelium

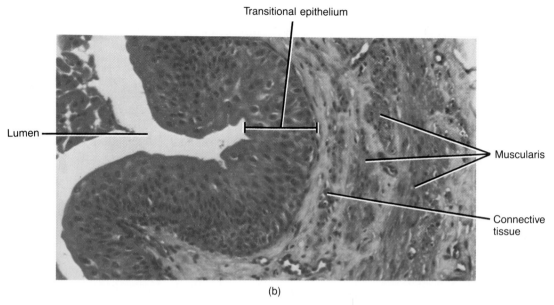

Lumen

Muscularis

Connective tissue

(b)

FIGURE 24-9 (*Continued*) Histology of the ureter. (b) Photomicrograph of an enlarged aspect of the mucosa of the ureter at a magnification of 160×. (© 1983 by Michael H. Ross. Used by permission.)

BLOOD AND NERVE SUPPLY

The arterial supply of the ureters is from the renal, testicular or ovarian, common iliac, and inferior vesical arteries. The veins terminate in the corresponding trunks.

The ureters are innervated by the renal and vesical plexuses.

PHYSIOLOGY

The principal function of the ureters is to transport urine from the renal pelvis into the urinary bladder. Urine is carried through the ureters primarily by peristaltic contractions of the muscular walls of the ureters, but hydrostatic pressure and gravity also contribute. Peristaltic waves pass from the kidney to the urinary bladder, varying in rate from 1 to 5/min depending on the amount of urine formation.

URINARY BLADDER

The **urinary bladder** is a hollow muscular organ situated in the pelvic cavity posterior to the symphysis pubis. In the male, it is directly anterior to the rectum. In the female, it is anterior to the vagina and inferior to the uterus. It is a freely movable organ held in position by folds of the peritoneum. The shape of the urinary bladder depends on how much urine it contains. When empty, it looks like a deflated balloon. It becomes spherical when slightly distended. As urine volume increases, it becomes pear-shaped and rises into the abdominal cavity.

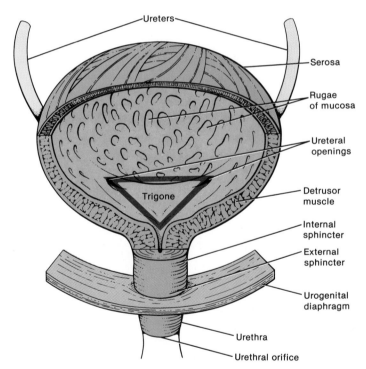

FIGURE 24-10 Urinary bladder and female urethra.

STRUCTURE

At the base of the urinary bladder is a small triangular area, the **trigone** (TRĪ-gōn), that points anteriorly (Figure 24-10). The opening to the urethra is found in the apex of this triangle. At the two points of the base, the ureters drain into the bladder. It is easily identified because the mucosa is firmly bound to the muscularis so that the trigone is typically smooth.

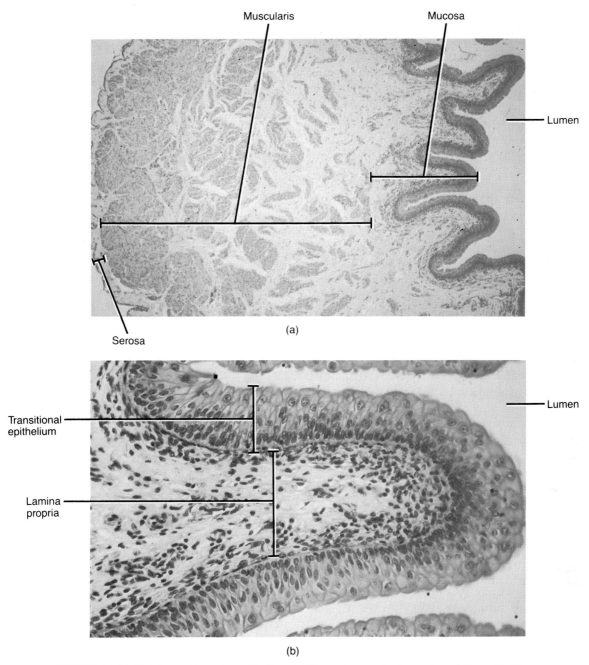

Muscularis   Mucosa

Lumen

(a)

Serosa

Transitional
epithelium

Lamina
propria

Lumen

(b)

FIGURE 24-11 Histology of the urinary bladder. (a) Photomicrograph of a portion of the wall of the urinary bladder at a magnification of 50×. (b) Photomicrograph of an enlarged aspect of the mucosa of the urinary bladder at a magnification of 200×. (© 1983 by Michael H. Ross. Used by permission.)

HISTOLOGY

Four coats make up the wall of the urinary bladder (Figure 24-11). The mucosa, the innermost coat, is a mucous membrane containing transitional epithelium. Transitional epithelium is able to stretch—a marked advantage for an organ that must continually inflate and deflate. Rugae (folds in the mucosa) are also present. The second coat, the submucosa, is a layer of dense connective tissue that connects the mucosa and muscular coats. The third coat—a muscular one called the **detrusor** (de-TROO-ser) muscle—consists of three layers: inner longitudinal, middle circular, and outer longitudinal muscles. In the area around the opening to the urethra, the circular fibers form an **internal sphincter** muscle. Below the internal sphincter is the **external sphincter,** which is composed of skeletal muscle. The outermost coat, the serous coat, is formed by the peritoneum and covers only the superior surface of the organ.

BLOOD AND NERVE SUPPLY

The arteries of the urinary bladder are the superior vesical, the middle vesical, and the inferior vesical. The veins from the urinary bladder pass to the internal iliac trunk.

The nerves are derived partly from the hypogastric sympathetic plexus and partly from the second and third sacral nerves. The fibers from the sacral nerves constitute the nervi erigentes.

PHYSIOLOGY

Urine is expelled from the urinary bladder by an act called **micturition** (mik'-too-RISH-un), commonly known as urination, or voiding. This response is brought about by a combination of involuntary and voluntary nervous impulses. The average capacity of the urinary bladder is 700 to 800 ml. When the amount of urine in the bladder exceeds 200 to 400 ml, stretch receptors in the urinary bladder wall transmit impulses to the lower portion of the spinal cord. These impulses initiate a conscious desire to expel urine and a subconscious reflex referred to as the **micturition reflex.** Parasympathetic impulses transmitted from the sacral area of the spinal cord reach the urinary bladder wall and internal urethral sphincter, bringing about contraction of the detrusor muscle of the urinary bladder and relaxation of the internal sphincter. Then the conscious portion of the brain sends impulses to the external sphincter, the sphincter relaxes, and urination takes place. Although emptying of the urinary bladder is controlled by reflex, it may be initiated voluntarily and stopped at will because of cerebral control of the external sphincter.

CLINICAL APPLICATION

A lack of voluntary control over micturition is referred to as **incontinence.** In infants about 2 years old and under, incontinence is normal because neurons to the external sphincter muscle are not completely developed. Infants void whenever the urinary bladder is sufficiently distended to arouse a reflex stimulus. Proper training overcomes incontinence if the latter is not caused by emotional stress or irritation of the urinary bladder.

Involuntary micturition in the adult may occur as a result of unconsciousness, injury to the spinal nerves controlling the urinary bladder, irritation due to abnormal constituents in urine, disease of the urinary bladder, and inability of the detrusor muscle to relax due to emotional stress.

Retention, a failure to void urine, may be due to an obstruction in the urethra or neck of the urinary bladder, nervous contraction of the urethra, or lack of sensation to urinate.

URETHRA

The **urethra** is a small tube leading from the floor of the urinary bladder to the exterior of the body (see Figure 24-10). In females, it lies directly posterior to the symphysis pubis and is embedded in the anterior wall of the vagina. Its undilated diameter is about 6 mm (¼ in), and its length is approximately 3.8 cm (1½ in). The female urethra is directed obliquely inferiorly and anteriorly. The opening of the urethra to the exterior, the **urethral orifice,** is located between the clitoris and vaginal opening.

In males, the urethra is about 20 cm (8 in) long. Immediately below the urinary bladder it passes vertically through the prostate gland, then pierces the urogenital diaphragm, and finally pierces the penis and takes a curved course through its body (see Figures 25-1 and 25-11).

HISTOLOGY

The wall of the female urethra consists of three coats: an inner mucous coat that is continuous externally with that of the vulva, an intermediate thin layer of spongy tissue containing a plexus of veins, and an outer muscular coat that is continuous with that of the urinary bladder and consists of circularly arranged fibers of smooth muscle.

The male urethra is composed of two coats. An inner mucous membrane is continuous with the mucous membrane of the urinary bladder. An outer submucous tissue connects the urethra with the structures through which it passes.

PHYSIOLOGY

The urethra is the terminal portion of the urinary system. It serves as the passageway for discharging urine from the body. The male urethra also serves as the duct through which reproductive fluid (semen) is discharged from the body.

AGING AND THE URINARY SYSTEM

The effectiveness of kidney function decreases with aging, and by age 70 the filtering mechanism is only about one-half what it was at age 40. Urinary incontinence and urinary tract infections are two major problems associated with aging of the urinary system. Other pathologies include polyuria (excessive urine production), nocturia (excessive urination at night), increased frequency of urination, dysuria (painful urination), retention of urine (failure to produce urine), and hematuria (blood in the urine). Changes and diseases in the kidney include acute and chronic kidney inflammations and renal calculi (kidney stones). The prostate gland is often implicated in various

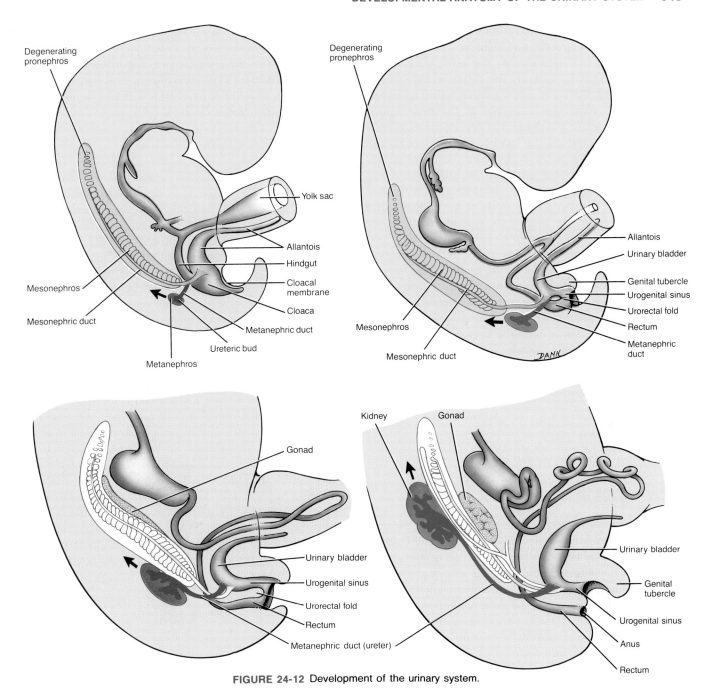

FIGURE 24-12 Development of the urinary system.

disorders of the urinary tract and cancer of the prostate is the most frequent malignancy in elderly males.

DEVELOPMENTAL ANATOMY OF THE URINARY SYSTEM

As early as the third week of development, a portion of the mesoderm along the posterior half of the dorsal side of the embryo, the **intermediate mesoderm,** differentiates into the kidneys. Three pairs of kidneys form within the intermediate mesoderm in successive time periods: pro-

nephros, mesonephros, and metanephros (Figure 24-12). Only the last remains as the functional kidneys of the adult.

The first kidney to form, the **pronephros,** is the superior of the three. Associated with its formation is a tube, the **pronephric duct.** This duct empties into the **cloaca,** which is the dilated caudal end of the gut derived from **endoderm.** The pronephros begins to degenerate during the fourth week and is completely gone by the sixth week. The pronephric ducts, however, remain.

The pronephros is replaced by the second kidney, the **mesonephros.** With its appearance, the retained portion of the pronephric duct, which connects to the mesonephros, becomes known as the **mesonephric duct.** The mesonephros begins to degenerate by the sixth week and is just about completely gone by the eighth week.

At about the fifth week, an outgrowth, called a **ureteric bud,** develops from the distal end of the mesonephric duct near the cloaca. This bud is the developing **metanephros.** As it grows toward the head of the embryo, its end widens to form the *pelvis* of the kidney with its *calyces* and associated *collecting tubules.* The unexpanded portion of the bud, the **metanephric duct,** becomes the *ureter.* The *nephrons,* the functional units of the kidney, arise from the intermediate mesoderm around each ureteric bud.

During development, the cloaca divides into a **urogenital sinus,** into which urinary and genital ducts empty, and a *rectum* that discharges into the anal canal. The *urinary bladder* develops from the urogenital sinus. In the female, the *urethra* develops from lengthening of the short duct that extends from the urinary bladder to the urogenital sinus. The *vestibule,* into which the urinary and genital ducts empty, is also derived from the urogenital sinus. In the male, the urethra is considerably longer and more complicated, but is also derived from the urogenital sinus.

APPLICATIONS TO HEALTH

RENAL CALCULI (KIDNEY STONES)

Occasionally, the crystals of salts found in urine may solidify into insoluble stones called **renal calculi (kidney stones).** They may be formed in any portion of the urinary tract from the kidney tubules to the external opening. Conditions leading to calculi formation include the ingestion of excessive mineral salts, a decrease in the amount of water, abnormally alkaline or acid urine, and overactivity of the parathyroid glands. Common constituents of stones are uric acid, calcium oxalate, and calcium phosphate crystals. The stones usually form in the pelvis of the kidney, where they cause pain, hematuria, and pyuria. A *staghorn calculus* may fill the entire collecting system; a *dendritic stone* involves some but not all of the calyces. Severe pain occurs when a stone passes through a ureter and stretches its walls. *Ureteral stones* are seldom completely obstructive because they are usually needle-shaped and urine can flow around them.

For kidney stones that become painful or obstructive, surgical removal is the typical alternative. However, there are now two new methods of treatment that do not involve conventional surgery. One procedure, called *extracorporeal shock wave lithotripsy* (*ESWL*), involves the use of ultrasound waves. In this procedure, a patient is placed in a water bath and is subjected to ultrasound waves. Once the stones are shattered, the fragments are elimi-

nated via the urine. The other procedure is called *percutaneous ultrasonic lithotripsy* (*PUL*). A tube resembling a cystoscope is inserted into the kidney through a small opening in the back. The surgeon can then pick out small stones whole or shatter large stones with ultrasound waves and remove the fragments by suction.

GOUT

Gout is a hereditary condition associated with an excessively high level of uric acid in the blood. When nucleic acids are catabolized, a certain amount of uric acid is produced as a waste. Some people seem to produce excessive amounts of uric acid, and others seem to have trouble excreting normal amounts. In either case, uric acid accumulates in the body and tends to solidify into crystals that are deposited in the joints and kidney tissue. When the crystals are deposited in the joints, the condition is called gouty arthritis. Gout is aggravated by excessive use of diuretics, dehydration, and starvation.

GLOMERULONEPHRITIS (BRIGHT'S DISEASE)

Glomerulonephritis (Bright's disease) is an inflammation of the kidney that involves the glomeruli. One of the most common causes of glomerulonephritis is an allergic reaction to the toxins given off by streptococci bacteria that have recently infected another part of the body, especially the throat. The glomeruli become so inflamed, swollen, and engorged with blood that the glomerular membranes become highly permeable and allow blood cells and proteins to enter the filtrate. Thus the urine contains many erythrocytes and much protein. The glomeruli may be permanently changed, leading to chronic renal disease and renal failure.

PYELITIS AND PYELONEPHRITIS

Pyelitis is an inflammation of the kidney pelvis and its calyces. **Pyelonephritis** is the interstitial inflammation of one or both kidneys. It usually involves both the parenchyma and the renal pelvis and is due to bacterial invasion from the middle and lower urinary tracts or the bloodstream.

CYSTITIS

Cystitis is an inflammation of the urinary bladder principally involving the mucosa and submucosa. It may be caused by bacterial infection, chemicals, or mechanical injury.

NEPHROSIS

Nephrosis is a condition in which the glomerular membrane leaks, allowing large amounts of protein to escape from the blood into the urine. Water and sodium then accumulate in the body creating edema, especially around the ankles and feet, abdomen, and eyes. Nephrosis is more common in children than adults, but occurs in all ages. Although it cannot always be cured, certain synthetic steroid hormones such as cortisone and prednisone, which are similar to the natural hormones secreted by the adrenal glands, can suppress some forms of it.

POLYCYSTIC DISEASE

Polycystic disease may be caused by a defect in the renal tubular system that deforms nephrons and results in cyst-like dilations along their course. It is the most common inherited disorder of the kidneys. The kidney tissue is riddled with cysts, small holes, and fluid-filled bubbles ranging in size from a pinhead to the diameter of an egg. These cysts gradually increase until they squeeze out the normal tissue, interfering with kidney function and causing uremia. The chief symptom is weight gain. The kidneys themselves may enlarge from their normal ¼ kg (½ lb) to as much as 14 kg (30 lb). Many people lead normal lives without ever knowing they have the disease, however, and the condition is sometimes discovered only after death. Although the disease is progressive, its advance can be slowed by diet, drugs, and fluid intake. Kidney failure as a result of polycystic disease seldom occurs before the mid-forties and frequently can be postponed until the sixties.

KEY MEDICAL TERMS ASSOCIATED WITH THE URINARY SYSTEM

Azotemia (*azo* = nitrogen-containing; *emia* = condition of blood) Pressure of urea or other nitrogenous elements in the blood.

Cystocele (*cyst* = bladder; *cele* = cyst) Hernia of the urinary bladder.

Cystoscope (*skopein* = to examine) Instrument used to examine the urinary bladder.

Dysuria (*dys* = painful; *uria* = urine) Painful urination.

Enuresis (*enourein* = to void urine) Bed-wetting; may be due to a hereditary delay in developing normal urinary bladder control, faulty toilet training, some psychological or emotional disturbance, or, rarely, to some physical disorder.

Intravenous pyelogram, or **IVP** (*intra* = within; *veno* = vein; *pyelo* = pelvis of kidney; *gram* = written or recorded) X-ray film of the kidneys after injection of a dye.

Nephroblastoma (*neph* = kidney; *blastos* = germ or forming; *oma* = tumor) Embryonal carcinosarcoma or Wilms's tumor.

Nephrocele Hernia of the kidney.

Oliguria (*olig* = scanty) Scanty urine.

Polyuria (*poly* = much) Excessive urine.

Stricture Narrowing of the lumen of a canal or hollow organ, as the ureter or urethra.

Uremia (*emia* = condition of blood) Toxic levels of urea in the blood resulting from severe malfunction of the kidneys.

Urethritis Inflammation of the urethra, caused by highly acid urine, the presence of bacteria, or constriction of the urethral passage.

STUDY OUTLINE

1. The primary function of the urinary system is to regulate the concentration and volume of blood by removing and restoring selected amounts of water and solutes. It also excretes wastes.
2. The organs of the urinary system are the kidneys, ureters, urinary bladder, and urethra.

Kidneys (p. 627)

1. The kidneys are retroperitoneal organs attached to the posterior abdominal wall.
2. Three layers of tissue surround the kidneys: renal capsule, adipose capsule, and renal fascia.
3. Internally, the kidneys consist of a cortex, medulla, pyramids, papillae, columns, calyces, and a pelvis.
4. Nephrons are the functional units of the kidneys. They help form urine and regulate blood composition.
5. Each juxtamedullary nephron consists of a glomerular (Bowman's) capsule, glomerulus, proximal convoluted tubule, descending limb of the loop of the nephron, loop of the nephron (loop of Henle), ascending limb of the loop of the nephron, distal convoluted tubule, and collecting duct. The limbs and loop of the nephron are absent in a cortical nephron.
6. The filtering unit of a nephron is the endothelial-capsular membrane. It consists of the glomerular endothelium, glomerular basement membrane, and epithelium (podocytes) of the visceral layer of the glomerular capsule.
7. The extensive flow of blood through the kidney begins in the renal artery and terminates in the renal vein.
8. The nerve supply to the kidney is derived from the renal plexus.
9. The juxtaglomerular apparatus consists of the juxtaglomerular cells of the afferent arteriole and the macula densa of the distal convoluted tubule.

Hemodialysis Therapy (p. 637)

1. Filtering blood through an artificial device is called hemodialysis.
2. The kidney machine filters the blood of wastes and adds

nutrients; a recent variation is called continuous ambulatory peritoneal dialysis (CAPD).

Ureters (p. 639)

1. The ureters are retroperitoneal and consist of a mucosa, muscularis, and fibrous coat.
2. The ureters transport urine from the renal pelvis to the urinary bladder, primarily by peristalsis.

Urinary Bladder (p. 640)

1. The urinary bladder is posterior to the symphysis pubis. Its function is to store urine prior to micturition.
2. Histologically, the urinary bladder consists of a mucosa (with rugae), a muscularis (detrusor muscle), and a serous coat.
3. A lack of control over micturition is called incontinence; failure to void urine is referred to as retention.

Urethra (p. 642)

1. The urethra is a tube leading from the floor of the urinary bladder to the exterior.
2. Its function is to discharge urine from the body.

Aging and the Urinary System (p. 642)

1. With age, kidney function decreases.

2. Common problems related to aging include incontinence, urinary tract infections, prostate disorders, and renal calculi.

Developmental Anatomy of the Urinary System (p. 643)

1. The kidneys develop from intermediate mesoderm.
2. They develop in the following sequence: pronephros, mesonephros, metanephros.

Applications to Health (p. 644)

1. Renal calculi (kidney stones) are composed of crystals of salts and may be formed in any part of the urinary tract.
2. Gout is a high level of uric acid in the blood.
3. Glomerulonephritis is an inflammation of the glomeruli of the kidney.
4. Pyelitis is an inflammation of the kidney pelvis and calyces; pyelonephritis is an interstitial inflammation of one or both kidneys.
5. Cystitis is an inflammation of the urinary bladder.
6. Nephrosis leads to protein in the urine due to glomerular membrane permeability.
7. Polycystic disease is an inherited kidney disease in which nephrons are deformed.

REVIEW QUESTIONS

1. What are the functions of the urinary system? What organs compose the system?
2. Describe the location of the kidneys. Why are they said to be retroperitoneal?
3. Prepare a labeled diagram that illustrates the principal external and internal features of the kidney.
4. What is a nephron? List and describe the parts of a nephron from the glomerular (Bowman's) capsule to the collecting duct.
5. Distinguish between cortical and juxtaglomerular nephrons.
6. How are nephrons supplied with blood?
7. Describe the structure and importance of the juxtaglomerular apparatus.
8. What is hemodialysis? Briefly describe the operation of an artificial kidney. What is continuous ambulatory peritoneal

dialysis (CAPD)?
9. Describe the structure, histology, and function of the ureters.
10. How is the urinary bladder adapted to its storage function? What is micturition?
11. Contrast the causes of incontinence and retention.
12. Compare the location of the urethra in the male and female.
13. Describe the effects of aging on the urinary system.
14. Describe the development of the urinary system.
15. Define each of the following: renal calculi (kidney stones), gout, glomerulonephritis, pyelitis, pyelonephritis, cystitis, nephrosis, and polycystic disease.
16. Refer to the glossary of key medical terms associated with the urinary system. Be sure that you can define each term.

25

The Reproductive Systems

Student Objectives

Define reproduction and classify the organs of reproduction by function.

Explain the structure, histology, and functions of the testes.

Define meiosis and explain the principal events of spermatogenesis.

Describe the seminiferous tubules, straight tubules, and rete testis as components of the duct system of the testes.

Describe the location, structure, histology, and functions of the ductus epididymis, ductus (vas) deferens, and ejaculatory duct.

Describe the structure and function of the ejaculatory duct.

Describe the three anatomical subdivisions of the male urethra.

Explain the location and functions of the seminal vesicles, prostate gland, and bulbourethral (Cowper's) glands, the accessory reproductive glands.

Explain the structure and functions of the penis.

Describe the location, histology, and functions of the ovaries.

Describe the principal events of oogenesis.

Explain the location, structure, histology, and functions of the uterine (Fallopian) tubes.

Describe the location and ligamentous attachments of the uterus.

Explain the histology and blood supply of the uterus.

Describe the location, structure, and functions of the vagina.

Describe the components of the vulva and explain their functions.

Describe the anatomical boundaries of the perineum.

Explain the structure and histology of the mammary glands.

Explain the roles of the male and female in sexual intercourse.

Describe the effects of aging on the reproductive systems.

Describe the development of the reproductive systems.

Explain the symptoms and causes of sexually transmitted diseases (STDs) such as gonorrhea, syphilis, genital herpes, trichomoniasis, and nongonococcal urethritis (NGU).

Describe the symptoms and causes of male disorders (prostate dysfunctions, impotence, and infertility) and female disorders [amenorrhea, dysmenorrhea, premenstrual syndrome (PMS), toxic shock syndrome (TSS), ovarian cysts, endometriosis, infertility, breast tumors, cervical cancer, and pelvic inflammatory disease or PID].

Define key medical terms associated with the reproductive systems.

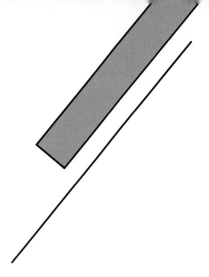

Reproduction is the mechanism by which the thread of life is sustained. It is the process by which a single cell duplicates its genetic material, allowing an organism to grow and repair itself. In this sense, reproduction maintains the life of the individual. But reproduction is also the process by which genetic material is passed from generation to generation. In this regard, reproduction maintains the continuation of the species.

The organs of the male and female reproductive systems may be grouped by function. The testes and ovaries, also called **gonads,** function in the production of gametes—sperm cells and ova, respectively. The gonads also secrete hormones. The production of gametes and their discharge into ducts classifies the gonads as exocrine glands, whereas their production of hormones classifies them as endocrine glands. The **ducts** transport, receive, and store gametes. Still other reproductive organs, called **accessory glands,** produce materials that support gametes.

The developmental anatomy of the reproductive system is considered later in the chapter.

MALE REPRODUCTIVE SYSTEM

The organs of the male reproductive system are the testes, or male gonads, which produce sperm; a number of ducts that either store or transport sperm to the exterior; accessory glands that add secretions constituting the semen; and several supporting structures, including the penis (Figure 25-1).

SCROTUM

The **scrotum** is a cutaneous outpouching of the abdomen consisting of loose skin and superficial fascia (Figure 25-1). It is the supporting structure for the testes. Externally, it looks like a single pouch of skin separated into lateral portions by a medium ridge called the **raphe** (RĀ-fē). Internally, it is divided by a septum into two sacs, each containing a single testis. The septum consists of superficial fascia and contractile tissue called the **dartos** (DAR-tōs), which consists of bundles of smooth muscle fibers. The dartos is also found in the subcutaneous tissue of the scrotum and is directly continuous with the subcutaneous tissue of the abdominal wall. The dartos causes wrinkling of the skin of the scrotum.

The location of the scrotum and the contraction of its muscle fibers regulate the temperature of the testes, organs that produce sperm and the hormones testosterone and inhibin. The production and survival of sperm require a temperature that is lower than body temperature. Because the scrotum is outside the body cavities, it supplies an environment about 3°F below body temperature. The **cremaster** (krē-MAS-ter) **muscle,** a small, circular band of skeletal muscle, elevates the testes on exposure to cold, moving them closer to the pelvic cavity where they can absorb body heat. Exposure to warmth reverses the process.

The blood supply of the scrotum is derived from the internal pudendal branch of the internal iliac, the cremasteric branch of the inferior epigastric artery, and the external pudendal artery from the femoral artery. The scrotal veins follow the arteries.

The scrotal nerves are derived from the pudendal, posterior cutaneous of the thigh, and ilioinguinal nerves.

TESTES

The **testes,** or **testicles,** are paired oval glands measuring about 5 cm (2 in) in length and 2.5 cm (1 in) in diameter (Figure 25-2). They weigh between 10 and 15 g. The testes develop high on the embryo's posterior abdominal wall, and usually begin their descent into the scrotum through the inguinal canals during the latter half of the seventh month (see Figure 25-9).

CLINICAL APPLICATION

When the testes do not descend, the condition is referred to as **cryptorchidism** (krip-TOR-ki-dizm). The condition occurs in about 3 percent of full-term infants

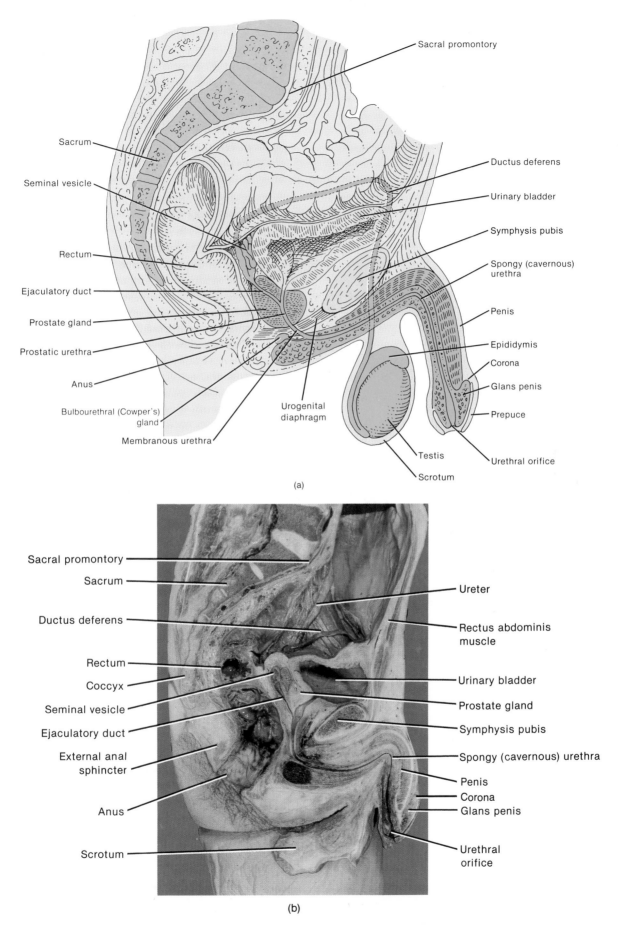

FIGURE 25-1 Male organs of reproduction and surrounding structures. (a) Diagram of structures in sagittal section. (b) Photograph of structures in sagittal section. (Courtesy of C. Yokochi and J. W. Rohen, *Photographic Anatomy of the Human Body*, 2nd ed., 1979, IGAKU-SHOIN, Ltd., Tokyo, New York.)

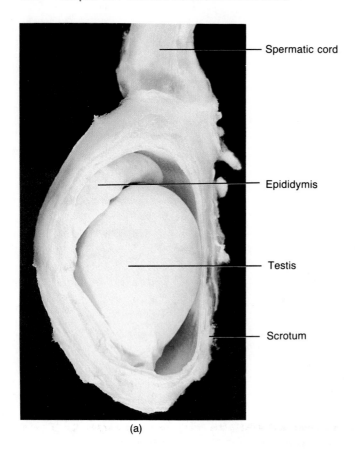

Spermatic cord

Epididymis

Testis

Scrotum

(a)

and about 30 percent of premature infants. Cryptorchidism results in sterility because the cells involved in the initial development of sperm cells are destroyed by the higher body temperature of the pelvic cavity. Undescended testes can be placed in the scrotum by administering hormones or by surgical means prior to puberty without ill effects.

The testes are covered by a dense layer of white fibrous tissue, the **tunica albuginea** (al'-byoo-JIN-ē-a), that extends inward and divides each testis into a series of internal compartments called **lobules.** Each of the 200 to 300 lobules contains one to three tightly coiled tubules, the convoluted **seminiferous tubules,** that produce sperm by a process called **spermatogenesis.** This process is considered shortly.

A cross section through a seminiferous tubule reveals that it is packed with sperm cells in various stages of development (Figure 25-3). The most immature cells, the **spermatogonia,** are located against the basement membrane. These cells collectively constitute the germinal epithelium. Toward the lumen of the tube, one can see layers of progressively more mature cells. In order of advancing maturity, these are primary spermatocytes, secondary

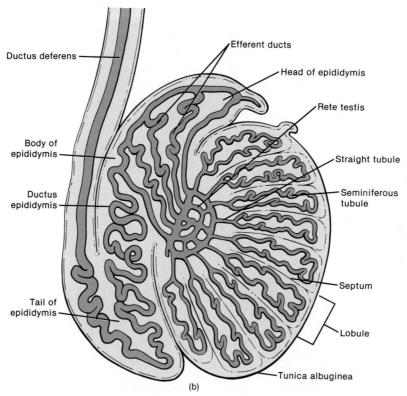

Ductus deferens

Body of epididymis

Ductus epididymis

Tail of epididymis

Efferent ducts

Head of epididymis

Rete testis

Straight tubule

Seminiferous tubule

Septum

Lobule

Tunica albuginea

(b)

FIGURE 25-2 Anatomy of the testes. (a) Photograph of external view in the scrotum. (Courtesy of C. Yokochi and J. W. Rohen, *Photographic Anatomy of the Human Body,* 2nd ed., 1979, IGAKU-SHOIN, Ltd., Tokyo, New York.) (b) Diagram of a sagittal section illustrating the internal anatomy.

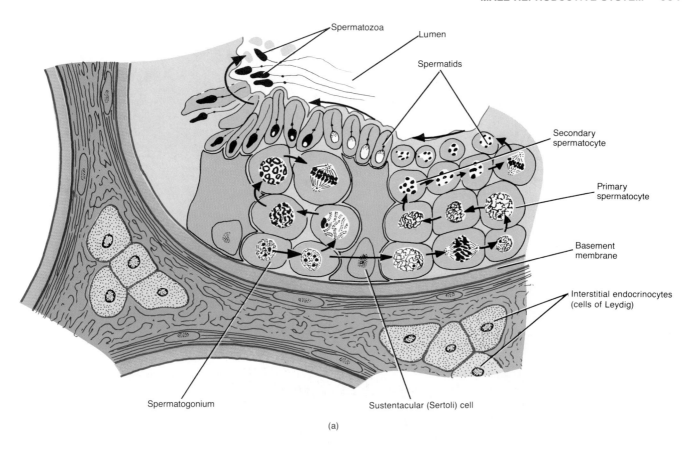

(a)

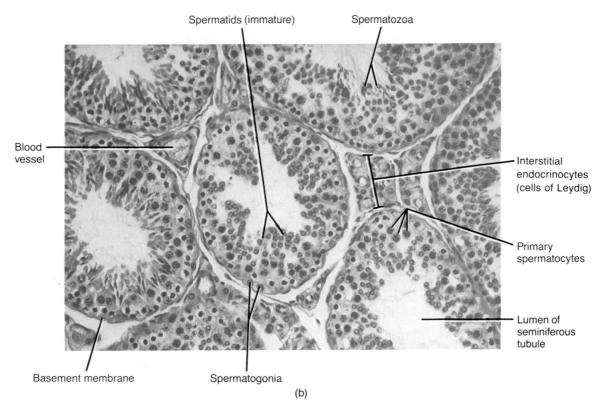

(b)

FIGURE 25-3 Histology of the testes. (a) Diagram of a cross section of a portion of a seminiferous tubule showing the stages of spermatogenesis. (b) Photomicrograph of an enlarged aspect of several seminiferous tubules at a magnification of 350×. (© 1983 by Michael H. Ross. Used by permission.)

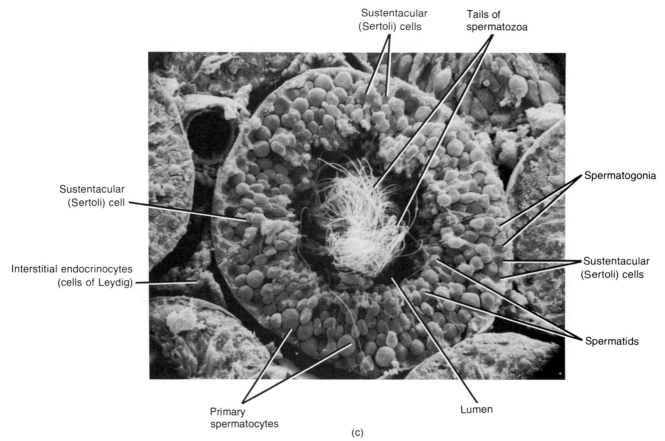

Sustentacular (Sertoli) cells

Tails of spermatozoa

Sustentacular (Sertoli) cell

Spermatogonia

Interstitial endocrinocytes (cells of Leydig)

Sustentacular (Sertoli) cells

Spermatids

Primary spermatocytes

Lumen

(c)

FIGURE 25-3 (*Continued*) Histology of the testes. (c) Scanning electron micrograph of a seminiferous tubule at a magnification of 663×. (Courtesy of Richard K. Kessel and Randy H. Kardon, *Tissues and Organs: A Text-Atlas of Scanning Electron Microscopy.* Copyright © 1979 by Scientific American, Inc.)

spermatocytes, and spermatids. By the time a **sperm cell,** or **spermatozoon** (sper'-ma-tō-ZŌ-on), has reached full maturity, it is in the lumen of the tubule and begins to be moved through a series of ducts. Embedded between the developing sperm cells in the tubules are **sustentacular** (sus'-ten-TAK-yoo-lar), or **Sertoli, cells.** These cells produce secretions that supply nutrients to the spermatozoa and secrete the hormone inhibin. Between the seminiferous tubules are clusters of **interstitial endocrinocytes (interstitial cells of Leydig).** These cells secrete the male hormone testosterone, the most important androgen.

The testes are supplied by the testicular arteries, which arise from the aorta immediately below the origin of the renal arteries. Blood from the testes is drained on the right side by the testicular vein, which enters the inferior vena cava, and on the left side by the testicular vein, which enters the left renal vein.

The nerve supply to the testes is from the testicular plexus, which contains vagal parasympathetic fibers and sympathetic fibers from the tenth and eleventh thoracic segments of the spinal cord.

Diploid and Haploid Cells

In sexual reproduction, each new organism is produced by the union and fusion of two different sex cells, one produced by each parent. The sex cells, called **gametes,** are the ovum produced in the female gonads (ovaries) and the sperm produced in the male gonads (testes). The union and fusion of gametes is called fertilization and the cell thus produced is known as a zygote. The zygote contains a mixture of chromosomes (DNA) from the two parents and, through its repeated mitotic division, develops into a new organism.

Gametes differ from all other body cells (somatic cells) with respect to the number of chromosomes in their nuclei. Somatic cells, such as brain cells, stomach cells, kidney cells, and all other uninucleated somatic cells, contain 46 chromosomes in their nuclei. Some somatic cells, such as skeletal muscle cells, are multinucleated and thus contain more than 46 chromosomes. However, since most somatic cells are uninucleated, these are the cells to which we will refer in the following discussion. Of the 46 chro-

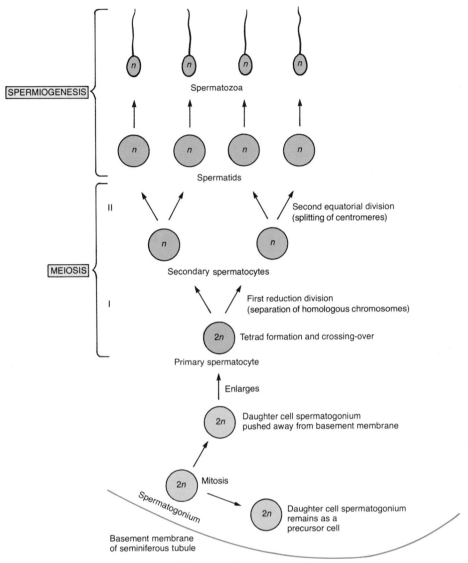

SPERMIOGENESIS

Spermatozoa

Spermatids

MEIOSIS

II

Second equatorial division
(splitting of centromeres)

Secondary spermatocytes

I

First reduction division
(separation of homologous chromosomes)

Tetrad formation and crossing-over

Primary spermatocyte

Enlarges

Daughter cell spermatogonium
pushed away from basement membrane

Mitosis

Spermatogonium

Daughter cell spermatogonium
remains as a
precursor cell

Basement membrane
of seminiferous tubule

FIGURE 25-4 Spermatogenesis.

mosomes, 23 are a complete set that contain all the genes necessary for carrying out the activities of the cell. In a sense, the other 23 chromosomes are a duplicate set. The symbol n is used to designate the number of different chromosomes within the nucleus. Since somatic cells contain two sets of chromosomes, they are referred to as **diploid** (DIP-loyd) **cells** (di = two), symbolized as $2n$. In a diploid cell, two chromosomes that belong to a pair are called **homologous** (ho-MOL-o-gus) **chromosomes,** or **homologues.**

If gametes had the same number of chromosomes as somatic cells, the zygote formed from their fusion would have double the number. The somatic cells of the resulting individual would have twice the number of chromosomes ($4n$) as the somatic cells of the parents, and with every succeeding generation, the number of chromosomes would double. The chromosome number does not double

with each generation because of a special nuclear division called **meiosis.** Meiosis occurs only in the production of gametes. It causes a developing sperm or ovum to relinquish its duplicate set of chromosomes so that the mature gamete has only 23. Thus gametes are **haploid** (HAP-loyd) **cells,** meaning "one-half," and are symbolized as n.

The formation in the testes of haploid spermatozoa by meiosis is called **spermatogenesis** (sper'-ma-tō-JEN-e-sis). The formation in the ovary of haploid ova by meiosis is referred to as **oogenesis** (ō'-ō-JEN-e-sis).

Spermatogenesis

In humans, spermatogenesis takes about two to three weeks. The seminiferous tubules are lined with immature cells called **spermatogonia** (sper'-ma-tō-GŌ-nē-a), or sperm mother cells (Figure 25-4). Spermatogonia contain

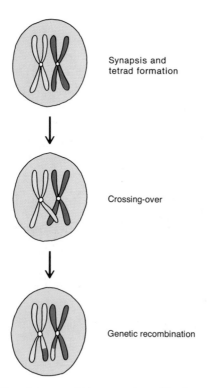

Synapsis and tetrad formation

Crossing-over

Genetic recombination

FIGURE 25-5 Crossing-over within a tetrad resulting in genetic recombination.

the diploid chromosome number and are the precursor cells for all the spermatozoa the male will produce. At puberty, when the anterior pituitary secrets FSH in response to GnRF from the hypothalamus, the spermatogonia embark on a lifetime of active mitotic division. Some of the daughter cells are pushed inward toward the lumen of the seminiferous tuble. These cells lose contact with the basement membrane of the seminiferous tubule, undergo certain developmental changes, and become known as **primary spermatocytes** (SPER-ma-tō-sītz'). Primary spermatocytes, like spermatogonia, are diploid, that is, they have 46 chromosomes. The other daughter cells formed by mitosis of the spermatogonia remain near the basement membrane and form a reservoir of precursor cells for future meiotic divisions.

● *Reduction Division*   Each primary spermatocyte enlarges before dividing. Then two nuclear divisions take place as part of meiosis. In the first, DNA is replicated and 46 chromosomes (each made up of two chromatids) form and move toward the equatorial plane of the nucleus. There they line up by homologous pairs so that there are 23 pairs of duplicated chromosomes in the center of the nucleus. This pairing of homologous chromosomes is called **synapsis.** The four chromatids of each homologous pair then twist around each other to form a **tetrad.** In a tetrad, portions of one chromatid may be exchanged with portions of another. This process, called **crossing-over,** permits an exchange of genes among chromatids

(Figure 25-5) that results in the combination of genes. Thus the spermatozoa eventually produced may be genetically unlike each other and unlike the cell that produced them—hence the great variation among humans. Next the mitotic spindle forms and the chromosomal microtubules produced by the centromeres of the paired chromosomes extend toward the poles of the cell. As the pairs separate, one member of each pair migrates to opposite poles of the dividing nucleus. The cells formed by the first nuclear division (reduction division) are called **secondary spermatocytes.** Each cell has 23 chromosomes—the haploid number. Each chromosome of the secondary spermatocytes, however, is made up of two chromatids. Moreover, the genes of the chromosomes of secondary spermatocytes may be rearranged as a result of crossing-over.

● *Equatorial Division*   The second nuclear division of meiosis is equatorial division. There is no replication of DNA. The chromosomes (each composed of two chromatids) line up in single file around the equatorial plane, and the chromatids of each chromosome separate from each other. The cells formed from the equatorial division are called **spermatids.** Each contains half the original chromosome number, or 23 chromosomes, and is haploid. Each primary spermatocyte therefore produces four spermatids by meiosis (reduction division and equatorial division). Spermatids lie close to the lumen of the seminiferous tubule.

● *Spermiogenesis*   The final stage of spermatogenesis, called **spermiogenesis** (sper'-mē-ō-JEN-e-sis), involves the maturation of spermatids into spermatozoa. Each spermatid embeds in a sustentacular cell and develops a head with an acrosome and a flagellum (tail). Sustentacular cells extend from the basement membrane to the interior of the seminiferous tubule where they nourish the developing spermatids. Since there is no cell division in spermiogenesis, each spermatid develops into a single **spermatozoon (sperm cell).**

Spermatozoa enter the lumen of the seminiferous tubule and migrate to the ductus epididymis, where in 18 hours to 10 days they complete their maturation and become capable of fertilizing an ovum. Spermatozoa are also stored in the ductus deferens. Here, they can retain their fertility for up to several months.

Spermatozoa

Spermatozoa are produced or matured at the rate of about 300 million per day and, once ejaculated, have a life expectancy of about 48 hours within the female reproductive tract. A spermatozoon is highly adapted for reaching and penetrating a female ovum. It is composed of a head, a midpiece, and a tail (Figure 25-6). Within the **head** are the nuclear material and the *acrosome,* which contains

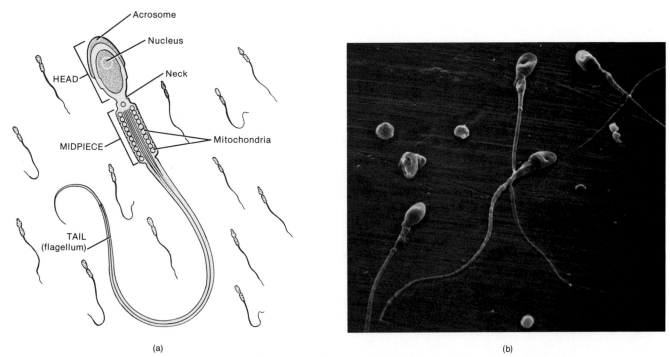

FIGURE 25-6 Spermatozoa. (a) Diagram of the parts of a spermatozoon. (b) Scanning electron micrograph of several spermatozoa at a magnification of 2,000×. (Courtesy of Fisher Scientific Company and S.T.E.M. Laboratories, Inc., Copyright 1975.)

enzymes (hyaluronidase and proteinases) that facilitate penetration of the sperm cell into the ovum. Numerous mitochondria in the **midpiece** carry on the metabolism that provides energy for locomotion. The **tail,** a typical flagellum, propels the sperm along its way.

DUCTS

Ducts of the Testis

When the sperm mature, they are moved through the convoluted seminiferous tubules to the **straight tubules** (see Figure 25-2b). The straight tubules lead to a network of ducts in the testis called the **rete** (RĒ-tē) **testis.** Some of the cells lining the rete testis possess cilia that probably help move the sperm along. The sperm are next transported out of the testis.

Epididymis

The sperm are transported out of the testis through a series of coiled **efferent ducts** in the epididymis that empty into a single tube called the ductus epididymis. At this point, the sperm are morphologically mature.

The two **epididymides** (ep'-i-DID-i-mi-dēz') are comma-shaped organs. Each lies along the posterior border of the testis (see Figures 25-1 and 25-2) and consists mostly of a tightly coiled tube, the **ductus epididymis,** (ep'-i-DID-i-mis). The larger, superior portion of the epi-

didymis is known as the **head.** It consists of the efferent ducts that empty into the ductus epididymis. The **body** of the epididymis contains the ductus epididymis. The **tail** is the smaller, inferior portion. Within the tail, the ductus epididymis continues as the ductus deferens.

The ductus epididymis measures about 6 m (20 ft) in length and 1 mm in diameter. It is tightly packed within the epididymis, which measures only about 3.8 cm (1.5 in). The ductus epididymis is lined with pseudostratified columnar epithelium, and its wall contains smooth muscle. The free surfaces of the columnar cells contain long, branching microvilli called *stereocilia* (Figure 25-7).

Functionally, the ductus epididymis is the site of sperm maturation. They require between 18 hours and 10 days to complete their maturation, that is, to become capable of fertilizing an ovum. The ductus epididymis also stores spermatozoa and propels them toward the urethra during ejaculation by peristaltic contraction of its smooth muscle. Spermatozoa may remain in storage in the ductus epididymis for up to 4 weeks. After that, they are reabsorbed.

Ductus (Vas) Deferens

Within the tail of the epididymis, the ductus epididymis becomes less convoluted, its diameter increases, and at this point it is referred to as the **ductus (vas) deferens,** or **seminal duct** (see Figure 25-2). The ductus deferens, about 45 cm long (18 in), ascends along the posterior border of the testis, penetrates the inguinal canal, and enters the pelvic cavity, where it loops over the side and down the posterior surface of the urinary bladder (see Figure 25-1). The dilated terminal portion of the ductus deferens is known as the **ampulla** (am-POOL-la). The ductus deferens is lined with pseudostratified epithelium and contains a heavy coat of three layers of muscle (Figure 25-8). Functionally, the ductus deferens stores sperm for up to several months and propels them toward the urethra during ejaculation by peristaltic contractions of the muscular coat.

CLINICAL APPLICATION

One method of sterilization of males is called **vasectomy,** in which a portion of each ductus deferens is removed. In the procedure, an incision is made in the scrotum, the ducts are located, each is tied in two places, and the portion between the ties is excised. Although sperm production continues in the testes, the sperm cannot reach the exterior because the ducts are cut and the sperm degenerate.

Traveling with the ductus deferens as it ascends in the scrotum are the testicular artery, autonomic nerves, veins that drain the testes (pampiniform plexus), lymphatics, and the cremaster muscle. These structures constitute the **spermatic cord,** a supporting structure of the male reproductive system (Figure 25-9a). The cremaster muscle, which also surrounds the testes, elevates the testes during sexual stimulation and exposure to cold. The spermatic cord and ilioinguinal nerve pass through the **inguinal** (IN-gwin-al) **canal** in the male (Figure 25-9b). The canal is an oblique passageway in the anterior abdominal wall just superior and parallel to the medial half of the inguinal ligament. The canal is about 4 to 5 cm (1.6 to 2.0 in) in length. It originates at the *deep (abdominal) inguinal ring,* a slitlike opening in the aponeurosis of the transversus abdominis muscle. The canal terminates at the *superficial (subcutaneous) inguinal ring,* a triangular opening in the aponeurosis of the external oblique muscle. In the female the round ligament of the uterus and ilioinguinal nerve pass through the inguinal canal.

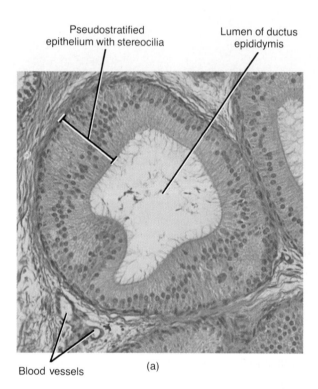

Pseudostratified epithelium with stereocilia

Lumen of ductus epididymis

Blood vessels

(a)

FIGURE 25-7 Histology of the ductus epididymis. (a) Photomicrograph of the ductus epididymis seen in cross section at a magnification of 160×. (b) Photomicrograph of an enlarged aspect of the mucosa of the ductus epididymis at a magnification of 350×. (© 1983 by Michael H. Ross. Used by permission.)

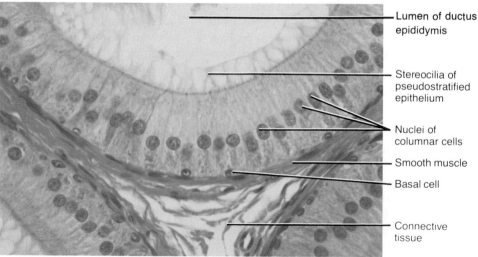

Lumen of ductus epididymis

Stereocilia of pseudostratified epithelium

Nuclei of columnar cells

Smooth muscle

Basal cell

Connective tissue

(b)

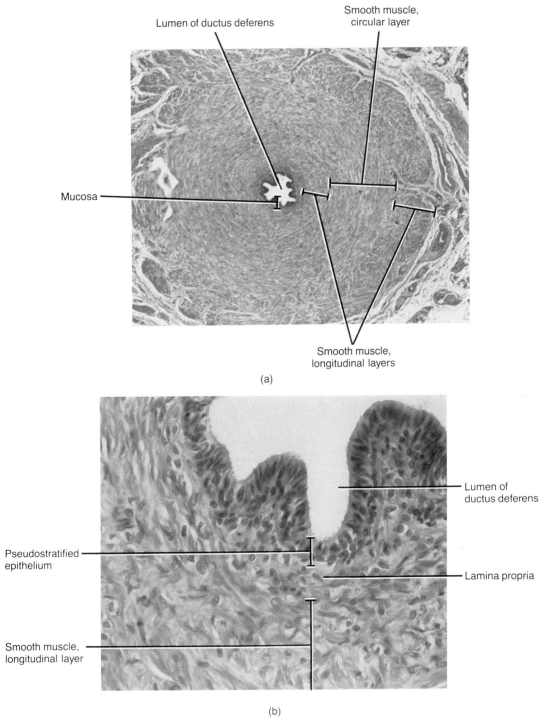

Lumen of ductus deferens

Smooth muscle, circular layer

Mucosa

Smooth muscle, longitudinal layers

(a)

Pseudostratified epithelium

Smooth muscle, longitudinal layer

Lumen of ductus deferens

Lamina propria

(b)

FIGURE 25-8 Histology of the ductus (vas) deferens. (a) Photomicrograph of the ductus (vas) deferens seen in cross section at a magnification of 40×. (b) Photomicrograph of an enlarged aspect of the mucosa of the ductus (vas) deferens at a magnification of 160×. (© 1983 by Michael H. Ross. Used by permission.)

CLINICAL APPLICATION

The inguinal canal represents a weak spot in the abdominal wall. It is frequently the site of an **inguinal hernia**—a rupture or separation of a portion of the abdominal wall resulting in the protrusion of a part of an organ. Since the inguinal canal is smaller in females, inguinal hernias occur much less frequently in females.

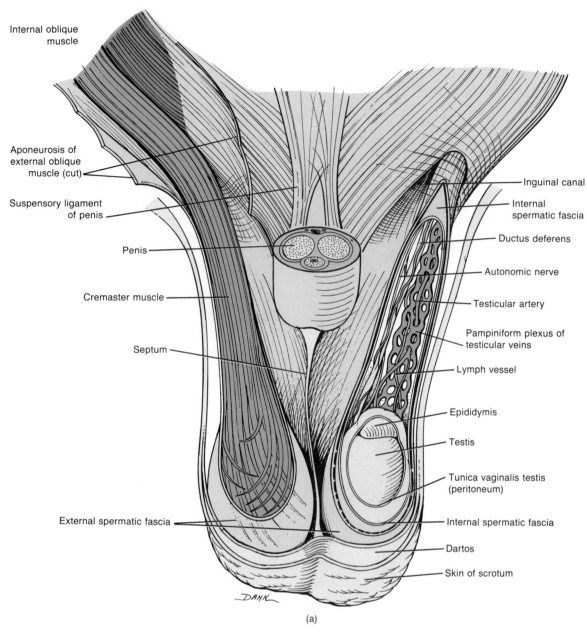

Internal oblique muscle

Aponeurosis of external oblique muscle (cut)

Suspensory ligament of penis

Penis

Cremaster muscle

Septum

External spermatic fascia

Inguinal canal

Internal spermatic fascia

Ductus deferens

Autonomic nerve

Testicular artery

Pampiniform plexus of testicular veins

Lymph vessel

Epididymis

Testis

Tunica vaginalis testis (peritoneum)

Internal spermatic fascia

Dartos

Skin of scrotum

(a)

FIGURE 25-9 Spermatic cord and inguinal canal. (a) The left spermatic cord has been opened to expose its contents.

Ejaculatory Duct

Posterior to the urinary bladder are the **ejaculatory** (ē-JAK-yoo-la-tō'-rē) **ducts** (Figure 25-10). Each duct is about 2 cm (1 in) long and is formed by the union of the duct from the seminal vesicle and ductus deferens. The ejaculatory ducts eject spermatozoa into the prostatic urethra.

Urethra

The **urethra** is the terminal duct of the system, serving as a passageway for spermatozoa or urine. In the male,

the urethra passes through the prostate (PROS-tāt) gland, the urogenital diaphragm, and the penis. It measures about 20 cm (8 in) in length and is subdivided into three parts (see Figure 24-1). The **prostatic urethra** is 2 to 3 cm (1 in) long and passes through the prostate gland. It continues inferiorly and as it passes through the urogenital diaphragm, a muscular partition between the two ischiopubic rami, it is known as the **membranous** (MEM-bra-nus) **urethra.** The membranous portion is about 1 cm (½ in) in length. As it passes through the corpus spongiosum of the penis, it is known as the spongy (cavernous) urethra. This portion is about 15 cm (6 in) long. The spongy urethra enters the bulb of the penis and terminates at the external **urethral orifice** (see Figure 25-11).

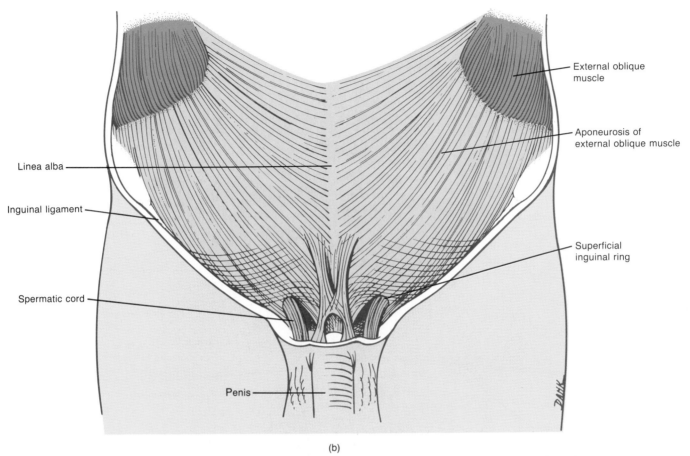

Linea alba

Inguinal ligament

Spermatic cord

Penis

External oblique muscle

Aponeurosis of external oblique muscle

Superficial inguinal ring

(b)

FIGURE 25-9 (*Continued*) Spermatic cord and inguinal canal. (b) Location of the inguinal canal.

ACCESSORY SEX GLANDS

Whereas the ducts of the male reproductive system store and transport sperm cells, the **accessory sex glands** secrete the liquid portion of semen. The paired **seminal vesicles** (VES-i-kuls) are convoluted pouchlike structures, about 5 cm (2 in) in length, lying posterior to and at the base of the urinary bladder in front of the rectum (see Figure 25-10). They secrete an alkaline, viscous fluid, rich in the sugar fructose, and pass it into the ejaculatory duct. This secretion contributes to sperm viability. It constitutes about 60 percent of the volume of semen.

The **prostate gland** is a single, doughnut-shaped gland about the size of a chestnut (Figure 25-10). It is inferior to the urinary bladder and surrounds the superior portion of the urethra. The prostate secretes an alkaline fluid into the prostatic urethra. The prostatic secretion constitutes 13 to 33 percent of the volume of semen and contributes to sperm motility.

The paired **bulbourethral** (bul'-bō-yoo-RĒ-thral), or **Cowper's, glands** are about the size of peas. They are located beneath the prostate on either side of the membranous urethra (Figure 25-10). The bulbourethral glands secrete mucus for lubrication and a substance that neutralizes urine. Their ducts open into the spongy urethra.

SEMEN (SEMINAL FLUID)

Semen (seminal fluid) is a mixture of sperm and the secretions of the seminal vesicles, the prostate gland, and the bulbourethral glands. The average volume of semen for each ejaculation is 2.5 to 5 ml, and the average range of spermatozoa ejaculated is 50 to 100 million/ml. When the number of spermatozoa falls below 20 million/ml, the male is likely to be sterile. The very large number is required because only a small percentage eventually reach the ovum. And, although only a single spermatozoon fertilizes an ovum, fertilization seems to require the combined action at the ovum of a larger number of them. The intercellular material of the cells covering the ovum presents a barrier to the sperm. This barrier is digested by the hyaluronidase and proteinases secreted by the acrosomes of sperm. However, it appears that a single sperm does not produce enough of these enzymes to dissolve the barrier. A passageway through which one may enter can be created only by the action of many.

Semen has a slightly alkaline pH of 7.20 to 7.60. The prostatic secretion gives semen a milky appearance, and fluids from the seminal vesicles and bulbourethral glands give it a mucoid consistency. Semen provides spermatozoa with a transportation medium and nutrients. It neutralizes

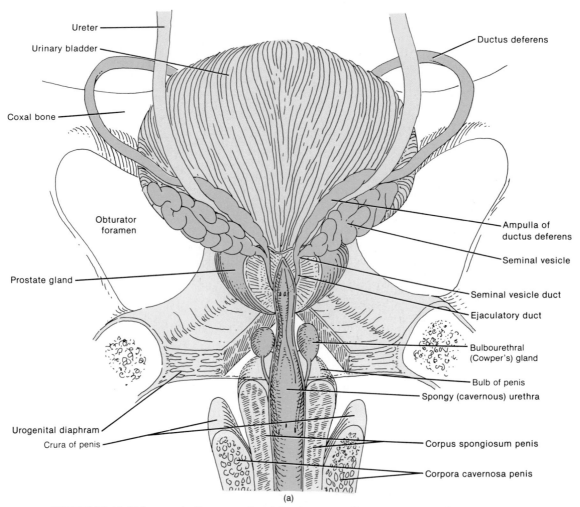

FIGURE 25-10 Male reproductive organs in relation to surrounding structures seen in posterior view. (a) Diagram.

the acid environment of the male urethra and the female vagina. It also contains enzymes that activate sperm after ejaculation.

Semen has been shown to contain an antibiotic, called *seminalplasmin,* which has the ability to destroy a number of bacteria. Its antimicrobial activity has been described as similar to that currently exerted by penicillin, strepto-mycin, and tetracyclines. Since both semen and the lower female reproductive tract contain bacteria, seminalplas-min may keep these bacteria under control to help ensure fertilization.

Once ejaculated into the vagina, liquid semen coagu-lates rapidly because of a clotting enzyme produced by the prostate that acts on a substance produced by the seminal vesicle. This clot liquefies in about 5 to 20 minutes because of another enzyme produced by the prostate gland. Abnormal or delayed liquefaction of coagulated semen may cause complete or partial immobilization of spermatozoa, thus inhibiting their movement through the cervix of the uterus.

CLINICAL APPLICATION

Semen analysis is the most valuable test in evaluating sterility. Among the criteria analyzed are the following:

1. Volume. A low volume might suggest an anatom-ical or functional destruction or inflammation.

2. Motility. This refers to the percentage of motile spermatozoa (40 to 60 percent) and quality of move-ment (forward and progressive).

3. Count. Sperm counts below 20 million/ml could indicate sterility.

4. Liquefaction. Delayed liquefaction of more than 2 hours suggests inflammation of accessory sex glands or enzyme defects in the glands.

5. Morphology. No more than about 35 percent of spermatozoa should have abnormal morphology.

6. Autoagglutination. Agglutination does not occur normally.

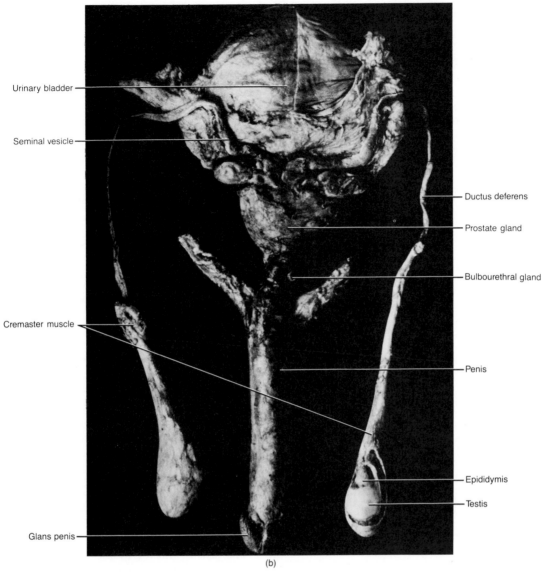

Urinary bladder

Seminal vesicle

Ductus deferens

Prostate gland

Bulbourethral gland

Cremaster muscle

Penis

Epididymis

Testis

Glans penis

(b)

FIGURE 25-10 (*Continued*) Male reproductive organs in relation to surrounding structures seen in posterior view. (b) Photograph. (Courtesy of C. Yokochi and J. W. Rohen, *Photographic Anatomy of the Human Body,* 2nd ed., 1979, IGAKU-SHOIN, Ltd., Tokyo, New York.)

7. pH. A rise in pH could indicate prostatitis.

8. Fructose. This sugar is present in a normal ejaculate. Its absence indicates obstruction or congenital absence of the ejaculatory ducts or seminal vesicles.

A normal semen analysis does not guarantee fertility; the absence of spermatozoa or zero motility are the only definitive signs of sterility.

PENIS

The **penis** is used to introduce spermatozoa into the vagina (Figure 25-11). The penis is cylindrical in shape and consists of a body, root, and glans penis. The **body** of the penis is composed of three cylindrical masses of tissue, each bound by fibrous tissue (**tunica albuginea**). The two dorsolateral masses are called the **corpora cavernosa penis.** The smaller midventral mass, the **corpus spongiosum penis,** contains the spongy urethra. All three masses are enclosed by fascia and loose-fitting skin and consist of erectile tissue permeated by blood sinuses. Under the influence of sexual stimulation, the arteries supplying the penis dilate, and large quantities of blood enter the blood sinuses. Expansion of these spaces compresses the veins draining the penis so most entering blood is retained. These vascular changes result in an **erection,** a parasympathetic reflex. The penis returns to its flaccid state when the arteries constrict and pressure on the veins is relieved. During ejaculation, a sympathetic reflex, the smooth mus-

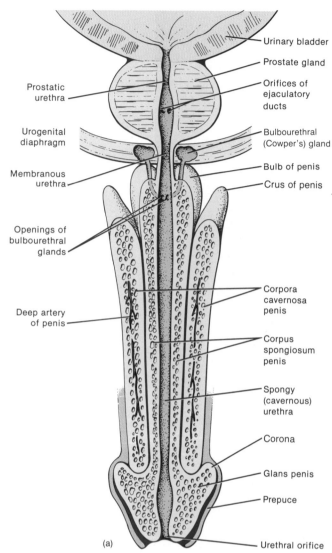

(a)

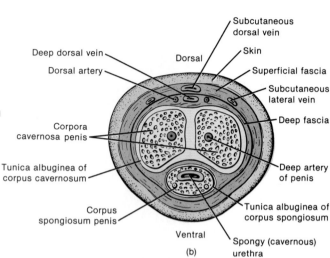

(b)

FIGURE 25-11 Internal structure of the penis. (a) Coronal section. (b) Cross section.

CLINICAL APPLICATION

Circumcision (*circumcido* = to cut around) is a surgical procedure in which part or all of the prepuce is removed. It is usually performed by the third or fourth day after birth (or on the eighth day as part of a Jewish religious rite).

The penis has a very rich blood supply from the internal pudendal artery and the femoral artery. The veins drain into corresponding vessels.

The sensory nerves to the penis are branches from the pudendal and ilioinguinal nerves. The corpora have a sympathetic and parasympathetic supply. As a result of parasympathetic stimulation, the blood vessels dilate and the muscles contract. The result is that blood is trapped within the penis and erection is maintained. The musculature of the penis is supplied by the pudendal nerve. The muscles include the bulbocavernosus muscle, which overlies the bulb of the penis, the ischiocavernosus muscles on either side of the penis, and the superficial transverse perineus muscles on either side of the bulb of the penis (see Figures 10-14 and 25-20a).

FEMALE REPRODUCTIVE SYSTEM

The female organs of reproduction include the ovaries, which produce ova (eggs); the uterine (Fallopian) tubes, which transport the ova to the uterus (womb); the vagina; and external organs that constitute the vulva, or pudendum (Figure 25-12). The mammary glands also are considered part of the female reproductive system.

cle sphincter at the base of the urinary bladder is closed because of the higher pressure in the urethra caused by expansion of the corpus spongiosum penis. Thus urine is not expelled during ejaculation and semen does not enter the urinary bladder.

The **root** of the penis is the attached portion and consists of the **bulb of the penis,** the expanded portion of the base of the corpus spongiosum penis, and the **crura of the penis,** the separated and tapered portion of the corpora cavernosa penis. The bulb of the penis is attached to the inferior surface of the urogenital diaphragm and enclosed by the bulbocavernosus muscle. Each crus of the penis is attached to the ischial and pubic rami and surrounded by the ischiocavernosus muscle.

The distal end of the corpus spongiosum penis is a slightly enlarged region called the **glans penis,** which means shaped like an acorn. The margin of the glans penis is referred to as the **corona.** Covering the glans is the loosely fitting **prepuce** (PRĒ-pyoos), or **foreskin.**

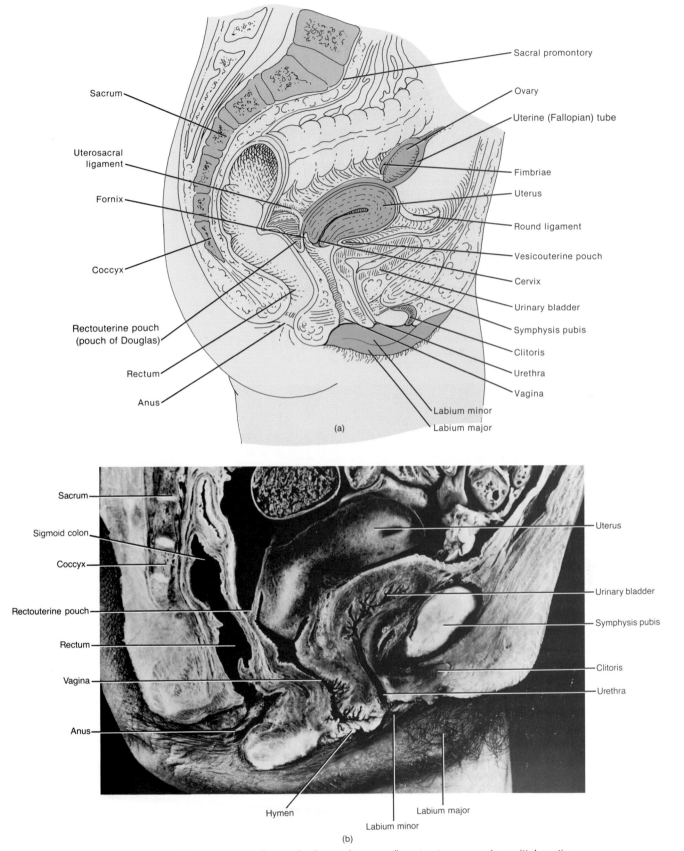

FIGURE 25-12 Female organs of reproduction and surrounding structures seen in sagittal section. (a) Diagram. (b) Photograph. (Courtesy of C. Yokochi and J. W. Rohen, *Photographic Anatomy of the Human Body,* 2nd ed., 1979, IGAKU-SHOIN, Ltd., Tokyo, New York.

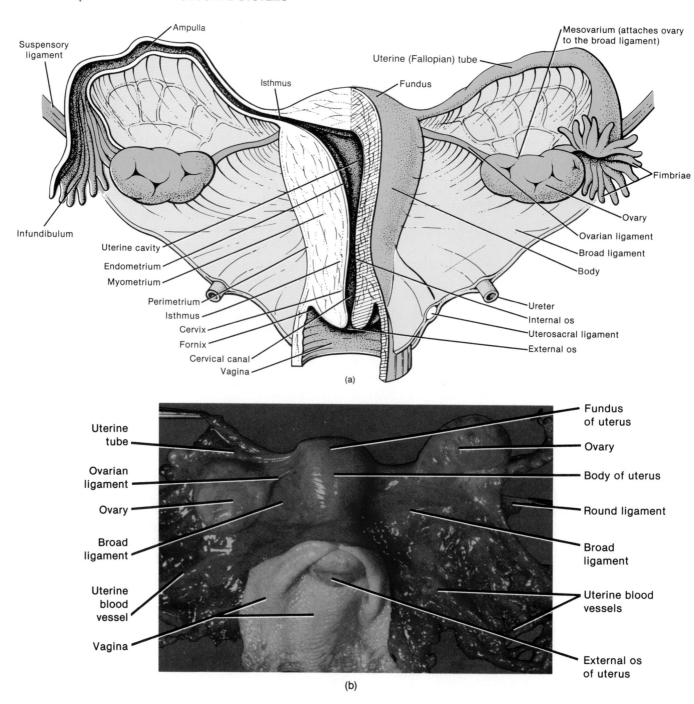

FIGURE 25-13 Uterus and associated structures seen in anterior view. (a) Diagram. The left side of the figure has been sectioned to show internal structures. (b) Photograph. (Courtesy of C. Yokochi and J. W. Rohen, *Photographic Anatomy of the Human Body,* 2nd ed., 1979, IGAKU-SHOIN, Ltd., Tokyo, New York.)

The specialized branch of medicine that deals with the diagnosis and treatment of diseases of the female reproductive system is called **gynecology** (gi'-ne-KOL-ō-jē; *gyneco* = woman).

OVARIES

The **ovaries,** or female gonads, are paired glands resembling unshelled almonds in size and shape. They descend to the brim of the pelvis during the third month of development. They are positioned in the upper pelvic cavity, one on each side of the uterus. The ovaries are maintained in position by a series of ligaments (Figure 25-13). They are attached to the broad ligament of the uterus, which is itself part of the parietal peritoneum, by a double-layered fold of peritoneum called the **mesovarium,** which

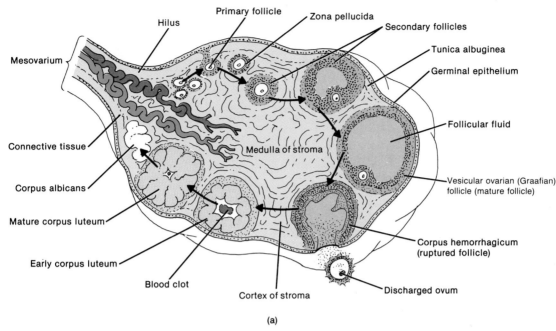

(a)

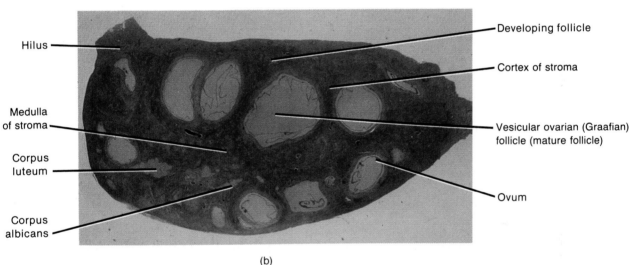

(b)

FIGURE 25-14 Histology of the ovary. (a) Diagram of the parts of an ovary seen in sectional view. The arrows indicate the sequence of developmental stages that occur as part of the ovarian cycle. (b) Photograph of an ovary in sectional view. (Courtesy of C. Yokochi and J. W. Rohen, *Photographic Anatomy of the Human Body,* 2nd ed., 1979, IGAKU-SHOIN, Ltd., Tokyo, New York.)

surrounds the ovary and ovarian ligament. The ovaries are anchored to the uterus by the **ovarian ligament** and are attached to the pelvic wall by the **suspensory ligament.** Each ovary also contains a **hilus,** the point of entrance for blood vessels and nerves.

The microscope reveals that each ovary consists of the following parts (Figure 25-14):

1. Germinal epithelium. A layer of simple cuboidal epithelium that covers the free surface of the ovary and serves as a source of ovarian follicles.

2. Tunica albuginea. A capsule of collagenous connective tissue immediately deep to the germinal epithelium.

3. Stroma. A region of connective tissue deep to the tunica albuginea and composed of an outer, dense layer called the *cortex* and an inner, loose layer known as the *medulla.* The cortex contains ovarian follicles.

4. Ovarian follicles. Ova and their surrounding tissues in various stages of development.

5. Vesicular ovarian (Graafian) follicle. An endocrine gland made up of a mature ovum and its surrounding tissues. The follicle secretes hormones called estrogens.

6. Corpus luteum. Glandular body that develops from a vesicular ovarian follicle after extrusion of an ovum (ovulation). The corpus luteum produces the hormones progesterone, estrogens, and relaxin.

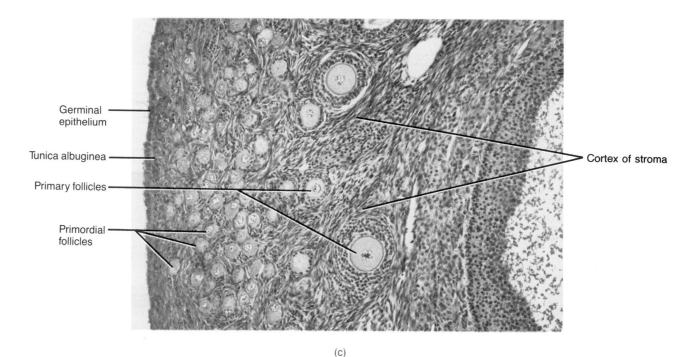

Germinal epithelium

Tunica albuginea

Primary follicles

Primordial follicles

Cortex of stroma

(c)

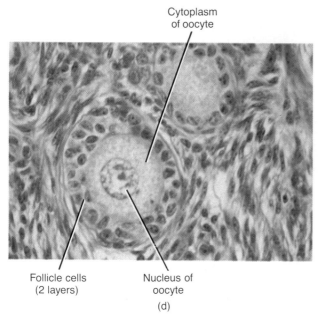

Cytoplasm of oocyte

Follicle cells (2 layers)

Nucleus of oocyte

(d)

FIGURE 25-14 (*Continued*) Histology of the ovary. (c) Photomicrograph of the cortex of the stroma of an ovary at a magnification of 60×. (d) Photomicrograph of an enlarged aspect of a primary follicle at a magnification of 300×.

The ovarian blood supply is furnished by the ovarian arteries, which anastomose with branches of the uterine arteries. The ovaries are drained by the ovarian veins. On the right side, they drain into the inferior vena cava, and on the left side, they drain into the renal vein.

Sympathetic and parasympathetic nerve fibers to the ovaries are said to terminate on the blood vessels and not enter the substance of the ovaries.

The ovaries produce ova, discharge ova (ovulation), and secrete the female sex hormones progesterone, estrogens, and relaxin. They are analogous to the testes of the male reproductive system.

Oogenesis

The formation of a haploid ovum by meiosis in the ovary is referred to as oogenesis. With some exceptions, oogenesis occurs in essentially the same manner as spermatogenesis. It involves meiosis and maturation.

• *Reduction Division*   The precursor cell in oogenesis is a diploid cell called the **oogonium** (ō′-o-GŌ-nē-um), or egg mother cell (Figure 25-15). Early in the prenatal development of the ovaries, oogonia proliferate by mitosis. But, as development continues, oogonia in the primary follicles lose their ability to carry on mitosis. At about the third month of prenatal development, oogonia develop into larger cells also containing the diploid number of chromosomes and called **primary oocytes** (Ō-o-sītz). They remain in this stage until their follicular cells respond to FSH from the anterior pituitary, which in turn has responded to GnRF from the hypothalamus.

Starting with puberty, several follicles respond each month to the rising level of FSH. As the preovulatory phase of the menstrual cycle proceeds and LH is secreted from the anterior pituitary, one of the follicles reaches a stage in which the diploid primary oocyte undergoes reduction division. Synapsis, tetrad formation, and crossing-over occur, and two cells of unequal size, both with 23 chromosomes of two chromatids each, are produced. The smaller cell, called the **first polar body,** is essentially a packet of discarded nuclear material. The larger cell,

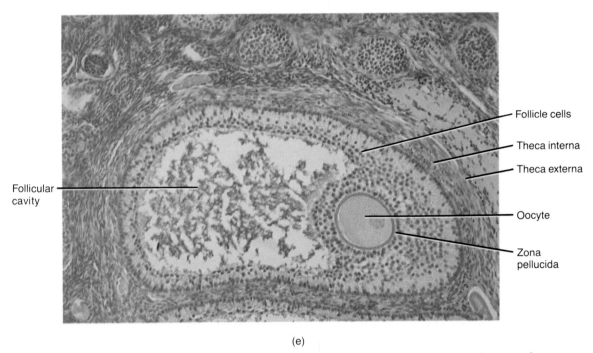

Follicle cells

Theca interna

Theca externa

Follicular cavity

Oocyte

Zona pellucida

(e)

FIGURE 25-14 (*Continued*) Histology of the ovary. (e) Photomicrograph of an enlarged aspect of a secondary follicle at a magnification of 160×. The theca interna and theca externa are connective tissue coverings around the secondary follicle. (Photomicrographs © 1983 by Michael H. Ross. Used by permission.)

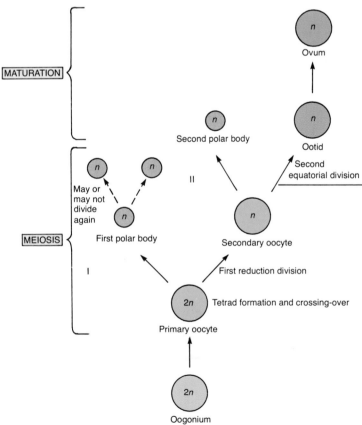

MATURATION

MEIOSIS

n
Ovum

n
Second polar body

n
Ootid

Second equatorial division

Fertilization

Ovulation

n   *n*

May or may not divide again

n
First polar body

II

n
Secondary oocyte

I

First reduction division

2*n*   Tetrad formation and crossing-over

Primary oocyte

2*n*
Oogonium

FIGURE 25-15 Oogenesis.

known as the **secondary oocyte,** receives most of the cytoplasm.

• *Equatorial Division*   At ovulation, the secondary oocyte with its polar body and surrounding supporting cells is discharged. The discharged secondary oocyte enters the uterine tube and, if spermatozoa are present and fertilization occurs, the second division takes place: the equatorial division.

• *Maturation*   The secondary oocyte produces two cells of unequal size, both of them haploid. The larger cell is called an **ootid** and eventually develops into an **ovum,** or mature egg; the smaller is the **second polar body.**

The first polar body may undergo another division. If it does, meiosis of the primary oocyte results in a single haploid ovum and three polar bodies. In any event, all polar bodies disintegrate. Thus each oogonium produces a single ovum, whereas each spermatocyte produces four spermatozoa.

UTERINE TUBES

The female body contains two **uterine (Fallopian) tubes** that extend laterally from the uterus and transport the ova from the ovaries to the uterus (see Figure 25-13). Measuring about 10 cm (4 in) long, the tubes are positioned between the folds of the broad ligaments of the uterus. The funnel-shaped open distal end of each tube, called the **infundibulum,** lies close to the ovary but is not attached to it and is surrounded by a fringe of finger-like projections called **fimbriae** (FIM-brē-ē). From the infundibulum the uterine tube extends medially and inferiorly and attaches to the superior lateral angle of the uterus. The **ampulla** of the uterine tube is the widest, longest portion, making up about two-thirds of its length. The **isthmus** of the uterine tube is the short, narrow, thick-walled portion that joins the uterus.

Histologically, the uterine tubes are composed of three layers. The internal **mucosa** contains ciliated columnar cells and secretory cells, which are believed to aid the movement and nutrition of the ovum. The middle layer, the **muscularis,** is composed of a thick, circular region of smooth muscle and an outer, thin, longitudinal region of smooth muscle. Peristaltic contractions of the muscularis and the ciliated action of the mucosa help move the ovum down into the uterus. The outer layer of the uterine tubes is a serous membrane, the **serosa.**

About once a month an immature ovum ruptures from the surface of the ovary near the infundibulum of the uterine tube, a process called **ovulation.** The ovum is swept into the tube by the ciliary action of the epithelium of the infundibulum. It is then moved along the tube by ciliary action supplemented by the peristaltic contractions of the muscularis. If the ovum is fertilized by a sperm cell, it usually occurs in the ampulla of the uterine tube. Fertilization may occur at any time up to about 24 hours following ovulation. The fertilized ovum, now referred to as a blastocyst, descends into the uterus within 7 days. An unfertilized ovum disintegrates.

CLINICAL APPLICATION

Ectopic (*ektopos* = displaced) **pregnancy (EP)** refers to the development of an embryo or fetus outside the uterine cavity. The majority occur in the uterine tube, usually in the ampullar and infundibular portions. Some occur in the ovaries, abdomen, uterine cervix, and broad ligaments. The basic cause of a tubal pregnancy is impaired passage of the fertilized ovum through the uterine tube related to factors such as pelvic inflammatory disease (PID), previous uterine tube surgery, previous ectopic pregnancy, repeated elective abortions, pelvic tumors, and developmental abnormalities. Ectopic pregnancy may be characterized by one or two missed periods, followed by vaginal bleeding and acute pelvic pain.

The uterine tubes are supplied by branches of the uterine and ovarian arteries. Venous return is via the uterine veins.

The uterine tubes are supplied with sympathetic and parasympathetic nerve fibers from the hypogastric plexus and the pelvic splanchnic nerves. The fibers are distributed to the muscular coat of the tubes and their blood vessels.

UTERUS

The site of menstruation, implantation of a fertilized ovum, development of the fetus during pregnancy, and labor is the **uterus.** Situated between the urinary bladder and the rectum, the uterus is shaped like an inverted pear (see Figures 25-12 and 25-13). Before the first pregnancy, the adult uterus measures approximately 7.5 cm (3 in) long, 5 cm (2 in) wide, and 2.5 cm (1 in) thick.

Anatomical subdivisions of the uterus include the dome-shaped portion above the uterine tubes called the **fundus,** the major tapering central portion called the **body,** and the inferior narrow portion opening into the vagina called the **cervix.** Between the body and the cervix is the **isthmus** (IS-mus), a constricted region about 1 cm (½ in) long. The interior of the body of the uterus is called the **uterine cavity,** and the interior of the narrow cervix is called the **cervical canal.** The junction of the isthmus with the cervical canal is the **internal os.** The **external os** is the place where the cervix opens into the vagina.

CLINICAL APPLICATION

Early diagnosis of **cancer of the uterus** is accomplished by the *Papanicolaou* (pap'-a-NIK-ō-la-oo) *test,* or Pap smear. In this generally painless procedure, a few cells from the part of the vagina surrounding the cervix and the cervix are removed with a swab and examined microscopically. Malignant cells have a characteristic appearance and indicate an early stage of cancer, even before symptoms occur. Estimates indicate that the Pap smear is more than 90 percent reliable in detecting cancer of the cervix.

To rule out invasive carcinoma, a *cone biopsy* of the cervix is performed. A cone biopsy is a hospital procedure in which an inverted cone of tissue is excised.

It requires an anesthetic and is usually done only when abnormal cells have been detected. In another procedure, *punch biopsy* is combined with an *endocervical curettage* (ku-re-TAZH), or *ECC;* this combination has a high degree of diagnostic accuracy. In a punch biopsy, a disc or segment of tissue is excised. Curetage is a procedure in which the cervix is dilated and the endometrium (lining) of the uterus is scraped with a spoon-shaped instrument called a curette. This procedure is commonly called a *D and C.* If the carcinoma has spread beyond the lining, treatment may involve complete or partial removal of the uterus, called a *hysterectomy,* or radiation treatment.

Normally the uterus is flexed between the uterine body and the cervix. This is called **anteflexion.** In this position, the body of the uterus projects anteriorly and slightly superiorly over the urinary bladder, and the cervix projects inferiorly and posteriorly and enters the anterior wall of the vagina at nearly a right angle. Several structures that are either extensions of the parietal peritoneum or fibromuscular cords, referred to as ligaments, maintain the position of the uterus. The paired **broad ligaments** are double folds of parietal peritoneum attaching the uterus to either side of the pelvic cavity. Uterine blood vessels and nerves pass through the broad ligaments. The paired **uterosacral ligaments,** also peritoneal extensions, lie on either side of the rectum and connect the uterus to the sacrum. The **cardinal (lateral cervical) ligaments** extend below the bases of the broad ligaments between the pelvic wall and the cervix and vagina. These ligaments contain smooth muscle, uterine blood vessels, and nerves and are the chief ligaments that maintain the position of the uterus and help to keep it from dropping down into the vagina. The **round ligaments** are bands of fibrous connective tissue between the layers of the broad ligament. They extend from a point on the uterus just below the uterine tubes to a portion of the external genitalia.

CLINICAL APPLICATION

Although the ligaments normally maintain the anteflexed position of the uterus, they also afford the uterine body some movement. As a result, the uterus may become malpositioned. A posterior tilting of the uterus is called **retroflexion.**

Histologically, the uterus consists of three layers of tissue. The outer layer, the **perimetrium,** or **serosa,** is part of the visceral peritoneum. Laterally, it becomes the broad ligament. Anteriorly, it is reflected over the urinary bladder and forms a shallow pouch, the **vesicouterine** (ves'-i-kō-YOO-ter-in) **pouch** (see Figure 25-12). Posteriorly, it is reflected onto the rectum and forms a deep pouch, the **rectouterine** (rek-tō-YOO-ter-in) **pouch (pouch of Douglas)**—the lowest point in the pelvic cavity.

The middle layer of the uterus, the **myometrium,** forms the bulk of the uterine wall (Figure 25-16). This layer consists of three layers of smooth muscle fibers and is thickest in the fundus and thinnest in the cervix. During childbirth, coordinated contractions of the muscles help expel the fetus from the body of the uterus.

The inner layer of the uterus, the **endometrium,** is a mucous membrane composed of two principal layers. The *stratum functionalis,* the layer closer to the uterine cavity, is shed during menstruation. The second layer, the *stratum basalis* (bā-SAL-is), is permanent and produces a new functionalis following menstruation. The endometrium contains numerous glands.

CLINICAL APPLICATION

More and more physicians are beginning to use **colposcopy** (kol-POS-ko-pē) to evaluate the status of the mucosa of the vagina and cervix. Colposcopy is the direct examination of the vaginal and cervical mucosae with a magnifying device (a colposcope) similar to a low-power binocular microscope. Various instruments that magnify the mucous membrane from about 6 to 40 times its actual size are commercially available. The application of a 3 percent solution of acetic acid removes mucus and enhances the appearance of mucosal columnar epithelium.

Blood is supplied to the uterus by branches of the internal iliac artery called *uterine arteries.* Branches called *arcuate arteries* are arranged in a circular fashion in the myometrium and give off *radial arteries* that penetrate deeply into the myometrium (Figure 25-17). Just before the branches enter the endometrium, they divide into two kinds of arterioles. The *straight arteriole* terminates in the basalis and supplies it with the materials necessary to regenerate the functionalis. The *spiral arteriole* penetrates the functionalis and changes markedly during the menstrual cycle. The uterus is drained by the *uterine veins.*

Sympathetic and parasympathetic fibers are supplied to the uterus via the hypogastric and pelvic plexuses. Both sets of nerves terminate on the uterine vessels. The myometrium is believed to be innervated by the sympathetic fibers alone.

VAGINA

The **vagina** serves as a passageway for the menstrual flow. It is also the receptacle for the penis during coitus, or sexual intercourse, and the lower portion of the birth canal. It is a muscular, tubular organ lined with mucous membrane and measures about 10 cm (4 in) in length, extending from the cervix to the vestibule (see Figures

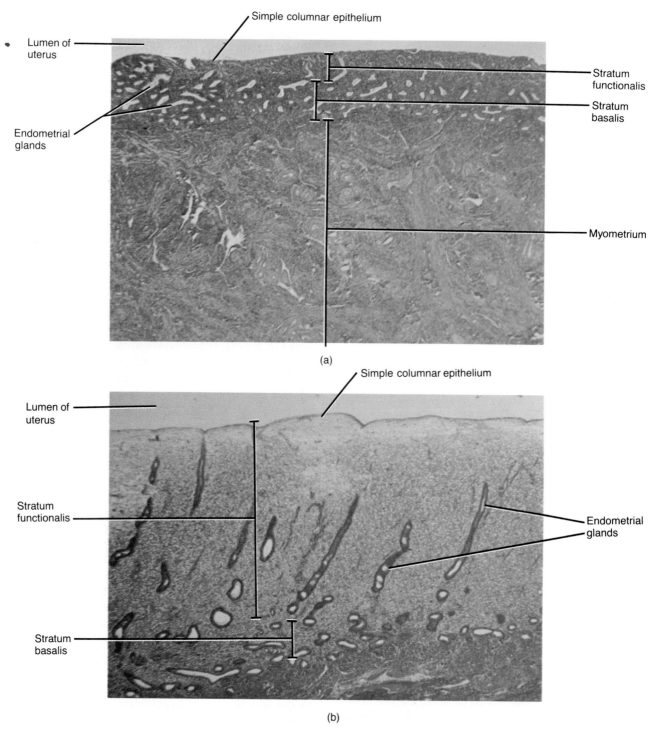

FIGURE 25-16 Histology of the uterus. (a) Photomicrograph of a portion of the uterine wall in which the stratum functionalis is in an early stage of proliferation at a magnification of 25×. (b) Photomicrograph of a portion of the uterine wall in which the stratum functionalis is in an advanced stage of proliferation at a magnification of 25×. Note the increased thickness of the stratum functionalis and the long, straight nature of the glands. (© 1983 by Michael H. Ross. Used by permission.)

25-12 and 25-13). Situated between the urinary bladder and the rectum, it is directed superiorly and posteriorly, where it attaches to the uterus. A recess, called the **fornix,** surrounds the vaginal attachment to the cervix. The dorsal recess, called the posterior fornix, is deeper than the ven-

tral and two lateral fornices. The fornices make possible the use of contraceptive diaphragms.

Histologically, the mucosa of the vagina is continuous with that of the uterus and consists of stratified squamous epithelium and connective tissue that lies in a series of

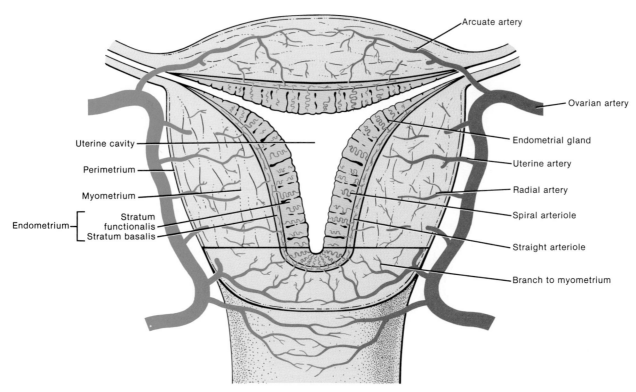

FIGURE 25-17 Blood supply of the uterus.

transverse folds, the **rugae** (Figure 25–18). Most of the lubrication during sexual intercourse is the result of a mucoid secretion produced by the vaginal epithelium. The muscularis is composed of longitudinal smooth muscle that can stretch considerably. This distension is important because the vagina receives the penis during sexual intercourse and serves as the lower portion of the birth canal. At the lower end of the vaginal opening, the **vaginal orifice,** is a thin fold of vascularized mucous membrane called the **hymen,** which forms a border around the orifice, partially closing it (see Figure 25-19).

CLINICAL APPLICATION

Sometimes the hymen completely covers the orifice, a condition called **imperforate** (im-PER-fō-rāt) **hymen.** Surgery is required to open the orifice to permit the discharge of the menstrual flow.

The mucosa of the vagina contains large amounts of glycogen, which on decomposition produces organic acids. These acids create a low pH environment that retards microbial growth. However, the acidity is also injurious to sperm cells. Semen neutralizes the acidity of the vagina to ensure survival of the sperm.

The blood supply of the vagina is derived from branches of the internal iliac arteries: vaginal, vaginal branch of the uterine, internal pudendal, and vaginal branches of the middle rectal arteries. The veins form vaginal venous plexuses along the sides of the vagina and are drained through their vaginal veins into the internal iliac veins.

The nerves of the vagina are derived from the uterovaginal plexus at the base of the broad ligament. Lower nerve fibers from the plexus supply the cervix and superior vagina; vaginal nerves supply the wall of the vagina.

VULVA

The term **vulva** (VUL-va), or **pudendum** (pyoo-DEN-dum), is a collective designation for the external genitalia of the female (Figure 25-19). Its components are as follows.

The **mons pubis (veneris),** an elevation of adipose tissue covered by coarse pubic hair, is situated over the symphysis pubis. It lies anterior to the vaginal and urethral openings. From the mons pubis, two longitudinal folds of skin, the **labia majora** (LĀ-bē-a ma-JŌ-ra), extend inferiorly and posteriorly. The labia majora, the female homologue of the scrotum, contain an abundance of adipose tissue and sebaceous (oil) and sudoriferous (sweat) glands; they are covered by pubic hair on their upper outer surfaces. Medial to the labia majora are two folds of mucous membrane called the **labia minora** (MĪ-nō-ra). Unlike the labia majora, the labia minora are devoid of pubic hair and fat and have few sudoriferous glands. They do, however, contain numerous sebaceous glands.

The **clitoris** (KLI-to-ris) is a small, cylindrical mass of erectile tissue and nerves. It is located at the anterior

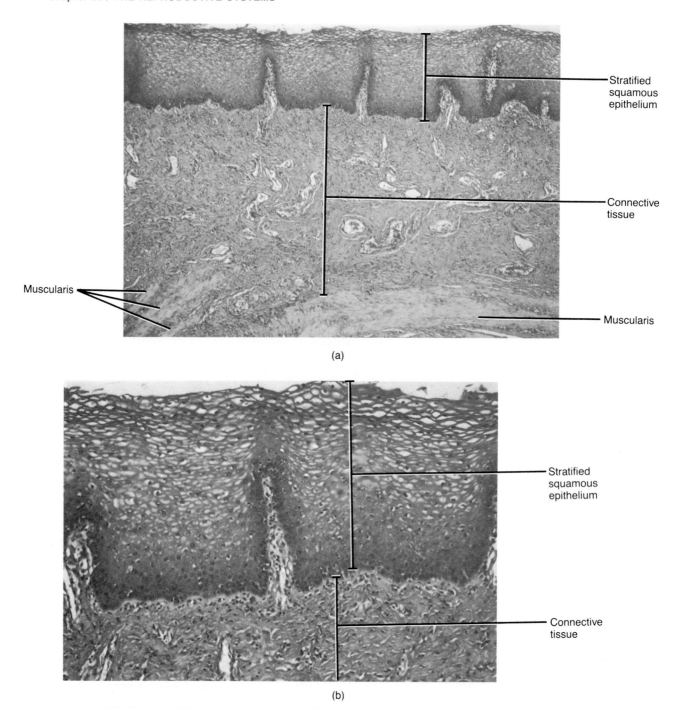

(a)

(b)

FIGURE 25-18 Histology of the vagina. (a) Photomicrograph of a portion of the wall of the vagina at a magnification of 50×. (b) Photomicrograph of an enlarged aspect of the mucosa of the vagina at a magnification of 125×. (© 1983 by Michael H. Ross. Used by permission.)

junction of the labia minora. A layer of skin called the **prepuce** (foreskin) is formed at the point where the labia minora unite and covers the body of the clitoris. The exposed portion of the clitoris is the **glans.** The clitoris is homologous to the penis of the male. Homologous means that two organs correspond in structure, position, and origin. Like the penis, the clitoris is capable of en-

largement upon tactile stimulation and assumes a role in sexual excitement of the female.

The cleft between the labia minora is called the **vestibule.** Within the vestibule are the hymen, vaginal orifice, urethral orifice, and the openings of several ducts. The **vaginal orifice** occupies the greater portion of the vestibule and is bordered by the hymen. Anterior to the vaginal

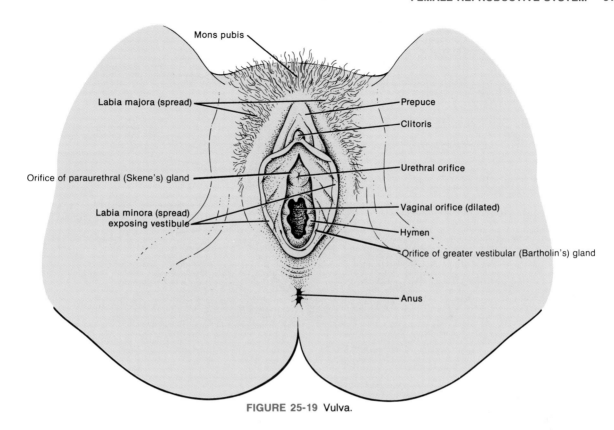

FIGURE 25-19 Vulva.

orifice and posterior to the clitoris is the **urethral orifice.** On either side of the urethral orifice are the openings of the ducts of the **paraurethral (Skene's) glands.** These glands secrete mucus. On either side of the vaginal orifice itself are the **greater vestibular (Bartholin's) glands.** These glands open by ducts into a groove between the hymen and labia minora and produce a mucoid secretion that supplements lubrication during sexual intercourse. A number of **lesser vestibular glands,** whose orifices are microscopic, open into the vestibule. The paraurethral glands are homologous to the male prostate. The greater vestibular glands are homologous to the male bulbourethral glands.

CLINICAL APPLICATION

One important sign in the **diagnosis of pregnancy** is a bluish discoloration of the vulva and vagina due to venous congestion. The discoloration appears at about the eighth to twelfth week and increases in intensity as the pregnancy progresses.

PERINEUM

The **perineum** (per'-i-NĒ-um) is the diamond-shaped area at the inferior end of the trunk between the thighs and buttocks of both males and females (Figure 25-20). It is bounded anteriorly by the symphysis pubis, laterally by the ischial tuberosites, and posteriorly by the coccyx. A transverse line drawn between the ischial tuberosities divides the perineum into an anterior **urogenital** (yoo'-rō-JEN-i-tal) **triangle** that contains the external genitalia and a posterior **anal triangle** that contains the anus.

CLINICAL APPLICATION

In the female, the region between the vagina and anus is known as the **clinical perineum.** If the vagina is too small to accommodate the head of an emerging fetus, the skin, vaginal epithelium, subcutaneous fat, and superficial transverse perineal muscle of the clinical perineum may tear. Moreover, the tissues of the rectum may be damaged. To avoid this, a small incision called an **episiotomy** (e-piz'-ē-OT-ō-mē) is made in the perineal skin and underlying tissues just prior to delivery. After delivery the episiotomy is sutured in layers.

MAMMARY GLANDS

The **mammary glands** are modified sweat glands (branched tubuloalveolar) that lie over the pectoralis major and serratus anterior muscles and are attached to them by a layer of connective tissue (Figure 25-21). Internally, each mammary gland consists of 15 to 20 **lobes,** or compartments, separated by adipose tissue. The amount of adipose tissue determines the size of the breasts.

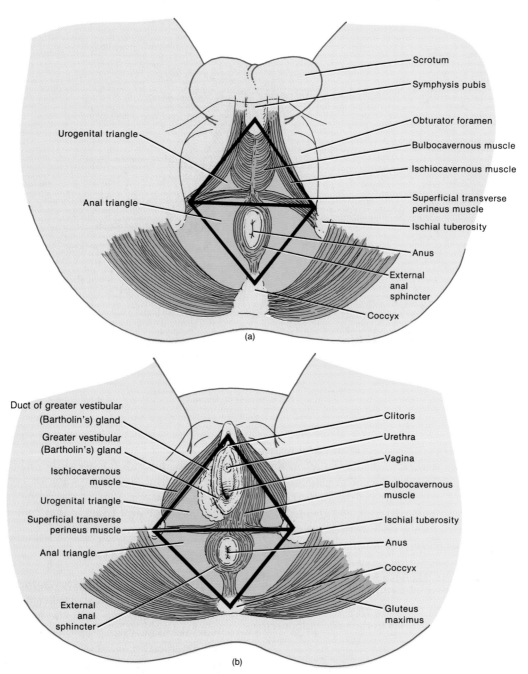

FIGURE 25-20 Perineum. Borders seen in the (a) male and (b) female.

However, breast size has nothing to do with the amount of milk produced. In each lobe are several smaller compartments called **lobules,** composed of connective tissue in which milk-secreting cells referred to as **alveoli** are embedded (Figure 25-22). Alveoli are arranged in grapelike clusters. Between the lobules are strands of connective tissue called the **suspensory ligaments of the breast (Cooper's ligaments).** These ligaments run between the skin and deep fascia and support the breast. Alveoli convey the milk into a series of **secondary tubules.** From here the milk passes into the **mammary ducts.** As the mammary ducts approach the nipple, they expand to form sinuses called **ampullae** (am-POOL-ē), where milk may be stored.

The ampullae continue as **lactiferous ducts** that terminate in the **nipple.** Each lactiferous duct conveys milk from one of the lobes to the exterior, although some may join before reaching the surface. The circular pigmented area of skin surrounding the nipple is called the **areola** (a-RĒ-ō-la). It appears rough because it contains modified sebaceous glands.

The essential function of the mammary glands is milk secretion and ejection, together called **lactation.** The secretion of milk is due largely to the hormone prolactin (PRL), with contributions from progesterone and estrogens. The ejection of milk occurs in the presence of oxytocin (OT).

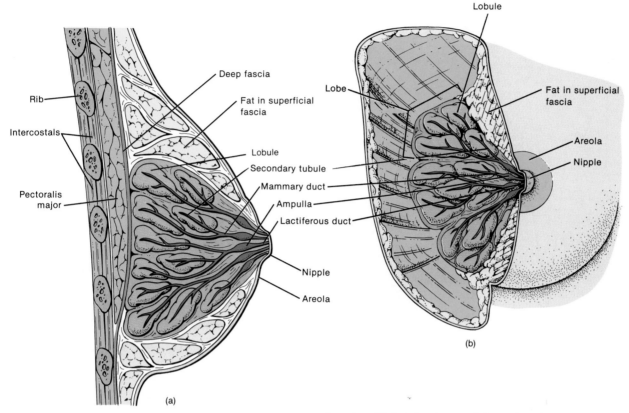

FIGURE 25-21 Mammary glands. (a) Sagittal section. (b) Anterior view, partially sectioned.

CLINICAL APPLICATION

Early detection—especially by breast self-examination and mammography—is still the most promising method to increase the survival rate for **breast cancer.** It is estimated that 95 percent of breast cancer is first detected by the women themselves. Each month after the menstrual period the breasts should be thoroughly examined for lumps, puckering of the skin, or discharge.

The most effective screening technique for routinely detecting tumors less than 1.27 cm (½ in) in diameter is *mammography.* One reason that mammography is so useful is that it can detect small calcium deposits, called microcalcifications, in breast tissue. Such calcifications frequently indicate the presence of a tumor. The mammographic image, called a mammogram, can be obtained in either of two ways. In one procedure, x rays are beamed onto a specially coated metal plate and the blue-on-white image produced provides the physician with detail of thicker as well as thinner portions of the breast. This procedure is called *xeroradiography.* In the other procedure, called *film-screen mammography,* the x rays are beamed onto a fluorescent screen, which exposes a film in contact with it. It uses a very low dose of radiation and can be adjusted to emphasize different kinds of tissue within the breast.

One of the most recent breast cancer detecting procedures is *ultrasound.* The procedure is performed while the patient lies on her stomach in a specially designed hospital bed with her breasts immersed in a tank of water. Ultrasound produces images using a device that first emits a pulse of high-frequency sound and then records the echo on a monitor. Although ultrasound can neither detect microcalcifications or tumors less than 1 cm in diameter, it can be used to determine whether a lump is a benign cyst or a malignant tumor.

Another technique combines computed tomography with mammography (*CT/M*) and appears to overcome some of the limitations of mammography. The procedure is based upon the fact that breast carcinoma has an abnormal affinity for iodide. CT scans are made before and after the rapid intravenous infusion of an iodide contrast material. Comparison of the initial density of a suspected lesion with the density following infusion of the iodide gives an indication of the status of the tumor. CT/M affords definitive diagnostic help in instances where the mammographic and physical examinations are inconclusive and appears to be a significantly improved method of breast cancer diagnosis.

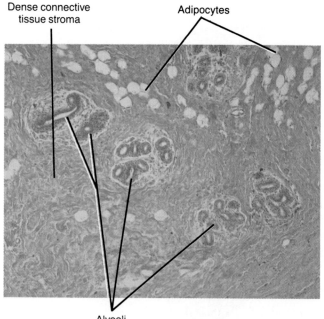

Dense connective tissue stroma

Adipocytes

Alveoli

(a)

FIGURE 25-22 Histology of the mammary glands. (a) Photomicrograph of several alveoli in a resting (inactive) mammary gland at a magnification of 60×. (b) Photomicrograph of several alveoli during late pregnancy at a magnification of 160×. (© 1983 by Michael H. Ross. Used by permission.)

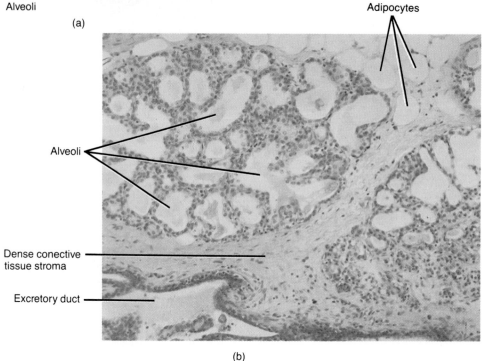

Adipocytes

Alveoli

Dense conective tissue stroma

Excretory duct

(b)

Long-term studies are now underway to determine the effectiveness of an experimental home device for breast cancer detection. It is called a *breast cancer screening indicator.* It is composed of two plastic discs that contain heat-sensitive chemicals. The discs are worn inside a female's brassiere for 15 minutes each month so that she can observe color changes that could indicate a breast abnormality. The screening device is said to detect tumors as small as 2 mm in diameter, their size about 2 to 10 years before they become palpable or can be seen by mammogram.

SEXUAL INTERCOURSE

Sexual intercourse, or **copulation** (in humans, called **coitus**), is the process by which spermatozoa are deposited in the vagina.

CLINICAL APPLICATION

Artificial insemination (*in* = into; *seminatus* = sown; *semen* = seed) refers to the deposition of seminal fluid within the vagina or cervix by artificial means. If the husband's seminal fluid is used, the process is known as **homologous insemination.** Such a procedure might be carried out if there is a developmental anomaly that prevents placement of the penis in the vagina or normal ejaculation. A small volume of seminal fluid is also an indication for homologous insemination. If seminal fluid from a donor is used, the process is called **heterologous insemination.** In such cases, an anonymous donor is selected on the basis of race, blood type compatibility, physical appearance, general health, and genetic background.

MALE SEXUAL ACT

Erection

The male role in the sexual act starts with **erection,** the enlargement and stiffening of the penis. An erection may be initiated in the cerebrum by stimuli such as anticipation, memory, and visual sensation, or it may be a reflex brought on by stimulation of the touch receptors in the penis, especially in the glans. In any case, parasympathetic impulses that pass from the sacral portion of the spinal cord to the penis cause dilation of the arteries of the penis, allowing blood to fill the cavernous spaces of the spongy bodies.

Lubrication

Parasympathetic impulses from the sacral cord also cause the bulbourethral (Cowper's) glands to secrete mucus, which affords some **lubrication** for intercourse. The mucus flows through the urethra. The major portion of lubricating fluid is produced by the female. Without satisfactory lubrication, the male sexual act is difficult since unlubricated intercourse causes pain impulses that inhibit rather than promote coitus.

Orgasm

Tactile stimulation of the penis brings about emission and ejaculation. When sexual stimulation becomes intense, rhythmic sympathetic impulses leave the spinal cord at the levels of the first and second lumbar vertebrae and pass to the genital organs. These impulses cause peristaltic contractions of the ducts in the testes, epididymides, and ductus (vas) deferens that propel spermatozoa into the urethra—a process called **emission.** Simultaneously, peristaltic contractions of the seminal vesicles and prostate expel seminal and prostatic fluid along with the spermatozoa. All these mix with the mucus of the bulbourethral glands, resulting in the fluid called semen. Other rhythmic impulses sent from the spinal cord at the levels of the first and second sacral vertebrae reach the skeletal muscles at the base of the penis, and the penis expels the semen from the urethra to the exterior. The propulsion of semen from the urethra to the exterior constitutes an **ejaculation.** A number of sensory and motor activities accompany ejaculation, including a rapid heart rate, an increase in blood pressure, an increase in respiration, and pleasurable sensations. These activities, together with the muscular events involved in ejaculation, are referred to as an **orgasm.**

FEMALE SEXUAL ACT

Erection

The female role in the sex act, like that of the male, also involves erection, lubrication, and orgasm. Stimulation of the female, as in the male, depends on both psychic and tactile responses. Under appropriate conditions, stimulation of the female genitalia, especially the clitoris, results in **erection** and widespread sexual arousal. This response is controlled by parasympathetic impulses sent from the sacral spinal cord to the external genitalia.

Lubrication

Parasympathetic impulses from the sacral spinal cord also cause the bulk of **lubrication** of the vagina. The impulses result in the production of a mucoid fluid from the epithelium of the vaginal mucosa. Some mucus is also produced by the greater vestibular (Bartholin's) glands. As was noted earlier, lack of sufficient lubrication results in pain impulses that inhibit rather than promote coitus.

Orgasm (Climax)

When tactile stimulation of the genitalia reaches maximum intensity, reflexes are initiated that cause the female **orgasm (climax).** Female orgasm is analogous to male ejaculation and may assume a role in fertilization of an ovum. The perineal muscles contract rhythmically from spinal reflexes similar to those that occur in the male ejaculation.

BIRTH CONTROL

Methods of **birth control** include removal of the gonads and uterus, sterilization, and mechanical and chemical contraception.

REMOVAL OF GONADS AND UTERUS

Castration (removal of the testes), **hysterectomy** (removal of the uterus), and **oophorectomy** (ō'-of-ō-REK-tō-mē; removal of the ovaries) are all absolute preventive methods. Once performed, these operations cannot be reversed and it is impossible to produce offspring. However, removal of the testes or ovaries has adverse effects because of the importance of these organs in the endocrine system. Generally these operations are performed only if the organs are diseased. Castration before puberty prevents the development of secondary sex characteristics.

STERILIZATION

One means of **sterilization** of males is **vasectomy**—a simple operation in which a portion of each ductus (vas) deferens is removed. An incision is made in the scrotum, the tubes are located, and each is tied in two places. Then the portion between the ties is cut out. Sperm production can continue in the testes, but the sperm cannot reach the exterior.

Sterilization in females generally is achieved by performing a **tubal ligation** (lī-GĀ-shun). An incision is made into the abdominal cavity, the uterine tubes are squeezed, and a small loop called a knuckle is made. A suture is tied tightly at the base of the knuckle and the knuckle is then cut. After 4 or 5 days the suture is digested by body fluids and the two severed ends of the tubes separate. The ovum thus is prevented from passing to the uterus, and the sperm cannot reach the ovum. Sterilization normally does not affect sexual performance or enjoyment.

Another method of sterilizing women is the **laparoscopic** (lap'-a-rō-SKŌ-pik) **technique.** After a woman receives local or general anesthesia, a harmless gas is introduced into her abdomen to create a gas bubble. The bubble expands the abdominal cavity and pushes the intestines away from the pelvic organs, permitting safe, easy access to the uterine tubes. The doctor makes a small incision at the lower rim of the umbilicus and inserts a laparoscope to view the inside of the abdominal cavity and the uterine tubes. The tubes can be closed with this instrument or a second incision can be made at the pubic hairline to insert a cautery forceps. Once the uterine tubes are sealed, the instrument is removed, the gas is released, and the incision is covered with a bandage. After a few hours the patient can usually go home.

CONTRACEPTION

Contraception is the prevention of fertilization without destroying fertility by natural, mechanical, or chemical means.

Natural

The **natural methods** include complete or periodic abstinence. An example of periodic abstinence is the rhythm method, which takes advantage of the fact that a fertilizable ovum is available only during a period of three to five days in each menstrual cycle. During this time the couple refrains from intercourse. Its effectiveness is limited by the fact that few women have absolutely regular cycles. Moreover, some women occasionally ovulate during the "safe" times of the month, such as during menstruation.

Mechanical

Mechanical methods of contraception include the use of the condom by the male and the diaphragm by the female. The **condom** is a nonporous, elastic (rubber or similar material) covering placed over the penis that prevents deposition of sperm in the female reproductive tract. The **diaphragm** is a dome-shaped structure that fits over the cervix and is generally used in conjunction with a sperm-killing chemical. The diaphragm stops the sperm from passing into the cervix. The chemical kills the sperm cells. It has recently been learned that toxic shock syndrome (TSS) and recurrent urinary tract infections are associated with diaphragm use in some females.

Another mechanical method of contraception is an **intrauterine device (IUD).** The device is a small object made of plastic, copper, or stainless steel and shaped like a loop, coil, T, or 7. It is inserted into the cavity of the uterus. It is not clear how IUDs operate. Some investigators believe they cause changes in the uterine lining that, in turn, produce a substance which destroys either the sperm or the fertilized ovum. Some IUDs release contraceptive agents in minute amounts; one new device secretes progesterone.

Chemical

Chemical methods of contraception include spermicidal and hormonal methods. Various foams, creams, jellies, suppositories, and douches make the vagina and cervix unfavorable for sperm survival. A recent spermicidal development is a **contraceptive sponge.** It is a nonprescription, polyurethane sponge that contains a spermicide called nonoxynol-9, also used in a variety of contraceptive foams and jellies. The sponge is placed in the vagina where it releases spermicide for up to 24 hours and also acts as a physical barrier to sperm. The sponge is equal to the diaphragm in effectiveness. Some cases of toxic shock syndrome (TSS) have been reported among users of the contraceptive sponge.

The hormonal method, **oral contraception** (the pill), has found rapid and widespread use. Although several pills are available, the one most commonly used contains a high concentration of progesterone and a low concentration of estrogens. These two hormones act on the anterior pituitary to decrease the secretion of FSH and LH by inhibiting GnRF by the hypothalamus. The low levels of FSH and LH are not adequate to initiate follicle maturation and ovulation. In the absence of a mature ovum, pregnancy cannot occur.

Women for whom all oral contraceptives are contraindicated include those with a history of thromboembolic disorders (predisposition to blood clotting), cerebral blood vessel damage, hypertension, liver malfunction, heart disease, or cancer of the breast or reproductive system. About 40 percent of all pill users experience side effects—generally minor problems such as nausea, weight gain, headache, irregular menses, spotting between periods, and amenorrhea. The statistics on the life-threatening conditions associated with the pill such as blood clots, heart attacks, liver tumors, and gallbladder disease are somewhat more reassuring. For all the problems combined, fewer than 2 deaths occurred per 100,000 users under age 30, 4 among those 30 to 35, 10 among those 35 to 39, and 18 among women over 40. The major exception is that women who take the pill and smoke face far higher odds of developing heart attack and stroke than do nonsmoking pill users.

A summary of birth control methods is presented in Exhibit 25-1.

CLINICAL APPLICATION

The quest for an efficient male oral contraceptive has been disappointing until recently. An oral contraceptive, **gossypol,** which is derived from the cotton plant, has achieved a high efficacy (power to produce effects) rate of 99.9 percent in clinical tests in China. Its action is to inhibit an enzyme required for spermatogenesis. Levels of blood luteinizing hormone (LH) and testosterone remained unchanged and potency was not impaired. The number and morphology of spermatozoa gradually recover and fertility is restored to normal within 3 months after termination of therapy in most individuals. In some cases, fertility is not regained when gossypol is stopped. Gossypol also lowers potassium levels in the blood.

EXHIBIT 25-1   BIRTH CONTROL METHODS

METHOD	COMMENTS
Removal of gonads and uterus	Irreversible sterility. Generally performed if organs are diseased rather than as contraceptive method because of importance of hormones produced by gonads.
Sterilization	Procedure involving severing ductus (vas) deferens in males (vasectomy) and uterine tubes in females (tubal ligation and laparoscopic technique).
Natural contraception	Abstinence from intercourse during time of month woman is fertile. Under ideal circumstances, effectiveness in women with regular menstrual cycles may approach that of mechanical and chemical contraceptives. Extremely difficult to determine fertile period. Effectiveness can be increased by recording body temperature each morning before getting up; a small rise in temperature indicates that ovulation has occurred one or two days earlier and ovum can no longer be fertilized.
Mechanical contraception **Condom**	Thin, strong sheath of rubber or similar material worn by male to prevent sperm from entering vagina. Failures caused by sheath tearing or slipping off after climax or not putting the sheath on soon enough. If used correctly and consistently, effectiveness similar to that of diaphragm.
Diaphragm	Flexible rubber dome inserted into vagina to cover cervix, providing barrier to sperm. Usually used with spermicidal cream or jelly. Must be left in place at least 6 hours after intercourse and may be left in place as long as 24 hours. Must be fitted by physician or other trained personnel and refitted every two years and after each pregnancy. Offers high level of protection if used with spermicide; rate of 2 to 3 pregnancies per 100 women per year is estimated for consistent users. If not used consistently, much higher pregnancy rates must be expected. Occasional failures caused by improper insertion or displacement during sexual intercourse.
Intrauterine device (IUD)	Small object (loop, coil, T, or 7) made of plastic, copper, or stainless steel and inserted into uterus by physician. May be left in place for long periods of time (some must be changed every two to three years). Does not require continued attention by user. Some women cannot use them because of expulsion, bleeding, or discomfort. Not recommended for women who have not had children because uterus is too small and cervical canal too narrow. Use of IUDs may lead to pelvic inflammatory disease.
Chemical contraception **Foams, creams, jellies, suppositories, vaginal douches, contraceptive sponge**	Sperm-killing chemicals inserted into vagina to coat vaginal surfaces and cervical opening. Provide protection for about 1 hour. Effective when used alone, but significantly more effective when used with diaphragm or condom. The contraceptive sponge is made of polyurethane and releases a spermicide for up to 24 hours. It rates with the diaphragm in effectiveness. A few cases of toxic shock syndrome (TSS) have been reported among users of the contraceptive sponge.
Oral contraceptives (OC)	Except for total abstinence or surgical sterilization, most effective contraceptive known. Side effects include nausea, occasional light bleeding between periods, breast tenderness or enlargement, fluid retention, and weight gain. Should not be used by women who have cardiovascular conditions (thromboembolic disorders, cerebrovascular disease, heart disease, hypertension), liver malfunction, cancer or neoplasia of breast or reproductive organs, or by women who smoke.

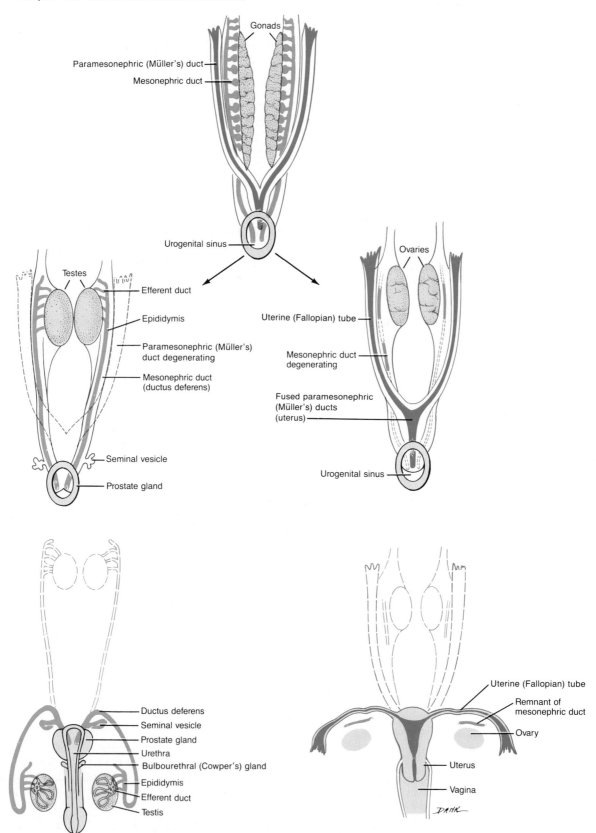

FIGURE 25-23 Development of the internal reproductive system.

AGING AND THE REPRODUCTIVE SYSTEMS

Although there are major age-specific physical changes in structure and function, few age-specific disorders are associated with the reproductive system. In the male, the decreasing production of testosterone produces less muscle strength, fewer viable sperm, and decreased sexual desire. However, abundant spermatozoa may be found even in old age. Most of the pathologies produce uneventful recoveries, except for prostate problems that could become serious and fatal.

The female reproductive system becomes less efficient, possibly as a result of less frequent ovulation and the declining ability of the uterine tubes and uterus to support the young embryo. There is a decrease in the production of the progesterone and estrogens. Menopause is only one of a series of phases that leads to reduced fertility, irregular or absent menstruation, and a variety of physical changes. Uterine cancer peaks at about 65 years of age, but cervical cancer is more common in younger women, and breast cancer is the leading cause of death among women between the ages of 40 and 60. Prolapse (falling down or sinking) of the uterus is possibly the commonest complaint among female geriatric patients.

DEVELOPMENTAL ANATOMY OF THE REPRODUCTIVE SYSTEMS

The *gonads* develop from the **intermediate mesoderm.** By the sixth week, they appear as bulges that protrude into the coelom (Figure 25-23). The gonads develop near the mesonephric ducts. A second pair of ducts, the **paramesonephric (Müller's) ducts,** develop lateral to the mesonephric ducts. Both sets of ducts empty into the urogenital sinus. At about the eighth week, the gonads are clearly differentiated into ovaries or testes.

In the male embryo, the *testes* connect to the mesonephric duct through a series of tubules. These tubules become the *seminiferous tubules.* Continued development of the mesonephric duct produces the *efferent ducts, ductus epididymis, ductus (vas) deferens, ejaculatory ducts,* and *seminal vesicle.* The *prostate* and *bulbourethral (Cowper's) glands* are **endodermal** outgrowths of the urethra. Shortly after the gonads differentiate into testes, the paramesonephric ducts degenerate without contributing any functional structures to the male reproductive system.

In the female embryo, the gonads develop into *ovaries.* At about the same time, the distal ends of the paramesonephric ducts fuse to form the *uterus* and *vagina.* The unfused portions become the *uterine (Fallopian) tubes.* The *greater (Bartholin's)* and *lesser vestibular glands* develop from **endodermal** outgrowths of the vestibule. The mesonephric ducts in the female degenerate without contributing any functional structures to the female reproductive system.

The *external genitals* of both male and female embryos also remain undifferentiated until about the eighth week. Before differentiation, all embryos have an elevated region, the **genital tubercle,** a point between the tail (future coccyx) and the umbilical cord where the mesonephric and paramesonephric ducts open to the exterior (Figure 25-24). The tubercle consists of a **urethral groove** (opening into the urogenital sinus), paired **urethral folds,** and paired **labioscrotal swellings.**

In the male embryo, the genital tubercle elongates and develops into a *penis.* Fusion of the urethral folds forms the *spongy (cavernous) urethra* and leaves an opening only at the distal end of the penis, the *urethral orifice.* The labioscrotal swellings develop into the *testes.* In the female, the genital tubercle gives rise to the *clitoris.* The urethral folds remain open as the *labia minora* and the labioscrotal swellings become the *labia majora.* The urethral groove becomes the *vestibule.*

APPLICATIONS TO HEALTH

SEXUALLY TRANSMITTED DISEASES (STDs)

The general term **sexually transmitted disease (STD)** is applied to any of the large group of diseases that can be spread by sexual contact. The group includes conditions traditionally specified as **venereal diseases (VD),** from Venus, goddess of love, such as gonorrhea, syphilis, and genital herpes, and several other conditions that are contracted sexually or may be contracted otherwise but are then transmitted to a sexual partner.

Gonorrhea

Gonorrhea is an infectious sexually transmitted disease that affects primarily the mucous membrane of the urogenital tract, the rectum, and occasionally the eyes. The disease is caused by the bacterium *Neisseria gonorrhoeae.* Discharges from the involved mucous membranes are the source of infection, and the bacteria are transmitted by direct contact, usually sexual or during passage of a newborn through the birth canal.

Males usually suffer inflammation of the urethra with pus and painful urination. Fibrosis sometimes occurs in an advanced stage, causing stricture of the urethra. There also may be involvement of the epididymis and prostate gland. In females, infection may occur in the urethra, vagina, and cervix, and there may be a discharge of pus. However, infected females often harbor the disease without any symptoms until it has progressed to a more advanced stage. If the uterine tubes become involved, pelvic inflammation may follow. Peritonitis, or inflammation of the peritoneum, is a very serious disorder. The infection should be treated and controlled immediately because, if neglected, sterility or death may result. Although antibi-

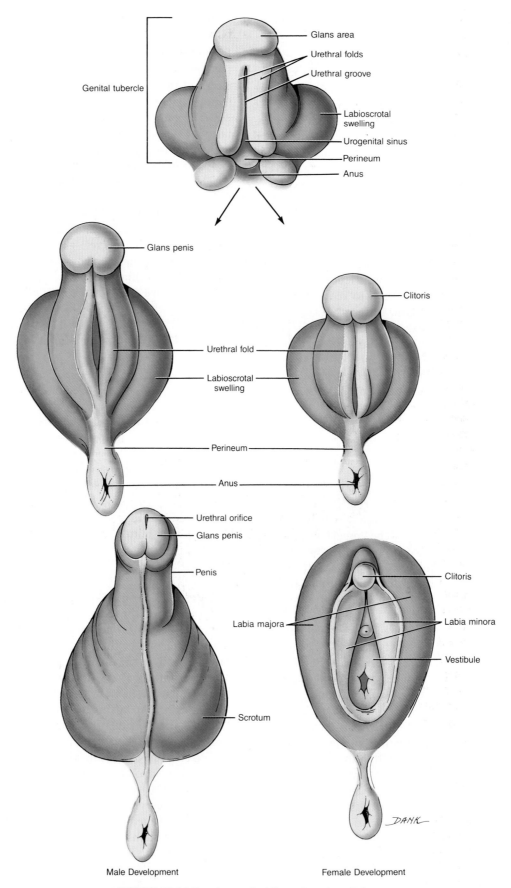

FIGURE 25-24 Development of the external genitals.

otics have greatly reduced the mortality rate of acute peritonitis, it is estimated that between 50,000 and 80,000 women are made sterile by gonorrhea every year as a result of scar tissue formation that closes the uterine tubes. If the bacteria are transmitted to the eyes of a newborn in the birth canal, blindness can result.

Administration of a 1 percent silver nitrate solution or penicillin in the infant's eyes prevents infection. Penicillin or tetracycline are the drugs of choice for the treatment of gonorrhea in adults.

Syphilis

Syphilis is a sexually transmitted disease caused by the bacterium *Treponema pallidum.* It is acquired through sexual contact or transmitted through the placenta to a fetus. The disease progresses through several stages: primary, secondary, latent, and sometimes tertiary. During the *primary stage,* the chief symptom is an open sore, called a *chancre,* at the point of contact. The chancre heals within 1 to 5 weeks. From 6 to 24 weeks later, symptoms such as a skin rash, fever, and aches in the joints and muscles usher in the *secondary stage.* These symptoms also eventually disappear (in about 4 to 12 weeks) and the disease ceases to be infectious, but a blood test for the presence of the bacteria generally remains positive. During this "symptomless" period, called the *latent stage,* the bacteria may invade body organs. When signs of organ degeneration appear, the disease is said to be in the *tertiary stage.*

If the syphilis bacteria attack the organs of the nervous system, the tertiary stage is called *neurosyphilis.* Neurosyphilis may take different forms, depending on the tissue involved. For instance, about 2 years after the onset of the disease, the bacteria may attack the meninges, producing meningitis. The blood vessels that supply the brain may also become infected. In this case, symptoms depend on the parts of the brain destroyed by oxygen and glucose starvation. Cerebellar damage is manifested by uncoordinated movements in such activities as writing. As the motor areas become extensively damaged, victims may be unable to control urine and bowel movements. Eventually, they may become bedridden, unable even to feed themselves. Damage to the cerebral cortex produces memory loss and personality changes that range from irritability to hallucinations.

Infection of the fetus with syphillis can occur after the fifth month. Infection of the mother is not necessarily followed by fetal infection, provided that the placenta remains intact. But once the bacteria gain access to fetal circulation, there is nothing to impede their growth and multiplication. As many as 80 percent of children born to untreated syphilitic mothers will be infected in the uterus if the fetus is exposed at the onset or in the early stages of the disease. About 25 percent of the fetuses will die within the uterus. Most of the survivors will arrive prematurely, but 30 percent will die shortly after birth. Of the infected and untreated children surviving infancy, about 40 percent will develop symptomatic syphilis during their lifetimes.

Syphilis can be treated with antibiotics (penicillin) during the primary, secondary, and latent periods. Certain forms of neurosyphilis may also be successfully treated, but the prognosis for others is very poor. Noticeable symptoms do not always appear during the first two stages of the disease. Syphilis, however, is usually diagnosed through a blood test whether noticeable symptoms appear or not. The importance of these blood tests and follow-up treatments cannot be overemphasized.

Genital Herpes

Another sexually transmitted disease, **genital herpes,** is now common in the United States. The sexual transmission of the herpes simplex virus is well established. Unlike syphilis and gonorrhea, genital herpes is incurable. Type I herpes simplex virus is the virus that causes the majority of infections above the waist such as cold sores. Type II herpes simplex virus causes most infections below the waist such as painful genital blisters on the prepuce, glans penis, and penile shaft in males and on the vulva or sometimes high up in the vagina in females. The blisters disappear and reappear in most patients, but the disease itself remains in the body.

Genital herpesvirus infection causes considerable discomfort, and there is an extraordinarily high rate of recurrence of the disease. The infection is usually characterized by fever, lymphadenopathy, and numerous clusters of genital blisters. Because genital herpes is statistically associated with later cervical carcinoma, all women with this infection should be reminded of the importance of annual Papanicolaou smears. For pregnant women with genital herpes, a cesarean section will usually prevent complications in the child. Complications range from a mild asymptomatic infection to CNS damage to death.

Treatment of the symptoms involves pain medication, saline compresses, sexual abstinence for the duration of the eruption, and use of a drug called acyclovir. This drug interferes with viral DNA replication but not with host cell DNA replication. Acyclovir speeds the healing and sometimes reduces the pain of initial genital herpes infections and shortens the duration of lesions in patients with recurrent genital herpes.

Trichomoniasis

The microorganism *Trichomonas vaginalis,* a flagellated protozoan (one-celled animal), causes **trichomoniasis,** an inflammation of the mucous membrane of the vagina in females and the urethra in males. Symptoms include vaginal discharge and severe vaginal itch in women. Men can have it without symptoms, but can transmit it to women nonetheless. Sexual partners must be treated simultaneously.

Nongonococcal Urethritis (NGU)

The term **nongonococcal urethritis (NGU),** also known as **nonspecific urethritis (NSU),** refers to an inflammation of the urethra not caused by the bacterium *Neisseria gonorrhoeae.* The bacterium *Chlamydia trachomatis* is known to cause some cases of NGU. NGU affects both males and females. The urethra swells and narrows, impeding the flow of urine. Both urination and the urgency to urinate increase. Urination is accompanied by burning pain, and there may be a purulent (pus-containing) discharge.

Although nonmicrobial factors such as trauma (passage of a catheter) or chemical agents (alcohol and certain chemotherapeutic agents) can cause this condition, it is estimated that at least 40 percent of the cases of NGU are acquired sexually. In fact, NGU may be the most common sexually transmitted disease in the United States today. Although it is not a disease which must be reported to state departments of health and exact data are lacking, the Centers for Disease Control estimates that 4 to 9 million Americans have NGU. Since the symptoms are often mild in males and females are usually asymptomatic, many cases of infection go untreated. Complications are not common, but can be serious. Males may develop inflammation of the epididymis. In females, inflammation may cause sterility by blocking the cervix or uterine tubes. As in gonorrhea, the bacteria can be passed from mother to infant during birth, infecting the eyes. NGU of chlamydial origin responds to treatment with tetracycline.

MALE DISORDERS

Prostate

The prostate gland is susceptible to infection, enlargement, and benign and malignant tumors. Because the prostate surrounds the urethra, any of these disorders can obstruct the flow of urine. Prolonged obstruction may result in serious changes in the urinary bladder, ureters, and kidneys and may perpetuate urinary tract infections. Therefore, if the obstruction cannot be relieved by other means, surgical removal of part of or the entire gland is indicated. The surgical procedure is called **prostatectomy** (pros'-ta-TEK-tō-mē).

Acute and chronic infections of the prostate gland are common in postpubescent males, often in association with inflammation of the urethra. In **acute prostatitis,** the prostate gland becomes swollen and tender. Appropriate antibiotic therapy, bed rest, and above-normal fluid intake are effective treatment.

Chronic prostatitis is one of the most common chronic infections in men of the middle and later years. On exami-

nation, the prostate gland feels enlarged, soft, and extremely tender. The surface outline is irregular and may be hard. This disease frequently produces no symptoms, but the prostate is believed to harbor infectious microorganisms responsible for some allergic conditions, arthritis, and inflammation of nerves (neuritis), muscles (myositis), and the iris (iritis).

An **enlarged prostate** gland, increasing to two to four times larger than normal, occurs in approximately one-third of all males over age 60. The cause is unknown, and the enlarged condition usually can be detected by rectal examination.

Tumors of the male reproductive system usually involve the prostate gland. Carcinoma of the prostate is the second leading cause of death from cancer in men in the United States, and it is responsible for approximately 19,000 deaths annually. Its incidence is related to age, race, occupation, geography, and ethnic origin. Both benign and malignant growths are common in elderly men. Both types of tumors put pressure on the urethra, making urination painful and difficult. At times, the excessive back pressure destroys kidney tissue and gives rise to an increased susceptibility to infection. Therefore, even when the tumor is benign, surgery is indicated.

Sexual Functional Abnormalities

Impotence is the inability of an adult male to attain or hold an erection long enough for normal intercourse. Impotence could be the result of physical abnormalities of the penis, systemic disorders such as syphilis, vascular disturbances, neurological disorders, testosterone deficiency, or psychic factors such as fear of causing pregnancy, fear of sexually transmitted diseases, religious inhibitions, and emotional immaturity.

Infertility (sterility) is an inability to fertilize the ovum. It does not imply impotence. Male fertility requires production of adequate amounts of viable, normal spermatozoa by the testes, unobstructed transportation of sperm through the seminal tract, and satisfactory deposition in the vagina. The tubules of the testes are sensitive to many factors—x rays, infections, toxins, malnutrition—that may cause degenerative changes and produce male sterility. If inadequate spermatozoa production is suspected, a sperm analysis should be performed. At least one type of fertility may be improved by the administration of vitamin C.

FEMALE DISORDERS

Menstrual Abnormalities

Because menstruation reflects not only the health of the uterus but also the health of the endocrine glands that control it, the ovaries and the pituitary gland, disorders of the female reproductive system frequently involve menstrual disorders.

Amenorrhea is the absence of menstruation. If a woman has never menstruated, the condition is called **primary amenorrhea.** Primary amenorrhea can be caused by endocrine disorders, most often in the pituitary gland and hypothalamus, or by genetically caused abnormal development of the ovaries or uterus. **Secondary amenorrhea,** the skipping of one or more periods, is commonly experienced by women at some time during their lives. Changes in body weight, either gains or losses, often cause amenorrhea. Obesity may disturb ovarian function, and similarly, the extreme weight loss that characterizes anorexia nervosa often leads to a suspension of menstrual flow. When amenorrhea is unrelated to weight, analysis of levels of estrogens often reveals deficiencies of pituitary and ovarian hormones. Amenorrhea may also be caused by continuous involvement in rigorous athletic training.

Dysmenorrhea is painful menstruation caused by forceful contraction of the uterus. It is often accompanied by nausea, vomiting, diarrhea, headache, fatigue, and nervousness. Some cases are caused by pathological conditions such as uterine tumors, ovarian cysts, endometriosis, and pelvic inflammatory disease (PID). However, other cases of dysmenorrhea are not related to any pathologies. Although the cause of these cases is unknown, it appears that they may be triggered by an overproduction of prostaglandins by the uterus. Prostaglandins are known to stimulate uterine contractions, but they cannot do so in the presence of high levels of progesterone. As we have noted earlier, progesterone levels are high during the last half of the menstrual cycle. During this time, prostaglandins are apparently inhibited by progesterone from producing uterine contractions. However, if pregnancy does not occur, progesterone levels drop rapidly and prostaglandin production increases. This causes the uterus to contract and slough off its lining and may result in dysmenorrhea. The other symptoms of dysmenorrhea—nausea, vomiting, diarrhea, and headache—may be due to prostaglandin-stimulated contractions of the smooth muscle of the stomach, intestines, and blood vessels in the brain. Drugs that inhibit prostaglandin synthesis (naproxen and ibuprofen) are used to treat dysmenorrhea.

Abnormal uterine bleeding includes menstruation of excessive duration or excessive amount, too frequent menstruation, intermenstrual bleeding, and postmenopausal bleeding. These abnormalities may be caused by disordered hormonal regulation, emotional factors, fibroid tumors of the uterus, and systemic diseases.

Premenstrual syndrome (PMS) is a term usually reserved for severe physical and emotional distress occurring late in the postovulatory phase of the menstrual cycle and sometimes overlapping with menstruation. Signs and symptoms include edema, breast swelling and tenderness, abdominal distension, backache, joint pain, constipation, skin eruptions, fatigue and lethargy, greater need for sleep, depression or anxiety, irritability, headache, poor coordination and clumsiness, and cravings for sweet or salty foods. The basic cause of PMS is unknown: One postulated mechanism involves excessive levels of estrogens or inadequate levels of progesterone or dopamine (DA). Other proposed causes are vitamin B_6 deficiency and altered glucose metabolism (hypoglycemia).

Toxic Shock Syndrome (TSS)

Toxic shock syndrome (TSS) is primarily a disease of previously healthy, young, menstruating females who use tampons. It is also recognized in males, children, and nonmenstruating females. Clinically, TSS is characterized by high fever up to 40.6°C (105°F), sore throat or very tender mouth, headache, fatigue, irritability, muscle soreness and tenderness, conjunctivitis, diarrhea and vomiting, abdominal pain, vaginal irritation, and erythematous rash. Other symptoms include lethargy, unresponsiveness, memory loss, hypotension, peripheral vasoconstriction, respiratory distress syndrome, intravascular coagulation, decreased platelet count, renal failure, circulatory shock, and liver involvement.

It is now clear that toxin-producing strains of the bacterium *Staphylococcus aureus* are necessary for development of the disease. Actually, it appears that a virus has become incorporated into *S. aureus,* causing the bacterium to produce the toxins. Apparently, certain high-absorbency tampons provide a substrate upon which the bacteria grow and produce toxins. Initial therapy is directed at correcting all homeostatic imbalances as quickly as possible. Antistaphylococcal antibiotics, such as penicillin or clindamycin, are also administered. In severe cases, high doses of corticosteroids are also administered.

Ovarian Cysts

Ovarian cysts are fluid-containing tumors of the ovary. Follicular cysts may occur in the ovaries of elderly women, in ovaries that have inflammatory diseases, and in menstruating females. They have thin walls and contain a serous albuminous material. Cysts may also arise from the corpus luteum or the endometrium. The endometrium is the inner lining of the uterus that is sloughed off in menstruation.

Endometriosis

Endometriosis is the growth of endometrial tissue outside the uterus. The tissue enters the pelvic cavity via the open uterine tubes and may be found in any of several sites—on the ovaries, rectouterine pouch, surface of the uterus, sigmoid colon, pelvic and abdominal lymph nodes, cervix, abdominal wall, kidneys, and urinary bladder. One theory for the development of endometriosis is that there is regurgitation of menstrual flow through the uterine tubes. Endometriosis is common in women 30 to 40 years of age. Symptoms include premenstrual pain or unusual menstrual pain. The unusual pain is caused by the displaced tissue sloughing off at the same time the normal uterine endometrium is being shed during menstruation. Infertility can be a consequence. Treatment usually consists of hormone therapy or surgery. Endometriosis disappears at menopause or when the ovaries are removed.

Infertility

Female infertility, or the inability to conceive, occurs in about 10 percent of married females in the United States. Once it is established that ovulation occurs regularly, the reproductive tract is examined for functional and anatomical disorders to determine the possibility of union of the sperm and the ovum in the uterine tube.

Disorders Involving the Breasts

The breasts of females are highly susceptible to cysts and tumors. Men are also susceptible to breast tumors, but certain breast cancers are 100 times more common in women.

In the female, the benign **fibroadenoma** is a common tumor of the breast. It occurs most frequently in young women. Fibroadenomas have a firm rubbery consistency and are easily moved about within the mammary tissue. The usual treatment is excision of the growth. The breast itself is not removed.

Breast cancer has one of the highest fatality rates of all cancers affecting women, but it is rare in men. In the female, breast cancer is rarely seen before age 30, and its occurrence rises rapidly after menopause. Breast cancer is generally not painful until it becomes quite advanced, so often it is not discovered early or, if noted, is ignored. Any lump, no matter how small, should be reported to a doctor at once. Treatment for breast cancer may involve hormone therapy, chemotherapy, a modified or radical mastectomy, or a combination of these. A *radical mastectomy* involves removal of the affected breast along with the underlying pectoral muscles and the axillary lymph nodes. Metastasis of cancerous cells is usually through the lymphatics or blood. Radiation treatment and chemotherapy may follow the surgery to ensure the destruction of any stray cancer cells.

Among the factors that clearly increase the risk of breast cancer development are: (1) a family history of breast cancer, especially in a mother or sister; (2) never having a child or having a first child after age 34; (3) previous cancer in one breast; (4) exposure to ionizing radiation; and (5) an abnormal mammogram.

Cervical Cancer

Another common disorder of the female reproductive system is cancer of the uterine cervix. It ranks third in frequency after breast and skin cancers. **Cervical cancer** starts with a change in the shape, growth, and number of the cervical cells called *cervical dysplasia* (dis-PLĀ-sē-a). Cervical dysplasia is not a cancer in itself, but the abnormal cells tend to become malignant.

Cervical cancer may be a sexually transmitted disease with a long incubation period. Inciting factors are not known, but type II herpes simplex virus has recently become suspect. Smegma and the DNA of spermatozoa have also been implicated. Cancer of the cervix (except for adenocarcinoma) rarely occurs in celibate women, and for unknown reasons it is rare in Jewish women.

Pelvic Inflammatory Disease (PID)

Pelvic inflammatory disease (PID) is a collective term for any extensive bacterial infection of the pelvic organs, especially the uterus, uterine tubes, or ovaries. A vaginal or uterine infection may spread into the uterine tube (*salpingitis*) or even farther into the abdominal cavity, where it infects the peritoneum (*peritonitis*). PID is most commonly caused by gonorrhea, but any bacterium can trigger infection. Often the early symptoms of PID, which include increased vaginal discharge and pelvic pain, occur just after menstruation. As infection spreads, fever may develop in advanced cases along with painful abscesses of the reproductive organs. Early treatment with antibiotics (tetracycline or penicillin) can stop the spread of PID.

KEY MEDICAL TERMS ASSOCIATED WITH THE REPRODUCTIVE SYSTEMS

Leukorrhea (*leuko* = white; *rrhea* = discharge) A non-bloody vaginal discharge that may occur at any age and affects most women at some time.

Neoplasia (*neo* = new; *plas* = form, grow) A condition characterized by the presence of new growths (tumors).

Oophorectomy (*oophoro* = ovary; *ectomy* = removal of) Excision of an ovary. Bilateral oophorectomy refers to the removal of both ovaries.

Pruritus Itching.

Salpingectomy (*salpingo* = tube) Excision of a uterine tube.

Smegma (*smegma* = soap) The secretion, consisting principally of desquamated epithelial cells, found chiefly about the external genitalia and especially under the foreskin of the male.

Vaginitis Inflammation of the vagina characterized by an increased vaginal discharge, pruritus, and occasional pain upon urination. Typically caused by bacteria (*Gardnerella vaginalis*), a protozoon (*Trichomonas vaginalis*), or fungi (*Candida albicans*).

STUDY OUTLINE

Male Reproductive System (p. 648)

1. Reproduction is the process by which genetic material is passed on from one generation to the next.
2. The organs of reproduction are grouped as: gonads (produce gametes), ducts (transport and store gametes), and accessory glands (produce materials that support gametes).
3. The male structures of reproduction include the testes, ductus epididymis, ductus deferens, ejaculatory duct, urethra, seminal vesicles, prostate gland, bulbourethral (Cowper's) glands, and penis.

Scrotum

1. The scrotum is a cutaneous outpouching of the abdomen that supports the testes.
2. It regulates the temperature of the testes by contraction of the dartos to elevate them closer to the pelvic cavity.

Testes

1. The testes are oval-shaped glands (gonads) in the scrotum containing seminiferous tubules, in which sperm cells are made; sustentacular cells, which nourish sperm cells; and interstitial endocrinocytes, which produce the male sex hormone testosterone.
2. Failure of the testes to descend is called cryptorchidism.
3. Ova and sperm are collectively called gametes, or sex cells, and are produced in gonads.
4. Uninucleated somatic cells divide by mitosis, the process in which each daughter cell receives the full complement of 23 chromosome pairs (46 chromosomes). Somatic cells are said to be diploid ($2n$).
5. Immature gametes divide by meiosis in which the pairs of chromosomes are split so that the mature gamete has only 23 chromosomes. It is said to be haploid (n).
6. Spermatogenesis occurs in the testes. It results in the formation of four haploid spermatozoa.
7. The spermatogenesis sequence consists of reduction division, equatorial division, and spermiogenesis.
8. Mature spermatozoa consist of a head, midpiece, and tail. Their function is to fertilize an ovum.
9. Spermatozoa are moved through the testes through the seminiferous tubules, straight tubules, rete testis, and efferent ducts.

Ducts

1. The duct system of the testes includes the seminiferous tubules, straight tubules, and rete testis.
2. Sperm are transported out of the testes through the efferent ducts.
3. The ductus epididymis is lined by stereocilia and is the site of sperm maturation and storage.
4. The ductus (vas) deferens stores sperm and propels them toward the urethra during ejaculation.
5. Alteration of the ductus (vas) deferens to prevent fertilization is called vasectomy.
6. The ejaculatory ducts are formed by the union of the ducts from the seminal vesicles and ductus (vas) deferens and eject spermatozoa into the prostatic urethra.
7. The male urethra is subdivided into three portions: prostatic, membranous, and spongy (cavernous).

Accessory Glands

1. The seminal vesicles secrete an alkaline, viscous fluid that constitutes about 60 percent of the volume of semen and contributes to sperm viability.
2. The prostate gland secretes an alkaline fluid that constitutes about 13 to 33 percent of the volume of semen and contributes to sperm motility.
3. The bulbourethral (Cowper's) glands secrete mucus for lubrication and a substance that neutralizes urine.
4. Semen (seminal fluid) is a mixture of spermatozoa and accessory gland secretions that provide the fluid in which spermatozoa are transported, provide nutrients, and neutralize the acidity of the male urethra and female vagina.

Penis

1. The penis is the male organ of copulation that consists of a root, body, and glans penis.
2. Expansion of its blood sinuses under the influence of sexual excitation is called erection.

Female Reproduction System (p. 662)

1. The female organs of reproduction include the ovaries (gonads), uterine tubes, uterus, vagina, and vulva.
2. The mammary glands are considered part of the reproductive system.

Ovaries

1. The ovaries are female gonads located in the upper pelvic cavity, on either side of the uterus.
2. They produce ova, discharge ova (ovulation), and secrete estrogens, progesterone, and relaxin.
3. Oogenesis occurs in the ovaries. It results in the formation of a single haploid ovum.
4. The oogenesis sequence consists of reduction division, equatorial division, and maturation.

Uterine Tubes

1. The uterine (Fallopian) tubes transport ova from the ovaries to the uterus and are the normal sites of fertilization.
2. Implantation outside the uterus (pelvic or tubular) is called an ectopic pregnancy.

Uterus

1. The uterus is an inverted, pear-shaped organ that functions in menstruation, implantation of a fertilized ovum, development of a fetus during pregnancy, and labor.
2. The uterus is normally held in position by a series of ligaments.
3. Histologically, the uterus consists of an outer perimetrium, middle myometrium, and inner endometrium.

Vagina

1. The vagina is a passageway for the menstrual flow, the receptacle for the penis during sexual intercourse, and the lower portion of the birth canal.
2. It is capable of considerable distension to accomplish its functions.

Vulva

1. The vulva is a collective term for the external genitals of the female.
2. It consists of the mons veneris, labia majora, labia minora, clitoris, vestibule, vaginal and urethral orifices, and the paraurethral (Skene's), greater vestibular (Bartholin's), and lesser vestibular glands.

Perineum

1. The perineum is a diamond-shaped area at the inferior end of the trunk between the thighs and buttocks.
2. An incision in the perineal skin prior to delivery is called an episiotomy.

Mammary Glands

1. The mammary glands are modified sweat glands (branched tubuloalveolar) over the pectoralis major muscles. Their function is to secrete and eject milk (lactation).
2. Milk secretion is due to mainly PRL and milk ejection is stimulated by OT.

Sexual Intercourse (p. 676)

1. The role of the male in the sex act involves erection, lubrication, and orgasm.
2. The female role also involves erection, lubrication, and orgasm (climax).

Birth Control (p. 677)

1. Methods include removal of gonads and uterus, sterilization (vasectomy, tubal ligation, laparoscopic technique), and contraception (natural, mechanical, and chemical).
2. Contraceptive pills of the combination type contain estrogens and progesterone in concentrations that decrease the secretion of FSH and LH and thereby inhibit ovulation.

Aging and the Reproductive Systems (p. 681)

1. In the male, decreased levels of testosterone decrease muscle strength, sexual desire, and viable sperm; prostate disorders are common.
2. In the female, levels of progesterone and estrogens decrease resulting in changes in menstruation; uterine and breast cancer increase in incidence.

Developmental Anatomy of the Reproductive Systems (p. 681)

1. The gonads develop from intermediate mesoderm and are differentiated into ovaries or testes by about the eighth week.
2. The external genitals develop from the genital tubercle.

Applications to Health (p. 681)

1. Sexually transmitted diseases (STDs) are diseases spread by sexual contact and include gonorrhea, syphilis, genital herpes, trichomoniasis, and nongonococcal urethritis (NGU).
2. Conditions that affect the prostate are prostatitis, enlarged prostate, and tumors.
3. Impotence is the inability of the male to attain or hold an erection long enough for intercourse.
4. Infertility is the inability of a male's sperm to fertilize an ovum.
5. Menstrual disorders include amenorrhea, dysmenorrhea, abnormal bleeding, and premenstrual syndrome (PMS).
6. Toxic shock syndrome (TSS) includes widespread homeostatic imbalances and is a reaction to toxins produced by *Staphylococcus aureus*.
7. Ovarian cysts are tumors that contain fluid.
8. Endometriosis refers to the growth of uterine tissue outside the uterus.
9. Female infertility is the inability of the female to conceive.
10. The mammary glands are susceptible to benign fibroadenomas and malignant tumors. The removal of a malignant breast, pectoral muscles, and lymph nodes is called a radical mastectomy.
11. Cervical cancer can be diagnosed by a Pap test.
12. Pelvic inflammatory disease (PID) refers to bacterial infection of pelvic organs.

REVIEW QUESTIONS

1. Define reproduction. Describe how the reproductive organs are classified and list the male and female organs of reproduction.
2. Describe the function of the scrotum in protecting the testes from temperature fluctuations. What is cryptorchidism?
3. Describe the internal structure of a testis. Where are the sperm cells made?
4. Describe the principal events of spermatogenesis. Why is meiosis important?
5. Identify the principal parts of a spermatozoon. List the function of each.
6. Which ducts are involved in transporting sperm within the testes?
7. Describe the location, structure, and functions of the ductus epididymis, ductus (vas) deferens, and ejaculatory duct.
8. What is the spermatic cord? What is an inguinal hernia?
9. Give the location of the three subdivisions of the male urethra.
10. Trace the course of spermatozoa through the male system of ducts from the seminiferous tubules through the urethra.

11. Briefly explain the locations and functions of the seminal vesicles, prostate gland, and bulbourethral (Cowper's) glands.
12. How is the penis structurally adapted as an organ of copulation? What is circumcision?
13. How are the ovaries held in position in the pelvic cavity?
14. Describe the microscopic structure of an ovary. What are the functions of the ovaries?
15. Describe the principal events of oogenesis.
16. Where are the uterine (Fallopian) tubes located? What is their function? What is an ectopic pregnancy?
17. Diagram the principal parts of the uterus.
18. Describe the arrangement of ligaments that hold the uterus in its normal position. What is retroflexion?
19. Discuss the blood supply to the uterus. Why is an abundant blood supply important?
20. Describe the histology of the uterus.
21. What is the function of the vagina? Describe its histology.
22. List the parts of the vulva and the functions of each part.
23. What is the perineum? Define episiotomy.
24. Describe the structure of the mammary glands. How are they supported?
25. Describe the passage of milk from the areolar cells of the mammary gland to the nipple. Define lactation.
26. Explain the role of the male's erection, lubrication, and orgasm in the sex act. How do the female's erection, lubrication, and orgasm (climax) contribute to the sex act?
27. Briefly describe the various methods of birth control and the effectiveness of each.
28. Explain the effects of aging on the reproductive systems.
29. Describe the development of the reproductive systems.
30. Define a sexually transmitted disease (STD). Describe the cause, clinical symptoms, and treatment of gonorrhea, syphilis, genital herpes, trichomoniasis, and nongonococcal urethritis (NGU).
31. Describe several disorders that affect the prostate gland.
32. What are some of the causes of amenorrhea, dysmenorrhea, and abnormal uterine bleeding?
33. Describe the clinical symptoms of premenstrual syndrome (PMS) and toxic shock syndrome (TSS).
34. What are ovarian cysts? Define endometriosis.
35. What is a radical mastectomy?
36. Define pelvic inflammatory disease (PID).
37. Refer to the glossary of key medical terms at the end of the chapter. Be sure that you can define each term.

26

Developmental Anatomy

Student Objectives

Explain the activities associated with fertilization, morula formation, blastocyst development, and implantation.

Describe how external human fertilization and embryo transfer are accomplished.

Discuss the formation of the primary germ layers, embryonic membranes, placenta, and umbilical cord as the principal events of the embryonic period.

Describe the principal events associated with fetal growth.

Explain the events associated with the three stages of labor.

Explain amniocentesis and chorionic villi sampling (CVS) as procedures for diagnosing diseases in the newborn.

Define key medical terms associated with developmental anatomy.

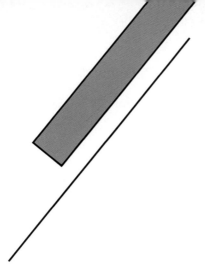

Developmental anatomy is the study of the sequence of events from the fertilization of an egg to the formation of an adult organism. As we look at the sequence from fertilization to birth, we will consider fertilization, implantation, embryonic development, fetal growth, gestation, parturition, and labor.

PREGNANCY

Once spermatozoa and ova are developed through meiosis and maturation and the spermatozoa are deposited in the vagina, pregnancy can occur. **Pregnancy** is a sequence of events that normally includes fertilization, implantation, embryonic growth, and fetal growth that terminates in birth.

FERTILIZATION AND IMPLANTATION

Fertilization

The term **fertilization** is applied to the penetration of an ovum by a spermatozoon and the subsequent union of the sperm nucleus and the nucleus of the ovum. Of the hundreds of millions of sperm cells introduced into the vagina, only about 2,000 arrive in the vicinity of the ovum. Fertilization normally occurs in the uterine tube when the ovum is about one-third of the way down the tube, usually within 24 hours after ovulation. Peristaltic contractions and the action of cilia transport the ovum through the uterine tube. The mechanism by which sperm reach the uterine tube is still unclear. Some believe that sperm swim up the female tract by means of whiplike movements of their flagella; others believe sperm are transported by muscular contractions of the uterus. Their motility is probably a combination of both.

Sperm must remain in the female genital tract 4 to 6 hours before they are capable of fertilizing an ovum. During this time, the enzymes hyaluronidase and proteinases are secreted by the acrosomes of the spermatozoa. They help dissolve the intercullular materials covering the ovum. The ovum is surrounded by a gelatinous covering called the **zona pellucida** (pe-LOO-si-da) and several layers of cells, the innermost of which are follicle cells, known as the **corona radiata** (Figure 26-1a). Normally, only one spermatozoon fertilizes an ovum, because once union is achieved, the electrical changes in the surface of the ovum block the entry of other sperm, and enzymes produced by the fertilized ovum alter receptor sites so that sperm already bound are detached and others are prevented from binding. In this way, polyspermy, the fertilization of an ovum by more than one spermatozoon, is prevented.

When a spermatozoon has entered the ovum, the tail is shed and the nucleus in the head develops into a structure called the **male pronucleus.** The nucleus of the ovum develops into a **female pronucleus** (Figure 26-1b). After the pronuclei are formed, they fuse to produce a **segmentation nucleus.** The segmentation nucleus contains 23 chromosomes from the male pronucleus and 23 chromosomes from the female pronucleus. Thus the fusion of the haploid pronuclei restores the diploid number. The fertilized ovum, consisting of a segmentation nucleus, cytoplasm, and enveloping membrane, is called a **zygote.**

CLINICAL APPLICATION

Dizygotic (fraternal) twins are produced from the independent release of two ova and the subsequent fertilization of each by different spermatozoa. They are the same age and are in the uterus at the same time, but they are genetically as dissimilar as any other siblings. They may or may not be the same sex. **Monozygotic (identical) twins** are derived from a single fertilized ovum that splits at an early stage in development. They contain the same genetic material and are always the same sex.

Monozygotic triplets are developed from a single fertilized ovum that splits twice; **dizygotic triplets** are derived from two ova from one or both ovaries, one of which splits after fertilization; and **trizygotic triplets** result from the fertilization of three ova from one or both ovaries.

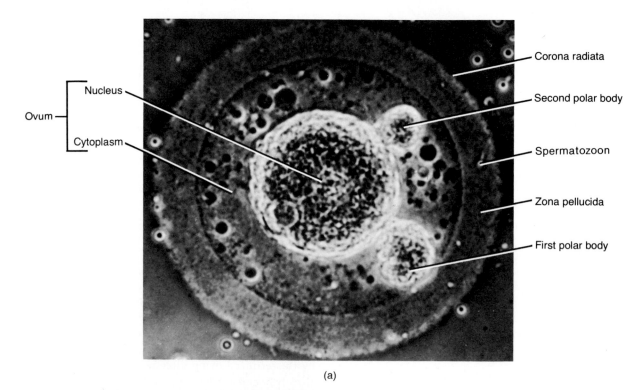

(a)

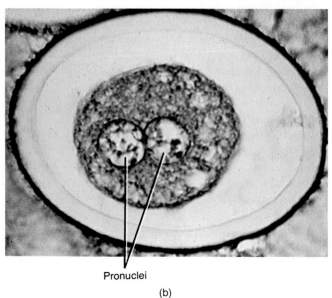

Pronuclei

(b)

FIGURE 26-1 Fertilization. (a) Photomicrograph of a spermatozoon moving through the zona pellucida on its way to reach the nucleus of the ovum. (Courtesy of *The Rand McNally Atlas of the Body and Mind,* Rand McNally and Company, New York, Chicago, San Francisco, in association with Mitchell Beazley Publishers Limited, London, 1976.) (b) Photomicrograph showing male and female pronuclei. (Courtesy of Carolina Biological Supply Company.)

cleavage is completed. By the end of the third day there are 16 cells. The progressively smaller cells produced are called **blastomeres** (BLAS-tō-mērz). A few days after fertilization the successive cleavages have produced a solid mass of cells, the **morula** (MOR-yoo-la) or mulberry, which is about the same size as the orignal zygote.

Development of the Blastocyst

As the number of cells in the morula increases, it moves from the original site of fertilization down through the ciliated uterine tube toward the uterus and enters the uterine cavity. By this time, the dense cluster of cells is altered to form a hollow ball of cells. The mass is now referred to as a **blastocyst** (Figure 26-3).

The blastocyst is differentiated into an outer covering of cells called the **trophectoderm** (trō-FEK-tō-derm), an **inner cell mass,** and an internal fluid-filled cavity called the **blastocoel** (BLAS-tō-sēl). The trophectoderm ultimately forms part of the membranes composing the fetal portion of the placenta; the inner cell mass develops into the embryo.

Formation of the Morula

Immediately after fertilization, rapid cell division of the zygote takes place. This early division of the zygote is called **cleavage.** During this time the dividing cells are contained by the zona pellucida. Although cleavage increases the number of cells, it does not result in an increase in the size of the developing organism.

The first cleavage is completed after about 36 hours, and each succeeding division takes slightly less time (Figure 26-2). By the second day after conception, the second

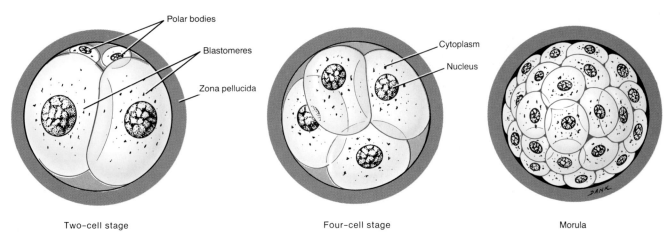

FIGURE 26-2 Formation of the morula.

Implantation

The attachment of the blastocyst to the endometrium occurs 7 to 8 days after fertilization and is called **implantation** (Figure 26-4). At this time, the endometrium is in its postovulatory phase. During implantation, the cells of the trophectoderm secrete an enzyme that enables the blastocyst to penetrate the uterine lining and become burried in the endometrium, usually on the posterior wall of the fundus or body of the uterus. The blastocyst, now only about $\frac{1}{100}$ of an inch in diameter, becomes oriented so that the inner cell mass is toward the endometrium. Implantation enables the blastocyst to absorb nutrients from the glands and blood vessels of the endometrium for its subsequent growth and development.

A summary of the principal events associated with fertilization and implantation is shown in Figure 26-5.

CLINICAL APPLICATION

In the early months of pregnancy, some females develop **hyperemesis gravidarum (morning sickness),** characterized by nausea and vomiting. Although the exact cause is unknown, it is possible that the degenerative products of digested portions of the endometrium during implantation may be responsible. Another possible cause is high levels of estrogens secreted by the placenta.

EXTERNAL HUMAN FERTILIZATION

On July 12, 1978, Louise Joy Brown was born near Manchester, England. Her birth was the first recorded case of **external human fertilization (in vitro fertilization)**—fertilization in a glass dish. The procedure developed for external human fertilization is carried out as follows. The female is given follicle-stimulating hormone (FSH) soon after menstruation, so that several ova, rather than the typical single one, will be produced (superovulation). Administration of luteinizing hormone (LH) may also ensure the maturation of the ova. Next, a small incision is made near the umbilicus, and the ova are aspirated from the follicles and placed in a medium that simulates the fluids in the female reproductive tract. The ova are then transferred to a solution of the male's sperm. Once fertilization has taken place, the fertilized ovum is put in another medium and is observed for cleavage. When the fertilized ovum reaches the 8-cell or 16-cell stage, it is introduced into the uterus for implantation and subsequent growth. The growth and development sequences that occur are similar to those in internal fertilization. It is also possible

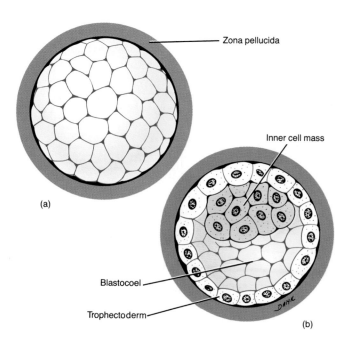

FIGURE 26-3 Blastocyst. (a) External view. (b) Internal view.

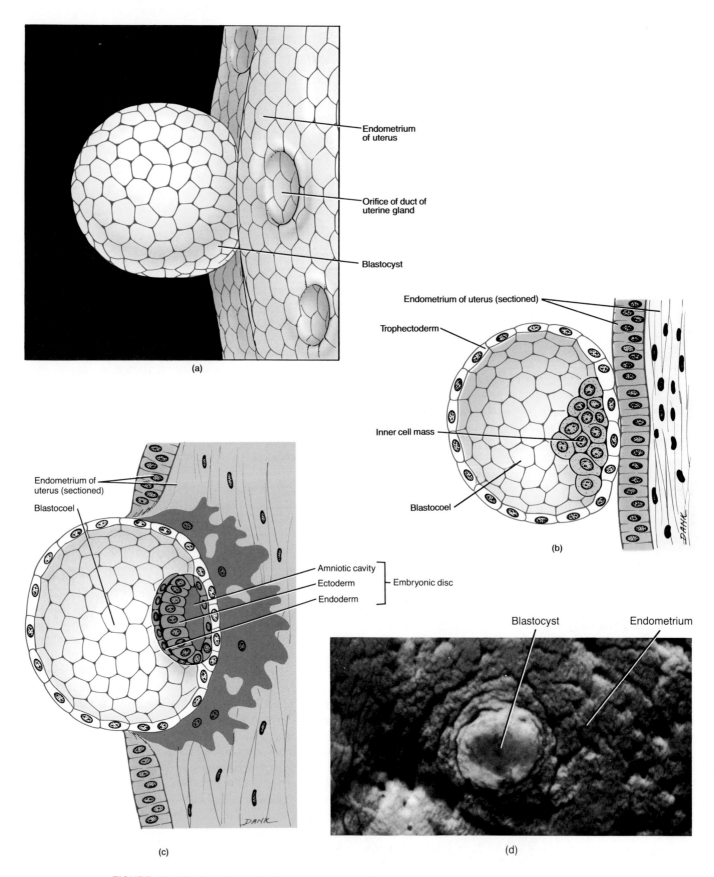

FIGURE 26-4 Implantation. (a) External view of the blastocyst in relation to the endometrium of the uterus about 5 days after fertilization. (b) Internal view of the blastocyst in relation to the endometrium about 6 days after fertilization. (c) Internal view of the blastocyst at implantation about 7 days after fertilization. (d) Photomicrograph of implantation. (Courtesy of Roberts Rugh and Landrum B. Shettles, M.D., with Richard N. Einhorn, from *From Conception to Birth: The Drama of Life's Beginnings,* Harper & Row, Publishers, Inc., New York, 1971.)

to freeze unused embryos (cryopreservation) to permit parents a successive pregnancy several years later or allow a second attempt at implantation if the first attempt is unsuccessful.

EMBRYO TRANSFER

Embryo transfer is an alternate to external human fertilization. It is a procedure in which a husband's seminal fluid is used to artificially inseminate a fertile ovum donor and, following fertilization, the morula or blastocyst is transferred from the donor to the infertile wife who carries it to term. Embryo transfer is indicated for females who are infertile, who have surgically untreatable blocked uterine tubes; or who are afraid of passing on their own genes because they are carriers of a serious genetic disorder.

In the procedure, the donor is monitored to ascertain the time of ovulation by checking her blood levels of luteinizing hormone (LH) and by ultrasound. The wife is also monitored to make sure that her ovarian cycle is synchronized with that of the donor. Once ovulation occurs in the donor, she is artificially inseminated with the husband's seminal fluid. Four days later, a morula or blastocyst is flushed from the donor's uterus through a soft plastic catheter and transferred to the uterus of the wife where it grows and develops until the time of birth. Embryo transfer is an office procedure that requires no anesthesia and may be performed in a few minutes.

EMBRYONIC DEVELOPMENT

The first two months of development are generally considered the **embryonic period.** During this period the developing human is called an **embryo.** The study of development from the fertilized egg through the eighth week in utero is referred to as **embryology.** The months of development after the second month are considered the **fetal period,** and during this time the developing human is called a **fetus.** By the end of the embryonic period the rudiments of all the principal adult organs are present, the embryonic membranes are developed, and the placenta is functioning.

BEGINNINGS OF ORGAN SYSTEMS

Following implantation, the inner cell mass of the blastocyst begins to differentiate into the three **primary germ layers:** ectoderm, endoderm, and mesoderm. They are the embryonic tissues from which all tissues and organs of the body will develop. The various movements of cell groups leading to the establishment of the primary germ layers are referred to as **gastrulation.**

In the human, the germ layers form so quickly that it is difficult to determine the exact sequence of events.

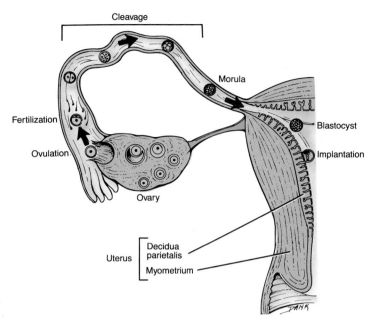

FIGURE 26-5 Summary of events associated with fertilization and implantation.

Before implantation, a layer of **ectoderm** (the trophectoderm) already has formed around the blastocoel (see Figure 26-4b). The trophectoderm will become part of the chorion, one of the fetal membranes. Within 8 days after fertilization, the inner cell mass moves downward so a space called the **amniotic cavity** lies between the inner cell mass and the trophectoderm. The bottom layer of the inner cell mass develops into the **endoderm.**

About the twelfth day after fertilization, striking changes appear (Figure 26-6a). A layer of cells from the inner cell mass has grown around the top of the amniotic cavity. These cells will become the amnion, another fetal membrane. The cells below the cavity are called the **embryonic disc.** They will form the embryo. At this stage, the embryonic disc contains ectodermal and endodermal cells; the mesodermal cells are scattered external to the disc. The cells of the endodermal layer have been dividing rapidly, so that groups of them now extend downward in a circle, forming the yolk sac, another fetal membrane. The cells of the **mesoderm** also have been dividing, and many have left the area of the embryonic disc and can be seen around the structures that are becoming fetal membranes.

About the fourteenth day, the cells of the embryonic disc differentiate into three distinct layers: the upper ectoderm, the middle mesoderm, and the lower endoderm (Figure 26-6b). At this time the two ends of the embryonic disc draw together, squeezing off the yolk sac. The resulting cavity inside the disc is the endoderm-lined **primitive**

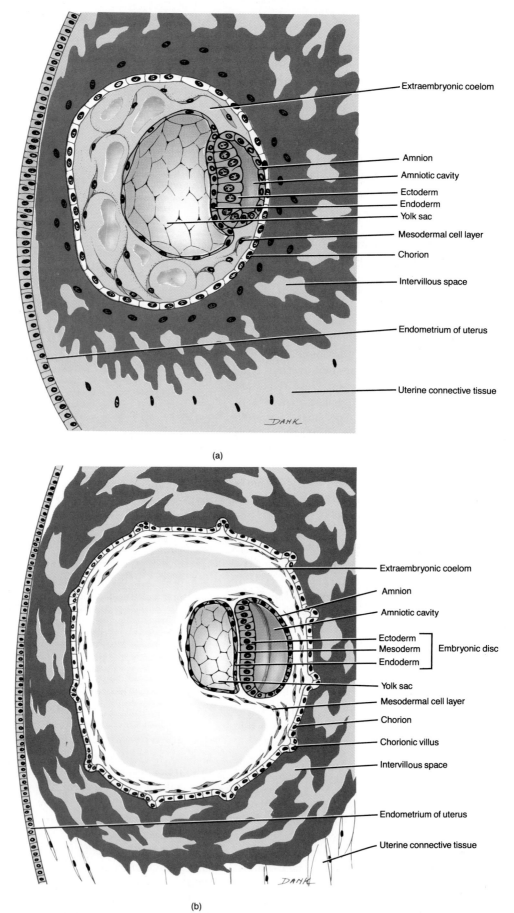

Extraembryonic coelom

Amnion
Amniotic cavity
Ectoderm
Endoderm
Yolk sac
Mesodermal cell layer
Chorion

Intervillous space

Endometrium of uterus

Uterine connective tissue

(a)

Extraembryonic coelom

Amnion

Amniotic cavity

Ectoderm ⎤
Mesoderm ⎬ Embryonic disc
Endoderm ⎦

Yolk sac
Mesodermal cell layer
Chorion
Chorionic villus
Intervillous space

Endometrium of uterus

Uterine connective tissue

(b)

FIGURE 26-6 Formation of the primary germ layers and associated structures. (a) Internal view of the developing embryo about 12 days after fertilization. (b) Internal view of the developing embryo about 14 days after fertilization.

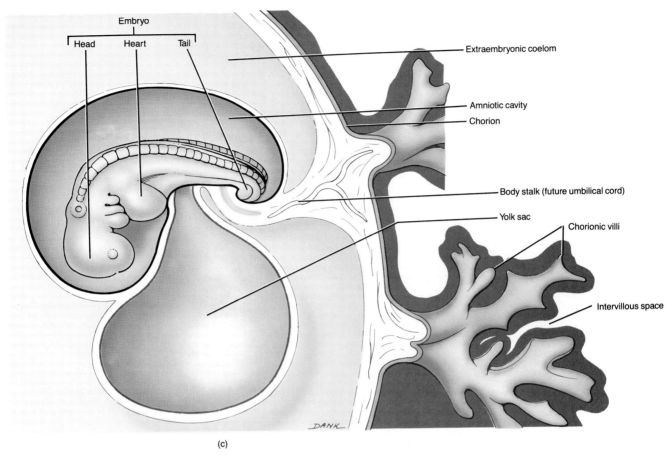

Embryo

Head | Heart | Tail

Extraembryonic coelom

Amniotic cavity

Chorion

Body stalk (future umbilical cord)

Yolk sac

Chorionic villi

Intervillous space

DANK

(c)

FIGURE 26-6 (Continued) Formation of the primary germ layers and associated structures. (c) External view of the developing embryo about 25 days after fertilization.

gut. The mesoderm in the disc soon splits into two layers, and the space between the layers becomes the **extraembryonic coelom.**

As the embryo develops (Figure 26-6c), the endoderm becomes the epithelial lining of the digestive tract and respiratory tract and a number of other organs. The mesoderm forms the peritoneum, muscle, bone, and other connective tissue. The ectoderm develops into the skin and nervous system. Exhibit 26-1 provides more details about the fates of these primary germ layers.

EXHIBIT 26-1   STRUCTURES PRODUCED BY THE THREE PRIMARY GERM LAYERS

ENDODERM	MESODERM	ECTODERM
Epithelium of digestive tract (except the oral cavity and anal canal) and the epithelium of its glands.	All skeletal, most smooth, and all cardiac muscle.	All nervous tissue.
	Cartilage, bone, and other connective tissues.	Epidermis of skin.
Epithelium of urinary bladder, gallbladder, and liver.		Hair follicles, arrector pili muscles, nails, and epithelium of skin glands (sebaceous and sudoriferous).
	Blood, bone marrow, and lymphoid tissue.	
Epithelium of pharynx, auditory tube, tonsils, larynx, trachea, bronchi, and lungs.	Endothelium of blood vessels and lymphatics.	Lens, cornea, and optic nerve of eye and internal eye muscles.
Epithelium of thyroid, parathyroid, pancreas, and thymus glands.	Dermis of skin.	Internal and external ear.
	Fibrous tunic and vascular tunic of eye.	Neuroepithelium of sense organs.
Epithelium of prostate and bulbourethral glands, vagina, vestibule, urethra, and associated glands such as the greater vestibular and lesser vestibular glands.	Middle ear.	Epithelium of oral cavity, nasal cavity, paranasal sinuses, salivary glands, and anal canal.
	Mesothelium of coelomic and joint cavities.	
	Epithelium of kidneys and ureters.	
	Epithelium of adrenal cortex.	Epithelium of pineal gland, hypophysis, and adrenal medulla.
	Epithelium of gonads and genital ducts.	

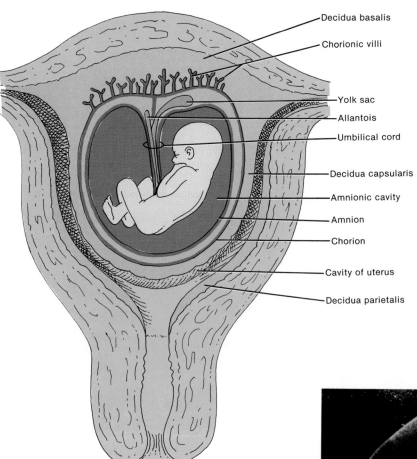

FIGURE 26-7 Embryonic membranes.

Embryonic Membranes

During the embryonic period, the **embryonic membranes** form (Figure 26-7). These membranes lie outside the embryo and protect and nourish the embryo and, later, the fetus. The membranes are the yolk sac, amnion, chorion, and allantois.

The **yolk sac** is an endoderm-lined membrane that, in many species, provides the primary or exclusive nutrient for the embryo (Figure 26-8). However, the human embryo receives its nourishment from the endometrium and the yolk sac remains small. During an early stage of development it becomes a nonfunctional part of the umbilical cord.

The **amnion** is a thin, protective membrane that initially overlies the embryonic disc and is formed by the eighth day following fertilization. As the embryo grows, the amnion entirely surrounds the embryo and becomes filled with **amniotic fluid** (Figure 26-8). Amniotic fluid serves as a shock absorber for the fetus. The amnion usually ruptures just before birth and with its fluid constitutes the "bag of waters."

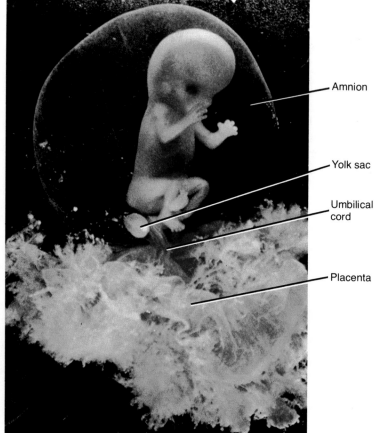

FIGURE 26-8 Ten-week fetus in which the amnion, yolk sac, umbilical cord, and placenta are clearly visible. (Courtesy of Roberts Rugh and Landrum B. Shettles, M.D., with Richard N. Einhorn, from *From Conception to Birth: The Drama of Life's Beginnings,* Harper & Row, Publishers, Inc., New York, 1971.)

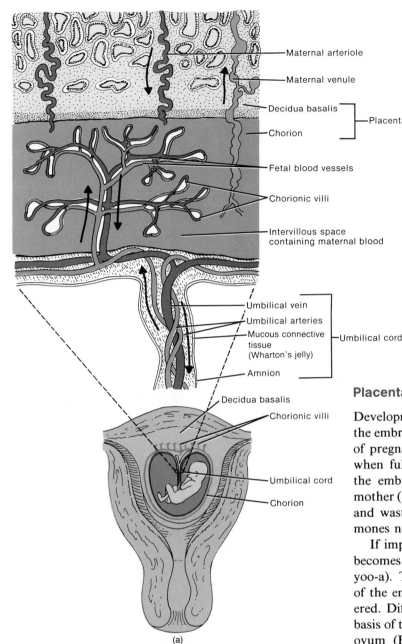

Maternal arteriole

Maternal venule

Decidua basalis ⎤
                 ⎦ Placenta
Chorion

Fetal blood vessels

Chorionic villi

Intervillous space containing maternal blood

Umbilical vein
Umbilical arteries
Mucous connective tissue (Wharton's jelly) ⎤ Umbilical cord
Amnion

Decidua basalis
Chorionic villi

Umbilical cord

Chorion

(a)

FIGURE 26-9 Placenta and umbilical cord. (a) Diagram of the structure of the placenta and umbilical cord.

The **chorion** (KŌ-rē-on) is derived from the trophectoderm of the blastocyst and its associated mesoderm. It surrounds the embryo and, later, the fetus. Eventually the chorion becomes the principal part of the placenta, the structure through which materials are exchanged between mother and fetus. The amnion also surrounds the fetus and eventually fuses to the inner layer of the chorion.

The **allantois** (a-LAN-tō-is) is a small vascularized membrane. Later its blood vessels serve as connections in the placenta between mother and fetus. This connection is the umbilical cord.

Placenta and Umbilical Cord

Development of the placenta, the third major event of the embryonic period, is accomplished by the third month of pregnancy. The **placenta** has the shape of a flat cake when fully developed and is formed by the chorion of the embryo and a portion of the endometrium of the mother (Figure 26-9). It provides an exchange of nutrients and wastes between fetus and mother and secretes hormones necessary to maintain pregnancy.

If implantation occurs, a portion of the endometrium becomes modified and is known as the **decidua** (dē-SID-yoo-a). The decidua includes all but the deepest layer of the endometrium and is shed when the fetus is delivered. Different regions of the decidua are named on the basis of their positions relative to the site of the implanted ovum (Figure 26-10). The *decidua parietalis* (pa-rī-e-TAL-is) is the portion of the modified endometrium that lines the entire pregnant uterus, except for the area where the placenta is forming. The *decidua capsularis* is the portion of the endometrium between the embryo and the uterine cavity. The *decidua basalis* is the portion of the endometrium between the chorion and the muscularis of the uterus. The decidua basalis becomes the maternal part of the placenta.

During embryonic life, fingerlike projections of the chorion, called **chorionic villi** (kō'-rē-ON-ik VIL-ī), grow into the decidua basalis of the endometrium (see Figure 26-8a). These will contain fetal blood vessels of the allantois. They continue growing until they are bathed in maternal blood sinuses called **intervillous** (in-ter-VIL-us) **spaces.** Thus, maternal and fetal blood vessels are brought into proximity. It should be noted, however, that maternal

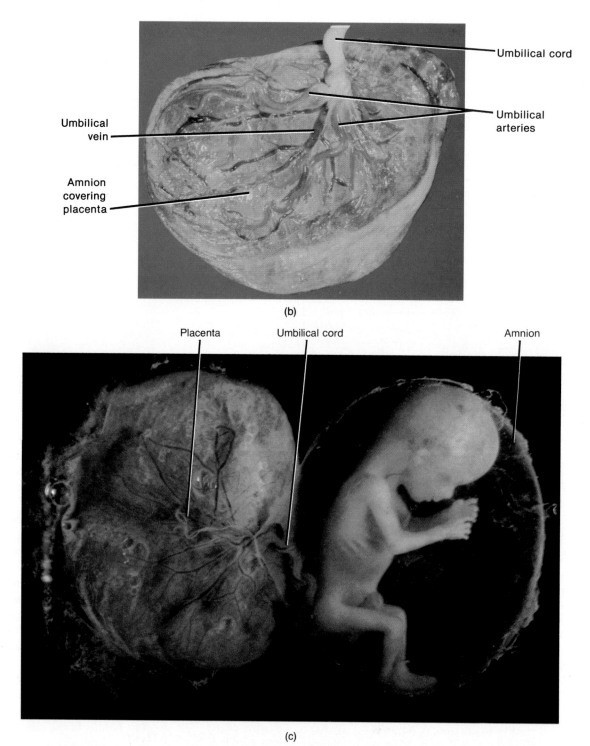

Umbilical cord

Umbilical arteries

Umbilical vein

Amnion covering placenta

(b)

Placenta

Umbilical cord

Amnion

(c)

FIGURE 26-9 (*Continued*) Placenta and umbilical cord. (b) Photograph of the fetal aspect of the placenta. (Photograph courtesy of C. Yokochi and J. W. Rohen, *Photographic Anatomy of the Human Body,* 2nd ed., 1979, IGAKU-SHOIN, Ltd., Tokyo, New York.) (c) Twelve-week fetus showing the relationship of the placenta to the umbilical cord. (Courtesy of Roberts Rugh and Landrum B. Shettles, M.D., with Richard N. Einhorn, from *From Conception to Birth: The Drama of Life's Beginnings,* Harper & Row, Publishers, Inc., New York, 1971.)

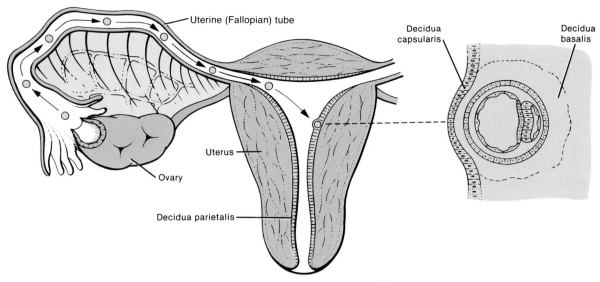

Uterine (Fallopian) tube

Decidua capsularis

Decidua basalis

Uterus

Ovary

Decidua parietalis

FIGURE 26-10 Regions of the decidua.

and fetal blood do not normally mix. Oxygen and nutrients from the mother's blood diffuse into the capillaries of the villi. From the capillaries the nutrients circulate into the umbilical vein. Wastes leave the fetus through the umbilical arteries, pass into the capillaries of the villi, and diffuse into the maternal blood. The **umbilical cord** consists of an outer layer of amnion containing the umbilical arteries and umbilical vein, supported internally by mucous connective tissue (Wharton's jelly) from the allantois (see Figure 26-8a).

CLINICAL APPLICATION

At delivery, the placenta detaches from the uterus and is referred to as the **afterbirth.** At this time, the umbilical cord is severed, leaving the baby on its own. The scar that marks the site of the entry of the fetal umbilical cord into the abdomen is the **umbilicus (navel).**

Pharmaceutical houses use human placenta as a source of hormones, drugs, and blood. Portions of placentas are also used for burn coverage. The placental and umbilical cord veins are used in blood vessel grafts.

FETAL GROWTH

During the **fetal period,** organs established by the primary germ layers grow rapidly. The organism takes on a human appearance. A summary of changes associated with the fetal period is presented in Exhibit 26-2.

GESTATION

The time the zygote, embryo, or fetus is carried in the female reproductive tract is called **gestation** (jes-TĀ-shun). The total human gestation period is about 280 days from the beginning of the last menstrual period. The specialized branch of medicine that deals with pregnancy, labor, and the period of time immediately following delivery is called **obstetrics** (ob-STET-riks; *obstetrix* = midwife).

By about the end of the third month of gestation, the uterus occupies most of the pelvic cavity, and as the fetus continues to grow, the uterus extends higher and higher into the abdominal cavity. In fact, toward the end of a full-term pregnancy, the uterus occupies practically all of the abdominal cavity, reaching above the costal margin nearly to the xiphoid process of the sternum (Figure 26-11), causing displacement of the maternal intestines, liver, and stomach upward, elevation of the diaphragm, and widening of the thoracic cavity. In addition, the breasts enlarge in anticipation of lactation and the areolae around the nipples become darkly pigmented.

CLINICAL APPLICATION

Physicians can see into the uteri of pregnant women without exposing them to the known dangers of x rays and without pain or intrusion. One such technique, called **ultrasound** (also **sonography** or **ultrasonography**), uses high-frequency, inaudible sound waves, which are directed into the abdomen of the mother-to-be and then reflected back to a receiver. The reflected waves give a visual "echo" of what's inside

EXHIBIT 26-2 CHANGES ASSOCIATED WITH FETAL GROWTH

END OF MONTH	APPROXIMATE SIZE AND WEIGHT	REPRESENTATIVE CHANGES
1	0.6 cm (³⁄₁₆ in)	Eyes, nose, and ears not yet visible. Backbone and vertebral canal form. Small buds that will develop into arms and legs form. Heart forms and starts beating. Body systems begin to form.
2	3 cm (1¼ in) 1 g (¹⁄₃₀ oz)	Eyes far apart, eyelids fused, nose flat. Ossification begins. Limbs become distinct as arms and legs. Digits are well formed. Major blood vessels form. Many internal organs continue to develop.
3	7.5 cm (3 in) 28 g (1 oz)	Eyes almost fully developed but eyelids still fused, nose develops bridge, and external ears are present. Ossification continues. Appendages are fully formed and nails develop. Heartbeat can be detected. Body systems continue to develop.
4	18 cm (6½–7 in) 113 g (4 oz)	Head large in proportion to rest of body. Face takes on human features and hair appears on head. Skin bright pink. Many bones ossified, and joints begin to form. Continued development of body systems.
5	25–30 cm (10–12 in) 227–454 g (½–1 lb)	Head less disproportionate to rest of body. Fine hair (lanugo) covers body. Skin still bright pink. Rapid development of body systems.
6	27–35 cm (11–14 in) 567–681 g (1¼–1½ lb)	Head becomes even less disproportionate to rest of body. Eyelids separate and eyelashes form. Skin wrinkled and pink.
7	32–42 cm (13–17 in) 1,135–1,362 g (2½–3 lb)	Head and body become more proportionate. Skin wrinkled and pink. Seven-month fetus (premature baby) is capable of survival.
8	41–45 cm (16½–18 in) 2,043–2,270 g (4½–5 lb)	Subcutaneous fat deposited. Skin less wrinkled. Testes descend into scrotum. Bones of head are soft. Chances of survival much greater at end of eighth month.
9	50 cm (20 in) 3,178–3,405 g (7–7½ lb)	Additional subcutaneous fat accumulates. Lanugo shed. Nails extend to tips of fingers and maybe even beyond.

the uterus. This echo is transformed electronically into an image on a screen. The potential risks of ultrasonography, such as decreased immune response, changes in plasma membrane functions, and breakdown of macromolecules, are currently under investigation. Another technique is the **Doppler detector,** a fetal monitor that registers the baby's heart rate. It is used routinely in many hospitals during labor.

In the United States today, most obstetricians prescribe ultrasound examinations only when there is some clinical question about the normal progress of the pregnancy. By far the most common use of diagnostic ultrasound is to determine true fetal age when the date of conception is unknown or mistaken by the mother. It is also used to evaluate fetal growth, determine fetal position, determine the time of ovulation in infertile females to enhance the chances of pregnancy, ascertain the cause of vaginal bleeding, diagnose ectopic and multiple pregnancies, and as an adjunct to special procedures such as amniocentesis. Ultrasound is not used routinely to determine the sex of a fetus; it is only performed for a specific medical indication.

Ultrasound has gained a wide application beyond obstetrics, including the detection of tumors, gallstones, and other abnormal internal masses.

PARTURITION AND LABOR

The term **parturition** (par'-too-RISH-un) refers to birth. Parturition is accompanied by a sequence of events commonly called **labor.** The onset of labor is apparently related to a complex interaction of many factors. Just prior to birth, the muscles of the uterus contract rhythmically and forcefully. Both placental and ovarian hormones seem to play a role in these contractions. Since progesterone inhibits uterine contractions, labor cannot take place until its effects are diminished. At the end of gestation, the level of estrogens in the mother's blood is sufficient to overcome the inhibiting effects of progesterone and labor commences. It has been suggested that some factor released by the placenta, fetus, or mother rather suddenly overcomes the inhibiting effects of progesterone so that estrogens can exert their effect. Prostaglandins may also play a role in labor. Oxytocin (OT) from the posterior pituitary gland also stimulates uterine contractions. And relaxin assists by relaxing the symphysis pubis and helping to dilate the uterine cervix.

Uterine contractions occur in waves, quite similar to peristaltic waves, that start at the top of the uterus and move downward. These waves expel the fetus. **True labor** begins when pains occur at regular intervals. The pains correspond to uterine contractions. As the interval be-

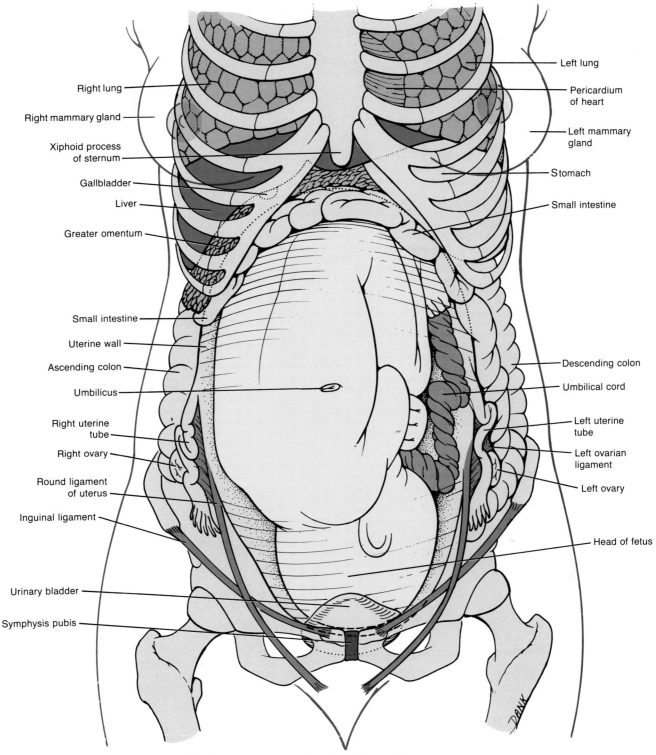

Right lung

Right mammary gland

Xiphoid process
of sternum

Gallbladder

Liver

Greater omentum

Small intestine

Uterine wall

Ascending colon

Umbilicus

Right uterine
tube

Right ovary

Round ligament
of uterus

Inguinal ligament

Urinary bladder

Symphysis pubis

Left lung

Pericardium
of heart

Left mammary
gland

Stomach

Small intestine

Descending colon

Umbilical cord

Left uterine
tube

Left ovarian
ligament

Left ovary

Head of fetus

FIGURE 26-11 Normal fetal position during a full-term pregnancy.

tween contractions shortens, the contractions intensify. Another sign of true labor in some females is localization of pain in the back, which is intensified by walking. A reliable indication of true labor is the "show" and dilation of the cervix. The "show" is a discharge of a blood-containing mucus that accumulates in the cervical canal during pregnancy. In **false labor,** pain is felt in the abdomen at irregular intervals. The pain does not intensify and is

not altered significantly by walking. There is no "show" and no cervical dilation.

Labor can be divided into three stages (Figure 26-12).

1. The **stage of dilation** is the time from the onset of labor to the complete dilation of the cervix. During this stage there are regular contractions of the uterus, usually a rupturing of the amniotic sac, and complete

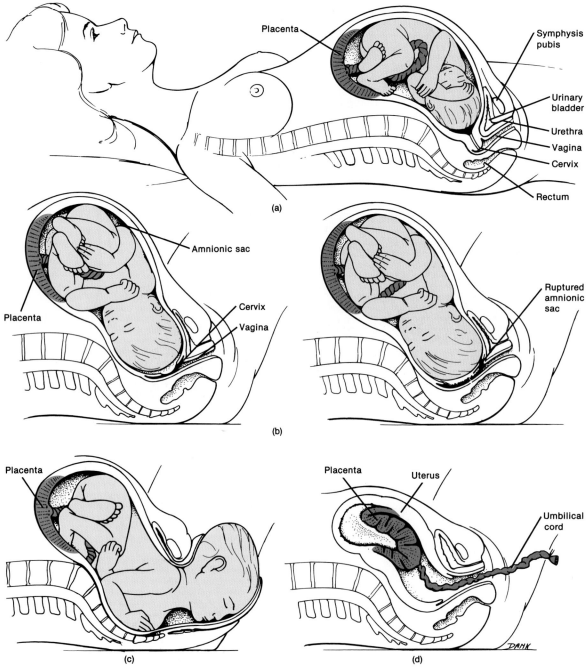

FIGURE 26-12 Parturition. (a) Fetal position prior to birth. (b) Dilation. Protrusion of amniotic sac through partly dilated cervix (left). Amniotic sac ruptured and complete dilation of cervix (right). (c) Stage of expulsion. (d) Placental stage.

dilation (10 cm) of the cervix. If the amniotic sac does not rupture spontaneously, it is done artificially.

2. The **stage of expulsion** is the time from complete cervical dilation to delivery.

3. The **placental stage** is the time after delivery until the placenta or "afterbirth" is expelled by powerful uterine contractions. These contractions also constrict blood vessels that were torn during delivery. In this way, the possibility of hemorrhage is reduced.

CLINICAL APPLICATION

Pudendal (pyoo-DEN-dal) **nerve block** is used for procedures such as episiotomy. The primary innervation to the skin and muscles of the perineum is the pudendal nerve. In the transvaginal approach, the needle is passed through the lateral vaginal wall to a point just medial to the ischial spine. The anesthesia results in

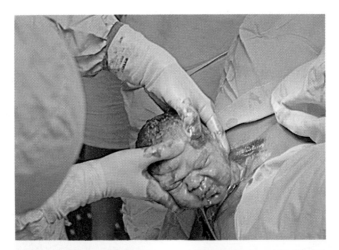

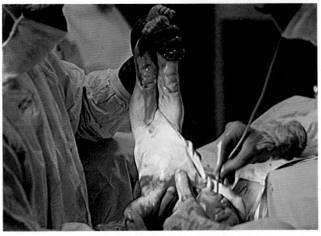
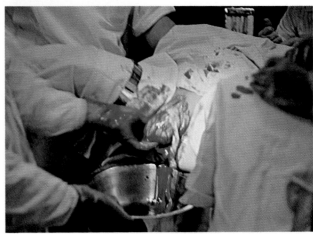

(e)

FIGURE 26-12 (*Continued*) Parturition. (e) Sequential photographs of parturition. (Top) Emergence of the infant's head. (Courtesy of Kinne, Photo Researchers.) (Bottom left) Delivery of the infant. (Courtesy of Thomas, Photo Researchers.) (Bottom right) Delivery of the placenta. (Courtesy of McCartney, Photo Researchers.)

loss of the anal reflex, relaxation of the muscles of the floor of the pelvis, and loss of sensation to the vulva and lower one-third of the vagina.

Various deformities of the female pelvis may be responsible for **dystocia** (dis-TŌ-sē-a), that is, difficult labor. Pelvic deformities may be congenital or acquired from disease, fractures, or poor posture. Among other conditions associated with difficult labor are malposition of the fetus, malpresentation of the fetus, and premature rupture of the fetal membranes.

CLINICAL APPLICATION

If either dystocia or prolonged labor are indicated, it may be necessary to deliver the baby via a **cesarean** (*caedere* = to cut) **section**. In this procedure, a low,

horizontal incision is made through the abdominal wall and uterus, through which the baby and placenta are removed.

PRENATAL DIAGNOSTIC TECHNIQUES

AMNIOCENTESIS

Amniocentesis is a technique of withdrawing some of the amniotic fluid that bathes the developing fetus to diagnose genetic disorders or to determine fetal maturity or wellbeing. Using ultrasound, the position of the fetus and placenta are first determined. About 10 to 20 ml of fluid is removed by hypodermic needle puncture of the uterus, usually 16 to 20 weeks after conception (Figure 26-13). Cells and fluid are subjected to microscopic examination

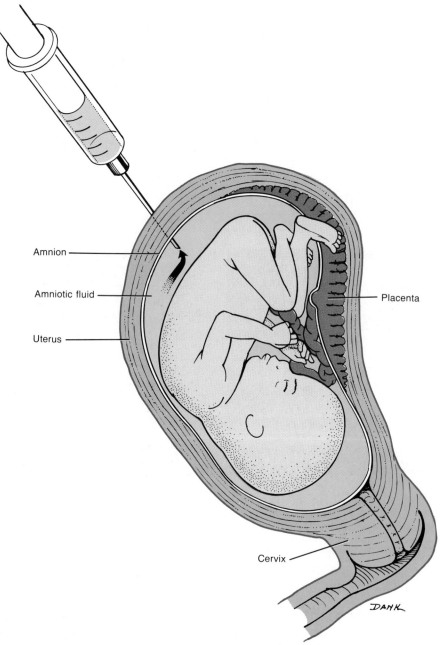

Amnion

Amniotic fluid

Uterus

Placenta

Cervix

DANK

FIGURE 26-13 Amniocentesis.

and biochemical testing to determine abnormalities in chromosome number or structure and biochemical defects. Close to 300 chromosomal disorders and over 50 inheritable biochemical defects can be detected through amniocentesis, including hemophilia, certain muscular dystrophies, Tay-Sachs disease, myelocytic leukemia, Klinefelter's and Turner's syndromes, sickle cell anemia, thalassemia, and cystic fibrosis. When both parents are known or suspected to be genetic carriers of any one of these disorders, amniocentesis is advised.

CLINICAL APPLICATION

One chromosome disorder that may be diagnosed through amniocentesis is **Down's syndrome.** This disorder is characterized by mental retardation, retarded physical development (short stature and stubby fingers), distinctive facial structures (large tongue, broad skull, slanting eyes, and round head), and malformation of the heart, ears, hands, and feet. Sexual maturity is rarely attained. Individuals with the disorder usually have 47 chromosomes instead of the normal 46 (an extra chromosome in the twenty-first pair).

CHORIONIC VILLI SAMPLING (CVS)

A new test used to detect prenatal genetic defects is now available and is referred to as **chorionic villi sampling (CVS).** Although the test picks up the same defects as amniocentesis, it has several advantages. First of all, it can be performed during the first trimester of pregnancy, usually at 8 to 10 weeks gestation. Also, results are available within 24 hours. Moreover, the procedure does not require penetration of the abdominal wall, uterine wall, or amniotic cavity.

CVS is performed as follows. A catheter is placed through the vagina into the uterus and then to the chorionic villi under ultrasound guidance. About 30 mg of tissue is suctioned out and prepared for chromosomal analysis. Chorion cells and fetal cells contain identical genetic information. The safety of the procedure is believed to compare to that for amniocentesis.

KEY MEDICAL TERMS ASSOCIATED WITH DEVELOPMENTAL ANATOMY

Abortion   Premature expulsion from the uterus of the products of conception—embryo or nonviable fetus.

Cautery   Application of a caustic (burning) substance or instrument for the destruction of tissue.

Colpotomy (*colp* = vagina; *tome* = cutting)   Incision of the vagina.

Culdoscopy (*skopein* = to examine)   A procedure in which a culdoscope (endoscope) is used to view the pelvic cavity. The approach is through the vagina.

Hermaphroditism   Presence of both male and female sex organs in one individual.

Karyotype (*karyon* = nucleus)   The chromosomal element typical of a cell, drawn in their true proportions, based on the average of measurements determined in a number of cells. Useful in judging whether or not chromosomes are normal in number and structure.

Lethal gene (*lethum* = death)   A gene that, when expressed, results in death either in the embryonic state or shortly after birth.

Lochia   The discharge from the birth canal consisting initially of blood and later of serous fluid occurring after childbirth. The discharge is derived from the former placental site and may last up to about a week and a half.

Mutation (*mutare* = change)   A permanent, heritable change in a gene that causes it to have a different effect than it had previously.

Puerperal (*puer* = child, *parere* = to bring forth) **fever**   Infectious disease of childbirth, also called puerperal sepsis and childbed fever. The disease results from an infection originating in the birth canal and affects the endometrium. It may spread to other pelvic structures and lead to septicemia.

STUDY OUTLINE

Pregnancy (p. 691)
1. Pregnancy is a sequence of events that includes fertilization, implantation, embryonic growth, fetal growth, and birth.
2. Its various events are hormonally controlled.

Fertilization and Implantation
1. Fertilization refers to the penetration of the ovum by a sperm cell and the subsequent union of the sperm and ovum nuclei to form a zygote.
2. Penetration is facilitated by hyaluronidase and proteinases produced by sperm.
3. Normally only one sperm fertilizes an ovum.
4. Early rapid cell division of a zygote is called cleavage, and the cells produced by cleavage are called blastomeres.
5. The solid mass of cells produced by cleavage is a morula.
6. The morula develops into a blastocyst, a hollow ball of cells differentiated into a trophectoderm (future embryonic membranes) and inner cell mass (future embryo).
7. The attachment of a bastocyst to the endometrium is called implantation.
8. It occurs by enzymatic degradation of the endometrium.

Embryonic Development (p. 695)
1. During embryonic growth, the primary germ layers and embryonic membranes are formed and the placenta is functioning.
2. The primary germ layers—ectoderm, mesoderm, and endoderm—form all tissues of the developing organism.
3. Embryonic membranes include the yolk sac, amnion, chorion, and allantois.
4. Fetal and maternal materials are exchanged through the placenta.

Fetal Growth (p. 701)
1. During the fetal period, organs established by the primary germ layers grow rapidly.
2. The principal changes associated with fetal growth are summarized in Exhibit 26-2.

Gestation (p. 701)
1. The time an embryo or fetus is carried in the uterus is called gestation.

2. Human gestation lasts about 280 days from the beginning of the last menstrual period.

Parturition and Labor (p. 702)

1. Parturition refers to birth and is accompanied by a sequence of events called labor.
2. The birth of a baby involves dilation of the cervix, expulsion of the fetus, and delivery of the placenta.

Prenatal Diagnostic Techniques (p. 705)

1. Amniocentesis is the withdrawal of amniotic fluid. It can be used to diagnose inherited biochemical defects and chromosomal disorders, such as hemophilia, Tay-Sachs disease, sickle cell anemia, and Down's syndrome.
2. Down's syndrome is a chromosomal abnormality characterized by mental retardation and retarded physical development.
3. Chorionic villi sampling (CVS) involves withdrawal of chorionic villi for chromosomal analysis.
4. CVS can be done sooner than amniocentesis and the results are available sooner.

REVIEW QUESTIONS

1. Define developmental anatomy.
2. Define fertilization. Where does it normally occur? How is a morula formed?
3. Describe the components of a blastocyst.
4. What is implantation? How does the fertilized ovum implant itself? What causes morning sickness?
5. Describe the procedure for external human fertilization and embryo transfer.
6. Define the embryonic period and the fetal period.
7. List several body structures formed by the endoderm, mesoderm, and ectoderm.
8. What is an embryonic membrane? Describe the functions of the four embryonic membranes.
9. Explain the importance of the placenta and umbilical cord to fetal growth.
10. Outline some of the major developmental changes during fetal growth.
11. Define gestation and parturition.
12. Distinguish between false and true labor. Describe what happens during the stage of dilation, the stage of expulsion, and the placental stage of delivery.
13. What is amniocentesis? What is its value?
14. Describe chorionic villi sampling (CVS). What are its advantages over amniocentesis?
15. Refer to the glossary of key medical terms associated with developmental anatomy. Be sure that you can define each term.

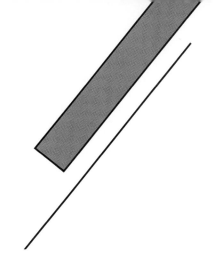

Appendix A: Abbreviations

Many terms, especially medical terms, are commonly expressed in abbreviated form. To familiarize you with some of these abbreviations, an alphabetical list has been prepared. Most of the terms listed have been used in the book; some terms not referred to in the book have been included because of their frequent use.

AID   automatic implantable defibrillation; artificial insemination by donor
AIDS   acquired immune deficiency syndrome
AMI   acute myocardial infarction
ALS   amyotrophic lateral sclerosis
ANS   autonomic nervous system
AV   atrioventricular
BBB   blood–brain barrier
BP   blood pressure
BS   blood sugar
CABG   coronary artery bypass grafting
CAC   cardioacceleratory center
CAD   coronary artery disease
CAPD   continuous ambulatory peritoneal dialysis
CBC   complete blood count
CCCC   closed-chest cardiac compression
CF   cystic fibrosis
CHD   coronary heart disease
CHF   congestive heart failure
CIC   cardioinhibitory center
CNS   central nervous system
CO   cardiac output; carbon monoxide
COAD   chronic obstructive airways disease
COPD   chronic obstructive pulmonary disease
CPR   cardiopulmonary resuscitation
CP   cor pulmonale
C-section   cesarean section
CSF   cerebrospinal fluid
CT (CAT)   computed tomography (computed axial tomography)
CVA   cerebrovascular accident
CVS   chorionic villi sampling
D & C   dilation and curettage
DBP   diastolic blood pressure
DES   diethylstilbestrol
DMD   Duchenne muscular dystrophy
DMSO   dimethyl sulfoxide
DNR   do not resuscitate

DSR   dynamic spatial reconstructor
DVT   deep vein thrombosis
EB virus   Epstein-Barr virus
ECF   extracellular fluid
ECG (EKG)   electrocardiogram
EEG   electroencephalogram
EM   electron micrograph
EMG   electromyogram
EOG   electrooculogram
EP   ectopic pregnancy
ER   endoplasmic reticulum
ESR   erythrocyte sedimentation rate
ESRD   end-stage renal disease
ESWL   extracorporeal shock wave lithotripsy
FAS   fetal alcohol syndrome
GI   gastrointestinal
Hb   hemoglobin
HBO   hyperbaric oxygenation
HBV   hepatitis B virus
HDL   high-density lipoprotein
HF   heart failure
HR   heart rate
HSV   herpes simplex virus
ICF   intracellular fluid
ID   intradermal
IM   intramuscular
INF   interferon
IUD   intrauterine device
IV   intravenous
LDL   low-density lipoprotein
LLQ   left lower quadrant
LUQ   left upper quadrant
LV   left ventricular
MG   myasthenia gravis
MI   myocardial infarction
mm³   cubic millimeter
MRI   magnetic resonance imaging
MS   multiple sclerosis
MVP   mitral valve prolapse
NGU   nongonococcal urethritis
NLMC   nocturnal leg muscle cramping
OC   oral contraceptive
OTC   over-the-counter
PCT   positron-computed tomography
PE   pulmonary embolism

PEMFs pulsating electromagnetic fields
PG prostaglandin
PID pelvic inflammatory disease
PKU phenylketonuria
PMNs polymorphonuclear leucocytes
PMS premenstrual syndrome
PNS peripheral nervous system
PPNG penicillinase-producing *Neisseria gonorrhea*
PTA percutaneous transluminal angioplasty
PUL percutaneous ultrasonic lithotripsy
RA rheumatoid arthritis
RAS reticular activating system
RBC red blood cell
RDS respiratory distress syndrome
RES reticuloendothelial system
Rh *Rhesus*
RLQ right lower quadrant
RUQ right upper quadrant
SA sinoatrial (sinuatrial)
SCA sickle cell anemia

SCIB severe combined immunodeficiency disease
SIDS sudden infant death syndrome
SLE systemic lupus erythematosus
SMD senile macular degeneration
SNS somatic nervous system
SPF sun protection factor
STD sexually transmitted disease
TIA transient ischemic attack
TM transcendental meditation
TM joint temporomandibular joint
TSS toxic shock syndrome
UA urinalysis
URI upper respiratory infection
UTI urinary tract infection
UV ultraviolet
VD venereal disease
VF ventricular fibrillation
VSD ventricular septal defect
VT ventricular tachycardia
WBC white blood cell

Appendix B: Eponyms Used in This Text

Eponymous terms are those named after a person. In general, eponyms should be avoided where possible, since they are totally nondescriptive, often vague, and do not necessarily indicate that the person whose name is used actually contributed anything very original. However, since eponyms are still in frequent use, this glossary has been prepared to indicate which current terms

have been used to replace eponyms in this book. In the body of the text eponyms are cited in parentheses, immediately following the current terms where they are used for the first time in a chapter or later in the book. In addition, although eponyms are included in the index, they have been cross referenced to their current terminology.

EPONYM	CURRENT TERMINOLOGY
Adam's apple	thyroid cartilage
ampulla of Vater (VA-ter)	hepatopancreatic ampulla
Bartholin's (BAR-tō-linz) gland	greater vestibular gland
Billroth's (BIL-rōtz) cord	splenic cord
Bowman's (BŌ-manz) capsule	glomerular capsule
Broca's (BRŌ-kaz) area	motor speech area
Brunner's (BRUN-erz) gland	duodenal gland
bundle of His (HISS)	atrioventricular (AV) bundle
canal of Schlemm (SHLEM)	scleral venous sinus
circle of Willis (WIL-is)	cerebral arterial circle
Cooper's (KOO-perz) ligament	suspensory ligament of the breast
Cowper's (KOW-perz) gland	bulbourethral gland
crypt of Lieberkühn (LE-ber-kyoon)	intestinal gland
duct of Rivinus (re-VĒ-nus)	lesser sublingual duct
duct of Santorini (san'-tō-RĒ-nē)	accessory duct
duct of Wirsung (VĒR-sung)	pancreatic duct
end organ of Ruffini (roo-FĒ-nē)	type II cutaneous mechanoreceptor
Eustachian (ū-STĀ-kē-an) tube	auditory tube
Fallopian (fal-LO-pē-an) tube	uterine tube
gland of Zeis (ZĪS)	sebaceous ciliary gland
Golgi (GOL-jē) tendon organ	tendon organ
Graafian (GRAF-ē-an) follicle	vesicular ovarian follicle
Hassall's (HAS-alz) corpuscle	thymic corpuscle
Haversian (ha-VĒR-shun) canal	central canal
Haversian (ha-VĒR-shun) system	osteon
Henle's (HEN-lēz) layer	pallid layer
Huxley's (HUKS-lēz) layer	granular layer
interstitial cell of Leydig (LĪ-dig)	interstitial endocrinocyte
islet of Langerhans (LANG-er-hanz)	pancreatic islet
Kupffer (KOOP-fer) cell	stellate reticuloendothelial cell
loop of Henle (HEN-lē)	loop of the nephron
Malpighian (mal-PIG-ē-an) corpuscle	splenic nodule
Meibomian (mi-BŌ-mē-an) gland	tarsal gland
Meissner's (MĪS-nerz) corpuscle	corpuscle of touch
Merkel's (MER-kelz) disc	tactile disc

Müller's (MIL-erz) duct	paramesonephric duct
Nissl (NIS-l) bodies	chromatophilic substance
node of Ranvier (ron-VĒ-ā)	neurofibral node
organ of Corti (KOR-tē)	spiral organ
Pacinian (pa-SIN-ē-an) corpuscle	lamellated corpuscle
Peyer's (PĪ-erz) patches	aggregated lymphatic follicles
plexus of Auerbach (OW-er-bak)	myenteric plexus
plexus of Meissner (MĪS-ner)	submucous plexus
pouch of Douglas	rectouterine pouch
Purkinje (pur-KIN-jē) fiber	conduction myofiber
Rathke's (rath-KEZ) pouch	hypophyseal pouch
Schwann (SCHVONZ) cell	neurolemmocyte
Sertoli (ser-TŌ-lē) cell	sustentacular cell
Skene's (SKĒNZ) gland	paraurethral gland
sphincter of Oddi (OD-dē)	sphincter of the hepatopancreatic ampulla
Stensen's (STEN-senz) duct	parotid duct
Volkmann's (FŌLK-manz) canal	perforating canal
Wharton's (HWAR-tunz) duct	submandibular duct
Wharton's (HWAR-tunz) jelly	mucous connective tissue
Wormian (WER-mē-an) bone	sutural bone

Selected Readings

Anderson, J. E., *Grant's Atlas of Anatomy,* 8th ed. Baltimore; Williams & Wilkins, 1983.

Bacon, R. L., and N. R. Niles, *Medical Histology,* New York: Springer-Verlag, 1983.

Barr, M. L., and J. A. Kiernan, *The Human Nervous System,* 4th ed. New York: Harper & Row, 1983.

Basmajian, J. V., *Grant's Method of Anatomy,* 10th ed. Baltimore: Williams & Wilkins, 1980.

Basmajian, J. V., *Primary Anatomy,* 8th ed. Baltimore: Williams & Wilkins, 1982.

Basmajian, J. V., *Surface Anatomy,* Baltimore: Williams & Wilkins, 1977.

Carlson, B. M., *Patten's Foundations of Human Embryology,* 4th ed. New York: McGraw-Hill Book Co., 1981.

Carpenter, M. B., *Human Neuroanatomy,* 7th ed. Baltimore: Williams & Wilkins, 1976.

Christensen, J. B., and I. R. Telford, *Synopsis of Gross Anatomy,* 4th ed. New York: Harper & Row, 1982.

Clemente, C. D., *Anatomy: A Regional Atlas of the Human Body,* 2d ed. Baltimore: Urban and Schwarzenberg, 1981.

Cunningham's Textbook of Anatomy, 12th ed. Edited by G. J. Romanes. London: Oxford University Press, 1981.

DiFiore, M. S. H., *An Atlas of Human Histology,* 5th ed. Philadelphia: Lea & Febiger, 1981.

Ellis, H., *Clinical Anatomy,* 6th ed. Philadelphia: Lippincott, 1977.

Ezrin, C., J. O. Godden, and R. Volpé, *Systematic Endocrinology,* 2d ed. New York: Harper & Row, 1979.

Gluhbegovic, N., and T. H. Williams, *The Human Brain: A Photographic Guide,* New York: Harper & Row, 1980.

Gray, H., *Anatomy of the Human Body,* 29th ed. Edited by C. M. Goss, Philadelphia: Lea & Febiger, 1973.

Hafez, E. S. E., *Human Reproduction,* 2d ed. New York: Harper & Row, 1980.

Ham, A. W., and D. H. Cormack, *Histology,* 8th ed., Philadelphia: Lippincott, 1979.

Hamilton, W. J., G. Simon, and S. G. I. Hamilton, *Surface and Radiological Anatomy,* 5th ed. Cambridge: Heffer & Sons, 1971.

Hollinshead, W. H., *Textbook of Anatomy,* 3d ed. New York: Harper & Row, 1974.

Kelly, D. E., R. L. Wood, and A. C. Enders, *Bailey's Textbook of Microscopic Anatomy,* 18th ed. Baltimore: Williams & Wilkins, 1984.

Kessel, R. G., and R. H. Kardon, *Tissues and Organs: A Text-Atlas of Scanning Electron Microscopy.* San Francisco: Freeman, 1979.

Kieffer, S. A., and E. R. Heitzman, *An Atlas of Cross-Sectional Anatomy.* New York: Harper & Row, 1979.

Langman, J., *Medical Embryology,* 4th ed. Baltimore: Williams & Wilkins, 1981.

Langman, J., and M. W. Woerdeman, *Atlas of Medical Anatomy.* Philadelphia: Saunders, 1982.

Lockhart, R. D., *Living Anatomy,* 7th ed. London: Faber & Faber, 1974.

Lockhart, R. D., G. F. Hamilton, and F. W. Fyfe, *Anatomy of the Human Body,* 2d ed. Philadelphia: Lippincott, 1969.

McMinn, R. M. H., and R. T. Hutchings, *Color Atlas of Human Anatomy.* Chicago: Year Book Medical Publishers, 1977.

Montgomery, R. L., *Basic Anatomy for the Allied Health Professions.* Baltimore: Urban & Schwarzenberg, 1981.

Moore, K. L., *Clinically Oriented Anatomy.* Baltimore: Williams & Wilkins, 1980.

Moyer, K. E., *Neuroanatomy.* New York: Harper & Row, 1980.

Netter, F. H., *Ciba Collection of Medical Illustrations,* Vols. 1–7. Summit, N.J.: CIBA, 1962–1983.

O'Rahilly, R., *Basic Human Anatomy.* Philadelphia: Saunders, 1983.

Pernkopf, E., *Atlas of Topographical and Applied Human Anatomy.* Edited by H. Ferner; translated by H. Monsen. Philadelphia: Saunders, Vol. 1, 1963; Vol. 2, 1964.

Ranson, S. W., and S. L. Clark, *Anatomy of the Nervous System, Its Development and Function,* 10th ed. Philadelphia: Saunders, 1959.

Reith, E. J., and M. N. Ross, *Histology: A Text and Atlas.* New York: Harper & Row, 1985.

Rohen, J. W., and C. Yokochi, *Color Atlas of Anatomy.* Tokyo, New York: Igaku-Shoin, Ltd., 1983.

Royce, J., *Surface Anatomy.* Philadelphia: Davis, 1965.

Snell, R. S., *Clinical Anatomy for Medical Students.* Boston: Little, Brown & Co., 1981.

Snell, R. S., *Clinical and Functional Histology for Medical Students.* Boston: Little, Brown & Co., 1984.

Sobotta, J., and F. Hammersen, *Histology.* Philadelphia: Lea & Febiger, 1976.

Sobotta Atlas of Human Anatomy, Vols. 1 and 2, 10th English ed. Edited by H. Ferner and J. Staubesand. Baltimore: Urban & Schwarzenberg, 1983.

Williams, R. H., *Textbook of Endocrinology,* 6th ed. Philadelphia: Saunders, 1981.

Woodburne, R. T., *Essentials of Human Anatomy,* 7th ed. New York: Oxford University Press, 1983.

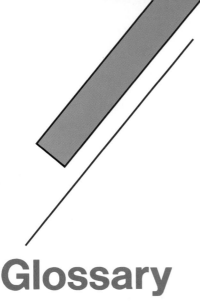

Glossary

PRONUNCIATION KEY

1. The strongest accented syllable appears in capital letters, for example, bilateral (bī-LAT-er-al) and diagnosis (dī-ag-NŌ-sis).
2. If there is a secondary accent, it is noted by a single quote mark ('), for example, constitution (kon'-sti-TOO-shun) and physiology (fiz'-ē-OL-ō-jē). Any additional secondary accents are also noted by a single quote mark, for example, decarboxylation (dē'-kar-bok'-si-LA-shun).
3. Vowels marked with a line above the letter are pronounced with the long sound as in the following common words:

   ā as in māke
   ē as in bē
   ī as in īvy
   ō as in pōle

4. Vowels not so marked are pronounced with the short sound, as in the following words:

   e as in bet
   i as in sip
   o as in not
   u as in bud

5. Other phonetic symbols are used to indicate the following sounds:

   a as in above
   oo as in sue
   yoo as in cute
   oy as in oil

Abatement (a-BĀT-ment) A decrease in the seriousness of a disorder or in the severity of pain or other symptoms.

Abdomen (ab-DŌ-men) The area between the diaphragm and pelvis.

Abdominal (ab-DŌM-i-nal) **cavity** Superior portion of the abdominopelvic cavity that contains the stomach, spleen, liver, gallbladder, pancreas, small intestine, and most of the large intestine.

Abdominal thrust maneuver A first-aid procedure for choking. Employs a quick, upward thrust against the diaphragm that forces air out of the lungs with sufficient force to eject any lodged material. Also called the **Heimlich maneuver.**

Abdominopelvic (ab-dō-men-ō'-PEL-vic) **cavity** Inferior component of the ventral body cavity that is subdivided into an upper abdominal cavity and a lower pelvic cavity.

Abduction (ab-DUK-shun) Movement away from the axis or midline of the body or one of its parts.

Abortion (a-BOR-shun) The premature loss or removal of the embryo or nonviable fetus; any failure in the normal process of developing or maturing.

Abscess (AB-ses) A localized collection of pus and liquefied tissue in a cavity.

Absorption (ab-SORP-shun) The taking up of liquids by solids or of gasses by solids or liquids; intake of fluids or other substances by cells of the skin or mucous membranes; the passage of digested foods from the GI tract into blood or lymph.

Accessory duct A duct of the pancreas that empties into the duodenum about 2.5 cm (1 in) superior to the hepatopancreatic ampulla. Also called the **duct of Santorini.**

Accretion (a-KRĒ-shun) A mass of material that has accumulated in a space or cavity; the adhesion of parts.

Acetabulum (as'-e-TAB-yoo-lum) The rounded cavity on the external surface of the coxal bone that receives the head of the femur.

Acetylcholine (as'-ē-til-KŌ-lēn) **(ACh)** A neurotransmitter liberated at synapses in the central nervous system that stimulates skeletal muscle contraction.

Achlorhydria (ā-klōr-HĪ-drē-a) Absence of hydrochloric acid in the gastric juice.

Acid (AS-id) A proton donor, or substance that dissociates into hydrogen ions (H^+) and anions, characterized by an excess of hydrogen ions and a pH less than 7.

Acinar (AS-i-nar) Flasklike.

Acini (AS-i-nē) Masses of cells in the pancreas that secrete digestive enzymes.

Acinus (AS-i-nus) A small sac; a grapelike dilation.

Acoustic (a-KOOS-tik) Pertaining to sound or the sense of hearing.

Acquired immune deficiency syndrome (AIDS) A deficiency of T cells and a reversed ratio of helper T cells to suppressor T cells that results in fever or night sweats, coughing, sore throat, fatigue, body aches, weight loss, and enlarged lymph nodes. Caused by a virus called human T-cell leukemia virus-3 (HTLV-3).

Acromegaly (ak'-rō-MEG-a-lē) Condition caused by hypersecretion of growth hormone (GH) during adulthood characterized by thickened bones and enlargement of other tissues.

Actin (AK-tin) The contractile protein that makes up thin myofilaments in muscle fiber.

Active transport The movement of substances, usually ions, across cell membranes, against a concentration gradient, requiring the expenditure of energy.

Acuity (a-KYOO-i-tē) Clearness or sharpness, usually of vision.

Acupuncture (AK-ū-punk'-chur) The insertion of a needle into a tissue for the purpose of drawing fluid or relieving pain. It is also an ancient Chinese practice employed to cure illnesses by inserting needles into specific locations of the skin.

Acute (a-KYOOT) Having a rapid onset, severe symptoms, and a short course; not chronic.

Adaptation (ad'-ap-TĀ-shun) The adjustment of the pupil of the eye to light variations. The property by which a neuron relays a decreased frequency of action potentials from a receptor even though the strength of the stimulus remains constant. The decrease in perception of a sensation over time while the stimulus is still present.

Addison's (AD-i-sonz) **disease** Disorder caused by hyposecretion of glucocorticoids characterized by muscular weakness, mental lethargy, weight loss, low blood pressure, and dehydration.

Adduction (ad-DUK-shun) Movement toward the axis or midline of the body or one of its parts.

Adenohypophysis (ad'-e-nō-hī-POF-i-sis) The anterior portion of the pituitary gland.

Adenoids (AD-e-noyds) The pharyngeal tonsils.

Adenosine triphosphate (a-DEN-ō-sēn trī-FOS-fāt) (**ATP**) The universal energy-carrying molecule manufactured in all living cells as a means of capturing and storing energy. It consists of the purine base *adenine* and the five-carbon sugar *ribose,* to which are added, in linear array, three *phosphate* molecules.

Adherence (ad-HEAR-ens) Firm contact between the plasma membrane of a phagocyte and an antigen.

Adhesion (ad-HĒ-zhun) Abnormal joining of parts to each other.

Adipocyte (AD-i-pō-sīt) Fat cell, derived from a fibroblast.

Adrenal cortex (a-DRĒ-nal KOR-teks) The outer portion of an adrenal (suprarenal) gland, divided into three zones, each of which has a different cellular arrangement and secretes different hormones.

Adrenal glands Two glands located superior to each kidney. Also called the **suprarenal glands.**

Adrenal medulla (me-DUL-a) The inner portion of an adrenal (suprarenal) gland, consisting of cells that secrete epinephrine and norepinephrine (NE) in response to the stimulation of preganglionic sympathetic neurons.

Adrenergic (ad'-ren-ER-jik) **fiber** A nerve fiber that, when stimulated, releases norepinephrine (noradrenaline) at a synapse.

Adrenocorticotropic (ad-rē'-nō-kor-ti-kō-TRŌP-ik) **hormone**

(**ACTH**) A hormone produced by the adenohypophysis (anterior lobe) of the pituitary gland that influences the production and secretion of certain hormones of the adrenal cortex.

Adrenoglomerulotropin (a-drē'-nō-glō-mer'-yoo-lō-TRŌ-pin) A hormone secreted by the pineal gland that may stimulate aldosterone secretion.

Adventitia (ad-ven-TISH-yah) The outermost covering of a structure or organ.

Aerobic (air-Ō-bik) Requiring molecular oxygen.

Afferent arteriole (AF-er-ent ar-TĒ-rē-ōl) A blood vessel of a kidney that breaks up into the capillary network called a glomerulus; there is one afferent arteriole for each glomerulus.

Afferent neuron (NOO-ron) A neuron that carries an impulse toward the central nervous system. Also called a *sensory neuron.*

Afterimage Persistence of a sensation even though the stimulus has been removed.

Agglutination (a-gloot'-i-NĀ-shun) Clumping of microorganisms, blood corpuscles or particles; an immune response; an antigen-antibody reaction.

Agglutinin (a-GLOO-ti-nin) A specific principle or antibody in blood serum capable of causing the clumping of bacteria, blood corpuscles, or particles. Also called **isoantibody.**

Agglutinogen (a-GLOOT-in-ō-gen) A genetically determined antigen located on the surface of erythrocytes; basis for the ABO grouping and Rh system of blood classification. Also called **isoantigen.**

Aggregated lymphatic follicles Aggregated lymph nodules that are most numerous in the ileum. Also called **Peyer's patches.**

Aging Progressive failure of the body's homeostatic adaptive responses.

Agnosia (ag-NŌ-zē-a) A loss of the ability to recognize the meaning of stimuli from the various senses (visual, auditory, touch).

Agonist (AG-ō-nist) The prime mover—the muscle directly engaged in contraction as distinguished from muscles that are relaxing at the same time.

Agraphia (a-GRAF-ē-a) An inability to write.

Albinism (AL-bin-ism) Abnormal, nonpathological, partial or total absence of pigment in skin, hair, and eyes.

Albumin (al-BYOO-min) The most abundant (60 percent) and smallest of the plasma proteins, which functions primarily to regulate osmotic pressure of plasma.

Albuminuria (al-byoo'-min-UR-ēa) Presence of albumin in the urine.

Aldosterone (al-do-STER-ōn) A mineralocorticoid produced by the adrenal cortex that brings about sodium and water reabsorption and potassium excretion.

Aldosteronism (al'-do-STER-ōn-izm') Condition caused by hypersecretion of aldosterone characterized by muscular paralysis, high blood pressure, and edema.

Alimentary (al-i-MEN-ta-rē) Pertaining to nutrition.

Alkaline (AL-ka-līn) Containing more hydroxyl ions than hydrogen ions to produce a pH of more than 7.

Allantois (a-LAN-tō-is) A small, vascularized membrane between the chorion and amnion of the fetus.

Allergen (AL-er-jen) An antigen that evokes a hypersensitivity reaction.

Allergic (a-LER-jik) Pertaining to or sensitive to an allergen.

All-or-none principle In muscle physiology, muscle fibers of a motor unit contract to their fullest extent or not at all. In neuron physiology, if a stimulus is strong enough to initiate an action potential, an impulse is transmitted along the entire neuron at a constant and maximum strength.

Alpha (AL-fa) **cell** A cell in the pancreatic islets (islets of Langerhans) in the pancreas that secretes glucagon.

Alveolar-capillary (al-VĒ-ō-lar) **membrane** Structure in the lungs consisting of the alveolar wall and basement membrane and a capillary endothelium and basement membrane through which the diffusion of respiratory gases occurs. Also called the **respiratory membrane.**

Alveolar duct Branch of a respiratory bronchiole around which alveoli and alveolar sacs are arranged.

Alveolar macrophage (MAK-rō-fāj) Cell found in the alveolar walls of the lungs that is highly phagocytic. Also called a **dust cell.**

Alveolar sac A collection or cluster of alveoli that share a common opening.

Alveolus (al-VĒ-ō-lus) A small hollow or cavity; an air sac in the lungs; milk-secreting portion of a mammary gland.

Alzheimer's (ALTZ-hī-merz) **disease** Disabling neurological disorder characterized by dysfunction and death of specific cerebral neurons resulting in widespread intellectual impairment, personality changes, and fluctuations in alertness.

Amenorrhea (ā-men-ō-RĒ-a) Absence of menstruation.

Amnesia (am-NĒ-zē-a) A lack or loss of memory.

Amniocentesis (am'-nē-ō-sen-TĒ-sis) Removal of amniotic fluid by inserting a needle transabdominally into the amniotic cavity.

Amnion (AM-nē-on) The innermost fetal membrane; a thin transparent sac that holds the fetus suspended in amniotic fluid. Also called the **"bag of waters."**

Amorphous (a-MOR-fus) Without definite shape or differentiation in structure; pertains to solids without crystalline structure.

Amphiarthrosis (am'-fē-ar-THRŌ-sis) Articulation midway between diarthrosis and synarthrosis, in which the articulating bony surfaces are separated by an elastic substance to which both are attached, so that the mobility is slight but may be exerted in all directions.

Ampulla (am-POOL-la) A saclike dilation of a canal.

Amyotrophic (a-mē-ō-TROF-ik) **lateral sclerosis (ALS)** Progressive neuromuscular disease characterized by degeneration of motor cells in the spinal cord that leads to muscular weakness. Also called **Lou Gehrig's disease.**

Anaerobic (an-AIR-ō-bik) Not requiring molecular oxygen.

Anal (Ā-nal) **canal** The terminal 2 or 3 cm of the rectum; opens to the exterior at the anus.

Anal column A longitudinal fold in the mucous membrane of the anal canal that contains a network of arteries and veins.

Analgesia (an-al-JĒ-zē-a) Absence of normal sense of pain.

Anal triangle The subdivision of the male or female perineum that contains the anus.

Anaphase (AN-a-fāz) The third stage of mitosis in which the chromatids that have separated at the centromeres move to opposite poles.

Anaphylaxis (an'-a-fi-LAK-sis) A hypersensitivity reaction involving IgE antibodies, mast cells, and basophils; examples are hay fever, bronchial asthma, hives, and anaphylactic shock.

Anastomosis (a-nas-tō-MŌ-sis) An end-to-end union or joining together of blood vessels, lymphatics, or nerves.

Anatomical (an'-a-TOM-i-kal) **position** A position of the body universally used in anatomical descriptions in which the body is erect, facing the observer, the upper extremities are at the sides, the palms of the hands are facing forward.

Anatomy (a-NAT-ō-mē) The structure or study of structure of the body and the relationship of its parts to each other.

Androgen (AN-drō-jen) Substance producing or stimulating male characteristics, such as the male hormone testosterone.

Anemia (a-NĒ-mē-a) Condition of the blood in which the number of functional red blood cells or their hemoglobin content is below normal.

Anesthesia (an-es-THĒ-zē-a) A total or partial loss of feeling or sensation, usually defined with respect to loss of pain sensation.

Aneurysm (AN-yoo-rizm) A saclike enlargement of a blood vessel caused by a weakening of the wall.

Angina pectoris (an-JĪ-na *or* AN-ji-na PEK-tō-ris) A pain in the chest related to reduced coronary circulation that may or may not involve heart or artery disease.

Angiography (an-jē-OG-ra-fē) X-ray examination of blood vessels after injection of a radiopaque substance into the common carotid or vertebral artery; used to demonstrate cerebral blood vessels and may detect brain tumors with specific vascular patterns.

Angstrom (Å) (ANG-strum) One-tenth of a millimicron or about 1/250,000,000 inch.

Ankyloglossia (ang'-ki-lō-GLOSS-ē-a) "Tongue-tied"; restriction of tongue movements by a short lingual frenulum.

Ankylose (ANG-ke-lōs) Immobilization of a joint by pathological or surgical process.

Anomaly (a-NOM-a-lē) An abnormality that may be a developmental (congenital) defect; a variant from the usual standard.

Anopsia (an-OP-sē-a) A defect of vision.

Anorexia nervosa (an-ō-REK-sē-a ner-VŌ-sa) A disorder characterized by loss of appetite and bizarre patterns of eating.

Anosmia (an-OZ-mē-a) Loss of the sense of smell.

Anoxia (an-OK-sē-a) Deficiency of oxygen.

Antagonist (an-TAG-ō-nist) A muscle that has an action opposite that of the agonist and yields to the movement of the agonist.

Antepartum (an-tē-PAR-tum) Before delivery of the child; occurring (to the mother) before childbirth.

Anterior (an-TĒR-ē-or) Nearer to or at the front of the body. Also called **ventral.**

Anterior root The structure composed of axons of motor or efferent fibers that emerges from the anterior aspect of the spinal cord and extends laterally to join a posterior root, forming a spinal nerve. Also called a **ventral root.**

Antibiotic (an'-ti-bī-OT-ik) Literally, "antilife"; a chemical produced by a microorganism that is able to inhibit the growth of or kill other microorganisms.

Antibody (AN-ti-bod'-ē) A substance produced by certain cells in the presence of a specific antigen that combines with that antigen to neutralize, inhibit, or destroy it.

Anticoagulant (an-tī-cō-AG-yoo-lant) A substance that is able to delay, suppress, or prevent the clotting of blood.

Antidiuretic (an'-ti-dī-yoo-RET-ik) Substance that inhibits urine formation.

Antidiuretic hormone (ADH) Hormone produced by neurosecretory cells in the paraventricular nucleus of the hypothalamus that stimulates water reabsorption from kidney cells into the blood and vasoconstriction of arterioles.

Antigen (AN-ti-jen) Any substance that when introduced into the tissues or blood induces the formation of antibodies or reacts with them.

Antrum (AN-trum) Any nearly closed cavity or chamber, especially one within a bone, such as a sinus.

Anulus fibrosus (AN-yoo-lus fī-BRŌ-sus) A ring of fibrous tissue and fibrocartilage that encircles the pulpy substance (nucleus pulposus) of an intervertebral disc.

Anuria (a-NOO-rē-a) Absence of urine formation.

Anus (Ā-nus) The distal end and outlet of the rectum.

Aorta (ā-OR-ta) The main systemic trunk of the arterial system of the body; emerges from the left ventricle.

Aperture (AP-er-chur) An opening or orifice.

Apex (Ā-peks) The pointed end of a conical structure.

Aphasia (a-FĀ-zē-a) Loss of ability to express oneself properly through speech or loss of verbal comprehension.

Apheresis (a-FER-ē-sis) Procedure in which blood is removed from the body, components are selectively separated, the undesired component is removed, and the remainder is returned to the body.

Apnea (ap-NĒ-a) Temporary cessation of breathing.

Apneustic (ap-NOO-stik) Pertaining to sustained inspiratory effort, even to the point of asphyxia.

Apneustic area Portion of the respiratory center in the pons that helps coordinate the transition between inspiration and expiration.

Apocrine (AP-ō-krin) **gland** A type of gland in which the secretory products gather at the free end of the secreting cell and are pinched off, along with some of the cytoplasm, to become the secretion, as in mammary glands.

Aponeurosis (ap'-ō-nyoo-RŌ-sis) A sheetlike layer of dense, regularly arranged connective tissue joining one muscle with another or with bone.

Appendage (a-PEN-dij) A part or thing attached to the body.

Appendicitis (a-pen-di-SĪ-tis) Inflammation of the vermiform appendix.

Aqueduct (AK-we-duct) A canal or passage, especially for the conduction of a liquid.

Aqueous humor (ĀK-we-us HYOO-mor) The watery fluid that fills the anterior cavity of the eye.

Arachnoid (a-RAK-noyd) The middle of the three coverings (meninges) of the brain

Arachnoid villus (VIL-us) Berrylike tuft of arachnoid that protrudes into the superior sagittal sinus and through which the cerebrospinal fluid enters the bloodstream.

Arbor vitae (AR-bōr VĒ-tē) The treelike appearance of the white matter tracts of the cerebellum when seen in midsagittal section. A series of branching ridges within the cervix of the uterus.

Arch of the aorta (ā-OR-ta) The most superior portion of the aorta, lying between the ascending and descending segments of the aorta.

Areflexia (a'-rē-FLEK-sē-a) Absence of reflexes.

Areola (a-RĒ-ō-la) Any tiny space in a tissue. The pigmented ring around the nipple of the breast.

Arm The portion of the upper extremity from the shoulder to the elbow.

Arrector pili (a-REK-tor PI-lē) Smooth muscles attached to hairs; contraction pulls the hairs into a more vertical position, resulting in "goose bumps."

Arrhythmia (a-RITH-mē-a) Irregular heart rhythm. Also called a **dysrhythmia.**

Arteriogram (ar-TĒR-ē-ō-gram) A roentgenogram of an artery after injection of a radiopaque substance into the blood.

Arteriole (ar-TĒ-rē-ōl) A small arterial branch that delivers blood to a capillary.

Artery (AR-ter-ē) A blood vessel that carries blood away from the heart.

Arthritis (ar-THRĪ-tis) Inflammation of a joint.

Arthrology (ar-THROL-ō-jē) The study or description of joints.

Arthroscopy (ar-THROS-co-pē) Surgical technique in which an arthroscope is inserted into a small incision, usually in the knee, to repair torn cartilage.

Arthrosis (ar-THRŌ-sis) A joint or articulation.

Articular (ar-TIK-yoo-lar) **capsule** Sleevelike structure around a synovial joint composed of a fibrous capsule and a synovial membrane.

Articular cartilage (KAR-ti-lij) Hyaline cartilage attached to articular bone surfaces.

Articular disc Fibrocartilage pad between articular surfaces of bones of some synovial joints. Also called a **meniscus** (men-IS-cus).

Articulate (ar-TIK-yoo-lāt) To join together as a joint to permit motion between parts.

Articulation (ar-tik'-yoo-LĀ-shun) A joint.

Artificial insemination (in-sem'-i-NĀ-shun) The deposition of seminal fluid within the vagina or cervix by artificial means.

Artificial pacemaker A device that generates and delivers electrical signals to the heart to maintain a regular heart rhythm.

Arytenoid (ar'-i-TĒ-noyd) Ladle-shaped.

Arytenoid cartilages A pair of small cartilages of the larynx that articulates with the cricoid cartilage.

Ascending colon (KŌ-lon) The portion of the large intestine that passes upward from the cecum to the lower edge of the liver where it bends at the right colic (hepatic) flexure to become the transverse colon.

Ascites (as-SĪ-tēz) Serous fluid in the peritoneal cavity.

Aseptic (ā-SEP-tik) Free from any infectious or septic material.

Asphyxia (as-FIX-ē-a) Unconsciousness due to interference with the oxygen supply of the blood.

Aspirate (AS-pir-āt) To remove by suction.

Association area A portion of the cerebral cortex connected by many motor and sensory fibers to other parts of the cortex. The association areas are concerned with motor patterns, memory, concepts of word-hearing and word-seeing, reasoning, will, judgment, and personality traits.

Association neuron (NOO-ron) A nerve cell lying completely within the central nervous system that carries impulses from sensory neurons to motor neurons. Also called an **internuncial** or **connecting neuron.**

Astereognosis (as-ter'-ē-ōg-NŌ-sis) Inability to recognize objects or forms by touch.

Asthenia (as-THĒ-nē-a) Lack or loss of strength; debility.

Astigmatism (a-STIG-ma-tizm) An irregularity of the lens or cornea of the eye causing the image to be out of focus and producing faulty vision.

Astrocyte (AS-trō-sīt) A neuroglial cell having a star shape.

Ataxia (a-TAK-sē-a) A lack of muscular coordination, lack of precision.

Atelectasis (at'-ē-LEK-ta-sis) A collapsed or airless state of all or part of the lung, which may be acute or chronic.

Atherosclerosis (ath'-er-ō-skle-RŌ-sis) A process in which fatty substances (cholesterol and triglycerides) are deposited in the walls of medium and large arteries in response to certain stimuli (hypertension, carbon monoxide, dietary cholesterol). Following endothelial damage, platelets release chemicals that stimulate uptake of cholesterol and proliferation of smooth muscle cells in the tunica media, resulting in formation of an atherosclerotic plaque that decreases the size of the arterial lumen.

Atresia (a-TRĒ-zē-a) Abnormal closure of a passage, or absence of a normal body opening.

Atrial fibrillation (Ā-trē-al fib-ri-LĀ-shun) Asynchronous contraction of the atria that results in the cessation of atrial pumping.

Atrioventricular (AV) (ā'-trē-ō-ven-TRIK-yoo-lar) **bundle** The portion of the conduction system of the heart that begins at the atrioventricular (AV) node and passes through the cardiac skeleton separating the atria and the ventricles, then runs a short distance down the interventricular septum before splitting into right and left bundle branches. Also called the **bundle of His.**

Atrioventricular (AV) node The portion of the conduction system of the heart made up of a compact mass of conducting cells located near the orifice of the coronary sinus in the right atrial wall.

Atrioventricular (AV) valve A structure made up of membranous flaps or cusps that allows blood to flow in one direction only, from an atrium into a ventricle.

Atrium (Ā-trē-um) A superior chamber of the heart.

Atrophy (AT-rō-fē) Wasting away or decrease in size of a part, due to a failure, abnormality of nutrition, or lack of use.

Auditory ossicle (AW-di-tō-rē OS-si-kul) One of the three small bones of the middle ear called the malleus, incus, and stapes.

Auditory tube The tube that connects the middle ear with the nose and nasopharynx of the throat. Also called the **Eustachian tube.**

Aura (OR-a) A feeling or sensation that precedes an epileptic seizure or any paroxysmal attack (like those of bronchial asthma).

Auricle (OR-i-kul) The flap or pinna of the ear. An appendage of an atrium of the heart.

Auscultation (aws-kul-TĀ-shun) Examination by listening to sounds in the body.

Autoimmunity An immunologic response against a person's own tissue antigens.

Autolysis (aw-TOL-i-sis) Spontaneous self-destruction of cells by their own digestive enzymes upon death or a pathological process.

Autonomic ganglion (aw'-tō-NOM-ik GANG-lē-on) A cluster of sympathetic or parasympathetic cell bodies located outside the central nervous system.

Autonomic nervous system (ANS) Visceral efferent neurons, both sympathetic and parasympathetic, that transmit impulses from the central nervous system to smooth muscle, cardiac muscle, and glands; so named because this portion of the nervous system was thought to be self-governing or spontaneous.

Autonomic plexus (PLEK-sus) An extensive network of sympathetic and parasympathetic fibers; the cardiac, celiac, and pelvic plexuses are located in the thorax, abdomen, and pelvis, respectively.

Autopsy (AW-top-sē) The examination of the body after death.

Autosome Any chromosome other than the pair of sex chromosomes.

Axilla (ak-SIL-a) The small hollow beneath the arm where it joins the body at the shoulders. Also called the **armpit.**

Axon (AK-son) The process of a nerve cell that carries an impulse away from the cell body.

Axon terminal Terminal branch of an axon and its collateral. Also called a **telodendrium.**

Azygos (AZ-i-gos) An anatomical structure that is not paired; occurring singly.

Babinski (ba-BIN-skē) **sign** Extension of the great toe, with or without fanning of the other toes, in response to stimulation of the outer margin of the sole of the foot; normal up to 1½ years of age.

Back The posterior part of the body; the dorsum.

Ball-and-socket joint A synovial joint in which the rounded surface of one bone moves within a cup-shaped depression or fossa of another bone, as in the shoulder or hip joint. Also called a **spheroid joint.**

Baroreceptor (bar'-ō-rē-SEP-tor) Receptor stimulated by pressure change.

Basal metabolic (BĀ-sal met'-a-BOL-ik) **rate (BMR)** The rate of metabolism measured under standard or basal conditions.

Basal ganglia (GANG-glē-a) Paired clusters of cell bodies that make up the central gray matter in each cerebral hemisphere, including the caudate nucleus, lentiform nucleus, claustrum, and amygdaloid body. Also called **cerebral nuclei.**

Base The broadest part of a pyramidal structure. A nonacid, or a proton acceptor, characterized by excess of hydroxide ions and a pH greater than 7. A ring-shaped, nitrogen-containing organic molecule which is one of the components of a nucleotide, for example, adenine, guanine, cytosine, thymine, and uracil.

Basement membrane Thin, extracellular layer consisting of basal lamina secreted by epithelial cells and reticular lamina secreted by connective tissue cells.

Basilar (BAS-i-lar) **membrane** A membrane in the cochlea of the inner ear that separates the cochlear duct from the scala tympani and on which the spiral organ (organ of Corti) rests.

Basophil (BĀ-sō-fil) A type of white blood cell characterized by a pale nucleus and large granules that stain readily with basic dyes.

B cell A lymphocyte that develops into a plasma cell that produces antibodies or a memory cell.

Belly The abdomen. The gaster or prominent, fleshy part of a skeletal muscle.

Benign (be-NĪN) Not malignant.

Beta (BĀ-ta) **cell** A cell in the pancreatic islets (islets of Langerhans) in the pancreas that secretes insulin.

Bicuspid (bī-KUS-pid) **valve**   Atrioventricular (AV) valve on the left side of the heart. Also called the **mitral valve.**

Bifurcate (bī-FUR-kāt)   Having two branches or divisions; forked.

Bilateral (bī-LAT-er-al)   Pertaining to two sides of the body.

Bile (bīl)   A secretion of the liver.

Biliary (BIL-ē-er-ē)   Relating to bile, the gallbladder, or the bile ducts.

Biliary calculi (CAL-kyoo-lē)   Gallstones formed by the crystallization of cholesterol in bile.

Bilirubin (bil-ē-ROO-bin)   A red pigment that is one of the end products of hemoglobin breakdown by the liver cells and is excreted as a waste material in the bile.

Bilirubinuria (bil-ē-roo-bi-NOO-rē-a)   The presence of above-normal levels of bilirubin in urine.

Biliverdin (bil-ē-VER-din)   A green pigment that is one of the first products of hemoglobin breakdown in the liver cells and is converted to bilirubin or excreted as a waste material in bile.

Biofeedback   Process by which an individual gets constant signals (feedback) about various visceral body functions.

Biopsy (BĪ-op-sē)   Removal of tissue or other material from the living body for examination, usually microscopic.

Blastocoel (BLAS-tō-sēl)   The fluid-filled cavity within the blastocyst.

Blastocyst (BLAS-tō-sist)   In the development of an embryo, a hollow ball of cells that consists of a blastocoel (the internal cavity), trophectoderm (outer cells), and inner cell mass.

Blastomere (BLAS-tō-mēr)   One of the cells resulting from the cleavage of a fertilized ovum.

Blastula (BLAS-tyoo-la)   An early stage in the development of a zygote.

Blepharism (BLEF-a-rizm)   Spasm of the eyelids; continuous blinking.

Blind spot   Area in the retina at the end of the optic nerve in which there are no light receptor cells.

Blood   The fluid that circulates through the heart, arteries, capillaries, and veins and that constitutes the chief means of transport within the body.

Blood–brain barrier (BBB)   A special mechanism that prevents the passage of materials from the blood to the cerebrospinal fluid and brain.

Blood reservoir (REZ-er-vwar)   Systemic veins that contain large amounts of blood that can be moved quickly to parts of the body requiring the blood.

Body cavity   A space within the body that contains various internal organs.

Bolus (BŌ-lus)   A soft, rounded mass, usually food, that is swallowed.

Bony labyrinth (LAB-i-rinth)   A series of cavities within the petrous portion of the temporal bone, forming the vestibule, cochlea, and semicircular canals of the inner ear.

Brachial plexus (BRĀ-kē-al PLEK-sus)   A network of nerve fibers of the anterior rami of spinal nerves C5, C6, C7, C8, and T1. The nerves that emerge from the brachial plexus supply the upper extremity.

Brain   A mass of nerve tissue located in the cranial cavity.

Brain electrical activity mapping (BEAM)   Noninvasive procedure that measures and displays the electrical activity of the brain; used primarily to diagnose epilepsy.

Brain sand   Calcium deposits in the pineal gland that are laid down starting at puberty.

Brain stem   The portion of the brain immediately superior to the spinal cord, made up of the medulla oblongata, pons, and midbrain.

Broad ligament   A double fold of parietal peritoneum attaching the uterus to the side of the pelvic cavity.

Broca's (BRŌ-kaz) **area**   Motor area of the brain in the frontal lobe that translates thoughts into speech. Also called the **motor speech area.**

Bronchi (BRONG-kē)   Branches of the respiratory passageway including primary bronchi (the two divisions of the trachea), secondary or lobar bronchi (divisions of the primary that are distributed to the lobes of the lung), and tertiary or segmental bronchi (divisions of the secondary that are distributed to bronchopulmonary segments of the lung).

Bronchial asthma (BRONG-kē-al AZ-ma)   Usually allergic reaction characterized by smooth muscle spasms in bronchi resulting in wheezing and difficult breathing.

Bronchial tree   The trachea, bronchi, and their branching structures.

Bronchiectasis (brong'-kē-EK-tā-sis)   A chonic disorder in which there is a loss of the normal tissue and expansion of lung air passages; characterized by difficult breathing, coughing, expectoration of pus, and foul breath.

Bronchiole (BRONG-kē-ōl)   Branch of a tertiary bronchus further dividing into terminal bronchioles (distributed to lobules of the lung), which divide into respiratory bronchioles (distributed to alveolar sacs).

Bronchitis (BRONG-kī-tis)   Inflammation of the bronchi characterized by a productive cough.

Bronchogenic carcinoma (brong'-kō-JEN-ik kar'-si-NŌ-ma)   Cancer originating in the bronchi.

Bronchogram (BRONG-kō-gram)   A roentgenogram of the lungs and bronchi.

Bronchopulmonary (brong'-kō-PUL-mō-ner'-ē) **segment**   One of the smaller divisions of a lobe of a lung supplied by its own branches of a bronchus.

Bronchoscope (BRONG-kō-skōp)   An instrument used to examine the interior of the bronchi of the lungs.

Bronchus (BRONG-kus)   One of the two large branches of the trachea. *Plural*, **bronchi** (BRONG-kē).

Buccal (BUK-al)   Pertaining to the cheek or mouth.

Bulb of penis   Expanded portion of the base of the corpus spongiosum penis.

Bulbourethral (bul'-bō-yoo-RĒ-thral) **gland**   One of a pair of glands located inferior to the prostate gland on either side of the urethra that secretes an alkaline fluid into the cavernous urethra. Also called a **Cowpers' gland.**

Bulimia (boo-LIM-ē-a)   A disorder characterized by uncontrollable overeating followed by forced vomiting or overdoses of laxatives.

Bullae (BYOOL-ē)   Blisters beneath or within the epidermis.

Bundle branch   One of the two branches of the atrioventricular (AV) bundle, made up of specialized muscle fibers that transmit electrical impulses to the ventricles.

Bunion (BUN-yun)   Inflammation and thickening of the bursa of a toe joint due to displacement of the great toe.

Burn   An injury caused by heat (fire, steam), chemicals, electricity, or the ultraviolet rays of the sun.

Bursa (BUR-sa)   A sac or pouch of synovial fluid located at friction points, especially about joints.

Bursitis (bur-SĪ-tis)   Inflammation of a bursa.

Buttocks (BUT-oks)   The two fleshy masses on the posterior aspect of the lower trunk, formed by the gluteal muscles.

Cachexia (kah-KEK-sē-ah)   A state of ill health, malnutrition, and wasting.

Calcaneal (Achilles) tendon   The tendon of the soleus and gastrocnemius muscles at the back of the heel.

Calcify (KAL-si-fī)   To harden by deposits of calcium salts.

Calcitonin (kal-si-TŌ-nin) **(CT)**   A hormone produced by the thyroid gland that lowers the calcium and phosphate levels of the blood by inhibiting bone breakdown and accelerating calcium absorption by bones.

Calculus (KAL-kyoo-lus)   A stone, or insoluble mass of crystallized salts or other material, formed within the body, as in the gallbladder, kidney, or urinary bladder.

Callus (KAL-lus)   A growth of new bone tissue in and around a fractured area, ultimately replaced by mature bone. An acquired, localized thickening.

Calyx (KĀL-iks)   Any cuplike division of the kidney pelvis. *Plural,* **calyces** (KĀ-li-sēz).

Canal (ka-NAL)   A narrow tube, channel, or passageway.

Canaliculus (kan'-a-LIK-yoo-lus)   A small channel or canal, as in bones, where they connect the lacunae. *Plural,* **canaliculi** (kan'-a-LIK-yoo-lī).

Cancellous (KAN-sel-us)   Having a reticular or latticework structure, as in spongy tissue of bone.

Cancer (KAN-ser)   A malignant tumor of epithelial origin tending to infiltrate and give rise to new growths or metastases. Also called **carcinoma** (kar'-si-NŌ-ma).

Capillary (KAP-i-lar'-ē)   A microscopic blood vessel located between an arteriole and venule through which materials are exchanged between blood and body cells.

Carbohydrate (kar'-bō-HĪ-drāt)   An organic compound containing carbon, hydrogen, and oxygen in a particular amount and arrangement and comprised of sugar subunits; usually has the formula $(CH_2O)_n$.

Carbon monoxide poisoning   Hypoxia due to increased levels of carbon monoxide as a result of its preferential and tenacious combination with hemoglobin compared to oxygen.

Carbuncle (KAR-bung-kal)   A hard, pus-filled, and painful inflammation involving the skin and underlying connective tissues; similar to boils but larger and more deeply rooted with several openings.

Carcinogen (kar-SIN-ō-jen)   Any substance that causes cancer.

Cardiac muscle   An organ specialized for contraction, composed of striated muscle cells, forming the wall of the heart, and stimulated by an intrinsic conduction system and visceral efferent neurons.

Cardiac notch   An angular notch in the anterior border of the left lung.

Cardiac tamponade (tam'-pon-ĀD)   Compression of the heart due to excessive fluid or blood in the pericardial sac that could result in cardiac failure.

Cardinal ligament   A ligament of the uterus, extending laterally from the cervix and vagina as a continuation of the broad ligament.

Cardioacceleratory (kar-dē-ō-ak-SEL-er-a-tō-rē) **center**   A group of neurons in the medulla from which cardiac nerves (sympathetic) arise; impulses along the nerves release epinephrine that increases the rate and force of heartbeat.

Cardioinhibitory (kar-dē-ō-in-HIB-i-tō'-rē) **center**   A group of neurons in the medulla from which parasympathetic fibers that reach the heart via the vagus (X) nerve arise; impulses along the nerves release acetylcholine that decreases the rate

and force of heartbeat.

Cardiology (kar-de-OL-ō-jē)   The study of the heart and diseases associated with it.

Cardiopulmonary resuscitation (rē-sus-i-TĀ-shun) **(CPR)**   A technique employed to restore life or consciousness to a person apparently dead or dying; includes external respiration (exhaled air respiration) and external cardiac massage.

Caries (KĀ-rēz)   Decay of a tooth or bone.

Carina (ka-RĪ-na)   A ridge on the inside of the division of the right and left primary bronchi.

Carotid (ka-ROT-id) **body**   Receptor in the carotid sinus that responds to alterations in blood levels of oxygen, carbon dioxide, and hydrogen ions.

Carotid sinus   A dilated region of the internal carotid artery immediately above the bifurcation of the common carotid artery that contains receptors that monitor blood pressure.

Carpus (KAR-pus)   A collective term for the eight bones of the wrist.

Cartilage (KAR-ti-lij)   A type of connective tissue consisting of chondrocytes in lacunae embedded in a dense network of collagenous and elastic fibers and a matrix of chondroitin sulfate.

Cartilaginous (kar'-ti-LAJ-i-nus) **joint**   A joint without a joint cavity where the articulating bones are held tightly together by cartilage, allowing little or no movement.

Caruncle (KAR-ung-kul)   A small fleshy eminence, often abnormal.

Cast   A small mass of hardened material formed within a cavity in the body and then discharged from the body; can originate in different areas and be composed of various materials.

Castration (kas-TRĀ-shun)   The removal of the testes.

Cataract (KAT-a-rakt)   Loss of transparency of the crystalline lens of the eye or its capsule or both.

Catheter (KATH-i-ter)   A tube that can be inserted into a body cavity through a canal or into a blood vessel; used to remove fluids, such as urine or blood, and to introduce diagnostic materials or medication.

Cauda equina (KAW-da ē-KWĪ-na)   A tail-like collection of roots of spinal nerves at the inferior end of the spinal canal.

Caudal (KAW-dal)   Pertaining to any tail-like structure; inferior in position.

Cecum (SĒ-kum)   A blind pouch at the proximal end of the large intestine to which the ileum is attached.

Celiac (SĒ-lē-ak)   Pertaining to the abdomen.

Celiac plexus (PLEK-sus)   A large mass of ganglia and nerve fibers located at the level of the upper part of the first lumbar vertebra; the solar plexus.

Cell   The basic structural and functional unit of all organisms; the smallest structure capable of performing all the activities vital to life.

Cell inclusion   A lifeless, often temporary, constituent in the cytoplasm of a cell as opposed to an organelle.

Cementum (se-MEN-tum)   Calcified tissue covering the root of a tooth.

Center   An area in the brain where a particular function is localized.

Center of ossification (os'-i-fi-KĀ-shun)   An area in the cartilage model of a future bone where the cartilage cells hypertrophy, secrete enzymes that result in the calcification of their matrix resulting in the death of the cartilage cells, followed by the invasion of the area by osteoblasts that then lay down bone.

Central canal A circular channel running longitudinally in the center of an osteon (Haversian system) of mature compact bone, containing blood and lymph vessels and nerves. Also called a **Haversian canal.** A microscopic tube running the length of the spinal cord in the gray commissure.

Central fovea (FŌ-vē-a) A cuplike depression in the center of the macula lutea of the retina, containing cones only; the area of clearest vision.

Central nervous system (CNS) Brain and spinal cord.

Centrioles (SEN-trē-ōlz) Paired, cylindrical organelles within a centrosome, each consisting of a ring of microtubules and arranged at right angles to each other; function in cell division.

Centromere (SEN-trō-mēr) The clear, constricted portion of a chromosome where the two chromatids are joined; serves as the point of attachment for the chromosomal microtubules.

Centrosome (SEN-trō-sōm) A rather dense area of cytoplasm, near the nucleus of a cell, containing a pair of centrioles.

Cephalic (se-FAL-ik) Pertaining to the head; superior in position.

Cerebellar peduncle (ser-e-BEL-ar pe-DUNG-kul) A bundle of nerve fibers connecting the cerebellum with the brain stem.

Cerebellum (ser-e-BEL-um) The portion of the hindbrain lying posterior to the medulla and pons, concerned with coordination of movements.

Cerebral aqueduct (SER-e-bral AK-we-dukt) A channel through the midbrain connecting the third and fourth ventricles and containing cerebrospinal fluid.

Cerebral arterial circle A ring of arteries forming an anastomosis at the base of the brain between the internal carotid and vertebral arteries and arteries supplying the brain. Also called the **circle of Willis.**

Cerebral cortex The surface of the cerebral hemispheres, 2 to 4 mm thick, consisting of six layers of nerve cell bodies (gray matter) in most areas.

Cerebral palsy (PAL-zē) A group of nonprogressive motor disorders caused by damage to motor areas of the brain (cerebral cortex, basal ganglia, and cerebellum) during fetal life, birth, or infancy.

Cerebral peduncle (pe-DUNG-kul) One of a pair of nerve fiber bundles located on the ventral surface of the midbrain, conducting impulses between the pons and the cerebral hemispheres.

Cerebrospinal (se-rē'-brō-SPĪ-nal) **fluid (CSF)** A fluid produced in the choroid plexuses of the ventricles of the brain that circulates in the ventricles and the subarachnoid space around the brain and spinal cord.

Cerebrovascular (se-rē'-brō-VAS-kyoo-lar) **accident (CVA)** Destruction of brain tissue (infarction) resulting from disorders of blood vessels that supply the brain. Also called a **stroke.**

Cerebrum (SER-ē-brum) The two hemispheres of the forebrain, making up the largest part of the brain.

Ceruminous (se-ROO-mi-nus) **gland** A modified sudoriferous (sweat) gland in the external auditory canal that secretes cerumen (ear wax).

Cervical ganglion (SER-vi-kul GANG-glē-on) A cluster of nerve cell bodies of postganglionic sympathetic neurons, located in the neck, near the vertebral column.

Cervical plexus (PLEK-sus) A network of neuron fibers formed by the anterior rami of the first four cervical nerves.

Cervix (SER-viks) Neck; any constricted portion of an organ, especially the lower cylindrical part of the uterus.

Cesarean (se-SA-rē-an) **section** Procedure in which a low, horizontal incision is made through the abdominal wall and uterus for removal of the baby and placenta.

Chalazion (ka-LĀ-zē-on) A small tumor of the eyelid.

Chemonucleolysis (kē'-mō-noo'-klē-OL-i-sis) Dissolution of the nucleus pulposus of an intervertebral disc by injection of a proteolytic enzyme (chymopapain) to relieve the pressure and pain associated with a herniated (slipped) disc.

Chemoreceptor (kē'-mō-rē-SEP-tor) Receptor that detects the presence of chemicals.

Chemotaxis (kē'-mō-TAK-sis) Attraction of phagocytes to a chemical stimulus.

Chemotherapy (kē'-mō-THER-a-pē) The treatment of illness or disease by chemicals.

Chiasma (kī-AZ-ma) A crossing; especially the crossing of the optic (II) nerve fibers.

Chiropractic (kī-rō-PRAK-tik) A system of treating disease by using one's hands to manipulate body parts, mostly the vertebral column.

Choana (KŌ-a-na) A funnel-shaped structure; the posterior opening of the nasal fossa, or internal naris.

Cholecystectomy (kō'-lē-sis-TEK-tō-mē) Surgical removal of the gallbladder.

Cholesterol (kō-LES-te-rol) Classified as a lipid, the most abundant steroid in animal tissues; located in cell membranes and used for the synthesis of steroid hormones and bile salts.

Cholinergic (kō'-lin-ER-jik) **fiber** A nerve ending that liberates acetylcholine at a synapse.

Cholinesterase (kō'-lin-ES-ter-ās) An enzyme that hydrolyzes acetylcholine.

Chondrocyte (KON-drō-sīt) Cell of mature cartilage.

Chondroitin (kon-DROY-tin) **sulfate** An amorphous matrix material found outside the cell.

Chordae tendineae (KOR-dē TEN-di-nē) Cords that connect the heart valves with the papillary muscles.

Chorion (KŌ-rē-on) The outermost fetal membrane; serves a protective and nutritive function.

Chorionic villi sampling (CVS) The removal of a sample of chorionic villi tissue by means of a catheter to analyze the tissue for prenatal genetic defects.

Choroid (KŌ-royd) One of the vascular coats of the eyeball.

Choroid plexus (PLEX-sus) A vascular structure located in the roof of each of the four ventricles of the brain; produces cerebrospinal fluid.

Chromaffin (krō-MAF-in) **cell** Cell that has an affinity for chrome salts, due in part to the presence of the precursors of the neurotransmitter epinephrine; found, among other places, in the adrenal medulla.

Chromatid (KRŌ-ma-tid) One of a pair of identical connected nucleoprotein strands that are joined at the centromere and separate during cell division, each becoming a chromosome of one of the two daughter cells.

Chromatin (KRŌ-ma-tin) The threadlike mass of the genetic material consisting principally of DNA, which is present in the nucleus of a nondividing or interphase cell.

Chromatolysis (krō'-ma-TOL-i-sis) The disintegration and disappearance of the chromophil granules of a cell.

Chromatophilic substance Rough endoplasmic reticulum in the cell bodies of neurons that functions in protein synthesis. Also called **Nissl bodies.**

Chromosomal microtubules (mī-krō-TOOB-yool) Microtubules formed during prophase of mitosis that originate from centromeres, extend from a centromere to a pole of the cell, and assist in chromosomal movement; constitute a part of the mitotic spindle.

Chromosome (KRŌ-mō-sōm) One of the 46 small, dark-staining bodies that appear in the nucleus of a human diploid cell during cell division.

Chronic (KRON-ik) Long-term or frequently recurring; applied to a disease that is not acute.

Chyle (kīl) The milky fluid found in the lacteals of the small intestine after digestion.

Chyme (kīm) The semifluid mixture of partly digested food and digestive secretions found in the stomach and small intestine during digestion of a meal.

Cicatrix (SIK-a-triks) A scar left by a healed wound.

Ciliary (SIL-ē-ar'-ē) Pertaining to any hairlike processes; an eyelid, an eyelash.

Ciliary body One of the three portions of the vascular tunic of the eyeball, the others being the choroid and the iris; includes the ciliary muscle and the ciliary processes.

Ciliary ganglion (GANG-glē-on) A very small parasympathetic ganglion whose preganglionic fibers come from the oculomotor (III) nerve and whose postganglionic fibers carry impulses to the ciliary muscle and the sphincter muscle of the iris.

Cilium (SIL-ē-um) A hair or hairlike process projecting from a cell that may be used to move the entire cell or to move substances along the surface of the cell.

Circumcision (ser'-kum-SIZH-un) Removal of the foreskin, the fold over the glans penis.

Circumduction (ser'-kum-DUK-shun) A movement at a synovial joint in which the distal end of a bone moves in a circle while the proximal end remains relatively stable.

Circumvallate papilla (ser'-kum-VAL-āt pa-PIL-a) One of the circular projections that are arranged in an inverted V-shaped row at the posterior portion of the tongue; the largest of the elevations on the upper surface of the tongue; contains taste buds.

Cirrhosis (si-RŌ-sis) A liver disorder in which the parenchymal cells are destroyed and replaced by connective tissue.

Cisterna chyli (sis-TER-na KĪ-lē) The origin of the thoracic duct.

Cleavage The rapid mitotic divisions following the fertilization of an ovum, resulting in an increased number of progressively smaller cells, called blastomeres, so that the overall size of the zygote remains the same.

Cleft palate Condition in which the palatine processes of the maxilla do not unite before birth; cleft lip, a split in the upper lip, is often associated with cleft palate.

Climacteric (klī-mak-TER-ik) Cessation of the reproductive function in the female or diminution of testicular activity in the male.

Climax The peak period or moments of greatest intensity during sexual excitement.

Clitoris (KLI-to-ris) An erectile organ of the female that is homologous to the male penis.

Clot The end result of a series of biochemical reactions that changes liquid plasma into a gelatinous mass; specifically, the conversion of fibrinogen into a tangle of polymerized fibrin molecules.

Coarctation (kō'-ark-TĀ-shun) **of the aorta** Congenital condition in which the aorta is too narrow and results in reduced blood supply, increased ventricular pumping, and high blood pressure.

Coccygeal (kok-SIJ-ē-al) Pertaining to the coccyx bone.

Coccygeal plexus (PLEK-sus) A network of nerves formed by the anterior rami of the coccygeal and the fourth and fifth sacral nerves. Fibers from the plexus supply the skin in the region of the coccyx.

Coccyx (KOK-six) The fused bones at the end of the vertebral column.

Cochlea (KŌK-lē-a) A winding, cone-shaped tube forming a portion of the inner ear and containing the spiral organ (organ of Corti).

Cochlear duct The membranous cochlea consisting of a spirally arranged tube enclosed in the bony cochlea and lying along its outer wall. Also called the **scala media.**

Coitus (KŌ-i-tus) Sexual intercourse. Also called **copulation.**

Collagen (KOL-a-jen) A protein that is the main organic constituent of connective tissue.

Collateral circulation The alternate route taken by blood through an anastomosis.

Colliculus (ko-LIK-yoo-lus) A small elevation.

Colon The division of the large intestine consisting of ascending, transverse, descending, and sigmoid portions.

Color blindness Any deviation in the normal perception of colors, resulting from the lack of one or more of the photopigments of the cones.

Colostomy (kō-LOS-tō-mē) The surgical creation of a new opening from the colon to the body surface.

Colostrum (kō-LOS-trum) A thin, cloudy fluid secreted by the mammary glands a few days prior to or after delivery before true milk is secreted.

Colposcopy (kol-POS-kō-pē) Direct examination of the vaginal and cervical mucosa using a magnifying device.

Coma (KŌ-ma) Profound unconsciousness from which one cannot be roused.

Commissure The angular junction of the eyelids at either corner of the eyes.

Common bile duct A tube formed by the union of the common hepatic duct and the cystic duct that empties bile into the duodenum at the hepatopancreatic ampulla (ampulla of Vater).

Compact (dense) bone Bone tissue with no apparent spaces in which the layers or lamellae are fitted tightly together. Compact bone is found immediately deep to the periosteum and external to spongy bone.

Complete blood count (CBC) Hematology test that usually includes hemoglobin determination, hematocrit, red and white blood cell count, differential count, and comments about blood cell morphology.

Computed tomography (tō-MOG-ra-fē) **(CT)** X-ray technique that provides a cross-sectional picture of any area of the body. Also called **computed axial tomography (CAT).**

Concha (KONG-ka) A scroll-like bone found in the skull. *Plural,* **conchae** (KONG-kē).

Concussion (kon-KUSH-un) Traumatic injury to the brain that produces no visible bruising, but may result in abrupt, temporary loss of consciousness.

Conduction myofiber Muscle fiber in the subendocardial tissue of the heart specialized for conducting an impulse to the myocardium; part of the conduction system of the heart. Also called a **Purkinje fiber.**

Cone   The light-sensitive receptor in the retina concerned with color vision.

Congenital (kon-JEN-i-tal)   Present at the time of birth.

Congestive heart failure (CHF)   Chronic or acute state that results when the heart is not capable of supplying the oxygen demands of the body.

Conjunctiva (kon'-junk-TĪ-va)   The delicate membrane covering the eyeball and lining the eyelids.

Conjunctivitis (kon-junk'-ti-VĪ-tis)   Inflammation of the delicate membrane covering the eyeball and lining of the eyelids.

Connective tissue   The most abundant of the four tissue types in the body, performing the functions of binding and supporting; consists of relatively few cells in a great deal of intercellular substance.

Constipation (con-sti-PĀ-shun)   Infrequent or difficult defecation caused by decreased motility of the intestines.

Contact inhibition   Phenomenon by which migration of a growing cell is stopped when it makes contact with another cell of its own kind.

Continuous microtubules (mī-krō-TOOB-yoolz)   Microtubules formed during prophase of mitosis that originate from the vicinity of the centrioles, grow toward each other, extend from one pole of the cell to another, and assist in chromosomal movement; constitute a part of the mitotic spindle.

Contraception (kon'-tra-SEP-shun)   The prevention of conception or impregnation without destroying fertility.

Contractility (kon'-trak-TIL-i-tē)   The ability of muscle tissue to shorten.

Contralateral (kon'-tra-LAT-er-al)   On the opposite side; affecting the opposite side of the body.

Conus medullaris (KŌ-nus med-yoo-LAR-is)   The tapered portion of the spinal cord below the lumbar enlargement.

Convergence (con-VER-jens)   An anatomical arrangement in which the synaptic knobs of several presynaptic neurons terminate on one postsynaptic neuron. The medial movement of the two eyeballs so that both are directed toward a near object being viewed in order to produce a single image.

Convulsion (con-VUL-shun)   Violent, involuntary, tetanic contractions of an entire group of muscles.

Cornea (KŌR-nē-a)   The transparent fibrous coat that covers the iris of the eye.

Corona (kō-RŌ-na)   Margin of the glans penis.

Coronal (kō-RŌ-nal) **plane**   A plane that runs vertical to the ground and divides the body into anterior and posterior portions. Also called **frontal plane.**

Corona radiata   Innermost layer of follicle cells surrounding an ovum.

Coronary (KOR-ō-na-rē) **artery disease (CAD)**   A condition in which the heart muscle receives inadequate blood due to an interruption of its blood supply.

Coronary artery spasm   A condition in which the smooth muscle of a coronary artery undergoes a sudden contraction, resulting in vasoconstriction.

Coronary circulation   The pathway followed by the blood from the ascending aorta through the blood vessels supplying the heart and returning to the right atrium. Also called **cardiac circulation.**

Coronary sinus (SĪ-nus)   A wide venous channel on the posterior surface of the heart that collects the blood from the coronary circulation and returns it to the right atrium.

Corpora quadrigemina (KOR-por-a kwad-ri-JEM-in-a)   Four small elevations on the dorsal region of the midbrain concerned with visual and auditory functions.

Cor pulmonale (kor pul-mōn-ALE) **(CP)**   Right ventricular hypertrophy from disorders that bring about hypertension in pulmonary circulation.

Corpus (KOR-pus)   The principal part of any organ; any mass or body.

Corpus albicans (AL-bi-kanz)   A white fibrous patch in the ovary that forms after the corpus luteum regresses.

Corpus callosum (ka-LŌ-sum)   The great commissure of the brain between the cerebral hemispheres.

Corpuscle of touch   The sensory receptor for the sensation of touch; found in the dermal papillae, especially in palms and soles. Also called a **Meissner's corpuscle.**

Corpus luteum (LOO-tē-um)   A yellow endocrine gland in the ovary formed when a follicle has discharged its ovum; secretes estrogens, progesterone, and relaxin.

Corpus striatum (strī-Ā-tum)   An area in the interior of each cerebral hemisphere composed of the caudate and lentiform nuclei of the basal ganglia and white matter of the internal capsule, arranged in a striated manner.

Cortex (KOR-teks)   An outer layer of an organ. The convoluted layer of gray matter covering each cerebral hemisphere.

Costal (KOS-tal)   Pertaining to a rib.

Costal cartilage (KAR-ti-lij)   Hyaline cartilage that attaches a rib to the sternum.

Cramp   A spasmodic, especially a tonic, contraction of one or many muscles, usually painful.

Cranial (KRĀ-nē-al) **cavity**   A subdivision of the dorsal body cavity formed by the cranial bones and containing the brain.

Cranial nerve   One of 12 pairs of nerves that leave the brain, pass through foramina in the skull, and supply the head, neck, and part of the trunk; each is designated by a Roman numeral and a name.

Craniosacral (krā-nē-ō-SĀ-kral) **outflow**   The fibers of parasympathetic preganglionic neurons, which have their cell bodies located in nuclei in the brain stem and in the lateral gray matter of the sacral portion of the spinal cord.

Craniotomy (krā'-nē-OT-ō-mē)   Any operation on the skull, as for surgery on the brain or decompression of the fetal head in difficult labor.

Cranium (KRĀ-nē-um)   The skeleton of the skull that protects the brain and the organs of sight, hearing, and balance; includes the frontal, parietal, temporal, occipital, sphenoid, and ethmoid bones.

Crenation (krē-NĀ-shun)   The shrinkage of red blood cells into knobbed, starry forms when placed in a hypertonic solution.

Cretinism (KRĒ-tin-izm)   Severe congenital thyroid deficiency during childhood leading to physical and mental retardation.

Crista (KRIS-ta)   A crest or ridged structure. A small elevation in the ampulla of each semicircular duct that serves as a receptor for dynamic equilibrium.

Crus (krus) **of penis**   Separated, tapered portion of the corpora cavernosa penis.

Cryosurgery (KRĪ-ō-ser-jer-ē)   The destruction of tissue by application of extreme cold.

Cryptorchidism (krip-TOR-ki-dizm)   The condition of undescended testes.

Cupula (KUP-yoo-la)   A mass of gelatinous material covering the hair cells of a crista; a receptor in the ampulla of a semicircular canal stimulated when the head moves.

Curvature (KUR-va-tūr)   A nonangular deviation of a straight line, as in the greater and lesser curvatures of the stomach.

Abnormal curvatures of the vertebral column include kyphosis, lordosis, and scoliosis.

Cushing's syndrome Condition caused by a hypersecretion of glucocorticoids characterized by spindly legs, "moon face," "buffalo hump," pendulous abdomen, flushed facial skin, and poor wound healing.

Cutaneous (kyoo-TĀ-nē-us) Pertaining to the skin.

Cyanosis (sī'-a-NŌ-sis) Slightly bluish or dark purple discoloration of the skin and the mucous membrane due to an oxygen deficiency.

Cystic (SIS-tik) **duct** The duct that transports bile from the gallbladder to the common bile duct.

Cystitis (sis-TĪ-tis) Inflammation of the urinary bladder.

Cystoscope (SIS-ti-skōp) An instrument used to examine the inside of the urinary bladder.

Cytokinesis (sī'-tō-ki-NĒ-sis) Division of the cytoplasm.

Cytology (sī-TOL-o-jē) The study of cells.

Cytoplasm (SĪ-tō-plazm) Substance within a cell's plasma membrane and external to its nucleus. Also called **protoplasm.**

Cytoskeleton Complex internal structure of cytoplasm consisting of microfilaments and microtubules.

Dartos (DAR-tōs) The contractile tissue under the skin of the scrotum.

Dead air volume The volume of air that is inhaled but remains in spaces in the upper respiratory system and does not reach the alveoli for participation in gaseous exchange; about 150 ml.

Deafness Lack of the sense of hearing or a significant hearing loss.

Debility (dē-BIL-i-tē) Weakness of tonicity in functions or organs of the body.

Decidua (dē-SID-yoo-a) That portion of the endometrium of the uterus (all but the deepest layer) that is modified for pregnancy and shed after childbirth.

Deciduous (dē-SID-yoo-us) Falling off or being shed seasonally or at a particular stage of development. In the body, referring to the first set of teeth.

Decubitus (dē-KYOO-bi-tus) **ulcer** Tissue destruction due to a constant deficiency of blood to tissues overlying a bony projection that has been subjected to prolonged pressure against an object such as a bed, cast, or splint. Also called **bedsore, pressure sore,** or **trophic ulcer.**

Decussation (dē'-ku-SĀ-shun) A crossing over; usually refers to the crossing of most of the fibers in the large motor tracts to opposite sides in the medullary pyramids.

Deep Away from the surface of the body.

Deep fascia (FASH-ē-a) A sheet of connective tissue wrapped around a muscle to hold it in place.

Deep inguinal (IN-gwi-nal) **ring** A slitlike opening in the aponeurosis of the transversus abdominis muscle that represents the origin of the inguinal canal.

Defecation (def-e-KĀ-shun) The discharge of feces from the rectum.

Defibrillation (dē-fib-ri-LĀ-shun) Delivery of a very strong electrical current to the heart in an attempt to stop ventricular fibrillation.

Deglutition (dē-gloo-TISH-un) The act of swallowing.

Dehydration (dē-hī-DRĀ-shun) Excessive loss of water from the body or its parts.

Delta cell A cell in the pancreatic islets (islets of Langerhans) in the pancreas that secretes somatostatin.

Demineralization (de-min'-er-al-i-ZĀ-shun) Loss of calcium and phosphorus from bones.

Dendrite (DEN-drīt) A nerve cell process carrying an impulse toward the cell body.

Dens (denz) Tooth.

Denticulate (den-TIK-yoo-lāt) Finely toothed or serrated; characterized by a series of small, pointed projections.

Dentin (DEN-tin) The osseous tissues of a tooth enclosing the pulp cavity.

Dentition (den-TI-shun) The eruption of teeth. The number, shape, and arrangement of teeth.

Deoxyribonucleic (dē-ok'-sē-ri'-bō-nyoo-KLĒ-ik) **acid (DNA)** A nucleic acid in the shape of a double helix constructed of nucleotides consisting of one of four nitrogen bases (adenine, cytosine, guanine, or thymine), deoxyribose, and a phosphate group; encoded in the nucleotides is genetic information.

Depression (dē-PRESS-shun) Movement of a part of the body downward.

Dermal papilla (pa-PILL-a) Fingerlike projection of the papillary region of the dermis that may contain blood capillaries or corpuscles of touch (Meissner's corpuscles).

Dermatology (der-ma-TOL-ō-jē) The medical specialty dealing with diseases of the skin.

Dermatome (DER-ma-tōm) An instrument for incising the skin or cutting thin transplants of skin. The cutaneous area developed from one embryonic spinal cord segment and receiving most of its innervation from one spinal nerve.

Dermis (DER-mis) A layer of dense connective tissue lying deep to the epidermis; the true skin or corium.

Descending colon (KŌ-lon) The part of the large intestine descending from the left colic (splenic) flexure to the level of the left iliac crest.

Detritus (de-TRI-tus) Any broken-down or degenerative tissue or carious matter.

Detrusor (de-TROO-ser) **muscle** Muscle in the wall of the urinary bladder.

Developmental anatomy The study of development from the fertilized egg to the adult form. The branch of anatomy called embryology is generally restricted to the study of development from the fertilized egg through the eighth week in utero.

Diabetes insipidus (dī-a-BĒ-tēz in-SIP-i-dus) Condition caused by hyposecretion of antidiuretic hormone (ADH) and characterized by excretion of large amounts of urine and thirst.

Diabetes mellitus (MEL-i-tus) Condition caused by hyposecretion of insulin and characterized by hyperglycemia, increased urine production, excessive thirst, and excessive eating.

Diagnosis (dī-ag-NŌ-sis) Recognition of disease states from signs and symptoms by inspection, palpation, laboratory tests, and other means.

Dialysis (dī-AL-i-sis) The process of separating crystalloids (smaller particles) from colloids (larger particles) by the difference in their rates of diffusion through selectively permeable membrane.

Diaphragm (DĪ-a-fram) Any partition that separates one area from another, especially the dome-shaped skeletal muscle between the thoracic and abdominal cavities.

Diaphysis (dī-AF-i-sis) The shaft of a long bone.

Diarrhea (dī-a-RĒ-a) Frequent defecation of liquid feces caused by increased motility of the intestines.

Diarthrosis (dī-'ar-THRŌ-sis)   Articulation in which opposing bones move freely, as in a hinge joint.

Diastole (dī-AS-tō-lē)   In the cardiac cycle, the phase of relaxation or dilation of the heart muscle, especially of the ventricles.

Diencephalon (dī-'en-SEF-a-lon)   A part of the brain consisting primarily of the thalamus and the hypothalamus.

Differential (dif-fer-EN-shal) **count**   Determination of the number of each kind of white blood cell in a sample of 100 cells for diagnostic purposes.

Differentiation (dif'-e-ren'-shē-Ā-shun)   Acquisition of specific functions different from those of the original general type.

Diffusion (dif-YOO-zhun)   A passive process in which there is a net or greater movement of molecules or ions from a region of high concentration to a region of low concentration until equilibrium is reached.

Digestion (di-JES-chun)   The mechanical and chemical breakdown of food to simple molecules that can be absorbed by the body.

Dilate (DĪ-lāte)   To expand or swell.

Diploid (DIP-loyd)   Having the number of chromosomes characteristically found in the somatic cells of an organism. Symbolized as *2n*.

Diplopia (di-PLŌ-pē-a)   Double vision.

Dislocation (dis-lō-KĀ-shun)   Displacement of a bone from a joint with tearing of ligaments, tendons, and articular capsules. Also called **luxation** (luks-Ā-shun).

Dissect (DĪ-sekt)   To separate tissues and parts of a cadaver (corpse) or an organ for anatomical study.

Distal (DIS-tal)   Farther from the attachment of an extremity to the trunk or a structure; farther from the point of origin.

Diuretic (dī-yoo-RET-ik)   A chemical that increases urine volume by inhibiting facultative reabsorption of water.

Diurnal (dī-UR-nal)   Daily.

Divergence (di-VER-jens)   An anatomical arrangement in which the synaptic end knobs of one presynaptic neuron terminate on several postsynaptic neurons.

Diverticulitis (dī-ver-tik-yoo-LĪ-tis)   Inflammation of diverticula; saclike outpouchings of the colonic wall when the muscularis becomes weak.

Diverticulum (dī'-ver-TIK-yoo-lum)   A sac or pouch in the walls of a canal or organ, especially in the colon.

Dorsal (DOR-sal)   Nearer to, located at, or referring to the back of the body; posterior.

Dorsal body cavity   Cavity near the dorsal surface of the body that consists of a cranial cavity and vertebral canal.

Dorsal ramus (RĀ-mus)   A branch of a spinal nerve containing motor and sensory fibers supplying the muscles, skin, and bones of the posterior part of the head, neck, and trunk.

Dorsiflexion (dor'-si-FLEK-shun)   Flexion of the foot at the ankle.

Down's syndrome   An inherited defect due to an extra copy of chromosome 21. Symptoms include mental retardation; a small skull, flattened front to back; a short, flat nose; short fingers; and a widened space between first two digits of the hand and foot. Also called **trisomy 21.**

Dropsy (DROP-sē)   A condition in which there is an abnormal accumulation of water in the tissues and cavities.

Ductus arteriosus (DUK-tus ar-tē-rē-Ō-sus)   A small vessel in the fetus that connects the pulmonary trunk with the aorta.

Ductus deferens (DEF-er-ens)   The duct that conducts sperma-

tozoa from the epididymis to the ejaculatory duct. Also called the **vas deferens,** or **seminal duct.**

Ductus venosus (ve-NŌ-sus)   A small vessel in the fetus that helps the circulation bypass the liver.

Duodenal gland   Gland in the submucosa of the duodenum that secretes intestinal juice. Also called **Brunner's gland.**

Duodenal papilla (doo-ō-DĒ-nal pa-PILL-a)   An elevation on the duodenal mucosa that receives the hepatopancreatic ampulla (ampulla of Vater).

Duodenum (doo'-ō-DĒ-num)   The first portion of the small intestine.

Dura mater (DYOO-ra MĀ-ter)   The outer membrane (meninx) covering the brain and spinal cord.

Dynamic equilibrium (ē-kwi-LIB-rē-um)   The maintenance of body position, mainly the head, in response to sudden movements such as rotation, acceleration, and deceleration.

Dynamic spatial reconstructor (DSR)   An x-ray machine that has the ability to construct moving, three-dimensional, life-size images of all or part of an internal organ from any view desired.

Dysfunction (dis-FUNK-shun)   Absence of complete normal function.

Dyslexia (dis-LEK-sē-a)   Impairment of ability to comprehend written language.

Dysmenorrhea (dis'-men-ō-RĒ-a)   Painful menstruation.

Dystrophia (dis-TRŌ-fē-a)   Progressive weakening of a muscle.

Ectoderm   The outermost of the three primary germ layers that gives rise to the nervous system and the epidermis of skin and its derivatives.

Ectopic (ek-TOP-ik)   Out of the normal location.

Eczema (EK-ze-ma)   A skin rash characterized by itching, swelling, blistering, oozing, and scaling of the skin.

Edema (e-DĒ-ma)   An abnormal accumulation of fluid in body tissues.

Effector (e-FEK-tor)   The organ of the body, either a muscle or a gland, that responds to a motor neuron impulse.

Efferent arteriole (EF-er-ent ar-TĒ-rē-ōl)   A vessel of the renal vascular system that transports blood from the glomerulus to the peritubular capillary.

Efferent ducts   A series of coiled tubes that transport spermatozoa from the rete testis to the epididymis.

Efferent neuron (NOO-ron)   A neuron that conveys impulses from the brain and spinal cord to effectors that may be either muscles or glands. Also called a **motor neuron.**

Effusion (e-FYOO-zhun)   The escape of fluid from the lymphatics or blood vessels into a cavity or into tissues.

Ejaculation (ē-jak-yoo-LĀ-shun)   The reflex ejection or expulsion of semen from the penis.

Ejaculatory (e-JAK-yoo-la-tor'-ē) **duct**   A tube that transports spermatozoa from the ductus (vas) deferens to the prostatic urethra.

Elasticity (e-las-TIS-i-tē)   The ability of tissue to return to its original shape after contraction or extension.

Electrocardiogram (e-lek'-trō-KAR-dē-ō-gram) **(ECG,** or **EKG)**   A recording of the electrical changes that accompany the cardiac cycle.

Electroencephalogram (e-lek'-trō-en-SEF-a-lō-gram) **(EEG)**   A recording of the electrical impulses of the brain.

Eleidin (el-Ē-i-din)   A translucent substance found in the skin.

Elevation (el-e-VĀ-shun)   Movement of a part of the body upward.

Ellipsoidal (e-lip-SOY-dal) **joint**   A synovial joint structured so that an oval-shaped condyle of one bone fits into an elliptical cavity of another bone, permitting side-to-side and back-and-forth movements, as at the joint at the wrist between the radius and carpals. Also called a **condyloid joint.**

Embolus (EM-bō-lus) A blood clot, bubble of air, fat from broken bones, mass of bacteria, or other debris or foreign material transported by the blood.

Embryo (EM-brē-ō) The young of any organism in an early stage of development; in humans, the developing organism from fertilization to the end of the eighth week in utero.

Embryology (em'-brē-OL-ō-jē) The study of development from the fertilized egg to the end of the eighth week in utero.

Embryo transfer Procedure in which a husband's seminal fluid is used to artifically inseminate a fertile ovum donor and, following fertilization, the morula or blastocyst is transferred from the donor to the infertile wife who carries it to term.

Emesis (EM-e-sis) Vomiting.

Emmetropia (em'-e-TRŌ-pē-a) The ideal optical condition of the eyes.

Emphysema (em'-fi-SĒ-ma) A swelling or inflation of air passages due to loss of elasticity in the alveoli.

Emulsification (ē-mul'-si-fi-KĀ-shun) The dispersion of large fat globules to smaller uniformly distributed particles.

Enamel (e-NAM-el) The hard, white substance covering the crown of a tooth.

Endocardium (en-dō-KAR-dē-um) The layer of the heart wall, composed of endothelium and smooth muscle, that lines the inside of the heart and covers the valves and tendons that hold the valves open.

Endochondral ossification (en'-dō-KON-dral os'-i-fi-KĀ-shun) The replacement of cartilage by bone. Also called **intracartilaginous ossification.**

Endocrine (EN-dō-krin) **gland** A gland that secretes hormones into the blood; a ductless gland.

Endocrinology (en'-dō-kri-NOL-ō-jē) The science concerned with the structure, function, and disorders of endocrine glands.

Endocytosis (en'-dō-sī-TŌ-sis) The uptake into a cell of large molecules and particles in which a segment of plasma membrane surrounds the substance, encloses it, and brings it in; includes phagocytosis, pinocytosis, and receptor-mediated endocytosis.

Endoderm The innermost of the three primary germ layers of the developing embryo that gives rise to the digestive tract, urinary bladder and urethra, and respiratory tract.

Endodontics (en'-dō-DON-tiks) The branch of dentistry concerned with the prevention, diagnosis, and treatment of diseases that affect the pulp, root, periodontal ligament, and alveolar bone.

Endogenous (en-DOJ-e-nus) Growing from or beginning within the organism.

Endolymph (EN-dō-lymf') The fluid within the membranous labyrinth of the inner ear.

Endometriosis (en'-dō-MĒ-trē-ō-sis) The growth of endometrial tissue outside the uterus.

Endometrium (en-dō-MĒ-trē-um) The mucous membrane lining the uterus.

Endomysium (en'-dō-MIZ-ē-um) Invaginations of the perimysium separating each individual muscle fiber (cell).

Endoneurium (en'-dō-NYOO-rē-um) Connective tissue wrapping around individual nerve fibers.

Endoplasmic reticulum (en'-dō-PLAZ-mik re-TIK-yoo-lum) **(ER)** A network of channels running through the cytoplasm of a cell that serves in intracellular transportation, support, storage, synthesis, and packaging of molecules. Portions of ER where ribosomes are attached to the outer surface are called granular or rough reticulum; portions that have no ribosomes are called agranular or smooth reticulum.

Endorphin (en-DOR-fin) A peptide in the central nervous system that acts as a painkiller.

Endoscope (EN-dō-skōp') An instrument used to look inside hollow organs such as the stomach (gastroscope) or urinary bladder (cystoscope).

Endosteum (en-DOS-tē-um) The membrane that lines the medullary cavity of bones.

Endothelial-capsular (en-dō-THĒ-lē-al) **membrane** A filtration membrane in a nephron of a kidney consisting of the endothelium and basement membrane of the glomerulus and the epithelium of the visceral layer of the glomerular capsule.

Endothelium (en'-dō-THĒ-lē-um) The layer of simple squamous epithelium that lines the cavities of the heart and blood and lymphatic vessels.

Enkephalin (en-KEF-a-lin) A peptide found in the central nervous system that acts as a painkiller.

Enteroendocrine (en-ter-ō-ENDŌ-krin) **cell** A stomach cell that secretes the hormone stomach gastrin. Also called an **argentaffin cell.**

Enuresis (en'-yoo-RĒ-sis) Involuntary discharge of urine, complete or partial, after age 3.

Enzyme (EN-zīm) A substance that affects the speed of chemical changes; an organic catalyst, usually a protein.

Eosinophil (ē'-ō-SIN-ō-fil) A type of white blood cell characterized by granular cytoplasm readily stained by eosin.

Ependyma (e-PEN-de-ma) Neuroglial cells that line ventricles of the brain.

Epicardium (ep'-i-KAR-dē-um) The thin outer layer of the heart wall, composed of serous tissue and mesothelium. Also called the **visceral pericardium.**

Epidemiology (ep'-i-dē-mē-OL-ō-jē) Medical science concerned with the occurrence and distribution of disease in human populations.

Epidermis (ep'-i-DERM-is) The outermost layer of skin, composed of stratified squamous epithelium.

Epididymis (ep'-i-DID-i-mis) An elongated tube along the posterior bladder of the testis in which sperm are stored. *Plural,* **epididymides** (ep'-i-DID-i-mi-dēz).

Epidural (ep'-i-DOO-ral) **space** A space between the spinal dura mater and the vertebral canal, containing loose connective tissue and a plexus of veins.

Epiglottis (ep'-i-GLOT-is) A large, leaf-shaped piece of cartilage lying on top of the larynx, with its "stem" attached to the thyroid cartilage and its "leaf" portion unattached and free to move up and down to cover the glottis.

Epilepsy (EP-i-lep'-sē) Neurological disorder characterized by short, periodic attacks of motor, sensory, or psychological malfunction.

Epimysium (ep'-i-MIZ-ē-um) Fibrous connective tissue around muscles.

Epinephrine (ep-ē-NEF-rin) Hormone secreted by the adrenal medulla that produces actions similar to those that result from sympathetic stimulation. Also called **adrenaline.**

Epineurium (ep'-i-NYOO-rē-um) The outermost covering around the entire nerve.

Epiphyseal (ep'-i-FIZ-ē-al) **line** The remnant of the epiphyseal plate in a long bone.

Epiphyseal plate The plate between the epiphysis and diaphysis in a long bone.

Epiphysis (ē-PIF-i-sis) The end of a long bone, usually larger in diameter than the shaft (the diaphysis).

Epiphysis cerebri (se-RĒ-brī) Pineal gland.

Episiotomy (ē-piz'-ē-OT-ō-mē) Incision of the clinical perineum at the end of second stage of labor to avoid tearing.

Epistaxis (ep'-i-STAK-sis) Hemorrhage from the nose; nosebleed.

Epithelial (ep'-i-THĒ-lē-al) **tissue** The tissue that forms glands or the outer part of the skin and lines blood vessels, hollow organs, and passages that lead externally from the body.

Eponychium (ep'-ō-NIK-ē-um) The horny embryonic structure from which the nail develops.

Erection (ē-REK-shun) The enlarged and stiff state of the penis (or clitoris) resulting from the engorgement of the spongy erectile tissue with blood.

Eructation (e-ruk'-TĀ-shun) The forceful expulsion of gas from the stomach. Also called **belching.**

Erythema (er'-e-THĒ-ma) Skin redness usually caused by engorgement of the capillaries in the lower layers of the skin.

Erythematosus (er-i'-them-a-TŌ-sis) Pertaining to redness.

Erythrocyte (e-RITH-rō-sīt) Red blood cell.

Erythropoiesis (e-rith'-rō-poy-Ē-sis) The process by which red blood cells are formed.

Erythropoietin (ē-rith'-rō-POY-ē-tin) A hormone formed from a plasma protein that stimulates red blood cell production.

Esophagus (e-SOF-a-gus) A hollow muscular tube connecting the pharynx and the stomach.

Estrogens (ES-tro-jens) Female sex hormones produced by the ovaries concerned with the development and maintenance of female reproductive structures and secondary sex characteristics, fluid and electrolyte balance, and protein anabolism. Examples are β-estradiol, estrone, and estriol.

Etiology (ē'-tē-OL-ō-jē) The study of the causes of disease, including theories of origin and the organisms, if any, involved.

Euphoria (yoo-FŌR-ē-a) A subjectively pleasant feeling of well-being marked by confidence and assurance.

Eupnea (yoop-NĒ-a) Normal quiet breathing.

Euthanasia (yoo'-tha-NĀ-zēā) The practice of ending a life in case of incurable disease.

Eversion (ē-VER-zhun) The movement of the sole outward at the ankle joint.

Exacerbation (eg-zas'-er-BĀ-shun) An increase in the severity of symptoms or of disease.

Excitability (ek-sīt'-a-BIL-i-tē) The ability of muscle tissue to receive and respond to stimuli; the ability of nerve cells to respond to stimuli and convert them into impulses.

Excrement (EKS-kre-ment) Material cast out from the body as waste, especially fecal matter.

Excretion (eks-KRĒ-shun) The process of eliminating waste products from a cell, tissue, or the entire body; or the products excreted.

Exocrine (EK-sō-krin) **gland** A gland that secretes substances into ducts that empty at covering or lining epithelium or directly onto a free surface.

Exocytosis (ex'-ō-sī-TŌ-sis) A process of discharging cellular products too big to go through the membrane. Particles for export are enclosed by Golgi membranes when they are synthesized. Vesicles pinch off from the Golgi complex and carry the enclosed particles to the interior surface of the cell membrane where the vesicle membrane and cell membrane fuse and the contents of the vesicle are discharged.

Exogenous (ex-SOJ-e-nus) Originating outside an organ or part.

Exophthalmic goiter (ek'-sof-THAL-mik GOY-ter) Condition caused by hypersecretion of thyroid hormones characterized by protrusion of the eyeballs and an enlarged thyroid.

Exophthalmos (ek'-sof-THAL-mus) An abnormal protrusion or bulging of the eyeball.

Expiration (ek-spi-RĀ-shun) Breathing out; expelling air from the lungs into the atmosphere. Also called **exhalation.**

Expiratory (eks-PĪ-ra-tō-rē) **reserve volume** The volume of air in excess of tidal volume that can be forcibly exhaled; about 1,200 ml.

Extensibility (ek-sten'-si-BIL-i-tē) The ability of tissue to be stretched when pulled.

Extension (ek-STEN-shun) An increase in the anterior angle between two bones, except in extension of the knee and toes, in which the posterior angle is involved; restoring a body part to its anatomical position after flexion.

External Located on or near the surface.

External auditory (AW-di-tōr-ē) **canal** or **meatus** (mē-Ā-tus) A canal in the temporal bone that leads to the middle ear.

External ear The outer ear, consisting of the pinna, the external auditory canal, and the tympanic membrane (eardrum).

External nares (NA-rēz) The external nostrils, or the openings into the nasal cavity on the exterior of the body.

External respiration The exchange of respiratory gases between the lungs and blood.

Exteroceptor (eks'-ter-ō-SEP-tor) A receptor adapted for the reception of stimuli from outside the body.

Extracorporeal (eks'-tra-kor-PŌ-rē-al) The circulation of blood outside of the body.

Extravasation (eks-trav-a-SĀ-shun) The escape of fluid from a vessel into the tissues, especially blood, lymph, or serum.

Extrinsic (ek-STRIN-sik) Of external origin.

Exudate (EKS-yoo-dāt) Escaping fluid or semifluid material that oozes from a space that may contain serum, pus, and cellular debris.

Eyebrow The hairy ridge above the eye.

Face The anterior aspect of the head.

Facilitated diffusion (fa-SIL-i-tā-ted dif-YOO-zhun) Diffusion in which a substance not soluble by itself in lipids is transported across a selectively permeable membrane by combining with a carrier substance.

Falciform ligament (FAL-si-form LIG-a-ment) A sheet of parietal peritoneum between the two principal lobes of the liver. The ligamentum teres, or remnant of the umbilical vein, lies within its fold.

Falx cerebelli (falks ser'-e-BEL-lē) A small triangular process of the dura mater attached to the occipital bone in the posterior cranial fossa and projecting inward between the two cerebellar hemispheres.

Falx cerebri (SER-e-brē) A fold of the dura mater extending down into the longitudinal fissure between the two cerebral hemispheres.

Fascia (FASH-ē-a) A fibrous membrane covering, supporting, and separating muscles.

Fascicle (FAS-i-kul) A small bundle or cluster, especially of nerve or muscle fibers. Also called a **fasciculus** (fa-SIK-yoo-lus); *plural,* **fasciculi** (fa-SIK-yoo-lī).

Fat   A lipid compound formed from one molecule of glycerol and three molecules of fatty acids; the body's most highly concentrated source of energy. Adipose tissue, composed of loose connective tissue and cells specialized for fat storage and present in the form of soft pads between various organs for support, protection, and insulation.

Fauces (FAW-sēs)   The opening from the mouth into the pharynx.

Febrile (FĒ-bril)   Feverish; pertaining to a fever.

Feces (FĒ-sēz)   Material discharged from the rectum and made up of bacteria, excretions, and food residue. Also called **stool.**

Fenestra cochlea (fe-NES-tra KŌK-lē-a)   A small opening between the middle and inner ear, directly below the oval window, covered by the second tympanic membrane. Also called the **round window.**

Fenestration (fen-e-STRĀ-shun)   Surgical opening made into the labyrinth of the ear for some conditions of deafness.

Fenestra vestibuli (ves-TIB-yoo-lī)   A small opening between the middle and inner ear into which the footplate of the stapes fits. Also called the **oval window.**

Fertilization (fer'-ti-li-ZĀ-shun)   Penetration of an ovum by a spermatozoon and subsequent union of the nuclei of the cells.

Fetal circulation   The circulatory system of the fetus, including the placenta and special blood vessels involved in the exchange of materials between fetus and mother.

Fetus (FĒ-tus)   The latter stages of the developing young of an animal; in humans, the developing organism in utero from the beginning of the third month to birth.

Fever   An elevation in body temperature above its normal temperature of 37°C (98.6°F).

Fibrillation (fi-bre-LĀ-shun)   Irregular twitching of individual muscle fibers (cells) or small groups of muscle fibers preventing effective action by an organ or muscle.

Fibroblast (FĪ-brō-blast)   A large, flat cell that forms collagenous and elastic fibers and the viscous ground substance of loose connective tissue.

Fibrocyte (FĪ-brō-sīt)   A mature fibroblast that no longer produces fibers or matrix in connective tissue.

Fibromyositis (fi'-brō-mī-ō'-SĪ-tis)   A group of symptoms including pain, tenderness, and stiffness of joints, muscles, or adjacent structures. Called **"charleyhorse"** when it involves the thigh.

Fibrosis (fī-BRŌ-sis)   Abnormal formation of fibrous tissue.

Fibrous (FĪ-brus) **joint**   A joint that allows little or no movement, such as a suture and syndesmosis.

Fibrous tunic (TOO-nik)   The outer coat of the eyeball, made up of the posterior sclera and the anterior cornea.

Fight-or-flight response   The effect of the stimulation of the sympathetic division of the autonomic nervous system.

Filiform papilla (FIL-i-form pa-PIL-a)   One of the conical projections that are distributed in parallel rows over the anterior two-thirds of the tongue and contain no taste buds.

Filum terminale (FĪ-lum ter-mi-NAL-ē)   Nonnervous fibrous tissue of the spinal cord that extends inferiorly from the conus medullaris to the coccyx.

Fimbriae (FIM-brē-ē)   Fingerlike structures, especially the lateral ends of the uterine (Fallopian) tubes.

Fissure (FISH-ur)   A groove, fold, or slit that may be normal or abnormal.

Fistula (FIS-choo-la)   An abnormal passage between two organs or between an organ cavity and the outside.

Fixator   A muscle that stabilizes the origin of the prime mover so that the prime mover can act more efficiently.

Fixed macrophage (MAK-rō-fāj)   Stationary phagocytic cell found in the liver, lungs, brain, spleen, lymph nodes, subcutaneous tissue, and bone marrow. Also called a **histiocyte.**

Flaccid (FLAK-sid)   Relaxed, flabby, or soft; lacking muscle tone.

Flagellum (fla-JEL-um)   A hairlike, motile process on the extremity of a cell. *Plural,* **flagella** (fla-JEL-a).

Flatfoot   A condition in which the ligaments and tendons of the arches of the foot are weakened and the height of the longitudinal arch decreases.

Flatus (FLĀ-tus)   Gas or air in the digestive tract; commonly used to denote passage of gas rectally.

Flexion (FLEK-shun)   A folding movement in which there is a decrease in the angle between two bones anteriorly, except in flexion of the knee and toes, in which the bones are approximated posteriorly.

Fluoroscope (FLOOR-ō-skōp)   An instrument for visual observation of the body by means of x ray.

Follicle (FOL-i-kul)   A small secretory sac or cavity.

Follicle-stimulating hormone (FSH)   Hormone secreted by the adenohypophysis (anterior lobe) of the pituitary gland that initiates development of ova and stimulates the ovaries to secrete estrogens in females and initiates sperm production in males.

Fontanel (fon'-ta-NEL)   A membrane-covered spot where bone formation is not yet complete, especially between the cranial bones of an infant's skull.

Foot   The terminal part of the lower extremity.

Foramen (fo-RĀ-men)   A passage or opening; a communication between two cavities of an organ or a hole in a bone for passage of vessels or nerves.

Foramen ovale (ō-VAL-ē)   An opening in the fetal heart septum between the right and left atria. A hole in the greater wing of the sphenoid bone that transmits the mandibular branch of the trigeminal (V) nerve.

Forearm (FOR-arm)   The part of the upper extremity between the elbow and the wrist.

Fornix (FOR-niks)   An arch or fold; a tract in the brain made up of association fibers, connecting the hippocampus with the mammillary bodies; a recess around the cervix of the uterus where it protrudes into the vagina.

Fossa (FOS-a)   A furrow or shallow depression.

Fourth ventricle (VEN-tri-kul)   A cavity within the brain lying between the cerebellum and the medulla and pons.

Fovea (FŌ-vē-a)   A pit or cuplike depression.

Fracture (FRAK-chur)   Any break in a bone.

Frenulum (FREN-yoo-lum)   A small fold of mucous membrane that connects two parts and limits movement.

Frontal plane   A plane at a right angle to a midsagittal plane that divides the body or organs into anterior and posterior portions. Also called **coronal plane.**

Fulminate (FUL-mi-nāt')   To occur suddenly with great intensity.

Functional residual (re-ZID-yoo-al) **volume**   The sum of residual volume plus expiratory reserve volume; about 2,400 ml.

Fundus (FUN-dus)   The part of a hollow organ farthest from the opening.

Fungiform papilla (FUN-ji-form pa-PIL-a) A mushroomlike elevation on the upper surface of the tongue appearing as a red dot; most contain taste buds.

Furuncle (FYOOR-ung-kul) A boil; painful nodule caused by bacteria infection and inflammation of a hair follicle or sebaceous (oil) gland.

Gallbladder A small pouch that stores bile, located under the liver, and filled and emptied of bile via the cystic duct.

Gamete (GAM-ēt) A male or female reproductive cell; the spermatozoon or ovum.

Ganglion (GANG-glē-on) A group of nerve cell bodies that lie outside the central nervous system. *Plural,* **ganglia** (GANG-glē-a).

Gangrene (GANG-rēn) Death and rotting of a considerable mass of tissue that usually is caused by interruption of blood supply followed by bacterial invasion (*Clostridium*).

Gastroenterology (gas-trō-en'-ter-OL-ō-jē) The medical specialty that deals with the structure, function, diagnosis, and treatment of diseases of the stomach and intestines.

Gastrointestinal (gas-trō-in-TES-ti-nal) **(GI) tract** A continuous tube running through the ventral body cavity extending from the mouth to the anus. Also called the **alimentary canal.**

Gavage (ga-VAZH) Feeding through a tube passed through the esophagus and into the stomach.

Gene (jēn) Biological unit of heredity; an ultramicroscopic, self-reproducing DNA particle located in a definite position on a particular chromosome.

Genetics The study of heredity.

Genital herpes (JEN-i-tal HER-pēz) A sexually transmitted disease caused by type II herpes simplex virus.

Genitalia (jen'-i-TĀL-ya) Reproductive organs.

Geriatrics (jer'-ē-AT-riks) The branch of medicine devoted to the medical problems and care of elderly persons.

Germinal (JER-mi-nal) **epithelium** A layer of epithelial cells that covers the ovaries and produces ova and lines the seminiferous tubules of the testes and produces spermatozoa.

Germinativum (jer'-mi-na-TĒ-vum) Skin layers where new cells are germinated.

Gerontology (jer'-on-TOL-o-jē) The study of old age.

Gestation (jes-TĀ-shun) The period of intrauterine fetal development.

Giantism (GĪ-an-tizm) Condition caused by hypersecretion of growth hormone (GH) during childhood characterized by excessive bone growth and body size.

Gingivae (jin-JI-vē) Gums. They cover the alveolar processes of the mandible and maxilla and extend slightly into each alveolus (socket).

Gingivitis (jin'-je-VĪ-tis) Inflammation of the gums.

Gland Single or group of specialized epithelial cells that secrete substances.

Glans penis (glanz PĒ-nis) The slightly enlarged region at the distal end of the penis.

Glaucoma (glaw-KŌ-ma) An eye disorder in which there is increased pressure due to an excess of fluid within the eye.

Gliding joint A synovial joint having articulating surfaces that are usually flat, permitting only side-to-side and back-and-forth movements, as between carpal bones, tarsal bones, and the scapula and clavicle. Also called an **arthrodial joint.**

Glomerular (glō-MER-yoo-lar) **capsule** A double-walled globe at the proximal end of a nephron that encloses the glomerulus. Also called **Bowman's capsule.**

Glomerulonephritis (glō-mer-yoo-lō-nef-RĪ-tis) Inflammation of the glomeruli of the kidney. Also called **Bright's disease.**

Glomerulus (glō-MER-yoo-lus) A rounded mass of nerves or blood vessels, especially the microscopic tuft of capillaries that is surrounded by the glomerular (Bowman's) capsule of each kidney tubule.

Glottis (GLOT-is) The air passageway between the vocal folds in the larynx.

Glucagon (GLOO-ka-gon) A hormone produced by the pancreas that increases the blood glucose level.

Glucocorticoids (gloo-kō-KOR-ti-koyds) A group of hormones of the adrenal cortex.

Glucose (GLOO-kōs) A six-carbon sugar, $C_6H_{12}O_6$; the major energy source for every cell type in the body. Its metabolism is possible by every known living cell for the production of ATP.

Glycogen (GLĪ-ko-jen) A highly branched polymer of glucose containing thousands of subunits; functions as a compact store of glucose molecules in liver and muscle cells.

Glycosuria (glī'-kō-SOO-rē-a) The presence of glucose in the urine.

Gnostic (NOS-tik) Pertaining to the faculties of perceiving and recognizing.

Gnostic area Sensory area of the cerebral cortex that receives and integrates sensory input from various parts of the brain so that a common thought can be formed.

Goblet cell A goblet-shaped unicellular gland that secretes mucus. Also called a **mucus cell.**

Goiter (GOY-ter) An enlargement of the thyroid gland.

Golgi (GOL-jē) **complex** An organelle in the cytoplasm of cells consisting of four to eight flattened channels, stacked on one another, with expanded areas (vesicles) at their ends; functions in packaging secreted proteins, lipid secretion, and carbohydrate synthesis.

Gomphosis (gom-FŌ-sis) A fibrous joint in which a cone-shaped peg fits into a socket.

Gonad (GŌ-nad) A gland that produces gametes and hormones; the ovary in the female and the testis in the male.

Gonadocorticoids (gō-na-dō-KOR-ti-kodyz) Sex hormones secreted by the adrenal cortex.

Gonadotropic (go'-nad-ō-TRŌ-pik) **hormone** A hormone that regulates the functions of the gonads.

Gonorrhea (gon'-ō-RĒ-a) Infectious, sexually transmitted disease caused by the bacterium *Neisseria gonorrhoeae* and characterized by inflammation of the urogenital mucosa, discharge of pus, and painful urination.

Gout (gowt) Hereditary condition associated with excessive uric acid in the blood; the acid crystallizes and deposits in joints and the kidneys.

Gradient (GRĀ-dē-ent) A slope or gradation in the body; difference in concentration or electrical charges across a selectively permeable membrane.

Gray commissure (KOM-i-shur) A narrow strip of gray matter connecting the two lateral gray masses within the spinal cord.

Gray matter Area in the central nervous system consisting of nonmyelinated nervous tissue.

Gray ramus communicans (RĀ-mus ko-MYOO-ni-kans) A short nerve containing postganglionic sympathetic fibers; the cell bodies of the fibers are in a sympathetic chain ganglion, and the nonmyelinated axons run by way of the gray ramus to a spinal nerve and then to the periphery to supply smooth muscle in blood vessels, arrector pili muscles, and sweat glands. *Plural,* **rami communicantes** (RĀ-mē kō-myoo-ni-KAN-tēz).

Greater omentum (ō-MEN-tum)   A large fold in the serosa of the stomach that hangs down like an apron over the front of the intestines.

Greater vestibular (ves-TIB-yoo-lar) **glands**   A pair of glands on either side of the vaginal orifice that open by a duct into the space between the hymen and the labia minora. Also called **Bartholin's glands.**

Groin (groyn)   The depression between the thigh and the trunk; the inguinal region.

Gross anatomy   The branch of anatomy that deals with structures that can be studied without using a microscope. Also called **macroscopic anatomy.**

Growth hormone (GH)   Hormone secreted by the adenohypophysis (anterior lobe) of the pituitary gland that brings about growth of body tissues, especially skeletal and muscular. Also known as **somatotropin** and **somatotropic hormone (STH).**

Gustatory (GUS-ta-tō'-rē)   Pertaining to taste.

Gynecology (gī'-ne-KOL-ō-jē)   The branch of medicine dealing with the study and treatment of disorders of the female reproductive system.

Gyrus (JĪ-rus)   One of the convolutions of the cerebral cortex of the brain. *Plural,* **gyri** (JĪ-rī).

Hair   A threadlike structure produced by hair follicles that develops in the dermis. Also called **pilus.**

Hair follicle (FOL-li-kul)   Structure composed of epithelium surrounding the root of a hair from which hair develops.

Hallucination (ha-loo'-sī-NĀ-shun)   A sensory perception of something that does not really exist in the world, that is, a sensory experience created from within the brain.

Hand   The terminal portion of an upper extremity, including the carpus, metacarpus, and phalanges.

Haploid (HAP-loyd)   Having half the number of chromosomes characteristically found in the somatic cells of an organism; characteristic of mature gametes. Symbolized as *n.*

Hard palate (PAL-at)   The anterior portion of the roof of the mouth, formed by the maxillae and palatine bones and lined by mucous membrane.

Haustra (HAWS-tra)   The sacculated elevations of the colon.

Head   The superior part of a human, cephalic to the neck. The superior or proximal part of a structure.

Heart   A hollow muscular organ lying slightly to the left of the midline of the chest that pumps the blood through the cardiovascular system.

Heart block   An arrhythmia (dysrhythmia) of the heart in which the atria and ventricles contract independently because of a blocking of electrical impulses through the heart at a critical point in the conduction system.

Heart-lung machine   A device that pumps blood, functioning as a heart, and removes carbon dioxide from blood and oxygenates it, functioning as lungs; used during heart transplantation, open heart surgery, and coronary artery bypass surgery.

Hematocrit (hē'-MAT-ō-krit)   An expression of the percentage of the volume occupied by blood cells. Usually calculated by centrifuging a blood sample in a graduated tube and then reading off the volumes of cells and total blood.

Hematology (hē'-ma-TOL-ō-jē)   The branch of science concerned with the study of blood and blood-forming tissues and disorders associated with them.

Hematoma (hē'-ma-TŌ-ma)   A tumor or swelling filled with blood.

Hematopoiesis (hem'-a-tō-poy-Ē-sis)   Blood cell production occurring in the red marrow of bones. Also called **hemopoiesis** (hē-mō-poy-Ē-sis).

Hematuria (hē'-ma-TOOR-ē-a)   Blood in the urine.

Hemiballismus (hem'-i-ba-LIZ-mus)   Violent muscular restlessness of half of the body, especially of the upper extremity.

Hemocytoblast (hē'-mō-SĪ-tō-blast)   Immature stem cell in bone marrow that develops along different lines into all the different mature blood cells.

Hemodialysis (hē'-mō-dī-AL-i-sis)   Filtering of the blood by means of an artificial device so that certain substances are removed from the blood as a result of the difference in rates of their diffusion through a semipermeable membrane while the blood is being circulated outside the body.

Hemoglobin (hē'-mō-GLŌ-bin) **(Hb)**   A substance in erythrocytes consisting of the protein globin and the iron-containing, red pigment heme and constituting about 33 percent of the cell volume; involved in the transport of oxygen and carbon dioxide.

Hemolysis (hē-MOL-i-sis)   The escape of hemoglobin from the interior of the red blood cell into the surrounding medium; results from disruption of the integrity of the plasma membrane by toxins or drugs, freezing or thawing, or hypotonic solutions.

Hemolytic disease of the newborn   A hemolytic anemia of a newborn child that results from the destruction of the infant's rbc's by antibodies produced by the mother; usually the antibodies are due to an Rh blood type incompatibility. Also called **erythroblastosis fetalis.**

Hemophilia (hē'-mō-FĒL-ē-a)   A hereditary blood disorder where there is a deficient production of certain factors involved in blood clotting, resulting in excessive bleeding into joints, deep tissues, and elsewhere.

Hemorrhage (HEM-or-rij)   Bleeding; the escape of blood from blood vessels, especially when it is profuse.

Hemorrhoids (HEM-ō-royds)   Dilated or varicosed blood vessels (usually veins) in the anal region. Also called **piles.**

Hemostasis (hē-MŌS-ta-sis)   The stoppage of bleeding.

Hemostat (HĒ-mō-stat)   An agent or instrument used to prevent the flow or escape of blood.

Hepatic (he-PAT-ik)   Refers to the liver.

Hepatic duct   A duct that receives bile from the bile capillaries. Small hepatic ducts merge to form the larger right and left hepatic ducts that unite to leave the liver as the common hepatic duct.

Hepatic portal circulation   The flow of blood from the digestive organs to the liver before returning to the heart.

Hepatitis (hep-a-TĪ-tis)   Inflammation of the liver due to a virus, drugs, and chemicals.

Hepatopancreatic (hep'-a-tō-pan'-krē-A-tik) **ampulla**   A small, raised area in the duodenum where the combined common bile duct and main pancreatic duct empty into the duodenum. Also called the **ampulla of Vater.**

Hernia (HER-nē-a)   The protrusion or projection of an organ or part of an organ through the wall of the cavity containing it.

Herniated (her'-nē-Ā-ted) **disc** A rupture of an intervertebral disc so that the nucleus pulposus protrudes into the vertebral cavity. Also called a **slipped disc.**

Hiatus (hī-Ā-tus) An opening; a foramen.

Hilus (HĪ-lus) An area, depression, or pit where blood vessels and nerves enter or leave the organ. Also called a **hilum.**

Hinge joint A synovial joint in which a convex surface of one bone fits into a concave surface of another bone, such as the elbow, knee, ankle, and interphalangeal joints. Also called a **ginglymus joint.**

Histology (his-TOL-ō-jē) Microscopic study of the structure of tissues.

Holocrine (HŌL-ō-krin) **gland** A type of gland in which the entire secreting cell, along with its accumulated secretions, makes up the secretory product of the gland, as in the sebaceous (oil) glands.

Holter monitor Electrocardiograph worn by a person while going about everyday routines.

Homeostasis (hō'-mē-ō-STĀ-sis) The condition in which the body's internal environment remains relatively constant, within limits.

Homologous (hō-MOL-ō-gus) Similar in structure and origin but not necessarily in function.

Horizontal plane A plane that runs parallel to the ground and divides the body or organs into superior and inferior portions. Also called **transverse plane.**

Hormone (HOR-mōn) A secretion of an endocrine gland that alters the physiological activity of target cells of the body.

Horn Principal area of gray matter in the spinal cord.

Human chorionic gonadotropin (kō-rē-ON-ik gō-nad-ō-TRŌ-pin) **(HCG)** A hormone produced by the developing placenta that maintains the corpus luteum.

Human chorionic somatomammotropin (sō-mat-ō-mam-ō-TRŌ-pin) **(HCS)** A hormone produced by the chorion of the placenta that may stimulate breast tissue for lactation, enhance body growth, and regulate metabolism.

Hyaluronic (hī'-a-loo-RON-ik) **acid** An amorphous matrix material found outside the cell.

Hyaluronidase (hī'-a-loo-RON-i-dās) An enzyme that breaks down hyaluronic acid, increasing the permeability of connective tissues by dissolving the substances that hold body cells together.

Hydrocele (HĪ-drō-sēl) A fluid-containing sac or tumor. Specifically, a collection of fluid formed in the space along the spermatic cord and in the scrotum.

Hydrocephalus (hī-drō-SEF-a-lus) Abnormal accumulation of cerebrospinal fluid on the brain.

Hydrophobia (hī'-drō-FŌ-bē-a) Rabies; a condition characterized by severe muscle spasms when attempting to drink water. Also, an abnormal fear of water.

Hymen (HĪ-men) A thin fold of vascularized mucous membrane at the vaginal orifice.

Hyperemia (hī'-per-Ē-mē-a) An excess of blood in an area or part of the body.

Hyperextension (hī'-per-ek-STEN-shun) Continuation of extension beyond the anatomical position, as in bending the head backward.

Hyperglycemia (hī'-per-glī-SĒ-mē-a) An elevated blood sugar level.

Hypermetropia (hī'-per-mē-TRŌ-pē-a) A condition in which visual images are focused behind the retina with resulting defective vision of near objects; farsightedness.

Hyperplasia (hī'-per-PLĀ-zē-a) An abnormal increase in the number of normal cells in a tissue or organ, increasing its size.

Hypersecretion (hī'-per-se-KRĒ-shun) Overactivity of glands resulting in excessive secretion.

Hypersensitivity (hī'-per-sen-si-TI-vi-tē) Reaction to an antigen that results in pathological changes. Also called **allergy.**

Hypertension (hī'-per-TEN-shun) High blood pressure.

Hyperthermia (hī'-per-THERM-ē-a) An elevated body temperature.

Hypertonic (hī'-per-TON-ik) Having an osmotic pressure greater than that of a solution with which it is compared.

Hypertrophy (hī-PER-trō-fē) An excessive enlargement or overgrowth of tissue without cell division.

Hyperventilation (hī'-per-ven-ti-LĀ-shun) A rate of respiration higher than that required to maintain a normal level of plasma pCO_2.

Hyponychium (hī'-pō-NIK-ē-um) Free edge of the fingernail.

Hypophysis (hī-POF-i-sis) Pituitary gland.

Hyposecretion (hī'-pō-se-KRĒ-shun) Underactivity of glands resulting in diminished secretion.

Hypothalamic-hypophyseal (hī'-pō-thal-AM-ik hī'-po-FIZ-ē-al) **tract** A bundle of nerve processes made up of fibers that have their cell bodies in the hypothalamus but release their neurosecretions in the posterior pituitary gland (neurohypophysis).

Hypothalamus (hī'-pō-THAL-a-mus) A portion of the diencephalon lying beneath the thalamus and forming the floor and part of the wall of the third ventricle.

Hypothermia (hī'-pō-THER-mē-a) A low body temperature.

Hypotonic (hī'-pō-TON-ik) Having an osmotic pressure lower than that of a solution with which it is compared.

Hypoxia (hī-POKS-ē-a) Lack of adequate oxygen. Also called **anoxia.**

Hysterectomy (his-te-REK-to-mē) The surgical removal of the uterus.

Ileocecal (il'-ē-ō-SĒ-kal) **sphincter** A fold of mucous membrane that guards the opening from the ileum into the large intestine.

Ileum (IL-ē-um) The terminal portion of the small intestine.

Immunity (i-MYOON-i-tē) The state of being resistant to injury, particularly by poisons, foreign proteins, and invading parasites, due to the presence of antibodies.

Immunoglobulin (im-yoo-nō-GLOB-yoo-lin) **(Ig)** An antibody synthesized by special lymphocytes (plasma cells) in response to the introduction of antigen. Immunoglobulins are divided into five kinds (IgG, IgM, IgA, IgD, IgE) based primarily on the larger protein component present in the immunoglobulin.

Immunosuppression (im'-yoo-nō-su-PRESH-un) Inhibition of the immune response.

Imperforate (im-PER-fō-rāt) Abnormally closed.

Impetigo (im'-pe-TĪ-gō) A contagious skin disorder characterized by pustular eruptions.

Implantation (im-plan-TĀ-shun) The insertion of a tissue or a part into the body. The attachment of the blastocyst to the lining of the uterus seven to eight days after fertilization.

Impotence (IM-pō-tens) Weakness; inability to copulate; failure to maintain an erection.

Incontinence (in-KON-ti-nens) Inability to retain urine, semen, or feces, through loss of sphincter control.

Infarction (in-FARK-shun) The presence of a localized area of necrotic tissue, produced by inadequate oxygenation of the tissue.

Infectious mononucleosis (mon-ō-nook'-lē-Ō-sis) **(IM)** Contagious disease caused by a virus and characterized by an elevated mononucleocyte and lymphocyte count, fever, sore throat, stiff neck, cough, and malaise.

Inferior (in-FĒR-ē-or) Away from the head or toward the lower part of a structure. Also called **caudad.**

Inferior vena cava (VĒ-na CĀ-va) Large vein that collects blood from parts of the body inferior to the heart and returns it to the right atrium.

Infertility Inability to conceive or to cause conception. Also called **sterility.**

Inflammation (in'-fla-MĀ-shun) Localized, protective response to tissue injury designed to destroy, dilute, or wall off the infecting agent or injured tissue; characterized by redness, pain, heat, swelling, and sometimes loss of function.

Infraspinatous (in'-fra-SPĪ-na-tus) Bony process found below the spine of the scapula used for muscle attachment.

Infundibulum (in'-fun-DIB-yoo-lum) The stalklike structure that attaches the pituitary gland (hypophysis) to the hypothalamus of the brain. The funnel-shaped, open, distal end of the uterine (Fallopian) tube.

Ingestion (in-JES-chun) The taking in of food, liquids, or drugs, by mouth.

Inguinal (IN-gwi-nal) Pertaining to the groin.

Inguinal canal An oblique passageway in the anterior abdominal wall just superior and parallel to the medial half of the inguinal ligament that transmits the spermatic cord and ilioinguinal nerve in the male and round ligament of the uterus and ilioinguinal nerve in the female.

Inheritance The acquisition of body characteristics and qualities by transmission of genetic information from parents to offspring.

Inhibin A male sex hormone secreted by sustentacular (Sertoli) cells that inhibits FSH release by the adenohypophysis (anterior pituitary) and thus spermatogenesis.

Insertion (in-SER-shun) The manner or place of attachment of a muscle to the bone that it moves.

In situ (in SĪ-too) In position.

Inspiration (in-spi-RĀ-shun) The act of drawing air into the lungs.

Inspiratory (in-SPĪ-ra-tō-rē) **capacity** The total inspiratory ability of the lungs; the sum of tidal volume and inspiratory reserve; about 3,600 ml.

Inspiratory reserve volume The volume of air in excess of tidal volume that can be inhaled by forced inspiration; about 3,100 ml.

Insula (IN-su-la) A triangular area of cerebral cortex that lies deep within the lateral cerebral fissure, under the parietal, frontal, and temporal lobes, and cannot be seen in an external view of the brain. Also called the **island** or **isle of Reil.**

Insulin (IN-su-lin) A hormone produced by the pancreas that decreases the blood glucose level.

Integumentary (in-teg'-yoo-MEN-rat-ē) Relating to the skin.

Intercalated (in-TER-ka-lāt-ed) **disc** An irregular transverse thickening of sarcolemma that separates cardiac muscle fibers (cells) from each other.

Intercellular (in'-ter-SEL-yoo-lar) Between the cells of a structure.

Intercostal (in'-ter-KOS-tal) **nerve** A nerve supplying a muscle located between the ribs.

Intermediate Between two structures, one of which is medial and one of which is lateral.

Internal Away from the surface of the body.

Internal capsule A thick sheet of white matter made up of myelinated fibers connecting various parts of the cerebral cortex and lying between the thalamus and the caudate and lentiform nuclei of the basal ganglia.

Internal ear The inner ear or labyrinth, lying inside the temporal bone, containing the organs of hearing and balance.

Internal nares (NA-rēz) The two openings posterior to the nasal cavities opening into the nasopharynx. Also called the **choanae.**

Internal respiration The exchange of respiratory gases between the blood and body cells.

Interphase (IN-ter-fāz) The period during its life cycle when a cell is carrying on every life process except division; the stage between two mitotic divisions.

Interstitial (in'-ter-STISH-al) **endocrinocyte** A cell located in the connective tissue between seminiferous tubules in a mature testis that secretes testosterone. Also called an **interstitial cell of Leydig.**

Interstitial fluid The fluid filling the microscopic spaces between the cells of tissues.

Interventricular foramen (in'-ter-ven-TRIK-yoo-lar) A narrow, oval opening through which the lateral ventricles of the brain communicate with the third ventricle. Also called the **foramen of Monro.**

Intervertebral (in'-ter-VER-te-bral) **disc** A pad of fibrocartilage located between the bodies of two vertebrae.

Intestinal gland Simple tubular gland that opens onto the surface of the intestinal mucosa and secretes digestive enzymes. Also called a **crypt of Lieberkühn.**

Intracellular (in'-tra-SEL-yoo-lar) Within cells.

Intrafusal (in'-tra-FYOO-zal) **fibers** Three to ten specialized muscle fibers (cells), partially enclosed in a connective tissue capsule that is filled with lymph; the fibers compose muscle spindles.

Intramembranous ossification (in'-tra-MEM-bra-nus os'-i-fi-KĀ-shun) The method of bone formation in which the bone is formed directly in membranous tissue.

Intraocular (in-tra-OC-yoo-lar) **pressure** Pressure in the eyeball, produced mainly by aqueous humor.

Intrauterine device (IUD) A small metal or plastic object inserted into the uterus for the purpose of preventing pregnancy.

Intrinsic (in-TRIN-sik) Of internal origin; for example, the intrinsic factor, a mucoprotein formed by the gastric mucosa that is necessary for the absorption of vitamin B_{12}.

Intrinsic factor A glycoprotein synthesized and secreted by the gastric mucosa that facilitates vitamin B_{12} absorption.

Intubation (in'-too-BĀ-shun) Insertion of a tube through the nose or mouth into the larynx and trachea for entrance of air or to dilate a stricture.

Intussusception (in'-ta-sa-SEP-shun) The infolding (invagination) of one part of the intestine within another segment.

In utero (YOO-ter-ō) Within the uterus.

Invagination (in-vaj'-i-NĀ-shun) The pushing of the wall of a cavity into the cavity itself.

Inversion (in-VER-zhun) The movement of the sole inward at the ankle joint.

In vitro (VĒ-trō) Literally, in glass; outside the living body and in an artificial environment such as a laboratory test tube.

In vivo (VĒ-vō) In the living body.

Ipsilateral (ip'-si-LAT-er-al) On the same side; affecting the same side of the body.

Ischemia (is-KĒ-mē-a) A lack of sufficient blood flow to a part due to obstruction of circulation.

Isotonic (Ī-sō-TON-ik) Having equal tension or tone. Having equal osmotic pressure between two different solutions or between two elements in a solution.

Isthmus (IS-mus) A narrow strip of tissue or narrow passage connecting two larger parts.

Jaundice (JAWN-dis) A condition characterized by yellowness of the skin, white of the eyes, mucous membranes, and body fluids.

Jejunum (jē-JOO-num) The middle portion of the small intestine.

Joint kinesthetic (kin'-es-THET-ik) **receptor** A proprioceptive receptor located in a joint, stimulated by joint movement.

Juxtaglomerular (juks-ta-glō-MER-yoo-lar) **apparatus** Consists of the macula densa (cells of the distal convoluted tubule adjacent to the afferent arteriole) and juxtaglomerular cells (modified cells of the afferent arteriole); secretes renin when blood pressure starts to fall.

Keratin (KER-a-tin) An insoluble protein found in the hair, nails, and other keratinized tissues of the epidermis.

Keratohyalin (ker'-a-tō-HĪ-a-lin) A compound involved in the formation of keratin.

Kidney (KID-nē) One of the paired reddish organs located in the lumbar region that regulates the composition and volume of blood and produces urine.

Kinesiology (ki-nē'-sē-OL-ō-jē) The study of the movement of body parts.

Kinesthesia (kin-is-THĒ-szē-a) Ability to perceive extent, direction, or weight of movement; muscle sense.

Kyphosis (kī-FŌ-sis) An exaggeration of the thoracic curve of the vertebral column, resulting in a "round-shouldered" or hunchback appearance.

Labial frenulum (LĀ-bē-al FREN-yoo-lum) A medial fold of mucous membrane between the inner surface of the lip and the gums.

Labia majora (LĀ-bē-a ma-JOR-a) Two longitudinal folds of skin extending downward and backward from the mons pubis of the female.

Labia minora (min-OR-a) Two small folds of mucous membrane lying medial to the labia majora of the female.

Labium (LĀ-bē-um) A lip. A liplike structure. *Plural,* **labia** (LĀ-bē-a).

Labyrinth (LAB-i-rinth) Intricate communicating passageways, especially the internal ear.

Labyrinthine (lab-i-RIN-thēn) **disease** Malfunction of the internal ear characterized by deafness, tinnitus, vertigo, nausea, and vomiting.

Laceration (las'-er-Ā-shun) A torn, ragged, or mangled wound due to trauma.

Lacrimal (LAK-ri-mal) Pertaining to tears.

Lacrimal canal A duct, one on each eyelid, commencing at the punctum at the medial margin of an eyelid and conveying the tears medially into the nasolacrimal sac.

Lacrimal gland Secretory cells located at the superior lateral portion of each orbit that secrete tears into excretory lacrimal ducts that open onto the surface of the conjunctiva.

Lacrimal sac The superior expanded portion of the nasolacrimal duct that receives the tears from a lacrimal canal.

Lactation (lak-TĀ-shun) The secretion and ejection of milk by the mammary glands.

Lacteal (LAK-tē-al) One of many intestinal lymph vessels in villi that absorb fat from digested food.

Lacuna (la-KOO-na) A small, hollow space, such as that found in bones in which the osteoblasts lie. *Plural,* **lacunae** (la-KOO-nē).

Lambdoidal (lam-DOY-dal) **suture** The line of union in the skull between the parietal bones and the occipital bone.

Lamellae (la-MEL-ē) Concentric rings found in compact bone.

Lamellated corpuscle Oval pressure receptor located in subcutaneous tissue and consisting of concentric layers of connective tissue wrapped around an afferent nerve fiber. Also called a **Pacinian corpuscle.**

Lamina (LAM-i-na) A thin, flat layer or membrane, as the flattened part of either side of the arch of a vertebra. *Plural,* **laminae** (LAM-i-nē).

Lamina propria (PRO-prē-a) The connective tissue layer of a mucous membrane.

Lanugo (lan-YOO-gō) Fine downy hairs that cover the fetus.

Large intestine The portion of the digestive tract extending from the ileum of the small intestine to the anus, divided structurally into the cecum, colon, rectum, and anal canal.

Laryngitis (la-rin-JĪ-tis) Inflammation of the mucous membrane lining the larynx.

Laryngopharynx (la-rin'-gō-FAR-inks) The inferior portion of the pharynx, extending downward from the level of the hyoid bone to divide posteriorly into the esophagus and anteriorly into the larynx.

Laryngoscope (la-RINJ-ō-skōp) An instrument for examining the larynx.

Larynx (LAR-inks) The voice box, a short passageway that connects the pharynx with the trachea.

Lateral (LAT-er-al) Farther from the midline of the body or a structure.

Lateral apertures (AP-er-choorz) Two of the three openings in the roof of the fourth ventricle through which cerebrospinal fluid enters the subarachnoid space of the brain and spinal cord. Also known as the **foramina of Luschka.**

Lateral ventricle (VEN-tri-kul) A cavity within a cerebral hemisphere that communicates with the lateral ventricle in the other cerebral hemisphere and with the third ventricle by way of the interventricular foramen (of Monro).

Leg The part of the lower extremity between the knee and the ankle.

Lens A transparent structure lying posterior to the pupil and iris of the eyeball and anterior to the vitreous humor.

Lesion (LĒ-zhun) Any localized, abnormal change in tissue formation.

Lesser omentum (ō-MEM-tum) A fold of the peritoneum that extends from the liver to the lesser curvature of the stomach and the commencement of the duodenum.

Lesser vestibular (ves-TIB-yoo-lar) **gland** One of the paired mucus-secreting glands that have ducts that open on either side of the urethral orifice in the vestibule of the female.

Leucocyte (LOO-kō-sīt) A white blood cell.

Leucocytosis (loo-kō-sī-TŌ-sis) An increase in the number of white blood cells, characteristic of many infections and other disorders.

Leucopenia (loo-kō-PĒ-nē-a) A decrease of the number of white blood cells below 5,000/mm³.

Leukemia (loo-KĒ-mē-a)   A cancerlike disease of the blood-forming organs characterized by a rapid and abnormal increase in the number of white blood cells plus many immature cells in the circulating blood.

Leukoplakia (loo-kō-PLĀ-kē-a)   A disorder in which there are white patches in the mucous membranes of the tongue, gums, and cheeks.

Libido (li-BĒ-dō)   The sexual drive, conscious or unconscious.

Ligament (LIG-a-ment)   Dense, regularly arranged connective tissue that attaches bone to bone.

Limbic system   A portion of the forebrain, sometimes termed the visceral brain, concerned with various aspects of emotion and behavior, that includes the limbic lobe (hippocampus and associated areas of gray matter plus the cingulate gyrus), certain parts of the temporal and frontal cortex, some thalamic and hypothalamic nuclei, and parts of the basal ganglia.

Lingual frenulum (LIN-gwal FREN-yoo-lum)   A fold of mucous membrane that connects the tongue to the floor of the mouth.

Lipase (LĪ-pās)   A fat-splitting enzyme.

Lipid   An organic compound composed of carbon, hydrogen, and oxygen that is usually insoluble in water but soluble in alcohol, ether, and chloroform; examples include fats, phospholipids, steroids, and prostaglandins.

Lipoma (li-PŌ-ma)   A fatty tissue tumor, usually benign.

Liver   Large gland under the diaphragm that occupies most of the right hypochondriac region and part of the epigastric region; functionally, it produces bile salts, heparin, and plasma proteins; converts one nutrient into another; detoxifies substances; stores glycogen, minerals, and vitamins; carries on phagocytosis of blood cells and bacteria; and helps activate vitamin D.

Lobe (lōb)   A curve or rounded projection.

Local   Pertaining to or restricted to one spot or part.

Locus coeruleus (LŌ-kus sē-ROO-lē-us)   A group of neurons in the brain stem where norepinephrine (NE) is concentrated.

Lordosis (lor-DŌ-sis)   An exaggeration of the lumbar curve of the vertebral column.

Lower extremity   The appendage attached at the pelvic girdle, consisting of the thigh, knee, leg, ankle, foot, and toes.

Lumbar (LUM-bar)   Region of the back and side between the ribs and pelvis; loin.

Lumbar plexus (PLEK-sus)   A network formed by the anterior branches of spinal nerves L1 through L4.

Lumen (LOO-men)   The space within an artery, vein, intestine, or tube.

Lung   One of the two main organs of respiration, lying on either side of the heart in the thoracic cavity.

Lunula (LOO-nyoo-la)   The moon-shaped white area at the base of a nail.

Luteinizing (LOO-tē-in'-īz-ing) **hormone (LH)**   A hormone secreted by the adenohypophysis (anterior lobe) of the pituitary gland that stimulates ovulation, progesterone secretion by the corpus luteum, and readies the mammary glands for milk secretion in females and stimulates testosterone secretion by the testes in males.

Lymph (limf)   Fluid confined in lymphatic vessels and flowing through the lymphatic system to be returned to the blood.

Lymphangiogram (lim-FAN-jē-ō-gram)   An x ray of lymphatic vessels and lymphatic organs.

Lymphangiography (lim-fan'-jē-OG-ra-fē)   A procedure by which lymphatic vessels and lymph organs are filled with a radiopaque substance in order to be x-rayed.

Lymphatic (lim-FAT-ik)   Pertaining to lymph. A large vessel that collects lymph from lymph capillaries and converges with other lymphatics to form the thoracic and right lymphatic ducts.

Lymph capillary   Blind-ended microscopic lymph vessel that begins in spaces between cells and converges with other lymph capillaries to form lymphatics.

Lymph node   An oval or bean-shaped structure located along the lymphatic vessels.

Lymphocyte (LIM-fō-sīt)   A type of white blood cell, found in lymph nodes, associated with the immune system.

Lysosome (LĪ-sō-sōm)   An organelle in the cytoplasm of a cell, enclosed by a single membrane and containing powerful digestive enzymes.

Macula (MAK-yoo-la)   A discolored spot or a colored area. A small, flat region on the wall of the utricle and saccule that serves as a receptor for static equilibrium.

Macula lutea (LOO-tē-a)   The yellow spot in the center of the retina.

Magnetic resonance imaging (MRI)   A diagnostic procedure that focuses on the nuclei of atoms of a single element in a tissue, usually hydrogen, to determine if they behave normally in the presence of an external magnetic force; used to indicate the biochemical activity of a tissue. Formerly called **nuclear magnetic resonance (NMR).**

Malaise (ma-LĀYZ)   Discomfort, uneasiness, and indisposition, often indicative of infection.

Malignant (ma-LIG-nant)   Referring to diseases that tend to become worse and cause death; especially the invasion and spreading of cancer.

Mammary (MAM-ar-ē) **gland**   Modified sudoriferous (sweat) gland of the female that secretes milk for the nourishment of the young.

Marrow (MAR-ō)   Soft, spongelike material in the cavities of bone. Red marrow produces blood cells; yellow marrow, formed mainly of fatty tissue, has no blood-producing function.

Mast cell   A cell found in loose connective tissue along blood vessels that produces heparin, an anticoagulant. The name given to a basophil after it has left the bloodstream and entered the tissues.

Mastectomy (mas-TEK-tō-mē)   Surgical removal of breast tissue.

Mastication (mas'-ti-KĀ-shun)   Chewing.

Meatus (mē-Ā-tus)   A passage or opening, especially the external portion of a canal.

Mechanoreceptor (me-KAN-ō-rē'-sep-tor)   Receptor that detects mechanical deformation of the receptor itself or adjacent cells; stimuli so detected include touch, pressure, proprioception, hearing, and equilibrium.

Medial (MĒ-dē-al)   Nearer the midline of the body or a structure.

Medial lemniscus (lem-NIS-kus)   A flat band of myelinated nerve fibers extending through the medulla, pons, and midbrain and terminating in the thalamus on the opposite side. Second-order sensory neurons in this tract transmit impulses for discriminating touch, pressure, and vibration sensations.

Median aperture (AP-er-choor)   One of the three openings in the roof of the fourth ventricle through which cerebrospinal fluid enters the subarachnoid space of the brain and cord. Also called the **foramen of Magendie.**

Median plane A vertical plane dividing the body into right and left halves. Situated in the middle.

Mediastinum (mē'-dē-as-TĪ-num) A mass of tissue found between the pleurae of the lungs that extends from the sternum to the vertebral column.

Medulla (me-DUL-la) An inner layer of an organ, such as the medulla of the kidneys.

Medulla oblongata (ob'-long-GA-ta) The most inferior part of the brain stem.

Medullary (MED-yoo-lar'-ē) **cavity** The space within the diaphysis of a bone that contains yellow marrow.

Medullary rhythmicity (rith-MIS-i-tē) **area** Portion of the respiratory center in the medulla that controls the basic rhythm of respiration.

Meiosis (mē-Ō-sis) A type of cell division restricted to sex-cell production involving two successive nuclear divisions which result in daughter cells with the haploid number of chromosomes.

Melanin (MEL-a-nin) A dark black, brown, or yellow pigment found in some parts of the body such as the skin.

Melanocyte (MEL-a-nō-sīt') A pigmented cell located between or beneath cells of the deepest layer of the epidermis that synthesizes melanin.

Melanocyte-stimulating hormone (MSH) A hormone secreted by the adenohypophysis (anterior lobe) of the pituitary gland that stimulates the dispersion of melanin granules in melanocytes.

Melanoma (mel'-a-NŌ-ma) A usually dark, malignant tumor of the skin containing melanin.

Melatonin (mel-a-TŌN-in) A hormone secreted by the pineal gland that may inhibit reproductive activities.

Membrane A thin, flexible sheet of tissue composed of an epithelial layer and an underlying connective tissue layer, as in an epithelial membrane, or of loose connective tissue only, as in a synovial membrane.

Membranous labyrinth (mem-BRA-nus LAB-i-rinth) The portion of the labyrinth of the inner ear that is located inside the bony labyrinth and separated from it by the perilymph; made up of the membranous semicircular canals, the saccule and utricle, and the cochlear duct.

Menarche (me-NAR-kē) Beginning of the menstrual function.

Ménière's (men-YAIRZ) **syndrome** A type of labyrinthine disease characterized by fluctuating loss of hearing, vertigo, and tinnitus.

Meninges (me-NIN-jēz) Three membranes covering the brain and spinal cord, called the dura mater, arachnoid, and pia mater. *Singular,* **meninx** (MEN-inks).

Meningitis (men-in-JĪ-tis) Inflammation of the meninges, most commonly the pia mater and arachnoid.

Menopause (MEN-ō-pawz) The termination of the menstrual cycles.

Menstrual (MEN-stroo-al) **cycle** A series of changes in the endometrium of a nonpregnant female that prepares the lining of the uterus to receive a fertilized ovum.

Menstruation (men'-stroo-Ā-shun) Periodic discharge of blood, tissue fluid, mucus, and epithelial cells that usually lasts for 5 days; caused by a sudden reduction in estrogens and progesterone. Also called the **menstrual phase,** or **menses.**

Merocrine (MER-ō-krin) **gland** A secretory cell that remains intact throughout the process of formation and discharge of the secretory product, as in the salivary and pancreatic glands.

Mesenchyme (MEZ-en-kīm) An embryonic connective tissue from which all other connective tissues arise.

Mesentery (MEZ-en-ter'-ē) A fold of peritoneum attaching the small intestine to the posterior abdominal wall.

Mesocolon (mez'-ō-KŌ-lon) A fold of peritoneum attaching the colon to the posterior abdominal wall.

Mesoderm The middle of the three primary germ layers that gives rise to connective tissues, blood and blood vessels, and muscles.

Mesothelium (mez'-ō-THĒ-lē-um) The layer of simple squamous epithelium that lines serous cavities.

Mesovarium (mez'-ō-VAR-ēum) A short fold of peritoneum that attaches an ovary to the broad ligament of the uterus.

Metabolism (me-TAB-ō-lizm) The sum of all the biochemical reactions that occur within an organism, including the synthetic (anabolic) reactions and decomposition (catabolic) reactions.

Metacarpus (met'-a-KAR-pus) A collective term for the five bones that make up the palm of the hand.

Metaphase (MET-a-phāz) The second stage of mitosis in which chromatid pairs line up on the equatorial plane of the cell.

Metaphysis (me-TAF-i-sis) Growing portion of a bone.

Metarteriole (met'-ar-TĒ-rē-ōl) A blood vessel that emerges from an arteriole, traverses a capillary network, and empties into a venule.

Metastasis (me-TAS-ta-sis) The transfer of disease from one organ or part of the body to another.

Metatarsus (met'-a-TAR-sus) A collective term for the five bones located in the foot between the tarsals and the phalanges.

Microcephalus (mi-kro-SEF-a-lus) An abnormally small head; premature closing of the anterior fontanel so that the brain has insufficient room for growth, resulting in mental retardation.

Microfilament (mī-krō-FIL-a-ment) Rodlike cytoplasmic structure, ranging in diameter from 30 Å to 120 Å; comprises contractile units in muscle fibers (cells) and provides support, shape, and movement in nonmuscle cells.

Microglia (mi-krō-GLĒ-a) Neuroglial cells that carry on phagocytosis. Also called **brain macrophages.**

Microphage (MĪK-rō-fāj) Granular leucocyte that carries on phagocytosis, especially neutrophils and eosinophils.

Microtubule (mī-krō-TOOB-yool') Cylindrical cytoplasmic structure, ranging in diameter from 180 Å to 300 Å, consisting of the protein tubulin; provides support, structure, and transportation.

Microvilli (mī'-krō-VIL-ē) Microscopic, fingerlike projections of the plasma membranes of small intestinal cells that increase surface area for absorption.

Micturition (mik'-too-RISH-un) The act of expelling urine from the bladder. Also called **urination.**

Midbrain The part of the brain between the pons and the diencephalon. Also called the **mesencephalon.**

Middle ear A small, epithelial-lined cavity hollowed out of the temporal bone, separated from the external ear by the eardrum and from the internal ear by a thin bony partition containing the oval and round windows; extending across the middle ear are the three auditory ossicles. Also called the **tympanic cavity.**

Midsagittal (mid-SAJ-i-tal) Median or relating to the midline.

Midsagittal plane A plane through the midline of the body running vertical to the ground and dividing the body or organs into equal right and left sides. Also called a **median plane.**

Milk letdown Contraction of alveolar cells to force milk into ducts of mammary glands, stimulated by oxytocin (OT) that is released from the posterior pituitary in response to sucking action.

Mineralocorticoids (min'-er-al-ō-KOR-ti-koyds) A group of hormones of the adrenal cortex.

Minimal volume The volume of air in the lungs even after the thoracic cavity has forced out some of the residual volume.

Minute volume of respiration (MVR) Total volume of air taken into the lungs per minute; about 6,000 ml.

Mitochondrion (mi'-tō-KON-drē-on) A double-membraned organelle that plays a central role in the production of ATP; known as the "powerhouse of the cell."

Mitosis (mī-TŌ-sis) The orderly division of the nucleus of a cell that ensures that each new daughter nucleus has the same number and kind of chromosomes as the original parent nucleus. The process includes the replication of chromosomes and the distribution of the two sets of chromosomes into two separate and equal nuclei.

Mitotic apparatus Collective term for continuous and chromosomal microtubules and centrioles; involved in cell division.

Mitotic spindle The combination of continuous and chromosomal microtubules, involved in chromosomal movement during mitosis.

Mittelschmerz (MIT-el-shmerz) Abdominopelvic pain that supposedly indicates the release of an egg from the ovary.

Modality (mō-DAL-i-tē) Any of the specific sensory entities, such as vision, smell, or taste.

Modiolus (mō-DĪ-ō'-lus) The central pillar or column of the cochlea.

Monoclonal antibody Antibody produced by in vitro clones of B cells hydridized with cancerous cells.

Monocyte (MON-ō-sīt') A type of white blood cell characterized by agranular cytoplasm; the largest of the leucocytes.

Mons pubis (monz PYOO-bis) The rounded, fatty prominence over the symphysis pubis, covered by coarse pubic hair.

Morbid (MOR-bid) Diseased; pertaining to disease.

Morula (MOR-yoo-la) A solid mass of cells produced by successive cleavages of a fertilized ovum a few days after fertilization.

Motor area The region of the cerebral cortex that governs muscular movement, particularly the precentral gyrus of the frontal lobe.

Motor unit A motor neuron together with the muscle fibers (cells) it stimulates.

Mucin (MYOO-sin) A protein found in mucus.

Mucous (MYOO-kus) **cell** A unicellular gland that secretes mucus. Also called a **goblet cell.**

Mucous membrane A membrane that lines a body cavity that opens to the exterior. Also called the **mucosa.**

Mucus The thick fluid secretion of the mucous glands and mucous membranes.

Multiple sclerosis (skler-Ō-sis) Progressive destruction of myelin sheaths of neurons in the central nervous system, short-circuiting conduction pathways.

Mumps Inflammation and enlargement of the parotid glands accompanied by fever and extreme pain during swallowing.

Murmur An unusual heart sound; may indicate a disorder such as a malfunctioning bicuspid (mitral) valve or may have no clinical significance.

Muscle An organ composed of one of three types of muscle tissue (skeletal, cardiac, or smooth), specialized for contraction to produce voluntary or involuntary movement of parts of the body.

Muscle spindle An encapsulated receptor in a skeletal muscle, consisting of specialized muscle cells and nerve endings, stimulated by changes in length or tension of muscle cells; a proprioceptor. Also called a **neuromuscular spindle.**

Muscle tissue A tissue specialized to produce motion in response to nerve impulses by its qualities of contractility, extensibility, elasticity, and excitability.

Muscular dystrophy (DIS-trō-fē) Inherited myopathy characterized by degeneration of muscle fibers (cells) that leads to progressive atrophy.

Muscularis (MUS-kyoo-la'-ris) A muscular layer or tunic of an organ.

Muscularis mucosae (myoo-KŌ-sē) A thin layer of smooth muscle fibers (cells) located in the outermost layer of the mucosa of the alimentary canal, underlying the lamina propria of the mucosa.

Musculi pectinati (MUS-kyoo-lī pek-ti-NA-tē) Projecting muscle bundles of the anterior atrial walls and the lining of the auricles.

Myasthenia (mī-as-THĒ-nē-a) **gravis** Weakness of skeletal muscles caused by antibodies directed against acetylcholine receptors that inhibit muscle contraction.

Myelin (MĪ-e-lin) **sheath** A white, phospholipid, segmented covering, formed by neurolemmocytes (Schwann cells), around the axons and dendrites of many peripheral neurons.

Myenteric plexus A network of nerve fibers from both autonomic divisions located in the muscularis coat or tunic of the small intestine. Also called the **plexus of Auerbach.**

Myocardial infarction (mī'-ō-KAR-dē-al in-FARK-shun) Gross necrosis of myocardial tissue due to interrupted blood supply. Also called a **heart attack.**

Myocardium (mī-ō-KAR-dē-um) The middle layer of the heart wall, made up of cardiac muscle, comprising the bulk of the heart, and lying between the epicardium and the endocardium.

Myofibril (mī'-ō-FĪ-bril) A threadlike structure, running longitudinally through a muscle fiber (cell), consisting mainly of thick myofilaments (myosin) and thin myofilaments (actin).

Myoglobin (mī-ō-GLŌ-bin) The oxygen-binding, iron-containing conjugated protein complex present in the sarcoplasm of muscle fibers (cells); contributes the red color to muscle.

Myogram (MĪ-ō-gram) The record or tracing produced by the myograph, the apparatus that measures and records the effects of muscular contractions.

Myology (mī-OL-ō-jē) The study of the muscles and their parts.

Myometrium (mī'-ō-MĒ-trē-um) The smooth muscle coat or tunic of the uterus.

Myopia (mī-Ō-pē-a) Defect in vision so that objects can be seen distinctly only when very close to the eyes; nearsightedness.

Myosin (MĪ-ō-sin) The contractile protein that makes up the thick myofilaments of muscle fibers (cells).

Myxedema (mix-e-DĒ-ma) Condition caused by hypothyroidism during the adult years characterized by swelling of facial tissues.

Nail A hard plate, composed largely of keratin, that develops from the epidermis of the skin to form a protective covering on the dorsal surface of the distal phalanges of the fingers and toes.

Nail matrix (MĀ-triks) The part of the nail beneath the body and root from which the nail is produced.

Narcosis (nar-KŌ-sis) Unconscious state due to narcotics.

Nasal (NĀ-zal) **cavity** A mucosa-lined cavity on either side of the nasal septum that opens onto the face at an external naris and into the nasopharynx at an internal naris.

Nasal septum (SEP-tum) A vertical partition composed of bone and cartilage, covered with a mucous membrane, separating the nasal cavity into left and right sides.

Nasolacrimal (nā'-zō-LAK-ri-mal) **duct** A canal that transports the lacrimal secretion from the nasolacrimal sac into the nose.

Nasopharynx (nā'-zō-FAR-inks) The uppermost portion of the pharynx, lying posterior to the nose and extending down to the soft palate.

Nebulization (neb'-yoo-li-ZĀ-shun) Treatment with medication by spray method.

Neck The part of the body connecting the head and the trunk. A constricted portion of an organ such as the neck of a femur or the neck of the uterus.

Necrosis (ne-KRŌ-sis) Death of a cell or group of cells as a result of disease or injury.

Neonatal (nē'-ō-NĀ-tal) Pertaining to the first four weeks after birth.

Neoplasm (NĒ-ō-plazm) A mass of new, abnormal tissue; a tumor.

Nephritis (ne-FRĪT-is) Inflammation of the kidney.

Nephron (NEF-ron) The functional unit of the kidney.

Nephrosis (nef-RŌ-sis) A degenerative disease of the kidney in which glomerular damage results in leakage of protein into urine.

Nerve A cordlike bundle of nerve fibers and their associated connective tissue coursing together outside the central nervous system.

Nerve impulse A wave of negativity that sweeps along the outside of the membrane of a neuron. Also called an **action potential.**

Nervous tissue Tissue that initiates and transmits nerve impulses to coordinate homeostasis.

Neurilemma (noo-ri-LEM-a) A delicate sheath around a nerve fiber composed of the cytoplasm, nucleus, and enclosing cell membrane of a neurolemmocyte (Schwann cell). There may or may not be a layer of myelin between the neurilemma and the axis cylinder (axon or dendrite) of the neuron.

Neuritis (noo-RĪ-tis) Inflammation of a nerve.

Neuroeffector (noo-rō-e-FEK-tor) **junction** Cumulative term for neuromuscular and neuroglandular junctions.

Neurofibral (noo-rō-FĪ-bral) **node** A space, along a myelinated nerve fiber, between the individual neurolemmocytes (Schwann cells) that form the myelin sheath and the neurilemma. Also called **node of Ranvier.**

Neurofibril (noo-rō-FĪ-bril) One of the delicate threads that forms a complicated network in the cytoplasm of the cell body and processes of a neuron.

Neuroglandular (noo-rō-GLAND-yoo-lar) **junction** Area of contact between a motor neuron and a gland.

Neuroglia (noo-ROG-lē-a) Cells of the nervous system that are specialized to perform the functions of connective tissue. The neuroglia of the central nervous system are the astrocytes, oligodendrocytes, microglia, and ependyma; neuroglia of the peripheral nervous system include the neurolemmocytes (Schwann cells) and the ganglion satellite cells. Also called **glial cells.**

Neurohypophysis (noo-rō-hī-POF-i-sis) The posterior lobe of the pituitary gland.

Neurolemmocyte A neuroglial cell of the peripheral nervous system that forms the myelin sheath and neurilemma of a nerve fiber by wrapping around a nerve fiber in a jelly-roll fashion. Also called a **Schwann cell.**

Neurology (noo-ROL-ō-jē) The branch of science that deals with the normal functioning and disorders of the nervous system.

Neuromuscular (noo-rō-MUS-kyoo-lar) **junction** The area of contact between a motor neuron and a muscle fiber (cell). Also called a **myoneural junction,** or **motor end plate.**

Neuron (NOO-ron) A nerve cell, consisting of a cell body, dendrites, and an axon.

Neurosecretory (noo-rō-SEC-re-tō-rē) **cell** A cell in a nucleus in the hypothalamus that produces oxytocin (OT) or antidiuretic hormone (ADH), hormones stored in the neurohypophysis of the pituitary gland.

Neurosyphilis (noo-rō-SIF-i-lis) A form of the tertiary stage of syphilis in which various types of nervous tissue are attacked by the bacteria and degenerate.

Neurotransmitter One of a variety of molecules synthesized within the nerve axon terminals, released into the synaptic cleft in response to an action potential, and affecting the membrane potential of the postsynaptic neuron.

Neutrophil (NOO-trō-fil) A type of white blood cell characterized by granular cytoplasm that stains as readily with acid or basic dyes. Also called a **polymorph.**

Night blindness Poor or no vision in dim light or at night, although good vision present during bright illumination; frequently caused by a deficiency of vitamin A.

Nipple A pigmented, wrinkled projection on the surface of the mammary gland that is the location of the openings of the lactiferous ducts for milk release.

Nociceptor (nō'-sē-SEP-tor) A receptor that detects pain.

Nongonococcal urethritis (NGU) A sexually transmitted disease characterized by inflammation of the urethra caused by a microorganism other than *Neisseria gonorrhoeae* or by nonmicrobial factors.

Norepinephrine (nor'-ep-ē-NEF-rin) **(NE)** A hormone secreted by the adrenal medulla that produces actions similar to those that result from sympathetic stimulation. Also called **noradrenaline.**

Nuclear medicine The branch of medicine concerned with the use of radioisotopes in the diagnosis of disease and therapy.

Nuclease (NOO-klē-ās) An enzyme that breaks nucleotides into pentases and nitrogen bases; examples are ribonuclease and deoxyribonuclease.

Nucleic (noo-KLĒ-ic) **acid** An organic compound that is a long polymer of nucleotides, with each nucleotide containing a pentose sugar, a phosphate group, and one of four possible nitrogen bases (adenine, cytosine, guanine, and thymine or uracil).

Nucleosome (NOO-klē-ō-sōm) Elementary structural subunit of a chromosome consisting of histones and DNA.

Nucleus (NOO-klē-us) A spherical or oval organelle of a cell that contains the hereditary factors of the cell, called genes. A cluster of nerve cell bodies in the central nervous system. The central portion of an atom made up of protons and neutrons.

Nucleus cuneatus (kyoo-nē-Ā-tus) A group of nerve cells in the inferior portion of the medulla oblongata in which fibers of the fasciculus cuneatus terminate.

Nucleus gracilis (gras-I-lis) A group of nerve cells in the medulla oblongata in which fibers of the fasciculus gracilis terminate.

Nucleus pulposus (pul-PŌ-sus) A soft, pulpy, highly elastic substance in the center of an intervertebral disc, a remnant of the notochord.

Nystagmus (nis-TAG-mus) Constant, involuntary, rhythmic movement of the eyeballs; horizontal, rotary, or vertical.

Obesity (ō-BĒS-i-tē) Body weight 10 to 20 percent over a desirable standard as a result of excessive accumulation of fat.

Obstetrics (ob-STET-riks) The specialized branch of medicine that deals with pregnancy, labor, and the period of time immediately following delivery.

Obturator (OB-tyoo-rā'-ter) Anything that obstructs or closes a cavity or opening.

Occlusion (o-KLOO-zhun) The act of closure or state of being closed.

Odontoid (ō-DON-toyd) Toothlike.

Olfactory (ōl-FAK-tō-rē) Pertaining to smell.

Olfactory bulb A mass of gray matter at the termination of an olfactory (I) nerve, lying beneath the frontal lobe of the cerebrum on either side of the crista galli of the ethmoid bone.

Olfactory cell A bipolar neuron with its cell body lying between supporting cells located in the mucous membrane lining the upper portion of each nasal cavity.

Olfactory tract A bundle of axons that extends from the olfactory bulb posteriorly to the olfactory portion of the cortex.

Oligodendrocyte (ol'-i-gō-DEN-drō-sīt) A neuroglial cell that supports neurons and produces a myelin sheath around axons of neurons of the central nervous system.

Oligospermia (ol'-i-gō-SPER-mē-a) A deficiency of spermatozoa in the semen.

Olive A prominent oval mass on each lateral surface of the superior part of the medulla.

Oncogene (ONG-kō-jēn) Gene that has the ability to transform a normal cell into a cancerous cell.

Oncology (ong-KOL-ō-jē) The study of tumors.

Oogenesis (ō'-ō-JEN-e-sis) Formation and development of the ovum.

Oophorectomy (ō'-of-ō-REK-tō-mē) The surgical removal of the ovaries.

Ophthalmic (of-THAL-mik) Pertaining to the eye.

Ophthalmology (of'-thal-MOL-ō-jē) The study of the structure, function, and diseases of the eye.

Optic (OP-tik) Refers to the eye, vision, or properties of light.

Optic chiasma (kī-AZ-ma) A crossing point of the optic (II) nerves, anterior to the pituitary gland.

Optic disc A small area of the retina containing openings through which the fibers of the ganglion neurons emerge as the optic (II) nerve. Also called the **blind spot.**

Optic tract A bundle of axons that transmits impulses from the retina of the eye between the optic chiasma and the thalamus.

Oral contraceptive (OC) A hormonal compound that is swallowed and that prevents ovulation, and thus pregnancy. Also called **the pill.**

Ora serrata (Ō-ra ser-RĀ-ta) **of the retina** The notched anterior edge of sensory portion of the nervous tunic (retina).

Orbit (OR-bit) The bony, pyramid-shaped cavity of the skull that holds the eyeball.

Organ A structure of definite form and function composed of two or more different kinds of tissues.

Organelle (or-gan-EL) A permanent structure within a cell with characteristic morphology that is specialized to serve a specific function in cellular activities.

Organism (OR-ga-nizm) A total living form; one individual.

Orgasm (OR-gazm) Sensory and motor events involved in ejaculation for the male and involuntary contraction of the perineal muscles in the female at the climax of sexual intercourse.

Orifice (OR-i-fis) Any aperture or opening.

Origin (OR-i-jin) The place of attachment of a muscle to the more stationary bone, or the end opposite the insertion.

Oropharynx (or'-ō-FAR-inks) The second portion of the pharynx, lying posterior to the mouth and extending from the soft palate down to the hyoid bone.

Orthopedics (or'-thō-PE-diks) The branch of medicine that deals with the preservation and restoration of the skeletal system, articulations, and associated structures.

Osmosis (os-MŌ-sis) The net movement of water molecules through a selectively permeable membrane from an area of high water concentration to an area of lower water concentration until an equilibrium is reached.

Osmotic pressure The force under which a solvent (water) moves through a selectively permeable membrane from higher to lower water concentration.

Osseous (OS-ē-us) Bony.

Ossicle (OS-si-kul) Small bone, as in the middle ear (malleus, incus, stapes).

Ossification (os'-i-fi-KĀ-shun) Formation of bone. Also called **osteogenesis.**

Osteoblast (OS-tē-ō-blast') A bone-forming cell.

Osteoclast (OS-tē-ō-clast') A large, multinuclear cell that destroys or resorbs bone tissue.

Osteocyte (OS-tē-ō-sīt') A mature osteoblast that has lost its ability to produce new bone tissue.

Osteogenic (os'-tē-ō-JEN-ik) **layer** The inner layer of the periosteum.

Osteology (os'-tē-OL-ō-jē) The study of bones.

Osteomalacia (os'-tē-ō-ma-LA-shē-a) A deficiency of vitamin D causing demineralization and softening of bone.

Osteomyelitis (os'-tē-ō-mī-i-LĪ-tis) Inflammation of bone marrow or of the bone and marrow.

Osteon The basic unit of structure in adult compact bone, consisting of a central (Haversian) canal with its concentrically arranged lamellae, lacunae, osteocytes, and canaliculi. Also called a **Haversian system.**

Osteoporosis (os'-tē-ō-pō-RŌ-sis) Increased porosity of bone.

Ostium (OS-tē-um) Any small opening, especially an entrance into a hollow organ or canal.

Otic (Ō-tik) Pertaining to the ear.

Otic ganglion (GANG-glē-on) A ganglion of the parasympathetic subdivision of the autonomic nervous system, made up of cell bodies of postganglionic parasympathetics to the parotid gland.

Otitis media (ō-TĪ-tus MĒ-dē-a) Acute infection of the middle ear cavity characterized by an inflamed tympanic membrane, subject to rupture.

Otolith (Ō-tō-lith) A particle of calcium carbonate embedded in the otolithic membrane that functions in maintaining static equilibrium.

Otolithic (ō-tō-LITH-ik) **membrane** Thick, gelatinous, glycoprotein layer located directly over hair cells of the macula in the saccule and utricle of the inner ear.

Otorhinolaryngology (ō'-tō-rī-nō-lar'-in-GOL-ō-jē) The branch of medicine that deals with the diagnosis and treatment of diseases of the ears, nose, and throat.

Oval window A small opening between the middle and inner ear into which the footplate of the stapes fits. Also called the **fenestra vestibuli.**

Ovarian (ō-VAR-ē-an) **cycle** A monthly series of events in the ovary designed to produce a mature ovum.

Ovarian follicle (FOL-i-kul) A general name for an ovum in any stage of development, along with its surrounding group of epithelial cells.

Ovarian ligament (LIG-a-ment) A rounded cord of connective tissue that attaches the ovary to the uterus.

Ovary (Ō-var-ē) Female gonad that produces ova and the hormones estrogens, progesterone, and relaxin.

Ovulation (ō-vyoo-LĀ-shun) The rupture of a vesicular (Graafian) ovarian follicle with discharge of an ovum into the pelvic cavity.

Ovum (Ō-vum) The female reproductive or germ cell; an egg cell.

Oxyhemoglobin (ok'-sē-HĒ-mō-glō-bin) **(HbO₂)** Hemoglobin combined with oxygen.

Oxyphil cell A cell found in the parathyroid gland that secretes parathyroid hormone (PTH).

Oxytocin (ok'-sē-TŌ-sin) **(OT)** A hormone secreted by neurosecretory cells in the paraventricular nucleus of the hypothalamus that stimulates contraction of the smooth muscle fibers (cells) in the pregnant uterus and contractile cells around the ducts of mammary glands.

Paget's (PAJ-ets) **disease** A disorder characterized by an irregular thickening and softening of the bones as a result of a greatly accelerated remodeling process. The cause is unknown.

Palate (PAL-at) The horizontal structure separating the oral and the nasal cavities; the roof of the mouth.

Palliative (PAL-ē-a-tiv) Serving to relieve or alleviate without curing.

Palpate (PAL-pāt) To examine by touch; to feel.

Palpebra (PAL-pe-bra) Eyelid.

Pancreas (PAN-krē-as) A soft, oblong organ lying along the greater curvature of the stomach and connected by a duct to the duodenum. It is both exocrine (secreting pancreatic juice) and endocrine (secreting insulin, glucagon, and somatostatin).

Pancreatic (pan'-krē-AT-ik) **duct** A single, large tube that drains pancreatic juice into the duodenum at the hepatopancreatic ampulla (ampulla of Vater).

Pancreatic islet A cluster of endocrine gland cells in the pancreas that secretes insulin, glucagon, and somatostatin. Also called an **islet of Langerhans.**

Papanicolaou (pap'-a-NIK-ō-la-oo) **test** A cytological staining test for the detection and diagnosis of premalignant and malignant conditions of the female genital tract. Cells scraped from the genital epithelium are smeared, fixed, stained, and examined microscopically. Also called a **Pap smear.**

Papilla (pa-PIL-a) Any small projection or elevation. *Plural,* **papillae** (pa-PIL-ē).

Paralysis (pa-RAL-a-sis) Loss or impairment of motor function due to a lesion of nervous or muscular origin.

Paranasal sinus (par'-a-NĀ-zal SĪ-nus) A mucus-lined air cavity in a skull bone that communicates with the nasal cavity. Paranasal sinuses are located in the frontal, maxillary, ethmoid, and sphenoid bones.

Paraplegia (par-a-PLĒ-jē-a) Paralysis of the lower limbs only.

Parasympathetic (par'-a-sim-pa-THET-ik) **division** One of the two subdivisions of the autonomic nervous system, having cell bodies of preganglionic neurons in nuclei in the brain stem and in the lateral gray matter of the sacral portion of the spinal cord; primarily concerned with activities that restore and conserve body energy. Also called the **craniosacral division.**

Parathyroid (par'-a-THĪ-royd) **gland** One of four small endocrine glands embedded on the posterior surfaces of the lateral lobes of the thyroid gland.

Parathyroid hormone (PTH) A hormone secreted by the parathyroid glands that decreases blood phosphate level and increased blood calcium level.

Paraurethral (par'-a-yoo-RĒ-thral) **glands** Glands whose ducts open on either side of the urethral orifice and secrete mucus. Also called **Skene's glands.**

Parenchyma (par-EN-ki-ma) The functional parts of any organ, as opposed to tissue that forms its ground substance or framework.

Parenteral (par-EN-ter-al) Situated or occurring outside the intestines; referring to introduction of substances into the body other than by way of the intestines such as intradermal, subcutaneous, intramuscular, intravenous, or intraspinal.

Parietal (pa-RĪ-e-tal) Pertaining to or forming the outer wall of a body cavity.

Parietal cell The secreting cell of the gastric glands that produces hydrochloric acid and intrinsic factor.

Parietal pleura (PLOO-ra) The outer layer of the serous pleural membrane that encloses and protects the lungs; the layer that is attached to the wall of the pleural cavity.

Parkinson's disease Progressive degeneration of basal ganglia of the cerebrum resulting in decreased production of dopamine (DA) that leads to tremor, slowing of voluntary movements, and muscle weakness.

Parotid (pa-ROT-id) **gland** One of the paired salivary glands located inferior and anterior to the ears connected to the oral cavity via a duct that opens into the inside of the cheek opposite the upper second molar tooth.

Paroxysm (PAR-ok-sizm) A sudden periodic attack or recurrence of symptoms of a disease.

Pars intermedia A small avascular zone between the adenohypophysis and neurohypophysis of the pituitary gland.

Parturition (par'-too-RISH-un) Act of giving birth to young; childbirth, delivery.

Patent (PĀ-tent) **ductus arteriosus** Congenital anatomical heart defect in which the fetal connection between the aorta and pulmonary trunk remains open instead of closing completely after birth.

Pathogen (PATH-ō-jen)   A disease-producing organism.

Pathogenesis (path'-ō-JEN-e-sis)   The development of disease or a morbid or pathological state.

Pathological (path'-ō-LOJ-i-kal)   Pertaining to or caused by disease.

Pathological anatomy   The study of structural changes caused by disease.

Pectoral (PEK-tō-ral)   Pertaining to the chest or breast.

Pediatrician (pē'-dē-a-TRISH-un)   A physician who specializes in the care and treatment of children and their illnesses.

Pedicel (PED-i-sel)   Footlike structure, as on podocytes of a glomerulus.

Pedicle (PED-i-kul)   A short, thick process found on vertebrae.

Pelvic (PEL-vik) **cavity**   Inferior portion of the abdominopelvic cavity that contains the urinary bladder, sigmoid colon, rectum, and internal female and male reproductive structures.

Pelvic inflammatory disease (PID)   Collective term for any extensive bacterial infection of the pelvic organs, especially the uterus, uterine (Fallopian) tubes, and ovaries.

Pelvic splanchnic (PEL-vik SPLANGK-nik) **nerves**   Preganglionic parasympathetic fibers from the levels of S2, S3, and S4 that supply the urinary bladder, reproductive organs, and the descending and sigmoid colon and rectum.

Pelvimetry (pel-VIM-e-trē)   Measurement of the size of the inlet and outlet of the birth canal.

Pelvis   The basinlike structure formed by the two pelvic bones, the sacrum, and the coccyx. The expanded, proximal portion of the ureter, lying within the kidney and into which the major calyces open.

Penis (PĒ-nis)   The male copulatory organ, used to introduce spermatozoa into the female vagina.

Pepsin   Protein-digesting enzyme secreted by zymogenic (chief) cells of the stomach as the inactive form pepsinogen that is converted to active pepsin by hydrochloric acid.

Peptic ulcer   An ulcer that develops in areas of the gastrointestinal tract exposed to hydrochloric acid; classified as a gastric ulcer if in the lesser curvature of the stomach and as a duodenal ulcer if in the first part of the duodenum.

Perforating canal   A minute passageway by means of which blood vessels and nerves from the periosteum penetrate into compact bone. Also called **Volkmann's canal.**

Pericardial (per'-i-KAR-dē-al) **cavity**   Small potential space between the visceral and parietal pericardium.

Pericardium (per'-i-KAR-dē-um)   A loose-fitting serous membrane that encloses the heart, consisting of an outer fibrous layer and an internal serous layer. Also called **parietal pericardium,** or **pericardial sac.**

Perichondrium (per'-i-KON-drē-um)   The membrane that covers cartilage.

Perikaryon (per'-i-KAR-ē-on)   The nerve cell body that contains the nucleus and other organelles.

Perilymph (PER-i-lymf')   The fluid contained between the bony and membranous labyrinths of the inner ear.

Perimetrium (per'-i-MĒ-trē-um)   The serosa of the uterus.

Perimysium (per'-i-MIZ-ē-um)   Invagination of the epimysium that divides muscles into bundles.

Perineum (per'-i-NĒ-um)   The pelvic floor; the space between the anus and the scrotum in the male and between the anus and the vulva in the female.

Perineurium (per'-i-NYOO-rē-um)   Connective tissue wrapping each muscle bundle.

Periodontal (per'-ē-ō-DON-tal) **disease**   A collective term for conditions characterized by degeneration of gingivae, alveolar bone, periodontal ligament, and cementum.

Periodontal membrane   The periosteum lining the alveoli (sockets) for the teeth in the alveolar processes of the mandible and maxillae.

Periosteum (per'-ē-OS-tē-um)   The membrane that covers bone and is essential for bone growth, repair, and nutrition.

Peripheral (pe-RIF-er-al)   Located on the outer part or a surface of the body.

Peripheral nervous system (PNS)   The part of the nervous system that lies outside the central nervous system—nerves and ganglia.

Periphery (pe-RIF-er-ē)   Outer part or a surface of the body; part away from the center.

Peristalsis (per'-i-STAL-sis)   Successive muscular contractions along the wall of a hollow muscular structure.

Peritoneum (per'-i-tō-NĒ-um)   The largest serous membrane of the body that lines the abdominal cavity and covers the viscera.

Peritonitis (per'-i-tō-NĪ-tis)   Inflammation of the peritoneum.

Pernicious (per-NISH-us)   Fatal.

Peroxisome (pe-ROKS-ī-sōm)   Organelle similar in structure to a lysosome that contains enzymes related to hydrogen peroxide metabolism; abundant in liver cells.

Perspiration   Substance produced by sudoriferous (sweat) glands containing water, salts, urea, uric acid, amino acids, ammonia, sugar, lactic acid, and ascorbic acid; helps maintain body temperature and eliminate wastes.

pH   A symbol of the measure of the concentration of hydrogen ions in a solution. The pH scale extends from 0 to 14 with a value of 7 expressing neutrality, values lower than 7 expressing increasing acidity, and values higher than 7 expressing increased alkalinity.

Phagocytosis (fag'-ō-si-TŌ-sis)   The process by which cells ingest particulate matter; especially the ingestion and destruction of microbes, cell debris, and other foreign matter.

Phalanx (FĀ-lanks)   The bone of a finger or toe. *Plural,* **phalanges** (fa-LAN-jēz).

Phantom pain   A sensation of pain as originating in a limb that has been amputated.

Pharmacology (far'-ma-KOL-ō-jē)   The science that deals with the effects and uses of drugs in the treatment of disease.

Pharynx (FAR-inks)   The throat; a tube that starts at the internal nares and runs partway down the neck where it opens into the esophagus posteriorly and into the larynx anteriorly.

Phenylketonuria (fen'-il-kē'-tō-NOO-rē-a) **(PKU)**   A disorder characterized by an elevation of the amino acid phenylalanine in the blood.

Pheochromocytoma (kē-ō-krō'-mō-sī-TŌ-ma)   Tumor of the chromaffin cells of the adrenal medulla.

Phlebotomy (fle-BOT-ō-me)   The cutting of a vein to allow the escape of blood.

Photoreceptor   Receptor that detects light on the retina of the eye. Also called **electromagnetic receptor.**

Physiology (fiz'-ē-OL-ō-jē)   Science that deals with the functions of an organism or its parts.

Pia mater (PĒ-a MĀ-ter)   The inner membrane (meninx) covering the brain and spinal cord.

Pilonidal (pī'-lō-NĪ-dal)   Containing hairs resembling a tuft inside a cyst or sinus.

Pineal (PIN-ē-al) **gland** The cone-shaped gland located in the roof of the third ventricle. Also called the **epiphysis cerebri.**

Pinealocyte (pin-ē-AL-o-sīt) Secretory cell of the pineal gland that produces hormones.

Pinna (PIN-na) The projecting part of the external ear. Also called the **auricle.**

Pinocytosis (pi'-nō-sī-TŌ-sis) The process by which cells ingest liquid.

Pituicyte (pi-TOO-i-sīt) Supporting cell of the posterior lobe of the pituitary gland.

Pituitary (pi-TOO-i-tar'-ē) **dwarfism** Condition caused by hyposecretion of growth hormone (GH) during the growth years and characterized by childlike physical traits in an adult.

Pituitary gland A small endocrine gland lying in the sella turcica of the sphenoid bone and attached to the hypothalamus by the infundibulum; nicknamed the "master gland." Also called the **hypophysis.**

Pivot joint A synovial joint in which a rounded, pointed, or conical surface of one bone articulates with a shallow depression of another bone, as in the joint between the atlas and axis and between the proximal ends of the radius and ulna. Also called a **trochoid joint.**

Placenta (pla-SEN-ta) The special structure through which the exchange of materials between fetal and maternal circulations occurs. Also called the **afterbirth.**

Plantar flexion (PLAN-tar FLEK-shun) Extension of the foot at the ankle joint.

Plaque (plak) A cholesterol-containing mass in the tunica media of arteries. A mass of bacterial cells, dextran (polysaccharide), and other debris that adheres to teeth.

Plasma (PLAZ-ma) The extracellular fluid found in blood vessels; blood minus the formed elements.

Plasma cell Cell that produces antibodies and develops from a B cell (lymphocyte).

Plasma (cell) membrane Outer, limiting membrane that separates the cell's internal parts from extracellular fluid and the external environment.

Pleura (PLOOR-a) The serous membrane that enfolds the lungs and lines the chest wall and diaphragm.

Pleural cavity Small potential space between the visceral and parietal pleurae.

Plexus (PLEK-sus) A network of nerves, veins, or lymphatic vessels.

Plicae circulares (PLĪ-kē ser-kyoo-LAR-es) Permanent, deep, transverse folds in the mucosa and submucosa of the small intestine that increase the surface area for absorption.

Pneumonia (noo-MŌ-nē-a) Acute infection or inflammation of the alveoli of the lungs.

Pneumotaxis (noo-mō-TAK-sik) **area** Portion of the respiratory center in the pons that helps coordinate the transition between inspiration and expiration.

Podiatry (pō-DĪ-a-trē) The diagnosis and treatment of foot disorders.

Polar body The smaller cell resulting from the unequal division of cytoplasm during the meiotic divisions of an oocyte. The polar body has no function and is resorbed.

Poliomyelitis (pō'-lē-ō-mī-e-LĪ-tis) Viral infection marked by fever, headache, stiff neck and back, deep muscle pain and weakness, and loss of certain somatic reflexes; a serious form of the disease, **bulbar polio,** results in destruction of motor neurons in anterior horns of spinal nerves that leads to paralysis.

Polycythemia (pol'-ē-sī-THĒ-mē-a) An abnormal increase in the number of red blood cells.

Polysaccharides (pol'-ē-SAK-a-rīds) Three or more monosaccharides joined chemically.

Polyuria (pol'-ē-YOO-rē-a) An excessive production of urine.

Pons (ponz) The portion of the brain stem that forms a "bridge" between the medulla and the midbrain, anterior to the cerebellum.

Positron emission tomography (PET) A type of radioactive scanning based on the release of gamma rays when positrons collide with negatively charged electrons in body tissues; it indicates where radioisotopes are used on the body.

Postcentral gyrus A gyrus immediately posterior to the central sulcus that contains the general sensory area of the cerebral cortex.

Posterior (pos-TĒR-ē-or) Nearer to or at the back of the body. Also called **dorsal.**

Posterior root The structure composed of afferent (sensory) fibers lying between a spinal nerve and the dorsolateral aspect of the spinal cord. Also called the **dorsal root,** or **sensory root.**

Posterior root ganglion A group of cell bodies of sensory (afferent) neurons and their supporting cells located along the posterior root of a spinal nerve. Also called a **dorsal,** or **sensory, root ganglion.**

Postganglionic neuron (pōst'-gang-lē-ON-ik NOO-ron) The second visceral efferent neuron in an autonomic pathway, having its cell body and dendrites located in an autonomic ganglion and its unmyelinated axon ending at cardiac muscle, smooth muscle, or a gland.

Postpartum (pōst-PAR-tum) After parturition; occurring after the delivery of a baby.

Postsynaptic (pōst-sin-AP-tik) **neuron** The nerve cell that is activated by the release of a neurotransmitter substance from another neuron and carries action potentials away from the synapse.

Precentral gyrus A gyrus immediately anterior to the central sulcus that contains the primary motor area of the cerebral cortex.

Preganglionic (prē'-gang-lē-ON-ik) **neuron** The first visceral efferent neuron in an autonomic pathway, with its cell body and dendrites in the brain or spinal cord and its myelinated axon ending at an autonomic ganglion where it synapses with a postganglionic neuron.

Pregnancy Sequence of events including fertilization, implantation, embryonic growth, and fetal growth that normally terminates in birth.

Premenstrual syndrome (PMS) Severe physical and emotional stress occurring late in the postovulatory phase of the menstrual cycle and sometimes overlapping with menstruation.

Premonitory (prē-MON-i-tō-rē) Giving previous warning; as premonitory symptoms.

Prepuce (PRĒ-pyoos) The loosely fitting skin covering the glans of the penis and clitoris. Also called the **foreskin.**

Presbyopia (prez-bē-Ō-pē-a) A loss of elasticity of the lens of the eye due to advancing age with resulting inability to focus clearly on near objects.

Pressor (PRES-or) Stimulating the activity of a function, especially vasomotor, usually accompanied by an increase in blood pressure.

Pressoreceptor (press-ō-rē-SEP-tor) Nerve cells capable of responding to changes in blood pressure. Also called a **baroreceptor.**

Presynaptic (prē-sin-AP-tik) **neuron** A nerve cell that carries action potentials toward a synapse.

Prevertebral ganglion (prē-VERT-e-bral GANG-lē-on) A cluster of cell bodies of postganglionic sympathetic neurons anterior to the spinal column and close to large abdominal arteries. Also called a **collateral ganglion.**

Primary germ layer One of three layers of embryonic tissue, called ectoderm, mesoderm, and endoderm, that gives rise to all tissues and organs of the organism.

Primigravida (prī-mi-GRAV-i-da) A woman pregnant for the first time.

Primordial (prī-MŌR-dē-al) Existing first; especially primordial egg cells in the ovary.

Principal cell Cell found in the parathyroid glands that secretes parathyroid hormone (PTH). Also called a **chief cell.**

Proctology (prok-TOL-ō-jē) The medical specialty that deals with the diagnosis and treatment of disorders of the rectum and anus.

Progeny (PROJ-e-nē) Refers to offspring or descendants.

Progesterone (prō-JES-te-rōn) A female sex hormone produced by the ovaries that helps prepare the endometrium for implantation of a fertilized ovum and the mammary glands for milk secretion.

Prognosis (prog-NŌ-sis) A forecast of the probable results of a disorder; the outlook for recovery.

Projection (prō-JEK-shun) The process by which the brain refers sensations to their point of stimulation.

Prolactin (prō-LAK-tin) **(PRL)** A hormone secreted by the adenohypophysis (anterior lobe) of the pituitary gland that initiates and maintains milk secretion by the mammary glands.

Prolapse (PRŌ-laps) A dropping or falling down of an organ, especially the uterus or rectum.

Proliferation (pro-lif'-er-Ā-shun) Rapid and repeated reproduction of new parts, especially cells.

Promontory (PROM-on-tor'-ē) A projecting process or part.

Pronation (prō-NĀ-shun) A movement of the flexed forearm in which the palm of the hand is turned posteriorly.

Prophase (PRŌ-fāz) The first stage in mitosis during which chromatid pairs are formed and aggregate around the equatorial plane region of the cell.

Proprioception (prō'-prē-ō-SEP-shun) The receipt of information from muscles, tendons, and the labyrinth that enables the brain to determine movements and position of the body and its parts. Also called **kinesthesia.**

Proprioceptor (prō'-prē-ō-SEP-tor) A receptor located in muscles, tendons, or joints that provides information about body position and movements.

Prostaglandin (pros-ta-GLAN-din) **(PG)** A membrane-associated lipid composed of 20-carbon fatty acids with 5 carbon atoms joined to form a cyclopentane ring; synthesized in small quantities and basically mimics hormones in activities.

Prostatectomy (pros'-ta-TEK-tō-mē) The surgical removal of part of or the entire prostate gland.

Prostate (PROS-tāt) **gland** A muscular, doughnut-shaped gland inferior to the urinary bladder that surrounds the su-

perior portion of the male urethra.

Prosthesis (pros-THĒ-sis) An artificial device to replace a missing body part.

Protein An organic compound consisting of carbon, hydrogen, oxygen, nitrogen, and sometimes sulfur and phosphorus and made up of amino acids linked by peptide bonds.

Protraction (prō-TRAK-shun) The movement of the mandible or clavicle forward on a plane parallel with the ground.

Proximal (PROK-si-mal) Nearer the attachment of an extremity to the trunk or a structure; nearer to the point of origin.

Pruritus (proo'-RĪ-tus) Itching.

Pseudopodia (soo'-dō-PŌ-dē-a) Temporary, protruding projections of cytoplasm.

Psoriasis (sō-RĪ-a-sis) Chronic skin disease characterized by reddish plaques or papules covered with scales.

Psychosomatic (sī'-kō-sō-MAT-ik) Pertaining to the relationship between mind and body. Commonly used to refer to those physiological disorders thought to be caused entirely or partly by emotional disturbances.

Pterygopalatine ganglion (ter'-i-gō-PAL-a-tin GANG-glē-on) A cluster of cell bodies of parasympathetic postganglionic neurons ending at the lacrimal and nasal glands.

Ptosis (TŌ-sis) Drooping, as of the eyelid or the kidney.

Puberty (PYOO-ber-tē) The time of life during which the secondary sex characteristics begin to appear and the capability for sexual reproduction is possible; usually between the ages of 10 and 15.

Pudendum (pyoo-DEN-dum) A collective designation for the external genitalia of the female.

Puerperium (pyoo'-er-PER-ē-um) The state immediately after childbirth, usually 4 to 6 weeks.

Pulmonary (PUL-mo-ner'-ē) Concerning or affected by the lungs.

Pulmonary circulation The flow of deoxygenated blood from the right ventricle to the lungs and the return of oxygenated blood from the lungs to the left atrium.

Pulmonary edema (e-DĒ-ma) An abnormal accumulation of interstitial fluid in the tissue spaces and alveoli of the lungs due to increased pulmonary capillary permeability or increased pulmonary capillary pressure.

Pulmonary emoblism (EM-bō-lizm) **(PE)** The presence of a blood clot or other foreign substance in a pulmonary arterial blood vessel that obstructs circulation to lung tissue.

Pulmonary ventilation The inflow (inspiration) and outflow (expiration) of air between the atmosphere and the lungs. Also called **breathing.**

Pulp cavity A cavity within the crown and neck of a tooth, filled with pulp, a connective tissue containing blood vessels, nerves, and lymphatics.

Pulsating electromagnetic fields (PEMFs) A procedure that uses electrotherapy to treat improperly healing fractures.

Pupil The hole in the center of the iris; the area through which light enters the posterior cavity of the eyeball.

Pus The liquid product of inflammation that contains leucocytes or their remains and debris of dead cells.

Pyemia (pī-Ē-mē-a) Infection of the blood, with multiple abscesses, caused by pus-forming microorganisms.

Pyloric (pī-LOR-ik) **sphincter** A thickened ring of smooth muscle through which the pylorus of the stomach communicates with the duodenum. Also called the **pyloric valve.**

Pyogenesis (pi'-ō-JEN-e-sis) Formation of pus.

Pyorrhea (pī-ō-RĒ-a) A discharge or flow of pus, especially in the alveoli (sockets) and the tissues of the gums.

Pyramid (PIR-a-mid) A pointed or cone-shaped structure; one of two roughly triangular structures on the ventral side of the medulla composed of the largest motor tracts that run from the cerebral cortex to the spinal cord; a triangular-shaped structure in the renal medulla composed of the straight segments of renal tubules.

Pyramidal (pi-RAM-i-dal) **pathways** Collections of motor nerve fibers arising in the brain and passing down through the spinal cord to motor cells in the anterior horns.

Pyrexia (pī-REK-sē-a) A condition in which the temperature is above normal.

Pyuria (pī-YOO-rē-a) The presence of leucocytes and other components of pus in urine.

Quadrant (KWOD-rant) One of four parts.

Quadriplegia (kwod'-ri-PLĒ-jē-a) Paralysis of all four limbs.

Radiographic (rā'-dē-ō-GRAF-ic) **anatomy** Diagnostic branch of anatomy that includes the use of x rays.

Radiography (rā'-dē-OG-ra-fē) The making of x-ray pictures.

Rami communicantes (RĀ-mē kō-myoo-ni-KAN-tēz) Branches of a spinal nerve. *Singular,* **ramus communicans** (RĀ-mus ko-MYOO-ni-kans).

Ramus (RĀ-mus) A branch, especially a nerve or blood vessel.

Raphe (RĀ-fē) A crease, ridge, or seam noting union of the halves of a part, as the division of the halves of the scrotum.

Receptor A specialized cell or a nerve cell terminal modified to respond to some specific sensory modality, such as touch, pressure, cold, light, or sound. A specific molecule or arrangement of molecules organized to accept only molecules with a complementary shape.

Reciprocal innervation (re-SIP-rō-kal in-ner-VĀ-shun) The activation of those nerve fibers that specifically inhibit contractions of the antagonistic muscles in any muscle action.

Recombinant DNA Synthetic DNA, formed by joining a fragment of DNA from one source to a portion of DNA from another.

Rectouterine pouch A pocket formed by the parietal peritoneum as it moves posteriorly from the surface of the uterus and is reflected onto the rectum; the lowest point in the pelvic cavity. Also called the **pouch** or **cul de sac of Douglas.**

Rectum (REK-tum) The last 20 cm (8 in) of the gastrointestinal tract, from the sigmoid colon to the anus.

Recumbent (re-KUM-bent) Lying down.

Red nucleus A cluster of cell bodies in the midbrain, occupying a large portion of the tegmentum and sending fibers into the rubroreticular and rubrospinal tracts.

Referred pain Pain that is felt at a site remote from the place of origin.

Reflex arc The most basic conduction pathway through the nervous system, connecting a receptor and an effector and consisting of a receptor, a sensory neuron, a center in the central nervous system for a synapse, a motor neuron, and an effector.

Regimen (REJ-i-men) A strictly regulated scheme of diet, exercise, or activity designed to achieve certain ends.

Regional anatomy The division of anatomy dealing with a specific region of the body, such as the head, neck, chest, or abdomen.

Regulating factor Chemical secretion of the hypothalamus that can either stimulate or inhibit secretion of hormones of the adenohypophysis (anterior pituitary).

Regurgitation (rē-gur'-ji-TĀ-shun) Return of solids or fluids to the mouth from the stomach; flowing backward of blood through incompletely closed heart valves.

Relapse (rē-LAPS) The return of a disease weeks or months after its apparent cessation.

Relaxin A female hormone produced by the ovaries that relaxes the symphysis pubis and helps dilate the uterine cervix to facilitate delivery.

Remodeling Replacement of old bone by new bone tissue.

Renal (RĒ-nal) Pertaining to the kidney.

Renal corpuscle (KOR-pus'-l) A glomerular capsule and its enclosed glomerulus.

Renal erythropoietic (ē-rith'-rō-poy-Ē-tik) **factor** An enzyme released by the kidneys and liver in response to hypoxia that acts on a plasma protein to bring about the production of erythropoietin, which stimulates red blood cell production.

Renal pelvis A cavity in the center of the kidney formed by the expanded, proximal portion of the ureter, lying within the kidney, and into which the major calyces open.

Renal pyramid A triangular structure in the renal medulla composed of the straight segments of renal tubules.

Renal suppression The sudden stoppage of urine production and excretion. Also called **anuria.**

Residual (re-ZID-yoo-al) **volume** The volume of air still contained in the lungs after a maximal expiration; about 1,200 ml.

Resistance Ability to ward off disease. The hindrance encountered by an electrical charge as it moves through a substance from one point to another. The hindrance encountered by blood as it flows through the vascular system or by air through respiratory passageways.

Respiration (res-pi-RĀ-shun) Overall exchange of gases between the atmosphere, blood, and body cells consisting of pulmonary ventilation, external respiration, and internal respiration.

Respiratory center Neurons in the reticular formation of the brain stem that regulate the rate of respiration.

Respiratory distress syndrome (RDS) of the newborn A disease of newborn infants, especially premature ones, in which insufficient amounts of surfactant are produced and breathing is labored. Also called **hyaline membrane disease (HMD).**

Resuscitation (rē-sus'-i-TĀ-shun) Act of bringing a person back to full consciousness.

Retention (rē-TEN-shun) A failure to void urine due to obstruction, nervous contraction of urethra, or absence of sensation of desire to urinate.

Rete (RĒ-tē) **testis** The network of ducts in the testes.

Reticular (re-TIK-yoo-lar) **activating system (RAS)** An extensive network of branched nerve cells running through the core of the brain stem. When these cells are activated, a generalized alert or arousal behavior results.

Reticular formation A network of small groups of nerve cells scattered among bundles of fibers beginning in the medulla as a continuation of the spinal cord and extending upward through the central part of the brain stem.

Reticulocyte (rē-TIK-yoo-lō-sīt) An immature red blood cell.

Reticulum (re-TIK-yoo-lum) A network.

Retina (RET-i-na) The inner coat of the eyeball, lying only in the posterior portion of the eye and consisting of nervous tissue and a pigmented layer comprised of epithelial cells lying in contact with the choroid. Also called the **nervous tunic.**

Retraction (rē-TRAK-shun) The movement of a protracted part of the body backward on a plane parallel to the ground, as in pulling the lower jaw back in line with the upper jaw.

Retroflexion (re-trō-FLEK-shun) A malposition of the uterus in which it is tilted posteriorly.

Retroperitoneal (re'-trō-per-i-tō-NĒ-al) External to the peritoneal lining of the abdominal cavity.

Rheumatism (ROO-ma-tizm') Any painful state of the supporting structures of the body—bones, ligaments, joints, tendons, or muscles.

Rhinology (rī-NOL-ō-jē) The study of the nose and its disorders.

Ribonucleic (rī'-bō-nyoo-KLĒ-ik) **acid (RNA)** A single-stranded nucleic acid constructed of nucleotides consisting of one of four possible nitrogen bases (adenine, cytosine, guanine, or uracil), ribose, and a phosphate group; three types are messenger RNA (mRNA), transfer RNA (tRNA), and ribosomal RNA (rRNA), each of which cooperates with DNA for protein synthesis.

Ribosome (RĪ-bō-sōm) An organelle in the cytoplasm of cells, composed of ribosomal RNA and ribosomal proteins, that synthesizes proteins; nicknamed the "protein factory."

Rickets (RIK-ets) Condition affecting children characterized by soft and deformed bones resulting from inadequate calcium metabolism due to a vitamin D deficiency.

Right lymphatic (lim-FAT-ik) **duct** A vessel of the lymphatic system that drains lymph from the upper right side of the body and empties it into the right subclavian vein.

Rigor mortis State of partial contraction of muscles following death due to lack of ATP that causes cross bridges of thick myofilaments to remain attached to thin myofilaments, thus preventing relaxation.

Rod A visual receptor in the retina of the eye that is specialized for vision in dim light.

Roentgen (RENT-gen) The international unit of radiation; a standard quantity of x or gamma radiation.

Roentgenogram (RENT-gen-ō-gram) A photographic image made with x rays.

Root canal A narrow extension of the pulp cavity lying within the root of a tooth.

Root hair plexus (PLEK-sus) A network of dendrites arranged around the root of a hair as free or naked nerve endings that are stimulated when a hair shaft is moved.

Root of penis Attached portion of penis that consists of the bulb and crura.

Rotation (rō-TĀ-shun) Moving a bone around its own axis, with no other movement.

Round ligament (LIG-a-ment) A band of fibrous connective tissue enclosed between the folds of the broad ligament of the uterus, emerging from a point on the uterus just below the uterine (Fallopian) tube, extending laterally along the pelvic wall, and penetrating the abdominal wall through the deep inguinal ring to end in the labia majora.

Round window A small opening between the middle and inner ear, directly below the oval window, covered by the second tympanic membrane. Also called the **fenestra cochleae.**

Rugae (ROO-gē) Large folds in the mucosa of an empty hollow organ, such as the stomach and vagina.

Saccule (SAK-yool) The lower and smaller of the two chambers in the membranous labyrinth inside the vestibule of the inner ear containing a receptor organ for static equilibrium.

Sacral plexus (PLEK-sus) A network formed by the anterior branches of spinal nerves L4 through S3.

Sacral promontory (PROM-on-tor'-ē) The superior surface of the body of the first sacral vertebra that projects anteriorly into the pelvic cavity; a line from the sacral promontory to the superior border of the symphysis pubis divides the abdominal and pelvic cavities.

Saddle joint A synovial joint in which the articular surfaces of both of the bones are saddle-shaped or concave in one direction and convex in the other direction, as in the joint between the trapezium and the metacarpal of the thumb. Also called a **sellaris joint.**

Sagittal plane A plane parallel to a midsagittal plane that divides the body or organs into unequal left and right portions. Also called a **parasagittal plane.**

Saliva (sa-LĪ-va) A clear, alkaline, somewhat viscous secretion produced by the three pairs of salivary glands; contains various salts, mucin, lysozyme, and salivary amylase.

Salivary amylase (SAL-i-ver-ē AM-ilās) An enzyme in saliva that initiates the chemical breakdown of starch, mostly in the mouth.

Salivary gland One of three pairs of glands that lie outside the mouth and pour their secretory product (called saliva) into ducts that empty into the oral cavity; the parotid, submandibular, and sublingual glands.

Salpingitis (sal'-pin-JĪ-tis) Inflammation of the uterine (Fallopian) or auditory (Eustachian) tube.

Sarcolemma (sar'-kō-LEM-ma) The plasma membrane of a muscle fiber (cell), especially of a skeletal muscle fiber (cell).

Sarcoma (sar-KŌ-ma) A connective tissue tumor, often highly malignant.

Sarcomere (SAR-kō-mēr) A contractile unit in a striated muscle fiber (cell) extending from one Z line to the next Z line.

Sarcoplasm (SAR-kō-plazm) The cytoplasm of a muscle fiber (cell).

Sarcoplasmic reticulum (sar'-kō-PLAZ-mik re-TIK-yoo-lum) A network of saccules and tubes surrounding myofibrils of a muscle fiber (cell), comparable to endoplasmic reticulum; functions to reabsorb calcium ions during relaxation and to release them to cause contraction.

Scala tympani (SKA-la TIM-pan-ē) The lower spiral-shaped channel of the bony cochlea, filled with perilymph.

Scala vestibuli (ves-TIB-u-lē) The upper spiral-shaped channel of the bony cochlea, filled with perilymph.

Sciatic (sī-AT-ik) Pertaining to the ischium.

Sciatica (sī-AT-i-ka) Inflammation and pain along the sciatic nerve, felt at the back of the thigh running down the inside of the leg.

Sclera (SKLĒ-ra) The white coat of fibrous tissue that forms the outer protective covering over the eyeball except in the area of the anterior cornea; the posterior portion of the fibrous tunic.

Scleral venous sinus A circular venous sinus located at the junction of the sclera and the cornea through which aqueous humor drains from the anterior chamber of the eyeball into the blood. Also called the **canal of Schlemm.**

Sclerosis (skle-RŌ-sis)   A hardening with loss of elasticity of the tissues.

Scoliosis (skō'-lē-Ō-sis)   An abnormal lateral curvature from the normal vertical line of the spine.

Scotoma (skō-TŌ-ma)   A blind spot or area of depressed vision within the visual field.

Scrotum (SKRŌ-tum)   A skin-covered pouch that contains the testes and their accessory organs.

Sebaceous (se-BĀ-shus)   Secreting oil.

Sebaceous gland   An exocrine gland in the dermis of the skin, almost always associated with a hair follicle, that secretes sebum. Also called an **oil gland.**

Sebum (SĒ-bum)   Secretion of sebaceous (oil) glands.

Secondary sex characteristic   A feature characteristic of the male or female body that develops at puberty under the stimulation of sex hormones, but is not directly involved in sexual reproduction, such as distribution of body hair, voice pitch, body shape, and muscle development.

Secretion (se-KRĒ-shun)   Production and release from a gland cell of a fluid, especially a functionally useful product as opposed to a waste product.

Selectively permeable membrane   A membrane that permits the passage of certain substances, but restricts the passage of others. Also called a **semipermeable membrane.**

Sella turcica (SEL-a TUR-si-ka)   A saddlelike depression on the middle upper surface of the sphenoid bone enclosing the pituitary gland.

Semen (SĒ-men)   A fluid discharge at ejaculation by a male that consists of a mixture of spermatozoa and the secretions of the seminal vesicles, the prostate gland, and the bulbourethral (Cowper's) glands. Also called **seminal fluid.**

Semicircular canals   Three bony channels projecting superiorly and posteriorly from the vestibule of the inner ear, filled with perilymph, in which lie the membranous semicircular canals filled with endolymph. They contain receptors for equilibrium.

Semicircular ducts   The membranous semicircular canals filled with endolymph and floating in the perilymph of the bony semicircular canals. They contain cristae that are concerned with dynamic equilibrium.

Semilunar (sem'-ē-LOO-nar) **valve**   A valve guarding the entrance into the aorta or the pulmonary trunk from a ventricle of the heart.

Seminal vesicle (SEM-i-nal VES-i-kul)   One of a pair of convoluted, pouchlike structures, lying posterior and inferior to the urinary bladder and anterior to the rectum, that secrete a component of semen into the ejaculatory ducts.

Seminiferous tubule (sem'-i-NI-fer-us TOO-byool)   A tightly coiled duct, located in a lobule of the testis, where spermatozoa are produced.

Senescence (se-NES-ens)   The process of growing old; the period of old age.

Senile macular (MAK-yoo-lar) **degeneration (SMD)**   A disease in which blood vessels grow over the macula lutea.

Senility (se-NIL-i-tē)   A loss of mental or physical ability due to old age.

Sensation   A state of awareness of external or internal conditions of the body.

Sensory area   A region of the cerebral cortex concerned with the interpretation of sensory impulses.

Sepsis (SEP-sis)   A morbid condition resulting from the presence of pathogenic bacteria and their products.

Septal defect   An opening in the septum between the left and right sides of the heart.

Septum (SEP-tum)   A wall dividing two cavities.

Serosa (ser-Ō-sa)   Any serous membrane. The outermost layer or tunic of an organ formed by a serous membrane. The membrane that lines the pleural, pericardial, and peritoneal cavities.

Serous (SIR-us) **membrane**   A membrane that lines body cavities that does not open to the exterior. Also called the **serosa.**

Sesamoid (SES-a-moyd) **bones**   Small bones usually found in tendons.

Sex chromosomes   The twenty-third pair of chromosomes, designated X and Y, which determine the genetic sex of an individual; in males, the pair is XY; in females, XX.

Sexual intercourse   The insertion of the erect penis of a male into the vagina of a female. Also called **coitus.**

Sexually transmitted disease (STD)   General term for any of a large number of diseases spread by sexual contact.

Shingles   Acute infection of the peripheral nervous system caused by a virus.

Shinsplints   Soreness or pain along the tibia probably caused by inflammation of the periosteum brought on by repeated tugging of the muscles and tendons attached to the periosteum. Also called **tibia stress syndrome.**

Shoulder   A synovial or diarthrotic joint where the humerus joins the scapula.

Sigmoid (SIG-moyd)   S-shaped, from the Greek letter sigma.

Sigmoid colon (KŌ-lon)   The S-shaped portion of the large intestine that begins at the level of the left iliac crest, projects inward to the midline, and terminates at the rectum at about the level of the third sacral vertebra.

Sign   Any objective evidence of disease such as a lesion, swelling, or fever.

Sinoatrial (si-nō-Ā-trē-al) **(SA) node**   A compact mass of cardiac muscle fibers (cells) specialized for conduction, located in the right atrium beneath the opening of the superior vena cava. Also called the **sinuatrial node,** or **pacemaker.**

Sinus (SĪ-nus)   A hollow in a bone (paranasal sinus) or other tissue; a channel for blood; any cavity having a narrow opening.

Sinusitis (sīn-yoo-SĪT-is)   Inflammation of the mucous membrane of a paranasal sinus.

Sinusoid (SĪN-yoo-soyd)   A microscopic space or passage for blood in certain organs such as the liver or slpeen.

Skeletal muscle   An organ specialized for contraction, composed of striated muscle fibers (cells), supported by connective tissue, attached to a bone by a tendon or an aponeurosis, and stimulated by somatic efferent neurons.

Skull   The skeleton of the head consisting of the cranial and facial bones.

Sliding-filament theory   The most commonly accepted explanation for muscle contraction in which actin and myosin myofilaments move into interdigitation with each other, decreasing the length of the sarcomeres.

Small intestine   A long tube of the gastrointestinal tract that begins at the pyloric sphincter of the stomach, coils through the central and lower part of the abdominal cavity, and ends at the large intestine; divided into three segments: duodenum, jejunum, and ileum.

Smooth muscle   An organ specialized for contraction, composed of smooth muscle fibers (cells), located in the walls of hollow internal structures and stimulated by visceral efferent neurons.

Soft palate (PAL-at)   The posterior portion of the roof of the mouth, extending posteriorly from the palatine bones and ending at the uvula. It is a muscular partition lined with mucous membrane.

Somatic (sō-MAT-ik) **nervous system (SNS)**   The portion of the peripheral nervous system made up of the somatic efferent fibers that run between the central nervous system and the skeletal muscles and skin.

Somesthetic (sō'-mes-THET-ik)   Pertaining to sensations and sensory structures of the body.

Spasm (spazm)   An involuntary, convulsive, muscular contraction.

S period   Period of interphase during which chromosomes are replicated preceded by a G_1 period and followed by a G_2 period, when cells grow, metabolize, and produce substances required for division.

Spermatic (sper-MAT-ik) **cord**   A supporting structure of the male reproductive system, extending from a testis to the deep inguinal ring, that includes the ductus (vas) deferens, arteries, veins, lymphatics, nerves, cremaster muscle, and connective tissue.

Spermatogenesis (sper'-ma-tō-JEN-e-sis)   The formation and development of the spermatozoa.

Spermatozoon (sper'-ma-tō-ZŌ-on)   A mature sperm cell.

Spermicide (SPER-mi-sīd')   An agent that kills spermatozoa.

Spermiogenesis (sper'-mē-ō-JEN-e-sis)   The maturation of spermatids into spermatozoa.

Sphincter (SFINGK-ter)   A circular muscle constricting an orifice.

Sphincter of the hepatopancreatic ampulla   A circular muscle at the opening of the common bile and main pancreatic ducts in the duodenum. Also called the **sphincter of Oddi.**

Sphygmomanometer (sfig'-mō-ma-NOM-e-ter)   An instrument for measuring arterial blood pressure.

Spina bifida (SPĪ-na BIF-i-da)   A congenital defect of the vertebral column in which the two sides of the neural arch of a vertebra fail to fuse in midline.

Spinal (SPĪ-nal) **cord**   A mass of nerve tissue located in the vertebral canal from which 31 pairs of spinal nerves originate.

Spinal (lumbar) puncture   Withdrawal of some of the cerebrospinal fluid from the subarachnoid space in the lumbar region.

Spinal nerve   One of the 31 pairs of nerves that originate on the spinal cord from posterior and anterior roots.

Spinous (SPĪ-nus) **process**   A sharp or thornlike process or projection. Also called a **spine.** A sharp ridge running diagonally across the posterior surface of the scapula.

Spiral organ   The organ of hearing, consisting of supporting cells and hair cells that rest on the basilar membrane and extend into the endolymph of the cochlear duct. Also called the **organ of Corti.**

Spirometer (spī-ROM-e-ter)   An apparatus used to measure air capacity of the lungs. Also called a **respirometer.**

Splanchnic (SPLANK-nik)   Pertaining to the viscera.

Spleen (splēn)   Large mass of lymphatic tissue between the fundus of the stomach and the diaphragm that functions in phagocytosis, production of lymphocytes, and blood storage.

Sprain   Forcible wrenching or twisting of a joint with partial rupture or other injury to its attachments without disloca-

tion.

Sputum (SPYOO-tum)   Substance ejected from the mouth containing saliva and mucus.

Squamous (SKWĀ-mus)   Scalelike.

Stasis (STĀ-sis)   Stagnation or halt of normal flow of fluids, as blood, urine, or of the intestinal mechanism.

Static equilibrium (ē-kwi-LIB-rē-um)   The maintenance of posture in response to changes in the orientation of the body, mainly the head, relative to the ground.

Stellate reticuloendothelial (STEL-āte re-tik'-yoo-lō-en'-dō-THE-lē-al) **cell**   Phagocytic cell that lines a sinusoid of the liver. Also called a **Kupffer cell.**

Stenosis (sten-Ō-sis)   An abnormal narrowing or constriction of a duct or opening.

Stereocilia (ste'-rē-ō-SIL-ē-a)   Groups of extremely long, slender, nonmotile microvilli projecting from epithelial cells lining the epididymis.

Stereognosis (ste'-rē-og-NŌ-sis)   The ability to recognize the size, shape, and texture of an object by touch.

Sterile (STE-rīl)   Free from any living microorganisms. Unable to conceive or produce offspring.

Sterilization (ster'-i-li-ZĀ-shun)   Elimination of all living microorganisms. The rendering of an individual incapable of reproduction (e.g., castration, vasectomy, hysterectomy).

Sternal puncture   Introduction of a wide-bore needle into the marrow cavity of the sternum for aspiration of a sample of red bone marrow.

Stimulus   Any change in the environment, capable of initiating a response by the nervous system.

Stomach   The J-shaped enlargement of the gastrointestinal tract directly under the diaphragm in the epigastric, umbilical, and left hypochondriac regions of the abdomen, between the esophagus and small intestine.

Strabismus (stra-BIZ-mus)   A condition in which the visual axes of the two eyes differ, so that they do not both fix on the same object.

Straight tubule (TOO-byool)   A duct in a testis leading from a convoluted seminiferous tubule to the rete testis.

Stratum (STRĀ-tum)   A layer.

Stratum basalis (ba-SAL-is)   The outer layer of the endometrium, next to the myometrium, that is maintained during menstruation and gestation and produces a new functionalis following menstruation or parturition.

Stratum functionalis (funk'-shun-AL-is)   The inner layer of the endometrium, the layer next to the uterine cavity, that is shed during menstruation and that forms the maternal portion of the placenta during gestation.

Stretch reflex   A monosynaptic reflex triggered by a sudden stretch of a muscle and ending with a contraction of that same muscle.

Stricture (STRIK-cher)   A local contraction of a tubular structure.

Stroma (STRŌ-ma)   The tissue that forms the ground substance, foundation, or framework of an organ, as opposed to its functional parts.

Subarachnoid (sub'-a-RAK-noyd) **space**   A space between the arachnoid and the pia mater that surrounds the brain and spinal cord and through which cerebrospinal fluid circulates.

Subcutaneous (sub'-kyoo-TĀ-nē-us)   Beneath the skin. Also called **hypodermic.**

Subcutaneous layer   A continuous sheet of loose connective tissue and adipose tissue between the dermis of the skin and the deep fascia of the muscles. Also called the **superficial fascia.**

Subdural (sub-DOO-ral) **space** A space between the dura mater and the arachnoid of the brain and spinal cord that contains a small amount of fluid.

Sublingual (sub-LING-gwal) **gland** One of a pair of salivary glands situated in the floor of the mouth under the mucous membrane and to the side of the lingual frenulum, with a duct that opens into the floor of the mouth.

Submandibular ganglion (sub-man-DIB-yoo-lar GANG-lē-on) A cluster of cell bodies of postganglionic parasympathetic neurons, located above the submandibular gland with fibers ending at the submandibular and sublingual salivary glands and other small salivary glands in the floor of the mouth.

Submandibular gland One of a pair of salivary glands found beneath the base of the tongue under the mucous membrane in the posterior part of the floor of the mouth, posterior to the sublingual glands, with a duct situated to the side of the lingual frenulum. Also called the **submaxillary gland.**

Submucosa (sub-myoo-KŌ-sa) A layer of connective tissue located beneath a mucous membrane, as in the gastrointestinal tract or the urinary bladder, where a submucosa connects the mucosa to the muscularis tunic.

Submucous plexus A network of autonomic nerve fibers located in the outer portion of the submucous layer or tunic of the small intestine. Also called the **plexus of Meissner.**

Subserous fascia (sub-SE-rus FASH-ē-a) A layer of connective tissue internal to the deep fascia, lying between the deep fascia and the serous membrane that lines the body cavities.

Subthreshold stimulus A stimulus of such weak intensity that it cannot initiate muscle contraction. Also called a **subliminal stimulus.**

Sudoriferous (soo'-dor-IF-er-us) **gland** An apocrine or eccrine exocrine gland in the dermis or subcutaneous layer that produces perspiration. Also called a **sweat gland.**

Sulcus (SUL-kus) A groove or depression between parts, especially between the convolutions of the brain. *Plural,* **sulci.**

Superficial (soo'-per-FISH-al) Located on or near the surface.

Superficial fascia (FASH-ē-a) A continuous sheet of fibrous connective tissue between the dermis of the skin and the deep fascia of the muscles. Also called **subcutaneous layer.**

Superficial inguinal (IN-gwi-nal) **ring** A triangular opening in the aponeurosis of the external oblique muscle that represents the termination of the inguinal canal.

Superior (soo-PĒR-ē-or) Toward the head or upper part of a structure. Also called **cephalad,** or **craniad.**

Superior vena cava (VĒ-na CĀ-va) Large vein that collects blood from parts of the body superior to the heart and returns it to the right atrium.

Supination (soo-pī-NĀ-shun) A movement of the forearm in which the palm of the hand is turned anteriorly.

Suppuration (sup'-yoo-RĀ-shun) Pus formation and discharge.

Surface anatomy The study of the structures that can be identified from the outside of the body.

Surfactant (sur-FAK-tant) A phospholipid substance produced by the lungs that decreases surface tension.

Susceptibility (sus-sep'-ti-BIL-i-tē) Lack of resistance of the body to the deleterious or other effects of an agent such as pathogenic microorganisms.

Suspensory ligament (sus-PEN-so-rē LIG-a-ment) A fold of

peritoneum extending laterally from the surface of the ovary to the pelvic wall.

Sustentacular (sus'-ten-TAK-yoo-lar) **cell** A supporting cell of seminiferous tubules that produces secretions for supplying nutrients to spermatozoa and the hormone inhibin. Also called a **Sertoli cell.**

Sutural (SOO-cher-al) **bone** A small bone located within a suture of certain cranial bones. Also called **Wormian bone.**

Suture (SOO-cher) A fibrous joint, especially in the skull, where bone surfaces are closely united.

Sympathetic (sim'-pa-THET-ik) **division** One of the two subdivisions of the autonomic nervous system, having cell bodies of preganglionic neurons in the lateral gray columns of the thoracic segment and first two or three lumbar segments of the spinal cord; primarily concerned with processes involving the expenditure of energy. Also called the **thoracolumbar division.**

Sympathetic trunk ganglion (GANG-glē-on) A cluster of cell bodies of postganglionic sympathetic neurons lateral to the vertebral column, close to the body of a vertebra. These ganglia extend downward through the neck, thorax, and abdomen to the coccyx on both sides of the vertebral column and are connected to one another to form a chain on each side of the vertebral column. Also called **lateral,** or **sympathetic, chain** or **vertebral chain ganglia.**

Sympathomimetic (sim'-pa-thō-mi-MET-ik) Producing effects that mimic those brought about by the sympathetic division of the autonomic nervous system.

Symphysis (SIM-fi-sis) A line of union. A slightly movable cartilaginous joint such as the symphysis pubis between the anterior surfaces of the pubic bones.

Symphysis pubis (PYOO-bis) A slightly movable cartilaginous joint between the anterior surfaces of the pelvic bones.

Symptom (SIMP-tum) An observable abnormality that indicates the presence of a disease or disorder of the body.

Synapse (SIN-aps) The junction between the processes of two adjacent neurons; the place where the activity of one neuron affects the activity of another.

Synaptic (sin-AP-tik) **cleft** The narrow gap that separates the axon terminal of one nerve cell from another nerve cell or muscle fiber (cell) and across which a neurotransmitter diffuses to affect the postsynaptic cell.

Synaptic end bulb Expanded distal end of an axon terminal that contains synaptic vesicles.

Synaptic gutter Invaginated portion of a sarcolemma under an axon terminal. Also called a **synaptic trough.**

Synaptic vesicle Membrane-enclosed sac in a synaptic end bulb that stores neurotransmitter substances.

Synarthrosis (sin'-ar-THRŌ-sis) An immovable joint.

Synchondrosis (sin'-kon-DRŌ-sis) A cartilaginous joint in which the connecting material is hyaline cartilage.

Syncope (SIN-kō-pē) A temporary cessation of consciousness due to cerebral ischemia; a faint.

Syndesmosis (sin'-dez-MŌ-sis) A fibrous joint in which articulating bones are united by dense fibrous tissue.

Syndrome (SIN-drōm) A group of signs and symptoms that occur together in a pattern characteristic of a particular disease or abnormal condition.

Synergist (SIN-er-jist) A muscle that assists the prime mover by steadying a movement. Thus, unwanted movements are prevented and the prime mover functions more efficiently.

Synostosis (sin'-os-TŌ-sis)   A joint in which the dense fibrous connective tissue that unites bones at a suture has been replaced by bone, resulting in a complete fusion across the suture line.

Synovial (si-NŌ-vē-al) **cavity**   The space between the articulating bones of a synovial or diarthrotic joint, filled with synovial fluid.

Synovial fluid   Secretion of synovial membranes that lubricates joints and nourishes articular cartilage.

Synovial joint   A fully movable or diarthrotic joint in which a joint or synovial cavity is present between the two articulating bones.

Synovial membrane   The inner of the two layers of the articular capsule of a synovial joint, composed of loose connective tissue covered with epithelium that secretes synovial fluid into the joint cavity.

Syphilis (SIF-i-lis)   A sexually transmitted disease caused by the bacterium *Treponema pallidum.*

System   An association of organs that have a common function.

Systemic (sis-TEM-ik)   Affecting the whole body; generalized.

Systemic anatomy   The study of particular systems of the body, such as the skeletal, muscular, nervous, cardiovascular, or urinary systems.

Systemic circulation   The routes through which oxygenated blood flows from the left ventricle through the aorta to all the organs of the body and deoxygenated blood returns to the right atrium.

Systemic lupus erythematosus (er-i-them-a-TŌ-sus) **(SLE)**   An autoimmune, inflammatory disease that may affect every tissue of the body.

Systole (SIS-tō-lē)   In the cardiac cycle, the phase of contraction of the heart muscle, especially of the ventricles.

Tactile (TAK-tīl)   Pertaining to the sense of touch.

Tactile disc   An encapsulated, cutaneous receptor for touch located in deeper layers of epidermal cells. Also called Merkel's disc.

Taenia coli (TĒ-nē-a KŌ-lī)   One of three flat bands of muscles running the length of the large intestine.

Target cell   A cell whose activity is affected by a particular hormone.

Tarsal (TAR-sal) **gland**   Sebaceous (oil) gland that opens on the edge of each eyelid. Also called a **Meibomian gland.**

Tarsal plate   A thin, elongated sheet of connective tissue, one in each eyelid, giving the eyelid form and support. The aponeurosis of the levator palpebrae superioris is attached to the tarsal plate of the superior eyelid.

Tarsus (TAR-sus)   A collective term for the seven bones of the ankle.

T cell   A lymphocyte that can differentiate into one of four kinds of cells—killer, helper, suppressor, or memory—all of which function in cellular immunity.

Tectorial (tek-TŌ-rē-al) **membrane**   A gelatinous membrane projecting over and in contact with the hair cells of the spiral organ (organ of Corti) in the cochlear duct.

Telophase (TEL-ō-fāz)   The final stage of mitosis in which the daughter nuclei become established.

Tendinitis (ten'-din-Ī-tis)   Inflammation of a tendon and synovial membrane at a joint. Also called **tenosynovitis.**

Tendon (TEN-don)   A white fibrous cord of dense, regularly arranged connective tissue that attaches muscle to bone.

Tendon organ   A proprioceptive receptor found chiefly near the junction of tendons and muscles. Also called a **Golgi tendon organ.**

Tentorium cerebelli (ten-TŌ-rē-um ser'-e-BEL-ē)   A transverse shelf of dura mater that forms a partition between the occipital part of the cerebral hemispheres and the cerebellum and that covers the cerebellum.

Teratogen (TER-a-tō-jen)   Any agent or factor that causes physical defects in a developing embryo.

Terminal ganglion (TER-min-al GANG-lē-on)   A cluster of cell bodies of postganglionic parasympathetic neurons either lying very close to the visceral effectors or located within the walls of the visceral effectors supplied by the postganglionic fibers.

Testis (TES-tis)   Male gonad that produces sperm and the hormones testosterone and inhibin. Also called a **testicle.**

Testosterone (tes-TOS-te-rōn)   A male sex hormone (androgen) secreted by interstitial endocrinocytes (Sertoli cells) of a mature testis; controls the growth and development of male sex organs, secondary sex characteristics, spermatozoa, and body growth.

Tetanus (TET-a-nus)   An infectious disease caused by the toxin of *Clostridium tetani,* characterized by tonic muscle spasms and exaggerated reflexes, lockjaw, and arching of the back. A smooth, sustained contraction produced by a series of very rapid stimuli to a muscle.

Tetany (TET-a-nē)   A nervous condition caused by hypoparathyroidism and characterized by intermittent or continuous tonic muscular contractions of the extremities.

Tetralogy of Fallot (tet-RAL-ō-jē of fal-Ō)   A combination of four congenital heart defects: (1) constricted pulmonary semilunar valve, (2) interventricular septal opening, (3) emergence of aorta from both ventricles instead of from the left only, and (4) enlarged right ventricle.

Thalamus (THAL-a-mus)   A large, oval structure located above the midbrain, consisting of two masses of gray matter covered by a thin layer of white matter.

Thalassemia (thal'-a-SĒ-mē-a)   A group of hereditary hemolytic anemias.

Therapy (THER-a-pē)   The treatment of a disease or disorder.

Thermoreceptor (THER-mō-rē-sep-tor)   Receptor that detects changes in temperature.

Thigh   The portion of the lower extremity between the hip and the knee.

Third ventricle (VEN-tri-kul)   A slitlike cavity between the right and left halves of the thalamus and between the lateral ventricles.

Thoracic cavity   Superior component of the ventral body cavity that contains two pleural cavities, the mediastinum, and the pericardial cavity.

Thoracic duct   A lymphatic vessel that begins as a dilation called the cisterna chyli, receives lymph from the left side of the head, neck, and chest, the left arm, and the entire body below the ribs, and empties into the left subclavian vein. Also called the **left lymphatic duct.**

Thoracolumbar outflow   The fibers of the sympathetic preganglionic neurons, which have their cell bodies in the lateral gray columns of the thoracic segment and first two or three lumbar segments of the spinal cord.

Thorax (THŌ-raks)   The chest.

Thrombocyte (THROM-bō-sīt)   A fragment of cytoplasm enclosed in a plasma membrane and lacking a nucleus; found in the circulating blood; plays a role in blood clotting. Also called a **platelet.**

Thrombophlebitis (throm'-bō-fle-BĪ-tis) A disorder in which inflammation of a vein wall is followed by the formation of a blood clot (thrombus).

Thrombus A clot formed in an unbroken blood vessel.

Thymectomy (thī-MEK-tō-mē) Surgical removal of the thymus.

Thymus (THĪ-mus) **gland** A bilobed organ, located in the upper mediastinum posterior to the sternum and between the lungs, that plays a role in the immunity mechanism of the body.

Thyroglobulin (thī-rō-GLŌ-byoo-lin) **(TGB)** A large glycoprotein molecule secreted by follicle cells of the thyroid gland in which iodine is combined with tyrosine to form thyroid hormones.

Thyroid cartilage (THĪ-royd KAR-ti-lij) The largest single cartilage of the larnyx, consisting of two fused plates that form the anterior wall of the larynx. Also called the **Adam's apple.**

Thyroid colloid (KOL-loyd) A complex in thyroid follicles consisting of thyroglobulin and stored thyroid hormones.

Thyroid follicle (FOL-i-kul) Spherical sac that forms the parenchyma of the thyroid gland and consists of follicular cells that produce thyroxine (T_4) and triiodothyronine (T_3) and parafollicular cells that produce calcitonin (CT).

Thyroid gland An endocrine gland with the right and left lateral lobes on either side of the trachea connected by an isthmus located in front of the trachea just below the cricoid cartilage.

Thyroid-stimulating hormone (TSH) A hormone secreted by the adenohypophysis (anterior lobe) of the pituitary gland that stimulates the synthesis and secretion of hormones produced by the thyroid gland.

Thyroxine (thī-ROK-sin) **(T_4)** A hormone secreted by the thyroid that regulates metabolism, growth and development, and the activity of the nervous system.

Tic Spasmodic, involuntary twitching by muscles that are ordinarily under voluntary control.

Tidal volume The volume of air breathed in and out in any one breath; about 500 ml in quite, resting conditions.

Tinnitus (ti-NĪ-tus) A ringing or tingling sound in the ears.

Tissue A group of similar cells and their intercellular substance joined together to perform a specific function.

Tissue rejection Phenomenon by which the body recognizes the proteins in transplanted tissues or organs as foreign and produces antibodies against them.

Tongue A large skeletal muscle on the floor of the oral cavity.

Tonsil (TON-sil) A mass of lymphoid tissue embedded in mucous membrane.

Torn cartilage A tearing of an articular disc (meniscus) in the knee.

Total lung capacity The sum of tidal volume, inspiratory reserve volume, expiratory reserve volume, and residual volume; about 6,000 ml.

Toxic (TOK-sik) Pertaining to poison; poisonous.

Toxic shock syndrome (TSS) A disease caused by toxin-producing strains of *Staphylococcus aureus* and characterized by sore throat, headache, fatigue, irritability, abdominal pain, lethargy, respiratory distress, and erythematous rash.

Trabecula (tra-BEK-yoo-la) Irregular latticework of thin plate of spongy bone. Fibrous cord of connective tissue serving as supporting fiber by forming a septum extending into an organ from its wall or capsule. *Plural,* **trabeculae** (tra-BEK-yoo-lē).

Trabeculae carnae (tra-BEK-yoo-lē KAR-nē) Ridges and folds of the myocardium in the ventricles.

Trachea (TRĀ-kē-a) Tubular air passageway extending from the larynx to the fifth thoracic vertebra. Also called the **windpipe.**

Tracheostomy (trā-kē-OS-tō-mē) Creation of an opening into the trachea through the neck, with insertion of a tube to facilitate passage of air or evacuation of secretions.

Trachoma (tra-KŌ-ma) A chronic infectious disease of the conjunctiva and cornea of the eye caused by a strain of the bacterium *Chlamydia trachomatis.*

Tract A bundle of nerve fibers in the central nervous system.

Transplantation (trans-plan-TĀ-shun) The replacement of injured or diseased tissues or organs with natural ones.

Transverse colon (trans-VERS KŌ-lon) The portion of the large intestine extending across the abdomen from the right colic (hepatic) flexure to the left colic (splenic) flexure.

Transverse fissue (FISH-er) The deep cleft that separates the cerebrum from the cerebellum.

Trauma (TRAW-ma) An injury, either a physical wound or psychic disorder, caused by an external agent or force, such as a physical blow or emotional shock; the agent or force that causes the injury.

Triad (TRĪ-ad) A complex of three units in a muscle fiber (cell) composed of a T tubule and the segments of sarcoplasmic reticulum on both sides of it.

Tricuspid (trī-KUS-pid) **valve** Atrioventricular (AV) valve on the right side of the heart.

Trigeminal neuralgia (trī-JEM-i-nal noo-RAL-jē-a) Pain in one or more of the branches of the trigeminal (V) nerve. Also called **tic douloureux.**

Trigone (TRĪ-gon) A triangular area at the base of the urinary bladder.

Triiodothyronine (trī-ī-od-ō-THĪ-rō-nēn) **(T_3)** A hormone produced by the thyroid gland that regulates metabolism, growth and development, and the activity of the nervous system.

Trochlea (TROK-lē-a) A pulleylike surface on the humerus.

Trophectoderm (trō-FEK-tō-derm) The outer covering of cells of the blastocyst.

Tropic (TRŌ-pik) **hormone** A hormone whose target is another endocrine gland.

Trunk The part of the body to which the upper and lower extremities are attached.

T tubules (TOOB-yools) **(transverse tubules)** Minute, cylindrical invaginations of the plasma membrane of a muscle fiber (cell) that carry the surface action potentials deep into the muscle fiber (cell).

Tubal ligation (li-GĀ-shun) A sterilization procedure in which the uterine (Fallopian) tubes are tied and cut.

Tuberculosis (too-berk-yoo-LŌ-sis) An inflammation of the lungs and pleurae caused by *Mycobacterium tuberculosis* resulting in destruction of lung tissue and its replacement by fibrous connective tissue.

Tumor (TOO-mor) A growth of excess tissue due to an unusually rapid division of cells.

Tunica albuginea (TOO-ni-ka al'-byoo-JIN-ē-a) A dense layer of white fibrous tissue covering a testis or deep to the surface of an ovary.

Tunica externa (eks-TER-na)   The outer coat of an artery or vein, composed mostly of elastic and collagenous fibers. Also called the **adventitia.**

Tunica interna (in-TER-na)   The inner coat of an artery or vein, consisting of a lining of endothelium and its supporting layer of connective tissue. Also called the **tunica intima.**

Tunica media (MĒ-dē-a)   The middle coat of an artery or vein, composed of smooth muscle and elastic fibers.

Twitch   Rapid, jerky contraction of a muscle in response to a single stimulus.

Tympanic antrum (tim-PAN-ik AN-trum)   An air space in the posterior wall of the middle ear that leads into the mastoid air cells or sinus.

Tympanic membrane   A thin, semitransparent partition of fibrous connective tissue between the external auditory canal and the middle ear. Also called the **eardrum.**

Ulcer (UL-ser)   An open lesion of the skin or a mucous membrane of the body with loss of substance and necrosis of the tissue.

Umbilical (um-BIL-i-kal)   Pertaining to the umbilicus or navel.

Umbilical cord   The long, ropelike structure, containing the umbilical arteries and vein, which connects the fetus to the placenta.

Umbilicus (um-BIL-i-kus)   A small scar on the abdomen that marks the former attachment of the umbilical cord to the fetus. Also called the **navel.**

Upper extremity   The appendage attached at the shoulder girdle, consisting of the arm, forearm, wrist, hand, and fingers.

Uremia (yoo-RĒ-mē-a)   Accumulation of toxic levels of urea and other nitrogenous waste products in the blood, usually resulting from severe kidney malfunction.

Ureter (yoo-RĒ-ter)   One of two tubes that connect the kidney with the urinary bladder.

Urethra (yoo-RĒ-thra)   The duct from the urinary bladder to the exterior of the body that conveys urine in females and urine and semen in males.

Urinalysis   The physical, chemical, and microscopic analysis or examination of urine.

Urinary (YOO-ri-ner-ē) **bladder**   A hollow, muscular organ situated in the pelvic cavity posterior to the symphysis pubis.

Urine   The fluid produced by the kidneys that contains wastes or excess materials and excreted from the body through the urethra.

Urobilinogenuria   The presence of urobilinogen in urine.

Urogenital (YOO'-rō-JEN-i-tal) **triangle**   The region of the pelvic floor below the symphysis pubis, bounded by the symphysis pubis and the ischial tuberosities and containing the external genitalia.

Urology (yoo-ROL-ō-jē)   The specialized branch of medicine that deals with the structure, function, and diseases of the male and female urinary systems and the male reproductive system.

Urticaria (ur'-ti-KĀ-rē-a)   A skin reaction to certain foods, drugs, or other substances to which a person may be allergic; hives.

Uterine (YOO-ter-in) **tube**   Duct that transports ova from the ovary to the uterus. Also called the **Fallopian tube** or **oviduct.**

Uterosacral ligament (yoo'-ter-ō-SĀ-kral LIG-a-ment)   A fibrous band of tissue extending from the cervix of the uterus laterally to attach to the sacrum.

Uterovesical (yoo'-ter-ō-VES-î-kal) **pouch**   A shallow pouch formed by the reflection of the peritoneum from the anterior surface of the uterus, at the junction of the cervix and the body, to the posterior surface of the urinary bladder.

Uterus (YOO-te-rus)   The hollow, muscular organ in females that is the site of menstruation, implantation, development of the fetus, and labor. Also called the **womb.**

Utricle (YOO-tri-kul)   The larger of the two divisions of the membranous labyrinth located inside the vestibule of the inner ear, containing a receptor organ for static equilibrium.

Uvea (YOO-vē-a)   The three structures that make up the vascular tunic of the eye.

Uvula (YOO-vyoo-la)   A soft, fleshy mass, especially the V-shaped pendant part, descending from the soft palate.

Vagina (va-JĪ-na)   A muscular, tubular organ that leads from the uterus to the vestibule, situated between the urinary bladder and the rectum of the female.

Valvular stenosis (VAL-vyoo-lar STEN-Ō-sis)   A narrowing of a heart valve, usually the bicuspid (mitral) valve.

Varicocele (VAR-i-kō-sēl)   A twisted vein; especially, the accumulation of blood in the veins of the spermatic cord.

Varicose (VAR-i-kōs)   Pertaining to an unnatural swelling, as in the case of a varicose vein.

Vas   A vessel or duct.

Vasa vasorum (VĀ-sa va-SŌ-rum)   Blood vessels supplying nutrients to the larger arteries and veins.

Vascular (VAS-kyoo-lar)   Pertaining to or containing many blood vessels.

Vascular (venous) sinus   A vein with a thin endothelial wall that lacks a tunica media and externa and is supported by surrounding tissue.

Vascular tunic (TOO-nik)   The middle layer of the eyeball, composed of the choroid, the ciliary body, and the iris.

Vasectomy (va-SEK-tō-mē)   A means of sterilization of males in which a portion of each ductus (vas) deferens is removed.

Vasoconstriction (vāz-ō-kon-STRIK-shun)   A decrease in the size of the lumen of a blood vessel caused by contraction of the smooth muscle in the wall of the vessel.

Vasodilation (vās'-ō-DĪ-lā-shun)   An increase in the size of the lumen of a blood vessel caused by relaxation of the smooth muscle in the wall of the vessel.

Vasomotor (vā-sō-MŌ-tor) **center**   A cluster of neurons in the medulla that controls the diameter of blood vessels, especially arteries.

Vein   A blood vessel that conveys blood from tissues back to the heart.

Vena cava (VĒ-na KĀ-va)   One of two large veins that open into the right atrium, returning to the heart all of the deoxygenated blood from the systemic circulation except from the coronary circulation.

Ventral (VEN-tral)   Pertaining to the anterior or front side of the body; opposite of dorsal.

Ventral body cavity   Cavity near the ventral aspect of the body that contains viscera and consists of a superior thoracic cavity and an inferior abdominopelvic cavity.

Ventral ramus (RĀ-mus)   The anterior branch of a spinal nerve, containing sensory and motor fibers to the muscles and skin of the anterior surface of the head, neck, trunk, and the extremities.

Ventricle (VEN-tri-kul)   A cavity in the brain or an inferior chamber of the heart.

Ventricular fibrillation (ven-TRIK-yoo-lar fib-ri-LĀ-shun) Asynchronous ventricular contractions that result in circulatory failure.

Venule (VEN-yool) A small vein that collects blood from capillaries and delivers it to a vein.

Vermiform appendix (VER-mi-form a-PEN-diks) A twisted, coiled tube attached to the cecum.

Vermilion (ver-MIL-yon) The area of the mouth where the skin on the outside meets the mucous membrane on the inside.

Vermis (VER-mis) The central constricted area of the cerebellum that separates the two cerebellar hemispheres.

Vertebral (VER-te-bral) **canal** The cavity formed by the vertebral foramina of all vertebrae together; the inferior portion of the dorsal cavity, containing the spinal cord. Also called the **spinal canal.**

Vertebral column The 26 vertebrae; encloses and protects the spinal cord and serves as a point of attachment for the ribs and back muscles. Also called the **spine, spinal column,** or **backbone.**

Vertigo (VER-ti-go) Sensation of dizziness.

Vesicle (VES-i-kul) A small bladder or sac containing liquid.

Vesicular ovarian follicle An endocrine gland consisting of a mature ovum and its surrounding tissue that secretes estrogens. Also called a **Graafian follicle.**

Vestibular (ves-TIB-yoo-lar) **membrane** The membrane that separates the cochlear duct from the scala vestibuli.

Vestibule (VES-ti-byool) A small space or cavity at the beginning of a canal, especially the inner ear, larynx, mouth, nose, and vagina.

Villus (VIL-lus) A projection of the intestinal mucosal cells containing connective tissue, blood vessels, and a lymphatic vessel; functions in the absorption of food. *Plural,* **villi** (VIL-ē).

Viscera (VIS-er-a) The organs inside the ventral body cavity. *Singular,* **viscus.**

Visceral (VIS-er-al) Pertaining to the organs or to the covering of an organ.

Visceral effector (e-FEK-tor) Cardiac muscle, smooth muscle, and glandular epithelium.

Visceral pleura (PLOO-ra) The inner layer of the serous membrane that covers the lungs.

Visceroceptor (vis'-er-ō-SEP-tor) Receptor that provides information about the body's internal environment.

Viscosity (vis-KOS-i-tē) The state of being sticky or thick.

Vital capacity The sum of inspiratory reserve volume, tidal volume, and expiratory reserve volume; about 4,800 ml.

Vitreous humor (VIT-rē-us HYOO-mor) A soft, jellylike substance that fills the posterior cavity of the eyeball, lying between the lens and the retina.

Vocal folds Pair of mucous membrane folds below the ventricular folds that function in voice production. Also called **true vocal cords.**

Vomiting Forcible expulsion of the contents of the upper gastrointestinal tract through the mouth.

Vulva Collective designation for the external genitalia of the female. Also called **pudendum.**

Wallerian (wal-LE-rē-an) **degeneration** Degeneration of the distal portion of the axon and myelin sheath.

Wandering macrophage (MAK-rō-fāj) Phagocytic cell that develops from a monocyte, leaves the blood, and migrates to infected tissues.

Wart Generally benign tumor of epithelial skin cells caused by a virus.

Wheal (hwēl) Elevated lesion of the skin.

White matter Aggregations or bundles of myelinated axons located in the brain and spinal cord.

White pulp The portion of the spleen composed of ovoid masses of lymphoid tissue called lymph follicles that contain germinal centers and where lymphocytes are produced.

White ramus communicans (RĀ-mus co-MYOO-ni-kans) The portion of a preganglionic sympathetic nerve fiber that branches away from the anterior ramus of a spinal nerve to enter the nearest sympathetic trunk ganglion.

Xiphoid (ZĪ-foyd) Sword-shaped. The lowest portion of the sternum.

Yolk sac An extraembryonic membrane that connects with the midgut during early embryonic development but is nonfunctional in humans.

Zona fasciculata (ZŌ-na fa-sik'-yoo-LA-ta) The middle zone of the adrenal cortex that consists of cells arranged in long, straight cords and that secretes glucocorticoid hormones.

Zona glomerulosa (glo-mer'-yoo-LŌ-sa) The outer zone of the adrenal cortex, directly under the connective tissue covering, that consists of cells arranged in arched loops or round balls and that secretes mineralocorticoid hormones.

Zona pellucida (pe-LOO-si-da) Gelatinous covering that surrounds an ovum.

Zona reticularis (ret-ik'yoo-LAR-is) The inner zone of the adrenal cortex, consisting of cords of branching cells that secrete sex hormones, chiefly androgens.

Zygote (ZĪ-gōt) The single cell resulting from the union of a male and female gamete; the fertilized ovum.

Zymogenic (zī'-mō-JEN-ik) **cell** A cell that secretes enzymes, for example, the zymogenic (chief) cells of the gastric glands that secrete pepsinogen.

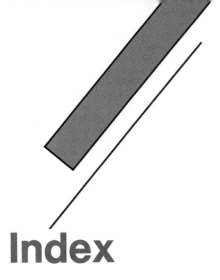

Index

Page numbers followed by the letter *E* indicate terms to be found in exhibits.

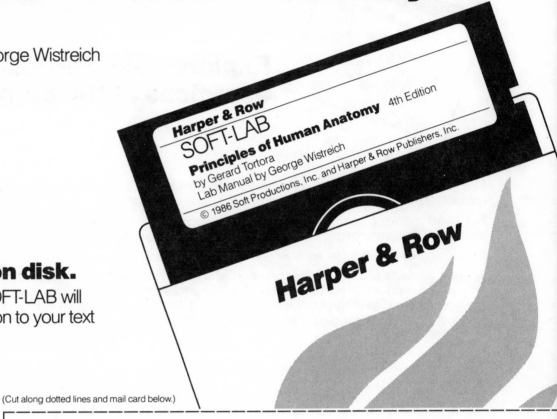

Harper & Row
SOFT-LAB

Explore selected laboratory exercises at the computer.

SOFT-LAB is a new computer program for the Apple, the Macintosh, and the IBM PC (and most IBM PC compatible) computers keyed directly to your text and lab manual.

SOFT-LAB enables you to manipulate variables, observe the results, and produce a lab report for selected exercises. Through exciting graphics, SOFT-LAB brings the laboratory experience to you.

If you do not own a computer, ask your instructor where you can use one on campus.

To order, complete and mail the card below or call Soft Productions, Inc. at (219) 255-3911. Be sure to indicate the type of computer disk you want and to include your method of payment. If the order card has been removed, mail your order to Soft Productions, Inc., P.O. Box 1003, Notre Dame, IN 46556. Be sure to indicate the title and author of the text and the type of computer disk you want (Apple, Macintosh, or IBM) and to include your method of payment.

(Cut along dotted lines and mail card below.)

Offer available in the
U.S. and Canada only.

NO POSTAGE
NECESSARY
IF MAILED
IN THE
UNITED STATES

BUSINESS REPLY MAIL
FIRST CLASS   PERMIT NO. 4501   NOTRE DAME, IN

POSTAGE WILL BE PAID BY ADDRESSEE

Soft Productions, Inc.
P.O. Box 1003
Notre Dame, IN 46556